T0360821

INTRODUCTION TO GEOCHEMISTRY

Introduction to Geochemistry

Principles and Applications

Kula C. Misra

Emeritus Professor, Department of Earth and Planetary Sciences,
The University of Tennessee, Knoxville, Tennessee, USA

WILEY-BLACKWELL

A John Wiley & Sons, Ltd., Publication

Blackwell Publishing was acquired by John Wiley & Sons in February 2007. Blackwell's publishing program has been merged with Wiley's global Scientific, Technical and Medical business to form Wiley-Blackwell.

Registered Office
John Wiley & Sons, Ltd, The Atrium, Southern Gate, Chichester, West Sussex, PO19 8SQ, UK

Editorial Offices
9600 Garsington Road, Oxford, OX4 2DQ, UK
The Atrium, Southern Gate, Chichester, West Sussex, PO19 8SQ, UK
111 River Street, Hoboken, NJ 07030-5774, USA

For details of our global editorial offices, for customer services and for information about how to apply for permission to reuse the copyright material in this book please see our website at www.wiley.com/wiley-blackwell.

Library of Congress Cataloging-in-Publication Data

Misra, Kula C.
Introduction to geochemistry : principles and applications / by Kula C. Misra.
p. cm.
Includes bibliographical references and index.
ISBN 978-1-4443-5095-1 (cloth) – ISBN 978-1-4051-2142-2 (pbk.)
1. Geochemistry–Textbooks. I. Title.
QE515.M567 2012
551.9–dc23

2011046009

A catalogue record for this book is available from the British Library.

Wiley also publishes its books in a variety of electronic formats. Some content that appears in print may not be available in electronic books

Set in 9.5/11.5pt Sabon by SPi Publisher Services, Pondicherry, India

1 2012

To

Anand, Lolly, and Tom

Brief Contents

Contents

COMPANION WEBSITE

This book has a companion website

www.wiley.com/go/misra/geochemistry

with Figures and Tables from the book for downloading.

Preface

Geochemistry deals essentially with the processes and consequences of distribution of elements in minerals and rocks in different physical–chemical environments and, as such, permeates all branches of geology to varying degrees. An adequate background in geochemistry is, therefore, an imperative for earth science students. This book is an attempt to cater to that need. It covers a wide variety of topics, ranging from atomic structures that determine the chemical behavior of elements to modern biogeochemical cycles that control the global–scale distribution of elements. It is intended to serve as a text for an introductory undergraduate/graduate level course in geochemistry, and it should also provide the necessary background for more advanced courses in mineralogy, petrology, and geochemistry.

The organization of the book is logical and quite different from the geochemistry texts in the market. Excluding the "Introduction", the 12 chapters of the book are divided into four interrelated parts. Part I (Crystal Chemistry – Chapters 2 and 3) provides a brief review of the electronic structure of atoms and of different kinds of chemical bonds. Part II (Chemical Reactions – Chapters 4 through 9) discusses the thermodynamic basis of chemical reactions involving phases of constant and variable composition, including reactions relevant to aqueous systems and reactions useful for geothermometry and geobarometry. A substantial portion of the chapter on oxidation–reduction reactions (Chapter 8) is devoted to a discussion of the role of bacteria in such reactions. The last chapter of Part II is a brief introduction to the kinetic aspects of chemical reactions. Part III (Isotope Geochemistry – Chapters 10 and 11) introduces the students to radiogenic and stable isotopes, and their applications to geologic problems, ranging from dating of rocks and minerals to the interpretation of an anoxic atmosphere during the Hadean and Archean eras. Part IV (The Earth Supersystem – Chapters 12 and 13) is an overview of the origin and evolution of the solid Earth (core, mantle, and crust), and of the atmosphere and hydrosphere. A brief discussion of some important biogeochemical cycles provides a capstone to the introductory course.

The treatment in this book recognizes the welcome fact that geochemistry has become increasingly more quantitative, and assumes that the students have taken the usual selection of elementary courses in earth sciences, chemistry, and mathematics. Nevertheless, most relevant chemical concepts and mathematical relations are developed from first principles. It is my experience that the derivation of an equation enhances the appreciation for its applications and limitations. To maintain the flow of the text, however, some derivations and tangential material are separated from the text in the form of "boxes." Supplementary data and explanations are presented in 10 appendixes.

Quantitative aspects of geochemistry are emphasized throughout the book to the extent they are, in my judgment, appropriate at an introductory level.

Each chapter in the book contains many solved examples illustrating the application of geochemistry to real-life geological and environmental problems. At the end of each chapter is a list of computational techniques the students are expected to have learned and a set of questions to reinforce the importance of solving problems. It is an integral part of the learning process that the students solve every one of these problems. To help the students in this endeavor, answers to selected problems are included as an appendix (Appendix 10).

I owe a debt of gratitude to all my peers who took the time to review selected parts of the manuscript: D. Sherman, University of Bristol; D.G. Pearson, Durham University; Hilary Downes, University College (London); Harry McSween, Jr., University of Tennessee (Knoxville); and Harold Rowe, University of Texas (Arlington). Their constructive critiques resulted in significant improvement of the book, but I take full responsibility for all shortcomings of the book. Thanks are also due to many of my colleagues in the Department of Earth and Planetary Sciences, University of Tennessee – Christopher Fedo, Robert Hatcher, Linda Kah, Theodre Labotka, Colin Sumrall, and Lawrence Taylor – who in course of many discussions patiently shared with me their expertise on selected topics covered in the book. I am particularly grateful to Harry McSween for many prolonged discussions regarding the origin and early history of the Earth, and to Ian Francis, Senior Commissioning Editor, Earth and Environmental Sciences, Wiley-Blackwell Publishers, for his sustained encouragement throughout this endeavor. I am also indebted to the many publishers and individuals who have kindly allowed me to include copyrighted figures in the book. Lastly, and most importantly, this book could not have been completed without the patience of my wife, children, and grandchildren, who had to endure my preoccupation with the book for long stretches.

Kula C. Misra
Department of Earth and Planetary Sciences
University of Tennessee
Knoxville, TN 37996
May 2011

1 Introduction

Geochemistry, as the name suggests, is the bridge between geology and chemistry and, thus, in essence encompasses the study of all chemical aspects of the Earth and their interpretation utilizing the principles of chemistry.

Rankama and Sahama (1950)

1.1 Units of measurement

A unit of measurement is a definite magnitude of a physical quantity, defined and adopted by convention and/or by law, that is used as a standard or measurement of the same physical quantity. Any other value of the physical quantity can be expressed as a simple multiple of the unit of measurement.

The original metric system of measurement was adopted in France in 1791. Over the years it developed into two somewhat different systems of metric units: (a) the *MKS system*, based on the meter, kilogram, and second for length, mass, and time, respectively; and (b) the *CGS system* (which was introduced formally by the British Association for the Advancement of Science in 1874), based on the centimeter, gram, and second. There are other traditional differences between the two systems, for example, in the measurements of electric and magnetic fields. The recurring need for conversion from units in one of the two systems to units of the other, however, defeated the metric ideal of a universal measuring system, and a choice had to be made between the two systems for international usage.

In 1954, the Tenth General Conference on Weights and Measures adopted the meter, kilogram, second, ampere, degree Kelvin, and candela as the basic units for all international weights and measures. Soon afterwards, in 1960, the Eleventh General Conference adopted the name *International System of Units* (abbreviated to *SI* from the French "Système International d'Unitès") for this collection of units. The "degree Kelvin" was renamed the "kelvin" in 1967.

1.1.1 *The SI system of units*

In the SI system, the modern form of the MKS system, there are seven base units from which all other units of measurement can be derived (Table 1.1). The International Union of Pure and Applied Chemistry (IUPAC) has recommended the use of SI units in all scientific communications. This is certainly desirable from the perspective of standardization of data, but a lot of the available chemical data were collected prior to 1960 and thus are not necessarily in SI units. It is therefore necessary for geochemists to be familiar with both SI and non-SI units. Equivalence between SI and non-SI units, and some of the commonly used physical constants are given in Appendix 1.

Most chemists, physicists, and engineers now use SI system of units, but the use of CGS (centimeter–gram–second) and other non-SI units is still widespread in geologic literature. In this book we will use SI units, but with two exceptions. As pointed out by Powell (1978) and Nordstrom and Munoz (1994), the SI unit pascal (Pa) is unwieldy for reporting geological pressures. For example, many geochemical measurements have been done at 1 atmosphere (atm) ambient pressure (the pressure exerted by the atmosphere at sea level), which translates into 101,325 pascals or 1.01 megapascals

Introduction to Geochemistry: Principles and Applications, First Edition. Kula C. Misra.

Table 1.1 The SI base units and examples of SI derived units.

SI base units			Examples of SI derived units			
Physical quantity	**Name**	**Symbol**	**Physical quantity**	**Name**	**Symbol**	**Definition in terms of the SI base units**
Length	Meter	m	Force	Newton	N	m kg s^{-2}
Mass	Kilogram	kg	Pressure	Pascal	Pa	m^{-1} kg s^{-2}=Nm^{-2}
Time	Second	s	Energy, work, heat	Joule	J	m^2 kg s^{-2}=Nm
Temperature	Kelvin	K	Electric charge	Coulomb	C	sA
Amount of substance	Mole	mol	Electric potential difference	Volt	V	m^2 kg s^{-3} A^{-1}
Electric current	Ampere	A	Volume	Liter	L	m^3 10^{-3}
Luminous intensity	Candela	cd	Electric conductance	Siemens	S	m^{-2} kg^{-1} s^3 A^2

Newton=the force that will accelerate a mass of 1 kg by 1 m s^{-2}.
Pascal=the pressure exerted when a force of 1 N acts uniformly over an area of 1 m^2.
Joule=work done when a force of 1 N produces a displacement of 1 m in the direction of the force.

(Mpa), a rather cumbersome number to use. Most geochemists prefer to use bar as the unit of pressure, which can easily be converted into pascals (1 bar=10^5 pascals or 0.1 MPa) and which is close enough to pressure expressed in atmosphere (1 bar=0.987 atm) for the difference to be ignored in most cases without introducing significant error. A similar problem exists in the use of the SI unit joule (J), instead of the more familiar non-SI unit calorie (cal). The calorie, defined as the quantity of heat required to raise 1 gram (g) of water from 14.5 to 15.5°C, has a physical meaning that is easy to understand. Moreover, tables of thermodynamic data, especially the older ones, use calories instead of joules. Thus, we may use calories in the calculations and report the final results in joules (1 cal=4.184 J).

The familiar scale of temperature is the Celsius scale (°C), which is based on two reference points for temperature: the *ice point*, the temperature at which ice is in equilibrium with liquid water at 1 atm pressure; and the *steam point*, the temperature at which steam is in equilibrium with liquid water at 1 atm pressure. The Celsius scale arbitrarily assigns a temperature of zero to the ice point and a temperature of 100 to the steam point. The SI unit of temperature is kelvin (K), which is the temperature used in all thermodynamic calculations. If pressure–temperature (°C) plots at different volumes are constructed for any gas, the extrapolated lines all intersect at a point representing zero pressure at a temperature around −273°C (Fig. 1.1). This temperature, which is not physically attainable (although it has been approached very closely), is called the *absolute zero of temperature*. It is the temperature at which the molecules of a gas have no translational, rotational, or vibrational motion and therefore no thermal energy. The temperature scale with absolute zero as the starting point is the *kelvin temperature scale* and the unit of temperature on this scale is kelvin (K, not °K), so named after Lord Kelvin who proposed it in 1848. The kelvin unit of temperature is defined as the 1/273.16 fraction of the so-called *triple point*

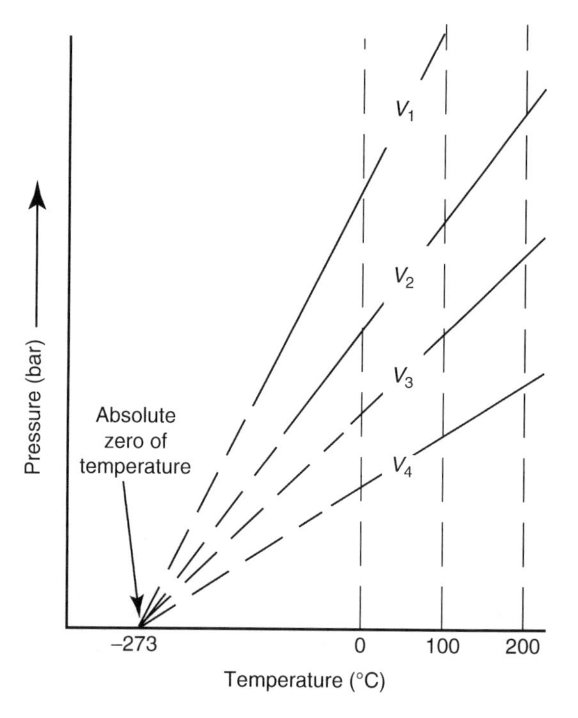

Fig. 1.1 The definition of absolute zero of temperature. The lines V_1–V_4 show the variation of different volumes of a gas as a function of temperature and pressure. When extrapolated, the lines intersect at a point representing zero pressure at a temperature around −273°C. This temperature, which is not physically attainable, is called the absolute zero of temperature.

for H_2O (the temperature at which ice, liquid water, and steam coexist in equilibrium at 1 atm pressure), which is 0.01 K greater than the ice point. Thus, the ice point, which is defined as 0°C, corresponds to 273.15 K (see Fig. 4.3) and the relationship between kelvin and Celsius scales of temperature is given by:

$$T\,(\mathrm{K}) = t\,(°\mathrm{C}) + 273.15 \tag{1.1}$$

Evidently, the steam point (100°C) corresponds to 373.15 K. It follows from equation (1.1) that the degree Celsius is equal in magnitude to the kelvin, which in turn implies that the numerical value of a given temperature difference or temperature interval is the same whether it is expressed in the unit degree Celsius (°C) or in the unit kelvin. (In the USA, temperatures are often measured in the Fahrenheit scale (F). The expression relating temperatures in the Celsius and Fahrenheit scales is: F=9/5°C+32.)

1.1.2 *Concentration units for solutions*

Concentrations of solutes (dissolved substances) in solutions (solids, liquids, or gases) are commonly expressed either as mass concentrations (parts per million, or milligrams per liter, or equivalent weights per liter) or as molar concentrations (*molality*, *molarity*, or *mole fraction*; Table 1.2).

To obtain the number of moles (abbreviated *mol*) of a substance, the amount of the substance (in grams) is divided by its gram-molecular weight; to obtain the mole fraction of a substance, the number of moles of the substance is divided by the total number of moles in the solution (see section 2.2 for further elaboration). For example, the mole fraction of NaCl (gram-molecular weight=58.44) in a solution of 100 g of NaCl in 2 kg of H_2O (gram-molecular weight=18.0) can be calculated as follows:

Number of moles of NaCl=100/58.44=1.7112
Number of moles of H_2O=2 (1000)/18.0=111.1111
Total number of moles in the solution= 1.7112 + 111.1111
= 112.8223
Mole fraction of NaCl in solution=1.7112 /112.8223=0.0152

Note that the mole fraction of a pure substance (solid, liquid, or gas) is unity.

The concentration units mg/L and ppm, as well as molality and molarity, are related through the density of the solution (ρ):

Table 1.2 Concentration units for a solute.

Concentration unit	Definition
Milligrams per liter (mg/L)	Mass of solute (mg) / volume of solution (L)
Parts per million (ppm)	Mass of solute (mg) / mass of solution (kg)
Mole fraction (*X*)	Moles of solute / total moles of solution[1]
Molarity (*M*)	Moles of solute / volume of solution (L)
Molality (*m*)	Moles of solute / mass of solvent (kg)
Normality (*N*)	Equivalent weight of solute (g) / volume of solution (L)

[1]Moles of a substance=weight of the substance (g)/gram-molecular weight of the substance.

$$\text{concentration (ppm)} = \frac{\text{concentration of solute (g L}^{-1})}{\rho\ (\text{g mL}^{-1})} \tag{1.2}$$

$$m = M\left(\frac{\text{weight of solution (g)}}{\text{weight of solution (g)} - \text{total weight of solutes (g)}}\right)\left(\frac{1}{\rho\ (\text{g mL}^{-1})}\right) \tag{1.3}$$

Concentrations expressed in molality or mole fraction have the advantage that their values are independent of temperature and pressure; molarity, on the other hand, is dependent on the volume of the solution, which varies with temperature and pressure. The advantage of using molarity is that it is often easier to measure the volume of a liquid than its weight. For dilute aqueous solutions at 25°C, however, the density of the solution is very close to that of pure water, ρ=(1 kg)/(1 L), so that little error is introduced if the difference between mg/L and ppm or molality and molarity is ignored for such a solution.

The strength of an acid or a base is commonly expressed in terms of *normality*, the number of equivalent weights of the acid or base per liter of the solution, the *equivalent weight* being defined as the gram-molecular weight per number of Hs or OHs in the formula unit. For example, the equivalent weight of H_2SO_4 (gram-molecular weight=98) is 98/2=49, and the normality of a solution of 45 g of H_2SO_4 in 2 L of solution is 45/(49×2)=0.46.

1.2 The Geologic Time Scale

Discussions of events require a timeframe for reference. The Geologic Time Scale provides such a reference for past geologic events. Forerunners of the current version of the time scale were developed in small increments during the 19th century, long before the advent of radiometric dating, using techniques applicable to determining the relative order of events. These techniques are based on the principles of *original horizontality* (sediments are deposited in horizontal layers), *superposition* (in a normal sequence of sedimentary rocks or lava flows, the layer above is younger than the layer below), and *faunal succession* (fossil assemblages occur in rocks in a definite and determinable order). Although the time scale evolved haphazardly, with units being added or modified in different parts of the world at different times, it has been organized into a universally accepted workable scheme of classification of geologic time.

The Geologic Time Scale spans the entire interval from the birth of the Earth (t=4.55 Ga, i.e., 4.55 billion years before the present) to the present (t=0), and is broken up into a hierarchical set of relative time units based on the occurrence of distinguishing geologic events. Generally accepted divisions for increasingly smaller units of time are *eon*, *era*, *period*, and *epoch* (Fig. 1.2). Different spans of time on the time scale are usually delimited by major tectonic or paleontological events

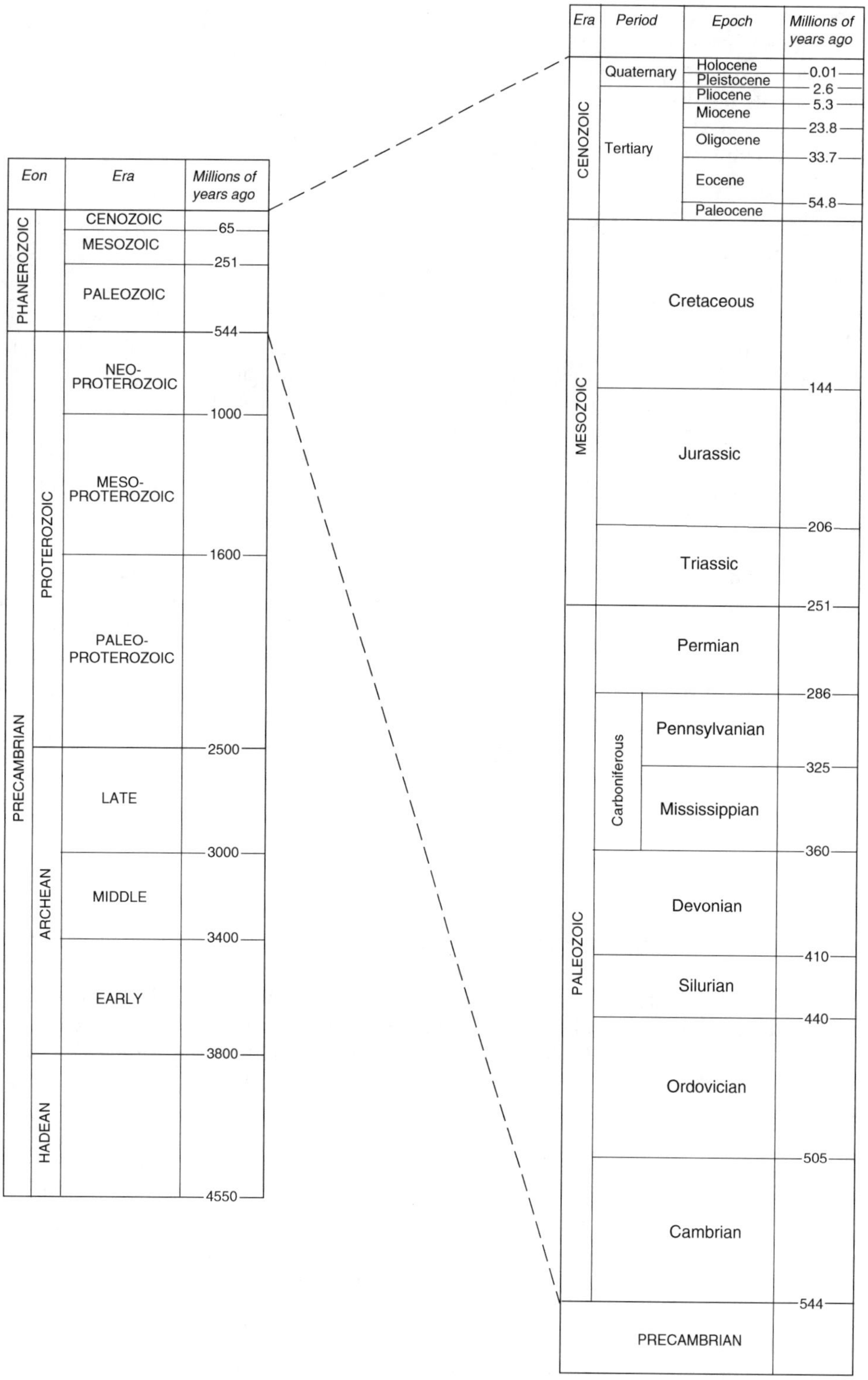

Eon	Era	Millions of years ago
PHANEROZOIC	CENOZOIC	65
	MESOZOIC	251
	PALEOZOIC	544
PRECAMBRIAN – PROTEROZOIC	NEO-PROTEROZOIC	1000
	MESO-PROTEROZOIC	1600
	PALEO-PROTEROZOIC	2500
PRECAMBRIAN – ARCHEAN	LATE	3000
	MIDDLE	3400
	EARLY	3800
PRECAMBRIAN – HADEAN		4550

Era	Period	Epoch	Millions of years ago
CENOZOIC	Quaternary	Holocene	0.01
		Pleistocene	2.6
	Tertiary	Pliocene	5.3
		Miocene	23.8
		Oligocene	33.7
		Eocene	54.8
		Paleocene	
MESOZOIC	Cretaceous		144
	Jurassic		206
	Triassic		251
PALEOZOIC	Permian		286
	Carboniferous – Pennsylvanian		325
	Carboniferous – Mississippian		360
	Devonian		410
	Silurian		440
	Ordovician		505
	Cambrian		544
PRECAMBRIAN			

Fig. 1.2 The Geologic Time Scale. The age of the Earth, based on the age of meteorites, is 4.55 ± 0.05 Ga according to Patterson (1956) and 4.55–4.57 Ga according to Allègre *et al.* (1995).

such as orogenesis (mountain–building activity) or mass extinctions. For example, the Cretaceous–Tertiary boundary is defined by a major mass extinction event that marked the disappearance of dinosaurs and many marine species.

Absolute dates for the boundaries between the divisions were added later on, after the development of techniques for dating rocks using radioactive isotopes (see Chapter 10). The time scale shown in Fig. 1.2 includes these dates, producing an integrated geologic time scale. Time units that are older than the Cambrian Period (that is, units in the Precambrian Eon) pre-date reliable fossil records and are defined by absolute dates.

1.3 Recapitulation

Terms and concepts

Absolute zero of temperature
Celsius scale (temperature)
CGS system of units
Eon
Epoch
Era
Equivalent weight
Fahrenheit scale (temperature)
Faunal succession
Geologic Time Scale
Fahrenheit scale (temperature)
Geologic time scale
Gram–molecular weight
Kelvin scale (temperature)
Mass concentration
Mass extinction
MKS system of units
Molality
Molarity
Mole
Mole fraction
Original horizontality
Period
SI units
Superposition

Computation techniques

- Conversion of SI units to non-SI units.
- Conversion of °C to °F and K.
- Calculations of number of moles, mole fraction, molarity, molality, ppm.

1.4 Questions

Gram atomic weights: H=1.0; C=12.01; O=16.00; Na=22.99; Al=26.98; Si=28.09; S=32.07; Cl=35.45; K=39.10; Ca=40.08.

1. The gas constant, R, has the value 1.987 cal K^{-1} mol^{-1}. Show that

   R=8.317 Joules K^{-1} mol^{-1}=8.317×10^7 ergs K^{-1} mol^{-1}
   =83.176 cm^3 bar K^{-1} mol^{-1}

2. Show that (a) 1 calorie bar^{-1}=41.84 cm^3, and (b) 1 m^3= 1 joule $pascal^{-1}$
3. What is the molarity of one molal NaCl solution (at 25°C and 1 bar)? Density of the NaCl solution (at 25°C and 1 bar) is 1.0405 kg L^{-1}.
4. What are the mole fractions of C_2H_5OH (ethanol) and H_2O (water) in a solution prepared by mixing 70.0 g of ethanol with 30.0 g of water?
5. What are the mole fractions of C_2H_5OH (ethanol) and H_2O (water) in a solution prepared by mixing 70.0 mL of ethanol with 30.0 mL of water at 25°C? The density of ethanol is 0.789 g mL^{-1}, and that of water is 1.00 g mL^{-1}.
6. When dissolved in water, NaCl dissociates into Na^+ and Cl^- ions ($NaCl=Na^++Cl^-$). What is the molality of Na^+ in a solution of 1.35 g of NaCl dissolved in 2.4 kg of water? What is the concentration of Na^+ the solution in ppm?
7. The density of an aqueous solution containing 12.5 g K_2SO_4 in 100.00 g solution is 1.083 g mL^{-1}. Calculate the concentration of K_2SO_4 in the solution in terms of molality and molarity. What is the mole fraction of the solvent in the solution?
8. The ideal chemical formula of the mineral albite is $NaAlSi_3O_8$. How many moles of $NaAlSi_3O_8$ do 5 g of the mineral contain? How many moles of Si?
9. A solution made by dissolving 16.0 g of $CaCl_2$ in 64.0 g of water has a density of 1.180 g mL^{-1} at 25°C. Express the concentration of Ca in the solution in terms of molality and molarity.

Part I
Crystal Chemistry

The task of crystal chemistry is to find systematic relationships between the chemical composition and physical properties of crystalline substances, and in particular to find how crystal structure, i.e., the arrangement of atoms or ions in crystals, depend on chemical composition.

Goldschmidt (1954)

2 Atomic Structure

I knew of [Heisenberg's) theory, of course, but I felt discouraged, not to say repelled, by the methods of transcendental algebra, which appeared difficult to me, and by the lack of visualizability (Edwin Schrödinger, 1926).

The more I think about the physical portion of Schrödinger's theory, the more repulsive I find it.... What Schrödinger writes about the visualizability of his theory is probably not quite right, in other words it's crap (Werner Heisenberg, 1926).

[Extracted from the worldwide web]

The three major physical states in which a system (part of the universe that we have chosen for consideration) occurs are solid, liquid, and gaseous. Although we commonly think of distinctions among the three states in terms of physical properties such as bounding surface, hardness, viscosity, etc., the fundamental difference lies in the arrangement of atoms in the three states. *Crystalline* solids, which can occur spontaneously in a form bounded by planar surfaces, are characterized by *long-range order*, a regularity in the arrangement of atoms or molecules that is repeated along parallel lines. *Amorphous* solids consist of extremely small solid units with a very small number of atoms per unit, so that "the forces which lead to the planar surfaces of solids and their internal order are destroyed by the enormous number of atoms in surface position" (Fyfe, 1964). Gases have no unique volume or boundaries, except those imposed by the chosen container; they lack ordering of their atoms or molecules. The liquid state may be viewed as being somewhere in between the solid and gaseous states. The atoms and molecules of liquids show only *short-range order*, i.e., ordering of the atoms and molecules extend only over a few molecular diameters. From the atomic perspective, *glass* may be regarded as a very viscous fluid. Our focus here is on crystalline solids because minerals, the main constituents of rocks, are by definition crystalline solids. We begin with a discussion of the general features of atomic structures because the physical and chemical properties of minerals are determined by the structure and arrangement of their constituent atoms.

2.1 Historical development

2.1.1 *Discovery of the electron*

The Greek philosopher Democritus (5th century BC) believed in the existence of four elementary substances – air, water, stone, and fire – all formed by a very large number of very small particles called *atoms*, the word for "indivisibles" in Greek (Gamow, 1961). Our understanding of the atom was not much better until about the beginning of the 20th century. Sir Isaac Newton (1642–1727) described atoms as "solid, massy, hard, impenetrable, moveable particles" and this was the generally accepted view throughout the 19th century.

In 1897, the British scientist J. J. Thompson (1856–1940), one of the pioneers in the investigation of atomic structure, proved by direct experiments that atoms are complex systems composed of positively and negatively charged parts. The experimental set-up, a primitive version of the modern television tube, was rather simple. It consisted of a glass tube containing highly rarified gas with a cathode (–) placed at one end, an anode (+) in the middle part, and a fluorescent screen at the other end (Fig. 2.1). When an electric current was passed through the gas in the tube, the fluorescent screen became luminous because of bombardment by fast moving particles resulting from the gas. When a piece of metal was placed in the path of the moving particles, it cast a shadow on the fluorescent screen, indicating a straight-line path for the particles

Introduction to Geochemistry: Principles and Applications, First Edition. Kula C. Misra.

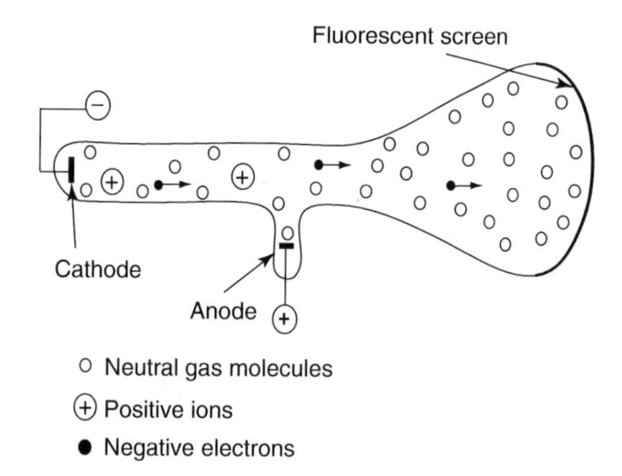

Fig. 2.1 Thompson's experimental set-up to study the effect of electric current on rarefied gas.

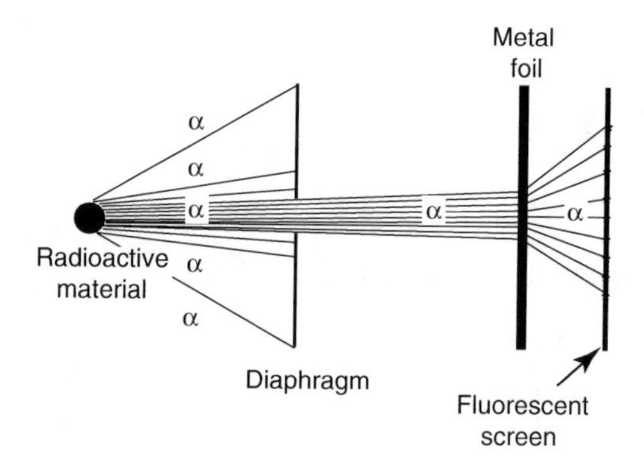

Fig. 2.2 Rutherford's experimental set-up for studying the scattering of α-particles emitted from a radioactive source.

in question, similar to light rays. Thompson visualized the atom as a complex system consisting of a positively charged substance (positive electric fluid) distributed uniformly over the entire body of the atom, with negatively charged particles (*electrons*) embedded in this continuously positive charge like seeds in a watermelon (Gamow, 1961). The model, however, was soon found to be unsatisfactory as the theoretically calculated optical line spectra (a set of characteristic light frequencies emitted by an "excited" atom) of different elements based on this model could not be matched with the observed optical spectra. Thompson also conducted experiments to determine the charge : mass ratio (*e/m*) of an electron (1.76×10^{-8} coulomb g^{-1}). A few years later, Robert A. Millikan experimentally measured the charge of an electron (1.602×10^{-19} coulomb) and computed its mass (about $9.109\times10^{-28}\,g^{-1}$).

2.1.2 *The Rutherford–Bohr atom*

In 1911, Ernest Rutherford (1871–1937), a New Zealand-born physicist, advanced the concept that the mass of an atom is concentrated at its center, which he named the *nucleus*. The experimental set-up that led to this discovery was quite simple (Fig. 2.2). A small amount of radioactive material emitting α-particles (positively charged helium ions that are ejected from the nucleus of an atom if it undergoes radioactive decay) was put on a pinhead and placed at a certain distance from a thin foil made from the metal to be investigated. The beam of α-particles was collimated by passing it through a lead diaphragm. A fluorescent screen was placed behind the foil to record the α-particles that would pass through the foil, each α-particle producing a little spark (scintillation) on the screen at the point of impact that could be viewed with the help of a microscope. In his experiments, Rutherford noticed that the majority of the α-particles passed through the foil almost without deflection, but some were deflected considerably and in a few cases (with a somewhat different experimental arrangement) some α-particles bounced back toward the source. Rutherford reasoned that collisions between the α-particles of the beam and the atoms of the target could not possibly deflect the incident particles by more than a few degrees; the observed large deflections required strong electrostatic repulsion between the positive charge of the bombarded atom and the positive charge of the incident α-particles. He concluded that the positive charge of the atom (associated with most of its mass) was not distributed throughout its body, as Thompson had envisaged, but had to be concentrated in a small central region of the atom, which he called the "atomic nucleus." It followed that the rest of the atom must be composed of a bunch of negatively charged electrons, rotating around the nucleus at high velocities so as not to fall into the positively charged nucleus because of electrostatic attraction. The positive charge of the nucleus was attributed to subatomic particles called *protons*. Rutherford speculated that the nucleus might also contain electrically neutral particles, although such elementary particles, called *neutrons*, were discovered only in 1932. Thus, the atomic model proposed by Rutherford consisted of negatively charged, light electrons whirling at very high velocities in circular paths around a positively charged, heavy nucleus at the center, so that the outward centrifugal force associated with such a motion would balance the electrostatic attraction between the electrons and the nucleus.

Rutherford's model, however, faced a serious problem because of the inherent instability of an electron orbiting around a positively charged nucleus. According to the laws of classical physics, such an electron would lose energy by emitting an electromagnetic wave, resulting in an increase in the velocity of the electron and a decrease in the radius of its orbit until the electron falls into the nucleus. Consider the hydrogen atom consisting of a bare proton and a single electron of mass m_e and charge e orbiting the nucleus of circular path of radius r at velocity v_e (Fig. 2.3). For this system, the energy of the atom (E_{atom}) is inversely proportional to the radius of the orbit (see Box 2.1):

$$E_{atom} = -\frac{1}{2}\left(\frac{e^2}{r}\right) = -\frac{1}{2}m_e v_e^2 \tag{2.1}$$

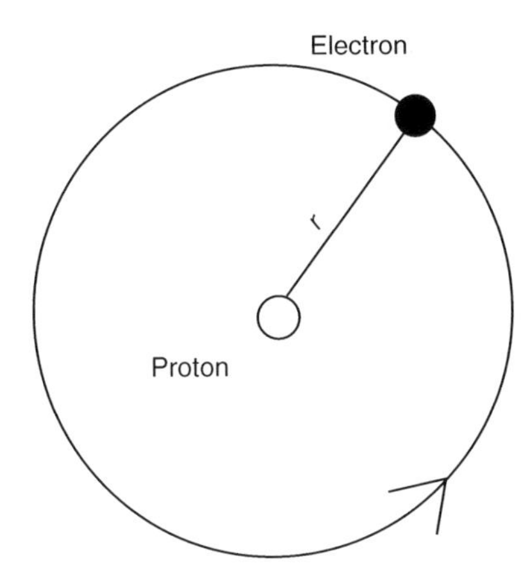

Fig. 2.3 Model of a hydrogen atom with a single electron moving in a circular orbit of radius r around a nucleus consisting of one proton.

Box 2.1 Derivation of equation for E_{atom}

Let us consider an electron of mass m_e and charge e orbiting around the nucleus of an atom in a circular path of radius r at velocity v_e (Fig. 2.3). The condition of stability for such an atom is that the force of attraction between the proton and electron (e^2/r^2) must be balanced by the centrifugal force($m_e v_e^2/r^2$):

$$\frac{e^2}{r^2} = \frac{m_e v_e^2}{r} \qquad (2.2)$$

The energy of the system is the sum of the kinetic energy ($1/2\, m_e v_e^2$) and potential energy ($-e^2/r$):

$$E_{atom} = \frac{1}{2}\, m_e v_e^2 - \frac{e^2}{r} \qquad (2.3)$$

The potential energy term has a negative sign because the force between the proton in the nucleus and the electron is due to electrostatic attraction, which by convention is assigned a negative sign. Substituting the value of m_e from equation (2.2), $m_e = e^2/rv_e^2$ we get

$$E_{atom} = \frac{1}{2}\left(\frac{e^2}{r}\right) - \frac{e^2}{r} = -\frac{1}{2}\left(\frac{e^2}{r}\right) = -\frac{1}{2} m_e v_e^2 \qquad (2.1)$$

Thus, the energy of the atom is negative and is inversely related to the radius of the orbit. The atom should become more stable as r decreases, and the electron should gradually fall into the nucleus. Calculations using classical physics predicted that electrons orbiting around a positively charged nucleus would lose all their energy in the form of electromagnetic waves within about one-hundred-millionth of a second and collapse into the nucleus.

The Danish physicist Neils Bohr (1885–1962), who had joined Rutherford as a postdoctoral fellow after a falling out with Thompson at the Cavendish Laboratory, Cambridge, provided the answer by applying Max Planck's revolutionary theory of quantization of electromagnetic energy. Speaking at the meeting of the German Physical Society on December 14, 1900, Max Planck (1858–1947) had proposed that light energy can exist only in the form of discrete packages, which he called "light quanta," and that the amount of energy of a light quantum (E) is directly proportional to the frequency (ϑ) of the radiation and inversely proportional to its wave length (λ). Since wavelength and frequency for light waves are related by the equation $\lambda\vartheta = c$, we have (see section 13.1.2)

$$E = h\vartheta = \frac{hc}{\lambda} \qquad (2.4)$$

where h is the proportionality constant known as the Planck's Constant ($h = 6.62517 \times 10^{-34}$ J s), c is the speed of light "($c = 3 \times 10^{10}$ cm s^{-1}), λ is expressed in angstrom units (1 Å $= 10^{-8}$ cm), and E is in units of kiloelectron volts (keV). (Electron and X-ray energies are expressed in electron volts, 1 eV being the kinetic energy gained by a single unbound electron when it accelerates through an electric potential difference of 1 volt.) In his first article on the *Theory of Relativity* in 1905, Albert Einstein (1879–1955) had also used the quantum theory to explain empirical laws of the photoelectric effect, the emission of electrons from metallic surfaces irradiated by violet or ultraviolet rays. Bohr reasoned that, if the electromagnetic energy is *quantized* – i.e., permitted to have only certain discrete values – mechanical energy must be quantized too, although perhaps in a somewhat different way. After struggling with the idea for almost two years, he finally published it in 1913 (Bohr, 1913a,b). Bohr retained Rutherford's concept of electron motion in circular orbits (but rejected the classical law that moving charged particles radiate energy), and postulated that electrons moving around the atomic nucleus can reside only in a few permitted circular orbits or "shells," each with a specific level of energy (E_n) and a radius (r_n) given by (for a hydrogen atom, which has only one electron):

$$E_n = -\frac{2\pi^2 m_e e^4}{h^2 n^2} \qquad (2.5)$$

$$r_n = \frac{h^2 n^2}{4\pi^2 m_e e^4} \qquad (2.6)$$

where m_e is the mass of the electron, e is the magnitude of its charge, h is Planck's Constant, and n is the *principal quantum number* (see section 2.2.1) that can assume only integral values (1, 2, 3, ...), each value of n defining a particular energy level for the electron. Note that the closer an orbit is to the nucleus (i.e., as the value of n gets smaller), the larger is E_n in absolute value but actually smaller in arithmetic value because of the negative sign (an arbitrary convention for attractive forces). The smallest radius permitted by Bohr's theory, the so-called *Bohr radius* (r_0), is obtained from equation (2.6) by setting $n = 1$. This is the radius of the orbit ($r_0 = 0.529 \times 10^{-8}$ cm) with the lowest permitted energy and thus represents the most stable state. The orbits corresponding to $n=1, 2, 3, \ldots, 7$ are sometimes referred to as $K, L, M, \ldots, Q$ "shells", respectively. An essential feature of the model is that an orbit of principal quantum number n can accept no more that $2n^2$ electrons. Bohr postulated further that an electron moving around the nucleus in a particular orbit is prohibited from emitting any electromagnetic radiation, but it does emit a quantum of monochromatic radiation when it jumps from an orbit of higher energy, say E_1, to an orbit of lower energy, say E_2, according to the relation derived earlier by Einstein:

$$E_1 - E_2 = \Delta E = h\vartheta \qquad (2.7)$$

where ϑ is the frequency of the radiation and h is the Planck's Constant. Bohr's model of the atom provided a convincing explanation for the emission of X-rays (which had been discovered by William Konrad Röntgen in1895) from target elements bombarded with a stream of electrons, and for the characteristic X-ray spectra of elements (Charles G. Barkla, 1911), which constitute the theoretical basis of modern electron microprobe and X-ray fluorescence analytical techniques.

Bohr's notion of circular quantum orbits worked very well for the hydrogen atom, the simplest of all atoms containing a single electron. By this time, the optical line spectra for hydrogen was well known from spectroscopic studies, and the spectra predicted from Bohr's model was a perfect match. The model, however, broke down almost completely for atoms having two or more electrons. Soon after, to allow more freedom in choosing the "permitted" orbits for multi-electron atoms, Arnold Sommerfeld (1868–1951) introduced the idea of elliptical orbits, which had different geometrical shapes but corresponded to almost the same energy levels as Bohr's circular orbits. According to Sommerfeld's postulate, the orbit closest to the nucleus ($n=1$) is circular and corresponds to the lowest energy of the electron. The next four orbits ($n=2$), one circular and the other three energetically equivalent but elliptical, have higher energy than that associated with the first orbit; the next nine orbits ($n=3$), one circular and the rest eight energetically equivalent but elliptical, correspond to a still higher level of energy, and so on (see Table 2.2 and Fig. 2.7).

2.1.3 *Wave mechanics*

In a doctoral thesis presented in 1925, Louis Victor de Broglie (1892–1987) proposed a new interpretation of Bohr's quantum orbits. He postulated that each electron moving along a given orbit is accompanied by some mysterious "pilot waves" (now known as *de Broglie waves*), whose propagation velocity and wavelength depend on the velocity of the electron in question. He deduced that the wavelength λ of an electron of mass m_e and velocity v_e is inversely proportional to its momentum ($m_e v_e$) and related to Planck's Constant (h) by the equation:

$$\lambda = \frac{h}{m_e v_e} \qquad (2.8)$$

The validity of this relationship was confirmed later by experiments demonstrating diffraction effects for electrons similar to those of X-rays. A year later, in 1926, de Broglie's ideas were brought into more exact mathematical form by Werner Heisenberg (1901–1976) and Edwin Schrödinger (1887–1961). The two scientists used entirely different formulations but arrived at the same results concerning atomic structure and optical spectra. Heisenberg developed the *Uncertainty Principle*, which states that the position and velocity of an electron in motion, whether in a circular or in an elliptical orbit, cannot be measured simultaneously with high precision. The mathematical formulation, however, was abstract and relied on matrix algebra. Most physicists of the time were slow to accept "matrix mechanics" because of its abstract nature and its unfamiliar mathematics. They gladly embraced Schrödinger's alternative wave mechanics, since it entailed more familiar concepts and equations, and it seemed to do away with quantum jumps and discontinuities. However, Schrödinger soon published a proof that matrix mechanics and wave mechanics gave equivalent results: mathematically they were the same theory, although he argued for the superiority of wave mechanics over matrix mechanics.

The recognition of the wave-like nature of the electron forced a fundamental change in ideas regarding the distribution of electrons in an atom. The concept of electrons as physical particles moving in orbits of definite geometrical form was replaced by a probability distribution of electron density (the number of electrons per unit volume) around the nucleus, rendering it possible to calculate the probability of finding the position of the electron at any point around the nucleus. From classical equations governing the behavior of waves, Schrödinger developed a general equation for de Broglie waves and proved its validity for all kinds of electron motion in three-dimensional space. Schrödinger's theory, which has now become known as *wave mechanics* (or *quantum mechanics*), explains not only all the atomic phenomena for which Bohr's model works, but also those phenomena (such as intensities of optical spectral lines) for which Bohr's model does not. In its most commonly used form (Fyfe, 1964),

$$\frac{\partial^2\psi}{\partial x^2} + \frac{\partial^2\psi}{\partial y^2} + \frac{\partial^2\psi}{\partial z^2} + \frac{8\pi^2 m_e}{h^2}(E-V)\psi = 0 \qquad (2.9)$$

the Schrödinger's wave equation, in essence, is a differential equation that relates a quantity ψ, the "wave function" of the system, to its total energy E and potential energy V. As defined earlier, m_e is the mass of the electron, h is Planck's Constant, and n is the principal quantum number. Such an equation can be satisfied by an infinite number of values of ψ, which lead to separate (not continuous) values of E for a given potential V. Of greatest interest are those solutions that yield the lowest possible values of E, the stable stationary state. The significance of ψ for our purpose lies in the fact that the value of ψ^2 at any point in space is a measure of the probability of finding the electron in an infinitely small volume at that point; it is thus a measure of electron density. The larger the value of ψ^2, the greater the probability of finding the electron there (although there is also a very small but finite probability of finding the electron far away from the nucleus). A region of space in which the probability of finding an electron is high is called an *(atomic) orbital* (to distinguish it from "orbits" of the Bohr–Sommerfeld model). An orbital may be occupied (fully or partially) or empty. The probability interpretation is consistent with the idea that the electron is a particle, although described by a wave function, if we visualize an electron as a diffuse cloud of negative charge rather than a small discrete entity.

The simplest solutions of Schrödinger's equation are the ones that predict spherical probability distributions, which give the same energies as Bohr's model for circular orbits of different principal quantum numbers. For example, let us consider the spherical solution for a hydrogen atom. The electron distribution around a nucleus can be described by the radial distribution function $4\pi r^2\psi^2$, which is a measure of the probability of finding the electron in an orbital at a certain distance r from the nucleus. For a hydrogen atom in ground state (the most stable state of an atom, with the lowest permitted energy), this function when plotted against r passes through a maximum (Fig. 2.4) that can be shown for a spherical solution ($n=1$) to be identical in magnitude to the radius permitted by Bohr's model. Thus, there is a striking correlation between the results of the two treatments despite a vast difference in the physical significance.

The Schrödinger equation has been solved exactly only for one-electron systems such as the hydrogen atom and the ions He^+ and Li^{2+}. Even in a two-electron system such as helium, the repulsion between the two electrons makes the potential energy term V very complicated and requires simplifying assumptions to solve the equation. However, experimental observations justify the extrapolation of the one-orbital results to bigger, multi-orbital atoms (Companion, 1964).

2.2 The working model

For our purpose, the atom (also referred to as *nuclide*), which is about 10^{-8} cm in diameter, may be considered to be composed of three elementary particles: neutron, proton, and electron. Each electron carries one unit of negative charge and has a mass so small that it can be ignored for simplicity. Each proton carries one unit of positive charge (measured in terms of the charge of the electron as the unit) and one unit of mass. Each neutron carries one unit of mass but no charge. The number of neutrons (N) in the nucleus, the *neutron number*, affects the atomic mass but has no direct bearing on chemical properties of the element. The atom consists of a nucleus (about 10^{-13} cm in diameter), which contains all its positive charge and practically all of its mass, and negatively charged electrons that orbit around the nucleus (Fig. 2.5). The atom in ground state is electrically neutral, so that the number of protons is balanced by an equal number of electrons. Each element is uniquely identified by its *atomic number* (Z), the number of protons in its nucleus. The *mass number* or *atomic weight* (A) of an element is defined as $A=Z+N$, and the chemical notation for an element includes both its Z and A numbers. In reality, the mass of a nuclide is slightly less than the combined mass of its neutrons and protons. The "missing" mass is expressed as the nuclear binding energy, which represents the amount of energy required to break up the nucleus into its constituent *nucleons* (protons and neutrons). The mass number is not unique in the sense that the same element may have different

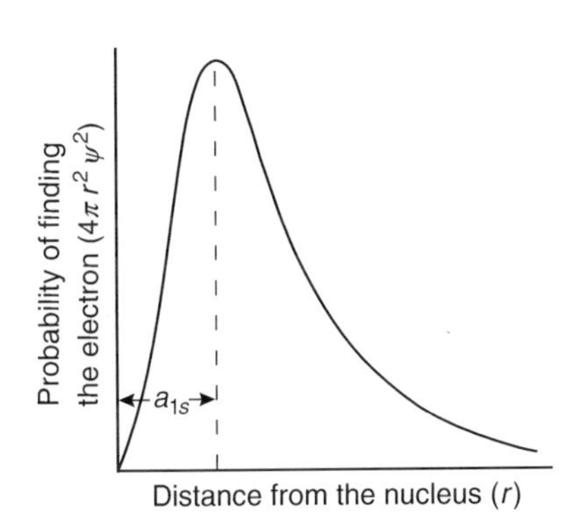

Fig. 2.4 Variation of the radial distribution function for the ground state hydrogen atom with increasing distance (r) from the nucleus. The magnitude of the function at a given value of r represents the electron density in a thin spherical orbital at distance r from the nucleus. In this case, the maximum probability occurs at a distance a_{1s} that can be shown to be exactly the same as the radius of the smallest orbit ($n=1$) permitted by the Bohr model.

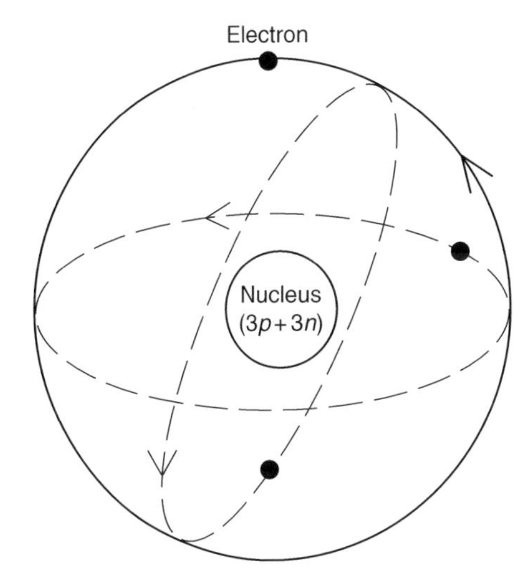

Fig. 2.5 Schematic representation of the Bohr-type model of a lithium atom (Z=3), which has a centrally located nucleus consisting of three protons (p) and three neutrons (n), and three electrons that orbit around the nucleus. Not to scale. The electronic structure of the ^3Li atom is represented as $1s^22s^1$.

Table 2.1 Isotopes, isobars, and isotones.

	Relationship among Z, N, and A	Examples
Isotopes	Same Z, different N and A	${}^{16}_{8}O$, ${}^{17}_{8}O$, ${}^{18}_{8}O$ (Z=8)
Isobars	Same A, different N and Z	${}^{12}_{7}N$, ${}^{12}_{6}C$, ${}^{12}_{5}B$, ${}^{12}_{4}Be$ (A=12)
Isotones	Same N, different A and Z	${}^{16}_{6}C$, ${}^{17}_{7}N$, ${}^{18}_{8}O$, ${}^{19}_{9}F$ (N=10)

values of N, and different elements may have the same value of A. In terms of Z, N, and A, atoms are classified as *isotopes*, *isobars*, and *isotones* (Table 2.1).

The chemical notation for an element includes both its Z and A numbers. For example, ${}^{238}_{92}U$ denotes an isotope of uranium having Z=92 and A=238, whereas ${}^{235}_{92}U$ denotes another isotope of uranium with the same Z but with A=235. Evidently, the nucleus of ${}^{238}U$ contains three more neutrons than that of ${}^{235}U$, and the two isotopes respond quite differently in nuclear reactions involving their nuclei. The characteristic chemical properties of the isotopes of an element, however, are essentially the same because they have the same number and distribution of electrons in their atoms. This is the reason why the ratio of various isotopes in a mass of a naturally occurring element is nearly fixed. The atomic weight (or mass number) of an element is the sum of the masses of its naturally occurring isotopes weighted by the fractional abundance of each isotope, and it is expressed in atomic mass units. The *atomic mass unit* (amu) is defined as one-twelfth of the mass of the carbon isotope ${}^{12}_{6}C$, which is arbitrarily fixed at 12.000, and the atomic weights of all other elements are obtained by comparison to the mass of ${}^{12}_{6}C$. On this scale, the atomic weight of hydrogen (H) is 1.00794 amu (usually approximated as 1 amu), that of Na atom is 22.989768amu, and that of Mg is 24.3050 amu. Thus, a Na atom has nearly 23 times the mass of an H atom, and a Mg atom has nearly 24 times the mass of an H atom.

Example 2–1: Calculation of atomic weights from abundances of naturally occurring isotopes

Calculate the atomic weights (in amu) of the elements K and Ar from the given data on the abundances of their isotopes:

K			Ar		
Isotope	Mass (amu)	Abundance (atom%)	Isotope	Mass (amu)	Abundance (atom%)
${}^{39}K$	38.9637074	93.2581	${}^{36}Ar$	35.96754552	0.3365
${}^{40}K$	39.9639992	0.01167	${}^{38}Ar$	37.9627325	0.0632
${}^{41}K$	40.9618254	6.7302	${}^{40}Ar$	39.9623837	99.6003

$$\text{Atomic weight of K} = (38.9637074 \times 0.932581) + (39.9639992 \times 0.0001167) + (40.9618254 \times 0.067302) = 39.0983$$

$$\text{Atomic weight of Ar} = (35.96754552 \times 0.003365) + (37.9627325 \times 0.000632) + (39.9623837 \times 00.9960003) = 39.9477$$

The *gram-atomic weight* of an element is numerically equal to its atomic weight expressed in grams. Similarly, the *gram-molecular weight* or the *gram-formula weight* (gfw) of a compound, which is the sum of the gram-atomic weights of its constituent atoms, is numerically equal to its molecular weight or formula weight expressed in grams. Both the gram-atomic weight and gram-molecular weight are referred to as *mole* (mol). In other words, the *molar mass* of an element (or a compound) expressed in units of (g mol^{-1}) is numerically the same as the atomic weight of the element (or molecular weight of the compound) in amu. The mole is defined as *the amount of substance that contains as many entities (atoms, molecules, ions, or other particles) as there are atoms in exactly 0.012 kg of pure carbon-12 atoms*. The currently accepted value is: 1 mole=6.0221367×10^{23} entities. This number, which is often rounded off to 6.022×10^{23}, is known as *Avogadro's number* in honor of Amedeo Avogadro (1776–1856). It is a consequence of the hypothesis he proposed in 1811 that equal volumes of all gases at the same temperature and pressure contain the same number of molecules. (The molar volume of an ideal gas is taken to be 22.414 L per mole at *standard temperature and pressure*, 273.15 K and 1 atm). The number of moles of an element or a compound in a given mass of an element (or a compound) is calculated by dividing the mass by the corresponding molar mass (see Chapter 1).

2.2.1 *Quantum numbers*

As stated earlier, a variable that is allowed by the properties of the system to have only certain discrete values is said to be quantized. The integer that enumerates these permitted values is called a *quantum number*. To describe the motion of an electron in the space around the nucleus of an atom, solution of the Schrödinger equation requires three characteristic interrelated quantum numbers: the *principal quantum number* (n), the *azimuthal* (or *subsidiary*, or *angular momentum) quantum number* (l), and the *magnetic quantum number* (m_l) (Table 2.2).

Table 2.2 Some principal quantum numbers and corresponding electron orbitals.

Principal quantum number, n	Maximum permissible electrons for given n ($2n^2$)	Azimuthal quantum number, l (l=0, 1, 2, 3, ... n – 1)	Names of sub-shells	Magnetic quantum number, m_l (maximum number of orbitals=1, 3, 5, 7, ...2l+1)	Designation of atomic orbitals for electrons (see Fig. 2.6)
1 (*K* shell)	2	0	1*s*	0 [1]	1*s*
2 (*L* shell)	8	0	2*s*	0 [1]	2*s*
		1	2*p*	–1, 0, +1 [3]	$2p_x$, $2p_y$, $2p_z$
3 (*M* shell)	18	0	3*s*	0 [1]	3*s*
		1	3*p*	–1, 0, +1 [3]	$3p_x$, $3p_y$, $3p_z$
		2	3*d*	–2, –1, 0, +1, +2 [5]	$3d_{z^2}$, $3d_{x^2-y^2}$, $3d_{xy}$, $3d_{xz}$, $3d_{yz}$

The principal quantum number n, which is analogous to Bohr's principal quantum number, determines the size of the orbital and also governs the allowed energy levels in the atom; we may view n as defining the "shell" in which the orbitals will occur. The value of n must be an integer (e.g., 1, 2, 3, 4, ...) and the corresponding shells, from the nucleus outward, are sometimes designated as *K, L, M, N*, ... A shell of principal quantum number n can accommodate a maximum of $2n^2$ electrons. Thus, the innermost or *K* shell (n=1) can hold 2 electrons, the *L* shell (n=2) 8 electrons, the *M* shell (n=3) 18 electrons, and so on. Not all electrons in a given shell, however, are identical because those in a shell of principal quantum number n are distributed over n "subshells" of different energy levels characterized by an azimuthal quantum number l. The azimuthal quantum number l determines the shape of the subshell, and for a given value of n may assume any of the values 0, 1, 2, 3, 4, 5, ..., n – 1 (the maximum possible value), and the corresponding sub-shells are designated as *s, p, d, f, g, h*, ... (Table 2.2). Thus, an electron corresponding to n=1 and l=0 is symbolized as 1*s*, one corresponding to n =3 and l=1 as 3*p*, one corresponding to n = 3 and l=2 as 3*d*, and so on. Further, the number of electrons in a sub-shell may be indicated by a superscript attached to the symbol for the sub-shell. For example, $1s^2$ represents 2 electrons in the *s* sub-shell of the *K* shell, $2p^6$ represents 6 electrons in the *p* sub-shell of the *L* shell, and $3d^5$ represents 5 electrons in the *d* sub-shell of the *M* shell. This scheme provides a convenient way to represent the electronic configuration of an atom. For example, the electronic configuration of Ni (Z=28) is written as $1s^22s^22p^63s^23p^63d^84s^2$.

The electrons in each sub-shell are distributed in one or more *atomic orbitals* (AOs), the number of orbitals being determined by the magnetic quantum number m_l, which has no effect on the size or shape of the orbitals but is related to the orientation of an orbital in space. For a given n and l, there are 2l+1 different possible values of m_l: –l, 0, +l (although the actual numerical values allowed for m_l are not important for our purpose).

Now let us consider how the quantum number l determines the distribution of electron clouds around the nucleus (Fig. 2.6). For l=0, the number of possible values of m_l is 1 for all values of n, and the corresponding orbitals are *s* orbitals such as 1*s*, 2*s*, 3*s*, etc. Electron clouds in *s* orbitals have a spherical symmetry in the sense that there is the same probability of finding an electron at a given distance from the nucleus in any direction in space, and the sphere gets larger as

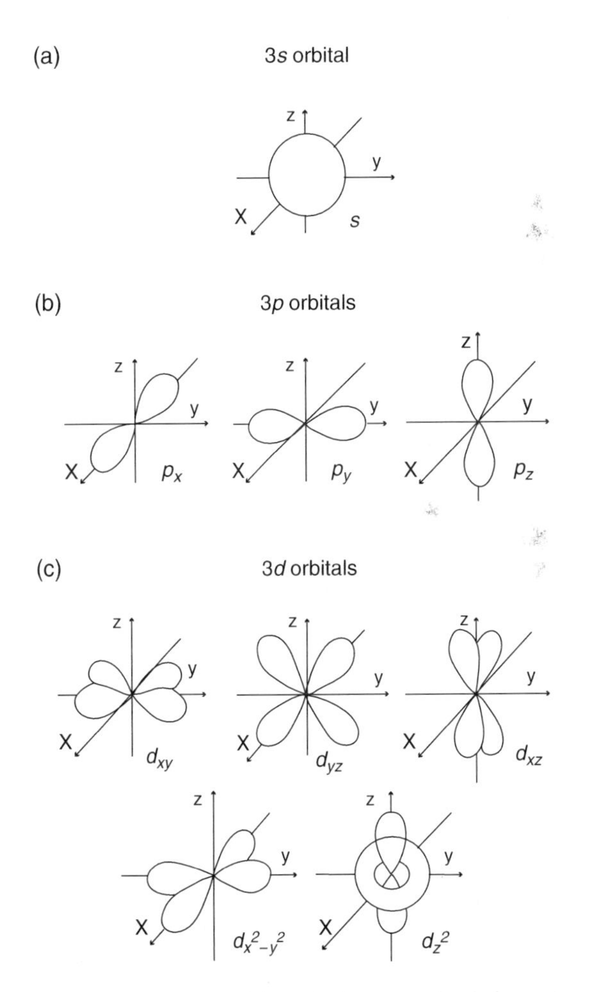

Fig. 2.6 Representation of the "shapes" of electron clouds for various types of atomic orbitals for n (principal quantum number)=3: (a) 3*s* orbital with spherical symmetry; (b) threefold degenerate 3*p* orbitals; and (c) fivefold degenerate 3*d* orbitals. Note that the boundary surfaces represent the regions likely to contain most (perhaps 95%) of the electrons (meaning that there is a 95% probability of finding the electrons within the region) and that the electron density is not the same everywhere within a given orbital lobe.

n gets larger. Thus, a 1s electron is more strongly attracted by the nucleus than a 2s electron because the former spends more time closer to the nucleus. When l=1, m_l (= 2l+1) has three possible values (m_l=−1, 0, or +1), giving rise to three spatially distinct but equally probable p orbitals, all having the same energy, for a given n (i.e., the p solutions of ψ are threefold degenerate). The three p orbitals are named p_x, p_y, and p_z to remind us that their dumbbell-shaped lobes of maximum electron density lie along arbitrarily defined x, y, and z orthogonal axes in space, respectively. An electron in p_x orbital, for example, exists (at least 95% of the time) somewhere in the dumbbell-shaped space along the x axis, with equal probability for each lobe. Unlike s orbitals, the p orbitals have a plane of zero electron density, a so-called *nodal plane* separating the lobes, a plane that is perpendicular to the long axis of the orbital and contains the nucleus. In the case of the p_z orbital, for example, it is the xy plane.

For l=2, m_l has five possible values (i.e., the d solutions of ψ are fivefold degenerate) and, therefore, there are five spatially distinct d orbitals of equal energy for a given n: d_{xy}, d_{yz}, d_{xz}, $d_{x^2-y^2}$, and d_{z^2}. The $d_{x^2-y^2}$ orbital lies in the xy plane with its four lobes coinciding with the x and y axes; d_{xy} also lies in the xy plane, but with its lobes pointed between the axes; d_{xz} and d_{yz} lie in the xz and yz planes, respectively, and like d_{xy} have their lobes of electron density pointed between the axes. The d_{z^2} orbital has a very different shape; most of its electron density is concentrated around the z-axis as shown in Fig. 2.6. The more complex shapes of f, g, and higher orbitals will not be discussed here because they are not particularly important in geochemistry.

For a discussion of the orbital distribution of electrons for various elements, we need to define one more quantum number, the *spin quantum number* (m_s). Even before the Schrödinger equation came into play, experimentalists had found that a great deal of spectroscopic data could be explained if it were postulated that the electron is able to spin in one of two possible directions about an arbitrary axis through its center. The spin quantum number describes the spin of an electron and the direction of the magnetic field produced by the spin. For every set of n, l, and m_l values, m_s can take the value of either +1/2 (α spin, commonly denoted by the symbol ↑) or −1/2 (β spin, commonly denoted by the symbol ↓) depending on the direction of the spin of the electron. The existence of spin requires us to consider another postulate known as the *Pauli's exclusion principle*, which states that any orbital (defined by a set of n, l, m_l) can accommodate a maximum of two electrons, and then only if they have spins in opposite directions ($\alpha\beta$ pair, ↑↓). Thus, the quantum numbers n, l, m_l, and m_s uniquely define the state of an electron in an atom; two or more electrons cannot exist in the same state at the same time.

2.2.2 *Energy levels of the atomic orbitals*

The energy of an electron is determined by how strongly it is attracted by the nucleus (i.e., by the closeness of its orbital to the nucleus). It is, therefore, necessary to consider the relative energy levels of the different atomic orbitals (Fig. 2.7), which have been determined from a combination of the wave equation and study of atomic spectra. An examination of Fig. 2.7 leads to the following general conclusions (Evans, 1966):

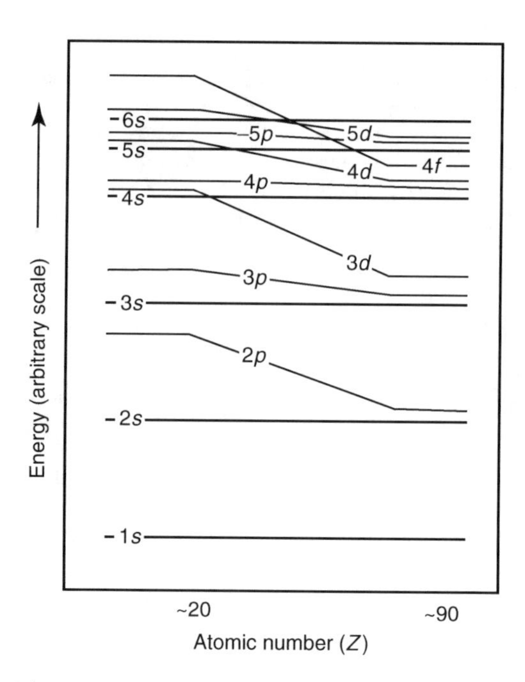

Fig. 2.7 Schematic representation of the variations in the energy levels of atomic orbitals in the ground state of atoms as a function of atomic number. (From *An Introduction to Crystal Chemistry*, 2nd edition, by R.C. Evans, Figure 2.02, p. 19; Copyright 1966, Cambridge University Press. Reproduced with permission of the publisher.)

(1) For a given n and l, the energy level is the same for all the possible atomic orbitals. Thus, for n=2, the three 2p orbitals have the same energy; for n=3, the three 3p orbitals have the same energy and so do the five 3d orbitals.

(2) For a given l, the orbital energy for any value of atomic number (Z) increases with increasing n. Thus, the sequences of increasing energy are:
1s < 2s < 3s < 4s < 5s < 6s < 7s
2p < 3p < 4p < 5p < 6p
3d < 4d < 5d < 6d
4f < 5f

(3) For a given n and any value of Z less than about 20, orbital energy increases with increasing l. Thus, the sequences of increasing energy are:
2s < 2p
3s < 3p < 3d
4s < 4p < 4d < 4f, etc.
For Z > 20, the energy level of the 4s orbitals is lower than that of the three 3d orbitals, and the relationships between 5s and 4d orbitals and between 6s and 5d orbitals are similar.

Table 2.3 The electronic configuration of the elements of the first three periods.

Element	Z	K-shell (n=1)	L-shell (n=2)				M-shell (n=3)				Notation for electronic configuration
		$1s$	$2s$	$2p_x$	$2p_y$	$2p_z$	$3s$	$3p_x$	$3p_y$	$3p_z$	
H	1	↑									$1s^1$
He	2	↑↓									$1s^2$
Li	3	↑↓	↑								$1s^2 2s^1$
Be	4	↑↓	↑								$1s^2 2s^2$
B	5	↑↓	↑	↑							$1s^2 2s^2 2p^1$
C	6	↑↓	↑↓	↑	↑						$1s^2 2s^2 2p^2$
N	7	↑↓	↑↓	↑	↑	↑					$1s^2 2s^2 2p^3$
O	8	↑↓	↑↓	↑↓	↑	↑					$1s^2 2s^2 2p^4$
F	9	↑↓	↑↓	↑↓	↑↓	↑					$1s^2 2s^2 2p^5$
Ne	10	↑↓	↑↓	↑↓	↑↓	↑↓					$1s^2 2s^2 2p^6$
Na	11	↑↓	↑↓	↑↓	↑↓	↑↓	↑				$1s^2 2s^2 2p^6 3s^1$
Mg	12	↑↓	↑↓	↑↓	↑↓	↑↓	↑				$1s^2 2s^2 2p^6 3s^2$
Al	13	↑↓	↑↓	↑↓	↑↓	↑↓	↑	↑			$1s^2 2s^2 2p^6 3s^2 3p^1$
Si	14	↑↓	↑↓	↑↓	↑↓	↑↓	↑↓	↑	↑		$1s^2 2s^2 2p^6 3s^2 3p^2$
P	15	↑↓	↑↓	↑↓	↑↓	↑↓	↑↓	↑	↑	↑	$1s^2 2s^2 2p^6 3s^2 3p^3$
S	16	↑↓	↑↓	↑↓	↑↓	↑↓	↑↓	↑↓	↑	↑	$1s^2 2s^2 2p^6 3s^2 3p^4$
Cl	17	↑↓	↑↓	↑↓	↑↓	↑↓	↑↓	↑↓	↑↓	↑	$1s^2 2s^2 2p^6 3s^2 3p^5$
Ar	18	↑↓	↑↓	↑↓	↑↓	↑↓	↑↓	↑↓	↑↓	↑↓	$1s^2 2s^2 2p^6 3s^2 3p^6$

2.3 The ground state electron configuration of elements

The *ground state* of an isolated atom is its quantum state of lowest permissible energy. The distribution of electrons among various electron "shells" of an atom corresponding to the quantum state of minimum energy is called the *ground state electron configuration*. An atom is said to be in an *excited state* if it has a higher energy than the ground state because of one or more of its electrons occupying one or more "shells" of higher energy compared to the ground state.

2.3.1 *Filling atomic orbitals with electrons: the Aufbau principle*

The single electron in an H atom (Z=1) enters the orbital with the lowest energy ($1s^1$). Subject to the Pauli's exclusion principle, the electronic configuration of atoms of increasing atomic number is constructed by adding appropriate number of electrons (depending on the atomic number) to the possible orbitals in a way that minimizes the energy of the atom, because the lowest energy state of an atom is its most stable (or ground) state. This filling-up procedure is called the *Aufbau* (meaning "build-up") *principle*, which is governed by the following set of guidelines (Companion, 1964; Evans, 1966):

(1) Because of unsystematic variations in energy levels of atomic orbitals as a function of quantum numbers n and l (Fig. 2.7), (i) electrons are assigned to orbitals in order of increasing value of $(n+l)$, and (ii) for subshells with the same value of $(n+l)$, electrons are assigned first to the sub-shell with lower n. For example, the $5s$ orbital $(n+l=5+0=5)$ would fill before the $4d$ orbital $(n+l=4+2 = 6)$, and the $4d$ orbital before the $5p$ orbital $(n+l=5+1=6)$ because $4d$ has a lower value of n. The sequence in which orbitals are filled is:

$1s, 2s, 2p, 3s, 3p, 4s, 3d, 4p, 5s, 4d, 5p, 6s, 4f, 5d, 6p, 7s, 6d$

(2) No two electrons may have identical sets of the four quantum numbers (Pauli's exclusion principle).

(3) As many of the orbitals as possible are occupied by a single electron before any pairing of electrons takes place (see Table 2.3); the unpaired electrons have parallel spins, and the paired electrons have opposite spins (*Hund's rule of maximum multiplicity*). This is because, even with pairing of spins, two electrons that are in the same orbital repel each other more strongly than two electrons in different orbitals.

The rules listed above should be considered only as a guide to predicting electron distribution in atoms. The experimentally determined electron configurations of lowest total energy do not always match those predicted by these guidelines, especially for the B group elements of the Periodic Table (see section 2.3.2).

The electronic configurations of isolated atoms in the ground state are presented in Appendix 2. Alternative electronic configurations may have to be considered for atoms in excited states or when they are not isolated (e.g., when involved in chemical reactions).

2.3.2 *The Periodic Table*

The *Periodic Table* reflects an attempt at a systematic organization of the elements from the perspective of their atomic structures. In 1869, two very similar arrangements of the known elements, much like the modern Periodic Table (Fig. 2.8), were published independently, one by the Russian chemist Dimitri Ivanovich Mendeleev (1834–1907) and the other by the German chemist Lothar Meyer (1830–1895). Both were based on regular periodic repetition of properties with increasing atomic weight of the elements, Mendeleev's largely on chemical properties and Meyer's on physical properties. The modern version of the Periodic Table is organized on the basis of atomic number of the elements, a concept that was developed some 50 years later than Mendeleev's work and is more fundamental to the identity of each element.

The vertical columns in the Periodic Table are referred to as *groups*, and the horizontal rows are referred to as *periods*, which are numbered in accordance with the first quantum number of the orbitals that are being filled with increasing atomic number. For example, electrons fill the 1*s* orbital in elements belonging to the 1st period, 2*s* and 2*p* orbitals in elements of the 2nd period, and 3*s* and 3*p* orbitals in elements of the 3rd period. The electron distribution in elements included in the 4th to 7th periods becomes more complicated because of filling of *d* and *f* orbitals (3*d*, 4*d*, 5*d*, 6*d*, 4*f*, 5*f*) as illustrated in Fig. 2.8. Elements within a period have properties that change progressively with increasing atomic number because of addition of electrons. In contrast, elements in any group have similar physical and chemical properties because of similar electronic configuration. The groups are either designated as A and B, and numbered from left to right in accordance with the highest possible positive valence of the elements in that group (American system) or numbered continuously from left to right as 1 through 18 (International system). Some of the groups are commonly referred to by special names: alkali metals (Group IA, except H); alkaline earth metals (Group IIA); halogens (Group VIIA), and noble (or inert) gases (Group VIIIA).

The Periodic Table can also be viewed as a systematic representation of the electronic configurations of the elements. The elements are arranged in blocks based on the kinds of atomic orbitals (*s*, *p*, *d*, or *f*) being filled. The A groups comprise elements in which *s* and *p* orbitals are being filled. The B groups include elements in which the *s* orbital of the outermost occupied shell contains one or two electrons (e.g., $5s^1$ and $5s^2$ for ^{37}Rb and ^{38}Sr, respectively), and the *d* orbitals, one shell smaller, are being filled. The electronic configurations of the A group elements, including the noble gases (Group VIIIA), are quite systematic and can be predicted from their positions in the Periodic Table, but some pronounced irregularities exist for elements of the B group and of the 5th and higher periods.

2.3.3 *Transition elements*

Application of the guidelines discussed earlier to the filling of successive atomic orbitals with electrons is quite straightforward for atoms from ^{2}He($1s^2$) to ^{18}Ar ($1s^2 2s^2 2p^6 3s^2 3p^6$), as illustrated in Table 2.3. The *M*-shell, which can contain up to 18 electrons, has room for 10 more electrons in the five 3*d* orbitals, but in ^{19}K [(Ar core)$^{18}4s^1$] and ^{20}Ca [(Ar core)$^{18}4s^2$], the additional electrons are accommodated in the energetically more favorable 4*s* orbital of the *N*-shell compared with the 3*d* orbital (Fig. 2.7). After ^{20}Ca, the 3*d* orbital is more stable than the 4*s* orbital, so that from ^{21}Sc to ^{30}Zn, electrons enter the 3*d* orbital of the *M*-shell in preference to the 4*p* orbital of the *N*-shell. The 3*d* orbitals become completely filled in ^{30}Zn, and filling of the 4*p* orbitals starts with the element ^{31}Ga and continues progressively to the element ^{36}Kr. The elements from ^{21}Sc to ^{30}Zn are called the *3d transition elements* or the *first transition series*. Analogous schemes of filling the *d* orbitals give rise to *4d transition series* (^{39}Y through ^{48}Cd), *5d transition series* (^{57}La and ^{72}Hf through ^{80}Hg), and *6d transition series* (^{89}Ac and ^{104}Rf through element 112) (see Fig. 2.8). The elements of these four transition series are all metals, and they contain electrons in both *ns* and $(n-1)d$ orbitals, but not in the *np* orbitals. Two additional transition series exist between groups IIIB and IVB of the Periodic Table: the *4f transition series* (^{58}Ce through ^{71}Li) and the *5f transition series* (^{90}Th through ^{103}Lr). These transition elements are also metals, and are characterized by the progressive filling of 4*f* and 5*f* orbitals, respectively (Fig. 2.8).

2.4 Chemical behavior of elements

The chemical behavior of an element is governed by its electronic configuration because the energy level of the atom is determined by the spatial distribution of its electron cloud. It is only the most loosely bound electrons in the outermost orbitals that take part in chemical interaction with other atoms. For example, the alkali elements (group IA of the Periodic Table), all of which have one electron in the outermost orbital, exhibit similar chemical properties; so do the alkaline earth metals (group IIA of the Periodic Table), all of which have two electrons in the outermost orbital.

2.4.1 *Ionization potential and electron affinity*

Two concepts are useful in predicting the chemical behavior of elements: *ionization potential (or ionization energy)*; and *electron affinity*. Ions are produced by the removal of electron(s) from or the addition of electron(s) to a neutral atom. *The energy that must be supplied to a neutral atom (M) in the gas phase to remove an electron to an infinite distance is called the ionization potential (I).* In other words, the ionization potential is the difference in potential between the initial state, in which the electron is bound, and the final state, in which it is at rest at infinity; the lower the ionization potential, the easier it is to convert the atom into a cation. This is the reason why the ionization potential generally increases from left to right in a given period and from top to bottom within a given group of the Periodic Table (Fig. 2.9). The *first ionization potential* (I_1) refers to the energy required to remove the first (the least tightly bound) electron, the

PERIODIC TABLE

Group / Period	1A (1)	IIA (2)	IIIB (3)	IVB (4)	VB (5)	VIB (6)	VIIB (7)	VIIIB (8)	VIIIB (9)	VIIIB (10)	1B (11)	IIB (12)	IIIA (13)	IVA (14)	VA (15)	VIA (16)	VIIA (17)	VIIIA (18)
1	1 H																	2 He
	1s																	**1s**
2	3 Li	4 Be											5 B	6 C	7 N	8 O	9 F	10 Ne
	2s												**2p**					
3	11 Na	12 Mg											13 Al	14 Si	15 P	16 S	17 Cl	18 Ar
	3s												**3p**					
4	19 K	20 Ca	21 Sc	22 Ti	23 V	24 Cr	25 Mn	26 Fe	27 Co	28 Ni	29 Cu	30 Zn	31 Ga	32 Ge	33 As	34 Se	35 Br	36 Kr
	4s		**3d** (3*d* transition elements)										**4p**					
5	37 Rb	38 Sr	39 Y	40 Zr	41 Nb	42 Mo	43 Te	44 Ru	45 Rh	46 Pd	47 Ag	48 Cd	49 In	50 Sn	51 Sb	52 Te	53 I	54 Xe
	5s		**4d** (4*d* transition elements)										**6p**					
6	55 Cs	56 Ba	57 La	72 Hf	73 Ta	74 W	75 Re	76 Os	77 Ir	78 Pt	79 Au	80 Hg	81 Tl	82 Pb	83 Bi	84 Po	85 At	86 Rn
	6s		**5d** (5*d* transition elements)										**6p**					
7	87 Fr	88 Ra	89 Ac	104 Rf	105 Db	106 Sg	107 Bh	108 Hs	109 Mt	110	111	112						
	7s		**6d** (6*d* transition elements)															

6	58 Ce	59 Pr	60 Nd	61 Pm	62 Sm	63 Eu	64 Gd	65 Tb	66 Dy	67 Ho	68 Er	69 Tm	70 Yb	71 Lu
	4f (4*f* transition elements) — Lanthanide series													
7	90 Th	91 Pa	92 U	93 Np	94 Pu	95 Am	96 Cm	97 Bk	98 Cf	99 Es	100 Fm	101 Md	102 No	103 Lr
	5f (5*f* transition elements) — Actinide series													

Fig. 2.8 The Periodic Table showing the filling of atomic orbitals of elements with increasing atomic number. The electron configurations of the A group elements, including the noble gases (group VIIIA), are quite systematic and can be predicted from their positions in the periodic table, but some pronounced irregularities exist for B group elements of the fifth and higher periods. The proposed names for elements with atomic numbers 104 to 109 – dubnium (Db), joliotium (Jl), rutherfordium (Rf), bohrium (Bh), hahnium (Hn), and meitnerium (Mt) – have not yet been formally approved.

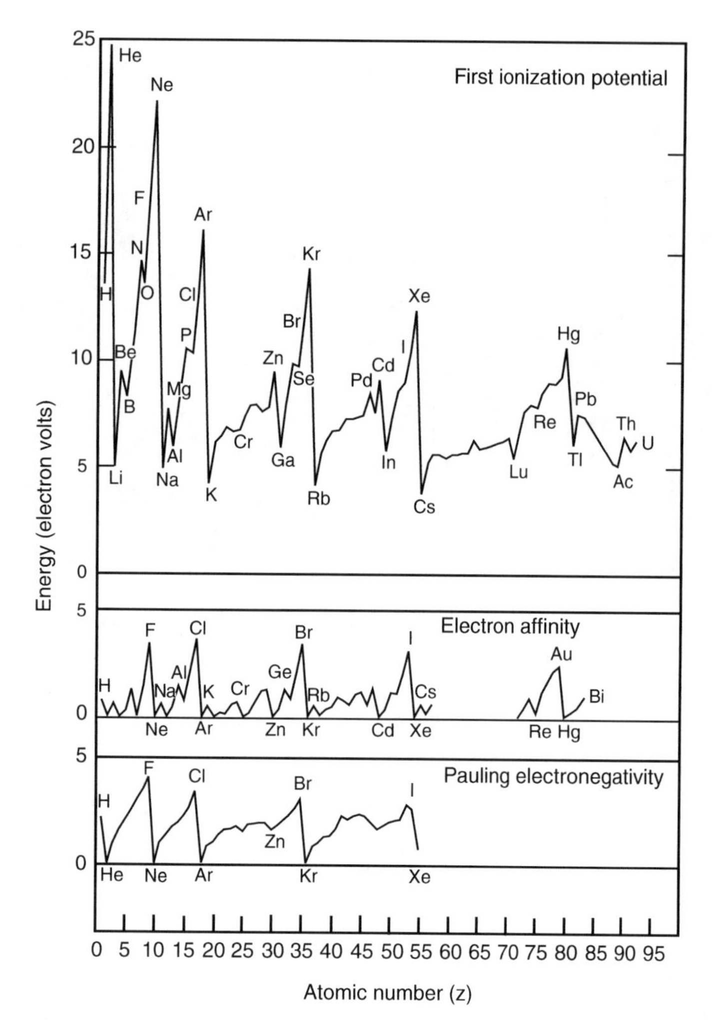

Fig. 2.9 Variation of first ionization potential, electron affinity, and Pauling electronegativity (see section 3.6.5) of atoms with increasing atomic number. Note that the noble gases (He, Ne, etc.) appear at peak values of first ionization potential, reflecting their chemical inertness, whereas the alkali metals (Li, Na, K, Rb, Cs) appear at minimum values, consistent with their reactivity and ease of cation formation. The peak values of electron affinity, on the other hand, are marked by halogens (F, Cl, Br, I) and the minimum values by the noble gases. $1\,eV = 96.48532\,kJ\,mol^{-1}$. (Source of data: compilations in Lide, 1998.)

second ionization potential (I_2) to the energy corresponding to the removal of the second electron, and so on:

$$M \Rightarrow M^+ + e^- \quad (I_1) \tag{2.10}$$

$$M^+ \Rightarrow M^{2+} + e^- \quad (I_2) \tag{2.11}$$

Electron affinity (E_{ea}) *is defined as the energy required for detaching an electron from a singly charged anion* (Lide, 2001):

$$M^- \Rightarrow M + e^- \quad (E_{ea}) \tag{2.12}$$

In other words, electron affinity is the energy difference between the lowest (ground) state of a neutral atom and the lowest state of its corresponding negative ion. Note that it is not exactly the reverse of the ionization process. The sign convention for E_{ea} is opposite to most thermodynamic quantities (see Chapter 4); a positive E_{ea} indicates that energy is released when the electron goes from an atom to an anion. All atoms have positive values of E_{ea}; a high value of E_{ea} indicates strong attraction for extra electrons. Electron affinity is a precise quantitative term like ionization potential, but it is difficult to measure. Values of I_1 and E_{ea} for selected elements are listed in Appendix 3, and how they vary with atomic number is presented graphically in Fig. 2.9.

Why is there so much variation in ionization potential and electron affinity of elements? The explanation lies in the "*screening effect*" (or "*shielding effect*") of the electrons. For example, consider the Na atom ($1s^22s^22p^63s^1$). The difficulty of removing the outer 3*s* electron of the Na atom is due to the electrostatic attraction of the positive nucleus. However, the effective nuclear charge (Z_{eff}) is somewhat less than what would be exerted by the 11+ nuclear charge because of the electrostatic repulsion or shielding effect ($S_{electron\ screening}$) of the 10 inner-shell electrons on the 3*s* electron. For an atom with a known distribution of the inner electrons, it is possible to calculate ($S_{electron\ screening}$) and, thus, Z_{eff} ($Z_{eff} = Z_{atomic\ number} - S_{electron\ screening}$) (Fyfe, 1964).

2.4.2 *Classification of elements*

On the basis of first ionization potentials and electron affinities, the elements may be separated into three classes (Table 2.4):

(1) *Electron donors* (*metals*), such as elements of Groups 1A and IIA in the Periodic Table, which easily lose one or more *valence electrons* (electrons in the outermost occupied *s* and *p* orbitals) and become positively charged ions (*cations*); these elements have relatively low values of I_1 and E_{ea}. For any period in the Periodic Table, the metallic character of the elements generally decreases with increasing atomic number (i.e., with progressive filling of the atomic orbitals), as the first ionization potential increases in the same direction.
(2) *Electron acceptors* (*nonmetals*), such as elements of Groups VIA and VIIA in the Periodic Table, which readily acquire one or more added electrons and become negatively charged ions (*anions*); these elements are characterized by relatively high values of I_1 and E_{ea}. For any period, E_{ea} generally increases with increasing atomic number.
(3) *Noble elements* (inert gases), such as the elements of Group VIIIA in the Periodic Table, which do not easily lose or gain electrons. These elements are characterized by a complete octet of electrons in their outermost *s* and *p* orbitals (ns^2np^6) (except for ^2He, which has only 2 electrons), very

Table 2.4 Examples of the three classes of elements in terms of ionization potentials and electron affinities.

Element type	Example	Electron distribution 1s 2s 2p 3s	First ionization potential* (eV)	Second ionization potential* (eV)	Electron affinity[1] (eV)
Electron acceptors	F (Z=9)	↑↓ ↑↓ ↑↓ ↑↓ ↑	17.42	34.97	3.40
Noble elements	Ne (Z=10)	↑↓ ↑↓ ↑↓ ↑↓ ↑↓	21.56	40.96	~0
Electron donors	Na (Z=11)	↑↓ ↑↓ ↑↓ ↑↓ ↑↓ ↑	5.14	47.29	0.55

[1]Source of data: Lide (1998).
The values of these parameters are often expressed in molar equivalent of electronvolt (eV), which is the kinetic energy that would be gained by a mole of electrons passing through a potential difference of one volt. 1 eV = 96.48532 kJ mol^{-1}.

high values of I_1, and near-zero values of E_{ea}; they do not normally occur in the ionized state. By losing or gaining electrons, the cation- and anion-forming elements achieve the stable noble element configurations characterized by 8 electrons in their outermost shells (the *octet rule*).

To the above list, we may add a fourth class of elements (such as B, Si, Ge, As, Sb, Te, Po, and At), called *metalloids* (or *semimetals*), which show certain properties that are characteristic of metals and other properties that are characteristic of nonmetals. Many of the metalloids (especially Si and Ge) are used as *semiconductors* in solid-state electronic circuits. Whereas metals become less conductive with increasing temperature, semiconductors are insulators at low temperatures, but become conductors at higher temperatures.

2.5 Summary

1. The atom consists of a nucleus (about 10^{-13} cm in diameter), which contains all its positive charge in the form of protons and practically all of its mass (neutrons and protons), and negatively charged electrons that orbit around the nucleus. Each element is uniquely identified by its atomic number (Z), the number of protons in its nucleus. The mass number or atomic weight (A) of an element is the sum of its protons (Z) and neutrons (N). Atoms with the same atomic number but different mass numbers are called isotopes.
2. The state of an electron in the structure of an atom can be described by four quantum numbers, not all of which can be the same for any two electrons: the principal quantum number (n), which determines the size of the orbital; the azimuthal quantum number (l), which determines the shape of the orbital for any given value of n; the magnetic quantum number (m_l), which determines the number of atomic orbitals ($2l+1$) accommodating the electrons of the orbital; and the spin quantum number (m_s), which can take the value of +1/2 (α spin, commonly denoted by the symbol ↑) or −1/2 (β spin, commonly denoted by the symbol ↓) depending on the direction of the spin of the electron. An atomic orbital (defined by a set of n, l, m_l) can accommodate a maximum of two electrons, and then only if they have spins in opposite directions ($\alpha\beta$ pair, ↑↓), a postulate known as Pauli's exclusion principle.
3. The electronic configuration of atoms of increasing atomic number is constructed by adding the appropriate number of electrons (depending on the atomic number) to the possible orbitals in a way that minimizes the energy of the atom.
4. The Periodic Table can be viewed as a systematic representation of the electronic configurations of the elements.
5. Two concepts useful in predicting the chemical behavior of elements are: ionization potential (I), the energy that must be supplied to a neutral atom (M) in the gas phase to remove an electron to an infinite distance ($M \Rightarrow M^+ + e^-$); and electron affinity (E_{ea}), the energy required to detach an electron from a singly charged anion ($M^- \Rightarrow M + e^-$). On the basis of these two parameters, the elements may be divided into three broad categories: electron donors (metals); electron acceptors (nonmetals), and noble elements (inert gases).

2.6 Recapitulation

Terms and concepts

Anions
Aufbau principle
Avogadro's number
Azimuthal quantum number
Bohr radius
Cations
Electron affinity
Electron acceptors (nonmetals)
Electron donors (metals)
Excited state

Ground state
Hund's rule
Ionization potential
Inert elements
Isobars
Isotones
Isotopes
Magnetic quantum number
Metalloids (semimetals)
Molar mass
Mole
Nodal plane
Octet rule
Pauli's exclusion principle
Periodic Table
Principal quantum number
Schrödinger equation
Screening effect
Spin quantum number
Standard temperature and pressure (STP)
Transition elements
Valence electrons

Computation techniques

- The atomic weight (in amu) of an element from the data on the abundances of its isotopes.
- Ionization potential, electron affinity.

2.7 Questions

1. Show that 1 amu = 1.6606×10^{-24} g
   [Hint: 1 amu = 1/12 of the mass of a $^{12}_{6}C$ atom; 1 mole of $^{12}_{6}C$ contains 6.022×10^{23} atoms.]
2. Calculate the atomic weights (in amu) of the elements U and Pb from the abundances of their isotopes given below:

U			Pb		
Isotope	Mass (amu)	Abundance (atom %)	Isotope	Mass (amu)	Abundance (atom%)
^{234}U	234.0409468	0.0055	^{204}Pb	203.973020	1.4
^{234}U	235.0439242	0.7200	^{204}Pb	205.974440	24.1
^{234}U	238.0507847	99.2745	^{204}Pb	206.975872	22.1
			^{204}Pb	207.976627	52.4

3. A saturated solution of AgCl in water contains 1.3×10^{-5} mole of AgCl per liter of the solution at 25°C temperature and 1 atm pressure. Calculate the mass of dissolved AgCl in 1 L of the solution. Gram atomic weights of the elements: Ag = 107.87; Cl = 35.453.
4. Calculate the number of H_2O molecules that will evaporate per second, given that one drop of water weighing 0.05 g evaporates in 1 h. Avogadro's number = 6.022×10^{23}.
5. A geologist has identified two iron deposits. One of them contains 100 million tons of magnetite (Fe_3O_4), and the other contains 20 million tons of magnetite (Fe_3O_4), and 80 million tons of hematite (Fe_2O_3). Based on the iron content alone, which of the two deposits should be recommended for mining? Gram atomic weights of the elements: Fe = 55.85; O = 16.00.
6. What is the maximum number of electrons that can be accommodated in the following atomic orbitals?
   (a) all the 6*g* orbitals; (b) all the 7*s* orbitals;
   (c) all the 8*f* orbitals; and (d) all the orbitals with $n = 5$
7. Write down the ground–state electronic configuration of the following elements:
   Fe (Z=26); Rubidium (Z=37); Xenon (Z=54); and Uranium (Z=92).
8. Using the same format as for Table 2.3, prepare a table illustrating the gradual filling with electrons of the atomic orbitals of the 4*d* transition series.
9. Light near the middle of the ultraviolet region of the electromagnetic radiation spectrum has a frequency of $2.73\times10^{16}\,s^{-1}$, whereas yellow light within the visible spectrum has a frequency of $5.26\times10^{14}\,s^{-1}$.
   (a) Calculate the wavelength corresponding to each of these two frequencies of light.
   (b) Calculate how much more is the energy associated with a photon of ultraviolet light compared to that of yellow light.
10. What is the wavelength associated with a neutron of mass 1.675×10^{-24}g moving with a speed of 2360 m s^{-1}.

3 Chemical Bonding

The forces responsible for chemical combination of atoms are of two main types. There are those forces which arise between electrically charged species, repulsive if the charges are similar, attractive if dissimilar. ... The second type of force may be called an exchange force and these can be described in terms of the Schrödinger wave equation.... A little reflection will reveal that we understand neither – all we have is two ways of describing different situations, and in all probability the origin of both forces is the same.

Fyfe (1964)

Chemical compounds are formed by the combination of two or more atoms (or ions), and the formation of a stable compound occurs when the combination results in a lower energy than the total energy of the separated atoms (or ions). Interatomic (or interionic) net attractive forces that hold atoms (or ions) in solids together are called *chemical bonds*.

Chemical bonds usually involve only the *valence electrons* (s and p electrons in the outermost orbitals) of an atom. Physical and chemical properties of all substances depend on the character of the chemical bonds that hold them together.

Much of the bonding in solids of geochemical interest can be described in terms of two end-member types: (i) *ionic* (or *electrovalent*) *bonds* that exist because of electrostatic attraction between cations and anions formed by transfer of one or more electrons between atoms; and (ii) *covalent bonds* that arise because of sharing of electrons between atoms that results from overlap of orbitals from the two atoms. For example, the ionic bonding in NaCl results from the electrostatic attraction between a Na^+ cation formed by the loss of a valence electron from the $3s$ orbital of the Na atom ($1s^22s^22p^63s^1$) and a Cl^- anion formed by incorporation of that electron into the $3p$ orbital of the Cl atom ($1s^22s^22p^63s^23p^5$) (Fig. 3.1a):

$$Na\ (1s^22s^22p^63s^1) \Rightarrow Na^+\ (1s^22s^22p^6) + e^-$$
$$Cl\ (1s^22s^22p^63s^23p^5) + e^- \Rightarrow Cl^-\ (1s^22s^22p^63s^23p^6)$$
$$2Na_{(s)} + Cl_{2(g)} \Rightarrow 2Na^+Cl^-_{(s)}$$

The NaCl molecule itself is electrically neutral because its structure contains a Na^+ cation for every Cl^- anion. Sodium and chlorine atoms combine readily because of the large difference in their first ionization potential and electron affinity (see section 2.4). Covalent bonding, on the other hand, arises from sharing of electrons. The covalent bonding in Cl_2, for example, may be viewed as resulting from the sharing of a pair of electrons between two Cl atoms, each of which contributes one electron to the shared pair (Fig. 3.1b). Atoms can share one, two, or three electron pairs, forming, respectively, single, double, and triple covalent bonds. All bonds between atoms of different elements have some degree of both ionic and covalent character.

Compounds containing predominantly ionic bonding are called *ionic compounds*, and those that are held together mainly by covalent bonds are called *covalent compounds*. This difference in bonding accounts for the differences in some properties associated with simple ionic and covalent compounds (Table 3.1). Other types of bonds that will be discussed briefly in this chapter include *metallic bonds, Van der Waals bonds*, and *hydrogen bonds*.

Introduction to Geochemistry: Principles and Applications, First Edition. Kula C. Misra.

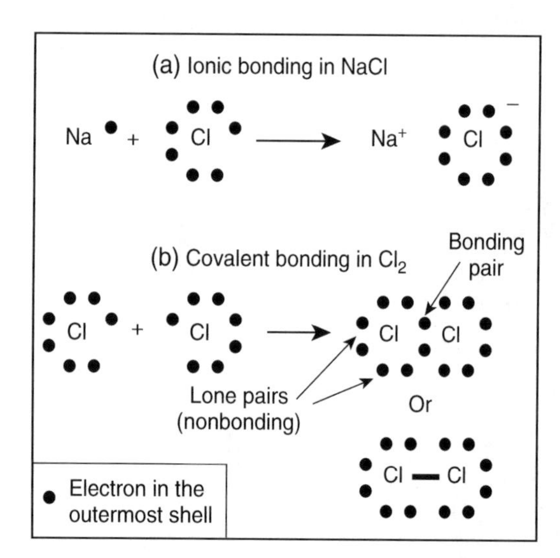

Fig. 3.1 Lewis dot representation of (a) ionic bonding (solid NaCl) and (b) covalent bonding (Cl_2 gas). In this kind of illustration, the chemical symbol of the element includes the inner complete shells of electrons, and the valence electrons (i.e., electrons in the outermost occupied *s* and *p* orbitals) are represented by dots. The single covalent bond in Cl_2 is represented by the two dots of the "bonding pair" or by a single line representing that pair; any pair of unshared electrons in the same orbital, which does not participate in the formation of covalent bonds, are referred to as a "lone pair."

3.1 Ionic bonding

3.1.1 *Ionic radii*

The potential energy of a system (E_p) comprised of two oppositely charged ions (e.g., Na^+ and Cl^-), each with its electron cloud around the nucleus, approaching each other is given by (Fyfe, 1964):

$$E_p = -\frac{e^2}{R} + \frac{be^2}{R^n} \quad (3.1)$$

where R is the interionic distance (i.e., distance between the centers of the two ions, which are assumed to be hard spheres), b is a constant, and n is an integer with values between 8 and 12. The term $-e^2/R$ represents the coulombic attraction between the opposite net charges ($\pm e$), and the term e^2/R^n arises out of the repulsion caused by interpenetration of the electron clouds and by the repulsion between the nuclei of the ions. As R gets smaller, the attraction term (which, by convention, is assigned a negative sign) becomes more negative, indicating lower potential energy and thus increased stability. The repulsion (which, by convention, is assigned a positive sign) contributes little to E_p for large values of R, but its contribution increases very rapidly when R becomes smaller than a critical value R_0, which is the equilibrium interionic distance at which the isolated ion pair is most stable (Fig. 3.2). When the two ions are separated by the distance R_0, we have a stable *ionic bond* formed between them. The value of R_0, the *bond length*, can be determined from the fact that it corresponds to the minimum value of E_p and occurs when $dE/dR = 0$. (When the cation is associated with more than one anion, as in a crystal, repulsion between the anions makes R_0 somewhat larger.) The curve for E_p in Fig. 3.2 is typical of all atomic–molecular systems, and it is the basis of the concepts of interionic distance and *ionic radius*, and the premise that to a first approximation ions have a more or less constant ionic size. Strictly speaking, the electron density distribution around a nucleus does not have a spherical symmetry, as implied by the term "ionic radius." A more appropriate term, according to Gibbs *et al.* (1992), is "bonded radius," which refers to the distance between the center of one atom to the point of minimum electron density in the direction between two nuclei. The outer extent of an atom in other directions is usually different because of a different distribution of electrons (i.e., the atom is not spherical). The bonded radius can be measured from electron distribution maps. However, we continue to rely on the concept of ionic radii because the approach has been quite successful in explaining most of the ionic crystal structures.

Table 3.1 Some properties of ionic and covalent compounds.

Property	Ionic compounds	Covalent compounds
Participating elements	Commonly between two elements with quite different electronegativities[1], usually a metal and a nonmetal	Commonly between two elements with similar electronegativities[1], usually nonmetals. Homonuclear molecules (such as Cl_2 comprised of only one element) are covalent
Melting point	They are solids with high melting points (typically > 400°C). Ionic compounds do not exist as gases in nature	They are gases, liquids, or solids with low melting points (typically < 300°C)
Solubility	Many are soluble in polar solvents such as water, and most are insoluble in nonpolar solvents such as carbon tetrachloride (CCl_4).	Many are insoluble in polar solvents, and most are soluble in nonpolar solvents such as carbon tetrachloride (CCl_4)
Electrical conductivity	Molten compounds and aqueous solutions are good conductors of electricity because they contain charged particles (ions)	Due to lack of charged particles, liquid and molten compounds do not conduct electricity, and aqueous solutions are usually poor conductors of electricity

[1]See section 3.6.5.

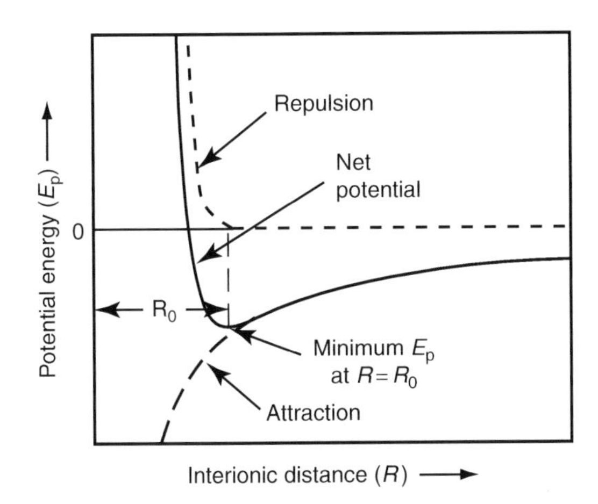

Fig. 3.2 Variation of the potential energy (E_p) of a system consisting of a singly charged cation and a singly charged anion as a function of interionic distance. The equilibrium interionic distance (R_0) is marked by the minimum value of E_p, when $dE/dR = 0$. For $R > R_0$, E_p is essentially determined by the coulombic attraction between the opposite charges and for $R < R_0$ by the repulsion between the nuclei and the electron clouds.

For the discussion below, we assume a model of pure ionic bonding arising out of a geometric framework of ions represented by hard spheres of constant radius. But how do we determine the radius of each ion? Actually, it is not possible to measure the radius of individual ions in a solid, but we can measure the interionic distance between centers of two ions in a solid from its cell dimensions determined with X-ray diffraction techniques, and then determine the radius of individual ions through some manipulation (Companion, 1964). For the purpose of illustration, let us suppose that Fig. 3.3 represents the packing in LiCl and KCl crystals as revealed by X-ray diffraction data. It is reasonable to expect that Li^+, with only two electrons, is a very small cation and assume that the packing in LiCl be largely determined by the much larger Cl^- anions (each containing18 electrons) touching each other. In this case, the radius of the Cl^- ion (r_{Cl^-}) is one-half of the measurable interionic distance d_1. We can now determine the radius of K^+ ion from the measured interionic distance d_2 in a KCl crystal: $r_{K^+} = d_2 - r_{Cl^-}$. It turns out that the ionic radii calculated by this strategy are reasonably constant from compound

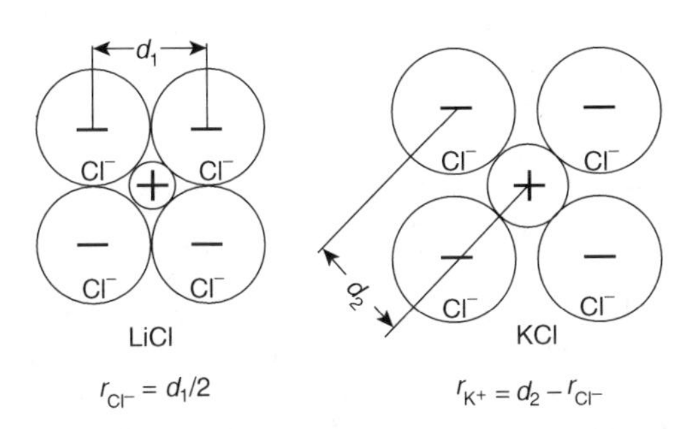

Fig. 3.3 Strategy for determining ionic radii from packing of ions (assumed to be hard spheres) in LiCl and KCl.

to compound and, carried over the entire Periodic Table, this has enabled the setting up a self-consistent set of average ionic radii (Fig. 3.4).

As expected, ionic radii of cations and anions vary with atomic number. The radius of a given ion is also a function of the *coordination number*, the number of nearest neighbors of the ion in a crystalline structure (see section 3.1.2). Some general trends for ionic radii (expressed in Å) with octahedral (or six-fold) coordination (Fig. 3.4), the most common kind of coordination for most ions in silicate minerals, are summarized below:

(1) Cations are smaller than anions, the only exceptions being the five largest cations (Rb^+, Cs^+, Fr^+, Ba^{2+}, and Ra^{2+}), which are larger than F^-, the smallest anion.

(2) Within an *isoelectronic series* – a series of ions with the same number of electrons – ionic radius decreases with increasing atomic number because of increased nuclear attraction for the electron cloud. For example,

$$r_{Si^{4+}}(0.48) < r_{Al^{3+}}(0.54) < r_{Mg^{2+}}(0.72) < r_{Na^+}(1.02) < r_{F^-}(1.33) < r_{O^{2-}}(1.40)$$

(3) On the other hand, in the lanthanide (or rare-earth) series characterized by cations with 3+ charge, the ionic radius decreases with increasing atomic number, from 1.13 for La^{3+} to 0.94 for Lu^{3+}. This so-called *lanthanide contraction* can be attributed to the influence of the increasing nuclear charge.

(4) Within a family of ions, such as the alkali metals or the halogens, the ionic size increases as we go down the Periodic Table. For example,

$$r_{Li^+}(0.76) < r_{Na^+}(1.02) < r_{K^+}(1.38) < r_{Rb^+}(1.52) < r_{Cs^+}(1.67) < r_{Fr^+}(1.80)$$

$$r_{F^-}(1.33) < r_{Cl^-}(1.81) < r_{Br^-}(1.96) < r_{I^-}(2.20)$$

This variation is a consequence of adding electrons with their most probable distance farther from the nucleus.

(5) In the case of cations of the same element, ionic radius decreases with increase in ionic charge because of a decrease in the number of electrons. For example,

$$r_{Fe^{3+}}(0.73) < r_{Fe^{2+}}(0.78)$$

$$r_{Mn^{4+}}(0.62) < r_{Mn^{3+}}(0.73) < r_{Mn^{2+}}(0.83)$$

$$r_{Ti^{4+}}(0.61) < r_{Ti^{3+}}(0.72) < r_{Ti^{2+}}(0.87)$$

$$r_{U^{6+}}(0.81) < r_{U^{4+}}(0.97)$$

The opposite is true for anions, although anions with variable charge are not common.

3.1.2 *Coordination number and radius ratio*

How do ions fit together to produce different crystal structures? The fundamental constraint is that, for a given set of ions, the most stable arrangement is the one that has the

IONIC RADII (Å) (octahedral coordination)

1	2	3	4	5	6	7	8	9	10	11	12	13	14	15	16	17	18
1 H Very small																	2 He
3 Li^+ 0.76	4 Be^{2+} 0.27*											5 B^{3+} 0.11*	6 C^{4+} 0.15*	7 N^{5+} 0.13	8 O^{2-} 1.40	9 F^- 1.33	10 Ne
11 Na^+ 1.02	12 Mg^{2+} 0.72											13 Al^{3+} 0.54	14 Si^{4+} 0.26*	15 P^{5+} 0.17*	16 S^{-2} 1.84	17 Cl^- 1.81	18 Ar
19 K^+ 1.38	20 Ca^{2+} 1.00	21 Sc^{3+} 0.75	22 Ti^{4+} 0.61	23 V^{5+} 0.54	24 Cr^{3+} 0.62	25 Mn^{2+} 0.83	26 Fe^{2+} 0.78	27 Co^{2+} 0.75	28 Ni^{2+} 0.69	29 Cu^{2+} 0.73	30 Zn^{2+} 0.74	31 Ga^{3+} 0.62	32 Ge^{4+} 0.73	33 As^{3+} 0.58	34 Se^{2-} 1.98	35 Br^- 1.96	36 Kr
37 Rb^+ 1.52	38 Sr^{2+} 1.18	39 Y^{3+} 0.90	40 Zr^{4+} 0.72	41 Nb^{5+} 0.64	42 Mo^{4+} 0.65	43 Te^{6+} 0.56	44 Ru^{3+} 0.76	45 Rh^{3+} 0.75	46 Pd^{2+} 0.86	47 Ag^+ 0.94	48 Cd^{2+} 0.95	49 In^{3+} 0.80	50 Sn^{4+} 0.69	51 Sb^{5+} 0.60	52 Te^{6+} 0.56	53 I^- 2.20	54 Xe
55 Cs^+ 1.67	56 Ba^{2+} 1.35	57–71 Lanth	72 Hf^{4+} 0.71	73 Ta^{5+} 0.64	74 W^{6+} 0.60	75 Re^{4+} 0.63	76 Os^{4+} 0.71	77 Ir^{4+} 0.71	78 Pt^{4+} 0.71	79 Au^+ 1.37	80 Hg^{2+} 1.02	81 Tl^{3+} 0.67	82 Pb^{2+} 1.19	83 Bi^{3+} 1.03	84 Po^{6+} 0.67	85 At^{7+} 0.62	86 Rn
87 Fr^+ 1.80	88 Ra^{2+} 1.43	89–103 Actin	104 Rf	105 Db	106 Sg	107 Bh	108 Hs	109 Mt	110	111	112						

Lanthanides	57 La^{3+} 1.13	58 Ce^{3+} 1.09	59 Pr^{3+} 1.08	60 Nd^{3+} 1.06	61 Pm^{3+} 1.04	62 Sm^{3+} 1.04	63 Eu^{3+} 1.03	64 Gd^{3+} 1.02	65 Tb^{3+} 1.00	66 Dy^{3+} 0.99	67 Ho^{3+} 0.98	68 Er^{3+} 0.97	69 Tm^{3+} 0.96	70 Yb^{3+} 0.95	71 Lu^{3+} 0.94
Actinides	89 Ac^{3+} 1.18	90 Th^{4+} 1.08	91 Pa^{4+} 0.98	92 U^{4+} 0.97	93 Np^{4+} 0.95	94 Pu^{4+} 0.88	95 Am^{4+}	96 Cm^{4+}	97 Bk^{4+}	98 Cf^{4+}	99 Es	100 Fm	101 Md	102 No	103 Lr

* Tetrahedral coordination

Fig. 3.4 Ionic radii of ions in octahedral coordination. Å = 10^{-10} m. Sources of data: compilations by Krauskopf and Bird (1995), and Faure (1991).

lowest potential energy. The general rules that need to be observed for attaining maximum stabilization of a crystal structure are as follows:

(1) The crystal structure must be electrically neutral, that is, the cation: anion ratio must be such that the positive charges are exactly balanced by the negative charges.

(2) The cation–anion separation must be close to the equilibrium interionic distance (R_0 in Fig. 3.2) for the compound under consideration.

(3) The arrangement of the ions must be in a regular pattern, with as many cations around anions as possible and as far away from each other as possible; analogous restrictions apply to the anions. In other words, we may treat the ions as spherical balls and pack them as closely as possible, subject to the constraints of electrical neutrality of the structure and minimum interionic distance. In a given three-dimensional close packing of spheres, the number of oppositely charged nearest neighbors surrounding an ion is called its *coordination number* (*CN*). If an ion *A*, for example, is surrounded by four ions of *B*, $CN_A = 4$ (tetrahedral coordination); if *A* is surrounded by six ions of *B*, $CN_A = 6$ (octahedral coordination), and so on. As discussed below, coordination number is an important consideration in crystal chemistry.

Generally, cations are smaller than anions, so the number of anions that can be packed around the smaller cations determines crystal structures. The combined influence of cations and anions on coordination number can be predicted from a consideration of the magnitudes of their radii

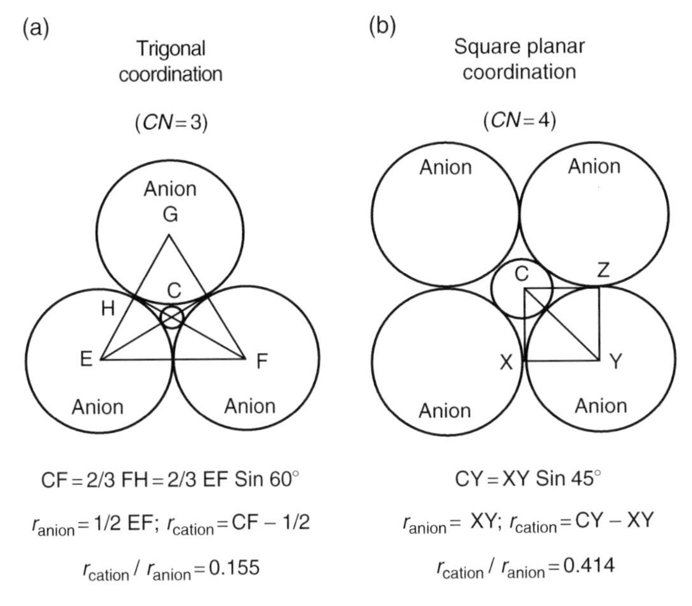

Fig. 3.5 Critical radius ratios for (a) threefold (trigonal) and (b) fourfold (square planar) coordinations.

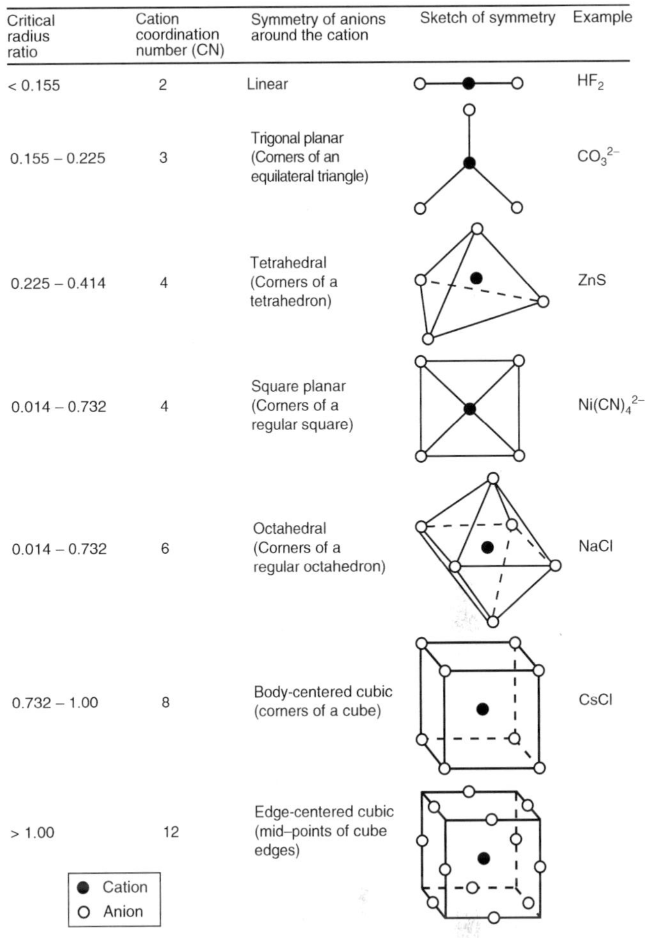

Critical radius ratio	Cation coordination number (CN)	Symmetry of anions around the cation	Sketch of symmetry	Example
< 0.155	2	Linear		HF_2
0.155 – 0.225	3	Trigonal planar (Corners of an equilateral triangle)		CO_3^{2-}
0.225 – 0.414	4	Tetrahedral (Corners of a tetrahedron)		ZnS
0.014 – 0.732	4	Square planar (Corners of a regular square)		$Ni(CN)_4^{2-}$
0.014 – 0.732	6	Octahedral (Corners of a regular octahedron)		NaCl
0.732 – 1.00	8	Body-centered cubic (corners of a cube)		CsCl
> 1.00	12	Edge-centered cubic (mid-points of cube edges)		

● Cation
○ Anion

Fig. 3.6 The effect of critical radius ratios on coordination number and possible geometrical arrangements of ions in ionic crystals. ZnS (sphalerite) itself is not an ionic compound but its name is given to the structure because it is the most common compound in which this geometrical arrangement occurs (Evans, 1966).

expressed as the *radius radio (RR)*. For a cation in a binary ionic solid, *RR* is defined as

$$\text{Radius ratio } (RR) = \frac{r_c}{r_a} \tag{3.2}$$

where r_c and r_a are ionic radius of the cation and the anion, respectively. Evidently, as the cation becomes larger relative to the anion, a larger number of anions may fit around the cation. In other words, the *CN* of the cation is likely to increase as *RR* increases.

Accepting a model based on close packing of spheres, we can easily calculate the *critical radius ratios* (i.e., the limiting values of the radius ratio) for different geometrical arrangements of the spheres. The smallest value of *CN* is 2, which represents the situation when the cation is so small that it is possible to pack only two anions around it if anion–cation contact is to be maintained. As the size of the cation increases relative to that of the anion, it becomes possible to place three anions in mutual contact around the cation (i.e., $CN = 3$) when *RR* reaches a critical value of 0.155 (Fig. 3.5a). With increasing size of the cation relative to that of the anion, the *CN* changes to higher values. The calculated critical radius ratios for different possible symmetries are: 0.155–0.225 for $CN = 3$ (trigonal coordination); 0.225–0.414 (Fig. 3.5b) for $CN = 4$ (tetrahedral or square planar coordination); 0.414–0.732 for $CN = 6$ (octahedral coordination); 0.732–1.0 for $CN = 8$ (body-centered cubic coordination); and > 1.0 for $CN = 12$ (edge-centered cubic coordination) (Fig. 3.6). Other coordination numbers, such as 5, 7, 9, 10, and 11, do exist but are quite uncommon because such coordination polyhedra cannot be extended into infinite, regular three-dimensional arrays (Greenwood, 1970). In mineral structures, the most common anion is O^{2-}, which has an ionic radius of 1.40 Å, and the ionic radii of most common cations are between 0.60 and 1.10 Å. Thus, the radius ratios with oxygen in minerals mostly lie between 0.43 and 0.79, suggesting that the most frequent coordination number in minerals is 6. This is why Fig. 3.4 lists ionic radii for octahedral coordination rather than for tetrahedral or cubic coordination. Examples of some ionic crystal structures characterized by different coordinations are presented in Fig. 3.7.

Many cations in silicate minerals occur exclusively in a particular coordination with oxygen, but some occur in more than one coordination, to some extent controlled by the temperature and pressure of crystallization. For example, the radius ratio of Al^{3+} bonded to O^{2-} is 0.54 Å/1.40 Å = 0.386, which is very close to the theoretical boundary of 0.414 between $CN = 4$ and $CN = 6$. Thus, in silicate minerals formed

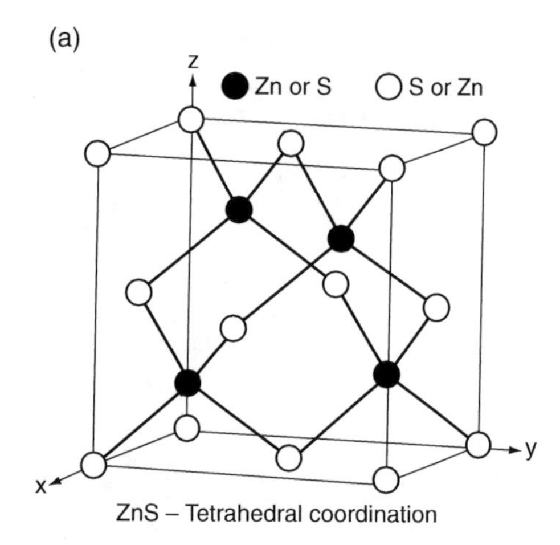

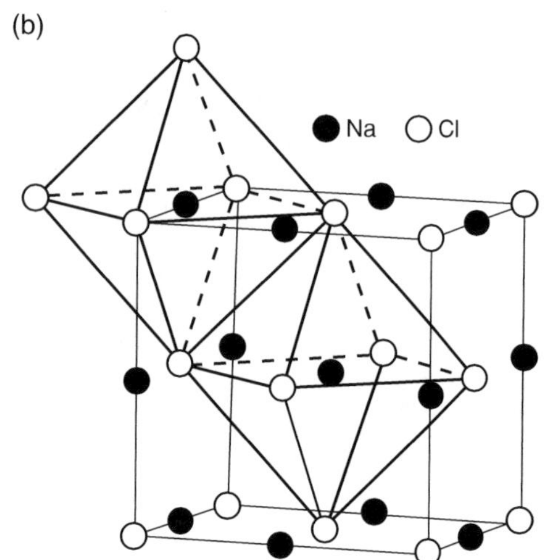

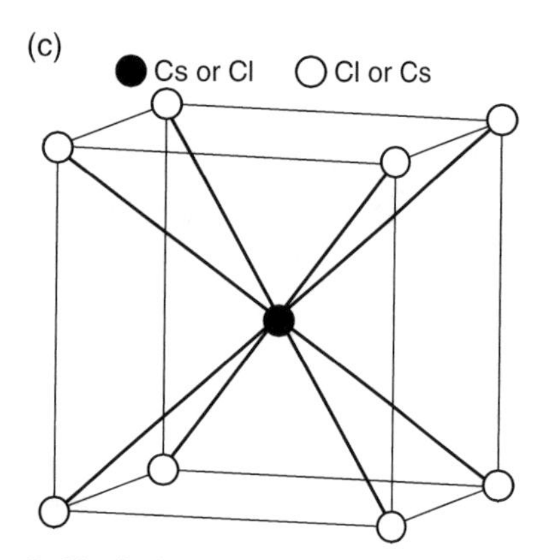

Fig. 3.7 Examples of some ionic crystal structures characterized by different coordinations: (a) ZnS (tetrahedral); (b) NaCl (octahedral); and (c) CsCl (body-centered cubic). It should be stressed that such illustrations of crystal structures represent only the geometry of arrangement of the centers of the ions or the mean positions of the vibrating nuclei; the electron density is concentrated near the nuclei and along directions to near neighbors. (From *An Introduction to Crystal Chemistry*, 2nd edition, by R.C. Evans, Figure 3.04, p. 35, Figure 3.02, p. 33, and Figure 3.03, p. 34; Copyright 1966, Cambridge University Press. Reproduced with permission of the publisher.)

at high temperatures or low pressures, Al^{3+} tends to assume tetrahedral coordination and substitute for Si^{4+}, whereas in minerals formed at low temperatures or high pressures, Al^{3+} tends to occur in octahedral coordination.

Crystal structures are often more complicated than what can be predicted on the basis of radius ratios. An important complicating factor is the degree of covalence (electron sharing) in many dominantly ionic minerals, resulting in distortion of the electronic charge density around ions (*polarization*) (see section 3.6.5).

3.1.3 *Lattice energy of ideal ionic crystals*

The stability of ionic compounds is optimized by close packing of oppositely charged ions together in extended arrays. The three-dimensional arrangement of atoms or ions in a crystal is commonly referred to as (crystal) *lattice*. Every lattice is associated with a certain amount of stabilization energy, which is called the (crystal) *lattice energy*. *The lattice energy* (U_L) *of a perfectly ionic crystal is defined as the amount of energy required at absolute zero (i.e., −273°C) to convert one mole of the solid into its constituent ions at infinite separation in the gas phase*:

$$M_aX_b(s) \Rightarrow aM^{b+}(g) + bX^{a-}(g)\ (U_L) \qquad (3.3)$$

The lattice energy for NaCl, for example, is 786 kJ mol^{-1}; that is, it would require 786 kJ of energy per mole of NaCl to produce infinite separation of Na^+ and Cl^- ions in the gas phase. We can also conclude that one mole of solid NaCl compound is 786 kJ lower in energy, and thus more stable, than a mixture of one mole each of the constituent ions.

For 1 gram-mole of a binary compound, the lattice energy can be calculated directly from the properties of the ions by means of the following equation, originally derived by Born and Landé (see Box 3.1):

$$U_L = M_c A \frac{z_c\, z_a\, e^2}{R}\left(1 - \frac{1}{n}\right) \qquad (3.4)$$

where z_c and z_a are the charges on cations and anions (expressed as a multiple of the electron charge e), R is the cation–anion interionic distance, n is a constant for the particular crystal structure, A is the Avogadro's number, and M_c is a numerical quantity called the *Madelung constant*, the value of which depends on the crystal structure (Table 3.2). From experimental studies of compressibility, it is known that for most ionic crystals n has a value between 8 and 12. Suggested values of n for some common solids are: LiF, 5.9; LiCl, 8.0; LiBr, 8.7; NaCl, 9.1; NaBr, 9.5. By convention, the lattice energy is considered positive because energy is consumed for separation of the ions. The energy released during

Box 3.1 Expression for the lattice energy of an ionic crystal

The electrostatic potential energy (E_p) associated with a cation–anion pair in an ionic crystal can be described by a relation of the form (Fyfe, 1964)

$$E_p = -\frac{z_c\, z_a\, e^2}{R} + \frac{be^2}{R^n} \qquad (3.5)$$

where z_c and z_a are the charges on cations and anions (expressed as a multiple of the electron charge e), R is the interionic distance separating the ions, and b and n are constants characteristic of the crystal structure. As in equation (3.1), the first term in equation (3.5) represents normal coulombic attraction and the second term repulsion. The potential energy of an entire crystal, which is obtained by adding together all such interactions over the three-dimensional lattice of the entire crystal, is (for derivation see Fyfe, 1964, pp. 48–49):

$$E_{p\,(crystal)} = -M_c \frac{z_c\, z_a\, e^2}{R}\left(1-\frac{1}{n}\right) \qquad (3.6)$$

where M_c is a numerical quantity called the *Madelung constant*, the value of which depends on the crystal structure (Table 3.2). To obtain the lattice energy of 1 gram-mole of a crystal, this energy must be multiplied by Avogadro's number (A), the number of molecules in one mole of a substance. The lattice energy (U_L) of a crystal is defined as $-AE_{p(crystal)}$, so that

$$U_L = -AE_p = M_c A \frac{z_c\, z_a\, e^2}{R}\left(1-\frac{1}{n}\right) \qquad (3.4)$$

Note that U_L decreases with increasing R, and approaches zero as R approaches infinity.

Table 3.2 Madelung constants for selected types of crystal structures.

Structure type	Madelung constant	Coordination (anion) : Coordination (cation)	Solid compound
NaCl type	1.747	6:6	Halite
CsCl type	1.747	8:8	CsCl
ZnS type	1.638	4:4	Sphalerite
CaF_2 type	2.519	8:4	Flourite
TiO_2 type	2.408	6:3	Rutile
Al_2O_3 type	4.172	6:4	Corundum

Source of data: Lide (1998).

the reverse process, the formation of a crystal from its widely separated constituent ions, called the *energy of crystallization* (E_{cryst}), is assigned a negative sign (i.e., $U_L = -E_{cryst}$). Equation (3.4) does not include the contribution due to the van der Waals forces of attraction between the ions (see section 3.8), but the correction arising out of this weak force is very small – for example, less than about 12 kJ mol^{-1} for the alkali halides (Evans, 1966).

The Born–Landé equation, strictly speaking, applies only to binary compounds, but it enables us to make some qualitative statements regarding the lattice energy of more complex substances. For a particular structure type, lattice energies are greater the higher the charge on the ions, the smaller the ions, and the closer the packing (Mason, 1966).

For crystals of the same structure, with ions of the same charge, the lattice energy varies as the interionic distance: $U_{LiCl} < U_{NaCl} < U_{KCl}$.

Example 3–1: Calculation of the lattice energy and the energy of crystallization of one mole of NaCl crystals using equation (3.5)

Given: radius of Na^+ (r_{Na^+}) = 1.02 Å; radius of Cl^- (r_{Cl^-}) = 1.81 Å; $A = 6.02 \times 10^{23}$; $e = 4.80 \times 10^{-10}$ coulombs; $n = 9$; and $M = 1.747$

R = 1.02 Å + 1.81 Å = 2.83 Å = 2.83×10^{-8} cm

$$U_L(\text{NaCl}) = M_c A \frac{z_c z_a e^2}{R}\left(1-\frac{1}{n}\right)$$
$$= (1.747)(6.02 \times 10^{23}) \frac{(1)(1)(4.80 \times 10^{-10})^2}{2.83 \times 10^{-8}}\left(1-\frac{1}{9.1}\right)$$
$$= 10.517 \times 8.141 \times 0.890 \times 10^{-11}$$
$$= 76.201 \times 10^{11} \text{ ergs mol}^{-1}$$
$$= (76.201 \times 10^{11})(2.389 \times 10^{-11}) \text{ kcal mol}^{-1}$$
$$= 182 \text{ kcal mol}^{-1} = 761 \text{ kJ mol}^{-1}$$

$E_{cryst}(\text{NaCl}) = -U_L(\text{NaCl}) = -761$ kJ mol^{-1}

It is impossible to measure precisely the lattice energy of a crystal directly, but it can be estimated from more readily measurable quantities by applying *Hess's Law of Heat Summation*, which says that the heat of a reaction (exothermic or endothermic) is the same whether it occurs in one step or a series of steps. The energy cycle used in this approach to calculate lattice energy is known as the *Born–Haber cycle*, which is illustrated in Example 3–2 for NaCl. Lattice energies of selected solid halides, estimated by application of the Born–Haber cycle, are listed in Table 3.3.

Example 3–2: Estimation of the lattice energy of 1 mole of NaCl crystal using the Born–Haber cycle approach

The estimate of lattice energy obtained by the application of Born–Haber cycle will vary somewhat, depending on the reactions chosen for the cycle; the cycle depicted in Fig. 3.8 will suffice to illustrate the concept.

Table 3.3 Estimated lattice energies (U_L) of selected solid compounds (halides).

Solid	(U_L) kJ mol^{-1}	Solid	(U_L) kJ mol^{-1}	Solid	(U_L) kJ mol^{-1}	Solid	(U_L) kJ mol^{-1}
LiF	1036	LiCl	853	LiBr	807	LiI	757
NaF	923	NaCl	786	NaBr	747	NaI	704
KF	821	KCl	715	KBr	682	KI	649
RbF	785	RbCl	689	RbBr	660	RbI	630
CsF	740	CsCl	659	CsBr	631	CsI	604
MgF_2	2957	$MgCl_2$	2526	$MgBr_2$	2440	MgI_2	2327
AlF_3	5215	$AlCl_3$	5492	$AlBr_3$	5361	AlI_3	5218

Source of data: Lide (1998).

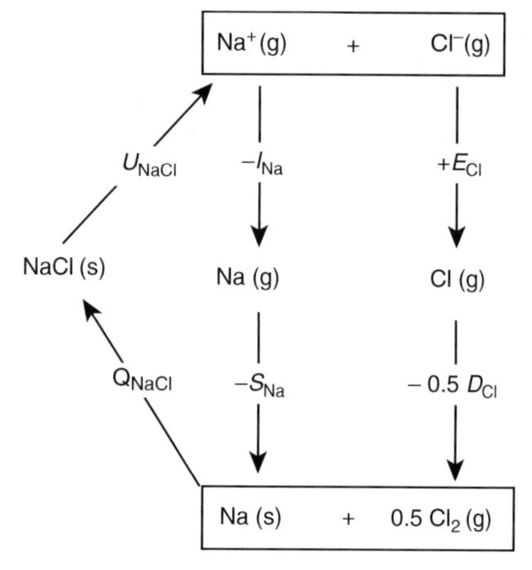

Fig. 3.8 Born–Haber cycle for estimation of the lattice energy of solid NaCl. Sources of data: Fyfe (1964); Evans (1966).

The reactions included in this Born–Haber cycle are:

Reaction	Heat of reaction (kJ mol^{-1})
$Na^+(g) + e^- = Na\ (g)$	$(-I_{Na}) = 495.9$
$Cl^-(g) = Cl\ (g) + e^-$	$(+E_{Cl}) = 348.3$
$Na\ (g) = Na\ (s)$	$(-S_{Na}) = 108.8$
$Cl\ (g) = 0.5Cl_2(g)$	$(-0.5D_{Cl}) = 236.0$
$Na\ (s) + 0.5\ Cl_2\ (g) = NaCl\ (s)$	$(+Q_{NaCl}) = -411.3$

Addition of the above equations gives

$$Na^+(g) + Cl^-(g) = NaCl\ (s) \qquad [-U_L(NaCl)\ \text{or}\ E_{cryst}(NaCl)]$$

Rearranging and substituting the given energy values, we get the lattice energy of NaCl:

$$U_{L(NaCl)} = I_{Na} + S_{Na} + 0.5D_{Cl} - Q_{NaCl} - E_{Cl}$$
$$= 495.9 + 108.8 + 118.0 + 411.3 - 348.3 = 785.7\ \text{kJ mol}^{-1}$$

Note that the estimated value of $U_{L\ (NaCl)}$ is very close to that obtained in Example 3–1.

To a first approximation, the lattice energy of a crystal may be viewed as representing the binding energy of the ions in the crystal. The magnitude of the lattice energy, therefore, has significant influence on certain physical properties, such as melting point and solubility, of a solid. Melting points of solids with similar crystal structures tend to increase with increasing lattice energy. For example, the melting point of NaCl ($U_L \approx 760$ kJ mol^{-1}) is 801°C, but that of MgO ($U_L \approx 3790$ kJ mol^{-1}), which also has a simple cubic crystal structure, is 2800°C. The higher melting point is partly due to a smaller cation–anion distance in the MgO structure, but primarily due to its much higher lattice energy (Companion, 1964). This is to be expected as $E_{p\ (crystal)}$ (and consequently U_L) increases fourfold when the ionic charge increases from 1 to 2 (equations 3.4 and 3.6).

Lattice energy is one of many factors that determine the solubility of a salt in water (see Chapter 7). When a salt, such as NaCl, dissolves in water, it dissociates into Na^+ and Cl^- ions, both of which become dispersed in the solution:

$$NaCl\ (s) \xrightarrow{H_2O} Na^+_{aq} + Cl^-_{aq} \qquad (3.7)$$

The lattice energy of a salt gives a rough indication of the solubility of a salt in water because it reflects the energy that must be supplied to separate the cations in the salt from its anions. It follows that solubility of salts should decrease with increasing lattice energy. Thus, from the data in Table 3.3, we may predict that halides of alkaline earth metals are less soluble than those of alkali metals, and aluminum halides are even less so. Similarly, the solubility of NaOH (U_L = 900 kJ mol^{-1})

is very high (420 g L^{-1}), but that of $Mg(OH)_2$ (U_L = 3006 kJ mol^{-1}) is very low (0.009 g L^{-1}), and $Al(OH)_3$ (U_L = 5627 kJ mol^{-1}) is essentially insoluble in water.

3.2 Crystal structures of silicate minerals

Assuming a model of pure ionic bonding, the basic unit for building crystal structures of silicate minerals is considered to be the *silicon–oxygen tetrahedron* (SiO_4^{4-}), which is comprised of four O^{2-} anions at the four corners of a tetrahedron and the much smaller Si^{4+} cation filling the interstitial space at the center of the tetrahedron, giving the unit a net charge of −4 (Fig. 3.9). In essence, silicate structures (except those formed at extremely high pressures) consist of linked silicon–oxygen tetrahedra and cations as can be accommodated by the structure to achieve electrical neutrality. The different structural classes of silicate minerals result from the various ways in which the silicon–oxygen tetrahedra are linked to each other (Table 3.4). The tetrahedra may exist as isolated units, with no shared oxygen, or may be linked by sharing one, two, three, or four oxygen ions, as illustrated in Fig. 3.10. The bonds between silicon and oxygen within SiO_4^{4-} units are so strong that the dimensions and shape of the tetrahedra remain nearly constant irrespective of what the rest of a silicate structure may be.

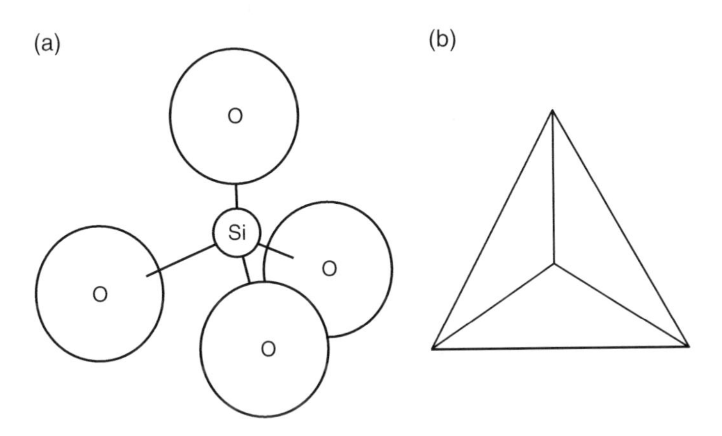

Fig. 3.9 The silicon–oxygen tetrahedron, the basic building block for crystal structures of silicate minerals: (a) A model of the tetrahedron using rods to depict the ionic bonds that connect the central Si^{4+} cation to the four O^{2-} anions positioned at the corners; (b) A commonly used graphical representation of the tetrahedron.

3.3 Ionic substitution in crystals

Most rock-forming minerals are dominantly ionic compounds, and they are seldom pure phases. Deviation of the chemical composition of a mineral from its ideal chemical formula occurs due to incorporation of minor amounts of foreign ions into the lattice, primarily by ionic substitutions governed by similarities in size and charge of the ions involved. The ability of different elements to occupy the same lattice position in a particular crystal structure is called *diadochy*. For example, Mg, Fe, Mn, and Sr are diadochic in the structure of calcite because they can substitute for Ca in this structure. The concept of diadochy, if used rigorously, always applies to a particular crystal structure. Two elements may be diadochic in one mineral and not in another (Mason, 1966).

3.3.1 *Goldschmidt's rules*

It is well known from experience that there exists a preferential association, or *geochemical coherence*, between certain elements in natural assemblages because ions of such elements substitute easily for each other in minerals. The Norwegian

Table 3.4 Structural classification of silicate minerals.

Silicate class	Linkage of tetrahedra	Repeat unit	Si:O ratio	Example of silicate mineral
Nesosilicates	Independent tetrahedra (no sharing of oxygen ions)	SiO_4^{4-}	1:4	Forsterite (olivine) [Mg_2SiO_4]
Sorosilicates	Two tetrahedra sharing one oxygen ion	$Si_2O_7^{6-}$	2:7	Åkermanite [$Ca_2MgSi_2O_7$]
Cyclosilicates (ring structures)	Closed rings of tetrahedra, each sharing two oxygen ions	SiO_3^{2-}	1:3	Beryl [$Al_2Be_3Si_6O_{18}$]
Inosilicates (single chains)	Continuous single chains of tetrahedra, each sharing two oxygen ions	SiO_3^{2-}	1:3	Enstatite (pyroxene) [$MgSiO_3$]
Inosilicates (double chains)	Continuous double chains of tetrahedra, alternately sharing two and three oxygen ions	$Si_4O_{11}^{6-}$	4:11	Anthophyllite (amphibole) [$Mg_7(Si_4O_{11})_2(OH)_2$]
Phylosilicates (sheet structures)	Continuous sheets of tetrahedra, each sharing three oxygen ions	$Si_2O_5^{2-}$	2:5	Phlogopite (mica) [$KMg_3(AlSi_3O_{10})(OH)_2$]
Tektosilicates (framework structures)	Continuous framework of tetrahedra, each sharing four oxygen ions	SiO_2	1:2	Albite (plagioclase feldspar) [$Na(AlSi_3)O_8$]

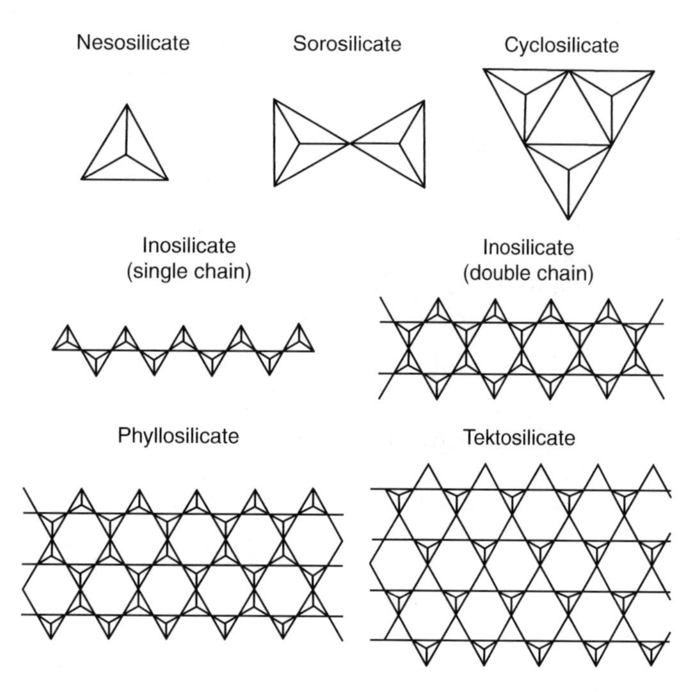

Fig. 3.10 Different patterns of silicon–oxygen tetrahedra linkages for the various structural classes of silicate minerals listed in Table 3.4. The sketch is a two-dimensional representation of crystal structures that are three-dimensional. In all cases, except the tektosilicate, the apexes of all the tetrahedra are pointing upward. The tektosilicate structure consists of layers of hexagonal rings of tetrahedra with alternate rows of apexes pointing in opposite directions, and the upward-pointing apexes of a given layer coincide with the downward-pointing apexes of the layer above it. Thus, all four oxygens of each tetrahedron of the tektosilicate are shared with oxygens of other tetrahedra in the structure.

geochemist V.M. Goldschmidt (1888–1947) was the first to propose a series of rules governing the mutual replacement (diadochy) of ions in magmatic minerals, on the assumption that the bonding in these minerals was purely ionic. He considered that between two ions capable of diadochy, the one that makes the larger contribution to the energy of the crystal structure is incorporated preferentially. *Goldschmidt's rules of substitution* may be summarized as follows:

(1) If two ions have the same radius and the same charge, they would enter into solid solution in a given mineral with equal ease, in amounts roughly proportional to their abundances. The ionic radii must not differ by more than 15%; substitution is limited or rare if the radii differ by 15% to 30%, and nonexistent if the difference is more than 30%. Examples are the common substitution of Ta^{5+} (0.64 Å) for Nb^{5+} (0.64 Å), Hf^{4+} (0.71 Å) for Zr^{4+} (0.72 Å), Ga^{3+} (0.62 Å) for Al^{3+} (0.54 Å), and Fe^{2+} (0.78 Å) for Mg^{2+} (0.72 Å) in many silicate minerals. In such cases, the minor element is said to be *camouflaged* in the crystal structure (Shaw, 1953).

(2) When two ions possessing the same charge but different radii compete for a particular lattice site, the ion with the smaller radius would be incorporated preferentially because the smaller ion forms a stronger ionic bond. During magmatic crystallization, for example, the earlier formed olivine tends to be enriched in Mg^{2+} (0.72 Å) relative to Fe^{2+} (0.78 Å).

(3) When two ions having similar radii but different charges compete for a particular lattice site, the ion with the higher charge would be incorporated preferentially because the ion with higher charge forms a stronger ionic bond. If the substituting ion has a higher charge than the ion in the lattice being substituted, it is said to be *captured* by the crystal structure; if the substituting ion has a lower charge, it is said to be *admitted* by the crystal structure. For example, the K-feldspar structure captures Ba^{2+} (1.35 Å) for replacement of K^+ (1.38 Å), and the biotite structure admits Li^+ (0.76 Å) for replacement of Mg^{2+} (0.72 Å).

(4) Ions whose charges differ by one unit may substitute for one another provided electrical neutrality of the crystal is maintained by *coupled (or compensatory) substitution*. Example: concurrent substitution of Na^+ (1.02 Å) by Ca^{2+} (1.00 Å) and of Si^{4+} (0.26) Å) by Al^{3+} (0.47 Å) in plagioclase feldspars:

$$NaAlSi_3O_8 + Ca^{2+} + Al^{3+} \Rightarrow CaAl_2Si_2O_8 + Na^+ + Si^{4+} \quad (3.8)$$

In general, very little or no substitution takes place when the difference in charge on the ions is > 1, even if the size is appropriate. Zr^{4+} (0.72 Å) does not substitute for Mg^{2+} (0.72 Å) nor Cr^{3+} (0.62 Å) for Li^+ (0.76 Å); this is probably because of the difficulty in achieving charge balance by compensatory substitutions.

3.3.2 *Ringwood's rule*

Over the years the generalizations embodied in the Goldschmidt's rules have been found to be fraught with many exceptions. Moreover, some ion pairs of similar size and charge, such as Mg^{2+} (0.72 Å) and Ni^{2+} (0.69 Å), show only a moderately close association, whereas a few ion pairs, for example Sr^{2+} (1.18 Å) and Hg^{2+} (1.02 Å), show virtually no geochemical coherence. Obviously, there are other properties that are also important in predicting geochemical associations.

A major limitation in the application of Goldschmidt's rules is the assumption of pure ionic bonding because most minerals have a significant component of covalent bonding. Ringwood (1955) suggested that substitution may be limited, even when the size and charge criteria are satisfied, if the competing ions have different electronegativities (see section 3.6.5) and, therefore, form bonds of different strengths. For example, Cu^{2+} (0.73 Å) rarely substitutes for Mg^{2+} (0.72 Å) because of the large difference in electronegativity (1.90 versus 1.31), although it should from the consideration of charge and size. Si^{4+} (0.26 Å; electronegativity = 1.9) and

Ge^{4+} (0.73 Å; electronegativity = 2.01), on the other hand, show strong geochemical coherence because of almost identical values of electronegativity, despite a large difference in ionic radius. Ringwood's rule states that between two cations satisfying the criteria of charge and size for diadochic substitution in a crystal structure, preference will be for the one with lower electronegativity because it forms a more ionic, stronger bond. In most cases, the difference in electronegativity required before elements obey this rule is about 0.1. For example, Pb^{2+} (1.19 Å; electronegativity = 1.8) does not substitute easily for K^+ (1.38 Å; electronegativity = 0.82) in potash–bearing minerals and hence becomes concentrated in the residual magma.

It should be evident that the extent of ionic substitution is determined by the nature of the crystal structure and how similar the ions are in terms of size, charge, and electronegativity. Other factors that affect the extent of substitution are the temperature and pressure at which the substitution takes place, elevated temperatures and lower pressures favoring increased substitution. Higher temperatures promote greater atomic vibration and open structures, which are easier to distort locally to accommodate cations of different sizes. Thus, the concentration of minor and trace elements in minerals (e.g., of Fe in sphalerite, ZnS) provides a potential means of determining the temperatures of mineral formation (geothermometry; see section 6.5). Pressure at the time of substitution has the opposite effect, but is much less important (except at very high pressures) because most minerals are quite incompressible.

Although the rules discussed above can be applied to explain to a certain extent the distribution of elements during many geochemical processes (especially magmatic crystallization and metamorphic recrystallization), their application is limited because of numerous exceptions, especially when the transition elements are involved (Burns and Fyfe, 1967; Burns, 1973). The order of uptake of the transition-metal ions by minerals formed by magmatic crystallization is better explained by the crystal-field theory.

3.4 Crystal-field theory

Crystal-field theory was developed originally by the physicists Hans Bethe and John Hasbrouck van Vleck in the 1930s to explain the absorption spectra of the transition metals such as Ni, Co, Fe, Ti, etc. At its present state of development, the theory can be applied to account for some magnetic properties, colors, hydration enthalpies, and spinel structures ("normal" versus "inverse" spinels) of transition metal complexes (Burns and Fyfe, 1967; Burns, 1973, 1993).

Crystal-field theory describes the effects of electrostatic fields on the energy levels of the valence electrons (electrons in the outermost orbitals) of a transition-metal when it is surrounded by negatively charged ligands in a crystal structure. (A *ligand* is an ion, a molecule, or a molecular group that binds to another chemical entity to form a larger complex.) The ligands are assumed to be point negative charges sited on the Cartesian axes, and the bonding entirely ionic. The more comprehensive *ligand-field theory*, which is too complicated to be discussed here, treats the metal-ligand interaction as a covalent bonding interaction involving overlap between the *d*-orbitals of the metals and the ligand-donor orbitals.

3.4.1 *Crystal-field stabilization energy*

Transition elements are characterized by incompletely filled inner *d* or *f* orbitals. Let us consider the first transition series involving 3*d* orbitals. Electronic structures of this transition series are of the general form [Ar core $(1s^22s^22p^63s^23p^6)3d^{1\text{ to }10}$ $4s^{1\text{ or }2}$], and ions are formed when the 4*s* electrons, and in some cases 3*d* electrons, are removed from the metal atom. When such an ion is surrounded by ligands, for example, in octahedral coordination (i.e., coordinated to six identical ligands), the increased repulsion between the anions and the electrons of the 3*s* orbital, which has a spherical symmetry, results simply in raising the energy level of 3*s* electrons. The 3*p* orbitals of the metal pointing directly towards the point charges of anions are also raised to a higher energy level because of increased repulsion between the ligands and 3*p* electrons, but remain degenerate.

The main effect of the ligands on the transitional-metal ion arises from interaction with 3*d* electrons. In a free (isolated) transition-metal ion, the 3*d* orbitals ($3d_{z^2}$, $3d_{x^2-y^2}$, $3d_{xy}$, $3d_{xz}$, $3d_{yz}$; see Table 2.2 and Fig. 2.6) are fivefold degenerate – i.e., they are energetically equivalent and the *d* electrons have equal probability of being located in any of the five 3*d* orbitals. When the same ion is placed in a crystal, for example in an octahedral coordination, the five 3*d* orbitals do not experience exactly the same kind of interaction because they do not have the same spatial configuration relative to the ligands. The $3d_{z^2}$ and $3d_{x^2-y^2}$ orbitals (designated as e_g symmetry group) have lobes that point directly towards the point charges of the ligands, whereas the $3d_{xy}$, $3d_{xz}$, $3d_{yz}$ orbitals (designated as t_{2g} symmetry group) have lobes that point between the negative charges. This results in a greater electrostatic repulsion for the e_g electrons than the t_{2g} electrons. The 3*d* orbitals can no longer remain degenerate, and they split into two groups that have different levels of energy (Fig. 3.11a). The energy separation between t_{2g} and e_g orbitals is termed the *crystal-field splitting parameter* and denoted by Δ_o (the "o" stands for an octahedral coordination). Energy is lowered by 2/5 Δ_o for the three t_{2g} orbitals and raised by 3/5 Δ_o for the two e_g orbitals relative to the mean energy of an unperturbed ion (Fig. 3.11a). Each electron in a t_{2g} orbital lowers the energy of the transition-metal ion, and thus increases its stability, by 2/5 Δ_o, whereas each electron in the e_g orbital diminishes the stability by 3/5 Δ_o relative to a hypothetical nontransitional-metal ion of the same size and charge. The resultant net energy, which depends on the number of electrons and how they fill the

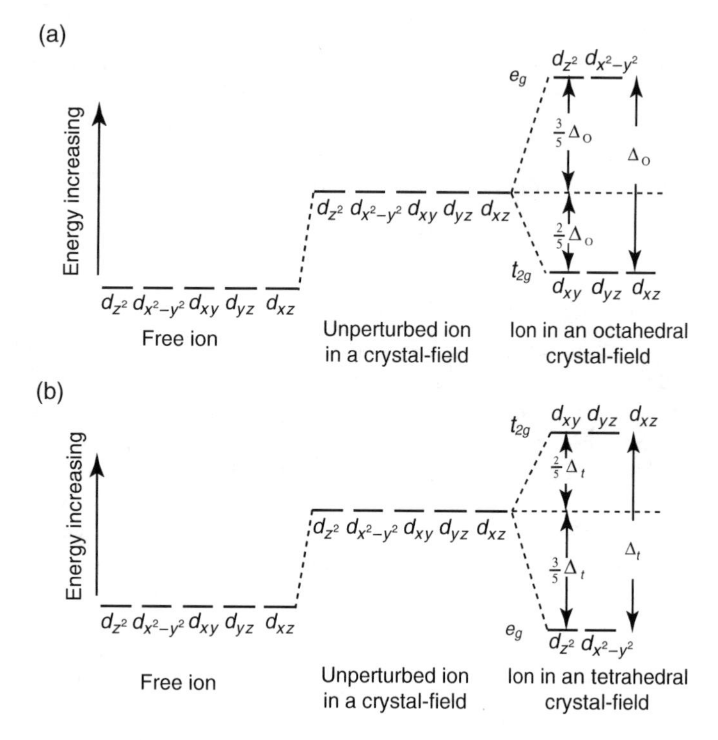

Fig. 3.11 Schematic energy level diagram for $3d$ orbitals of a transition-metal ion (not to scale): (a) octahedral coordination in a crystal; (b) tetrahedral coordination in a crystal. Δ_o = crystal-field splitting parameter for octahedral coordination; Δ_t = crystal-field splitting parameter for tetrahedral coordination; $\Delta_t = 4/9\ \Delta_o$.

orbitals, is called *crystal-field stabilization energy* (CFSE), the magnitude of which can be estimated from absorption spectra measurements.

Let us examine how the $3d$ orbitals are filled by electrons in an octahedral field (Fig. 3.12), the most common type of coordination between transition-metal ions and ligands in silicate minerals. As per Hund's rule of maximum multiplicity (see Section. 2.3.1), ions possessing one, two, or three $3d$ electrons (e.g., Ti^{3+}, V^{3+}, and Cr^{3+}, respectively) can have only one electronic configuration, each with $3d$ electrons occupying different t_{2g} orbitals, with parallel orientation of their spins. The CFSE for these three cases, respectively, are: $2/5\ \Delta_o$; $2/5\ \Delta_o + 2/5\ \Delta_o = 4/5\ \Delta_o$; and $4/5\ \Delta_o + 2/5\ \Delta_o = 6/5\ \Delta_o$. Ions carrying four, five, six, or seven $3d$ electrons (e.g., Mn^{3+}, Fe^{3+}, Fe^{2+}, and Co^{2+}, respectively) present a choice between two states: a weak-field (or high-spin) state; and a strong-field (or low-spin) state. In the high-spin state, two electrons occupy e_g orbitals without pairing before the rest are paired in t_{2g} orbitals. No energy is expended by pairing of electrons in t_{2g} orbitals already filled with unpaired electrons, but the CFSE is reduced by $3/5\ \Delta_o$ for every electron in e_g orbitals. Thus, the calculated CFSEs for this state with four, five, six, and seven $3d$ electrons, respectively, are: $6/5\ \Delta_o - 3/5\ \Delta_o = 3/5\ \Delta_o$; $3/5\ \Delta_o - 3/5\ \Delta_o = 0$; $0 + 2/5\ \Delta_o = 2/5\ \Delta_o$; and $2/5\ \Delta_o + 2/5\ \Delta_o = 4/5\ \Delta_o$. In the low-spin state, it is energetically more favorable for the electrons to fill

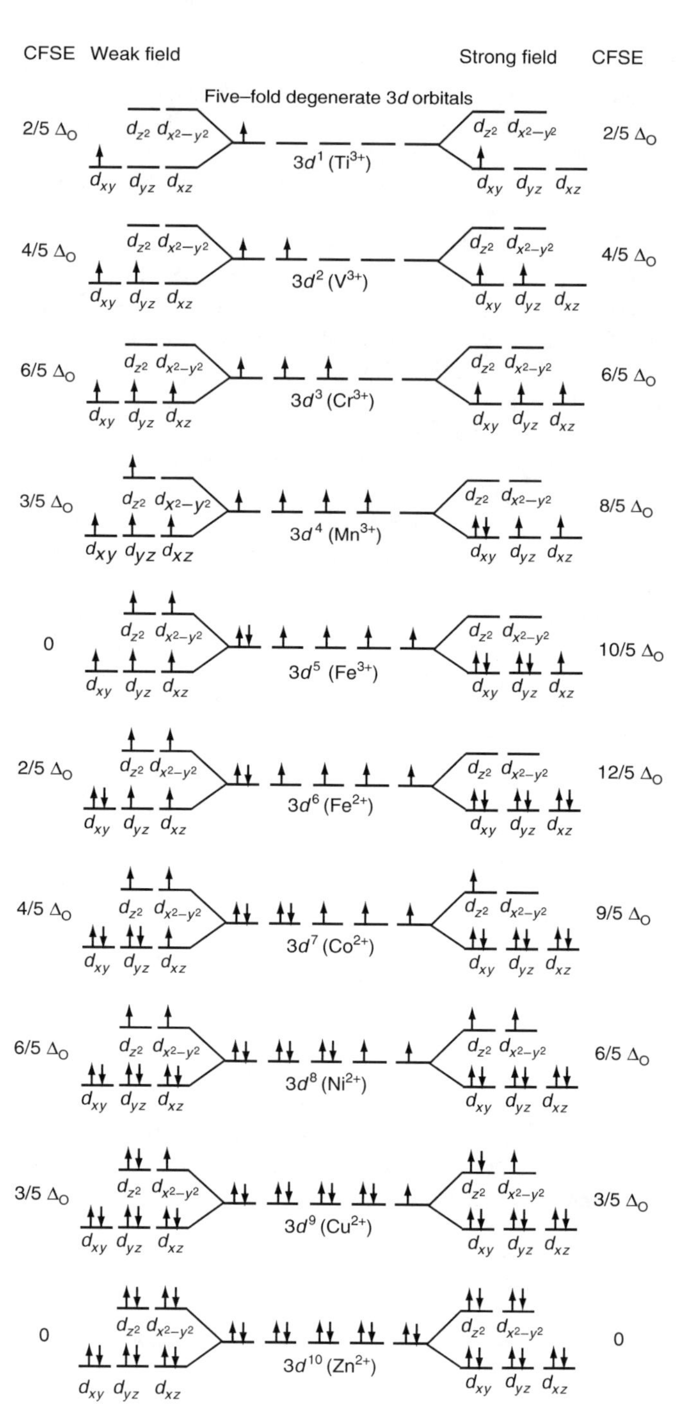

Fig. 3.12 Filling of electrons in $3d$ orbitals of $3d$ transition elements in octahedral coordination in a crystal and the resulting values of CFSE for strong-field (low-spin) and weak-field (high-spin) configurations (not to scale). Δ_o = octahedral crystal-field splitting parameter; CFSE = crystal-field stabilization energy. The calculated CFSE must be reduced by the energy required to pair two electrons in a t_{2g} orbital. Source of data: Burns and Fyfe (1967).

lower-energy t_{2g} orbitals before e_g orbitals, and the calculated CFSEs for this state with four, five, six, and seven $3d$ electrons, respectively, are: $6/5\ \Delta_o + 2/5\ \Delta_o = 8/5\ \Delta_o$, $8/5\ \Delta_o + 2/5\ \Delta_o = 10/5\ \Delta_o$, $10/5\ \Delta_o + 2/5\ \Delta_o = 12/5\ \Delta_o$, and $12/5\ \Delta_o - 3/5\ \Delta_o = 9/5\ \Delta_o$. Finally, each ion with eight, nine, or ten electrons (e.g., Ni^{2+}, Cu^{2+}, and Zn^{2+}, respectively) possesses only one electronic configuration in which t_{2g} and e_g orbitals are filled as shown in Fig. 3.12, along with calculated CFSEs that are identical for both low-spin and high-spin states. The distinction between low-spin and high-spin states is important in understanding magnetic properties of transition-metal compounds. For instance, pyrite is diamagnetic (i.e., weakly repelled by magnetic fields), indicating a low-spin configuration, $(t_{2g})^6$, of Fe^{2+}; ferromagnesian silicates, on the other hand, are paramagnetic (i.e., weakly attracted to magnetic fields), indicating a high-spin configuration, $(t_{2g})^4(e_g)^2$ (Burns and Fyfe, 1967).

The electronic configurations of $3d$ transition-metals in tetrahedral coordination, and the corresponding CFSEs, can be worked out the same way as explained above for octahedral coordination, remembering that in a tetrahedral field the $3d_{z^2}$ and $3d_{x^2-y^2}$ orbitals are more stable (i.e., have a lower energy) than the $3d_{xy}$, $3d_{xz}$, $3d_{yz}$ orbitals (Fig. 3.11b). The tetrahedral crystal-field splitting parameter, Δ_t, is smaller than Δ_o. For the same transition-metal cation in a tetrahedral coordination with identical ligands and identical metal-ligand interatomic distances, $\Delta_t = 4/9\ \Delta_o$. The difference between Δ_o and Δ_t, referred to as the *octahedral site-preference energy* parameter, is useful for the interpretation of cation distributions in mineral-forming processes and crystal structures.

3.4.2 *Nickel enrichment in early-formed magmatic olivine*

It is well documented that Ni in mafic and ultramafic igneous rocks is contained mainly in olivine, the earliest silicate mineral to crystallize from a basaltic magma, followed next in importance by orthopyroxene and then by clinopyroxene. Furthermore, there is a positive correlation between the Mg : Fe ratio and Ni concentration in these minerals, the forsterite-rich members of the olivine series, for example, being richer in Ni than fayalite-rich members. In the older literature, this correlation has been attributed to the substitution of Ni^{2+} (0.69 Å) for Mg^{2+} (0.72 Å) according to Goldschmidt's rules for ionic substitution (camouflage), but this is not consistent with Ringwood's rule because of the much higher electronegativity of Ni (1.88 compared to 1.31 for Mg). Ringwood (1955) argued that the dominant geochemical characteristic of Ni is its camouflage by Fe^{2+} (0.78 Å) in silicate crystals, as should be expected from their almost identical electronegativities (1.83 for Fe compared to 1.88 for Mg), but the smaller size of Ni^{2+} facilitates its preferential entry into early crystals.

Burns and Fyfe (1964, 1966, 1967) invoked the crystal-field theory to explain the Ni enrichment of early magmatic crystals. Absorption spectra of transition-metal compounds and their melts are similar, indicating that ions receive comparable CFSE in the two phases. Moreover, heats of fusion are generally small, one to three orders of magnitude lower than lattice energies, indicating that bond energies in solid and liquid phases are comparable. In silicate melts of basaltic and granitic compositions, transition-metal ions occupy both tetrahedral and octahedral sites, but almost exclusively octahedral sites in silicate minerals crystallizing from such magmas, and the magnitude of the octahedral site-preference energy parameter gives an indication of the relative affinity of an ion in a magma for a silicate crystal. The predicted orders of uptake arranged in terms of the octahedral site-preference energy (in kcal mol^{-1}) are:

Divalent cations: Ni (20.6) > Cu (15.2) > Co (7.4) > Fe (4.0) > Mn (0)

Trivalent cations: Cr (37.7) > Co (19.0) > V (12.8) > Ti (6.9) > Fe (0)

Thus, Ni^{2+} and Cr^{3+}, which have the largest values of the octahedral site-preference energy, should be expected to be readily accommodated by early-formed olivine and spinel. Cu^{2+} in the above list is an exception; presumably it prefers more deformable sites in a magma and is taken up by late-stage minerals (Burns and Fyfe, 1967).

3.4.3 *Colors of transition-metal complexes*

Crystal-field effects are the most common origin of color in transition-metal compounds and many minerals. If the $3d$ orbitals of a transition-metal ion in a molecule have been split into two sets as described above, absorption of a photon of energy in the visible or infrared region of the electromagnetic spectrum (see Fig. 13.1) can cause one or more electrons to jump momentarily from a lower energy orbital to a higher energy orbital, thus creating an excited transient-metal ion. The difference in energy between the ground state and the excited state is equal to the energy of the absorbed photon. Only specific wavelengths of light (or colors) are absorbed, and the substance takes on the color of the transmitted light, which is the complementary color of the absorbed light (Table 3.5).

Table 3.5 Wavelength absorbed and color observed.

Wavelength (λ) absorbed	Color observed
400 nm (violet)	Green-yellow (λ = 560 nm)
450 nm (blue)	Yellow (λ = 600 nm)
490 nm (blue-green)	Red (λ = 620 nm)
570 nm (yellow-green)	Violet (λ = 410 nm)
580 nm (yellow)	Dark blue (λ = 430 nm)
600 nm (orange)	Blue (λ = 450 nm)
650 nm (red)	Green (λ = 520 nm)

$1\ nm = 10^{-9}\ m = 10$ Å.

3.5 Isomorphism, polymorphism, and solid solutions

3.5.1 *Isomorphism*

The term *isomorphism*, which was introduced by Mitscherlich in 1819, means "equal form." In practice, *two substances are called isomorphous if they have similar crystal structures but different chemical formulas*. A typical pair is sodium nitrate ($NaNO_3$) and calcium carbonate ($CaCO_3$), which have almost identical crystal structure but very different physical properties such as hardness and solubility. Isomorphism is widespread among minerals of the spinel group, the garnet group, the pyroxene group, and the amphibole group. The basis of isomorphism is the similarity in ionic size relations of the different ions, leading to the same coordination and the same structure type. Replacement of cations or anions in a crystal lattice should be termed "substitution" or "solution" rather than isomorphous replacement (Fyfe, 1964).

3.5.2 *Polymorphism*

The term *polymorphism*, meaning "many forms", applies to *two or more substances (elements or compounds) that have the same or closely similar chemical formulas but different crystal structures*. The difference among the various structural forms may involve cation coordination, arrangement of the ions, or the nature of bonding. Polymorphism is quite common among minerals. Some familiar examples are the polymorphs of $CaCO_3$ (calcite, trigonal; aragonite, orthorhombic), FeS_2 (pyrite, cubic; marcasite, orthorhombic), C (diamond, cubic; graphite, hexagonal), and Al_2SiO_5 (andalusite, orthorhombic; sillimanite, orthorhombic; kyanite, triclinic). The polymorphs have discernibly different physical and chemical properties, and different pressure–temperature stability fields (see Figs 4.12, 6.4 and 7.10), the high-temperature modifications generally showing a higher crystallographic symmetry.

The transformation from one polymorph to another may be *displacive* or *reconstructive*, or of the *order-disorder* type. Transformations, such as from low-quartz to high-quartz ("low" and "high" denoting, respectively, lower and higher crystallographic symmetry) are displacive because there is very little difference in energy of the two polymorphs so that the change can be accomplished quite readily by minor displacement of the ions. The transformation of graphite to diamond, on the other hand, is an example of reconstructive transformation; it is much slower and more difficult, because it involves the formation of a new crystal structure with reconstructed bonds. In general, higher pressure favors polymorphs with high densities and large coordination numbers (e.g., calcite = aragonite; sillimanite = kyanite), whereas higher temperature favors polymorphs with low densities and small coordination numbers (e.g., quartz = tridymite; kyanite = sillimanite).

A crystal structure is said to be "disordered" if the constituent ions (or atoms) are randomly distributed among crystallographically equivalent sites (sites that are the same in terms of crystallographic symmetry), and "ordered" if the distribution is not random. Perfect ordering occurs only at absolute zero temperature; the degree of ordering gradually decreases with increasing temperature, and the crystal structure becomes completely disordered above a certain temperature characteristic of the structure and composition of the crystal. Between these two end-states of perfect order and complete disorder, there may be stable states of varying degrees of disorder, each of which may be considered a separate polymorph. The three polymorphs of $KAlSi_3O_8$ (potassium feldspar) – microcline, orthoclase, and sanidine – are an important example of order–disorder relationship with respect to the three Si^{4+} ions and one Al^{3+} ion, which occupy four tetrahedral sites that differ slightly in size and nearest neighbor configuration. Microcline is the low-temperature, ordered (triclinic) form of $KAlSi_3O_8$ in which Al^{3+} ions occupy one kind of site and three Si^{4+} ions the other three; orthoclase is the medium-temperature, partially ordered form (pseudomonoclinic); and sanidine is the high-temperature, disordered form (monoclinic) characterized by more random distribution of the Si^{4+} and Al^{3+} ions.

3.5.3 *Solid solutions*

A *solid solution* is a solution in the solid state of one or more solutes in a solvent whose crystal structure remains unchanged by addition of the solutes. Almost all minerals are solid solutions to varying degrees. The range of compositions produced by solid solution in a given mineral is known as a *solid solution series* and its compositional extremes as *end members*. A solid solution series may be *continuous*, in which case all intermediate members are possible (e.g., the olivine solid solution series with Mg_2SiO_4 and Fe_2SiO_4 as end members) or *discontinuous*, in which case only a restricted range of composition between the end members is found (e.g., the limited solid solution between ZnS and FeS).

Solid solution should not be confused with isomorphism, because they are distinct concepts (Mason, 1966). Isomorphism is neither necessary to, nor sufficient for, solid-solution formation. Many isomorphous substances show little solid solution (e.g., calcite, $CaCO_3$ and smithsonite, $ZnCO_3$), and extensive solid solution may occur between components that are not isomorphous (e.g., FeS and ZnS, which have very different crystal structures).

There are three kinds of solid solutions based on the mechanism that causes their chemical composition to vary (Fig. 3.13):

(1) *Substitutional solid solution* (Fig. 3.13a), in which one or more kinds of ions or atoms are substituted by other kinds. This is the most common type in the realm of minerals. A very good example is the binary olivine solid solution series, $(MgFe)_2SiO_4$, that spans all intermediate compositions ranging from the magnesium end-member Mg_2SiO_4 (forsterite) to the iron end-member Fe_2SiO_4 (fayalite) because of diadochic substitution of Mg^{2+} by Fe^{2+}, which carry the same charge and have similar ionic size. Another familiar

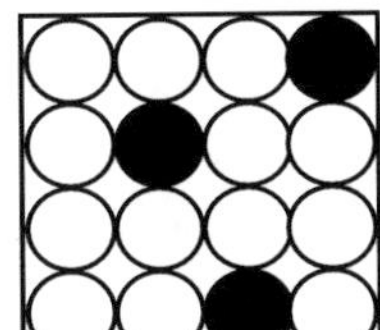

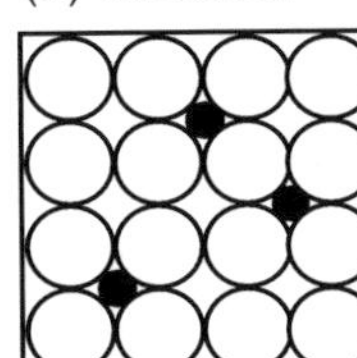
(c) Omission
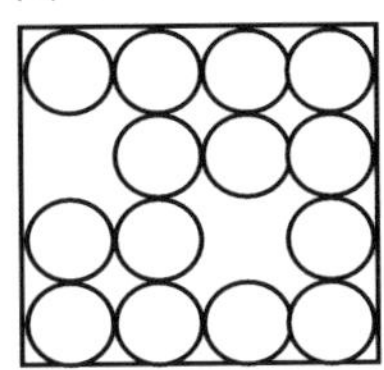

Fig. 3.13 Schematic representation of the three kinds of solid solutions: (a) substitutional solid solution; (b) interstitial solid solution; and (c) omission solid solution.

example is the plagioclase solid solution series characterized by progressive coupled substitution of Na^+ and Si^{4+} by Ca^{2+} and Al^{3+} (equation 3.8).

(2) *Interstitial solid solution* (Fig. 3.13b), in which foreign ions or atoms are added to fill unoccupied interstitial crystal sites (❑) that exist between ions or ion groups. An example of interstitial solid solution is the incorporation of Na^+ into the structure of cristobalite (a high-temperature polymorph of quartz) to compensate for the charge imbalance created by the replacement of a small amount of Si^{4+} by Al^{3+}. Minerals such as beryl ($Be_3Al_2Si_6O_{18}$), whose structures provide large openings, are particularly susceptible to interstitial substitution. Beryl contains large channel-like cavities that can be occupied by relatively large monovalent cations, the charge balance being maintained by coupled substitution of Al^{3+} or Be^{2+} for Si^{4+} in tetrahedral sites:

❑ + $Si^{4+} \Leftrightarrow Al^{3+}$ + (K^+, Rb^+, Cs^+) or ❑ + $Si^{4+} \Leftrightarrow Be^{2+}$ + 2(K^+, Rb^+, Cs^+)

This type of solid solution is very common in metals, which easily accommodate small atoms such as H, C, B, and N.

(3) *Omission solid solution* (Fig. 3.13c), in which some ion sites that are normally occupied remain vacant. The best example of such a solid solution – actually a type of crystal defect – is the monosulfide mineral pyrrhotite whose chemical analysis always shows more sulfur than its theoretical proportion in FeS. It is now well established that this discrepancy is due to a deficiency of Fe, not an excess of S, in the crystal structure, so that the generalized chemical formula of pyrrhotite is written as $Fe_{1-x}S$, where x varies between 0 and 0.125. The electrical neutrality in this structure is maintained by the replacement of three Fe^{2+} by only two Fe^{3+} leaving one site vacant: $3Fe^{2+} \Rightarrow 2Fe^{3+}$ + ❑.

The extent of solid solution between given solutes and solvent depend on the factors that affect the degree of substitution (see section 3.3). The favorable factors include: similar ionic size, charge, and electronegativity; flexibility of the solvent crystal structure to accommodate local strains (by bending bonds rather than by stretching or compressing them); and high temperature and low pressure. Thermodynamic aspects of solutions, including solid solutions, will be discussed in Chapter 5.

3.6 Covalent bonding

3.6.1 *Valence bond theory versus molecular orbital theory*

Soon after the application of quantum mechanics to the hydrogen atom, scientists began applying quantum mechanics to molecules, starting with the H_2 molecule. The treatment of covalent bonding, based on the principles of quantum mechanics, has developed in two distinct, but mutually consistent, forms: the *valence bond theory*, and the *molecular orbital theory*. In the valence bond approach, covalent bonds between two atoms A and B are formed by the sharing of valence electrons of opposite spin, resulting from interaction between atomic orbitals (AOs) that contain these electrons. In the molecular orbital approach, we start with the nuclei of the two atoms and feed all the electrons of the molecule into *molecular orbitals* (MOs) that are constructed by appropriate combination of atomic orbitals of about the same energy. Thus, atomic orbitals are associated with a single atom (or ionic species), whereas molecular orbitals are associated with the molecule as a whole. It is assumed that electrons would fill molecular orbitals following the same principles as followed by electrons for filling atomic orbitals:

(1) The molecular orbitals are filled in a way that yields the lowest potential energy for the molecule.
(2) Each molecular orbital can accommodate a maximum of two electrons of opposite spin (Pauli's exclusion principle).
(3) Orbitals of equal energy are half filled with electrons having parallel spin before they begin to pair up with electrons of opposite spin (Hund's rule).

The molecular orbital approach is more powerful because the orbitals reflect the geometry of the molecule to which they are applied. The downside is that it is a more difficult concept to visualize.

The essence of covalent bonding can be explained by the formation of the simplest of all molecules, the homonuclear diatomic H_2 molecule. An isolated H atom has the ground-state electron configuration $1s^1$, and the probability density for this one electron is distributed spherically about the H nucleus. When two hydrogen atoms approach each other, the electron charge clouds represented by the two $1s$ atomic orbitals begin to merge (which is referred to as *overlap* of atomic orbitals), and the electron charge density begins to shift. Due to the attraction between the electron of one H atom and the positively charged nucleus of the other H atom, the overlap gradually increases with decrease in the internuclear distance (R) between the two atoms. How close to each other the two atoms can get is determined by the short-range repulsion between the two positively charged nuclei. The merger reaches its limit when the potential energy of the system (E_p) reaches a minimum; this is the position of closest approach of the two atoms and it defines the equilibrium internuclear distance (R_0). At this point, we have formed a H_2 molecule consisting of two

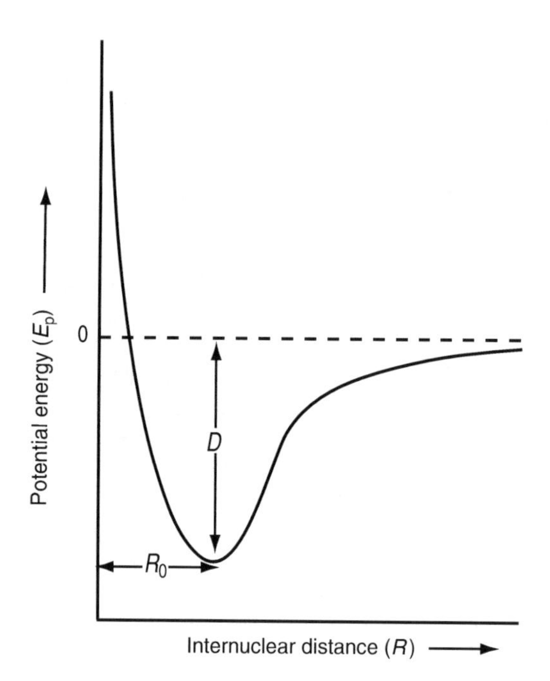

Fig. 3.14 Potential energy (E_p) diagram for the diatomic molecule H_2. For large values of internuclear distance (R), the potential energy of the system, which consists of two isolated hydrogen atoms, is arbitrarily assigned a value of zero. R_0 represents the equilibrium internuclear distance at which the potential energy reaches a minimum and the system attains maximum stability; at smaller values of R, the strong repulsion between the nuclei causes the potential energy to rise sharply. For the formation of a covalent H_2 molecule, $R_0 = 0.74$ Å, and the corresponding minimum potential energy is -435 kJ mol^{-1} (which corresponds to -7.23×10^{-19}kJ per H_2 molecule). The energy D closely approximates the experimental dissociation energy of the molecule into its component atoms in their ground states.

nuclei held together by a covalent bond comprised of two shared electrons with opposite spins. This system has a lower potential energy than the system comprised of the two isolated H atoms (Fig. 3.14), and thus is more stable. For each such covalent bond, there is a condition of optimal overlap that results in maximum bond strength (*bond energy*) at a particular internuclear distance (*bond length*).

The formation of a covalent bond between the two H atoms can be rationalized in two ways. Viewed from the perspective of the valence bond theory, the two electrons occupy the region of highest density of electron negative charge between the two nuclei, the region where the two orbitals overlap (Fig. 3.15a). Both electrons are now in the orbitals of both H atoms, and each H atom may be considered to have the stable electron configuration of He ($1s^2$). In terms of the molecular orbital theory, the two 1*s* atomic orbitals merge into a bigger electron cloud, a *molecular orbital* (MO). The MO may be visualized as the volume within which we should find a high percentage of the negative charge generated by the electrons (Fig. 3.15b). The two electrons are shared equally between the nuclei and are identified with the entire H_2 molecule, not with either of the nuclei. This sharing of the electron pair forms a covalent bond.

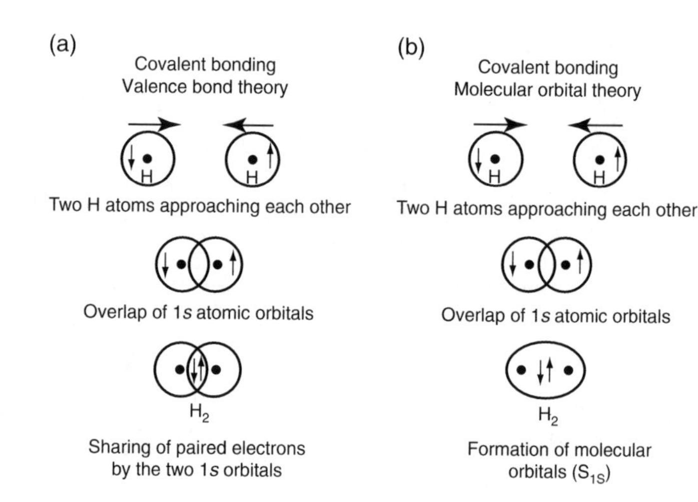

Fig. 3.15 Formation of a covalent bond in a H_2 molecule by sharing of electrons: (a) Overlap of two 1*s* atomic orbitals (valence bond theory); (b) Formation of a molecular orbital by merger of two 1*s* orbitals (molecular orbital theory).

3.6.2 *Covalent radii*

As in the case of ionic bonding (see Fig. 3.2), R_0 in Fig. 3.14 represents the sum of the covalent radii of the two H atoms, the bond length, from which the covalent radius of each atom can be determined. In the case of the H_2 molecule, the covalent radius will just be the one-half of R_0, and it is applicable only to crystal structures in which H is covalently bonded. The covalent radius of an atom varies with the number of bonds the atom has with its neighbors and the distortion of its atomic orbitals as a result of hybridization (see section 3.6.3). Tabulated values of covalent radii represent either average or idealized values.

The covalent radius of an atom, unlike its ionic radius, should not be visualized as the radius of a spherical atom; the concept of covalent radius is applicable only to interatomic distances between atoms joined by covalent bonds and not to distances between atoms of the same kind when not so joined (Evans, 1966). For example, the single-bond covalent radius of Cl atom is 0.99 Å, which is one-half of the measured Cl–Cl interatomic distance of 1.99 Å in the covalent Cl_2 molecule. In the crystal structure of solid chlorine, however, the interatomic distance between neighboring Cl atoms of different molecules, which are bound together only by weak van der Waals forces (see section 3.8), is about 3.6 Å, giving about 1.8 Å as the van der Waals radius of the Cl atom, the same as the ionic radius of Cl^- in octahedral coordination. In general, the covalent radius of an element is much smaller than its ionic radius because of greater penetration of unpaired electron orbitals in covalent bonding.

3.6.3 *Hybridization of atomic orbitals*

The formation of molecular orbitals is not all that happens to atomic orbitals when atoms approach each other. Before combining across atoms, atomic orbitals that are close to each other in energy within the same atom have the ability to combine in an additive way, forming a new set of orbitals that is at a lower total energy in the presence of the other atoms than the pure atomic orbitals would be. This process of blending of atomic orbitals is called *hybridization*, and the newly formed orbitals are called *hybrid orbitals*. (1s and 2s orbitals do not hybridize because of the large difference in their energy levels.) The number of hybrid orbitals formed always equals the number of atomic orbitals involved in the hybridization (Companion, 1964, p. 59). The hybrid orbitals of an atom (or ionic species) can overlap with orbitals on other atoms (or ions) to share electrons and form covalent bonds. The importance of hybrid orbitals lies in the fact that they usually provide a better description of the experimentally observed geometry of the molecule (or ionic species), especially for molecules formed with carbon, nitrogen, or

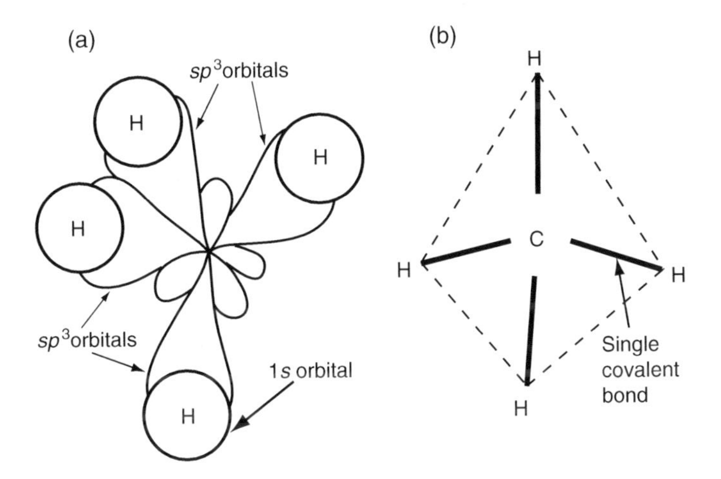

Fig. 3.16 Hybridization of atomic orbitals in the formation of a CH_4 molecule: (a) overlap between a carbon atom with sp^3 hybrid orbitals and four 1*s* orbitals of four H atoms; (b) the tetrahedral symmetry of the CH_4 molecule with four equivalent σ bonds.

oxygen (and to a lesser extent phosphorous and sulfur). This is why the main application of hybridization lies in the field of organic chemistry.

To illustrate the concept of hybridization, let us consider the covalent bonding in CH_4 (methane gas). Experimental results tell us that the C in this compound is bonded to the four H atoms by four equivalent covalent bonds, but the ground-state electronic configuration of C ($Z = 6$), $1s^2\ 2s^2\ 2p_x^{\ 1}\ 2p_y^{\ 1}$, suggests that the two *p* orbitals with unpaired electrons should result in the formation of only two covalent bonds. This discrepancy can be resolved by invoking hybridization of the 2*s* orbital with the three 2*p* orbitals (whose energy level is not that different from that of the 2*s* orbital) to form four sp^3 hybrid orbitals, each with one unpaired electron:

$$\underset{1s}{\underline{\uparrow\downarrow}}\ \underset{2s}{\underline{\uparrow\downarrow}}\ \underset{2p}{\underline{\uparrow}}\ \underset{2p}{\underline{\uparrow}}\ \underset{2p}{\underline{\ \ }}\quad \underset{\text{hybridization}}{\underline{\Rightarrow}}\quad \underset{1s}{\underline{\uparrow\downarrow}}\ \underset{sp^3}{\underline{\uparrow}}\ \underset{sp^3}{\underline{\uparrow}}\ \underset{sp^3}{\underline{\uparrow}}\ \underset{sp^3}{\underline{\uparrow}}$$

Ground-state C atom — Four sp^3 hybrid orbitals

The four sp^3 orbitals are directed in space toward the four corners of a regular tetrahedron. When four H atoms, each with an unpaired 1*s* electron, approach such a C atom, the overlap of the four 1*s* AOs of H and the four hybrid AOs of C form the CH_4 molecule with four equivalent covalent bonds. The shape of the CH_4 molecule would consequently be like that of a tetrahedron (Fig. 3.16); experiments have confirmed that is the case.

There are many types of hybridization that give rise to characteristic molecular configurations; some common configurations are listed in Table 3.6 and illustrated in Fig. 3.17.

3.6.4 *Sigma (σ), pi (π), and delta (δ) molecular orbitals*

There are three kinds of molecular orbitals (MOs) that are of interest to us: (i) sigma (σ) MOs; (ii) pi (π) MOs; and (c) delta (δ) MOs. Only σ MOs can be constructed from *s* AOs; only σ and

Table 3.6 Some common hybrid orbital configurations.

Hybrid[1]	Number of bonds	Orbitals on the hybridized atom	Configuration of bonds	Examples
sp	2	$s + p$	Linear	CO_2, C_2H_2
sp^2	3	s + two ps	Trigonal-planar (to corners of an equilateral triangle)	C_2H_4, C (graphite)
sp^3	4	s + three ps	Tetrahedral (to corners of a regular tetrahedron)	CH_4, C (diamond)
dsp^2	4	s + two ps + d	Square planar (to corners of a square)	$Ni(CN)_4^{2-}$
d^2sp^3	6	s + three ps + two ds	Octahedral (to corners of a regular octahedron)	SF_6
d^4sp	6	$s + p$ + four ds	Trigonal prismatic (to corners of a trigonal prism)	MoS_2

[1]Hybridization at the central atom (e.g., C in CH_4 molecule).

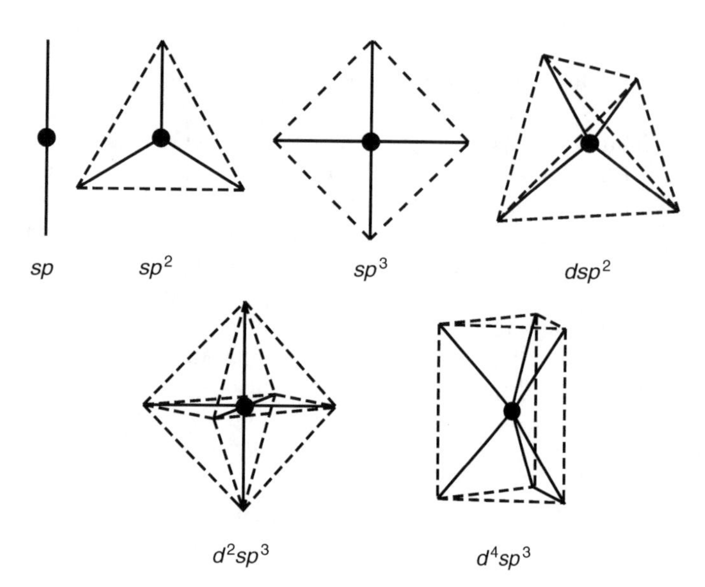

Fig. 3.17 The bond configurations corresponding to some simple hybrid orbitals. The solid dot in each represents the central atom (e.g., C in a CH_4 molecule) for the bonds. *sp*, linear; sp^2, to corners of an equilateral triangle; dsp^2, to corners of a square; sp^3, to corners of a regular tetrahedron; d^2sp^3, to corners of a regular octahedron; d^4sp, to corners of a trigonal prism. (From *An Introduction to Crystal Chemistry*, 2nd edition, by R.C. Evans, Figure 4.01, p. 60; Copyright 1966, Cambridge University Press. Reproduced with permission of the publisher.)

π MOs can be constructed from *p* AOs; and σ, π, and δ MOs can be constructed from *d* AOs (Companion, 1964). Some typical examples of molecular orbital formation are described below.

The MO formed by the overlap of *s* orbitals is designated as a σ MO (Fig. 3.18a), and for the H_2 molecule as $\sigma_{1s}{}^2$ since it is occupied by two electrons due to the merger of two 1*s* AOs. The distinguishing criterion of σ MOs is that it has no nodal plane (a plane of zero electron density) containing the internuclear axis. The strongest kinds of covalent bonds are associated with σ MOs. Actually, the covalent bonding is more complicated because of the formation of complementary *bonding* and *antibonding* molecular orbitals (see Box 3.2); for simplicity only bonding MOs are shown in Fig. 3.18.

When larger atoms are involved in the formation of covalent molecules, σ MOs can also form when two *p* AOs overlap end-on (e.g., in a diatomic molecule such as O_2, F_2, or N_2), or when a *p* orbital in one atom interacts with an *s* orbital in another atom (e.g., in a molecule of HF). As can be seen in Fig. 3.18, the σ MO formed by end-on overlap of two p_z orbitals has two nodal planes (the *xy* and *xz* planes), but neither contains the internuclear axis; the π MO formed by sideways overlap of two p_z orbitals has only one nodal plane (the *xy* plane) and it contains the internuclear axis; and the δ MO formed from face-to face overlap of two $d_{x^2-y^2}$ orbitals has two nodal planes (the *xz* and *yz* planes) each of which contains the internuclear axis.

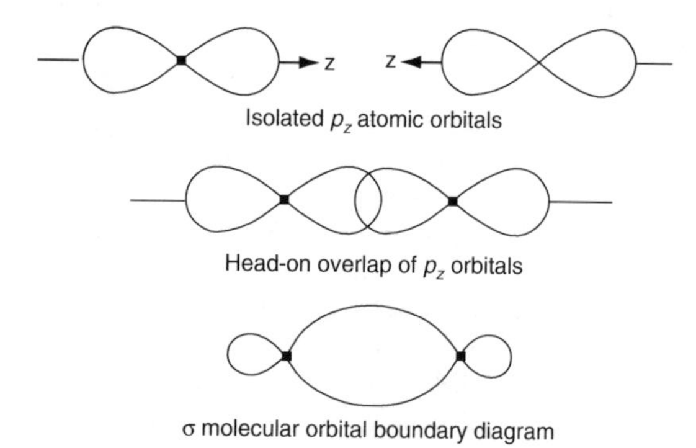

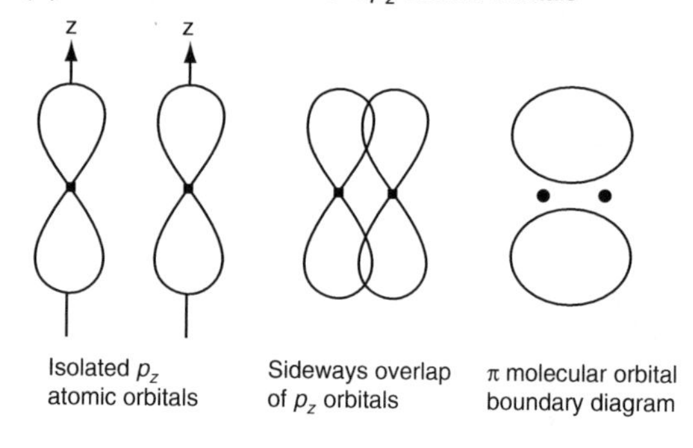

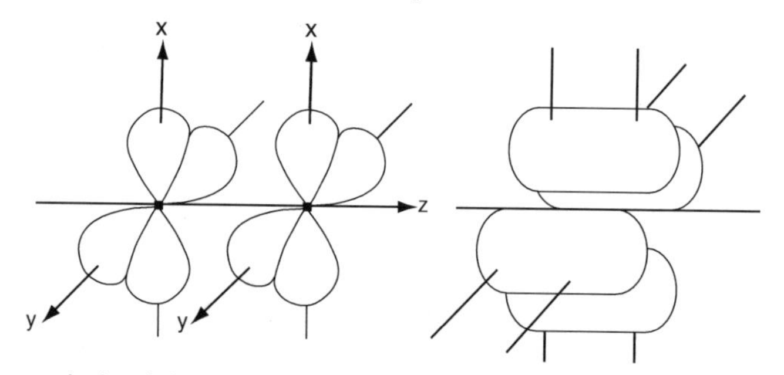

Fig. 3.18 The formation of bonding molecular orbitals: (a) σ molecular orbital by end-on overlap of two $2p_z$ atomic orbitals; (b) π molecular orbital by sideways overlap of two $2p_z$ atomic orbitals; and (c) δ molecular orbital by face-to-face overlap of two $3d_{x^2-y^2}$ atomic orbitals. For molecules larger than H_2, the *z* direction is assumed to be the internuclear axis, but the labeling of *x* axis and *y* axis is arbitrary. (After Companion, 1964.)

3.6.5 *The degree of ionic character of a chemical bond: Electronegativity*

In pure ionic bonding the transferred electron(s) should be associated solely with, and distributed symmetrically around, the nucleus of the anion so formed. Pure ionic bonds, however, do not exist in nature. In reality, the coulombic attraction exerted by

Box 3.2 Bonding and antibonding molecular orbitals

The wave properties of the electron cause variation in the intensity of negative charge generated by the electron. Just as in-phase superposition of light waves leads to an increase in the light intensity and out-of-phase superposition a decrease, the intensity of the negative charge is enhanced by in-phase interaction of electron waves and decreased by out-of-phase interaction.

Let us consider the molecular orbital (MO) formation from two 1*s* atomic orbitals (AOs). The model of MO formation assumes that two 1*s* AOs can overlap in two extreme ways – in-phase and out-of-phase interaction – to form two MOs. The in-phase interaction creates an increase in the negative charge between the two nuclei, leading to an increase in the attraction between the electron and the atoms in the bond, and thus to lower potential energy, which makes it energetically preferable to the two separate 1*s* AOs. This orbital is called a *bonding molecular orbital*, and is designated as σ_{1s} in which the symbol σ (sigma) stands for the fact that the orbital is cylindrically symmetrical about the internuclear (or bond) axis (Fig. 3.19).

The out-of-phase interaction creates exactly the opposite situation, a MO of higher potential energy and, therefore, energetically less favorable compared to the separate 1*s* AOs. A molecular orbital of this type, in which the electrons tend to destabilize the bond between atoms, is called an *antibonding molecular orbital*. It is also symmetrical about the bond axis, and to distinguish it from a bonding orbital is designated with an asterisk as σ^*_{1s} (Fig. 3. 19).

Bonding and antibonding orbitals are also formed by the overlap of *p* and *d* AOs. For example, when two O atoms combine to form a O_2 molecule, the end-on overlap of the two $2p_z$ AOs generates the $\sigma_{2p(z)}$ (bonding) and $\sigma^*_{2p(z)}$ (antibonding) MOs (the *z* direction is assumed to be the internuclear axis). If the remaining *p* orbitals overlap, they must do so sideways, forming what are designated as π MOs. Thus, the sideways overlap of the two $2p_x$ AOs would generate $\pi_{2p(x)}$ (bonding) and $\pi^*_{2p(x)}$ (antibonding) MOs, and the sideways overlap of the two $2p_y$ AOs would generate another pair of $\pi_{2p(y)}$ (bonding) and $\pi^*_{2p(y)}$ (antibonding) MOs of the same potential energy.

Bonding and antibonding configuration of molecular orbitals are commonly depicted in diagrams such as Fig. 3.19 and Fig. 3.20. Figure 3.19 is for the H_2 molecule, the simplest case; Fig. 3.20 represents a generalized framework of the expected molecular orbital diagram resulting from the overlap of 1*s*, 2*s*, and 2*p* AOs, and is applicable to molecules such as O_2, F_2, CO, and NO. Because they meet head-on, the interaction between $2p_z$ orbitals is stronger than the interaction between $2p_x$ or $2p_y$ orbitals, which meet edge-on. As a result, the σ_{2p} orbital has a lower energy than the π_{2p} orbitals, and the σ^*_{2p} orbital has a higher energy than the π^*_{2p} orbitals.

The procedure for filling electrons in the MOs is as follows:

(1) Find out (or work out) the electronic configuration of the atoms involved (for the F atom, for example, it is $1s^2\ 2s^2\ 2p^5$).
(2) Fill the molecular orbitals from bottom to top until all the electrons are added, remembering that the number of MOs generated must equal the number of AOs being merged (because we must have the same number of places to put electrons in the molecule that we had in the atoms) and that MOs of equal energy are half-filled with parallel spin (↑) before they are paired with opposite spin (↑↓).

The number of electrons in the bonding and antibonding MOs can be used to calculate the *bond order* (BO) and predict the stability of the covalent molecule. The bond order is defined as

$$\text{Bond order (BO)} = 0.5\ (\text{number of electrons in bonding MOs} - \text{number of electrons in antibonding MOs}) \qquad (3.9)$$

Bond order values of 0, 1, 2, and 3 correspond to classical no bond, single, double, and triple bonds. If for a molecule BO = 0, the molecule is unstable; if BO > 0, the molecule is stable. The higher the bond order, the more stable is the covalent bond. We can also use the molecular orbital diagram to predict whether the molecule is paramagnetic (i.e., weakly attracted to magnetic fields) or diamagnetic (i.e., weakly repelled by magnetic fields). Molecules that contain unpaired electrons are paramagnetic; those that contain only paired electrons are diamagnetic. The H_2 molecule (see Fig. 3.19), for example, is diamagnetic and its bond order is 0.5 (2–0) = 1, indicating that it is stable.

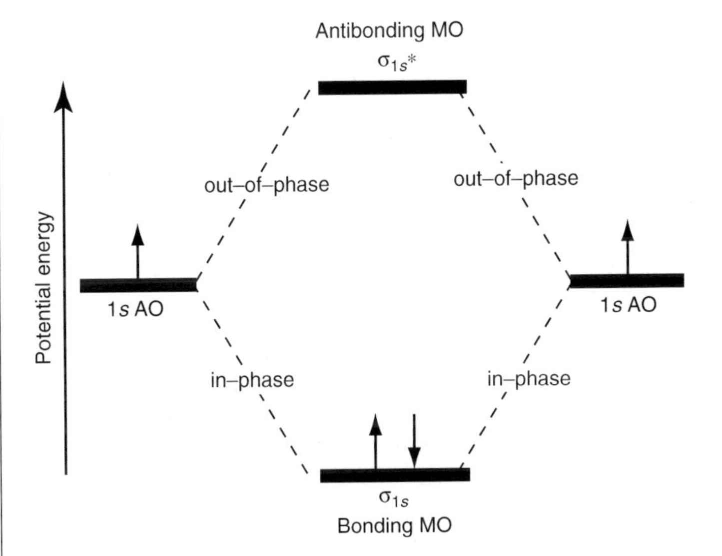

Fig. 3.19 Molecular orbital diagram for the H_2 molecule (Z = 1) showing sigma bonding and antibonding molecular orbitals (MO) formed from the overlap of 1*s* atomic orbitals (AO).

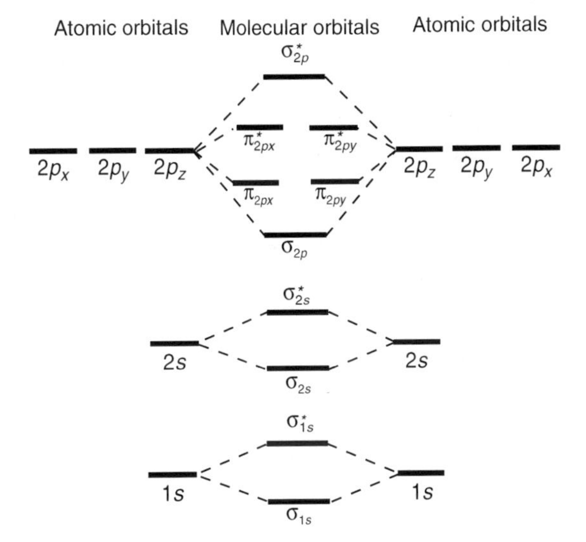

Fig. 3.20 Generalized framework of expected molecular orbital diagram from the overlap of 1*s*, 2*s*, and 2*p* orbitals for covalent molecules such as O_2, F_2, CO, and NO.

the neighboring cations forces the electron density of the anion to be concentrated to a small extent in the region between the nuclei. This distortion, referred to as *polarization* of the anion, results in a degree of electron sharing, or partial covalency, in any real ionic bond. The larger the *ionic potential* (charge to radius ratio) of a cation, the stronger is the polarization.

Polarization of a covalent bond occurs when the electron density becomes asymmetric around the atoms because one of the atoms exerts greater attraction on the shared electrons than the other, giving that atom a slight negative net charge (and leaving a complementary small positive charge on the other atom). This, in effect, amounts to transferring a fraction of an electron from one atom to another and imparting a degree of ionic character to the predominantly covalent bond. Such bonds, as in the *heteronuclear* diatomic molecule HF, formed by unequal sharing of electron pairs, are called *polar covalent bonds*, as opposed to *nonpolar covalent bonds*, in all *homonuclear* diatomic molecules (such as H_2, O_2, N_2, F_2, Cl_2, etc.) in which the bonding electron pairs are shared equally between the nuclei. In the case of HF, for example, the F atom (which has a higher electronegativity) attracts the shared electron pair more strongly than does the H atom. This causes a distortion of the electron density, and its small shift toward the F atom leaves the H-end of the HF molecule slightly positive and the F-end slightly negative. The polar HF molecule is commonly represented as

$$\overset{\delta+}{\mathrm{H}} - \overset{\delta-}{\mathrm{F}}$$

where the $\delta+$ over the H atom and the $\delta-$ over the F atom indicate that the "H-end" of the molecule is more positive relative to the "F-end," and *vice versa* (not that H has a charge of +1 and F a charge of −1). The separation of charge in a polar covalent bond creates an *electric dipole*, whose strength is expressed in terms of its *electric dipole moment* (μ), which is defined as

$$\mu = d\delta \qquad (3.10)$$

where d is the distance separating the charges of equal magnitude and opposite sign, and δ is the magnitude of the charge. All molecules whose positive-charge and negative-charge centers do not coincide possess an electric dipole moment.

Realizing that there must exist a continuous progression between purely covalent bonds and dominantly ionic bonds, the American Nobel laureate Linus Pauling (1901–1974) introduced the concept of *electronegativity* in 1932 to quantitatively express the degree of ionic character of a mixed ionic–covalent bond. Electronegativity (χ) is a parameter that describes *the tendency of an atom in a molecule to attract electrons towards itself when chemically combined with another atom.* (The equivalent property of a free atom is its electron affinity – see section 2.4.1.) Thus, electronegativity, which is related to ionization potential and electron affinity, is a measure of the ability of an atom to compete for electrons with other atoms to which it is bonded. For example, the electronegativity of chlorine is higher than that of hydrogen in an HCl molecule, which means that the chlorine atom displays greater attraction for electrons than does the hydrogen atom.

Whereas the chemical applications of ionization potential are limited to elements that lose electrons to form cations, the concept of electronegativity is applicable to all kinds of elements and the bonds they form. However, unlike ionization potential, electronegativity cannot be measured and has to be calculated from other atomic or molecular properties.

Several methods of calculation have been proposed, but the most commonly used method is the one originally proposed by Pauling based on measured electric dipole moments. Pauling's electronegativity is a dimensionless quantity that is estimated on an arbitrary scale ranging from 0 (helium) to 4.1 (fluorine) (see Appendix 3). In general, electronegative values increase within a period of the Periodic Table with increasing atomic number – e.g., from $\chi = 0.98$ for Li ($Z = 3$) to $\chi = 3.98$ for F($Z = 9$) – and decrease from top to bottom within a group – e.g., from $\chi = 3.98$ for F ($Z = 9$) to $\chi = 2.2$ for At ($Z = 85$). The inert gases (He, Ne, etc.) have zero (or close to zero) electronegativity values, and the peak values belong to the halogens (F, Cl, Br, I; see Fig. 2.9). Cation-forming elements (electron donors) with low values of electronegativity, such as alkali and alkaline earth metals, are called *electropositive*; anion-forming elements (electron acceptors) characterized by high values of electronegativity, such as the halogens and oxygen ($\chi = 3.44$), are called *electronegative*.

Although electronegativity is not a precisely defined atomic property, the electronegative difference between two atoms provides a useful measure of the polarity and ionic character of the bond between them in solid compounds. The larger the electronegativity difference between two atoms, the greater is the probability of them forming an ionic bond in which the more electronegative element represents an anion and the less electronegative element a cation. Between elements with similar electronegativity values, neither has a stronger preference for electrons, and they tend to share electrons by covalent bonding. Typically, a bond is considered ionic if the electronegativity difference between the two bond-forming elements is > 2.1. Thus, the Na–Cl bond ($|\chi_{Na} - \chi_{Cl}| = |0.93–3.16| = 2.23$) should be ionic whereas the Fe–S bond ($|\chi_{Fe} - \chi_S| = |1.83–2.58| = 0.75$) should be covalent. The Mg–O bond ($\Delta\chi = 2.13$) and Ca–O bond ($\Delta\chi = 2.44$), which are common in rock-forming silicate minerals, are also ionic. However, Si–O and Al–O bonds ($\Delta\chi = 1.54$ and 1.83, respectively) in the same minerals should be considered covalent by this definition, although they are generally regarded as ionic for the purpose of silicate mineral crystal structures.

Based on the premise that elements having different electronegativities form bonds whose ionic character is proportional to the magnitude of electronegativity difference (Fig. 3.21), Linus Pauling suggested the following empirical relation to calculate the percentage ionic character of a bond (PIC) between two atoms A and B:

$$\mathrm{PIC}_{A\text{–}B} = 16|\chi_A - \chi_B| + 3.5|\chi_A - \chi_B|^2 \qquad (3.11)$$

where χ_A and χ_B are the electronegativities of the two atoms (see Appendix 3). According to this formulation, the Na–Cl bond (PIC = 53.1%) is only slightly more ionic than covalent

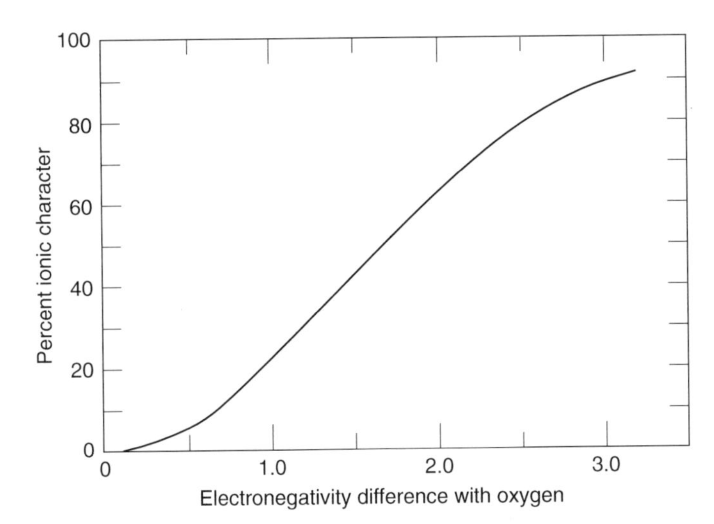

Fig. 3.21 The degree of ionic character of a single chemical bond between an element (A) and oxygen (B) as a function of the electronegativity difference $|\chi_A - \chi_B|$. An electronegativity difference of about 1.7 corresponds to a bond that is 50% ionic; bonds with larger electronegativity difference are primarily ionic, those with a smaller difference are primarily covalent, but there are exceptions. Source of data: Faure (1998).

(although NaCl is often cited as a typical example of an ionic compound), the Ca–O bond (PIC = 59.9%) is mostly ionic, the Mg–O bond (PIC = 50%) is as much ionic as covalent, and the Si–O bond (PIC = 32.9%) is dominantly covalent. Thus, electronegativity alone does not provide accurate predictions about the bonding character and coordination numbers in all kinds of compounds.

Example 3.3: Calculation of the percentage ionic character of the bond between H and F in the HF molecule, using the electronegativity values and empirical relation given by Pauling

$\chi_H = 2.20$; $\chi_F = 3.98$ (Appendix 3)

$$\begin{aligned} PIC_{H-F} &= 16|\chi_H - \chi_F| + 3.5|\chi_H - \chi_F|^2 \\ &= 16|2.20 - 3.98| + 3.5\,|2.20 - 3.98|^2 \\ &= (16 \times 1.78) + (3.5 \times (1.78)^2) = 28.48 + 11.09 = 39.57 \end{aligned}$$

So, the H–F bond in the HF molecule is about 40% ionic and 60% covalent.

3.7 Metallic bonds

Most metallic elements display *close-packed structures* (spheres, representing atoms, touching all their immediate neighbors), which represent the geometrically most compact arrangement of spheres in space (see Evans, 1966, fig. 5.01, p. 81). The close-packed arrangement results in crystal structures with hexagonal or cubic (face-centered and body-centered) symmetry (Fig. 3.22).

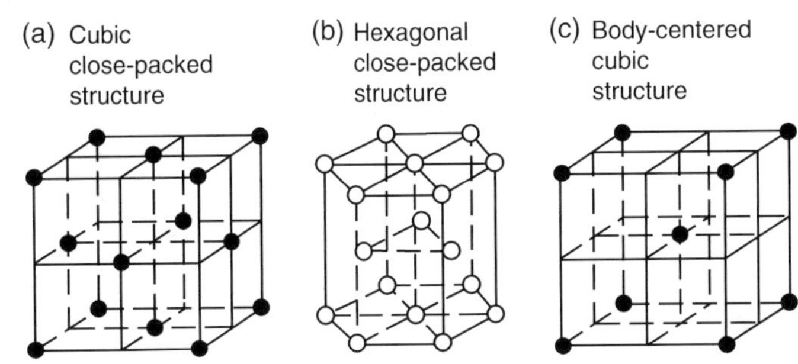

Fig. 3.22 Crystal structures of metallic elements. (a) Cubic close-packed structure (face-centered cubic), in which each atom is surrounded by 12 neighbors (e.g., Cu, Ni, Au). (b) Hexagonal close-packed structure, in which each atom is surrounded by 12 neighbors, 6 at a slightly greater distance than the other 6 (e.g., Mg, Ti, Zn). (c) Body-centered cubic structure, in which each atom is surrounded by 8 neighbors at the corners of a cube (e.g., Li, Na, Mo). Assuming the atoms to be perfect spheres, the packing efficiency (volume of the spheres occupied by the sphere/total volume) is 74.05% for both cubic and hexagonal close-packed structures; the body-centered cubic structure is not a close-packed array and its packing efficiency drops to 68.02%.

Metallic bonding is the bonding that exists among the atoms within the crystal structure of a metallic element. The bonding cannot be ionic because a metal is not composed of cations and anions. Van der Waals bonds (see section 3.8) are too weak to account for the high melting points of metals, and localized covalent bonds are unlikely in view of the fact that each atom in the crystal structure of metals has 12 to 14 near neighbors. Moreover, the bonding in metals must explain their special properties, such as electrical and thermal conductivity and malleability, which are not characteristics of ionic or covalent compounds.

Most metals have *s* electrons external to filled or partially filled inner shells – for example, the two 4*s* electrons in Ni ($1s^2$ $2s^2$ $2p^6$ $3s^2$ $3p^6$ $3d^8$ $4s^2$). These *s* electrons are "delocalized," partly because they are most easily ionized and partly because they occupy a large volume in space relative to other electrons and present an overall low electron density to the positive nucleus (Companion, 1964). For a simple model to explain metallic bonding, it is assumed that the lattice sites in the metal crystal are occupied by the relatively small positive ionic cores (e.g., Ni^{2+}), which are surrounded by a "sea" of delocalized (loosely bound) electrons. This is why atoms or layers of atoms are allowed to slide past each other, resulting in the characteristic properties of malleability and ductility.

Metallic bonding is the electrostatic attraction between the ionic cores and the delocalized electrons. The latter are not associated with a single atom or a covalent bond; they are contained within a molecular orbital that is shared with all the ionic cores in the crystal and extends over the entire crystal in all directions with equal probability. These electrons are free

Table. 3.7 Van der Waals radius (Å) of some elements.

Element	Radius	Element	Radius	Element	Radius	Element	Radius
H (Z = 1)	1.20	N (Z = 7)	1.55	F (Z = 9)	1.35	S (Z = 16)	1.85
C (Z = 6)	1.7	O (Z = 8)	1.52	P (Z = 13)	1.9	Cl (Z = 17)	1.8

to move around throughout the crystal structure and thus are capable of conducting electricity. Without any particular direction for a net movement of electrons in the crystal structure, there is no flow of electricity in an isolated piece of metal, but a current will flow through the metal piece if it is connected to the terminals of a battery, confirming the presence of delocalized electrons. The thermal conductivity of metals arises from the fact that their loosely bound electrons can transfer heat energy at a faster rate than the tightly bound electrons in substances with covalent bonding. (There are a few nonmetals that can conduct electricity: graphite because, like metals, it has free electrons, and molten and aqueous ionic compounds because they have moving ions.)

Atomic radii of metals can be derived from measurements of their cell dimensions. In the case of metals with close-packed structures, the atomic radius is half the distance of the interatomic distance. In structures of lower coordination, such as the cubic body-centered structure of the alkali metals, a similar definition of atomic radius is applicable, but the values so obtained are not immediately comparable with those derived from close-packed structures (Evans, 1966). The radius of an isolated atom is not a meaningful concept unless the coordination is specified because a small but systematic decrease in the atomic radius occurs with decreasing coordination. In general, atomic radii of metals are comparable to their covalent radii; the corresponding ionic radii are smaller because outer electrons are stripped away during the formation of an ion and those that remain are bound closely to the nucleus.

3.8 Van der Waals bonds

As early as 1873, Johannes Diderik van der Waals (1837–1923), a Nobel laureate from the Netherlands, postulated the existence of weak attractive and repulsive forces among the molecules of a gas and attributed the observed deviations from the ideal gas law to these forces (see section 4.1.1). Although the application of his postulate was limited to correcting the gas law through empirically derived constants, it is now recognized that van der Waals bonds operate between all atoms, ions, and molecules in all solids, and add a small contribution to the binding forces in ionic and covalent solids.

The van der Waals bonds are associated with energies of only about 10–50 kJ mol^{-1}, and their effect is largely masked in any crystal structure held together by other interatomic bonds, which are much stronger (see Table 3.7). The only solids in which the properties of the van der Waals bonds can be studied in isolation are the inert gases in the solid state to which they condense at sufficiently low temperatures. Such solids have a spatially undirected, cubic and hexagonal close-packed arrangement of atoms, similar to those found in metals (see Fig. 3.22).

The *van der Waals radius* of an atom is the radius of an imaginary hard sphere that can be used to model the atom for many purposes. Van der Waals radii are determined from measurements of atomic spacing between pairs of unbonded but touching atoms in crystals. Some examples are listed in Table 3.7.

3.9 Hydrogen bond

The *hydrogen bond* is a special kind of intermolecular bond that forms between the hydrogen atom in a polar bond such as O–H and an electronegative atom such as oxygen, hydrogen, or fluorine in a neighboring molecule. The fact that the hydrogen bond is found only between the atoms of strongly electronegative elements such as fluorine, oxygen, nitrogen, and (occasionally) chlorine suggests that it must be essentially ionic in character (Evans, 1966). Hydrogen is the only element capable of forming such bonds because the H^+ ion is unique in terms of both its very small size and its lack of extranuclear electrons. The typical hydrogen bond, with energy in the order of about 20 kJ mol^{-1}, is considerably weaker than covalent or ionic bonds, but stronger than van der Waals bonds.

The most ubiquitous, although not completely understood, example of a hydrogen bond is found between molecules of water, which has many anomalous physical and chemical properties (such as a high boiling point compared to other covalent molecules of similar molecular weights such as CO and NO) because of this kind of bonding (see section 7.1). Let us examine how a hydrogen bond forms between water molecules.

As shown below, there are eight valence electrons involved in a molecule of water, six associated with oxygen atoms and one with each of the hydrogen atoms:

↑↓	↑↓	↑	↑		↑	↑
2*s*	2*p*	2*p*	2*p*		1*s*	1*s*

Ground-state O atom — Ground state H atoms

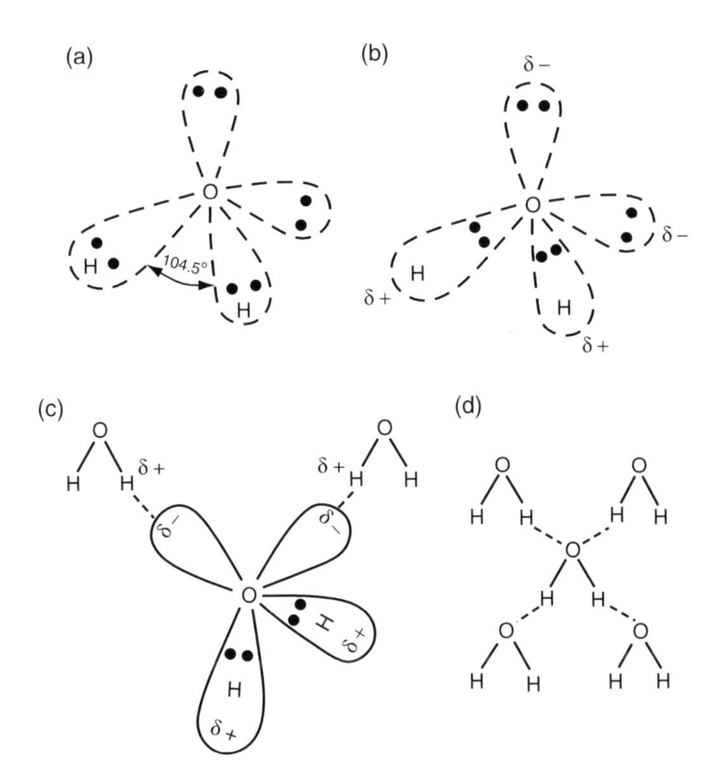

Fig. 3.23 Structure of the water molecule and the linkage of water molecules through hydrogen bonds. (a) Tetrahedral distribution of the four pairs of electrons (two bonding pairs shared with hydrogens and two lone pairs) surrounding the oxygen in a water molecule. (b) Charge distribution in a water molecule (δ+ with the hydrogens and δ– with the lone pairs). (c) Formation of hydrogen bonds between lone pairs on the oxygen and the hydrogens of neighboring molecules; (d) Linkage of a water molecule to four neighboring water molecules through hydrogen bonds. (After O'Neill, 1985.)

Two pairs of electrons are shared between the oxygen atom and two hydrogen atoms forming covalent bonds (H–O–H); the other two pairs are nonbonding or *lone pairs* because they are not shared with another atom (Fig. 3.23a). Because of repulsion between negative charges, the pairs of electrons would move as far apart as possible. The maximum separation is achieved if the four pairs are arranged in a tetrahedral configuration within the sphere of influence of the oxygen atom, one pair at each corner of the tetrahedron. The two lone pairs of electrons occupy a smaller volume and have a greater repulsive effect than the two bonding pairs, which causes the H–O–H bond angle to be reduced from the theoretical value of 109.5° for tetrahedral coordination to 104.5°.

Being more electronegative than hydrogen, oxygen attracts the shared bonding pair of electrons more strongly than does hydrogen. The result is a slight negative charge on the oxygen atom and a slight positive charge on each of the two bonding hydrogen atoms; the extra negative charge is concentrated on the two lone pairs of electrons (Fig. 3.23b). In other words, the two corners of the tetrahedron occupied by hydrogen atoms carry a positive charge, whereas the two remaining corners carry a negative charge. We can thus picture the water molecule behaving like a dipolar molecule, with a net positive charge on one side and a net negative charge on the other. The polarity of the water molecules is responsible for its remarkable solvent properties (see section 7.1).

A hydrogen bond forms by dipole–dipole interaction when the positively charged end of a water molecule is electrostatically attracted to the negatively charged end of a neighboring water molecule (Fig. 3.23c). This electrostatic bond is called a hydrogen bond. Each water molecule can be involved in four hydrogen bonds: two hydrogen bonds to the oxygen via its lone pairs, and a hydrogen bond between each hydrogen and the oxygen in a neighboring water molecule (Fig. 3.23d). This kind of linkage creates an open structure by holding the water molecules apart in fixed positions (O'Neill, 1985).

3.10 Comparison of bond types

A comparison of the bond types associated with solids is summarized in Table 3.8, with a comparison of dissociation energies (E_D) for various bond types in Table 3.9.

3.11 Goldschmidt's classification of elements

The distribution of the elements in a gravitational field, such as that of the Earth, is controlled not by their densities or atomic weights, but by their affinities for the three major groups of phases that can be formed — metals, sulfides, and silicates. Early in the 19th century, V.M. Goldschmidt (1888–1947), the famous Norwegian geochemist, reached this conclusion on the basis of three kinds of observations: (i) the composition of chondritic meteorites (see section 12.1), which are believed to have an average composition similar to that of the primordial Earth and to have undergone similar differentiation; (ii) analyses of metal, slag (silicates), and matte (sulfides) produced during smelting of sulfide ores for metal extraction; and (iii) the compositions of naturally occurring silicate rocks, sulfide ores, and native metals. The conclusion can also be supported from a comparison of free energies of reactions involved in the formation of various alloys, sulfide compounds, and silicate compounds. The concept of free energy and the calculation of free energy change for reactions will be discussed in Chapter 4.

Based on the way in which the elements distribute themselves among iron liquid, sulfide liquid, silicate liquid, and a gas phase, Goldschmidt (1937) classified the elements into four groups (Table 3.10): (i) *siderophile* (iron-loving) elements, which prefer to form metallic alloys (e.g., Fe–Ni alloy); (ii) *chalcophile* (sulfur-loving) elements, which most easily form sulfides; (iii) *lithophile* (rock-loving) elements, which readily combine with oxygen to form silicate and oxide minerals; and (iv) *atmophile* elements, which are commonly found in a gas phase. The geochemical character of an element is

Table 3.8 Comparison of properties of solids with different kinds of bonding.

Property	Ionic	Covalent	Metallic	Van der Waals
Formation	Commonly by combination of two elements with quite different electronegativities, usually a metal and a nonmetal	Commonly by combination of two elements with similar electronegativities, usually nonmetals	Within the crystal structures of metallic elements	Present in all solids as a contributing binding force
Crystal structure	Spatially nondirected; structures of high coordination and symmetry	Spatially directed; structures of low coordination and low symmetry	Spatially nondirected; structures of high coordination and symmetry	Formally analogous to metallic bond
Mechanical strength	Strong, giving hard crystals	Strong, giving hard crystals	Variable; gliding common	Weak, giving soft crystals
Dissociation energy (see Table 3.9)	Similar values (400–1300 kJ mol^{-1})			Much smaller values (< 50 kJ mol^{-1})
Melting point	Fairly high (typically 400 to 3000°C)	High (1200–4000°C)	Variable (–39° to 3400°C)	Low
Electrical conductivity	Molten compounds and solutions in polar solvents are good conductors (because they contain mobile ions)	Molten compounds and aqueous solutions are poor conductors (because they do not contain charged particles)[1]	Good conductors (because they contain delocalized electrons)	Insulators in solid and in melt

[1]Exceptions: diamond is a good conductor of heat; graphite is a good conductor of electricity.

Table 3.9 Comparison of dissociation energies (E_D) for various bond types.

Bond type	Reaction	E_D (kJ mol^{-1})	Bond type	Reaction	E_D (kJ mol^{-1})
Ionic	$NaCl_{(s)} \Rightarrow Na^+_{(g)} + Cl^-_{(g)}$	761*	Metallic	$Na_{(s)} \Rightarrow Na_{(g)}$	109
Ionic	$NaCl_{(s)} \Rightarrow Na_{(g)} + Cl_{(g)}$	640	Metallic	$Pb_{(s)} \Rightarrow Pb_{(g)}$	194
Ionic	$CaO_{(s)} \Rightarrow Ca_{(g)} + O_{(g)}$	1075	Metallic	$Ni_{(s)} \Rightarrow Ni_{(g)}$	423
Covalent	$C_{(diamond)} \Rightarrow C_{(g)}$	720	van der Waals	$S_{(s)} \Rightarrow S_{(g)}$	13
Covalent	$H_2O_{(g)} \Rightarrow 2H_{(g)} + O_{(g)}$	932	van der Waals	$Ar_{(s)} \Rightarrow Ar_{(g)}$	6
Covalent	$SiC_{(s)} \Rightarrow Si_{(g)} + C_{(g)}$	1197	Hydrogen bond	$H_2O_{(ice)} \Rightarrow H_2O_{(g)}$	50

* The lattice energy of NaCl crystal.
Sources of data: Compilations in Fyfe (1964) and Gill (1996).

Table 3.10 Goldschmidt's geochemical classification of elements.[1]

Siderophile	Chalcophile	Lithophile	Atmophile
Fe, Co, Ni	Cu, Ag, (Au)	Li, Na, K, Rb, Cs	H, N, (C), (O)
Ru, Rh, Pd	Zn, Cd, Hg	Be, Mg, Ca, Sr, Ba, (Pb)	(F), (Cl), (Br), (I)
Re, Os, Ir, Pt, Au	Ga, In, Tl	B, Al, Sc, Y, REE	He, Ne, Ar, Kr, Xe
Mo, Ge, Sn, C, P	(Ge), (Sn), Pb	(C), Si, Ti, Zr, Hf, Th	
(Pb), (As), (W)	As, Sb, Bi	(P), V, Nb, Ta	
	S, Se, Te	O, Cr, W, U	
	(Fe), (Mo), (Re)	(Fe), Mn	
		F, Cl, Br, I	
		(H), (Tl), (Ga), (Ge), (N)	

[1]Elements in parentheses belong primarily in another class. For example, Fe is dominantly siderophile, but also behaves as a chalcophile element as well as a lithophile element.
Source of data: Goldschmidt (1937).

largely governed by the electronic configuration of its atoms and hence is closely related to its position in the Periodic Table. Comparison of Table 3.10 with the Periodic Table (Fig. 2.8) shows that in general siderophile elements are concentrated at the center of the Periodic Table, lithophile elements to the left of the center, chalcophile elements to the right, and atmophile elements to the extreme right (Krauskopf and Bird, 1955) It can also be correlated with electrode potentials (see Chapter 8): siderophile elements are mostly noble metals with low electrode potentials; lithophile elements have high electrode potentials; and chalcophile elements have intermediate values. Goldschmidt's classification also included a fifth group named *biophile* elements (such as C, H, O, N, P, S, Cl, I), which are concentrated by living organisms.

It is clear from the many overlaps among the groups in Table 3.10 that the classification scheme is not perfect. Some elements show affinity for more than one group, because the distribution of any element is dependent to some extent on temperature, pressure, and chemical environment of the system (e.g., oxygen fugacity, competing atoms, etc.). It is, however, a useful qualitative guide to the behavior of elements during geochemical processes.

3.12 Summary

1. Chemical bonds hold the atoms or ions in a crystal structure together and determine the physical and chemical properties of the crystal. There are five kinds of chemical bonds: ionic bonds, covalent bonds, metallic bonds, van der Waals bonds, and hydrogen bonds.
2. Ionic bonds arise from electrostatic attraction between cations and anions, covalent bonds from sharing of electrons between atoms, and metallic bonds from the electrostatic attraction between the ionic cores of atoms at lattice positions and the delocalized electrons. Van der Waals bonds are weak attractive and repulsive forces that operate between all atoms, ions, and molecules in all solids. The hydrogen bond is a special kind of intermolecular bond that forms between the hydrogen atom in a polar bond such as O–H and an electronegative element such as oxygen on a neighboring molecule.
3. In ionic compounds, the number of oppositely charged nearest neighbors surrounding an ion is called its coordination number (*CN*). The coordination number and, therefore, the geometry of packing in ionic crystals can be predicted from the radius ratio r_c/r_a, where r_c and r_a are the cation and anion radius, respectively.
4. The lattice energy (U_L) of a perfectly ionic crystal is defined as the amount of energy required at absolute zero (i.e., –273°C) to convert one mole of the solid into its constituent ions at infinite separation in the gas phase. For 1 gram-mole of a binary ionic compound it can be calculated using the equation

$$U_L = MA\frac{z_c z_a e^2}{R}\left(1 - \frac{1}{n}\right)$$

where z_c and z_a are the charges on cations and anions (expressed as a multiple of the electron charge e), R is the interionic distance separating the ions, n is a constant for the particular crystal structure, A is the Avogadro's number, and M is a numerical quantity called the Madelung constant, the value of which depends on the crystal structure.
5. The crystal structures of silicate minerals can be rationalized in terms of different patterns of linkages among silicon–oxygen tetrahedra (SiO_4^{4-}).
6. The extent of ionic substitution in minerals is governed by the nature of the crystal structure and how similar the ions are in terms of size, charge, and electronegativity. Elevated temperatures and lower pressures favor increased substitution.
7. Crystal-field theory, which describes the effects of electrostatic fields on the energy levels of the valence electrons (electrons in the outermost orbitals) of a transition-metal when it is surrounded by negatively charged ligands in a crystal structure, provides reasonable explanations for some magnetic properties, colors, hydration enthalpies, and spinel structures of transition metal complexes.
8. Isomorphism refers to substances having similar crystal structures but different chemical formulas, whereas polymorphism refers to substances having similar chemical formulas but different crystal structures.
9. A solid solution is a solution in the solid state of one or more solutes in a solvent whose crystal structure remains unchanged by addition of the solutes. Solid solutions form: by substitution of ions of one kind by another in the lattice, by interstitial accommodation of ions, or by omission of ions from the lattice leaving vacant sites.
10. Covalent bonding is best explained through the formation of bonding and antibonding molecular orbitals (σ, π, and δ) from the overlap of atomic orbitals on the atoms in the molecule.
11. Before combining across atoms, atomic orbitals that are close to each other in energy within the same atom have the ability to combine with one another to form hybrid covalent bonds. The concept of orbital hybridization is useful for explaining the bonding in some molecules such as CH_4.
12. Anomalous physical and chemical properties of water, such as its high boiling point and high solvent capacity, are due to hydrogen bonding within and between water molecules.
13. Goldschmidt classified the elements into four groups: (i) siderophile (iron-loving) elements, which prefer to form metallic alloys (e.g., Fe–Ni); (ii) chalcophile (sulfur-loving) elements, which most easily form sulfides; (iii) lithophile (rock-loving) elements, which readily combine with oxygen to form silicate and oxide minerals; and (iv) atmophile elements, which are commonly found in a gas phase. This classification scheme is not perfect, but it is a useful qualitative guide to the behavior of elements during geochemical processes.

3.13 Recapitulation

Terms and concepts

Antibonding molecular orbital
Atomic orbital
Atomic radius
Avogadro's number
Bond length
Bond order
Bonding molecular orbital
Born–Haber cycle
Close-packed crystal structures
Coordination number
Coupled substitution
Covalent bond (polar and nonpolar)
Covalent radius
Crystal-field theory
Crystal-field stabilization energy (CFSE)
Delocalized electrons
Diadochy
Diamagnetism
Dipole moment
Dissociation energy
Electric dipole moment
Electronegativity
Energy of crystallization
Geochemical coherence
Goldschmidt's classification of elements
Goldschmidt's rules of ionic substitution
Hybridization
Hydrogen bond
Ionic bond
Ionic potential
Ionic radius
Isomorphism
Lanthanide contraction
Lattice energy
Molecular orbitals (σ, π, and δ)
Madelung constant
Metallic bond
Ordered structure
Paramagnetism
Polarization
Polymorphism
Radius ratio
Rare earth elements
Silicon–oxygen tetrahedron
Solid solution
Silicate crystal structures
Valence bond theory
Van der Waals bond
Van der Waals radius

Computation techniques

- Lattice energy using the Born–Haber cycle.
- Percent ionic character of a bond.

3.14 Questions

1. Which of the following series of ions are isoelectronic? Justify your answer.
   (a) Au^+, Hg^{2+}, Tl^{3+}, and Pb^{4+}
   (b) Mn^{2+}, Fe^{2+}, Co^{2+}, Ni^{2+}, Cu^{2+}, and Zn^{2+}
   (c) Li^+, Be^{2+}, B^{3+}, C^{4+}, and N^{5+}
2. From the data given in Table 3.3, what generalizations can you make about the trends in variation of lattice energy?
3. Calculate the lattice energy and the energy of crystallization of 1 mole of KCl crystals using equation (3.4). Given: $n = 9$; $r_{K^+} = 1.38$, $r_{Cl^-} = 1.81$, $A = 6.02 \times 10^{23}$, $e = 4.80 \times 10^{-10}$ Coulomb, and $M_{KCl} = 1.747$.
4. Calculate the lattice energy of one mole of solid LiCl (U_{LiCl}) by constructing a Born–Haber cycle of appropriate reactions based on the heat data given below:
   First ionization potential of lithium (I_{Li}) = 520 kJ mol^{-1}
   Electron affinity of chlorine (E_{Cl}) = 348 kJ mol^{-1}
   Heat of sublimation of lithium (S_{Li}) = 159 kJ mol^{-1}
   Heat of dissociation of chlorine (D_{Cl}) = 236 kJ mol^{-1}
   Heat of formation of LiCl (Q_{LiCl}) = −409 kJ mol^{-1}
   [Hint: see Fig. 3.2 and Example 3-2.]
5. Applying the principles of molecular orbital formation, explain why oxygen forms a stable O_2 molecule whereas helium does not form a stable He_2 molecule. Is oxygen paramagnetic or diamagnetic?
6. Determine the percent ionic character of the following bonds between: (i) Na and Cl; (ii) Cl and O; and (iii) Zn and O.
7. Explain why the following substitutions are not common in minerals: Cu^+ for Na^+: Cl^- for F^-, and C^{4+} for Si^{4+}.
8. Name three elements you would expect to find in trace quantities in calcite ($CaCO_3$) substituting for Ca. Justify your answer.
9. Explain why both lithium and cesium are concentrated in the late mica minerals of pegmatites, although their ionic radii are very different.
10. Explain why boron minerals are more common than vanadium minerals in igneous and sedimentary rocks, although vanadium is a more abundant element than boron.

Part II
Chemical Reactions

As a science grows more exact, it becomes possible to employ more extensively `the accurate and concise methods and notations of mathematics. At the same time it becomes desirable, and indeed necessary, to use words in a more precise sense.

Lewis and Randall (1961)

4 Basic Thermodynamic Concepts

Thermodynamics, like other sciences, has a theoretical side, expressed in mathematical language, and a practical side, in which experiments are performed to produce physical data required and interpreted by the theoretical side. The mathematical side of thermodynamics is simple and elegant and is easily derived from first principles. The difficulty in understanding and using thermodynamics is conceptual, not mathematical.

Anderson and Crerar (1993)

Thermodynamics is the science that deals with differences in and possible transfers of energy, and in some cases of matter, among systems. It is an integral part of geochemistry because it provides a universally applicable framework for characterizing chemical reactions that ultimately control the distribution of elements and minerals in the Earth. The beauty of *classical thermodynamics* lies in the fact that it is based on two empirical laws (and a few consistent conventions) that are adequate to quantitatively relate a host of *macroscopic* measurements such as temperature, pressure, volume, etc., without requiring knowledge of the underlying molecular structure of the materials involved in chemical reactions. *Statistical thermodynamics* (or *statistical mechanics*), on the other hand, is the science that relates the properties of individual molecules and their interactions to the empirical results of classical thermodynamics. The laws of classical and quantum mechanics are applied to molecules; then, by suitable statistical averaging methods, the rules of macroscopic behavior that would be expected from an assembly of a large number of molecules are formulated (Klotz and Rosenberg, 1994, p. 5).

A limitation of thermodynamics is that it does not provide any information about the rate or mechanism of a thermodynamically predicted chemical reaction. Some reactions, especially at low temperatures, are so sluggish that, for all practical purposes, they do not seem to proceed at all. If this were not the case, we would have no igneous or metamorphic rocks surviving on the Earth's surface. The study of the rates and mechanisms (pathways) of reactions belong to the field of *kinetics*. Thus, a comprehensive understanding of chemical reactions requires an integration of thermodynamics and kinetics.

In this chapter we will restrict our attention to the basic concepts and applications of classical *equilibrium thermodynamics* related to equilibrium states and reversible processes (processes that proceed in infinitely small steps so that equilibrium can be maintained at each step), with the primary objective of establishing criteria for determining the feasibility or spontaneity of given chemical transformation. *Nonequilibrium thermodynamics* dealing with irreversible processes – processes that proceed at finite rates – will not be discussed in this book. More detailed discussions on many aspects of thermodynamics, especially for earth scientists, can be found in books by Kern and Weisbrod (1967), Wood and Fraser (1976), Powell (1978), Fletcher (1993), Anderson and Crerar (1993), Klotz and Rosenberg, (1994), Nordstrom and Munoz (1994), White (2002), and McSween *et al.* (2003).

4.1 Chemical equilibrium

4.1.1 *Law of Mass Action – equilibrium constant* (K_{eq})

Suppose, we mix two substances X and Y in a closed container and at some appropriate temperature so that they would chemically react to form two other substances C and D. This rearrangement of atoms from one configuration to another is called a

Introduction to Geochemistry: Principles and Applications, First Edition. Kula C. Misra.

chemical reaction. It is possible that the forward reaction $X+Y \rightarrow C+D$ will result in complete conversion of X and Y into C and D. Many chemical reactions, however, remain incomplete because of the reverse reaction $C+D \rightarrow X+Y$. The reaction is considered to have attained *chemical equilibrium* if and when the rate of the forward reaction becomes equal to the rate of the reverse reaction. If conditions of the experiment remain unchanged, then at equilibrium all the four substances would coexist without any further change in their concentrations with time. For a chemical reaction, the terms reversibility and equilibrium are often used interchangeably because a reversible process is conceptualized as a process that proceeds in such infinitely small steps that the system is at equilibrium for every step. Real geochemical systems seldom, if ever, attain a state of equilibrium, but the equilibrium model, because of its simple mathematical relationships, serves as a useful reference for evaluating chemical reactions. An irreversible chemical reaction is unidirectional and so can never achieve equilibrium. A reaction may also not attain equilibrium either because the rate of the forward or the reverse reaction is too slow or because one or more of the products are removed from the system.

If a chemical reaction occurs as written at the molecular level (i.e., without involving intermediate steps), then its rate is proportional to the concentrations of the reacting substances (see Chapter 9 for a more elaborate discussion of rates of chemical reactions). Assuming that the reaction $X+Y=C+D$ is such a reaction, the rates of the forward reaction ($R_{forward}$) and the reverse reaction ($R_{reverse}$) at any instant may be expressed as

$$R_{forward} = k_{forward}\,[X]\,[Y] \qquad (4.1)$$

$$R_{reverse} = k_{reverse}\,[C]\,[D] \qquad (4.2)$$

where [X], [Y], [C], and [D] represent concentrations, and $k_{forward}$ and $k_{reverse}$ are proportionality constants, which are commonly referred to as rate constants (see Chapter 9). As the forward reaction progresses, [X] and [Y] decrease with a corresponding increase in [C] and [D]. It follows that $R_{forward}$ decreases with time whereas $R_{reverse}$ increases correspondingly, until $R_{forward} = R_{reverse}$ at equilibrium (Fig. 4.1). Thus, the condition at equilibrium may be stated as

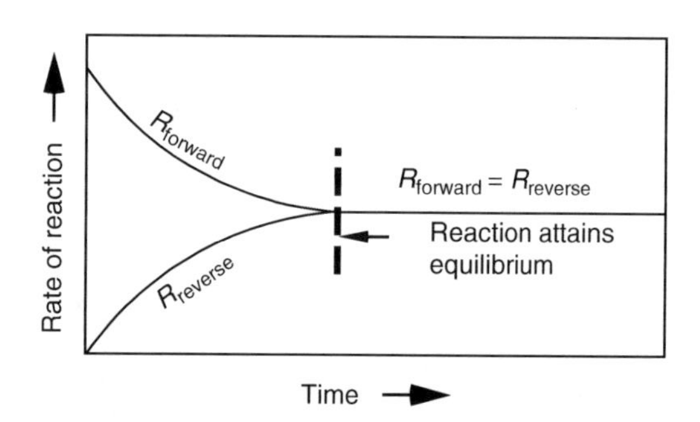

Fig. 4.1 A reaction, such as $X+Y=C+D$, is to have attained equilibrium when the rate of the forward reaction $X+Y \rightarrow C+D$ ($R_{forward}$) becomes equal to the rate of the reverse reaction $C+D \rightarrow X+Y$ ($R_{reverse}$). It is assumed that only X and Y were present at the start of the reaction. The properties of the system at equilibrium will not change with time.

$$k_{forward}\,[X]\,[Y] = k_{reverse}\,[C]\,[D] \qquad (4.3)$$

which can be rearranged as:

$$\frac{k_{forward}}{k_{reverse}} = \frac{[C]\,[D]}{[X]\,[Y]} = K \qquad (4.4)$$

where K is a constant because $k_{forward}$ and $k_{reverse}$ are constants. Equation (4.4) is a statement of the *Law of Mass Action*. Note that the concentration terms for the reaction products are placed in the numerator and those for the reactants in the denominator.

For a balanced, reversible reaction (indicated by the symbol $\Leftrightarrow$) of the general form

$$\text{x X} + \text{y Y} + \ldots \Leftrightarrow \text{c C} + \text{d D} + \ldots \qquad (4.5)$$

reactants *products*

that attains equilibrium at pressure P and temperature T, the Law of Mass Action takes the form

$$\boxed{\frac{[C]^{c}_{eq}\,[D]^{d}_{eq}\ldots}{[X]^{x}_{eq}\,[Y]^{y}_{eq}\ldots} = K^{app}} \qquad (4.6)$$

where x, y, c, d,... are the stoichiometric coefficients required to balance the reaction, $[C]_{eq}$, $[D]_{eq}$, etc., are concentrations at equilibrium (for simplicity, we shall omit the subscript "eq" in subsequent discussions related to equilibrium constants), and K^{app} is the *apparent equilibrium constant* for the particular reaction at the specified P and T. The true equilibrium constant for the reaction is defined by replacing the concentrations by activities at equilibrium:

$$\boxed{\frac{a_C^{c}\,a_D^{d}\ldots}{a_X^{x}\,a_Y^{y}\ldots} = K_{eq}} \qquad (4.7)$$

For a species i, the activity (a_i) is related to, but not necessarily equal to, the concentration of i. The concept of activity will be developed more rigorously later (Chapter 5). For the present, we will use the following approximations for a_i: (i) for gaseous species in a mixture of gases, a_i = partial pressure of i (P_i) (see Box 4.1); (ii) for solutes in a very dilute solution, a_i = molality of i (m_i) or molarity of i (M_i), or occasionally the mole fraction of i (so that for a pure liquid, $a_i = 1$); (iii) for components in the solid phase, a_i = mole fraction of i (so that for a pure solid, $a_i = 1$). Activities are dimensionless numbers, so equilibrium constant, a ratio of activity products, is also a dimensionless number.

Box 4.1 Dalton's Law of Partial Pressures

Consider n moles of a gas mixture (total volume V, total pressure P, temperature T) that consists of m different species of ideal gases and behaves ideally, i.e., it obeys the *Ideal Gas Law*,

$$\boxed{P\,V = n\,R\,T} \qquad (4.8)$$

where R is the empirically determined Gas Constant = 8.314 J mol^{-1} K^{-1}. Let $n_1, n_2, n_i, \ldots, n_m$ be the number of moles of the constituent gas species and $P_1, P_2\, P_i, \ldots, P_m$, the respective partial pressures. According to *Dalton's Law of Partial Pressures*, which was deduced by Dalton in 1801, the partial pressure of the ith constituent in the gas mixture is

$$P_i = n_i \frac{RT}{V}$$

The sum of the partial pressures, $\sum_i P_i$, is equal to the total pressure (P):

$$\sum_i P_i = \frac{RT}{V} \sum_i n_i = n \frac{RT}{V} = P$$

It follows that the partial pressure of each constituent in an ideal mixture of ideal gases is equal to its mole fraction times the total pressure of the gas mixture. That is,

$$P_i = n_i \frac{RT}{V} = n_i \frac{P}{n} = X_i P \qquad (4.9)$$

where X_i = mole fraction of $i = n_i/n$.

Some useful points to remember about equilibrium constants are as follows: (i) if K and K' are the equilibrium constants for a reaction in the forward and reverse directions, respectively, then $K' = 1/K$; (ii) adding two equilibrium reactions, with equilibrium constants K_1 and K_2, results in a new equilibrium reaction, with equilibrium constant K_3, such that $K_3 = (K_1)(K_2)$; and (iii) subtracting a reaction with equilibrium constant K_1 from one with equilibrium constant K_2 results in a new equilibrium constant K_4 such that $K_4 = (K_1)/(K_2)$. Applying these relations, it can be shown that equations (4.6) and (4.7) apply even if reaction (4.5) represents an overall reaction consisting of one or more intermediate steps. The Law of Mass Action cannot be derived from kinetics because the intermediate steps of an overall reaction occur at different rates, and the overall rate of the reaction is not necessarily a linear function of the molar concentrations of reactants and products. A more rigorous derivation of the Law of Mass Action from the laws of thermodynamics will be discussed in Chapter 5.

The magnitude of K_{eq} provides useful information about the likely tendency of the reaction. A large value of K_{eq} indicates a favorable forward reaction (a product-favored reaction) that would result in higher concentrations of the products relative to the reactants in the equilibrium assemblage; a small magnitude of K_{eq} indicates the opposite, a reactant-favored reaction. When equilibrium constants are given without specifying temperature and pressure, it is to be understood that they refer to 298.15 K (or 25°C) and 1 bar (or 10^5 Pa). We will discuss later (Chapter 6) how variation of K with temperature and pressure can be utilized for estimating the equilibration temperature and pressure of mineral assemblages.

Example 4–1: Concentration of carbonic acid in water in equilibrium with the atmosphere

A geologically important chemical reaction involves the combination of CO_2 with water to form carbonic acid (H_2CO_3), which plays a key role in controlling the pH of natural waters (Box 4.2). When CO_2 is brought into contact with water, the CO_2 dissolves in water until equilibrium is established. This equilibrium, ignoring the equilibrium between $CO_{2(air)}$ and $CO_{2(aq)}$ (i.e., CO_2 dissolved in water), can be represented as

$$CO_{2(g)} + H_2O_{(l)} \Leftrightarrow H_2CO_{3(aq)} \qquad (4.10)$$

$$K_{CO_2} = \frac{a_{H_2CO_3\,(aq)}}{a_{CO_2\,(g)}\; a_{H_2O\,(l)}} \qquad (4.11)$$

where $a_{H_2CO_3\,(aq)}$ represents the activity of dissolved but undissociated acid molecules. The convention adopted for the above formulation is that all the dissolved CO_2 exists as H_2CO_3. Actually, CO_2 in the solution exists predominantly as $CO_{2(aq)}$ (the concentration of $H_2CO_{3(aq)}$ in the solution at 25°C being slightly less than 0.3% of the $CO_{2(aq)}$ present; Langmuir, 1997), but this does not affect calculations as long as the convention is applied consistently because the equilibrium constants used are consistent with this convention. It is also the convention to represent the activity of water by its mole fraction ($X_{H_2O(l)}$) in the solution. For a very dilute solution, as in the present case, $a_{H_2O(l)} = X_{H_2O(l)} = 1$, so that equation (4.11) reduces to

$$K_{CO_2} = \frac{a_{H_2CO_3\,(aq)}}{a_{CO_2\,(g)}} \qquad (4.12)$$

The published value of K_{CO_2} at 1 bar and 298.15 K (25°C) is $10^{-1.47}$ (Plummer and Busenberg, 1982), but we can also calculate it from the solubility data for CO_2. At such a low pressure, CO_2 behaves as an ideal gas so that $a_{CO_2} = P_{CO_2}$. Let us choose $P_{CO_2} = 1$ bar. The solubility of CO_2 in water at 25°C and 1 atm pressure is 0.145 g of CO_2 per 100 g of water, which works out to 0.032 moles of CO_2 per 1000 g of water at 1 atm or 0.032 moles of CO_2 per 1000 g of water at 1 bar (1 atm = 1.013 bar). According to reaction (4.10), 1 mole of CO_2 produces 1 mole of $H_2CO_{3(aq)}$. Thus, $m_{H_2CO_3(aq)} = 0.032$, and for such a dilute solution we can assume that $a_{H_2CO_3\,(aq)} = m_{H_2CO_3\,(aq)} = 0.032$. Substituting these values in equation (4.12),

$$K_{CO_2} = \frac{a_{H_2CO_3\,(aq)}}{a_{CO_2(g)}} = \frac{m_{H_2CO_3\,(aq)}}{P_{CO_2}} = \frac{0.032}{1} = 0.032 = 10^{-1.49} \qquad (4.13)$$

which is almost the same as the value given by Plummer and Busenberg (1982). Equation (4.13) establishes that in the water–CO_2 system at equilibrium, for every P_{CO_2} there is a corresponding $m_{H_2CO_3\,(aq)}$ and vice versa. In fact, it is a common practice to report $m_{H_2CO_3\,(aq)}$ as the corresponding P_{CO_2}, even when no gas phase is present, because equation (4.13) can be used to calculate $m_{H_2CO_3\,(aq)}$ for different values of P_{CO_2} at 298.15 K, provided CO_2 behaves as an ideal gas.

As a simple exercise, let us calculate $m_{H_2CO_3(aq)}$ in surface water (at 25°C=298.15 K) exposed to the atmosphere ($P \approx 1$ bar$=10^5$ Pa). Atmospheric air contains about 0.03% CO_2 by volume, or its volume fraction in air is 0.0003. Assuming the volume fraction of CO_2 to be approximately equal to its mole fraction,

$$P_{CO_2} = P \cdot X_{CO_2} = 1 \times 0.0003 = 0.0003 \text{ bar} = 10^{-3.5} \text{bar}$$

and

$$m_{H_2CO_3\,(aq)} = K_{CO_2} \cdot P_{CO_2} = 10^{-1.47} \times 10^{-3.5} = 10^{-4.97} \qquad (4.14)$$

With such a small amount of dissolved carbonic acid, this water would be only mildly acidic (see section 7.3.1 for calculation of pH for the $CO_2 - H_2O$ system), but acidic enough to make natural surface waters a powerful agent of chemical weathering. As $m_{H_2CO_3(aq)}$ is directly proportional to P_{CO_2}, it is logical to conclude that chemical weathering must have been more intense during the Precambrian when the atmosphere was much richer in CO_2 and so had a much higher P_{CO_2}.

Box 4.2 The pH of an aqueous solution

The pH of a solution is defined as the negative logarithm of hydrogen-ion activity in the solution:

$$pH = -\log_{10} a_{H^+} \qquad (4.15)$$

This is a very useful parameter of aqueous solutions and it can be measured by electrical methods (Stumm and Morgan, 1981). Note that the pH value decreases with an increase in the hydrogen-ion activity, so that a solution of lower pH is more acidic compared with that of a higher pH.

The concept of pH arises from the dissociation of water, which can be represented by the reaction

$$H_2O \Leftrightarrow H^+ + OH^{\pm} \qquad (4.16)$$

and the equilibrium constant for this reaction, K_w, is 10^{-14} at 298.15 K and 1 bar. (Actually $K_w = 10^{-13.995}$ at 25°C and $K_w = 10^{-14}$ at 24°C, but the difference is ignored for simplicity). Since a_{H_2O} (by convention),

$$K_w = \frac{a_{H^+}\, a_{OH^-}}{a_{H_2O}} = a_{H^+}\, a_{OH^-} = 10^{-14} \qquad (4.17)$$

Although equation (4.17) was obtained for pure water, it is also valid for dilute aqueous solutions at 298.15 K (25°C) and 1 bar, irrespective of other solutes in the solution.

Careful measurements show that, for pure water at 298.15 K and 1 bar,

$$a_{H^+} = a_{OH^-} = 10^{-7}\ ; \quad pH = 7 \qquad (4.18)$$

At 298.15 K and 1 bar, any aqueous solution with pH=7 is, by definition, neutral, one with pH<7 acidic, and one with pH>7 alkaline. The neutral pH value decreases with an increase of temperature or pressure because of a corresponding increase in K_w (Helgeson and Kirkham, 1974). The pH scale ranges from 0 (i.e., $a_{H^+} = 10^0$) to 14 (i.e., $a_{H^+} = 10^{-14}$), but pH values of natural waters lie mostly in the range 5 to 9. In a geochemical system, pH is theoretically determined by all the equilibria reactions in the system.

4.1.2 *Le Chatelier's principle*

Let us consider reaction (4.10) in equilibrium. If the equilibrium is disturbed, for example, by subjecting the system to a small increase in P_{CO_2}, the reaction will proceed in the forward direction forming more $H_2CO_{3(aq)}$ until equilibrium is restored. Similarly, a small decrease in P_{CO_2} will promote the reverse reaction, the dissociation of $H_2CO_{3(aq)}$, until equilibrium is re-established. In both cases, the system responds to counter the effect of the imposed perturbation and establish a new equilibrium. A generalization of this predictable behavior of a system in equilibrium, which actually follows from the Law of Mass Action, is known as *Le Chatelier's principle* after the French chemist Henri Louis Le Chatelier (1850–1936) who first stated it explicitly. We can apply this principle also to predict qualitatively the effect of temperature or pressure on a system at equilibrium. For example, an increase in temperature should favor a reaction, such as melting of rocks, that is accompanied by absorption of heat (an *endothermic* reaction), whereas a decrease in temperature should favor the reverse process, such as crystallization of magma, that involves release of heat (an *exothermic* reaction). The effect of pressure changes on equilibria is accommodated by changes in volume. Thus, metamorphism at high pressure favors the reaction

	$NaAlSi_3O_8$ +	$NaAlSiO_4$ =	2 $NaAlSi_2O_6$	
	albite	nepheline	jadeite	
molar volume (cal/bar)	2.3917	1.2944	1.4435	(4.19)

because the formation of jadeite results in a significant reduction in volume relative to the combined volume of the reactants.

4.2 Thermodynamic systems

4.2.1 *Attributes of a thermodynamic system*

A thermodynamic *system* is some part of the universe we choose to study for addressing a given problem. The system we define can be as small as a rock chip or as large as the

State of the system

Defined by various properties
Extensive – Mass (Σn_i), Volume (V), Enthalpy (H), Entropy (S), Gibbs free energy (G)
Intensive – Temperature (T), Pressure (P), Chemical potential of i (μ_i)

Contents of the system

Phases (p), Components (c)
Phase rule: $f = c - p + 2$

Functions of state

Functions whose values are determined entirely by the initial and final states of the system, independent of the path taken for the change,
Examples: Volume, Enthalpy, Entropy, Gibbs free energy

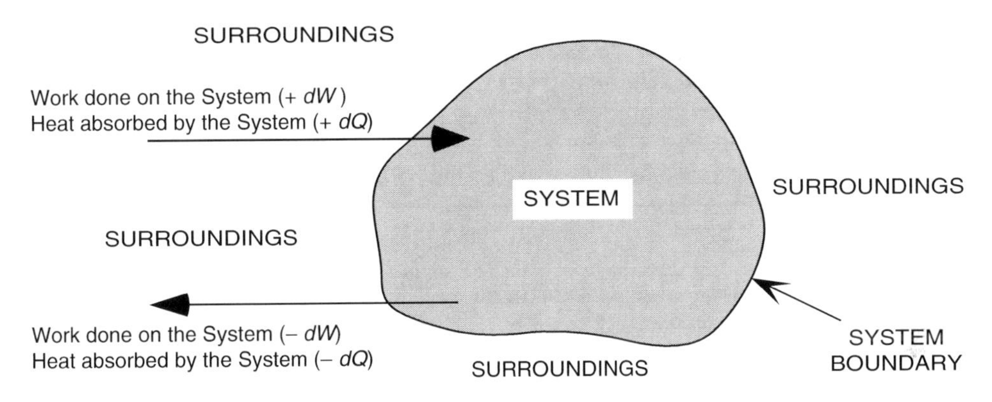

Kinds of Systems (relative to surroundings)

	Mass exchange	Energy exchange
Isolated	no	no
Closed	no	yes
Open	yes	yes

Kinds of processes for change of state

Isothermal (constant temperature; $dT = 0$)
Isobaric (constant pressure; $dP = 0$)
Isochoric (constant volume; $dV = 0$)
Adiabatic (constant heat content; $dQ = 0$)

Fig. 4.2 Attributes of a thermodynamic system, a well-defined portion of the universe that is of interest in a particular investigation. The system consists of one or more phases, the chemical compositions of which are described by a number of components. The system is characterized by a series of extensive and intensive properties and classified in terms of its ability to exchange matter and energy with the surroundings.

entire Earth. The rest of the universe beyond the chosen system constitutes the *surroundings*. As discussed below, a system is characterized by its contents and by its state (Fig. 4.2).

The matter contained in a system can be described completely in terms of phases and components. A *phase* is defined as a physically distinct, homogeneous body of matter that is, at least in principle, mechanically separable from the rest of the system. Each mineral in a rock is a single phase even if it is dispersed inhomogeneously throughout the system. A mixture of gases, on the other hand, is one phase because gases are miscible in all proportions. A mixture of miscible liquids is also one phase, but immiscible liquids, such as water and oil, must be counted as separate phases. A system consisting of a single phase is called *homogeneous*; a system consisting of more than one phase is called *heterogeneous*. Thus, a hypothetical system consisting of calcite ($CaCO_3$) crystals, barite ($BaSO_4$) crystals, liquid water ($H_2O_{(l)}$), water vapor ($H_2O_{(g)}$), and $CO_{2(g)}$, is a four-phase (two solids, one liquid, and one gas) heterogeneous system.

The chemical composition of a system can be described in terms of chemical species, which may be ions, gaseous (g) or liquid (l) molecules, solid phases (s), etc. Some possible chemical species in the $CaCO_3 - H_2O$ system, for example, are $H_2O_{(g)}$, $H_2O_{(l)}$, solid $CaCO_3$, and aqueous ions such as H^+, OH^-, Ca^{2+}, CO_3^{2-}, and $CaCO_3^0$. In thermodynamics, however, we use *components*, instead of species, for a more systematic and precise description of the chemical composition of a system.

The components of a system are the minimum number of chemical species required to describe the composition of all the phases in the system (although every component need not be present in every phase), and the number is fixed for a given system in equilibrium. For example, the number of *system components* in the three-phase system andalusite (solid)–kyanite (solid)–sillimanite (solid) is only one, because the single component Al_2SiO_5 completely describes the chemical composition of each of the three polymorphs. The number of components is also fixed in a multicomponent system, but the choice of the chemical species is based on convenience. For example, in the three-phase system $CaCO_{3(s)} - CaO_{(s)} - CO_{2(g)}$, the number of components is two but they are not unique. An obvious choice perhaps is CaO and CO_2 ($CaCO_3$ is simply $CaO + CO_2$), but $CaCO_3$ and CaO ($CO_2 = CaCO_3 - CaO$), or $CaCO_3$ and CO_2 ($CaO = CaCO_3 - CO_2$) are equally valid choices. As a general rule, the number of system components (c) is determined by identifying the number of chemical species in the system (m) and subtracting the number of independent chemical reactions (r) that can occur among those species: $c = m - r$.

The thermodynamic *state* of a system is described by certain measurable *extensive* and *intensive* physical properties (or variables). An extensive property is dependent on the quantity of matter being considered, and is additive. Mass (number of moles), volume, and energy (and some others such as enthalpy and entropy, which will be defined later) are extensive properties. For example, if a system is comprised of two solid phases, A and B, that do not react with each other and their volumes are V_A and V_B, respectively, then the total volume of the system is $(V_A + V_B)$. An intensive property has the same value at each point in a system at equilibrium; its value does not depend on the quantity of matter, and it is not additive. Examples are pressure, temperature, density, and concentration (and some others such as chemical potential, which will be defined later). Note that the quotient of two extensive properties is an intensive property. For example, density, which is equal to mass divided by volume, is an intensive property. The extensive and intensive properties have fixed values for a system at a particular state of equilibrium.

For easy reference, a list of symbols for thermodynamic parameters used in this book is included as Appendix 4.

4.2.2 *State functions*

Properties of a system that depend only on the final state of the system, irrespective of the path taken to reach that state, are called *state functions* (or *state variables*). The mathematical test of a function ϕ being a state function is that its total differential $d\phi$ must be an exact differential (Box 4.3). For example, the molar volume (or volume per mole, $\bar{V}$) of an ideal gas is a function of both temperature and pressure (equation 4.8), and the total change in molar volume, $d\bar{V}$, is

$$d\bar{V}_{(P,T)} = \left(\frac{\partial \bar{V}}{\partial P}\right)_T dP + \left(\frac{\partial \bar{V}}{\partial T}\right)_P dT = -\frac{RT}{P^2}\, dP + \frac{R}{P}\, dT \quad (4.20)$$

Box 4.3 Exact differential – state function

Consider a function $\phi\,(x, y)$. Its total differential, $d\phi\,(x, y)$, can be written as

$$d\phi\,(x,y) = \left(\frac{\partial \phi}{\partial x}\right)_y dx + \left(\frac{\partial \phi}{\partial y}\right)_x dy$$

If $d\phi\,(x, y)$ is an exact differential, then

$$\left[\frac{\partial}{\partial y}\left(\frac{\partial \phi}{\partial x}\right)_y\right]_x = \left[\frac{\partial}{\partial x}\left(\frac{\partial \phi}{\partial y}\right)_x\right]_y$$

This is the mathematical test for $\phi\,(x, y)$ to qualify as a state function. The change in the value of ϕ when a system is taken from state 1 to state 2 is given by

$$\Delta\phi = \phi_2 - \phi_1 = \int_{\text{state 1}}^{\text{state 2}} d\phi$$

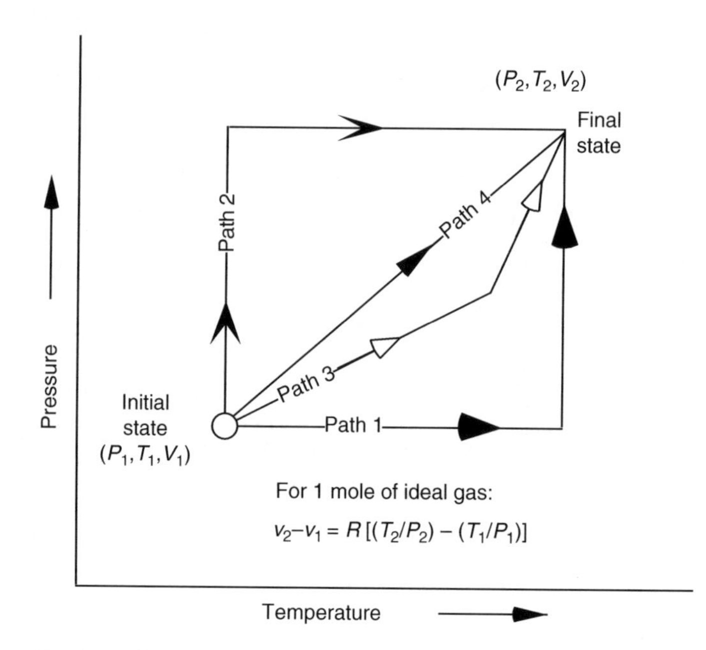

Fig. 4.3 Volume (V) is a state function because it is an exact differential. The volume at the final state (V_2) is independent of whether that state was reached by path 1, path 2, path 3, path 4, or any other path from an initial state (V_1).

The molar volume $\bar{V}$ is a state function because $d\bar{V}_{(P,T)}$ is an exact differential, i.e.,

$$\left[\frac{\partial}{\partial P}\left(\frac{\partial \bar{V}}{\partial T}\right)_P\right]_T = \left[\frac{\partial}{\partial T}\left(\frac{\partial \bar{V}}{\partial P}\right)_T\right]_P = -\frac{R}{P^2} \quad (4.21)$$

For instance, if a gas is taken from an initial state (P_1, T_1, V_1) to a final state (P_2, T_2, V_2), the molar volume change $(\Delta\bar{V} = \bar{V}_2 - \bar{V}_1)$ will be the same, independent of the path taken to reach the final state (Fig. 4.3) because $\bar{V}_2$ depends only on the state of the system. It follows that in a closed cycle (e.g., state 1 → state 2 → state 3 → state 2 → state 1), the net change in a state function, such as $\bar{V}$, is always zero:

$$\oint d\bar{V} = 0 \quad (4.22)$$

4.2.3 *The Gibbs phase rule*

Experience has shown that when a certain number of intensive properties of a system in equilibrium have been specified, all other properties of the system have fixed values. For example, at a specified state of temperature and pressure, the volume of an ideal gas is fixed by the ideal gas law (equation 4.8). The minimum number of intensive thermodynamic variables that must be specified to fix all other intensive variables of a system at equilibrium is called the *variance* or *degrees of freedom* of the system (f). The variance of a system is related to the number of phases (p) and the number of components (c) in the system at equilibrium and can be predicted by a simple rule, the *Gibbs phase rule*:

$$f = c - p + 2 \quad (4.23)$$

The "2" in this equation comes from the fact that the stable equilibrium state of a system of fixed composition is determined by fixing a maximum of two state variables (*Duhem's law*; Anderson and Crerar, 1993, p. 91). For this purpose, only the usual thermodynamic variables are included (i.e., magnetic, electrical, and gravitational properties of systems are ignored). For geochemical systems, these two variables are usually taken to be pressure and temperature.

The phase rule, originally formulated by the American chemist J. Willard Gibbs (1839–1903) toward the end of the 19th century, can be derived from first principles of thermodynamics (see Box 5.1). The rule does not predict the existence of certain states of matter at specified pressures and temperatures, but it does tell us what is possible or impossible at equilibrium. As an illustration, the possible equilibrium assemblages in the low-pressure portion of the one-component (unary) system H_2O (Fig. 4.4) are summarized in Table 4.1.

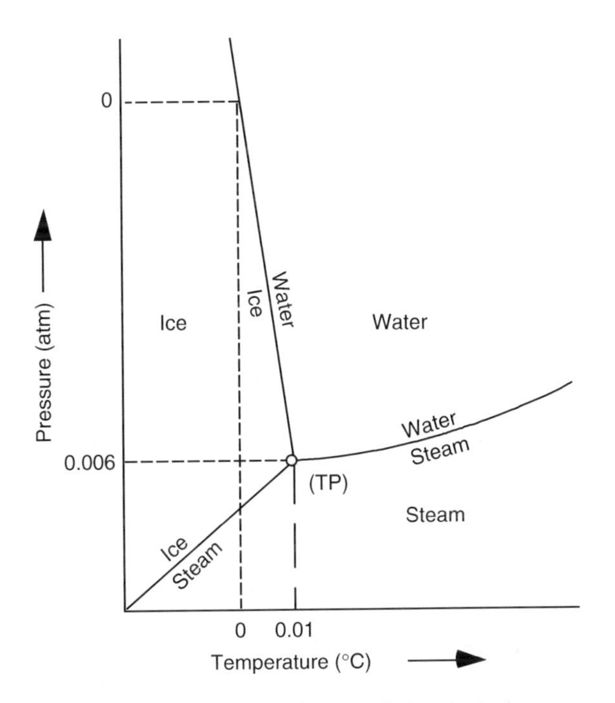

Fig. 4.4 Schematic representation of phase relations in the low-pressure portion of the unary H_2O system. The pressure and temperature scales are in atm and Celsius (°C) units, respectively, to illustrate that zero degree Celsius is defined as the freezing point of water (temperature achieved by a mixture of pure water and ice in equilibrium with air at 1 atm pressure). The temperature of the invariant point, commonly referred to as the triple point (TP) of H_2O, is slightly higher (0.01°C) because of the effect of dissolved gases on the freezing point of water and the effect of pressure on the melting point of ice. T (K) $= t$ (°C) $+ 273.15$.

Table 4.1 Phase equilibria in the unary system H_2O ($c=1$) shown in Fig. 4.4.

Phases in equilibrium	Number of phases in equilibrium (p)	Variance (f)	Description
Ice or water or steam	1	2 (P and T)	Divariant areas
Ice ⇔ water (melting) or Water ⇔ steam (boiling) or Steam ⇔ ice (sublimation)	2	1 (P or T)	Univariant curves
Ice+water+steam	3	0 (both P and T are fixed)	Invariant point

The phase rule predicts that an equilibrium assemblage in a unary system cannot have more than three phases; when $p > 3$, the variance becomes a negative number, which indicates a disequilibrium assemblage.

A system may be studied under some externally imposed restriction such as constant P or constant T or both, which reduces the variance of the system accordingly. At constant P or constant T, the phase rule becomes $f = c - p + 2 - 1 = c - p + 1$; at constant P and constant T, $f = c - p + 2 - 2 = c - p$. Thus the general form of the phase rule may be stated as

$$f = c - p + 2 - \psi \quad (4.24)$$

where ψ is the number of restrictions imposed on the system.

The phase rule can be applied as a test of equilibrium in mineral assemblages and thus is very useful in the construction and interpretation of phase diagrams. For an equilibrium mineral assemblage that crystallized over a range of both temperature and pressure, as usually is the case with metamorphic mineral assemblages, $f=2$, so that $c=p$. This simple relationship, termed the *mineralogical phase rule* by Goldschmidt (1911), states that the number of minerals (phases) should equal the number of components in a mineral assemblage in divariant equilibrium. There is nothing in this statement that is unique to minerals; it applies to any heterogeneous system in divariant equilibrium. It does, however, explain why most rocks are composed of a relatively small number of minerals.

4.2.4 *Equations of state*

An *equation of state* describes a quantitative relationship that exists between intensive (mass-independent) and extensive (mass-dependent) variables in a thermodynamic system.

The only well established equation of state is for an *ideal gas*, which is conceptualized as consisting of vanishingly small particles that do not interact in any way with each other. The *ideal gas equation* that applies to all gases at very low pressures and relatively high temperatures is:

$$\boxed{PV = nRT} \tag{4.8}$$

where P is the pressure arising out of the particles bouncing off the walls or boundaries of the container, V is the volume occupied by the particles, T is the temperature (K) related to the movement and individual energies of the particles, n is the number of moles of the gas, and R is the Gas Constant. For 1 mole of ideal gas (i.e., $n = 1$), the equation reduces to:

$$\boxed{P\bar{V} = RT} \tag{4.25}$$

where $\bar{V}$ is the molar volume. Most gases deviate appreciably from ideal behavior at even moderate pressures and low temperatures because of the finite volume of molecules (assumed to be infinitely small in the case of ideal gases) and because of intermolecular attractive and repulsive forces (assumed to be nonexistent in the case of ideal gases). Many alternative equations of state have been suggested for such nonideal (or "real") gases. The best known is the *van der Waals* (VDW) *equation*:

$$P = \frac{RT}{\bar{V} - b} - \frac{a}{\bar{V}^2} \tag{4.26}$$

where a and b are empirical constants specific to the gas under consideration (Table 4.2), the term $a/\bar{V}^2$ takes care of the intermolecular attractive forces, and the term $\bar{V} - b$ represents the volume available for movement of the molecules and thus is a measure of the repulsive forces.

The VDW equation is useful only over limited ranges of pressure and temperature. To obtain a better match with the experimental P–V–T data for real gases at moderate to high pressures, a modification of the VDW equation was introduced by Redlich and Kwong (1949):

$$P = \frac{RT}{\bar{V} - b} - \frac{a}{\sqrt{T}\ (\bar{V} + b)\ \bar{V}} \tag{4.27}$$

Table 4.2 Van der Waals constants for selected gases.

Gas	a L² atm mol⁻²	b L mol⁻¹	Gas	a L² atm mol⁻²	b L mol⁻¹
Helium	0.0346	0.0238	Chlorine	6.343	0.0542
Argon	1.355	0.3220	CO_2	3.658	0.0429
Hydrogen	0.245	0.0265	Methane	2.303	0.0431
Oxygen	1.382	0.0319	Water	5.537	0.0305
Nitrogen	1.370	0.0387	Ammonia	4.225	0.0371

Source of data: *CRC Handbook of Chemistry and Physics*, 81st edn. (2000)

The *Redlich–Kwong* (RK) *equation*, which expresses the attractive forces as a more complex function compared to the VDW equation, showed better agreement with experimental data for pure gases, such as N_2, O_2, CO, and CH_4, composed of nonpolar molecules. In the original RK equation, the parameters a and b were constants, independent of T and P. To obtain a better fit with the experimental data, many of the modified versions of the Redlich–Kwong equation (Kerrick and Jacobs, 1981) treat the parameters a and b as a function of T or P, or both T and P.

4.2.5 *Kinds of thermodynamic systems and processes*

Thermodynamic systems are classified into various kinds (Fig. 4.2) depending on the exchange of energy (in the form of heat or work) and matter between the system and the surroundings:

Isolated system – a system that permits no exchange of either matter or energy (in the form of heat or its mechanical equivalent, work) with the surroundings; such a system maintains a constant mass, a constant bulk chemical composition, and a constant level of energy.

Closed system – a system that permits exchange of energy with the surroundings but not matter; such a system has a constant mass and a constant bulk chemical composition but variable energy levels.

Open system – a system that permits exchange of both matter and energy with the surroundings; such a system has variable mass, bulk chemical composition, and energy levels.

Most geological systems are open, but some may be treated as closed because of very slow rate of mass exchange with the surroundings. Isolated systems do not exist in nature.

In addition to reversible and irreversible processes defined earlier, a system may undergo change by special processes that are named according to the constraint imposed on the system: *isothermal* (constant temperature); *isobaric* (constant pressure); *isochoric* (constant volume); and *adiabatic* (constant heat content, although the system can exchange energy with the surroundings in the form of work).

4.3 Laws of thermodynamics

The laws of thermodynamics discussed below cannot be derived mathematically or proven formally. They are simply postulates based on laboratory experiments and experience that continue to provide consistent explanations for many natural processes.

4.3.1 *The first law: conservation of energy*

The *first law of thermodynamics*, based mainly on the series of experiments carried out by J.P. Joule between 1843 and 1848, is a statement of the *principle of conservation of energy* – that is, energy is neither created nor destroyed in chemical reactions and physical changes. For a mathematical formulation of the law, we postulate that the system has a state function called *internal energy* (U). For a mineral, for example, it is the sum of the potential energy stored in the interatomic bonding and the kinetic energy of the atomic vibrations. For a given state, the internal energy of a system is difficult to measure, but we are interested in the difference in internal energy between two states, and the first law defines a way to measure it.

In classical thermodynamics, only two ways of changing the energy content of a system are considered: addition or subtraction of *heat* or *work* (the mechanical equivalent of heat) during the transformation of the system from one state to the other. The first law may be stated as:

> *The change in the internal energy of a system (ΔU) between two states is equal to the heat flowing into the system from the surroundings (Q) plus the work done on the system (W).*

The corresponding mathematical expression is

$$\boxed{\Delta U = Q + W} \tag{4.28}$$

It can be shown that neither Q nor W is a state function, but U is (i.e., dU is an exact differential, whereas dQ and dW are not). Commonly, the first law is written in the differential form for an infinitesimal change as

$$dU = dQ + dW \tag{4.29}$$

where dU, dQ, and dW refer to infinitesimal increments. According to the convention used by scientists (see Fig. 4.2), the sign of energy change in a system is considered from the point of view of the system, not the surroundings. Thus, heat absorbed by a system is positive (whereas heat given off by a system is negative), and work done on a system by some agency in the surroundings is positive (whereas work done by a system on its surroundings is negative). This convention is consistent with the idea that from the system's point of view a decrease in energy is negative and an increase in energy is positive.

How do we define heat and temperature? Heat is the energy that flows across a system boundary in response to a temperature gradient (i.e., from a body at a higher temperature to a body at a lower temperature). The concept of temperature is empirically defined by what is known as the *zeroth law of thermodynamics*: *two bodies that are in thermal equilibrium with a third body are in thermal equilibrium with each other.* A consequence of the zeroth law is that two or more bodies in thermal equilibrium possess a property in common. This is the property we call temperature, which is an intensive property because it does not depend on the amount of substance being considered. Suppose one of these bodies is a thermometer that has been calibrated according to some defined scale (e.g., the Celsius or the Kelvin scale we discussed in Chapter 1). The "hotness" as recorded by the thermometer is the temperature of all the bodies in thermal equilibrium with the thermometer. Thus, the temperature of a body is recorded by a thermometer with which the body is in thermal equilibrium.

Work is done when an object is displaced by some distance in response to an applied force. Consider a gas that is held under an applied force (F) by a piston in a vertical cylinder of cross-sectional area A. If the piston is released, the gas will expand. For an infinitesimal amount of work done by the gas (dW) corresponding to an infinitesimal displacement of the piston (dx),

$$dW = -F\,dx \tag{4.30}$$

The only form of work that is of appreciable magnitude in most geological environments is "PV work" involving a volume change (dV) against a constant external pressure (P). We can express dW in terms of P and dV as follows:

$$P = \frac{F}{A} \qquad \text{and} \qquad A\,dx = dV \tag{4.31}$$

$$dW = \frac{-F}{A}\,A\,dx = -P\,dV \tag{4.32}$$

Note that the SI unit of pressure is (mass) (distance^{-1}) (time^{-2}) and that of volume is distance^3, so that the product of pressure and volume is (mass) (distance^2) (time^{-2}), the units of energy (1 joule $=$ 1 kg m^2 s^{-2}). Substituting for dW,

$$dU = dQ - P\,dV \tag{4.33}$$

The first law makes no distinction between reversible and irreversible processes and no prediction about the direction of an irreversible reaction. The second law of thermodynamics provides these kinds of information.

4.3.2 *The second law: the concept and definition of entropy (S)*

The ideas that led to what we now refer to as the *second law of thermodynamics* and *entropy* were formulated by the German physicist Rudolf Clausius (1822–1888) in an article published in 1850. Similar ideas were published a year later by William Thompson (Lord Kelvin), who coined the term "entropy." The second law of thermodynamics can be stated in a number of different ways (see, e.g., Anderson and Crerar, 1993; Nordstrom and Munoz, 1994); perhaps the easiest one to understand is as follows (White, 2002):

> *There is a natural direction in which reactions will tend to proceed. This direction is inevitably that of higher entropy of the system and its surroundings.*

Consider, for example, a well-insulated cylinder C_A filled with a gas and connected to another well insulated but evacuated cylinder C_B by a stopcock. When the stopcock is opened, the gas will spontaneously expand to fill both cylinders, but the reverse process, in which all gas molecules diffuse back into cylinder C_A, will not happen spontaneously whether the stopcock is left open or closed. Neither the spontaneous expansion of the gas nor the lack of its spontaneous unmixing can be explained by the first law because we have neither added energy to nor removed energy from the system. It appears that there is some other system property that dictates the direction of spontaneous change. This property is *entropy* (S) as defined by the second law. As explained below, the direction of a spontaneous (irreversible) change is inevitably toward a state of higher entropy. Like V, S is an extensive property and also a state function. The unit of entropy is J mol^{-1} K^{-1}.

According to the second law, the infinitesimal change in entropy (dS) of a *closed* system due to a reversible process at temperature T (K) is equal to the infinitesimal amount of heat absorbed (dQ) from the surroundings during the process divided by the temperature T (K):

$$dS = \frac{dQ}{T} \quad \text{Reversible process} \tag{4.34}$$

For an irreversible process, dS cannot be determined precisely, but it will always be greater than the value calculated by using equation (4.34):

$$dS > \frac{dQ}{T} \quad \text{Irreversible process} \tag{4.35}$$

Thus, a complete expression of the second law is written as

$$dS \geq \frac{dQ}{T} \tag{4.36}$$

Changes that take place spontaneously in a system, such as mixing of two gases, are called *natural processes*, and it is known from experience that the system in question cannot be restored to its original condition only by transferring a quantity of heat elsewhere. Thus, natural processes are irreversible and they always result in increasing the entropy of an adiabatic system. In the gas expansion experiment mentioned above, dQ=0 as the system is assumed to be well insulated (or adiabatic), so that dS > 0. We may rationalize that the gas expansion occurred because it enabled the system to attain a state of higher entropy.

Entropy is commonly interpreted as a measure of the disorder or randomness of the molecules or atoms (or ions) in a system: entropy increasing with increasing disorder. For example, the entropy of a silicate mineral with a completely ordered crystal structure in which each crystallographic site is occupied by a specific kind of atom (or ion) is less than the entropy of the same mineral with partly disordered crystal structure in which random substitutions occur in one or more of the crystallographic sites. In general, for the same substance occurring in gaseous, liquid, and solid state, $\overline{S}_{gas} > \overline{S}_{liquid} > \overline{S}_{solid}$, where $\overline{S}$ represents molar entropy (entropy per mole) of the substance. For example, $\overline{S}_{H_2O\,(steam)} = 188.8$, $\overline{S}_{H_2O\,(water)} = 70.0$, and $\overline{S}_{H_2O\,(ice)} = 44.7$ J mol^{-1} K^{-1}. Because all natural processes are irreversible to varying degrees, the entropy of the Earth has been increasing since its birth!

4.3.3 *The fundamental equation: the first and second laws combined*

The first and second laws can be combined by substituting the expression for dQ from equation (4.36) into equation (4.33). The result is:

$$\boxed{dU \leq T\,dS - P\,dV} \tag{4.37}$$

This is often referred to as the fundamental equation for either reversible or irreversible changes in a closed system of constant composition because it includes all the primary variables (P, T, V, U, and S) necessary to describe the system. The restrictions are that volume change is the only form of work, and the system is in internal equilibrium. Changes in most geological systems occur very slowly and a close approach to reversible process would be a reasonable assumption. We can then use the equation

$$\boxed{dU = T\,dS - P\,dV} \tag{4.38}$$

4.3.4 *The third law: the entropy scale*

Experimental evidence indicates that entropies of crystalline substances might either vanish or reach a minimum at absolute zero temperature (0 K), although this temperature itself is unattainable. This postulate has evolved into what is now known as the *third law of thermodynamics*. The third law has been stated in many different ways; the following formal statement of the law by Lewis and Randall (1925, p. 448) is the most convenient for our purpose:

> *If the entropy of each element in some crystalline state be taken as zero at the absolute zero of temperature: every substance has a finite positive entropy, but at the absolute zero of temperature the entropy may become zero, and does so become in the case of perfectly crystalline substances.*

A perfectly crystalline substance is one that is characterized by a single quantum state at absolute zero temperature, that is, it has only one possible arrangement (and only one possible energetic configuration) of its constituent atoms at absolute zero.

The third law does not have same status as the first or second law, but it has an important practical significance in the calculation of the *absolute entropy* (also referred to as the *third-law entropy* or *conventional entropy*) of a substance. The stipulation that entropy of a pure (i.e., it cannot be separated into fractions of different properties), perfectly crystalline substance is zero at absolute zero temperature provides the reference point for a scale to measure absolute values of entropy of the substance at different states. Thus, the absolute entropy of a substance is its entropy relative to its entropy in a perfectly ordered crystalline form at 0 K, where its entropy is zero. The absolute entropy of a substance at 298.15 K, for exampde, is given by

$$\int_0^{298.15} dS = S_{298.15} - S_0 = S_{298.15} - 0 = S_{298.15}$$

and $S_{298.15}$ can be determined by measuring the heat capacity of the substance from temperatures as near absolute zero as possible to ambient conditions (see section 4.4.2). In this sense, entropy is unique among thermodynamic functions. As we will discuss in following sections, other thermodynamic functions are measured relative to an arbitrary reference state and are assigned relative rather than absolute values.

Various types of disorders may exist in an imperfect crystal at absolute zero. Entropy derived from such disorders is called *residual entropy* and commonly denoted by the symbol S_o. Contributions to the residual entropy, which are calculated using quantum mechanics, include *configurational entropy* that arises because of mixing of two or more kinds of atoms in crystallographically equivalent sites, crystal imperfections, and magnetic spin ordering in paramagnetic substances. The total entropy of a substance also includes the entropy associated with any phase changes between absolute zero and the temperature of interest (equation 4.69).

4.4 Auxiliary thermodynamic functions

We now introduce three additional extensive properties of systems – *enthalpy* (H), *Hemholtz free energy* (A), and *Gibbs free energy* (G) – as defined below:

Enthalpy $\quad H = U + PV \quad$ (4.39)

Hemholtz free energy $\quad A = U - TS \quad$ (4.40)

Gibbs free energy $\quad G = H - TS \quad$ (4.41)

H, A, and G are state functions because they are defined in terms of variables that are state functions. The values of these functions (commonly expressed in units of kJ mol^{-1}), unlike that of entropy, are measured or computed as changes relative to an arbitrary reference state and are assigned relative rather than absolute values. These three functions are called *auxiliary thermodynamic functions* because they do not add any new information beyond that provided by the first and second laws. Their usefulness lies in the fact that they are more convenient to use in certain applications. As the Hemholtz function is seldom used in geochemistry, we will not discuss it any further, and refer to Gibbs free energy simply as free energy.

4.4.1 *Enthalpy (H)*

Enthalpy of a substance is its heat content at constant pressure and it is defined by equation (4.39). We can measure only the change in enthalpy of a substance between two states, not its absolute value in any state. Consider a process carried out at constant pressure. Taking the total derivative of equation (4.39) and substituting for dU as per equation (4.33), we get

$$dH = dU + P\,dV + V\,dP = dQ - P\,dV + P\,dV + V\,dP = dQ + V\,dP \quad (4.42)$$

At constant pressure P (i.e., $dP = 0$),

$$\boxed{(dH)_P = (dQ)_P} \quad (4.43)$$

The change in enthalpy (ΔH) corresponding to a change from an initial state (temperature T_1) to a final state (temperature T_2) at constant pressure P is given by

$$\int_{T_1}^{T_2} (dH)_P = \int_{T_1}^{T_2} (dQ)_P \quad \text{or} \quad (H_{T_2})_P - (H_{T_1})_P = (Q_{T_2})_P - (Q_{T_1})_P$$

which can be written as

$$\boxed{(\Delta H)_P = (\Delta Q)_P} \quad (4.44)$$

Thus, the change in enthalpy due to a reversible or an irreversible change in the state of a system at constant pressure can be determined by measuring the heat absorbed (or released) by the system during that change. Also, note that for a reaction at constant P and constant V, $(dH)_{P,V} = dU$, i.e., the entire change in internal energy is represented by the change in enthalpy.

4.4.2 *Heat capacity (C_p, C_v)*

The enthalpy and entropy of a substance are quantitatively related to a function called *heat capacity* (C), the capacity of a substance to absorb heat, which can be measured directly. Heat capacity, an intrinsic property of any substance, is defined as *the amount of heat (in joules) required to raise the temperature of 1 mole of a substance by 1* and expressed mathematically as

$$\boxed{C = \frac{dQ}{dT}} \quad (4.45)$$

The commonly used units of *(molar) heat capacity* are J mol^{-1} K^{-1}. (Heat capacity is also expressed in units of J g^{-1} K^{-1}, and is referred to as *specific heat capacity*.)

A substance has two values of heat capacity, one for heat being added at constant pressure (C_p) and the other for heat being added at constant volume (C_V). Since $(\Delta Q)_P = (\Delta H)_P$,

$$\boxed{C_P = \left(\frac{\partial H}{\partial T}\right)_P \quad \text{or} \quad (dH)_P = C_P dT} \quad (4.46)$$

When the volume is held constant (i.e., $dV=0$), $(\Delta Q)_P=(\Delta U)_V$ from equation (4.33), and

$$C_V = \left(\frac{\partial U}{\partial T}\right)_V \quad \text{or} \quad (dU)_V = C_V dT \tag{4.47}$$

C_p is always larger than C_V for the same substance, because at constant pressure part of the heat added to a substance will be consumed as work if the volume changes by expansion or contraction, whereas at constant volume all the added heat will produce a rise in temperature. However, for relatively incompressible substances, such as minerals, the difference between C_p and C_V is generally small. In any case, C_P is of greater interest to us than C_V because the constraint of constant volume does not usually apply to geochemical reactions.

For a substance of constant composition, the enthalpy difference for a reversible or irreversible change between an initial state (temperature T_1) and a final state (temperature T_2) at constant pressure P, assuming no phase transformation between T_1 and T_2, is obtained by integration:

$$\Delta H^P_{T_1 \to T_2} = \int_{T_1}^{T_2} (dH)_P = H^P_{T_2} - H^P_{T_1} = \int_{T_1}^{T_2} C_P\, dT \tag{4.48}$$

$$H^P_{T_2} = H^P_{T_1} + \int_{T_1}^{T_2} C_P\, dT \tag{4.49}$$

For the calculation of entropy change for a reversible reaction at constant pressure P, we turn to equations (4.34) and (4.44):

$$dS = \frac{dQ}{T} = \frac{(dH)_P}{T} = \frac{C_P}{T}\, dT \tag{4.50}$$

By analogy with equation (4.48), for a substance of constant composition the entropy difference for a reversible change between state 1 (temperature T_1) and state 2 (temperature T_2) at constant pressure is given by

$$\Delta S^P_{T_1 \to T_2} = \int_{T_1}^{T_2} (dS)_P = S^P_{T_2} - S^P_{T_1} = \int_{T_1}^{T_2} \frac{C_P}{T}\, dT \tag{4.51}$$

$$S^P_{T_2} = S^P_{T_1} + \int_{T_1}^{T_2} \frac{C_P}{T}\, dT \tag{4.52}$$

If there are phase transformations between temperature T_1 and temperature T_2, then

$$H^P_{T_2} = H^P_{T_1} + \int_{T_1}^{T_2} C_P\, dT + \sum \Delta H_{\text{trans}} \tag{4.53}$$

$$S^P_{T_2} = S^P_{T_1} + \int_{T_1}^{T_2} \frac{C_P}{T}\, dT + \sum \frac{\Delta H_{\text{trans}}}{T_{\text{trans}}} \tag{4.54}$$

where T_{trans} is the temperature at which a particular phase transition occurs, ΔH_{trans} is the enthalpy change associated with the transition, and the summation represents the combined effect of all transitions occurring between T_1 and T_2.

The integrations in equations (4.49) and (4.52) are a little cumbersome because C_p itself is a function of temperature. (The variation of C_p with pressure is very small and can be ignored for our purpose.) The integrals can be evaluated graphically from C_p versus T and C_p/T versus T graphs, as illustrated in Fig. 4.5 for crystalline quartz as an example. The common practice, however, is to integrate a polynomial function that empirically describes the variation of molar heat capacity, $\bar{C}_p$, with temperature in accordance with experimental measurements of C_p for the substance under consideration. Many forms of empirical equations have been proposed for

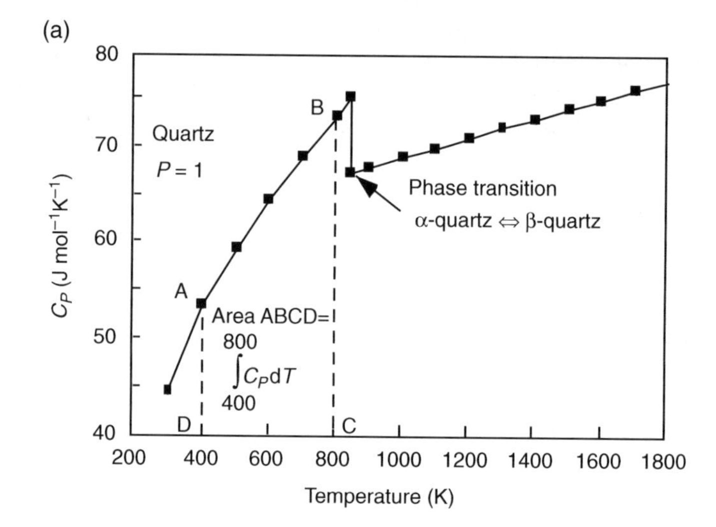

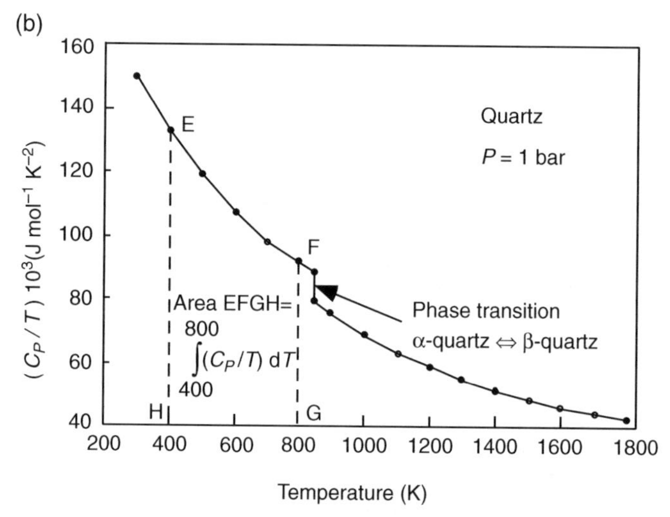

Fig. 4.5 Variation of (a) C_p (heat capacity) and (b) C_p/T with temperature (K) for crystalline quartz at 1 bar constant pressure. For a reversible change from 400 K to 800 K, $\Delta H = H^1_{800} - H^1_{400} \equiv$ area ABCD under the curve in (a), and $\Delta S = S^1_{800} - S^1_{400} \equiv$ area EFGH under the curve in (b). ΔH and ΔS corresponding to a temperature interval encompassing the phase transition at 844 K must include, in addition to the area under the curve between the appropriate temperatures, a term to account for the discontinuity. The additional term is ΔH_T for enthalpy change and $\Delta H_T/T$ for entropy change, where T is the phase transition temperature (844 K). Source of C_p data: Robie and Hemingway (1995).

the fitting of heat-capacity data (see Nordstrom and Munoz, 1994, p. 383–390, for an excellent summary). The compilation of thermodynamic data for substances of geochemical interest by Robie and Hemmingway (1995) includes analytical expressions for $\bar{C}_p$ of the form

$$\bar{C}_p = A_1 + A_2 T + A_3 T^{-2} + A_4 T^{-0.5} + A_5 T^2 \tag{4.55}$$

where A_1, A_2, A_3, A_4, and A_5 are constants whose values vary with the substance. For example, for α-quartz the expression for molar heat capacity at 1 bar constant pressure, $\bar{C}^1_p$, is given by Robie and Hemmingway (1995) as

$$\bar{C}^1_p = 8.1145 \times 10^1 + (1.828 \times 10^{-2})T - (1.810 \times 10^5)T^{-2} - (6.985 \times 10^2)T^{-0.5} + (5.406 \times 10^{-6})T^2 \tag{4.56}$$

Thus, for α-quartz, analytical solutions equivalent to the graphical solutions for a temperature interval from 400 to 800 K illustrated in Fig. 4.5 can be written as

$$\bar{H}^1_{800} - \bar{H}^1_{400} = \int_{400}^{800} [8.1145 \times 10^1 + (1.828 \times 10^{-2})T - (1.810 \times 10^5)T^{-2} - (6.985 \times 10^2)T^{-0.5} + (5.406 \times 10^{-6})T^2]\, dT$$

$$\bar{S}^1_{800} - \bar{S}^1_{400} = \int_{400}^{800} \{[8.1145 \times 10^1 + (1.828 \times 10^{-2})T - (1.810 \times 10^5)T^{-2} - (6.985 \times 10^2)T^{-0.5} + (5.406 \times 10^{-6})T^2] / T\}\, dT$$

where $\bar{H}$ and $\bar{S}$ denote molar enthalpy (enthalpy per mole) and molar entropy (entropy per mole), respectively.

4.4.3 *Gibbs free energy (G)*

The Gibbs free energy auxiliary function is named in honor of Josiah Willard Gibbs (1839–1903), who lectured at Yale University from 1871 to 1903 and published a series of papers that laid the theoretical foundation for all subsequent work on the subject. The Gibbs free energy of a system indicates the amount of energy available for the system to do useful work at constant pressure and temperature. It is one of the most useful functions in thermodynamics as it defines the criteria for both equilibrium (or reversible) and spontaneous (or irreversible) change at constant temperature and pressure, the most frequently encountered condition for geochemical reactions in a closed system. Let us examine how it works for a reversible reaction. Substituting equation (4.39) in equation (4.41), and taking the total derivative, we get

$$G = H - TS = U + PV - TS$$

$$dG = dU + P\,dV + V\,dP - T\,dS - S\,dT$$

Substituting for dU (equation 4.33) and $T\,dS$ (equation 4.34),

$$\boxed{dG = V\,dP - S\,dT} \tag{4.57}$$

It follows that for an infinitesimal reversible change in a closed system,

at constant pressure $$\left(\frac{\partial G}{\partial T}\right)_P = -S \tag{4.58}$$

at constant temperature $$\left(\frac{\partial G}{\partial P}\right)_T = V \tag{4.59}$$

at constant temperature and constant pressure $$\boxed{(dG)_{T,P} = 0} \tag{4.60}$$

Thus, G decreases with increasing temperature at constant pressure (in contrast to H that increases with increasing temperature at constant pressure, equation (4.46)) but increases with increasing pressure at constant temperature (Fig. 4.6);

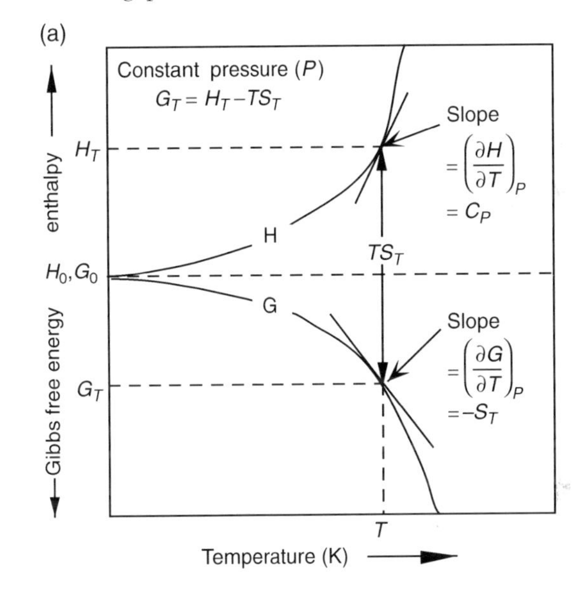

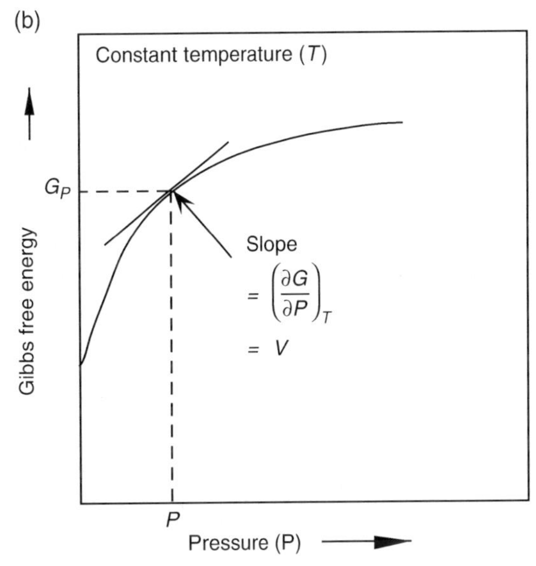

Fig. 4.6 (a) Variation of enthalpy (H) and Gibbs free energy (G) of a pure substance with temperature at constant pressure; (b) Variation of Gibbs free energy (G) with pressure at constant temperature.

and a reversible change at constant temperature and pressure is characterized by no change in the free energy. *The latter is the thermodynamic condition for chemical equilibrium in a closed system.*

The free energy is not directly measurable for any system, but its value can be computed from its enthalpy and entropy. Differentiation of the defining equation for G, $G=H-TS$, at constant temperature gives

$$dG = dH - T\,dS \tag{4.61}$$

which on integration results in the equation that is commonly used for computation of ΔG:

$$\boxed{\Delta G = \Delta H - T\,\Delta S} \tag{4.62}$$

4.4.4 *Computation of the molar free energy of a substance at T and P ($\bar{G}_T^P$)*

We adopt the following strategy for the computation of the molar free energy, $\bar{G}_T^P$, of a substance, with T and P being the temperature and pressure of interest:

(1) The values of $\bar{H}$, $\bar{S}$, and $\bar{G}$ for a substance are calculated as changes ($\Delta\bar{H}$, $\Delta\bar{S}$, and $\Delta\bar{G}$) from some convenient, common *reference state* of temperature (T_{ref}) and pressure (P_{ref}). The choice of the reference state is immaterial as long as we know how to calculate $\bar{H}$, $\bar{S}$, and $\bar{G}$ of the substance at the reference state, which we denote as $\bar{H}_{T_{ref}}^{P_{ref}}$, $\bar{S}_{T_{ref}}^{P_{ref}}$, and $\bar{G}_{T_{ref}}^{P_{ref}}$, respectively.

(2) As $\bar{H}$, $\bar{S}$, and $\bar{G}$ are state functions, it does not matter what path is taken for the change from the reference state (T_{ref}, P_{ref}) to the final state (T, P). For convenience of calculation, we consider the path to consist of two steps (Fig. 4.7): an isobaric step in which the temperature is changed from T_{ref} to T at constant pressure P_{ref} to obtain $\bar{G}_T^{P_{ref}}$ (the molar free energy at T and reference pressure); and an isothermal step in which the pressure is changed from P_{ref} to P at constant temperature T to obtain $\bar{G}_T^P$. Thus,

$$\bar{G}_T^P = \bar{G}_T^{P_{ref}} + \int_{P_{ref}}^{P} \left(\frac{\partial \bar{G}}{\partial P}\right)_T dP \tag{4.63}$$

(3) We use equation (4.41) to calculate $\bar{G}_T^{P_{ref}}$ and equation (4.59) to evaluate the effect of pressure:

$$\bar{G}_T^P = \bar{G}_T^{P_{ref}} + \int_{P_{ref}}^{P} \bar{V}\,dP = \left(\bar{H}_T^{P_{ref}} - T\,\bar{S}_T^{P_{ref}}\right) + \int_{P_{ref}}^{P} \bar{V}_T\,dP \tag{4.64}$$

Note that in equation 4.64, $\bar{V}$ represents the molar volume at temperature T, not T_{ref}.

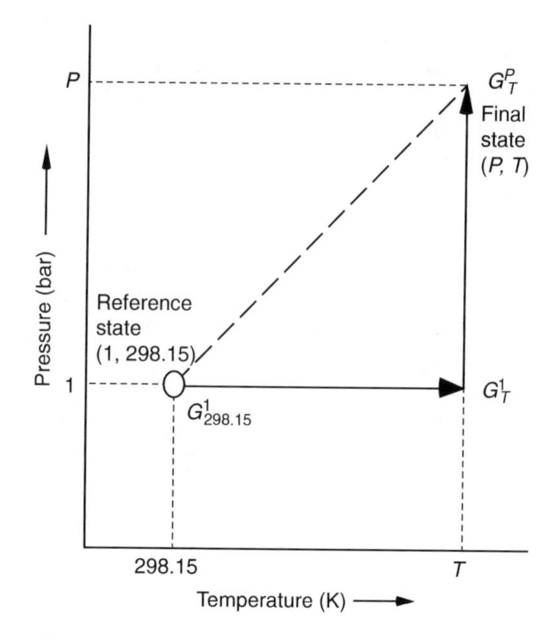

Fig. 4.7 The strategy we employ for computing Gibbs free energy of a substance at temperature T and pressure P, G_T^P, consists of calculating G_T^1 (at 1 bar constant reference pressure and temperature T) and then calculating G_T^P (at pressure P and constant temperature T).

(4) $\bar{H}_T^{P_{ref}}$ and $\bar{S}_T^{P_{ref}}$ are calculated from the following relations that are analogous to equations (4.49) and (4.52):

$$\bar{H}_T^{P_{ref}} = \bar{H}_{T_{ref}}^{P_{ref}} + \int_{T_{ref}}^{T} \bar{C}_P^{P_{ref}}\,dT \tag{4.65}$$

$$\bar{S}_T^{P_{ref}} = \bar{S}_{T_{ref}}^{P_{ref}} + \int_{T_{ref}}^{T} \frac{\bar{C}_P^{P_{ref}}}{T}\,dT \tag{4.66}$$

Let us choose the reference state as T_{ref}=298.15 K (often denoted as 298 for the sake of brevity) and P_{ref}=1 bar. The reference pressure is denoted by "o" as a superscript in many publications, but we will denote it more explicitly by "1" as a superscript. This reference state is a convenient choice, because thermodynamic data for substances commonly are available for this state. With this stipulation, we obtain the following expressions for $\bar{G}_T^1$ and $\bar{G}_T^P$:

$$\boxed{\bar{G}_T^1 = \left(\bar{H}_{298.15}^1 + \int_{298.15}^{T} \bar{C}_P^1\,dT\right) - T\left(\bar{S}_{298.15}^1 + \int_{298.15}^{T} \frac{\bar{C}_P^1}{T}\,dT\right)} \tag{4.67}$$

$$\boxed{\bar{G}_T^P = \left(\bar{H}_{298.15}^1 + \int_{298.15}^{T} \bar{C}_P^1\,dT\right) - T\left(\bar{S}_{298.15}^1 + \int_{298.15}^{T} \frac{\bar{C}_P^1}{T}\,dT\right) + \int_1^P \bar{V}_T\,dP} \tag{4.68}$$

Box 4.4 Standard state enthalpy and free energy of formation of α-quartz

The most stable form of SiO_2 at the chosen standard state (1 bar, 298.15 K) is α-quartz. The formation of 1 mole of α-quartz at 1 bar and 298.15 K from its constituent elements in their standard states and the associated, directly measured enthalpy change, $\Delta H^1_{f,\,298.15}$ (α-quartz) (Robie and Hemmingway, 1995) can be written as:

$$\text{Si (crystals)} + O_2 \text{ (ideal diatomic gas)} = SiO_2(\alpha\text{-quartz}) \quad \Delta H^1_{f,\,298.15} = -910.7\,\text{kJ mol}^{-1}$$
$$\Delta G^1_{f,\,298.15} = -856.3\,\text{kJ mol}^{-1}$$

$$\begin{aligned}\Delta H^1_{f,\,298.15}(\alpha\text{-quartz}) &= \bar{H}^1_{298.15}(\alpha\text{-quartz}) - [\bar{H}^1_{298.15}(\text{Si}) + \bar{H}^1_{298.15}(O_2)]\\ &= \bar{H}^1_{298.15}(\alpha\text{-quartz}) - 0 - 0\\ &= \bar{H}^1_{298.15}(\alpha\text{-quartz}) = -910.7\,\text{kJ mol}^{-1}\end{aligned}$$

$$\begin{aligned}\Delta G^1_{f,\,298.15}(\alpha\text{-quartz}) &= \bar{G}^1_{298.15}(\alpha\text{-quartz}) - [\bar{G}^1_{298.15}(\text{Si}) + \bar{G}^1_{298.15}(O_2)]\\ &= \bar{G}^1_{298.15}(\alpha\text{-quartz}) - 0 - 0\\ &= \bar{G}^1_{298.15}(\alpha\text{-quartz}) = -856.3\,\text{kJ mol}^{-1}\end{aligned}$$

We can also calculate $\Delta G^1_{f,\,298.15}$ (α-quartz) from $\bar{S}^1_{298.15}$ (total entropy) values given in Robie and Hemingway (1995) to first calculate $\Delta S^1_{f,\,298.15}$ (the standard state entropy of formation from constituent elements in their standard states) and then calculate $\Delta G^1_{f,\,298.15}$ using the relation

$$\Delta G^1_{f,\,298.15} = \Delta H^1_{f,\,298.15} - T\,\Delta S^1_{f,\,298.15} \quad (T = 298.15\,\text{K})$$

$$\begin{aligned}\Delta S^1_{f,\,298.15}(\alpha\text{-quartz}) &= S^1_{298.15}(\alpha\text{-quartz}) - [S^1_{298.15}(\text{Si}) + S^1_{298.15}(O_2)]\\ &= 41.46 - (18.81 + 205.15) = -182.50\,\text{J mol}^{-1}\,\text{K}^{-1}\\ &= -0.1825\,\text{kJ mol}^{-1}\,\text{K}^{-1}\end{aligned}$$

$$\begin{aligned}\Delta G^1_{f,\,298.15}(\alpha\text{-quartz}) &= \Delta H^1_{f,\,298.15}(\alpha\text{-quartz}) - 298.15\,\Delta S^1_{f,\,298.15}(\alpha\text{-quartz})\\ &= -910.7 - 298.15\,(-0.1825) = -856.3\,\text{kJ mol}^{-1}\,\text{K}^{-1}\end{aligned}$$

Standard state enthalpy of formation

We have already discussed how to evaluate the integrals involving C_P in equations (4.58) and (4.59), but what are the values of $H^1_{298.15}$ and $S^1_{298.15}$? We can measure only the change in enthalpy resulting from a change in state, not the intrinsic enthalpy of a substance. We get around this limitation by assigning $H^1_{298.15}$ a value called the *standard state enthalpy of formation* of the substance at 298.15 K (often abbreviated as *standard enthalpy*).

> The standard state enthalpy of formation of a substance is the enthalpy change that results when 1 mole of the pure substance is formed from the constituent elements in their standard states, the standard state enthalpy of each element being arbitrarily assigned a value of zero.

The standard state of a substance refers to the most stable form of the pure substance at a specified temperature and pressure. The standard state pressure is usually fixed at 1 bar and the standard state temperature is the stated temperature, often chosen as 298.15 K. For example, at 1 bar and 298.15 K, the standard state of silica is α-quartz and of oxygen is the diatomic O_2 because these are the most stable forms of these substances at the specified conditions. The enthalpy change associated with the exothermic reaction $H_{2(g)} + 0.5\ O_{2(g)} \Rightarrow H_2O_{(l)}$ at 1 bar and 298.15 K represents the standard state enthalpy of H_2O, $\Delta H^1_{f,\,298.15}(H_2O)$, because liquid water is the most stable form of H_2O at the chosen standard state pressure and temperature. $\Delta H^1_{f,\,T}$ values of many substances, including minerals, are available in the published literature. As an example, the calculation of $H^1_{298.15}$ $(= \Delta H^1_{f,\,298.15})$ for the mineral α-quartz is shown in Box 4.4. (The significance and usefulness of standard state will become clearer when we revisit the topic in connection with activities in Chapter 5.)

The total molar entropy of a substance in the chosen reference state, $\bar{S}^1_{298.15}$, can be calculated using the following equation:

$$\bar{S}^1_{298.15} = \bar{S}_0 + \int_0^{298.15} \frac{\bar{C}^1_P}{T}\,dT + \Delta\bar{S}_\phi \tag{4.69}$$

where $\bar{S}_0$ is the molar residual entropy at 0 K, the integrand represents the calorimetric contribution, and $\Delta\bar{S}_\phi$ is the molar entropy change associated with phase transformations, if any, within the temperature interval from 0 K to 298.15 K. However, compilations of $\bar{S}^1_{298.15}$ available in the published literature (e.g., Robie and Hemingway, 1995) will suffice for most geochemical applications.

Standard state free energy of formation

The $\bar{G}^1_T$ of a substance can also be assigned the value of its *standard state free energy of formation* at temperature T (often abbreviated as *standard free energy*), which is defined the same way as the standard state enthalpy of formation:

> free energy change that results when 1 mole of the pure substance in its standard state is formed from the constituent elements in their standard states, the standard state free energy of each element being arbitrarily assigned a value of zero.

The standard-state pressure is usually fixed at 1 bar and the standard state temperature is the stated temperature, often chosen as 298.15 K. For the purpose of illustration, the calculation of $\bar{G}^1_{298.15}$ (= $\Delta G^1_{f,\,298.15}$) for the mineral α-quartz is included in Box 4.4. Values of standard free energy of formation of many substances, including minerals, at various temperatures ($\Delta G^1_{f,\,T}$) are available in the published literature (e.g., Robie and Hemingway, 1995).

Example 4–2: Calculation of H, S, and G of a solid phase at T and 1 bar using C_p data

Let us calculate the molar free energy of the mineral forsterite (Mg_2SiO_4) at 500 K and 1 bar, given that $\Delta H^1_{f,298.15}$ (Mg_2SiO_4)=−2173 kJ mol^{-1}, $S^1_{298.15}$(Mg_2SiO_4)=94.1 J mol^{-1}K^{-1}, and

$\bar{C}_P$ (J mol^{-1} K^{-1})=$A_1+A_2T+A_3T^{-2}+A_4T^{-0.5}+A_5T^2$

The values of the constants for forsterite are:

$A_1 = 8.735\times10^1$, $A_2=8.717\times10^{-2}$, $A_3=-3.699\times10^6$,
$A_4 = 8.436\times10^2$, $A_5=-2.237\times10^{-5}$

$$\begin{aligned}\int_{298.15}^{T} \bar{C}^1_P\, dT\ (\text{fo}) &= \int_{298.15}^{500} [A_1 + A_2T + A_3T^{-2} + A_4T^{-0.5} + A_5T^2]\, dT \\ &= [(A_1)T + (A_2/2)T^2 - (A_3)T^{-1} + (A_4/0.5)T^{0.5} \\ &\quad + (A_5/3)T^3]^{500}_{298.15} \\ &= [17{,}631.6 + 7021.83 - 5008.45 + 8589.53 \\ &\quad + 734.8] \\ &= 27{,}499.7 \text{ J mol}^{-1} = 27.5 \text{ kJ mol}^{-1}\text{ K}^{-1}\end{aligned}$$

$$\begin{aligned}\int_{298.15}^{T} \frac{\bar{C}^1_P}{T}\, dT\ (\text{fo}) &= \int_{298.15}^{500} \frac{1}{T}[A_1 + A_2T + A_3T^{-2} + A_4T^{-0.5} \\ &\quad + A_5T^2]\, dT \\ &= [(A_1)\ln T + (A_2)T - (A_3/2)T^{-2} \\ &\quad - (A_4/0.5)T^{-0.5} + (A_5/2)T^2]^{500}_{298.15} \\ &= [45.16 + 17.59 - 13.41 + 22.27 - 1.802] \\ &= 69.808 \text{ J mol}^{-1} = 0.0698 \text{ kJ mol}^{-1}\end{aligned}$$

Substituting values in equations (4.65) and (4.56),

$$\begin{aligned}\bar{H}^1_{500}(\text{fo}) &= \bar{H}^1_{298.15} + \int_{298.15}^{500} \bar{C}^1_p\, dT = -2173.0 + 27.5 \\ &= -2145.5 \text{ kJ mol}^{-1}\end{aligned}$$

$$\begin{aligned}\bar{S}^1_{500}(\text{fo}) &= \bar{S}^1_{298.15} + \int_{298.15}^{500} \frac{\bar{C}^1_P}{T}\, dT = 94.11 + 69.81 \\ &= 163.92 \text{ J mol}^{-1}\text{K}^{-1} = 0.164 \text{ kJ mol}^{-1}\text{ K}^{-1}\end{aligned}$$

$$\begin{aligned}\bar{G}_{500}{}^1(\text{fo}) &= \bar{H}^1_{500}(\text{fo}) - 500\ \bar{S}^1_{500}(\text{fo}) = -2145.5 - (500 \times 0.164) \\ &= -2145.5 - 81.96 = -2227.46 \text{ kJ mol}^{-1}\end{aligned}$$

Note that H^1_{500} (fo) and G^1_{500} (fo) calculated above from the C_P data are not the same as the enthalpy of formation and the free energy of formation ($\Delta H^1_{f,\,500}$(fo)=−2172.6 kJ mol^{-1}, and $\Delta G^1_{f,\,500}$ (fo)=−1972.7 kJ mol^{-1}; Robie and Hemingway, 1995). However, for a reaction, both approaches yield identical values for the change in enthalpy ($\Delta H^1_{r,\,T}$) and free energy ($\Delta G^1_{r,\,T}$) (see section 4.9).

The volume intregral

To evaluate the volume integral in equation (4.68) we need to know $\bar{V}$ at T, the temperature of interest. For solids of constant composition, the combined effect of temperature and pressure on the volume can be calculated using equation (4.72; Box 4.5), but the effect is very small (except under mantle conditions) and very little error is introduced if the molar volumes are assumed to be constant. The evaluation of the volume integral, then, becomes quite simple:

$$\begin{aligned}\int_1^P \bar{V}_T\ (\text{solid})\ dP &= \bar{V}_T\ (\text{solid})\ (P-1) \\ &= \bar{V}_{T_{ref}}(\text{solid})\ (P-1)\end{aligned} \tag{4.70}$$

The effects of temperature and pressure on a fluid phase, on the other hand, can be significant, and its molar volumes cannot be assumed to be a constant for the purpose of integration. In the special case where the reaction involves a single fluid phase composed of a pure ideal gas,

$$\int_1^P \bar{V}_T\ (\text{fluid})\ dP = \int_1^P \frac{RT}{P}\, dP = RT \ln P \tag{4.71}$$

Reactions involving phases of variable composition will be discussed in Chapter 5.

Box 4.5 Effects of temperature and pressure on the volume of solids

The variation in the volume of a solid phase with temperature and pressure is most commonly measured, respectively, as the *coefficient of thermal expansion at constant pressure* $P(\alpha_p)$ and the *coefficient of compressibility at constant temperature* T (β_T), which are defined as:

$$\alpha_P = \frac{1}{\bar{V}}\left(\frac{\partial \bar{V}}{\partial T}\right)_P \quad \text{and} \quad \beta_T = -\frac{1}{\bar{V}}\left(\frac{\partial \bar{V}}{\partial P}\right)_T \tag{4.72}$$

The coefficients α_p and β_T reflect the response of the bond angles and bond lengths in the crystal structure to changes in temperature and pressure.

Assuming that α_p and β_T are constants in the temperature and pressure range under consideration, a reasonable assumption for moderate temperatures (< 1400°C) and pressures (< 50 kbar), and taking the reference state as 1 bar and 298.15 K, the definitions of α_p and β_T can be combined to obtain

$$d\bar{V} = \alpha_P\ \bar{V}\ dT - \beta_T\ \bar{V}\ dP$$

$$\int_{V_{298.15,\,1}}^{V_{T,\,P}} \frac{d\bar{V}}{\bar{V}} = \int_{298.15}^{T} \alpha_P\ dT - \int_1^P \beta_T\ dP \tag{4.73}$$

> **Box 4.5 (cont'd)**
>
> This can be further reduced to obtain the molar volume at the pressure (P) and temperature (T) of interest (McSween *et al.*, 2003):
>
> $$\bar{V}^P_T = \bar{V}^{P_{ref}}_{T_{ref}}\left[1+\alpha_P\,(T-T_{ref})-\beta_T\,(P-P_{ref})\right]$$
> $$= \bar{V}^1_{298.15}\left[1+\alpha_P\,(T-298.15)-\beta_T\,(P-1)\right] \quad (4.74)$$
>
> For most solids the difference between $\bar{V}^1_{298.15}$ and $\bar{V}^P_T$ is very small for two reasons. First, the values of α_p and β_T are very small. For example, for the mineral enstatite ($Mg_2Si_2O_6$), α_p=33 × $10^{-6}K^{-1}$ (at 800°C) and β_T=1.01 × 10^{-6} bar^{-1} (Birch, 1966; Skinner, 1966). The temperature dependence of α_p is very large near 0K, where it goes to zero, but becomes rather small at high temperatures. The effect of pressure on volume becomes significant only at pressures greater than 20 to 30kbar (and thus is relevant for the interpretation of seismic data on the constitution of the mantle and core). Second, the effect of increased temperature (an increase in volume in most cases) is just the opposite of that of pressure (a decrease in volume in all cases).

4.5 Free energy change of a reaction at T and P ($\Delta G^P_{r,T}$)

Consider an extensive property Z (e.g., H, S, G, or V) of the constituents in a balanced chemical reaction such as

$$\underset{\text{reactants}}{a\,A+b\,B+\ldots} \Rightarrow \underset{\text{products}}{c\,C+d\,D+\ldots} \quad (4.75)$$

The change in Z for the reaction (denoted by the suffix r), ΔZ_r, is calculated as

$$\Delta Z_r=[c\,\bar{Z}\,(C)+d\,\bar{Z}\,(D)+\ldots]-[a\,\bar{Z}\,(A)+b\,\bar{Z}\,(B)+\ldots] \quad (4.76)$$

where $\bar{Z}$ (A), $\bar{Z}$ (B), $\bar{Z}$ (C), $\bar{Z}$ (D), ... are molar quantities. The general expression for ΔZ_r is:

$$\Delta Z_r = \sum_j Z_j\ (\text{products}) - \sum_k Z_k\ (\text{reactants}) \quad (4.77)$$

$$= \sum_j \nu_j \bar{Z}_j\ (\text{products}) - \sum_k \nu_k \bar{Z}_k\ (\text{reactants}) \quad (4.78)$$

where ν_j and ν_k are stoichiometric coefficients of the jth product and kth reactant, respectively, in the corresponding balanced reaction. Note that ΔZ_r is not necessarily a molar quantity. The generalized form of equation (4.78) is also written as

$$\Delta Z_r = \sum_i \nu_i\,\bar{Z}_i \quad (4.79)$$

with the convention that the stoichiometric coefficient ν_i for a product is positive and that for a reactant is negative. For example, the change in enthalpy (i.e., the heat transfer between the system and its surroundings) at constant pressure for the reaction

$$\underset{\text{serpentine}}{5\,Mg_3Si_2O_5(OH)_4}=\underset{\text{forsterite}}{6\,Mg_2SiO_4}+\underset{\text{talc}}{1\,Mg_3Si_4O_{10}(OH)_2}+9\,H_2O \quad (4.80)$$

will be calculated from the molar enthalpies of the products and the reactants as

$$\Delta H_r=6\,\bar{H}\,(Mg_2SiO_4)+1\,\bar{H}\,(Mg_3Si_4O_{10}(OH)_2$$
$$+\,9\,\bar{H}\,(H_2O)+(-5)\,\bar{H}\,(Mg_3Si_2O_5(OH)_4 \quad (4.81)$$

Similar equations can be set up for computing $(\Delta C_P)_r$, ΔS_r, ΔV_r, and ΔG_r.

The strategy for calculating the *free energy change of a reaction* (also referred to as *free energy of a reaction*) at any specified T and P, $\Delta G^P_{r,T}$, is the same as for $\bar{G}^P_T$ discussed earlier (equation 4.64; Fig. 4.6). Choosing P_{ref}=1 bar,

$$\Delta G^P_{r,T} = \Delta G^1_{r,T}+\int_1^P \Delta V_{r,T}\ dP = \Delta H^1_{r,T} - T\,\Delta S^1_{r,T} + \int_1^P \Delta V_{r,T}\ dP \quad (4.82)$$

4.5.1 Computation of $\Delta G^1_{r,T}$

Depending on the thermodynamic data available in the published literature, $\Delta G^1_{r,T}$ can be computed in different ways, all converging to the same answer within the limits of rounding errors (see example in section 4.5.3).

(1) The simplest procedure is to compute $\Delta G^1_{r,T}$ from $\Delta G^1_{f,T}$ values of the substances at the temperature of interest:

$$\Delta G^1_{r,T}=\Sigma\Delta G^1_{f,T}\,(\text{products}) - \Sigma\Delta G^1_{f,T}\,(\text{reactants}) \quad (4.83)$$

(2) As $\Delta G=\Delta H - T\,\Delta S$, an equivalent procedure is to compute $\Delta G^1_{r,T}$ from $\bar{H}^1_T$ (= $\Delta H^1_{f,T}$) and $\bar{S}^1_T$ (the absolute entropy) of the substances at the temperature of interest:

$$\Delta G^1_{r,T}=\Delta H^1_{r,T} - T\,\Delta S^1_{r,T}$$
$$= [\Sigma \bar{H}^1_T\,(\text{products}) - \Sigma \bar{H}^1_T\,(\text{reactants})]$$
$$-T[\Sigma \bar{S}^1_T\,(\text{products}) - \Sigma \bar{S}^1_T\,(\text{reactants})] \quad (4.84)$$

(3) A more elaborate procedure for computing $\Delta G^1_{r,T}$ is to use equation (4.68) for calculating $\bar{G}^1_T$, which yields the following general equation for $\Delta G^1_{r,T}$:

$$\Delta G^1_{r,T} = \Delta H^1_{r,298.15} + \int_{298.15}^{T} (\Delta C^1_P)_r\ dT - T\left[\Delta S^1_{r,298.15} + \int_{298.15}^{T} \frac{(\Delta C^1_P)_r}{T}\ dT\right] \quad (4.85)$$

Equation (4.85) is considerably simplified in two special cases, which work quite well for many reactions involving solid phases only:

(1) $(\Delta C_P^1)_r$=constant (i.e., independent of temperature)

$$\int_{298.15}^{T} (\Delta C_P^1)_r \, dT = (\Delta C_P^1)_r \, (T - 298.15) \quad (4.86)$$

$$\int_{298.15}^{T} \frac{(\Delta C_P^1)_r}{T} \, dT = (\Delta C_P^1)_r \, (\ln T - \ln 298.15) \quad (4.87)$$

$$\Delta G^1_{r,T} = \Delta H^1_{r,298.15} + (\Delta C_P^1)_r \, (T - 298.15) - T[\Delta S^1_{r,298.15} + (\Delta C_P^1)_r \, (\ln T - \ln 298.15)] \quad (4.88)$$

(2) $(\Delta C_P^1)_r=0$ (i.e., ΔH_r and ΔS_r are independent of temperature)

$$\int_{298.15}^{T} (\Delta C_P^1)_r \, dT = 0 \quad (4.89)$$

$$\int_{298.15}^{T} \frac{(\Delta C_P^1)_r}{T} \, dT = 0 \quad (4.90)$$

$$\Delta G^1_{r,T} = \Delta H^1_{r,298.15} - T\,\Delta S^1_{r,298.15} \quad (4.91)$$

4.5.2 *Evaluation of the volume integral*

$$\int_1^P \Delta V_{r,T} \, dP = \int_1^P [\Delta V_{r,T} \text{ (solids)} + \Delta V_{r,T} \text{ (liquid)} + \Delta V_{r,T} \text{ (gas)}] \, dP \quad (4.92)$$

For reasons discussed earlier, ΔV_r(solids) can be taken as constant, so that for T_{ref}=298.15 K and P_{ref}=1 bar,

$$\int_1^P \Delta V_{r,T} \text{ (solids) } dP = (P-1)\, \Delta V^1_{r,298.15} \text{ (solids)} \quad (4.93)$$

which may be further approximated to $P\,\Delta V^1_{r,298.15}$(solids) if $P >> 1$ bar. If the gas phase is pure and behaves as an ideal gas, then

$$\int_1^P \Delta V_{r,T} \text{ (ideal gas) } dP = \int_1^P \nu \bar{V}_{r,T} \text{ (ideal) } dP = \int_1^P \frac{\nu RT}{P} \, dP = \nu \, RT \ln P \quad (4.94)$$

where ν is the stoichiometric coefficient of the gas phase (positive if it is a product and negative if it is a reactant) in the balanced reaction. We will deal with evaluation of the volume integral involving solid and fluid phases of variable composition in Chapter 5.

4.5.3 *General equation for $\Delta G^P_{r,T}$*

Substituting for $\Delta G^1_{r,T}$ (equation 4.84) and $\int_1^P \Delta V_{r,T} \, dP$ (equation 4.92) in equation (4.82), we get the general equation for the free energy change of a reaction at T and P, $\Delta G^P_{r,T}$, as

$$\Delta G^P_{r,T} = \Delta H^1_{r,298.15} + \int_{298.15}^{T} (\Delta C_P^1)_r \, dT - T\left[\Delta S^1_{r,298.15} + \int_{298.15}^{T} \frac{(\Delta C_P^1)_r}{T} \, dT\right] + \int_1^P [\Delta V_{r,T} \text{ (solids)} + \Delta V_{r,T} \text{ (liquid)} + \Delta V_{r,T} \text{ (gas)}] \, dP \quad (4.95)$$

4.6 Conditions for thermodynamic equilibrium and spontaneity in a closed system

A chemical reaction, reversible or irreversible, is accompanied by either release of heat ($-Q$) to the surroundings (*exothermic* reaction) or absorption of heat ($+Q$) from the surroundings (*endothermic* reaction). This exchange of heat associated with a chemical reaction at constant pressure, commonly known as the *heat of the reaction* (which, for a particular type of reaction, may be described as latent heat of melting, latent heat of crystallization, latent heat of vaporization, etc.), is the same as the enthalpy of the reaction (ΔH_r). Thus ΔH_r values are negative for exothermic reactions and positive for endothermic reactions. In general, the enthalpy of a reaction is not a measure of its tendency to occur.

The free energy change of a chemical reaction at the temperature and pressure of interest, $\Delta G^P_{r,T}$, which takes into account changes in both enthalpy and entropy, allows us to answer two important questions about chemical reactions: (i) are the reactants and products in equilibrium? and (ii) if not in equilibrium, in which direction would the spontaneous reaction proceed to achieve equilibrium? The rules may be summarized as follows (Fig. 4.8):

(1) $\Delta G^P_{r,T}=0$; the reactants and products are in equilibrium at the P and T of interest.
(2) $\Delta G^P_{r,T} < 0$; the reaction should be expected to proceed spontaneously in the direction of the product assemblage, which has a lower free energy and thus is more stable than the reactant assemblage at the P and T of interest.
(3) $\Delta G^P_{r,T} > 0$; the reaction should be expected to proceed spontaneously in the direction of the reactant assemblage, which has a lower free energy and thus is more stable than the product assemblage at the P and T of interest.

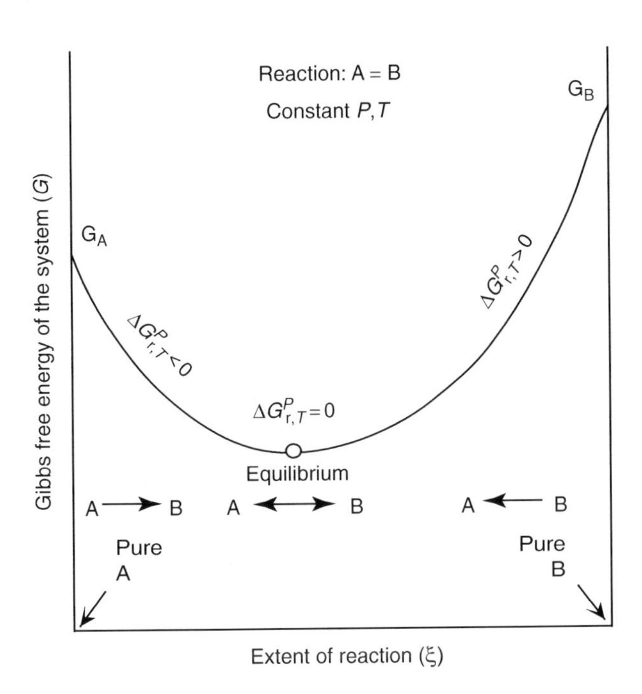

Fig. 4.8 Schematic illustration of the change in Gibbs free energy of a system (G) consisting of two pure phases, A and B, as a function of the extent of reaction (ξ) between A and B at constant temperature (T) and pressure (P). Starting with either pure A or pure B, G of the system decreases with increasing ξ and reaches a minimum at equilibrium. At equilibrium, $\Delta G^P_{r,T}=0$.

The magnitude of $\Delta G^P_{r,T}$ for a reaction may be looked upon as a measure of the "driving force" for the reaction, but $\Delta G^P_{r,T}$ does not give any information about the rate of the reaction. Another limitation of the free energy approach to geologic systems is that its direct application is restricted to closed systems in which the phases are of constant composition. Nevertheless, computed free energy changes of reactions provide a powerful tool for evaluating which reactions are possible in a given closed geologic system and which are not.

As discussed earlier (section 4.1), we can also use the equilibrium constant of a reaction to predict in what direction the reaction must proceed to attain equilibrium. This is because, as we will show later (Chapter 5), the equilibrium constant of a reaction and its free energy change are related.

Example 4–3: Free energy change of a reaction involving solid phases of fixed composition

What is the free energy change of the reaction

$$\underset{\text{forsterite}}{Mg_2SiO_4}+\underset{\alpha\text{-quartz}}{SiO_2}=\underset{\text{clinoenstatite}}{2\ MgSiO_3} \tag{4.96}$$

at 1 bar and 500 K? How will the reaction be affected if the pressure is raised to 1 kbar?

Relevant thermodynamic data at 1 bar for the phases involved are given in Table 4.3. Analytical expressions for $\bar{C}^1_p$ (J mol^{-1} K^{-1}) are of the form:

$$\bar{C}_p=A_1+A_2T+A_3T^{-2}+A_4T^{-0.5}+A_5T^2$$

and the constants are listed in Table 4.4. For simplicity, let us assume ΔV_r to be independent of temperature and pressure.

Let us compute $\Delta G^1_{r,500}$ by the three methods listed in section 4.5.1 and see whether we actually get congruent results, as we should.

(1) $\Delta G^1_{r,500}=2\ \Delta G^1_{f,500}\ (\text{clinoenstatite}) - \{(\Delta G^1_{f,500}\ (\text{forsterite}) + \Delta G^1_{f,500}\ (\alpha\text{-quartz})\}$
$= 2\ (-1399.2) - (-1972.7) - (-819.4) = -6.3\ \text{kJ}$

(2) $\Delta G^1_{r,500}=\Delta H^1_{r,500} - 500\ \Delta S^1_{r,500}$

$\Delta H^1_{r,500}=2\ \Delta\bar{H}^1_{f,500}(\text{clinoenstatite}) - \{\Delta\bar{H}^1_{f,500}(\text{forsterite}) + \Delta\bar{H}^1_{f,500}(\alpha\text{-quartz})\}$
$=2\ (-1545.1) - (-2172.6) - (-910.5) = -7.1\ \text{kJ}$

$\Delta S^1_{r,500} = 2\ \bar{S}^1_{500}(\text{clinoenstatite}) - \{\bar{S}^1_{500}\ (\text{forsterite}) + \bar{S}^1_{500}(\alpha\text{-quartz})\}$
$=2\ (115.27) - 163.92 - 68.45 = -1.83\ \text{J K}^{-1}$
$=-0.00183\ \text{kJ K}^{-1}$

$\Delta G^1_{r,500}=-7.1 - 500(-0.00183) = -6.145\ \text{kJ}$

(3) $$\Delta G^1_{r,500} = \Delta H^1_{r,298.15} + \int_{298.15}^{500} (\Delta C^1_p)_r\ dT - T\left[\Delta S^1_{r,298.15} + \int_{298.15}^{500} \frac{(\Delta C^1_p)_r}{T}\ dT\right]$$

$\Delta H^1_{r,298.15}=2\ \Delta\bar{H}^1_{f,298.15}(\text{clinoenstatite}) - \{\Delta\bar{H}^1_{f,298.15}(\text{forsterite}) + \Delta\bar{H}^1_{f,298}(\alpha\text{-quartz})\}$
$= 2\ (-1545.0) - (-2173.0) - (-910.7) = -6.3\ \text{kJ}$

$\Delta S^1_{r,298.15} = 2\ \bar{S}^1_{298.15}(\text{clinoenstatite}) - \{\bar{S}^1_{298.15}(\text{forsterite}) + \bar{S}^1_{298.15}(\alpha\text{-quartz})\}$
$= 2\ (67.86) - 94.11 - 41.46 = 0.15\ \text{J K}^{-1}$
$=0.00015\ \text{kJ K}^{-1}$

$(\Delta A_1)_r=2(2.056\times 10^2) - (8.736\times 10^1) - (8.1145\times 10^1) = 242.695$

$(\Delta A_2)_r=2(-1.280\times 10^{-2}) - (8.717\times 10^{-2}) - (1.828\times 10^{-2}) = -0.13105$

$(\Delta A_3)_r=2(1.193\times 10^6) - (-3.699\times 10^6) - (-1.810\times 10^5) = 6266000$

$(\Delta A_4)_r=2(-2.298\times 10^3) - (8.436\times 10^2) - (-6.985\times 10^2) = -4741.1$

$(\Delta A_5)_r=-(-2.237\times 10^{-5}) - (5.406\times 10^{-6}) = 0.000016964$

Table 4.3 Thermodynamic data for the forsterite–clinoenstatite reaction (Robie and Hemingway, 1995).

	T (K)	$\bar{C}_P^1$ (J mol^{-1} K^{-1})	$\Delta\bar{H}^1_{f,T}$ (kJ mol^{-1})	$\bar{S}^1_T$ (J mol^{-1} K^{-1})	$\Delta\bar{G}^1_{f,T}$ (kJ mol^{-1})	$\bar{V}^1_{298.15}$ (J bar^{-1}mol^{-1})
Forsterite	298.15	118.61	−2173.0	94.11	−2053.6	4.365
	500	148.28	−2172.6	163.92	−1972.7	
α-Quartz	298.15	44.59	−910.7	41.46	−856.3	2.269
	500	59.68	−910.5	68.45	−819.4	
Clinoenstatite	298.15	82.12	−1545.0	67.86	−1458.1	3.128
	500	101.20	−1545.1	115.27	−1399.2	

Table 4.4 Constants for the heat capacity equation (Robie and Hemingway, 1995).

	A_1	A_2	A_3	A_4	A_5
Forsterite (298–1800 K)	8.736×10^1	8.717×10^{-2}	-3.699×10^6	8.436×10^2	-2.237×10^{-5}
α-Quartz (298–844 K)	8.1145×10^1	1.828×10^{-2}	-1.810×10^5	-6.985×10^2	5.406×10^{-6}
Clinoenstatite (298–1600 K)	2.056×10^2	-1.280×10^{-2}	1.193×10^6	-2.298×10^3	0

$$\int_{298.15}^{500} (\Delta C_P^1)_r \, dT = \int_{298.15}^{500} [(\Delta A_1)_r + (\Delta A_2)_r T + (\Delta A_3)_r T^{-2} + (\Delta A_4)_r T^{-0.5} + (\Delta A_5)_r T^2] \, dT$$
$$= -826.386 \text{ J.K}^{-1} = -0.8264 \text{ kJ.K}^{-1}$$

$$\int_{298.15}^{500} \frac{(\Delta C_P^1)_r}{T} dT = \int_{298.15}^{500} \left[\frac{1}{T}\left((\Delta A_1)_r + (\Delta A_2)_r T + (\Delta A_3)_r T^{-2} + (\Delta A_4)_r T^{-0.5} + (\Delta A_5)_r T^2\right)\right] dT$$
$$= -1.9914 \text{ J.K}^{-1} = -0.00199 \text{ kJ.K}^{-2}$$

$$\Delta G^1_{r,500} = -6.3 - 0.8264 - 500\,(0.00015 - 0.00199) = -6.206 \text{ kJ}$$

Thus, the three methods yield identical values (within rounding errors) of free energy change for the reaction.

As $\Delta G^1_{r,500} < 0$, the reaction at (500 K, 1 bar) should spontaneously proceed toward the formation of clinoenstatite at the expense of forsterite + α-quartz.

To calculate the free energy change of the reaction at (500 K, 1 kbar), we assume ΔV_r to be independent of temperature and pressure and calculate it as $\Delta V^1_{r,298.15}$:

$$\Delta V^1_{r,298.15} = 2\,\bar{V}^1_{298.15}(\text{clinoenstatite}) - \{\bar{V}^1_{298.15}(\text{forsterite}) + \bar{V}^1_{298.15}(\alpha\text{-quartz})\}$$
$$= 2\,(3.128) - 4.365 - 2.269 = -0.378 \text{ J bar}^{-1}$$
$$= -0.0004 \text{ kJ bar}^{-1}$$

$$\int_1^{1000} \Delta V^1_{r,298.15}\, dP = \Delta V^1_{r,298.15}\,(1000 - 1)$$
$$= (-0.0004)\,999 = -0.3996 \text{ kJ}$$

We can now calculate $\Delta G^{1000}_{r,500}$ using equation (4.70):

$$\Delta G^{1000}_{r,500} = \Delta G^1_{r,500} + \int_1^{1000} \Delta V^1_{r,298.15}\, dP$$
$$= -\ 6.206 - 0.3996 - 6.606 \text{ kJ}$$

We conclude that clinoenstatite is stable relative to forsterite + α-quartz even at 1 kbar pressure if the temperature of the reaction is held at 500 K.

The computation of $\Delta G^1_{r,500}$ becomes much simpler if we assume $(\Delta C_P)_r = 0$ (i.e., ΔH_r and ΔS_r are independent of temperature):

$$\Delta G^1_{r,500} = \Delta H^1_{r,298.15} - 500\,\Delta S^1_{r,298.15}$$
$$= \Delta G^P_{r,T} - 6.3 - 500\,(-0.00015) = -6.225 \text{ kJ}$$

Note that this simplification introduces a very small error for the free energy of the reaction relative to the uncertainty in ΔH values. This is the case for almost all solid–solid reactions. The same assumption is also adequate for reactions involving gases and liquids, but generally over a relatively small temperature range, 200 or 300 K or so (Wood and Fraser, 1976).

4.7 Metastability

Metastability refers to the persistence of a mineral or a mineral assemblage in a pressure–temperature environment where thermodynamically it is not stable. A familiar example of metastability is diamond. Diamond is stable only under the high pressure–temperature conditions of the mantle, yet it occurs at or near the Earth's surface in kimberlites and placers instead of being oxidized to CO_2 by atmospheric oxygen:

$$\text{C (diamond)} + O_2(g) \Rightarrow CO_2(g),\ \Delta G^1_{r,298} = -394\,\text{kJ mol}^{-1} \quad (4.97)$$

Another example is the metastable occurrence of the mineral olivine in rocks exposed at the Earth's surface. It is well known that olivine typically occurs in mafic and ultramafic igneous rocks and is a mineral that crystallizes from magmas at relatively high temperatures. It is, therefore, very susceptible to chemical weathering when such rocks are exposed on the Earth's surface. In the presence of water, olivine alters to serpentine, which may be represented by the reaction

$$\underset{\text{forsterite}}{2Mg_2SiO_4} + \underset{\text{water}}{3H_2O} = \underset{\text{serpentine}}{Mg_3Si_2O_5(OH)_4} + \underset{\text{brucite}}{Mg(OH)_2} \quad (4.98)$$

Using the $\Delta G^1_{f,298}$ (in kJ mol^{-1}) values of the phases from Robie and Hemmingway (1995),

$$\begin{aligned}\Delta G^1_{r,298.15} &= \Delta G^1_{f,298.15}(\text{serp}) + \Delta G^1_{f,298.15}(\text{brucite}) \\ &\quad - 2\Delta G^1_{f,298.15}(\text{fo}) - 3\Delta G^1_{f,298.15}(\text{water}) \\ &= -4032.4 - 883.5 - 2\,(-2053.6) - 3(-237.1) \\ &= -47.4\,\text{kJ}\end{aligned}$$

Thus, from a free energy consideration the stable assemblage at and near the Earth's surface (298.15 K, 1 bar) is serpentine+brucite, but in many cases olivine continues to coexist with serpentine as a partially altered metastable phase because of the slow rate of the alteration reaction. The occurrence of metastable phases in rocks due to kinetic constraints is a common phenomenon. The mere persistence of a mineral assemblage through time, therefore, does not guarantee that it is a stable or an equilibrium assemblage. Some minerals and mineral assemblages continue to exist in a metastable state, because the reactions that would convert them into more stable states are too slow at the ambient P–T condition. We will discuss the kinetics of chemical reactions in much more detail in Chapter 9.

4.8 Computation of simple *P–T* phase diagrams

Phase diagrams depicting stability fields of minerals and mineral assemblages in P–T space is widely used for predicting which mineral or mineral assemblage is stable at any particular T and P. The preferred way to construct such phase diagrams is to conduct experiments under controlled P–T conditions. Subject to the availability of relevant thermodynamic data, simple phase diagrams can also be constructed quite easily by application of the concepts and computational techniques developed in the previous sections. The task involves computation of reaction boundaries that delineate the stability fields of minerals and mineral assemblages of interest in P–T space. This approach is particularly useful for determining phase relations at low temperatures where achievement of equilibrium in experimental runs is often a problem.

4.8.1 Procedure

Let us assume that the curve in Fig. 4.9a represents the reaction boundary in P–T space for a reaction of the type X+Y=C+D. For any point on this curve, say E, the reactants and products are in equilibrium at a particular combination of T and P – in this case, T_E and P_E. The reaction boundary is a univariant curve, because for a chosen value of temperature,

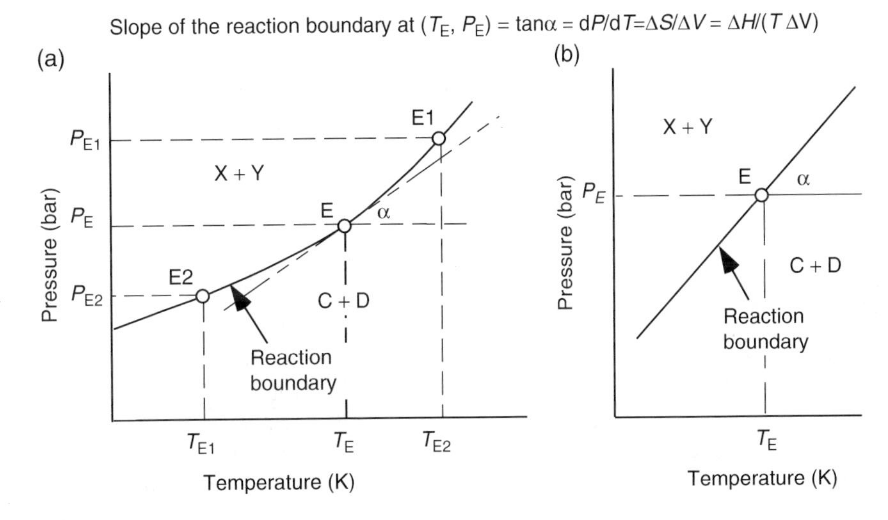

Fig. 4.9 Schematic illustration of the procedure for constructing a univariant reaction boundary in P–T space. A curved boundary requires the calculation of equilibrium pressures corresponding to several arbitrarily chosen temperatures. A straight-line boundary can be drawn by calculating one point (T_E, P_E) on the boundary and its slope using equation (4.105) or (4.106).

T_E, there is only one pressure, P_E, at which the reaction can be in equilibrium, and vice versa.

Also, for any point such as (T_E, P_E) on the reaction boundary, the free energy change of the reaction, $\Delta G^{P_E}_{r, T_E}$, is equal to zero. Now let us see how we can utilize this constraint to calculate the pressure P_E corresponding to some chosen temperature T_E. In terms of (T_E, P_E), equation (4.82) should be written as

$$\Delta G^{P_E}_{r, T_E} = \Delta G^1_{r, T_E} + \int_1^{P_E} \Delta V_{r, T_E} \, dP = 0 \tag{4.99}$$

For the general case of a reaction involving both solid and fluid phases,

$$\Delta G^1_{r, T_E} + \int_1^{P_E} \Delta V_{r, T_E} \text{ (solids) } dP + \int_1^{P_E} \Delta V_{r, T_E} \text{ (fluid) } dP = 0 \tag{4.100}$$

We know how to calculate $\Delta G^1_{r, T_E}$, so we can solve equation (4.100) for P_E if we can evaluate the volume integrals. If a reaction involves only solid phases of constant composition and ΔV_r (solids) can be considered a constant, equation 4.100 simplifies to

$$\Delta G^1_{r, T_E} + (P_E - 1) \, \Delta V_{r, 298.15} \text{ (solids)} = 0 \tag{4.101}$$

which can be solved for P_E:

$$\boxed{P_E = 1 - \frac{\Delta G^1_{r, T_E}}{\Delta V_{r, 298.15} \text{ (solids)}}} \tag{4.102}$$

Equation 4.102 can also be easily solved for P_E if the fluid phase involved in the reaction is a pure phase and it behaves as an ideal gas. In this case, we apply equation (4.94):

$$\int_1^{P_E} \Delta V_{r, T_E} \text{ (fluid) } dP = \int_1^{P_E} \nu \, \bar{V} \text{ (fluid) } dP = \int_1^{P_E} \frac{\nu R T_E}{P} \, dP$$
$$= \nu R T_E \ln \left(\frac{P_E}{1} \right) \tag{4.103}$$

Solids and fluids of variable composition will be discussed in Chapter 5.

To plot a curved reaction boundary, we need to fix a few more points on the boundary. This is accomplished by using the procedure outlined above to calculate pressures (P_{E1}, P_{E2}, ...) corresponding to arbitrarily chosen temperatures (T_{E1}, T_{E2}, ...), or *vice versa* (Fig. 4.9a).

4.8.2 *The Clapeyron equation*

Modifying equation (4.57), $\Delta G = V \, dP - S \, dT$, for a reaction, we can write:

$$d(\Delta G)_r = \Delta V_r \, dP - \Delta S_r \, dT \tag{4.104}$$

Since $d(\Delta G_r) = 0$ at equilibrium, for any point such as (T_E, P_E) on the reaction boundary in Fig. 4.10, the slope of the boundary is given by what is known as the *Clapeyron equation*:

$$\boxed{\frac{dP}{dT} = \frac{\Delta S_r}{\Delta V_r}} \tag{4.105}$$

An alternative form of the Clapeyron equation, which can easily be derived from first principles, is sometimes referred to as the *Clausius–Clapeyron equation*:

$$\boxed{\frac{dP}{dT} = \frac{\Delta H_r}{T \, \Delta V_r}} \tag{4.106}$$

When the reaction boundary is a curved line, the slope at any point on the boundary is the slope of the tangent to the curve at that point (Fig. 4.9a). If ΔH_r, ΔS_r, and ΔV_r can be assumed to be constants (i.e., independent of temperature and pressure), the slope will be a constant, that is, the reaction boundary will be a straight line (Fig. 4.9b). For many reactions involving only solids of constant composition, this is a reasonable assumption over a limited *P–T* range. In such cases, all that is needed to construct the reaction boundary is one point

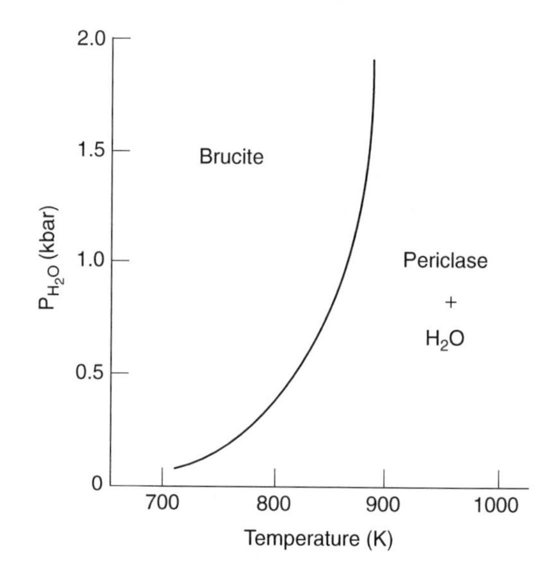

Fig. 4.10 Univariant reaction boundary in P–T space for the dehydration reaction $Mg(OH)_2$ (brucite) = MgO (periclase) + H_2O. Note the gentle slope of the boundary at low pressures and the steep slope at high pressures, which is typical of reactions involving a fluid phase. (From *Geochemical Thermodynamics* by D.K.Nordstrom and J.L.Munoz, 2nd edition, Figure 8.4, p. 230; Copyright 1994, Blackwell Scientific Publications. Used with permission of the publisher.)

on the boundary and its slope at that point, or two points on the boundary.

A reaction involving a fluid phase (e.g., H_2O or CO_2) is strongly affected by pressure and the slope of the reaction boundary varies markedly with pressure, producing a curved line (Fig. 4.10). At low pressures, ΔV_r is relatively large because of the large molar volume of the fluid, and the slope of the boundary, $\Delta S_r/\Delta V_r$, is gentle. At higher pressure ΔV_r becomes smaller as the fluid becomes denser, and the slope of the reaction boundary becomes steeper. In fact, at very high pressures, ΔV_r may even become negative (whereas ΔS_r continues to be positive), and the negative slope will be manifested by the reaction boundary bending back on itself. We will learn more about reactions involving fluid phases in the next chapter.

Example 4–4: Calculation of P–T reaction boundary involving solid phases of fixed composition

Let us construct the *P–T* phase diagram for the two polymorphs of $CaCO_3$, calcite and aragonite (thermodynamic data for these minerals are given in Table 4.5

$$CaCO_3\ (\text{calcite}) \Leftrightarrow CaCO_3\ (\text{aragonite}) \quad (4.107)$$

The topology of the calcite–aragonite phase diagram can be predicted even without any calculation. From the molar entropy and molar volume data, it is evident that the reaction boundary will have a positive slope (dP/dT) because both ΔS and ΔV for the reaction are negative numbers. Also, aragonite has a smaller molar volume relative to calcite. As higher pressure favors a smaller molar volume (Le Chatelier's principle), the stability field of aragonite must be on the higher-pressure side of the reaction boundary (Fig. 4.11).

Now, we proceed to fix the position of the reaction boundary in *P–T* space. As only solid phases of constant composition are involved in this reaction, it is reasonable to assume that ΔV_r is independent of temperature and pressure. So we use equation (4.102) to calculate values of *P* corresponding to chosen values of *T*. To start with, let us choose $T=298.15$ K.

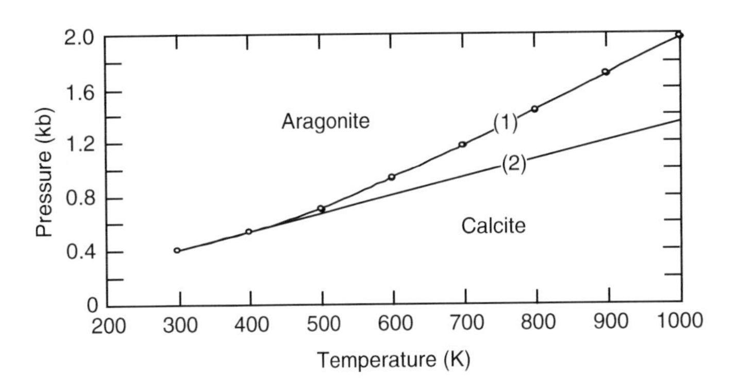

Fig. 4.11 *P–T* phase diagram for calcite–aragonite. Reaction boundaries (1) and (2) were constructed by calculating the equilibrium pressures for various chosen values of temperature using equation (4.102), and for (2) with the assumption that $(\Delta C_P^1)_r=0$ (i.e., ΔH_r^1 and ΔS_r^1 are independent of temperature).

$$\Delta G^1_{r,298.15}=\Delta\bar{G}^1_{f,298.15}(\text{aragonite})-\Delta\bar{G}^1_{f,298.15}(\text{calcite})$$
$$=-1127.793-(-1128.842)=1.049\ \text{kJ}=1049\ \text{J}$$

$$\Delta V^1_{r,298.15}=\bar{V}^1_{298.15}(\text{aragonite})-\bar{V}^1_{298.15}(\text{calcite})$$
$$=3.4150-3.6934=-0.2784\ \text{J bar}^{-1}$$

$$P=1-\frac{\Delta G^1_{r,298.15}}{\Delta V^1_{r,298.15}}=1-\frac{1049}{-0.2784}$$
$$=1+3767.96\approx 3769\ \text{bar}$$

Thus, one point on the calcite–aragonite reaction boundary is (298.15 K, 3769 bar). Repeating the calculation for other chosen temperatures, we obtain several points on the reaction boundary:

T (K)	*P* (bar)	*T* (K)	*P* (bar)	*T* (K)	*P* (bar)	*T* (K)	*P* (bar)
298.15	3769	500	7120	700	11736	900	16980
400	5260	600	9307	800	14322	1000	19635

Table 4.5 Thermodynamic data for calcite and aragonite (Robie *et al.*, 1978).

	Calcite			Aragonite		
T (K)	$\Delta\bar{G}^1_{f,T}$ (kJ mol^{-1})	$\bar{S}^1_T$ (J mol^{-1} K^{-1})	$\Delta\bar{H}^1_{f,T}$ (kJ mol^{-1})	$\Delta\bar{G}^1_{f,T}$ (kJ mol^{-1})	$\bar{S}^1_T$ (J mol^{-1} K^{-1})	$\Delta\bar{H}^1_{f,T}$ (kJ mol^{-1})
298.15	−1128.842	91.71	−1207.370	−1127.793	87.99	−1207.430
400	−1102.155	118.36	−1206.301	−1100.691	113.75	−1206.681
500	−1076.292	140.88	−1204.822	−1074.310	135.23	−1205.665
600	−1050.723	160.43	−1203.276	−1048.132	153.96	−1204.567
700	−1025.427	177.70	−1201.791	−1022.160	170.68	−1203.438
800	−1000.238	193.19	−1201.145	−996.251	185.86	−1203.022
900	−975.195	207.27	−1199.789	−970.468	199.84	−1201.749
1000	−950.299	220.22	−1198.689	−944.833	212.87	−1200.573

$\bar{V}^1_{298.15}$ (calcite)=3.6934 J bar^{-1} mol^{-1}; $\bar{V}^1_{298.15}$ (aragonite)=3.4150 J bar^{-1} mol^{-1} (1 J=10 cm^3 bar; 3.6934 J bar^{-1} mol^{-1}=36.934 cm^3 mol^{-1}).

The resulting phase diagram is presented in Fig. 4.11. For the sake of comparison, Fig. 4.11 also includes the reaction boundary, a line with constant slope, based on the assumption that $(\Delta C_p^1)_r=0$ (i.e., ΔH_r^1 and ΔS_r^1 are independent of temperature). The significant departure between the two graphs, progressively increasing with increasing temperature, is not a surprise. We could have guessed it by looking at the enthalpy and entropy data and realizing that neither ΔH_r^1 nor ΔS_r^1 is, strictly speaking, independent of temperature and that their values decrease with increasing temperature.

4.9 Thermodynamic data tables

A prerequisite for performing thermodynamic calculations is the availability of appropriate thermodynamic data. Many textbooks dealing with geochemical thermodynamics (e.g. Faure, 1986; Nordstrom and Munoz, 1994; Krauskopf and Bird, 1995) include an appendix listing standard state thermodynamic data at 298.15 K temperature and 1 bar or 1 atm pressure for pure substances of interest in geochemistry. Some of the important sources of data for these compilations include Helgeson *et al.* (1978), Wagman *et al.* (1982), Weast *et al.* (1986), Garvin *et al.* (1987), Cox *et al.* (1989), and Robie and Hemingway (1995). Nordstrom and Munoz (1994) provide a good discussion (Chapter 11) of how thermodynamic properties are measured and estimated, and of the potential problems in dealing with multiple data sources.

Published thermodynamic data sets for minerals are of two types: (i) compilations based largely on calorimetrically derived data (Robie *et al.*, 1978; Robie and Hemingway, 1995); and (ii) tabulations developed using statistical or mathematical programming methods applied to all the experimental data available for numerous related compositional systems (Helgeson *et al.*, 1978; Berman, 1988; Gottschalk, 1997; Holland and Powell, 1998). The latter approach generates self-consistent data sets, but may introduce significant systematic errors in the data sets. To the extent possible, all the standard thermodynamic data used in this book are from Robie and Hemingway (1995), because the most consistent results are obtained if all the data are taken from the same compilation. The entropy and heat capacity values given in Robie and Hemingway (1995) are based almost exclusively on calorimetric measurements, whereas the values for enthalpy or Gibbs free energy of formation were obtained from both equilibrium and calorimetric measurements.

The thermodynamic data tables in Robie and Hemingway (1995) are for 1 bar reference pressure (denoted by the superscript "o"). As mentioned earlier, the equation for temperature dependence of heat capacity is of the form

$$\bar{C}_P = A_1 + A_2T + A_3T^{-2} + A_4T^{-0.5} + A_5T^2 \quad (4.53)$$

and the constants for various substances are listed in a separate table. In the data table for a given substance, S_T^o denotes the total entropy (not the entropy of formation), $\Delta_f H^o$ and $\Delta_f G^o$ denote the standard enthalpy and standard free energy of formation from elements in their standard states, and log K_f represents the equilibrium constant for the reaction that defines the formation of the substance from its elements.

Their data tables include two other functions at 100 K intervals that allow calculation of high temperature ΔH and ΔG are the "*heat content*" *function*, $(H_T^o - H_{298}^o)/T$, and the *Gibbs free energy function*, $(G_T^o - H_{298}^o)/T$. Both of these functions refer to the compound itself and not formation from the elements, and they vary gradually and smoothly between transition temperatures, permitting more accurate interpolation. It should be noted that H_T^o and G_T^o values extracted from the said functions are not the same as $\Delta_f H^o$ and $\Delta_f G^o$. For example, for forsterite,

$$(H_{500}^o - H_{298}^o)/T = 55.01\ \text{J mol}^{-1}\ \text{K}^{-1}$$
$$H_{500}^o = H_{298}^o + (55.01 \times 500/1000) = -2173.0 + 27.505\ \text{kJ mol}^{-1} = -2145.5\ \text{kJ mol}^{-1}$$
$$(G_T^o - H_{298}^o)/T = -108.91\ \text{J mol}^{-1}\ \text{K}^{-1}$$
$$G_{500}^o = H_{298}^o + (-108.91 \times 500/1000) = -2173.0 - 54.455\ \text{kJ mol}^{-1} = 2227.46\ \text{kJ mol}^{-1}$$

Thus, the H_{500}^o(fo) and G_{500}^o(fo) values are identical to what we calculated in Example 4–2, whereas $\Delta_f H_{500}^o$ (fo) = −2173.0 kJ mol^{-1}, and $\Delta_f G_{500}^o$ (fo) = −2053.6 kJ mol^{-1}. The two sets are related by the following equations:

$$\left(\frac{G_T^o - H_{298}^o}{T}\right) = \left(\frac{H_T^o - H_{298}^o}{T}\right) - S_T^o \quad (4.108)$$

$$\Delta H_{f,T}^o = \Delta H_{f,298}^o + \Delta(H_T^o - H_{298}^o) \quad (4.109)$$

$$\Delta G_{f,T}^o = \Delta H_{f,298}^o + T\Delta\left(\frac{G_T^o - H_{298}^o}{T}\right) \quad (4.110)$$

Example 4–5: Calculation of the enthalpy and free energy change of a reaction using the "heat content" and free energy functions (Table 4.6)

Let us revisit the reaction we considered in Example 4–3:

$$\underset{\text{forsterite}}{Mg_2SiO_4} + \underset{\text{quartz}}{SiO_2} = 2\ \underset{\text{clinoenstatite}}{MgSiO_3} \quad (4.96)$$

$$\Delta H_{r,500}^1 = 2H_{500}^1\ (\text{clino}) - H_{500}^1\ (\text{fo}) - H_{500}^1\ (\text{qz})$$
$$= 2(-1526.325) - (-2145.495) - (-900.03)$$
$$= -3052.65 + 2145.495 + 900.03 = -7.125\ \text{kJ}$$
$$\Delta G_{r,500}^1 = 2G_{500}^1\ (\text{clino}) - G_{500}^1\ (\text{fo}) - G_{500}^1\ (\text{qz})$$
$$= 2(-1583.96) - (-2227.455) - (-934.26)$$
$$= -3167.92 + 2227.455 + 934.6 = -6.205\ \text{kJ}$$

Table 4.6 Relevant standard state ($T=500$ K, $P=1$ bar) thermodynamic data.

	$H^{1}_{f,298}$ kJ mol^{-1}	$(H^{1}_{T}-H^{1}_{298})/T$ J mol^{-1} K^{-1}	H^{1}_{T} kJ mol^{-1}	$(G^{1}_{T}-H^{1}_{298})/T$ J mol^{-1} K^{-1}	G^{1}_{T} kJ mol^{-1}
	(a)	(b)	(c)	(d)	(e)
Forsterite	−2173.0	55.01	−2145.495	−108.91	−2227.455
Quartz	−910.7	21.34	−900.03	−47.11	−934.255
Clinoenstatite	−1545.0	37.35	−1526.325	−77.92	−1583.96

Source of thermodynamic data: Robie and Hemingway (1995).
(c)=(a)+(b/1000) T; (e)=(a)+(d/1000) T.

The $\Delta H^{1}_{r,500}$ and $\Delta G^{1}_{r,500}$ values calculated above are identical to those in Example 4–3.

A listing of the standard state (298.15 K, 1 bar) data for selected substances, extracted mainly from Robie and Hemingway (1995), is given in Appendix 5.

4.10 Summary

1. The equilibrium constant for a reaction such as

   $$\mathrm{x\,X+y\,Y+\ldots \Leftrightarrow c\,C+d\,D+\ldots}$$

   at T and P is given by the ratio of activities as

   $$K_{T,P}=\frac{a_{C}^{c}\,a_{D}^{d}}{a_{X}^{x}\,a_{Y}^{y}}$$

   Every chemical reaction in equilibrium at specified P and T has a unique value of equilibrium constant.
2. The matter contained in a thermodynamic system can be described completely in terms of phases (p) and components (c), which are related to the degrees of freedom of the system (f) at equilibrium by the Gibbs phase rule:

   $$f=c-p+2$$

3. The thermodynamic state of a system is described by certain extensive (mass-dependent) and intensive (mass-independent) properties. Properties of a system that depend only on the final state of the system, irrespective of the path taken to reach that state, are called state functions (or state variables). If ϕ is a state function, the $d\phi$ is an exact differential.
4. The first law of thermodynamics is a statement of the principle of conservation of energy. Mathematically, it is expressed, in the differential form:

   $$dU=dQ+dW=dQ-P\,dV$$

   where dU is the change in internal energy of the system, dQ is the heat absorbed by the system, and dW is the work done on the system. The work in most geochemical systems is "PV work," so for a change in volume (dV) against constant external pressure (P), $dW=-P\,dV$.
5. The second law of thermodynamics introduces a state function, entropy (S), which for reversible reactions is defined as

   $$dS=\frac{dQ}{T},$$

   and for irreversible reactions as

   $$dS>\frac{dQ}{T}$$

6. The third law of thermodynamics assigns a value of zero to perfectly crystalline substances at absolute zero temperature (0 K=−273°C) and thereby provides a reference point for absolute entropies at other temperatures.
7. Two auxiliary functions, enthalpy ($H=U+PV$)) and Gibbs free energy ($G=H-TS$), both of which are extensive properties and state functions, are commonly used in thermodynamic calculations pertaining to geochemical systems.
8. The molar enthalpy and molar entropy of a substance at T (the temperature of interest) and some reference pressure (e.g., 1 bar) can be calculated using molar heat capacity data at the reference pressure:

   $$\bar{H}^{1}_{T}=\bar{H}^{1}_{T_{ref}}+\int_{T_{ref}}^{T}\bar{C}^{1}_{P}\,dT \quad \text{and} \quad \bar{S}^{1}_{T}=\bar{S}^{1}_{T_{ref}}+\int_{T_{ref}}^{T}\frac{\bar{C}^{1}_{P}}{T}\,dT$$

9. The Gibbs free energy change of a reaction at P and T is given by

   $$\Delta G^{P}_{r,T}=\Delta G^{1}_{r,T}+\int_{1}^{P}\Delta V_{r}\,dP=\Delta H^{1}_{r,T}-T\Delta S^{1}_{r,T}+\int_{1}^{P}\Delta V_{r,T}\,dP$$

   $$=\Delta H^{1}_{r,298.15}+\int_{298.15}^{T}(\Delta C^{1}_{P})_{r}\,dT$$

   $$-T\left[\Delta S^{1}_{r,298.15}+\int_{298.15}^{T}\frac{(\Delta C^{1}_{P})_{r}}{T}\,dT\right]$$

   $$+\int_{1}^{P}[\Delta V_{r,T}(\text{solids})+\Delta V_{r,T}(\text{liquid})+\Delta V_{r,T}(\text{gas})]dP$$

10. The free energy change of a reaction at P and T, $\Delta G^P_{r,T}$, indicates whether the reaction is at equilibrium and, if not, in which direction should it proceed. For a reaction in equilibrium, $\Delta G^P_{r,T}=0$. If $\Delta G^P_{r,T}<0$, the reaction should proceed spontaneously in the direction of the product assemblage; if $\Delta G^P_{r,T}>0$, the reaction should proceed spontaneously in the direction of the reactant assemblage. $\Delta G^P_{r,T}$ does not give any information about the rate of a reaction.
11. At any point on a reaction boundary in P–T space, $\Delta G^P_{r,T}=0$, and the slope of the boundary is given by

$$\frac{dP}{dT}=\frac{\Delta S_r}{\Delta V_r}=\frac{\Delta H_r}{T\,\Delta V_r}$$

4.11 Recapitulation

Terms and concepts

Adiabatic process
Closed system
Component
Coefficient of thermal expansion
Dalton's Law of Partial Pressure
Endothermic reaction
Enthalpy
Entropy
Equation of state
Equilibrium
Equilibrium constant
Exothermic reaction
Extensive property
Equilibrium
Equilibrium constant
Gibbs free energy
Gibbs phase rule
Heat capacity
Hemholtz free energy
Ideal Gas Law
Intensive property
Internal energy
Irreversible process
Isolated system
Isothermal process
Laws of thermodynamics
Mole fraction
Le Chatelier's principle
Metastability
Open system
Partial pressure
pH
Phase
Reaction boundary
Real gas
Reversible process
Standard enthalpy of formation
Standard free energy of formation
State function
Thermodynamic system

Computation techniques

- Partial pressure of a gas.
- Enthalpy, entropy, and free energy of a substance at the temperature and pressure of interest.
- Enthalpy, entropy, and free energy change of a reaction at the temperature and pressure of interest.
- Reaction boundaries in P–T space.

4.12 Questions

[Unless stated otherwise, the thermodynamic data are from Robie and Hemingway (1995).]

1. The reaction between nitrogen and oxygen to form $NO_{(g)}$ is represented by the reaction

$$N_{2(g)}+O_{2(g)} \Leftrightarrow 2NO_{(g)}$$

Equilibrium concentrations of the gases at 1500 K are 1.7×10^{-3} mol l^{-1} for $O_{2(g)}$, 6.4×10^{-3} mol l^{-1} for $N_{2(g)}$, and 1.1×10^{-5} mol l^{-1} for $NO_{(g)}$. Calculate the equilibrium constant for the reaction at 1500 K. Assume the gases to be ideal.

2. From the reactions and equilibrium constants (at 298 K, 1 bar) given below,

$$H_2O \Leftrightarrow H^+ + OH^- \qquad K_w=10^{-14}$$

$$HCO_3^- \Leftrightarrow H^+ + CO_3^{2-} \qquad K_{HCO_3^-}=10^{-10.3}$$

compute the equilibrium constant of the reaction $CO_3^{2-}+H_2O \Leftarrow HCO_3^-+OH^-$ at 298.15 K and 1 bar.

3. The equilibrium constant for the reaction

$$\underset{\text{magnetite}}{2Fe_3O_{4\,(s)}}+0.5\,O_2 \Leftrightarrow \underset{\text{hematite}}{3Fe_2O_{3\,(s)}}$$

is 5×10^{43} at 298.15 K and 1 atm. Calculate the pressure of oxygen in equilibrium with a mixture of pure hematite and pure magnetite at 298.15 K and 1 atm. How many molecules per liter does this represent? Assume the oxygen gas to be ideal. At 298.15 K and 1 atm, the volume of 1 mole of an ideal gas is 24.47 L and it contains 6.022×10^{23} molecules per liter. (Note that it does not require a very oxidizing condition to oxidize magnetite into hematite.)

4. Calculate the amount of heat required to raise the temperature of 1 kg of pure water at 25°C to its boiling point (100°C). Assume that no heat goes to heat the container. The specific heat capacity of water is 4.18 J g^{-1} °C^{-1}.

5. When 100 g of a metal at 75°C is added to 175 g of water at 15°C, the temperature of the water rises to 18°C as the water reaches thermal equilibrium with the metal. C_p (water)=4.18 J g^{-1} deg^{-1}. Assume that no heat is lost to the surroundings. What is the specific heat capacity of the metal?
6. The standard enthalpy of formation of $CH_{4(g)}$ cannot be measured directly as it cannot be synthesized directly from C (graphite) and $H_{2(g)}$. Calculate the value of ΔH_f^1 ($CH_{4(g)}$) from the following reactions:

   (1) C (graphite)+$O_{2(g)}$=$CO_{2(g)}$,
   $\Delta H^1_{r,\,298.15} = -393.5$ kJ mol^{-1}

   (2) $2H_{2(g)}+O_2\,(g)=2H_2O_{(l)}$,
   $\Delta H^1_{r,\,298.15} = 2(-285.8)$ kJ mol^{-1}

   (3) $CO_{2(g)}+2H_2O_{(l)}=CH_{4(g)}+2O_{2(g)}$,
   $\Delta H^1_{r,\,298.15} = +890.3$ kJ mol^{-1}

   (This is an application of the so-called *Hess's Law of Summation*, which states that the enthalpy change for a reaction is the same whether it occurs in one step or in a series of steps.)
7. Given that the solubility of H_2S gas in water is 2.3 L per liter of solution at 298.15 K and 1 atm, calculate the equilibrium constant for the reaction

   H_2S (gas) ⇔ H_2S (aq)

   Assume that both H_2S gas and the solution behave ideally. The volume of 1 mole of ideal gas at 298.15 K and 1 atm is 24.47 L.
8. Compute the work done when one mole of an ideal gas is reversibly taken from state 1 (P_1, T_1) to state 2 (P_2, T_2) along each of the following paths:

   (a) $P_1 \Rightarrow P_2$ at constant temperature T_1; then
   $T_1 \Rightarrow T_2$ at constant pressure P_2

   (b) $T_1 \Rightarrow T_2$ at constant pressure P_1; then
   $P_1 \Rightarrow P_2$ at const. Temperature T_2

   Is the work done the same for both the paths?

   [Hint: W (work done by the ideal gas)= $\int_{\text{State 1}}^{\text{State 2}} -P\,dV$;
   $d\bar{V} = \left(\frac{\partial \bar{V}}{\partial P}\right)_T dP + \left(\frac{\partial \bar{V}}{\partial T}\right)_P dT$.]
9. Show that for any pure substance:

   $(dS)_T = -\alpha_P\, V\, dP$
   $(dH)_T = V(1-\alpha_P T)\, dP$
   $C_P - C_V = T\,\bar{V}(\alpha_P^2/\beta_T)$
10. Calculate the volume change for the following reaction at 298.15 K and 1 bar, using the given molar volume data (in cm^3/mole):

    $Ca_3Al_2Si_3O_{12}+SiO_2=CaAl_2Si_2O_8+2CaSiO_3$

    grossular quartz anorthite wollastonite

    $\bar{V}$(grossular)=125.30; $\bar{V}$(quartz)=22.69; $\bar{V}$(anorthite) =100.79; $\bar{V}$(wollastonite)=39.93
    Express the answer in J bar^{-1}.
11. Calculate molar H^1_{800}, S^1_{800}, and G^1_{800} for albite, given that (Robie and Hemingway, 1995):

    (a) $\Delta\bar{H}^1_{f,\,298.15}=-3935.0$ kJ mol^{-1}, $\bar{S}^1_{298.15}=207.40$ J mol^{-1} K^{-1}, and the expression for $\bar{C}^1_P$ (J mol^{-1} K^{-1}) as a function of temperature is

    $\bar{C}_P{}^1=5.839\times10^2-9.285\times10^{-2}T+1.678\times10^6T^{-2}-6.424\times10^3T^{-0.5}+2.272\times10^{-5}T^2$

    (b) the "heat content function" and the "Gibbs free energy function" for albite at 500 K are as follows:

    $$\frac{H^1_T-H^1_{298}}{T}=166.38\ \text{J mol}^{-1}\ \text{K}^{-1}$$
    $$\frac{G^1_T-H^1_{298}}{T}=-295.30\ \text{kJ mol}^{-1}$$

    Do the two calculations yield the same numbers?
12. Modify the calculation of the free energy change of the reaction (4.96) at 500 K and 1 kbar in the worked out Example 4–2 to include the volume change for the reaction, assuming that α_p and β_T are constants in the temperature and pressure range under consideration. Use equation (4.74) and the following data:

	Forsterite	α-Quartz	Clinoenstatite
α_p(deg^{-1})	44×10^{-6}	69×10^{-6}	33×10^{-6}
β_T(bar^{-1})	0.79×10^{-6}	26.71×10^{-6}	1.01×10^{-6}

    Comment on the difference between your results and that of the worked out example.
13. Consider the reaction

    $CaTiO_3$ (perovskite)+SiO_2 (quartz)=$CaTiSiO_5$ (titanite)

    at 298.15 K and 1 bar. Using the thermodynamic data given below (Robie and Hemingway, 1995), answer the following:

    (a) Is the reaction exothermic or endothermic?
    (b) Which way should the reaction proceed to attain equilibrium?
    (c) Will the product or the reactants be favored by (i) increase in temperature or (ii) increase in pressure?

    [Hint: $\left(\frac{\partial(\Delta G)_r}{\partial T}\right)_P = -\Delta S_r$; $\left(\frac{\partial(\Delta G)_r}{\partial P}\right)_T = \Delta V_r$.]

Mineral	$\Delta H^1_{f,\,298}$ (kJ mol^{-1})	S^1_{298} (J mol^{-1} K^{-1})	V^1_{298} (J bar^{-1} mol^{-1})
Perovskite	−1660.6	93.64	3.363
Quartz	−910.7	41.46	2.269
Titanite	−2596.6	129.20	5.574

14. Derive the Clausius–Clapeyron equation: $\frac{dP}{dT}=\frac{\Delta H_r}{T\,\Delta V_r}$.
15. Construct the phase diagram for the three Al_2SiO_5 polymorphs (andalusite, sillimanite, and kyanite) in P–T

	Andalusite		Kyanite		Sillimanite	
T (K)	$\Delta H^1_{f,T}$ (kJ mol^{-1})	$\bar{S}^1_T$ (J mol^{-1} K^{-1})	$\Delta H^1_{f,T}$ (kJ mol^{-1})	$\bar{S}^1_T$ (J mol^{-1} K^{-1})	$\Delta H^1_{f,T}$ (kJ mol^{-1})	$\bar{S}^1_T$ (J mol^{-1} K^{-1})
298.15	−2589.9	91.39	−2593.8	82.80	−2586.1	95.40
400	−2590.7	131.57	−2594.7	122.81	−2586.9	135.63
500	−2590.1	166.71	−2594.1	157.97	−2586.4	170.58
600	−2588.8	197.79	−2592.7	189.10	−2585.2	201.44
700	−2587.2	225.45	−2591.0	216.84	−2583.7	228.91
800	−2585.4	250.30	−2589.2	241.77	−2582.0	253.59
900	−2583.6	272.83	−2587.3	264.37	−2580.4	275.97

$V^1_{298.15}$(J bar^{-1}mol^{-1}): andalusite=5.152; kyanite=4.415; sillimanite=4.986.

space, using the thermodynamic data (Robie and Hemingway, 1995) given above. Assume the ΔV of the reactions to be independent of temperature and pressure.

(a) Label the stability fields of the three minerals. Explain how you deduced this.
(b) What are the pressure and temperature of the triple point?
(c) Compare the calculated phase diagram with the experimentally determined phase relations (Fig. 6.4) and comment on the discrepancies between the two, if any.

16. For the equilibrium graphite ⇔ diamond, construct a phase diagram in P–T space showing the stability fields of diamond and graphite for the following conditions: (i) $(\Delta C_p)_r = 0$, (ii) $(\Delta C_p)_r$ = constant, and (iii) $(\Delta C_p)_r$ is a function of temperature. Thermodynamic data required for the problem are given below:

	$\Delta H^1_{f,298.15}$ (kJ mol^{-1})	$S^1_{298.15}$ (J mol^{-1} K^{-1})	$V^1_{298.15}$ (J bar^{-1})	α_P (K^{-1})	β_T (bar^{-1})
Diamond	1.9	2.38	0.3417	4×10^{-6}	2×10^{-7}
Graphite	0.0	5.74	0.5298	25×10^{-6}	30×10^{-7}

Analytical expressions for C_p^1 (J mol^{-1} K^{-1}):

$$C_p^1\,(\text{diamond}) = 9.845 - 3.655 \times 10^{-2}T + 1.217 \times 10^{6}T^{-2} - 1.659 \times 10^{3}T^{-0.5} + 1.098 \times 10^{-5}T^{2}$$

(temperature range 298.15–1800 K)

$$C_p^1\,(\text{graphite}) = 6.086 - 1.024 \times 10^{-2}T + 7.139 \times 10^{5}T^{-2} - 9.922 \times 10^{2}T^{-0.5} + 1.669 \times 10^{-6}T^{2}$$

(temperature range 298.15–2500 K)

(a) Which is the stable polymorph at the surface of the Earth?
(b) What is the minimum depth at which diamond is the stable polymorph in a region where the temperature increases uniformly at the rate of 15°C km^{-1} depth? Assume that the pressure increases with depth at the rate of 0.03 GPa km^{-1}.

5 Thermodynamics of Solutions

At the outset it may be remarked that the chief difficulty in the study of solutions is that thermodynamics provides no detailed information concerning the dependence of the chemical potential (or other thermodynamic functions) on the composition.

Denbigh (1971)

The relations we developed in the previous chapter apply only to closed systems (i.e., systems of constant bulk composition) consisting of phases of constant composition. In the natural world, however, most geologic materials are solutions of variable composition, a *solution* being defined as a homogeneous phase formed by dissolving one or more substances (solid, liquid, or gas) in another substance (solid, liquid, or gas). A solution may be a solid (e.g., a solid solution, such as olivine or plagioclase), a liquid (e.g., an aqueous solution resulting from dissolution of NaCl crystals or CO_2 gas in water, or a silicate melt generated by partial melting of rocks), or a gas (e.g., a mixture of O_2 and CO_2). A solution may be ideal or nonideal, and in many cases the value of a thermodynamic function at the same pressure and temperature may be significantly different, depending on whether the solution is treated as ideal or nonideal.

Among the solutions, perhaps the best understood from a thermodynamic perspective are the gaseous solutions (mixtures). Complications in aqueous solutions arise from electrostatic and nonelectrostatic behavior involving solute–solute and solute–solvent interactions. Silicate melts, which occur only at very high temperatures and are characterized by short-range ordering of discrete structural units for some compositions, are not well understood, and will not be discussed here. The thermodynamics of crystalline materials, such as minerals, present some unique problems because of ordering that is characteristic of the crystalline state. In this chapter we will learn how to evaluate quantitatively the thermodynamic properties of solutions, excluding aqueous electrolyte solutions, which will be discussed in Chapter 7.

We will see that gas mixtures are treated somewhat differently than liquid and solid solutions. For nonideal gas mixtures, the strategy commonly employed is to modify the ideal gas equation to fit the available P–V–T data. In the absence of equations of state comparable to the ideal gas equation, nonideality in liquid and solid solutions commonly are described as deviation from ideality through *excess functions*, which are then used to derive P–V–T equations of state (Anderson and Crerar, 1993).

To describe the composition of a solution, we need to know what are the substances present in the solution and in what quantities. In some cases the substances in a solution may be specified in terms of "components" in the sense of the Gibbs phase rule, as defined in section 4.2.3, but many reactions cannot be represented in terms of "components" in the true sense. We will use the less restrictive term *constituents* as defined by Anderson and Crerar (1993) – a constituent of a phase or a system is "any combination of elements in the system in any stoichiometry." Thus, constituents may be "components", actual chemical species present in a solution (e.g., HCO_3^- in a H_2O–CO_2 solution, which are referred to as *phase components*), end-members of a solid solution series (e.g., $NaAlSi_3O_8$ and $CaAl_2Si_2O_8$ for plagioclase), or even hypothetical species. The quantity of any constituent i in a solution is expressed either as number of moles or *mole number* (n_i), or as *mole fraction* (X_i).

Introduction to Geochemistry: Principles and Applications, First Edition. Kula C. Misra.

5.1 Chemical potential

5.1.1 *Partial molar properties*

In general, thermodynamic properties of a solution cannot be calculated by treating it as a mechanical mixture of its constituents. For example, if we add 1 mole of NaCl to 1 kg of water, the dissolved NaCl will dissociate into Na^+ and Cl^-, and the volume of the solution will not be equal to the sum of the two volumes; actually it will be slightly less because of electrostatic interactions between the ions and H_2O molecules. Similarly, the entropy and the free energy of a plagioclase solid solution will not be simply the weighted sum of the entropies and free energies of its end-members, $NaAlSi_3O_8$ and $CaAl_2Si_2O_8$, because of the entropy and free energy associated with the mixing of the two end-members.

In order to address the effect of composition of a system on its extensive thermodynamic properties, we introduce the concept of partial molar properties (also referred to as partial molal properties by some authors). Consider an extensive property Y (such as volume, enthalpy, entropy, or free energy) of a homogeneous phase α that consists of k different constituents. Let the amounts (mole numbers) of the constituents in the whole phase be denoted as $n_1^\alpha, n_2^\alpha, \ldots, n_k^\alpha$. Mathematically, the corresponding *partial molar property* of the ith constituent in the phase α, $\bar{y}_i^\alpha$, is defined as a partial derivative of Y^α with respect to n_i^α at constant temperature, pressure, and n_j^α:

$$\bar{y}_i^\alpha = \left(\frac{\partial Y^\alpha}{\partial n_i^\alpha}\right)_{P,T,n_j^\alpha} \tag{5.1}$$

where n_j^α refers to all constituents in the phase α other than i (i.e., $j \neq i$). In other words, $\bar{y}_i^\alpha$ represents the change in Y^α resulting from an infinitesimal addition of the ith constituent (so as not to change its overall composition significantly) when the pressure, temperature, and mole numbers of all other constituents in the phase are held constant. The partial molar quantity $\bar{y}_i^\alpha$ is an intensive property.

As Y is a function of temperature, pressure, and composition, the total differential of Y, dY, is

$$\mathrm{d}Y^\alpha = \left(\frac{\partial Y^\alpha}{\partial T}\right)_{P,n_i^\alpha} \mathrm{d}T + \left(\frac{\partial Y^\alpha}{\partial P}\right)_{T,n_i^\alpha} \mathrm{d}P + \sum_{i=1}^{i=k} \left(\frac{\partial Y^\alpha}{\partial n_i^\alpha}\right)_{P,T,n_j^\alpha} \mathrm{d}n_i^\alpha \tag{5.2}$$

which can be integrated at constant temperature and pressure (ie., dT=0 and dP=0) to obtain (see, e.g., Fletcher, 1993)

$$Y^\alpha = \sum_i n_i^\alpha \left(\frac{\partial Y^\alpha}{\partial n_i^\alpha}\right)_{P,T,n_j^\alpha} = \sum_i n_i^\alpha\, \bar{y}_i^\alpha \tag{5.3}$$

Thus, at constant temperature and pressure, an extensive property of a multiconstituent phase is the sum of the corresponding partial molar property of all its constituents multiplied by their respective mole numbers.

Dividing both sides of equation (5.3) by the total number of moles, $\sum_i n_i^\alpha$, we obtain

$$\bar{Y}^\alpha = \sum_i X_i^\alpha\, \bar{y}_i^\alpha \tag{5.4}$$

where $\bar{Y}^\alpha$ is the corresponding molar quantity for the phase α, and X_i^α is the mole fraction of the ith constituent in the phase. For example, the molar volume of a phase equals the sum of the partial molar volumes of its constituents multiplied by their respective mole fractions:

$$\bar{V}^\alpha = \sum_i X_i^\alpha\, \bar{v}_i^\alpha$$

Note that for a pure phase (i.e., a phase composed of one constituent only so that i=1 and X_i^α=1), the partial molar quantity is the same as the corresponding molar quantity ($\bar{y}_i^\alpha$):

$$\bar{Y}^\alpha = \sum_i X_i^\alpha\, \bar{y}_i^\alpha = \bar{y}^\alpha_{i\,(X_i=1)} = \bar{Y}^\alpha_{i\,(X_i=1)} \tag{5.5}$$

If a system consists of several multiconstituent phases ($\alpha, \beta, \gamma, \ldots$), we can write an equation similar to equation (5.3) for each phase, and add them to obtain the value of Y for the whole system at constant temperature and pressure:

$$Y^{\text{system}} = \sum_\alpha \sum_i n_i^\alpha \left(\frac{\partial Y^\alpha}{\partial n_i^\alpha}\right)_{P,T,n_j^\alpha} = \sum_\alpha \sum_i n_i^\alpha\, \bar{y}_i^\alpha \tag{5.6}$$

Example 5–1: Estimation of partial molar quantities of dissolved constituents in a solution

Let us consider a simple binary solution of n_1 moles of NaCl (solute) in n_2 moles of H_2O (solvent) at constant temperature and pressure. What are the partial molar volumes of NaCl and H_2O in the solution?

The volume of the NaCl – H_2O solution, according to equation (5.3), is

$$V^{\text{solution}} = n_1 \bar{v}_{\text{NaCl}}^{\text{solution}} + n_2 \bar{v}_{\text{H}_2\text{O}}^{\text{solution}} \tag{5.7}$$

where $\bar{v}_{\text{NaCl}}^{\text{solution}}$ and $\bar{v}_{\text{H}_2\text{O}}^{\text{solution}}$ are the partial molar volumes. To express equation (5.7) in terms of mole fractions of the constituents, $X_{\text{NaCl}} = n_1 / (n_1+n_2)$ and $X_{\text{H}_2\text{O}} = n_2 / (n_1+n_2)$, we divide each term in the equation by (n_1+n_2):

$$\bar{V}^{\text{solution}} = X_{\text{NaCl}}\, \bar{v}_{\text{NaCl}}^{\text{solution}} + X_{\text{H}_2\text{O}}\, \bar{v}_{\text{H}_2\text{O}}^{\text{solution}} \tag{5.8}$$

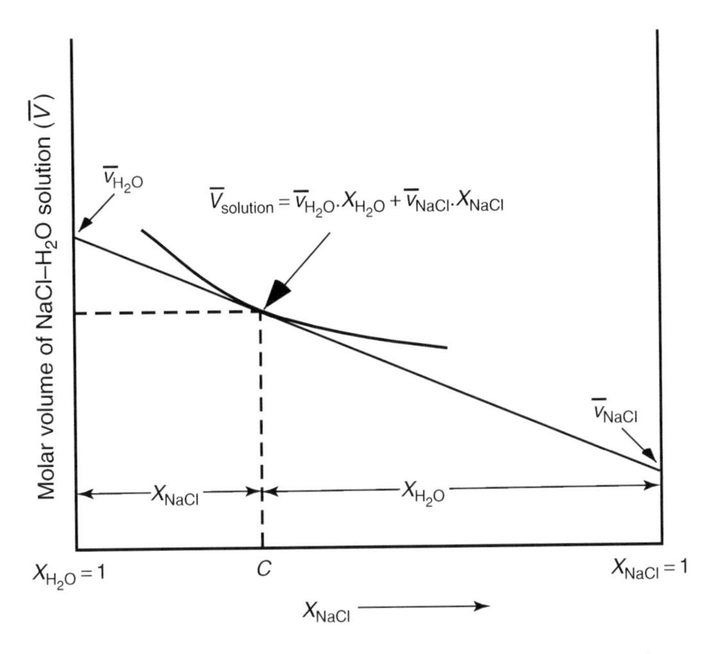

Fig. 5.1 Illustration of a graphical method used for determination of partial molar volumes. The curve is a schematic representation of the variation of the molar volume of NaCl–water solution as a function of the mole fraction of NaCl (X_{NaCl}) in the solution. For a solution of composition C, consisting of n_1 moles of NaCl and n_2 moles of water, the partial molar volumes are given by the y axis intercepts at $X_{NaCl}=1$ and $X_{water}=1$ of the tangent to the curve drawn at the specified composition $[X_{NaCl}=n_1/(n_1+n_2)]$. See Lewis and Randall (1961, p. 208) for the proof of this method.

where $\bar{V}^{solution}$ is the molar volume of the solution. We can determine graphically $\bar{v}_{NaCl}$ and $\bar{v}_{H_2O}$ corresponding to the given composition of the NaCl – H_2O solution if we know from experimental data how $\bar{V}^{solution}$ varies with composition (Fig. 5.1). Expressions similar to equations (5.7) and (5.8) can be written for other extensive properties, such as enthalpy and free energy, of the solution in terms of corresponding partial molar quantities.

5.1.2 *Definition of chemical potential* (μ)

In order to extend the application of the auxiliary thermodynamic functions (equations 4.39–4.41) to systems that may undergo a change in composition, Gibbs in 1876–1878 introduced the term *chemical potential* (μ). The chemical potential of the ith constituent in phase α, μ_i^α, can be defined mathematically as a partial derivative of U^α (at constant S, V, n_j), H^α (at constant S, P, n_j), A^α (at constant T, V, n_j), or G^α (at constant P, T, n_j), but the most useful definition for our purpose is with respect to G^α:

$$\mu_i^\alpha = \left(\frac{\partial G^\alpha}{\partial n_i^\alpha}\right)_{P,\,T,\,n_j^\alpha} \tag{5.9}$$

This is because μ_i^α in equation (5.9) is defined at constant temperature and pressure, and so represents *partial molar free energy*. Thus, *the chemical potential of a constituent* i *is its partial molar free energy*, and like free energy, is a state function. Note that for a pure phase, the partial molar free energy is equal to the molar free energy (i.e., $\bar{G}_i^\alpha=\mu_i^\alpha$) at constant temperature and pressure. It follows that for a reaction (such as equation 4.96) involving only pure phases, the change in chemical potential, $\Delta\mu_r$, is the same as the change in free energy, ΔG_r, at constant temperature and pressure:

$$\begin{aligned}\Delta\mu_r(P,T) &= 2\,\mu_{MgSiO_3}^{clinoenstatite} - \mu_{Mg_2SiO_4}^{forsterite} - \mu_{SiO_2}^{\alpha\text{-quartz}}\\ &= 2\,\bar{G}(\text{clinoenstatite}) - \bar{G}(\text{forsterite})\\ &\quad - \bar{G}(\alpha\text{-quartz}) = \Delta G_r(P,T)\end{aligned}$$

5.1.3 *Expression for free energy in terms of chemical potentials*

Now let us see how we can express the free energy of a multiconstituent phase α in terms of the chemical potentials of its constituents. Expressing G^α as a function of P, T, and composition, the total differential dG^α can be written as (see equation 5.2):

$$dG^\alpha = \left(\frac{\partial G^\alpha}{\partial T}\right)_{P,\,n_i^\alpha,\,n_j^\alpha} dT + \left(\frac{\partial G^\alpha}{\partial P}\right)_{T,\,n_i^\alpha,\,n_j^\alpha} dP + \sum_i \left(\frac{\partial G^\alpha}{\partial n_i^\alpha}\right)_{P,\,T,\,n_j^\alpha} dn_i^\alpha \tag{5.10}$$

Substituting for the temperature and pressure partial derivatives (equations 4.58 and 4.59),

$$dG^\alpha = -S^\alpha\, dT + V^\alpha\, dP + \sum_i \mu_i^\alpha\, dn_i^\alpha \tag{5.11}$$

At constant pressure and temperature ($dP=0$, $dT=0$), equation (5.11) reduces to

$$\left(dG^\alpha\right)_{P,T} = \sum_i \mu_i^\alpha\, dn_i^\alpha \tag{5.12}$$

which can be integrated to yield

$$\left(G^\alpha\right)_{P,T} = \sum_i \mu_i^\alpha\, n_i^\alpha \tag{5.13}$$

It follows that for a system composed of several multiconstituent phases (α, β, γ, …), at constant P and T,

$$\left(G^{system}\right)_{P,T} = \sum_\alpha\sum_i \mu_i^\alpha\, n_i^\alpha \tag{5.14}$$

Note that for 1 mole of a pure phase α, $\bar{G}^\alpha=\mu^\alpha$ and $G^\alpha=\mu^\alpha n^\alpha$, and for a system comprised of multiple pure phases,

$$G^{system} = \sum_\alpha G^\alpha = \sum_\alpha \bar{G}^\alpha n^\alpha = \sum_\alpha \mu^\alpha n^\alpha.$$

Equations (5.10)–(5.14) are valid for both closed systems, involving changes in composition due to reactive combination of existing constituents within the system, and for open systems, in which changes in composition may occur that are also due to mass transfer between the system and the surroundings. Note that these equations are consistent with those derived for closed systems of constant composition in Chapter 4. For example, equation (5.11) reduces to equation (4.57) and equation (5.12) to equation (4.60) when the term accounting for compositional variation vanishes in the case of constant composition (i.e., $dn_i^\alpha=0$).

5.1.4 *Criteria for equilibrium and spontaneous change among phases of variable composition*

Consider a hypothetical system composed of two phases, α and β, with a common chemical constituent i. If dn_i moles of i are transferred from α to β at constant temperature and pressure, then conservation of mass requires that

$$-dn_i^\alpha=dn_i^\beta$$

The change in the free energy of the system is given by

$$dG=\mu_i^\alpha\, dn_i^\alpha+\mu_i^\beta\, dn_i^\beta=\mu_i^\alpha\, dn_i^\alpha+\mu_i^\beta\,(-dn_i^\alpha)=(\mu_i^\alpha-\mu_i^\beta)\, dn_i^\alpha$$

For a reversible process at constant temperature and pressure, $dG=0$ (equation 4.60) and, since $dn_i^\alpha\neq 0$,

$$\mu_i^\alpha=\mu_i^\beta$$

Extending the above argument to pairs of phases in a system containing several phases (α, β, γ, ...), it can be shown that at equilibrium (constant temperature and pressure),

$$\boxed{\mu_i^\alpha=\mu_i^\beta=\mu_i^\gamma=\ldots} \tag{5.15}$$

Thus, in a system at equilibrium (constant temperature and pressure), the chemical potential of a constituent must be the same in all phases in the system among which this constituent can freely pass. If two phases are separated by a semi-permeable membrane that is permeable to some constituents but not to others, equation (5.15) would apply only to those constituents that can pass freely through the membrane.

A spontaneous reaction should occur (at constant temperature and pressure) if either $dG<0$ or $dG>0$. For a reaction $\alpha=\beta$, if $dG<0$

$$\boxed{\mu_i^\alpha>\mu_i^\beta} \tag{5.16}$$

and there should be spontaneous transfer of i from α to β; If $dG>0$

$$\boxed{\mu_i^\beta>\mu_i^\alpha} \tag{5.17}$$

and there should be spontaneous transfer of the constituent i from β to α. In either case, to minimize the free energy of the system, spontaneous transfer of i would occur from a phase in which its chemical potential is higher to another phase in which its chemical potential is lower, until the chemical potentials become equal and the two phases are in equilibrium. This is why μ_i is called a "potential," a sort of energy level of a constituent in one phase of a system. Chemical potential gradients due to differences in chemical potentials are the cause of mass transfer by diffusion. The direction of decreasing chemical potential usually, but not always, coincides with the direction of decreasing concentration.

Example 5–2: Equilibrium and mass transfer between calcite and aragonite

Let us take another look at the calcite–aragonite equilibrium (Fig. 4.11):

$$CaCO_3\ \text{(calcite)} \Leftrightarrow CaCO_3\ \text{(aragonite)} \tag{4.102}$$

that we discussed in the previous chapter, and understand how chemical potentials of the constituent $CaCO_3$ in these phases are related to equilibrium or mass transfer between them (Fig. 5.2).

Consider the calcite–aragonite system at pressure and temperature defined by point X in Fig. 5.2. At this condition of pressure and temperature, which lies in the stability field of aragonite, $\mu_{CaCO_3}^{aragonite}<\mu_{CaCO_3}^{calcite}$, and material should flow from the calcite phase to the aragonite phase until all the calcite disappears. The opposite should happen if the system is placed in the stability field of calcite, such as the point Y. In this case, $\mu_{CaCO_3}^{calcite}<\mu_{CaCO_3}^{aragonite}$, and material should flow from the aragonite phase to the calcite phase until all the aragonite disappears. At any point, such as Z, on the reaction boundary, calcite and aragonite are in equilibrium, $\mu_{CaCO_3}^{calcite}=\mu_{CaCO_3}^{aragonite}$, and there

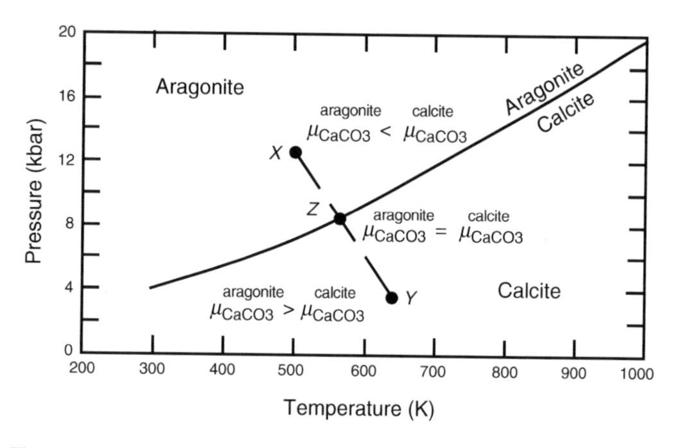

Fig. 5.2 Pressure–temperature phase diagram for the calcite–aragonite system illustrating conditions for equilibrium and spontaneous reaction in terms of the chemical potentials of $CaCO_3$ in the two polymorphs. At the pressure and temperature represented by point X, calcite should spontaneously change to aragonite; at point Y, aragonite should spontaneously change to calcite; and at point Z, calcite and aragonite should coexist in equilibrium.

should be no net flow of material between the two phases once equilibrium has been established.

5.1.5 *Criteria for equilibrium and spontaneous change for a reaction*

To find the condition of equilibrium at constant temperature and pressure for a reaction in terms of chemical potentials, we define a quantity called *the extent of reaction* (ξ) as

$$d\xi = \frac{dn_i}{v_i} \quad (5.18)$$

where v_i is the stoichiometric coefficient of the *i*th constituent in the reaction. Substituting for dn_i in equation (5.12), at constant temperature and pressure,

$$(dG)_{T,P} = \sum_i (v_i \ \mu_i) \ d\xi_i$$

$$\left(\frac{\partial G}{\partial \xi}\right)_{T,P} = \sum_i v_i \ \mu_i \quad (5.19)$$

As stated earlier, the convention is that v_i is positive if it refers to a product species and negative if it refers to a reactant species. At equilibrium, *G* must be a minimum with respect to any displacement of the system. So,

$$\left(\frac{\partial G}{\partial \xi}\right)_{T,P} = 0 \quad (5.20)$$

Thus, the most general condition of equilibrium for a reaction at constant temperature and pressure, irrespective of composition and physical state (solid, liquid or gas) of the substances involved, is

$$\boxed{\sum_i v_i \ \mu_i = 0} \quad (5.21)$$

For example, the breakdown of plagioclase (plag) to garnet (gt), sillimanite (sil), and quartz (qz) during metamorphism of rocks can be represented in terms of phase-components as

$$\underset{\text{anorthite}}{\overset{\text{plag}}{3CaAl_2Si_2O_8}} = \underset{\text{grossular}}{\overset{\text{gt}}{Ca_3Al_2Si_3O_{12}}} + \underset{\text{sillimanite}}{\overset{\text{sil}}{2Al_2SiO_5}} + \underset{\text{quartz}}{\overset{\text{qz}}{SiO_2}} \quad (5.22)$$

The condition of equilibrium for this reaction at the *P* and *T* of interest (taking v_i as positive for products and negative for reactants) is:

$$\Delta\mu^P_{r,T} = \mu^{gt}_{Ca_3Al_2Si_3O_{12}} + 2\mu^{sil}_{Al_2SiO_5} + \mu^{qz}_{SiO_2} - 3\mu^{plag}_{CaAl_2Si_2O_8} = 0 \quad (5.23)$$

The evaluation of chemical potentials will be discussed in section 5.2.

5.1.6 *The Gibbs–Duhem equation*

Differentiating equation (5.13), we get

$$dG^\alpha = \sum_i \mu_i^\alpha \ dn_i^\alpha + \sum_i n_i^\alpha \ d\mu_i^\alpha \quad (5.24)$$

Subtracting the above equation from equation (5.11), we obtain what is known as the *Gibbs–Duhem equation* for a single phase:

$$\boxed{-S^\alpha \ dT + V^\alpha \ dP - \sum_i n_i^\alpha \ d\mu_i^\alpha = 0} \quad (5.25)$$

The Gibbs–Duhem equation describes the interdependence among the intensive variables (pressure, temperature, and chemical potentials) in a single phase in a state of internal equilibrium. In a multiphase system, a Gibbs–Duhem equation can be written for each phase. The equations impose the restriction that in a closed system at equilibrium, net changes in chemical potential will occur only as a result of changes in temperature or pressure or both.

The Gibbs–Duhem equation is particularly useful for closed systems at constant pressure and constant temperature, in which case equation (5.25) reduces to

$$\boxed{\sum_i n_i^\alpha \ d\mu_i^\alpha = 0} \quad (5.26)$$

Equation (5.26) tells us that there can be no net change in chemical potential at equilibrium if pressure and temperature are held constant. In other words, if the phase α contains *k* constituents, the chemical potential of only (*k* – 1) constituents can vary independently. To illustrate the significance of

Box 5.1 Derivation of the Gibbs phase rule

Consider a heterogeneous system in equilibrium, consisting of *c* components and *p* phases as defined in section 4.2. In order to derive the phase rule, we need to count the number of independent variables for the system and the number of equations relating these variables; the difference between these two numbers is the variance of the system (*f*).

Each component adds a variable in the form of its chemical potential, but for *c* components only (*c* – 1) of these are independent because of the restriction imposed by equation (5.26) for each phase, giving *p* (*c* – 1) independent compositional variables for the system. Including *P* and *T*, the total number of independent variables for the system is *p* (– 1)+2.

At equilibrium (constant temperature and pressure), the chemical potential of each component must conform to equation (5.15), which gives (*p* – 1) equations for *p* phases. Thus, the total number of equations relating the *p* phases and *c* components is *c* (*p* – 1). We can now compute the variance of the system as

$$f = [p\ (c-1)+2] - [c\ (p-1)] = c - p + 2$$

This is the Gibbs phase rule (equation 4.23).

this statement, let us consider the NaCl – H_2O solution we mentioned earlier. For this phase in equilibrium at constant pressure and temperature,

$$n_{NaCl}^{solution}\ d\mu_{NaCl}^{solution} + n_{H_2O}^{solution}\ d\mu_{H_2O}^{solution} = 0$$

In this closed binary system we can change independently the chemical potential of either H_2O or NaCl, but for the equilibrium to be maintained the corresponding change in the chemical potential of the other constituent must satisfy the above relation. This is an important constraint for binary systems as it is useful in the calculation of phase equilibria and in the design of phase equilibria experiments.

5.2 Variation of chemical potential (μ_i^α) with temperature, pressure, and composition

5.2.1 *Temperature dependence of chemical potential*

From equation (5.10), at constant pressure,

$$dG^\alpha = \left(\frac{\partial G^\alpha}{\partial T}\right)_{P,\,n_i^\alpha,\,n_j^\alpha} dT + \sum_i \left(\frac{\partial G^\alpha}{\partial n_i^\alpha}\right)_{P,\,T,\,n_j^\alpha} dn_i^\alpha$$

Since G is a state function, dG is an exact differential (see Box 4.3):

$$\left[\frac{\partial}{\partial T}\left(\frac{\partial G^\alpha}{\partial n_i^\alpha}\right)_{P,\,T,\,n_j^\alpha}\right]_{n_i^\alpha} = \left[\frac{\partial}{\partial n_i^\alpha}\left(\frac{\partial G^\alpha}{\partial T}\right)_{P,\,n_i^\alpha,\,n_j^\alpha}\right]_T$$

Using the definition of μ_i^α (equation 5.9) and equation (4.58) for the partial derivative of G with respect to temperature, we obtain the variation of μ_i^α with temperature:

$$\left(\frac{\partial \mu_i^\alpha}{\partial T}\right)_{P,\,n_i^\alpha,\,n_j^\alpha} = \left(\frac{\partial S^\alpha}{\partial n_i^\alpha}\right)_{P,\,T,\,n_j^\alpha} = -\bar{s}_i^\alpha \qquad (5.27)$$

where $\bar{s}_i^\alpha$ is the partial molar entropy of the constituent i.

5.2.2 *Pressure dependence of chemical potential*

An analogous derivation, using equation (4.59) for the partial derivative of G with respect to pressure, yields the variation of μ_i^α with pressure:

$$\left(\frac{\partial \mu_i^\alpha}{\partial P}\right)_{T,\,n_i^\alpha,\,n_j^\alpha} = \left(\frac{\partial V^\alpha}{\partial n_i^\alpha}\right)_{P,\,T,\,n_j^\alpha} = \bar{v}_i^\alpha \qquad (5.28)$$

where $\bar{v}_i^\alpha$ is the partial molar volume of the constituent i.

In view of the definition of μ_i, it is not surprising that the above two equations are analogous to equations (4.58) and (4.59), with the molar property replaced by the corresponding partial molar property.

5.2.3 *Dependence of chemical potential on composition: the concept of activity*

We have shown that the free energy of a multiconstituent phase at the pressure and temperature of interest can be calculated using equation (5.13) and that of a multiphase system using equation (5.14). In order to perform such calculations, we must first learn how to calculate the chemical potential of a given constituent in a particular phase, μ_i^α, at the temperature and pressure of interest.

We employ the following strategy for calculation of μ_i^α at constant temperature and pressure. We first determine the chemical potential of the pure phase composed entirely of pure i at some chosen *standard state* pressure and temperature (P^0, T^0), and then add a composition-dependent term to account for the deviation from the standard state. The general algebraic form of the resulting equation, as will become evident in the course of subsequent discussions, is

$$\mu_i^\alpha (P, T) = \mu_i^{\alpha\,(pure)} (P^0, T^0) + RT \ln a_i^\alpha(P, T) \qquad (5.29)$$

where P and T are the pressure and temperature of interest, R is the Gas Constant, and a_i^α is the *activity* of i in phase α. The activity of a constituent in a solution is related to, but generally not equal to, its concentration; it may be viewed as the "thermodynamic concentration" or "effective concentration" of a constituent in a solution. The term $\mu_i^{\alpha\,(pure)}$ (P^0, T^0) represents the chemical potential of pure α at the chosen standard state (P^0, T^0) and is called the *standard state chemical potential* of i; it is independent of the actual composition of phase α. Note that for a pure phase α, $\mu^\alpha = \bar{G}^\alpha$, and $\bar{G}^\alpha$ can be calculated as discussed in Chapter 4.

The standard state of a chemical species serves the same purpose as the *reference state* we used in Chapter 4 to calculate thermodynamic parameters at P and T, but it is a more useful concept. A standard state has four attributes: temperature, pressure, composition, and physical state (solid, liquid, solution, etc.). The choice of the standard state is a matter of convenience in a given case, and it may be different for different substances in a single reaction (whereas the reference state temperature and pressure are the same for all substances in a reaction). The standard state may even be a hypothetical state. The only requirement is that we know or can determine the values of the thermodynamic parameters of a substance in the chosen standard state. The values of $\mu_i^{\alpha\,(pure)}$ (P^0, T^0) and a_i^α depend on the standard state chosen, but the value of μ_i^α (P, T) is independent of the standard state chosen. Note that a_i^α has no unique value; its value depends on the chosen standard state (see Box 5.2). It is always a function of composition and it may or may not be a function temperature or pressure, depending on the chosen standard state temperature and pressure. If the standard state is chosen as the pure substance at

Box 5.2 Activities and chemical potentials based on two commonly chosen standard states

Let us consider the simple case of a pure phase α consisting of a single constituent i, so that the molar quantities are numerically equal to their corresponding partial molar quantities at the pressure (P) and temperature (T) of interest.

(1) *Standard state: pure substance, P (bar), T (K)*

$$\bar{G}_T^P = \mu_i^\alpha (P, T) = \mu_i^{0(\alpha)} (P, T) \tag{5.31}$$

Comparison with equation (5.29) shows that

$$\boxed{RT \ln a_i^\alpha = 0 \quad \text{or} \quad a_i^\alpha = 1} \tag{5.32}$$

Thus, *the activity of the constituent of a pure phase at P (bar) and T (K) is unity if the standard state is taken to be the pure phase at P and T*. This is the reason why in many petrologic problems it is advantageous to choose pure phases at the pressure and temperature of interest as the standard state. For 1 mole of a substance with molar volume $\bar{V}_i^i$ (J bar^{-1}),

$$\bar{G}_T^P = \bar{G}_T^1 + \int_1^P \bar{V}_i^\alpha \, dP \tag{4.64}$$

For a reaction,

$$\begin{aligned} \Delta G_{r,T}^P &= \Delta\mu_r (P, T) = \Delta G_{r,T}^1 + \int_1^P \Delta V_r \, dP \\ &= \Delta H_{r,T}^1 - T \Delta S_{r,T}^1 + \int_1^P \Delta V_r \, dP \end{aligned} \tag{5.33}$$

(2) *Standard state: pure substance, 1 (bar), T (K)*

$$\begin{aligned} \bar{G}_T^P &= \mu_i^\alpha (P, T) = \mu_i^{0(\alpha)} (1, T) + RT \ln a_i^\alpha \\ &= \bar{G}_T^1 + RT \ln a_i^\alpha \end{aligned} \tag{5.34}$$

Comparison with equation (4.64) shows that the activity a_i^α corresponding to the standard state of 1 bar and T (K) is given by

$$\boxed{RT \ln a_i^\alpha = \int_1^P \bar{V}_i^\alpha dP} \tag{5.35}$$

For a reaction at P and T,

$$\begin{aligned} \Delta G_{r,T}^P &= \Delta\mu_r (P, T) = \Delta G_{r,T}^1 + RT \left[\sum_i \ln a_i \text{ (products)} - \sum_i \ln a_i \text{ (reactants)} \right] \\ &= \Delta H_{r,T}^1 - T \Delta S_{r,T}^1 + \int_1^P \Delta V_r \, dP \end{aligned} \tag{5.36}$$

Thus, for the standard state of 1 bar and T (K), the activity a_i^α at P and T is a function of the volume equation (5.35). Evidently, both the standard states will yield identical values for $\Delta G_{r,T}^P$.

It is permissible to choose different standard states for the different substances involved in a reaction. In fact, for a reaction involving both solid and gas phases, it is a common practice to choose pure solids at P and T of interest as the standard state for solid phases and pure, ideal gas at 1 bar and T as the standard state for the gas phase.

pressure and temperature of interest (P, T), then a_i^α is a function of composition only. Note that for a pure phase (i.e. $X_i^\alpha = 1$), $\mu^\alpha = \mu^{\alpha \text{(pure)}}$ if the standard state P and T are chosen as the P and T of interest, a convenient choice for solid phases in many petrologic problems, then $RT \ln a^\alpha = 0$, or $a^\alpha = 1$.; i.e., the activity of a pure substance in its standard state is unity if the standard state is chosen as the pure substance at the temperature and pressure of interest.

For the sake of brevity, we often omit the superscript denoting the phase to which the ith constituent refers and the standard state information, and write equation (5.29) as

$$\mu_i = \mu_i^0 + RT \ln a_i \tag{5.30}$$

in which the standard state chemical potential is identified by the subscript "0." For almost all applications it is logical to choose the temperature of interest (T) as the standard state temperature (i.e., $T^0 = T$), in which case μ_i^0 becomes a function of temperature, and a_i a function of pressure and composition. This is called a variable temperature standard state. For such a choice we will have a series of standard states corresponding to a series of equilibrium states at different temperatures, one for each temperature. In principle, we could choose a fixed temperature standard state (e.g., T^0=298.15 K), but this may render the computations unnecessarily more cumbersome. The standard state pressure is commonly chosen as 1 bar or the pressure of interest (P). We will include the standard state temperature and pressure information in the equations where doing so would be a helpful reminder.

Equation (5.30) is the basis for calculation of chemical potentials of constituents in all multiconstituent solutions – solids, liquids, or gases. Assuming that the standard chemical potential of a constituent is known, the critical factor in this calculation is the value of its activity. As will be elaborated in the following sections, the deviation of the activity of i (a_i) from its concentration (commonly expressed as its mole fraction, X_i) depends on whether the solution is ideal or nonideal and, if nonideal, on the assumed model of nonideality.

5.3 Relationship between Gibbs free energy change and equilibrium constant for a reaction

Since the chemical potential of a substance is related to its free energy and the equilibrium constant of a reaction is defined as a ratio of the activities of substances taking part in the reaction, there must be a relationship between the free energy change and equilibrium constant of the reaction. For a reaction in equilibrium at P and T, it can be shown that (see Box 5.3)

$$\Delta G_r^0 = -RT \ln K_{eq} = -2.3026\, RT \log K_{eq} \qquad (5.37)$$

where ΔG_r^0 is the *standard state Gibbs free energy change* for the reaction, with each reactant and product in its standard state, and K_{eq} is the equilibrium constant of the reaction at the P and T of interest. This relation is universally valid and one of the most useful in chemical thermodynamics because it allows us to compute numerical values of equilibrium constants from standard state free energy data and vice versa. The relation, however, is a peculiar one in the sense that the two sides of equation (5.37) refer to completely different physical situations; it is only a numerical equality. The left-hand side refers to free energy change of a reaction involving substances at chosen standard states (e.g., P bar, T K; 1 bar, T K; 1 bar, 298.15 K, etc), without any connotation of an equilibrium state. The right-hand side, on the other hand, refers to a system in which the constituents have achieved mutual equilibrium under the pressure–temperature–composition conditions of interest, or more exactly, to the activity product that would be observed if the system reached equilibrium. The chosen standard states must be the same for both sides of the equation. If we choose a fixed-temperature standard state (e.g., 298.15 K), equation (5.37) is valid only at that standard state temperature; if we choose a variable-temperature standard state, equation (5.37) is valid for all temperatures, but ΔG_r^0 then is a function of temperature. Similar consideration applies to pressure.

Box 5.3 Derivation of the relation $\Delta G_r^0 = -RT \ln K_{eq}$

Let us consider the following balanced reaction in equilibrium at P and T:

x X+y Y+...=c C+d D+...
reactants *products*

Following equation (5.30) for the chemical potential of a constituent at P and T,

$\mu_X=\mu_X^0+RT \ln a_X$; $\mu_Y=\mu_Y^0+RT \ln a_Y$; $\mu_C=\mu_C^0+RT \ln a_C$; $\mu_D = \mu_D^0+RT \ln a_D$

The change in chemical potential for the reaction at P and T, can be calculated as

$$\begin{aligned}\Delta\mu_r(P,T) &= (c\mu_C + d\mu_D) - (x\mu_X + y\mu_Y)\\ &= c\,(\mu_C^0 + RT \ln a_C) + d\,(\mu_D^0 + RT \ln a_D)\\ &\quad - x\,(\mu_X^0 + RT \ln a_X) - y\,(\mu_Y^0 + RT \ln a_Y)\\ &= (c\mu_C^0 + d\mu_D^0 - x\mu_X^0 - y\mu_Y^0)\\ &\quad + RT\,(\ln a_C + \ln a_D - \ln a_X - \ln a_Y)\\ &= \Delta\mu_r^0 + RT \ln\left(\frac{a_C^c\, a_D^d}{a_X^x\, a_Y^y}\right) = \Delta G_{r,T}^P\end{aligned}$$

The general equation for $\Delta\mu_r(P,T)$ can be written as

$$\Delta\mu_r(P,T) = \Delta\mu_r^0 + RT \ln \prod_i a_i^{\nu_i} = \Delta G_{r,T}^P \qquad (5.38)$$

where ν_i is the stoichiometric coefficient of the ith constituent in the reaction, positive if it refers to a product and negative if it refers to a reactant, $\prod_i a_i^{\nu_i}$ is the *reaction quotient*, and $\Delta G_{r,T}^P$ is calculated using methods discussed in section 4.5.1. If the *standard Gibbs free energy change* for the reaction (i.e., the free energy change with each of the reactants and products in its standard state) is denoted by ΔG_r^0, then $\Delta\mu_r^0=\Delta G_r^0$, and

$$\Delta G_{r,T}^P = \Delta G_r^0 + RT \ln \prod_i a_i^{\nu_i} \qquad (5.39)$$

If a_i represents the activity at equilibrium, then $\prod_i a_i^{\nu_i} = K_{eq}$ and $\Delta G_{r,T}^P=0$. Substituting and replacing natural logarithm by $\log_{10}$ ($\ln X = 2.3026 \log_{10} X$), equation (5.39) reduces to

$$\Delta G_r^0 = -RT \ln K_{eq} = -2.3026\, RT \log K_{eq} \qquad (5.37)$$

It follows from equation (5.37) that when two balanced reactions, rx1 and rx2, at T and P are added to obtain a third balancd reaction, rx3, at the same temperature and pressure, $\Delta G_{rx3}=\Delta G_{rx1}+\Delta G_{rx2}$, but $K_{rx3}=(K_{rx1})\,(K_{rx2})$.

Example 5–3: Calculation of K_{eq} from ΔG_r^0

Consider the formation of the mineral andalusite (Al_2SiO_5) by the reaction

$$SiO_{2\,(s)} + Al_2O_{3\,(s)} = Al_2SiO_{5\,(s)}$$

quartz corundum andalusite

The free energy of formation of the phases at 298.15 K and 1 bar are (in kJ mol^{-1}): andalusite=2441.8; corundum=−1582.3; and quartz=−856.3 (Robie and Hemingway, 1995). What is the value of K_{eq} for this reaction at 298.15 K and 1 bar?

Taking the standard state as pure phases at 298.15 K and 1 bar,

$$\Delta G_r^0 = \Delta G_f^0(\text{andalusite}) - \Delta G_f^0(\text{corundm}) - \Delta G_f^0(\text{quartz})$$
$$= -2441.8 - (-1582.3) - (-856.3) = -3.2\,\text{kJ mol}^{-1}$$
$$= -3200\,\text{J mol}^{-1}$$

From equation (5.37),

$$\log K_{eq} = -\Delta G_r^0/2.3026\,RT$$
$$= -(-3200)/(2.3026)\,(8.314)\,(298.15) = 0.56$$

$$K_{eq}(1\text{ bar}, 298.15\text{ K}) = 10^{0.56} = 3.63$$

5.4 Gases

5.4.1 *Pure ideal gases and ideal gas mixtures*

An ideal gas is conceptualized as consisting of vanishingly small particles that do not interact in any way with each other, i.e., there are no forces or energies of attraction or repulsion among the particles. We start with pure ideal gases, which are the easiest to model, because the equation of state for ideal gases is well established (equation 4.8). For a pure ideal gas, the molar volume is the same as the partial molar volume (equation 5.5). From equation (5.28), at constant temperature T,

$$\left(\frac{\partial \mu}{\partial P}\right)_T = \bar{v} = \bar{V} = \frac{RT}{P} \tag{5.40}$$

Choosing T as the standard state temperature and integrating between the limits of an arbitrarily chosen standard pressure P^0 and the pressure of interest P, we obtain

$$\int_{P_0}^{P}\left(\frac{\partial \mu}{\partial P}\right)_T = \mu^{\text{ideal gas}}(P,T) - \mu^0(P^0,T) = RT\ln\frac{P}{P^0} \tag{5.41}$$

which can be rearranged to yield

$$\boxed{\mu^{\text{ideal gas}}(P,T) = \mu^0(P^0,T) + RT\ln\frac{P}{P^0}} \tag{5.42}$$

Comparison with equation (5.30) shows that the activity of a pure ideal gas is P/P^0 and it is a dimensionless number. In many cases, a convenient choice for standard-state pressure is 1 bar.

A mixture of two ideal gases behaves as an ideal gas because the particles of each constituent gas are considered to have no interaction with any other particle in the mixture; i.e., there is no change in enthalpy or volume due to mixing. (Ideality in a liquid solution, on the other hand, assumes a complete uniformity of intermolecular forces irrespective of composition.) For a constituent i in a mixture of ideal gases, the chemical potential μ_i at P and T may be expressed in terms of its partial pressure P_i:

$$\boxed{\mu_i^{\text{ideal gas mixture}}(P,T) = \mu_i^0(P^0,T) + RT\ln\frac{P_i}{P^0}} \tag{5.43}$$

Comparison with equation (5.30) shows that the activity in this case is P_i/P^0 and it is a dimensionless number. According to Dalton's Law of partial pressures (Box 4.1), $P_i = P\,X_i$, where X_i is the mole fraction of i and P is the total pressure in bars.

Example 5–4: Calculation of partial pressures of constituents in a gas mixture

A gas mixture consisting of CO, CO_2 and O_2 is at equilibrium at 600 K (327°C). What are the partial pressures of CO_2 and CO if the total pressure is 1 bar and the partial pressure of O_2 is 0.8 bar. Assume that each gas behaves ideally. The standard free energies of formation at 1 bar and 600 K are (in kJ mol^{-1}): $\Delta G_f^1(CO_2) = -395.2$; $\Delta G_f^1(CO) = -164.2$; and $\Delta G_f^1(O_2) = 0.0$ (Robie and Hemingway, 1995).

The first step in this exercise is to recognize that partial pressures of the constituents in the gas mixture are controlled by the reaction

$$CO + 0.5\,O_2 \Leftrightarrow CO_2 \tag{5.44}$$

Taking the standard state for the gases as 1 bar and 600 K, and $a_i = P_i$ (ideal gases), equation (5.37) gives

$$\Delta G^1_{r,600} = -RT\ln K_{eq}$$

where

$$\Delta G^1_{r,600} = \Delta G^1_{f,600}(CO_2) - \Delta G^1_{f,600}(CO) - 0.5\Delta G^1_{f,600}(O_2)$$
$$= -395.2 - (-164.2) - 0.0 = -231.0\,\text{kJ} = -231{,}000\,\text{J}$$

and

$$K_{eq} = \frac{a_{CO_2}^{gas}}{a_{CO}^{gas}\,(a_{O_2}^{gas})^{0.5}} = \frac{P_{CO_2}}{P_{CO}\,P_{O_2}^{0.5}}$$

Substituting,

$$K_{eq} = \exp\left(\frac{-\Delta G^1_{r,600}}{RT}\right) = \exp\left(\frac{231{,}000}{(8.314)\,(600)}\right) = 1.29\times10^{20}$$

Since $P_{CO_2} + P_{CO} + P_{O_2} = P_{\text{total}} = 1$ bar, $P_{CO_2} = P_{CO}\,P_{O_2}^{0.5}\,K_{eq}$, and $P_{O_2} = 0.8$ bar,

$$P_{CO} = \frac{1 - P_{O_2}}{1 + K_{eq}\,(P_{O_2})^{0.5}} = 1.734\times10^{-21}\text{bar}$$

$$P_{CO_2} = P_{CO}\,P_{O_2}^{0.5}\,K_{eq} = (1.734\times10^{-21})\,(0.8)^{0.5}\,(1.29\times10^{20}) = 2.000\times10^{-1} = 0.2\text{ bar}$$

Since $P_i = P_{total} X_i$ (equation 4.9), at equilibrium the gas would be a mixture essentially of CO_2 and O_2, with only a trace of CO.

5.4.2 *Pure nonideal gases: fugacity and fugacity coefficient*

In the high-temperature–high-pressure environments encountered in igneous and metamorphic petrology, the gas phase commonly behaves as a nonideal (or *real*) gas. An equation of the same algebraic form as equation (4.8) can be applied to a pure nonideal gas by replacing pressure (P) with a new variable called *fugacity* (f), which is related to pressure by a *fugacity coefficient* (χ), which is defined as

$$\chi = \frac{f}{P} \text{ such that } \frac{f}{P} \rightarrow 1 \text{ as } P \rightarrow 0 \qquad (5.45)$$

That is, as pressure P tends to zero, the value of χ approaches unity and the gas approaches ideality. Thus, $\chi=1$ for an ideal gas, and the fugacity coefficient is a measure of the deviation of a nonideal gas from ideal gas behavior.

Now we can derive a relation for a nonideal gas analogous to equation (5.41), with T as the standard state temperature and the pressure replaced by fugacity:

$$\mu^{\text{non-ideal gas}}(P, T) = \mu^0 (P^0, T) + RT \ln \frac{f}{f^0} \qquad (5.46)$$

where f is the fugacity corresponding to pressure P of interest ($f=\chi P$) and f^0 the fugacity corresponding to the chosen standard state pressure P^0. Comparison with equation (5.30) shows that the activity in this case is f/f^0, again a dimensionless number.

What should we chose as the standard state for a real gas? A standard state of unit fugacity is not a good choice because real gases may have states of unit fugacity in which the pressure is not equal to unity. This leads to different real gases having different standard states. For convenience of computation, we choose a standard state for which $f^0=1$ at all temperatures (including the temperature of interest). The only substance for which this is true is an ideal gas (i.e., $\chi=1$) at a pressure of 1 bar (i.e., $P^0=1$). Note that the chosen standard state is hypothetical, not an actual state of any real gas (Fig. 5.3). The advantage of this choice is that nonideality of real gases can be compared in terms of their fugacity coefficients because the fugacity coefficient for any real gas fully accounts for the deviation from ideality at the specified temperature and pressure.

Fugacity coefficient is a function of both temperature and pressure. The classical method of obtaining the fugacity and fugacity coefficient of a pure, real gas at specified P and T is by experimental measurement of its molar volume ($\bar{V}_{real}$) at several pressures up to P (at constant T) followed by calculation using the equation (see Box 5.4):

$$RT \ln f = RT \ln P + \int_0^P (\bar{V}_{real} - \bar{V}_{ideal}) \, dP \qquad (5.47)$$

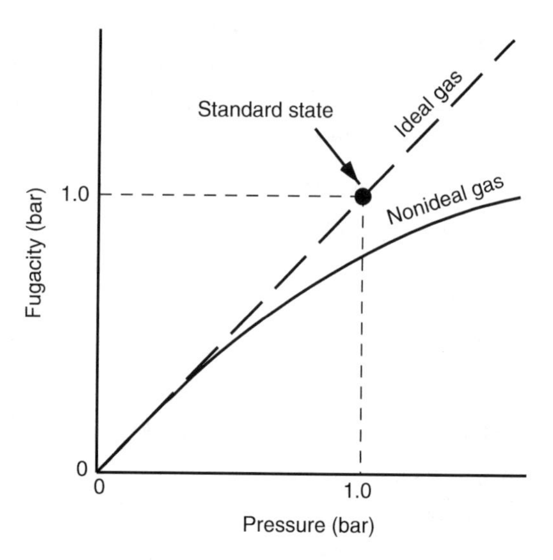

Fig. 5.3 Schematic illustration of the hypothetical standard state commonly chosen for nonideal (or real) gases. If the standard state is chosen as an ideal gas with unit fugacity, $f^0=P^0=1$ bar. For a nonideal gas, the deviation from ideality is accounted for by the fugacity coefficient (χ) of the gas at the pressure and temperature of interest. Note that f is numerically equal to P for an ideal gas (i.e., $\chi=1$).

As illustrated in Fig. 5.4 for water as an example, the integrand in equation (5.47) can be evaluated graphically. The P–V–T method, however, is difficult, expensive, and time consuming when experiments are conducted above 1 kbar pressure and 500°C. As such, very few experimental P–V–T data are available for gases of geologic interest beyond about 10 kbar and 1000°C (Holloway, 1977).

For a pure gas, thermodynamic properties, including fugacities, within and beyond the range of P–V–T measurements are commonly obtained by formulating equations that conform to the available P–V–T data for the gas. One approach involves generating a polynomial, without the constraint of an equation of state, for the sole purpose of obtaining the best fit (i.e., lowest residuals) with the experimental data. This was the approach used by Burnham *et al.* (1969) for calculating fugacity coefficients of H_2O up to 1000°C and 10,000 bars, and by Holland and Powell (1990) for calculating fugacities of H_2O and CO_2 over the range 1–50 kbar and 100–1600°C. Another approach is to derive an appropriate equation of state, based on molecular thermodynamics, that can not only account for the existing P–V–T data but also is likely to yield satisfactory results on extrapolation. For example, Kerrick and Jacobs (1981) used a modified version of the Redlich–Kwong (MRK) equation of state (equation 4.27) to obtain fugacity coefficients for H_2O and CO_2 gases at a series of

Box 5.4 Calculation of fugacity of a real gas at P and T, the pressure and temperature of interest

The fugacity of a pure, real gas at constant temperature T can be calculated from its measured volumes at various pressures up to the pressure of interest (P) and its theoretical volume at the same pressures if it behaved as an ideal gas. To derive an appropriate mathematical relation for this calculation (Nordstrom and Munoz, 1994), let us define a quantity ϕ at the P and T of interest as:

$$\phi = \bar{V}_{\text{ideal gas}} - \bar{V}_{\text{real gas}} = \frac{RT}{P} - \bar{V}_{\text{real}} \quad (5.48)$$

Rearranging and integrating between the limits P^0 (standard state pressure) and P (pressure of interest), we obtain

$$\int_{P^0}^{P} \bar{V}_{\text{real}}\, dP = \int_{P^0}^{P} \left(\frac{RT}{P} - \phi\right) dP = RT \ln \frac{P}{P^0} - \int_{P^0}^{P} \phi\, dP$$

Since for a pure gas, molar volume is the same as partial molar volume, using equations (5.28) and (5.46), we get

$$\int_{P^0}^{P} \bar{V}_{\text{real}}\, dP = \int_{P^0}^{P} d\mu = \mu\,(P, T) - \mu^0(P^0, T) = RT \ln \frac{f}{f^0}$$

We can now write

$$RT \ln \frac{f}{f^0} = RT \ln \frac{P}{P^0} - \int_{P^0}^{P} \phi\, dP$$

which on rearrangement gives

$$RT \ln \frac{f}{P} = RT \ln \frac{f^0}{P^0} - \int_{P^0}^{P} \phi\, dP$$

Considering equation (5.45) in terms of P^0 and f^0,

$$\text{when } P^0 \to 0, \ \frac{f^0}{P^0} \to 1, \text{ so that } RT \ln \frac{f^0}{P^0} \to 0$$

Substituting for ϕ, we get the following equation for calculation of the fugacity of a gas:

$$\ln \chi = \ln \frac{f}{P} = -\frac{1}{RT} \int_0^P (\bar{V}_{\text{ideal}} - \bar{V}_{\text{real}})\, dP = -\frac{1}{RT} \int_0^P \left(\frac{RT}{P} - \bar{V}_{\text{real}}\right) dP \quad (5.49)$$

which on rearrangement gives

$$\boxed{RT \ln f = RT \ln P + \int_0^P (\bar{V}_{\text{real}} - \bar{V}_{\text{ideal}})\, dP} \quad (5.47)$$

temperatures and pressures up to 1000°C and 10 kbar, whereas Holland and Powell (1998) used what they called Compensated Redlich–Kwong (CORK) equations to calculate fugacities of H_2O and CO_2 over the range 0.5–10 kbar and 200–1000°C (Appendix 6). Mäder and Berman (1991) used a van der Waals type of equation and claimed that it yielded thermodynamic properties for mineral equilibrium calculations reliably in the

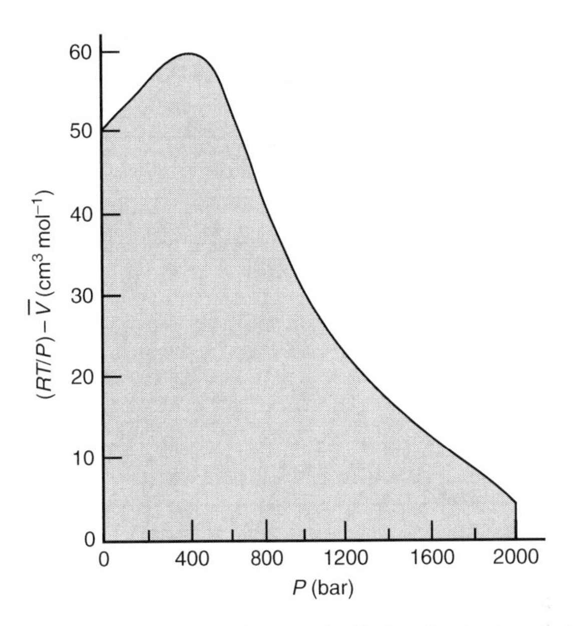

Fig. 5.4 A plot of the difference between the ideal and real volumes of H_2O at 500°C as a function of pressure, using data from Burnham *et al.* (1969). The shaded area under the curve is equivalent to the integrand in equation (5.49) with P=2000 bars, and can be used to solve for the fugacity of H_2O at 500°C and 2000 bars. The data for the plot are from Burnham *et al.* (1969), who give the calculated value of the fugacity as 721 bars, which translates into a fugacity coefficient of 0.361 (equation 5.45).

range 400–1800 K and 1–42 kbar with good extrapolation properties to higher pressures and temperatures.

5.4.3 *Nonideal gas mixtures*

For a constituent i in a mixture of nonideal (or real) gases, the chemical potential μ_i at P and T may be expressed in terms of its corresponding fugacity f_i by an equation analogous to equation (5.46):

$$\boxed{\mu_i^{\text{nonideal gas mixture}}(P, T) = \mu_i^0\ (P^0, T) + RT \ln \frac{f_i}{f_i^0}} \quad (5.50)$$

where $f_i = \chi_i\, P_i$, such that $\frac{f_i}{P_i} \to 1$ as $P \to 0$

and f_i^0 is the fugacity of pure i at the standard-state pressure P^0, and χ_i is the fugacity coefficient of i. In this case, activity of i, a dimensionless number, is given by

$$\boxed{a_i = \frac{f_i}{f_i^0}} \quad (5.51)$$

As in the case of pure nonideal gases, if the standard state is chosen as the state in which the pure gas has unit fugacity (i.e., f_i^0=1) and behaves as an ideal gas (i.e., χ_i=1), then P^0=1.

If the mixing of the gases can be assumed to be ideal, f_i can be calculated using the Lewis fugacity rule (see section 5.5.4):

$$\boxed{f_i = f_i^{\text{pure}}\, X_i^{\text{solution}}} \quad (5.52)$$

Box 5.5 Activity coefficients of H_2O and CO_2 in binary $H_2O - CO_2$ fluids based on the experimental data of Aranovich and Newton (1999)

Aranovich and Newton (1999) defined activity coefficient of each of the components (γ_i) in a $CO_2 - H_2O$ gas mixture by the relation

$$a_i = \frac{f_i}{f_i^0} = \gamma_i X_i \tag{5.53}$$

where a_i is the activity of the component i, f_i is the fugacity of i in the mixture, f_i^0 is the fugacity of the pure i at the same P–T conditions, X_i is the mole fraction of i, *and* γ_i is the activity coefficient.

Activity–composition relations in $CO_2 - H_2O$ solutions were determined by Aranovich and Newton (1999) at 6, 10, and 14 kbar over broad temperature –fluid composition ranges, and activity values of CO_2 and H_2O were retrieved based on the thermodynamic data set of Holland and Powell (1991). Equations that are consistent with their experimental data and can be used to calulate the activity coefficients of CO_2 (γ_{CO_2}) and H_2O (γ_{H_2O}) are:

$$RT \ln \gamma_{H_2O} = (X_{CO_2})^2 W \left[\frac{V^o_{H_2O} (V^o_{CO_2})^2}{(V^o_{H_2O} + V^o_{CO_2})(X_{H_2O} V^o_{H_2O} + X_{CO_2} V^o_{CO_2})^2} \right]$$

$$RT \ln \gamma_{CO_2} = (X_{H_2O})^2 W \left[\frac{V^o_{CO_2} (V^o_{H_2O})^2}{(V^o_{H_2O} + V^o_{CO_2})(X_{H_2O} V^o_{H_2O} + X_{CO_2} V^o_{CO_2})^2} \right] \tag{5.54}$$

where $V^o_{H_2O}$ and $V^o_{CO_2}$ are the specific volumes of pure H_2O and CO_2 at a given P (in kbar) and T (in K), X_{CO_2} and X_{H_2O} are the mole fractions ($X_{CO_2} + X_{H_2O} = 1$), and $W = (A + BT\,[1 - \exp(-20P)] + CPT$, with T in K and P in kbar. The experimentally determined best-fit values of the constants are: $A = 12893$ J, $B = -6.501$ J K^{-1}, and $C = 1.0112$ J K^{-1} kbar^{-1}. The volumes of H_2O and CO_2 were obtained from the CORK equations of Holland and Powell (1991). The experiments were conducted in the range 600–1000° C and 6–14 kbar, but calculations with these equations down to 1 kbar or below at temperatures as low as 500°C are likely to be reasonably accurate.

regardless of the proportion of other constituents in the gas mixture. For example, the fugacity coefficient of pure H_2O gas at 800°C and 4 kbar is 0.843 (Kerrick and Jacobs, 1981), which translates to $f^{pure}_{H_2O} = 4.0 \times 0.843 = 3.372$ kbar. In an ideal gas mixture with $X_{H_2O} = 0.2$, $f^{gas\ mixture}_{H_2O} = 0.2 \times 3.372 = 0.674$ kbarat the pressure and temperature of interest, irrespective of other constituents in the gas mixture. The rule works best at high temperatures, at low pressures, and for dilute solutions of i.

In principle, the ideal mixing model should not be applicable to real gas mixtures – such as $H_2 - H_2O$, $CO_2 - H_2O$ – composed of unlike molecules. This is because interactions between unlike molecules generally result in volumes, and hence fugacities, that are not the same as obtained by the addition of the end-member volumes. $CO_2 - H_2O$ fluids are of particular importance in geochemistry, because they play vital roles in crystallization of magmas and in metamorphism of rocks. A number of mixing models have been proposed for nonideal C–O–H fluids (e.g., Holloway, 1981; Kerrick and Jacobs, 1981; Saxena and Fei, 1988; Duan *et al.*, 1996; Aranovich and Newton, 1999; Huizenga, 2001). A commonly used model is that proposed by Kerrick and Jacobs (1981), which uses a modified Redlich–Kwong (MRK) equation of state for the pure gases and simple mixing rules for binary $H_2O - CO_2$ fluids. Equations proposed by Aranovich and Newton (1999) for the fugacity coefficients of CO_2 and H_2O (Box 5.5) are simpler in analytical form and not tightly linked to any particular equation of state for the end-member gases. Calculated activity–concentration relations according to both models are similar and show significant positive nonideality of mixing that increases with decreasing temperature at constant pressure and with increasing pressure at constant temperature (Fig. 5.5).

Equation (5.52) is commonly used as the general definition of activity (a_i):

> *the activity of a constituent i in a phase is equal to the fugacity of i in the phase divided by the standard-state fugacity of i in the phase.*

The activities of all the other ideal and nonideal gas systems discussed above are special cases that can be derived from this definition. Moreover, this definition of activity applies equally well to liquids and solids because all constituents of a phase have a fugacity, whether it is measurable or not. In a gas mixture the fugacity of a constituent is related to its partial pressure, which is a function of its mole fraction in the mixture; in a liquid or solid solution the fugacity of a constituent is related to its vapor pressure, which is a function of its mole fraction in the solution. Note that a_i depends on the choice of standard-state fugacity, and does not have a unique value even if temperature, pressure, and composition are fixed.

If fugacity values are available, equation (5.50) is sufficient for the calculation of chemical potentials and the activity concept is redundant. For most constituents of liquid and solid solutions, however, fugacities are too small to measure. In such cases, the activity concept is useful as it can be determined in other ways.

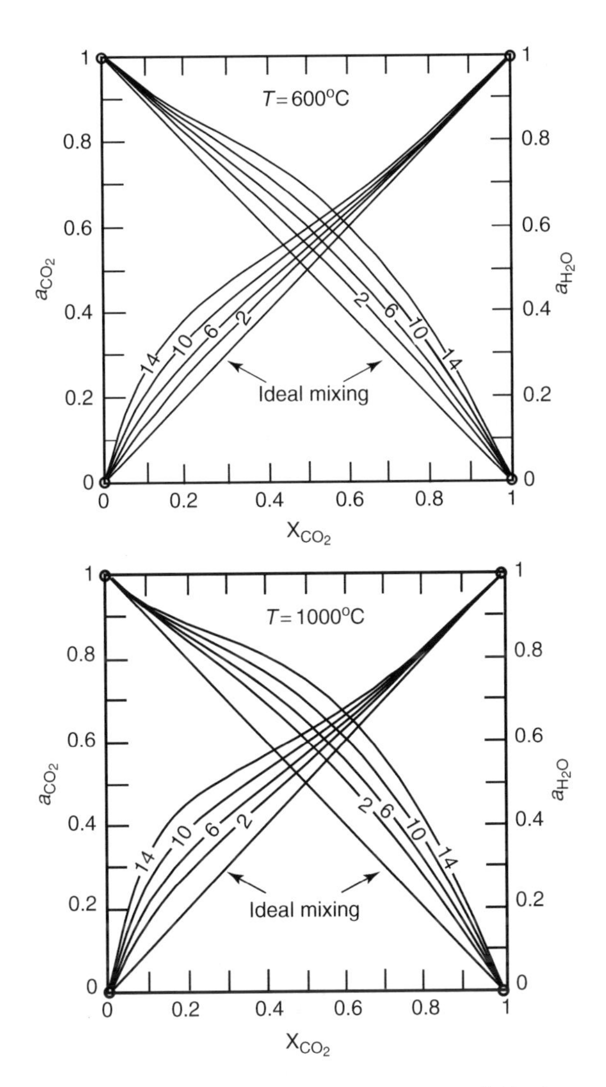

Fig. 5.5 Activity–composition relationship in CO_2–H_2O fluid mixture ($X_{CO_2} + X_{H_2O} = 1$) at 600°C and 1000°C at total pressures of 2, 6, 10, and 14 kbars, as calculated by Aranovich and Newton (1999). The mixture becomes closer to an ideal mixture with decreasing pressure at constant temperature and with increasing temperature at constant pressure. (After Aranovich and Newton, 1999, Figure 8, p. 1330.)

Example 5–5: P–T relationship for a dehydration reaction

We have now learned enough to calculate the *P–T* relationships for reactions involving fluid phases. The example discussed below is a dehydration reaction that is capable of producing corundum during metamorphism of aluminum-rich rocks:

$$\underset{\text{91muscovite}}{KAl_3Si_3O_{10}(OH)_2} \Leftrightarrow \underset{\text{sanidine}}{KAlSi_3O_8} + \underset{\text{corundum}}{Al_2O_3} + H_2O \tag{5.55}$$

For the purpose of illustration, let us calculate the equilibrium temperature corresponding to $P_{total} = 3$ kbar. For simplicity, we will assume that all the solid phases are pure, the fluid phase consists of only H_2O and behaves as a nonideal gas, $P_{H_2O} = P_{total}$, $(\Delta C_P)_r = 0$, and ΔV_r(solids) is independent of temperature and pressure.

At equilibrium, using equation (5.37),

$$\Delta G_r^0 = -RT \ln K_{eq} = -RT \ln \frac{a^{san}_{KAlSi_3O_8}\, a^{cor}_{Al_2O_3}\, a^{fluid}_{H_2O}}{a^{mus}_{KAl_3Si_3O_{10}(OH)_2}} \tag{5.56}$$

To evaluate the activities at the pressure and temperature of interest (*P, T*), we need to choose a standard state for each species involved in the reaction. As is the common practice in addressing petrologic problems, we choose a mixed standard state of pure substances at *P* (bar) and *T* (K) for the solid species and pure ideal gas at 1 (bar) and *T* (K) for the fluid phase. This choice has the advantage that a_i(solid) = 1 and a_i(gas) = $f_i/f_i^o = f_i/1 = f_i$, so that

$$\ln K_{eq} = \ln f_{H_2O} \tag{5.57}$$

It follows from our choice of mixed standard state that

$$\begin{aligned}\Delta G_r^0 &= \Delta G^P_{r,T}\text{(solid phases in the reaction)} \\ &\quad + \Delta G^1_{r,T}\text{(fluids in the reaction)} \\ &= [\Delta H^1_{r,T}\text{(solids)} - T\,\Delta S^1_{r,T}\text{(solids)} \\ &\quad + \int_1^P \Delta V_r\text{(solids)}\, dP] + [\Delta H^1_{r,T}(H_2O) - T\,\Delta S^1_{r,T}(H_2O)] \\ &= \Delta H^1_{r,T}(\text{solids} + H_2O) - T\,\Delta S^1_{r,T}(\text{solids} + H_2O) \\ &\quad + (P-1)\,\Delta V_r\text{(solids)} \\ &= \Delta H^1_{r,T} - T\,\Delta S^1_{r,T} + (P-1)\,\Delta V_r\text{(solids)} \\ &= -RT \ln f_{H_2O}\end{aligned}$$

Rearranging,

$$\ln f_{H_2O} = \frac{-\Delta H^1_{r,T}}{RT} + \frac{\Delta S^1_{r,T}}{R} - \frac{(P-1)\,\Delta V_r\text{(solids)}}{RT} \tag{5.58}$$

Chatterjee and Johannes (1974) determined that in the temperature range 600–800°C (873–1073 K) the enthalpy, entropy, and volume changes for the reaction at 1 bar, within experimental error, could be assumed to have the following constant values:

$$\Delta H^1_{r,T} = 23460\ \text{cal} = 98157\ \text{J}$$

$$\Delta S_{r,T} = 39.2\ \text{cal K}^{-1} = 164\ \text{J K}^{-1}$$

$$\Delta V^1_{r,298.15} = -0.1515\ \text{cal bar}^{-1} = -0.6339\ \text{J bar}^{-1}$$

Substituting these values in equation (5.58),

$$\begin{aligned}\ln f_{H_2O} &= \frac{1}{R}\left[\frac{-98157}{T} + 164 + \frac{(2999)\,(0.6339)}{T}\right] \\ &= \frac{1}{R}\left[\frac{-96256}{T} + 164\right]\end{aligned}$$

This equation cannot be solved directly for T because we do not know f_{H_2O} (it is not equal to P_{H_2O}), but it can be solved by successive approximations of the equilibrium temperature (Wood and Fraser, 1976). If we choose $T=973$ K (700°C), f_{H_2O} works out to be 2507 bar, which corresponds approximately to $P_{H_2O} \approx 3650$ bar in the tables of Burnham *et al.* (1969), higher than the pressure of our interest. Since f_{H_2O} becomes smaller with decreasing temperature, the equilibrium temperature must be less than 973 K. Repeating the calculation for a few lower values of T, we find that the temperature for $P_{H_2O}=3000$ bar must lie between 953 K and 933 K:

for $T = 953$ K (680°C), $f_{H_2O} = 1953$ bar, and $P_{H_2O} \approx 3050$ bar

for $T = 933$ K (660°C), $f_{H_2O} = 1505$ bar, and $P_{H_2O} \approx 2500$ bar

Let T_E=the equilibrium temperature corresponding to $P_{H_2O}=3000$ bar. Assuming the variation between these two sets of T–P_{H_2O} values to be approximately linear, we can apply the lever rule to calculate T_E from the equation

$$\frac{953 - T_E}{953 - 933} = \frac{3050 - 3000}{3050 - 2500}$$

Solving the equation, we get $T_E \approx 951$ K ≈ 678°C at $P_{H_2O}=3$ kbar.

5.5 Ideal solutions involving condensed phases

5.5.1 Mixing properties of ideal solutions

For any extensive property Z of a solution, the molar value $\bar{Z}$ can be written as

$$\bar{Z} = \sum_i X_i \bar{Z}_i + \Delta Z_{mixing} \tag{5.59}$$

where $\bar{Z}_i$ is the molar value of the pure ith constituent and X_i its mole fraction in the solution, $\sum_i X_i \bar{Z}_i$ represents the contribution arising from a hypothetical mechanical mixture of the constituents (i.e., mixing without any interaction among the constituents), and ΔZ_{mixing} is the change in Z that accompanies formation of the solution.

In terms of mixing properties, solutions involving condensed phases (liquids and solids are referred to as condensed phases because they have much higher densities than gases) can be grouped into three main categories (Table 5.1): ideal, regular (which are only or slightly nonideal; Anderson and Crerar, 1993), and nonideal (or real). Ideality in a solution is defined by complete uniformity of intermolecular forces among the constituents of the solution. Ideal solutions, therefore, are characterized by no change in volume or enthalpy due to mixing of the constituents, but the change in entropy due to ideal mixing, $\Delta S_{ideal\ mixing}$, is not zero as the state of ordering in a solution is different from that in the constituents being mixed. In contrast, a nonideal solution lacks complete uniformity of intermolecular forces among its constituents and is characterized by a change in volume and enthalpy, in addition to entropy, due to mixing of its constituents.

It can be shown that (see Anderson and Crerar (1993, pp. 231–233) or Nordstrom and Munoz (1994, pp. 127–129) for derivations)

$$\Delta S_{ideal\ mixing} = -R \sum_i X_i \ln X_i \tag{5.60}$$

so that

$$\bar{S}^{ideal\ solution} = \sum_i X_i \bar{S}_i - R \sum_i X_i \ln X_i \tag{5.61}$$

where $\bar{S}_i$ is the molar entropy of a pure ith constituent in the solution and X_i its mole fraction. It follows that the change in Gibbs free energy that accompanies ideal mixing is

$$\begin{aligned} \Delta G_{ideal\ mixing} &= \Delta H_{ideal\ mixing} - T\,\Delta S_{ideal\ mixing} \\ &= RT \sum_i X_i \ln X_i \end{aligned} \tag{5.62}$$

The molar free energy of an ideal solution, taking the standard state of each constituent as the pure substance at P and T, is the sum of the free energies of the pure constituents being mixed plus the change in free energy due to ideal mixing of the constituents:

$$\begin{aligned} \bar{G}^{ideal\ solution}(P,T) &= G_{mechanical\ mixture} + G_{ideal\ mixing} \\ &= \sum_i X_i \bar{G}_i (P,T) + RT \sum_i X_i \ln X_i \\ &= \sum_i X_i \mu_i^0 (P,T) + RT \sum_i X_i \ln X_i \end{aligned} \tag{5.63}$$

Table 5.1 Types of solutions in terms of mixing properties.

Type of solution	ΔV_{mixing}	ΔH_{mixing}	ΔS_{mixing}	$\Delta G_{mix}=\Delta H_{mix} - T\,\Delta S_{mix}$
Ideal	0	0	$\Delta S_{ideal\ mix} = -R \sum_i X_i \ln X_i$	$RT \sum_i X_i \ln X_i$
Regular	≈0	≠0	$\Delta S_{reg\ mix} = -R \sum_i X_i \ln X_i$	$\Delta H_{reg\ mix} + RT \sum_i X_i \ln X_i$
Nonideal	≠0	≠0	$\Delta S_{nonideal\ mix} = -R \sum_i X_i \ln a_i$	$\Delta H_{nonideal} + RT \sum_i X_i \ln a_i$

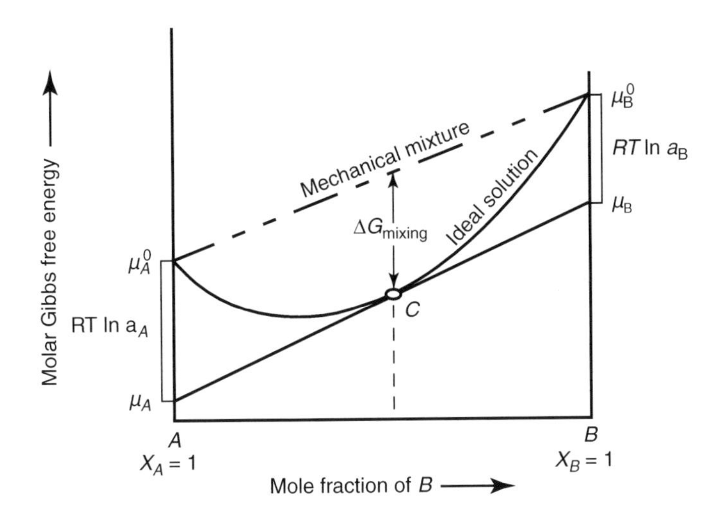

Fig. 5.6 Molar Gibbs free energy–composition (X) diagram for a hypothetical binary solution or mechanical mixture of pure phases A and B at the P and T of interest. μ_A^0 and μ_B^0 are the standard chemical potentials of A and B, respectively, the chosen standard state being the pure constituents at P and T. The straight line joining μ_A^0 and μ_B^0 represents the molar free energy of a mechanical mixture of A and B as a function of composition, and the parabola that of the ideal solution. For any particular composition C, the difference between the two is the $\Delta G_{\text{ideal mixing}}$, and the tangent to the parabola at that composition gives μ_A and μ_B, the chemical potentials of A and B in the ideal solution.

where μ_i^0 is the chemical potential of the pure *i*th constituent at the chosen standard state. Since $X_i < 1$, the term $RT \sum_i X_i \ln X_i$ would always yield a negative number. Therefore, the molar free energy of an ideal solution at a specified temperature and pressure condition would always be lower than that of a mechanical mixture of the same constituents (Fig. 5.6); that is, thermodynamically an ideal solution would be more stable than the corresponding mechanical mixture. The free energy of a nonideal solution may be more or less than that of the mechanical mixture, depending on the relative magnitudes of ΔH_{mixing} and ΔS_{mixing}; the solution would be less stable than the mechanical mixture in the former case and more stable in the latter case. Figure 5.7a illustrates the general form of the $\bar{G} - X$ graph of a hypothetical binary solution A– B, relative to the corresponding mechanical mixture, for ideal mixing and the two cases of nonideal mixing mentioned above.

The $\bar{G} - X$ graph of each of the hypothetical binary ideal solutions shown in Fig. 5.7a is marked by one minimum, but for a highly nonideal solution the $G - X$ graph may have two minima (Fig. 5.7b). In the latter case, the solution may separate into two phases coexisting in equilibrium. This is the phenomenon of *immiscibility* or *exsolution*. For the hypothetical example shown in Fig. 5.7b, the solution, which exists as a single phase at high temperatures, will separate at lower temperatures into two solutions – solution *a* of composition x_a and solution *b* of composition x_b – in equilibrium with each other, i.e., $\mu_A^a = \mu_A^b$ and $\mu_B^a = \mu_B^b$. The compositions of the immiscible phases, *a* and *b*, depend on the temperature and, therefore, can be used for geothermometry (see Chapter 7). For many systems, increasing the temperature will cause the compositions of *a* and *b* to become closer together, finally coalescing to a single point at some temperature. The solution will be completely miscible above this temperature, which is called the *consolute temperature* and has the same significance as the *critical temperature* for liquid–gas separation. For example, for the phenol–water binary solution, the consolute temperature is 66.8°C and the corresponding composition is 34.5% phenol.

An example of exsolution familiar to petrologists is the formation of perthite (intergrowths of Na-rich feldspar in a K-rich feldspar host; Fig. 5.8a) because of the breakdown of high-temperature alkali feldspar (single phase) into two feldspar phases. Other examples of exsolution textures include intergrowths of K-rich feldspar in a plagioclase host (antiperthite), intergrowths of clinopyroxene blebs or lamallae in orthopyroxene (Fig. 5.8b), Fe-Ni sulfide globules in peridotites and basalts, and pentlandite rims around pyrrhotite crystals.

5.5.2 *Raoult's Law*

An important property of a solution is the partial vapor pressures of its constituents. At equilibrium (at constant P, T), the chemical potential of a common constituent must be the same in all of the coexisting phases (equation 5.15). Thus, the chemical potential of any constituent in a condensed phase (solid or liquid) must be equal to its chemical potential in the coexisting vapor phase:

$$\mu_i^{\text{solution}}(P, T) = \mu_i^{\text{vapor}}(P, T) \quad (5.64)$$

This relationship enables us to address the properties of a condensed phase from a consideration of its coexisting vapor phase. A quantitative formulation of this relationship in a liquid solution–vapor equilibrium system at constant P and T is known as *Raoult's Law*:

$$\boxed{P_i = P_i^0 X_i^{\text{solution}}} \quad (5.65)$$

where P_i is the partial pressure of the *i*th constituent in the vapor, X_i the mole fraction of *i* in the solution, and P_i^0 the vapor pressure of *i* in equilibrium with pure *i*. This law was established through the experimental work by the French chemist François Marie Raoult (1830–1901) at relatively low total vapor pressures so that the vapor phase could be regarded as an ideal gas mixture.

Applying Raoult's Law to a binary solution of A and B (Fig. 5.9), the partial pressures of A and B in the vapor phase, as long as it behaves as an ideal gas, can be calculated as:

$$P_A = P_A^0 X_A, \quad P_B = P_B^0 X_B, \quad \text{and} \quad P_{\text{vapor}} = P_A + P_B$$

For example, given a solution of $X_A^{\text{solution}} = 0.3$, $X_B^{\text{solution}} = 1 - 0.3 = 0.7$, $P_A^0 = 2.2$, and $P_B^0 = 1.5$, we will have

$P_A = 2.2 \times 0.3 = 0.66$; $P_B = 1.5 \times 0.7 = 1.05$; and $P_{\text{vapor}} = 0.66 + 1.05 = 1.71$.

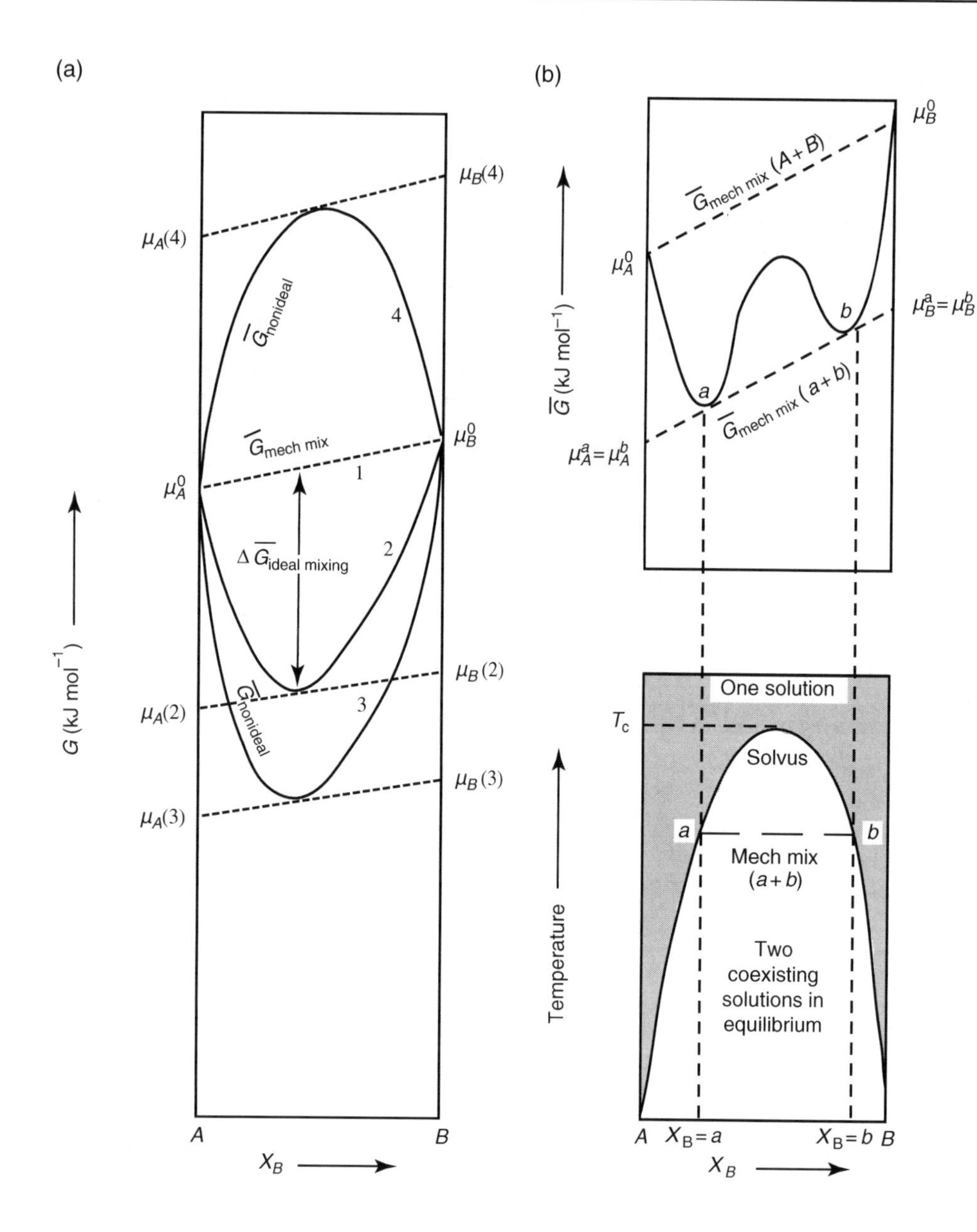

Fig. 5.7 (a) Molar Gibbs-free-energy ($\overline{G}$)–composition (X) diagram for a hypothetical binary solution or mechanical mixture of A and B similar to Fig. 5.6: 1 = mechanical mixture; 2 = ideal solution; 3 and 4 = nonideal solutions, depending on the magnitude of ΔS_{mixing}. (b) Separation (or exsolution) of two phases of composition $X_B = a$ and $X_B = b$ in equilibrium at a particular temperature from a nonideal solution, so that $\mu_A^a = \mu_A^b$ and $\mu_B^a = \mu_B^b$. The boundary between the one-phase region and the two-phase region is called a *solvus*. As shown in the T–X diagram, no immiscibility can occur above T_C, the consolute temperature. Note that the nonideal solution is more stable than a mechanical mixture of A and B, but less stable than a mechanical mixture of a and b. (After Anderson and Crerar, 1993.)

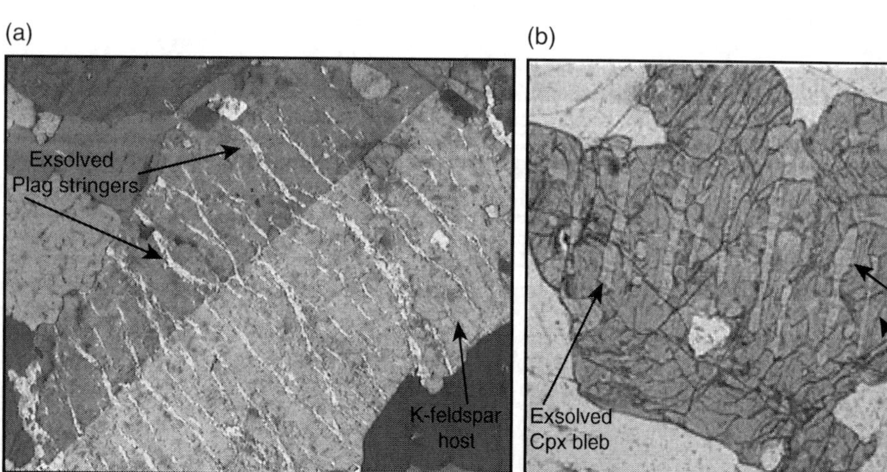

Fig. 5.8 Photomicrographs (plane polarized light) showing examples of exsolution textures formed due to unmixing of high-temperature solid solutions: (a) A large, twinned crystal of K-feldspar with oriented stringers of exsolved plagioclase (Plag) (http://www.univ-lille1.fr/geosciences/cours/cours_minerralo/cours_mineralo_3.html); (b) A large crystal of orthopyroxene (Opx) with exsolved blebs of clinopyroxene (Cpx) (http://www.ucl.ac.uk/%7Eucfbrxs/PLM/opx.html).

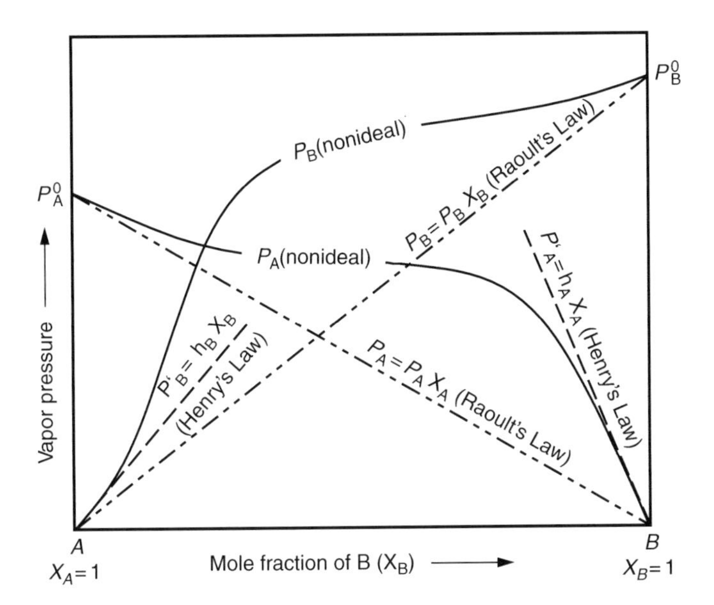

Fig. 5.9 Vapor pressure versus composition diagram for a binary solution *A–B* at constant temperature, assuming ideal behavior of the constituents according to Raoult's Law and Henry's Law, and positive deviation from ideality for a nonideal solution. P^o_A and P^o_B represent vapor pressures of pure *A* and pure *B*, respectively. In reality, only the solvent compositions close to pure *A* and pure *B* obey Raoult's Law and the corresponding solute compositions obey Henry's Law.

The mole fractions of *A* and *B* in the vapor phase, applying Dalton's Law of Partial Pressures ($P_i=X_i\, P_{total}$), can be calculated as:

$$X_A^{vapor}=\frac{P_A}{P_{vapor}}=\frac{0.66}{1.71}=0.39 \text{ and } X_B^{vapor}=\frac{P_B}{P_{vapor}}=\frac{1.05}{1.71}=0.61$$

Raoult's Law holds only for ideal solutions. In other words, a solution that obeys Raoult's Law over the entire composition range is an ideal solution, a condition seldom satisfied by solutions in the real world. Vapor pressure curves of constituents in a nonideal solution show either positive (Fig. 5.9) or negative deviation from ideality.

5.5.3 Henry's Law

In a very dilute solution, a dissolved constituent (solute) is overwhelmed by the solvent, and this presents a different situation. Any solute molecule is completely surrounded by a uniform environment of the solvent molecules, and the solute–solvent interactions remain independent of composition. In such a solution (i.e., with $X_i << 1$), the partial pressure of the solute vapor is given by *Henry's Law*, which states that the pressure of a gas above a solution is proportional to the concentration of the gas dissolved in the solution:

$$\boxed{P_i=h_i\, X_i^{solution}} \qquad (5.66)$$

where h_i is the proportionality constant called the *Henry's Law constant*. It is a pressure- and temperature-dependent constant with the dimensions of pressure. The constant may be viewed as the activity coefficient of species *i* in a phase at high dilution. At a given pressure and temperature condition, h_i varies with the nature of the solute and the solvent, but is independent of X_i. This relationship was documented by the British chemist William Henry (1775–1836) in 1803 to describe the dependence of the solubility of gases in liquids on pressure, but it is not restricted to gas–liquid systems. Henry's Law is applicable to a wide variety of fairly dilute solutions (i.e., $a_i=h_i X_i^{solution}$), and is particularly useful for characterizing trace element behavior in solid solutions (see Box 5.6).

Box 5.6 Nernst distribution coefficients for trace elements

Elements that occur in rocks or minerals in concentrations of a few tenths of a percent or less by weight (typically < 0.1 wt%) are commonly called *trace elements*. The low concentrations (and activities) of trace elements generally are not conducive to the formation of phases for which they are major constituents, and they are accommodated as minor constituents in mineral solid solutions, melts, and other fluid phases.

At very low concentrations of a trace solute constituent, the thermodynamic behavior of the solution may be expected to approach that of an ideal dilute solution and obey Henry's Law. The concentration range over which Henry's Law is obeyed by a given trace constituent cannot as yet be predicted; it must be determined by experiments.

Consider a trace element *i* partitioned between two phases, α and β, coexisting at equilibrium. If the distribution obeys Henry's Law (i.e., $a_i=h_i X_i$), then

$$\mu_i^\alpha=(\mu_i^\alpha)^0+RT\ln a_i^\alpha=(\mu_i^\alpha)^0+RT\ln(h_i^\alpha X_i^\alpha) \qquad (5.68)$$

$$\mu_i^\beta=(\mu_i^\beta)^0+RT\ln a_i^\beta=(\mu_i^\beta)^0+RT\ln(h_i^\beta X_i^\beta) \qquad (5.69)$$

where h_i^α, h_i^β are the Henry's Law constants, and X_i^α and X_i^β are the mole fractions.

Since $\mu_i^\alpha=\mu_i^\beta$ at equilibrium (equation 5.15),

$$\exp\left[\frac{(\mu_i^\alpha)^0-(\mu_i^\beta)^0}{RT}\right]=\left(\frac{h_i^\beta X_i^\beta}{h_i^\alpha X_i^\alpha}\right) \qquad (5.70)$$

If the standard state is chosen as the pure phase at *P* and *T* of interest, the left side of equation (5.70) becomes independent of composition and has a constant value at a given *P–T* condition. Designating this constant as k^*, equation (5.70) can be rearranged to define what is known as the *Nernst distribution coefficient* (K_D):

$$K_{D(i)}^{\alpha-\beta}=\left(\frac{h_i^\beta}{h_i^\alpha}\right)\frac{1}{k*}=\left(\frac{X_i^\alpha}{X_i^\beta}\right)=\left(\frac{C_i^\alpha}{C_i^\beta}\right) \qquad (5.67)$$

where C_i^α and C_i^β represent concentrations (e.g., in wt% or ppm) of *i* in the two phases. $K_{D(i)}^{\alpha-\beta} > 1$ indicates that the element would be enriched in phase α, where as $K_{D(i)}^{\alpha-\beta} < 1$ indicates that the element would be enriched in phase β. For example, the distribution coefficient for the partitioning of Ni between olivine (ol) and parent magma (melt),

Box 5.6 (cont'd)

$$K_{D(\mathrm{Ni})}^{\mathrm{ol-melt}} = \left(\frac{X_{\mathrm{Ni}}^{\mathrm{ol}}}{X_{\mathrm{Ni}}^{\mathrm{melt}}}\right) = \left(\frac{C_{\mathrm{Ni}}^{\mathrm{ol}}}{C_{\mathrm{Ni}}^{\mathrm{melt}}}\right)$$

is > 1. That is why Ni commonly is enriched in the olivine relative to the parent magma.

The Nernst distribution coefficient, which was originally formulated by W. Nernst to account for the distribution of trace elements between crystals and aqueous solutions, varies with temperature (see, e.g., Green and Pearson, 1985; Dunn and Sen, 1994), but its potential use for estimating equilibration temperatures in mineral assemblages is diminished by the fact that it also is a function of pressure, composition, and structure of the phases of interest, and oxygen fugacity. Nernst distribution coefficients, however, are very useful in modeling the progress of partial melting and magmatic crytallization processes (see section 12.3.2).

As the mole fraction of the solute in a binary solution tends to zero, the mole fraction of the solvent approaches unity and the behavior of the solvent approaches that predicted by Raoult's Law. Thus, for a solution in which the solutes obey Henry's Law, the solvent obeys Raoult's Law (Fig. 5.9). For a gaseous species obeying the ideal gas law, the form of Henry's Law that can be used to calculate its concentration in solution (C_i) is $C_i = K_H\ P_i$, where P_i is the partial pressure of the gaseous species, and K_H is the Henery's Law constant. Henry's Law constants for some gases at 1 bar total pressure are listed in Table 5.2.

It may be a little confusing that a solution that obeys Henry's Law is also called an ideal solution, but the ideality in this case is clearly different from that of a solution obeying Raoult's Law. It is, therefore, necessary to specify if the ideality in a given case conforms to Raoult's Law or to Henry's Law.

5.5.4 *The Lewis Fugacity Rule*

It can be shown that for ideal solutions (solutions for which $\Delta V_{\mathrm{mixing}} = 0$ and $\Delta H_{\mathrm{mixing}} = 0$) a form of Raoult's Law, called the *Lewis Fugacity Rule*, is obtained by replacing pressures with fugacities in equation (5.65):

$$f_i = f_i^{\mathrm{pure}}\ X_i^{\mathrm{solution}} \tag{5.71}$$

where f_i^{pure} is the fugacity of pure i and f_i the fugacity of i in the solution at pressure and temperature of interest. The rule is generally used for estimating fugacities in gas or supercritical solutions, but it is also applicable to liquid and solid solutions. Thus, the example of the *A–B* solution shown in Fig. 5.9 can also be represented in terms of fugacities (Fig. 5.10). The rule is very useful because it allows us to calculate the fugacity of a constituent in a multiconstituent, ideal solution if we know the fugacity of the pure constituent. The limitation of this rule is that it applies only to ideal solutions, that is, solutions in which there are no molecular interactions. Ignoring such interactions, except in dilute solutions, can lead to significant errors in calculated fugacities.

Table 5.2 Henry's Law constants (K_H) for selected gases (in mol L^{-1} bar^{-1}) at 1 bar total pressure.

T (°C)	O_2 ($\times10^{-3}$)	N_2 ($\times10^{-4}$)	CO_2 ($\times10^{-2}$)	H_2S ($\times10^{-2}$)	SO_2
0	2.18	10.5	7.64	20.8	3.56
5	1.91	9.31	6.35	17.7	3.0115
10	1.70	8.30	5.33	15.2	2.53
15	1.52	7.52	4.55	13.1	2.11
20	1.38	6.89	3.92	11.5	1.76
25	1.26	6.40	3.39	10.2	1.46
30	1.16	5.99	2.97	9.09	1.21
35	1.09	5.60	2.64	8.17	1.00
40	1.03	5.28	2.36	7.41	0.837
50	0.932	4.85	1.95	6.21	–

Source of data: Pagenkopf (1978).

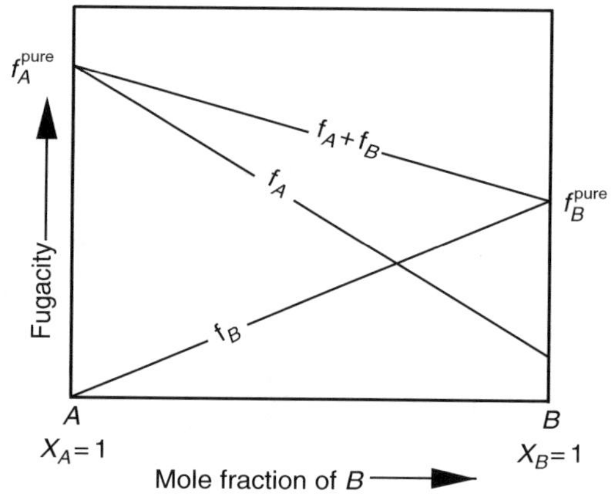

Fig. 5.10 Schematic representation of the variation of fugacity of the constituents of a binary solution, which obeys the Lewis Fugacity Rule, as a function of composition. f_A^{pure} and f_B^{pure} are fugacities of pure *A* and pure *B*, respectively.

5.5.5 *Activities of constituents in ideal solutions*

As f_i^0 in equation (5.52) represents the fugacity of pure i at the temperature of interest (T) and behaving ideally, $f_i^0 = f_i^{\mathrm{pure}}$. Combining equations (5.52) and (5.71), we get

$$a_i^{\mathrm{ideal,\ Raoult's}} = \frac{f_i}{f_i^0} = \frac{f_i}{f_i^{\mathrm{pure}}} = X_i \tag{5.72}$$

Thus in ideal solutions, the activity of a constituent is equal to its mole fraction. Substituting for a_i in equation (5.29), we get the equation for ideal solutions based on Raoult's Law:

$$\mu_i^{\mathrm{ideal,\ Raoult's}}(P, T) = \mu_i^0(P, T) + RT \ln a_i = \mu_i^0(P, T) + RT \ln X_i \tag{5.73}$$

The corresponding equation for ideal solutions based on Henry's Law is

$$\begin{aligned}\mu_i^{\text{ideal, Henry's}}(P,T) &= \mu_i^0(P,T) + RT\ln a_i \\ &= \mu_i^0(P,T) + RT\ln h_i X_i \\ &= \mu_i^0(P,T) + RT\ln h_i + RT\ln X_i\end{aligned} \tag{5.74}$$

Collecting the two terms that are independent of composition and denoting

$$\mu_i^* = \mu_i^0 + RT\ln h_i \tag{5.75}$$

equation (5.74) is often written as

$$\mu_i^{\text{ideal, Henry's}}(P,T) = \mu_i^*(P,T) + RT\ln X_i \tag{5.76}$$

Comparison of equation (5.74) with equation (5.29) shows that:

$$a_i^{\text{ideal, Henry's}} = X_i h_i \tag{5.77}$$

if the standard state is defined as μ_i^* (equation 5.75).

5.6 Nonideal solutions involving condensed phases

For the activity of the *i*th constituent in a homogeneous (i.e., single-phase) nonideal solution ($\Delta V_{\text{nonideal mixing}} \neq 0$ and $\Delta H_{\text{nonideal mixing}} \neq 0$), the deviation from ideality is taken into account by introducing a function called the *rational activity coefficient*, λ_i, which embodies interaction effects between the constituent *i* and all the other constituents in the solution, and is defined by the equation

$$a_i^{\text{nonideal}} = a_i^{\text{ideal}}\,\lambda_i = \lambda_i X_i \tag{5.78}$$

such that as $\lambda_i \rightarrow 1$ the solution approaches ideal behavior. In other words, for an ideal solution, $\lambda_i = 1$ and $a_i = X_i$. Note that for the same value of X_i in a solution, λ_i will have different values depending on whether the nonideality is relative to Raoult's Law or Henry's Law. In either case, λ_i is a dimensionless number and a function of temperature, pressure, and composition of the solution. The value of λ_i in a given solution is determined experimentally or theoretically.

As discussed by Powell (1978) and Nordstrom and Munoz (1994), the behavior of a constituent *i* over the entire compositional range of a solution (from $X_i = 1$ to $X_i = 0$) may be broken into three regions in a plot of μ_i versus $\ln X_i$ (Fig. 5.11): (1) Raoult's Law region of ideality at one end; (2) Henry's Law region of ideality at the other end; and (3) an intermediate region of nonideality where the solution obeys neither Raoult's Law nor Henry's Law.

The Raoult's Law region extends from pure end-member *i* (i.e., $X_i = 1$, $\ln X_i = 0$) to some composition where *i* is diluted somewhat but not enough to be significantly affected by other constituents in the solution. In this region, where *i* may be thought of as a solvent, μ_i is given by equation (5.73), which plots as a straight line (a–b in Fig. 5.11), with a slope equal to *RT* and a *y* axis intercept of μ_i^0(at $X_i = 1$), the standard chemical potential

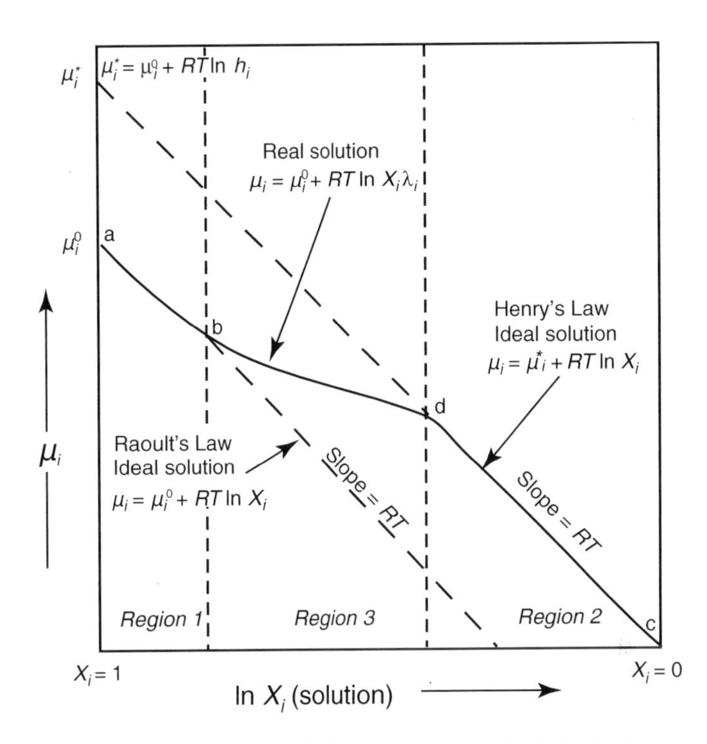

Fig. 5.11 Separation of three distinct regions of solution behavior in a schematic plot of chemical potential of a constituent *i* (μ_i) as a function of its mole fraction (X_i) in the solution. μ_i^0 is the standard chemical potential of *i*, the standard state being pure *i* at the *P* and *T* of interest. See text for details. (From *Geochemical Thermodynamics*, 2nd edition, by D.K. Nordstrom and J.L. Munoz, Fig. 5.10, p. 139; Copyright 1994, Blackwell Scientific Publications. Used with permission of the publisher.)

of *i* when the standard state is chosen as pure *i* at the pressure and temperature of interest. In this region, $a_i = X_i$ (equation 5.72).

The Henry's Law region extends from infinite dilution (in practice, very small values of X_i), in which case the constituent *i* is so dispersed that it can be regarded as being surrounded by a uniform environment, to some composition (which will vary from system to system) in which the dilution is not so excessive. The Henry's Law region is of particular interest because concentrations of dissolved neutral species in H_2O at Earth's surface conditions typically fall in this region. In this region, where *i* may be thought of as a solute, μ_i is given by equation (5.76) that plots as a straight line (c–d in Fig. 5.11), with a slope equal to *RT* and a *y* axis intercept of μ_i^* at $X_i = 1$. In this region, a_i is a function not only of X_i, but also of h_i, and $a_i = h_i X_i$ (equation 5.77). μ_i^* may be looked upon as the chemical potential of *i* for a hypothetical standard state in which the environment of the molecules of *i* is assumed to be the same as in an infinitely dilute solution.

In the intermediate region, the region of nonideality, μ_i is represented by a curved line (b–d in Fig. 5.11) and is given by

$$\begin{aligned}\mu_i^{\text{nonideal}}(P,T) &= \mu_i^0(P,T) + RT\ln a_i \\ &= \mu_i^0(P,T) + RT\ln(\lambda_i X_i) \\ &= \mu_i^0(P,T) + RT\ln X_i + RT\ln\lambda_i\end{aligned} \tag{5.79}$$

Note that for Raoult's Law ideality, $\lambda_i=1$, so that $a_i=X_i$; for Henry's Law ideality, $\lambda_i=h_i$, so that $a_i=h_iX_i$.

5.7 Excess functions

An expression for the molar Gibbs free energy of a nonideal solution can be obtained simply by replacing X_i in equation (5.63) with $\lambda_i X_i$:

$$\bar{G}^{\text{nonideal solution}}(P,T)=\sum_i X_i \mu_i^0(P,T)+RT\sum_i X_i \ln(\lambda_i X_i)$$
$$=\sum_i X_i \mu_i^0(P,T) + RT\sum_i X_i \ln X_i + RT\sum_i X_i \ln \lambda_i \quad (5.80)$$

The difference between the molar free energy of a nonideal solution and that of its conceptual ideal equivalent is called the *excess free energy* function ($\bar{G}^{\text{EX}}$), and it is given by

$$\bar{G}^{\text{EX}} = \bar{G}^{\text{nonideal}} - \bar{G}^{\text{ideal}} = RT\sum_i X_i \ln \lambda_i \quad (5.81)$$

As $\bar{G}^{\text{EX}}$ incorporates the mole fraction and activity coefficient of each constituent in the solution, it adequately describes the nonideality of a multiconstituent solution. Similar expressions can be derived for other excess thermodynamic functions, such as $\bar{H}^{\text{EX}}$, $\bar{S}^{\text{EX}}$, and $\bar{V}^{\text{EX}}$. Molar excess properties for petrologically important solutions (both solid and liquid) are commonly represented by Margules-type equations (see section 5.9), which have evolved over the past century from a study first published in 1885 by Margules and from subsequent works of many authors such as, to mention a few, Thompson (1967), Grover (1977), Ganguly and Saxena (1984), Fei *et al.* (1986), Berman (1990), Mukhopadhyay *et al.* (1993), and Cheng and Ganguly (1994).

5.8 Ideal crystalline solutions

A crystalline solid differs from a liquid or a gas in that a crystal has a distinctive atomic structure, a three-dimensional lattice produced by the regular arrangement of atoms or ions, each atom or ion being in a state of thermal motion about an average position in the lattice. The mixing of constituents in the formation of a solid solution, unlike the mixing in the case of a liquid or gas solution, involves substitution of atoms or ions on one kind of atomic site or on several energetically nonequivalent atomic sites in the crystal structure, with the crystal structure usually remaining essentially unchanged. Such a substitutional solid solution model introduces additional considerations in the evaluation of thermodynamic properties of solid solutions, including activities of the constituents. In this section we will emphasize the calculation of activities of constituents of some common rock-forming silicate minerals. The necessity for this exercise will be evident in Chapter 6.

Ideal solid solutions, like all ideal solutions, obey Raoult's Law, Henry's Law, and Lewis Fugacity Rule (assuming that the vapor pressures of the constituents can be measured), and are characterized by no change in volume or enthalpy due to mixing (i.e., $\Delta V_{\text{ideal mixing}}=0$ and $\Delta H_{\text{ideal mixing}}=0$). It is known from experience that binary substitutional oxide and silicate solid solutions tend to behave ideally if there is no discontinuity in the crystallography in proceeding from one end-member to the other and the molar volumes of the two end-members are nearly the same – no more than 5% difference according to Kerrick and Darken (1975). Most solid solutions, however, do not strictly conform to the criteria for ideal mixing over their entire compositional range, but the assumption of ideality is a useful approximation in many petrologic problems.

The thermodynamic treatment of a solid solution requires knowledge of the distribution pattern of atoms or ions in the crystal structure obtained by spectroscopic and X-ray diffraction methods and from our understanding of crystal chemistry. In crystal structures of silicate minerals, cations occupy atomic sites of specific coordination number (see section 3.12), the most common of which are tetrahedral (fourfold coordination), octahedral (sixfold coordination), and cubic (eightfold coordination). Mixing of ions to form a solid solution may be conceptualized to occur with two very different sets of constraints (Helgeson *et al.*, 1978; Nordstrom sand Munoz, 1994). One is the *local charge balance (LCB) model* in which charge balance is maintained over short distances in the lattice by coupled substitution. An example of the LCB model is simultaneous substitution of a divalent cation (Mg^{2+} or Fe^{2+}) by a trivalent cation (Fe^{3+} or Al^{3+}) for every substitution of a divalent ion by a monovalent ion (e.g., Ca^{2+} by Na^{+}) in an adjacent site. The other is the more commonly used *mixing-on-sites (MOS) model,* which assumes that all ions occupying a given crystallographic site mix independently and randomly on that site, i.e., there is no ordering of cations on that site and no charge balance constraint. The MOS model of mixing may produce local charge imbalances, but is assumed to satisfy the average stoichiometry of the solid solution.

5.8.1 *Application of the mixing-on-sites model to some silicate minerals*

The calculation of MOS-model activities is based on the entropy change associated with random mixing of cations on the same atom site. From statistical thermodynamics it can be shown that the ΔS for ideal mixing associated with the formation of 1 mole of a solid solution (ss) according to the MOS model is:

$$\Delta\bar{S}_{\text{ideal mixing}} = -R\sum_j\left(n_j \sum_i X_{i,j}^{\text{ss}} \ln X_{i,j}^{\text{ss}}\right) \quad (5.82)$$

where i refers to ions (or atoms) in the solid solution, j refers to the nonequivalent mixing sites in its structural formula, n_j is site multiplicity for the jth site (i.e., the number of positions in the jth site per formula unit on which mixing takes place) and $X_{i,j}$ is the mole fraction (or the fractional occupancy) of the ith ion (or atom) on the jth site.

It can be shown (see Nordstrom and Munoz, 1994, pp. 426–428) that, taking the standard state as the pure substance at the pressure and temperature of interest, the activity of an individual cation i in a particular atomic site j in the crystal structure can be calculated as:

$$a_{i,j}^{\text{ideal ss}} = \left(X_{i,j}^{\text{ss}}\right)^{n_j} \tag{5.83}$$

The activity of a component c in a solid solution is the product of the activity of the individual cations in each mixing site in the solid solution:

$$a_c^{\text{ideal ss}} = X_c^{\text{ss}} = \prod a_{i,j}^{\text{ss}} = \prod \left(X_{i,j}^{\text{ss}}\right)^{n_j} \tag{5.84}$$

The chemical potential of the component c, by analogy with equation (5.73), is

$$\begin{aligned}\mu_c^{\text{ideal ss}}(P,T) &= \mu_c^0(P,T) + RT \ln a_c \\ &= \mu_c^0(P,T) + RT \ln X_c^{\text{ss}}\end{aligned} \tag{5.85}$$

and the molar free energy of an ideal solid solution, by analogy with equation (5.67), is

$$\bar{G}^{\text{ideal ss}}(P,T) = \sum_c X_c^{\text{ss}}\, \mu_c^0\,(P,T) + RT \sum_j \left(n_j \sum_i X_{i,j}^{\text{ss}} \ln X_{i,j}^{\text{ss}}\right) \tag{5.86}$$

It follows from equation (5.84) that the activity of a pure phase (i.e., comprised of one phase component only of fixed composition), such as $Mg_3Al_2Si_3O_{12}$ (pyrope garnet), is unity, whereas the activity of the same phase component in an ideal garnet solid solution, such as $(Mg, Ca)_3^{\text{cubic}}(Al, Fe)_2^{\text{oct}}Si_3^{\text{tet}}O_{12}$, is calculated as:

$$\begin{aligned}a_{Mg_3Al_2Si_3O_{12}}^{\text{ss}} &= X_{Mg_3Al_2Si_3O_{12}}^{\text{ss}} = a_{Mg^{2+}\,(\text{cubic})}\; a_{Al^{3+}\,(\text{oct})} \\ &= a_{Si^{4+}\,(\text{tet})}\; X^3_{Mg^{2+}\,(\text{cubic})}\; X^2_{Al^{3+}\,(\text{oct})}\; X^3_{Si^{4+}\,(\text{tet})}\end{aligned} \tag{5.87}$$

where $X_{Mg^{2+}} = Mg^{2+}/(Mg^{2+} + Ca^{2+})$, $X_{Al^{3+}} = Al^{3+}/(Al^{3+} + Fe^{3+})$, and $X_{Si^{4+}\,(\text{tet})} = 1$. The calculation of $X_{Mg_3Al_2Si_3O_{12}}^{\text{ss}}$ as a product of the mole fractions of cations at the various sites follows the basic principle of conditional probability. If the probabilities of finding a Mg^{2+} on the cubic site, a Al^{3+} on the octahedral site, and a Si^{4+} on the tetrahedral site are p_1, p_2, and p_3, respectively, then the combined probability of finding all of them in the structure is $p_1\, p_2\, p_3$. The chemical potential of the component $Mg_3Al_2Si_3O_{12}$ with P and T as the standard state is:

$$\begin{aligned}\mu_{Mg_3Al_2Si_3O_{12}}^{\text{ideal gt ss}}(P,T) &= \mu_{Mg_3Al_2Si_3O_{12}}^0 + RT \ln a_{Mg_3Al_2Si_3O_{12}} \\ &= \mu_{Mg_3Al_2Si_3O_{12}}^0 + RT \ln X_{Mg_3Al_2Si_3O_{12}}^{\text{gt}} \\ &= \mu_{Mg_3Al_2Si_3O_{12}}^0 + RT \ln\left[X^3_{Mg^{2+}(\text{cubic})}\, X^2_{Al^{3+}\,(\text{oct})}\right] \\ &= \mu_{Mg_3Al_2Si_3O_{12}}^0 + 3RT \ln X_{Mg^{2+}(\text{cubic})} \\ &\quad + 2RT \ln X_{Al^{3+}\,(\text{oct})}\end{aligned} \tag{5.88}$$

Note that Si makes no contribution to $\mu_{Mg_3Al_2Si_3O_{12}}$ because $X_{Si^{4+}(\text{tet})} = 1$.

Now let us apply equations (5.83) and (5.84) to calculate activities in some common rock-forming silicate minerals, assuming them to be ideal solid solutions.

Single-site mixing

In the silicate mineral olivine (ol), $Z_2^{2+}\, Si^{4+}\, O_4$, divalent cations reside on two nonequivalent octahedral sites, $M1$ and $M2$, but for Mg-Fe olivines we can consider them as equivalent sites because cation ordering between Mg^{2+} and Fe^{2+} is weak (Motoyama and Matsumoto, 1989). Si^{4+} ions occupy the tetrahedral positions in the crystal lattice. The activities Mg^{2+} and Fe^{2+} in $(Mg, Fe)_2SiO_4$ olivine, as per equation (5.83), are:

$$a_{Mg^{2+}}^{\text{ol}} = X^2_{Mg^{2+}(\text{oct})} \quad \text{and} \quad a_{Fe^{2+}}^{\text{ol}} = X^2_{Fe^{2+}(\text{oct})} \tag{5.89}$$

and from equation (5.84) activities of the two end-member components of the olivine solid solution, Mg_2SiO_4 (forsterite) and Fe_2SiO_4 (fayalite), are calculated as:

$$a_{Mg_2SiO_4\,(\text{fo})}^{\text{ol}} = X_{\text{fo}}^{\text{ol}} = X^2_{Mg^{2+}(\text{oct})} \quad \text{and} \quad a_{Fe_2SiO_4\,(\text{fa})}^{\text{ol}} = X_{\text{fa}}^{\text{ol}} = X^2_{Fe^{2+}(\text{oct})} \tag{5.90}$$

There is no contribution to the activities from the tetrahedral site, because we have assumed no mixing on this site (i.e., $X_{Si^{4+}\,(\text{tet})} = 1$). For a hypothetical Mg-Fe olivine of composition $Mg_{1.6}Fe_{0.4}SiO_4$, for example, the mole fractions of Mg^{2+} and Fe^{2+} on the octahedral site are:

$$\begin{aligned}X^{\text{ol}}_{Mg^{2+}(\text{oct})} &= 1.6/(1.6+0.4) = 1.6/2 = 0.8;\; X^{\text{ol}}_{Fe^{2+}(\text{oct})} \\ &= 0.4/(1.6+0.4) = 0.4/2 = 0.2\end{aligned}$$

Note that the mole fraction of a cation on a particular site is calculated relative to the total number of cations on that site. Now we can calculate the MOS-model activities of the cations and components as:

$$\begin{aligned}a_{Mg}^{\text{ol}} &= X^2_{Mg^{2+}(\text{oct})} = (0.8)^2 = 0.64;\; a_{Fe}^{\text{ol}} = X^2_{Fe^{2+}(\text{oct})} = (0.2)^2 = 0.04 \\ a_{fo}^{\text{ol}} &= X^2_{Mg^{2+}(\text{oct})} = (0.8)^2 = 0.64;\; a_{fa}^{\text{ol}} = X^2_{Fe^{2+}(\text{oct})} = (0.2)^2 = 0.04\end{aligned} \tag{5.91}$$

Multisite mixing

When mixing occurs randomly on more than one crystallographically equivalent site, the calculation of activity requires a separate summation for each site. For example, let us consider the silicate mineral garnet (gt), $Z_3Y_2Si_3O_{12}$, which has three different kinds of atomic sites that are occupied by different ions: eight-coordinated cubic sites (Z) by Mg^{2+}, Fe^{2+}, Ca^{2+}, and Mn^{2+}; six-coordinated octahedral sites (Y) mostly by Fe^{3+}, Al^{3+}, Ti^{3+}, and Cr^{3+}; and four-coordinated tetrahedral sites by Si^{4+} (and in some cases also by a very small proportion of Al^{3+}). So, we can write the general formula of garnet as $(Mg, Fe^{2+}, Ca, Mn)_3{}^{cubic}(Al, Fe^{3+}, Ti, Cr)_2{}^{oct}(Si, Al)_3{}^{tet}O_{12}$. Note that for the cubic site ($j=1$), $n_j=3$; for the octahedral site ($j=2$), $n_j=2$; and for the tetrahedral site ($j=3$), $n_j=3$. Ignoring Mn and Cr for simplicity, a garnet solid solution may be described by four end-member phase components: $Mg_3Al_2Si_3O_{12}$, pyrope (py); $Fe_3{}^{2+}Al_2Si_3O_{12}$, almandine (alm); $Ca_3Al_2Si_3O_{12}$, grossular (gr); and $Ca_3Fe_2{}^{3+}Si_3O_{12}$, andradite (and).

For the purpose of illustration, let us consider an ideal garnet solid solution of composition $(Mg_{0.352}Fe^{2+}_{2.413}Ca_{0.235})(Al_{1.926}Fe^{3+}_{0.074})(Si_{2.985}Al_{0.015})O_{12}$. The MOS-model activities of the various cations in this garnet solid solution, using equation (5.83), are:

$$
\begin{aligned}
a^{gt}_{Mg^{2+}} &= X^3_{Mg^{2+}(cubic)} = (0.352/(0.352+2.413+0.235))^3 \\
&= (0.352/3)^3 = 1.615\times10^{-3} \\
a^{gt}_{Fe^{2+}} &= X^3_{Fe^{2+}(cubic)} = (2.413/(0.352+2.413+0.235))^3 \\
&= (2.413/3)^3 = 0.520 \\
a^{gt}_{Ca^{2+}} &= X^3_{Ca^{2+}(cubic)} = (0.235/(0.352+2.413+0.235))^3 \\
&= 4.807\times10^{-4} \\
a^{gt}_{Fe^{3+}} &= X^2_{Fe^{3+}(oct)} = (0.074/(1.926+0.074))^2 = (0.074/2)^2 \\
&= 1.369\times10^{-3} \\
a^{gt}_{Al^{3+}} &= X^2_{Al^{3+}(oct)} = (1.926/(1.926+0.074))^2 = (1.926/2)^2 \\
&= 0.927 \\
a^{gt}_{Si^{4+}} &= X^2_{Si^{4+}(tet)} = (2.985/(2.985+0.015))^2 = (2.985/3)^3 \\
&= 0.985
\end{aligned}
\tag{5.92}
$$

The activities of the four end-member components, using equation (5.84), are:

$$
\begin{aligned}
a^{gt}_{py} &= X^{gt}_{py} = X^3_{Mg^{2+}(cubic)}X^2_{Al^{3+}(oct)}X^3_{Si^{4+}(tet)} \\
&= (1.615\times10^{-3})(0.927)(0.985) = 1.475\times10^{-3} \\
a^{gt}_{alm} &= X^{gt}_{alm} = X^3_{Fe^{2+}(cubic)}X^2_{Al^{3+}(oct)}X^3_{Si^{4+}(tet)} \\
&= (0.520)(0.927)(0.985) = 0.475 \\
a^{gt}_{gr} &= X^{gt}_{gr} = X^3_{Ca^{2+}(cubic)}X^2_{Al^{3+}(oct)}X^3_{Si^{4+}(tet)} \\
&= (4.807\times10^{-4})(0.927)(0.985) = 4.389\times10^{-4} \\
a^{gt}_{and} &= X^{gt}_{and} = X^3_{Ca^{2+}(cubic)}X^2_{Fe^{3+}(oct)}X^3_{Si^{4+}(tet)} \\
&= (4.807\times10^{-4})(1.369\times10^{-3})(0.985) = 6.482\times10^{-7}
\end{aligned}
\tag{5.93}
$$

The procedure described above is applicable to any solid solution in which there is mixing on more than one type of site and in which there is complete disorder of ions on each site. A slightly different procedure is required for calculation of the activity of a phase component, such as the phlogopite component in biotite, which contains two ions occupying the same site in fixed stoichiometric proportions (Nordstrom and Munoz, 1994).

5.8.2 *Application of the local charge balance model to some silicate minerals*

For some solid solutions, such as clinopyroxene and plagioclase, the local charge balance (LCB) model may be more appropriate, because complete cation disordering assumed in the MOS model results in unacceptable local charge imbalances in the crystal structure. Coupled substitution necessary to maintain local charge balance in the structure increases cation ordering (resulting in a decrease in the molar entropy of mixing than predicted by equation 5.78) and thus alters the activity–composition relations of constituents in the solid solution. We shall illustrate the LCB model with a few of simple examples.

Clinopyroxene

The generalized structural formula of the silicate mineral clinopyroxene (cpx) is $(Ca, Mg, Fe^{2+}, Na)_{M2}(Mg, Fe^{2+}, Al, Cr, Fe^{3+})_{M1}(Si, Al)_2O_6$. The crystal structure contains three energetically nonequivalent sites: a cubic site designated as *M*2 ($j=1$, $n_j=1$) that is occupied mostly by relatively large cations such as Ca^{2+} and Na^+; an octahedral site designated as *M*1 ($j=2$, $n_j=1$) that is occupied mostly by relatively small cations such as Mg^{2+} and Al^{3+}; and a tetrahedral site ($j=3$, $n_j=2$) that is occupied by Si^{4+} and Al^{3+}. To calculate the mole fractions of components in a cpx solid solution, we have to know the proportional distribution of Fe and Mg between the *M*1 and *M*2 sites. This information can be obtained from spectroscopic and X-ray diffraction analysis of the phase but is often not available, and we have to be content with simplifying assumptions.

For the present purpose, let us consider a simpler clinopyroxene composition, $(Ca^{2+}, Na^+)_{M2}(Al^{3+}, Mg^{2+})_{M1}(Si, Al)_{2(tet)}O_6$, whose end-member components are $CaMgSi_2O_6$ (diopside) and $NaAlSi_2O_6$ (jadeite). If there is complete disorder on *M*1 and *M*2 sites, the MOS-model activities of the components are:

$$
\begin{aligned}
a^{cpx}_{CaMgSi_2O_6(diop)} &= X^{cpx}_{diop} = X_{Ca\,(M2)}\,X_{Mg\,(M1)}\,X^2_{Si\,(tet)} \\
a^{cpx}_{NaAlSi_2O_6(jd)} &= X^{cpx}_{jd} = X_{Na\,(M2)}\,X_{Al\,(M1)}\,X^2_{Si\,(tet)}
\end{aligned}
\tag{5.94}
$$

where the mole fractions are calculated as

$$X_{Ca\,(M2)} = [Ca/(Ca+Na)]_{M2};\; X_{Al\,(M1)} = [Al/(Al+Mg)]_{M1}, \text{etc.}$$

To calculate LCB-model activities for the same clinopyroxene, let us suppose that the Al-Mg mixing on *M*1 is completely random, but for every Al^{3+} on *M*1 there is a Na^+ in one of the three nearest-neighbor *M*2 positions to preserve local charge balance. In the case of complete ordering between Al^{3+} on *M*1 and Na^+ on *M*2, the positions of Na^+ ions are fixed by the positions of Al^{3+} ions, so that the entropy of mixing in the solid solution arises entirely due to disorder on *M*1, with no contribution from *M*2 positions. The activity–composition relations for such a clinopyroxene are analogous to those of the one-site mixing in Mg-Fe olivine we discussed earlier: (Ganguly, 1973):

$$a^{cpx}_{CaMgSi_2O_6} = X_{Mg\,(M1)} \quad \text{and} \quad a^{cpx}_{NaAlSi_2O_6} = X_{Al\,(M1)} \qquad (5.95)$$

Plagioclase

Calculation of activities in a plagioclase solid solution is controversial because of its complex crystal structure. The general structural formula of plagioclase is $Y(Si, Al)_4O_8$, where the Y positions are occupied by (Na^+, Ca^{2+}), and (Si^{4+}, Al^{3+}) fill the four tetrahedral positions per formula unit. Most of the iron reported in feldspar analyses is considered to be Fe^{3+} replacing Al^{3+} in the structure. The small amount of Fe^{2+} reported in some analyses is either replacing Ca^{2+} in the structure or is an impurity. The solid solution may be viewed as a mixture of the two end-member components, $NaAlSi_3O_8$ (albite) and $CaAl_2Si_2O_8$ (anorthite), the local charge balance being maintained by the coupled substitution $Na^++Si^{4+}=Ca^{2+}+Al^{3+}$. It is well known that throughout most of its stability range, anorthite has a fully ordered structure in which the equal numbers of Si and Al atoms alternate with strict regularity. Albite, on the other hand, displays Al–Si order–disorder, so that the ordering pattern of Al_2Si_2 in stoichiometric anorthite cannot be derived by simple modification of the Al_1Si_3 pattern in stoichiometric albite but requires complete reconstruction. This, combined with the coupled substitution, the constraint imposed by the "aluminum avoidance principle" (the exclusion of Al–O–Al bonds that are believed to be energetically unfavorable; Saxena and Ribbe, 1973), and the varying distortion of the tetrahedral (Si, Al)–O framework to suit chemical composition and temperature render the crystal structure of plagioclase quite complex.

Kerrick and Darken (1975) discussed four possible models with different restrictions for calculation of the activities of albite and anorthite components of plagioclase solid solutions, none of which would apply to the entire range of plagioclase compositions. One of these models (model number 4 of Kerrick and Darken, 1975) assumes complete local charge balance and complete Al–Si ordering in the tetrahedral positions adjacent to each (Na, Ca) site in both albite and anorthite domains without adherence to the aluminum avoidance principle. Under these restrictions, mixing of Al and Si on the tetrahedral sites was not considered to contribute to the entropy of mixing and the activities of the albite and anorthite components were shown to be equal, respectively, to the mole fractions of Na and Ca:

$$\begin{aligned} a^{plag}_{NaAlSi_3O_8\,(ab)} &= X^{plag}_{ab} = X^{plag}_{Na} \\ a^{plag}_{CaAl_2Si_2O_8\,(an)} &= X^{plag}_{an} = X^{plag}_{Ca} = 1 - X_{ab} \end{aligned} \qquad (5.96)$$

In another model that is constrained by the aluminum avoidance principle (model number 2 of Kerrick and Darken, 1975), the activities were calculated as:

$$\begin{aligned} a^{plag}_{NaAlSi_3O_8\,(ab)} &= (X^{plag}_{ab})^2(2 - X^{plag}_{ab}) \\ a^{plag}_{CaAl_2Si_2O_8\,(an)} &= (1/4)X^{plag}_{an}(1 + X^{plag}_{an})^2 \end{aligned} \qquad (5.97)$$

Activity–composition relations in the ternary $CaAl_2Si_2O_8$ – $NaAlSi_3O_8$ – $KAlSi_3O_8$ system are more complex, as discussed, for example, by Ghiorso (1984), Fuhrman and Lindsley (1988), and Elkins and Grove (1990).

5.9 Nonideal crystalline solutions

Ideal solution relationship in minerals is the exception rather than the rule. Most solid solutions deviate to some extent from ideality (i.e., ΔH_{mixing} or ΔV_{mixing} is not exactly zero), as evidenced by partitioning of elements inconsistent with ideal behavior and immiscibility at lower temperatures. For many solid solutions the nonideality is large enough to make a significant difference in the activity–composition relationships.

5.9.1 General expressions

For a nonideal solid solution, $X_{i,j}$ in equations (5.83), (5.84), and (5.86) is replaced by ($X_{i,j}\,\lambda_{i,j}$), where $\lambda_{i,j}$ is the *rational activity coefficient*:

$$a^{\alpha\,(nonideal\ ss)}_{i,j} = \left(X^{\alpha}_{i,j}\,\lambda^{\alpha}_{i,j}\right)^{n_j} \qquad (5.98)$$

$$a^{ss\,(nonideal)}_{c} = X^{ss}_{c} = \prod\left(X^{ss}_{i,j}\,\lambda_{i,j}\right)^{n_j} = \prod a^{ss}_{i,j} \qquad (5.99)$$

$$\begin{aligned} \bar{G}^{ss\,(nonideal)}(P, T) &= \sum_c X^{ss}_c\,\mu^0_c\,(P, T) \\ &\quad + RT\,\sum_j\left(n_j\,\sum_i X^{ss}_{i,j}\,\ln\left(X^{ss}_{i,j}\,\lambda_{i,j}\right)\right) \end{aligned} \qquad (5.100)$$

The expression for $\bar{G}^{EX}$, by analogy with equation (5.81), is

$$\bar{G}^{EX} = \bar{G}^{nonideal\ ss} - \bar{G}^{ideal\ ss} = RT\sum_j\left(n_j\,\sum_i X^{ss}_{i,j}\,\ln\lambda_{i,j}\right) \qquad (5.101)$$

For example, in olivine, $(Mg, Fe)_2SiO_4$, $j=1$ and $n_j=2$, and equation (5.101) gives

$$\bar{G}^{EX} = 2RT\,(X^{ol}_{Mg}\ln\lambda_{Mg} + X^{ol}_{Fe}\ln\lambda_{Fe}) \qquad (5.102)$$

The evaluation of $\lambda_{i,j}$, however, is not an easy task, because it is a function of the composition of the solid solution as well as of pressure and temperature. A number of equations based on different mixing models, but all obeying the constraints of the Raoult's and Henry's ideality, have been proposed for the evaluation of activity coefficients. The set of equations most commonly used in geochemistry involves Margules-type formulation of *interaction parameters* (or *Margules parameters*) that reflect the energy of interactions among the molecules of the components that are being mixed to form the solid solution. The expression for the free energy interaction parameter (W^G), which is a function of temperature and pressure (but not of composition), is of the same form as that for Gibbs free energy at P and T:

$$W^G = W^H - TW^S + PW^V \tag{5.103}$$

where W^H, W^S, and W^V are the interaction parameters for enthalpy, entropy, and volume, respectively. As explained by Anderson and Crerar (1993), a free energy interaction parameter W^G_{12} (in kJ mol^{-1}), may be looked upon as the energy necessary to interchange 1 mole of component 1 with 1 mole of component 2 in a binary (or two-component) solution (i.e., $X_1+X_2=1$), without changing its composition. If W^G_{12} is positive, molecules of 1 and 2 repel each other, and this may lead to immiscibility. If W^G_{12} is negative, the molecules of components1 and 2 prefer to stick together, and this may lead to the formation of intermediate compounds. The interaction parameters are determined from experimental data and theoretical principles.

5.9.2 *Regular solution*

A mixing model commonly used for crystalline solutions in petrologic applications is a binary *regular solution*, which may be further classified into two types, *symmetric* (or *strictly regular*) and *asymmetric* (or *subregular*), depending respectively on whether one or two adjustable interaction parameters are needed to represent the excess thermodynamic functions along the binary join. If a binary solid solution of two end-member components 1 and 2 held at some pressure P and temperature T is modeled as an asymmetric regular solution, the excess molar free energy due to nonideal mixing, $\bar{G}^{EX}$, is (Thompson, 1967; Mukhopadhaya *et al.*, 1993):

$$\bar{G}^{EX} = X_1X_2\,(X_2W^G_{12} + X_1W^G_{21}) \tag{5.104}$$

where $W_{12}{}^G$ and $W_{21}{}^G$ are the two adjustable free-energy interaction parameters (in kJ mol^{-1}) that depend on temperature and pressure but not on composition. The corresponding equations for the activity coefficients are:

$$\begin{aligned} RT\ln\lambda_1 &= 2X_1X_2W^G_{21} + X_2{}^2W^G_{12} - 2\bar{G}^{EX} \\ &= (X_2)^2\,[W^G_{12} + 2X_1(W^G_{21} - W^G_{12})] \\ RT\ln\lambda_2 &= 2X_1X_2W^G_{12} + X_1{}^2W^G_{21} - 2\bar{G}^{EX} \\ &= (X_1)^2\,[W^G_{21} + 2X_2(W^G_{12} - W^G_{21})] \end{aligned} \tag{5.105}$$

If the solid solution is modeled as a symmetric regular solution, $\bar{G}^{EX}$ (and all other nonideal thermodynamic functions) are symmetrical about the midpoint composition (i.e., $X_1=X_2=0.5$) and $W_{ij}{}^G = W_{ji}{}^G$. Equation (5.104) then reduces to a one-parameter Margules equation:

$$\bar{G}^{EX\,(ss)} = X_1X_2W^G_{12} = W^G_{12}X_2(1-X_1) \tag{5.106}$$

The corresponding equations for activity coefficients are:

$$\begin{aligned} RT\ln\lambda_1 &= X_2W^G_{12} - \bar{G}^{EX} = (X_2)^2W^G_{12} \\ RT\ln\lambda_2 &= X_1W^G_{12} - \bar{G}^{EX} = (X_1)^2W^G_{12} \end{aligned} \tag{5.107}$$

Equations for regular solutions comprised of more than two end-member components are of the same form as described above, but involve a greater number of interaction parameters. Many different formulations for $\bar{G}^{EX}$ and $RT\ln\lambda_i$ have been discussed in the literature; expressions for binary, ternary, and quaternary symmetric and asymmetric regular solutions, as formulated by Mukhopadhaya *et al.* (1993), are summarized in Appendix 7.

We will revisit the issue of activity–composition relationship in nonideal solid solutions in the next chapter.

Example 5–6: Calculation of activities in a nonideal olivine solid solution of composition $(Mg_{1.6}Fe_{0.4})SiO_4$ at 1400K and 1 bar

Considering olivine as a binary solution of phase components Mg_2SiO_4 (fo) and Fe_2SiO_4 (fa), and ignoring the weak ordering between $M1$ and $M2$ sites, the activities of the components are

$$a^{ol}_{fo} = X^{ol}_{fo}\,\lambda^{ol}_{fo} = (X^{ol}_{Mg^{2+}}\lambda^{ol}_{Mg^{2+}})^2 \quad\text{and}\quad a^{ol}_{fa} = X^{ol}_{fa}\,\lambda^{ol}_{fo} = (X^{ol}_{Fe^{2+}}\lambda^{ol}_{Fe^{2+}})^2 \tag{5.108}$$

Experimental determinations of activity coefficients in Fe-Mg olivine at high temperatures have indicated that this phase shows small positive deviations from ideality (i.e., λ values > 1) and its activity–composition relations are compatible with a symmetrical regular solution model (Hackler and Wood, 1989). Using equation (5.107), we have for the olivine components with two mixing sites for Mg^{2+} or Fe^{2+},

$$RT\ln\lambda^{ol}_{Mg^{2+}} = \left(X^{ol}_{Fe^{2+}}\right)^2 W^G_{MgFe} \quad\text{and}\quad RT\ln\lambda^{ol}_{Fe^{2+}} = \left(X^{ol}_{Mg^{2+}}\right)^2 W^G_{MgFe} \tag{5.109}$$

There is a considerable spread in the published values of W^G_{MgFe} for olivine, but let us calculate W^G_{MgFe} using the equation

$$W^G_{MgFe} = W^H_{MgFe} - T\,W^S_{MgFe} + P\,W^V_{MgFe} \tag{5.110}$$

and the values given by Berman and Aranovich (1996) on the basis of two oxygen atoms (or one cation):

$W^H_{MgFe} = 10{,}366.20$ J mol^{-1}; W^S_{MgFe} $= 4.0$ J mol^{-1} K^{-1}; $W^V_{MgFe} = 0.011$ J bar^{-1} mol^{-1}

Given $X^{ol}_{Mg}=0.8$ and $X^{ol}_{Fe}=0.2$, $T=1{,}400$ K, and $P=1$ bar, $W^G_{MgFe} = 10{,}366.20 - (1{,}400)(4) + (1)(0.011) = 4{,}766.21$ J mol^{-1}
Substituting in equation (5.109),

$\lambda_{Mg}=1.0165$, and $\lambda_{Fe}=1.2996$.

The calculated activities, using equation (5.108), are:

$$a_{fo}^{ol} = (X_{Mg^{2+}}^{ol} \lambda_{Mg^{2+}}^{ol})^2 = (0.8)^2(1.0165)^2 = 0.66$$
$$a_{fa}^{ol} = (X_{Fe^{2+}}^{ol} \lambda_{Fe^{2+}}^{ol})^2 = (0.2)^2(1.2996)^2 = 0.07$$

For the same olivine composition, the activities would work out to be 0.64 and 0.04 if it were treated as an ideal solid solution.

5.10 Summary

1. A solution is defined as a homogeneous phase formed by dissolving one or more substances (solid, liquid, or gas) in another substance (solid, liquid, or gas).
2. A partial molar property of the ith constituent in the phase α, $\bar{y}_i^{\alpha}$, is defined as:

$$\bar{y}_i^{\alpha} = \left(\frac{\partial Y^{\alpha}}{\partial n_i^{\alpha}}\right)_{P,T,n_j^{\alpha}}$$

where Y is an extensive property of the phase α, and n_j^{α} refers to all constituents in the phase α other than i. Partial molar properties are intensive. For any phase α,

$$Y^{\alpha} = \sum_i n_i^{\alpha}\, \bar{y}_i^{\alpha}$$

3. The chemical potential of the ith constituent in phase α, μ_i^{α}, is its partial molar Gibbs free energy and is defined as:

$$\mu_i^{\alpha} = \left(\frac{\partial G^{\alpha}}{\partial n_i^{\alpha}}\right)_{P,T,n_j^{\alpha}}$$

so that at constant temperature and pressure the free energy of the phase α is

$$G^{\alpha} = \sum_i \mu_i^{\alpha}\, n_i^{\alpha}$$

and for the pure phase, $\bar{G}_i^{\alpha} = \mu_i^{\alpha}$.
4. The chemical potential of a common constituent is the same in all the phases (α, β, γ, ...) coexisting in equilibrium (at constant temperature and pressure):

$$\mu_i^{\alpha} = \mu_i^{\beta} = \mu_i^{\gamma} = \ldots$$

Chemical potential gradients due to differences in chemical potentials are the cause of all processes of diffusion.
5. The general condition of equilibrium for a reaction at constant temperature and pressure, irrespective of composition and physical state (solid, liquid or gas) of the substances involved, is:

$$\sum_i \nu_i\, \mu_i = 0$$

where ν_i, the stoichiometric coefficient of the ith constituent in the reaction, is positive if it refers to a product and negative if it refers to a reactant.
6. For a reaction in equilibrium at P and T,

$$\Delta G_r^{0} = -RT \ln K_{eq}$$

where ΔG_r^{0} is the standard Gibbs free energy change for the reaction, with each reactant and product in its standard state, and K_{eq} is the equilibrium constant of the reaction at the P and T of interest. This relationship allows us to compute equilibrium constants from free energy data and vice versa.
7. The general equation relating chemical potential at specified (P, T) and composition is

$$\mu_i\,(P, T) = \mu_i^{0}\,(P^{0}, T^{0}) + RT \ln a_i$$

where μ_i^{0} is the standard state chemical potential of i, and a_i its activity at P and T.
8. For any species i in the standard state, $a_i = 1$. For a pure ideal gas, $a = P/P^0$, where P^0 is the standard state pressure; for a nonideal gas mixture of ideal gases $a_i = f/f^0$, where f is the fugacity corresponding to pressure P of interest and f^0 the fugacity corresponding to the chosen standard state pressure P^0. Equations for evaluating activities in solutions, including gas mixtures, are summarized in Table 5.3
9. The molar excess free energy function of a nonideal solid solution, $\bar{G}^{EX}$, a measure of the deviation from ideality, is given by

$$\bar{G}^{EX} = \bar{G}^{\text{nonideal ss}} - \bar{G}^{\text{ideal ss}}$$
$$= \left[\sum_c X_c^{ss}\, \mu_c^0 + RT \sum_j \left(n_j \sum_i X_{i,j}^{ss} \ln\left(X_{i,j}^{ss}\, \lambda_{i,j}\right)\right)\right]$$
$$- \left[\sum_c X_c^{ss}\, \mu_c^0 + RT \sum_j \left(n_j \sum_i X_{i,j}^{ss} \ln X_{i,j}^{ss}\right)\right]$$
$$= RT \sum_j \left(n_j \sum_i X_{i,j}^{ss} \ln \lambda_{i,j}\right)$$

10. In many petrologic applications, nonideal solid solutions may be modeled as regular solutions ($\Delta H_{mixing} \neq 0$, $\Delta V_{mixing} \approx 0$). For a binary symmetric regular solution,

$$RT \ln \lambda_1 = (X_2)^2 W_{12}^G$$
$$RT \ln \lambda_2 = (X_1)^2 W_{12}^G$$

and for a binary asymmetric regular solution,

$$RT \ln \lambda_1 = (X_2)^2[W_{12}^G + 2X_1(W_{12}^G - W_{21}^G)]$$
$$RT \ln \lambda_2 = (X_1)^2[W_{21}^G + 2X_2(W_{12}^G - W_{21}^G)]$$

where W^Gs denote interaction free-energy parameters along the binary join and $X_1 + X_2 = 1$.

Table 5.3 Activity of constituent i in ideal and nonideal solutions.

Phase	Ideal solution ($\Delta H_{mixing}=0$; $\Delta V_{mixing}=0$)	Nonideal solution ($\Delta H_{mixing}\neq 0$; $\Delta V_{mixing}\neq 0$)
Pure gas	$a=P/P^0$	$a=f/f^0$
Mixture of ideal gases	$a_i=P_i/P^0$	$a_i=f_i/f_i^0$
Mixture of nonideal gases	$a_i=f_i/f_i^0$	$a_i=f_i/f_i^0$
	$f_i=f_i^{pure}X_i$ (Lewis fugacity rule)	$f_i=P_i\,\chi_i$
Solution involving condensed phases	$a_i=X_i$ (as per Raoult's Law)	$a_i=X_i\,\lambda_i$
	$a_i=h_iX_i$ (as per Henry's Law)	
Crystalline solid solution	MOS model	MOS model
	$a_{i,j}^{ss}=(X_{i,j}^{ss})^{nj}$	$a_{i,j}^{ss}=(X_{i,j}^{ss}\,\lambda_{i,j}^{ss})^{nj}$
	$a_c^{ss}=\prod(X_{i,j}^{ss})^{nj}=\prod a_{i,j}^{ss}$	$a_c^{ss}=\prod(X_{i,j}^{ss}\,\lambda_{i,j}^{ss})^{nj}=\prod a_{i,j}^{ss}$

P=pressure, f=fugacity, χ=fugacity coefficient, a=activity, λ=rational activity coefficient, h=Henry's Law constant, j=mixing site, n_j=number of positions in the jth site per formula unit, X=mole fraction.

5.11 Recapitulation

Terms and concepts

Activity
Activity coefficient
Aluminum avoidance
Chemical potential
Consolute temperature
Dehydration reaction
Extent of reaction
Excess functions
Fugacity
Fugacity coefficient
Gibbs–Duhem equation
Henry's Law
Ideal solutions
Immiscibility (exsolution)
Interaction parameters
LCB mixing model
Lewis fugacity rule
Margules parameters
Mixing properties of solutions
Mole fraction
MOS mixing model
Nernst distribution coeficient
Nonideal solutions
Partial molar quantity
Raoult's Law
Rational activity coefficient
Regular solution (symmetric and asymmetric)
Solvus
Standard states
Standard state chemical potential
Vapor pressure

Computation techniques

- Mole fractions.
- Activities of constituents in ideal and nonideal solutions at the temperature and pressure of interest.
- Chemical potentials of constituents in ideal and nonideal solutions at the temperature and pressure of interest.
- *P–T* reaction boundary of a dehydration reaction.

5.12 Questions

1. A mineral assemblage consisting of hematite (Fe_2O_3), magnetite (Fe_3O_4), and chromite ($FeCr_2O_4$) is interpreted to have attained thermodynamic equilibrium at temperature T (K) and pressure P (bar). What thermodynamic condition(s) must be satisfied if the assemblage attained equilibrium (i) in a closed system or (ii) in an open system?
2. Calculate the Gibbs free energy change and equilibrium constant for the reaction

$$H_{2\,(g)}+0.5O_{2\,(g)} \Leftrightarrow H_2O_{(l)}$$

at 298.15 K and 1 bar, using the standard data given below (Robie and Hemingway, 1995):

$\Delta H^1_{f,\,298.15}$ (kJ mol^{-1})			$S^1_{298.15}$ J mol^{-1} K^{-1}		
$H_{2\,(g)}$	$O_{2\,(g)}$	$H_2O_{(l)}$	$H_{2\,(g)}$	$O_{2\,(g)}$	$H_2O_{(l)}$
0	0	−285.8	130.68	205.15	70.0

What can you infer about the reaction from your calculations?

3. Show that the chemical potential of a pure phase α consisting of a single constituent i at pressure P (bars) and temperature T (K), $\mu_i^{\alpha}(P, T)$, has the same value for the following choice of standard states: (i) pure substance, P

(bar), T (K); (ii) pure substance, 1 (bar), T (K); and (iii) pure substance, 1 (bar), 298.15 (K). What is the expression for activity in each case?

4. For the dehydration reaction

$$\underset{\text{brucite}}{Mg(OH)_2} = \underset{\text{periclase}}{MgO} + \underset{\text{fluid}}{H_2O}$$

(a) Calculate the approximate equilibrium temperature at P_{H_2O}=1 kbar. Assume that brucite and periclase are pure phases, $(\Delta C_P)_r=0$, ΔV_r(solids) is independent of temperature and pressure, the fluid phase is composed of H_2O only, H_2O behaves as a nonideal gas, and $P_{H_2O}=P_{total}$. Thermodynamic data relevant to the problem are as follows:

For the reaction at 1 bar and 800 K (data from Robie and Hemingway, 1995)	P_{H_2O} versus f_{H_2O} relationship (Selected data from Burnham *et al.*, 1969)	
	f_{H_2O} (bar)	P_{H_2O} (bar)
$\Delta H^1_{r,800}$=72 kJ	459	700
	498	800
$\Delta S^1_{r,800}$=135.8 J.K^{-1}	562	900
	601	1000
$\Delta V^1_{r,298}$(solids)=–1.338 J.bar^{-1}	713	1200
	754	1300

[Hint: a spreadsheet for the computation of f_{H_2O} for assumed values of temperature would be useful.]

(b) What would be the error in the calculated equilibrium temperature if we had assumed H_2O to behave as an ideal gas?

(c) If the fluid phase is not pure H_2O, will the equilibrium temperature increase or decrease with decreasing mole fraction of H_2O in the fluid? What does your conclusion imply about the role of water in the fluid phase during metamorphism?

5. Calculate activities of end-member components (using P and T of interest as the standard state) in the solid solutions as specified below, assuming ideal behavior and mixing-on-sites (MOS) model:

(a) Components Mg_2SiO_4 (forsterite) and Fe_2SiO_4 (fayalite) in olivine,

$$(Mg_{1.63}Fe_{0.35}Mn_{0.02})_2(Si_{0.97}Al_{0.03})_1O_4$$

(b) Components $NaAlSi_2O_6$ (jadeite), $CaMgSi_2O_6$ (diopside), and $NaFe^{3+}Si_2O_6$ (aegirine) in clinopyroxene,

$$(Ca_{0.452}Mg_{0.003}Fe^{2+}_{0.003}Na_{0.525})^{M2}_{0.983}(Mg_{0.420}Fe^{2+}_{0.028}Al_{0.517}Ti_{0.012}Fe^{3+}_{0.023})^{M1}_1(Si_{1.985}Al_{0.015})_2O_6$$

(c) Components $Fe_3^{2+}Al_2Si_3O_{12}$ (almandine), $Ca_3Al_2Si_3O_{12}$ (grossular), $Mg_3Al_2Si_3O_{12}$ (pyrope), and $Mn_3Al_2Si_3O_{12}$ (spessartine) in garnet,

$$(Mg_{0.108}Fe^{2+}_{2.004}Ca_{0.775}Mn_{0.078})_{2.965}(Al_{1.963}Ti_{0.045})_{2.008}(Si_{2.941}Al_{0.059})_3O_{12}$$

6. (a) Calculate the activity of $Ca_3Al_2Si_3O_{12}$ component in pure grossular garnet ($Ca_3Al_2Si_3O_{12}$) at 10 kbar and 1000 K, taking the standard state as pure grossular at 1 bar and 1000 K. $\bar{V}^{grossular}$=12.528 J bar^{-1}. Assume that $\bar{V}^{grossular}$ is independent of temperature and pressure. [Hint: see Box 5.2.]

(b) What will be the activity if the standard state is taken as pure grossular at the pressure and temperature of interest? Is this result consistent with equations (5.79) and (5.80)?

7. Plot on a graph paper the activity of anorthite (a^{plag}_{an}) against the mole fraction of anorthite in plagioclase (X^{plag}_{an}) for the following cases:

(a) The plagioclase is an ideal solid solution (i.e., it obeys Raoult's Law; $a^{plag}_{an}=X^{plag}_{an}$).

(b) The plagioclase is a nonideal solid solution (i.e., it obeys Henry's Law; $a^{plag}_{an}=\gamma^{plag}_{an}X^{plag}_{an}$ and can be modeled using the activity model of Orville (1972):

X_{an}^{plag}	0.00–0.50	0.60	0.70	0.80	0.90	1.00
γ_{an}^{plag}	1.267	1.213	1.112	1.042	1.000	1.000

(c) The plagioclase is a non–ideal solid solution and can be modeled using the aluminum avoidance model of Kerrick and Darken (1975):

$$RT \ln \gamma^{plag}_{an}=(X^{plag}_{ab})^2\,[W^{plag}_{an}+2X^{plag}_{an}(W^{plag}_{ab}-W^{plag}_{an})]$$

where the Margules parameters (at 1000°C) are W_{an}^{plag}=2025 cal and W_{ab}^{plag}=6746 cal, and the activity is calculated as

$$a^{plag}_{an}=\gamma^{plag}_{an}\,[X^{plag}_{an}(1-X^{plag}_{an})^2/4].$$

Comment on the results.

8. We wish to model the mixing in biotite as a Mg–Fe–Al–Ti symmetric regular solution. Using the formulation of Mukhopadhyay *et al.* (1993) given in Appendix 7, show that

$$RT \ln\left(\frac{\lambda^{bi}_{Mg}}{\lambda^{bi}_{Fe}}\right) = W_{MgFe}(X_{Fe}-X_{Mg}) + X_{Al}(W_{MgAl}-W_{FeAl}) + X_{Ti}(W_{MgTi}-W_{FeTi})$$

where λ_i and X_i denote, respectively, the rational activity coefficient and mole fraction of the ith constituent in the biotite solid solution, and W_{ij} the various free-energy interaction parameters. Remember that for a symmetric regular solution, $W_{ij}=W_{ji}$.

9. Write a general expression for the free energy of an ideal garnet solid solution comprised of the following four end–member components: $Mg_3Al_2Si_3O_{12}$, pyrope (py); $Fe_3^{2+}Al_2Si_3O_{12}$, almandine (alm); $Ca_3Al_2Si_3O_{12}$, grossular (gr); and $Ca_3Fe_2^{3+}Si_3O_{12}$, andradite (and). For each phase component the standard state is pure substance at the pressure (P) and temperature (T) of interest.

10. (a) Draw $\bar{G}-X$ diagrams for a binary system at $T=800\,K$ that is: (i) an ideal solution (i.e., $W^G_{12}=0$); (ii) a symmetric regular solution with $W^G_{12}=7\,kJ/mol$; and (iii) a symmetric regular solution with $W^G_{12}=18\,kJ/mol$. Explain how the solution behavior changes with increasing W^G_{12}. Use $\bar{G}_1=\bar{G}_2=0$, and the same horizontal and vertical scales in these diagrams for ease of comparison. (After Powell, 1978.)
    (b) Draw a $\bar{G}-X$ diagram for a binary symmetric regular solution with $W^G_{12}=18\,kJ/mol$ at various temperatures (700, 800, 900, 1000, 1100, and 1200 K), and retrieve data from this diagram to plot a $T-X$ diagram (as shown in Figure 5.7b). Explain the behavior of this solution with $X_1=0.2$ as it cools from 1100 K to 800 K. (After Powell, 1978.)
11. If the standard state of each constituent in a mixture of gases is chosen as the pure constituent at the pressure and temperature of interest (P, T), show that if the gas mixture is ideal

$$\mu_i\,(P, T)=\mu_i^0\,(P, T)+RT\ln X_i$$

but if the gas mixture is nonideal

$$\mu_i\,(P, T)=\mu_i^0\,(P, T)+RT\ln (X_i\chi_i)$$

where X_i=mole fraction of the ith constituent and χ_i=its fugacity coefficient. What is the value of a_i in each case?
12. Calculate the equilibrium constant for the reaction

$$CO_{2\,(g)}+H_2O_{(l)}=H_2CO_{3\,(aq)}$$

at 298.15 K and 1 bar. The $\Delta G^1_{f,\,298.15}$ values (in kJ mol^{-1}) are as follows (Robie and Hemingway, 1995): $CO_{2\,(g)}$ = −394.4; $H_2O_{(l)}$=−237.1; $H_2CO_{3\,(aq)}$=−623.2.
13. Calculate the solubility of oxygen in water in equilibrium with the atmosphere at 25°C. Assume that the volume fraction of oxygen in the atmosphere is equal to its mole fraction. The Henry's Law constant for O_2 at 25°C is 1.26×10^{-3} mol l^{-1} bar^{-1}.
14. Using the thermodynamic data given below, and assuming $(\Delta C_P)_r=0$ and ΔV_r(solids) to be independent of temperature and pressure, calculate the equilibrium pressure at 1200 K for the reaction

$$\underset{\text{anorthite}}{\overset{\text{plag}}{3CaAl_2Si_2O_8}} = \underset{\text{grossular}}{\overset{\text{gt}}{Ca_3Al_2Si_3O_{12}}} + \underset{\text{sillimanite}}{\overset{\text{sil}}{2Al_2SiO_5}} + \underset{\text{quartz}}{\overset{\text{qz}}{SiO_2}}$$

if (i) all the phases are pure, and (ii) quartz and sillimanite are pure phases, but garnet ($(Ca, Fe^{2+}, Mg, Mn)_3Al_2Si_3O_{12}$) and plagioclase ($NaAlSi_3O_8 - CaAl_2Si_2O_8$) are ideal solid solutions in which

$$X^{plag}_{CaAl_2Si_2O_8}=0.2 \text{ and } X^{gt}_{Ca}=Ca/(Ca+Fe^{2+}+Mg+Mn)=0.05$$

Mineral	$\Delta H^1_{f,\,298.15}$ (kJ mol^{-1})	$S^1_{298.15}$ (J mol^{-1} K^{-1})	$V^1_{298.15}$ (J bar^{-1})
Anorthite	−4234.0	199.30	10.079
Grossular	−6640.0	260.12	12.528
Sillimanite	−2586.1	95.40	4.986
Quartz	−910.7	41.46	2.269

15. From the Henry's Law constants given in Table 5.2 for $N_{2(g)}$ at 20°C and 50°C, calculate the solubility of $N_{2(g)}$ at these temperatures for a $N_{2(g)}$ partial pressure of 0.78 bar above the solution (at a total pressure of 1 bar). Does the solubility increase or decrease with increasing temperature?

6 Geothermometry and Geobarometry

The intersecting curves of the two classes (univariant curves for equilibria involving solids only and for equilibria involving a gas phase) will thus cut the general P–T diagram into a grid which we may call a petrogenetic grid. With the necessary data determined by experiment we might be able to locate very closely on the grid both the temperature and the pressure of formation of those rocks and mineral deposits of any terrane that were formed at a definite stage of its history, provided always that a sufficient variety of composition of materials occurred in the terrane to permit adequate cross reference. The determinations necessary for the production of such a grid constitute a task of colossal magnitude, but the data will be gradually acquired, and we shall thus slowly proceed toward an adequate knowledge of the conditions of formation of rocks and mineral deposits.

Bowen (1940)

In the previous two chapters we learned how to evaluate the thermodynamic properties of phases of fixed or variable compositions, and calculate changes in the thermodynamic properties of such phases involved in a chemical reaction. In this chapter we will learn how this knowledge may be applied for the estimation of temperature (*geothermometry*) and pressure (*geobarometry*) of equilibration of a given mineral assemblage at some point in its history, an application of prime interest to geologists. Such information provides quantitative constraints on *P–T* environments of geologic events such as magmatic crystallization, prograde and retrograde metamorphism, and formation of ore deposits. A mineral or mineral assemblage useful for geothermometry is called a *geothermometer*; one useful for geobarometry, a *geobarometer*. As most reactions are sensitive to both temperature and pressure, many authors refer to them as *geothermobarometers* or simply as *thermobarometers*.

6.1 Tools for geothermobarometry

Except for the use of fluid inclusions in minerals to obtain temperature (and, in some cases, pressure) information about the host minerals, the techniques employed for quantitative geothermometry and geobarometry rely on chemical reactions that have been calibrated as a function of temperature, or pressure, or both. A *petrogenetic grid*, as was conceived by Bowen (1940) several decades ago, is merely a collection of such calibrated reactions relevant to a portion of the *P–T* space. Petrogenetic grids are useful for estimating *P–T* limits of assemblages if the minerals are well approximated by reactions included in the grid.

The tools that have commonly been used for geothermobarometry are (Essene, 1982, 1989; Bohlen and Lindsley, 1987):

(1) univariant reactions, in which all phases essentially have fixed compositions, and *displaced equilibria*, in which one or more phases are solid solutions of variable compositions;
(2) exchange reactions, in which exchange of one element for another between two coexisting phases occurs primarily as a function of temperature;
(3) solvus equilibria, in which the solubility of a component in a phase varies as a function of temperature (or, less commonly as a function of pressure);
(4) study of cogenetic fluid inclusions in minerals;
(5) fractionation (or partitioning) of oxygen or sulfur isotopes between coexisting phases, which is a function of temperature only except at very high pressures (see Chapter 11).

Reactions used for thermobarometry generally involve minerals of high variance in natural compositions (e.g., garnet,

Introduction to Geochemistry: Principles and Applications, First Edition. Kula C. Misra.

Table 6.1 Some commonly used geothermometers and geobarometers.

	Mineral assemblage	Reaction/basis	Selected references
Intercrystalline exchange reactions			
Mg^{2+}–Fe^{2+} exchange	Garnet–clinopyroxene	1/3$Mg_3Al_2Si_3O_{12}$ (py) + $CaFeSi_2O_6$ (hd) = 1/3 $Fe_3Al_2Si_3O_{12}$ (alm) + $CaMgSi_2O_6$ (diop)	Råheim and Green (1974); Ellis and Green (1979); Dahl (1980); Hodges and Spear (1982); Krogh (1988); Pattison and Newton (1989); Kitano *et al.* (1994); Berman *et al.* (1995); Ganguly *et al.* (1996); Holdaway *et al.* (1997); Holdaway (2000)
	Garnet–orthopyroxene	$Mg_3Al_2Si_3O_3$ (py) + 3$FeSiO_3$ (fs) = $Fe_3Al_2Si_3O_3$ (alm) + 3$MgSiO_3$ (en)	Mori and Green (1978); Harley (1984); Sen and Bhattacharya (1984); Aranovich and Berman (1997)
	Garnet–biotite (GABI)	$KMg_3Si_3AlO_{10}(OH)_2$ (phl) + $Fe_3Al_2Si_3O_{12}$ (alm) = $KFe_3Si_3AlO_{10}(OH)_2$ (ann) + $Mg_3Al_2Si_3O_{12}$ (py)	Ferry and Spear (1978); Perchuk and Lavrent'eva (1983); Hodges and Spear (1982); Indares and Martignole (1985); Berman (1990); Dasgupta *et al.* (1991); Bhattacharya *et al.* (1992); Kleeman and Reinhardt (1994); Berman and Aranovich (1996); Ganguly *et al.* (1996); Gessmann *et al.* (1997); Mukhopadhyay *et al.* (1997); Holdaway (2000)
	Garnet–olivine	1/3$Mg_3Al_2Si_3O_{12}$ (py) + 1/2Fe_2SiO_4 (fa) = 1/3$Fe_3Al_2Si_3O_{12}$ (alm) + 1/2Mg_2SiO_4 (fo)	O'Neill and Wood (1979); Kawasaki and Matsui (1983); Hackler and Wood (1989); Ryan *et al.* (1996)
	Garnet–cordierite	2$Fe_3Al_2Si_3O_{12}$ (alm) + 3$Mg_2Al_4Si_5O_{18}$ (Mg-cord) = 1/3$Mg_3Al_2Si_3O_{12}$ (py) + 3$Fe_2Al_4Si_5O_{18}$ (Fe-cord)	Currie (1971); Hensen and Green (1971); Thompson (1976); Holdaway and Lee (1977); Perchuk and Lavrent'eva (1983)
	Garnet–ilmenite	$Fe_3Al_2Si_3O_{12}$ (alm) + 3$MnTiO_3$ = $Mn_3Al_2Si_3O_{12}$ (spess) + 3$FeTiO_3$ (ilm)	Pownceby *et al.* (1987); Anderson and Lindsley (1981)
	Olivine–orthopyroxene	$Mg_2Si_2O_6$ (en) + Fe_2SiO_4 (fa) = $Fe_2Si_2O_6$ (fs) + Mg_2SiO_4 (fo)	Von Seckendorff and O'Neill (1993)
	Olivine–clinopyroxene (calcic)	1/2 Mg_2SiO_4 (fo) + $CaFeSi_2O_6$ (hd) = 1/2Fe_2SiO_4 (fa) + $CaMgSi_2O_6$ (diop)	Powell and Powell (1974); Perkins and Vielzeuf (1992); Kawasaki and Ito (1994); Loucks (1996)
	Olivine–spinel	$MgSi_{0.5}O_2$ (fo)+ Fe(Y)O_4 (sp) = $FeSi_{0.5}O_2$ (fa)+ Mg(Y)O_4 (sp) [Y = Al, Cr]	Engi (1983); Jameison and Roeder (1984); Sack and Ghiorso (1991)
	Orthopyroxene–biotite	1/3$KFe_3Si_3AlO_{10}(OH)_2$ (ann) + 1/2 $Mg_2Si_2O_6$ (en) = 1/3$KMg_3Si_3AlO_{10}(OH)_2$ (phl) + 1/2 $Mg_2Si_2O_6$	Sengupta *et al.* (1990)
	Magnetite–ilmenite (FeO–TiO_2–O_2 system)	Fe_3O_4 (mt) + $FeTiO_3$ (ilm) = Fe_2TiO_4 (usp) + Fe_2O_3 (hem) 4Fe_3O_4 (mt) + O_2 = 6Fe_2O_3 (hem)	Buddington and Lindsley (1964); Spencer and Lindsley (1981); Bohlen and Lindsley (1987); Anderson and Lindsley (1988); Brown and Navrotsky (1994); Woodland and Wood (1994)
Mg^{2+}–Ni^{2+} exchange	Olivine–orthopyroxene	1/2Mg_2SiO_4 (fo) + $NiSiO_3$ = 1/2Ni_2SiO_4 + $MgSiO_3$ (en)	Medaris (1968); Podvin (1988); von Seckendorff and O'Neill (1993); Ganguly and and Tazzoli (1994)
	Olivine–basaltic melt	1/2Mg_2SiO_4 (fo) + NiO (melt)= 1/2Ni_2SiO_4 + MgO (melt)	Duke (1976); Hart and Davis (1978); Hirschman (1991); Roeder andEmslie (1970); Irvine (1975)
Univariant reactions and displaced equilibria			
	Andalusite–sillimanite–kyanite	andalusite = sillimanite; sillimanite = kyanite; kyanite = andalusite	Holdaway (1971); Hodges and Spear, 1982; Bohlen *et al.* (1991); Hemingway *et al.* (1991); Holdaway and Mukhopadhyay (1993) Harlov and Newton (1993); Whitney (2002)
	Orthopyroxene–olivine–quartz	2$FeSiO_3$ (fs) = Fe_2SiO_4 (fo) + SiO_2 (qz)	Bohlen *et al.* (1980a, 1980b); Davison and Lindsley (1989)

Clinopyroxene–plagioclase–quartz	$NaAlSi_3O_8$ (alb/analbite) = $NaAlSi_2O_6$ (jd) + SiO_2 (qz)	Ganguly (1973); Holland (1980); Newton (1983)
Garnet–orthopyroxene (Al content of opx in equilibrium with garnet)	$Mg_2Si_2O_6$ (en) + $MgAl_2Si_2O_6$ (Mg-Tschermak, MgTs) = $Mg_3Al_2Si_3O_{12}$ (py)	Wood and Banno (1973); MacGregor (1974); Harley (1984); Sen and Bhattacharya (1984); Lee and Ganguly (1988); Brey and Köhler (1990); Aranovich and Berman (1997); Taylor (1998)
Garnet–plagioclase–sillimanite /kyanite–quartz (GASP)	$3CaAl_2Si_2O_8$ (an) = $Ca_3Al_2Si_3O_{12}$ (gr) + $2Al_2SiO_5$ (sil/ky) + SiO_2 (qz)	Ghent (1976); Ghent *et al.* (1979); Newton and Haselton (1981); Hodges and Spear (1982); Ganguly and Saxena (1984); Koziol and Newton (1988); Koziol (1989); Spear (1993); Holdaway (2001)
Garnet–plagioclase–orthopyroxene–quartz (GAPES)	$CaAl_2Si_2O_8$ (an) + $Mg_2Si_2O_6$ (en) = $2/3Mg_3Al_2Si_3O_{12}$ (py) + $1/3Ca_3Al_2Si_3O_{12}$ (gr) + SiO_2 (qz)	Newton and Perkins (1982); Bohlen *et al.* (1983a); Perkins and Chipera (1985); Eckert *et al.* (1991)
Garnet–plagioclase–clinopyroxene–quartz (GADS)	$CaAl_2Si_2O_8$ (an) + $CaMgSi_2O_6$ (diop) = $2/3Ca_3Al_2Si_3O_{12}$ (gr) + $1/3Mg_3Al_2Si_3O_{12}$ (py) + SiO_2 (qz)	Newton and Perkins (1982); Perkins and Chipera (1985); Moecher *et al.* (1988); Eckert *et al.* (1991)
Garnet–rutile–Al_2SiO_5 polymorph–ilmenite–quartz (GRAIL)	$Fe_3Al_2Si_3O_{12}$ (alm) + $3TiO_2$ (rut) = $3FeTiO_3$ (ilm) + Al_2SiO_5 (sil) + $2SiO_2$ (qz)	Bohlen *et al* (1983b)
Garnet–rutile–ilmenite–plagioclase–quartz (GRIPS)	$CaFe_2Al_2Si_3O_{12}$ (gr_1alm_2) + 2 TiO_2 (rut) = $2FeTiO_3$ (ilm) + $CaAl_2Si_2O_8$ (an) + SiO_2 (qz)	Bohlen and Liotta (1986)
Garnet–plagioclase–amphibole–quartz	$6CaAl_2Si_2O_8$ (an) + $3NaAlSi_3O_8$ (ab) + $3Ca_2Mg_5Si_8O_{22}(OH)_2$ (trem) = $2Ca_3Al_2Si_3O_{12}$ (gr) + $3NaCa_2Mg_4Al_3Si_6O_{22}(OH)_2$ (par) + $Mg_3Al_2Si_3O_{12}$ (py) + 18 SiO_2 (qz)	Kohn and Spear (1989)
Garnet–plagioclase–muscovite–biotite	$Mg_3Al_2Si_3O_{12}$ (py) + $Ca_3Al_2Si_3O_{12}$ (and) + $KAl_3Si_3O_{10}(OH)_2$ (mus) = $3CaAl_2Si_2O_8$ (an) + $KMg_3AlSi_3O_{10}(OH)_2$ (bi)	Ghent and Stout (1981); Hodges and Spear (1982); Hodges and Crowley (1985); Hoisch (1991)
Solvus equilibria		
Alkali feldspar–albite (two-feldspar thermometry)	The distribution of K and Na between alkali feldspar and albite coexisting in equilibrium	Stormer (1975); Whitney and Stormer (1977); Green and Udansky (1986); Fuhrman and Lindsley (1988); Holland and Powell (1992)
Alkali feldspar–plagioclase (ternary feldspar)	Limits of miscibility gap in the ternary system (or–ab–an)	Whitney and Stormer (1977); Ghiorso (1984); Fuhrman and Lindsley (1988); Elkins and Grove (1990)
Calcite–dolomite	The distribution of Mg and Ca between calcite and dolomite coexisting in equilibrium	Harker and Tuttle (1955); Goldsmith and Newton (1969); Powell *et al.* (1984); Anovitz and Essene (1987)
Clinopyroxene–orthopyroxene	The distribution of Ca and Mg between clinopyroxene and orthopyroxene coexisting in equilibrium	Lindsley (1983); Gasparik (1984b); Bertrand and Mercier (1985); Davidson and Lindsley (1985); Wood (1987); Carlson and Lindsley (1988); Brey and Köhler (1990)

Mineral abbreviations: act = actinolite, ab = albite, alm = almandine, and = andalusite, an = anorthite, ann = annite, bi = biotite; cal = calcite, cord = cordierite diop = diopside, en = enstatite, fs = ferrosilite, gr = grossular, hc = hercynite, hd = hedenbergite, hem = hematite, ilm = ilmenite, jd = jadeite, ksp = K-feldspar, ky = kyanite, mt = magnetite, mus = muscovite, or = orthoclase, par= pargasite, phl = phlogopite, py = pyrope, qz = quartz, rut = rutile, sil = sillimanite, sp = spinel, trem = tremolite, usp = ulvospinel.

pyroxene, feldspar), so that the corresponding thermodynamic equations are sliding scale or continuous *P–T* indicators. Garnet, a common mineral in medium- and high-grade metamorphic rocks, is a particularly useful mineral for both geothermometry and geobarometry. The equilibrium constants for Fe–Mg exchange reactions between garnet and most other silicates (such as clinopyroxene, biotite, etc.) have large values because of the strong preference of Fe^{2+} (relative to Mg^{2+}) for the garnet structure. This, in turn, results in significantly temperature-dependent Fe–Mg distribution coefficients, a favorable condition for geothermometry. The usefulness of garnet in geobarometry arises from the fact that it is a dense mineral favored at high pressures. The downside of using garnet is the uncertainty associated with the activity models formulated for this mineral, which can incorporate a variety of elements such as Mn, Ti, and Cr.

As discussed in several excellent reviews (Ferry, 1980: Essene, 1982, 1989; Newton, 1983; Finnerty and Boyd, 1987; Bohlen and Lindsley, 1987; Spear, 1989; Smith, 1999), the list of promising thermobarometers is a long one. Essene (1989), for example, listed 60 equilibria that may be useful for thermobarometry, some more than others for a given metamorphic facies. A selection from the more commonly used thermobarometers is summarized in Table 6.1; we will discuss a few examples from this list to illustrate the principles involved. Many computer programs are available for calculations relevant to many thermobarometers included in the list; two widely used programs, both of which can be accessed online, are: THERMOBAROMETRY (version 2.1, May 1999) by Frank Spear and Mathew Kohn (http:ees2.geo.rpi.edu/MetaPetaRen/Software/GTB_Prog/), and THERMOCALC (updated 03–24–2011) by Tim Holland, Roger Powell, and Richard White (http://www.metamorph.geo.uni-mainz.de/thermocalc/).

6.2 Selection of reactions for thermobarometry

All thermobarometric studies based on mineral assemblages embody two inherent assumptions: (i) the mineral assemblage under consideration was in equilibrium at some former *P–T* condition, which we wish to determine; and (ii) the assemblage either has remained essentially unchanged since then or can be reconstructed from textural and other information preserved in the rock. It is not easy to verify either of these assumptions. There is no way to prove equilibrium; an assemblage is assumed to be in equilibrium if such an assumption is compatible with a lack of textural and chemical evidence indicative of disequilibrium. A commonly employed test of equilibrium is the concordance of *P* and *T* calculated from a number of independent reactions constructed from a multimineral assemblage, at least within the limits of uncertainties of the thermobarometers used. The assumption of no subsequent change may be tested by using a particular geothermometer or geobarometer for different samples of the same rock collected from an area of fairly uniform *P–T* regime as inferred from other lines of evidence. One can always calculate a pressure or a temperature from an appropriate geobarometer or geothermometer, but the numbers may or may not actually pertain to the geologic event being investigated. For example, calculations based on the compositions of only the outer rims of adjacent phases in a metamorphic assemblage, a common practice in thermobarometric applications, may yield satisfactory estimates of *P–T* conditions during cooling or uplift, but not of peak metamorphism.

For a given mineral assemblage it is generally possible to write a number of mass-balanced chemical reactions involving components of the phases (minerals) present in the assemblage. Of these only some may be suitable for thermobarometry in a particular case. The selection of reactions is based on the following considerations:

(1) An accurate thermodynamic database must be available for the end-member components included in the reaction. It is advisable to use self-consistent data sets to avoid introducing systematic errors, although a self-consistent data set is not necessarily an accurate one.
(2) The reaction must be well calibrated based either on empirical data or on data obtained from reversed experiments (from which standard enthalpies and entropies can be extracted). Empirical calibration is derived from the composition of minerals in natural assemblages for which the equilibrium pressure and temperature have been, or can be, estimated by some other thermobarometer(s) such as a metamorphic isograd (Ferry, 1980), an invariant point (Hodges and Spear, 1982), or other equilibria (Ghent and Stout, 1981; Hoisch, 1990). Empirical calibration is much simpler than its experimental counterpart, and the results may be more readily applicable to complex natural systems. Such a calibration, however, may be saddled with large uncertainties in the independent *P–T* estimate, and it may not apply to rocks with significantly different bulk compositions. Carefully conducted experiments can provide more accurate calibration, but experiments are almost always performed on simple systems composed of end-member components. Thermodynamic properties of phases are obtained either directly from the experiments involving the reaction being considered (e.g., Holdaway, 2000), or from a self-consistent thermodynamic database developed using statistical methods (such as multiple regression) or mathematical programming methods applied to all experimental data available for numerous related compositional systems (e.g., Berman, 1990; Holland and Powell, 1998).
(3) In natural assemblages we often have to deal with *displaced equilibria*, which refer to variations in temperature and pressure of a reaction that results from one or

more phases being solid solutions of variable composition. Displaced equilibria increase the P–T stability range of an assemblage and, therefore, are more useful than univariant reactions involving only pure components. However, the extension of experimental data to solid solution phases in natural assemblage(s) requires suitable solution models for determining activity–composition relations. Unfortunately, there is still no consensus regarding the correct solution model even for most anhydrous mineral groups (including those frequently used in thermobarometry, such as olivine, pyroxene, plagioclase, and garnet). The task is more difficult for hydrous silicates (e.g., amphiboles and micas), because mixing models for such minerals are based largely on unreversed experimental data or on simple ionic mixing models that are generally not supported by reversed experiments (Essene, 1989, p. 3). In many cases, the formulation of appropriate mixing models for solid solutions poses the most difficult challenge in thermobarometry.

(4) In general, reactions involving a fluid phase are not suitable for thermobarometry because of a lack of accurate information about the fluid composition and the ratio of fluid pressure (P_f) to total pressure (P_{total}).

A recurring issue in connection with the composition–activity relationship in Fe-bearing minerals is the recasting of the total Fe obtained with electron microprobe analysis into Fe^{2+} and Fe^{3+} using some established procedure (e.g., Deer *et al.*, 1966; Stormer, 1983; Spear and Kimball, 1984; Droop, 1987; Spear, 1993). For some minerals, such as olivine and orthopyroxene, all the Fe may be treated as Fe^{2+} without introducing much error; for others, such as clinopyroxene and spinel, Fe^{2+} can be estimated by normalizing the analysis to cations and calculating the Fe^{2+}: Fe^{3+} that will satisfy the oxygen stoichiometry (see example in section 6.5.3). This procedure does not produce unique solutions for micas and amphiboles because of different possible substitutions, nor can it be applied to minerals with cation vacancies in their crystal structures. In such cases, Fe^{2+}–Fe^{3+} should be measured directly by wet-chemical analysis or by Mossbaüer spectroscopy. In some cases, all that may be feasible is to assume a Fe^{2+}: Fe^{3+} ratio based on experience and the published literature.

6.3 Dependence of equilibrium constant on temperature and pressure

The ultimate theoretical basis of most geothermometers and geobarometers is the temperature and pressure dependence of the equilibrium constant of a reaction (K_{eq}). This is because the compositional variations in minerals and coexisting fluids are already incorporated in the calculation of K_{eq} at a specified condition of temperature and pressure. The total variation of K_{eq} with respect to temperature and pressure can be represented by the total derivative of $\ln K_{eq}$:

$$\mathrm{d}\ln K_{eq} = \left(\frac{\partial \ln K_{eq}}{\partial T}\right)_P \mathrm{d}T + \left(\frac{\partial \ln K_{eq}}{\partial P}\right)_T \mathrm{d}P \quad (6.1)$$

From the relation $-RT \ln K_{eq} = \Delta G_r^0 = \Delta H_r^0 - T\Delta S_r^0$ (equations 4.62 and 5.37),

$$\ln K_{eq} = \frac{-\Delta G_r^0}{RT} = \frac{-\Delta H_r^0}{RT} + \frac{\Delta S_r^0}{R} \quad (6.2)$$

where the superscript "0" refers to the chosen standard state. It follows from equation (6.2) that a plot of $\ln K_{eq}$ against $1/T$ (at constant pressure) will have a slope of $-\Delta H_r^0/R$ and an intercept of $\Delta S_r^0/R$ (Fig. 6.1a). The magnitudes of $\Delta H_r^0/R$ and $\Delta S_r^0/R$ will vary with temperature, unless they are assumed to be independent of temperature [i.e., $(\Delta C_P^0)_r = 0$], in which case the plot will be a straight line with a uniform slope (Fig. 6.1b). In the latter case, the enthalpy and entropy change of a reaction can be estimated if the value of K_{eq} for the reaction is known at several different temperatures (at constant pressure).

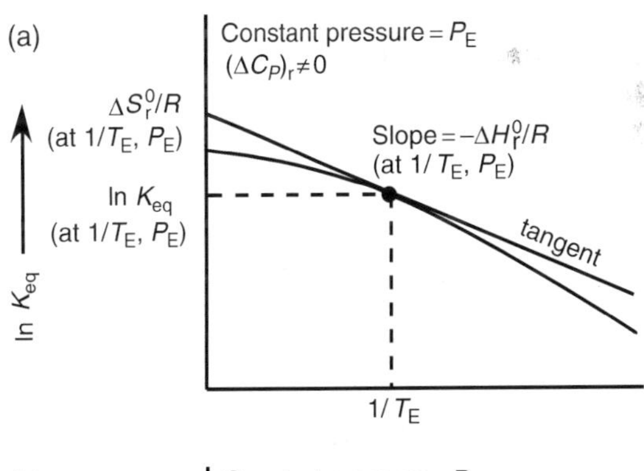

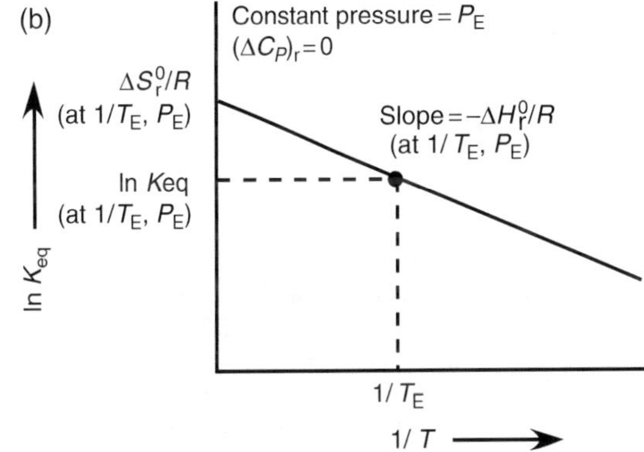

Fig. 6.1 Plot (schematic) of $\ln K_{eq}$ versus $1/T$ (K) at constant pressure with (a) $\Delta C_P \neq 0$ and (b) $\Delta C_P \neq 0$.

It can be shown (see Box 6.1) that the equation describing the dependence of K_{eq} on temperature at constant pressure, assuming ΔH_r^0 to be independent of temperature, is

$$\ln K_T = \ln K_{T\text{ref}} - \frac{\Delta H_r^0}{R}\left(\frac{1}{T} - \frac{1}{T_{\text{ref}}}\right) \tag{6.3}$$

The equation describing the dependence of K_{eq} on pressure at constant temperature, assuming ΔV_r^0 to be independent of temperature and pressure, is

$$\ln K_P = \ln K_{P\text{ref}} - \frac{\Delta V_r^0}{RT}(P - P_{\text{ref}}) \tag{6.4}$$

Box 6.1 Expressions for K_{eq} as a function of temperature and pressure (Krauskopf and Bird, 1995, pp. 203–210; Langmuir, 1997, pp. 20–23 and 28–30)

From equation (6.2), the variations of K_{eq} with temperature (at constant pressure) and with pressure (at constant temperature) are given by

$$\left(\frac{\partial \ln K_{eq}}{\partial T}\right)_P = -\frac{1}{R}\left(\frac{\partial(\Delta G_r^0/T)}{\partial T}\right)_P = \frac{\Delta H_r^0}{RT^2} \quad \text{van't Hoff equation} \tag{6.5}$$

$$\left(\frac{\partial \ln K_{eq}}{\partial P}\right)_T = -\frac{1}{RT}\left(\frac{\partial \Delta G_r^0}{\partial P}\right)_T = -\frac{\Delta V_r^0}{RT} \tag{6.6}$$

The total change in K_{eq} with temperature at constant pressure is obtained by integrating equation (6.5) from some reference temperature (T_{ref}, usually taken as 298.15 K), for which ΔH_r and are known, to the temperature of interest, T:

$$\int_{T_{\text{ref}}}^{T} d(\ln K) = \ln K_T - \ln K_{T_{\text{ref}}} = \int_{T_{\text{ref}}}^{T} \frac{\Delta H_r^0}{RT^2} dT \tag{6.7}$$

Assuming ΔH_r^0 to be independent of temperature [i.e., $(\Delta C_p^0)_r = 0$],

$$\ln K_T = \ln K_{T\text{ref}} - \frac{\Delta H_r^0}{R}\left(\frac{1}{T} - \frac{1}{T_{\text{ref}}}\right) \tag{6.3}$$

If this assumption is valid, a plot of $\ln K_T$ versus $1/T$ would yield a straight line with

$$\text{slope} = \frac{-\Delta H_r^0}{R} \quad \text{and} \quad \text{intercept} = -\ln K_{T_{\text{ref}}} + \frac{\Delta H_r^0}{R}\frac{1}{T_{\text{ref}}}$$

The linearity of such a plot would provide evidence that the assumption of constant ΔH_r^0 is justified. The value of ΔH_r^0 can then be evaluated from the slope of the plot. This method is used for obtaining enthalpy changes (at constant pressure) for reactions involving only condensed phases that would be difficult to measure directly.

If ΔH_r^0 varies with temperature, then its evaluation must take into account the variation of $(\Delta C_p^0)_r$ as a function of temperature, using analytical expressions, such as equation (4.53), for the C_p of each phase involved in the reaction. In the special case where $(\Delta C_p^0)_r$ = constant, , and the van't Hoff expression can be integrated to give

$$\ln K_T = \ln K_{T\text{ref}} - \frac{\Delta H_r^0}{R}\left(\frac{1}{T} - \frac{1}{T_{\text{ref}}}\right) - \frac{(\Delta C_P^0)_r}{R}\left[\left(\frac{T_{\text{ref}}}{T} - 1\right) - \ln\frac{T_{\text{ref}}}{T}\right] \tag{6.8}$$

The total change in K_{eq} with pressure at constant temperature is obtained by integrating equation (6.6) from some reference pressure (P_{ref}, usually taken as 1 bar), for which ΔV_r and are known, to the pressure of interest, P:

$$\int_{P_{\text{ref}}}^{P} d(\ln K) = \ln K_P - \ln K_{P_{\text{ref}}} = -\int_{P_{\text{ref}}}^{P} \frac{\Delta V_r^0}{RT} dP \tag{6.9}$$

If, for simplicity, ΔV_r^0 can be assumed to be independent of temperature and pressure,

$$\ln K_P = \ln K_{P\text{ref}} - \frac{\Delta V_r^0}{RT}(P - P_{\text{ref}}) \tag{6.4}$$

In this case, a plot of $\ln Kp$ versus P would yield a straight line with slope $= -\Delta V_r^0/RT$ and intercept $= \ln K_{P_{\text{ref}}} + \frac{\Delta V_r^0}{RT} P_{\text{ref}}$.

If ΔV_r^0 is a function of temperature and pressure, equation (6.4) has to be modified by taking into account the coefficient of thermal expansion at constant pressure (α_P) and the coefficient of compressibility at constant temperature (β_T) of the substances involved in the reaction (see Box 4.5).

For a reaction to qualify as a sensitive geothermometer, its equilibrium constant should be a function largely of temperature (i.e., almost independent of pressure); the opposite is the case for a potentially good geobarometer. It is evident from equation (6.5) that a sensitive geothermometer should have a large value of ΔH_r^0 (and a small value of ΔV_r^0). The larger the value of ΔH_r^0 (positive or negative), the more rapid is the change of K_{eq} with temperature. This means that an error in ΔH_r^0 or in the activity–composition relationships will produce a small error in the calculated T if ΔH_r^0 is large. Equation (6.6) indicates that a sensitive geobarometer, on the other hand, should have a large value of ΔV_r^0 (positive or negative). For rigorous thermobarometry it is advisable to consider heat-capacity data to calculate ΔH_r^0 and equation (4.74) or some other appropriate equation to calculate ΔV_r^0.

To determine the temperature or pressure from a univariant reaction, one of the two variables must be known from an independent source. For a given mineral assemblage, the intersection of two univariant reaction boundaries in P–T space, however, will fix both temperature and pressure, with the uncertainty of the estimates determined by the angle of intersection of the two reaction boundaries (Fig. 6.2).

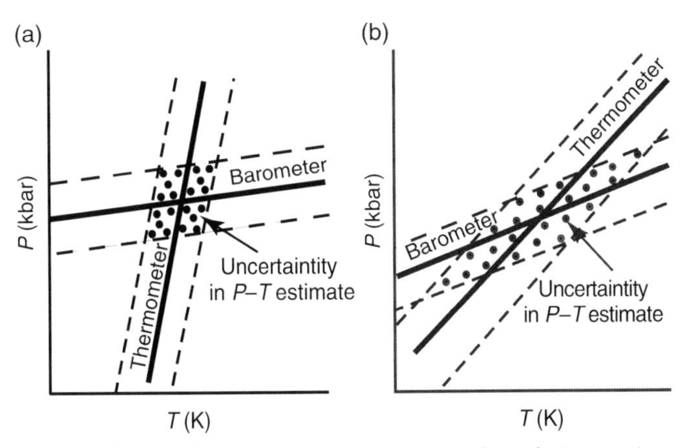

Fig. 6.2 Schematic illustration of thermobarometry from the intersection of two univariant reaction boundaries. The uncertainty in the position of each of the hypothetical reaction boundaries is shown by dashed lines and the uncertainties in the estimated pressure and temperature by the box with a pattern. Note that the uncertainties in the *P–T* estimate increase with a decrease in the angle of intersection, as in (a).

Example 6–1: Clinopyroxene–analbite equilibria as a function of pressure, temperature and the jadeite content of clinopyroxene

Consider the reaction that characterizes the equilibrium between omphacitic (Na-bearing) clinopyroxene (cpx) and disordered albite in blueschists, metamorphic rocks that are formed at high pressures and low temperatures in subduction zones. Representing the reaction in terms of end-member components,

$$\underset{\text{jadeite}}{\overset{\text{cpx}}{NaAlSi_2O_6}} + \overset{\text{quartz}}{SiO_2} \Leftrightarrow \overset{\text{analbite}}{NaAlSi_3O_8} \tag{6.10}$$

the equilibrium constant for the reaction at pressure P and temperature T is given by

$$\Delta G_r^0 = -RT \ln K_{eq} = -RT \ln \frac{a^{anb}_{NaAlSi_3O_8}}{a^{cpx}_{NaAlSi_2O_6}\, a^{qz}_{SiO_2}} \tag{6.11}$$

In blueschists, jadeitic pyroxene exhibits a wide range of composition because of the components diopside ($CaMgSi_2O_6$) and acmite ($NaFe^{3+}Si_2O_6$) in solid solution. In contrast, quartz and analbite are nearly pure phases. Let us choose the standard state for each phase as the pure substance at P and T, so that $a^{qz}_{SiO_2} = 1$ and $a^{anb}_{NaAlSi_3O_8} = 1$. Let us also stipulate, as a first approximation, that the cpx solid solution is ideal and that the activity of jadeite can be approximated on the basis of a one-site mixing model: $a^{cpx}_{jd} = (X^{cpx}_{Na})^{M2}$ (Holland, 1979). Substituting the activity values in equation (6.11), we get

$$K_{eq} = 1/a^{cpx}_{NaAlSi_2O_6} = 1/X^{cpx}_{NaAlSi_2O_6} \tag{6.12}$$

Thus, assuming that the pyroxene composition has not changed since its formation, the mole fraction of jadeite in the cpx solid solution will be a measure of the equilibrium constant at the pressure and temperature of metamorphism.

Assuming $(\Delta C_p)_r = 0$ and ΔV_r to be independent of temperature and pressure, ΔG_r at the chosen standard state (P, T) is given by (see equation 4.95)

$$\Delta G_r^0 = \Delta G^P_{r,T} = \Delta H^1_{r,T} - T\,\Delta S^1_{r,T} + (P-1)\,\Delta V_r$$

so that the equation relating X^{cpx}_{jd}, P, and T (equation 6.12) can be written as

$$\ln K_{eq} = \ln \frac{1}{X^{cpx}_{jd}} = \frac{-\Delta G_r^0}{RT} = \frac{-\Delta H^1_{r,T}}{RT} + \frac{\Delta S^1_{r,T}}{R} - \frac{(P-1)\,\Delta V_r}{RT} \tag{6.13}$$

We can use equation (6.13) to calculate P as a function of X^{cpx}_{jd} (or $\ln K_{eq}$) for any given temperature. For the purpose of illustration, the results for $T = 800$ K are presented in Table 6.2 and Fig. 6.3.

Similar calculations will yield a straight line in $P - \ln K_{eq}$ space for each chosen T, with the equilibration pressure being higher at higher temperatures for the same X_{jd}^{cpx} (Fig. 6.3). Thus, if an independent estimate of temperature is available, the jadeite content of cpx in a clinopyroxene–albite–quartz assemblage can be used as a geobarometer (e.g., Reinsch,

Table 6.2 Thermodynamic data (Robie and Hemingway, 1995) and calculation of pressure for the clinopyroxene–analbite equilibria at T = 800 K as a function of clinopyroxene composition.

	Thermodynamic data			**Calculation of equilibrium *P* for the reaction**					
	$\Delta H^1_{f,T}$ (kJ mol^{-1})	S^1_T (J mol^{-1} K^{-1})	$V^1_{298.15}$ (J bar^{-1})	X^{cpx}_{jd}	$\ln K_{eq}$	P (kbar)	X^{cpx}_{jd}	$\ln K_{eq}$	P (kbar)
Jadeite	−3027.8	332.22	6.04	0.1	2.30	4.80	0.6	0.51	11.67
				0.2	1.61	7.46	0.7	0.36	12.26
Analbite	−3920.2	480.35	10.043	0.3	1.20	9.02	0.8	0.22	12.78
				0.4	0.92	10.12	0.9	0.11	13.23
Quartz	−907.4	99.83	2.269	0.5	0.69	10.97	1.0	0	13.63

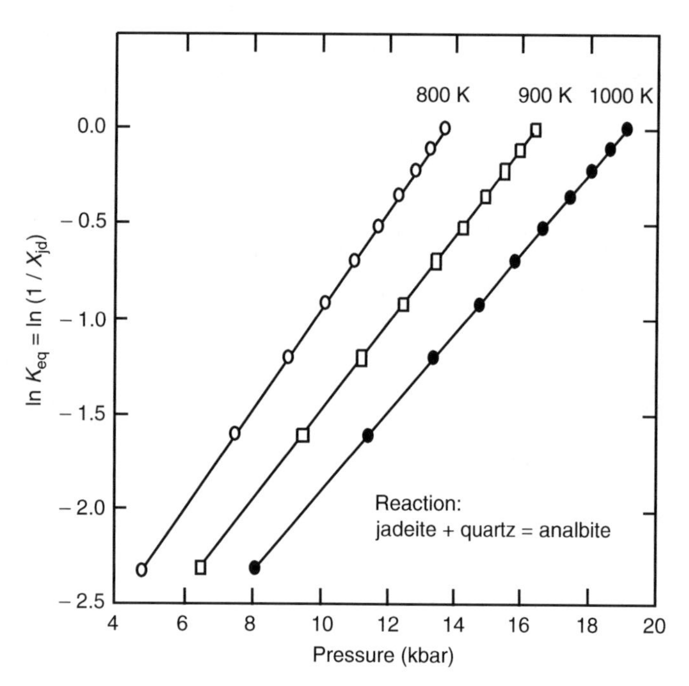

Fig. 6.3 Calculated relationship between ln K_{eq} and pressure for the reaction jadeite + quartz = analbite at constant temperatures 800 K, 900 K, and 1000 K. It is assumed that quartz and analbite are pure phases and clinopyroxene is an ideal solid solution with one mixing site.

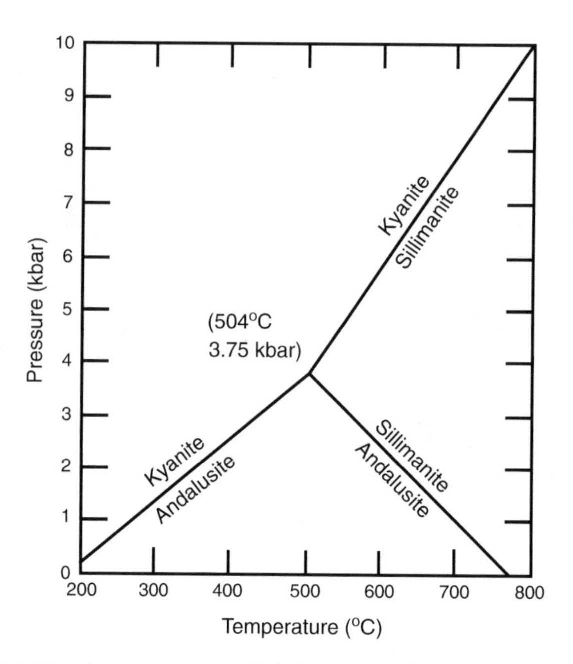

Fig. 6.4 The phase diagram for Al_2SiO_5 polymorphs (andalusite, kyanite, and sillimanite) as calculated by Holdaway and Mukhopadhyay (1993) from available thermodynamic data. The triple point is located at 504 ± 20°C and 3.75 ± 0.25 kbar. (After Holdaway and Mukhopadhyay, 1993, Fig. 5, p. 312.)

1977), provided the Na content of the cpx is measured accurately, for example, with an electron microprobe without any significant loss due to volatilization under the electron beam. The calculations, of course, will be a little more complicated for natural assemblages in which the plagioclase is not pure analbite or the cpx should not be assumed to behave as an ideal solid solution (see, e.g., Ganguly, 1973).

6.4 Univariant reactions and displaced equilibria

Univariant reactions and displaced equilibria involve consumption and production of phases and are referred to as *net transfer reactions*. The essence of thermobarometry based on such reactions is the determination of reaction boundaries as a function of temperature and pressure. The calculations, however, tend to get more cumbersome compared with the examples discussed above and in the previous two chapters. This is because most minerals in natural assemblages are solid solutions of variable composition, which generally do not conform to ideal mixing models.

6.4.1 *Al_2SiO_5 polymorphs*

One of the most commonly used thermobarometers for crustal rocks is based on the relative stabilities of the three Al_2SiO_5 polymorphs: andalusite, kyanite, and sillimanite. The phase diagram for this system consists of three univariant reaction boundaries (andalusite ⇔ sillimanite, sillimanite ⇔ kyanite, and kyanite ⇔ andalusite) that intersect at an invariant point (the triple point of the system) where all the three phases coexist in equilibrium at a specific set of *P–T* values. The phase relations in this system have been investigated experimentally by a number of workers (see reviews in Kerrick, 1990 (Chapter 3), and Holdaway and Mukhopadhyay, 1993); most petrologists now accept the reaction boundaries as determined by Holdaway and Mukhopadhyay (1993), who located the triple point at 504 ± 20°C and 3.75 ± 0.25 kbar (Fig. 6.4), not much different from some of the earlier determinations: 501 ± 20°C and 3.76 ± 0.3 kbar by Holdaway (1971); 503°C and 3.73 kbar by Berman (1988); and 511°C and 3.87 kbar by Hemingway *et al.* (1991).

Metamorphosed aluminous rocks, such as pelitic schists and micaceous quartzites, commonly contain one or more of these polymorphs. Assemblages containing only one of the polymorphs are not of much use in thermobarometry, but those containing any two of the three can be used as a geothermometer or a geobarometer, depending on whether the pressure or the temperature can be estimated from some other thermobarometer. Assemblages in which all three polymorphs coexist are not common, but have been reported (e.g., Grambling, 1981; Hiroi and Kobayashi, 1996; Whitney, 2002). In such a natural assemblage, the triple point of the Al_2SiO_5 phase diagram uniquely defines the equilibrium pressure and temperature, provided the Al_2SiO_5 phases are almost pure, and they formed contemporaneously.

A possible problem with the application of this thermobarometry is the Fe_2O_3 contents of andalusite and sillimanite, the effect of which is to increase the temperature of the andalusite ⇔ sillimanite reaction in proportion to the amount of Fe_2O_3 in the minerals undergoing reaction. Holdaway and Mukhopadhyay (1993) have cautioned that, with Fe_2O_3 as an impurity, the andalusite ⇔ sillimanite phase boundary may be up to 20°C higher than that in the pure aluminum silicate system. We should also remember that coexistence of the three polymorphs in a rock does not necessarily mean that they grew together at the *P–T* condition of the triple point. In some cases the textures clearly indicate a sequential growth of the polymorphs (Whitney, 2002).

6.4.2 *Garnet–rutile–Al_2SiO_5 polymorph–ilmenite–quartz (GRAIL) barometry*

A useful geobarometer is based on the natural assemblage garnet (gt)–rutile (rut)–Al_2SiO_5 polymorph (andalusite, kyanite, or sillimanite)–ilmenite (ilm)–quartz (qz) (commonly referred to as GRAIL) that is common in medium to high-grade, Al-rich metamorphic rocks. The equilibrium of interest in this case, with kyanite (ky) as the Al_2SiO_5 polymorph, can be written in terms of end-member components as

$$\underset{\text{almandine}}{\overset{\text{gt}}{Fe_3Al_2Si_3O_{12}}} + \overset{\text{rut}}{3TiO_2} \Leftrightarrow \overset{\text{ilm}}{3FeTiO_3} + \overset{\text{ky}}{Al_2SiO_5} + \overset{\text{qz}}{2SiO_2} \quad (6.14)$$

$$K_{eq} = \frac{\left(a_{FeTiO_3}^{ilm}\right)^3 \left(a_{Al_2SiO_5}^{ky}\right) \left(a_{SiO_2}^{qz}\right)^2}{\left(a_{alm}^{gt}\right) \left(a_{TiO_2}^{rut}\right)^3} \quad (6.15)$$

The experimentally based *P–T* calibration of this reaction by Bohlen *et al.* (1983b) is shown in Fig. 6.5. The figure also shows contours of $\log_{10}K_{eq}$ calculated from the relation

$$\Delta P \cong -\frac{RT \log_{10} K_{eq}}{\Delta V_r} \quad (6.16)$$

using available data on molar volumes, isobaric thermal expansion, and isothermal compressibility to calculate ΔV_r. Note that the slopes of the $\log_{10}K_{eq}$ contours change slightly when they pass from the stability field of one mineral to that of another. The very gentle d*P*/d*T* slopes of the $\log_{10}K_{eq}$ contours translate to a maximum error of only about 0.5 kbar in the inferred pressure corresponding to temperature uncertainties of ±50°C; this is the reason why the GRAIL equilibrium qualifies as a sensitive geobarometer. We can use the $\log_{10}K_{eq}$ contours in Fig. 6.5 to estimate the equilibration pressure of a given GRAIL assemblage, provided (i) the assemblage can be inferred to have coexisted in equilibrium, (ii) we can obtain a reasonable estimate of the equilibration temperature of the assemblage from some geothermometer (e.g., garnet–biotite exchange reaction), and (iii) we have appropriate solution models to calculate activities of the end-member components in impure minerals and thus K_{eq} for the assemblage. However, the $\log_{10}K_{eq}$ contours in Fig. 6.5, derived from experimental data, are not dependent on the mixing models for the solid solutions in the reaction, and potential users can make their own choice of solution models for geobarometry applications.

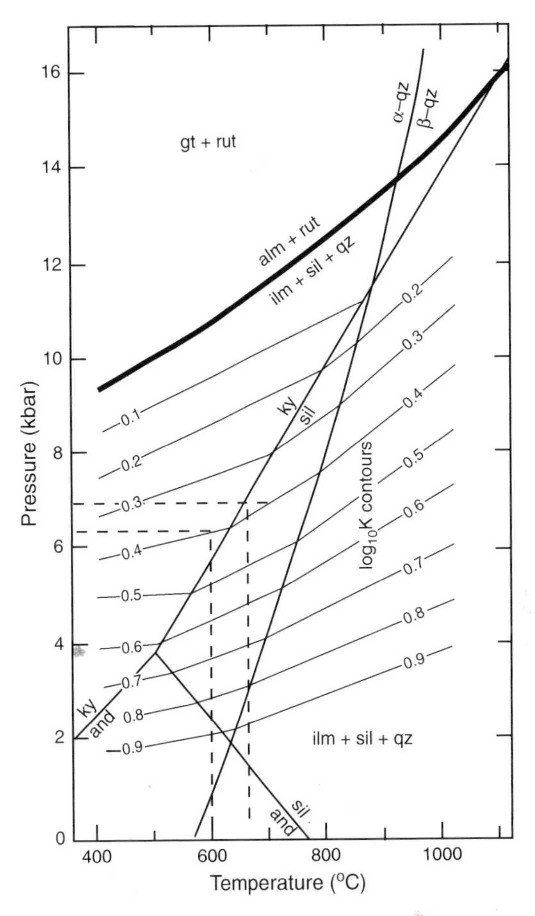

Fig. 6.5 Pressure–temperature diagram showing the experimentally based univariant reaction boundary for the GRAIL equilibrium (equation 6.9) involving only pure phases, and calculated contours of $\log_{10} K_{eq}$ for impure garnet and ilmenite compositions. The Al_2SiO_5 phase relations from Holdaway (1971) and the α-quartz–β-quartz transition from Cohen and Klement (1967) are shown for reference. Note that slopes of the contours change slightly when they cross reaction boundaries, reflecting changes in the Δ*V* of the reaction. Mineral abbreviations: and = andalusite, alm = almandine, gt = garnet, ilm = ilmenite, ky = kyanite, qz = quartz, rut = rutile, sil = sillimanite. (After Bohlen *et al.*, 1983b, Fig.3, p.1054.)

Equilibrium coexistence of the GRAIL assemblage minerals is sometimes difficult to demonstrate because of the very low (commonly < 1%) modal abundance of the Ti-bearing oxides. Bohlen *et al.* (1983b) recommended using assemblages in which ilmenite and rutile are present as inclusions in garnet and exhibit grain boundary relations indicative of chemical equilibrium.

An advantage of this geobarometer is the relatively less complicated calculation of its equilibrium constant. In natural GRAIL assemblages, compositions of three of the phases – quartz, rutile, and the Al_2SiO_5 polymorph – commonly differ little from the end-member components considered for reaction (6.9), so that their activities are equal to unity if the standard state of each phase is taken as the pure substance at the pressure and temperature of interest. Ilmenite in the GRAIL assemblage seldom contains more than 15% hematite (Fe_2O_3) and pyrophanite ($MnTiO_3$), and can be treated as an ideal solution or even as a pure phase in some cases. Thus, to a first approximation, the pressure inferred from a GRAIL assemblage in equilibrium is a function of the garnet composition (at constant temperature).

In order to calculate activities, Bohlen *et al.* (1983b) adopted the following symmetric ternary (Ca–Fe–Mg) mixing model for garnet (Perkins, 1979, as quoted in Bohlen *et al.*, 1983a):

$$RT \ln \lambda^{gt}_{alm} = W^G_{CaFe} X^2_{Ca} + W^G_{FeMg} X^2_{Mg} + (W^G_{CaFe} - W^G_{CaMg} + W^G_{FeMg}) X_{Ca} X_{Mg}$$

$$W^G_{FeMg} = 3480 - 1.2\, t\ (°C)\ \text{cal gm} - \text{atom}^{-1}$$

$$W^G_{CaFe} = 4180 - 1.2\, t\ (°C)\ \text{cal gm} - \text{atom}^{-1} \qquad (6.17)$$

$$W^G_{CaFe} = 1050 - 1.2\, t\ (°C)\ \text{cal gm} - \text{atom}^{-1}$$

$$a^{gt}_{alm} = (X_{Fe} \lambda^{gt}_{alm})^3$$

where W^Gs represent free-energy interaction parameters. The Perkins model, according to Bohlen *et al.* (1983b), is consistent with the bulk of the empirical data and yields pressures consistent with the appropriate Al_2SiO_5 mineral in the GRAIL assemblage. For garnets with significant Mn content, the authors assumed that Fe–Mn garnets mix ideally (Ganguly and Kennedy, 1974) and that $W^G_{CaMn} = W^G_{CaFe}$ and $W^G_{MgMn} = W^G_{MgFe}$.

Bohlen *et al.* (1983b) applied the GRAIL geobarometers to several metamorphic terranes and found the calculated pressures to be consistent with the appropriate Al_2SiO_5 polymorph(s) and generally in good agreement with other well-calibrated geobarometers.

Example 6–2: Application of GRAIL geobarometry to a sample from the Settler Schist, British Columbia, Canada

Let us estimate the equilibrium pressures for the Settler Schist corresponding to 600°C, using the analytical data of Pigage (1976), the solution model of Perkins (1979) for garnet, and the calibration of the GRAIL equilibrium by Bohlen *et al.* (1983b).

Electron microprobe analyses of minerals from the staurolite–kyanite–garnet–biotite–muscovite–quartz–plagioclase–ilmenite–rutile assemblage in the Settler Schist (Pigage, 1976) indicate that ilmenite and rutile are pure phases. Thus, taking the standard state as pure substances at the pressure and temperature of interest, and assuming kyanite and quartz to be a pure phases, $a^{qz}_{SiO_2} = 1$, $a^{ilm}_{FeTiO_3} = 1$, $a^{rut}_{TiO_2} = 1$, and $a^{ky}_{Al_2SiO_5} = 1$. Substituting the activity values in equation (6.15),

$$K_{eq} = \frac{1}{a^{gt}_{alm}} \qquad (6.18)$$

For the purpose of illustration, let us choose a typical composition of garnet in the Settler Schist: $(Mg_{0.4}Fe^{2+}_{2.15}Ca_{0.4}Mn_{0.05})Al_2Si_3O_{12}$. The calculated mole fractions for this composition are: $X_{Mg} = 0.133$, $X_{Fe^{2+}} = 0.717$, $X_{Ca} = 0.133$, and $X_{Mn} = 0.017$. Now we can calculate $a_{alm}{}^{gt}$ using equation (6.17) and K_{eq} using equation (6.18), and then obtain P from $\log_{10} K_{eq}$ contours in Fig. 6.5. The result, ignoring the very small contribution of Mn in the cpx solid solution, is summarized in Table 6.3.

The pressure calculated with the GRAIL geobarometer agrees well with the metamorphic conditions – 550 to 770°C and 6 to 8 kbar – inferred for the Settler Schist by Pigage (1976) using other methods.

6.4.3 *Garnet–plagioclase–pyroxene–quartz (GAPES and GADS) barometry*

Two of the potentially most useful geobarometers are based on the mineral assemblage garnet (gt)–plagioclase (plag)–orthopyroxene (opx) or clinopyroxene (cpx)–quartz (qz), which occurs widely in quartzofeldspathic and mafic lithologies of granulite metamorphic facies. The reactions considered for this purpose, commonly referred to as GAPES

Table 6.3 Results of the GRAIL geobarometry calculation for the Settler Schist.

Temperature (°C)	W^G_{FeMg}	W^G_{CaMg}	W^G_{CaFe}	$RT \ln \lambda^{gt}_{alm}$	λ^{gt}_{alm}	$a^{gt}_{alm} = (X_{Fe^{2+}} \lambda^{gt}_{alm})^3$	K_{eq}	$\log_{10} K_{eq}$	P (kbar)
600	2760	3460	330	47.999	1.205	0.646	2.494	0.397	≈ 6.2

W^Gs are in cal gm - atom^{-1}; $R = 1.987$ cal mol^{-1}K^{-1}.

(grossular–anorthite–pyrope–enstatite–silica) and GADS (garnet–anorthite–diopside–silica), are:

$$\underset{\text{anorthite (an)}}{\overset{\text{plag}}{CaAl_2Si_2O_8}} + \underset{\text{enstatite (en)}}{\overset{\text{opx}}{Mg_2Si_2O_6}} \Leftrightarrow 2/3\underset{\text{pyrope (py)}}{\overset{\text{gt}}{Mg_3Al_2Si_3O_{12}}} + 1/3\underset{\text{grossular (gr)}}{\overset{\text{gt}}{Ca_3Al_2Si_3O_{12}}} + \overset{\text{qz}}{SiO_2} \quad (6.19)$$

$$\underset{\text{anorthite (an)}}{\overset{\text{plag}}{CaAl_2Si_2O_8}} + \underset{\text{diopside (diop)}}{\overset{\text{opx}}{CaMgSi_2O_6}} \Leftrightarrow 1/3\underset{\text{pyrope (py)}}{\overset{\text{gt}}{Mg_3Al_2Si_3O_{12}}} + 2/3\underset{\text{grossular (gr)}}{\overset{\text{gt}}{Ca_3Al_2Si_3O_{12}}} + \overset{\text{qz}}{SiO_2} \quad (6.20)$$

These reactions are accompanied by large volume changes and thus are suitable, in principle, for geobarometry.

Experimental investigations of the two reactions at high temperature and high pressure have been hindered by apparent metastable persistence of anorthite + pyroxene assemblages, but the reactions were calibrated by Eckert *et al.* (1991) using measured values of the relevant thermodynamic parameters. For simplicity, let us assume that $(\Delta C_p)_r = 0$ (i.e., ΔH_r and ΔS_r are independent of temperature) and ΔV_r is independent of temperature and pressure. Choosing the standard state for each phase as the pure substance at P and T, the pressure and temperature of interest, and recognizing that $P >> 1$ bar, equation (6.13) reduces to

$$\Delta G^0_r = \Delta H^1_{r,T} - T\,\Delta S^1_{r,T} + P\,\Delta V_r = -RT \ln K_{eq} \quad (6.21)$$

Substituting values of ΔH^1_r, ΔS^1_r, and ΔV^1_r listed in Table 6.4, the thermobarometric equations, with uncertainties in the calculated P, are (Eckert *et al.*, 1991):

GADS... GAPES: $$P\ (\text{kbars}) = 3.47 + 0.01307\,T + 0.003503\,T \ln K_{eq}\ (\pm 1.55\,\text{kbar}) \quad (6.22)$$

GADS: $$P\ (\text{kbars}) = 2.60 + 0.01718\,T + 0.003596\,T \ln K_{eq}\ (\pm 1.90\,\text{kbar}) \quad (6.23)$$

To calculate P from equations (6.22) and (6.23) for a known or assumed value of T, we need to evaluate the corresponding values of K_{eq}:

$$K_{eq\,(GAPES)} = \frac{(a^{gt}_{py})^{\frac{2}{3}}(a^{gt}_{gr})^{\frac{1}{3}}\,a^{qz}_{SiO_2}}{a^{plag}_{an}\,a^{opx}_{en}} \quad (6.24)$$

$$K_{eq\,(GADS)} = \frac{(a^{gt}_{py})^{\frac{1}{3}}(a^{gt}_{gr})^{\frac{2}{3}}\,a^{qz}_{SiO_2}}{a^{plag}_{an}\,a^{cpx}_{diop}} \quad (6.25)$$

The activity of quartz, almost always a pure phase, was taken as unity. For the solid solutions in equation (6.19),

Table 6.4 Values used by Eckert et al. (1991) for calibration of GAPES and GADS geobarometers.

	ΔH^1_r (J)	ΔS^1_r (J K^{-1})	ΔV^1_r (J bar^{-1})
GAPES reaction	8230	−31.033	−2.373
GADS reaction	6020	−39.719	−2.312

Eckert *et al.* (1991) adopted the solution models discussed in Newton and Perkins (1982) and Perkins and Chipera (1985). The garnet was treated as a ternary (Ca–Fe–Mg) symmetric regular solution (Ganguly and Kennedy, 1974), with $W^G_{CaFe} \approx 0$, and the activity coefficients of the grossular and pyrope components in garnet solid solution were calculated using the following expressions:

$$\begin{aligned} RT \ln \lambda^{gt}_{Ca} &= W^G_{CaFe}X^2_{Fe} + W^G_{CaMg}X^2_{Mg} + (W^G_{CaFe} - W^G_{FeMg} + W^G_{CaMg})X_{Fe}X_{Mg} \\ &= W^G_{CaMg}(X^2_{Mg} + X_{Fe}X_{Mg}) \end{aligned} \quad (6.26)$$

$$\begin{aligned} RT \ln \lambda^{gt}_{Mg} &= W^G_{FeMg}X^2_{Fe} + W^G_{CaMg}X^2_{Ca} + (W^G_{FeMg} - W^G_{CaFe} + W^G_{CaMg})X_{Fe}X_{Ca} \\ &= W^G_{CaMg}(X^2_{Ca} + X_{Fe}X_{Ca}) \end{aligned}$$

Assuming no mixing on the octahedral and tetrahedral sites and taking into account the three positions of Ca–Fe–Mg mixing on the cubic site in the formula unit of garnet (12 O atoms basis),

$$a^{gt}_{py} = (X_{Mg}\lambda_{Mg})^3 = X^3_{Mg}\left\{\exp\left[\frac{W^G_{CaMg}}{RT}(X^2_{Ca} + X_{Ca}X_{Fe})\right]\right\}^3 \quad (6.27)$$

$$a^{gt}_{gr} = (X_{Ca}\lambda_{Ca})^3 = X^3_{Ca}\left\{\exp\left[\frac{W^G_{CaMg}}{RT}(X^2_{Ca} + X_{Mg}X_{Fe})\right]\right\}^3 \quad (6.28)$$

where $W^G_{CaMg} = 13807 - 6.3T$ J mol^{-1} (four O atoms basis), and $R = 8.314$ J mol^{-1} K^{-1}.

Expressions used for calculating the activities of enstatite and diopside components in pyroxene on the basis of an ideal two-site MOS (mixing-on-site) model were adopted from Wood and Banno (1973):

$$a^{opx}_{en} = X^{M2}_{Mg}\,X^{M1}_{Mg} \quad (6.29)$$

$$a^{cpx}_{diop} = X^{M2}_{Ca}\,X^{M1}_{Mg} \quad (6.30)$$

The cation assignments for calculation of mole fractions in clinopyroxene were as follows: Ca, Na, Mn, and Fe^{2+} to $M2$; and Mg, Ti, Fe^{3+}, Al^{VI}, and remaining Fe^{2+} to $M1$. The cation assignments were the same for orthopyroxene, except that Fe^{2+} and Mg were considered to be distributed randomly between $M1$ and $M2$ sites and partitioned between the two sites such that

$$\left(\frac{Mg}{Mg+Fe^{2+}}\right)^{M1} = \left(\frac{Mg}{Mg+Fe^{2+}}\right)^{M2} = \left(\frac{Mg}{Mg+Fe^{2+}}\right)^{mineral} \quad (6.31)$$

The above formulation assumes that there is no mixing on the tetrahedral positions.

The calculation of the activity coefficient of the anorthite component in plagioclase (λ_{an}^{plag}) was based on the "Al-avoidance" model of Kerrick and Darken (1975): that

$$RT \ln \lambda_{an}^{plag} = X_{ab}^2[W_{an}^G + 2X_{an}(W_{ab}^G - W_{an}^G)] \quad (6.32)$$

where $W_{an}^G = 2025\,\text{cal} = 8473\,\text{J}$ and $W_{ab}^G = 6746\,\text{cal} = 28225\,\text{J}$ (Newton *et al.*, 1980). The activity coefficient multiplied with the activity of anorthite in an ideal plagioclase solid solution (equation 5.88) led to the following expression for a_{an}^{plag}

$$\begin{aligned} a_{an}^{plag} &= \frac{X_{an}(1+X_{an})^2}{4}\lambda_{an}^{plag} \\ &= \frac{X_{an}(1+X_{an})^2}{4}\exp\left[\frac{X_{ab}^2(8473+39505X_{an})}{RT}\right] \end{aligned} \quad (6.33)$$

Eckert *et al.* (1992) calculated GAPES and GADS pressures for a granulite-facies core of Paleozoic metamorphism in the Blue Ridge province of the North Carolina Appalachians, at an inferred temperature of 750°C, using five different formulations for each barometer. The calculated GAPES pressures (7.2 to 8.9 kbars) showed reasonable to good agreement, all falling in the sillimanite field of the experimental Al_2SiO_5 phase diagram, consistent with the location of the samples relative to mapped kyanite and sillimanite isograds in the area and with the presence of peak-metamorphic sillimanite in nearby metapelites. The calculated GADS pressures agreed reasonably well with the GAPES pressures from the same rock or adjacent outcrops, but were systematically higher by 130–590 bars, suggesting that an empirical pressure adjustment of −350 bars to the GADS calculations would be necessary to force an agreement between the two geobarometers.

6.5 Exchange reactions

Exchange reactions are heterogeneous reactions that are written only in terms of the exchange of two similar cations (in respect of ionic radius and charge) between nonequivalent atomic sites in one mineral (*intracrystalline exchange*) or between atomic sites in two different coexisting minerals (*intercrystalline exchange*). Exchange reactions are accompanied by very small volume changes but relatively large entropy changes, so that they are largely independent of pressure and potentially good geothermometers (Essene, 1982). However, intracrystalline exchange, such as of Fe^{2+}–Mg^{2+} between $M1$ and $M2$ sites in orthopyroxene,

$$Fe^{2+}(M2) + Mg^{2+}(M1) = Fe^{2+}(M1) + Mg^{2+}(M2) \quad (6.34)$$

requires diffusion of Fe^{2+} and Mg^{2+} to be operative over very short distances, which makes the cation distribution prone to resetting. In general, intercrystalline cation distributions are relatively less vulnerable to resetting, as the diffusion has to occur over much larger distances.

The general form of an intercrystalline exchange equilibrium involving 1 mole each of two elements (or cations) A and B between the phases α and β may be written as

$$A(\alpha) + B(\beta) \Leftrightarrow A(\beta) + B(\alpha) \quad (6.35)$$

The equilibrium constant for the reaction at P and T is

$$\begin{aligned} K_{eq}(P,T) &= \frac{(a_A)^\beta\,(a_B)^\alpha}{(a_A)^\alpha\,(a_B)^\beta} = \frac{(X_A)^\beta\,(X_B)^\alpha}{(X_A)^\alpha\,(X_B)^\beta}\cdot\frac{(\lambda_A)^\beta\,(\lambda_B)^\alpha}{(\lambda_A)^\alpha\,(\lambda_B)^\beta} \\ &= \frac{(X_A/X_B)^\alpha}{(X_A/X_B)^\beta}\cdot\frac{(\lambda_A/\lambda_B)^\alpha}{(\lambda_A/\lambda_B)^\beta} \end{aligned} \quad (6.36)$$

where a_i, X_i, and λ_i represent, respectively, the activity, mole fraction, and rational activity coefficient of the constituent i. Denoting the ratio of mole fractions in equation (6.36) as K_D, the *empirical distribution coefficient* for the exchange (see section 3.7.2), and the ratio of activity coefficients as K_λ, then

$$K_{eq}(P,T) = K_D\,K_\lambda(P,T) \quad (6.37)$$

where K_λ is a function of P and T. Note that if α and β are ideal solutions, then $\lambda_i = 1$ and $K_{eq}(P,T) = K_D(P,T)$. In this case, a plot of $(X_A/X_B)^\alpha$ versus $(X_A/X_B)^\beta$ for a suite of rocks that equilibrated at the same temperature and pressure should define a straight line with a slope equal to K_D. As a general rule, the preference of a phase to exchange one element for another decreases with increasing temperature and K_D approaches a value of 1. This is because as temperature is increased, the energetic distinction between different elements becomes relatively smaller, so the crystal structures display less of a preference for one element over another (Spear, 1993).

Choosing the standard state of each phase as a pure substance at the P and T of interest, and assuming $(\Delta C_P)_r = 0$, ΔV_r (solids) to be independent of temperature and pressure, and $P >> 1$ bar, we have (see equation 6.21)

$$-RT \ln K_{eq}(P,T) = \Delta H_{r,T}^1 - T\,\Delta S_{r,T}^1 + P\,\Delta V_r \quad (6.38)$$

Application of this relationship to geothermometry requires the calibration of K_{eq} as a function of T for known or assumed values of P.

Many exchange geothermometers have been discussed in the literature, but here we will restrict our attention to Fe^{2+}–Mg^{2+} exchange reactions, the most widely applied thermometers in metamorphic rocks. This is because Fe^{2+}–Mg^{2+} substitute easily for each other in many common metamorphic minerals and the relatively high concentrations of Fe^{2+}–Mg^{2+} in these minerals permit their precise measurement with an electron microprobe. Microprobe analysis, however, measures the total Fe in a mineral, which then has to be recast into Fe^3 and Fe^2 by some established procedure.

6.5.1 *Garnet–clinopyroxene thermometry*

The temperature dependence of Fe^{2+}–Mg^{2+} exchange between the minerals garnet and clinopyroxene has long since been recognized as an important potential geothermometer, because the assemblage is common in amphibolites, granulites, and eclogites that span a broad temperature range. It is probably the most useful and consistent of the Fe–Mg exchange thermometers for high-grade metamorphic terranes (Pattison and Newton, 1989) and is particularly applicable to low-Na and low-Cr mineral compositions that are typical of eclogites and granulites. The pertinent exchange reaction in terms of end-member components is

$$\underset{\text{pyrope (py)}}{\overset{\text{gt}}{1/3Mg_3Al_2Si_3O_{12}}} + \underset{\text{hedenbergite (hd)}}{\overset{\text{cpx}}{CaFeSi_2O_6}} \Leftrightarrow \underset{\text{almandine (alm)}}{\overset{\text{gt}}{1/3Fe_3Al_2Si_3O_{12}}} + \underset{\text{diopside (diop)}}{\overset{\text{cpx}}{CaMgSi_2O_6}} \quad (6.39)$$

Taking into account the three positions for Fe–Mg mixing in the garnet (12 O atom basis) and only one position of Fe–Mg mixing in the clinopyroxene, the equilibrium constant of the reaction is:

$$K_{eq} = \frac{(a_{alm}^{gt})^{\frac{1}{3}}\, a_{diop}^{cpx}}{(a_{py}^{gt})^{\frac{1}{3}}\, a_{hd}^{cpx}} = \frac{X_{Fe}^{gt}\, X_{Mg}^{cpx}}{X_{Fe}^{cpx}\, X_{Mg}^{gt}}\ \frac{\lambda_{Fe}^{gt}\, \lambda_{Mg}^{cpx}}{\lambda_{Fe}^{cpx}\, \lambda_{Mg}^{gt}}$$
$$= \frac{\left(X_{Fe}/X_{Mg}\right)^{gt}}{\left(X_{Fe}/X_{Mg}\right)^{cpx}}\ \frac{\left(\lambda_{Fe}/\lambda_{Mg}\right)^{gt}}{\left(\lambda_{Fe}/\lambda_{Mg}\right)^{cpx}} = K_D\, K_\lambda \quad (6.40)$$

Several authors have proposed equations relating T to K_{eq} (Table 6.5), which can be used to calculate T from the compositions of coexisting garnet and clinopyroxene if P is estimated from some other source. The empirical calibration of Dahl

Table 6.5 Some explicit equations proposed for garnet–clinopyroxene geothermometry

Author(s)	Data for calibration	Equation for T (K) (P in kbar)
Råheim and Green (1974)	Experimental	$\dfrac{3686 + 28.35\,P}{\ln K_D + 2.33}$ (6.41)
Ellis and Green (1979)	Experimental	$\dfrac{3104\,(X_{Ca}^{gt}) + 3030 + 10.86\,P}{\ln K_D + 1.9034}$ (6.42)
Dahl (1980)	Empirical (based on a granulite terrane)	$\dfrac{2482 + 1509(X_{Fe}^{gt} - X_{Mg}^{gt}) + 2810(X_{Ca}^{gt}) + 2855(X_{Mn}^{gt})}{R\ln K_D}$ (6.43)
Krogh (1988)	Reinterpretation of existing experimental data	$\dfrac{-6137\,(X_{Ca}^{gt})^2 + 6731\,(X_{Ca}^{gt}) + 1879 + 10\,P}{\ln K_D + 1.393}$ (6.44)
Pattison and Newton (1989)	Experimental	$\dfrac{a'X^3 + b'X^2 + c'X + d'}{\ln K_D + a_oX^3 + b_oX^2 + c_oX + d_o} + 5.5\,(P - 15)$ (6.45) where a_o, b_o, c_o, d_o, a′, b′, c′, and d′ are parameters determined from experimental data
Ai (1994)	Experimental	$\dfrac{[-1629(X_{Ca}^{gt})^2 + 3648.55(X_{Ca}^{gt}) - 6.59\,\{mg\,\#\,(gt)\} + 1987.98 + 17.66P]}{\ln K_D + 1.076}$ (6.46)

mg# (magnesium number) = [molar Mg/(Mg + Fe)]100. The uncertainties in the temperatures estimated by these geothermometers are generally ±5% in the 800°–1200°C range, but may be somewhat larger toward the low temperature end of this range because of the lack of extensive experimental data or well-constrained thermochemical parameters at such temperatures.

(1980) (equation 6.43) gives widely scattered, erratic results and is not reliable. The experimental calibration of Råheim and Green (1974) (equation 6.41) was based on a series of natural basaltic rocks [Mg/(Mg + Fe) = 0.062–0.85] crystallized to eclogite at 20–40 kbar and 600°–1400°C. In order to test the effects of variations in the compositions (Na, Ca) of clinopyroxene and garnet on K_D, Ellis and Green (1979) obtained additional experimental data on a series of basaltic compositions in the range of 24–30 kbar and 750°–1300°C. The variation of K_D over a wide range of pressure, temperature, and rock composition was found to be a function of X^{gt}_{Ca} [where X^{gt}_{Ca} = Ca/(Ca + Mg + Fe^{2+} + Mn)] at any given *P, T* condition and could be accounted for by the linear relationship

$$\ln K_D = c\, X^{gt}_{Ca} + d$$

where the constants *c* and *d*, which incorporate the total non-ideal effects on K_D, were determined from the experimental data (equation 6.42). Using experimental data available at the time, Krogh (1988) proposed a new equation (equation 6.44), which replaced the linear relationship between X_{Ca}(gt) and ln K_D by a curvilinear relationship. Application of this calibration to a suite of samples of eclogites and associated omphacite–garnet-bearing gneisses showed that the calculated temperatures did not vary much with rather large variations in the Mg:(Mg + Fe) ratio of the garnet (0.17–0.54) and Na content of the clinopyroxene (0.11–0.44).

Pattison and Newton (1989) showed from experimental study that the magnesium number (mg #) of garnet had a significant effect on K_D. They fitted their experimental data on synthetic compositions (mg # 12.5 to 60) to a third-order polynomial,

$$\ln K_D = aY^3 + bY^2 + cY + d$$

where *Y* represents the Mg:(Mg + Fe) ratio of the garnet and the coefficients are functions of $1/T$ (e.g., $a = a_o + a_1(1/T)$, etc.), an apparent contradiction with the experimental observation of Råheim and Green (1974) that K_D increases with increasing temperature. The geothermometer (equation 6.45) requires the extraction of eight parameters (or coefficients) for each garnet composition and, as the authors have cautioned, it should not be applied outside the experimental range of garnet composition, X^{gt}_{Ca} = 0.125 – 0.600. Overall, temperatures calculated using equation (6.45) are 0 to 60°C lower than those obtained by Råheim and Green (1974) and 60° to150°C lower than those of Ellis and Green (1979). For high-pressure mantle rocks (*T* greater than about 1000° C) temperatures calculated with the calibration of Ai (1994) (equation 6.46), based on all the experimental data available at the time, are similar to those of Ellis and Green (1979), but may be lower by as much as about 100°C for crustal rocks.

Berman *et al.* (1995) proposed a "provisional" solution model to evaluate ln K_λ for the calibration of the garnet–clinopyroxene geothermometer. For this model, the *M*1 site in clinopyroxene is treated as a symmetric ternary regular solution of Mg, Fe, and Al atoms, and the *M*2 site as a quaternary solution of Na, Ca, Fe and Mg atoms; Mg:(Mg + Fe) is assumed to be equal in both *M*1 and *M*2. The garnet is considered to be a Ca–Mg–Fe ternary solution, with each of the three binary joins modeled as asymmetric regular solution.

The geothermometer of Ellis and Green (1979) is the one that is most commonly used for intermediate, mafic, and ultramafic rocks from metamorphic terranes and from xenolith assemblages as it agrees more closely with feldspar- and oxide-based temperature determinations than other formulations. From a comparison of the various formulations of the garnet–clinopyroxene thermometer, Green and Adam (1991) concluded that the Ellis and Green (1979) calibration may result in an overestimation of temperature when applied to crustal rocks formed at low to moderate pressures (10–20 kbar), but temperatures for such rocks calculated using the solution model of Berman *et al.* (1995) are quite similar to that produced with the formulation by Ellis and Green (1979).

6.5.2 *Garnet–biotite (GABI) thermometry*

The garnet (gt)–biotite (bi) Fe–Mg exchange geothermometer is one of the most widely used geothermometers for estimating the temperature of equilibration of medium- and high-grade metamorphic rocks. The cation exchange equilibrium can be expressed in terms of end-member components of garnet and biotite as:

$$\underset{\text{almandine (garnet)}}{Fe_3Al_2Si_3O_{12}} + \underset{\text{phlogopite (biotite)}}{KMg_3AlSi_3O_{10}(OH)_2} \Leftrightarrow \underset{\text{pyrope (garnet)}}{Mg_3Al_2Si_3O_{12}} + \underset{\text{annite (biotite)}}{KFe_3AlSi_3O_{10}(OH)_2} \qquad (6.47)$$

The equilibrium constant of the reaction, taking into account the Fe–Mg exchange on three crystallographic sites, is:

$$K_{eq} = \frac{\left(X^{gt}_{Mg}\right)^3 / \left(X^{gt}_{Fe}\right)^3}{\left(X^{bi}_{Mg}\right)^3 / \left(X^{bi}_{Fe}\right)^3} \cdot \frac{\left(\lambda^{gt}_{Mg}\right)^3 / \left(\lambda^{gt}_{Fe}\right)^3}{\left(\lambda^{bi}_{Mg}\right)^3 / \left(\lambda^{bi}_{Fe}\right)^3} = (K_D\, K_\lambda)^3 \qquad (6.48)$$

If ΔH_r and ΔS_r for reaction (6.47) are assumed to be independent of temperature and ΔV_r independent of temperature and pressure, we can combine equations (6.38) and (6.48) to get the following (see Box 6.2):

$$\ln (K_D\, K_\lambda)(P,T) = \frac{1}{T}\left[\frac{-\left(\Delta H^1_{r,\,298.15} + P\,\Delta V_r\right)}{3R}\right] + \frac{\Delta S^1_{r,\,298.15}}{3R} \qquad (6.49)$$

Conducting reversed experiments for calibration of the partitioning of Fe and Mg between a large amount of synthetic Fe–Mg garnet of known composition and a small amount of

Box 6.2 Derivation of equation (6.49) for GABI thermometry

Separating the mole fractions term from the activity coefficients term in equation (6.48),

$$K_{eq} = (K_D\, K_\lambda)^3$$

$$\ln K_{eq} = 3 \ln (K_D\, K_\lambda)$$

where

$$K_D = \frac{\left(X_{Mg}^{gt}\right)/\left(X_{Fe}^{gt}\right)}{\left(X_{Mg}^{bi}\right)/\left(X_{Fe}^{bi}\right)} \quad \text{and} \quad K_\lambda = \frac{\left(\lambda_{Mg}^{gt}\right)/\left(\lambda_{Fe}^{gt}\right)}{\left(\lambda_{Mg}^{bi}\right)/\left(\lambda_{Fe}^{bi}\right)}$$

Substituting for K_{eq} in equation (6.38),

$$-RT \ln K_{eq}(P, T) = \Delta H^1_{r,T} - T\,\Delta S^1_{r,T} + P\,\Delta V_r$$

$$-3RT \ln (K_D\, K_\lambda)\,(P, T) = \Delta H^1_{r,T} - T\,\Delta S^1_{r,T} + P\,\Delta V_r$$

$$\ln (K_D\, K_\lambda)\,(P, T) = \frac{-\Delta H^1_{r,T} - T\,\Delta S^1_{r,T} + P\,\Delta V_r}{3RT}$$

which can be rearranged to give equation (6.49)

$$\ln (K_D\, K_\lambda)(P,T) = \frac{1}{T}\left[\frac{-\left(\Delta H^1_{r,298.15} + P\,\Delta V_r\right)}{3R}\right] + \frac{\Delta S^1_{r,298.15}}{3R} \qquad (6.49)$$

Fe–Mg biotite of unknown composition at temperatures ranging from 550 to 800°C, Ferry and Spear (1978) established the relation between K_D and T (at $P = 2.07$ kbar, the pressure at which the experiment was conducted) as:

$$\ln KD = -2109/T(K) + 0.782 \qquad (6.50)$$

where $K_D = (Mg/Fe)_{gt}/(Mg/Fe)_{bi}$ (either on a weight or atomic basis). This equation is consistent with ideal mixing of Fe and Mg in biotite and garnet solid solutions (i.e., $a_i = X_i$ so that $K_\lambda = 1$ in equation (6.49)), at least in the compositional range studied in their experiments ($0.80 \le Fe/(Fe + Mg) \le 1.00$). The coefficients in this equation were determined by a linear least-squares fit of the experimental values of $\ln K_D$ versus $1/T$ (Fig. 6.6). The straight-line plot in Fig. 6.7 indicates that the assumptions $-\Delta C_r^0 = 0$ and $\Delta V_r = 0$ – are quite reasonable. Comparing equation (6.49) with equation (6.50), and substituting $\Delta V_r = 0.057\,\text{cal bar}^{-1} = 0.238\,\text{J bar}^{-1}$ (using molar volume data at 298.15 K and 1 bar from Robie *et al.*, 1978), they determined that at $P = 2070$ bars (the pressure at which the experiment was conducted),

$$\Delta S^1_{r,298.15} = 3R\,(0.782) = 4.662\,\text{cal K}^{-1} = 19.51\,\text{J K}^{-1}$$

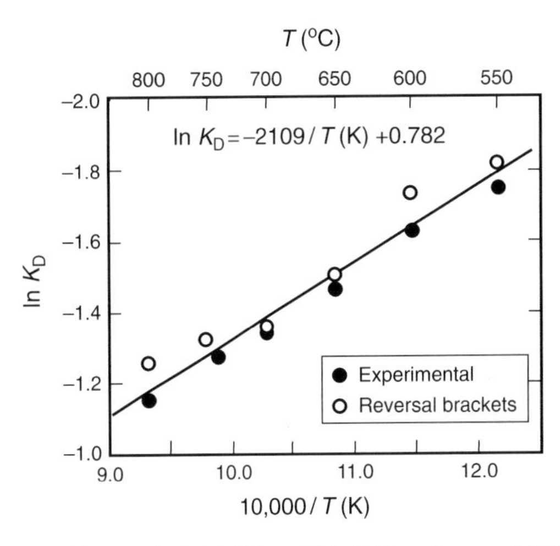

Fig. 6.6 Plot of $\ln K_D[= \ln \{(Mg/Fe)_{garnet}/(Mg/Fe)\}_{biotite}]$ versus 1/T (K) for experimentally equilibrated garnet–biotite pairs by Ferry and Spear (1978). The solid circle–open circle pairs represent reversal brackets, and the solid line the calculated least-squares fit to the data points. (After Ferry and Spear, 1978, Figure 3, p.115.)

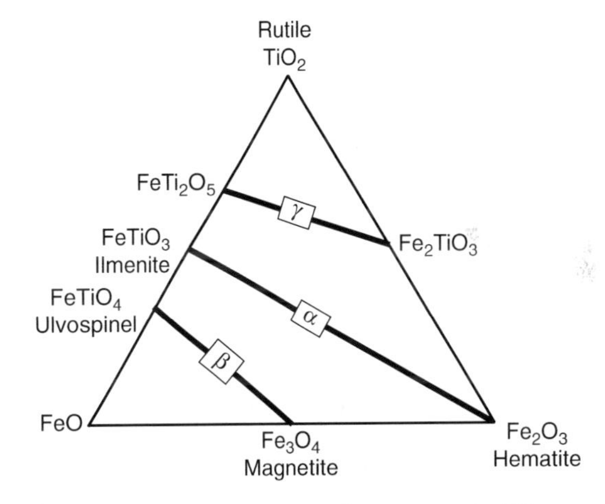

Fig. 6.7 The major solid solution series of the FeO–Fe_2O_3–TiO_2 system: ilmenite ($FeTiO_3$)–hematite (Fe_2O_3) solid solution [α phase, rhombohedral]; magnetite (Fe_3O_4)–ulvöspinel (Fe_2TiO_4) solid solution [β phase, cubic]; and ferropseudobrookite ($FeTi_2O_5$)–pseudobrookite (Fe_2TiO_5) solid solution [γ phase, orthorhombic]. Compositions are in mole percent.

$$\Delta H^1_{r,298.15} = 3R\,(2109) - 2070\,(\Delta V_r) = 12{,}454\,\text{cal} = 52{,}108\,\text{J}$$

Substituting the above values in equation (6.49) (and assuming $K_\lambda = 1$),

$$52,108\,(J) - 19.51\,T\,(K) + 0.238\,P\,(\text{bars}) + 3RT \ln K_D = 0 \qquad (6.51)$$

which can be rearranged as ($R = 8.314\,\text{J mol}^{-1}\,\text{K}^{-1}$):

$$\ln K_D = -(2089 + 0.0096P)/T + 0.782 \qquad (6.52)$$

Note that the effect of pressure on this exchange equilibrium is small (as it should be for a good geothermometer) so that the error introduced on account of uncertainty in the estimated pressure will be small. This reaction is appropriate for geothermometry because of the relatively large ΔH_r and quite small ΔV_r.

Equation (6.52) can be used to calculate T for a given rock first by calculating K_D from electron microprobe analysis of coexisting garnet and biotite, and then obtaining an independent estimate of P, provided that the garnet and biotite compositions lie close to the Mg–Fe binary and the Mg–Fe mixing is ideal in both the minerals. In reality, the activity–composition relations are more complicated because of dilution of the solid solutions – with Ca and Mn in garnets, and with Al^{VI} and Ti in biotites – and the nonideality of the solutions. In addition, natural and synthetic biotites and garnets contain Fe^{3+}, which was not been taken into account in the calibration of Ferry and Spear (1978). They considered it a useful thermometer (with a maximum practical resolution of approximately ±50°C) without correction for components other than Fe and Mg [(Ca + Mn)/(Ca + Mn + Fe + Mg) up to ~0.2 in garnet and $(Al^{VI} + Ti)/(Al^{VI} + Ti + Fe + Mg)$ up to ~0.15 in biotite].

Over the past two decades, a number of calibrations of the GABI geothermometer, incorporating different mixing models for garnet or biotite or both, have been discussed in the literature (see Table 6.1). Holdaway (2000) made a statistical comparison of several of these models and concluded that the models of Berman and Aranovich (1996), Ganguly *et al.* (1996), and Mukhopadhyay *et al.* (1997) can be used to produce reasonable garnet–biotite geothermometer calibrations.

6.5.3 *Magnetite–ilmenite thermometry and oxygen barometry*

The compositions of coexisting magnetite–ulvöspinel solid solution (mt_{ss} or β phase; cubic) and ilmenite–hematite solid solution (ilm_{ss} or α phase; rhombohedral) (Fig. 6.7), the two most common iron–titanium oxide minerals in rocks and mineral deposits, provide information about not only the temperature of equilibration but also the corresponding oxygen fugacity. The method is best suited to extrusive and hypabyssal igneous rocks that were subjected to relatively rapid cooling and are unmetamorphosed. Its application to plutonic igneous rocks and mineral deposits should be viewed with caution because of likely reequilibration subsequent to crystallization.

The crystal structure of both hematite ($Fe_2^{3+}O_3$) and ilmenite ($Fe^{2+}Ti^{4+}O_3$) may be described as a hexagonal close-packed array of oxygen atoms with octahedrally coordinated metal atoms in the interstices. In pure ilmenite, there are two cation sites, one (site A) containing Fe^{2+} and the other (site B) containing Ti^{4+}. In pure hematite the two sites are indistinguishable as both contain only Fe^{2+}. In the structure of the ilm_{ss}, Ti is assumed to be restricted to the B sites, as in pure ilmenite, where it mixes randomly with Fe^{3+}, and Fe^{2+} is restricted to the A sites where it mixes randomly with the rest of Fe^{3+} (Rumble, 1977). For the pure Fe–Ti–O system, a further constraint in this model is that the amount of Fe^{2+} and Fe^{3+} on the A and B sites must be equal in order to preserve charge balance. Thus, we may accept a structural model for ilm_{ss} that is ordered in terms of A and B sites but disordered in respect of Fe^{3+}.

In spinels ($A^{IV}B_2{}^{VI}O_4$) with "normal" structure, A cations occupy the tetrahedral sites and B cations only the octahedral sites. But both magnetite (Fe_3O_4) and ulvöspinel (Fe_2TiO_4) have "inverse" spinel structure – that is, one-half of the B cations occupy tetrahedral sites, and the other half of the B cations and the A cations occupy octahedral sites – and their structural formulas may be written as $(Fe^{3+})^{IV}[Fe^{2+}\,Fe^{3+}]^{VI}O_4$ and $(Fe^{2+})^{IV}[Fe^{2+}\,Ti^{4+}]^{VI}O_4$. Several models have been proposed to account for the cation distribution in mt_{ss}, and the activity–composition relationships vary depending on the model adopted.

Thermodynamic equilibrium between mt_{ss} and ilm_{ss} may be represented by the reaction among the end-member components as

$$\begin{array}{ccc} \beta \text{ phase} & & \alpha \text{ phase} \\ [y\mathrm{Fe_2TiO_4} + (1-y)\mathrm{Fe_3O_4}] & + \tfrac{1}{4}\mathrm{O_2} \Leftrightarrow & \left[y\mathrm{FeTiO_3} + \left(\tfrac{3}{2}-y\right)\mathrm{Fe_2O_3}\right] \end{array} \quad (6.53)$$

$$K_{eq} = \frac{\left(a^{\alpha}_{FeTiO_3}\right)^{y}\left(a^{\alpha}_{Fe_2O_3}\right)^{\frac{3}{2}-y}}{\left(a^{\beta}_{Fe_2TiO_4}\right)^{y}\left(a^{\beta}_{Fe_3O_4}\right)^{1-y}\left(f_{O_2}\right)^{\frac{1}{4}}} \quad (6.54)$$

As K_{eq} is a constant at given P, T and the effect of total pressure on this equilibrium is negligible (Buddington and Lindsley 1964), K_{eq} is essentially a function of temperature and f_{O_2}. Thus, the equilibrium temperature of the reaction and its associated f_{O_2} are defined simultaneously if the activities of all the solids are fixed. Buddington and Lindsley (1964) experimentally determined calibration curves for various compositions of the two solid solutions as a function of temperature and f_{O_2}, which could then be used to determine temperature and f_{O_2} for coexisting mt_{ss} and ilm_{ss} in equilibrium in natural assemblages. The calibration has subsequently been refined incorporating additional experimental data and solution models.

For thermodynamic modeling, the compositions of coexisting mt_{ss} and ilm_{ss} phases are considered in terms of two separate reactions involving end-member components:

(1) a Fe-Ti exchange reaction,

$$\underset{\text{ilm}}{\overset{\text{ilm}_{ss}}{\mathrm{FeTiO_3}}} + \underset{\text{mt}}{\overset{\text{mt}_{ss}}{\mathrm{Fe_3O_4}}} \Leftrightarrow \underset{\text{usp}}{\overset{\text{mt}_{ss}}{\mathrm{Fe_2TiO_4}}} + \underset{\text{hem}}{\overset{\text{ilm}_{ss}}{\mathrm{Fe_2O_3}}} \quad (6.55)$$

Since $\Delta G_r{}^0 = -RT \ln K_{eq}$ (equation 5.37),

$$-\frac{\Delta G^0_{exch}}{RT} = \ln K_{eq} = \ln \frac{\left(a^{mt_{ss}}_{usp}\right)^{\phi=1}\left(a^{ilm_{ss}}_{hem}\right)^{\phi=2}}{\left(a^{mt_{ss}}_{mt}\right)^{\phi=1}\left(a^{ilm_{ss}}_{ilm}\right)^{\phi=2}}$$
$$= \ln \frac{\left(X^{mt_{ss}}_{usp}\right)^{\phi=1}\left(X^{ilm_{ss}}_{hem}\right)^{\phi=2}}{\left(X^{mt_{ss}}_{mt}\right)^{\phi=1}\left(X^{ilm_{ss}}_{ilm}\right)^{\phi=2}} + \ln \frac{\left(\lambda^{mt\,ss}_{usp}\right)^{\phi=1}\left(\lambda^{ilm\,ss}_{hem}\right)^{\phi=2}}{\left(\lambda^{mt\,ss}_{mt}\right)^{\phi=1}\left(\lambda^{ilm\,ss}_{ilm}\right)^{\phi=2}} \quad (6.56)$$

(2) an oxidation reaction (oxidation of magnetite to hematite),

$$\underset{\text{mt}}{\overset{\text{mt}_{\text{ss}}}{4\text{Fe}_3\text{O}_4}} \Leftrightarrow \underset{\text{hem}}{\overset{\text{ilm}_{\text{ss}}}{6\text{Fe}_2\text{O}_3}} \tag{6.57}$$

$$-\frac{\Delta G^0_{\text{oxid}}}{RT} = \ln K_{\text{eq}} = \ln \frac{\left(a^{\text{ilm ss}}_{\text{hem}}\right)^{6\,.\,\phi=2}}{\left(a^{\text{mt ss}}_{\text{mt}}\right)^{4.\,\phi=1} f_{\text{O}_2}}$$
$$= \ln \frac{\left(X^{\text{ilm ss}}_{\text{hem}}\right)^{6\,.\,\phi=2}}{\left(X^{\text{mt ss}}_{\text{mt}}\right)^{4\,.\,\phi=1}} + \ln \frac{\left(\lambda^{\text{ilm ss}}_{\text{hem}}\right)^{6\,.\,\phi=2}}{\left(\lambda^{\text{mt ss}}_{\text{mt}}\right)^{4\,.\,\phi=1}} - \ln f_{\text{O}_2} \tag{6.58}$$

The variable ϕ in equations (6.56) and (6.58) is set to 2 for ilm_{ss} and to 1 for mt_{ss} in accordance with configurational entropy expressions (Rumble, 1977). Spencer and Lindsley (1981) used the above reactions to revise the calibration of Buddington and Lindsley (1964) with a solution model based on least squares fit of thermodynamic parameters to experimental data in the range 550–1200°C obtained by various workers. The assumptions for the model are as follows: (i) ilm_{ss} behaves as a binary asymmetric regular solution; (ii) mt_{ss} behaves as a binary asymmetric regular solution below 800°C and as a binary ideal solution above 800°C; (iii) the activities of the components can be approximated by a molecular mixing model for the mt_{ss} (with the proportions of the different end-member "molecules" calculated in a specified sequence) and by disorder of Fe^{3+} in ilm_{ss}. The equations derived by them for calculation of the temperature and f_{O_2} of equilibration of a mt_{ss}–ilm_{ss} pair are:

$$T\ (\text{K}) = \frac{-\text{A}_1 W^H_{\text{usp}} - \text{A}_2 W^H_{\text{mt}} + \text{A}_3 W^H_{\text{ilm}} + \text{A}_4 W^H_{\text{hem}} + \Delta H^0_{\text{exch}}}{-\text{A}_1 W^S_{\text{usp}} - \text{A}_2 W^S_{\text{mt}} + \text{A}_3 W^S_{\text{ilm}} + \text{A}_4 W^S_{\text{hem}} + \Delta S^0_{\text{exch}} - R \ln K_{\text{exch}}} \tag{6.59}$$

$$\log f_{\text{O}_2} = \log \text{MH} + \frac{1}{2.303}\left[\begin{array}{l} 12 \ln(1 - X_{\text{ilm}}) - 4 \ln(1 - X_{\text{usp}}) + \\ \dfrac{1}{RT}\left\{\begin{array}{l} 8X^2_{\text{usp}}\left(X_{\text{usp}} - 1\right)W^G_{\text{usp}} + 4X^2_{\text{usp}}\left(1 - 2X_{\text{usp}}\right)W^G_{\text{mt}} \\ +12X^2_{\text{ilm}}\left(1 - X_{\text{ilm}}\right)W^G_{\text{ilm}} - 6X^2_{\text{ilm}}\left(1 - 2X_{\text{ilm}}\right)W^G_{\text{hem}} \end{array}\right\} \end{array}\right] \tag{6.60}$$

where W^H and W^G are enthalpy and free-energy interaction parameters, respectively, and log MH is the value of log f_{O_2} of coexisting mt_{ss}–ilm_{ss} at T and is calculated as log MH = 13.996 – 24634/T.

The parameters for this model are listed in Table 6.6. Spencer and Lindsley (1981) suggested using the low-T parameters to calculate an initial temperature, and recalculating T by setting W^Gs = 0 for mt and usp (i.e., considering mt_{ss} to be ideal) if the initial temperature turned out to be greater than 800°C. Uncertainties in T and f_{O_2} according to the authors are approximately 40–80°C and 0.5–1.0 log units f_{O_2} (2σ), assuming 1% uncertainties in mt_{ss} and ilm_{ss} compositions. The T– f_{O_2} curves calculated from this model are shown in Fig. 6.8.

The calibration of Spencer and Lindsley (1981), like that of Buddington and Lindsley (1964), did not consider minor constituents (such as Mg, Mn, Ca, Al, Cr, V, Si, etc.) that may occur in the solid solutions due to ionic substitutions. Stormer (1983) recommended a scheme for calculation of what he called "apparent mole fractions" of the solid solution components that considered the effect of ionic substitutions while being consistent with the solution models of Spencer and Lindsley (1981). This scheme assumes that in ilm_{ss} the 2+ ions are confined to the A site with Fe^{2+}, the 4+ cations occupy the B sites with Ti^{4+}, and the 3+ cations and Fe^{3+} are randomly mixed in equal proportions on both sites. For mt_{ss} it is assumed that all the 4+ cations substitute for Ti^{4+} on the B sites, the two octahedral sites always contain one Fe^{2+} per formula unit, Fe^{2+} and Fe^{3+} substitute on the tetrahedral sites in ratios equal to the molar ratio of usp to mt, and the local charge balance is maintained by the substitution of 2+ cations for Fe^{2+} and of 3+ cations for Fe^{3+}. Incorporation of this scheme into the solution model of Spencer and Lindsley (1981) simply requires replacing the mole fractions (X_{ilm}, X_{hem}, X_{usp}, X_{mt}) used in equations (6.59) and (6.60) by corresponding apparent mole fractions (X^*_{ilm}, X^*_{hem}, X^*_{usp}, X^*_{mt}), which are calculated as follows:

$$X^*_{\text{ilm}} = \frac{\sqrt{n_{(\text{Ti+Si, F})}\, n_{(\text{Fe}^{2+},\,\text{F})}}}{0.5n_{(\text{Fe}^{3+},\,\text{F})} + \sqrt{n_{(\text{Ti+Si, F})}\, n_{(\text{Fe}^{2+},\,\text{F})}}} \tag{6.61}$$
$$X^*_{\text{hem}} = 1 — X^*_{\text{ilm}}$$

$$X^*_{\text{usp}} = \frac{n_{(\text{Ti+Si, F})}\, X_{(\text{Fe}^{2+},\,\text{S}^{2+})}}{0.5n_{(\text{Fe}^{3+},\,\text{F})}\, X_{(\text{Fe}^{3+},\,\text{S}^{3+})} + n_{(\text{Ti+Si, F})}\, X_{(\text{Fe}^{2+},\,\text{S}^{2+})}} \tag{6.62}$$
$$X^*_{\text{mt}} = 1 - X^*_{\text{usp}}$$

where n denotes the number of moles, F is the formula unit, $X_{\text{Fe}^{2+},\,\text{S}^{2+}}$ is the mole fraction of Fe^{2+} versus all other 2+ cations, and $X_{\text{Fe}^{3+},\,\text{S}^{3+}}$ is the mole fraction of Fe^{3+} versus all other 3+ cations.

Table 6.6 Parameters for the solution model of Spencer and Lindsley (1981).

	Interaction and other parameters*	
	kJ mol^{-1}	**J mol^{-1} K^{-1}**
$A_1 = -3X^2_{usp} + 4X_{usp} - 1$	$W^G_{usp}(T \geq 800°C) = 0$	$W^G_{mt}(T \geq 800°C) = 0$
$A_2 = 3X^2_{usp} - 2X_{usp}$	$W^H_{usp}(T < 800°C) = 64.835$	$W^S_{usp}(T < 800°C) = 60.296$
$A_3 = -3X^2_{ilm} + 4X_{ilm} - 1$	$W^H_{mt}(T < 800°C) = 20.798$	$W^S_{mt}(T < 800°C) = 19.652$
$A_4 = 3X^2_{ilm} - 2X_{ilm}$	$W^H_{ilm} = 102.374$	$W^S_{ilm} = 71.095$
$K_{exch} = (X_{usp}X^2_{hem})/(X_{mt}X_{ilm}{}^2)$	$W^H_{hem} = 36.818$	$W^S_{hem} = 7.7714$
$\Delta H_{exch}{}^0 = 27.799$ kJ mol^{-1}	$\Delta H^0_{usp} = -3.0731$**	$\Delta S^0_{usp} = 10.724$**
$\Delta S_{exch}{}^0 = 4.1920$ J mol^{-1} K^{-1}		

*$W^G = W^H - TW^S$, where W^G, W^H, and are free energy, enthalpy, and entropy exchange parameters, respectively.
**These values represent the intercept and slope of a line on a $G_f^0 - T$ plot, and are not values for 298 K.

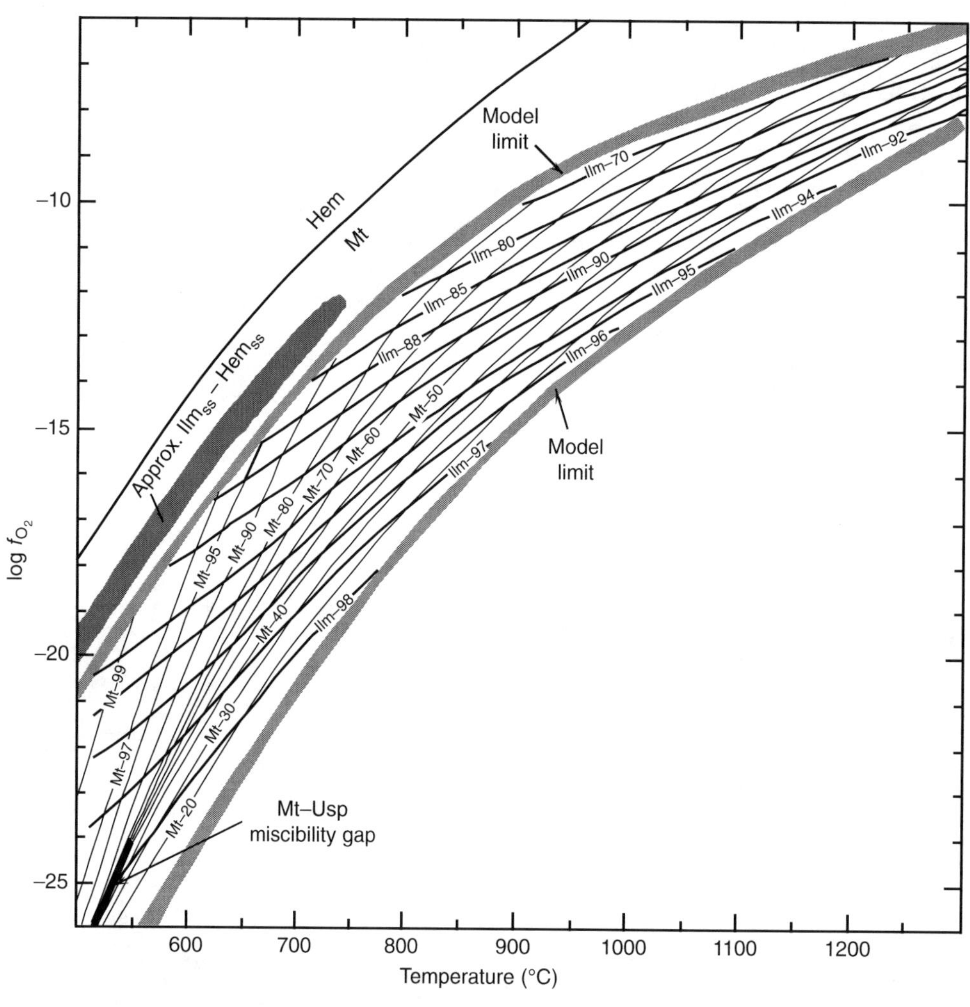

Fig. 6.8 Temperature (°C)– f_{O_2} grid for compositions, in mole percent, of coexisting magnetite (mt)–ulvöspinel (usp) solid solution (mt$_{ss}$) and ilmenite (ilm)–hematite (hem) solid solution (ilm$_{ss}$) pairs based on experimental data and the solution model of Spencer and Lindsley (1981). The compositional contours are effectively univariant calibration curves because of the nondetectable effect of pressure on the equilibrium. The intersection of contours for the compositions of these two phases provide a measure of both temperature and oxygen fugacity. The light grey shaded fields are estimates of the limits of the model. The mt–usp miscibility gap is calculated for the three-phase assemblage mt$_{ss}$ + usp$_{ss}$ + ilm$_{ss}$. The ilmenite–hematite miscibility gap (labeled "Approx. ilm$_{ss}$-hem$_{ss}$") is the best guess from experimental data (it is not calculated). The mt$_{ss}$ and ilm$_{ss}$ are assumed to be pure binary Fe–Ti oxides; no minor constituents are considered. (After Spencer and Lindsley 1981, Figure 4, p.1197.)

Example 6–3: Fe–Ti oxide geothermometry and oxygen barometry applied to a sample of Tertiary tholeiitic basalt from Iceland in which the coexisting mt_{ss} and ilm_{ss} have compositions as given in Table 6.7

The results of the first set of necessary calculations are presented in Table 6.8. The first task is to determine $Fe^{2+}$and Fe^{3+} in the mt_{ss} and ilm_{ss} from the total Fe given as FeO. To accomplish this we follow the steps recommended by Stormer (1983): (i) calculate the molar proportions of all cations in the analyses (columns 2 and 3); (ii) normalize the cations in mt_{ss} to a formula unit of 3 sites and the cations in ilm_{ss} to a formula unit of 2 sites (columns 4 and 5); (iii) calculate the sum of the cationic charges per formula unit, and subtract 8 for the mt_{ss} and 6 for the ilm_{ss} (columns 6 and 7) to get the cation charge deficiency (or excess) for each. In the present example, there is a charge deficiency in both cases, which turns out to be 0.3772 for mt_{ss} and 0.1199 for ilm_{ss}.

Table 6.7 Electron microprobe analysis of mtss and ilmss (Carmichael, 1967a).

Wt%	mt_{ss}	ilm_{ss}
SiO_2	0.23	0.17
TiO_2	27.1	49.5
Al_2O_3	1.22	0.13
V_2O_3	0.03	–
Cr_2O_3	0.71	0.10
FeO (t)	67.0	47.8
MnO	0.83	0.49
MgO	0.19	1.09
CaO	0.19	0.16
ZnO	0.12	–
Total	97.6	99.4

FeO (t) = total Fe calculated as FeO.

The charge deficiency numbers show that:

Fe^{3+} cations in the formula unit of mt_{ss} (3 cations total) = 0.3772

Fe^{3+} cations in the formula unit of ilmss. (2 cations total) = 0.1199

To get the number of moles of Fe^{3+} in the formula unit of each solid solution, we subtract the Fe^{3+} cations from its total number of Fe cations:

Fe^{2+} cations in mt_{ss} = 2.1027 – 0.3772 = 1.7255

Fe^{2+} cations in ilm_{ss} = 1.0017 – 0.1199 = 0.8818

We now have the number of moles of each cation per formula unit for both phases (columns 9 and 10) so that we can calculate the apparent mole fractions of usp in mt_{ss} and of ilm in ilm_{ss}, using equations (6.61) and (6.62).
For mt_{ss}:

$$n_{(Ti+Si,\,F)} = 0.7648 + 0.0086 = 0.7734$$

$$X_{(Fe^{2+},\,S^{2+})} = \frac{n_{(Fe^{2+},\,F)}}{n_{(Fe^{2+},\,F)} + n_{(Mn,\,F)} + n_{(Mg,\,F)} + n_{(Ca,\,F)} + n_{(Zn,\,F)}} = 0.9730$$

$$X_{(Fe^{3+},\,S^{3+})} = \frac{n_{(Fe^{3+},\,F)}}{n_{(Fe^{3+},\,F)} + n_{(Al,\,F)} + n_{(V,\,F)} + n_{(Cr,\,F)}} = 0.8324$$

$$X^{*}_{usp} = \frac{n_{(Ti+Si,\,F)}\,X_{(Fe^{2+},\,S^{2+})}}{0.5n_{(Fe^{3+},\,F)}\,X_{(Fe^{3+},\,S^{3+})} + n_{(Ti+Si,\,F)}\,X_{(Fe^{2+},\,S^{2+})}} = 0.8275$$

$$X^{*}_{mt} = 1 - X^{*}_{usp} = 0.1725$$

Table 6.8 First set of calculations for Fe–Ti oxide geothermometry and oxygen barometry of Tertiary tholeiitic basalt from Iceland.

	Molar proportion of cations		Normalized to 3 cations for mt_{ss} and 2 cations for ilm_s		Cationic charge			Recomputed molar proportion of cations	
(1)	(2) mt_{ss}	(3)ilm_{ss}	(4)mt_{ss}	(5) ilm_{ss}	(6) mt_{ss}	(7) ilm_{ss}	(8)	(9) mt_{ss}	(10) ilm_{ss}
Si^{4+}	0.0038	0.0028	0.0086	0.0043	0.0345	0.0171	Si^{4+}	0.0086	0.0043
Ti^{4+}	0.3392	0.6195	0.7648	0.9328	3.0591	3.7313	Ti^{4+}	0.7648	0.9329
Al^{3+}	0.0239	0.0025	0.0540	0.0038	0.1619	0.0115	Al^{3+}	0.0540	0.0038
V^{3+}	0.0004	0.0000	0.0009	0.0000	0.0027	0.0000	V^{3+}	0.0009	0.0000
Cr^{3+}	0.0093	0.0013	0.0211	0.0020	0.0632	0.0059	Cr^{3+}	0.0211	0.0020
Fe^{2+} (t)	0.9325	0.6653	2.1027	1.0017	4.2054	2.0035	Fe^{3+}	0.3770	0.1199
Mn^{2+}	0.0117	0.0069	0.0264	0.0104	0.0528	0.0208	Fe^{2+}	1.7256	0.8817
Mg^{2+}	0.0047	0.0270	0.0106	0.0407	0.0213	0.0814	Mn^{2+}	0.0264	0.0104
Ca^{2+}	0.0034	0.0029	0.0076	0.0043	0.0153	0.0086	Mg^{2+}	0.0106	0.0407
Zn^{2+}	0.0015	0.0000	0.0033	0.0000	0.0067	0.0000	Ca^{2+}	0.0076	0.0043
Total	1.3305	1.3283	3.0000	2.0001	7.6228	5.8801	Zn^{2+}	0.0033	0.0000
Cation deficiency					0.3772	0.1199	Total	3.0000	2.0000

Fe^{2+} (t) = total Fe expressed as Fe^{2+}.

For ilm_{ss}:

$$n_{(Ti+Si,\ F)} = 0.9329 + 0.0043 = 0.9372$$

$$X^*_{ilm} = \frac{\sqrt{n_{(Ti+Si,F)}\ n_{(Fe^{2+},\ F)}}}{0.5n_{(Fe^{3+},\ F)} + \sqrt{n_{(Ti+Si,\ F)}\ n_{(Fe^{2+},\ F)}}} = 0.9193$$

$$X^*_{hem} = 1 - X^*_{ilm} = 0.0807$$

Next we calculate the following parameters as defined in Table 6.6, replacing X_i values with X_i^* values:

$A_1 = -3X^2_{usp} + 4X_{usp} - 1 = 0.2557;$ $A_2 = 3X^2_{usp} - 2X_{usp} = 0.3992$

$A_3 = -3X^2_{ilm} + 4X_{ilm} - 1 = 0.1419;$ $A_4 = 3X^2_{ilm} - 2X_{ilm} = 0.6967$

$$K_{exch} = (X_{usp}\ X^2_{hem}/X_{mt}\ X^2_{ilm}) = 0.0369$$

Substituting the values calculated above and the low-T parameters (Table 6.6) in equation (6.59), we obtain T = 1622 K (1349°C). Since T is above 800°C, we recalculate T by setting $W^H_{usp} = 0$, $W^S_{usp} = 0$, $W^H_{mt} = 0$, and $W^S_{mt} = 0$, and then use the new value of T to calculate f_{O_2} using equation (6.60), remembering that $W^G_{usp} = 0$ and $W^G_{mt} = 0$ at $T < 800$°C.

The final result: T = 1443 K (1170°C); $\log f_{O_2} = -9.23$.

6.6 Solvus equilibria

As mentioned earlier (section 5.5.1), the unmixing of a non-ideal binary solid solution below the consolute temperature at a given pressure provides a method of geothermometry. For example, the calcite ($CaCO_3$)–dolomite ($CaMg(CO_3)_2$) solvus (Fig. 6.9) can be used to estimate the equilibration temperature from the compositions of calcite coexisting in equilibrium with dolomite. (Strictly speaking, the immiscibility between two phases of different crystal structures, such as calcite and dolomite, is not a "solvus", but for brevity we shall refer to such a miscibility gap as "solvus".). The solvus, however, is not applicable to natural calcites, which generally contain small amounts of $MgCO_3$ and $FeCO_3$ in solid solution.

Reliable experimental data on the ternary $CaCO_3 - MgCO_3 - FeCO_3$ system are not available. Anovitz and Essene (1987) constructed isothermal phase diagrams for the system from natural carbonate assemblages at various grades of metamorphism and used these to obtain approximate activity–composition relations for the phase components, assuming an asymmetric regular solution model for the ternary system. The equation they derived for the calculation of equilibration temperature from the mole fractions of $CaCO_3$, $MgCO_3$, and $FeCO_3$ in calcite (cal) is:

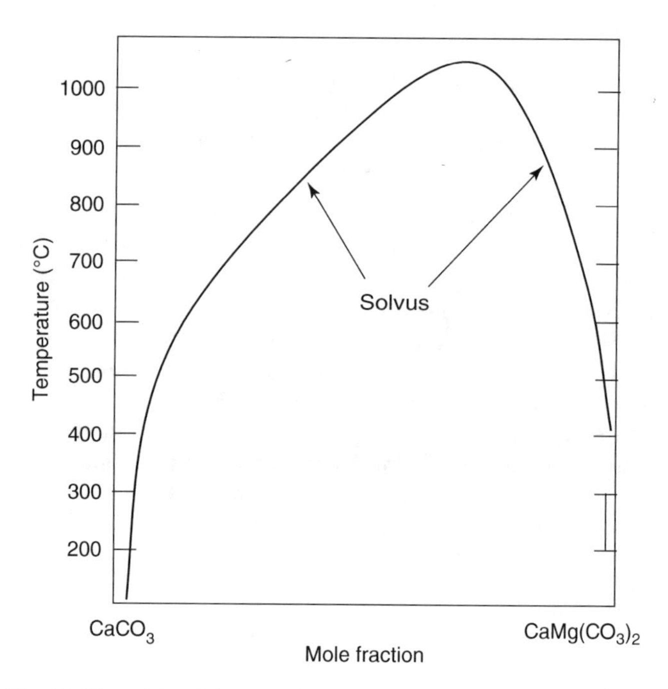

Fig. 6.9 The calcite–dolomite solvus constructed by Anovitz and Essene (1987). The position of the calcite limb of the solvus was based on reversed experiments of several earlier workers. Equilibration temperatures were calculated from the calcite limb of the solvus and natural dolomite compositions in equilibrium with calcites were used to locate the dolomite limb of the solvus. The solvus seems to be well calibrated in the temperature range from 400 to 800°C. (After Anovitz and Essene, 1987, Figure 3, p. 394.)

$$T\ (K) = T_{Fe-Mg} + a(X^{cal}_{FeCO_3}) + b(X^{cal}_{FeCO_3})^2 + c(X^{cal}_{FeCO_3}/X^{cal}_{MgCO_3}) + d(X^{cal}_{FeCO_3}.X^{cal}_{MgCO_3}) + e(X^{cal}_{FeCO_3}/X^{cal}_{MgCO_3})^2 + f(X^{cal}_{FeCO_3}.X^{cal}_{MgCO_3})^2 \quad (6.63)$$

where

$$T_{Fe-Mg}(K) = A(X^{cal}_{MgCO_3}) + B/(X^{cal}_{MgCO_3})^2 + C(X^{cal}_{MgCO_3})^2 + D(X^{cal}_{MgCO_3})^{0.5} + E \quad (6.64)$$

and the values of the constants (A, B, C, D, E, a, b, c, d, e, and f) are as given in Table 6.9. Evidently, for very small values of

Table 6.9 Coefficients for equations (6.63) and (6.64) (Anovitz and Essene, 1987).

Coefficient	Value	Coefficient	Value
A	–2360	a	1718
B	–0.01345	b	–10610
C	2620	c	22.49
D	2608	d	–26260
E	334	e	1.333
		f	0.32837×10^7

$X_{FeCO_3}^{cal}$, the difference in T calculated from equations (6.63) and (6.64) will be very small.

The formulation of Anovitz and Essene (1987) does not consider the possible effect of Mn in calcite, but the mole fraction of Mn in natural calcites is commonly too small to make a significant contribution to the temperature estimate. The two-carbonate thermometer is not recommended for high-temperature metamorphic rocks (above about 600°C) because of possible loss of Mg from the original high-Mg calcite due to exsolution (resulting in dolomite lamellae in calcite) or diffusion during retrogression.

6.7 Uncertainties in thermobarometric estimates

Realistic interpretation of thermobarometric results requires an appreciation of the uncertainties associated with the *P–T* estimates. The calculation of uncertainties is largely an exercise in statistics, which is beyond the scope of the present book. The purpose of this section is to provide an overview of the subject; the details are discussed in many recent publications (e.g., Hodges and McKenna, 1987; Kohn and Spear, 1991; Spear, 1993; Powell and Holland, 1994).

Sources of error in geothermobarometry include (modified from Spear, 1993):

(1) accuracy of calibration, experimental or calculated (in the latter case because of uncertainties of thermodynamic datasets);
(2) error in the ΔV_r;
(3) uncertainty associated with the compositions of standards and matrix correction factors used in electron microprobe analyses;
(4) analytical imprecision in electron microprobe analyses and in the estimation of Fe^{3+} from Fe_{total} such analyses;
(5) cross correlations between errors in temperature estimates (for geobarometers) and errors in pressure (for geothermometers) as all thermobarometers are dependent to some extent on both temperature and pressure;
(6) compositional heterogeneities in minerals; and
(7) uncertainties in activity–composition models for solid solutions.

Items (1) to (5) produce *random errors* that can be handled statistically by standard techniques of error propagation (see, e.g., Hodges and McKenna, 1987; Spear, 1993), whereas items (6) and (7) generate *systematic errors* that are handled differently (see, e.g., Kohn and Spear, 1991; Powell and Holland, 1994). Since the pressure term in the calculations (see, e.g., equation 6.4) is a difference between the pressure of interest and the reference pressure, many of the sources of uncertainty in pressure cancel out. Many of the uncertainties in the *P–T* estimates arise from sources that contribute to error in the temperature estimate, which also contributes to error in the pressure estimate.

Calculated *P–T* estimates are critically dependent on the mixing models adopted for the various solid solution phases involved in the chosen thermobarometric reactions. For example, as shown by Spear (1993, pp. 528–531), the *P–T* obtained for a single set of mineral analyses from Mount Moosilauke, New Hampshire (Sample 90A, Hodges and Spear, 1982), using the calibrations of different authors for garnet–biotite thermometry and GASP barometry, show a spread of approximately 80°C and as much as 3.5 kbar. Without an absolute reference against which the calculated pressures can be compared, it is not possible to make a judgment as to which of these calibrations is the most accurate, and an arithmetic average of the estimates will not be meaningful.

A practical approach for obtaining more consistent *P–T* estimates may lie in using an internally consistent set of thermometers and barometers, which is calibrated either with the same thermodynamic database and solution models for all phases or against the same set of empirical data. The results of such an approach for the same Mt Moosilauke sample 90A mentioned above are shown in Fig. 6.10. The internally consistent thermodynamic data set of Berman (1988) and Holland and Powell (1990) would also yield a consistent set of *P–T* estimates for this sample.

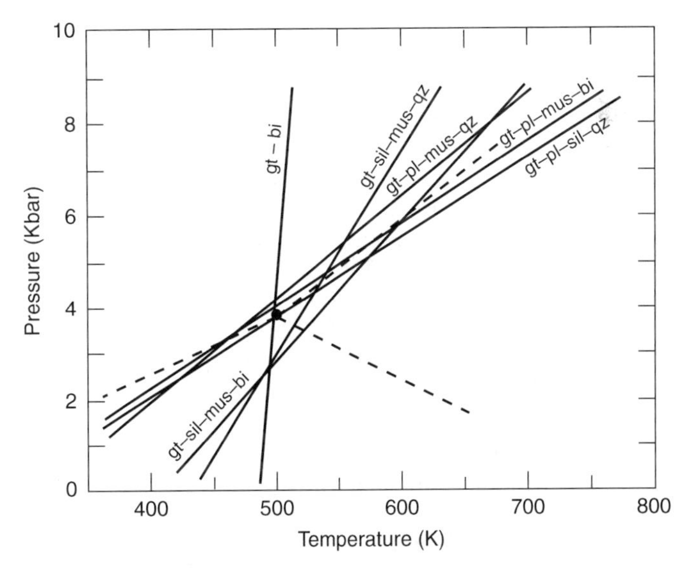

Fig. 6.10 Pressure–temperature diagram showing an internally consistent set of geothermobarometers applied to sample 90A from Mount Moosilauke, New Hampshire. The three garnet–plagioclase geobarometers (gt–pl–mus–qz, gt–pl–mus–bi, gt–pl–sil–qz) give pressures very close to 4 kbar at 500°C, whereas the two plagioclase-absent geobarometers (gt–mus–sil–qz, gt–mus–bi–sil) give slightly lower pressures, with a mean value of approximately 3.25 ±. 0.75 kbar. Calibration of the gt–bi geothermometer is from Hodges and Spear (1982) and calibrations of the five geobarometers are from Hodges and Crowley (1985). Phase relations among the Al_2SiO_5 polymorphs are shown for reference. (After Spear, 1993, Figure 15–10, p. 532.)

To obtain optimal results, Powell and Holland (1985, 1994) have proposed a further refinement of the internally consistent approach, which they call "the average *P–T* method". For a given sample (assemblage), this method considers equilibria, including uncertainties and enthalpies of formation of the end-members, pertaining to all possible independent thermobarometric reactions in the system, and employs an iterative least-squares regression technique to calculate an average *P–T* with an ellipse of uncertainty. The calculations are quite complex, but the results can be obtained readily with the computer program THERMOCALC (Powell and Holland, 2001).

It should be pointed out that the internally consistent approach provides a reasonably good test for equilibrium in the assemblage under consideration, but it does not guarantee that the *P–T* estimates are accurate. A consistent set of *P–T* estimates (i.e., with a narrow spread of values) may not be correct because of systematic errors in the calibrations (Spear, 1993).

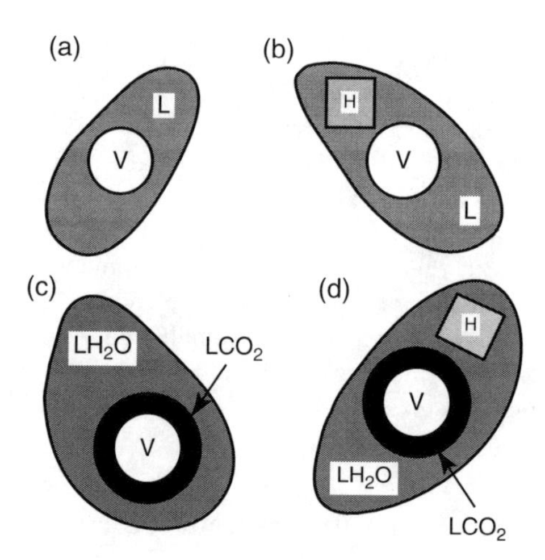

Fig. 6.11 Sketches of some common types of fluid inclusions at room temperature: (a) two-phase (L + V) aqueous inclusion; (b) three-phase (L + V + H) aqueous inclusion; (c) CO_2-bearing inclusion (LCO_2 + V + LH_2O); (d). CO_2-bearing inclusion with a salt crystal (LCO_2 + V + LH_2O + H). Abbreviations: L = liquid, V = vapor, H = halite.

6.8 Fluid inclusion thermobarometry

Fluid inclusions are minute aliquots of fluid that were trapped inside a crystal at some stage of its history. The trapped fluid may be liquid, vapor, or supercritical fluid. In terms of composition, the trapped fluid may be essentially pure water, brines of different salinity (the total dissolved salt content), gas or gas-bearing fluids, or even silicate, carbonate or sulfide melts. Subsequent to trapping at some temperature and pressure, a fluid inclusion may undergo various physical changes (e.g., necking down to form several small inclusions, leakage) and phase transformations (e.g., separation into immiscible gas and liquid phases, crystallization of daughter minerals). At surface temperature–pressure conditions, the inclusions usually consist of some combination of gaseous, liquid, and solid phases (Figs 6.11 and 6.12). The essence of fluid inclusion thermobarometry is to estimate the *trapping temperature* (T_t) and/or *trapping pressure* (P_t) from the present state of the inclusions in a mineral.

This reconstruction requires that the physical characteristics of the fluid inclusions and their behavior during microthermometric analysis (heating and cooling experiments) are consistent with the following assumptions (Roedder, 1984):

(1) the inclusion fluid was trapped and sealed as a single, homogeneous phase;
(2) the inclusion has behaved as an isochoric (constant density) system throughout its history;
(3) nothing has been added to or lost from the inclusion after sealing.

In terms of the timing of trapping of the inclusion relative to the crystallization of the host mineral, fluid inclusions may be classified as *primary*, *secondary*, or *pseudosecondary*. Primary inclusions result from fluids trapped in growth irregularities during the growth of the host crystal. Thus, inclusions trapped along growth surfaces of a crystal are the most obvious examples of primary inclusions. A crystal may be fractured one or more times after its formation and the fractures healed in the presence of fluids; fluids trapped in these fractures result in secondary fluid inclusions, which are most easily identified when they occur along a fracture that crosscuts the entire crystal. A crystal may contain secondary inclusions of several generations, each possibly representing a different fluid. Pseudosecondary inclusions originate from trapping of fluid in a fracture that formed during the growth of the crystal; such inclusions show a distribution similar to that of secondary inclusions but are primary in terms of their age relative to the host crystal, although not necessarily of the same composition as the primary inclusions. Distinction among the three types of fluid inclusions (and different generations of secondary inclusions) in a given sample is of fundamental importance in correlating the fluid inclusion data to geologic processes. The petrographic and morphological criteria listed by Roedder (1984, pp. 43–45) for categorization of fluid inclusions are very useful, but the task can be difficult and sometimes impossible. Nevertheless, for thermobarometry it is imperative that we focus on a *fluid inclusion assemblage* (FIA; Goldstein and Reynolds, 1994) – a group of fluid inclusions that were all trapped at the same time and thus from a fluid of about the same composition at approximately the same temperature and pressure – that represents the fluid event of interest.

The practical and interpretative aspects of fluid inclusion study are quite involved, as discussed by several authors (e.g., Hollister and Crawford, 1981; Roedder, 1984; Shepherd *et al.*, 1985; Bodnar, 2003). The basic principles of thermobarometric application, however, can be illustrated

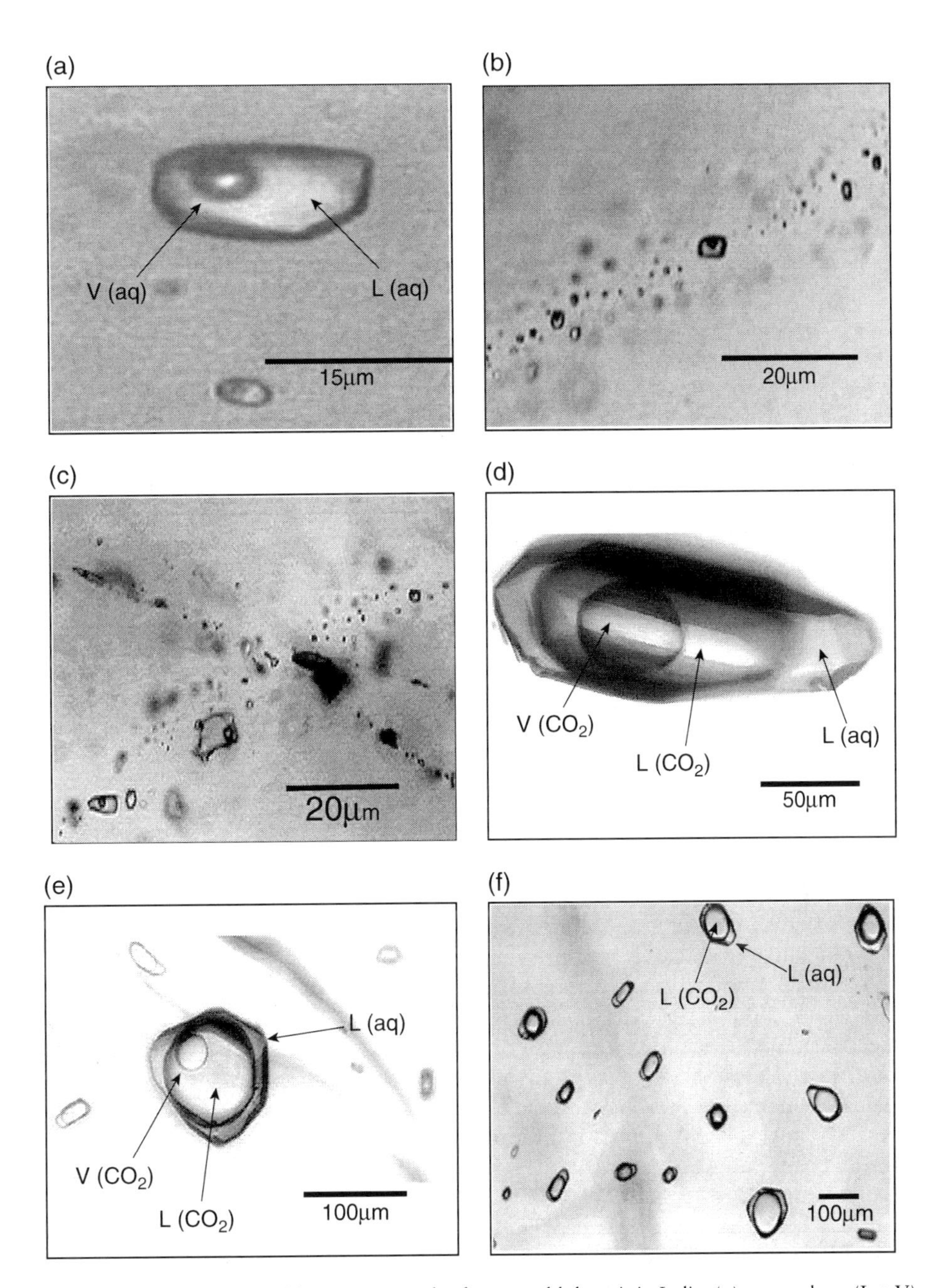

Fig. 6.12 Photomicrographs of fluid inclusions hosted by quartz samples from a gold deposit in India: (a) a two-phase (L + V) aqueous inclusion; (b) secondary aqueous inclusions trapped along a fracture; (c) aqueous inclusions trapped along intersecting fractures; (d) a CO_2–bearing three–phase inclusion; (e) a CO_2–bearing three–phase inclusion; (f) a cluster of CO_2-bearing inclusions. Abbreviations: L = liquid, V = vapor, H = halite. (Courtesy Dr M.K. Panigrahi, Indian Institute of Technology, Khragpur, India.)

by considering an assemblage of two-phase (L + V) aqueous inclusions that can be approximated by the NaCl – H_2O system with no detectable gases, the most common kind of fluid inclusions found in hydrothermal minerals. Suppose we have determined from laboratory experiments that such an assemblage of NaCl – H_2O inclusions have an average salinity (dissolved salt content) of 20 wt% NaCl and an average homogenization temperature (T_h) of 300°C, and that the calculated *P–T* phase relations in this system are as shown in Fig. 6.13. The *homogenization temperature*, the temperature at which an inclusion homogenizes to a single fluid phase, and the vapor pressure in the inclusion at homogenization (P_h; which is about 0.1 kbar in this case) represent T_t and P_t only if the inclusions were trapped in a boiling or immiscible fluid system. In all other cases they represent the minimum values of trapping temperature and trapping pressure,

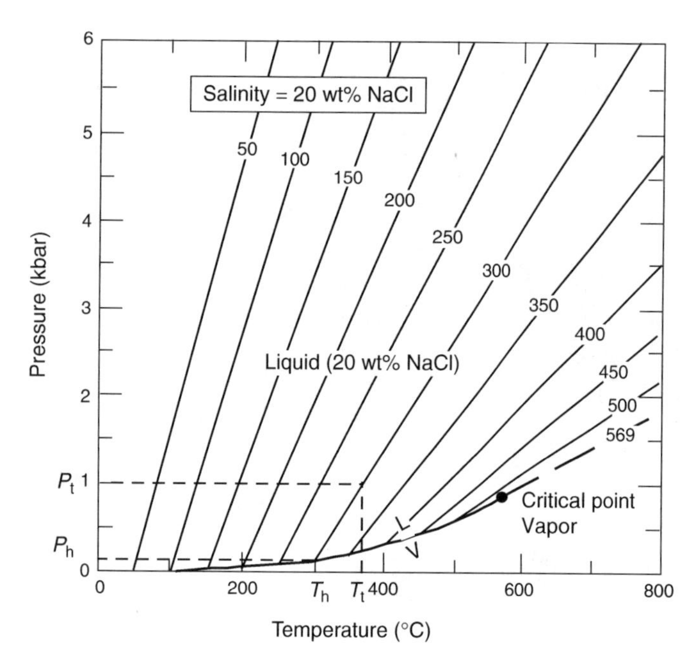

Fig. 6.13 Iso-T_h lines for NaCl-H_2O inclusions having a salinity of 20 wt% NaCl. The L–V curve was constructed using the empirical equations in Atkinson (2002) and the iso-T_h lines using data from Bodnar and Vityk (1994). (After Bodnar, 2003, Fig. 4–10, p. 90.)

irrespective of the fluid composition. Since the fluid inclusions are assumed to have evolved with constant density, the T_t and P_t for this fluid inclusion assemblage are defined by some point along the iso-T_h line (approximately the same as an isochore) marked as 300°C. The liquid (L)–vapor (V) curve in Fig. 6.13 defines the equilibrium between an aqueous solution containing 20 wt% NaCl and a vapor phase of lesser salinity; it was constructed using the empirical equation of Atkinson (2002), which describes the vapor pressure of NaCl – H_2O solutions as a function of salinity and temperature. The iso-T_h lines in this figure were calculated using the equation formulated by Bodnar and Vityk (1994), which calculates the slope of an iso-T_h line as a function of T_h and salinity. As shown in Fig. 6.13 we can use the 300°C iso-T_h line to estimate the T_h as ~370°C if we have an independent estimate of P_t as 1 kbar (e.g., from the thickness of the overlying rocks at the time of trapping or some other geobarometer), and vice versa.

A noteworthy limitation of the approach discussed above is that aqueous inclusion fluids in many geologic environments contain other dissolved salts, such as KCl and $CaCl_2$, in addition to NaCl. At present our ability to use such inclusions for thermobarometry is hindered by (i) the difficulty of observing and identifying phase changes in complex aqueous solutions during heating and cooling experiments (especially in small, natural inclusions) and (ii) the paucity of *P–V–T–X* data applicable to these more complex compositions (Bodnar, 2003).

6.9 Summary

1. Geothermometry and geobarometry refer, respectively, to the estimation of temperature and pressure at which a given mineral assemblage equilibrated during some event in its history (e.g., magmatic crystallization, prograde or retrograde metamorphism, hydrothermal alteration).
2. The tools that have commonly been used for geothermobarometry are: (i) univariant reactions and displaced equilibria; (ii) exchange reactions; (iii) solvus equilibria; (iv) study of cogenetic fluid inclusions in minerals; and (v) fractionation of oxygen and sulfur isotopes between coexisting phases.
3. The dependence of the equilibrium constant of a reaction on temperature and pressure is given by the following equations:

$$\left(\frac{\partial \ln K_{eq}}{\partial T}\right)_P = -\frac{1}{R}\left(\frac{\partial (\Delta G_r^0 / T)}{\partial T}\right)_P$$

$$= \frac{\Delta H_r^0}{RT^2} \quad \text{van't Hoff equation}$$

$$\left(\frac{\partial \ln K_{eq}}{\partial P}\right)_T = -\frac{1}{RT}\left(\frac{\partial \Delta G_r^0}{\partial P}\right)_T = -\frac{\Delta V_r^0}{RT}$$

Since the equilibrium constant is a function of both temperature and pressure, one of the variables must be known from an independent source in order to calculate the other. For a reaction to be a sensitive geothermometer, its equilibrium constant should largely be a function of *T*, that is the reaction should have a large value of ΔH_r and a small value of ΔV_r; for a reaction to be a sensitive geobarometers, it should have a large value of ΔV_r and a small value of ΔH_r.
4. In general, reactions involving a fluid phase are not suitable for thermobarometry because of a lack of information about the fluid composition and the ratio of fluid pressure (P_f) to total pressure (P_{total}).
5. The basic relation used for calculation of equilibrium temperature and pressure of reactions is

$$\ln K_{eq} = \frac{-\Delta G_r^0}{RT} = \frac{-\Delta H_r^0}{RT} + \frac{\Delta S_r^0}{R}$$

where the superscript "0" refers to the chosen standard state.
6. For univariant reactions and displaced equilibria (net transfer reactions) involving solid phases only,

$$\ln K_{eq} = \frac{-\Delta H^1_{r,T}}{RT} + \frac{\Delta S^1_{r,T}}{R} - \frac{(P-1)\,\Delta V_r}{RT}$$

if the standard state for each substance is taken as the pure substance at the temperature (T) and pressure (P) of interest, and ΔV_r is assumed to be independent of temperature and pressure.

7. For an exchange reaction involving 1 mole each of two elements (or cations) A and B between the phases α and β,

$$K_{eq}(P,T) = K_D\, K_\lambda(P,T)$$

where,

$$K_D = \frac{(X_A/X_B)^\alpha}{(X_A/X_B)^\beta} \quad \text{and} \quad K_\lambda = \frac{(\lambda_A/\lambda_B)^\alpha}{(\lambda_A/\lambda_B)^\beta}$$

8. Reactions selected for geothermobarometry commonly include one or more solid solutions. The activity–composition relations in such solid solutions depend on the choice of solution models for the particular problem.
9. Rigorous thermobarometry should include an estimate of uncertainties in the estimated temperature and pressure due to several possible sources of error, some of which are random and others systematic.
10. Study of fluid inclusion assemblages can provide reasonable estimates of trapping temperature and trapping pressure if P–V–T–X data are available for the inclusion fluid composition. The application of fluid inclusion thermobarometry to complex compositions is hindered at present by the lack of reliable P–V–T–X data for such systems.

6.10 Recapitulation

Terms and concepts

Al_2SiO_5 triple point
Displaced equilibria
Exchange reaction
Fluid inclusion assemblage
Fluid inclusion microthermometry
Geobarometer
Geobarometry
Geothermometer
Geothermometry
Homogenization temperature
Oxygen fugacity
Petrogenetic grid
Solid solutions
Solvus equilibria
Trapping pressure
Trapping temperature
Univariant reaction
van't Hoff equation

Computation techniques

- Activities of constituents of solid solutions.
- Variation of equilibrium constant as a function of temperature and pressure.
- Reaction boundaries in P–T space.
- Equilibration temperature and pressures of appropriate mineral assemblages.

6.11 Questions

1. Show that for a reaction at constant pressure, the difference in the equilibrium constant at temperatures T_1 and T_2, assuming $(\Delta C_P)_r = 0$, is related to the standard state enthalpy, ΔH_r^0, of the reaction by the relation

$$\ln K_{T_2} - \ln K_{T_1} = -\frac{\Delta H_r^0}{R}\left(\frac{1}{T_2} - \frac{1}{T_1}\right)$$

2. The ΔG^0 (in kJ mol^{-1}) of the reaction

$$\underset{\text{dolomite}}{3CaMg(CO_3)_2} + \underset{\text{quartz}}{4SiO_2} + \underset{\text{fluid}}{H_2O} = \underset{\text{talc}}{Mg_3Si_4O_{10}(OH)_2}$$
$$+ \underset{\text{calcite}}{3CaCO_3} + \underset{\text{fluid}}{3CO_2}$$

can be described by the relation $\Delta G_r^0 = 173.1 - 0.2275T$ at 2 kbar pressure, assuming $(\Delta C_P)_r = 0$ (Powell, 1978, p. 116). How does the equilibrium constant for the reaction change if the temperature of the reaction is raised from 300 K to 700 K? What is the slope of the line in a plot of ln K_{eq} versus $10^3/T$?

3. Calculate the equilibrium metamorphic pressure at 750°C for GAPES and GADS assemblages using the solid solution models discussed in the text. Assume quartz to be a pure phase, no mixing in the octahedral or tetrahedral sites in garnet, and no mixing in the tetrahedral sites in pyroxene. The cation and component mole fractions in the different minerals in the GAPES and GADS assemblages are given below:

Garnet	Plagioclase	Clinopyroxene	Orthopyroxene
$X_{Mg}^{cubic} = 0.181$	$X_{ab} = 0.613$	$X_{Ca}^{M2} = 0.870$	$X_{Mg}^{M2} = 0.482$
$X_{Fe^{2+}}^{cubic} = 0.598$	$X_{an} = 0.373$	$X_{Mg}^{M1} = 0.664$	$X_{Mg}^{M1} = 0.468$
$X_{Ca}^{cubic} = 0.191$			

4. Given below are the compositions of coexisting mt_{ss} and ilm_{ss} in a sample of basalt (Carmichael, 1967b). Calculate the temperature and oxygen fugacity of equilibration represented by the assemblage. Assume that the

Wt%	mt_{ss}	ilm_{ss}
SiO_2	0.09	0.06
TiO_2	28.80	50.32
Al_2O_3	1.18	0.02
V_2O_3	0.97	0.11
Cr_2O_3	0.71	–
Fe_2O_3	10.74	4.34
FeO	56.06	43.85
MnO	0.82	0.50
MgO	0.50	0.49
CaO	0.14	0.07
ZnO	0.12	–
Total	99.44	99.76

assemblage had achieved equilibrium and has experienced no subsequent change.

5. Given below are the electron microprobe analyses (in wt%) of coexisting garnet and clinopyroxene in an eclogite xenolith recovered from the Udachnaya kimberlite pipe, Siberia (Misra *et al.*, 2004):

	Garnet	Clinopyroxene
SiO_2	40.5	55.9
TiO_2	0.55	0.50
Al_2O_3	21.9	11.2
Cr_2O_3	0.06	0.07
MgO	12.2	9.01
FeO (total)	17.3	4.87
CaO	6.95	11.7
MnO	0.34	0.04
Na_2O	0.23	6.67
K_2O	0.04	0.12
Total	100.07	100.08

Calculate the temperature of equilibration using the calibrations of Ellis and Green and (1979) and Krogh (1988), assuming FeO (total) = FeO (i.e., no Fe^{3+} in the garnet or clinopyroxene) and an equilibrium pressure of 6.5 GPa. Comment on the discrepancy, if any, between the two estimates of temperature.

6. The compositions of coexisting garnet and biotite (expressed as mole ratios) in a sample from the sillimanite metamorphic zone in a pelitic schist unit are as follows (Ferry, 1980):

garnet: Fe/(Fe + Mg) = 0.872; and biotite: Fe/(Fe + Mg) = 0.507

Assuming equilibrium at a pressure of 3474 bars, calculate the temperature of metamorphic equilibration of the pelitic schist, using the calibration of Ferry and Spear (1978).

7. Consider the metamorphic assemblage garnet–biotite–plagioclase–kyanite–quartz. Microprobe analyses have shown that in the cubic garnet site $X_{Ca} = 0.04$, $X_{Mg} = 0.18$, and $X_{Fe^{2+}} = 0.78$. The biotite has the octahedral-site contents $X_{Mg} = 0.63$ and $X_{Fe^{2+}} = 0.37$. If the activity of the anorthite component of plagioclase has been calculated to be 0.3, calculate the equilibrium P and T values that are consistent with these mineral compositions. For biotite–garnet equilibrium,

$$\underset{\text{almandine}}{\overset{\text{garnet}}{Fe_3Al_2Si_3O_{12}}} + \overset{\text{phlogopite}}{KMg_3AlSi_3O_{10}(OH)_2} \Leftrightarrow \underset{\text{pyrope}}{\overset{\text{garnet}}{Mg_3Al_2Si_3O_{12}}} + \underset{\text{annite}}{\overset{\text{biotite}}{KFe_3AlSi_3O_{10}(OH)_2}} \quad (6.47)$$

use the following equation calculated from Ferry and Spear (1978):

$$\ln K = \ln\left(\frac{X_{Mg}}{X_{Fe}}\right)_{gt} - \ln\left(\frac{X_{Mg}}{X_{Fe}}\right)_{biot} = -\frac{2089}{T} - \frac{0.0096\,(P-1)}{T} + 0.782$$

For plagioclase–garnet–kyanite–quartz equilibrium,

$$\underset{\text{grossular}}{\overset{\text{gt}}{Ca_3Al_2Si_3O_{12}}} + \underset{\text{kyanite}}{\overset{\text{kyanite}}{2Al_2SiO_5}} + \underset{\text{quartz}}{\overset{\text{qz}}{SiO_2}} \Leftrightarrow \underset{\text{anorthite}}{\overset{\text{plag}}{3CaAl_2Si_2O_8}}$$

use the following equation calculated from Koziol and Newton (1988):

$$\ln K = \ln\left(\frac{(a_{an}^{plag})^3}{a_{gross}^{gt}}\right) = -\frac{5815}{T} - 0.795\frac{(P-1)}{T} + 18.12$$

In both equations, T is in Kelvin and P is in bars. Calculate the equilibrium temperature and pressure for the mineral assemblage, assuming all the solid solutions to be ideal. (After Nordstrom and Munoz, 1994.)

8. The metamorphic mineral assemblage garnet + plagioclase + orthopyroxene + quartz occurs within granulite facies rocks of pelitic bulk composition. The equilibrium among these minerals can be represented by the reaction

$$\underset{\text{grossular}}{\overset{\text{garnet}}{Ca_3Al_2Si_3O_{12}}} + \underset{\text{almandine}}{\overset{\text{garnet}}{2Fe_3Al_2Si_3O_{12}}} + \overset{\text{quartz}}{3SiO_2} \Leftrightarrow \underset{\text{anorthite}}{\overset{\text{plagioclase}}{3CaAl_2Si_2O_8}} + \underset{\text{ferosillite}}{\overset{\text{orthopyroxene}}{3Fe_2Si_2O_6}}$$

(a) Calculate the equilibrium constant for the reaction if the composition of the minerals at equilibrium are as follows:

garnet: $Ca_{1.04}Fe_{1.43}Mg_{0.46}Mn_{0.07}Al_{2.15}Si_{2.86}O_{12}$
plagioclase: $Ca_{0.91}Na_{0.07}K_{0.02}Al_{2.35}Si_{1.65}O_8$
orthopyroxene: $Ca_{0.09}Fe_{1.67}Mg_{0.19}Al_{0.07}Si_{1.98}O_6$

Assume that the solid solutions are ideal (i.e., $a = X^n$, where X = mole fraction and n = site multiplicity).

(b) From the P–T calibration of this reaction by Bohlen *et al.* (1983a) estimate the equilibrium pressure corresponding to a temperature of 850°C.

9. In Koziol and Newton (1988), as cited in the question 7, the experimental fit to the univariant line for the reaction

$$\underset{\text{grossular}}{Ca_3Al_2Si_3O_{12}} + \underset{\text{kyanite}}{2Al_2SiO_5} + \underset{\text{quartz}}{SiO_2} \Leftrightarrow \underset{\text{anorthite}}{3CaAl_2Si_2O_8}$$

is given by

P (bars) = 22.80 T (°C) – 1093

Derive the ln K equation used in the above problem from the P–T fit to the experimental data. Take $\Delta V_{(s)} = 6.608$ J bar^{-1}. Note that the temperature scales are different. State any assumptions that you need to make. (After Nordstrom and Munoz, 1994.)

7 Reactions Involving Aqueous Solutions

Aquatic chemistry is concerned with the chemical processes affecting the distribution and circulation of chemical compounds in natural waters; its aims include the formulation of an adequate theoretical basis for the chemical behavior of ocean waters, estuaries, rivers, lakes, groundwaters, and soil water systems, as well as the description of the processes involved in water treatment. Obviously, aquatic chemistry draws primarily on the fundamentals of chemistry, but it is also influenced by other sciences, especially geology and biology.

Stumm and Morgan (1981)

Water, the most abundant liquid on Earth, plays an important role in almost all geochemical and biochemical processes. These include chemical weathering, formation of chemical and biochemical sediments, diagenesis, transport and accumulation of contaminants, metamorphism, and even magmatic crystallization. Understanding the behavior of aqueous solutions, therefore, is an essential aspect of geochemistry.

Water is a remarkable naturally occurring compound with many unusual properties. Water has a much higher boiling point compared to other hydrides of higher molecular weight (e.g., H_2S) so that it is a liquid at room temperature. It has the highest heat capacity of all common liquids (except ammonia) and solids (a property of considerable climatic significance considering the extent of oceans on the Earth's surface), the highest latent heat of vaporization of all common substances, and the highest surface tension of all common liquids. Its maximum density occurs at 4°C, ice being less dense than liquid water. This is why aquatic life can be supported by water below a few meters of ice in a frozen body of water, and freezing of water in the cracks of rocks is an important mechanism of mechanical weathering in cold climates. Water is a very good solvent for ionic compounds; compounds that dissociate into positively and negatively charged ionic species when dissolved in water.

In this chapter, our main objective will be to learn concepts and techniques required for computing concentrations of dissolved species in aqueous solutions under specified conditions. We will apply this knowledge to predict dissolution/precipitation of carbonate minerals, especially calcite, in natural aqueous systems and alteration of silicate minerals in the course of chemical weathering. As in the previous three chapters, thermodynamic equilibrium, whether attainable or not, will serve as the reference for much of our discussions.

7.1 Water as a solvent

The effectiveness of water as a solvent is reflected in its high *dielectric constant* (ε_{H_2O}), higher than that of any other common liquid. The dielectric constant of a solvent is a measure of the capacity of its molecules to prevent recombination of the positively and negatively charged solute particles. For liquid water at 20°C and 1 bar, the dielectric constant is 80.4, which means that the oppositely charged solute particles in water attract each other with a force of only 1/80.4 as strong as in vacuum. The high value of ε_{H_2O} arises from the dipolar nature of the water molecules (see section 3.9), which causes the solute ions to be *hydrated* — that is, surrounded by a loose layer of water molecules, aligned according to the charge on the solute ions (Fig. 7.1). The hydration provides a sort of protective shield around the cations and anions, thus reducing the probability of their recombination.

Introduction to Geochemistry: Principles and Applications, First Edition. Kula C. Misra.

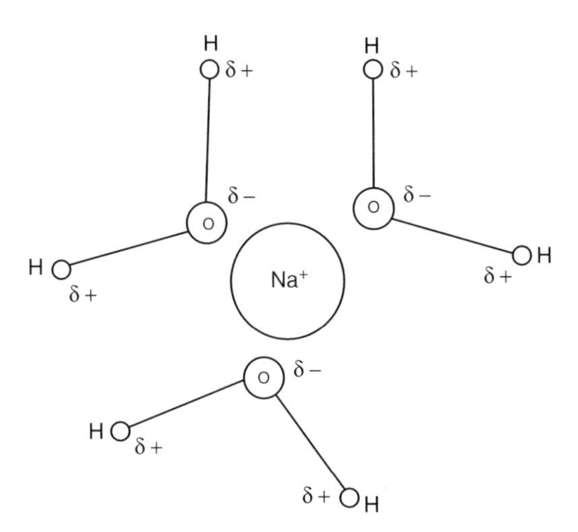

Fig. 7.1 When an ionic salt such as NaCl is dissolved in water, each Na^+ ion is surrounded mostly by the oxygen atoms of the polar water molecules. Similarly, each Cl^- ion is surrounded by the hydrogen atoms of the polar water molecules (not shown). Not to scale. (From *An Introduction to Environmental Chemistry*, 2nd edition, by J.E. Andrews, P. Brimblecombe, T.D. Jickells, P.S. Liss, and B. Reid, Box 6.4, Figure 1, p. 198; Copyright 2004, Blackwell Scientific Publications. Used with permission of the publisher.)

For example, when NaCl, an ionic compound, is dissolved in water, it dissociates into Na^+ and Cl^- ions:

$$NaCl_{(s)} \Leftrightarrow Na^+ + Cl^- \tag{7.1}$$

The dipolar water molecules align their positive and negative ends in such a way that the attraction between Na^+ and Cl^- is partially neutralized. Also, the sodium forms a hydrated cation (radius 0.95 Å) and the chlorine a hydrated anion (radius 1.81 Å), and the hydration weakens the attraction between them. This is the reason for the high solubility of NaCl in water.

The capacity of water to dissolve ionic compounds leads to some important consequences. Water occurring in nature is practically never pure; it invariably contains dissolved ionic species (and neutral molecules) and, therefore, is an aqueous solution. As the dissolved ionic species are charged particles, or *electrolytes*, that can conduct an electric current, aqueous solutions are *electrolyte solutions*. Nonelectrolytes exist as electrically neutral molecules in an aqueous solution, so a nonelectrolyte solution does not conduct an electric current.

7.2 Activity–concentration relationships in aqueous electrolyte solutions

The distribution of ions in an ionic aqueous solution is not completely random because there is a tendency for cations to be surrounded by a "cloud" of anions and for anions by a "cloud" of cations. This cloud is unrelated to the solvation shell. The result is a nonideal solution with a lower free energy compared to an ideal solution with a completely random distribution of ions. Thus, electrolyte solutions, in general, are nonideal, and the activity of a *solute* (any dissolved aqueous species) is not equal to its concentration in the solution, but is a function of both concentration and *activity coefficient*, the latter being a measure of the deviation of the solution from ideality. The activity of the solvent, on the other hand, remains close to unity even for solutions of quite high concentrations.

7.2.1 *Activity coefficient of a solute*

As discussed in Chapter 5, for a solute i in an ideal solution,

$$\mu_i(P, T) = \mu_i^0\,(P, T) + RT \ln a_i = \mu_i^0\,(P, T) + RT \ln X_i \tag{5.73}$$

where P and T are the pressure and temperature of interest, μ_i^0 (P, T) is the standard state chemical potential, and $a_i = X_i$. For a nonideal solution,

$$\mu_i(P, T) = \mu_i^0(P, T) + RT \ln a_i = \mu_i^0(P, T) + RT \ln (\lambda_i X_i) \tag{5.79}$$

and $a_i = \lambda_i\, X_i$, where γ_i is the rational activity coefficient. In the case of aqueous solutions, concentrations are commonly expressed as molalities rather than mole fractions, so that the expression for a_i is modified to

$$a_i = \gamma_i\, m_i \tag{7.2}$$

where m_i is the molality and γ_i is the *practical activity coefficient*. As $\gamma_i \rightarrow 1$, the solution approaches ideal behavior; for a very dilute solution, $\gamma_i \approx 1$ and $a_i \approx m_i$. A similar stipulation applies to aqueous solutions in which concentrations are expressed in molarity. In most cases, activity coefficients are less than unity for individual ions so that their activities are less than the actual concentrations.

The activity of a solute calculated using equation (7.2) has units of molality, instead of being a dimensionless number as per equation (5.72). To get around this discrepancy, the molality in equation (7.2) should be referenced to unit molality (Nordstrom and Munoz, 1994):

$$\gamma_i = \frac{a_i m_i^0}{m_i} \tag{7.3}$$

where m_i^{o} is 1 mol kg^{-1}. The inclusion of unit molality in equation (7.3) results in dimensionless γ_i and dimensionless a_i in equation (7.2), without affecting the numerical value of a_i or γ_i. For the sake of brevity, the unit molality reference is not explicitly mentioned in derivations and calculations, and activities are often expressed simply as molalities.

7.2.2 *Standard state of an aqueous solute*

We have emphasized earlier that the activity of a substance has no meaning unless it is referred to a defined standard state. What is the standard state for a solute in an aqueous solution? It is evident from equation (5.73) that

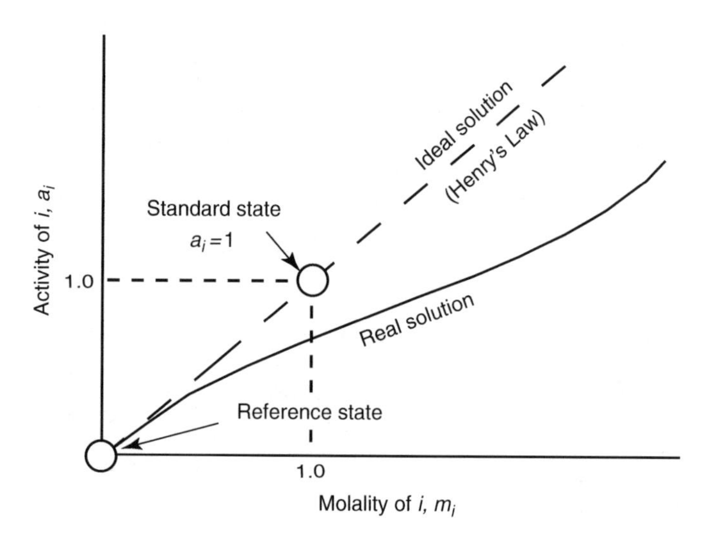

Fig. 7.2 Schematic illustration of the standard state and reference state for an aqueous solute (i). The dashed line represents the ideal solution (for which $a_i = m_i$ and $\gamma_i = 1$) that is really an expression of Henry's Law (equation 5.73), and the solid line the real solution (for which $a_i = \gamma_i m_i$ and $\gamma_i \neq 1$). In a real solution, $a_i = 1$ does not correspond to $m_i = 1$. Standard state (hypothetical) for aqueous solute: $a_i = m_i = 1$. Reference state for both solute and solvent: infinite dilution so that $a_i = m_i$. (From *Geochemical Thermodynamics*, 2nd edition, by D.K. Nordstrom and J.L. Munoz, Figure 7.3, p. 186; Copyright 1994, Blackwell Scientific Publications. Used with permission of the publisher.)

$$\mu_i\,(P, T) = \mu_i^0\,(P, T), \text{ if } RT \ln X_i = 0, \text{ i.e., } X_i = 1$$

However, $X_i = 1$ is not a reasonable choice for the standard state of electrolytes because most mineral–water solutions become saturated at much lower values of X_i (not much above $X_i = 0.1$ and for many of the more common minerals at $X_i = 0.01$).

> By convention, the standard state adopted for most solutes is a hypothetical 1 molal (1 mol kg^{-1}) solution at 298.15 K and 1 bar for which $a_i = m_i = 1$ (Nordstrom and Munoz, 1994).

The solution is a hypothetical one because for any real solution $a_i \neq 1$ when $m_i = 1$ since $\gamma_i \neq 1$ at any measurable concentration of the solute. A real solution would have $a_i = 1$ at some $m_i > 1$, and it should not be confused with standard state (Fig. 7.2).

As the standard state cannot be attained in reality, we need an attainable *reference state* for experimental measurements and extrapolation from such measurements.

> The reference state for an aqueous solute is the state of infinite dilution where the activity of the solute approaches its molal concentration (Fig. 7.2).

Thus, the reference state represents the limiting condition

$$\boxed{\lim_{m_i \to 0} \frac{a_i}{m_i} = 1} \qquad (7.4)$$

in which the solution is ideal (i.e., $a_i = m_i$). This reference state also has the advantage that it is the same for both the solute and the solvent because an aqueous solution is pure water at infinite dilution (Nordstrom and Munoz, 1994).

7.2.3 *Estimation of activity coefficients of solutes*

Ionic strength

Properties of individual ions in a solution cannot be measured independently. This is because a solution is electrically neutral, which demands that each ion is accompanied by an ion of the opposite charge. So, activity coefficients of aqueous ions have to be estimated indirectly.

The electrostatic forces between the charged solute species in an aqueous electrolyte solution depend on the charges of the species (the higher the charges, the greater the magnitude of electrostatic forces) and the total concentration of the species (because an increase in the total solute concentration decreases the mean distance between the ions and, thus, affects the magnitude of electrostatic forces). Both these factors are included in the *ionic strength* (I) of an aqueous solution, a concept introduced by G. N. Lewis and M. Randall in 1921 for calculating activity coefficients. The ionic strength is calculated by the formula

$$\boxed{I = \frac{1}{2}\sum_i m_i\, z_i^2} \qquad (7.5)$$

where m_i is the molality of the ith species (which is essentially the same as molarity, except in concentrated solutions with total dissolved solids in excess of about 7000 mg L^{-1}) and z_i is its charge (positive or negative). This formulation of ionic strength emphasizes the effect of higher charges of multivalent ions (positive and negative) and does not include any contribution from neutral molecules. Activity coefficients for solutes without charge commonly are close to 1.0. Also, H^+ and OH^- ions arising out of dissociation of water commonly are not included in the calculation of I (because their concentrations are very low, about 10^{-7} mol kg^{-1}) unless the water is highly acidic or highly alkaline.

For example, consider a hypothetical NaCl–$BaSO_4$ aqueous solution containing 0.1 mol kg^{-1} each of dissolved Na^+ and Cl^-, and 0.001 mol kg^{-1} each of dissolved Ba^{2+} and SO_4^{2-}. Ignoring m_{H^+} and m_{OH^-}, the ionic strength of the solution is

$$\begin{aligned} I &= 0.5\left[(m_{Na^+}\, z^2_{Na^+}) + (m_{Cl^-}\, z^2_{Cl^-}) + (m_{Ba^{2+}}\, z^2_{Ba^{2+}}) + (m_{SO_4^{2-}}\, z^2_{SO_4^{2-}})\right] \\ &= 0.5\left[(0.1)\,(1)^2 + (0.1)\,(-1)^2 + (0.001)\,(2)^2 + (0.001)(-2)^2\right] \\ &= 0.104 \text{ mol kg}^{-1} \end{aligned}$$

Ideally, the calculation of ionic strength would require a complete analysis of the solution; in practice, however, about 99% of the dissolved material in waters from streams, lakes, underground aquifers, and the oceans are accounted for by 10 elements: H, O, Na, K, Mg, Ca, Si, Cl, S, and C.

Table 7.1 Analysis and calculated ionic strengths of some natural waters*.

	Ganges River water[a]		Groundwater, limestone aquifer, Florida[b]		Surface seawater[c]	
Ions	mg L^{-1}	Molality (× 10^{-3})†	mg L^{-1}	Molality (× 10^{-3})†	mg L^{-1}	Molality (× 10^{-3})†
Ca^{2+}	25.4	0.634	56	1.397	406	10.13
Mg^{2+}	6.9	0.284	12	0.494	1271	52.29
Na^{+}	10.1	0.44	7.9	0.34	10570	460
K^{+}	2.7	0.069	1.0	0.026	380	9.719
Cl^{-}	5	0.14	12	0.34	19011	536
Br^{-}	–	–	–	–	66	0.826
SO_4^{2-}	8.5	0.088	53	0.552	2664	27.73
HCO_3^{-}	127	2.082	160	2.623	121	1.983
CO_3^{2-}	–	–	–	–	18	0.3
I (mol kg^{-1} water)		0.0034		0.0066		0.6852

[a]Sarin *et al.* (1989); [b]Goff (1971) as quoted in McSween *et al.* (2003); [c]Berner (1971).
*The concentration shown for any ionic species represents the sum of free ions plus ion pairs formed with ions of opposite charge. For example, the concentration of SO_4^{2-} in seawater is the sum of concentrations of the following species: SO_4^{2-}, $NaSO_4^{-}$, $CaSO_4^{0}$, and $MgSO_4^{0}$.
†To (simplify calculations it is assumed here that 1 L of water weighs 1 kg.

Table 7.2 Values of parameters for the Deby –Hückel equations.

Parameter $\mathring{a}$ of ions in aqueous solutions		Parameters at 1 bar		
Ions	$\mathring{A} \times 10^{-8}$	T (°C)	A	B × 10^{8}
Rb^+, Cs^+, NH_4^+, Tl^+, Ag^+	2.5	0	0.4883	0.3241
K^+, Cl^-, Br^-, I^-, NO_3^-	3	5	0.4921	0.3249
OH^-, F^-, HS^-, BrO_3^-, IO_4^-, MnO_4^-	3.5	10	0.4960	0.3258
Na^+, HCO_3^-, $H_2PO_4^-$, HSO_3^-, Hg_2^{2+},	4.0–4.5	15	0.5000	0.3262
SO_4^{2-}, SeO_4^{2-}, CrO_4^{2-}, CrO_4^{2-}, HPO_4^{2-}, PO_4^{3-}		20	0.5042	0.3273
Pb^{2+}, CO_3^{2-}, SO_3^{2-}, MoO_4^{2-}	4.5	25	0.5085	0.3281
Sr^{2+}, Ba^{2+}, Ra^{2+}, Cd^{2+}, Hg^{2+}, S^{2-}, WO_4^{2-}	5.0	30	0.5130	0.3290
Li^+, Ca^{2+}, Cu^{2+}, Zn^{2+}, Sn^{2+}, Mn^{2+}, Fe^{2+}, Ni^{2+}, Co^{2+}	6.0	35	0.5175	0.3297
Mg^{2+}, Be^{2+}	8.0	40	0.5221	0.3305
H^+, Al^{3+}, Cr^{3+}, REE^{3+}	9.0	45	0.5271	0.3314
Th^{4+}, Zr^{4+}, Ce^{4+}, Sn^{4+}	11.0	50	0.5319	0.3321
		55	0.5371	0.3329
		60	0.5425	0.3338

Source: Garrels and Christ (1965).

The ionic strength of most natural waters is less than 1 mol kg^{-1} water (Table 7.1).

Activity coefficients of individual ions

In 1923 the Dutch scientist Peter. J. W. Debye (1884–1966), a Nobel laureate in Chemistry, and the German scientist E. Hückel (1896–1980) theoretically derived an equation to calculate activity coefficients of individual ions in dilute aqueous solutions on the basis of the effect ionic interactions should have on the free energy of the solution. Assumptions incorporated in the derivation of the equation include the following: (i) the electrolytes are completely dissociated into ions; (ii) positive ions are surrounded by a cloud of negative charges and vice versa; (iii) the size of the ions do not vary with ionic strength; and (iv) interactions among the aqueous species are entirely electrostatic (with no interactions between ions of the same sign). For individual ions, the *Debye–Hückel equation* (commonly referred to as the *extended Debye–Hückel equation*), is

$$\log \gamma_i = \frac{-A\, z_i^2 \sqrt{I}}{1 + \mathring{a}_i B \sqrt{I}} \qquad (7.6)$$

where A and B are constants characteristic of the solvent (water in this case) at specified P and T, and $\mathring{a}_i$ is the *hydrated ionic radius*, or *"effective" ionic radius*, of the ith ion. The hydrated radius is significantly larger than the radius of the same ion in a crystal. In practice, the value of $\mathring{a}_i$ is chosen to give the best fit to the experimental data. Values of A, B, and $\mathring{a}_i$ for the common aqueous species are listed in Table 7.2. The effect of temperature on activity coefficients is largely dictated by changes in the value of A. With increasing temperature, the activity coefficients become smaller because the value of A becomes larger. The higher the charge on an ion, the greater is the effect of temperature on its activity coefficient. The extended Debye–Hückel equation is most useful for solutions with $I \le 0.1$ mol kg^{-1} and provides adequate approximations for ionic strengths up to about 1 mol kg^{-1}.

At $I \le 0.001$ mol kg^{-1}, the term $\mathring{a}_i B\sqrt{I}$ approaches zero, and the *Debye–Hückel equation* reduces to

$$\log \gamma_i = -A\, z_i^2 \sqrt{I} \qquad (7.7)$$

This is called the *Debye–Hückel limiting law* because it applies to the limiting case of very dilute solutions.

The form of the *Debye–Hückel (D–H) equation* is such that the calculated activity coefficient of an individual aqueous ion continuously decreases with increasing ionic strength of the solution (Fig. 7.3). Experimental data, however, show that in solutions of high ionic strength (I > about 0.8 to1 for most ions) activity coefficients actually increase with increasing ionic strength. To accommodate this behavior, Truesdell and Jones (1974) proposed a modification that simply added a term bI to the D–H equation:

$$\log \gamma_i = \frac{-A\, z_i^2 \sqrt{I}}{1 + \mathring{a}_i B \sqrt{I}} + bI \qquad (7.8)$$

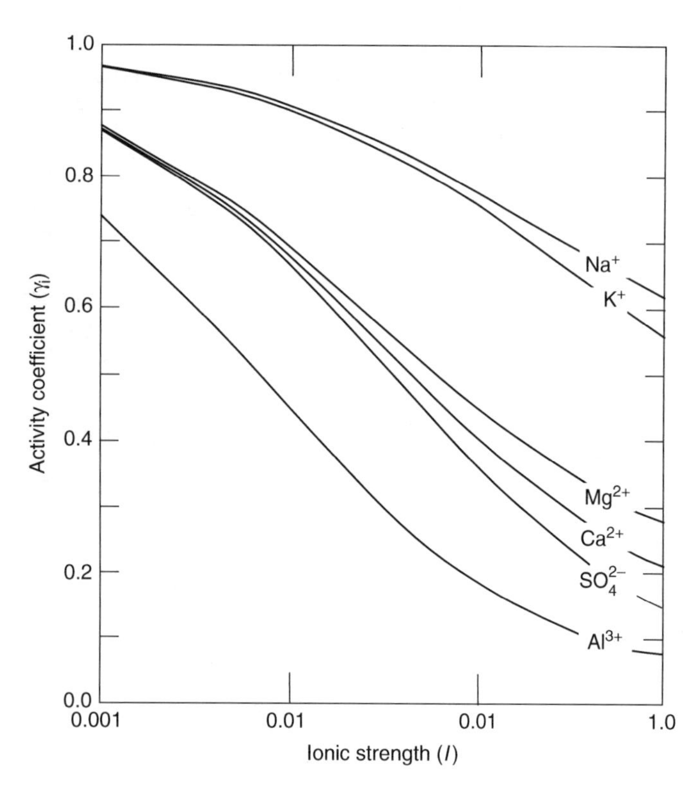

Fig. 7.3 Activity coefficients of some common aqueous species at 298.15 K (25°C) and 1 bar as a function of ionic strength, calculated using the extended Debye–Hückel equation (equation 7.6) and values of the parameters given in Table 7.2. Note that for a particular ion, the activity coefficient decreases with increasing ionic strength and for ions of equal charge the differences in activity coefficients due to different values of the å parameter become important at $I > 0.01\,m$.

Table 7.3 Parameters for the Truesdell–Jones equation.

Ion	å × 10⁻⁸	b	Ion	å × 10⁻⁸	b	Ion	å × 10⁻⁸	b
H^+	4.78	0.24	Mn^{2+}	7.04	0.22	Al^{3+}	6.65	0.10
Na^+	4.32	0.06	Fe^{2+}	5.08	0.16	OH^-	10.65	0.21
K^+	3.71	0.01	Co^{2+}	6.17	0.22	F^-	3.46	0.08
Mg^{2+}	5.46	0.22	Ni^{2+}	5.51	0.22	Cl^-	3.71	0.01
Ca^{2+}	4.86	0.15	Zn^{2+}	4.87	0.24	HCO_3^-	5.4	0
Sr^{2+}	5.48	0.11	Cd^{2+}	5.80	0.10	CO_3^{2-}	5.4	0
Ba^{2+}	4.55	0.09	Pb^{2+}	4.80	0.01	SO_4^{2-}	5.31	−0.07

Sources: Truesdell and Jones (1974); Parkhurst (1990).

where b is a constant specific to the individual ion. The additional term compensates for the lowering of the dielectric constant of water and increased ion pairing (see section 7.3.3) caused by the increased concentration of solutes. Values of å and b for selected ions are listed in Table 7.3; values of A and B are the same as given in Table 7.2. Equation (7.8) is applicable to solutions with ionic strength ranging from 0 to 2 mol kg⁻¹ (Langmuir, 1997). The calculations are easily

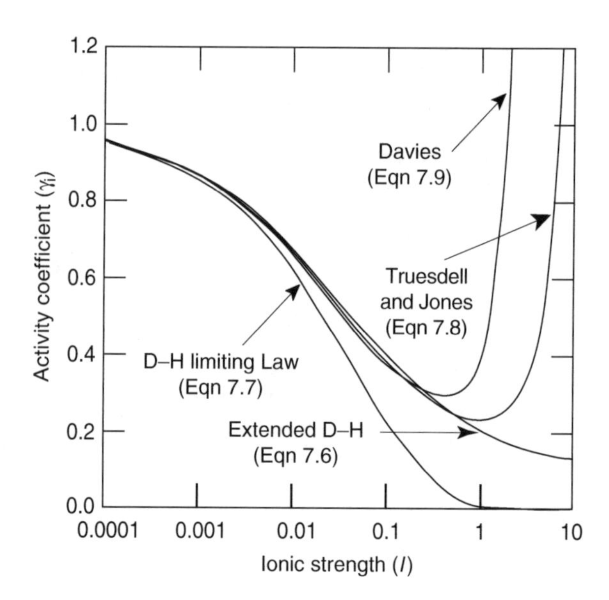

Fig. 7.4 Variation of the activity coefficient of Ca^{2+} as a function of ionic strength according to different equations discussed in the text.

accomplished with a spreadsheet, or with a computer code such as SOLMINEQ.88 (Kharaka *et al.*, 1988; Perkins *et al.*, 1990) or PHREEQE (Parkhurst *et al.*, 1980).

Experimentally measured values of å are not available for many ions; for such ions we may use the following empirical equation suggested by Davies (1962):

$$\log \gamma_i = -A\, z_i^2 \left[\frac{\sqrt{I}}{1+\sqrt{I}} - 0.3I \right] \tag{7.9}$$

The *Davies equation* yields reasonable results for ionic strength up to about 0.6 mol kg⁻¹, but the lack of an ion size parameter in the equation makes it less accurate than the D–H equation at low ionic strengths ($I < 0.1$). A comparison of the variation of activity coefficient as a function of ionic strength according to the above equations is presented in Fig. 7.4.

The high density of ions in concentrated solutions ($I > 2$ to 3.5) can lead to more complicated interactions such as binary interactions between species of like charge and ternary interactions among three or more ions. The most appropriate model for high ionic strength electrolyte solutions (I up to 6) is that of Pitzer (1973, 1979, 1980), which takes into account all specific ion interactions except the strongly bound, covalent association of ions (see Langmuir, 1997, pp. 138–142 for a concise discussion of the Pitzer model). The calculations, however, are complex and best carried out using an appropriate computer code such as SOLMINEQ.88 (Kharaka *et al.*, 1988; Perkins *et al.*, 1990), PHRQPITZ (Plummer *et al.*, 1988; Plummer and Parkhurst, 1990), or PHREEQC (Parkhurst, 1995) that uses the Pitzer model (see Appendix 8). The Pitzer model, however, is less accurate than the extended D–H equation for dilute solutions because it lacks an ion size parameter.

Example 7–1: Calculation of the activity coefficient and activity of Ca^{2+} in the Ganges River water (Table 7.1) using different equations and appropriate values of the parameters as given in Tables 7.2 and 7.3

For the Ganges water, $m_{Ca^{2+}} = 0.634 \times 10^{-3}$ mol kg^{-1} and $I = 0.0034$ (Table 7.1). The activity coefficient values of Ca^{2+}, $\gamma_{Ca^{2+}}$, using the different equations, are as follows:

(1) Limiting D–H equation ($A = 0.5085$)

$$\log \gamma_{Ca^{2+}} = -A\, z^2_{Ca^{2+}} \sqrt{I} = -(0.5085)(2^2)\sqrt{0.0034} = -0.1186$$

$$\gamma_{Ca^{2+}} = 0.761$$

(2) Extended D–H equation ($A = 0.5085$, $B = 0.328 \times 10^8$, $\mathring{a} = 6 \times 10^{-8}$)

$$\log \gamma_{Ca^{2+}} = \frac{-A\, z^2_{Ca^{2+}} \sqrt{I}}{1 + \mathring{a}_{Ca^{2+}} B\sqrt{I}} = \frac{-(0.5085)\,(2^2)\sqrt{0.0034}}{1 + (6)\,(0.3281)\sqrt{0.0034}} = 0.1064$$

$$\gamma_{Ca^{2+}} = 0.761$$

(3) Truesdell–Jones equation ($A = 0.5085$, $B = 0.328 \times 10^8$, $\mathring{a} = 4.86 \times 10^{-8}$, $b = 0.15$)

$$\log \gamma_{Ca^{2+}} = \frac{-A\, z^2_{Ca^{2+}} \sqrt{I}}{1 + \mathring{a}_{Ca^{2+}} B\sqrt{I}} + bI = \frac{-(0.5085)(2^2)\sqrt{0.0034}}{1 + (4.86)(0.3281)\sqrt{0.0034}} + (0.15)(0.0034) = -0.108$$

$$\gamma_{Ca^{2+}} = 0.780$$

(4) Davies equation ($A = 0.5085$)

$$\log \gamma_{Ca^{2+}} = -A\, z^2_{Ca^{2+}} \left[\frac{\sqrt{I}}{1+\sqrt{I}} - 0.3I\right]$$

$$= -(0.5085)(2^2)\left[\frac{\sqrt{0.0034}}{1+\sqrt{0.0034}} - 0.3(0.0034)\right] = -0.1100;$$

$$\gamma_{Ca^{2+}} = 0.776$$

Applying the extended D–H equation (since I < 0.1 m for this solution),

$$\gamma_{Ca^{2+}} = 0.783;\ a_{Ca^{2+}} = m_{Ca^{2+}} \gamma_{Ca^{2+}} = (0.634 \times 10^{-3})\,(0.783) = 0.496 \times 10^{-3} \text{ mol kg}^{-1}$$

Error in estimated $a_{Ca^{2+}}$ if $\gamma_{Ca^{2+}}$ is ignored = [(0.634 − 0.496)/0.634] × 100 ≈ 22%.

7.3 Dissociation of acids and bases

Towards the end of the 19th century, the Swedish chemist Svante August Arrhenius (1859–1927) defined an *acid* as a hydrogen-bearing compound that dissociates when dissolved in water, releasing H^+ ions (protons) and thus lowering the pH of the solution. This is a convenient definition for geochemical reactions, although H^+ ions do not exist in an isolated form in the presence of water; they interact with water molecules to form aqueous hydronium ions, $H_3O^+_{(aq)}$. Similarly, a *base* is defined as a substance containing the OH group that dissociates when dissolved in water, releasing OH^- (hydroxide ions) and thus raising the pH of the solution. As is the case with H^+ ions, OH^- ions do not occur as isolated ions in water; they actually occur as hydrated ions but we write them as though they were not hydrated.

Following the definitions stated above, the dissociation of an acid (H_mA_n) and a base (B_pOH_q) at equilibrium may be represented as

$$H_mA_n \Leftrightarrow mH^+ + nA^-; \qquad K_a = \frac{(a_{H^+})^m\,(a_{A^-})^n}{a_{H_mA_n}} \tag{7.10}$$

$$B_pOH_q \Leftrightarrow pB^+ + qOH^-; \qquad K_b = \frac{(a_{B^+})^p\,(a_{OH^-})^q}{a_{B_pOH_q}} \tag{7.11}$$

where a_i represents activity of the species i, $a_{H_mA_n}$ and $a_{B_pOH_q}$ refer to undissociated acid and base molecules in the solution, and K_a and K_b are the equilibrium constants, which are commonly referred to as *dissociation constants* to emphasize the nature of the reactions. The higher the value of a dissociation constant, the greater is the *degree of dissociation* (D) of the acid or the base, which is calculated as follows:

$$\%\ \text{Dissociation of an acid} = \frac{\text{amount of acid dissociated}}{\text{amount of acid originally present}} \times 100 \tag{7.12}$$

$$\%\ \text{Dissociation of a base} = \frac{\text{amount of base dissociated}}{\text{amount of base originally present}} \times 100 \tag{7.13}$$

Note that the percent dissociation of an acid or a base is defined as the ratio of the amount dissociated relative to the total amount originally present, not the ratio of the concentrations of dissociated and undissociated acid/base at equilibrium.

Some acids (e.g., HCl, HNO_3, H_2SO_4) are referred to as *strong acids* because they dissociate almost completely when dissolved in water (i.e., they release most or all of their H^+ ions). In the case of complete dissociation, the value of K_a is undefined because the activity of the undissociated acid is zero. *Weak acids* (e.g., CH_3COOH, H_2CO_3, H_3PO_4, H_4SIO_4) dissociate only to a small extent (i.e. they release only a small fraction of their H^+ ions into the solution). Similarly, bases are

Table 7.4 Dissociation constants of selected acids and bases at 298.15 K (25°C) and 1 bar.

Acid	Formula	Constants				Base (hydroxide)	Formula	Constants			
		pK_{a_1}	pK_{a_2}	pK_{a_3}	pK_{a_4}			pK_{a_1}	pK_{a_2}	pK_{a_3}	pK_{a_4}
Acetic	CH_3COOH	4.75	*	*	*	Ammonium hydroxide	NH_4OH	4.7	*	*	*
Carbonic	H_2CO_3	6.35	10.3	*	*	Copper hydroxide	$Cu(OH)_2$	13.0	6.3	*	*
Hydrochloric	HCl	~ −3	*	*	*	Ferrous hydroxide	$Fe(OH)_2$	10.6	4.5	*	*
Hydrofluoric	HF	3.18	*	*	*	Ferric hydroxide (amorphous)	$Fe(OH)_3$	16.5	10.5	11.8	*
Phosphoric	H_3PO_4	2.1	7.2	12.35	*	Gibbsite	$Al(OH)_3$	14.8	10.3	9.0	*
Silicic	H_4SiO_4	9.71	13.28	9.86	13.1	Magnesium hydroxide	$Mg(OH)_2$	8.6	2.6	*	*
Sulfuric	H_2SO_4	~ −3	1.99	*	*	Manganese hydroxide	$Mn(OH)_2$	9.4	3.4	*	*
						Zinc hydroxide (amorphous)	$Zn(OH)_2$	10.5	5.0	*	*

*Does not apply or value not known; $pK_a = -\log K_a$, $pK_b = -\log K_b$.
Sources of data: Compilations by Krauskopf and Bird (1985), Faure (1991), and Eby (2004).

classified as *strong* (e.g., hydroxides of alkali and alkaline-earth elements) or *weak* (e.g., NH_4OH, $Mg(OH)_2$, $Fe(OH)_3$) on the basis of their degree of dissociation when dissolved in water. Strong acids and strong bases are strong electrolytes.

Polyprotic acids (i.e., acids containing more than one H in their formula units) dissociate in a stepwise manner, with a different dissociation constant for each step. For example, phosphoric acid (H_3PO_4) dissociates in three steps, which may be written as:

$$H_3PO_4 \Leftrightarrow H^+ + H_2PO_4^-; \quad K_{a_1} = 10^{-2.1} \tag{7.14}$$

$$H_2PO_4^- \Leftrightarrow H^+ + HPO_4^{2-}; \quad K_{a_2} = 10^{-7.2} \tag{7.15}$$

$$HPO_4^{-2} \Leftrightarrow H^+ + PO_4^{3-}; \quad K_{a_3} = 10^{-12.35} \tag{7.16}$$

where K_{a1}, K_{a2}, and K_{a3}, respectively, are the first, second, and third dissociation constant. Similarly, dissociation of the base ferrous hydroxide ($Fe(OH)_2$) occurs in two steps, each step with its own dissociation constant:

$$Fe(OH)_2 \Leftrightarrow Fe(OH)^+ + (OH)^-; \quad K_{b_1} = 10^{-10.6} \tag{7.17}$$

$$Fe(OH)^+ \Leftrightarrow Fe^{2+} + (OH)^-; \quad K_{b_2} = 10^{-4.5} \tag{7.18}$$

Dissociation constants for a few common acids and bases are listed in Table 7.4.

Example 7–2: Calculation of the concentrations of dissociated species in a solution of 0.01 mole of acetic acid, CH_3COOH, in 1 kg of pure water

When acetic acid (CH_3COOH), a weak acid, is dissolved in water, it dissociates into CH_3COO^- and H^+. At equilibrium,

$$CH_3COOH_{(aq)} \Leftrightarrow CH_3COO^- + H^+ \tag{7.19}$$

$$K_{CH_3COOH} = \frac{a_{CH_3COO^-}\, a_{H^+}}{a_{CH_3COOH\,(aq)}} = 10^{-4.75} (298.15\ K, 1\ bar) \tag{7.20}$$

This is a very dilute solution, so we can assume $a_i = m_i$ (i.e., $\gamma_i = 1$).

Let $m_{H^+} = x$. Since each mole of CH_3COOH on dissociation produces 1 mole of H^+ and 1 mole of CH_3COO^-, $m_{CH_3COOH^-} = x$ and $m_{CH_3COOH(aq)} = 0.01 - x$. Substituting in equation (7.20),

$$\frac{(x)\,(x)}{0.01 - x} = 10^{-4.75} \quad \text{or} \quad x^2 + 10^{-4.75}x - 10^{-6.75} = 0$$

This is a quadratic equation of the form $ax^2 + bx + c = 0$, where $a = 1$, $b = 10^{-4.75}$, $c = 10^{-6.76}$, and the roots of the equation are;

$$x = \frac{-b \pm \sqrt{(b^2 - 4ac)}}{2a} \tag{7.21}$$

Solving for x and rejecting the negative value, we get $x = 4.31 \times 10^{-4}$. Thus, at equilibrium,

$$m_{H^+} = m_{CH_3COOH^-} = 4.31 \times 10^{-4} \text{mol kg}^{-1}$$

$$m_{CH_3COOH(aq)} = 0.01 - (4.31 \times 10^{-4}) = 0.0096 \text{ mol kg}^{-1}$$

$$\text{Degree of dissociation } (D) = \frac{4.31 \times 10^{-4}}{0.01} \times 100 = 4.31\%$$

7.4 Solubility of salts

7.4.1 The concept of solubility

A *salt* is a compound that contains a cation other than H^+ and an anion other than OH^- or O^{2-}. Reaction of an acid (source of anion) with a base (source of cation) produces water and a salt that is named after the acid that provided the anion. For example, the salt produced by the reaction of $Ca(OH)_2$ (calcium hydroxide) and H_2CO_3 (carbonic acid) is calcium carbonate ($CaCO_3$):

$$\underset{\text{acid}}{H_2CO_3} + \underset{\text{base}}{Ca(OH)_2} = \underset{\text{salt}}{CaCO_3} + 2H_2O \quad (7.22)$$

Similarly, reaction of a base with hydrochloric acid (HCl) produces a chloride, with sulfuric acid (H_2SO_4) a sulfate, with silicic acid (H_4SiO_4) a silicate, and so on. Minerals, except for native elements, oxides, and hydroxides, are salts.

A salt may remain in solution or precipitate as a solid depending on its solubility. *The solubility of a substance in water is the concentration (not the activity) of the substance in the water that is in equilibrium with the substance at the pressure and temperature of interest.* Such a solution is called a *saturated solution* of the substance under consideration. A similar definition also applies to solutions with solvents other than water. Thus, the ability of a solution to transport substances in a dissolved form (as solutes) depends on the solubilities of the solutes. The solubility is expressed in one of the following units: molality (m), molarity (M), g kg^{-1}, g L^{-1} parts per million (ppm), or parts per billion (ppb) (see Chapter 1).

Let us imagine an experiment in which a lump of a sparingly soluble salt, B_pA_n, is placed in a beaker of pure water and the concentrations of the solutes in the dilute aqueous solution are monitored continuously. A very small amount of the salt will dissociate into the anion of the acid (A^-) and the cation of the base (B^+) from which they were derived:

$$B_pA_{n(s)} \Rightarrow pB^+_{(aq)} + nA^-_{(aq)} \quad (7.23)$$

The reaction does not include H_2O because it does not take part in the reaction. We assume for the time being that the ions resulting from dissociation do not interact with water. As

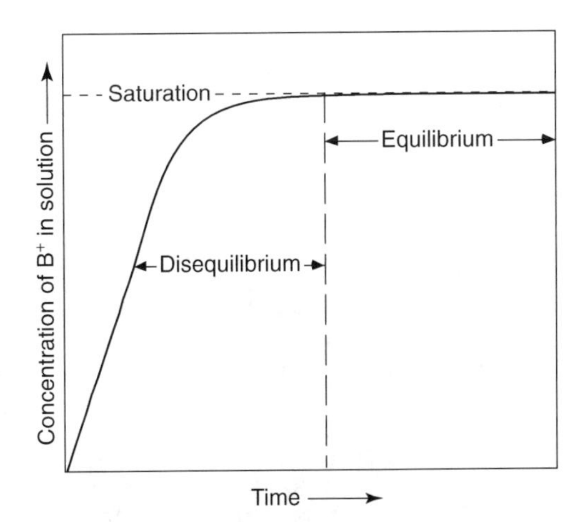

Fig. 7.5 Illustration of the concept of solubility of a salt BA in water.

shown schematically in Fig. 7.5, the concentration of B^+ (and A^-) in the solution will increase with time due to more and more of the salt being dissolved, and then stay constant regardless of how long the experiment is continued or how much of the solid salt is left undissociated. At the point where the concentration attains a constant value, the water becomes saturated with the salt and the dissociation reaction (equation 7.23) attains a state of equilibrium. As each mole of dissociated B_pA_n in this case produces p moles of B^+ (and n moles of A^-), the solubility of the salt B_pA_n is given by m_{B^+} / p (or m_{A^-} / q) in the saturated solution.

7.4.2 *Solubility product*

We can apply the Law of Mass Action to express solubility of a slightly soluble salt in terms of the equilibrium constant of its dissociation reaction. Since $a_{B_pA_n(s)} = 1$ (the amount of the salt is immaterial for equilibrium as long as it is present in excess of its solubility), the equilibrium constant for reaction (7.23) reduces to merely a product of the activities of the ions (each raised to the power that corresponds to its stoichiometric coefficient in the balanced dissociation reaction). In equilibria that involve slightly soluble compounds in water, the equilibrium constant is called the *solubility product constant*, or simply the *solubility product* (K_{sp}), although it actually represents a product of activities:

$$K_{eq} = \frac{(a_{B^+})^p\,(a_{A^-})^n}{a_{B_pA_n\,(s)}} = \frac{(a_{B^+})^p\,(a_{A^-})^n}{1} = (a_{B^+})^p(a_{A^-})^n = K_{sp\,(B_pA_n)} \quad (7.24)$$

As $a_i = m_i\gamma_i$, substitution for a_i in equation (7.24) gives

$$\boxed{K_{sp\,(B_pA_n)} = (a_{B^+})^p\,(a_{A^-})^n = (m_{B^+}\gamma_{B^+})^p\,(m_{A^-}\gamma_{A^-})^n} \quad (7.25)$$

For the mineral barite ($BaSO_4$), for example,

$$BaSO_{4\,(s)} \Leftrightarrow Ba^{2+} + SO_4^{2-};\; K_{sp\,(BaSO_4)} = a_{Ba^{2+}}\,a_{SO_4^{2-}}$$
$$= (m_{Ba^{2+}}\gamma_{Ba^{2+}})\,(m_{SO_4^{2-}}\gamma_{SO_4^{2-}})$$

Since dissociation of 1 mole of $BaSO_4$ produces 1 mole of Ca^{2+}

$$\text{Solubility of } BaSO_4 = m_{Ba^{2+}} = m_{SO_4^{2-}} = \left(\frac{K_{sp\,(BaSO_4)}}{\gamma_{Ba^{2+}}\,\gamma_{SO_4^{2-}}}\right)^{1/2} \quad (7.26)$$

If the solution can be assumed to be ideal, solubility of $BaSO_4 = (K_{sp\,(BaSO_4)})^{1/2}$.

For the mineral fluorite (CaF_2), to consider another example,

$$CaF_{2\,(s)} \Leftrightarrow Ca^{2+} + 2F^-;\; K_{sp\,(CaF_2)} = a_{Ca^{2+}}\,a^2_{F^-} = (m_{Ca^{2+}}\gamma_{Ca^{2+}})(m_{F^-}\gamma_{F^-})^2$$

Let x = solubility of CaF_2 (in moles kg^{-1} of water). Since the dissociation produces only one-half as many moles of Ca^{2+} as F^-,

$$x = m_{Ca^{2+}} = 0.5\,m_{F^-}$$

Substituting for m_{F^-},

$$K_{sp\,(CaF_2)} = (m_{Ca^{2+}}\gamma_{Ca^{2+}})\,(2m_{Ca^{2+}}\gamma_{F^-})^2 = 4m^3_{Ca^{2+}}\gamma_{Ca^{2+}}\gamma_{F^-}$$

$$\text{Solubility of } CaF_2 = x = m_{Ca^{2+}} = 0.5\,m_{F^-} = \left(\frac{1}{4}\,\frac{K_{sp\,(CaF_2)}}{\gamma_{Ca^{2+}}\,\gamma^2_{F^-}}\right)^{1/3} \quad (7.27)$$

If the solution can be assumed to be ideal,

$$\text{Solubility of } CaF_2 = \left(\frac{K_{sp\,(CaF_2)}}{4}\right)^{1/3} \quad (7.28)$$

The solubility of a substance is a function of temperature and pressure because the solubility product is an equilibrium constant that varies with temperature and pressure as discussed in section 6.2. For most salts, the solubility product – and, therefore, the solubility – increases with increasing temperature, but the opposite is true for some salts such as calcite and other carbonates. Some salts and some ions in concentrated solutions have activity coefficients greater than unity. This means that as a solution becomes more concentrated, the solubility of a slightly soluble salt may pass through a maximum and then decrease rapidly with further increase in the solute concentration (Krauskopf and Bird, 1995, p. 51). For all gases, solubility decreases as the solution temperature rises. Other major factors affecting solubilities, as will be illustrated by some of the examples below, include ionic strength, pH, and composition of the solution.

Example 7–3: Calculation of the solubility of the mineral gypsum ($CaSO_4{\cdot}2H_2O$) at 298.15 K and 1 bar, given $K_{sp\,(gypsum)} = 10^{-4.59}$

Dissociation of gypsum in pure water (i.e., without any other source of Ca^{2+} or SO_4^{2-}) at equilibrium can be represented by the reaction

$$CaSO_4\cdot 2H_2O_{(s)} \Leftrightarrow Ca^{2+} + SO_4^{2-} + 2H_2O;$$

$$K_{eq} = \frac{a_{Ca^{2+}}\,a_{SO_4^{2-}}\,a^2_{H_2O}}{a_{CaSO_4\cdot 2H_2O(s)}} \quad (7.29)$$

Let us choose the standard state as 298.15 K and 1 bar. Since $a_{CaSO_4\cdot 2H_2O} = 1$ (standard state activity) and $a_{H_2O} = 1$ (convention),

$$K_{sp\,(gypsum)} = a_{Ca^{2+}}\,a_{SO_4^{2-}} = (m_{Ca^{2+}}\gamma_{Ca^{2+}})\,(m_{SO_4^{2-}}\gamma_{SO_4^{2-}}) = 10^{-4.59}$$

As 1 mole of gypsum dissociates into 1 mole each of Ca^{2+} and SO_4^{2-}, $m_{Ca^{2+}} = m_{SO_4^{2-}}$,

$$\text{Solubility of gypsum (in pure water)} = m_{Ca^{2+}} = m_{SO_4^{2-}} = \left(\frac{K_{sp\,(gypsum)}}{\gamma_{Ca^{2+}}\gamma_{SO_4^{2-}}}\right)^{1/2} \quad (7.30)$$

Here we run into a problem: to compute the activity coefficients, we have to first calculate the ionic strength of the solution for which we need to know $m_{Ca^{2+}}$ and $m_{SO_4^{2-}}$, but we do not. In fact, that is what we want to find out. The problem can be addressed by an iteration procedure involving successive approximations as follows:

Step 1. To start, assume that the solution is ideal, i.e., $\gamma_{Ca^{2+}} = 1$ and $\gamma_{SO_4^{2-}} = 1$, and calculate the ionic strength of the solution:

$$m_{Ca^{2+}} = m_{SO_4^{2-}} = (K_{sp})^{1/2} = (10^{-4.59})^{1/2} = 10^{-2.29} = 5.13\times10^{-3}\text{mol kg}^{-1} \quad (7.31)$$

$$m_{Ca^{2+}} = m_{SO_4^{2-}} = (K_{sp})^{1/2} = (10^{-4.60})^{1/2} = 10^{-2.30} = 5.01\times10^{-3}\text{mol kg}^{-1}$$

$$I = 0.5\left[(m_{Ca^{2+}}\,z^2_{Ca^{2+}}) + (m_{SO_4^{2-}}\,z^2_{SO_4^{2-}})\right] = 0.5\left[(10^{-2.29})\,(2)^2 + (10^{-2.29})\,(-2)^2\right] = 0.0205$$

Step 2. Calculate $\gamma_{Ca^{2+}}$ and $\gamma_{SO_4^{2-}}$ from the extended D–H equation, using parameters in Table 7.1, and improved values of $m_{Ca^{2+}}$ and $m_{SO_4^{2-}}$ from equation (7.30).

Step 3. Calculate the improved value of I with the new values of $m_{Ca^{2+}}$ and $m_{SO_4^{2-}}$.

Step 4. Repeat steps 2 and 3 until the difference between the new and immediately preceding values of $m_{Ca^{+}}$ and $m_{SO_4^{2-}}$ are negligibly small (Table 7.5).

As shown in Table 7.5, it took seven iterations in this case to obtain the final answer:

$$\text{Solubility of gypsum} = m_{Ca^{2+}} = m_{SO_4^{2-}} = 10.25\times10^{-3} = 10^{-1.99}\ \text{mol kg}^{-1}$$

This value is about twice the value calculated without corrections for activity coefficients.

Table 7.5 Results of iterative calculations for gypsum solubility

Iteration	I (initial)	$\gamma_{Ca^{2+}}$	$\gamma_{SO_4^{2-}}$	$m_{Ca^{2+}}\times10^{-3}$	$m_{SO_4^{2-}}\times10^{-3}$	I (new)
1	Unknown	1	1	5.13	5.13	0.0205
2	0.0205	0.5926	0.5748	8.69	8.69	0.0347
3	0.0347	0.5280	0.5043	9.82	9.82	0.0393
4	0.0393	0.5128	0.4876	10.14	10.14	0.0406
5	0.0406	0.5089	0.4833	10.22	10.22	0.0409
6	0.0409	0.5079	0.4822	10.24	10.24	0.0410
7	0.0410	0.5077	0.4820	10.25	10.25	

Example 7–4: Calculation of the solubility of gypsum ($CaSO_4{\cdot}2H_2O$) at 298.15 K and 1 bar, in a 0.1 m $CaCl_2$ aqueous solution — the common-ion effect

Since the solution has another source of Ca^{2+} from the dissociation of $CaCl_2$, ($CaCl_2 \Leftrightarrow Ca^{2+} + 2\ Cl^-$), we can reasonably guess that gypsum will not supply all the Ca^{2+} necessary for saturation. In other words, the solubility of gypsum should be less in this solution compared to pure water. Let us see how we can quantify this guess.

Let the solubility of gypsum = x moles/kg of water

Moles of Ca^{2+} in 1 kg of the solution = x (from gypsum) + 0.1 (from $CaCl_2$)

Moles of SO_4^{2-} in 1 kg of solution = x (from gypsum)

Moles of Cl^- in 1 kg of solution = 0.2

As the presence of $CaCl_2$ in the solution does not affect $K_{sp\,(gypsum)}$,

$$K_{sp\,(gypsum)} = a_{Ca^{2+}}\ a_{SO_4^{2-}}$$
$$= [(x+0.1)\ \gamma_{Ca^{2+}}]\ [\ x\ \gamma_{SO_4^{2-}}] = 10^{-4.59} = 2.57\times10^{-5}$$

If we assume the solution to be ideal, $\gamma_{Ca^{2+}} = 1$ and $\gamma_{SO_4^{2-}} = 1$, and the equation is simplified to

$$x^2 + 0.1x = 10^{-4.59}$$

This equation can be solved for x by the familiar formula for a quadratic equation, but we will employ a shortcut. We can guess from Example 7–3 that x *will* be a small number and $x^2 \ll x$. Ignoring x^2 relative to x,

$$x = \frac{10^{-4.59}}{0.1} = 10^{-3.59} = 25.7\times10^{-5}\ m$$

Thus, the slubility of gypsum in a 0.1 *m* $CaCl_2$ aqueous solution works out to be much less than its solubility in pure water (Example 7–3).

Now let us find out whether it would make a difference if we took into account the activity coefficients. To estimate $\gamma_{Ca^{2+}}$ and $\gamma_{SO_4^{2-}}$ we have to first calculate the ionic strength, for which we need to know $m_{Ca^{2+}}$ and $m_{SO_4^{2-}}$, but we do not. To get around this impasse, we will employ the iteration procedure as outlined in Example 7–3:

Step 1. Assume $\gamma_{Ca^{2+}} = 1$ and $\gamma_{SO_4^{2-}} = 1$. We have already calculated that

$$m_{SO_4^{2-}} = 10^{-3.59} = 0.000257;\ m_{Ca^{2+}} = 0.1 + 10^{-3.59} = 0.100257;\ m_{Cl^-} = 0.2$$

With these values, the starting ionic strength of the solution is:

$$I\ (\text{initial}) = 0.5\left[(m_{Ca^{2+}} z^2_{Ca^{2+}}) + (m_{SO_4^{2-}} z^2_{SO_4^{2-}}) + (m_{Cl^-} \cdot z^2_{Cl^-})\right] = 0.301$$

Step 2. Calculate the activity coefficients of the ions, using the extended D–H equation:

$$\gamma_{Ca^{2+}} = 0.2907 \text{ and } \gamma_{SO_4^{2-}} = 0.2418$$

Using the approximation that $x^2 + 0.1x \approx 0.1x$ because $x^2 \ll 0.1$, new molalities for the second step of the iteration are:

$$m_{SO_4^{2-}} = \frac{K_{sp}(\text{gypsum})}{0.1\ \gamma_{Ca^{2+}}\ \gamma_{SO_4^{2-}}} = 0.0036$$
$$m_{Ca^{2+}} = m_{SO_4^{2-}} + 0.1 = 0.1036\ ;\ m_{Cl^-} = 0.2$$

Step 3. Calculate the improved value of I with the new values of $m_{Ca^{2+}}$ and $m_{SO_4^{2-}}$.

Step 4. Repeat steps 2 and 3 until the difference between the new and immediately preceding values of $m_{Ca^{2+}}$ and $m_{SO_4^{2-}}$ are negligibly small.

Table 7.6 Results of iterative calculations for gypsum solubility in 0.1 *m* $CaCl_2$ aqueous solution.

Iteration	I (initial)	$\gamma_{Ca^{2+}}$	$\gamma_{SO_4^{2-}}$	$m_{Ca^{2+}}$	$m_{SO_4^{2-}}$	I (new)
1	Unknown	1	1	0.100257	0.000257	0.3010
2	0.3010	0.2907	0.2418	0.103656	0.003656	0.3146
3	0.3146	0.2869	0.2367	0.103769	0.003769	0.3151
4	0.3151	0.2868	0.2375	0.103773	0.003773	0.3151
5	0.3151	0.2868	0.2375	0.103773	0.003773	0.3151

As shown in Table 7.6, four iterations of the above procedure yielded the following constant values of the ionic strength and molalities:

$$I = 0.3151$$
$$m_{SO_4^{2-}} = 0.0038$$
$$m_{Ca^{2+}} = m_{SO_4^{2-}} + 0.1 = 0.1038$$

Thus, the solubility of gypsum in a 0.1 *m* $CaCl_2$ aqueous solution is

$$m_{SO_4^{2-}} = 0.0038 = 3.7\times10^{-3} = 10^{-2.43}\ \text{mol kg}^{-1}$$

which is significantly less than its solubility in pure water (Example 7–3).

A check for our answer is provided by the fact that the calculated molalities satisfy the condition of electrical neutrality of the solution:

$$2m_{Ca^{2+}} = 2\times0.1038 = 0.2076;$$
$$2m_{SO_4^{2-}} + m_{Cl^-} = (2\times0.0038) + 0.2 = 0.2076$$

The decrease in the solubility of a salt due to the presence of its own ions in solution from some other source is called the *common-ion effect*. The presence of ions other than those produced by dissociation of the salt itself, on the other hand, generally increases the salt's solubility, but we could not predict this from simple equilibrium reasoning.

7.4.3 *Saturation index*

The *saturation index* (*SI*) of a salt in an aqueous solution is defined as

$$SI = \log\left(\frac{IAP_{solution}}{K_{sp}}\right)^{1/v} \quad (7.32)$$

where *IAP* is the *ion activity product* and K_{sp} the solubility product at the *P* and *T* of interest. The *SI* is independent of the way the dissociation reaction is written, and this is achieved by the exponent $1/v$, where v denotes the number of ions in the expression for the *IAP*. For example, when the dolomite dissociation reaction is written as

$$Ca_{0.5}Mg_{0.5}CO_3 = 0.5Ca^{2+} + 0.5Mg^{2+} + CO_3^{2-}$$

$v = 0.5 + 0.5 + 1 = 2$, but if the reaction is written as

$$CaMg(CO_3)_2 = Ca^{2+} + Mg^{2+} + 2CO_3^{2-}$$

$v = 1 + 1 + 2 = 4$. The IAP/K_{sp} ratio in the latter case, however, will be the square of what it would be in the first case so that the value of *SI* will be the same in the first case.

The *SI* enables us to determine if a solution is supersaturated, saturated, or undersaturated with respect to the salt:

(1) $SI = 0$. The solution is (just) saturated ($IAP = K_{sp}$), i.e., the solution should neither dissolve any more of the salt nor precipitate the salt.
(2) $SI > 0$. The solution is supersaturated ($IAP > K_{sp}$), i.e., the solution should precipitate some of the dissolved salt to restore equilibrium.
(3) $SI < 0$. The solution is undersaturated ($IAP < K_{sp}$), i.e., some more of the salt can be dissolved in the solution to achieve saturation (equilibrium).

The actual dissolution of a mineral in an undersaturated solution or its precipitation from a supersaturated solution, however, will depend on kinetic factors.

Example 7–5: Calculation of saturation index of $CaSO_4$ for the Ganges River water (Table 7.1), given $K_{sp\,(CaSO_4)} = 10^{-4.59}$ at 298.15 K and 1 bar

Applying the extended D–H equation ($A = 0.5085$, $B = 0.3281 \times 10^8$, $\mathring{a}_{Ca^{2+}} = 6 \times 10^{-8}$, $\mathring{a}_{SO_4^{2-}} = 4.5 \times 10^{-8}$) to the Ganges River water,

$$m_{Ca^{2+}} = 0.496 \times 10^{-3} \text{(Table 7.1)};\ \gamma_{Ca^{2+}} = 0.783 \text{ (Example 7-1)}$$

$$a_{Ca^{2+}} = m_{Ca^{2+}}\gamma_{Ca^{2+}} = (0.634 \times 10^{-3})\ (0.783) = 0.496 \times 10^{-3}$$

$$\log\ \gamma_{SO_4^{2-}} = \frac{-A\, z^2_{SO_4^{2-}}\sqrt{I}}{1+\mathring{a}_{SO_4^{2-}}B\sqrt{I}} = \frac{-(0.5085)\ (2^2)\sqrt{0.0034}}{1+(4.5)\ (0.3281)\sqrt{0.0034}} = -0.1092$$

$$m_{SO_4^{2-}} = 0.0885 \times 10^{-3} \text{ (Table 7.1)};\ \gamma_{SO_4^{2-}} = 0.7777$$

$$a_{SO_4^{2-}} = m_{SO_4^{2-}}\gamma_{SO_4^{2-}} = (0.0885 \times 10^{-3})\ (0.7777) = 0.0688 \times 10^{-3}$$

$$IAP\ (CaSO_4,\ \text{Ganges water}) = a_{Ca^{2+}}\ a_{SO_4^{2-}} = (0.496)(0.0688) \times 10^{-6} = 0.034 \times 10^{-6}$$

Since *IAP* ($CaSO_4$, Ganges water) < $K_{sp\,(CaSO_4)}$, the Ganges River water is undersaturated with respect to $CaSO_4$, and $CaSO_4$ should not precipitate from this water.

Example 7–6: Calculation of how much of a salt should precipitate from a solution oversaturated with the salt

Suppose, the concentrations of Ca^{2+} and SO_4^{2-} in a sample of water are 5×10^{-2} *m* and 7×10^{-3} *m*, respectively. Let us first find out the saturation state of the solution with respect to gypsum. For simplicity, we will assume that $a_{Ca^{2+}} = m_{Ca^{2+}}$ and $a_{SO_4^{2-}} = m_{SO_4^{2-}}$

$$IAP\ (\text{solution}) = a_{Ca^{2+}}\ a_{SO_4^{2-}} = (5 \times 10^{-2})\ (7 \times 10^{-3}) = 35 \times 10^{-5} = 10^{-3.5}$$

$$K_{sp\,(CaSO_4)} = 10^{-4.59}$$

Since *IAP* (solution) > $K_{sp\,(gypsum)}$, the water sample is oversaturated with respect to gypsum, and some gypsum should precipitate out of this solution to restore equilibrium. Let us calculate how much.

Let y = moles of gypsum that should precipitate. Then, the concentrations of the ions at equilibrium would be reduced to:

$$m_{Ca^{2+}} = (5 \times 10^{-2}) - y \quad \text{and} \quad m_{SO_4^{2-}} = (7 \times 10^{-3}) - y$$

At equilibrium,

$$[(5 \times 10^{-2}) - y]\ [(7 \times 10^{-3}) - y] = K_{sp\,(gypsum)} = 10^{-4.59},$$

which gives the quadratic equation

$$y^2 - (5.7 \times 10^{-2})\ y + (3.18 \times 10^{-4}) = 0$$

The two roots of this quadratic equation are $y_1 = 5.05 \times 10^{-2}$ and $y_2 = 6.45 \times 10^{-3}$. Root y_1 is rejected because it results in negative values for $m_{Ca^{2+}}$ and $m_{SO_4^{2-}}$. So we accept $y_2 = 6.45 \times 10^{-3}$. Thus, the amount of gypsum that should precipitate = 6.45×10^{-3} moles per kg of water. Note that, with the new values of concentrations,

$$IAP \text{ (solution)} = a_{Ca^{2+}}\, a_{SO_4^{2-}} = [(5\times10^{-2}) - (6.45\times10^{-3})]\,[(7\times10^{-3}) - (6.45\times10^{-3})] = 10^{-4.6} = K_{sp\ (gypsum)}$$

as it should be at equilibrium.

7.4.4 *Ion pairs*

The extended D–H equation works well for very dilute solutions, such as most river and lake waters, in which the major dissolved species may be considered to exist as individual free ions. In more concentrated solutions, such as seawater ($I \sim 0.7$), activity coefficients of individual ions calculated using the extended D–H equation deviate markedly from experimentally determined values. This deviation can be attributed to the formation of dissolved *complex ions* due to short-range interactions among oppositely charged ionic species. Complex ions represent a combination of two or more simpler ionic species and may be divided into three broad classes: *ion pairs*, neutral molecules composed of two ions of opposite charge (e.g., $CaSO_4^0$, $CaCO_3^0$); *coordination complexes*, composed of a central atom surrounded by anions (e.g., $ZnCl_4^{2-}$, $Cu(H_2O)_6^{2+}$, $UO_2(CO_3)_3^{4-}$); and *chelates* with dissolved organic matter. Except in dense brines, the only important complexes of the major dissolved species in natural waters (not contaminated with dissolved organic matter) at low temperatures are ion pairs (Garrels and Christ, 1965).

Ionic strength calculated using equation (7.5) in which m_i represents the total concentration of the dissolved species i without any correction for the existence of ion pairs, is called the *stoichiometric ionic strength* (I_s). The *effective ionic strength* (I_e) of the same solution is less than because formation of ion pairs reduces the concentration of free ions in the solution. For fresh waters, which typically have less than 1% of the ionic species in the form of ion pairs, $I_s = I_e$ for all practical purposes, but for seawater, in which a significant percentage of many major ionic species form ion pairs (see Andrews *et al.*, 2004; Berner and Berner, 1996), $I_e = 0.668\,m$ compared with $I_s = 0.718\,m$ (Langmuir, 1997).

A decrease in the ionic strength generally leads to smaller values of activity coefficients of individual ions (see Fig. 7.2). The activity of a dissolved ion, however, should be the same, regardless of how the ionic strength is calculated, i. e.,

$$a_i = m_i^s \gamma_i^s = m_i^e \gamma_i^e \tag{7.33}$$

where the superscripts s and e refer to I_s and I_e, respectively. Thus, a lower value of γ_i^e will be compensated by a higher value of m_i^e resulting from further dissolution of the salt. The increased dissolution of a salt does not change the relation $K_{sp\ (salt)} = IAP$ at saturation, but the total concentration of electrolytes in the solution is enhanced due to complex formation (see Example 7–7). In fact, it is the formation and stability of complexes that enable some hydrothermal solutions to transport the large amount of dissolved metals needed to form ore deposits under favorable conditions (see section 7.4.5).

The effect of ion pair formation on the solubility of a salt can be evaluated from the equilibrium constant of the reaction of the form (see Example 7–3)

$$B^+ + A^- \Leftrightarrow BA^0; \qquad K_{stab} = \frac{a_{BA^0}}{a_{B^+}\, a_{A^-}} \tag{7.34}$$

where the superscript "0" denotes an electrically neutral molecule. Note that reaction (7.34) is written in the opposite direction compared to the mineral dissociation reactions considered earlier. The equilibrium constant for such a reaction is commonly called the *stability constant* (K_{stab}) or *association constant* (K_{assoc}). The higher the stability constant, the more stable is the complex under consideration. Evidently, $K_{stab} = 1/K_{dissociation}$.

Example 7–7: Calculation of the effect of ion pair formation on the solubility of gypsum, given $K_{sp\ (gypsum)} = 10^{-4.59}$ and $K_{stab\ (CaSO_4^0)} = 10^{2.32}$ at 298.15 K and 1 bar

In Example 7–3, we assumed that all the dissolved calcium existed as free ions, i.e.,

$$\sum m_{Ca} \text{(solution)} = m_{Ca^{2+}free} = 10.25\times10^{-3} = 10^{-1.99} \text{ mol kg}^{-1}$$

[log (10.25×10^{-3}) = log (10.25) + log (10^{-3}) = 1.011 − 3 = 1.99; antilog (−1.99) = $10^{-1.99}$]

The relevance of the formation of the ion pair $CaSO_4^0$ is that it decreases the amount of free Ca^{2+} in solution, thereby causing dissolution of more gypsum to restore the solution to saturation with respect to gypsum. Now the calcium mass balance equation is:

$$\text{Solubility of gypsum} = \sum m_{Ca} = m_{Ca^{2+}free} + m_{CaSO_4^0}$$

The value of $m_{CaSO_4^0}$ can be determined from $K_{stab\ (CaSO_4^0)} = 10^{2.32}$, which is defined as

$$Ca^{2+} + SO_4^{2-} = CaSO_4^0; \quad K_{stab\ (CaSO_4^0)} = \frac{a_{CaSO_4^0}}{a_{Ca^{2+}}\, a_{SO_4^{2-}}} = 10^{2.32} \tag{7.35}$$

Rearranging,

$$a_{\mathrm{CaSO_4^0}} = K_{\mathrm{stab\,(CaSO_4^0)}}\, a_{\mathrm{Ca^{2+}}}\, a_{\mathrm{SO_4^{2-}}} = K_{\mathrm{stab\,(CaSO_4^0)}}\, K_{\mathrm{sp\,(gypsum)}}$$
$$= (10^{2.32})(10^{-4.59}) = 10^{-2.27} = 5.37 \times 10^{-3}$$

The activity coefficient of $CaSO_4^0$, a neutral molecule, can be assumed to be 1. So,

$$a_{\mathrm{CaSO_4^0}} = m_{\mathrm{CaSO_4^0}}\, \gamma_{\mathrm{CaSO_4^0}} = m_{\mathrm{CaSO_4^0}} = 5.37 \times 10^{-3}$$

Substituting,

$$\sum m_{\mathrm{Ca}} = m_{\mathrm{Ca^{2+}\,free}} + m_{\mathrm{CaSO_4^0}}$$
$$= (10.25 \times 10^{-3}) + (5.37 \times 10^{-3})$$
$$= 15.62 \times 10^{-3} = 10^{-1.81}\ \mathrm{mol\ kg^{-1}}$$

[log (15.62×10^{-3}) = log (15.62) + log (10^{-3}) = 1.194 − 3 = 1.81; antilog (−1.81) = $10^{-1.81}$]

The calculation shows that formation of $CaSO_4^0$ has resulted in an increase in the solubility of gypsum by about 50%, from 15.62×10^{-3} mol kg^{-1} to 15.62×10^{-3} mol kg^{-1}.

7.4.5 *Aqueous complexes of ore metals*

Most of the metals (such as lead, zinc, copper, silver, molybdenum, etc.) in ore deposits occur mainly as sulfide and sulfosalt minerals, and these were precipitated from aqueous solutions at moderate temperatures. This raises a serious issue about the formation of hydrothermal ore deposits because solubilities of metals as simple ions are extremely low at moderate temperatures and realistic pH values. The solubility issue is less severe if ore metals are transported as aqueous complexes, which generally have much greater stability than simple ions (Barnes, 1979).

Three kinds of aqueous complexes have been considered as likely candidates for transport of most ore mretals: (i) chloride complexes; (ii) sulfide and bisulfide complexes; and (iii) organometallic complexes. Organometallic complexes generally dissociate at moderate temperatures and are unlikely to be important carriers of ore metals at temperatures above about 300°C, and their importance at lower temperatures is limited by the low concentrations of organic acids in hydrothermal fluids (Gize and Barnes, 1987). For the temperature–pH range appropriate for ore-forming fluids, the dominant complexing ligand among sulfur species is more likely to be HS^- or H_2S rather than S^{2-} or SO_4^{2-}. Calculations show that bisulfide complexes – such as $Zn(HS)_3^-$, $Cu(HS)_2^-$, $HgS(HS)^-$, $Au(HS)_2^-$ – may be important for ore metal transport at temperatures below about 300°C and in moderately alkaline solutions. The major limitation is the high concentration of reduced sulfur (H_2S or HS^-) required to keep the bisulfide complexes stable, concentrations much higher than what have been observed in fluid inclusions in ore minerals or in the fluids of modern hydrothermal systems.

The high Cl^- contents of fluid inclusions and modern hydrothermal fluids point to the high probability of ore metal transport as chloride complexes (e.g., $PbCl^+$, $PbCl_4^{2-}$). The case for the dominance of chloride complexes is also supported by experimental studies and thermodynamic calculations (e.g., Barrett and Anderson, 1988; Ruaya, 1988; Hemley *et al.*, 1992), and the observed increase in the solubility of lead and zinc with increasing chlorite concentration in basinal fluids (Hanor, 1997). The requirements for chloride complexing, in addition to an abundance of Cl^-, are an acidic pH, moderately elevated temperatures, and low concentrations of reduced sulfur species (H_2S, HS^-, or S^{2-}) in the solution. For example, at 200°C and saturated water vapor pressure, $H_2S = 10^{-3}$ molal, and pH = 3.0, the solubility of galena due to chloride complexation (in the presence of 1.0 molal Cl^-) was found to be 1038 mg L^{-1}, which is nearly 4.5 orders of magnitude higher compared to its solubility (47.4 μg L^{-1}) as simple Pb^{2+} ion (Wood and Samson, 1998). Krauskopf and Bird (1995) calculated that addition of only 0.1 m Cl^- to an aqueous solution in equilibrium with sphalerite (ZnS) would increase the solubility of zinc from $10^{-8.86}$ mol kg^{-1} to $10^{-5.35}$ mol kg^{-1} at the same P–T condition. This increase, by more than three orders of magnitude, is due to the formation of zinc chloride complexes ($ZnCl^+$, $ZnCl_2^0$, $ZnCl_3^-$, and $ZnCl_4^{2-}$). Thermodynamic calculations and experimental studies indicate that the formation of platinum–palladium and gold deposits involves transport of the metals as chloride and bisulfide complexes (such as $PtCl_3^-$, $PtCl_3^-$, $PdCl_4^{2-}$, $Pd(HS)_2^0$, $AuCl_2^-$, and $HAu(HS)_2^0$) in ore-forming solutions (Hayashi and Ohmoto, 1991; Pan and Wood, 1994; Wood *et al.*, 1994).

7.5 Dissociation of H_2CO_3 acid – the carbonic acid system

Let us consider dissociation of weak carbonic acid (H_2CO_3), which plays an important role in regulating the pH of natural waters, and the dissolution/precipitation of carbonate minerals. Following the convention that all of the dissolved $CO_{2(g)}$ forms $H_2CO_{3(aq)}$, and assuming that $CO_{2(g)}$ behaves as an ideal gas (i.e., $a_{\mathrm{CO_2\,(g)}} = P_{\mathrm{CO_2\,(g)}}$), the equilibrium between $CO_{2(g)}$ and water can be represented by the reaction (see section 4.1.1):

$$CO_{2\,(g)} + H_2O \Leftrightarrow H_2CO_{3\,(aq)};$$
$$K_{\mathrm{CO_2}} = \frac{a_{\mathrm{H_2CO_3\,(aq)}}}{P_{\mathrm{CO_2}}} = 10^{-1.47}\ (298.15\ \mathrm{K},\ 1\ \mathrm{bar}) \qquad (7.36)$$

An aqueous solution of H_2CO_3 acid in water contains five dissolved species — $H_2CO_{3\,(aq)}$ (undissociated dissolved H_2CO_3), HCO_3^-, CO_3^{2-}, H^+, and OH^- — arising from the two-step dissociation of $H_2CO_{3\,(aq)}$ and the dissociation of H_2O, as listed below (along with equilibrium constants at 298.15 K and 1 bar):

$$H_2CO_{3\,(aq)} \Leftrightarrow H^+ + HCO_3^-\ ;$$
$$K_{\mathrm{H_2CO_3}} = \frac{a_{\mathrm{H^+}}\, a_{\mathrm{HCO_3^-}}}{a_{\mathrm{H_2CO_3\,(aq)}}} = 10^{-6.35} \qquad (7.37)$$

$$HCO_3^- \Leftrightarrow H^+ + CO_3^{2-}; \quad K_{HCO_3^-} = \frac{a_{H^+}\, a_{CO_3^{2-}}}{a_{HCO_3^-}} = 10^{-10.33} \quad (7.38)$$

$$H_2O \Leftrightarrow H^+ + OH^-; \quad K_w = \frac{a_{H^+}\, a_{OH^-}}{a_{H_2O}} = 10^{-14} \quad (7.39)$$

A further constraint on the system is provided by a fundamental principle of solution chemistry that positive and negative charges of the dissolved species in a solution must be equal to maintain electrical neutrality (i.e., $\Sigma m_i z_i = 0$). In the present case, the so-called *charge balance equation* is:

$$m_{H^+} = m_{HCO_3^-} + 2m_{CO_3^{2-}} + m_{OH^-} \quad (7.40)$$

Note that the charge balance equation is written in terms of concentrations (molalities), not activities (although in dilute solutions the distinction may be considered insignificant), and that x moles of a divalent anion, such as CO_3^{2-}, contribute $2x$ units of negative charge.

For estimation of concentrations of the five dissolved species, we need another equation relating one or more of these variables. The fifth equation is constructed from the relation of the carbonic acid system with the surroundings – whether the system is *open* or *closed*. In an open system the exchange of CO_2 between water and a gas phase (e.g., the atmosphere) in equilibrium keeps the externally imposed P_{CO_2} at a constant value; in a closed system (i.e., involving no exchange with a gas phase), the total carbonate content of the system remains fixed and the P_{CO_2} is determined by equilibrium within the system.

7.5.1 *Open system*

Consider a body of water in equilibrium with the prevailing atmosphere. Rearranging equation (7.37),

$$a_{H^+} = \frac{K_{H_2CO_3}\, a_{H_2CO_3\,(aq)}}{a_{HCO_3^-}} \quad (7.41)$$

Assuming the atmosphere to be an ideal gas mixture (i.e., $a_{CO_2\,(g)} = P_{CO_2\,(g)}$) and the atmospheric $P_{CO_2} = 10^{-3.5}$ bar, $a_{H_2CO_3\,(aq)}$ can be calculated using equation (7.36):

$$a_{H_2CO_3\,(aq)} = K_{CO_2} P_{CO_2} = (10^{-1.47})\,(10^{-3.5}) = 10^{-4.97} \quad (7.42)$$

To calculate $a_{HCO_3^-}$, we apply two reasonable approximations:

(1) Since $K_{H_2CO_3} >> K_{HCO_3^-}$ (by about four orders of magnitude), H^+ in the solution is almost entirely from equation (7.37) and $m_{CO_3^{2-}}$ is too small.
(2) We know from experience that the solution will be acidic so that $m_{OH^-} < m_{H^+}$ and may be neglected relative to $m_{HCO_3^-}$ and m_{H^+}.

Neglecting $m_{CO_3^{2-}}$ and m_{OH^-}, equation (7.40) simplifies to $m_{HCO_3^-} = m_{H^+}$ and equation (7.41), assuming $a_i = m_i$, gives

$$(a_{H^+})^2 = K_{H_2CO_3}\, a_{H_2CO_3\,(aq)} = (10^{-6.35})\,(10^{-4.97}) = 10^{-11.32}$$

$$a_{H^+} = 10^{-5.66} \quad \text{(i.e., pH = 5.66)} \quad (7.43)$$

Thus, rainwater in equilibrium with atmospheric CO_2 at a level of 300 ppm should be acidic. In fact, surface water in equilibrium with the Earth's atmosphere would acquire a pH near 5.7 unless it contains solutes that neutralize the dissolved CO_2. The rainwater would be even more acidic in response to a higher P_{CO_2} in the atmosphere or other acid solutes, such as sulfuric acid (H_2SO_4) and nitric acid (HNO_3), in the rainwater. Sulfuric acid and nitric acid are formed in the atmosphere by oxidation of SO_2 gas and nitrogen gases released primarily from anthropogenic sources. The term *acid rain* is applied to rainwater that has a pH of less than about 5.7, as is the case at present with rainwater in many regions around the world (Berner and Berner, 1996). Acidification of freshwater bodies can be toxic to aquatic life. The problem for fish is that the dissolved Al^{3+} in the acidic water precipitates as an insoluble $Al(OH)_3$ gel on the less acidic gill tissues, preventing normal uptake of oxygen and suffocating the fish (Andrews et al., 2004).

Concentrations of other ionic species in the solution can be calculated as follows (assuming $a_i = m_i$):

$$m_{HCO_3^-} = m_{H^+} = 10^{-5.66} \text{ mol kg}^{-1}$$

$$m_{OH^-} = \frac{K_w}{m_{H^+}} = \frac{10^{-14}}{10^{-5.66}} = 10^{-8.34} \text{ mol kg}^{-1}$$

$$m_{CO_3^{2-}} = \frac{K_{HCO_3^-}\, m_{HCO_3^-}}{m_{H^+}} = \frac{10^{-10.33}\; 10^{-5.66}}{10^{-5.66}} = 10^{-10.33} \text{ mol kg}^{-1}$$

The very small values of $m_{CO_3^{2-}}$ and m_{OH^-} relative to $m_{HCO_3^-}$ and m_{H^+} confirm that we were justified in neglecting $m_{CO_3^{2-}}$ and m_{OH^-} in equation (7.40).

7.5.2 *Closed system*

In a closed carbonic acid system, as is the case when it is not in equilibrium with the atmosphere, the total concentration of the dissolved carbonate species, carb (t) (also denoted by many authors as ΣCO_3 or ΣCO_2), remains constant:

$$m_{carb\,(t)} = m_{H_2CO_3\,(aq)} + m_{HCO_3^-} + m_{CO_3^{2-}} \quad (7.44)$$

Expressions for the activities of the dissolved species in the carbonic acid system can be derived as follows (Eby, 2004). From equation (7.37),

$$a_{HCO_3^-} = \frac{K_{H_2CO_3} a_{H_2CO_3\,(aq)}}{a_{H^+}} \tag{7.45}$$

Substituting for $a_{HCO_3^-}$ in equation (7.38), we get

$$a_{CO_3^{2-}} = \frac{K_{H_2CO_3} K_{HCO_3^-}\, a_{H_2CO_3(aq)}}{(a_{H^+})^2} \tag{7.46}$$

Rearrangement of equation (7.44) after substituting for $a_{HCO_3^-}$ and $a_{CO_3^{2-}}$ gives:

$$a_{H_2CO_3(aq)} = \frac{m_{carb\,(t)}}{\left(1 + \frac{K_{H_2CO_3}}{a_{H^+}} + \frac{K_{H_2CO_3} K_{HCO_3^-}}{(a_{H^+})^2}\right)} \tag{7.47}$$

Noting that the denominators in equations (7.45)–(7.47) are functions of a_{H^+} (i.e., of pH), let us define a variable F_H such that it is a function of a_{H^+} only:

$$F_H = \left(1 + \frac{K_{H_2CO_3}}{a_{H^+}} + \frac{K_{H_2CO_3} K_{HCO_3^-}}{(a_{H^+})^2}\right) \tag{7.48}$$

Some simple algebraic manipulations yield the following equations for calculating $a_{H_2CO_3\,(aq)}$, $a_{HCO_3^-}$, and $a_{CO_3^{2-}}$ as a function of pH in a closed carbonic acid system with known $m_{carb\,(t)}$:

$$a_{H_2CO_3\,(aq)} = \frac{m_{carb\,(t)}}{F_H} \tag{7.49}$$

$$a_{HCO_3^-} = \frac{m_{carb\,(t)} K_{H_2CO_3}}{a_{H^+} F_H} \tag{7.50}$$

$$a_{CO_3^{2-}} = \frac{m_{carb\,(t)} K_{H_2CO_3} K_{HCO_3^-}}{(a_{H^+})^2 F_H} \tag{7.51}$$

Example 7–8: Calculation of activities of dissolved species in a solution of 0.01 m H_2CO_3 at 298.15 K and a measured pH of 4.2 when the system is isolated from atmospheric CO_2, assuming $a_i = m_i$

"A solution of 0.01 m H_2CO_3" refers to the total amount of dissolved carbonate species in the solution (undissociated H_2CO_3, HCO_3^-, and CO_3^{2-}) that arise from the dissociation of 0.01 mole of H_2CO_3 formed by dissolving 0.01 mole of CO_2 gas in 1 kg of water. The mass balance equation is:

$$m_{carb\,(t)} = m_{H_2CO_3\,(aq)} + m_{HCO_3^-} + m_{CO_3^{2-}} = 0.01 \text{ mol kg}^{-1}\ H_2O$$

$$F_H = \left(1 + \frac{K_{H_2CO_3}}{a_{H^+}} + \frac{K_{H_2CO_3} K_{HCO_3^-}}{(a_{H^+})^2}\right)$$

$$= \left[1 + \frac{10^{-6.35}}{10^{-4.2}} + \frac{10^{-6.35}\,10^{-10.33}}{(10^{-4.2})^2}\right] = 1.007$$

$$a_{H_2CO_3\,(aq)} = \frac{m_{carb\,(t)}}{F_H} = \frac{0.01}{1.007} = 9.93 \times 10^{-3} = 10^{-2.0} \text{ mol kg}^{-1}$$

$$a_{HCO_3^-} = \frac{m_{carb\,(t)} K_{H_2CO_3}}{a_{H^+} F_H} = \frac{(0.01)\,(10^{-6.35})}{(10^{-4.2})(1.007)}$$

$$= 7.03 \times 10^{-5} = 10^{-4.15} \text{ mol kg}^{-1}$$

$$a_{CO_3^{2-}} = \frac{m_{carb\,(t)} K_{H_2CO_3} K_{HCO_3^-}}{(a_{H^+})^2 F_H} = \frac{(0.01)\,(10^{-6.35})(10^{-10.33})}{(10^{-4.2})^2(1.007)}$$

$$= 5.21 \times 10^{-11} = 10^{-10.28} \text{ mol kg}^{-1}$$

$$a_{OH^-} = \frac{K_w}{a_{H^+}} = \frac{10^{-14}}{10^{-4.2}} = 10^{-9.8} = 1.58 \times 10^{-10} \text{ mol kg}^{-1}$$

The above procedure can be repeated to calculate the activities of the dissolved species in carbonic acid system at any given pH and $m_{carb\,(t)}$. Calculations for the carbonic acid system under consideration indicate that different pH ranges are dominated by different dissolved carbonate species: $H_2CO_{3\,(aq)}$ at pH < 6.4; HCO_3^- at pH between 6.4 and 10.3; and CO_3^{2-} at pH > 10.3 (Fig. 7.6). Thus, in surface ocean water (pH ≈ 8),

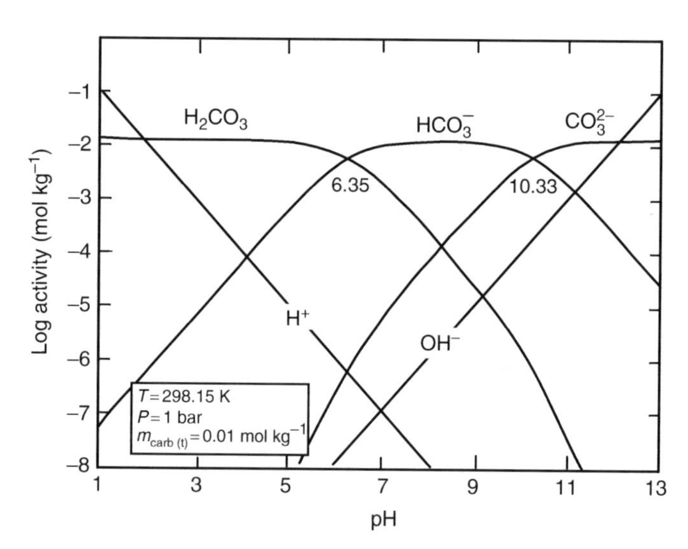

Fig. 7.6 Activities of dissolved species in the CO_2–H_2O system as a function of pH at 298.15 K and 1 bar, assuming $a_i = m_i$. Total dissolved carbonate ($m_{carb\,(t)}$) = 0.01 mol kg^{-1}. Note that the crossovers of the curves occur at pH = 6.35 ($a_{H_2CO_3(aq)} = a_{HCO_3^-}$) and at pH = 10.33 ($a_{HCO_3^-} = a_{CO_3^{2-}}$). This sort of diagram is commonly referred to as a Bjerrum plot.

HCO_3^- dominates over $H_2CO_{3\,(aq)}$ and CO_3^{2-}. Actually, we could have predicted these conclusions from equations (7.37) and (7.38).

7.6 Acidity and alkalinity of a solution

The *acidity* of a solution is its capacity to donate protons (or H^+) and it can be determined by titrating the solution with a strong base such as sodium hydroxide (NaOH). The titration results in OH^--consuming reactions such as

$$H^+ + OH^- = H_2O$$

$$H_2CO_{3\,(aq)} + OH^- = HCO_3^- + H_2O$$

$$HSO_4^- + OH^- = SO_4^{2-} + H_2O$$

Acidity is an important property because high acidity gives water a greater capacity to dissolve minerals and metals from rocks and waste dumps, and so is usually accompanied by high *total dissolved solids* (TDS), which may render the water unsuitable for drinking.

The *alkalinity* of a solution is its capacity to accept protons. It is equal to the stoichiometric sum of the bases in solution, and can be determined by titrating the solution with a strong acid such as HCl. The titration in this case results in H^+-consuming reactions such as

$$H^+ + OH^- = H_2O$$

$$HCO_3^- + H^+ = H_2CO_{3(aq)}$$

$$SO_4^{2-} + H^+ = HSO_4^-$$

The approximate charge balance equation for natural waters may be expressed as

$$\begin{aligned} 2m_{Ca^{2+}} + 2m_{Mg^{2+}} + m_{Na^+} + m_{K^+} + m_{H^+} \\ = m_{Cl^-} + 2m_{SO_4^{2-}} + m_{HCO_3^-} + 2m_{CO_3^{2-}} + m_{OH^-} \\ + m_{Br^-} + m_{B(OH)_4^-} + m_{H_3SiO_4^-} + m_{HS^-} + m_{\text{organic anions}} \end{aligned} \quad (7.52)$$

Some of the ionic species, such as Ca^{2+}, Mg^{2+}, Na^+, K^+, SO_4^{2-}, Cl^-, and Br^- are *conservative* because their concentrations will not be affected by changes in pH or some other intensive variable, such as pressure, or temperature (within the ranges normally encountered near the Earth's surface and assuming no precipitation, dissolution of solid phases, or biological transformations). The rest of the ionic species in equation (7.52) are *nonconservative* because the concentration of each will vary with pH or some other intensive variable (such as temperature or pressure). Virtually all the nonconservative ionic species are anions, the two major exceptions being H^+ and NH_4OH (which dissociates into NH_4^+ and OH^- at high pH). Variations in the concentrations of nonconservative ions result from reactions among them, and these reactions can occur without involving precipitation or dissolution (e.g., $H^+ + CO_3^{2-} = HCO_3^-$). An example of pH control on the concentrations of individual dissolved species in a carbonic acid system at 1 bar and 298.15 K (25°C) is shown in Fig. 7.6; the control of P and T on the concentrations arise from their effects on the equilibrium constants $K_{H_2CO_3(aq)}$, $K_{HCO_3^-}$, and K_w.

Rearrangement of equation (7.52) with conservative species on the left and nonconservative species on the right gives:

$$\begin{aligned} 2m_{Ca^{2+}} + 2m_{Mg^{2+}} + m_{Na^+} + m_{K^+} - m_{Cl^-} - 2m_{SO_4^{2-}} \\ = m_{HCO_3^-} + 2m_{CO_3^{2-}} + m_{OH^-} + m_{Br^-} + m_{B(OH)_4^-} \\ + m_{H_3SiO_4^-} + m_{HS^-} + m_{\text{organic anions}} - m_{H^+} \end{aligned}$$

in which the left represents the excess of conservative cations over conservative anions. The total alkalinity of most natural waters (C_{ALK}) is defined as (Drever, 1997):

$$\begin{aligned} C_{ALK} &= \sum \text{conservative cations (in gram-equivalent kg}^{-1}) \\ &- \sum \text{conservative anions (in gram-equivalent kg}^{-1}) \\ &= [m_{HCO_3^-} + 2m_{CO_3^{2-}} + m_{OH^-} + m_{B(OH)_4^-} + m_{H_3SiO_4^-} \\ &+ m_{HS^-} + m_{\text{organic anions}}] - [m_{H^+} \; m_{HSO_4^-}] \end{aligned} \quad (7.53)$$

In most natural waters, $m_{B(OH)_4^-}$, $m_{H_3SiO_4^-}$, m_{HS^-}, $m_{\text{org. anions}}$, m_{H^+}, and m_{OH^-} are very small quantities compared to $m_{HCO_3^-}$ and $m_{CO_3^{2-}}$ so that the total alkalinity is approximately equal to the *carbonate alkalinity* given by

$$\text{Carbonate alkalinity} = m_{HCO_3^-} + 2m_{CO_3^{2-}} \quad (7.54)$$

The expressions commonly used for calculation of acidity (C_{Acid}) and alkalinity (C_{Alk}) of waters in which the only acids and bases are species of carbonic acid and strong acids or bases (that dissociate almost completely) are as follows (Langmuir, 1997):

$$C_{Acid} = 2m_{H_2CO_3(aq)} + m_{HCO_3^-} + m_{H^+} - m_{OH^-} \quad (7.55)$$

$$C_{Alk} = m_{HCO_3^-} + 2m_{CO_3^{2-}} + m_{OH^-} - m_{H^+} \quad (7.56)$$

To describe the acidity or alkalinity of other systems, these equations are modified to reflect the combination of appropriate nonconservative species. For example, the acidity of acid mine waters (pH 2 to 3) may be defined as

$$\begin{aligned} C_{Acid} &= m_{H^+} + m_{HSO_4^-} + 2Fe^{2+} + 3Fe^{3+} \\ &+ 2Fe(OH)^{2+} + Fe(OH)_2^+ \end{aligned} \quad (7.57)$$

Acidity and alkalinity are commonly reported in equivalents per kilogram of water or liter of solution (gram-equivalent kg^{-1} or gram-equivalent L^{-1}) rather than as mol kg^{-1} or mol L^{-1}. The gram-equivalent weight of a species is its molecular or atomic weight divided by the valence, or in the case of acids and bases by the number of H^+ or OH^- that can be produced when the acid or base is dissolved in water. For example, an equivalent of 1 mole of HCl is (1.01 + 35.45)/1 = 36.46 and an equivalent of 1 mole of H_2CO_3 is (2.02 + 12.01 + 48)/2 = 31.01. This is the reason why $m_{H_2CO_3(aq)}$ in equation (7.55) and $m_{CO_3^{2-}}$ in equation (7.56) are multiplied by 2.

The addition or removal of $CO_{2\,(g)}$ does not change the alkalinity of a solution. This is because the net reaction – for example, $CO_2 + H_2O \Rightarrow HCO_3^- + H^+$ at neutral pHs, and $CO_2 + H_2O \Rightarrow CO_3^{2-} + 2H^+$ at high pHs – produces the same number of equivalents of species contributing positive charge (e.g., H^+) as species contributing negative charge (e.g., HCO_3^- or CO_3^{2-}). Thus, the alkalinity and the total dissolved CO_2 content of seawater can vary independently of each other, even though they are closely related. Addition of $CO_{2\,(g)}$ to a solution in contact with a solid, on the other hand, can affect the alkalinity, especially for groundwater or seaweater in contact with carbonate minerals. The dissolution or precipitation of carbonate minerals has a strong influence on alkalinity. This is because the dissociation of $CaCO_3$ would add Ca^{2+} and CO_3^{2-} into solution. Ca^{2+} would not influence alkalinity, but CO_3^{2-} would increase alkalinity by two units.

Example 7–9: Calculation of the acidity and the alkalinity of the carbonic acid system considered in Example 7–7

Substituting in equations (7.55) and (7.56) the molalities we calculated in Example 7–7 ($a_{H_2CO_3(aq)} = 10^{-2.0}\ m$; $a_{HCO_3^-} = 10^{-4.15}\ m$; $a_{CO_3^{2-}} = 10^{-10.28}\ m$; $m_{OH^-} = 10^{-9.8}$), the acidity and the alkalinity of the solution are:

$$C_{\text{Acid}} = 2m_{H_2CO_3(aq)} + m_{HCO_3^-} + m_{H^+} - m_{OH^-}$$
$$= 2(10^{-2.0}) + 10^{-4.15} + 10^{-4.2} - 10^{-9.8} = 2.01 \times 10^{-2}\text{eq kg}^{-1}$$

$$C_{\text{Alk}} = m_{HCO_3^-} + 2m_{CO_3^{2-}} + m_{OH^-} - m_{H^+}$$
$$= (10^{-4.15}) + 2(10^{-10.28}) + (10^{-9.8}) - (10^{-4.2})$$
$$= 1.34 \times 10^{-4}\text{eq kg}^{-1}$$

7.7 pH buffers

In general, the addition of an acid to a solution should decrease its pH due to an increase in a_{H^+} and the addition of a base increase its pH due to an increase in a_{OH^-}. Some solutions, however, resist such pH changes because of H^+-consuming or OH^--consuming reactions either among the dissolved species in the water or between water and mineral species with which it is in equilibrium. A solution in which the pH is maintained at a nearly constant level when an acid or a base is added is called a pH *buffer*. The concept helps us to understand chemical reactions in natural systems and control pH of chemical reactions in synthetic systems.

Suppose we add 0.0001 mole of hydrochloric acid (HCl) to 1 kg of pure water ($a_{H^+} = 10^{-7}$ or pH = 7). HCl is a strong acid, that is, it will dissociate almost completely into H^+ and Cl^- ($HCl \Leftrightarrow H^+ + Cl^-$). Assuming $a_i = m_i$, the dissociation of 0.0001 m HCl will increase m_{H^+} in the solution by $10^{-4}\,m$. In pure water, $m_{H^+} = 10^{-7}$, which is very small compared to $10^{-4}\,m$. Thus, m_{H^+} of the solution after addition of HCl will be very close to 10^{-4}; i.e., the pH of the solution will drop to about 4.

Now let us see what will be the pH of the solution if 0.0001 mole of HCl is added to a closed carbonic acid system with $m_{\text{Carb (t)}} = m_{H_2CO_3(aq)} + m_{HCO_3^-} + m_{CO_3^{2-}} = 0.01\,m$ and pH = 6. Using the procedure discussed in section 7.4.2 and assuming $a_i = m_i$, the concentrations of the dissolved species in this solution can be calculated as follows:

$$m_{H_2CO_3\ (aq)} = 10^{-2.16};\ m_{HCO_3^-} = 10^{-2.51};\ m_{OH^-} = 10^{-8}$$

Being a strong acid, 0.0001 mole of HCl will dissociate completely to give 0.0001 mole of H^+ and 0.0001 mole of Cl^-, and the H^+ so produced will react with HCO_3^- to form $H_2CO_{3\,(aq)}$:

$$H^+ + HCO_3^- \Rightarrow H_2CO_{3\,(aq)} \tag{7.58}$$

The result will be a decrease in $m_{HCO_3^-}$ by 10^{-4} moles and an increase in $m_{H_2CO_3(aq)}$ by 10^{-4} moles. Solving equation (7.42) with the new values of $m_{HCO_3^-}$ and $m_{H_2CO_3(aq)}$ gives

$$a_{H^+} = \frac{K_{H_2CO_3}\ a_{H_2CO_3\ (aq)}}{a_{HCO_3^-}} = \frac{10^{-6.35}(10^{-2.16} + 10^{-4})}{(10^{-2.51} - 10^{-4})} = 10^{-5.98}$$

Thus, the pH of the given carbonic acid system will change very little (only by 0.02 pH units) by the addition of $10^{-4}\,m$ HCl, i. e, the solution will act as a good pH buffer.

Employing similar calculations it can be shown that if 0.0001 mole of a strong base such as sodium hydroxide (NaOH) is added to 1 kg of pure water, the pH of the solution will increase to about 11, but the same amount of NaOH added to 1 kg of the given carbonic acid system (pH = 6) will increase the pH of the solution by only 0.35 pH units because of the reaction

$$H_2CO_{3\,(aq)} + OH^- \Rightarrow HCO_3^- + H_2O \tag{7.59}$$

The *buffering capacity* of a solution, a measure of the amount of H^+ or OH^- ions the solution can absorb without a significant change in pH, varies as a function of pH. For weak acids and bases, the maximum buffering capacity occurs at pH values that equal the dissociation constants of the weak acid or weak base. In the case of a CO_2–H_2O solution, for example, the maximum buffering capacity occurs at the two cross-over points corresponding to pH 6.4 and 10.3 (Fig. 7.6), which have the same numerical values as the dissociation constants $K_{H_2CO_3}$ and $K_{HCO_3^-}$, respectively.

A common pH buffer in geochemical systems, such as the seawater, is provided by solutions that contain significant amounts of a weak acid (e.g., HCO_3^-) and a salt of the acid (e.g., $CaCO_3$). The pH of water in equilibrium with calcite and atmospheric CO_2 can be shown to be 8.26 at 298.15 K and 1 bar (see Example 7–10). The buffering actions of $CaCO_3$ and HCO_3^- are largely responsible for maintaining the pH of seawater around 8 (mostly between 7.8 and 8.4). The oceans contain effectively an infinite amount of $CaCO_3$ particles suspended in surface waters and buried in bottom sediments, and the excess H^+ in seawater is depleted by the dissolution of $CaCO_3$,

$$CaCO_{3\,(s)} + H^+ \Rightarrow Ca^{2+} + HCO_3^- \tag{7.60}$$

and by other reactions such as

$$H^+ + HCO_3^- \Rightarrow H_2CO_{3(aq)} \quad \text{and} \quad H^+ + CO_3^{2-} \Rightarrow HCO_3^- \tag{7.61}$$

Excess OH^- in seawater is depleted by the precipitation of $CaCO_3$,

$$Ca^{2+} + HCO_3^- + OH^- \Rightarrow CaCO_3 + H_2O \tag{7.62}$$

and by other reactions such as

$$OH^- + HCO_3^- \Rightarrow CO_3^{2-} + H_2O \quad \text{and}$$
$$OH^- + H_2CO_{3(aq)} \Rightarrow HCO_3^- + H_2O \tag{7.63}$$

The seawater is a complex system containing many dissolved species and other reactions, such as those involving boron,

$$H^+ + H_2BO_3^- \Rightarrow H_3BO_3 \quad \text{and}$$
$$OH^- + H_3BO_3 \Rightarrow H_2BO_3^- + H_2O \tag{7.64}$$

also contribute to the buffering of seawater.

7.8 Dissolution and precipitation of calcium carbonate

In nature, $CaCO_3$ occurs as two common minerals: calcite (rhombohedral, density = 2.7 g cm^{-3}, $K_{sp} = 10^{-8.48}$ at 298.15 K and 1 bar); and aragonite (orthorhombic, density = 2.9 g cm^{-3}, $K_{sp} = 10^{-8.34}$ at 298.15 K and 1 bar). A third known polymorph, vaterite, forms only from solutions highly supersaturated with respect to calcite and aragonite, and is extremely rare in nature. We will focus our attention on calcite, the most stable and the most abundant of the $CaCO_3$ polymorphs in the sedimentary rock record. Equilibria involving other carbonate minerals in aqueous systems can be treated in a similar manner.

7.8.1 *Solubility of calcite in pure water*

When an excess of pure calcite ($CaCO_3$) is placed in pure water, it undergoes a very small degree of dissociation with a solubility product of $10^{-8.48}$ at 298.15 K and 1 bar:

$$CaCO_{3\,(s)} \Leftrightarrow Ca^{2+} + CO_3^{2-} \tag{7.65}$$

$$K_{sp\,(cal)} = a_{Ca^{2+}} a_{CO_3^{2-}} = (m_{Ca^{2+}} \gamma_{Ca^{2+}})(m_{CO_3^{2-}} \gamma_{CO_3^{2-}}) = 10^{-8.48} \tag{7.66}$$

Since each mole of $CaCO_3$ dissociates into 1 mole each of Ca^{2+} and CO_3^{2-}, $m_{Ca^{2+}} = m_{CO_3^{2-}}$, and the solubility of calcite is given by (see equation 7.26)

$$\boxed{m_{Ca^{2+}} = \left(\frac{K_{sp\,(cal)}}{\gamma_{Ca^{2+}}\, \gamma_{CO_3^{2-}}} \right)^{1/2}} \tag{7.67}$$

Calculation of $\gamma_{Ca^{2+}}$ and $\gamma_{CO_3^{2-}}$ by the iterative method employed in Example 7–3 will show that the activity coefficients are very close to 1 (i.e., $a_i \approx m_i$), a conclusion we could have predicted from the fact that this is a very dilute solution as judged from the very low value of $K_{sp(cal)}$. So, the solubility of calcite in pure water is

$$m_{Ca^{2+}} = \left[K_{sp\,(cal)}\right]^{1/2} = \left[10^{-8.48}\right]^{1/2} = 10^{-4.24} \text{mol kg}^{-1}\ H_2O \tag{7.68}$$

Now let us find out how the calcite solubility is affected by the presence of carbonic acid.

7.8.2 *Carbonate equilibria in the $CaCO_3$–CO_2–H_2O system*

Calcite solubility as calculated above does not represent the real situation in natural aqueous systems, which almost always involve equilibrium among solid calcite, an aqueous solution containing many dissolved species, and a gas phase containing CO_2. The treatment of equilibria in the $CaCO_3$–CO_2–H_2O system is very similar to that discussed in section 7.4 for the carbonic acid (CO_2–H_2O) system, except that the dissociation of $CaCO_3$ presents an additional variable, Ca^{2+}, as well as an additional equation to deal with it:

$$CaCO_{3\,(s)} \Leftrightarrow Ca^{2+} + CO_3^{2-}\ ;\ K_{sp(cal)} = a_{ca^{2+}}\, a_{co_3^{2-}} = 10^{-8.48} \tag{7.65}$$

Also, the charge balance equation for the $CaCO_3$–CO_2–H_2O system needs to be modified to take into account the contribution of Ca^{2+}:

$$m_{H^+} + 2m_{Ca^{2+}} = m_{HCO_3^-} + 2m_{CO_3^{2-}} + m_{OH^-} \tag{7.69}$$

Now we have five equations (7.37, 7.38, 7.39, 7.65, and 7.69), but six unknowns ($a_{H_2CO_3\,(aq)}$, $a_{HCO_3^-}$, $a_{CO_3^{2-}}$, a_{H^+}, a_{OH^-}, and $a_{Ca^{2+}}$). The sixth equation we need would be constructed from some other condition specified for the system: a fixed value of P_{CO_2} if the system is open to exchange of $CO_{2\,(g)}$ with a gas phase (e.g., water in a lake located in limestone), or a fixed value of total cabonate, carb (t) as defined by equation (7.44), if the system is closed to gas exchange (e.g., deep groundwater).

Open system

The sixth equation for an open system with a fixed value of P_{CO_2} is

$$a_{H_2CO_3\,(aq)} = K_{CO_2}\, P_{CO_2} \qquad (7.36)$$

We can solve the set of six simultaneous equations rigorously, but we will settle for an easier but a little less exact solution by making a reasonable assumption. The assumption is that m_{H^+}, m_{OH^-}, and $2m_{CO_3^{2-}}$ are very small quantities relative to $2m_{Ca^{2+}}$ and $m_{HCO_3^-}$, if we restrict our attention to pH < 9, and can be neglected in equation (7.69), which then reduces to:

$$\left(a_{HCO_3^-} = m_{HCO_3^-}\,\gamma_{HCO_3^-} = 2m_{Ca^{2+}}\,\gamma_{HCO_3^-}\right)$$
$$2m_{Ca^{2+}} = m_{HCO_3^-} \qquad (7.70)$$

Algebraic manipulations yield the following expressions for $m_{Ca^{2+}}$ (the solubility of calcite) and a_{H^+} (see Box 7.1):

$$m_{Ca^{2+}} = \left(\frac{K_{sp\,(cal)}\,K_{H_2CO_3}\,a_{H_2CO_3\,(aq)}}{4K_{HCO_3^-}\,\gamma^2_{HCO_3^-}\,\gamma_{Ca^{2+}}}\right)^{1/3} \qquad (7.71)$$

$$a_{H^+} = \left(\frac{K^2_{H_2CO_3}\,K_{HCO_3^-}\,a^2_{H_2CO_3\,(aq)}\,\gamma_{Ca^+}}{2K_{sp\,(cal)}\,\gamma_{HCO_3^-}}\right)^{1/3} \qquad (7.72)$$

We can compute $a_{H_2CO_3\,(aq)}$ using equation (7.36) for any given value of P_{CO_2}, and either estimate activity coefficients by the iterative procedure discussed earlier or assume them to be unity for a slightly approximate result.

Box 7.1 Expressions for $m_{Ca^{2+}}$ and a_{H^+} in the $CaCO_{3(s)}$–$CO_{2(g)}$–$H_2O_{(l)}$ system

Substituting for $a_{CO_3^{2-}}$ (equation 7.38) in equation (7.65),

$$m_{Ca^{2+}} = \frac{K_{sp\,(cal)}}{a_{CO_3^{2-}}\,\gamma_{Ca^{2+}}} = \frac{K_{sp\,(cal)}\,a_{H^+}}{K_{HCO_3^-}\,a_{HCO_3^-}\,\gamma_{Ca^{2+}}}$$

Substituting first for a_{H^+} (equation 7.37) and then for $a_{HCO_3^-}$ (equation 7.70)

$$m_{Ca^{2+}} = \frac{K_{sp\,(cal)}\,K_{H_2CO_3}\,a_{H_2CO_3\,(aq)}}{K_{HCO_3^-}\,a^2_{HCO_3^-}\,\gamma_{Ca^{2+}}} = \frac{K_{sp\,(cal)}\,K_{H_2CO_3}\,a_{H_2CO_3(aq)}}{4K_{HCO_3^-}\,m^2_{Ca^{2+}}\,\gamma^2_{HCO_3^-}\,\gamma_{Ca^{2+}}}$$

Rearranging,

$$m_{Ca^{2+}} = \left(\frac{K_{sp\,(cal)}\,K_{H_2CO_3}\,a_{H_2CO_3\,(aq)}}{4K_{HCO_3^-}\,\gamma^2_{HCO_3^-}\,\gamma_{Ca^{2+}}}\right)^{1/3} \qquad (7.71)$$

which can be used to calculate $m_{Ca^{2+}}$ as a function of P_{CO_2}. Substituting for $a_{CO_3^{2-}}$ (equation 7.65) in equation (7.38),

$$a_{H^+} = \frac{K_{HCO_3^-}\,a_{HCO_3^-}}{a_{CO_3^{2-}}} = \frac{K_{HCO_3^-}\,a_{HCO_3^-}\,m_{Ca^{2+}}\,\gamma_{Ca^+}}{K_{sp\,(cal)}}$$

Substituting first for $m_{Ca^{2+}}$ (equation 7.70) and then for $a_{CO_3^{2-}}$ (equation 7.65),

$$a_{H^+} = \frac{K_{HCO_3^-}\,a_{HCO_3^-}\,m_{HCO_3^-}\,\gamma_{Ca^+}}{2K_{sp\,(cal)}} = \frac{K_{HCO_3^-}\,\gamma_{Ca^+}}{2K_{sp\,(cal)}\,\gamma_{HCO_3^-}}\left(\frac{K_{H_2CO_3}\,a_{H_2CO_3\,(aq)}}{a_{H^+}}\right)^2$$
$$= \left(\frac{K^2_{H_2CO_3(aq)}\,K_{HCO_3^-}\,a^2_{H_2CO_3(aq)}\,\gamma_{Ca^{2+}}}{2K_{sp\,(cal)}\,\gamma_{HCO_3^-}}\right)^{1/3}$$

Since $a_{H_2CO_3\,(aq)} = P_{CO_2}\,K_{CO_2}$ (equation 7.36),

$$a_{H^+} = \left(P^2_{CO_2}\,\frac{K^2_{H_2CO_3(aq)}\,K_{HCO_3^-}\,K^2_{CO_2}\,\gamma_{Ca^{2+}}}{2K_{sp\,(cal)}\,\gamma_{HCO_3^-}}\right)^{1/3} \qquad (7.72)$$

Example 7–10: Calculation of $m_{Ca^{2+}}$ and a_{H^+} in water exposed to the atmosphere and in equilibrium with calcite at 25°C, assuming $a_i = m_i$

As discussed earlier (section 7.4.1), P_{CO_2}(atmosphere) = $10^{-3.5}$ bar. Since the system is open to the atmosphere, P_{CO_2} for calcite–water equilibrium remains constant at $10^{-3.5}$ bar, and $a_{H_2CO_3\,(aq)} = K_{CO_2}P_{CO_2} = 10^{-1.47}\;10^{-3.5} = 10^{-4.97}$.

Assuming $\gamma_{Ca^{2+}} = 1$ and $\gamma_{HCO_3^-} = 1$, and substituting for $K_{sp\,(cal)}$, $K_{H_2CO_3}$, $K_{HCO_3^-}$, and $a_{H_2CO_3\,(aq)}$, we can calculate $m_{Ca^{2+}}$ and a_{H^+} using equations (7.71) and (7.72):

$$m_{Ca^{2+}} = \left(\frac{K_{sp\,(cal)}\,K_{H_2CO_3}\,a_{H_2CO_3\,(aq)}}{4K_{HCO_3^-}\,\gamma^2_{HCO_3^-}\,\gamma_{Ca^{2+}}}\right)^{1/3}$$
$$= \left(\frac{(10^{-8.48})\,(10^{-6.35})\,(10^{-4.97})}{4\,(10^{-10.33})}\right)^{1/3} = 10^{-3.36}\ \text{mol kg}^{-1}$$

$$a_{H^+} = \left(\frac{K^2_{H_2CO_3}\,K_{HCO_3^-}\,a^2_{H_2CO_3(aq)}\,\gamma_{Ca^+}}{2K_{sp(cal)}\,\gamma_{HCO_3^-}}\right)^{1/3}$$
$$= \left(\frac{(10^{-6.35})^2\,(10^{-10.33})\,(10^{-4.97})^2}{2\,(10^{-8.48})}\right)^{1/3} = 10^{-8.26}$$

Thus, the final pH of the solution at equilibrium is 8.26, and the concentration of Ca^{2+} in the solution is $10^{-3.36}$ mol kg^{-1} = 4.36×10^{-4}mol kg^{-1}.

The concentration of Ca^{2+} can also be expressed in units of parts per million. The gram-molecular weight of $CaCO_3$ = 100.086; so the conentration of Ca^{2+} in the water in this case = $10^{-3.36} \times 100.086 \times 10^3$ mg kg^{-1} = 44 mg kg^{-1} = 44 ppm.

Note that the calcite solubility in this solution is significantly greater than in pure water, which is consistent with our experience that calcite is easily attacked by acidic solutions (including rain water). The neutralization of the acid by calcite, however, results in a solution that is alkaline, its pH being a function of P_{CO_2}. In general, surface waters and ground waters in carbonate terranes tend to be alkaline. The connection becomes evident if we consider hydrolysis reactions such as:

$$CaCO_3 + H_2O \Leftrightarrow Ca^{2+} + HCO_3^- + OH^- \quad (7.73)$$

Closed system

Suppose groundwater initially equilibrates in an open system with atmospheric P_{CO_2} and then equilibrates with calcite in a system closed to gas exchange. Such a closed system is characterized by a fixed value of total carbonate, and the conservation of carb (t), assuming zero initial concentration of Ca^{2+}, can be written as:

$$m_{\text{carb (t)}} = (m_{\text{carb (t)}})_{\text{initial}} + (m_{\text{carb (t)}})_{\text{from dissolution of calcite}} \quad (7.74)$$

where $(m_{\text{carb (t)}})_{\text{initial}}$ denotes the total carbonate before dissolution of any $CaCO_3$ Since dissolution of 1mole of $CaCO_3$ yields 1 mole of carb (t) for each mole of Ca^{2+},

$$m_{\text{carb (t)}} = (m_{\text{carb (t)}})_{\text{initial}} + m_{Ca^{2+}} \quad (7.75)$$

where

$$m_{\text{carb (t)}} = m_{H_2CO_3\,(aq)} + m_{HCO_3^-} + CO_3^{2-} \quad (7.76)$$

Neglecting the contribution of CO_3^{2-} (which is very small) to carb (t), equation (7.75) is simplified to:

$$m_{H_2CO_3\,(aq)} + m_{HCO_3^-} = (m_{H_2CO_3\,(aq)})_{\text{initial}} + m_{Ca^{2+}} \quad (7.77)$$

where $(m_{H_2CO_3\,(aq)})_{\text{initial}}$ denotes the concentration of $H_2CO_{3\,(aq)}$ calculated using equation (7.36) (assuming $a_i = m_i$):

$$(m_{H_2CO_3\,(aq)})_{\text{initial}} = K_{CO_2}\,(P_{CO_2})_{\text{initial}} \quad (7.78)$$

Algebraic manipulations involving $K_{H_2CO_3}$ (equation 7.37), $K_{HCO_3^-}$ (equation 7.38), and $K_{\text{sp (cal)}}$ (equation 7.65) yield the following equation, which can be solved for $m_{Ca^{2+}}$, the solubility of calcite (Drever, 1997, p. 62):

$$m^3_{Ca^{2+}} + \frac{K_{H_2CO_3}\,K_{\text{sp (cal)}}}{4K_{HCO_3^-}}\,m_{Ca^{2+}} - \frac{K_{H_2CO_3}\,K_{\text{sp (cal)}}\,K_{CO_2}}{4K_{HCO_3^-}}\,(P_{CO_2})_{\text{initial}} = 0 \quad (7.79)$$

Assuming $a_i = m_i$ and substituting for $a_{H_2CO_3}$ in equation (7.71), we get the following equation for calculating the final P_{CO_2} of the closed system:

$$P_{CO_2} = \frac{m^3_{Ca^{2+}}\,4K_{HCO_3^-}}{K_{\text{sp (cal)}}\,K_{H_2CO_3}K_{CO_2}} \quad (7.80)$$

7.8.3 *Factors affecting calcite solubility*

The dissolution and precipitation of calcite (and other carbonates) in natural aqueous systems are controlled by several factors as described below.

(1) *P_{CO_2} and pH.* The most important controls of calcite solubility are P_{CO_2} and pH, which are inversely correlated (Fig. 7.7a). At constant temperature and pressure, calcite solubility increases with increasing P_{CO_2} (equation 7.71; Fig. 7.7b) and, therefore, with increasing $a_{H_2CO_3(aq)}$ (equation 7.36). However, since an increase in $a_{H_2CO_3(aq)}$ results in a higher value of a_{H^+} (or a lower value of pH) (equation 7.37), calcite solubility decreses with increasing pH (Fig. 7.7c). In other words, precipitation of calcite is favored by a decrease in P_{CO_2} or an increase in pH. Thus, the current trend of increasing anthropogenic CO_2 in the atmosphere will tend to lower progressively the saturation state of surface seawater with respect to carbonate minerals, thereby reducing the precipitation and enhancing the dissolution of $CaCO_3$ (Kleypas *et al.*, 1999; Mackenzie *et al.*, 2000; Anderson *et al.*, 2003; Ridgwell and Zeebe, 2005).

(2) *Temperature.* Unlike most rock-forming minerals, the solubility of calcite, as well as as of a number of other carbonate minerals (and sulfates), decreases with increasing temperature (Fig. 7.7d). Actually, both $K_{H_2CO_3}$ and $K_{HCO_3^-}$ increase with increasing temperature, but this is more than compensated by decreases in K_{CO_2} and $K_{\text{sp (cal)}}$. Precipitation of carbonates in the warm surface waters of the ocean, especially in the tropics, is largely a consequence of the temperature dependence of $CaCO_3$ solubility.

(3) *Pressure.* The effect of total pressure, independent of its effect on P_{CO_2}, is to increase the solubility of $CaCO_3$, but the effect becomes appreciable only where the pressure is large, as is the case in deep parts of the oceans. Applying equation (6.4) to reaction (7.65), it can be shown that $K_{\text{sp (cal)}}$ will increase by ~ 0.2 log units for a pressure increase of 200 bars (Stumm and Morgan, 1981). Morse

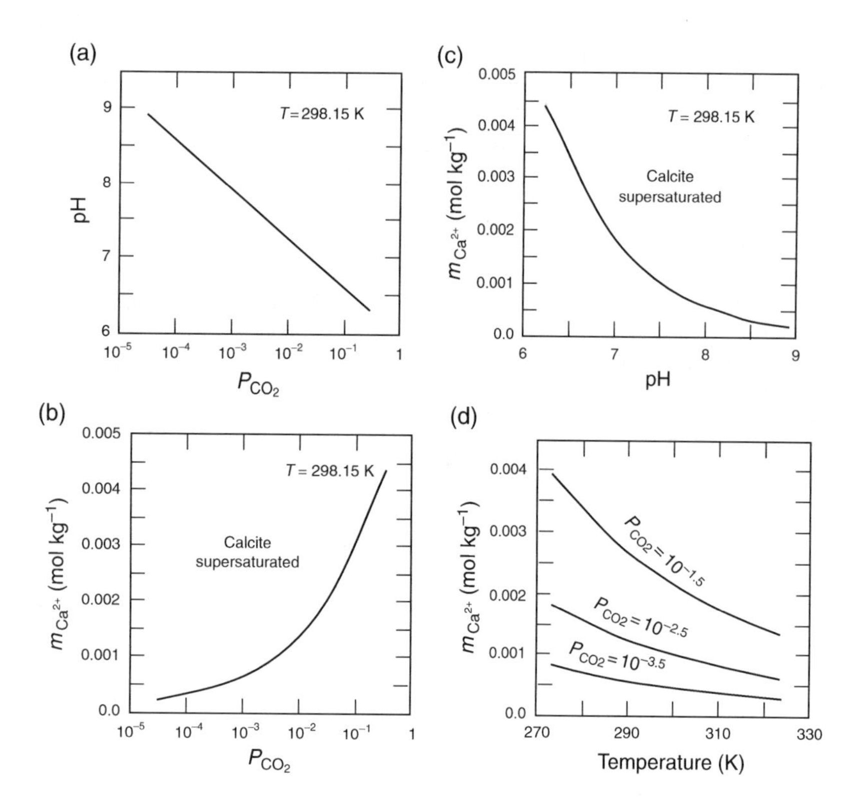

Fig. 7.7 Calcite solubility as a function of P_{CO_2}, pH, and temperature in the $CaCO_3$–CO_2–H_2O system at 1 bar. (a) Inverse relationship between P_{CO_2} and pH at 298.15 K. (b) Increase in calcite solubility with increasing P_{CO_2} at 298.15 K. (c) Decrease in calcite solubility with increasing pH at 298.15 K. (d) Decrease in calcite solubility with increasing temperature at two constant values of P_{CO_2}. Calcite solubilities for (d) were calculated using $K_{H_2CO_3}$, $K_{HCO_3^-}$, K_{CO_2}, and $K_{sp\,(cal)}$ values for the temperature interval 0 to 50°C compiled by Garrels and Christ (1965). Corresponding figures for aragonite will be very similar.

and Berner (1979) have discussed in detail the effects of temperature and pressure on calcite dissolution.

(4) *Evaporation.* Evaporation in the open ocean does not materially change the solute concentrations because of the large volume of water and its global-scale circulation. Evaporation in a restricted basin, however, may increase the Ca^{2+} concentration sufficiently to cause calcite precipitation by the reaction

$$Ca^{2+} + 2HCO_3^- \Rightarrow CaCO_3 + CO_{2\,(g)} \uparrow + H_2O \qquad (7.81)$$

This reaction is driven to the right because of removal of CO_2 from the system and because a_{H_2O} tends to become less than unity as the solute concentrations in the water increase with evaporation.

(5) *The common-ion effect.* Increases in the concentration of Ca^{2+}, HCO_3^-, or CO_3^{2-} due to supply from sources other than calcite dissolution may cause supersaturation with respect to calcite and lead to calcite precipitation.

(6) *Organic activity.* Microorganisms play a key role in the intracellular and extracellular precipitation of minerals, including $CaCO_3$. *Biomineralization*, the process by which organisms form minerals from simple compounds, may be either "biologically controlled" (BCM) or "biologically induced" (BIM) (Lowenstam, 1981; Mann, 1983; Lowenstam and Weiner, 1989). The BCM process usually involves deposition of minerals on or within organic matrices or vesicles within the cell, allowing the organism to exert a significant degree of control over the nucleation and growth of the minerals and thus over the composition, size, habit, and intracellular location of the minerals (Bazylinski and Frankel, 2000). Minerals formed by the BIM process generally nucleate and grow extracellularly as a result of metabolic activity of the organism and subsequent chemical reactions involving metabolic byproducts. Thus, biologically induced mineralization results in the minerals having crystal habits similar to those produced by precipitation from inorganic solutions (Lowenstam, 1981).

7.8.4 *Abiological precipitation of calcium carbonate in the oceans*

Calcite and aragonite saturation

Calcium carbonate, like most carbonate compounds, is relatively insoluble and should redily precipitate from solutions at relatively low activities of Ca^{2+} and CO_3^{2-}. Let us calculate whether the surface seawater (temperature 298.15 K and salinity 35‰) is saturated with respect to calcite or aragonite or both:

$m_{Ca^{2+}} = 10.13 \times 10^{-3}$ mol kg^{-1}; $\quad m_{CO_3^{2-}} = 0.3 \times 10^{-3}$ mol kg^{-1}

(see Table 7.1)

$\gamma_{Ca^{2+}} = 0.26$; $\gamma_{CO_3^{2-}} = 0.02$ (Berner, 1971)

$$IAP_{CaCO_3} = (m_{Ca^{2+}}\gamma_{Ca^{2+}})\,(m_{CO_3^{2-}}\gamma_{CO_3^{2-}})$$
$$= (10.13\times10^{-3}\times0.26)\,(0.3\times10^{-3}\times0.02)$$
$$= 15\times10^{-9} = 10^{-7.82}$$

$$\frac{IAP_{CaCO_3}}{K_{sp\ (calcite)}} = \frac{10^{-7.82}}{10^{-8.48}} = 10^{0.66} = 4.57$$

$$\frac{IAP_{CaCO_3}}{K_{sp\ (aragonite)}} = \frac{10^{-7.82}}{10^{-8.34}} = 10^{0.52} = 3.31$$

Thus, inspite of relatively low activities of Ca^{2+} and CO_3^{2-}, present–day surface seawater is oversaturated with respect to both calcite and aragonite. In fact, calculations by Riding and Liang (2005) indicate that seawater has been saturated with respect to both calcite and aragonite (and dolomite) at least for the past ~550 Ma, with calcite being consistently at a higher state of saturation compared to aragonite. Oversaturatioin of seawater should result in widespread spontaneous abiotic precipitation of $CaCO_3$, but this is not the case in modern oceans. Inorganic precipitation of $CaCO_3$ seems to be restricted to some special marine environments, such as the shallow banks off Florida and the Bahamas, where a combination of warm water temperatures and lower P_{CO_2} (because of loss of CO_2 due to photosynthesis and evaporation) decrease the solubility of $CaCO_3$. Even in these environments, it is not certain if the precipitation process is entirely abiotic. Perhaps, to overcome kinetic barriers, inorganic calcite precipitation requires a higher degree of supersaturation than we have in modern oceans (Berner, 1975; Bosak and Newman, 2003). In the experimental study reported by Morse *et al.* (2003), spontaneous nucleation of calcite in seawater solutions required an $IAP_{CaCO_3}:K_{sp\ (calcite)}$ ratio in excess of 20 to 25.

The calcite–aragonite puzzle: calcite sea versus aragonite sea

Another unresolved issue is the precipitation of aragonite and high-Mg calcite in preference to calcite (or low-Mg calcite) in modern oceans. Current estimates of the average pelagic aragonite to calcite production ratio range from 1 to 0.25 (Morse and Berner, 1979). The K_{sp} values ($10^{-8.48}$ for calcite and $10^{-8.34}$ for aragonite, at 298.15 K and 1 bar) suggest that aragonite should be slightly more soluble than calcite in surface seawater; that is, calcite should precipitate in preference to aragonite. The lack of authigenic low-Mg (< 5 mole % $MgCO_3$) calcite in modern marine sediments may be ascribed to non-equilibrium incorporation of Mg into the crystal structure during growth (Berner, 1975). The calcite versus aragonite precipitation has been alluded to various conditions of seawater (see reviews by Walter, 1986; Burton and Walter, 1987): CO_3^{2-} concentration, temperature, and the presence of other ions such as Mg^{2+}, Sr^{2+}, SO_4^{2-}, and PO_4^{3-} (Morse and Berner, 1979; Plummer *et al.*, 1979; Reddy, 1986). In general, precipitation of aragonite in preference to calcite is favored by: (i) high CO_3^{2-} concentration (Given and Wilkinson, 1985); (ii) relatively high temperature (Kitano *et al.*, 1962; Morse *et al.*, 1997), > 5°C according to the limited experimental study of Burton and Walter (1987); and (iii) high Mg : Ca ratio (Folk, 1974; Berner, 1975; Reddy *et al.*, 1981; Mucci and Morse, 1983; Hardie, 1996; Morse *et al.*, 1997; Drever, 1997; Stanley and Hardie, 1998; Lowenstein *et al.*, 2001) because of the inhibiting effect of Mg on calcite precipitation; (iv) high SO_4^{2-} concentration (Walter, 1986); (v) low PO_4^{3-} concentration (Walter, 1986); and (vi) the kinetics of calcite growth in marine environments (Morse and Berner, 1979). The kinetics of dissolution and precipitation of calcite in seawater is discussed further in section 9.6.2.

Most carbonate sedimentologists believe that whether calcite or aragonite would precipitate from seawater is determined mainly by its ambient mole Mg : Ca ratio. As summarized in Stanley *et al.* (2002), laboratory experiments have shown that in water at the ionic strength of modern seawater (~ 0.7) and temperatures typical of tropical seas (25–30°C), mineralogy of the nonskeletal $CaCO_3$ precipitate varies systematically with the ambient mole Mg : Ca ratio of the water: the precipitate is low-Mg calcite at Mg : Ca < ~1; high-Mg calcite (> 4 mol% $MgCO_3$) at Mg : Ca between about 1 and 2; aragonite along with high-Mg calcite at Mg : Ca between about 2 and slightly above 5; and only aragonite at Mg : Ca substantially above 5. According to many workers (e.g., Sandberg, 1983; Lowenstein *et al.*, 2001), the record of nonskeletal carbonates and brine inclusions in halite from evaporite deposits (salt deposits formed by evaporation of surface brines; see Box 13.6) indicates that the Phanerozoic chemistry of seawater oscillated between a "calcite sea," when the nonskeletal $CaCO_3$ precipitate was predominantly calcite (during Cambrian to middle Mississippian, and middle Jurassic to late Paleogene), and an "aragonite sea", when the $CaCO_3$ precipitate was predominantly aragonite (late Precambrian to early Cambrian, late Mississippian to middle Jurassic, and Neogene) (Fig. 7.8). Periods of "aragonite seas" were synchronous with the deposition of $MgSO_4$ evaporites, and periods of "calcite seas" with the deposition of KCl evaporites (Hardie, 1996) (see section 13.5.3).

Spencer and Hardie (1990) showed that the composition of modern seawater can be accounted for by steady-state mixing of the two major contributors to ocean chemistry – river water (RW) and mid–ocean ridge (MOR) hydrothermal brines – coupled with precipitation of solid $CaCO_3$ and SiO_2 phases. Their model for calculation of the chemistry of ancient ocean water based on the MOR/RW flux ratio predicts that relatively small changes in the MOR flux, as little as 10% of the modern value, can lead to changes in the Mg : Ca, Na : K, and Cl : SO_4 mole ratios of seawater that would drastically alter the primary mineralogy of marine evaporites and carbonate rocks (Hardie, 1996).

The secular shifts in the Mg : Ca ratio of seawater, together with synchronous oscillations in the mineralogy of marine evaporites, were probably driven largely by secular changes in

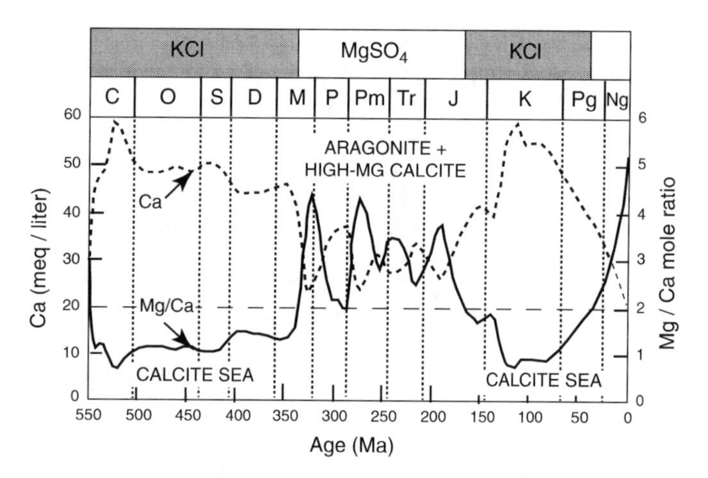

Fig. 7.8 Oscillation between "calcite sea" (when seawater should have precipitated calcite) and "aragonite sea" (when seawater should have precipitated aragonite + high-Mg calcite) during Phanerozoic time according to the model of Hardie (1996) for the secular variation in the Mg : Ca ratio and Ca concentration in seawater at 25°C. The boundary between the nucleation fields of (low-Mg) calcite and aragonite + high-Mg calcite is shown in the plot as a horizontal line at Mg : Ca = 2. Note the broad correlation between KCl evaporite deposits and "calcite sea," and between $MgSO_4$ evaporite deposits and "aragonite sea". (After Stanley and Hardie, 1998, Figure 1, p. 6.)

the hydrothermal brine flux along mid-ocean ridges due to secular changes in the rate of ocean crust formation (Hardie, 1996; Stanley and Hardie, 1998). Laboratory hydrothermal experiments indicate that high-temperature basalt–seawater reaction involves large-scale removal of Mg^{2+} and SO_4^{2-} from seawater and addition of Ca^{2+}, H_4SiO_4, and K^+ (Bischoff and Dickson, 1975; Mottl *et al.*, 1979). Thus, seawater–basalt interaction generates brines enriched in $CaCl_2$–KCl and impoverished in $MgSO_4$. High spreading rates would expand the mid-oceanic ridges, and greater Mg uptake at the ridges from increased reaction with hydrothermal brines would lower the Mg : Ca ratio of seawater. Many other authors have also argued in favor of a connection between plate tectonics and seawater chemistry (e.g., Kovalevich *et al.*, 1998; Lowenstein *et al.*, 2001; Horita *et al.*, 2002; Steuber and Veizer, 2002).

Hardie (1996) used the occurrence of nonskeletal $CaCO_3$ precipitates in the rock record to estimate the secular variation of the Mg : Ca ratio in seawater during the Phanerozoic Eon. His model also predicts that high-Mg calcite should precipitate along with aragonite, as is the case with the present-day aragonite sea. In a later article, Hardie (2003) extended his model to the Precambrian and argued that similar oscillations between calcite and aragonite seas occurred at least as far back as late Archean. Also, as the distribution coefficient of Sr is higher for argonite-seawater than for calcite-seawater, the Sr : Ca ratio was higher in calcite seas than in aragonite seas (Steuber and Veizer, 2002). Thus, despite widespread biological mediation of carbonate precipitation, seawater chemistry has generally exerted a primary control on the formation of shallow marine carbonates during the Phanerozoic and most, if not all, of the Precambrian. The paucity of aragonite in the Precambrian sedimentary rock record is generally attributed to transformation of aragonite to calcite, the more stable polymorph, during diagenesis (e.g., Grotzinger and James, 2000; Kah, 2000; Sumner and Grotzinger, 2000).

7.8.5 *Biological precipitation of calcium carbonate in the oceans*

Volumetrically, the most important mechanism for removal of dissolved $CaCO_3$ from seawater is by "biologically controlled" mineralization (*biomineralization*). The process involves secretion of skeletal hard parts by certain organisms such as foraminifera, mollusks, bryozoans, echinoderms, corals, petropods, and cocolithophores (Fig. 7. 9), the source of biochemical sediments that eventually accumulate on the seafloor. Foraminifera and cocolithophores are the most important sources of calcitic calcium carbonate; petropods are the most important sources of aragonitic calcium carbonate. In general, the organism constructs an organic framework into which the appropriate ions are actively introduced and then induced to crystallize and grow under genetic control. Apparently, such biochemical $CaCO_3$ deposition in cell structures occurs only in water that is saturated or nearly saturated with $CaCO_3$. Experimental studies (e.g., Langdon *et al.*, 2000) indicate that there is a strong correlation between the saturation state of the ocean and the rate of calcification of coralline algae and corals. From experimental studies on four types of marine invertebrates that secrete high-Mg calcite in modern seas, Reis (2004) concluded that the Mg : Ca ratio of unaltered fossils of calcareous organisms can be used to reconstruct the secular variation of the Mg : Ca ratio of seawater over the Phanerozoic Eon. Mg : Ca ratios of seawater calculated by Dickson (2002) from mole % $MgCO_3$ of fossil echinoderms (early Cambrian – 3.3; late Carboniferous – 3.4; late Triassic – 3.1; Jurassic – 1.4, and Cretaceous – 1.4) agree quite well with those shown in Fig. 7.8.

Precipitation of $CaCO_3$ in oceans can also be "biologically induced" by some bacteria, including cyanobacteria and fungi. Altermann *et al.* (2006) have proposed that cyanobacteria and heterotrophic bacteria were the principal contributors to the production of carbonate rocks during almost 70% of Earth's history (3.5 to 0.5 Ga). The biological involvement stems from the ability of the organisms to catalyze reactions that promote precipitation of $CaCO_3$ because of an increase in pH, or $a_{Ca^{2+}}$, or $a_{CO_3^{2-}}$. Favorable conditions for such calcite precipitation, as described in detail by Ehrlich (2002, pp. 193–198) are: (i) generation of CO_2 by aerobic and anaerobic oxidation of carbon compounds, such as organic acids and hydrocarbons, in well-buffered neutral or alkaline environments; (ii) increase in the pH of unbuffered environments by hydrolysis of ammonia ($NH_3 + H_2O = NH_4^+ + OH^-$) released by aerobic and anaerobic

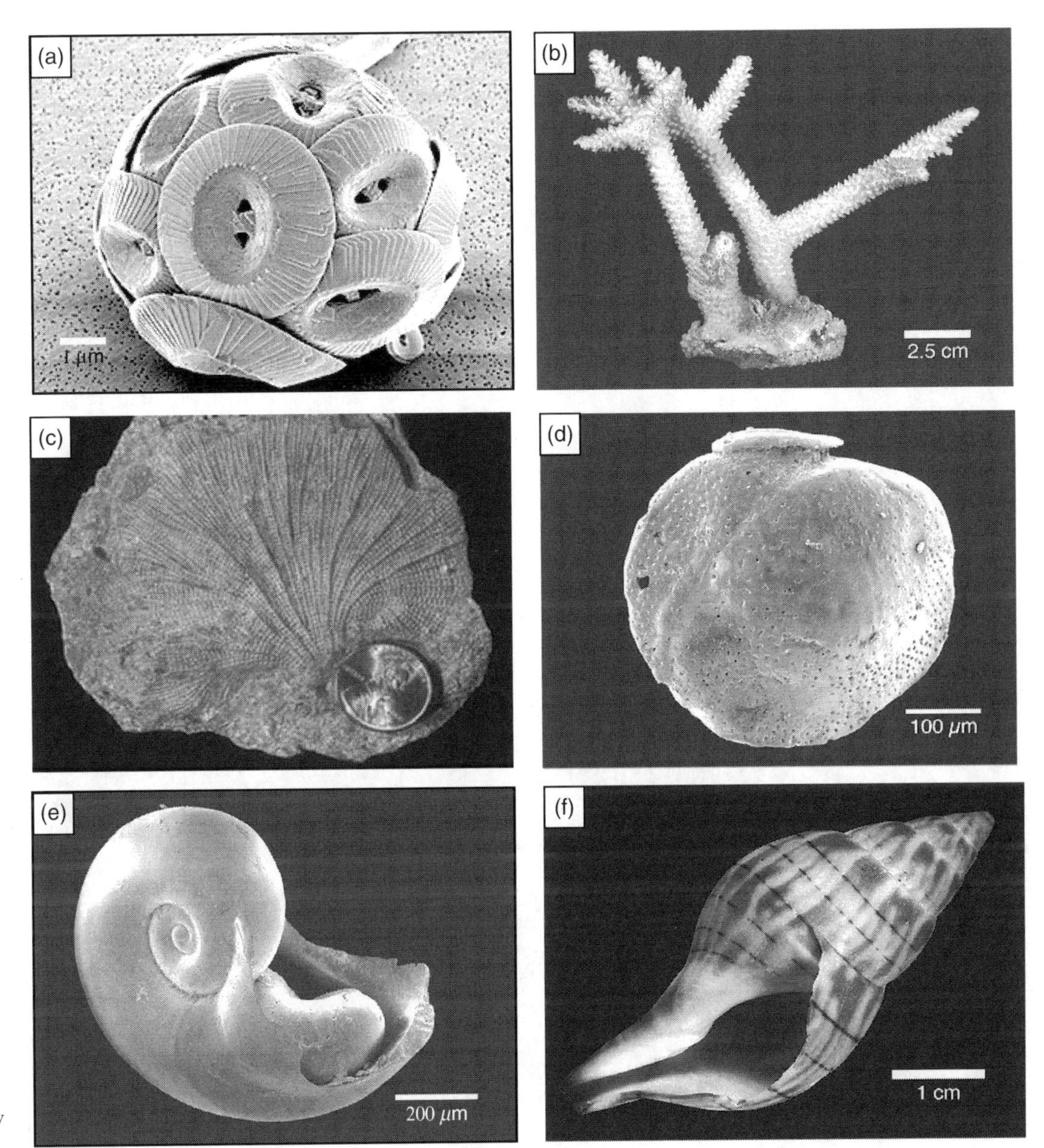

Fig. 7.9 Some marine organisms that extract $CaCO_3$ from seawater to form their skeletal hard parts. (a) Cocolithophore (a single-celled algae); (b) coral ; (c) bryozoan; (d) foraminifera; (e) petropod; and (f) mollusk (gastropod). (Sources of the images: <http://en.wikipedia.org/wiki/cocolithophore>; http://www.kgs.ku.edu/Extension/fossils/bryozoan.html; Dr Colin Sumrall, Department of Earth and Planetary Sciences, University of Tennessee, Knoxville.)

oxidation of organic nitrogen compounds; (iii) reduction of $CaSO_4$ to CaS by sulfate-reducing bacteria, with CO_2 as a byproduct; and (iv) increase in the concentration of CO_3^{2-} due to removal of CO_2 from a HCO_3^--containing solution by photosynthesis.

Biologically driven carbonate deposition provides a significant buffering of ocean chemistry and atmospheric CO_2 in the modern system. However, calcifying organisms that serve as a carbonate sink are being threatened by continued release of CO_2 to the atmosphere from burning of fossil fuels and increasing acidity of the surface ocean water. This poses a significant potential threat to the building of coral reefs.

7.8.6 Carbonate compensation depth

In most areas of the oceans, the abundance of $CaCO_3$ decreases with increasing water depth. This observation is consistent with the fact that the theoretical solubility of both calcite and aragonite in seawater, expressed as $m_{CO_3^{2-}}$ in equilibrium with the carbonate mineral, increases with depth (Fig. 7.10), primarily due to increasing pressure but also to a smaller extent due to decreasing temperature. The saturation state of seawater with respect to calcium carbonate minerals is largely determined by $a_{CO_3^{2-}}$ because $a_{Ca^{2+}}$ is almost conservative in seawater. The concentration of dissolved CO_3^{2-} in seawater, largely a result of decay and sinking of dead organisms,

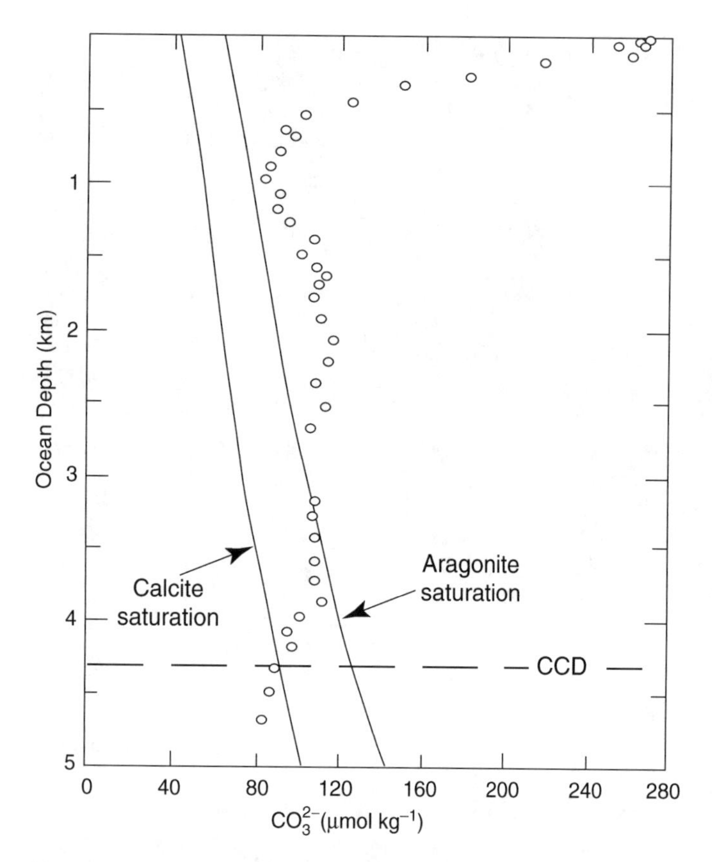

Fig. 7.10 Variation of measured dissolved CO_3^{2-} concentration in seawater measured at a station in the western South Atlantic Ocean relative to theoretical saturation curves for calcite (Cal) and aragonite (Arg). To the right of the calcite (or aragonite) saturation curve, seawater is supersaturated with respect to calcite (or aragonite); to the left it is undersaturated. The dashed lined marked CCD is the carbonate compensation depth at this location. (From Broecker and Peng, 1982, Figure 2–13, p. 74.)

is quite high at shallow depths, but decreases rapidly to lower values with increasing depth. The measured vertical $m_{CO_3^{2-}}$ profile shown in Fig. 7.10 is for a station in the western South Pacific Ocean. The disposition of the calcite and aragonite saturation curves relative to the $m_{Ca^{2+}}$ profile indicates that, at this location, there should be no aragonite precipitation below a depth of ~ 3300 m and no calcite precipitation below a depth of ~ 4300 m. The depth below which seawater is undersaturated with respect to calcite is referred to as the *carbonate (calcite) compensation depth* (CCD). In practice, it is difficult to determine the CCD at a particular location because the lack of calcite precipitation may be due to the extremely slow rate of reaction at the very low temperatures typical of such depths (~ 2°C; Berner and Berner, 1996) rather than calcite undersaturation. From a practical perspective, a more easily determined depth is the *lysocline*, the depth at which dissolution features indicative of $CaCO_3$ undersaturation are first observed in carbonate sediments. The CCD and the position of the lysocline vary from place to place in the oceans depending mainly on the $m_{CO_3^{2-}}$ profile. The *aragonite compensation depth* (ACD) commonly occurs at depths on the order of 3 km shallower than CCD (Morse and Berner, 1979).

In view of the increasing solubility of calcite and aragonite with depth, we should not expect $CaCO_3$ in deep-sea sediments. However, for kinetic reasons (see section 9.6.2) only 75–95% of the $CaCO_3$ being produced at present in the open oceans is subsequently dissolved, with the rest accumulating in the deep-sea sediments. The distribution of $CaCO_3$-rich sediments on the seafloor constitutes a composite record of the input of $CaCO_3$ (predominantly in the form of dead microorganisms) near the ocean surface, the rate of calcite dissolution, and the rate of sediment accumulation.

7.9 Chemical weathering of silicate minerals

Chemical weathering is the overall process by which minerals of the rocks exposed at or close to the Earth's surface interact with water (particularly acidic water) and the gases in the atmosphere (particularly O_2 and CO_2) to form mineral assemblages that are thermodynamically more stable under the ambient *P–T* conditions. The acid component of the water is provided by plant activity, bacterial metabolism, and even rainwater that normally is acidic (see section 7.4.1). *Physical weathering*, on the other hand, refers to fragmentation of rocks by physical processes, such as freezing of water in the cracks, and it aids chemical weathering by increasing the surface area exposed to weathering solutions. Among the most stable products of chemical weathering are clay minerals (e.g., kaolinite, smectite), iron oxides and hydroxides (e.g., hematite, goethite), and aluminum hydroxides (e.g., gibbsite). Chemical weathering is particularly effective in climatic conditions characterized by abundant rainfall and high temperatures and in environments of high organic decomposition. Notable consequences of chemical weathering include: soil formation and bioavailability of inorganic nutrients; production of sediments; pronounced changes in the compositions of surface waters because of dissolved species, including contaminants, contributed by chemical reactions; and, on time scales longer than a million years, buffering of atmospheric CO_2, thus moderating large fluctuations in global temperature and precipitation through the greenhouse effect (Berner *et al.*, 1983; Berner, 1995).

7.9.1 *Mechanisms of chemical weathering*

Carbonate rocks are very susceptible to chemical weathering because of dissolution of calcite and dolomite by reactions such as

$$CaCO_{3\,(s)} + H_2CO_{3\,(aq)} \Rightarrow Ca^{2+} + HCO_3^- \quad (7.82)$$

$$CaMg(CO_3)_{2\,(s)} + 2H_2CO_{3\,(aq)} \Rightarrow Ca^{2+} + Mg^{2+} + 4HCO_3^- \quad (7.83)$$

which are responsible for much of the dissolved Ca^{2+} and HCO_3^- in river waters and groundwaters, and in oceans. However, because of much greater abundance of silicate rocks in the Earth's crust, the overall effects of chemical weathering are dominated by what happens to silicate minerals.

Common rock-forming silicate minerals undergo chemical weathering by three main processes (Berner and Berner, 1996, pp. 149–150): *congruent dissolution*; *incongruent dissolution*; and *oxidation*. The reactions are caused by soil acids, dissolved oxygen, and water. Congruent dissolution is the process of simple dissolution of a mineral into dissolved species, such as the dissolution of water-soluble minerals such as halite producing Na^+ and Cl^- ions (reaction 7.1). Carbonate minerals typically weather by congruent dissolution (reactions 7.65, 7.74, 7.75). Most silicate minerals, however, weather by incongruent dissolution that produces one or more solid phases in addition to dissolved species. A typical example is the conversion of the aluminosilicate mineral Na-plagioclase ($NaAlSi_3O_8$) into the clay mineral kaolinite ($Al_2Si_2O_5(OH)_4$), with concomitant release of Na^+ and silica (in the form of silicic acid, H_4SiO_4) by reaction with acidic water:

$$2NaAlSi_3O_{8(s)} + 9H_2O_{(l)} + 2H^+ \Rightarrow Al_2Si_2O_5(OH)_{4\,(s)} + 2Na^+ + 4H_4SiO_{4\,(aq)} \qquad (7.84)$$

This reaction, which in essence controls the formation of most kaolinite deposits, would be favored by an abundant source of plagioclase, a steady supply of large amounts of acidic water (at least 4.5 moles of H_2O for every mole of plagioclase), and continual removal of one or more of the products from the system by good drainage. The reaction would also be favored at elevated temperatures because reaction rates generally double for every 10°C increase in temperature. (Balancing reactions describing incongruent dissolution of aluminosilicate minerals can be cumbersome; the reader may refer to guidelines listed in Faure (1991, p. 224)).

Oxidation produces oxide and hydroxide minerals of transition metals such as iron, copper, and manganese; most of the oxidation products are stable under a wide range of conditions in the surface environment. A more important role of oxidation lies in the generation of carbonic acid (H_2CO_3), sulfuric acid (H_2SO_4), and various organic acids, all of which release H^+ by dissociation and lower the pH of the environment. Oxidation of organic matter (represented for simplicity by the generalized formula for carbohydrate, CH_2O) produces CO_2, which then combines with water to form H_2CO_3:

$$CH_2O + O_2 \Rightarrow CO_2 + H_2O \Rightarrow H_2CO_{3\,(aq)} \qquad (7.85)$$

Oxidation of sulfide minerals, such as pyrite, produces sulfuric acid as well as some oxidized iron mineral, which may be designated as "$Fe(OH)_3$:"

$$4FeS_{2\,(s)} + 15O_2 + 14H_2O \Rightarrow 4Fe(OH)_{3\,(s)} + 8H_2SO_{4\,(aq)} \qquad (7.86)$$

Such oxidation reactions are almost always catalyzed by microorganisms. A more detailed discussion of oxidation reactions will be presented in Chapter 8.

7.9.2 *Solubility of Silica*

Before considering the behavior of more complex silicate minerals, let us examine the solubility of silica, which not only is chemically the simplest silicate mineral but also the most common mineral in a wide range of igneous and metamorphic rocks and in many sedimentary rocks.

Crystalline silica (SiO_2) occurs in nature as several polymorphs: α-quartz, β-quartz, tridymite, cristobalite, coesite, and stishovite. Coesite, which is found in rocks subjected to the impact of large meteorites and in xenoliths carried by kimberlite magmas, and stishovite, which has been identified in the impact-metamorphosed sandstone in the Meteor Crater, Arizona, are stable only at very high pressures. The *P–T* stability fields of the other polymorphs are shown in Fig. 7.11. *Chalcedony* is a group name for the compact varieties of silica composed of microcrystalline α-quartz with submicrosopic pores; chalcedony with color-banding, often in beautiful patterns, is given the special name *agate*. Opaque, dark colored or black chalcedony is commonly called *chert* when it occurs in stratified or massive form in rocks, and *flint* when it occurs in nodular form in a rock matrix (Deer *et al.*, 1992). Silica also occurs in two noncrystalline forms: *silica glass*, which is unstable below 1713°C (1986.15 K), and *opal*, which is amorphous silica with some water ($SiO_2 \cdot nH_2O$). Here we will focus only on α–quartz, the most common SiO_2 polymorph in the upper crust, and amorphous silica, the form in which silica precipitates from aqueous solutions; amorphous silica eventually transforms into chalcedony.

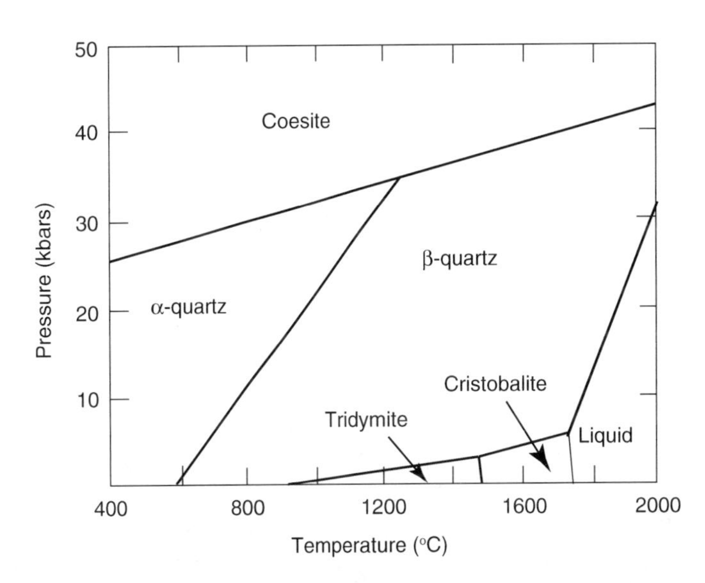

Fig. 7.11 Pressure–temperature stability fields of several polymorphs of SiO_2. Stishovite has been synthesized at 130 kbar and > 1200°C.

The dissolution of α–quartz and amorphous silica in water at equilibrium can be described by the following reactions (although in nature H_4SiO_4 acid is produced mainly by the dissolution of aluminosilicate minerals):

α-quartz:

$$SiO_2(\alpha - qz) + 2H_2O_{(l)} \Leftrightarrow H_4SiO_{4\,(aq)} \quad K_{SiO_2(\alpha\text{-}qz)} = 10^{-4} \quad (7.87)$$

$$a_{H_4SiO_4\,(aq)} = 10^{-4}\ \text{mol kg}^{-1}$$

Am silica:

$$SiO_2(am) + 2H_2O_{(l)} \Leftrightarrow H_4SiO_{4\,(aq)} \quad K_{SiO_2(am)} = 10^{-2.71} \quad (7.88)$$

$$a_{H_4SiO_4\,(aq)} = 10^{-2.71}\ \text{mol kg}^{-1}$$

where the equilibrium constants are for 298.15 K and 1 bar, which are consistent with the thermodynamic properties of α-quartz and amorphous silica determined by Richet *et al.* (1982). For the neutral molecule H_4SiO_4 in a dilute solution, $\gamma_{H_4SiO_4} \approx 1$, so that $a_{H_4SiO_4\,(aq)} = m_{H_4SiO_4\,(aq)}$. Since each mole of SiO_2 produces 1 mole of $H_4SiO_{4\,(aq)}$, the solubilities of α-quartz and amorphous silica at 298.15 K and 1 bar should be (gram-molecular weight of SiO_2 = 60.084):

$$m_{H_4SiO_4\,(aq)}(\alpha - qz)$$
$$= a_{H_4SiO_4\,(aq)}(\alpha - qz) = 10^{-4}\ \text{mol kg}^{-1}$$
$$= 60.084 \times 10^{-4}\text{g kg}^{-1} = 6.01\ \text{mg kg}^{-1} = 6.01\ \text{ppm}$$

$$m_{H_4SiO_4\,(aq)}(\text{am silica})$$
$$= a_{H_4SiO_4\,(aq)}(\text{am silica}) = 10^{-2.71}\ \text{mol kg}^{-1}$$
$$= 60.084 \times 10^{-2.71}\text{g kg}^{-1} = 117.1\ \text{mg kg}^{-1} = 117.1\ \text{ppm}$$

The concentrations calculated above, however, do not represent the actual silica solubilities in the solution because they do not take into account the contributions arising from the stepwise dissociation of the weak acid H_4SiO_4 (see Table 7.4). Incorporating these contributions and assuming $a_i = m_i$, the silica solubility, $m_{SiO_2\,(total)}$, is given by:

$$m_{SiO_2\,(total)} = m_{H_4SiO_4\,(aq)} + m_{H_3SiO_4^-} + m_{H_2SiO_4^{2-}} + m_{HSiO_4^{3-}} + m_{SiO_4^{4-}} \quad (7.89)$$

Substituting for activities in terms of the corresponding dissociation constants (Box 7.2),

$$m_{SiO_2\,(total)}$$
$$= K_{SiO_2}\left(1 + \frac{K_1}{a_{H^+}} + \frac{K_1\,K_2}{a_{H^+}^2} + \frac{K_1\,K_2\,K_3}{a_{H^+}^3} + \frac{K_1\,K_2\,K_3\,K_4}{a_{H^+}^4}\right) \quad (7.90)$$

where $K_{SiO_2} = K_{SiO_2(\alpha-qz)} = 10^{-4}$ in the case of α-quartz and $K_{SiO_2} = K_{SiO_2(am)} = 10^{-2.71}$ in the case of amorphous silica. Equation (7.90) can be used to calculate the solubility of α-quartz and amorphous silica at any given pH (see Example 7–10) and is the equation used for calculating the pH dependence of solubilities as presented in Fig. 7.12a. Up to a pH of

Box 7.2 Expressions for solubility of silica in dilute aqueous solutions

Silica solubility as a function of pH

Because of the stepwise dissociation of the polyprotic acid H_4SiO_4, the total silica solubility, $m_{SiO_2\,(total)}$, at 298.15 K and 1 bar is given by:

$$m_{SiO_2\,(total)} = m_{H_4SiO_4\,(aq)} + m_{H_3SiO_4^-} + m_{H_2SiO_4^{2-}} + m_{HSiO_4^{3-}} + m_{SiO_4^{4-}} \quad (7.89)$$

Assuming $a_i = m_i$, the molalities can be evaluated from the stepwise dissociation constants (see Table 7.4) as follows:

$$H_4SiO_4 \Leftrightarrow H_3SiO_4^- + H^+ \qquad K_1 = 10^{-9.71}$$

$$m_{H_3SiO_4^-} = a_{H_3SiO_4^-} = \frac{K_1\,a_{H_4SiO_4}}{a_{H^+}} = \frac{K_{SiO_2}\,K_1}{a_{H^+}} \quad (7.91)$$

$$H_2SiO_4^{2-} \Leftrightarrow HSiO_4^{3-} + H^+ \qquad K_3 = 10^{-9.86}$$

$$m_{H_2SiO_4^{2-}} = a_{H_2SiO_4^{2-}} = \frac{K_2\,a_{H_3SiO_4^-}}{a_{H^+}} = \frac{K_{SiO_2}\,K_1\,K_2}{a_{H^+}^2} \quad (7.92)$$

$$H_2SiO_4^{2-} \Leftrightarrow HSiO_4^{3-} + H^+ \qquad K_3 = 10^{-9.86}$$

$$m_{HSiO_4^{3-}} = a_{HSiO_4^{3-}} = \frac{K_3\,a_{H_2SiO_4^{2-}}}{a_{H^+}} = \frac{K_{SiO_2}\,K_1\,K_2\,K_3}{a_{H^+}^3} \quad (7.93)$$

$$HSiO_4^{3-} \Leftrightarrow SiO_4^{4-} + H^+ \qquad K_4 = 10^{-13.10}$$

$$m_{SiO_4^{4-}} = a_{SiO_4^{4-}} = \frac{K_4\,a_{HSiO_4^{3-}}}{a_{H^+}} = \frac{K_{SiO_2}\,K_1\,K_2\,K_3\,K_4}{a_{H^+}^4} \quad (7.94)$$

where $K_{SiO_2} = K_{SiO_2(\alpha-qz)} = 10^{-4}$ for α–quartz and $K_{SiO_2} = K_{SiO_2(am)} = 10^{-2.71}$ for amorphous silica. Substituting the expressions for the activities in equation (7.76), we get

$$m_{SiO_2\,(total)}$$
$$= K_{SiO_2}\left(1 + \frac{K_1}{a_{H^+}} + \frac{K_1\,K_2}{a_{H^+}^2} + \frac{K_1\,K_2\,K_3}{a_{H^+}^3} + \frac{K_1\,K_2\,K_3\,K_4}{a_{H^+}^4}\right) \quad (7.90)$$

Silica solubility as a function of temperature

Equations given by Gunnarsson and Arnórsson (1999) for the temperature dependence of solubilities, based on $a_{H_4SiO_4(aq)}$ only (i.e., for pH up to about 9), are:

α – quartz:

$$\log K_{SiO_2(\alpha-qz)} = -34.188 + 197.47\,T^{-1}$$
$$- 5.851 \times 10^{-6}\,T^2 + 12.245 \log T \quad (7.95)$$

Am silica:

$$\log K_{SiO_2(am)} = -15.433 - 151.60\,T^{-1}$$
$$- 2.977 \times 10^{-6}\,T^{-2} + 5.464 \log T \quad (7.96)$$

The equations are valid in the range from 0 to 350°C at 1 bar below 100°C and at $P_{saturation}$ for higher temperatures. They were obtained by multiple linear regressions of experimental solubility data.

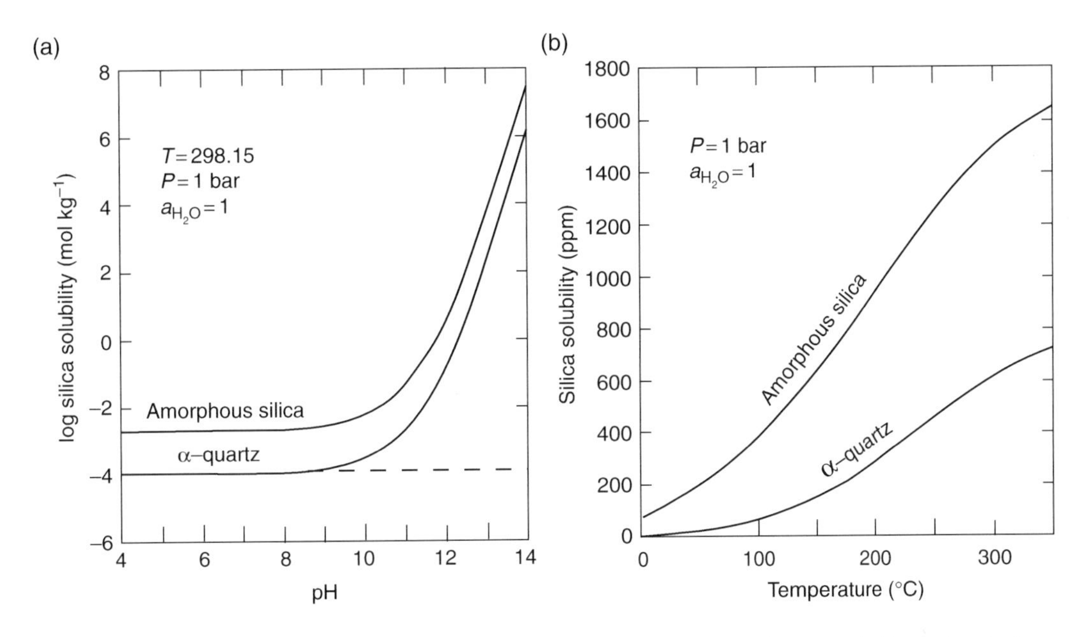

Fig. 7.12 (a) Calculated solubilities of crystalline α–quartz and amorphous silica in pure water as a function of pH at 298.15 K (25°C) and 1 bar. Note that the solubilities are constant up to about pH = 9 and are accounted for by equation (7.87) or equation (7.88); the exponential increase in the solubilities at higher pH vales is due to contributions from the stepwise dissociation of silicic acid, H_4SiO_4 (Box 7.2). (b) Calculated solubilities of crystalline α–quartz and amorphous silica in pure water as a function of temperature at pH < 9 according to equations (7.95) and (7.96) by Gunnarsson and Arnórsson (1999).

about 9, the pH limit of most natural aqueous environments, the solubility of α–quartz or amorphous silica is almost entirely due to the first term (K_{SiO_2}), which is independent of pH. At higher pH values, the solubility increases exponentially with increasing pH due to contributions from the other terms in equation (7.90).

The solubilities of α–quartz and amorphous silica are also strongly dependent on temperature, the solubilities increasing with increasing temperature (Fig. 7.12b). Equations for temperature dependence of the solubilities are given in Box 7.2.

Example 7–11: Calculation of silica solubility at 298.15 K (25°C), 1 bar, and a specified pH

For any given pH, the silica solubility at 298.15 K (25°C) and 1 bar can be calculated simply by substituting the values of a_{H^+} and the equilibrium constants in equation (7.82). For example, at pH = 10 (i.e., $a_{H^+} = 10^{-10}$), the solubility of α–quartz is

$$= 10^{-4}\left(1 + \frac{10^{-9.71}}{10^{-10}} + \frac{10^{-9.71}\,10^{-13.28}}{\left(10^{-10}\right)^2} + \frac{10^{-9.71}\,10^{-13.28}\,10^{-9.86}}{\left(10^{-10}\right)^3} + \frac{10^{-9.71}\,10^{-13.28}\,10^{-9.86}\,10^{-13.10}}{\left(10^{-10}\right)^4}\right)$$

$$= 10^{-4} + 10^{-3.71} + 10^{-6.99} + 10^{-6.85} + 10^{-9.95}$$

$$= 10^{-3.53}\text{mol kg}^{-1} = 17.74\text{ mg kg}^{-1}$$

Note that the last three terms contribute practically nothing to the solubility at this pH and, therefore, could have been ignored for the solubility calculation.

7.9.3 Equilibria in the system $K_2O–Al_2O_3–SiO_2–H_2O$

Chemical reactions that are likely to take place in the course of chemical weathering of a given parent material are controlled by the geochemical parameters of the environment and can be evaluated by application of thermodynamic principles. For simplicity we will assume the solid substances involved in the reactions to be pure phases. Taking the standard state of each phase as the pure substance at the pressure and temperature of interest, which we will assume to be 298.15 K (25°C) and 1 bar, the activity of each of the solid phases is unity, and the main control on the weathering reactions turns out to be the chemistry of the aqueous solution.

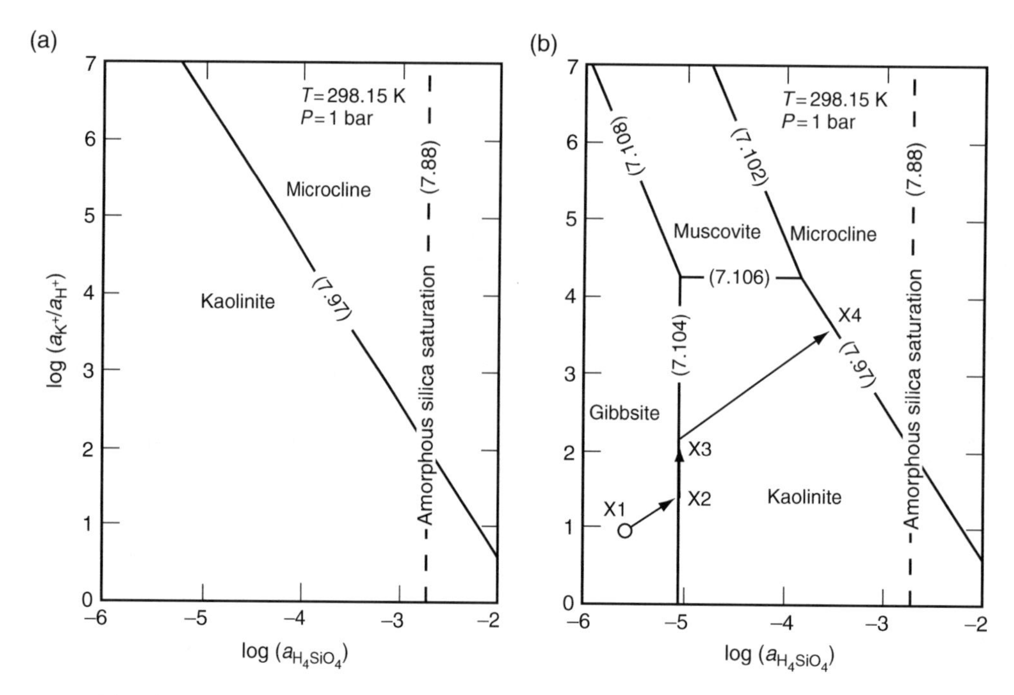

Fig. 7.13 Activity–activity stability diagrams (T = 298.15 K, P = 1 bar): (a) calculated reaction boundary between microcline and kaolinite in equilibrium with an aqueous solution (equation 7.97); (b) calculated reaction boundaries for aluminosilicate minerals in the K_2O–Al_2O_3–SiO_2–H_2O system in equilibrium with an aqueous solution. The saturation line for amorphous silica (equation 7.88) is shown for reference. The reaction boundaries, which were calculated using the thermodynamic data given in Table 7.7, are labeled according to the equation numbers they represent (Box 7.2). Stability fields of K-smectite and pyrophyllite are excluded from the figure to keep it simple. See text for explanation of the points X1, X2, X3, and X4.

For the purpose of illustration, let us consider the equilibrium between K-feldspar (represented by microcline, $KAlSi_3O_8$), a common aluminosilicate mineral in many igneous and metamorphic rocks, and kaolinite ($Al_2Si_2O_5(OH)_4$), a clay mineral of widespread occurrence in sediments and sedimentary rocks:

$$\underset{\text{microcline}}{2KAlSi_3O_8} + 9H_2O + 2H^+ \Rightarrow \underset{\text{kaolinite}}{Al_2Si_2O_5(OH)_4} + \underset{\text{solution}}{2K^+ + 4H_4SiO_{4\,(aq)}} \quad (7.97)$$

Assuming $a_{H_2O} = 1$, the equilibrium constant for the reaction (K_{eq}) is given by

$$K_{eq} = \frac{a_{kaol}\,(a_{K^+})^2\,(a_{H_4SiO_4(aq)})^4}{(a_{mic})^2\,(a_{H_2O})^9\,(a_{H^+})^2} = \frac{(a_{K^+})^2\,(a_{H_4SiO_4(aq)})^4}{(a_{H^+})^2} \quad (7.98)$$

Using logarithms to linearize the equation, we get:

$$\log K_{eq} = 2\log a_{K^+} + 4\log a_{H_4SiO_4\,(aq)} - 2\log a_{H^+} \quad (7.99)$$

For graphical representation of equation (7.99) in two-dimensional space, we reduce the three variables in the equation to two by rearranging the equation as:

$$\log\frac{a_{K^+}}{a_{H^+}} = -2\log a_{H_4SiO_4\,(aq)} + 0.5\log K_{eq} \quad (7.100)$$

which represents a straight line of the form $y = mx + b$, where $x = \log a_{H_4SiO_4\,(aq)}$, $y = \log\,(a_{K^+}/a_{H^+})$, m (slope) = −2, and b (y − axis intercept) = $0.5\log K_{eq}$.

We still have a little way to go for completing the exercise because we need to determine the value of $\log K_{eq}$ in order to plot this reaction boundary, as shown in Fig. 7.13a. Using ΔG_f values for the chosen standard states (298.15 K and 1 bar) as listed in Table 7.7, the free energy change of the reaction can be calculated as

$$\begin{aligned}\Delta G^1_{r,298.15} = {} & [\Delta G^1_{f,298.15}(\text{kaol}) + 2\Delta G^1_{f,298.15}(K^+) \\ & + 4\Delta G^1_{f,298.15}(H_4SiO_{4\,(aq)})] \\ & - [\Delta G^1_{f,298.15}(\text{mic}) + 9\Delta G^1_{f,298.15}(H_2O) \\ & - 2\Delta G^1_{f,298.15}(H^+)]\end{aligned}$$

Table 7.7 Standard free energy of formation of relevant substances (Robie and Hemingway, 1995).

Mineral	Formula	$\Delta G^1_{f,\,298.15}$ (kJ mol^{-1})	Substance	Formula	$\Delta G^1_{f,\,298.15}$ (kJ mol^{-1})
Gibbsite	$Al(OH)_3$	−1154.9	Silicic acid	H_4SiO_4	−1307.8
Kaolinite	$Al_2Si_2O_5(OH)_4$	−3797.5	Water	H_2O (l)	−237.1
Microcline	$KAlSi_3O_8$	−3749.3	K^+		−282.5
Muscovite	$Al_2Si_2O_5(OH)_4$	−5598.8	H^+		0

$$= [-3797.5 + 2(-282.5) + 4(-1307.8)] - [2(-3749.3) + 9(-237.1)] = +38.8\,\text{kJ} = +38.8 \times 10^3\,\text{J}$$

We can obtain $\log K_{eq}$ from the relation $\Delta G^0_{r,T} = -2.3026\ RT \log K_{eq}$ (equation 5.37):

$$\log K_{eq} = \frac{-\Delta G^1_{r,\,298.15}}{2.3026\ RT} = \frac{38.8 \times 10^3}{(2.3026)(8.314)(298.15)} = -6.798$$

Substituting for K_{eq} in equation (7.100), we get

$$\log \frac{a_{K^+}}{a_{H^+}} = -2 \log a_{H_4SiO_4\,(aq)} - 3.399 \qquad (7.101)$$

We can now plot the microcline–kaolinite reaction boundary representing equilibrium among microcline, kaolinite, and the aqueous solution, as shown in Fig. 7.13a. For any solution composition that plots on this boundary, kaolinite and microcline can coexist in equilibrium with the solution. Only one of the two minerals is stable if the solution composition does not plot on this line – microcline is the stable phase if $\log a_{H_4SiO_4\,(aq)}$ in the solution is more than is required for equilibrium and kaolinite is the stable phase if the opposite is the case.

Kaolinite is not the only possible weathering product of microcline. Depending on the condition of the aqueous environment, microcline may weather to produce muscovite (used here as a proxy for the clay mineral illite), and both kaolinite and muscovite may be converted into gibbsite (see Box 7.3). The calculation procedure outlined above for microcline–kaolinite can be used to draw reaction boundaries for other mineral pairs in the system $K_2O–Al_2O_3–SiO_2–H_2O$ and delineate the stability field of each of the minerals (Fig. 7.13b).

The procedure outlined above can be used to draw activity–activity diagrams for other silicate systems. Many such diagrams are available in the published literature (e.g., Wood, 1994; Faure, 1998).

What sort of information can we retrieve from a *mineral stability diagram* such as Fig. 7.13b? The diagram tells us the chemistry of the aqueous solution (in terms of $a_{H_4SiO_4\,(aq)}$ and a_{K^+} / a_{H^+} ratio) required for the stability of individual minerals and for equilibrium between pairs of minerals in the system, the conditions necessary for the alteration of one mineral into another if they are not in equilibrium, and the change in solution chemistry that should accompany such reactions in a closed system with a small water : rock ratio (so that a reaction can effectively change the $a_{H_4SiO_4\,(aq)}$ and a_{K^+} / a_{H^+} ratio in the aqueous solution in accordance with the stoichiometry of the reaction).

Let us consider a simple example to illustrate qualitatively how it works in a closed system with a low water : rock ratio,

Box 7.3 Reactions and equations for the reaction boundaries in Fig. 7. 13

Relevant thermodynamic data for the following reactions are listed in Table 7.6.

Microcline–kaolinite

$$2KAlSi_3O_8 + 9H_2O + 2H^+ \Leftrightarrow Al_2Si_2O_5(OH)_4 + 2K^+ + 4H_4SiO_{4\,(aq)} \qquad (7.97)$$

$$\log \frac{a_{K^+}}{a_{H^+}} = -2 \log a_{H_4SiO_4\,(aq)} - 3.399 \qquad (7.101)$$

Microcline–muscovite* (proxy for the clay mineral illite)

$$2KAlSi_3O_8 + 12H_2O + 2H^+ \Leftrightarrow KAl_3Si_3O_{10}(OH)_2 + 2K^+ + 6H_4SiO_{4\,(aq)} \qquad (7.102)$$

$$\log \frac{a_{K^+}}{a_{H^+}} = -3 \log a_{H_4SiO_4\,(aq)} - 7.227 \qquad (7.103)$$

Kaolinite–gibbsite

$$Al_2Si_2O_5(OH)_4 + 5H_2O \Leftrightarrow 2Al(OH)_3 + 2H_4SiO_{4\,(aq)} \qquad (7.104)$$

$$\log a_{H_4SiO_4\,(aq)} - 5.045 \qquad (7.105)$$

Muscovite*–kaolinite

$$2KAl_3Si_3O_{10}(OH)_2 + 3H_2O + 2H^+ \Leftrightarrow 3Al_2Si_2O_5(OH)_4 + 2K^+ \qquad (7.106)$$

$$\log \frac{a_{K^+}}{a_{H^+}} = 4.257 \qquad (7.107)$$

Muscovite*–gibbsite

$$2KAl_3Si_3O_{10}(OH)_2 + 9H_2O + H^+ \Leftrightarrow 3Al(OH)_3 + K^+ + 3H_4SiO_{4\,(aq)} \qquad (7.108)$$

$$\log \frac{a_{K^+}}{a_{H^+}} = -3 \log a_{H_4SiO_4\,(aq)} - 10.88 \qquad (7.109)$$

*Because of a lack of good thermodynamic data for illite, muscovite is used as a proxy.

assuming that the thermodynamically favored reactions are not inhibited by sluggish kinetics. Suppose a solution of composition X1 (Fig. 7.13b) comes into contact with a rock containing microcline. Since the solution composition falls within the stability field of gibbsite, microcline will react with the solution to form gibbsite, consuming H^+ and releasing K^+ and silicic acid:

$$\underset{\text{microcline}}{KAlSi_3O_8} + 7H_2O + H^+ \Rightarrow \underset{\text{gibbsite}}{Al(OH)_3} + K^+ + 3H_4SiO_{4\,(aq)} \quad (7.110)$$

The reaction will result in an increase in $a_{H_4SiO_{4(aq)}}$ and a_{K^+} / a_{H^+} ratio, and the solution will follow the diagonal path X1 → X2 (the slope of which can be calculated from the stoichiometry of the reaction (7.102)). When the solution composition has changed to X2 on the gibbsite–kaolinite boundary, gibbsite, which is not a stable mineral in contact with microcline, will be converted to kaolinite by the reaction

$$\underset{\text{microcline}}{2KAlSi_3O_8} + \underset{\text{gibbsite}}{4Al(OH)_3} + 2H^+ \Rightarrow \underset{\text{kaolinite}}{3Al_2Si_2O_5(OH)_4} + 2K^+ + H_2O \quad (7.111)$$

As long as gibbsite is present, the reaction will occur at constant $a_{H_4SiO_4\,(aq)}$ (because equation (7.111) does not involve H_4SiO_4) but will result in an increase in the a_{K^+} / a_{H^+} ratio of the solution. This process is represented by segment X2 → X3, X3 being the solution composition corresponding to conversion of the last bit of gibbsite to kaolinite. Now the solution will follow a diagonal path such as X3 → X4 within the stability field of kaolinite, producing progressively more kaolinite by the reaction

$$\underset{\text{microcline}}{2KAlSi_3O_8} + 9H_2O + 2H^+ \Leftrightarrow \underset{\text{kaolinite}}{Al_2Si_2O_5(OH)_4} + 2K^+ + 4H_4SiO_{4\,(aq)} \quad (7.97)$$

which we considered earlier for the microcline–kaolinite reaction boundary. Point X4 marks the evolved solution composition with which both microcline and kaolinite will be in equilibrium.

7.10 Summary

1. Water is an effective solvent for ionic compounds; natural waters, which invariably contain dissolved ions (electrolytes), are aqueous electrolyte solutions.
2. Because of the nonrandom distribution of ions, aqueous electrolyte solutions are, in principle, nonideal, and $a_i = m_i \gamma_i$, where γ_i is the practical activity coefficient of the *i*th solute species.
3. The standard state adopted for most solutes is a hypothetical 1 molal (1 mol kg^{-1}) solution at 298.15 K and 1 bar for which $a_i = m_i = 1$.

Table 7.8 Equations commonly used to calculate activity coefficients of solutes in aqueous solutions.

	Equation	Most useful range (Langmuir, 1997)
Debye–Hückel limiting	$\log \gamma_i = -A\, z_i^2 \sqrt{I}$	$I \le 0.01$
Debye–Hückel extended	$\log \gamma_i = \dfrac{-A\, z_i^2 \sqrt{I}}{1 + \mathring{a}_i B \sqrt{I}}$	$I \le 0.1$
Davies (1962)	$\log \gamma_i = -A\, z_i^2 \left[\dfrac{\sqrt{I}}{1+\sqrt{I}} - 0.3I\right]$	$I \le 0.6$
Truesdell and Jones (1974)	$\log \gamma_i = \dfrac{-A\, z_i^2 \sqrt{I}}{1 + \mathring{a}_i B \sqrt{I}} + bI$	$I \le 2.0$

4. Equations proposed (Table 7.8) for the calculation of activity coefficients of individual ions require the calculation of the ionic strength of the solution (I), which incorporates both the charge (Z_i) and molality of all the solutes:

   $$I = \frac{1}{2} \sum_i m_i\, z_i^2$$

5. The solubility of a substance in an aqueous solution is defined as its concentration in the solution when it attains saturation with respect to the substance. The solubility is a function of many factors: temperature, pressure, partial pressures of gases (e.g., P_{CO_2}) involved in the equilibria, activities of other dissolved constituents in the solution (especially a_{H^+}), complex formation, and bacterial activity. Ore-forming solutions appear to transport metals as chloride and bisulfide complexes.
6. Whether a given solution is supersaturated, (just) saturated, or undersaturated with respect to a particular compound can be determined from a comparison of its ion activity product (IAP) with its solubility product ($K_{sp\,(compound)}$).
7. The pH of natural waters is largely controlled by the H^+ released by dissociation of dissolved H_2CO_3 acid and the H^+-consuming reactions with carbonate and silicate minerals (chemical weathering). The pH of rainwater is acidic because of dissolved CO_2; the pH of seawater is maintained at ~ 8 because of reactions that buffer the pH.
8. Concentrations of the various dissolved species in the $CaCO_3$–CO_2–H_2O system can be calculated from equilibrium constants of appropriate reactions and the charge balance equation (i.e., $\Sigma m_i\, z_i = 0$), if either the P_{CO_2} (open system) or the total carbonate content (closed system) is known.
9. The surface seawater is saturated with respect to both calcite and aragonite, but $CaCO_3$ precipitation in the oceans is restricted to some special environments, the precipitation of aragonite being favored by a mole Mg : Ca ratio above about 2. The most important process of removal of

$CaCO_3$ from seawater, however, is through the secretion of skeletal hard parts by certain organisms.

10. Mineral stability diagrams are useful for predicting the likely path of an aqueous solution of given composition as it evolves toward an equilibrium assemblage.

7.11 Recapitulation

Terms and concepts

Acidity
Acids
Alkalinity
Aqueous solution
Bases
Carbonate alkalinity
Carbonate compensation depth
Charge balance equation
Chemical weathering
Closed system
Common-ion effect
Complex ions
Congruent dissolution
Conservative species
Degree of dissociation
Dissociation constant
Electrolyte solution
Incongruent dissolution
Ion pairs
Ionic strength
Lysocline
Mass balance equation
Mineral stability diagram
Nonconservative species
Open system
pH buffer
Practical activity coefficient
Saturation index
Solubility
Solubility product
Stability constant
Standard state (aqueous solutes)

Computation techniques

- Activity coefficients of individual ionic solutes.
- Solubility of minerals in aqueous solutions.
- Acidity and alkalinity of an aqueous solution.
- Concentrations of dissolved species in the CO_2–H_2O system.
- Concentrations of dissolved species in the $CaCO_3$–CO_2–H_2O system.
- Calcite solubility as a function of P_{CO_2}, pH, and temperature.
- Activity–activity diagrams showing stability fields of minerals.

7.12 Questions

1. Using the extended Debye–Huckel equation, plot a graph showing the variation of activity coefficients of Ca^{2+}, Mg^{2+}, and SO_4^{2-} as a function of ionic strength between 0.0 and 0.1. What general trends, if any, can you infer from this plot?
2. What is the concentration of HCO_3^- in groundwater with pH = 8.2 and $m_{Ca^{2+}} = 10^{-2}$ at 25°C? Assume that the groundwater is in equilibrium with calcite, and $a_i = m_i$.

   $K_{sp}(\text{calcite}) = 10^{-8.48}$, and $K_{HCO_3^-} = 10^{-10.33}$.

3. A saturated solution of silver chromate (Ag_2CrO_4) at 25°C contains 0.0435 g of (Ag_2CrO_4) per 1 kg of water. Calculate the solubility and the solubility product constant of (Ag_2CrO_4). Assume that the solution is ideal.
4. The solubility product constant for Ag_2SO_4 at 25°C and 1 bar is 1.7×10^{-5}. Calculate the solubility of Ag_2SO_4, assuming the aqueous solution to be ideal, and express it in grams per kg of water. Grammolecular weight of Ag_2SO_4 = 311.17.
5. (a) Is the seawater of composition given in in Table 7.1 undersaturated, saturated, or supersaturated with respect to halite (NaCl) at 25°C and 1 bar? Use the Truesdell–Jones equation, the parameters for which are given in Tables 7.2 and 7.3, to calculate activity coefficients of Na^+ and Cl^-. $\Delta G^1_{f,\,298.15}$ values (in kJ mol^{-1}) for the calculation of $K_{sp\,(NaCl)}$ are as follows (Robie and Hemingway, 1995): $Na^+ = -262.0$, $Cl^- = -131.2$, and NaCl = −384.2.
   (b) To what extent must the seawater evaporate to start precipitating halite, assuming that the relative ratios of dissolved species in the seawater stay the same?
6. (a) What is the alkalinity of seawater of the composition listed in Table 7.1? Assume pH (seawater) = 8.3.
   (b) How will the alkalinity of this seawater be affected by the addition of small amounts of the following substances: (i) $NaHCO_3$, (ii) NaCl, (iii) HCl, and (iv) $MgCO_3$?
7. A vein deposit contains some amounts of gypsum ($CaSO_4.2H_2O$) and barite ($BaSO_4$). From a consideration of solubilities, which is a more plausible interpretation: partial replacement of gypsum by barite or partial replacement of barite by gypsum? Assume $a_i = m_i$ and ignore ion–pair formation.

   $K_{sp\,(gypsum)} = 10^{-4.59}$ and $K_{sp\,(barite)} = 10^{-10.0}$.

   [Hint: the interaction between barite and gypsum may be represented by the reaction $Ba^{2+} + CaSO_4.2H_2O \Leftrightarrow Ca^{2+} + BaSO_4 + H_2O$.]
8. A solution not in contact with any CO_2-bearing gas phase contains 0.1 *m* total carbonate and has a pH of 6 at 25°C. If 1 ml of 6 *m* HCl acid is added to a liter of this solution,

what would be the final pH of the solution. Is this solution an effective pH buffer? Assume $a_i = m_i$ and that 1 L of the solution weighs 1 kg.

9. Using the standard state (298.15 K, 1 bar) enthalpy data given below, show that the formation of kaolinite from microcline (reaction 7.89) will be favored by an increase in temperature to 308.15 K (35°C). [Hint: Calculate K_{eq} using van't Hoff equation.]

	Microcline	Kaolinite	H_2O (l)	H_4SiO_4	K^+	H^+
$\Delta H^1_{f,\,298.15}$ (kJ.mol^{-1})	−3974.6	−4119.0	−285.8	−1460.0	−252.1	0.00

10. Show that the equilibrium constant for the hydrolysis reaction, $Fe^{3+} + H_2O \Leftrightarrow FeOH^{2+} + H^+$ ($K_{hydrolysis}$), is related to the dissociation constant of water (K_w) and the dissociation constant of $FeOH^{2+}$ ($K_{dissociation}$) by the equation

$$K_{hydrolysis} = \frac{K_w}{K_{dissociation}}$$

11. Explain, with appropriate reactions, why waters in carbonate terranes, even without dissolved CO_2, would generally tend to be alkaline.

12. Calculate the solubility of calcite in water at 10°C and 1 bar in equilibrium with $CO_{2\,(g)}$ at $10^{-2.5}$ bar partial pressure. What is the equilibrium pH of the solution? Assume $a_i = m_i$. Relevant equilibrium constants at 10°C and 1 bar are (Plummer and Busenberg, 1982): $K_{sp\,(calcite)} = 10^{-8.41}$; $K_{CO_2} = 10^{-1.27}$, $K_{H_2CO_3\,(aq)} = 10^{-6.46}$, $K_{HCO_3^-} = 10^{-10.49}$, $K_w = 10^{-14.53}$.

13. Calculate the total acidity and total alkalinity of a carbonic acid solution for which Carb (total) = 3×10^{-3} mol kg^{-1}, $T = 20$°C, and pH = 6.2. The relevant equilibrium constants at 20°C are as follows: $K_{H_2CO_3} = 10^{-6.38}$; $K_{HCO_3^-} = 10^{-10.38}$; and $K_w = 10^{-14.163}$. Assume $a_i = m_i$.

14. The bottom water of a lake has a pH of 7 and it contains 34 mg of H_2S per liter of water. Calculate the minimum concentration of Fe^{2+} in the lake water required to initiate the precipitation of amorphous FeS by the reaction $Fe^{2+} + S^{2-}$ = FeS (amorphous). The dissociation constants of FeS (amorphous), H_2S, and HS^- at 25°C are 10^{-19}, $10^{-6.96}$, and 10^{-15}, respectively. Assume that $a_i = M_i$, $a_{FeS\,(amorphous)} = 1$, and the temperature of the bottom water = 25°C.

15. (a) Modify Figure 7.13b by incorporating the stability field of pyrophyllite ($Al_2Si_4O_{10}(OH)_2$) and the saturation line for crystalline silica. Reactions for pyrophyllite–kaolinite and microcline–pyrophyllite equilibria are:

$$\underset{\text{pyrophyllite}}{0.5Al_2Si_4O_{10}(OH)_2} + 2.5H_2O \Leftrightarrow \underset{\text{kaolinite}}{0.5Al_2Si_2O_5(OH)_4} + H_4SiO_4$$

$$\underset{\text{microcline}}{KAlSi_3O_8} + H^+ + 2H_2O = \underset{\text{pyrophyllite}}{0.5Al_2Si_4O_{10}(OH)_2} + K^+ + H_4SiO_4.$$

$\Delta G^1_{f,\,298.15}$ (pyrophyllite) = −5266.1 kJ mol^{-1}. Values of the other species are listed in Table 7.6.

(b) Give a qualitative description of a likely fluid reaction path for chemical weathering of microcline by groundwater with $\log(a_{K^+}/a_{H^+}) = 0$ and $\log(a_{H_4SiO_4}) = -5$.

Assume a closed system and ignore the possible precipitation of amorphous silica. Show the fluid reaction path on your stability diagram.

(a) If the CO_2 concentration in the atmosphere rises to 700 ppm because of increased burning of coal, what would be the likely pH of rainwater? Assume that atmospheric CO_2 behaves as an ideal gas and that its mole fraction of CO_2 in the atmosphere can be approximated by its volume fraction. For all other carbonate species, $a_i = m_i$. Use dissociation constants for 25°C and 1 bar as given in the text.

(b) What would be the pH of this water if it equilibrated with calcite?

17. The experimental data of Morey *et al.* (1962) on the variation of quartz solubility in pure water as a function of temperature can be fitted to a linear regression equation,

$$y = -1.2014x + 2.0725$$

where y = log (concentration of $H_4SiO_{4\,(aq)}$ in the solution in mol kg^{-1}), and $x = 1000/T$ (K). Calculate the ΔH of the quartz solubility reaction at 298.15 K and 1 bar, assuming $(\Delta C_p)_r = 0$ and $a_i = m_i$.

[Hint: For the quartz solubility reaction, SiO_2 (qz) + $2H_2O = H_4SiO_4$, and $a_{H_4SiO_4\,(aq)} = K_{SiO_2\,(\alpha-qz)}$.]

8 Oxidation–Reduction Reactions

Although the ferrous/ferric ratio has long been used as a monitor of oxygen fugacity in rocks, this ratio is not a simple function of oxygen fugacity. It is true that this ratio monitors oxygen fugacity in melts or glasses, and it may be used as an indicator of oxygen fugacity in volcanic rocks that can be safely assumed to be free of cumulate minerals. In most rocks, however, oxygen fugacity is monitored by mineral equilibria and these equilibria are dependent more on mineral composition rather than simply on ferrous/ferric ratio.

Frost (1991)

Most of the chemical reactions we have considered so far, such as dissolution and precipitation of carbonate and silicate minerals, are acid–base processes involving transfer of protons (H^+). In contrast, *oxidation–reduction reactions* (or *redox reactions*, for brevity) involve transfer of electrons (e^-) between species having different valence states (or oxidation numbers), such as Cu^{+1}–Cu^{2+}, Fe^{2+}–Fe^{3+}, Mn^{2+}–Mn^{3+}–Mn^{4+}, U^{4+}–U^{6+}, etc. The valence state of an element significantly affects its geochemical behavior in seawater, surface water, groundwater, and hydrothermal solutions. For example, Fe^{2+} is much more soluble in aqueous solutions compared to Fe^{3+}, the higher valence species, but the opposite is the case between U^{4+} and U^{6+}. Thus, iron is commonly transported as dissolved Fe^{2+} species in aqueous solutions but precipitated as a Fe^{3+}-compound (e.g., hematite, Fe_2O_3; goethite, α–FeOOH). By contrast, uranium is transported mostly as dissolved U^{6+} species but precipitated as a U^{4+}-compound (e.g., uraninite, UO_2; coffinite, $USiO_4$). Oxidation–reduction reactions are important in all geologic processes: magmatic, diagenetic, metamorphic, chemical weathering (e.g., the familiar reddish-yellow rust on iron garden tools), and formation of mineral deposits. In the realm of industrial applications, oxidation–reduction reactions are the basis of electrochemistry (chemical changes produced by an electric current and the production of electricity by chemical reactions), recovery of metals from many types of ores and wastes, and treatment of municipal sewage and industrial effluents.

8.1 Definitions

Reactions resulting from addition of oxygen (e.g., $2Ni + O_2 = 2NiO$) evidently represent oxidation. However, a reaction does not have to involve oxygen to qualify as an oxidation or a reduction reaction.

> Oxidation is defined as a reaction involving the loss of one or more electrons and reduction as a reaction involving the gain of one or more electrons.

Thus, oxidation involves an increase in *oxidation state*, and reduction a decrease in oxidation state. For the purpose of illustration, let us consider the oxidation of magnetite ($Fe^{2+}O.Fe_2^{3+}O_3$) to hematite ($Fe_2^{3+}O_3$) in the presence of water. Following the convention used by Garrels and Christ (1965, p. 132) and subsequently by many other authors (e.g., Krauskopf and Bird, 1995; Faure, 1998; McSween *et al.*, 2003; Appelo and Postma, 2005; Walther, 2009), this reaction may be written as

$$2(Fe^{2+}O.Fe_2^{3+}O_3)_{(s)} + H_2O_{(l)} \Rightarrow 3Fe_2^{3+}O_{3(s)} + 2H^+ + 2e^- \tag{8.1}$$

Introduction to Geochemistry: Principles and Applications, First Edition. Kula C. Misra.

The essence of this reaction is the oxidation of ferrous ion (Fe^{2+}) to ferric ion (Fe^{3+}):

Oxidation: $Fe^{2+} \Rightarrow Fe^{3+} + e^-$
reducing agent (8.2)
electron donor

As electrons cannot exist in nature as free species, they must be consumed by some complementary reduction reaction such as:

Reduction: $2H^+ + 2e^- \Rightarrow H_2$
oxidizing agent (8.3)
electron acceptor

In reaction (8.1), an *oxidation half-reaction*, magnetite (or Fe^{2+}) is the *reducing agent* or *electron donor* (it loses electrons, reduces another species, and gets oxidized in the process); in reaction (8.3), a *reduction half-reaction*, H^+ is the *oxidizing agent* or *electron acceptor* (it gains electrons, oxidizes another species, and gets reduced in the process). In general, metals (elements with low values of electronegativity) are electron donors and nonmetals (elements with high values of electronegativity) are electron acceptors. Some elements, such as carbon and sulfur, can be either electron donors or electron acceptors. Oxygen is the most common electron acceptor, hence the term oxidation.

In natural environments, reactions involving loss or gain of electrons do not occur in isolation because electrons are not released unless they can be accepted by some other species in the environment. Thus, half-reactions are always coupled, with one species getting oxidized while another species is being reduced. The two can be added to represent the whole oxidation–reduction (or *redox*) reaction, which for the present example may be written either as

$$2(Fe^{2+}O.Fe_2^{3+}O_3)_{(s)} + H_2O_{(l)} \Rightarrow 3Fe_2^{3+}O_{3(s)} + H_{2\,(g)} \qquad (8.4)$$

or simply as

$$2Fe^{2+} + 2H^+ \Rightarrow 2Fe^{3+} + H_2 \qquad (8.5)$$

neither of which shows explicitly the transfer of electrons. It can be shown easily that the free energy change for the full reaction (8.4) is the sum of free energy changes for the half-reactions (8.1) and (8.3). As the coupled reaction does not contain any electrons, the value of $\Delta G_f(e^-)$ is immaterial; it is conveniently assigned a value of zero (Garrels and Christ, 1965). Many applications in geochemistry, however, do not require a consideration of the full oxidation–reduction reaction, and ΔG_r values of half-reactions are calculated with the assumption that $\Delta G_f(e^-) = 0$.

Oxidation–reduction reactions can be quite complex and difficult to balance. Guidelines for balancing such reactions are listed in Box 8.1.

Box 8.1 Guidelines for balancing oxidation–reduction reactions

(1) Write the reaction with known reactants on the left and known products on the right.
(2) Assign appropriate *valence number* (*oxidation number*) to each element involved in the reaction, making sure that the net charge of a neutral molecule works out to be zero and that of a complex ion equals the charge it should have. The valence number of an element is defined as the electrical charge an atom would acquire if it formed ions in solution. An increase in valence number corresponds to oxidation, and a decrease to reduction. For elements that form dominantly ionic bonds, valence number is the same as the actual state of the element in ionic form. For elements that dominantly or exclusively form covalent bonds, valence number is a hypothetical concept.
The valence number of all elements in pure form, including polyatomic elements (such as H_2, O_2) is 0; for hydrogen it is +1 in compounds, except in metal hydrides (e.g., NaH, CaH_2) where it is −1; for oxygen it is −2, except in peroxides, where it is −1.
(3) Adjust the stoichiometric coefficients to balance all elements, except oxygen and hydrogen, on both sides of the reaction.
(4) Balance the number of oxygen atoms by adding H_2O.
(5) Balance the number of hydrogen atoms by adding H^+.
(6) Balance the charge by adding electrons (e^-).

More detailed guidelines are discussed in Krauskopf and Bird (1995, pp. 614–617) and Faure (1998, pp. 226–227).

8.2 Voltaic cells

A *voltaic* (or *galvanic*) *cell* is a type of electrochemical cell in which spontaneous chemical reactions produce flow of electricity through an external circuit. The other type of electrochemical cell is called the *electrolytic cell* in which electrical energy from an external source causes nonspontaneous chemical reactions to occur, such as dissociation of molten NaCl to produce Na metal. Both types of cells contain electrodes that provide surfaces on which oxidation and reduction half-reactions can take place. The electrode at which oxidation occurs is called the *anode* and the electrode at which reduction occurs is called the *cathode*. Here we will focus on voltaic cells. The batteries commonly used in flashlights, watches and clocks, photographic equipment, toys, portable appliances, automobiles, etc., are different kinds of voltaic cells. To gain a better understanding of oxidation and reduction reactions, let us examine how a voltaic cell works.

8.2.1 *Zinc–hydrogen cell*

Consider a zinc–hydrogen voltaic cell consisting of two compartments (or half-cells) connected by a "salt bridge" (Fig. 8.1), the whole system being placed at 298.15 K (25°C) and 1 bar

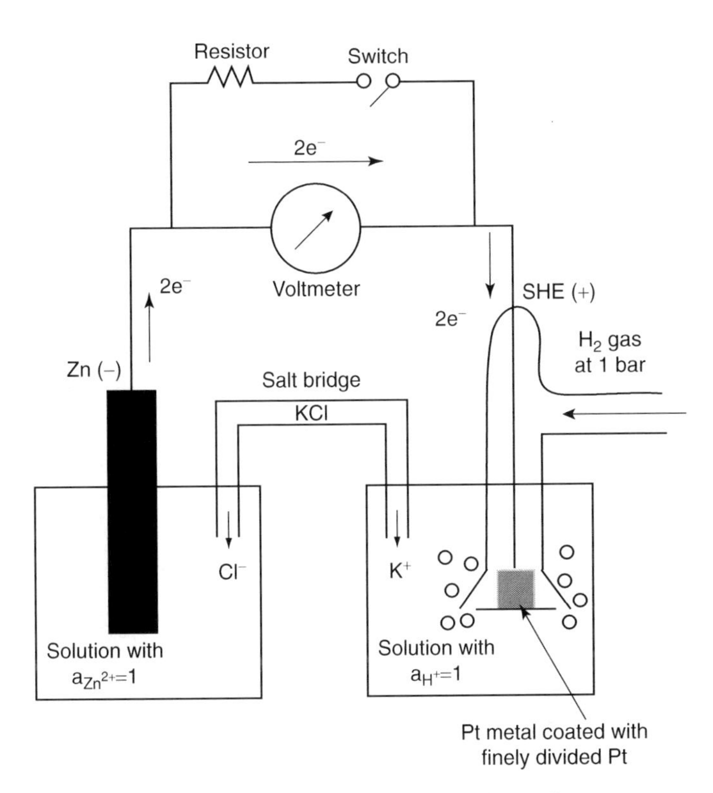

Fig. 8.1 Schematic diagram of a zinc–hydrogen voltaic cell. SHE = standard hydrogen electrode.

total pressure. The left compartment contains a piece of pure zinc metal (Zn electrode) immersed partly in a Zn^{2+} solution (say $ZnCl_2$). The right compartment contains a hydrogen electrode, which is created by bubbling H_2 gas at unit fugacity (i.e., $f_{H_2} = 1$ bar) over a piece of platinum metal coated with fine-grained platinum crystals and immersed partly in an acid such as HCl with $a_{H^+} = 1$. The platinum acts as a catalyst (see section 9.4) for the electrochemical reactions in the hydrogen half-cell. The salt bridge is a glass tube filled with a solution of a salt, such as KCl, which has a very low junction potential (potential across the junction of the two half-cells) because of nearly equal mobilities of K^+ and Cl^- ions. In addition to preventing physical mixing of solutions in the two compartments, a salt bridge allows electrical contact between the two cells, and maintains the electrical neutrality in each half-cell through diffusion of K^+ into the hydrogen half-cell and of Cl^- into the zinc half-cell.

If we connect the two electrodes of the cell by a conducting wire, a voltmeter in the circuit can be used to measure the electrical potential difference between the two electrodes. An ideal voltmeter is designed to measure potential differences without allowing electricity to pass through the meter, and modern voltmeters come very close to that ideal. The limiting value of the potential difference for zero current in the circuit (i.e., at the very beginning of the process, before any significant change in activities has occurred because of reactions in the cell), is called the *cell voltage* or the *electromotive force* (*EMF*) of the cell (E_{Zn-H_2}), which is the maximum voltage the cell is capable of delivering. In practice, the cell EMF is determined by measuring the voltage that counterbalances the potential difference between the electrodes, resulting in a stoppage of electron flow from the anode to the cathode. The EMF depends on factors that affect the activities of the constituents in the half-cell reactions. These include the purity of the electrode metals, temperature and pressure, concentration of Zn and HCl in the two solutions, and fugacity of the H_2 gas being bubbled through the hydrogen electrode compartment.

To permit flow of electricity between the two electrodes, let us insert a "load," such as a resistor, into the circuit (Fig. 8.1). If we now close the circuit by the switch, a current will flow through the circuit (because electrons will flow from the anode to the cathode). Whatever the value of E_{Zn-H_2}, it will slowly decrease as the cell progresses toward electrochemical equilibrium and eventually reach a value of 0.0 volt at equilibrium. The flow of electrons will stop at this point.

Our observations can be explained by the oxidation and reduction reactions occurring in the two half-cells. In the zinc half-cell, Zn spontaneously oxidizes to $Zn^{2+}_{(aq)}$, releasing electrons:

$$Zn_{(s)} \Rightarrow Zn^{2+}_{(aq)} + 2e^- \text{ (oxidation, anode)} \tag{8.6}$$

The excess electrons produced at the zinc electrode (*anode*) flow spontaneously through the wire to the hydrogen electrode (*cathode*); i.e., electric current flows from the cathode (+) to the anode (−). In the hydrogen half-cell, $H^+_{(aq)}$ is reduced to $H_{2\,(g)}$ by consuming the electrons:

$$2H^+_{(aq)} + 2e^- \Rightarrow H_{2\,(g)} \text{ (reduction, cathode)} \tag{8.7}$$

As the reactions proceeed, the flow of electrons from the anode to the cathode would produce a charge imbalance in both compartments if it were not for the salt bridge, which diffuses K^+ ions into the hydrogen half-cell and Cl^- ions into the zinc half-cell, and completes the electrical circuit.

The total or whole-cell reaction is the sum of the two half-cell (or electrode) reactions:

$$Zn_{(s)} + 2H^+_{(aq)} \Rightarrow Zn^{2+}_{(aq)} + H_{2\,(g)} \tag{8.8}$$

As the reactions progress, $a_{Zn^{2+}(aq)}$ will increase (at the expense of the zinc metal bar) in the solution around the zinc electrode, whereas $a_{H^+(aq)}$ will decrease in the hydrogen half-cell because of production of $H_{2\,(g)}$ at the hydrogen electrode, and the magnitude of the potential difference between the two electrodes will progressively decrease. Eventually, reaction (8.8) may reach equilibrium; at this state the potential difference will drop to 0.0 volt and there will be no flow of electrons through the circuit. As shown below, we can calculate the condition for this equilibrium from the equilibrium constant of the reaction.

The equilibrium constant of reaction (8.8) at standard state temperature and pressure (298.15 K and 1 bar total pressure) can be calculated from molar ΔG_f^1 values given in Robie and Hemingway (1995):

$$\Delta G^1_{r,\,298.15} = \Delta \bar{G}^1_{f,298.15}\,(Zn^{2+}_{(aq)}) + \Delta \bar{G}^1_{f,298.15}\,(H_{2\,(g)}) - \Delta \bar{G}^1_{f,298.15}\,(Zn_{(s)}) - 2\Delta \bar{G}^1_{f,298.15}\,(H^+_{(aq)})$$
$$= -147.3 + 0 - 0 - 0 = -147.3\,\text{kJ} = -147.3 \times 10^3\,\text{J}$$

$$\log K_{eq} = \frac{-\Delta G^1_{r,298.15}}{2.3026\,RT} = \frac{147.3 \times 10^3}{5707.78} = 25.81$$

Assuming the H_2 gas to be ideal (i.e., $f_{H_2} = 1$ bar),

$$K_{eq} = \frac{(a_{Zn^{2+}})\,(f_{H_2})}{(a_{Zn})\,(a_{H^+})^2} = \frac{(a_{Zn^{2+}})\,(1)}{(1)\,(a_{H^+})^2} = \frac{a_{Zn^{2+}}}{(a_{H^+})^2} = 10^{25.81}$$

The negative value of $\Delta G^1_{r,\,298.15}$ indicates that reaction (8.8) will proceed spontaneously to the right and the magnitude of K_{eq} indicates that the system will attain equilibrium (i.e., $E_{Zn\text{-}H_2} = 0$) when the activity ratio reaches a value of $10^{25.81}$.

8.2.2 *Standard hydrogen electrode and standard electrode potential*

There is no way to measure the potential of individual electrodes in isolation, but we can determine their potentials by following certain conventions and accepting the premise that the EMF of a cell is the sum of the potentials of its two half-cells (Anderson and Crerar, 1993). The EMF of the Zn–H_2 cell, $E_{Zn\text{-}H_2}$, then, is:

$$E_{Zn\text{-}H_2} = E_{Zn} + E_{H_2} \tag{8.9}$$

where E_{Zn} and E_{H_2} are the potentials of the Zn and H_2 half-cells, respectively. Let us stipulate that the EMF of the Zn–H_2 cell is being measured at the instant when the half-cell reactions are occurring with the constituents in their standard states, which we choose as follows: 298.15 K, 1 bar total pressure, 1 N HCl acid so that $a_{H^+} = 1$, 1 m $ZnCl_2$ solution so that $a_{Zn^{2+}} = 1$, pure zinc metal so that $a_{Zn\,(s)} = 1$, and $f_{H_2} = 1$ bar. The hydrogen electrode in the standard state is called the *standard hydrogen electrode* (SHE) and, by convention, its potential, denoted as E^0_{SHE} (the superscript "0" stands for standard conditions), is assigned a value of zero. The $E^0_{SHE} = 0$ convention is consistent with the fact that, for the hydrogen electrode reaction (reaction 8.7), $\Delta G^1_{r,298.15} = 0$ because, again by convention, $\Delta G^1_{f,\,298.15} = 0$ for $H_{2\,(g)}$, $H^+_{(aq)}$, and e^-. With the stated conventions, the zinc *electrode standard (state) potential*, denoted as E^0_{Zn}, is the same as the standard (state) EMF of the Zn–H_2 cell, $E^0_{Zn\text{-}H_2}$:

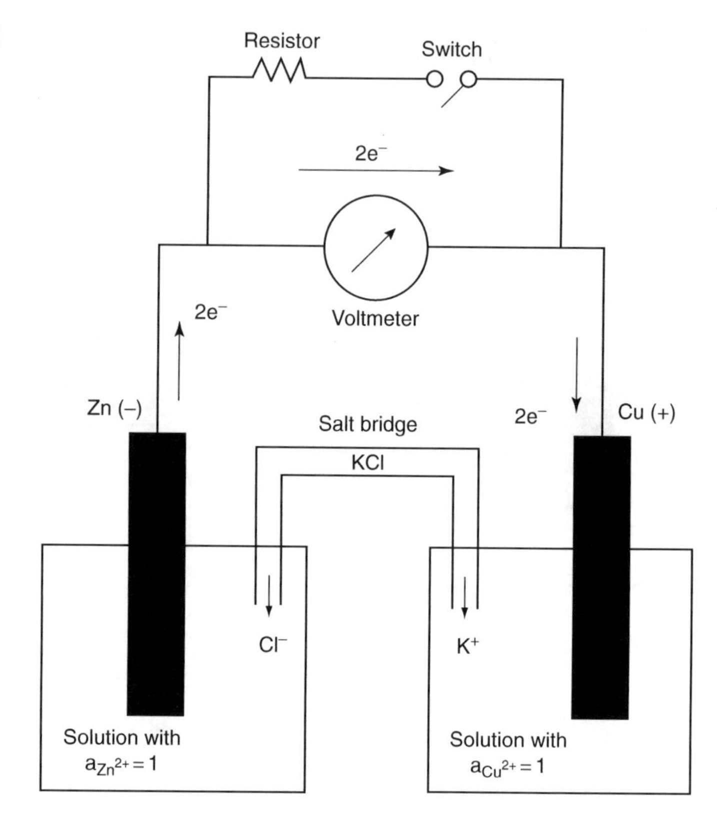

Fig. 8.2 Schematic representation of a Zn–Cu voltaic cell.

$$E^0_{Zn\text{-}H_2} = E^0_{Zn} + E^0_{SHE} = E^0_{Zn} + 0 = E^0_{Zn} \tag{8.10}$$

The measured magnitude of $E^0_{Zn\text{-}H_2}$ is 0.76 volts, which gives $E^0_{Zn} = -0.76$ volts. The negative sign for E^0_{Zn} originates from the negative sign of $\Delta G^1_{r,\,298.15}$ for reaction (8.8) as calculated above and reminds us that the zinc electrode standard potential is 0.76 volts more negative than that of the SHE. The prevalent practice of using the symbols E and E^0 for half-reactions as well as whole reactions can be a little confusing, so the context should be made clear with appropriate subscripts.

Equation (8.10) is the basis for determining the standard potential of any half-cell (or electrode) reaction.

8.2.3 *Zinc–copper cell*

To further elaborate on the application of the SHE concept to voltaic cells, let us consider the historical Daniel cell, a Zn–Cu voltaic cell (Fig. 8.2) consisting of two half-cells separated by a salt bridge. One of the half-cells contains a piece of pure zinc metal immersed partly in $ZnSO_4$ electrolyte solution and the other half-cell contains a piece of pure copper metal immersed partly in $CuSO_4$ electrolyte solution. Let us stipulate that the two half-cells are in standard state condition ($T = 298.15$ K, $P = 1$ bar, $a_{Zn^{2+}} = 1$, and $a_{Cu^{2+}} = 1$) at the start. If the EMF of this cell, $E^0_{Zn\text{-}Cu}$, is measured following the procedure described earlier for the zinc–hydrogen cell (section 8.2.1), it will turn out to be 1.10 volts, which will decrease with time and ultimately drop to 0.0 volt at equilibrium.

The potential difference between the two electrodes arises because of electrochemical reactions in the two half-cells. In the zinc half-cell, Zn metal (the reducing agent) goes into solution as oxidized Zn^{2+} cations (thereby increasing $a_{Zn^{2+}}$ in the solution):

$$Zn_{(s)} \Rightarrow Zn^{2+}_{(aq)} + 2e^- \quad \text{(oxidation, anode)} \qquad (8.11)$$

The excess electrons flow through the external circuit from the zinc electrode (anode) to the copper electrode (cathode). In the copper half-cell, Cu^{2+} cations (produced by the dissociation of $CuSO_4$, $CuSO_4 \Rightarrow Cu^{2+} + SO_4^{2-}$) combine with electrons to precipitate Cu metal (the reduced product) on the copper electrode (thereby decreasing $a_{Cu^{2+}}$ in the solution):

$$Cu^{2+}_{(aq)} + 2e^- \Rightarrow Cu_{(s)} \quad \text{(reduction, cathode)} \qquad (8.12)$$

To maintain electrical charge balance and complete the circuit, two Cl^- anions from the salt bridge migrate into the anode solution for every Zn^{2+} cation formed, and two K^+ cations migrate into the cathode solution to compensate for every Cu^{2+} cation reduced to Cu metal. Some Zn^{2+} cations from the anode half-cell and some SO_4^{2-} anions from the cathode half-cell also migrate into the salt bridge, but neither Cl^- nor K^+ ions are oxidized or reduced in preference to the Zn metal or Cu^{2+} cations. The whole-cell reaction is obtained simply by adding the two half-cell reactions (8.11 and 8.12):

$$Zn_{(s)} + Cu^{2+}_{(aq)} \Rightarrow Zn^{2+}_{(aq)} + Cu_{(s)} \qquad (8.13)$$

The published value of the standard electrode potential (relative to the SHE) of the zinc electrode, E^0_{Zn}, is –0.76 volts and that of the copper electrode, E^0_{Cu}, is –0.34 volts, so that the EMF of the Zn–Cu cell, E^0_{Zn-Cu}, can be calculated as:

$$E^0_{Zn-Cu} = E^0_{Zn} + E^0_{Cu} = (-0.76) + (-0.34) = -1.10 \text{ volts} \qquad (8.14)$$

which is the same in magnitude as the measured value of E^0_{Zn-Cu}, the negative sign of the calculated value originating from the negative sign of the free energy change for reaction (8.13) in the standard state (see Example 8–2). We will see a little later (section 8.3) that standard electrode potentials, such as E^0_{Zn} and E^0_{Cu}, can be calculated entirely from thermodynamic relations (equation 8.21), without having to set up voltaic cells.

Example 8–1: Calculation of the EMF of an electrochemical cell in the standard state

We can calculate the EMF of a voltaic cell in the standard state from E^0 values of the relevant electrode reactions (Table 8.1). For example, the standard state EMF of a Mn–Ag voltaic cell can be calculated as follows:

Table 8.1 An electromotive series listing calculated* standard electrode potentials (E^0) of selected half-reactions arranged in order of decreasing strengths as reducing agents (or of increasing strengths) as oxidizing agents.

Reducing agent	Half-reaction	E^0 (volts)
K	$K \Rightarrow K^+ + e^-$	–2.93
Ca	$Ca \Rightarrow Ca^{2+} + 2e^-$	–2.87
Na	$Na \Rightarrow Na^+ + e^-$	–2.71
Mg	$Mg \Rightarrow Mg^{2+} + 2e^-$	–2.36
Mn	$Mn \Rightarrow Mn^{2+} + 2e^-$	–1.18
Zn	$Zn \Rightarrow Zn^{2+} + 2e^-$	–0.76
Fe	$Fe \Rightarrow Fe^{2+} + 2e^-$	–0.47
Ni	$Ni \Rightarrow Ni^{2+} + 2e^-$	–0.24
Pb	$Pb \Rightarrow Pb^{2+} + 2e^-$	–0.13
H_2	$H_2 \Rightarrow 2H^+ + 2e^-$	0.00 (SHE)
Cu^+	$Cu^+ \Rightarrow Cu^{2+} + e^-$	+0.16
Cu	$Cu \Rightarrow Cu^{2+} + 2e^-$	+0.34
Cu	$Cu \Rightarrow Cu^+ + e^-$	+0.52
Fe^{2+}	$Fe^{2+} \Rightarrow Fe^{3+} + e^-$	+0.76
Ag	$Ag \Rightarrow Ag^+ + e^-$	+0.80
Hg	$Hg \Rightarrow Hg^{2+} + 2e^-$	+0.85
Cl^-	$2Cl^- \Rightarrow Cl_2 + 2e^-$	+1.36

*$E^0 = \Delta G^0_r / n\,F$ (equation 8.22); ΔG^0_r calculated using $\Delta G^1_{f,\,298.15}$ values in Robie and Hemmingway (1995) listed in Appendix 6.

Mn electrode half – reaction: $Mn \Rightarrow Mn^{2+} + 2e^-$
$E^0_{Mn} = -1.18$ volts

Ag electrode half – reaction: $2Ag^+ + 2e^- \Rightarrow 2Ag$
$E^0_{Ag} = -0.80$ volts

Whole – cell reaction: $E^0_{Mn-Ag} = E^0_{Mn} + E^0_{Ag} = (-1.18) + (-0.80) = -1.98$ volts

Note that the Ag electrode reaction is multiplied by a factor of two to balance the electrons released by the Mn electrode reaction, but not the corresponding E^0 value. This is because potential and, therefore, potential difference is not an extensive property. The effect of stoichiometric coefficients in an electrochemical reaction is included in the calculation of free energy change for the reaction (see equations 8.20 and 8.21).

8.2.4 *Electromotive series*

The negative sign of the free energy change for reaction (8.13) (see Example 8–2) indicates that the reaction will proceed spontaneously to the right. This means that Zn atoms are able to force Cu atoms to accept their two valence electrons rather than allow Cu atoms to force Zn atoms to accept their two valence electrons. In other words, Zn is capable of displacing Cu from a $CuSO_4$ solution, whereas there is no thermodynamic reason for Cu displacing Zn from a $ZnSO_4$ solution;

i.e., Zn is a stronger reducing agent than Cu. If we experimentally measured or thermodynamically calculated the EMF of a number of voltaic cells similar to the zinc–copper cell (such as zinc–lead, iron–copper, etc.), we would find that the metals can be arranged in a list according to their relative strengths as reducing agents (or their ability to displace one another from solution) as judged by the magnitude and sign of the EMF generated by different metal pairs. Such a list, ordered in terms of standard electrode potentials, is called an *electromotive series*.

A partial list, with calculated standard potentials, is presented in Table 8.1. For the sake of consistency, all elements in this list are considered to be reducing agents of varying strength; i.e., the corresponding electrode reactions are written as oxidation reactions (with electrons on the right side). If we choose to write any of these as a reduction reaction (with electrons on the left side), its standard potential will have the opposite sign. For example, the standard potential of the $Cu \Rightarrow Cu^{2+} + 2e^-$ electrode reaction is given as +0.34 volts in Table 8.1, but $E^0_{Cu} = -0.34$ volts, as we have considered earlier for reaction (8.12) written as a reduction reaction. Also, note that $E^0_{SHE} = 0$ provides a convenient reference for dividing the elements into two broad groups. All of the reducing agents above the H_2 electrode are stronger reducing agents than H_2, and their half-reactions have negative standard electrode potentials; those below the H_2 electrode are weaker reducing agents than H_2 and the related half-reactions have positive standard electrode potentials.

An important industrial application of the concept of the electromotive series is electroplating of metals for corrosion protection. For example, iron metal can be protected from oxidation by a coating of the relatively less easily oxidized copper metal (Table 8.1).

8.2.5 *Hydrogen–oxygen fuel cell*

Power generation by combustion of fossil fuels (petroleum, natural gas, or coal) involves the conversion of chemical energy stored in the fuels to heat energy, which is then converted partially into mechanical energy (e.g., a steam turbine), and finally into electrical energy by means of a generator. The process is inherently inefficient, and the combustion generates byproducts (e.g., CO_2, CO, SO_2, nitrogen oxides) that are deleterious to the environment. A fuel cell is a voltaic cell that is designed to produce electricity from oxygen gas and hydrogen gas through electrochemical reactions, without combustion and with much better efficiency.

In principle, a fuel cell is a simple device, consisting of an anode and a cathode, which are separated by an electrolyte (a specialized polymer or other substance that allows ions to pass but blocks electrons) and joined by an external circuit (Fig. 8.3). Pressurized hydrogen (the fuel) is fed into the anode compartment and pressurized oxygen (the oxidizing agent), commonly from air, into the cathode compartment. The anode and the cathode are made of porous carbon impregnated with

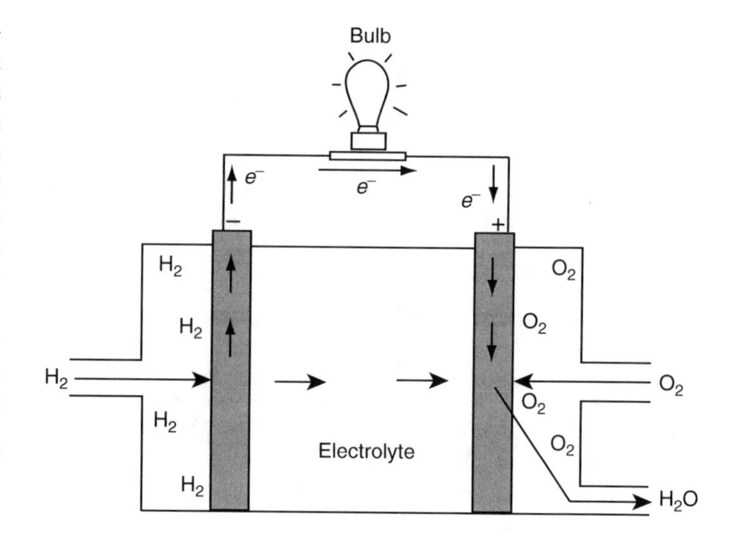

Fig. 8.3 Basic components of the design of a hydrogen–oxygen fuel cell. The two essential raw materials for a fuel cell are hydrogen and oxygen; the final products of a fuel cell are electricity and water.

fine particles of platinum or palladium that function as a catalyst. As platinum-group metals are expensive, other catalysts, such as nickel (for hydrogen) and silver (for oxygen), are used where possible. The very small catalyst particles provide a large number of active sites at which the normally very slow electrochemical reactions can take place at a fairly rapid rate.

The electrochemical reactions occurring at the electrodes vary with the nature of the electrolyte, but in principle they are as follows. At the anode, $H_{2\,(g)}$ oxidizes to H^+ ions, releasing electrons:

$$H_{2\,(g)} \Rightarrow 2H^+ + 2e^- \quad \text{(oxidation, anode)} \qquad (8.15)$$

When an electrical load (e.g., an electric bulb) is inserted in the external circuit, electrons travel along the connecting wire from the anode to the cathode. At the cathode, $O_{2\,(g)}$ combines with electrons and H^+ ions (that were formed at the anode) to produce water:

$$O_{2\,(g)} + 4H^+ + 4e^- \Rightarrow 2H_2O_{(l)} \quad \text{(reduction, cathode)} \qquad (8.16)$$

The net reaction in the hydrogen–oxygen fuel cell is obtained adding the two half-cell reactions (after adjusting the stoichiometric coefficients to balance the electrons):

$$2H_{2\,(g)} + O_{2\,(g)} \Rightarrow 2H_2O_{(l)} \quad \text{(oxidation–reduction)} \qquad (8.17)$$

Thus, the net product of the fuel cell is water, which is drawn off from the cell (and $CO_{2\,(g)}$ if the source of $H_{2\,(g)}$ is a hydrocarbon), and usable energy in the form of electricity.

The theoretical EMF of a fuel cell can be calculated from the free energy change for the reaction (8.17) (see section 8.3) – it is 1.23 volts at 25°C with the gases at atmospheric pressure. The discharge voltages observed in actual fuel cells are always less than the theoretical value, the difference increasing with increasing strength of the current drawn from the cell. For the moderate currents at which fuel cells normally operate, the output EMF from a single fuel cell is much lower, only 0.7 to 0.8 volts. To increase the output to a reasonable level, many separate fuel cells must be combined to form a fuel-cell stack.

Fuel cell technology has many advantages (Aaronson, 1971). Fuel cells are highly efficient in converting chemical energy into electrical energy; they are noiseless and can be operated at low temperatures; they remain efficient over a wide range of power levels and can be built in large or small units depending on the application; and they produce no potentially hazardous wastes. Oxygen needed for the fuel cell is extracted from air, almost an unlimited source. The main problems in the commercial application of fuel cells at present are the cost of hydrogen fuel and the potential danger associated with the handling of such a flammable gas. Substitutes for hydrogen fuel under consideration include natural gas, liquefied petroleum gases (propane and butane), methanol, and even carbon monoxide produced from coal. Various devices are used to clean up the hydrogen, but even then the hydrogen that comes out of the substitute fuels is not pure, and this lowers the efficiency of the fuel cell.

8.3 Relationship between free energy change (ΔG_r) and electrode potential *(E)* – the Nernst equation

Consider an oxidation half-reaction involving the transfer of n electrons:

$$\underset{\text{reduced assemblage}}{\text{a A} + \text{b B}} = \underset{\text{oxidized assemblage}}{\text{c C} + \text{d D}} + ne^- \tag{8.18}$$

In this book, following the convention of Garrels and Christ (1965), we write the half-reactions with electrons and the oxidized assemblage appearing on the right side and the reduced assemblage on the left side. Assuming pure phases, the free energy change for the reaction, $\Delta G^P_{r,T}$, is (see equation 5.38):

$$\Delta G^P_{r,T} = \Delta G^0_r + RT \ln \frac{a^c_C\, a^d_D}{a^a_A\, a^b_B} = \Delta G^0_r + RT \ln Q \tag{8.19}$$

where ΔG^0_r is the free energy change for the half-reaction in the chosen standard state, P and T are the pressure and temperature (in Kelvin) of interest, R is the gas constant (= 8.314 J mol^{-1} K^{-1}), and Q represents the *reaction quotient*, which is defined as the quotient of the activities (to be evaluated at P and T):

$$Q = \prod_i a_i^{\nu_i} \tag{8.20}$$

where ν_i is the stoichiometric coefficient of the ith constituent in the reaction, positive if it refers to a product and negative if it refers to a reactant (see Box 5.3). It is not necessary to include the activity of electrons that appear in reaction (8.18) because the electrode potential is measured against the standard hydrogen electrode, with activities of H_2 and H^+ assumed to be unity. Note that for a reaction at equilibrium, $Q = K_{eq}$.

The electrode potential E^P_T for the half-reaction at P and T is related to $\Delta G^P_{r,T}$ by the equation (for derivation of the equation, see Anderson and Crerar, 1993, p. 471):

$$\Delta G^P_{r,T} = n\,\Im\, E^P_T \quad \text{or} \quad E^P_T = \frac{\Delta G^P_{r,T}}{n\,\Im} \tag{8.21}$$

where n is the number of electrons being transferred and $\Im$ is the *Faraday constant* (= 96, 485 J volt^{-1} gram – equivalent^{-1}), which is the amount of electricity that corresponds to the gain or loss, and therefore the passage, of 1 mole (or 6.022×10^{23}) electricity.

An analogous relationship holds if the reaction occurs under standard state conditions (as denoted by the superscript "0" in the equation):

$$\Delta G^0_r = n\,\Im\, E^0 \quad \text{or} \quad E^0 = \frac{\Delta G^0_r}{n\,\Im} \tag{8.22}$$

where E^0 (in volts) is the standard potential (i.e., the potential with all substances present having unit activity at 298.15 K and 1 bar). The standard electrode potentials listed in Table 8.1 were calculated using equation (8.22).

Substituting for $\Delta G^P_{r,T}$ and ΔG^0_r in equation (8.19) (and remembering that $\ln X = 2.3026 \log_{10} X$),

$$E^P_T = E^0 + \frac{RT}{n\Im} \ln Q = E^0 + 2.3026 \frac{RT}{n\Im} \log Q \tag{8.23}$$

Thus, the general expression for E^P_T (in volts) for an oxidaton half-reaction may be written as:

$$E^P_T = E^0 + 2.3026 \frac{RT}{n\,\Im} \log \left(\frac{\text{activity product of species in the oxidized assemblage}}{\text{activity product of species in the reduced assemblage}} \right) \tag{8.24}$$

Equation (8.24) is called the *Nernst equation*, named after the physical chemist W. Nernst, who in 1897 derived a similar expression using concentrations instead of activities. Since E^0 can be calculated using equation (8.22) (if we know the ΔG^0_f values of the species involved in the reaction), the Nernst equation allows us to calculate electrode potentials of half-reactions occurring at nonstandard state conditions. The equation applies also to the calculation of the EMF for whole oxidation–reduction reactions, which do not show explicitly the loss and gain of electrons. The number of electrons transferred is immaterial for the calculation of ΔG^0_r because $\Delta G^0_f(e^-)=0$.

Example 8–2: Calculation of the standard state (298.15 K, 1 bar) EMF of the Zn–Cu electrochemical cell using standard free energy of formation data (Robie and Hemingway, 1995) listed in Appendix 6

Zn electrode half-reaction (oxidation):

$$Zn_{(s)} \Rightarrow Zn_{(aq)}^{2+} + 2e^- \text{(oxidation)} \quad (8.11)$$

$\Delta G^1_{r,298.15} = -147.3\,kJ$ (see section 8.2.1)

$$E^0_{Zn} = \frac{\Delta G^0_r}{n\,\Im} = \frac{\Delta G^1_{r,\,298.15}}{n\,\Im} = \frac{-147.3}{(2)\,(96.485)} = -0.76 \text{ volts}$$

Cu electrode half-reaction (reduction):

$$Cu_{(aq)}^{2+} + 2e^- \Rightarrow Cu_{(s)} \text{ (reduction)} \quad (8.12)$$

$$\Delta G^1_{r,\,298.15} = \Delta \bar{G}^1_{f,\,298.15}\,(Cu_{(s)}) - \Delta \bar{G}^1_{f,\,298.15}\,(Cu_{aq}^{2+}) = 0 - 65.1 = -65.1\,kJ$$

$$E^0_{Cu} = \frac{\Delta G^0_r}{n\,\Im} = \frac{\Delta G^1_{r,\,298.15}}{n\,\Im} = \frac{-65.1}{(2)\,(96.485)} = -0.34 \text{ volts}$$

Whole-cell reaction (oxidation–reduction):

$$Zn_{(s)} + Cu^{2+}_{(aq)} \Rightarrow Zn^{2+}_{(aq)} + Cu_{(s)} \quad (8.13)$$

$$\Delta G^1_{r,298.15} = \Delta \bar{G}^1_{f,298.15}\,(Zn_{aq}^{2+}) + \Delta \bar{G}^1_{f,298.15}\,(Cu_{(s)}) - \Delta \bar{G}^1_{f,298.15}\,(Zn_{(s)}) - \Delta \bar{G}^1_{f,298.15}\,(Cu_{aq}^{2+})$$

$$= -147.3 + 0 - 65.1 - 0 = -212.4\,kJ$$

$$E^0_{Zn-Cu} = \frac{\Delta G^0_r}{n\,\Im} = \frac{\Delta G^1_{r,\,298.15}}{n\,\Im} = \frac{-212.4}{(2)\,(96.485)} = -1.10 \text{ volts}$$

The same result is obtained by adding the standard state potentials of the two half-reactions:

$$E^0_{Zn-Cu} = E^0_{Zn} + E^0_{Cu} = (-0.76) + (-0.34) = -1.10 \text{ volts}$$

If our formulations are correct, then the potential difference for the Zn–Cu cell (reaction 8.13) at P and T, $E^P_{T\,(Zn-Cu)}$, should, as shown below, work out to be zero at eqilibrium. Substituting K_{eq} for Q in equation (8.23),

$$\log K_{eq} = \frac{-\Delta G^0_r}{2.3026\,RT} = \frac{-(-212.4)\times 10^3}{(2.3026)\,(8.314)\,(298.15)} = \frac{212.4 \times 10^3}{5716.2} = 37.21$$

Substituting values in equation (8.24),

$$E^P_{T\,(Zn-Cu)} = E^0_{Zn-Cu} + 2.3026\frac{RT}{n\,\Im}\log K_{eq} = -1.10 + \frac{(2.3026)\,(8.314)\,(298.15)\,(37.21)}{(2)\,(96485)} = -1.10 + 1.10 = 0.0 \text{ volt}$$

Since $a_{Zn(s)}=1$ and $a_{Cu(s)}=1$,

$$K_{eq} = \frac{a_{Zn^{2+}}}{a_{Cu^{2+}}} = 10^{37.2}$$

Thus, the EMF of the Zn–Cu cell will drop to 0.0 volt when the $a_{Zn^{2+}} : a_{Cu^{2+}}$ ratio reaches a value of $10^{37.2}$.

8.4 Oxidation potential (*Eh*)

In aqueous geochemistry, the electrode potential of a half-reaction at any state (E^P_T), measured relative to a standard hydrogen electrode (SHE), is called *oxidation potential* (or *redox potential* by some authors) and given a special symbol *Eh* to remind us that the standard EMF is referenced to the standard hydrogen electrode.

The *Eh* is measured with an electrode pair consisting of an inert electrode, usually platinum, immersed in the half-cell of interest, and a reference electrode such as the SHE. Because the hydrogen electrode is difficult to maintain and transport, measurements in the field are usually made with a calomel ($HgCl_2$) reference electrode ($HgCl_2 + 2e^- \Rightarrow 2Hg + 2Cl^-$). The measured EMF is then corrected for the potential of the reference electrode (Eby, 2004, p. 97):

$$Eh = E_{measured} + E_{reference} \quad (8.25)$$

The corrected *Eh* is the value that would be recorded if SHE were used as the reference.

There are, however, many practical difficulties pertaining to field measurements of *Eh*, and in many cases the *Eh* values measured with a platinum electrode differ significantly from values computed using the Nernst equation. The main reasons for such differences are (Garrels and Christ, 1965, pp. 135–136; Langmuir, 1997): (i) contamination of the electrodes;

(ii) irreversibility or slow kinetics of most redox couple reactions and resultant disequilibrium between different redox couples in the same water; and (iii) common existence of several redox couples in natural waters. For specific half-reactions, it is generally better to use the Nernst equation (equation 8.23), which we can write as:

$$Eh = E^0 + 2.3026\frac{RT}{nF}\log\left(\frac{\prod_i a_i^{\nu_i}\text{(oxidized assemblage)}}{\prod_j a_j^{\kappa_j}\text{(reduced assemblage)}}\right) \quad (8.26)$$

where a_i and a_j refer, respectively, to the reactant and product species in the reaction, and ν_i and κ_j refer to their stoichiometric coefficients.

For the Earth's surface environment we can assume T=298.15K (or 25°C). Substituting the values of T=298.15, $\Im$=96, 485J volt^{-1} gram – equivalent^{-1}, and R=8.314J mol^{-1} K^{-1}, equation (8.26) reduces to:

$$Eh = E^0 + \frac{0.0592}{n}\log\left(\frac{\prod_i a_i^{\nu_i}\text{(oxidized assemblage)}}{\prod_j a_j^{\kappa_j}\text{(reduced assemblage)}}\right) \quad (8.27)$$

The *Eh* value of an electrode reaction may be positive or negative. A positive *Eh* value implies that the environment is *less* reducing than the SHE and, thus, is oxidizing; a negative *Eh* value implies the converse, that the environment is more reducing than the SHE.

The *Eh* of a natural terrestrial environment is largely a function of its organic matter content. The shallow water on an open shelf presents an oxidizing environment because most organic material, irrespective of the abundance of organisms, decays to carbon dioxide and water before burial can take place. Sedimentary environments in which organic matter is accumulating or has accumulated in the past through burial (e.g., black shales) are highly reducing, with *Eh* values in the range of –0.1 to –0.5 volts (and pH reaching a minimum of about 4 where decay takes place under aerobic conditions).

Example 8–3: Calculation of the ratio of oxidized to reduced species in an environment of known Eh

If the measured *Eh* of the water in a lake is +0.65 volts, what is the dominant form of the dissolved copper in this water (Cu^+ or Cu^{2+})?

The standard potential ($E^0_{Cu^+}$) of the oxidation half-reaction

$$Cu^+ \Rightarrow Cu^{2+} + e^- \quad (8.28)$$

is +0.16 volts (Table 8.1). Since the measured *Eh* is greater than $E^0_{Cu^+}$, we should expect that $a_{Cu^{2+}}:a_{Cu^+} > 1$. Actually, we can calculate the activity ratio using equation (8.27):

$$Eh = E^0_{Cu^+} + \frac{0.0592}{n}\log\frac{a_{Cu^{2+}}}{a_{Cu^+}}$$

$$\log\frac{a_{Cu^{2+}}}{a_{Cu^+}} = \frac{Eh - E^0_{Cu^+}}{0.0592} = \frac{0.65\text{–}0.16}{0.0592} = 8.3$$

$$\frac{a_{Cu^{2+}}}{a_{Cu^+}} = 10^{8.3}$$

Thus, practically all the dissolved copper in this water should occur as Cu^{2+}.

8.5 The variable *pe*

An alternative way of characterizing the oxidation–reduction capability of an environment is by the variable *pe*, which is defined in terms of the activity of electrons:

$$pe = -\log a_{e^-} \quad (8.29)$$

As aqueous solutions do not contain free electrons, the activity of electrons, unlike normal solutes, does not correspond to a concentration; it is merely a mathematical variable used to characterize the redox properties of a system. It should be obvious that the definition of *pe* is analogous to that of pH,

$$\text{pH} = -\log_{10} a_{H^+} \quad (4.15)$$

and it is a dimensionless number like pH. In fact, the main advantage of using *pe* is that it allows redox reactions to be treated symmetrically with reactions in which pH is a controlling variable.

The variable *pe* should be thought of as a measure of the ability of a solution to donate electrons to an electron acceptor (oxidizing agent).

Just as a high value of a_{H^+} translates into a low pH and a high tendency to donate protons, a high value of a_{e^-} implies a low *pe*, and thus a relatively high tendency for oxidation. Also, by convention, the standard free energy of formation of an electron is zero, as is that of the H^+ ion (i.e., $\Delta G_f^0\ (e^-)=0$, and $\Delta G_f^0\ (H^+)=0$).

As elaborated in Box 8.2, an equation similar to the Nernst equation can be derived for *pe*:

$$pe = pe^0 + \frac{1}{n}\log Q \quad (8.30)$$

where

$$pe^0 = -\frac{1}{n}\log K_{eq} \tag{8.31}$$

Box 8.2 Equation for *pe* and equivalence between *pe* and *Eh*

Equation for *pe*

For the half-reaction (8.18) at equilibrium,

$$K_{eq} = \frac{a_C^c\, a_D^d\, a_e^n}{a_A^a\, a_B^b} \tag{8.32}$$

To separate a_{e^-} from the rest of the activity terms, we linearize the equation as

$$\log K_{eq} = \log a_{e^-}^n + \log \frac{a_C^c\, a_D^d}{a_A^a\, a_B^b} = n \log a_{e^-} + \log Q \tag{8.33}$$

If all the reactants and products of the reaction (except the electrons) are in their standard states, the activity of each constituent in the standard state, by convention, is unity, and $Q=1$. Denoting the corresponding value of pe as pe^0 (analogous to E^0), equation (8.33) reduces to:

$$pe^0 = -\frac{1}{n}\log K_{eq} \tag{8.31}$$

Substituting for pe and pe^0, equation (8.33) gives

$$pe = pe^0 + \frac{1}{n}\log \tag{8.30}$$

Equivalence between *pe* and *Eh*

From equation (8.26),

$$\log Q = \frac{(Eh - E^0)\, nF}{2.3026\, RT} \tag{8.34}$$

At equilibrium, since $Eh=0$ and $Q=K_{eq}$, equation (8.26) reduces to:

$$E^0 = -2.3026 \frac{RT}{nF} \log K_{eq}$$

Substituting for $\log K_{eq}$ (equation 8.31),

$$pe^0 = \frac{F}{2.3026\, RT} E^0 \tag{8.35}$$

Substituting for pe^0 and $\log Q$ (equation 8.34) in equation (8.30),

$$pe = \frac{E^0\, F}{2.3026\, RT} + \frac{Eh\, nF}{2.3026\, n\, RT} - \frac{E^0\, nF}{2.3026\, n\, RT}$$

Box 8.2 (cont'd)

$$pe = \frac{F}{2.3026\, RT} Eh = \frac{5040}{T} Eh \tag{8.36}$$

Thus, oxidation potential can be represented by either *Eh* or *pe*. The use of *pe* makes calculations simpler because every 10-fold change in the activity ratio causes a unit change in *pe*. Furthermore, because an electron can reduce a proton, the intensity parameter for oxidation should preferably be expressed in units equivalent to *pe* (Stumm and Morgan, 1981, p. 436). However, *Eh* has been used widely in the geologic literature and is an easier concept to grasp.

In the above equations, pe^0 refers to the situation where the activities of all the species taking part in the reaction, except the electrons, are in their standard states. A standard state for electrons is not meaningful because electrons will have different energy levels in different half-cells. Since $\log K_{eq}$ can be computed using equation (5.37) ($\Delta G_r^0 = -RT \ln K_{eq}$), equation (8.30) allows the calculation of *pe* for any half-reaction if we know the activities of the species involved in the reaction.

8.6 *Eh*–pH stability diagrams

For the Earth's surface environment, temperature and pressure can be assumed as fixed at 298.15 K (25°C) and 1 atm (1 atm = 1.013 bar ≈ 1 bar), so that *Eh (or pe)*, pH, and solute activities are the most important variables in oxidation and reduction reactions in such near-surface aqueous systems. It is, therefore, a common practice in geochemistry to depict phase equilibria in low-temperature aqueous systems on *Eh*–pH or *pe*–pH diagrams at specified solute activities. Such diagrams are useful for the interpretation of relative stabilities of minerals and aqueous species during chemical weathering, and solute transport in aqueous solutions, especially of elements with more than one valence state (e.g., Fe, Cu, Mn, Au, Hg, S, etc.).

A *Eh*–pH diagram looks exactly like a *pe*–pH diagram, except that the *y*-axis is shifted by a factor of $5040/T$ (equation 8.36). In this book we will deal with the more familiar *Eh*–pH diagrams.

8.6.1 *Stability limits of surface water*

The *Eh*–pH stability fields of solid compounds and dissolved species in natural aqueous environments are limited by the stability of water itself at the Earth's surface *P–T* conditions (approximated as $T=298.15$ K; $P=1$ bar). It is a common practice, therefore, to include the stability limits of water in a *Eh*–pH diagram.

In order to derive equations for the upper and lower stability limits of water in terms of *Eh* and pH, let us consider the oxidation of $H_2O_{(l)}$ to $O_{2(g)}$ at the *water electrode* of an imaginary $H_2O_{(l)}$–$H_{2(g)}$ electrochemical cell in which the $H_{2(g)}$ electrode is maintained in the standard state. The oxidation half-reaction (or the water electrode reaction) can be represented as:

$$H_2O_{(l)} \Rightarrow 2H^+_{(aq)} + 0.5\ O_{2\,(g)} + 2e^- \quad (8.37)$$

The *Eh* of the water electrode is given by (see equation 8.27)

$$Eh_{H_2O(l)} = E^0_{H_2O(l)} + \frac{0.0592}{n} \log \frac{(a_{H^+})^2\ (f_{O_2})^{0.5}}{a_{H_2O(l)}} \quad (8.38)$$

Since $a_{H_2O(l)} = 1$ (by convention), $n=2$ for the oxidation half-reaction, and $pH = -\log a_{H^+}$,

$$\begin{aligned} Eh_{H_2O(l)} &= E^0_{H_2O(l)} + \frac{0.0592}{2} \log \left((a_{H^+})^2\ (f_{O_2})^{0.5}\right) \\ &= E^0_{H_2O(l)} + 0.0148 \log f_{O_2} + 0.0592 \log a_{H^+} \quad (8.39) \\ &= E^0_{H_2O(l)} + 0.0148 \log f_{O_2} - 0.0592\ pH \end{aligned}$$

Using $\Delta G^1_{f,\,298.15}$ values from Robie and Hemingway (1995) in equation (8.22),

$$\begin{aligned} E^0_{H_2O(l)} &= \frac{2\ \Delta\bar{G}^1_f(H^+) + 0.5\ \Delta\bar{G}^1_f(O_2) + 2\ \Delta\bar{G}^1_f(e^-) - \Delta\bar{G}^1_f(H_2O)}{n\ \Im} \\ &= \frac{2\ (0) + 0.5\ (0) + 2\ (0) - (-237.1)}{(2)\ (96{,}485\ \times\ 10^{-3})} = \frac{237.1}{192.97} = +1.23\,\text{volts} \end{aligned}$$

Substituting for $E^0_{H_2O(l)}$ in equation (8.39) we get the *Eh–pH* equation for the water electrode:

$$Eh_{H_2O(l)} = 1.23 + 0.0148 \log f_{O_2} - 0.0592\ pH \quad (8.40)$$

Considering the limiting values of f_{O_2}, as elaborated in Box 8.3, the *Eh*–pH equation for the upper stability limit of surface liquid water in equilibrium with the Earth's atmosphere, is:

$$\boxed{Eh^{Up}_{H_2O(l)} = 1.23 - 0.0592\ pH} \quad (8.41)$$

and that for the lower stability limit is:

$$\boxed{Eh^{Lr}_{H_2O(l)} = -0.0592\ pH} \quad (8.42)$$

Equations (8.41) and (8.42) plot as parallel straight lines in *Eh* (*y*-axis)–pH (*x*-axis) space, each with a slope of –0.0592 volt per pH unit; the *y*-axis intercept for equation (8.41) is $E^0 = 1.23$ volt and for equation (8.42) is $E^0 = 0.0$ volt (Fig. 8.4). These two lines bound the *Eh*–pH stability field of surface water.

Box 8.3 Limits of f_{O_2} and f_{H_2} in the Earth's surface aqueous environment

The equilibrium between surface water and the gases O_2 and H_2 in the atmosphere is represented by

$$2\ H_2O_{(l)} \Leftrightarrow O_{2\,(g)} + 2H_{2\,(g)};\quad K_{eq} = \frac{f_{O_2}(f_{H_2})^2}{(a_{H_2O(l)})^2} = f_{O_2}(f_{H_2})^2 \quad (8.43)$$

Using $\Delta G^1_{f,\,298.15}$ values of Robie and Hemingway (1995), K_{eq} for reaction (8.43) at 298.15 K and 1 bar can be calculated from the relation $\Delta G^0_r = -2.3026\ RT \log K_{eq}$ (equation 5. 37):

$$\Delta G_r^{\,0} = \Delta\bar{G}^1_{f,298.15}(O_2) + 2\Delta\bar{G}^1_{f,298.15}(H_2) - 2\ \Delta\bar{G}^1_{f,298.15}(H_2O_{(l)})$$

$$= 0 + 0 - 2(-237.1) = 474.2\,\text{kJ}$$

$$\log K_{eq} = \frac{-\Delta G^0_r}{2.3026\ RT} = \frac{-474.2}{(2.3026)\ (8.314\ \times\ 10^{-3})\ (298.15)} = -83.08$$

$$K_{eq} = 10^{-83.08}$$

Therefore, at equilibrium (since $a_{H_2O} = 1$ by convention),

$$K_{eq} = f_{O_2}(f_{H_2})^2 = 10^{-83.08} \quad (8.44)$$

As the total pressure of the atmosphere on the Earth's surface is 1 bar (≈1 atm) or less, the maximum value of f_{O_2} or f_{H_2} in the atmosphere cannot exceed 1 bar. The upper stability limit of liquid water, therefore, is defined by

$$f_{O_2} = 1\ \text{bar or}\ f_{H_2} = \left(10^{-83.08}\right)^{0.5} = 10^{-41.54}\ \text{bar}$$

and its *Eh–pH* equation is obtained by substituting $f_{O_2} = 1$ bar (i.e., $\log f_{O_2} = 0$) in Equation (8.40):

$$\begin{aligned} Eh^{Up}_{H_2O(l)} &= 1.23 + 0.0148 \log f_{O_2} - 0.0592\ pH \\ &= 1.23 - 0.0592\ pH \quad (8.41) \end{aligned}$$

If we take $f_{O_2} = 0.2$ bar, considering that oxygen constitutes about 20% by volume of the present atmosphere, we get

$$Eh^{Up}_{H_2O(l)} = 1.22 - 0.0592\ pH \quad (8.45)$$

which is almost identical to equation 8.41.

The lower stability limit if liquid water is defined by

$$f_{O_2} = 10^{-83.08}\ \text{bar or}\ f_{H_2} = 1\ \text{bar}$$

and its *Eh*–pH equation is obtained by substituting $f_{O_2} = 10^{-83.08}$ bar in equation (8.40):

$$\begin{aligned} Eh^{Lr}_{H_2O(l)} &= 1.23 + 0.0148 \log f_{O_2} - 0.0592\ pH \\ &= 1.23 + 0.0148 \log (10^{-83.08}) - 0.0592\ pH \quad (8.42) \\ &= -0.0592\ pH \end{aligned}$$

Box 8.3 (cont'd)

Since the product of f_{O_2} and $(f_{H_2})^2$ at equilibrium must be $10^{-83.08}$, the same equations would be obtained if we chose to define the stability limits in terms of f_{H_2}, $f_{H_2} = 10^{-41.54}$ bar for the upper limit and $f_{H_2} = 1$ bar for the lower limit.

The values of f_{O_2} or f_{H_2} used for defining the upper and lower limits of water stability are conceptual and have little resemblance to physical reality. The *Eh*–pH fields of natural aqueous environments are largely controlled by (i) photosynthesis, respiration, and decay of organic matter; (ii) oxidation and reduction reactions involving iron, manganese, sulfur, nitrogen, and carbon; and (iii) carbonate equilibria (discussed in Chapter 7). The overall range of measured *Eh* and pH for natural environments and the ranges characteristic of some natural waters are shown in Fig. 8.5.

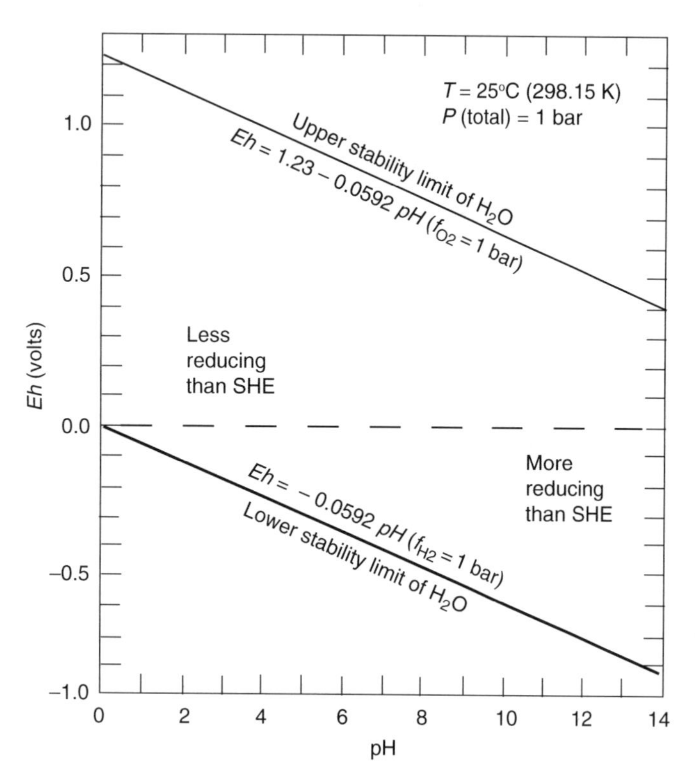

Fig. 8.4 *Eh*–pH diagram showing the upper and lower stability limits of water (USL and LSL) at 298.15 K and 1 bar total pressure (approximately 1 atm). The stability field of water can be contoured (not shown) by substituting different values of f_{O_2} in equation (8.40). The horizontal line at *Eh*=0.0 volts divides the field into two regions relative to SHE, a region of "oxidizing" environments (less reducing than SHE) characterized by positive *Eh* values and a region of "reducing" environments (more reducing than SHE) characterized by negative *Eh* values.

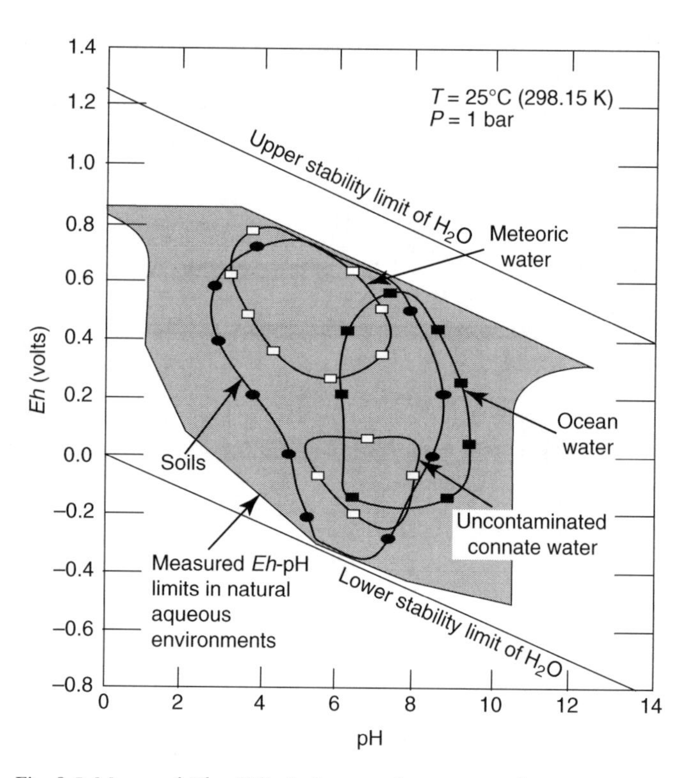

Fig. 8.5 Measured *Eh*–pH limits in natural aqueous environments. Also shown are approximate *Eh*–pH fields of meteoric water, ocean water, uncontaminated connate water, and soils. (After Baas Becking *et al.*, 1960.)

Example 8–4: Calculation of dissolved oxygen content in a body of surface water of known Eh and pH

Suppose the measured pH in the lake water (*Eh*=+0.65 volts) considered in Example 8–3 is 6.6. How much dissolved oxygen does this water contain?

From equation (8.37),

$$\log f_{O_{2\,(g)}} = \frac{Eh_{H_2O(l)} - 1.23 + 0.0592\ \text{pH}}{0.0148} = \frac{0.65 - 1.23 + (0.0592 \times 6.6)}{0.0148} = -12.8$$

$$f_{O_{2\,(g)}} = 10^{-12.8}\ \text{bar}$$

To compute the amount of oxygen dissolved in the water, we need to consider the equilibrium between aqueous oxygen and gaseous oxygen at 298.15 K and 1 bar:

$$O_{2\,(g)} \Leftrightarrow O_{2\,(aq)} \qquad K_{eq} = \frac{a_{O_2\,(aq)}}{fO_{2\,(g)}} = 10^{-2.9}$$

$$a_{O_2\,(aq)} = f_{O_2\,(g)} \times 10^{-2.9} = 10^{-12.8} \times 10^{-2.9} = 10^{-15.7} = 1.99 \times 10^{-16}$$

Assuming the aqueous solution to be ideal (i.e., $a_i = m_i$), the concentration of dissolved oxygen in the lake water is approximately 1.99×10^{-16} mol kg^{-1} or 6.46×10^{-12} ppm. This is very little oxygen indeed, enough to oxidize Cu^+ to Cu^{2+} but orders of magnitude less than the 5–7 ppm dissolved oxygen required for fish to survive (Eby, 2004, p. 103). Nevertheless, the lake water will be considered an oxidizing environment because its *Eh* is positive!

8.6.2 *Procedure for construction of Eh–pH diagrams*

The general procedure for plotting a half-reaction boundary in *Eh*–pH space (T=298.15 K, P (total)=1 bar) may be summarized as follows:

(1) For any pair of species to be included in the *Eh*–pH diagram, write the half-reaction in the form of an oxidation reaction:

$$\text{reduced assemblage} \Rightarrow \text{oxidized assemblage} + ne^-$$

Add water and hydrogen ions, as necessary, to balance the reaction.

(2) Calculate ΔG_r^0 for the half-reaction using $\Delta \bar{G}^1_{f,298.15}$ values from a well-accepted thermodynamic database (e.g., Robie and Hemingway 1995). According to convention, $\Delta \bar{G}^1_{f,298.15}(e^-) = 0$.

(3) Calculate E^0 for the half-reaction (equation 8.22).

(4) Using equation (8.27), establish the equation relating *Eh* and pH for the half-reaction under consideration. Activities of pure solids and pure liquids are taken to be unity, and the activity of a gas phase is taken to be equal to its fugacity or, if the gas is ideal, its partial pressure. The water is assumed to be almost pure so that $a_{H_2O(l)} = 1$.

(5) We follow certain conventions for constructing reaction boundaries in an *Eh*–pH diagram (Garrels and Christ, 1965). The limit of the stability field of a given solid compound or mineral in equilibrium with water is arbitrarily drawn where the sum of the activities of the ions in equilibrium with the solid exceeds some chosen value. The value commonly chosen for geochemical systems, largely based on experience, is 10^{-6} because at lower values the solid will behave as an immobile constituent in the environment. For ionic species in equilibrium with water, a given field is labeled with the ion that is preponderant within the field, and the boundary between two such adjacent fields (e.g., between the fields for Fe^{2+} and Fe^{3+}) is placed where the two ionic species have the same activity (e.g., $a_{Fe^{2+}} = a_{Fe^{3+}}$).

(6) Using the equation for the half-reaction calculate *Eh* for arbitrarily assumed values of pH. As the equations yield straight lines, two pairs of *Eh*–pH values should suffice to draw a reaction boundary. Only the pH value is required for a reaction that is independent of *Eh* and this yields a line parallel to the *Eh* axis. Similarly, only the *Eh* value is required for a reaction that is independent of pH and this yields a line parallel to the pH axis.

(7) Extract a final version of the diagram by deleting metastable extensions of the reaction boundaries, including extensions beyond the stability field of water.

Example 8–5: Eh–pH diagram for selected oxide minerals and ionic species in the Fe-O-H$_2$O system at 298.15 K and 1 bar

Eh(or *pe*)–pH diagrams for many geochemical systems are available in the published literature (e.g., Garrels and Christ, 1965; Brookins, 1978; Drever, 1997; Langmuir, 1997; Faure, 1998). We will discuss here one example, the Fe–O–H$_2$O system, to illustrate the strategy employed for the construction of such diagrams.

For this *Eh*–pH diagram we will compute the stability fields of the two most important iron oxide minerals, magnetite (Fe_3O_4) and hematite (Fe_2O_3), and the fields of preponderance of the two important dissolved ionic species, Fe^{2+} and Fe^{3+}. We will also take a look at the stability of metallic Fe, only to conclude that it is not a stable species in the surface aqueous environment. To keep the diagram simple, iron hydroxides, such as $Fe(OH)_2$ and $Fe(OH)_3$, and aqueous complexes, such as $HFeO_2^-$, $Fe(OH)^{2+}$, and $Fe(OH)_2^+$, will not be considered. The thermodynamic data needed for the computations are listed in Table 8.2. For the computations described below, $a_{Fe(s)} = 1$, $a_{Fe_2O_3\,(s)} = 1$, and $a_{Fe_3O_4\,(s)} = 1$ (pure solids in standard state), and $a_{H_2O(l)} = 1$ (almost pure phase in standard state). The computed boundaries are plotted in Fig. 8.6.

Table 8.2 Standard state (298.15 K, 1 bar) thermodynamic data (Robie and Hemmingway, 1995) for calculation of *Eh*–pH boundaries in the Fe-O-H$_2$O system.

	H^+	$Fe_{(s)}$	Fe^{2+}	Fe^{3+}	$H_2O_{(l)}$	$Fe_3O_{4\,(s)}$	$Fe_2O_{3\,(s)}$
$\Delta \bar{G}^1_{f,298.15}$ (kJ mol^{-1})	0.0	0.0	−90.0	−16.7	−237.1	−1012.7	−744.4

By convention, $\Delta G^1_{f,\,298.15}(e^-) = 0$.

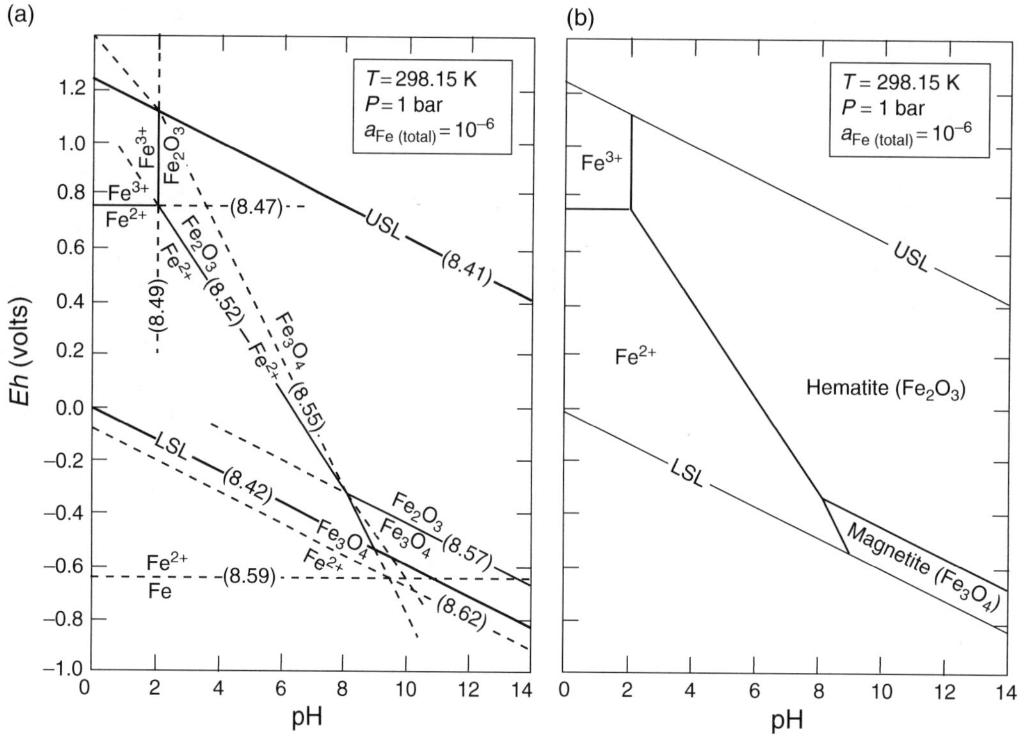

Fig. 8.6 *Eh*–pH diagram for the Fe–O–H_2O system at 298.15 K (25°C) and 1 bar. Fields labeled as magnetite and hematite are the stability fields of the respective oxide minerals in the presence of water. Fields labeled as Fe^{2+} (aq) and Fe^{3+} (aq) are dominated by the respective species and the boundary between the two fields is placed where $a_{Fe^{2+}} = a_{Fe^{3+}}$. The boundary between the fields of an aqueous species and a solid is placed where the activity of the dissolved species is 10^{-6}. (a) Reaction boundaries with dashed lines indicate metastable extensions. Each boundary is marked with the corresponding equation number in the text. (b) Final version of the *Eh*–pH diagram without the equation numbers and metastable extensions of the field boundaries. USL = upper stability limit of water; LSL = lower stability limit of water.

(1) Fe^{2+}–Fe^{3+} boundary:

$$Fe^{2+} \Rightarrow Fe^{3+} + e^- \qquad \Delta G_r^0 = +73.3\ \text{kJ} \tag{8.46}$$

$$E^0 = \frac{\Delta G_r^0}{n\Im} = \frac{+73.3}{(1)\,(96.485)} = +\,0.760\ \text{volts}$$

$$Eh = E^0 + \frac{0.0592}{n}\log\frac{a_{Fe^{3+}}}{a_{Fe^{2+}}}$$
$$= 0.760 + 0.0592\log a_{Fe^{3+}} - 0.0592\log a_{Fe^{2+}} \tag{8.47}$$

The reaction is independent of pH and equation (8.47) yields a straight line parallel to the pH-axis. Since at any point on the boundary between the fields of Fe^{2+} and Fe^{3+}, $a_{Fe^{2+}} = a_{Fe^{3+}}$ in the solution, the *Eh* coordinate of the boundary is +0.760 volts. At any point above this boundary, $a_{Fe^{3+}} > a_{Fe^{2+}}$, and at any point below, $a_{Fe^{3+}} < a_{Fe^{2+}}$.

(2) Fe^{3+}–$Fe_2O_{3\,(s)}$ boundary:

$$Fe_2O_{3(s)} + 6H^+ \Rightarrow 2Fe^{3+} + 3H_2O \qquad \Delta G_r^0 = -0.3\ \text{kJ} \tag{8.48}$$

This reaction is independent of *Eh* as it does not involve transfer of electrons. Therefore, the reaction boundary is parallel to the *Eh* axis at a partcular value of pH, which we can determine from the equilibrium constant of the reaction (see equation 5.37). Since $\text{pH} = -\log a_{H^+}$,

$$\log K_{eq} = \log\frac{(a_{Fe^{3+}})^2}{(a_{H^+})^6} = 2\log a_{Fe^{3+}} + 6\ \text{pH}$$

$$\log K_{eq} = \frac{-\Delta G_r^0}{2.3026\ RT} = \frac{0.3}{(2.3026)\,(8.314\times10^{-3})\,(298.15)}$$
$$= 0.0526$$

Substituting for $\log K_{eq}$,

$$\log a_{Fe^{3+}} = 0.0263 - 3\ \text{pH} \tag{8.49}$$

Assuming $a_{Fe^{3+}} = 10^{-6}$ (for reasons explained earlier), pH = 2.0009. It should be evident from equation (8.49) that the pH value would be lower for $a_{Fe^{3+}} > 10^{-6}$ and higher for $a_{Fe^{3+}} < 10^{-6}$. The same is true for $a_{Fe^{2+}}$, which needs to be fixed for the Fe^{2+}–$Fe_2O_{3\,(s)}$ and Fe^{2+}–$Fe_3O_{4\,(s)}$ boundaries calculated below. In other words, the oxide minerals become more soluble when the solution becomes more acidic, as we already know from experience.

(3) Fe^{2+}–$Fe_2O_{3\,(s)}$ boundary:

$$2Fe^{2+} + 3H_2O_{(l)} \Rightarrow Fe_2O_{3\,(s)} + 6H^+ + 2e^-$$

$$\Delta G_r^0 = +146.9\,\text{kJ} \quad (8.50)$$

$$E^0 = \frac{\Delta G_r^0}{n\Im} = \frac{+146.9}{(2)\,(96.485)} = +0.761 \text{ volts}$$

$$Eh = E^0 + \frac{0.0592}{n}\log\frac{(a_{H^+})^6}{(a_{Fe^{2+}})^2}$$
$$= 0.761 - 0.0592 \log a_{Fe^{2+}} - 0.1776 \text{ pH} \quad (8.51)$$

If we choose $a_{Fe^{2+}} = 10^{-6}$, as we did for the Fe^{3+}–$Fe_2O_{3\,(s)}$ boundary,

$$Eh = 0.761 - (0.0592)\,(-6) - 0.1776 \text{ pH}$$
$$= 1.1162 - 0.1776 \text{ pH} \quad (8.52)$$

The boundary, which is a function of both pH and *Eh*, is a straight line that can be plotted easily by calculating values of *Eh* for two arbitrarily chosen values of pH (e.g., pH=0 and pH=10).

(4) Fe^{2+}–$Fe_3O_{4\,(s)}$ boundary:

$$3Fe^{2+} + 4H_2O_{(l)} \Rightarrow Fe_3O_4 + 8H^+ + 2e^- \quad \Delta G_r^0 = +205.7\,\text{kJ} \quad (8.53)$$

$$E^0 = \frac{\Delta G_r^0}{n\Im} = \frac{+205.7}{(2)\,(96.485)} = +1.066 \text{ volts}$$

$$Eh = E^0 + \frac{0.0592}{n}\log\frac{(a_{H^+})^8}{(a_{Fe^{2+}})^3}$$
$$= 1.066 - 0.0888 \log a_{Fe^{2+}} - 0.2368 \text{ pH} \quad (8.54)$$

If we choose $a_{Fe^{2+}} = 10^{-6}$ as before,

$$Eh = 1.066 - (0.0888)\,(-6) - 0.2368 \text{ pH}$$
$$= 1.5988 - 0.2368 \text{ pH} \quad (8.55)$$

The boundary is a straight line that can be plotted easily by calculating *Eh* for two arbitrarily chosen values of pH (e.g., pH=0 and pH=10).

(5) $Fe_3O_{4\,(s)}$–$Fe_2O_{3\,(s)}$ boundary:

$$2Fe_3O_{4\,(s)} + H_2O_{(l)} \Rightarrow 3Fe_2O_{3\,(s)} + 2H^+ + 2e^-$$

$$\Delta G_r^0 = +29.3\,\text{kJ} \quad (8.56)$$

$$E^0 = \frac{\Delta G_r^0}{n\Im} = \frac{+29.3}{(2)\,(96.485)} = +0.152 \text{ volts}$$

$$Eh = E^0 + \frac{0.0592}{n}\log\frac{(a_{H^+})^2}{1} = 0.152 - 0.0592 \text{ pH} \quad (8.57)$$

This straight-line boundary, which separates the stability fields of magnetite and hematite in equilibrium with water, has the same slope as the lines defining the stability limits of water.

(6) $Fe_{(s)}$–Fe^{2+} boundary:

$$Fe \Rightarrow Fe^{2+} + 2e^- \qquad \Delta G_r^0 = -90.0\,\text{kJ} \quad (8.58)$$

$$E^0 = \frac{\Delta G_r^0}{n\Im} = \frac{-90.0}{(2)\,(96.485)} = -0.466 \text{ volts}$$

$$Eh = E^0 + \frac{0.0592}{n}\log\frac{(a_{Fe^{2+}})^2}{a_{Fe}} = -0.466 + 0.0296 \log a_{Fe^{2+}} \quad (8.59)$$

This reaction boundary is independent of pH and plots as a straight line parallel to the pH axis. For $a_{Fe^{2+}} = 10^{-6}$, the *Eh* coordinate for this boundary is

$$Eh = -0.466 + (0.0296)\,(-6) = -0.644 \text{ volts} \quad (8.60)$$

(7) $Fe_{(s)}$–$Fe_3O_{4\,(s)}$ boundary:

$$3Fe_{(s)} + 4H_2O_{(l)} \Rightarrow Fe_3O_{4\,(s)} + 8H^+ + 8e^- \quad \Delta G_r^0 = -64.3\,\text{kJ} \quad (8.61)$$

$$E^0 = \frac{\Delta G_r^0}{n\Im} = \frac{-64.3}{(8)\,(96.485)} = -0.083 \text{ volts}$$

$$Eh = E^0 + \frac{0.0592}{n}\log\frac{(a_{H^+})^8}{1} = -0.083 - 0.0592 \text{ pH} \quad (8.62)$$

The slope of this reaction boundary is parallel to the $Fe_3O_{4\,(s)}$–$Fe_2O_{3\,(s)}$ boundary.

The $Fe_{(s)}$–Fe^{2+} and $Fe_{(s)}$–$Fe_3O_{4\,(s)}$ boundaries lie below the lower stability limit of water, indicating that metallic iron cannot exist in equilibrium with Fe^{2+} or Fe_3O_4 in surface aqueous environments at any pH. Some striking aspects of the *Eh*–pH diagram for the Fe–O–H_2O system (Fig. 8.6) are the highly restricted field of Fe^{3+} predominance, the large stability field of hematite, and the requirement of reducing conditions and high pH for the stability of magnetite.

We can make Fig. 8.6 even more relevant to natural aqueous systems by adding stability fields of other common Fe-bearing minerals, such as siderite ($FeCO_3$), pyrite (FeS_2), and pyrrhotite ($Fe_{1-x}S$). Evidently, the construction of *Eh*–pH diagrams is tedious and repetitive, especially for complex systems involving a lot of reactions. Computer codes, however, are available for such tasks (see, e.g., discussion by Linkson *et al.*, 1979). Even a simple EXCEL spreadsheet can be of considerable help in performing the calculations and constructing an *Eh*–pH diagram.

8.6.3 *Geochemical classification of sedimentary redox environments*

Eh–pH diagrams are very helpful in characterizing the stability of minerals in sedimentary environments, but their application to the classification of sedimentary environments is limited by two factors. First, the range of pH in most marine and nonmarine subaqueous sediments is relatively narrow, commonly between 6 and 8. Second, measuring the *Eh* of natural environments is difficult because many of the species – such as SO_4^{2-}, NO_3^-, N_2, NH_4^+, HCO_3^-, CH_4—that are involved in important sedimentary oxidation–reduction reactions are not electroactive, i.e., they do not readily accept or release electrons at the surface of a platinum electrode used for *Eh* measurements. These limitations led Berner (1981) to propose a simplified classification of sedimentary redox environments (Table 8.3) based on the concentrations of dissolved oxygen (O_2) and total dissolved sulfide (H_2S+HS^-). In addition to being readily measurable, these two parameters are intimately involved in oxidation–reduction reactions, and they affect strongly both the ecology of organisms living in sediments and authigenic mineralogy.

The two major subdivisions in this classification scheme, *oxic* and *anoxic*, are based on the presence or absence of measurable dissolved O_2. In oxic environments, the organic matter is completely decomposed by aerobic degradation and there is not enough dissolved sulfide for precipitation of sulfide minerals. Anoxic environments (no measurable dissolved O_2) are classified as *sulfidic* or *nonsulfidic* depending on the presence or absence of measurable dissolved sulfide. Anoxic sulfidic conditions occur almost entirely due to biotic reduction of sulfate to H_2S and HS^- accompanying organic matter decomposition. Anoxic nonsulfidic environments are further subdivided into *post-oxic* and *methanic* based on the kind of reactions by which the organic matter is decomposed. In post-oxic environments, organic matter decomposition takes place, successively, by nitrate, manganese, and iron reduction (but not by sulfate reduction). Nonsulfidic methanic environments are highly reducing and are characterized by the formation of dissolved methane. A toxic environment of H_2S-laden bottom waters in a basin with restricted circulation is sometimes referred to as *euxinic*. The Black Sea is the best example of a euxinic basin.

Table 8.3 Geochemical classification of sedimentary environments proposed by Berner (1981).

Environment	Limits of dissolved O_2 or total sulfide (represented by H_2S) concentration	Characteristic authigenic phases (assuming that the sediments originally contain enough iron and manganese)
Oxic*	$O_2 \geq 1\,\mu M$	Hematite (α–Fe_2O_3), goethite (α–FeOOH), MnO_2-type minerals; no organic matter
Anoxic	$O_2 < 1\,\mu M$	
Sulfidic*	$H_2S \geq 1\,\mu M$	Pyrite (FeS_2), marcasite (FeS_2), rhodochrosite ($MnCO_3$), albandite (MnS); organic matter
Nonsulfidic	$H_2S < 1\,\mu M$	
Post-oxic		Glauconite** and other low-temperature Fe^{2+}–Fe^{3+} silicates, vivianite [$Fe_3(PO_4)_2.8H_2O$]), siderite ($FeCO_3$); no sulfide minerals; minor organic matter
Methanic		Siderite, vivianite, rhodochrosite, earlier formed sulfide minerals; organic matter

*An environment that is both anoxic and sulfidic is sometimes referred to as "euxinic."

**Glauconite, $K_{1.6}(Fe^{3+}, Fe^{2+}, Al, Mg)_{4.0}(Si, Al)_8O_{20}(OH)_4$, is a mica mineral that occurs almost exclusively in marine sediments, particularly in greensands.

8.7 Role of microorganisms in oxidation–reduction reactions

8.7.1 *Geochemically important microorganisms*

Microorganisms play an important role, as catalysts, in many geologic processes that involve solubilization, transport, and deposition of metals resulting from oxidation and reduction reactions (Ehrlich, 2002). They can manufacture or decompose organic matter, produce or consume oxygen, and oxidize or reduce sulfur, iron, nitrogen, and a variety of other substances. Microrganisms may also influence the rate of a geochemical reaction in which they are producers or consumers of any of the reactants or products. It is likely that microorganisms contributed significantly to the formation of some types of mineral deposits such as banded iron formations, placer gold deposits, roll-front uranium deposits, Mississippi Valley-type (MVT) Pb–Zn deposits, and sedimentary exhalative (SEDEX) Zn–Pb–Ag deposits (Southam and Saunders, 2005), and participated in the supergene enrichment of porphyry copper deposits (Enders *et al.*, 2006).

Archaea and Bacteria domains

In the past, microorganisms were classified into two megagroups – *procaryotes* and *eukaryotes* – depending on their cell structure. Prokaryotes, believed to have been the most primitive life forms on the Earth, have a simple cellular structure with rigid cell walls but lack a nucleus; eukaryotes, on the other hand, have a true nucleus and a more complex cell structure than prokaryotes. The prokaryotes, in turn, were divided into two domains: the *Archaea* and the *Bacteria*. More detailed studies of the cell structures, made possible with the advent of the electron-microscope, have shown that, despite similarity in shape and size, the evolutionary history and biochemistry of the Archaea and the Bacteria are quite distinctive. This has led to the currently popular classification of microorganisms into three domains – the *Archaea* (formerly known as *Archaebacteria*), the *Bacteria* (formerly designated as *Eubacteria*), and the *Eukarya* – as had been originally suggested by Dr Carl Woese in 1978. The three domains, it is believed, were derived from an unknown common ancestor way back in time, and they followed separate evolutionary paths.

Archea are found in a broad range of habitats: soils, wetlands, hot springs, salt lakes, guts of humans and ruminants where they aid in the digestion of food, and especially in oceans; the archaea in plankton may be one of the most abundant groups of organisms on the planet. Methanogenic archaea are part of the biotechnology used in the production of biogas and sewage treatment. Bacteria are ubiquitous in every habitat on Earth, ranging from soils to radioactive wastes to live bodies of plants and animals. Bacteria play an important role in many geochemical processes: cycling of nutrients such as nitrogen; oxidation–reduction reactions involving iron, manganese, or sulfur; nitrifying (oxidation of NH_4^+ ion to NO_3^- ion) and denitrifying (reduction of NO_3^- ion to N_2 gas, N_2O gas, or NH_4^+ ion) reactions; bioleaching; and biooxidation. Unfortunately, bacteria also cause infectious diseases such as cholera, leprosy, bubonic plague, tuberculosis, etc. Eukaryotic microorganisms include algae, fungi, actinomycetes (unicellular organisms that show similarities to both fungi and bacteria), and protozoa.

Assimilatory and dissimilatory metabolism

Metabolism (cellular biochemical activities) of bacteria is directed toward achieving two goals: growth and duplication. Bacterial metabolism has two components (Ehrlich, 2002): *dissimilatory metabolism* (or *catabolism*), the activity of obtaining energy; and *assimilatory metabolism* (or *anabolism*), the activity of obtaining building blocks. Dissimilatory metabolism provides the cell with needed energy through energy conservation, which is achieved by the oxidation of a suitable nutrient, and it may also yield to the cell some compounds that can serve as building blocks for polymers. Dissimilatory metabolism can occur in the presence or absence of oxygen, and take the form of aerobic respiration, anaerobic respiration, fermentation, or methanogenesis (see section 8.7.2). Large-scale oxidation of iron, manganese, and sulfur may involve dissimilatory metabolism.

Assimilatory metabolism deals with assimilation of nutrients for the formation of organic building blocks such as DNA, RNA, and proteins, and is an energy-requiring process. By contributing to an increase in cellular mass and duplicaion of vital molecules, assimilatory metabolism makes growth and reprodution possible. The two types of metabolism are tightly coupled in the sense that assimilatory metabolism depends on dissimilatory metabolism to provide energy and some of the building blocks.

Aerobes and anaerobes

In terms of oxygen requirement, microorganisms can be grouped as *aerobes* (oxygen-requiring microbes), and *anaerobes* (oxygen-shunning microbes). Aerobes can survive and grow in an oxygenated environment because they directly use molecular oxygen as the terminal electron acceptor. Anaerobes do not require molecular oxygen for their growth and may even die in its presence; they use other electron-poor species (such as SO_4^{2-}, NO_3^-) as the terminal electron acceptor. *Obligate anaerobes* can only function in the absence of oxygen; *facultative anaerobes* can use for growth either oxygen, when it is available, or other electron acceptors (e.g., NO_3^- Fe^{3+}, or organic compounds) when oxygen is not available. *Aerotolerant organisms* cannot use oxygen for growth, but can tolerate its presence.

Hyperthermophiles, thermophiles, mesophiles, and psychrophiles

From a consideration of preference for habitats, microorganisms are divided into several groups. *Hyperthrermophiles* (or *extreme thermophiles*) thrive in extremely hot environments (> 60°C; optimal temperature > 80°C). Most hyperthermophiles belong to the domain Archaea, although some bacteria are able to tolerate temperatures of around 100°C. *Thermophiles* thrive at relatively high temperatures (45° to 80°C); many thermophiles are archaea. *Mesophiles* grow best in moderate temperatures, typically between 15°C and 40°C, and are inacapable of growth above around 45°C. The optimal temperature of many pathogenic mesophiles is 37°C, the normal human body temperature. Mesophillic organisms have important uses in the preparation of cheese, yogurt, beer, and wine. *Psychrophiles* are capable of growth and reproduction in cold temperatures (−10 to 20°C), and they include both bacteria and archaea. They are widespread on the Earth, from arctic soils to high-latitude and deep ocean waters to continental and alpine glaciers, and are of particular interest in the investigations of possible extraterrestrial life and microbially mediated low-temperature geochemical processes. *Acidophiles* prefer to live in a relatively acidic environment, *halophiles* prefer high-saline environments, and *thermoacidophiles* flourish in high-temperature and low-pH environments.

Autotrophs and heterotrophs

Microorganisms need energy and carbon for their metabolism, and are classified as *autotrophs* and *heterotrophs* depending on the sources of energy and nutrients (Table 8.4). Autotrophs obtain carbon by assimilating inorganic nutrients (CO_2, CO, HCO_3^-, or CO_3^{2-}) and use external energy for the synthesis of complex organic compounds such as carbohydrates. Heterotrophs, on the other hand, obtain carbon to manufacture their own cell constituents by breaking down previously synthesized organic matter, and derive energy from the oxidation of organic compounds. Autotrophs comprise two broad groups: *photoautotrophs*, which acquire energy from sunlight by photosynthesis; and *chemoautotrophs*, for which the source of energy for organic carbon synthesis is chemical energy, derived from a change in the oxidation state of an energy-producing molecule such as Fe^{2+}, ammonia, or sulfide. Analogous to autotrophs, heterotrophs are also classified as *photoheterotrophs* and *chemoheterotrophs* depending on the source of energy (Table 8.4). Both autotrophs and heterotrophs can be either aerobic or anaerobic.

8.7.2 *Examples of oxidation–reduction reactions mediated by microorganism*

Aerobic degradation (Aerobic respiration)

Respiration, in a biological context, refers to oxidation utilizing an electron transport system that may operate with either oxygen or another external, reducible inorganic or organic compound as a terminal electron acceptor (Ehrlich, 2002). In natural environments, the major oxidizing agent is atmospheric oxygen, and the major reducing agent is organic carbon produced by photosynthesis, which may be represented in a simplified form as

$$CO_{2\,(g)} + H_2O_{(l)} + h\nu = CH_2O + O_{2\,(g)} \tag{8.63}$$

where $h\nu$ represents energy from sunlight (h=Planck's constant and ν=the frequency of light) and CH_2O (carbohydrate) the organic matter. Oxidation of the organic matter, in turn, by water produces $CO_{2\,(g)}$ and electrons:

$$CH_2O_{(aq)} + H_2O_{(l)} \Rightarrow CO_{2(g)} + 4H^+ + 4e^- \tag{8.64}$$

In aerobic environments oxygen serves as the terminal electron acceptor. As long as free molecular oxygen is available in the system, the H^+ ions and electrons are conveyed by an electron transport system (certain bacteria) to oxygen to produce water,

$$O_{2(g)} + 4H^+ + 4e^- \Rightarrow 2H_2O_{(l)} \tag{8.65}$$

Table 8.4 Classification of microorganisms in terms of their energy and carbon sources.

Organism classification	Sources of nutrients and energy	Examples
Autotroph	An organism that produces complex organic compounds by assimilating carbon as simple inorganic molecules (CO_2, HCO_3^-, or CO_3^{2-}) and using energy from external sources	
Photoautotroph	An autotroph that obtains energy from sunlight (by photosynthesis)	Green plants, most algae, cyanobacteria, some purple and green bacteria
Chemoautotroph	An autotroph that obtains energy from electron-donating molecules in the environment. These molecules may be organic (organotrophic autotroph) or inorganic (lithotrophic autotroph)	Some algae, most purple and green bacteria, some cyanobacteria
Heterotroph	An organism that acquires most or all of its carbon for growth from previously synthesized organic compounds	
Photoheterotroph	A heterotroph that uses sunlight for energy	Purple nonsulfur bacteria, green nonsulfur bacteria, heliobacteria
Chemoheterotroph	A heterotroph that derives energy from the breakdown of organic molecules such as glucose (organotrphic heterotroph) or inorganic molecules (lithotrophic heterotroph)	Prokaryotes, fungi, animals, and even some plants; all microorganisms that live the human body (e.g., *E. coli*)

and the net oxidation–reduction reaction (the sum of reactions 8.64 and 8.65) representing aerobic respiration is

$$CH_2O_{(aq)}+O_{2(g)} \Rightarrow CO_{2(g)}+H_2O_{(l)} \qquad \Delta G^1_{r,298.15}=-501.8\,\text{kJ mol}^{-1} \tag{8.66}$$

Other geochemically important examples of aerobic degradation (oxygen consumption), as listed in Table 8.5, include oxidation of Fe^{2+} to Fe^{3+}, which may be precipitated as FeOOH (equation 8.67), and of HS^- to SO_4^{2-} (equation 8.68), which may be precipitated as a sulfate mineral such as barite, $BaSO_4$ (Senko *et al.*, 2004).

Anaerobic degradation (Anaerobic respiration)

Anaerobic respiration (carbon dioxide consumption) generally occurs when all oxygen has been removed from the system by aerobic degradation. In the absence of oxygen, the role of the terminal electron acceptor is assumed by other reducible compounds such as $CO_{2\,(g)}$. The process is commonly mediated by certain anaerobic archaea and bacteria (the two domains of prokaryotes), including some cyanobacteria (Ehrlich, 2002). Some ecologically important examples of anaerobic degradation are (Table 8.5): ferric iron (Fe^{3+}) reduction (equation 8.69), nitrate (NO_3^-) reduction or denitrification (equation 8.70), and sulfate (SO_4^{2-}) reduction (equation 8.71). The terminal electron acceptor is a Fe^{3+}-compound such as FeOOH for ferric iron reduction, NO_3^- for nitrate reduction, and SO_4^{2-} for sulfate reduction. The metabolic products of these reductions (e.g., Fe^{2+} and HS^-) are reactive and they participate in subsequent formation of minerals such as siderite ($FeCO_3$) and base-metal sulfides.

Methanogenesis

Methanogenesis is the biologic production of methane, which occurs under even more reducing conditions than bacterial sulfate reduction. In environments (e.g., freshwater wetlands, water-logged soils, and deeply buried low-sulfate marine sediments) that lack external electron acceptors, certain microorganisms are capable of mediating reaction between organic carbon and water to produce $CO_{2\,(g)}$ and $CH_{4\,(g)}$ (methane) (equation 8.72; Table 8.5).

Table 8.5 Examples of oxidation–reduction reactions catalyzed by microorganisms.

Process		Reaction	$\Delta G^1_{r,\,298.15}$ (kJ mol^{-1})*
Organic C oxidation (8.66)	Ox	$CH_2O_{(aq)}+H_2O_{(l)} \Rightarrow CO_{2(g)}+4H^++4e^-$	
	Red	$O_{2(g)}+4H^++4e^- \Rightarrow 2H_2O_{(l)}$	
	Ox–Red	$CH_2O_{(aq)}+O_{2(g)} \Rightarrow CO_{2(g)}+H_2O_{(l)}$	−501.8
Fe^{2+} oxidation (8.67)	Ox	$FeCO_{3(s)}+2H_2O_{(l)} \Rightarrow FeOOH+HCO_3^-+2H^++2e^-$	
	Red	$O_{2(g)}+4H^++4e^- \Rightarrow 2H_2O_{(l)}$	
	Ox–Red	$2FeCO_{3(s)}+O_{2(g)}+2H_2O \Rightarrow 2FeOOH+2HCO_3^-$	−91.6
Sulfide oxidation (8.68)	Ox	$HS^-+4H_2O_{(l)} \Rightarrow SO_4^{2-}+9H^++8e^-$	
	Red	$O_{2(g)}+4H^++4e^- \Rightarrow 2H_2O_{(l)}$	
	Ox–Red	$HS^-+2O_{2(g)} \Rightarrow SO_4^{2-}+H^+$	−788.8
Fe^{3+} reduction (8.69)	Ox	$CH_2O_{(aq)}+H_2O_{(l)} \Rightarrow CO_{2(g)}+4H^++4e^-$	
	Red	$FeOOH_{(s)}+3H^++e^- \Rightarrow Fe^{2+}+2H_2O_{(l)}$	
	Ox–Red	$CH_2O_{(aq)}+4FeOOH_{(s)}+8H^+ \Rightarrow 4Fe^{2+}+CO_{2(g)}+7H_2O_{(l)}$	−498.8
Nitrate reduction (8.70)	Ox	$CH_2O_{(aq)}+H_2O_{(l)} \Rightarrow CO_{2(g)}+4H^++4e^-$	
	Red	$2NO_3^-+12H^++10e^- \Rightarrow N_{2(g)}+6H_2O_{(l)}$	
	Ox–Red	$2.5CH_2O_{(aq)}+2NO_3^-+2H^+ \Rightarrow 2.5CO_{2(g)}+N_{2(g)}+3.5H_2O_{(l)}$	−1270.0
Sulfate reduction (8.71)	Ox	$CH_2O_{(aq)}+H_2O_{(l)} \Rightarrow CO_{2(g)}+4H^++4e^-$	
	Red	$SO_4^{2-}+9H^++8e^- \Rightarrow HS^-+4H_2O_{(l)}$	
	Ox–Red	$2CH_2O_{(aq)}+SO_4^{2-}+H^+ \Rightarrow 2CO_{2(g)}+HS^-+2H_2O_{(l)}$	−214.8
Methane formation (8.72)	Ox	$CH_2O_{(aq)}+H_2O_{(l)} \Rightarrow CO_{2(g)}+4H^++4e^-$	
	Red	$CO_{2(g)}+8H^++8e^- \Rightarrow CH_{4(g)}+2H_2O_{(l)}$	
	Ox–Red	$2CH_2O_{(aq)} \Rightarrow CH_{4(g)}+CO_{2(g)}$	−185.7

*Calculated using $\Delta \bar{G}^1_{f,\,298.15}$ values listed in Appendix A-5.
Ox=oxidation, Red=reduction, Ox–Red=oxidation–reduction.

In fact, methanogenesis is a major contributor of methane, a greenhouse gas, to the atmosphere. In many landfills designated for domestic waste, the combustible gas methane is produced in such abundance that it must be collected and removed. Some landills produce sufficient amounts of methane to be harvested and burned with natural gas in order to generate electricity.

The free energy change for each oxidation–reduction reaction listed in Table 8.5 is negative, indicating that each reaction should proceed to the product side, which is thermodynamically more stable. Microorganisms merely act as catalysts for these oxidation–reduction reactions; i.e., they facilitate the transfer of electrons. Except for some minor overlaps, the reactions for organic matter decomposition occur more or less in the following succession (Stumm and Morgan, 1981): aerobic respiration, anaerobic respiration when all oxygen has been removed from the system, and finally methanogenesis when all sulfate and nitrate have been used up.

8.8 Oxidation of sulfide minerals

8.8.1 *Mediation by microorganisms*

The solubility of sulfide minerals in water is quite low, even at low pH conditions, but sulfides readily oxidize in the presence of water and dissolved oxygen. In nature, the oxidation of sulfide minerals commonly involves mediation by microorganisms (Singer and Stumm, 1970; Rawlings and Woods, 1995; Ehrlich, 1996, 2002; Nordstrom and Southam, 1997; Rawlings, 1997, 2002; Nordstrom and Alpers, 1999; Blowes *et al.*, 2003). The interaction may be direct or indirect. In *direct interaction,* oxidation of a metal sulfide (MS) to soluble sulfate occurs at the mineral surface with which the microorganisms are in contact:

$$M^{2+}S + 2O_2 \Rightarrow M^{2+} + SO_4^{2-} \qquad (8.73)$$

Cell attachment to the mineral particles takes place within minutes or hours, with the cells preferentially occupying irregularities on the surface structure. In *indirect interaction*, free-floating microorganisms catalyze the generation of an oxidant (e.g., Fe^{3+} from Fe^{2+} present in the mineral), which then oxidizes the sulfide mineral (and regenerates Fe^{2+}):

$$2M^{2+}SO_4 + H_2SO_4 + 0.5O_2 \Rightarrow M_2^{3+}(SO_4)_3 + H_2O \qquad (8.74)$$

$$M^{2+}S + 2M^{3+}(SO_4)_3 \Rightarrow 3M^{2+}SO_4 + S^0 \qquad (8.75)$$

The S^0 is oxidized by bacteria to regenerate H_2SO_4 (reaction 8.79).

Most studies to date in this context have focused on the reactivity and kinetics of two species of autotrophic and acidophilic bacteria: *Thiobacillus ferrooxidans* (recently renamed *Acidithiobacillus ferrooxidans*), an iron and sulfur oxidizer; and *Leptospirillum ferrooxidans*, an iron oxidizer. The species *T. ferrooxidans* is capable of utilizing the oxidation of Fe^{2+} ($Fe^{2+} \Rightarrow Fe^{3+} + e^-$), H_2S (e.g., $H_2S + 2O_2 \Rightarrow H_2SO_4$), S^o (e.g., $2S^0 + 3O_2 + 2H_2O \Rightarrow 2H_2SO_4$), or sulfide minerals (e.g., $CuS + 2O_2 \Rightarrow CuSO_4$) as a source of energy for metabolic activity. *Thiobacillus ferrooxidans* is also able to grow using Fe^{3+} as an electron acceptor provided reduced sulfur compounds are available to serve as an electron donor. The species *Leptospirillum ferrooxidans*, which is as ubiquitous as *T. ferrooxidans,* lacks the capacity to oxidize sulfur compounds and prefers the selective oxidation of Fe^{2+} (Blowes *et al*, 2003). Both *T. ferrooxidans* and *L. ferrooxidans* occur in similar environments, but *T. ferrooxidans* has a preferred pH range of 1.5–2.5, whereas *L. ferrooxidans* is more tolerant to a wider range of temperature and pH. Another important species is *Thiobacillus thiooxidans*, which is able to use only reduced sulfur compounds, but together with *L. ferrooxidans,* can rapidly degrade a variety of sulfide minerals. All of these bacteria are common in acidic environments where they play a fundamental role in the oxidation of pyrite.

8.8.2 *Oxidation of pyrite*

As an example of sulfide oxidation, let us consider the oxidation of pyrite (FeS_2), which is by far the most abundant sulfide mineral in rocks and base-metal sulfide deposits. The oxidation can occur when the surface of a pyrite crystal is exposed to an oxidant in an oxic or anoxic system. The process involves two important microbially mediated reactions (Hutchins *et al.*, 1986; Rawlings and Woods, 1995): (i) oxidation of sulfur (S^- to S^{6+}),

$$FeS_2 + 3.5O_2 + H_2O \Rightarrow FeSO_4 + H_2SO_4 \qquad (8.76)$$

producing soluble ferrous sulfate ($FeSO_4$) and an acidic environment conducive to the growth of iron-oxidizing bacteria; and (ii) oxidation of Fe^{2+} in $FeSO_4$ to Fe^{3+} in ferric sulfate ($Fe_2(SO_4)_3$),

$$2FeSO_4 + 0.5O_2 + H_2SO_4 \Rightarrow Fe_2(SO_4)_3 + H_2O \qquad (8.77)$$

which remains soluble at low pH (pH < 2.5). Ferric sulfate is a strong oxizing agent, capable of oxidizing pyrite (as well as a wide variety of other metal sulfides), even in the absence of oxygen or viable bacteria:

$$FeS_2 + Fe_2(SO_4)_3 \Rightarrow 3FeSO_4 + 2S^0 \qquad (8.78)$$

Similar oxidation reactions involving $Fe_2(SO_4)_3$ can be written for other metal sulfides such as chalcopyrite ($CuFeS_2$), chalcocite (Cu_2S), covellite (CuS), and bornite (Cu_5FeS_4).

The next step is microbially mediated oxidation of elemental sulfur to sulfuric acid:

$$2S^0 + 3O_2 + 2H_2O \Rightarrow 2H_2SO_4 \qquad (8.79)$$

which maintains the pH at levels favorable to bacteria and regenerates $Fe_2(SO_4)_3$ via reaction (8.77).

The ultimate product of pyrite oxidation is an insoluble chemical precipitate of ferric complexes and minerals formed by hydrolysis of ferric sulfate:

$$Fe_2(SO_4)_3 + 6H_2O \Rightarrow 2Fe(OH)_{3(s)} + 3H_2SO_4 \quad (8.80)$$

The solid product of reaction (8.80) is commonly referred to as "yellowboy" or "limonite." It is more likely to be a mixture of hydrous phases with variable stoichiometry such as goethite (α–FeOOH), ferrihydrite ($Fe_5(OH)_8.4H_2O$) and jarosite ($KFe_3(SO_4)_2(OH)_6$), rather than pure ferric hydroxide (Nordstrom and Southam, 1997). The reaction is pH dependent: under very acidic conditions (pH < about 3.5) the solid material does not form and Fe^{3+} remains in solution. Pyrite oxidation in an aqeous environment is easily recognized by the yellowish to reddish color of Fe^{3+}-precipitates and the reddish brown color of the water.

Pyrite oxidation by reaction (8.76) is mediated by bacteria capable of attacking the surface of pyrite crystals directly, but the microbial oxidation rate of pyrite by oxygen is very sluggish (on the order of 8.8×10^{-8} mol m^{-2} s^{-1}; Olson, 1991) and similar to the abiotic oxidation rate of pyrite by either oxygen or Fe^{3+} (0.3 to 3×10^{-9} mol m^{-2} s^{-1} and 1 to 2×10^{-8} mol m^{-2} s^{-1}, respectively; Nordstrom and Alpers, 1999). However, when catalyzed by free-floating bacteria, the rate of Fe^{2+} to Fe^{3+} oxidation (reaction 8.77), the rate-controlling step for the oxidation of pyrite by Fe^{3+} (reaction 8.78), is accelerated by about five orders of magnitude compared to the abiotic rate for the same reaction, from 3×10^{-12}mol l^{-1} s^{-1} to 5×10^{-7}mol l^{-1} s^{-1} (Singer and Stumm, 1970). Because significant concentrations of Fe^{3+} occur only at low pH values (see Fig. 8.6), it is believed that pyrite oxidation is initiated by reaction (8.76) at circumneutral pH. As the pyrite oxidation progresses, and the pH drops below about 4, the oxidation is accomplished mainly through microbial catalysis by the indirect mechanism (Nordstrom and Alpers, 1999).

8.8.3 *Acid mine drainage*

The net reaction for the oxidation of pyrite (sum of reactions (8.76), (8.77), and (8.80), and the dissociation of H_2SO_4), which may be written as

$$4FeS_2 + 15O_2 + 14H_2O_{(l)} \Rightarrow 4Fe(OH)_3 + 8SO_4^{2-} + 16H^+ \quad (8.81)$$

renders the water in the system more acidic than before. The greater the amount of pyrite oxidized, the lower would be the pH of the water. The oxidation of other sulfide minerals, such as chalcopyrite ($CuFeS_2$), in the system may also contribute to the acidity and dissolved metal content of the associated water:

$$2CuFeS_2 + 17.5O_2 + 5H_2O_{(l)} \Rightarrow 2Cu^+ + 2Fe(OH)_3 + 4SO_4^{2-} + 4H^+ \quad (8.82)$$

Similar reactions can be written for other common sulfide minerals such as pyrrhotite ($Fe_{1-x}S$), sphalerite (ZnS), galena (PbS), chalcocite (Cu_2S), covellite (CuS), etc.

Acid drainage typically is associated with the exploitation of base metal sulfide deposits (because mining and processing of the ore materials invariably generate large volumes of sulfide-bearing rock wastes and mill tailings that remain exposed to rain and surface drainage) and relatively sulfide-rich coal deposits (as in the southern Appalachians), and therefore is referred to as *acid mine drainage* (AMD). However, acid drainage can occur wherever rocks containing sulfide minerals are exposed to the atmosphere (e.g., in outcrops, road-cuts, quarries, abandoned mines, etc).

The drainage of acidic water into surface streams, rivers, and lakes is a potential environmental problem worldwide, because such waters typically have low pH values in the range of 2 to 4 and high concentrations of sulfate and metals (especially mercury, silver, and cadmium) known to be toxic to many organisms, including zooplankton, shellfish, and fish. In fact, in many major mining districts around the world, contamination of the surface drainage system by AMD has become a major environmental issue (Blowes *et al.*, 2003).

The acidity of water circulating within a pile of sulfidic material is reduced to some extent naturally by neutralizing chemical reactions with the gangue minerals within the pile, with concomitant reduction in the concentrations of some dissolved metals. The most significant neutralizing agents are carbonate, aluminum hydroxide, ferric oxyhydroxide, and aluminosilicate minerals. Natural attenuation, however, is commonly not adequate to eliminate AMD.

Meeting the challenge of AMD requires first to minimize the acid generation at the source by minimizing the access of the sulfidic material to oxygen, and then to prevent the acidic effluents from entering the surface drainage system by creating appropriate barriers. Controlling acid generation must involve steps to control one or more of the three essential components in the acid-generating process: sulfide minerals, water, and air (the source of oxygen). Steps that can be taken to achieve this objective include: blending of the sulfidic waste rock with enough waste rock of a net neutralizing potential to maintain a near-neutral pH of the water; adding materials (e.g., crushed limestone, lime, and soda ash) that can buffer acid-generating reactions; introducing bactericides, chemicals that can reduce the bacteria population responsible for catalyzing sulfide oxidation; and placing physical barriers, such as soil and synthetic cover, to minimize the infiltration of air and water into the acid-generating systems. A cover of oxygen-consuming materials, such as wood wastes, may also curtail the oxidation of underlying sulfide minerals. To be most effective, physical barriers must be applied soon after a waste pile reaches its planned size. To prevent discharge of the acid water into the local drainage system, a practice followed at many mine sites is to pool the effluent from the waste piles into a central treatment facility where it is treated chemically for acid neutralization and precipitation of the metals as a sludge. This preventive measure, however, does not eliminate the AMD issue because disposal of the potentially toxic sludge remains a problem.

For more details about the geochemistry of acid mine drainage as well as about preventive and remedial measures that can be taken to deal with the problem, the reader is referred to the recent article by Blowes *et al.* (2003).

8.8.4 *Bioleaching*

Many metals are recovered from ores that contain the metals as sulfide minerals. In essence, the recovery process consists of three steps: oxidation of the sulfides into oxides by roasting (heating below their melting points in the presence of oxygen from air; e.g., $2ZnS_{(s)}+3O_{2(g)} \Rightarrow 2ZnO_{(s)}+2SO_{2(g)}$); smelting of the oxides (reduction of the oxides to metals; e.g., $ZnO_{(s)}+C_{(s)} \Rightarrow Zn_{(s)}+CO_{(g)}$); and refining (purification) of the metals. This process, however, is not economic for very low-grade material. Oxidation of metal sulfides in aqueous environments, which creates environmental problems because of acid drainage, can be harnessed to extract desirable metals from mine wastes, low-grade ores, and even ore metal concentrates. The dissolution or leaching of metals from rocks was long believed to be a purely chemical reaction mediated by water and atmospheric oxygen, but it is now known to be primarily a biologically catalyzed process (Hutchins *et al.*, 1986).

Bioleaching is a process in which bacteria are utilized to leach a metal of value (e.g., copper, zinc, uranium, nickel, cobalt) from a mineral (usually a sulfide) and convert it into a soluble form (in most cases, a sulfate of the metal). The reactions occur under conditions that are close to atmospheric pressure and ambient temperature, thus requiring relatively little energy input. The process relies on microbially mediated solubilization through oxidation, reduction, and complexation, and recovery of the metals of intertest from the solution.

Iron-oxidizing bacteria may leach metal sulfides (e.g., MS_2) "directly" by oxidizing the sulfide into water-soluble sulfate (MSO_4) by a reaction similar to reaction (8.73):

$$MS_{2(s)}+3.5O_{2(g)}+H_2O_{(l)} \Rightarrow MSO_{4(aq)}+H_2SO_{4(aq)} \qquad (8.83)$$

The oxidation rate is vastly enhanced by the addition of Fe^{3+}, a strong oxidizing agent, to the system by reaction (8.74). This is easily accomplished by the fact that pyrite is ubiquitous in sulfidic rocks, ores, and wastes, and it is readily oxidized to $Fe_2(SO_4)_3$ by iron-oxidizing bacteria as discussed in section 8.8.2, resulting in "indirect" leaching of metals from their sulfide minerals by reactions such as:

$$MS_{2(s)}+Fe_2(SO_4)_{3(aq)} \Rightarrow M^{2+}_{(aq)}+MSO_{4(aq)}+2FeSO_{4(aq)}+2S^0_{(s)} \qquad (8.84)$$

Bacterially mediated oxidation of native sulfur regenerates H_2SO_4:

$$2S^0_{(s)}+3O_{2(g)}+2H_2O_{(l)} \Rightarrow 2H_2SO_{4(aq)} \qquad (8.79)$$

The replenishment of H_2SO_4 maintains the acidic environment favorable to the oxidizing bacteria and dissolution of minerals.

The low pH, metal-rich, inorganic mineral environments in which bioleaching reactions occur are populated by a group of bacteria that are highly adapted to growth under these conditions. The bacteria commonly isolated from inorganic mining environments are the acidophilic, mesophilic, iron-oxidizing bacteria *T. ferrooxidans* and *L. ferrooxidans*, which are the most important in most industrial bioleaching processes. As reported by Enders *et al.* (2006), samples from the weathering profile of the Morenci porphyry copper deposit (Arizona) contained up to 4×10^7 thiobacilli per gram, and bacterial iron oxidation at this site was estimated at 0.14 to 0.87 tons of Cu annually by a few kilograms of thiobacilli (Southam and Saunders, 2005). Other microorganisms that have been reported from bioleaching environments include the moderately thermophilic and autotrophic bacterium *Thiobacillus caldus*, and faculatively heterotrophic bacteria *Thiobacillus acidophilus* and *Thiobacillus cuprinus*, although heterotrophs do not seem to play a major role in most leaching operations.

Bioleaching has been employed *in situ* on deep ore deposits, resulting in large savings in the cost of bringing vast tonnages of ore and waste rock to the surface, but it is especially useful for the recovery of valuable metals from low-grade sulfide ores because of the lower cost compared to conventional extraction techniques. In addition, controlled microbial processing reduces the problem of acid mine drainage. The downside of bioleaching includes the generally sluggish rate of leaching and relatively low recovery (in the order of 6 months for ~60% recovery, and 9 months for ~ 85% recovery), and the potential risk of groundwater contamination through leakage of chemicals. Currently, bioleaching is being used to process ores of copper, cobalt, zinc, uranium, and gold in many countries such as Australia, Brazil, Chile, China, Ghana, Mexico, Namibia, Peru, South Africa, Tasmania, Uganda, and USA (Murr, 1980; Brierley, 1997; Schnell, 1997). Details of organisms and conditions used in commercial bioleach plants, however, are quite sparse in the published literature.

Copper ores are especially amenable to bioleaching because copper sulfate ($CuSO_4$) formed by the oxidation of copper sulfide minerals is very water-soluble. Typically, chalcopyrite ($CuFeS_2$) is the most abundant copper mineral in copper sulfide deposits and, therefore, also in low-grade ores and waste rock piles at the mines for these deposits. However, chalcopyrite has proved to be a stubborn candidate for obtaining copper recoveries greater than about 20% using normal bioleaching practices. At the present time, the commonly targeted copper minerals for bioleaching are chalcocite (Cu_2S) and covellite (CuS), which yield much better recovery (Songrong *et al.*, 2002). Inoculation of the ore heap with bacteria has been considered as an option to accelerate the bioleaching process, but bioleaching bacteria are ubiquitous and it is not clear to what extent inoculation would accelerate the process (Rawlings, 2002).

In essence, a bioleaching operation (Fig. 8.7), which includes adequate options for recycling of the solution through the different circuits for maximum metal recovery, starts with stacking the ore material into a heap, and sprinkling of the leach solution (lixiviant) at the top of the heap. The lixiviant can be water, water acidified with H_2SO_4, or spent acidic leach solution containing $Fe_2(SO_4)_3$ from a previous leaching cycle. In some operations aeration pipes are included in the pile to permit forced aeration and thus accelerate the bioleaching process. Oxidation by $Fe_2(SO_4)_3$ results in the leaching of copper from copper sulfide minerals in the form of soluble $CuSO_4$. The net reaction for bioleaching of copper from chalcocite may be represented as:

$$Cu_2S_{(s)} + 2Fe_2(SO_4)_{3(aq)} \Rightarrow 2CuSO_{4(aq)} + 4FeSO_{4(aq)} + S^0_{(s)} \quad (8.85)$$

Similar reactions can be written for other copper sulfide minerals such as covellite and chalcopyrite. The native sulfur is oxidized to sulfuric acid (reaction 8.77), which helps to maintain an acidic environment conducive to the growth of acidophilic bacteria. The leachate seeps to the bottom by gravity flow and is collected in special sumps. At the beginning of a leach operation, the leachate from the heap may be too low in copper for economic recovery, in which case it is fed back into the heap. Eventually, as the chemical and microbial activities continue, the leachate from the heap contains enough dissolved $CuSO_4$ to qualify as a *pregnant leach solution* (PLS).

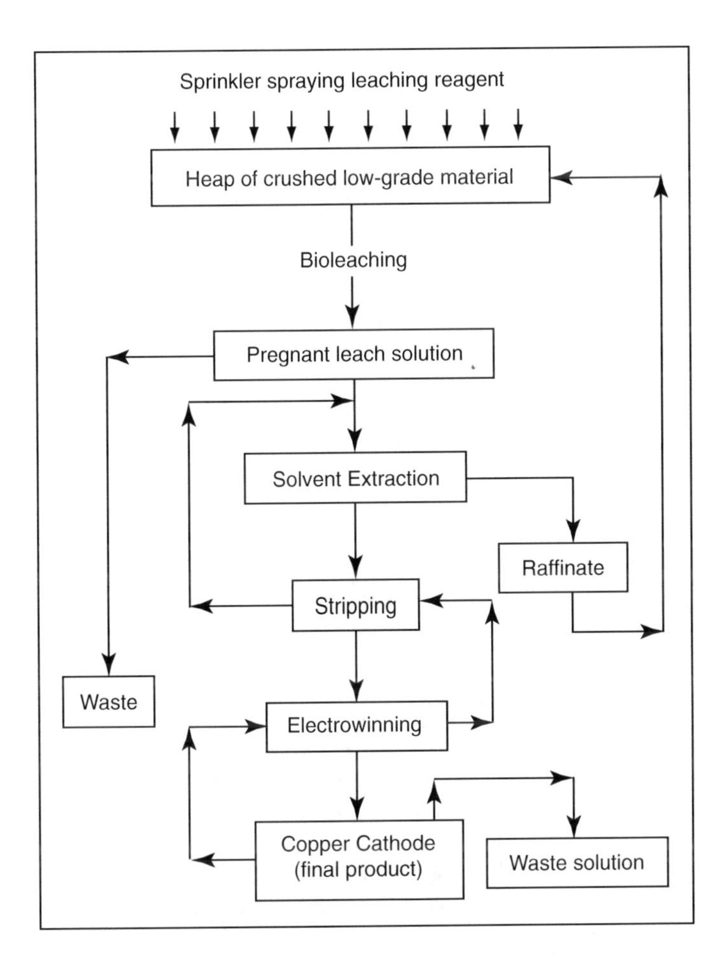

Fig. 8.7 Schematic flow sheet for bioleaching of low-grade copper ores.

Several techniques are available for stripping the copper metal from the PLS. A method widely used in the past involved the use of iron filings or sponge iron to precipitate copper by the oxidation–reduction reaction:

$$Cu^{2+}_{(aq)} + Fe_{(s)} \Rightarrow Cu_{(s)} + Fe^{2+}_{(aq)} \quad (8.86)$$

which is consistent with the relative positions of Fe^{2+} and Cu^{2+} in the electromotive series (Table 8.1). The process is simple and inexpensive, but the copper obtained has to be refined further by smelting to improve its purity. The modern method for copper metal recovery is by a process of solvent extraction followed by electrowinning. In the solvent extraction circuit, copper is extracted from the PLS by a selective organic solvent, which leaves other ions in the solution, and the barren solution (raffinate) is recycled for further leaching. The final step is the electrowinning of copper from the copper-containing solution by passing an electric current through it. High purity copper is deposited on copper cathode and the spent electrolyte is returned back to the stripping section.

Heap leaching has also been employed to recover gold from low-grade gold ores, the processing of which might not be economically viable otherwise. In these ores, gold is commonly contained within arsenopyrite (FeAsS) and pyrite (FeS_2). In the microbial leaching process, microorganisms (e.g., *T. ferrooxidans*) are able to attack and solubilize the gold-hosting minerals and release the trapped gold:

$$2FeAsS[Au]_{(s)} + 7O_{2(g)} + 2H_2O_{(l)} + H_2SO_{4(aq)} \Rightarrow Fe(SO_4)_{3(aq)} + 2H_3AsO_{4(s)} + [Au]_{(s)} \quad (8.87)$$

A dilute cyanide solution, such as NaCN, is used to solubilize gold as a soluble cyanide complex:

$$4Au_{(s)} + 8CN^-_{(aq)} + O_{2\,(g)} 2H_2O_{(l)} \Rightarrow 4[Au(CN)_2]^-_{(aq)} + 4OH^-_{(aq)} \quad (8.88)$$

This reaction proceeds very fast because of the very fine size of the gold particles, and a few hours of time usually are sufficient to dissolve the gold. Recovery of gold from the solution is achieved by electrolytic reduction or by a reduction reaction with zinc metal:

$$2Au(CN)^-_{2(aq)} + Zn_{(s)} \Rightarrow Zn(CN)^{2-}_{4\,(aq)} + 2Au_{(s)} \quad (8.89)$$

For further details about commercial bioleaching operations, the reader is referred to Ehrlich and Brierley (1990), Gaylarde and Videla (1995), and Rawlings (1997).

8.8.5 *Biooxidation*

Biooxidation is often used interchangeably with bioleaching, perhaps because both involve microbe-mediated oxidation by the same species of bacteria, but there are distinct differences between the two technologies. Biooxidation is mostly used at present for extraction of gold from refractory gold concentrates, which contain microparticles of the metal in very low concentrations and encapsulated by insoluble sulfides (e.g., arsenopyrite, FeAsS, or pyrite) that impede the solubilization of the gold. In bioleaching, the metal of interest is leached into a solution and then recovered from the solution; the solid residue is discarded as waste. In contrast, biooxidation causes selective dissolution of the undesired mineral matrix, and the solution is discarded; the concentration of the metal (in this case, gold) is enhanced in the solid phase, rendering the metal recovery by further treatment (e.g., cyanidation; reactions 8.88 and 8.89) more efficient. For example, the Yantai plant in China has achieved a biooxidation gold recovery of 96%, compared to only 10% without biooxidation (Songrong *et al.*, 2002). Large-scale commercial gold processing units are now in operation in South Africa, Brazil, Australia, Ghana, Peru, China, and USA. A schematic flowsheet for biooxidation of arsenical refractory gold ores is shown in Fig. 8.8.

8.8.6 *Biofiltration*

Biofiltration is a process used for removing undesirable volatile organic compounds (VOCs) and some inorganic compounds (such as H_2S and NH_3) from polluted air. It has been used extensively for more than 40 years in the USA and Europe for the control of odors from wastewater treatment facilities, composting facilities and other odor-producing operations, and is being used increasingly for treating high-volume, low-concentration air streams. Biofiltration is an attractive alternative to conventional air-pollution control technologies (e.g., thermal oxidizers, scrubbers) because of lower capital and operating costs, lower energy requirement, and the low volume of waste the process generates (Adler, 2001).

Biofilters are designed to convert gas-phase chemical compounds in an air stream to carbon dioxide, water, and inorganic salts. The contaminated air stream, after particulate removal, is sent to the bottom of a filter bed composed of soil, peat, composted organic material (such as wood or lawn waste), activated carbon, ceramic or plastic packing, or other inert or semi-inert media. The media provides a surface for microorganism attachment and growth. The filtration process is based on sorption of contaminants on organic media and their degradation by microorganisms in the filter bed. Food for the growth of microorganisms comes from the biooxidation of contaminants in the air stream or from the filter bed itself. Biological activiy in a filter bed eventually leads to degradation of a soil or compost media as organic matter is mineralized and the media particles are compacted. Degradable filter materials typically require replacement every 3–5 years.

Three parameters, in adition to the availability of oxygen for biooxidation and nutrients for microorganisms, determine the efficiency of a bioreactor. They are moisture, temperature, and pH (Boswell, 2002). A filter bed too dry would kill the microorganisms, and a filter bed too wet would drown or wash out some of the biomass. An appropriate level of moisture is maintained by passing the inflow air stream through a humidifier. Temperature controls the metabolism of microbes, typically doubling with each 10°C increase in temperature, as long as the organisms remain within their respective ranges of thermal tolerance. As a bioreactor contains hundreds and often thousands of species of microbes, the selection of operating temperature is a matter of experience and judgement in individual cases. A warm (approximately 25–30°C) biooxidation unit would generally support more organisms, both in terms of the number of species and the numbr of organisms. Microorganisms are also sensitive to pH, each species having its own survival and optimal range. Biofilters can function at pH values ranging from as low as 2–3 to as high as 8–9, with pH often changing (usually decreasing) after the operation begins. The choice of an operating pH range in a given case depends on the contaminants and the species of microorganism needed for their degradation.

For a detailed discussion of the design and operation of biofilters, the reader is referred to the review articles by Adler (2001) and Boswell (2002).

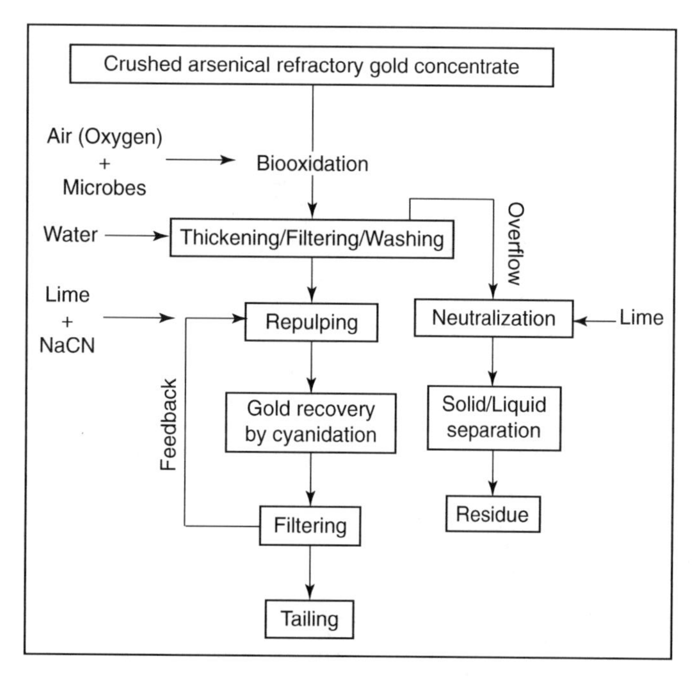

Fig. 8.8 Schematic flow sheet for biooxidation of low-grade arsenical gold ores.

8.9 Oxygen fugacity

Two methods are commonly used to measure and report the oxidation state of natural systems: the oxidation potential (*Eh*) or *pe*, as we have discussed in the earlier sections of this chapter, and the *oxygen fugacity* (fO_2). For conditions at or close to the Earth's surface, especially in aqueous systems, it is more convenient to use oxidation potential because it can be measured directly by an electrochemical method. At higher temperatures, especially in the absence of water, it is easier to use oxygen fugacity (fO_2), which can be measured directly. This, of course, works only for reactions involving molecular oxygen. For example, for the oxidation of magnetite (Fe_3O_4) to hematite (Fe_2O_3), the reaction may also be written as (instead of reaction (8.56), which involves water):

$$2Fe_3O_{4\,(s)}+0.5O_{2\,(g)} \Leftrightarrow 3Fe_2O_{3\,(s)} \tag{8.90}$$

If we choose the standard states as pure solids at *T* and *P* (the temperature and pressure of interest) for the solids, so that activity of each solid phase is unity, and ideal $O_{2\,(g)}$ at *T* K and 1 bar for the gaseous phase (see equation 5.52), then the activity of oxygen is numerically equal to its fugacity:

$$a_{O_2} = \frac{fO_2}{fO_2^0} = \frac{fO_2}{1} = fO_2 \tag{8.91}$$

The equilibrium constant for the reaction is:

$$K_{eq} = \frac{a^3_{Fe2O3(s)}}{a^2_{Fe3O4(s)}\, a^{0.5}_{O_2}} = \frac{1}{(fO_2)^{0.5}} \tag{8.92}$$

and the fO_2 for the reaction at equilibrium can be calculated using equation (5.37):

$$\log fO_2 = \frac{-\Delta G_r^0}{2.3026\, RT\, (-0.5)} \tag{8.93}$$

The fO_2 calculated using equation (8.93) and the *Eh* calculated using equation 8.57 are simply equivalent ways of expressing the same property, the oxidation state of chemical species within a given system.

It is evident from equation (8.93) that oxygen fugacity in a system is a function of temperature. As shown in Example 8–6, we can calculate the fO_2 for an oxidation reaction at any temperature (within the stability limit of the oxidized product) if we have the thermodynamic data to compute the free energy change for the reaction at that temperature. The effect of pressure on fO_2 can be calculated using equation (6.4).

Example 8–6: Calculation of oxygen fugacity as a function of temperature at 1 bar total pressure, assuming pure solid phases

Let us revisit the magnetite–hematite reaction:

$$2Fe_3O_{4\,(s)}+0.5O_{2\,(g)} \Leftrightarrow 3Fe_2O_{3\,(s)} \tag{8.90}$$

and calculate the fO_2 for this reaction at various temperatures.

This computation is a two-stage process. First, we calculate $\Delta G^0_{r,\,T}$ for the reaction at various arbitrarily chosen temperatures, using $\Delta \bar{G}_f$ values of the constituents at those temperatures (Table 8.6). We then calculate fO_2 at each temperature, using equation (8.93). For example, for the standard state T=298.15 K and P=1 bar,

$$\Delta G^0_r = 3\Delta \bar{G}^1_{f,\,298.15}\,(Fe_2O_3) - 2\Delta \bar{G}^1_{f,\,298.15}\,(Fe_3O_4)$$

$$= 3(-744.4) - 2(-1012.7) = -207.8\,\text{kJ} = -207800\,\text{J}$$

$$\log fO_2 = \frac{-207800}{(2.3026)\,(8.314)\,(298.15)(-0.5)} = -72.8\ \text{bar}$$

The calculated log fO_2 values at various temperatures are listed in Table 8.6 and plotted as a function of temperature in Fig. 8.9. Note that at low temperatures it does not take a whole lot of oxygen to oxidize magnetite to hematite!

8.9.1 Oxygen buffers

A plot of fO_2 as a function of temperature (at constant total pressure) for the magnetite–hematite assemblage yields a curve (labeled Fe_2O_3–Fe_3O_4 in Fig. 8.9) that is univariant according to the phase rule. In other words, for a magnetite–hematite assemblage in equilibrium at constant pressure, the value of fO_2 at a

Table 8.6 Thermodynamic data (Robie and Hemmingway, 1995) and calculations for the magnetite–hematite oxidation reaction at 1 bar total pressure.

T (K)	$\Delta G^1_{f,\,298.15}$ (Fe_3O_4) (kJ mol^{-1})	$\Delta G^1_{f,\,298.15}$ (Fe_2O_3) (kJ mol^{-1})	$\Delta G^0_{r,\,T}$ (kJ)	log fO_2 (bar)
298.15	−1012.7	−744.4	−207.8	−72.8
400	−977.9	−716.6	−194.0	−50.7
500	−944.5	−689.9	−180.7	−37.7
600	−912.0	−663.6	−166.8	−29.0
700	−880.3	−637.8	−152.8	−22.8
800	−849.4	−612.4	−138.4	−18.1
900	−819.5	−587.4	−123.2	−14.3

$\Delta G^1_{f,\,T}(O_2)$=0 for all values of *T*.

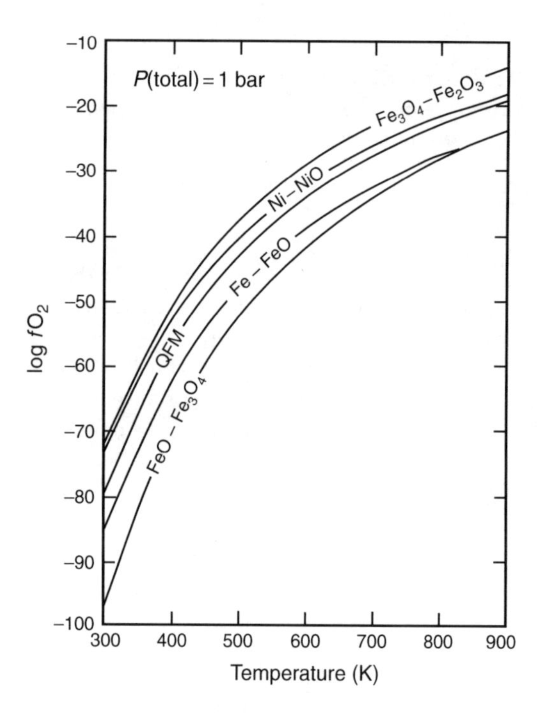

Fig. 8.9 Variation of oxygen fugacity as a function of temperature (at 1 bar total pressure) for selected oxidation reactions in equilibrium. Each of the curves represents an oxygen buffer. In a closed system, the fO_2 is regulated by the coexistence of the pure buffer components. The oxygen buffers have approximately the same slope because the enthalpy change associated with oxidation is approximately the same for all these reactions. The buffer curves will plot as straight lines in log fO_2 versus $1/T$ (K) space.

given temperature is fixed and predictable. Such an assemblage is referred to as an *oxygen buffer* because its presence in a rock may control the oxidation state at a given temperature (and pressure). This principle is widely used in experimental petrology to control the fO_2 of redox mineral reactions at high temperatures and pressures (e.g., Eugster and Wones, 1962; Huebner, 1971). The concept can also be applied to estimate fO_2 of natural mineral assemblages containing one or more oxygen buffers if the assemblage is in equilibrium and the temperature of equilibration can be obtained by some other technique.

In principle, any oxidation reaction involving molecular oxygen is a potential oxygen buffer. Figure 8.9 includes curves for several commonly used oxygen buffers: the corresponding phase assemblages and oxidation reactions are listed in Table 8.7.

The buffer curves in Fig. 8.9 are calculated using thermodynamic data at 1 bar total pressure. More commonly, the oxygen fugacity values of such reactions at different temperatures are measured directly in the laboratory. Oxygen fugacity for these equilibria also depends on the total pressure, increasing by more than an order of magnitude as pressure is increased to 10 kbar. The pressure dependence can be calculated using the thermodynamic principles discussed earlier. Also, for our

Table 8.7 Reactions for the oxygen buffer curves (Fig. 8.9).

Buffer curve	Phases for the buffer curve	Reaction for the buffer curve
IW buffer	Fe–FeO ("wustite")	$Fe+0.5O_2=FeO$ (8.94)
WM buffer	$FeO–Fe_3O_4$ (magnetite)	$3FeO+0.5O_2=Fe_3O_4$ (8.95)
QFM buffer	SiO_2 (quartz)–Fe_2SiO_4 (fayalite)–Fe_3O_4	$3Fe_2SiO_4+O_2=2Fe_3O_4+3SiO_2$ (8.96)
NNO buffer	Ni–NiO (bunsenite)	$Ni+0.5O_2=NiO$ (8.97)

calculations the solids were assumed to have pure end-member compositions; for buffer curves involving solid solutions, the calculations must consider activities of the components as discussed in Chapter 5.

Although fO_2 increases exponentially with increase in temperature, the numbers are very small even at high temperatures, which translates into too little oxygen to be physically significant. For example, an oxygen fugacity of 10^{-72} bar for the magnetite–hematite assemblage at 298.15 K and 1 bar total pressure corresponds to only about three molecules of O_2 gas in 10^{47} m^3 of space. The significance of the thermodynamically modeled oxygen fugacity lies in the fact that it is linked by equilibrium constants to all other constituent activities in the system, many of which are physically significant (Anderson and Crerar, 1993).

$FeO:Fe_3O_4$ ratios in silicic magmas with small amounts of iron and large amounts of water, and in mafic magmas with large amounts of iron and far less water, can be used to estimate the oxygen fugacity of the source regions of such magmas (Carmichael, 1991). The oxidation state of most geologic systems, including the Earth's upper mantle, lie between the buffer curves labeled Fe_2O_3–Fe_3O_4 (magnetite–hematite; MH) and Fe–FeO (iron–wüstite; IW); the oxidation state of the mantle is generally believed to be close to the QFM buffer.

8.9.2 Oxygen fugacity–sulfur fugacity diagrams

For assemblages containing both oxygen– and sulfur–bearing minerals, it is sometimes convenient to employ fO_2–fS_2 diagrams to represent the stability fields of various minerals. The calculations are analogous to those we performed for fO_2 in the previous section. As an example, let us calculate the boundary between the stability fields of hematite (Fe_2O_3) and pyrite (FeS_2) at 298.15 K and 1 bar total pressure, using the thermodynamic data listed in Table 8.8.

For the reaction defining this boundary at equilibrium,

$$Fe_2O_{3(s)} + 2S_{2(g)} \Leftrightarrow 2FeS_{2(s)} + 1.5O_{2(g)} \qquad (8.98)$$

Let us choose the standard states as pure solid phases at 298.15 K and 1 bar total pressure, the temperature and pressure of interest, so that activity of each solid phase is unity. For the reaction,

Table 8.8 Standard state (298.15 K, 1 bar) thermodynamic data (Robie and Hemmingway, 1995) for constituents in the Fe–O–S system.

	$O_{2(g)}$	$S_{2(g)}$	$FeS_{(s)}$	$FeS_{2(s)}$	$Fe_3O_{4(s)}$	$Fe_2O_{3(s)}$
$\Delta G^1_{f,\,298.15}$ (kJ mol^{-1})	0.0	79.7	−101.3	−160.1	−1012.7	−744.4

$$K_{eq}=\frac{a^2_{FeS2(s)}\, f^{1.5}_{O2(g)}}{a_{Fe2O3(s)}\, f^2_{S2(g)}}=\frac{f^{1.5}_{O2(g)}}{f^2_{S2(g)}}$$

$$\Delta G^1_{r,\,298.15}=2\,\Delta G^1_{f,\,298.15}(FeS_2)+1.5\,\Delta G^1_{f,\,298.15}(O_2)$$

$$-\Delta G^1_{f,\,298.15}(Fe_2O_3)-2\,\Delta G^1_{f,\,298.15}(S_2)$$

$$=(2)\,(-160)+0-(-744.4)-(2)\,(79.7)=264,800\,\mathrm{J}$$

$$\log K_{eq}=1.5\log fO_2-2\log fS_2=\frac{-\Delta G^1_{r,\,298.15}}{(2.3026)\,RT}$$

$$=\frac{-264{,}800}{(2.3026)\,(8.314)\,(298.15)}=-46.4$$

Thus, the equation for this reaction boundary is

$$\log fO_2=-30.9+1.3\log fS_2 \qquad (8.99)$$

which defines a straight line in $\log fO_2-\log fS_2$ space, with slope=+1.3 and intercept=−30.9 (Fig. 8.10). Equations for other boundaries included in Fig. 8.10 are listed in Table 8.9.

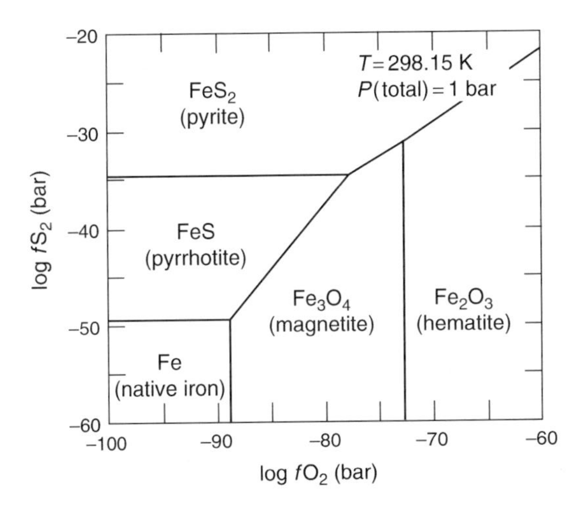

Fig. 8.10 $\log fO_2-\log fS_2$ diagram showing the stability fields of minerals in the Fe–O–S system at 298.15 K (25°C) and 1 bar total pressure. The equations for the various boundaries are listed in Table 8.9.

Table 8.9 Equations for stability field boundaries in fO_2-fS_2 diagram for the Fe –O –S system (Fig. 8.10).

Boundary	Reaction	Equation (fO_2 and fS_2 in bars)
$Fe_{(s)}-FeS_{(s)}$	$Fe_{(s)}+0.5S_{2(g)}\Leftrightarrow FeS_{(s)}$	$\log fS_2=-49.5$ (8.100)
$FeS_{(s)}-FeS_{2(s)}$	$FeS_{(s)}+0.5S_{2(g)}\Leftrightarrow FeS_{2(s)}$	$\log fS_2=-34.6$ (8.101)
$Fe_{(s)}-Fe_3O_{4(s)}$	$3Fe_{(s)}+2O_{2(g)}\Leftrightarrow Fe_3O_{4(s)}$	$\log fO_2=-88.7$ (8.102)
$Fe_3O_{4(s)}-Fe_2O_{3(s)}$	$2Fe_3O_{4(s)}+0.5O_{2(g)}\Leftrightarrow 3Fe_2O_{3(s)}$	$\log fO_2=-72.8$ (8.103)
$FeS_{(s)}-Fe_3O_{4(s)}$	$3FeS_{(s)}+2O_{2(g)}\Leftrightarrow Fe_3O_{4(s)}+1.5S_{2(g)}$	$\log fO_2=-51.6+0.75\log fS_2$ (8.104)
$Fe_3O_{4(s)}-FeS_{2(s)}$	$Fe_3O_{4(s)}+3S_{2(g)}\Leftrightarrow 3FeS_{2(s)}+2O_{2(g)}$	$\log fO_2=-25.7+1.5\log fS_2$ (8.105)
$Fe_2O_{3(s)}-FeS_{2(s)}$	$Fe_2O_{3(s)}+2S_{2(g)}\Leftrightarrow 2FeS_{2(s)}+1.5O_{2(g)}$	$\log fO_2=-30.9+1.3\log fS_2$ (8.106)

8.10 Summary

1. Oxidation is defined as a reaction involving the loss of one or more electrons and reduction as a reaction involving the gain of one or more electrons. A whole oxidation–reduction reaction consists of an oxidation half-reaction and a complementary reduction half-reaction.
2. A voltaic cell, essentially consisting of two electrodes and an electrolyte, is a device for generating usable electricity from spontaneous oxidation–reduction reactions at the electrodes. The EMF (E) of a voltaic cell is the sum of the potentials at the two electrodes, which are measured relative to a standard hydrogen electrode (SHE), following the convention that $E^0_{SHE}=0$.
3. The electrode potential (E^P_T) of a half-reaction at P and T, written as

   reduced state ⇒ oxidized state + ne^-

   is related to the free energy change for the reaction ($\Delta G^P_{r,T}$) by the equation

   $$\Delta G^P_{r,T}=n\,\mathfrak{F}\,E^P_T$$

   For a reaction occurring under standard state conditions,

   $$\Delta G^0_r=n\,\mathfrak{F}\,E^0$$

4. In aqueous geochemistry, the electrode potential of a half-reaction (written as oxidation reaction) relative to a standard hydrogen electrode (SHE) is called oxidation potential (Eh) and is calculated using the Nernst equation:

$$Eh = E^0 + 2.3026 \frac{RT}{n\Im} \log \left(\frac{\prod_i a_i^{\nu_i} \text{(oxidized assemblage)}}{\prod_j a_j^{\kappa_j} \text{(reduced assemblage)}} \right)$$

5. *Eh*–pH diagrams (at specified temperature, total pressure, and solute activities) are a convenient way of depicting relative stabilities of minerals and dissolved species in aqueous systems. For aqueous environments, the *Eh*–pH space of interest is bounded by the upper and lower stability limits of water, the equations for which at or near the Earth's surface ($T = 298.15\,K$, $P = 1$ bar) are

$$Eh^{Up}_{H_2O(l)} = 1.23 - 0.0592\ pH; \quad Eh^{Lr}_{H_2O(l)} = -0.0592\ pH$$

6. Oxidation and reduction reactions occurring in nature are commonly catalyzed by microorganisms.
7. Acid mine drainage (AMD) caused by the oxidation of sulfide minerals in ore and waste rock piles is a worldwide environmental issue. The two basic strategies for mitigating acid mine drainage are acid neutralization and minimizing the access of the sulfidic material to oxygen.
8. Bioleaching refers to the use of microorganisms to leach a metal of value (e.g., copper, zinc, uranium, nickel, cobalt) from a mineral (usually a sulfide) and convert it into a soluble form (in most cases, a sulfate of the metal). Biooxidation, on the other hand, enhances the concentration of the metal of interest in the solid phase by selective dissolution of the undesired mineral matrix. Such microbially mediated processes are especially useful for the recovery of valuable metals from low-grade ores because of the lower cost compared to conventional extraction techniques.
9. At higher temperatures, especially in nonaqueous systems, oxygen fugacity (fO_2) is commonly used as a measure of the oxidation state of a system.
10. An oxygen buffer curve represents the variation of fO_2 as a function of temperature at constant total pressure. The oxidation states of most geologic systems lie between the magnetite–hematite (MH) and iron–wüstite (IW) buffer curves.

8.11 Recapitulation

Terms and concepts

Acid mine drainage (AMD)
Acidophilic bacteria
Aerobic respiration
Anaerobic respiration
Anoxic
Archaea
Autotrophs
Bacteria
Biofiltration
Bioleaching
Biooxidation
Electrochemical cell
Electromotive force
Electromotive series
Fuel cell
Heterotrophs
Mesophilic bacteria
Nernst equation
Oxic
Oxidation
Oxidizing agent (oxidant)
Oxidation potential (*Eh*)
Oxygen buffer
Oxygen fugacity
Reducing agent (reductant)
Stability limits of water
Standard electrode potential
Standard hydrogen electrode
Sulfur fugacity
Voltaic cell

Computation techniques

- Calculation of EMF of half-cell and whole-cell reactions.
- Construction of *Eh*–pH diagrams showing stability fields of solid and aqueous species.
- Construction of oxygen buffer curves as a function of temperature.
- Construction of $f_{O_2} - f_{S_2}$ diagrams showing stability fields of minerals.

8.12 Questions

1. Balance the following chemical reactions:

(a) $FeS_{2\,(s)} + H_2O_{(l)} \Rightarrow Fe(OH)_{3\,(s)} + SO^{2-}_{4\,(aq)} + H^+_{(aq)} + e^-$
(b) $FeCO_{3\,(s)} + H_2O_{(l)} \Rightarrow Fe_3O_{4\,(s)} + H_2CO_{3\,(aq)} + H^+_{(aq)} + e^-$
(c) $[UO_2(CO_3)_{3\,(aq)}]^{4-} + HS^-_{(aq)} + H^+_{(aq)} \Rightarrow UO_{2\,(s)} + SO^{2-}_{4\,(aq)} + CO_{2\,(g)} + H_2O_{(l)}$

2. Determine which of the following reactions are oxidation–reduction reactions. For those that are, identify the oxidizing and reducing agents:

(a) $CaCO_{3\,(s)} + 2HNO_{3\,(aq)} \Rightarrow Ca(NO_3)_{2\,(aq)} + CO_{2\,(g)} + H_2O_{(l)}$
(b) $Fe_2O_{3\,(s)} + 3CO_{(g)} \Rightarrow 2Fe_{(s)} + 3CO_{2\,(g)}$
(c) $KFe_3(SO_4)_2(OH)_{6\,(s)} + 3OH^-_{(aq)} \Rightarrow 3Fe(OH)_{3\,(s)} + K^+_{(aq)} + 2SO^{2-}_{4\,(aq)}$
(d) $[UO_2(CO_3)_{3\,(aq)}]^{4-} + 2Fe^{2+}_{(aq)} + 3H_2O_{(l)} \Rightarrow UO_{2\,(s)} + 2Fe(OH)_{3\,(s)} + 3CO_{2\,(g)}$

3. Write down the oxidation and reduction half-reactions for the following whole oxidation–reduction reactions:

   (a) $2\,Fe^{2+} + 3\,Cl_2 = 2\,Fe^{3+} + 6\,Cl^-$
   (b) $4\,Fe^{2+}\,S_2^{-1} + 14\,H_2O + 15\,O_2 = 4\,Fe^{3+}\,(OH)_3 + 8\,SO_4^{2-} + 16\,H^+$

4. Using the standard electrode potentials given in Table 8.1, calculate the equilibrium constant of the reaction

$$Cu^{2+}_{(aq)} + Fe^{2+}_{(aq)} \Rightarrow Cu^{+}_{(aq)} + Fe^{3+}_{(aq)}$$

in the standard state condition.

5. What must be the *Eh* and *pe* of an aqueous environment for the activities of Fe^{2+} and Fe^{3+} to be equal? Free energy of formation (kJ mol^{-1}): $Fe^{2+} = -90.0$, $Fe^{3+} = -16.7$, $H^+ = 0$, and $H_{2\,(g)} = 0$.
6. The activities of Fe^{2+} and Fe^{3+} in a lake are $10^{-4.95}$ and $10^{-2.29}$ at pH 3.5 and 25°C. What would be the activity of Mn^{2+} in this water body if it were in equilibrium with MnO_2 in the sediment at the lake bottom? Given: E^0 of the oxidation half-reaction $Mn^{2+}_{(aq)} + 2H_2O_{(l)} \Rightarrow MnO_{2(s)} + 4H^+_{(aq)} + 2e^-$ is +1.23 volts.
7. What form of sulfur species (SO_4^{2-}, S^0, or H_2S) should dominate in bog water at 25°C, if the water has pH=4 and *pe*=−3? The relevant half-reactions are:

$$S^0_{(s)} + 2H^+ + 2e^- \Rightarrow H_2S_{(g)} \qquad \log K_{eq} = 4.8$$

$$SO_4^{2-} + 10H^+ + 8e^- \Rightarrow H_2S_{(g)} + 4H_2O_{(l)} \qquad \log K_{eq} = 41$$

Assume $a_i = m_i$.

8. Set up a voltaic cell based on the whole oxidation–reduction reaction

$$Zn_{(s)} + 2Ag^+_{(aq)} \Rightarrow Zn^{2+}_{(aq)} + 2Ag_{(s)}$$

What are the oxidation and reduction half-reactions? Calculate the EMF of the cell. The $\Delta G^1_{f,\,298.15}$ values (in kJ mol^{-1}) are as follows (Robie and Hemingway, 1995): $Ag^+ = +77.1$; $Zn^{2+} = -147.3$; $Ag_{(s)} = 0$; $Zn_{(s)} = 0$; $H^+ = 0$; and $H_{2\,(g)} = 0$.

9. Calculate the *Eh* for acid mine drainage (pH=4) and ocean water (pH=8.3) in equilibrium with atmospheric oxygen ($P_{O_2} = 0.2$ bar) at 25°C. Assume that the atmosphere is an ideal gas mixture:
10. (a) Construct an *Eh*–pH digram for the system Fe–O–H_2O–CO_2 at 298.15 K and 1 bar total pressure using the reactions discussed in the text for the Fe–O–H_2O system (Example 8–5) and the following reactions involving $FeCO_3$ (siderite):

$$3FeCO_3 + H_2O \Rightarrow Fe_3O_4 + 3CO_2 + 2H^+ + 2e^-$$

$$2FeCO_3 + H_2O \Rightarrow Fe_2O_3 + 2CO_2 + 2H^+ + 2e^-$$

$$FeCO_3 + 2H^+ \Rightarrow Fe^{2+} + CO_2 + H_2O$$

$\Delta G^1_{f,\,298.15}(FeCO_3) = -682.8$ kJ mol^{-1} and $\Delta G^1_{f,\,298.15}(CO_2) = -394.4$ kJ mol^{-1}.

Assume $a_{Fe^{3+}} = 10^{-6}$ mol kg^{-1}, $a_{Fe^{2+}} = 10^{-6}$ mol kg^{-1}, and $f_{CO_2} = 10^{-2}$ bar.

(Thermodynamic data from Robie and Heminhway, 1995.)

(b) At the level of CO_2 in the present atmosphere, which is approximately $f_{CO_2} = 10^{-3.5}$ bar, is siderite stable in the presence of water on the Earth's surface?

(c) How should the siderite–magnetite equation be modified for a closed system (i.e., not in contact with the atmosphere) containing a fixed concentration of carbonate species? [Hint: think of the control of pH on the abundance of different dissolved carbonate species.]

11. Construct an *Eh*–pH diagram for the H–S–O system at 298.15 K (25°C) and 1 bar total pressure. The species to be included are: $S_{(s)}$, SO_4^{2-}, HS^-, HSO_4^-, and $H_2S_{(aq)}$. ΣS (total dissolved sulfur)$=10^{-1}$ *m* (i.e., the activity of each aqueous species in its field of preponderance$=10^{-1}$ *m*). Include the stability limits of water in the diagram. The relevant reactions and thermodynamic data are listed below.

Reactions	Species	Standard free energy of formation $\Delta G^1_{f,\,298.15}$ (kJ mol^{-1})
(1) $H_2S_{(aq)} = HS^- + H^+$	$S_{(s)}$	0
(2) $H_2S_{(aq)} + 4\,H_2O_{(l)} = HSO_4^- + 9\,H^+ + 8\,e^-$	HS^-	12
(3) $H_2S_{(aq)} + 4\,H_2O_{(l)} = SO_4^{2-} + 10\,H^+ + 8\,e^-$	$H_2S_{(aq)}$	−27.7
(4) $HS^- + 4\,H_2O_{(l)} = SO_4^{2-} + 9\,H^+ + 8\,e^-$	SO_4^{2-}	−744.0
(5) $H_2S_{(aq)} = S\ \text{(rhombic)} + 2H^+ + 2e^-$	H^+	0
(6) $HS^- = S\ \text{(rhombic)} + H^+ + e^-$	$H_2O_{(l)}$	−237.1
(7) $S\ \text{(rhombic)} + 4\,H_2O_{(l)} = HSO_4^- + 7\,H^+ + 6\,e^-$		
(8) $S\ \text{(rhombic)} + 4\,H_2O_{(l)} = SO_4^{2-} + 8\,H^+ + 6\,e^-$		
(9) $SO_4^{2-} + H^+ = HSO_4^-$		

12. Nitrogen gas constitutes 89% of the Earth's atmosphere. Calculate the concentration of N_2 gas (in ppm) dissolved in water at 298.15 K and 1 bar total pressure if the water is in equilibrium with the atmosphere. The equilibrium constant for the $N_{2\,(g)} \Leftrightarrow N_{2\,(aq)}$ equilibrium is 6.40×10^{-4}.

Assume volume fraction of nitrogen ≈ mole fraction of nitrogen, $a_{i(aq)} = m_{i(aq)}$, and $f_{N_2\,(g)} = P_{N_2\,(g)}$.

13. At 1000 K and 5 kbar, the oxygen fugacity of the nickel–bunsenite (NiO) buffer

$$2Ni_{(s)} + O_{2\,(g)} = 2NiO_{(s)}$$

is $10^{-15.360}$ bar. Use this information to calculate the standard free energy of formation of bunsenite at 1 bar and 1000 K. The molar volumes are:

$$\bar{V}_{Ni} = 6.588\,cm^3\,mol^{-1} \text{ and } \bar{V}_{NiO} = 10.970\,cm^3\,mol^{-1}.$$

(After Nordstrom and Munoz, 1994.)

14. Construct a log fO_2 – log fS_2 phase diagram for the Cu–S–O system at 298.15 K (25°C) and 1 bar total pressure. The species to be included are: copper metal, $Cu_{(s)}$; tenorite, $CuO_{(s)}$; cuprite, $Cu_2O_{(s)}$; chalcocite, $Cu_2S_{(s)}$; and covellite, $CuS_{(s)}$. The relevant reactions and thermodynamic data (Robie and Hemingway, 1995) are listed below. (See Garrels and Christ, 1965, pp. 158–160.)

Reactions	Standard free energy of formation	
	Species	$\Delta G^1_{f,\,298.15}$ **(kJ mol⁻¹)**
(1) $2Cu_{(s)} + 0.5\,O_{2(g)} = Cu_2O_{(s)}$	$Cu_{(s)}$	0
(2) $Cu_2O_{(s)} + 0.5\,O_{2(g)} = 2CuO_{(s)}$	$Cu_2O_{(s)}$	−147.8
(3) $2Cu_{(s)} + 0.5\,S_{2(g)} = Cu_2S_{(s)}$	$CuO_{(s)}$	−128.3
(4) $Cu_2S_{(s)} + 0.5\,S_{2(g)} = 2CuS_{(s)}$	$Cu_2S_{(s)}$	−89.2
(5) $Cu_2O_{(s)} + 0.5\,S_{2(g)} = Cu_2S_{(s)} + 0.5O_{2\,(g)}$	$CuS_{(s)}$	−55.3
(6) $2CuO_{(s)} + 0.5\,S_{2(g)} = Cu_2S_{(s)} + O_{2\,(g)}$	$O_{2\,(g)}$	0
(7) $CuO_{(s)} + 0.5\,S_{2(g)} = CuS_{(s)} + 0.5O_{2\,(g)}$	$S_{2\,(g)}$	+79.7

9 Kinetics of Chemical Reactions

...it is interesting to note that geology has always stressed the element of time as a central concept in describing the earth. In this respect, the description of time-dependent phenomena (e.g., kinetics) is even more akin to geology than thermodynamics (time independent). The study of kinetics is inherently more difficult than that of thermodynamics because time-dependent processes are path dependent.

Lasaga (1998)

A chemical reaction may be homogeneous or heterogeneous. A *homogeneous reaction* occurs entirely within a single macroscopic phase (solid, liquid, or gas); a heterogeneous reaction occurs at the interface between two or more phases. For example, the dissociation of H_2CO_3 acid to aqueous H^+ and HCO_3^- (reaction 7.36) is a homogeneous reaction, whereas the dissociation of solid calcite in water to aqueous Ca^{2+} and CO_3^{2-} (reaction 7.65) is a heterogeneous reaction.

There are two fundamental questions about a homogeneous or heterogeneous chemical reaction at specified conditions: is the reaction possible and, if so, how fast is it likely to proceed? The first question – whether the reaction should occur or not – can be answered by the application of thermodynamic principles discussed in previous chapters. The answer to the second question – what would be the rate of the reaction if it is thermodynamically feasible – involves principles of chemical kinetics, which encompasses the study of rates of chemical reactions, the factors that affect reaction rates, and the mechanisms by which reactions occur.

The most obvious example of kinetic control on chemical reactions is the metastable persistence of high-temperature minerals, such as olivine, and high-pressure minerals, such as diamond, in rocks exposed on the Earth's surface (see Chapter 4). In reality, most geochemical processes such as nucleation and growth of minerals, chemical weathering, diagenesis, hydrothermal alteration, reactions at grain boundaries during metamorphism, and isotope exchange processes, are dominated by chemical kinetics.

There are two fundamental types of rate control in chemical reactions: transport, which refers to the physical movement of chemical species to and from the site of reaction; and the extent of the reaction itself, which involves the destruction and formation of chemical bonds. Reactions in geochemical systems are predominantly heterogeneous; they are extremely sensitive to the nature of the surface involved (e.g., type and density of crystal defects, kind and concentration of impurities, etc.) and much more difficult to quantify.

In this chapter, we will briefly review the basic principles of chemical kinetics, and discuss some applications of these principles to real geochemical systems.

9.1 Rates of chemical reactions ($\mathfrak{R}$): basic principles

9.1.1 *Elementary and overall reactions*

As we have discussed in earlier chapters, representation of a chemical reaction by a stoichiometrically balanced equation is an adequate description of the *overall reaction* for the purpose of thermodynamic evaluation. In thermodynamics, we are interested only in the overall reaction that describes the net

Introduction to Geochemistry: Principles and Applications, First Edition. Kula C. Misra.

result because the equilibrium state of a system is independent of the path taken to reach that state. Many overall reactions, whether homogeneous or heterogeneous, consist of several *elementary reactions* (or steps) – reactions that actually occur as written at the molecular level by collision of the reactant species – which usually take place at different rates. Generally, elementary steps are either unimolecular, involving only a single reacting species, or bimolecular, involving two reacting species. The description of an overall reaction in terms of each elementary reaction involved is called the *reaction mechanism*. The rate of an overall reaction depends on the reaction mechanism, which may be different under different circumstances, resulting in different rates for the overall reaction. The kinetics of an overall reaction can be predicted only if we know the reaction mechanism and the rates of the component elementary reactions at the conditions of interest. The goal of kinetic experiments is to derive a reaction mechanism and rate constants from the experimental data, which usually consist of measurements of reactant and product concentrations as a function of time during the course of the reaction with varying initial reaction conditions. Commonly, a likely reaction mechanism is assumed, which is then tested for agreement between the observed rate law and that derived for the assumed mechanism.

There are essentially unlimited possible combinations of elementary reactions that would yield complex reaction mechanisms for overall reactions. Here we will consider only a few simple examples; many more examples of reaction mechanisms have been discussed, for example, by Capellos and Bielski (1972), Stumm and Morgan (1981), Stumm and Wieland (1990), and Lasaga (1998).

9.1.2 *Rate-law expression*

The *rate of a chemical reaction* is the rate of change of the concentration of a reactant or a product with respect to time (t) divided by its stoichiometric coefficient. The rate is usually expressed in units of moles per volume (or surface area) per unit time. Consider, for example, a balanced overall reaction of the form

$$\text{a A} + \text{b B} \Rightarrow \text{c C} + \text{d D} \quad (9.1)$$

reactants *products*

For an irreversible reaction such as (9.1), the reaction continues until one of the reactants is completely consumed. Since a decrease in A by a moles simultaneously produces a decrease in B by b moles, but an increase in C and D by c and d moles, respectively, the average rate of the reaction ($\Re$) over a very short time period Δt can be written in terms of the rate of change in the concentration of any one of the four species in the reaction:

$$\Re = -\frac{1}{\text{a}}\left(\frac{\Delta[\text{A}]}{\Delta t}\right) = -\frac{1}{\text{b}}\left(\frac{\Delta[\text{B}]}{\Delta t}\right) = \frac{1}{\text{c}}\left(\frac{\Delta[\text{C}]}{\Delta t}\right) = \frac{1}{\text{d}}\left(\frac{\Delta[\text{D}]}{\Delta t}\right) \quad (9.2)$$

where $\Delta[i]$ represents the corresponding change in concentration of the ith species. The negative sign for the reactants ensures that the reaction rate is a positive number for both the reactants and the products; it also reminds us that the concentration of a reactant decreases with time. Since for most reactions the reaction rate gradually decreases as the reactants are consumed, the instantaneous rate of reaction is defined in terms of the derivatives of concentrations with respect to time:

$$\Re = -\frac{1}{\text{a}}\frac{\text{d}[\text{A}]}{\text{d}t} = -\frac{1}{\text{b}}\frac{\text{d}[\text{B}]}{\text{d}t} = \frac{1}{\text{c}}\frac{\text{d}[\text{C}]}{\text{d}t} = \frac{1}{\text{d}}\frac{\text{d}[\text{D}]}{\text{d}t} \quad (9.3)$$

In other words, the rate at any given time t is the slope of the concentration versus time curve at that point. The general expression for the rate of a reaction may be written as:

$$\boxed{\Re = \frac{1}{\nu_i}\frac{\text{d}[i]}{\text{d}t}} \quad (9.4)$$

where ν_i is the stoichiometric coefficient (positive for products and negative for reactants) and $[i]$ the concentration of the ith species. Thus, the rate of a reaction is simply the rate of consumption of a reactant or the rate of formation of a product divided by its stoichiometric coefficient.

In kinetics, concentrations are expressed as moles per volume or, in the case of reactions taking place at surfaces, as moles per area. For heterogeneous reactions, one or more of the concentrations should refer to the specific area (i.e., area per unit volume of solution) of the solids involved (Lasaga, 1981). It is, therefore, important to measure the total areas of reactive solids in heterogeneous reactions. In some cases, we should consider only the reactive surface area, which may be a fraction of the total area. The need to use concentration units, rather than activities, in writing rate laws distinguishes kinetics from thermodynamics. Whereas activities determine the equilibrium between thermodynamic components, spatial concentrations (e.g., moles per cm^3) of the colliding molecules determine the molecular collision rates, and hence the rates of reaction products (Lasaga, 1981).

Example 9–1: Calculation of the reaction rate for a homogeneous overall reaction

Consider the production of NO by oxidation of NH_3:

$$4NH_3 + 5O_2 \Rightarrow 4NO + 6H_2O$$

If at any given instant, NH_3 is reacting at the rate of 1.10 M per minute, what are the rates at which the other reactant and each of the products are changing at the same instant, and what is the rate of the overall reaction?

The rate of change of $[O_2]$

$$= -\frac{\Delta[O_2]}{\Delta t} = -\frac{5 \text{ moles } O_2}{4 \text{ moles } NH_3} \frac{1.10 \text{ moles } NH_3}{l \text{ . min}}$$
$$= -1.375 \text{ M min}^{-1}$$

The rate of change of [NO]

$$= \frac{\Delta[NO]}{\Delta t} = \frac{4 \text{ moles NO}}{4 \text{ moles } NH_3} \frac{1.10 \text{ moles } NH_3}{l \text{ . min}}$$
$$= 1.10 \text{ M min}^{-1}$$

The rate of change of $[H_2O]$

$$= \frac{\Delta[H_2O]}{\Delta t} = \frac{6 \text{ moles } H_2O}{4 \text{ moles } NH_3} \frac{1.10 \text{ moles}}{l \text{ . min}} = 1.65 \text{ M min}^{-1}$$

The reaction rate can be calculated from the rate of decrease of any reactant concentration or the rate of increase of any product concentration, and the answer is the same as shown below:

$$\text{Rate of reaction} = -\frac{1}{5}\left(\frac{\Delta[O_2]}{\Delta t}\right) = -\frac{1}{5}\left(-1.375 \text{ M min}^{-1}\right)$$
$$= 0.275 \text{ M min}^{-1}$$

$$\text{Rate of reaction} = \frac{1}{4}\left(\frac{\Delta[NO]}{\Delta t}\right) = \frac{1}{4}\left(1.10 \text{ M min}^{-1}\right)$$
$$= 0.275 \text{ M min}^{-1}$$

$$\text{Rate of reaction} = \frac{1}{4}\left(\frac{\Delta[H_2O]}{\Delta t}\right) = \frac{1}{6}\left(1.65 \text{ M min}^{-1}\right)$$
$$= 0.275 \text{ M min}^{-1}$$

Relationships describing the concentration dependence of reaction rates are called *rate law expressions* (or simply *rate laws*). The rate law for reaction (9.1) is of the form

$$\Re = k\ [A]^{n_A} [B]^{n_B} [C]^{n_C} [D]^{n_D} \qquad (9.5)$$

where $\Re$ is the rate of the reaction, k is the experimentally determined *rate constant* (also referred to as the *specific rate constant*) of the particular reaction (at a specified temperature and pressure), and [A], [B], etc., represent concentrations. The exponents n_A, n_B, etc., are real numbers (can be zero or fractions) that denote the *order of the reaction* with respect to the species A, B, etc., and represent how the rate of the reaction is related to the concentrations of the species involved. The overall order of the reaction ($n_{overall}$) is the sum of the exponents. In this case, $n_{overall} = n_A + n_B + n_C + n_D$; some real examples are listed in Table 9.1. Generally, values of the exponents bear no relationship to the stoichiometric coefficients in the balanced overall reaction and must be determined experimentally, although the values can be predicted if the reaction mechanism is known from experimental investigation. Only for elementary reactions, the orders are equal to the stoichiometric coefficients. The form of the rate law in equation (9.5) (i.e., as a product of concentration terms) has been validated in the majority of kinetic studies. The exact rate law may be more complicated, but in most cases can be reduced to the form in equation (9.5) by certain simplifications.

Table 9.1 Examples of orders of reactions from experimentally determined rate laws.

Reaction	Observed rate law for the reaction	Order of the reaction
$O_{3(g)} + NO_{(g)} \Rightarrow NO_{2(g)} + O_{2(g)}$	$\Re = k\ [O_3]\ [NO]$	First-order in O_3; first-order in NO; second-order overall
$H_2O_{2(aq)} + 3I^-_{(aq)} + 2H^+_{(aq)} \Rightarrow 2H_2O_{(l)} + I^{3-}_{(aq)}$	$\Re = k\ [H_2O_2]\ [I^-]$	First-order in H_2O_2; first-order in I^-; zero-order in H^+; second-order overall
$3NO_{(g)} \Rightarrow N_2O_{(g)} + NO_{2(g)}$	$\Re = k\ [NO]^2$	Second-order in NO; second-order overall
$2NO_{2(g)} \Rightarrow 2NO_{(g)} + O_{2(g)}$	$\Re = k\ [NO_2]^2$	Second-order in NO_2; second-order overall

The units of k depend on the form of the rate law. For reactions without gaseous reactants, the rate constants have the following (or comparable) units: mol cm^{-1} s^{-1} (zeroth–order); s^{-1} (first–order); and mol^{-1} cm^3 s^{-1} (second–order).

Equation (9.5) reduces to a much simpler form for elementary reactions. First, the exponent of each product is zero because the rate of an elementary reaction is independent of the concentrations of the products if the reaction only in the forward direction is considered (one of the tests for an elementary reaction). Second, the value of the exponent for a reactant is the same as the stoichiometric coefficient of the species in the reaction. Thus, if equation (9.1) represented an elementary reaction, its rate law would be of the form

$$\Re = k\ [A]^a [B]^b \qquad (9.6)$$

with a + b as the order of the reaction.

9.1.3 *Integrated rate equations for elementary reactions*

An integrated rate equation, obtained by integrating the corresponding rate law differential equation, expresses concentration as an explicit function of time. We will restrict our attention to zeroth-, first-, and second-order elementary reactions because in most geochemical systems of interest the reactions are second-order or lower. Remember that the rate of an elementary reaction in the forward direction, irrespective of the order of the reaction, is independent of the concentrations of the products.

Zeroth-order reaction

For a *zeroth-order reaction* of the form A ⇒ products, the reaction rate is independent of the reactant concentration, [A], and can be written as

$$\Re = -\frac{d[A]}{dt} = k \quad (9.7)$$

where k is the rate constant. Rearranging the equation and integrating, we get

$$\int d[A] = -k \int dt$$

$$[A] = -kt + C$$

where C is the constant of integration. If the concentrations at time $t = 0$ and time $t = t$ are denoted by $[A]_0$ and $[A]_t$, respectively, then $C = [A]_0$ and

$$\boxed{[A]_t - [A]_0 = -kt} \quad (9.8)$$

Equation (9.8) indicates that the plot of [A] versus time (t) would yield a straight line with slope = $-k$ and intercept = $[A]_0$ (Fig. 9.1a). The reduction of SO_4^{2-} by *Desulfovibro* bacteria to S^{2-} in marine environments seems to be a zeroth-order reaction with respect to SO_4^{2-} if the sulfate concentration is > 2 mM (Berner and Westrich, 1985). The rates of dissolution of silica polymorphs are zeroth-order but the precipitation reactions are first-order (Rimstidt and Barnes, 1980).

The decay of the reactant A with time can also be expressed in terms of *half-life* ($t_{1/2}$). The half-life of A is defined as the interval of time necessary for the concentration of A to be reduced to one-half of its original value. Substituting $[A]_t = 0.5\,[A]_0$ at $t = t_{1/2}$ in equation (9.8),

$$\frac{[A]_0}{2} - [A]_0 = -k\, t_{1/2}$$

$$\boxed{t_{1/2} = \frac{[A]_0}{2k}} \quad (9.9)$$

First-order reaction

For a *first-order reaction* of the form A → products, which is first-order in reactant A and first-order overall, the rate of the reaction is proportional to [A] and can be written as:

$$\Re = -\frac{d[A]}{dt} = k\,[A]^1 = k\,[A] \quad (9.10)$$

Following the same procedure as for the zeroth-order reaction, the integrated rate equation and the half-life for the first-order reaction can be derived as

$$\boxed{\ln [A]_t - \ln [A]_0 = -kt; \qquad [A]_t = [A]_0\, e^{-kt}} \quad (9.11)$$

$$\boxed{t_{1/2} = \frac{\ln 2}{k} = \frac{0.693}{k}} \quad (9.12)$$

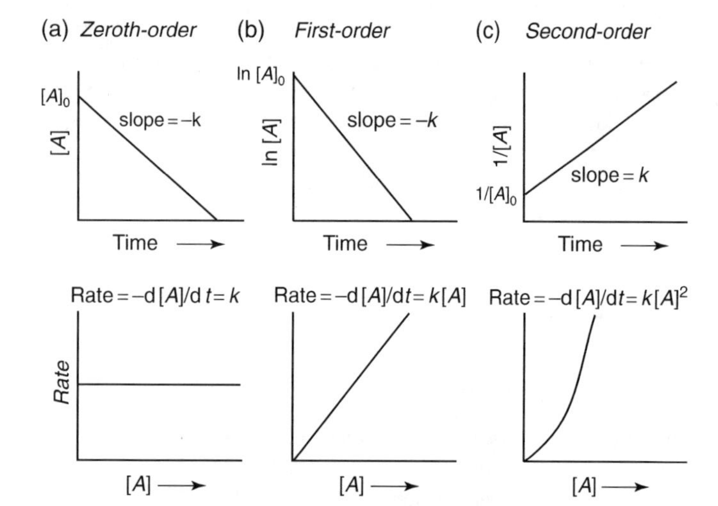

Fig. 9.1 Schematic illustration of the relationship between the concentration of a reactant A and time (at constant temperature) for: (a) a zeroth-order reaction; (b) a first-order reaction; and (c) a second-order reaction.

Equation (9.11) shows that a plot of $\ln [A]_t$ versus time (t) would yield a straight line with slope = $-k$ and intercept = $\ln [A]_0$ (Fig. 9.1b). Such a plot is the best way to prove that a reaction is first-order. Also, as the half-life of a first-order reaction is independent of the initial concentration of A (equation 9.12), a reaction that obeys first-order kinetics should yield the same half-life at various initial concentrations of A.

Radioactive decay is a good example of a first-order reaction (see Chapter 10). For example, the decay of the carbon isotope ^{14}C to the nitrogen isotope ^{14}N has a rate law given by

$$\frac{d[^{14}C]}{dt} = -k\,[^{14}C] \quad (9.13)$$

where $k = 1.209 \times 10^{-4}\,yr^{-1}$, which yields a half-life of 5732 yr for the decay of ^{14}C irrespective of its initial concentration (equation 9.12; Fig. 9.2).

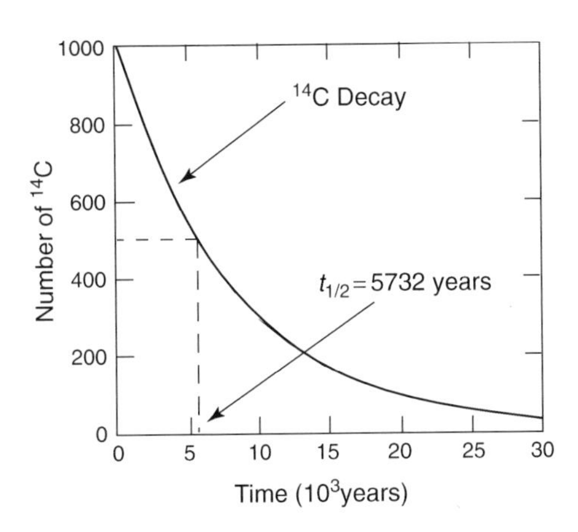

Fig. 9.2 The decay curve for radioactive isotope ^{14}C. The radioactive decay is a first-order reaction. The rate constant is $1.209 \times 10^{-4} yr^{-1}$ and the half-life of decay is 5732 yr.

Many higher-order reactions appear to follow *pseudo first-order* kinetics if one (or more than one) reactant is in such excess that its concentration can be considered to remain unchanged during the reaction (see Example 9–2).

Second-order reaction

For a *second-order reaction* of the form $2A \rightarrow$ products, which is second order in A and second order overall, the rate of the reaction is:

$$\Re = -\frac{d[A]}{dt} = k\,[A]^2 \tag{9.14}$$

Following the procedure above, the integrated rate equation and the half-life for the second-order reaction can be derived as

$$\boxed{\frac{1}{[A]_t} - \frac{1}{[A]_0} = kt} \tag{9.15}$$

$$\boxed{t_{1/2} = \frac{1}{k\,[A]_0}} \tag{9.16}$$

In this case (equation 9.15), a plot of $1/[A]_t$ versus time (t) would yield a straight line with slope $= k$ and intercept $= 1/[A]_0$ (Fig. 9.1c). Note that for a second-order reaction the half-life of [A] depends on its initial concentration.

The half-life of geochemical reactions can vary by orders of magnitude, from a few seconds to millions of years (Fig. 9.3).

9.1.4 *Principle of detailed balancing*

The *principle of detailed balancing* states that for an elementary reaction at equilibrium, the rate in the forward direction is equal to the rate in the reverse direction, a concept we actually used earlier to define chemical equilibrium (see section 4.1). In order to illustrate the principle, let us consider a reversible first-order elementary reaction for which k_+ and k_- are the rate constants associated with the forward and reverse directions, respectively, and $[A]_{eq}$ and $[B]_{eq}$ are the concentrations of the species at equilibrium:

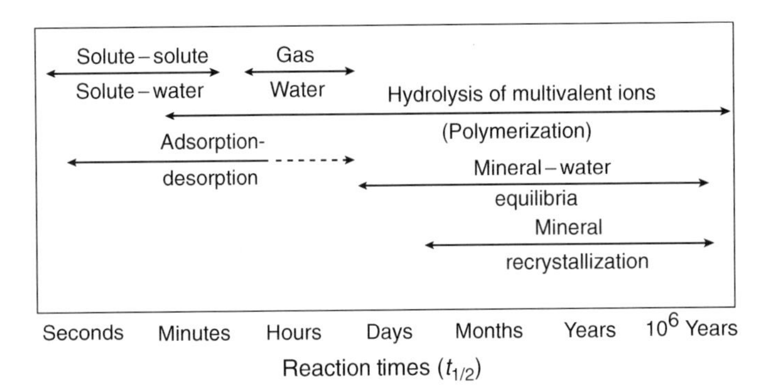

Fig. 9.3 A comparison of reaction times (in half-lives) for some common reactions in aqueous systems as estimated by Langmuir and Mahoney (1984). (Reproduced with permission from Proceedings of the First Canadian/American Conference on Hydrology, 69–95, by National Groundwater Association, Copyright 1984).

$$A \underset{k_-}{\overset{k_+}{\rightleftarrows}} B \quad \text{(rate constants } k_+ \text{ and } k_-) \tag{9.17}$$

For the fortward reaction, $-\frac{d[A]}{dt} = k_+[A]$; and for the reverse reaction, $-\frac{d[B]}{dt} = k_-[B]$

Since the rates in both directions of reaction (9.17) are equal at equilibrium in a closed system,

$$k_+\,[A]_{eq} = k_-\,[B]_{eq} \tag{9.18}$$

which can be rearranged as

$$\frac{k_+}{k_-} = \frac{[B]_{eq}}{[A]_{eq}} = \frac{a_B/\gamma_B}{a_A/\gamma_A} = K_{eq}\frac{\gamma_A}{\gamma_B} \tag{9.19}$$

where a_A and a_B are the activities, γ_A and γ_B the activity coefficients, and K_{eq} the equilibrium constant. Equation (9.19) provides a direct link between kinetic rate constants and thermodynamic equilibrium constants for the reaction under consideration. Such relationships can be used to determine one of the rate constants of an elementary reaction if the other rate constant and K_{eq} of the reaction are known. For example, Rimstidt and Barnes (1980) studied the reaction

$$SiO_{2(s)} + 2H_2O_{(l)} \underset{k_-}{\overset{k_+}{\rightleftarrows}} H_4SiO_{4(aq)} \tag{9.20}$$

in order to determine the rates of dissolution and precipitation of silica phases. They calculated the rate constant for the precipitation of a silica phase (k_-) from the measured rate constant for its dissolution (k_+) and the equilibrium constant (K_{eq}) for reaction (9.20), using the relation

$$k_- = k_+ \frac{\gamma_{H_4SiO_4}}{K_{eq}} \tag{9.21}$$

The principle of detailed balancing is also applicable to overall reactions (Lasaga, 1998).

9.1.5 *Sequential elementary reactions*

Elementary reactions (steps) comprising a reaction mechanism may take place sequentially (i.e., each must be completed before a subsequent reaction can occur), or concurrently in parallel paths. Each of these steps may be reversible (i.e., with appreciable rates in the forward and reverse directions) or essentially irreversible (i.e., with negligible rates in the reverse direction).

In the case of *sequential* (or *consecutive*) elementary reactions, the slowest of the elementary reactions is the *rate-determining step*; an overall reaction can never occur faster than its slowest elementary reaction. In other words, the rate of the slowest, rate-determining step is the rate of the overall reaction.

Consider an overall reaction in a closed system and at constant temperature:

$$A \Rightarrow C \tag{9.22}$$

A reaction mechanism frequently encountered for such overall reactions consists of a fast reversible reaction followed by a slow irreversible reaction:

$$\text{(a) } A \underset{k_{-(a)}}{\overset{k_{+(a)}}{\rightleftarrows}} B \quad \text{(fast, rate constants } k_{+(a)} \text{ and } k_{-(a)}) \tag{9.23}$$

$$\text{(b) } B \xrightarrow{k_{+(b)}} C \quad \text{(slow, rate constant } k_{+(b)}) \tag{9.24}$$

The two steps can be added to give the overall reaction $A \Rightarrow C$. Note that the intermediate product B formed in the first step is completely consumed in the second step, so that it does not appear in the overall reaction. Such a transient species is called a *reactive* (or *reaction*) *intermediate*, and the state of dynamic equilibrium is termed *steady state* (or *stationary state*). Dissolution of many silicate minerals in aqueous solutions involves steady state with respect to Al^{3+} as the reactive intermediate. A system in steady state is not necessarily in true thermodynamic equilibrium, but a system in equilibrium is also in steady state. For a system to reach equilibrium, all elementary steps must have equal forward and reverse rates and all species, not just reactive intermediates, must be at steady state (Lasaga, 1998).

The net rate of change of the reactive intermediate B in the closed system for this reaction mechanism is obtained by adding the contributions from each of the elementary reactions, in this case 9.23 and 9.24:

$$\frac{d[B]}{dt} = k_{+(a)}[A] - k_{-(a)}[B] - k_{+(b)}[B] \tag{9.25}$$

Note that the value of any of the rate constants ($k_{+(a)}$ or $k_{-(a)}$ or $k_{+(b)}$) is independent of the overall reaction; i.e., it will be the same for the same elementary reaction (at the same temperature) in any other reaction mechanism. Once reaction (9.23) attains steady state after a transient period, which is short compared to reaction times, the concentration of the reactive intermediate B changes so very slowly as to be equated to zero. In other words, we can assume

$$\frac{d[B]}{dt} \approx 0 \tag{9.26}$$

With this steady-state approximation,

$$\frac{d[B]}{dt} = k_{+(a)}[A] - k_{-(a)}[B]_{st} - k_{+(b)}[B]_{st} = k_{+(a)}[A] - [B]_{st}(k_{-(a)} + k_{+(b)}) = 0 \tag{9.27}$$

and the steady-state concentration of B, $[B]_{st}$, is given by

$$[B]_{st} = \frac{k_{+(a)}\,[A]}{k_{-(a)} + k_{+(b)}} \tag{9.28}$$

From the irreversible reaction (9.24), the rate equation for product C is

$$\frac{d[C]}{dt} = k_{+(b)}[B]_{st} \tag{9.29}$$

Substituting for $[B]_{st}$, we get

$$\frac{d[C]}{dt} = \frac{k_{+(a)}\,k_{+(b)}}{k_{-(a)} + k_{+(b)}}[A] \tag{9.30}$$

Equation (9.30) is also the rate equation for the overall reaction (9.22) because reaction (9.24) is the slower elementary reaction and thus the rate-determining step.

If $k_{-(a)} \gg k_{+(b)}$, equation (9.30) can be further simplified to

$$\frac{d[C]}{dt} = \frac{k_{+(a)}\,k_{+(b)}}{k_{-(a)}}[A] = K_a\,k_{+(b)}[A] \tag{9.31}$$

where $K_a = k_{+(a)}/k_{-(a)}$ for the equilibrium between A and B.

Example 9–2: Rate of conversion of ozone to oxygen

The conversion of ozone to oxygen described by Lasaga (1998) is an instructive example of how the steady state concept can be applied to the calculation of overall reaction rate. The overall reaction for this conversion is

$$2O_3 \xrightarrow{k_+} 3O_2 \tag{9.32}$$

and the overall reaction rate has been determined experimentally to be

$$-\frac{1}{2}\frac{d[O_3]}{dt} = \frac{1}{3}\frac{d[O_2]}{dt} = \frac{k_+[O_3]^2}{[O_2]} \tag{9.33}$$

where k_+ is the rate constant for the forward reaction. Since the reaction rate involves the product O_2, (9.32) is not an elementary reaction. A suggested mechanism for the overall reaction consists of two consecutive elementary reactions with oxygen atoms as the reactive intermediate:

$$\text{(a) } O_3 \underset{k_{-(a)}}{\overset{k_{+(a)}}{\rightleftarrows}} O_2 + O \qquad \text{(fast, reversible)} \tag{9.34}$$

$$\text{(b) } O + O_3 \xrightarrow{k_{+(b)}} 2O_2 \qquad \text{(slow, irreversible)} \tag{9.35}$$

Let us check if this reaction mechanism is consistent with the experimentally determined rate law (equation 9.33).

The total rate of change of oxygen atoms by the above reaction mechanism is the sum of contributions from the elementary reactions (9.34) and (9.35) (see equation 9.25):

$$\frac{d[O]}{dt} = k_{+(a)}[O_3] - k_{-(a)}[O_2][O] - k_{+(b)}[O][O_3] \tag{9.36}$$

Assuming steady state for the reactive intermediate oxygen atoms and applying the steady-state approximation (see equation 9.26),

$$\frac{d[O]}{dt} = 0 = k_{+(a)}[O_3] - k_{-(a)}[O_2][O]_{st} - k_{+(b)}[O]_{st}[O_3] \tag{9.37}$$

from which we obtain the steady-state concentration of oxygen atoms, $[O]_s$, as

$$[O]_s = \frac{k_{+(a)}[O_3]}{k_{-(a)}[O_2] + k_{+(b)}[O_3]} \tag{9.38}$$

From elementary reactions (9.34) and (9.35), the rate equation for the product O_2 is

$$\frac{d[O_2]}{dt} = k_{+(a)}[O_3] - k_{-(a)}[O_2][O] + 2k_{+(b)}[O][O_3] \tag{9.39}$$

Since we have already assumed steady state for oxygen atoms, we can use $[O]_s$ for [O] in equation (9.39) to obtain the rate equation for the overall reaction:

$$\begin{aligned}\frac{d[O_2]}{dt} &= k_{+(a)}[O_3] - \frac{k_{-(a)}[O_2]\,k_{+(a)}[O_3]}{k_{-(a)}[O_2] + k_{+(b)}[O_3]} + \frac{2k_{+(a)}\,k_{+(b)}[O_3]^2}{k_{-(a)}[O_2] + k_{+(b)}[O_3]} \\ &= \frac{k_{+(a)}[O_3]\,k_{+(b)}[O_3]}{k_{-(a)}[O_2] + k_{+(b)}[O_3]} + \frac{2k_{+(a)}\,k_{+(b)}[O_3]^2}{k_{-(a)}[O_2] + k_{+(b)}[O_3]} \\ &= \frac{3k_{+(a)}[O_3]\,k_{+(b)}[O_3]}{k_{-(a)}[O_2] + k_{+(b)}[O_3]}\end{aligned} \tag{9.40}$$

The above equation can be simplified by recognizing that reaction (9.35) is much slower than reaction (9.34) so that $k_{+(b)}$ is a very small quantity and $k_{-(a)}[O_2] \gg k_{+(b)}[O_3]$. Applying this approximation, equation (9.40) reduces to

$$\frac{d[O_2]}{dt} = \frac{3k_{+(a)}\,k_{+(b)}[O_3]^2}{k_{-(a)}[O_2]} \tag{9.41}$$

which is the same as the rate equation (9.33), where

$$k_+ = \frac{k_{+(a)}\,k_{+(b)}}{k_{-(a)}} \tag{9.42}$$

9.1.6 *Parallel elementary reactions*

Some overall reactions occur via two or more parallel paths (or *branches*) concurrently. For such *parallel* (or *concurrent* or *competitive*) reactions, the fastest path (or branch) determines the reaction mechanism, and the rate law for the overall reaction is the sum of the rates of its constituent elementary reactions.

Consider a simple reaction mechanism consisting of three parallel paths in which a single reactant gives different products:

$$\begin{aligned}&\text{(a) } A \xrightarrow{k_{+(a)}} B \\ &\text{(b) } A \xrightarrow{k_{+(b)}} C \\ &\text{(c) } A \xrightarrow{k_{+(c)}} D\end{aligned} \tag{9.43}$$

The rate law for this reaction mechanism is given by the overall rate of disappearance of A:

$$-\frac{d[A]}{dt} = k_{+(a)}[A] + k_{+(b)}[A] + k_{+(c)}[A] = (k_{+(a)} + k_{+(b)} + k_{+(c)})[A] \tag{9.44}$$

The rate law would be quite different if the reaction mechanism consisted of parallel paths that yielded the same product from different reactants:

$$\begin{aligned}&(a)\ A \xrightarrow{k_{+(a)}} D\\&(b)\ B \xrightarrow{k_{+(b)}} D\\&(c)\ C \xrightarrow{k_{+(c)}} D\end{aligned} \quad (9.45)$$

For this reaction mechanism,

$$-\frac{d[A]}{dt} = k_{+(a)}[A]; \quad -\frac{d[B]}{dt} = k_{+(b)}[B]; \quad -\frac{d[C]}{dt} = k_{+(c)}[C] \quad (9.46)$$

and the net change of the product D would be given by

$$\frac{d[D]}{dt} = -\frac{d[A]}{dt} - \frac{d[B]}{dt} - \frac{d[C]}{dt} = k_{+(a)}[A] + k_{+(b)}[B] + k_{+(c)}[C] \quad (9.47)$$

Example 9–3: The rate of hydration of carbon dioxide

In thermodynamic treatment of the CO_2–H_2O system in earlier chapters (see Example 4-1 and section 7.4), we made no distinction between $CO_{2\,(aq)}$ and $H_2CO_{3\,(aq)}$, and considered all the dissolved $CO_{2\,(g)}$ (i.e., $CO_{2\,(aq)} + H_2CO_{3\,(aq)}$) to be present as $H_2CO_{3\,(aq)}$, although less than 1% of the dissolved $CO_{2\,(g)}$ actually exists as $H_2CO_{3\,(aq)}$. For a kinetic treatment of the CO_2–H_2O system, we need to consider elementary reactions involving $CO_{2\,(aq)}$ and $H_2CO_{3\,(aq)}$ because rates of reactions involving $CO_{2\,(aq)}$ and $H_2CO_{3\,(aq)}$ with OH^- vary by orders of magnitude.

The suggested reaction mechanism for the hydration of $CO_{2(g)}$ consists of two parallel paths (Morel and Hering, 1993):

$$(a)\ CO_{2\,(aq)} + H_2O_{(l)} \xrightarrow{k_{+(a)}} H_2CO_{3\,(aq)} \quad (9.48)$$

$$(b)\ CO_{2\,(aq)} + H_2O_{(l)} \xrightarrow{k_{+(b)}} HCO_3^- + H^+ \quad (9.49)$$

The two paths are kinetically indistinguishable because of extremely rapid establishment of equilibrium between $H_2CO_{3\,(aq)}$ and HCO_3^-, and both are included in the experimentally determined rate law (Jones *et al.*, 1964):

$$-\frac{d[CO_{2\,(aq)}]}{dt} = k\,[CO_{2\,(aq)}] \quad (9.50)$$

where $k = 2 \times 10^{-3}\,s^{-1}$ at 0°C, which translates into a half-life of about 6 min (using equation 9.12). Note that the rate law is pseudo first-order instead of being second-order. This is because $[H_2O] >> [CO_{2\,(aq)}]$ so that the concentration of H_2O can be regarded as constant irrespective of the progress of the hydration reaction.

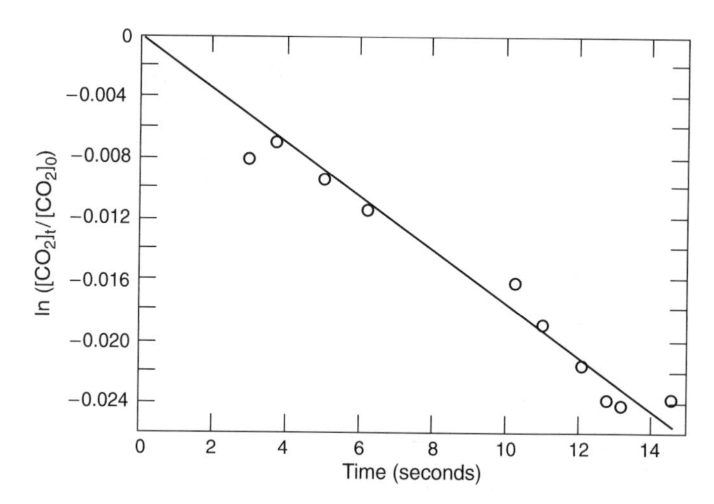

Fig. 9.4 Plot of ln ($[CO_{2(aq)}]_t/[CO_{2(aq)}]_0$) versus time (in seconds) representing the CO_2 hydration kinetics at 0°C. The slope of the graph is $0.002\,s^{-1}$, which corresponds to the observed rate constant (k) for the pseudo first-order hydration reaction. The rate constant was determined experimentally by Jones *et al.* (1964).

The integrated form of the rate law is (see equation 9.11)

$$\ln\left[\frac{[CO_{2\,(aq)}]_t}{[CO_{2\,(aq)}]_0}\right] = -\,kt \quad (9.51)$$

where $[CO_{2\,(aq)}]_0$ represents the initial concentration (at $t = 0$) and $[CO_{2\,(aq)}]_t$ the concentration at time t. Thus, a plot of ln ($[CO_{2\,(aq)}]_t/[CO_{2\,(aq)}]_0$) versus time is a straight line with slope = $-k$ (Fig. 9.4).

It should be pointed out that the above rate law applies at pH < 9 (i.e., at low concentrations of OH^-). At pH ≥ 9 an alternative reaction path

$$(c)\ CO_{2\,(aq)} + OH^- \xrightarrow{k_{+(c)}} HCO_3^- \quad (9.52)$$

becomes dominant. This path is pH-dependent and has a different rate constant.

9.2 Temperature dependence of rate constants

In addition to concentrations of the reactants (except for zero-order reactions), reaction mechanism, and surface area available for reaction (which is related in part to particle size), reaction rates are strongly influenced by temperature and the presence of catalysts. Of these temperature is by far the most important.

9.2.1 *The Arrhenius equation – activation energy*

We all know from experience that almost all chemical reactions occur at a faster rate when the temperature is raised. Lasaga (1984), for example, predicted that mineral dissolution rates would increase approximately an order of magnitude between 0°C and 25°C and another order of magnitude between 25°C and 55°C. The positive correlation between reaction rates and temperature arises from the fact that the average kinetic energy of a population of molecules is proportional to the absolute temperature, and increasing the temperature increases the number of molecules (or atoms or ions) that possess sufficient energy to undergo reaction. The temperature dependence of the reaction rate constant (k) is quantitatively expressed by the *Arrhenius equation*, so named after the Swedish chemist Svante August Arrhenius (1859–1927), who provided the theoretical justification for the equation originally proposed by the Dutch chemist J.H. van't Hoff in 1884. The Arrhenius equation, which is based on empirical observations, is:

$$\boxed{k = A\, e^{-E_a/RT}} \tag{9.53}$$

where A is a "frequency factor" (in s^{-1}), E_a the *activation energy* (in J mol^{-1}) of the reaction, R the gas constant (in J mol^{-1} K^{-1}), and T the temperature in Kelvin. The frequency factor is related to the frequency of collisions of the reactant molecules with proper orientations; it varies very slightly with temperature but is taken as a constant over small ranges of temperature. The activation energy is the minimum energy required for the reaction to occur, and the term $\exp(E_a/RT)$ represents the fraction of colliding molecules that possess kinetic energy equal to or in excess of the activation energy at temperature T. The equation indicates that the rate constant is very sensitive to temperature variations; the value of k increases exponentially with increasing temperature but becomes smaller at higher values of activation energy. Thus, the magnitude of E_a can sometimes be used to infer the nature of the rate-controlling process.

If A and E_a are assumed to be independent of temperature, then E_a can be determined by experimentally measuring the rate constant at different temperatures (but with the concentrations of the reactants kept constant) and using the logarithmic form of equation (9.53):

$$\boxed{\ln k = \ln A - \frac{E_a}{R}\left(\frac{1}{T}\right)} \tag{9.54}$$

A plot of measured ln k at various temperatures against $1/T$ will yield a straight line with slope = $-E_a/R$ and intercept = A (Fig. 9.5). Also, from equation (9.54) we can derive a relation between rate constants (k_1 and k_2) of a reaction at two different temperatures (T_1 and T_2) and E_a (see Fig. 9.5):

$$\boxed{\ln \frac{k_2}{k_1} = \frac{E_a}{R}\left(\frac{1}{T_1} - \frac{1}{T_2}\right)} \tag{9.55}$$

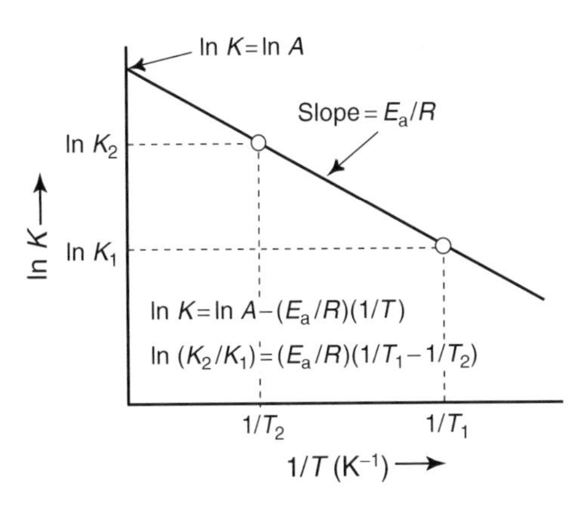

Fig. 9.5 A graphical method for determining activation energy (E_a) of a reaction using equation (9.54). A plot of ln k versus 1/T (K) gives a straight line with slope $-E_a/R$, from which E_a can be calculated. Both A and E_a are assumed to be independent of T over the temperature range of the plot. The graph can also be used to determine the ratio of the rate constants of the reaction at two different temperatures (T_1 and T_2), or one of the rate constants if the other one is known.

Example 9-4: Application of the Arrhenius equation to calculate the activation energy for a reaction from rate constants at two different temperatures

Consider the decomposition of $N_2O_{5\,(g)}$ by the reaction

$$N_2O_{5\,(g)} \Rightarrow 2NO_{2\,(g)} + 0.5O_{2\,(g)} \tag{9.56}$$

This reaction has been found to be first-order in $N_2O_{5\,(g)}$, with the following rate constants: k_1 (at 0°C) = $7.33 \times 10^{-7}\,s^{-1}$ and k_2 (at 65°C) = $4.87 \times 10^{-3}\,s^{-1}$. To calculate E_a for this reaction, we convert the natural log to $\log_{10}$, and substitute the values of k_1, k_2, T_1 = 0°C = 273 K, T_2 = 65°C = 338 K, and R = 8.314 J mol^{-1} K^{-1} in equation (9.55):

$$\log\left(\frac{4.87\times10^{-3}\ s^{-1}}{7.33\times10^{-7}\ s^{-1}}\right) = \frac{E_a}{(2.303)\ (8.314\ J\ mol^{-1}\ K^{-1})}\left(\frac{1}{273\ K} - \frac{1}{338\ K}\right) = \frac{E_a}{(2.303)\ (8.314)}\ \frac{(338-273)}{(273)\ (338)}\ J\ mol^{-1}$$

We can now solve for E_a:

$$E_a = \frac{(3.822)\ (1766783.381)}{65} = 1.04\times10^5\ J\ mol^{-1}$$

The calculated value of E_a indicates that decomposition of $N_2O_{5\,(g)}$ to $NO_{2\,(g)}$ and $O_{2\,(g)}$ requires activation of $N_2O_{5\,(g)}$ molecules by 1.04×10^5 J mol^{-1}. Note that given the value of E_a for a particular reaction, we can apply equation (9.55) to determine

Table 9.2 Activation energies (E_a) of some mineral surface dissolution reactions.

Mineral	Temperature °C	pH	E_a* kJ mol^{-1}	Mineral	Temperature °C	pH	E_a* kJ mol^{-1}
Albite[†]	5–300	Acid	60.0	Kaolinite[¶]	25–80	1	66.9
		Neutral	67.7			2	55.6
		Basic	50.1			3	43.1
Anorthite[†]	25–95	Acid	80.7			4	32.2
K-feldspar[†]	5–100	Acid	51.7			5	20.5
		Basic	57.8			6	9.6
Quartz[‡]	25–60	4	43.9			7	7.1
		6	46.0			8	14.2
		8	54.4			9	22.2
		10	92.0			10	29.3
		11	97.5			11	35.6
Calcite[§]	5–50		35.1			12	41.4

*E_a values should properly be called apparent activation energies because they represent the contribution from many terms in addition to the potential barrier of an elementary reaction (Lasaga, 1998).
[†]Blum and Stillings (1995).
[‡]Brady and Walther (1990).
[§]Sjoberg (1976).
[¶]Carroll and Walther (1990).

the ratio of rate constants at two different temperatures and even one of the rate constants if we know the other one.

The activation energies for many mineral–solution reactions commonly encountered in diagenetic and metamorphic processes are pH dependent and relatively low, mostly in the range of 40 to 80 kJ mol^{-1} at temperatures less than 100°C (Table 9.2).

9.2.2 *Transition states*

The energy associated with a chemical bond is a form of potential energy; thus, chemical reactions are accompanied by changes in potential energy. An elementary reaction involves molecular-level collision of reactant species for breaking the existing chemical bonds and making new bonds to form the specified product species. However, for the reaction to occur, the molecules must collide with enough kinetic energy to overcome the potential energy barrier. According to the *transition state theory*, the reactants pass through a transient, higher energy state, called a *transition state complex* or an *activated complex*, before reaching the final state. The activated complex represents an unstable molecular arrangement, in which bonds break and re-form either to generate the products or to degenerate back to the reactants. The transition state theory also assumes that thermodynamic equilibrium always exists among reactants, products, and the activated complex. The kinetic energy required to reach the transition state (i.e., to overcome the potential energy barrier) is the *activation energy* (E_a), which is different for the reaction in forward and reverse directions (Fig. 9.6). The transformation of an activated complex to product species is accompanied by release of energy.

For the purpose of illustration, let us consider a simple reaction

$$A + B \Leftrightarrow C^* \Leftrightarrow X + Y \qquad (9.57)$$

in which C* is the activated complex (actually a reactive intermediate). Let $E_{a\,(for)}$ and $E_{a\,(rev)}$ denote, respectively, the activation energies of the reaction in the forward and reverse

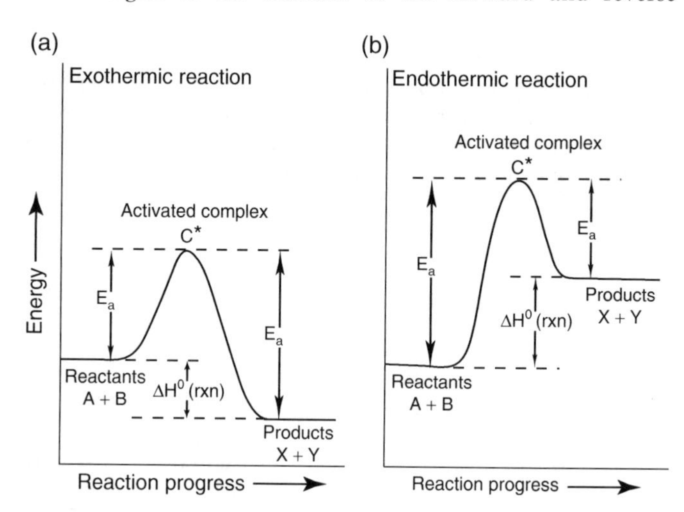

Fig. 9.6 Illustration of the concepts of activated complex (C*) and activation energy (E_a) for a reaction A + B ⇔ C* ⇔ X + Y: (a) exothermic reaction ($E_{a\,(for)} < E_{a\,(rev)}$) and (b) endothermic reaction ($E_{a\,(for)} > E_{a\,(rev)}$). In each case, the difference between $E_{a\,(for)}$ and $E_{a\,(rev)}$ can be shown to be approximately equal to ΔH_r^0, the enthalpy change of the overall reaction in the standard state (equation 9.72).

directions (i.e., A + B ⇒ C* and X + Y ⇒ C*, respectively). If the reactant molecules have sufficient kinetic energy to overcome the "energy barrier" ($E_{a\,(for)}$) to form the activated complex, the reaction can proceed. The transformation of the activated complex into the final product is accompanied by release of energy ($E_{a\,(rev)}$). If $E_{a\,(for)} < E_{a\,(rev)}$, the overall reaction is exothermic because it results in a net release of energy to the surroundings (Fig. 9.6a); if $E_{a\,(for)} > E_{a\,(rev)}$, the overall reaction is endothermic because it results in a net absorption of energy from the surroundings (Fig. 9.6b).

The transition state theory assumes that reaction C* ⇒ X + Y is spontaneous (fast), so that the slower reaction A + B ⇒ C* is the rate-determining step and $E_{a\,(for)}$ is the activation energy (E_a) for the overall reaction A + B ⇔ X + Y. It can be shown (see Box 9.1) that the terms E_a and A (see equation 9.53) for the reaction can be estimated from the relations

$$E_a = \Delta H^{0*} \tag{9.58}$$

and

$$A = \frac{k_B T}{h}\, e^{(\Delta S^{0*}/R)} \tag{9.59}$$

Box 9.1. Relationship between activation Energy (E_a) and enthalpy change of a reaction (ΔH_r)

Let us first consider the formation of the activated complex C* by the reaction A + B ⇔ C*. Assuming ideality (so that a_i = concentration $[i]$), the equilibrium constant for the reaction (K^*_{eq}) is:

$$K^*_{eq} = \frac{[C^*]}{[A]\,[B]} \tag{9.60}$$

In the present context, we use the superscript * to remind us that we are dealing with transition state equilibrium (involving an activated complex) of an activated complex. It can be shown from a statistical mechanical approach that the rate constant for the reaction (k^*) is given by

$$k^* = \frac{k_B T}{h} K^*_{eq} \tag{9.61}$$

where k_B is the Boltzmann constant, h the Planck constant, and T the temperature in Kelvin. The Gibbs free energy change for the activation reaction in the standard state can be written as

$$\Delta G^{0*} = -RT \ln K^*_{eq} \tag{9.62}$$

Substituting for K^*_{eq} in equation (9.61) we get

$$k^* = \frac{k_B T}{h}\, e^{-\Delta G_r^{0*}/RT} \tag{9.63}$$

Box 9.1. (cont'd)

Replacing free energy of activation (ΔG^{0*}) with enthalpy of activation (ΔH^{0*}) and entropy of activation (ΔS^{0*}) according to the realtion $\Delta G = \Delta H - T\Delta S$ we get what is commonly known as the *Eyring equation*:

$$k^* = \frac{k_B T}{h}\, e^{(-\Delta H^{0*}/RT + \Delta S^{0*}/R)} \tag{9.64}$$

The Eyring equation, like the Arrhenius equation, describes the trememperature dependence of reaction rate, but it is a theoretical construct based on transition state model. Taking the logarithm of equation (9.64), we get:

$$\ln k^* = \ln\left(\frac{k_B}{h}\right) + \ln T - \frac{\Delta H^{0*}}{RT} + \frac{\Delta S^{0*}}{R} \tag{9.65}$$

Differentiating, assuming ΔH^* and ΔS^* to be independent of temperature, we obtain

$$\frac{d(\ln k^*)}{dT} = \frac{1}{T} + \frac{\Delta H^{0*}}{RT^2} \tag{9.66}$$

Comparing equation (9.66) with the differential form of the Arrhenius equation (assuming A to be independent of temperature),

$$\frac{d(\ln k^*)}{dT} = \frac{E_a}{RT^2} \tag{9.67}$$

we find that

$$E_a = RT + \Delta H^{0*} \tag{9.68}$$

Because the value of RT is less than 5 kJ mol^{-1} up to 600 K, we can ignore the RT term in equation (9.68), which then simplifies to

$$E_a = \Delta H^{0*} \tag{9.69}$$

giving us an approximate relationship between activation energy and standard enthalpy of the activation complex. Applying equation (9.69) to the reaction in forward and reverse directions,

$$E_{a\,(for)} = \Delta H^{0*}_{for} \text{ (for the reaction A + B ⇔ C*)} \tag{9.70}$$

$$E_{a\,(rev)} = \Delta H^{0*}_{rev} \text{ (for the reaction X + Y ⇔ C*)} \tag{9.71}$$

and the difference in the activation energies is the standard enthalpy change for the whole reaction A + B ⇔ X + Y :

$$E_{a\,(for)} - E_{a\,(rev)} = \Delta H^{0*}_{for} - \Delta H^{0*}_{rev} = \Delta H_r^0 \text{ (for the whole reaction A + B ⇔ X + Y)} \tag{9.72}$$

As shown in Fig. 9.4,

if $E_{a\,(for)} < E_{a\,(rev)}$, ΔH_r^0 is negative, i.e., an exothermic reaction

if $E_{a\,(for)} > E_{a\,(rev)}$, ΔH_r^0 is positive, i.e., an endothermic reaction

where ΔH^{0*} and ΔS^{0*} denote, respectively, the standard state enthalpy and entropy of activation (i.e., the enthalpy and entropy change between the activated state and the initial ground state), which are assumed to be independent of temperature, k_B is the Boltzmann constant ($k_B \approx 1.3807 \times 10^{23}$ J K^{-1}), h is the Planck constant ($h = 6.626 \times 10^{-34}$ J s), R is the gas constant, and T is the temperature in Kelvin. The Boltzmann constant (named after the Austrian physicist Ludwig Boltzmann, 1844–1906) defines the relation between absolute temperature and the kinetic energy contained in each molecule of an ideas gas and is equal to the ratio of the ideal gas constant (R) and Avogadro's constant.

9.3 Relationship between rate and free energy change of an elementary reaction (ΔG_r)

It can be shown from transition state theory (see, e.g., Lasaga, 1981, pp. 145–146) that for a simple elementary reaction such as

$$\mathrm{A} + \mathrm{B} \underset{k_-}{\overset{k_+}{\rightleftarrows}} \mathrm{X} + \mathrm{Y} \tag{9.73}$$

the ratio of the forward and reverse reaction rates ($\Re_+$ and $\Re_-$, respectively) is given by

$$\frac{\Re_+}{\Re_-} = \frac{k_+\,[\mathrm{A}]\,[\mathrm{B}]}{k_-\,[\mathrm{X}]\,[\mathrm{Y}]} = \mathrm{e}^{-\Delta G_r/RT} \tag{9.74}$$

where T (K) is the temperature of interest and ΔG_r is the free energy change of the elementary reaction at T and the pressure of interest (i.e., $\Delta G^P_{r,T}$ according to the notation used in earlier chapters). ΔG_r can be calculated from the equilibrium constant (K_{eq}) of the reaction and its reaction quotient (Q) (see equation 8.19):

$$\begin{aligned}\Delta G_r &= \Delta G^0_{r,T} + RT \ln Q = -RT \ln K_{eq} + RT \ln Q \\ &= RT\,(\ln Q - \ln K_{eq}) \\ &= \mathrm{RT} \ln \frac{Q}{K_{eq}}\end{aligned} \tag{9.75}$$

The net rate of the reaction, $\Re_{net}$, is obtained as follows:

$$\begin{aligned}\Re_{net} &= \Re_+ - \Re_- = \Re_+ - \frac{\Re_+}{\mathrm{e}^{-\Delta G_r/RT}} \\ &= \Re_+ - \Re_+\,\mathrm{e}^{\Delta G_r/RT} \\ &= \Re_+\,(1 - \mathrm{e}^{\Delta G_r/RT})\end{aligned} \tag{9.76}$$

In this context, $-\Delta G_r$ is often called the *affinity of the reaction* (A_r). At equilibrium, $\Re_+ = \Re_-$, i.e., $\Re_{net} = 0$, and $\Delta G_r = 0$. Thus, $-\Delta G_r$ can be viewed as a measure of the departure from equilibrium or the driving force of the reaction toward equilibrium. If $-\Delta G_r$ is positive (i.e., $\Delta G_r < 0$), $\Re_{net}$ is positive and the reaction (9.73) proceeds from left to right. If $-\Delta G_r$ is negative (i.e., $\Delta G_r > 0$), $\Re_{net}$ is negative and the reaction (9.73) proceeds from right to left. For very small values of x, $\mathrm{e}^x \approx 1 + x$; thus, if the reaction is close enough to equilibrium so that $|\Delta G_r| \ll RT$, equation (9.76) simplifies to a linear relationship:

$$\boxed{\Re_{net} = -\frac{\Re_+\,\Delta G_r}{RT}} \tag{9.77}$$

Application of this rate law to heterogeneous reactions under high temperature metamorphic conditions is discussed by Walther and Wood (1984, 1986) and by Kerrick *et al.* (1991).

In equation (9.76), the term $\exp(\Delta G_r/T)$ represents the "reverse reaction." Because of its exponential form, the ratio $\Re_{net}/\Re_+$ increases rapidly as ΔG_r becomes more and more negative and, at 25°C, approaches unity (i.e., $\Re_{net} \approx \Re_+$) for ΔG_r values less than about -8 kJ mol^{-1} (Fig. 9.7) or for $\Delta G_r/RT$ values less than about -3. Thus, when ΔG_r is large and negative – i.e., the system is "far from equilibrium" – $\Re_{net}$ becomes practically independent of ΔG_r. Experimental studies, however, indicate much lower values of ΔG_r (e.g. about -38 kJ mol^{-1} for dissolution of albite at 80°C and pH 8.8; Burch *et al.*, 1993), below which the net reaction becomes independent of ΔG_r. As discussed by Burch *et al.* (1993), the dissolution kinetics of silicate minerals is more complicated than described here.

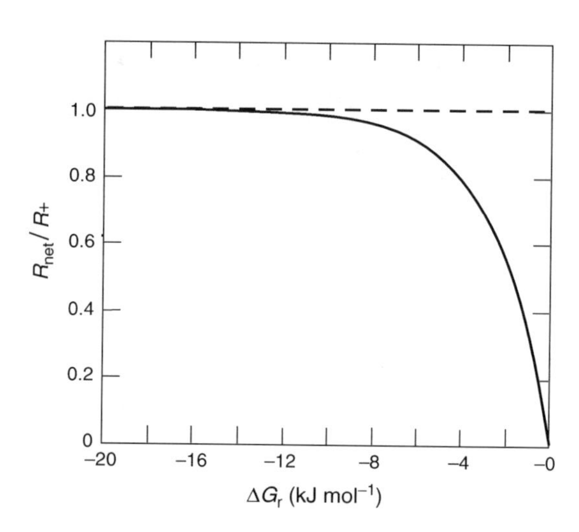

Fig. 9.7 Calculated trend of $\Re_{net}/\Re_+$ ratio as a function of the free energy change ΔG_r of a simple elementary reaction at 298.15 K (25°C) and 1 bar. For ΔG_r less than about 8 kJ mol^{-1}, $\Re_{net} \approx \Re_+$ and is independent of ΔG_r.

9.4 Catalysts

The term *catalysis* was coined in 1835 by the Swedish chemist Jons Jacob Berzelius, although ancient techniques such as fermenting wine to acetic acid and manufacturing soap from fats and bases used the principles of catalysis much earlier. It is estimated that at the present time about 60% of all commercial chemical products involve the use of catalysts at some stage of the manufacturing process.

A *catalyst* is a substance that accelerates the rate of a thermodynamically favorable chemical reaction without itself being transformed or consumed by the reaction. A catalyst generally enters into chemical combination with the reactants but is ultimately regenerated so that the amount of catalyst remains unchanged and the catalyst does not appear in the balanced equation for the net reaction. Since the catalyst is not consumed, each catalyst molecule is capable of affecting a large number of reactant molecules. An active catalyst may transform several million molecules of reactants per minute.

The rate acceleration occurs because a catalyst reduces the amount of energy needed to start a reaction by providing an alternative pathway that requires lower activation energy. All catalyzed reactions involve lowering of the activation energy, and it follows from the Arrhenius equation (equation 9.53) that, at constant temperature, the value of k would increase with decreasing value of E_a. A catalyst, however, cannot make a thermodynamically unfavorable reaction possible; it has no effect on the thermodynamic equilibrium of a reaction because it accelerates both the forward and the reverse reactions by the same factor. In the last chapter we discussed the catalytic role of microorganisms in many oxidation–reduction reactions. Here we discuss catalysis without the involvement of microorganisms.

9.4.1 *Homogeneous catalysis*

A catalytic reaction (or catalysis) may be homogeneous or heterogeneous. In *homogeneous catalysis*, the entire reaction occurs in a single phase, as in the case of a catalyst dissolved in a liquid reaction mixture or a gas in a gas reaction mixture. An example of homogeneous catalysis is the oxidation of ozone ($O_{3\,(g)}$) in the upper atmosphere to $O_{2\,(g)}$ catalyzed by chlorine radicals (denoted here by Cl•). (A radical is an atom containing an unpaired electron in an outer shell.) The reaction mechanism for the net oxidation reaction is believed to consist of two sequential elementary reactions:

$$\begin{array}{lll} Cl\bullet + O_{3\,(g)} \Rightarrow Cl-O + O_{2\,(g)} & \text{(first step)} & \\ Cl-O + O \Rightarrow Cl\bullet + O_{2\,(g)} & \text{(second step)} & (9.78) \\ \hline O_{3\,(g)} + O \Rightarrow 2O_{2\,(g)} & \text{(net reaction)} & \end{array}$$

The first step occurs because of the presence of Cl• in the upper atmosphere due to the breakdown of anthropogenic chlorofluorocarbon (CFC) compounds, such as CF_2Cl_2 (Freon-12), by ultraviolet radiation from the Sun, and the upper atmosphere has enough oxygen atoms available for the second step. Note that for every Cl• consumed in the first step, another Cl• is regenerated in the second step. Thus, the Cl• acts as a catalyst, and a single Cl• molecule is capable of destroying thousands of ozone molecules. The destruction of ozone in the upper atmosphere is a matter of serious concern because of the protection ozone offers against the potentially harmful ultraviolet radiation from the Sun reaching the Earth's surface (see Box 13.5).

9.4.2 *Heterogeneous catalysis*

In *heterogeneous catalysis* (also called *surface or contact catalysis*), the catalyst is present in a different phase than the reactants, as in the case of a solid catalyst in a liquid reaction mixture, and the reaction occurs at interfaces between phases. Typically, heterogeneous catalysis involves the interaction of a finely ground solid with liquid or gaseous reactants. The lowering of the activation energies in such catalytic reactions is accomplished by the large amount of surface area available for reactions. Most examples of heterogeneous catalysis may be viewed as progressing through a cycle of two steps: absorption of one or more reactants onto "active" sites on the surface of the catalyst (sites that are relatively more reactive because of some interaction between the surface and the reactant molecules); and desorption of the product molecules from the "active" sites so that these sites become available again for sorption. A good catalyst is one which absorbs the reactant molecules strongly enough for them to react but not so strongly that the product molecules cannot desorb. Noble metals (such as Pt, Pd, and Rh) and oxides of transition metals (such as NiO, V_2O_5, and Fe-oxides) are good heterogeneous catalysts.

An example of heterogeneous catalysis is the Haber–Bosch process for industrial production of ammonia (NH_3) by the exothermic reaction

$$N_{2\,(g)} + 3H_{2\,(g)} \Rightarrow 2NH_{3\,(g)} \qquad (9.79)$$

The very large value of the equilibrium constant for this reaction (3.6×10^8 at 298.15 K) indicates that nearly all of N_2 and H_2 mixed in a 1:3 mole ratio should be converted into NH_3 at 298.15 K. In reality, however, the reaction is so slow at 298.15 K that no measurable amount of NH_3 is formed within a reasonable time. Since the reaction results in a significant reduction in volume, it follows from Le Chatelier's Principle that the forward reaction will be favored by a decrease in temperature or an increase in pressure. However, if the pressure used is too high, the cost of generating ammonia exceeds the return obtained from the extra ammonia produced, and too

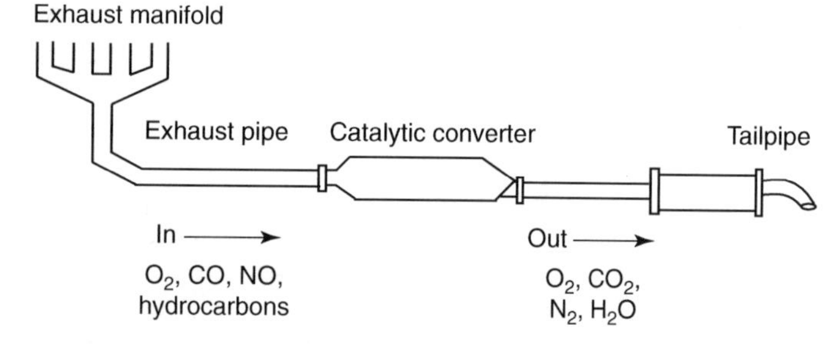

Fig. 9.8 Schematic diagram showing the position of catalytic converter in the exhaust system of an automobile.

Box 9.2. Catalytic converters in automobile exhaust systems

A catalytic converter is a fairly simple device that uses a catalyst (in the form of finely divided Pt, Pd, and Rh) to convert three kinds of air pollutants in automobile exhaust into harmless compounds:

(1) conversion of unburned hydrocarbon compounds in the gasoline (e.g., isooctane, C_8H_{18}) into $CO_{2\,(g)}$ and $H_2O_{(g)}$,

$$2C_8H_{18\,(g)} + 25O_{2\,(g)} \Rightarrow 16CO_{2\,(g)} + 18H_2O_{(g)} \quad (9.80)$$

(2) conversion of carbon monoxide ($CO_{(g)}$) formed by the combustion of gasoline into $CO_{2\,(g)}$,

$$2CO_{(g)} + O_{2\,(g)} \Rightarrow 2CO_{2\,(g)} \quad (9.81)$$

(3) conversion of nitrogen oxides formed by the combination of atmospheric $O_{2\,(g)}$ and $N_{2\,(g)}$ (because of the heat in the automobile engine) back into $O_{2\,(g)}$ and $N_{2\,(g)}$,

$$2NO_{(g)} \Rightarrow N_{2\,(g)} + O_{2\,(g)} \quad \text{and} \quad 2NO_{2\,(g)} \Rightarrow N_{2\,(g)} + 2O_{2\,(g)} \quad (9.82)$$

All of these reactions are thermodynamically favored and exothermic. A drawback of the catalytic converter is that it also catalyzes another thermodynamically favored reaction, the conversion of $SO_{2\,(g)}$ (which is formed by oxidation of the trace amounts of sulfur in gasoline) into $SO_{3\,(g)}$,

$$2SO_{2(g)} + O_{2\,(g)} \Rightarrow 2SO_{3\,(g)} \quad (9.83)$$

which probably is a worse pollutant than $SO_{2\,(g)}$.

low a temperature renders the forward reaction too slow. As a compromise, the system is commonly operated at about 450°C (~725 K) and about 200 bars. To increase the reaction rate, the process is carried out in the presence of a catalyst in the form of finely divided iron and small amounts of Fe-oxides. Nitrogen gas is obtained from liquefied air and hydrogen gas from coal gas or petroleum refining.

A more familiar example is the incorporation of catalytic converters in the design of automobile exhaust systems (Fig. 9.8) for the purpose of reducing the emission of harmful gases (such as CO, NO, NO_2) into the environment (see Box 9.2).

9.5 Mass transfer in aqueous solutions

For the discussion of reaction rates in section 9.1 we took it for granted that the reactants were available for the reactions under consideration and that the reactants and products continued to coexist in the case of reversible reactions. This requirement can be satisfied quite easily in experimental setups used for measurements of reaction rates, but the occurrence of chemical reactions in nature commonly involves transport of material into and out of the system. For example, the hydrothermal transformation of kaolinite ($Al_2Si_2O_5(OH)_4$) to muscovite ($KAl_3Si_3O_{10}(OH)_2$),

$$1.5 \text{ kaolinite} + K^+ \Rightarrow \text{muscovite} + H^+ + 1.5H_2O$$

is accomplished quite easily in laboratory experiments at moderate temperatures with KCl in the solution as a source of K^+ (Chermak and Rimstidt, 1990), but the rate of this reaction in nature depends on the rate of transport of K^+ to the solution–kaolinite interface.

Transport of a solute in a fluid occurs by two processes: *advection* and *diffusion*. Advection refers to the transport of solutes by the bulk movement of the fluid under the influence of a hydraulic gradient. It is the most important mass transfer process in geochemical systems, especially over long distances. Some examples are the transport of contaminants in groundwater, of dissolved constituents in magma, and of ore-forming constituents in hydrothermal solutions exsolved from a crystallizing magma. Diffusion refers to the process by which solutes move spontaneously along a chemical potential gradient in a fluid or solid medium. Diffusion is not as efficient as advection for long-distance transport, except in the case of gases, but it is an important mass transfer process for redistribution of constituents on a local scale (few meters at best) as in mineral zoning, exsolution, and cementation of sediments during diagenesis. Generally, mass transport of chemical constituents involves a combination of advection and diffusion, the former being important for large-scale transport and the latter for small-scale transport.

9.5.1 *Advection–diffusion equation*

Consider the flow of a fluid through a closed "elementary volume," dx cm (length) × 1 cm (width) × 1 cm (height) (Fig. 9.9). Generally, the composition of the output fluid would be

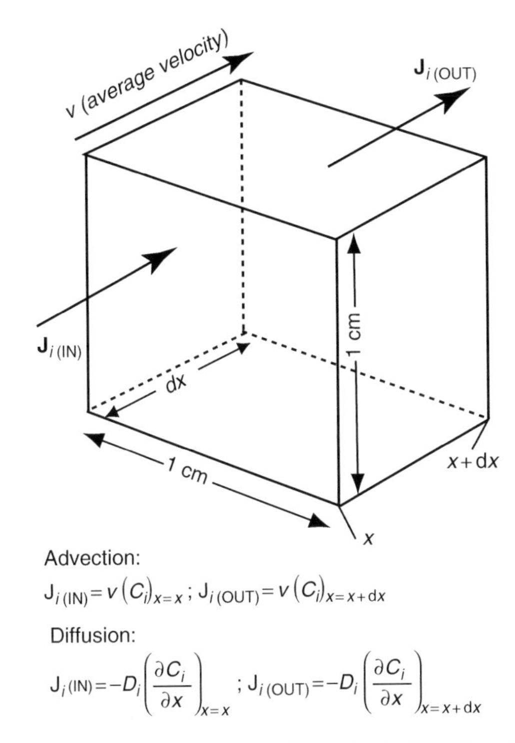

Fig. 9.9 Illustration of the concept of change in the flux of a solute relative to an "elementary volume."

different from that of the input fluid because of changes in the concentration of the solutes due to advection, diffusion, and reactions within the elementary volume. The one-dimensional change in the concentration of a solute i with respect to time (t) over the length (dx) of the elementary volume is given by the partial differential equation

$$\left(\frac{\partial C_i}{\partial t}\right)_{\text{fluid}} = -\left(\frac{\partial J_i}{\partial x}\right)_{\text{advection}} - \left(\frac{\partial J_i}{\partial x}\right)_{\text{diffusion}} \pm \left(\frac{\partial C_i}{\partial t}\right)_{\text{reaction}} \tag{9.84}$$

rate of change of i in the fluid with time	rate of mass transfer by advection	rate of mass transfer by diffusion	rate of mass transfer by chemical reaction

where C_i is the concentration of i expressed as moles per cm^3 of fluid (mol cm^{-3}), and J_i is the mass flux of i in the fluid measured in moles per unit cross-sectional area per second (mol cm^{-2} s^{-1}) entering the elementary volume. The negative signs of the advection and diffusion flux terms account for the fact that the flux decreases with distance along the flow direction. The reaction term is positive if i is added to the solution, negative if i is removed from the solution, and zero if i is a conservative species. The equation is written in terms of partial derivatives because it describes the change in C_i only along one direction (in this case, the x direction).

If v_x denotes the constant average linear velocity of fluid flow in the x direction (in centimeters per second, cm s^{-1}), then

$$J_{i\,(IN)}\,(\text{advection}) = v_x\,(C_i)_{x=x} \quad \text{and}$$
$$J_{i\,(OUT)}\,(\text{advection}) = v_x\,(C_i)_{x=x+dx} \tag{9.85}$$

and the rate of change of C_i with time, $\partial C_i/\partial t$, attributable to advection is the product of the velocity of the fluid entering the elementary volume (v_x) and the concentration gradient, ($\partial C_i/\partial x$):

$$\left(\frac{\partial C_i}{\partial t}\right)_{\text{fluid, advection}} = -\left(\frac{\partial J_i}{\partial x}\right)_{\text{advection}} = -v_x\left(\frac{\partial C_i}{\partial x}\right)_{\text{advection}} \tag{9.86}$$

For the construct described above, the gradient in fluid composition must originate outside the representative volume of rock under consideration. Possible sources for such concentration gradient in the fluid might include mineral dissolution or a groundwater contamination plume.

The flux for diffusion (in one dimension) is given by what is known as *Fick's first law of diffusion*, which states that, at steady state, the diffusion flux of a species i through a plane is proportional to its concentration gradient ($\partial C_i/\partial x$) normal to that plane:

$$J_i\,(\text{diffusion}) = -D_i\,\frac{\partial C_i}{\partial x} \tag{9.87}$$

where D_i, the proportionality constant, is called the *diffusion coefficient*, which has the dimensions of square centimeters per second (cm^2 s^{-1}). The negative sign in equation (9.87) takes care of the fact that the diffusion is from a region of high concentration to one of lower concentration. It is also evident from equation (9.87) that there will be no diffusion in the absence of a concentration gradient. Strictly speaking, we should use activity (a_i) instead of concentration (C_i) in equation (9.87), but for the sake of simplicity we will assume a dilute solution so that that $C_i = a_i$.

Equation (9.87) describes diffusion under steady-state conditions; that is, the flux and concentration gradients do not change with time. To calculate $\partial C_i/\partial x$ due to diffusion under transient conditions, we must take into account the change in $\partial C_i/\partial x$ with time by diffusion (see derivation in Drever, 1997, p. 355; Appelo and Postma, 1995, p. 351):

$$\left(\frac{\partial C_i}{\partial t}\right)_{\text{fluid, diffusion}} = -\left(\frac{\partial J_i}{\partial x}\right)_{\text{diffusion}} = \frac{\partial}{\partial x}\left(D_i\,\frac{\partial C_i}{\partial x}\right) \tag{9.88}$$

Assuming D_i to be constant at a specified temperature, equation (9.88) simplifies to

$$\left(\frac{\partial C_i}{\partial t}\right)_{\text{fluid, diffusion}} = D_i \left(\frac{\partial^2 C_i}{\partial x^2}\right) \quad (9.89)$$

which gives us the rate of change of the concentration of *i* with respect to time and is known as *Fick's second law of diffusion*. Solution of this partial differential equation is quite complicated (see, e.g., Lasaga, 1998) and will not be addressed here.

Equations (9.87) and (9.89) apply to diffusion in solid, liquid, and gas phases, but the diffusion process has unique characteristics in each of these phases. Diffusion coefficients at 25°C are usually in the order of 10^{-10} cm² s⁻¹ in the solid phase, in the order of 10^{-4} to 10^{-6} cm² s⁻¹ in the liquid phase (Table 9.3), and in the order of 10^{-1} cm² s⁻¹ in the gas phase. Diffusion in solids occurs by intracrystalline (or volume) diffusion, which is favored along crystallographic directions, and by intercrystalline (or grain-boundary) diffusion, which generally is more rapid than volume diffusion. Combining equations (9.84), (9.86), (9. 88), and (9.89) we get the *advection–diffusion equation* (in one dimension):

$$\left(\frac{\partial C_i}{\partial t}\right)_{\text{fluid}} = -v_x\left(\frac{\partial C_i}{\partial x}\right) - D_i\left(\frac{\partial^2 C_i}{\partial x^2}\right) \pm \left(\frac{\partial C_i}{\partial t}\right) \quad (9.90)$$

change due to advection — change due to diffusion — change due to reaction

9.5.2 *The temperature dependence of diffusion coefficient*

Diffusion coefficient is a function of temperature; in general it increases as temperature increases. The temperature dependence of D_i can be described by an equation of the Arrhenius form:

$$D_i = A_D\, e^{-E_D/RT} \quad (9.91)$$

where A_D (in cm² s⁻¹) is the *frequency factor*, which is usually a constant for a diffusing species in a particular medium, E_D is the *activation energy of diffusion* (in kJ mol⁻¹) for the species *i*, and *T* is the temperature in Kelvin. As illustrated in Fig. 9.10

Table 9.3 Diffusion coefficients (*D*) of some ionic species in water at 25°C.

Cation	*D* (10^{-6} cm² s⁻¹)	Cation	*D* (10^{-6} cm² s⁻¹)	Anion	*D* (10^{-6} cm² s⁻¹)
H^+	93.1	Sr^{2+}	7.94	OH^-	52.7
Na^+	13.3	Ba^{2+}	8.48	F^-	14.6
K^+	19.6	Ra^{2+}	8.89	Cl^-	20.3
Rb^+	20.6	Mn^{2+}	6.88	HS^-	17.3
Cs^+	20.7	Fe^{2+}	7.19	HCO_3^-	11.8
Mg^{2+}	7.05	Cr^{3+}	5.94	SO_4^{2-}	10.7
Ca^{2+}	7.93	Fe^{3+}	6.07	CO_3^{2-}	9.55

Source of data: Li and Gregory (1974).

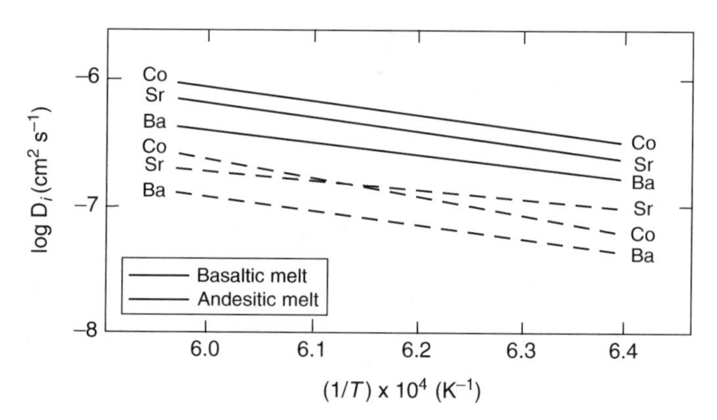

Fig. 9.10 Arrhenius plot of diffusion coefficients for the diffusion of cobalt ions in melts of basaltic and andesitic compositions. The activation energies (E_D) determined from the slopes of the two straight-line graphs are: basaltic melt, 220 kJ mol⁻¹; andesitic melt, 280 kJ mol⁻¹. (After Lowry, Henderson, and Nolan, 1982, Figure 6, p. 260.)

for diffusion of cobalt ions in silicate melts, E_D can be determined from a plot of ln D_i measured at various temperatures against 1/*T*. The plot will be a straight line with slope = $-E_D/R$ and intercept = ln A_D according to the equation

$$\ln D_i = \ln A_D - \frac{E_D}{R}\frac{1}{T} \quad (9.92)$$

Experimentally determined values of diffusion coefficients as a function of temperature have been applied to estimate cooling rates of rocks containing compositionally zoned minerals. For example, applying the experimentally determined diffusion coefficients of Fe^{2+} and Mg^{2+} in olivine (Buening and Buseck, 1973) to compositional profiles in zoned olivine crystals, Taylor *et al.* (1977) estimated the cooling rates of lunar rocks 12002 and 15555 as 10°C day⁻¹ and 5°C day⁻¹, respectively.

9.6 Kinetics of geochemical processes – some examples

A subject of prime interest in geochemistry is the dissolution and precipitation of minerals in contact with a fluid medium. Here we will discuss two examples involving aqueous solutions: the dissolution and growth of calcite, and the dissolution of silicate minerals.

9.6.1 *Diffusion-controlled and surface-controlled reaction mechanisms*

The reaction between a mineral and a solution usually involves several steps: diffusion of a reactant to the mineral surface; adsorption of the reactant on the surface; reaction with the mineral at a high-energy site (such as kinks and dislocations) forming a transition state complex; decomposition of the transition state complex; desorption of products; and diffusion of products to bulk solution. The slowest of these steps under a given set of conditions becomes the rate-controlling step. In most cases, different steps dominate the reaction kinetics at different degrees of disequilibrium.

Kinetic models proposed for the dissolution/precipitation of minerals fall into two general categories: *transport-controlled reactions* (often referred to as *diffusion-controlled reactions*); and *surface-controlled reactions*. For example, if dissolution of a silicate mineral is nonstoichiometric or if secondary precipitation occurs at the surface of the original mineral, then a protective layer could be built on the surface of the solid. The rate of the reaction in this case may be controlled by diffusion of reactants and products through this protective layer. On the other hand, stoichiometric dissolution is generally controlled by chemical reactions ocurring at the solid–fluid interface and, thus, are surface controlled. Commonly, surface-controlled reactions are most important near equilibrium and have relatively high activation energy (typically about 50 to 70 kJ mol^{-1}), whereas diffusion-controlled reactions are more important far from equilibrium (i.e., at low degrees of saturation) and have relatively low activation energy (typically about 20 kJ mol^{-1} or less).

9.6.2 *Dissolution and precipitation of calcite in aqueous solutions*

The dissolution and precipitation kinetics of calcite are quite complex, even in the simple $CaCO_3$–H_2O–CO_2 system, because such reactions are controlled by many variables. The variables include, in addition to temperature, total pressure, P_{CO_2}, pH, and activities of the constituents, the hydrodynamic mass transport properties of the mineral–water system and the kinetics of heterogeneous reactions at the calcite surface; above a pH of 4 the dissolution and crystal growth of calcite is largely controlled by surface reaction. In addition, "foreign ions", such as Mg^{2+} and PO_4^{3-}, present in natural waters inhibit the rate of dissolution and crystal growth of calcite (Berner and Morse, 1974; Berner, 1975; Morse and Burner, 1979; Mucci and Morse, 1983).

The fundamental reference point in this exercise is the congruent dissociation of $CaCO_{3\,(s)}$ to aqueous Ca^{2+} and CO_3^{2-}:

$$CaCO_{3\,(s)} \underset{\Re_p}{\overset{\Re_d}{\rightleftarrows}} Ca^{2+} + CO_3^{2-} \tag{9.93}$$

where $\Re_d$ represents the rate in the forward direction (dissolution) and $\Re_p$ the trate in the reverse direction (precipitation). At equilibrium, $\Re_d = \Re_p$.

The *saturation state* of an aqueous solution with respect to calcite, Ω_{cal}, can be defined as

$$\boxed{\Omega_{cal} = \frac{IAP_{sol}}{K_{cal}} = \frac{a_{Ca^{2+}}\, a_{CO_3^{2+}}}{K_{cal}}} \tag{9.94}$$

where K_{cal} represents the thermodynamic solubility constant of reaction (9.93) at the temperature and pressure of interest. Equation (9.94) is simply a different form of equation (7.32) with K_{sp} replaced by K_{cal} to account for P–T conditions other than 298.15 K and 1 bar. When $\Omega_{cal} < 1$, the solution is undersaturated and the net effect is calcite dissolution; when $\Omega_{cal} > 1$, the solution is supersaturated and the net effect is calcite precipitation; and the difference between Ω_{cal} and 1 is a measure of the degree of disequilibrium, one of the main factors controlling the rate of reaction. Although simple in concept, the determination of Ω_{cal} for seawater is not an easy task. As discussed by Morse and Berner (1979), the calculation of the activities is quite involved because of their dependence on pH, P_{CO_2}, total pressure, temperature, and seawater composition. Also, it is very difficult to measure solubility of calcite in seawater because supersaturated seawater precipitates Mg-enriched calcite rather than pure calcite (Berner, 1976).

Two general approaches have been used in treating $CaCO_3$ dissolution and precipitation rate data (Morse, 1983). The one used most commonly in recent years is to treat the reaction in an empirical manner, i.e., without the basis of a mechanistic model. The other type of approach attempts to interpret the observed relation between reaction rate and saturation state in terms of reaction mechanisms. The discussion below is a brief summary extracted from many published articles (e.g., Morse and Berner, 1979; Plummer *et al.*, 1978, 1979; Morse, 1983; Sjöberg and Rickard, 1984; Busenberg and Plummer, 1986), which treat the topic in considerable detail.

Empirical approach

The rate equations for dissolution and precipitation of calcite are written as functions of $(1 - \Omega)$ and $(\Omega - 1)$, respectively, instead of just Ω, because near equilibrium both dissolution and precipitation reactions are important simultaneously and quantitatively. The general form of the empirical rate laws that have been found to fit most of the relatively *near-equilibrium* data are (Morse, 1978, 1983):

$$\begin{aligned} &\text{Dissolution} \quad \Re_d = k_d(1-\Omega_{cal})^{n_d} \\ &\text{Precipitation} \quad \Re_p = k_p(\Omega_{cal}-1)^{n_p} \end{aligned} \tag{9.95}$$

where the subscripts d and p refer, respectively, to dissolution and precipitation, k is the rate constant, and n is the empirical reaction order. Note that $\Re_d = 0$ and $\Re_p = 0$ when $\Omega = 1$. We can extract values of n and k by fitting the rate data to logarithmic forms of these equations:

$$\begin{aligned} \log \Re_d &= n_d \log(1 - \Omega_{cal}) + \log k_d \\ \log \Re_p &= n_p \log(\Omega_{cal} - 1) + \log k_p \end{aligned} \tag{9.96}$$

A plot of $\Re_d$ versus $\log(1 - \Omega_{cal})$ would be a straight line with slope $= n_d$ and intercept $= \log k_d$, and the straight-line plot of

$\Re_d$ versus log $(\Omega_{cal} - 1)$ would have slope = n_p and intercept = log k_p. The values of n and k depend on the composition of the solution and nature of the carbonate sample (e.g., particle size, density of crystal defects, etc.). For example, dissolution experiments by Keir (1980) on shells of individual species of cocoliths and foraminifera, various size fractions of sediment from the Ontong–Java Plateau and the Rio Grande Rise, and synthetic calcite gave $n = 4.5 \pm 0.7$ (and ~4.2 for synthetic aragonite and petropods) but a widely variable rate constant among samples. On the other hand, near-equilibrium dissolution experiments by Morse (1978) yielded distinctly different values of n and k for $CaCO_3$-rich deep-sea sediments from different ocean basins:

$$\Re_d(\%\ day^{-1})_{Indian\ ocean} = 10^{4.3}(1 - \Omega_{cal})^{5.2}$$

$$\Re_d(\%\ day^{-1})_{Pacific\ ocean} = 10^{2.7}(1 - \Omega_{cal})^{3.0} \quad (9.97)$$

$$\Re_d(\%\ day^{-1})_{Atlantic\ ocean} = 10^{3.1}(1 - \Omega_{cal})^{4.5}$$

The reasons for the differences among the oceans are not known. Suggested explanations by Morse (1978) include different particle size distributions or different surface histories (e.g., deposition of differing amounts of metal oxides or phosphate on the surface of calcium carbonate) of the sediment samples from different locations. In any case, the above rates should not be taken as representative of these ocean basins, as variability within a given basin could probably be at least as large as that found in samples from different ocean basins. Zhong and Mucci (1989) showed that for seeded calcite and aragonite precipitated from artificial seawater solutions of various salinities (5 to 44‰) at 25°C and $P_{CO_2} = 10^{-2.5}$ atm, n varied from 2.5 to 3.3 for calcite and from 1.8 to 2.4 for aragonite.

Mechanistic model approach

Plummer *et al.* (1978) performed a series of experiments to study the dissolution kinetics of Iceland spar in the calcite $-CO_2-H_2O$ system (pH = 2 to 7; P_{CO_2} = 0.0 to 1.0 atm) at both far from equilibrium and near equilibrium conditions. In a plot of log-rate versus pH (Fig. 9.11), the combined results can be divided into three regions: (1) a "pH-dependent region" (pH < 3.5 at 25°C) in which the dissolution rate is independent of P_{CO_2} and the log-rate is a linear function of pH with slope = ~1; (2) a "transition region" (pH = 3.5 to 5.5 at P_{CO_2} = 1.0 atm and 25°C) of decreasing pH dependence in which the dissolution rate depends on both pH and P_{CO_2}; and (3) a "pH-independent region" (pH > 5.5 at P_{CO_2} = 1.0 atm and 25°C) in which the dissolution rate decreases rapidly with increasing pH because of significant backward reaction at near-equilibrium condition. Plummer *et al.* (1978) concluded that far from equilibrium (regions 1 and 2), the rate of calcite dissolution is at least in part a function of a_{H^+} in the

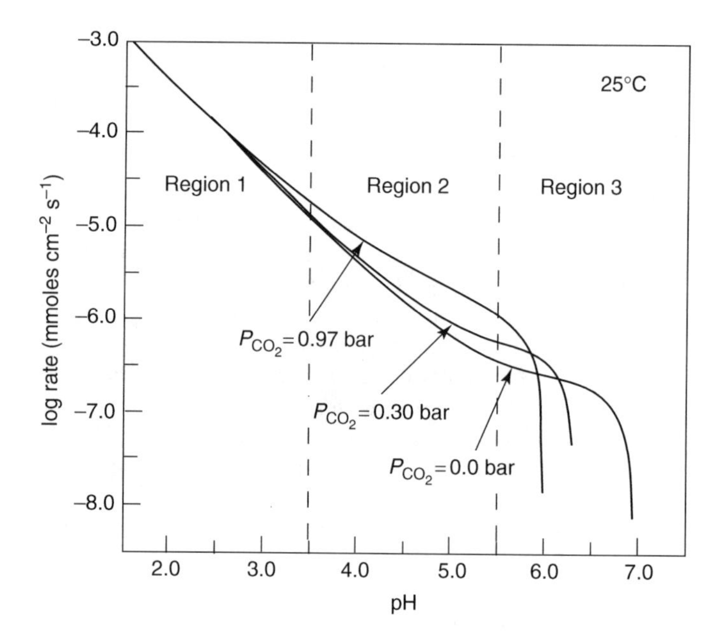

Fig. 9.11 Experimentally determined dissolution rates of calcite (Iceland spar) as a function of bulk fluid pH and P_{CO_2} at 25°C. The results can be divided into three regions: a pH-dependent regime (Region 1), a transition region (Region 2), and a pH-independent region (Region 3) (see text for explanation). The boundaries between the regimes are shown for dissolution at P_{CO_2} = 1 atm but, in general, the boundaries are a function of P_{CO_2} and reaction progress. (After Plummer *et al.*, 1978, Figure 1, p. 186.)

solution and hence of hydrodynamic transport, whereas at near-equilibrium condition (Region 3) the dissolution rate seems to be largely a function of surface reaction alone. Although covering a narrow range in pH (only ~0.5 pH units), Region 3 represents the final 75% of the reaction in terms of calcite solubility.

At fixed temperature and P_{CO_2}, the equation for net rate of calcite dissolution/precipitation that satisfies these experimental observations is of the form (Plummer *et al.*, 1978)

$$\Re = k_1\, a_{H^+} + k_2\, a_{H_2CO_3^*} + k_3\, a_{H_2O} - k_4\, a_{Ca^{2+}}\, a_{HCO_3^-} \quad (9.98)$$

where $\Re$ is in mmol cm^{-2} s^{-1}, $H_2CO_3^* = H_2CO_{3\,(aq)} + CO_{2\,(aq)}$, a_i represents activity in the bulk fluid, and k_1, k_2, and k_3 are temperature-dependent, first-order rate constants corresponding to the following three elementary reactions occurring concurrently on the calcite surface:

$$CaCO_3 + H^+ \xrightarrow{k_1} Ca^{2+} + HCO_3^- \quad (9.99)$$

$$CaCO_3 + H_2CO_3^* \xrightarrow{k_2} Ca^{2+} + 2HCO_3^- \quad (9.100)$$

$$CaCO_3 + H_2O \xrightarrow{k_3} Ca^{2+} + HCO_3^- + OH^- \quad (9.101)$$

Expressions for the rate constants and the corresponding Arrhenius activation energies (in kcal mol^{-1}) derived from their data are:

$\log k_1 = 0.198 - 444/T$	$E_a(k_1) = 2.0$
$\log k_2 = 2.84 - 2177/T$	$E_a(k_2) = 10.0$
$\log k_3\ (T < 298.15\,K) = -5.86 - 317/T$	$E_a(k_3) = 1.5\ (T < 298.15\,K)$
$\log k_3\ (T > 298.15\,K) = 1.10 - 1737/T$	$E_a(k_3) = 7.9\ (T > 298.15\,K)$

where T is in Kelvin. The rate constant k_4 is a function of both temperature and P_{CO_2}. The expression for k_4 in equation (9.98), theoretically derived by Plummer *et al.* (1978), is rather complex; a simpler but approximate expression derived by Langmuir (1997) from the $\log P_{CO_2}$ versus $\log k_4$ curves for different temperatures given in Plummer *et al.* (1978) is

$$\boxed{\log k_4 = -7.56 + 0.016\ T - 0.64 \log P_{CO_2}} \tag{9.102}$$

where T is in Kelvin and P_{CO_2} in bars. This function permits rough calculation of k_4 values for $P_{CO_2} < 10^{-1.5}$ bar.

The three positive terms on the right-hand side of equation (9.98) essentially describe the forward reaction rate far from equilibrium; the fourth, negative term (containing k_4) defines the reverse (precipitation) rate as the water approaches saturation with respect to calcite (and the net rate of reaction approaches zero). Furthermore, the low activation energy associated with k_1 indicates that at low pH values, where calcite dissolution is dominated by the stoichiometry of reaction (9.99), the dissolution rate is controlled by H^+ ion diffusion. Sjöberg and Rickard (1984) have suggested that the rate of calcite dissolution in the "pH-dependent region" (which is between pH=4 and pH=5.5 according to Sjöberg and Rickard, 1984), is given by

$$\boxed{\Re = k_1\ (a_{H^+})^{0.90}} \tag{9.103}$$

A fractional-order dependence of the rate on bulk phase a_{H^+} indicates a surface reaction-controlled dissolution (Sjöberg and Rickard, 1984; Schnoor, 1990) in the sense that the rate is controlled by diffusion of H^+ ions to the surface of calcite as well as the diffusion of reaction-product away from the surface. If the dissolution reaction were controlled by H^+ ion diffusion through a thin liquid film or residue layer surrounding the calcite, one would expect a first-order dependence on a_{H^+}; if the reaction were controlled by some other factor such as surface area alone, then the dependence on a_{H^+} should be zeroth-order.

At higher pH conditions, the forward rate becomes increasingly independent of pH, and the dissolution is both diffusion-controlled and surface-controlled (Sjöberg and Rickard, 1984). Based on equation (9.98), it can be computed that at pH>6 and P_{CO_2}<0.1 bar, conditions typical of shallow groundwaters, the rate of calcite dissolution far from equilibrium is approximated by

$$\boxed{\Re = k_3\ a_{H_2O}} \tag{9.104}$$

and for near equilibrium by

$$\boxed{\Re = k_3\ a_{H_2O} - k_4\ a_{Ca^{2+}} a_{CO_3^{2-}}} \tag{9.105}$$

as long as the surface area of calcite remains constant (Langmuir, 1997). Equation (9.104) can be further simplified to $\Re = k_3$ and equation (9.105) to $\Re = k_3 - k_4\ a_{Ca^{2+}} a_{CO_3^{2-}}$ if the solution is dilute so that a_{H_2O}=1.

The rate equations and reaction mechanisms for other one-component carbonate minerals, such as aragonite, witherite ($BaCO_3$) and magnesite ($MgCO_3$), are similar to those of calcite but, in solutions of the same composition, the rate of magnesite dissolution is approximately four orders of magnitude lower than that of calcite (Chou *et al.*, 1989). The rate of crystal growth of aragonite, however, is significantly slower than that of calcite in supersaturated $Ca(HCO_3)_2$ solutions, which may explain why aragonite is almost absent in freshwater environments. The dissolution kinetics and reaction mechanisms of two-component carbonate minerals, such as dolomite ($CaMg(CO_3)_2$), are more complicated (Busenberg and Plummer, 1986; Chou *et al.*, 1989; Wollast, 1990). The rate of dolomite dissolution is about two orders of magnitude slower than that of calcite and aragonite, and relatively small concentrations of HCO_3^- can almost completely suppress dolomite dissolution far from equilibrium (Busenberg and Plummer, 1986).

The role of magnesium in the precipitation of calcite and aragonite

Berner (1975) performed a series of closed-system experiments to study the seeded precipitation (overgrowth) of aragonite and calcite from "artificial seawater" (salinity 34–35‰) and "Mg-free seawater" (the same solution to which no Mg^{2+} was added and the deficiency of ionic strength was made up by adding excess NaCl) with varying state of calcite saturation (Ω_{cal}) or aragonite saturation (Ω_{arg}) and P_{CO_2}. The results presented in Fig. 9.12 show that: (i) the rate of calcite and aragonite precipitation increases exponentially with increase in Ω; (ii) at a fixed value of Ω, P_{CO_2} has no measurably consistent effect on the rate of precipitation of either calcite or aragonite;

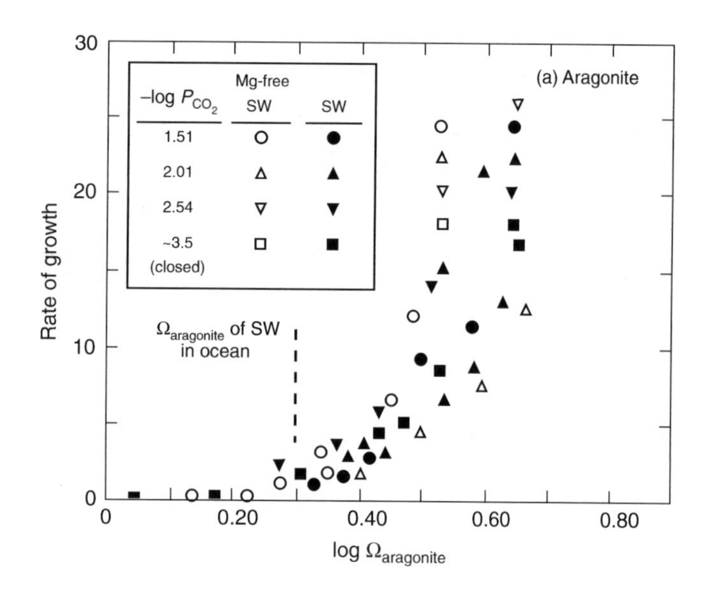

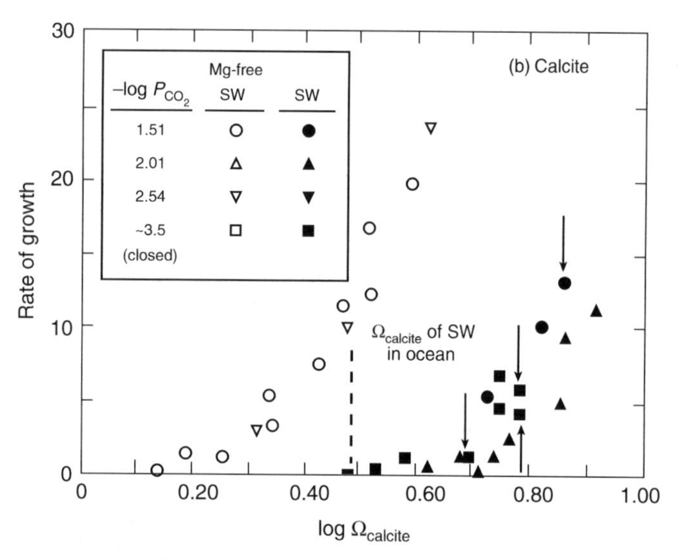

Fig. 9.12 Rate of seeded precipitation (in arbitrary units) versus (a) state of saturation with respect to calcite(Ω_{cal}) and (b) state of saturation with respect to aragonite(Ω_{arg}). SW = "artificial seawater;" Mg-free SW = "artificial seawater" of the same ionic strength but with no magnesium. The seeds used were Fisher reagent calcite and synthetic aragonite. T = 298.15 K (25°C). Salinity = 34 to 35‰. Arrows in (b) point to extended runs where calcite precipitate was identified. (After Berner, 1975, Figures 1 and 2, p. 492.)

and (iii) dissolved Mg^{2+} in seawater has virtually no effect on the seeded precipitation of aragonite, but severely retards the rate of seeded precipitation of calcite. According to Berner (1975, p. 501), aragonite growth is relatively unaffected by the presence of dissolved Mg^{2+} because Mg^{2+} is neither readily adsorbed onto the surface of aragonite nor incorporated to any extent in the aragonite crystal lattice. In contrast, Mg^{2+} is readily adsorbed onto the surface of calcite and incorporated into its crystal structure, which causes magnesian calcite to be considerably more soluble than pure calcite. Zhong and Mucci (1989) showed that calcite precipitation rates in seawater solutions did not vary appreciably over the salinity range from 5 to 44‰, but aragonite precipitation rates were higher at lower salinities (5, 15, and 20‰).

9.6.3 *Dissolution of silicate minerals*

Kinetics of the dissolution of silicate minerals are important to the understanding of many common geochemical processes such as chemical weathering, sediment formation, and acid neutralization in the environment. The mechanisms and rates of dissolution of silicate minerals differ considerably from those of carbonate and evaporite minerals: the latter are more soluble, and they dissolve rapidly (on a geologic time scale) and, in general, congruently. In contrast, the dissolution of silicate minerals is a much slower process and at least in part incongruent (i.e., the reactions are of the form mineral + H_2O ⇒ aqueous solution + secondary solid). A collection of instructive articles on chemical weathering rates of silicate minerals, including a succinct overview by the editors, can be found in White and Brantley (1995).

Kinetic experimental studies indicate that the dissolution of silicate minerals, such as albite, in CO_2-free aqueous solutions far from equilibrium typically involves three successive steps (Wollast and Chou, 1984; Walther and Wood, 1986). The first step is a rapid exchange of alkaline and alkaline earth metals with H^+. For albite ($NaAlSi_3O_8$), the exchange reaction leads to the formation of hydrogen feldspar ($HAlSi_3O_8$),

$$NaAlSi_3O_8 + H^+ \Rightarrow HAlSi_3O_8 + Na^+ \tag{9.106}$$

and at 25°C the total amount of Na exchanged within a few minutes corresponds to that initially present in a layer of the feldspar about one unit cell thick (Wollast and Chou, 1984). The second step is one of rapid dissolution that follows a parabolic rate law:

$$\boxed{\mathfrak{R}'\ (\text{mol cm}^{-2}\ \text{s}^{-1}) = \frac{dm}{dt} = \frac{1}{2} k'\ t^{-1/2}} \tag{9.107}$$

where k' is a constant at constant temperature, pressure, and solution composition. Holdren and Berner (1979) attributed this step to initial rapid dissolution of fine particles that stick to the surfaces of coarser grains in crushed samples. Wollast and Chou (1984), however, have argued that this step corresponds to a strong preferential leaching of Na and, to a smaller extent, of Al relative to Si, and the incongruent dissolution results in the formation of a depleted layer of about 30 Å in average thickness on the albite surface. The thickness of the layer differs for Si and Al, and is a function of pH of the solution.

After a few tens of days or less at 25°C, a steady state is reached when the rate of diffusion of the constituents of the mineral through the depleted layer equals the rate of retreat of the surface at the solid–solution interface. Under these conditions, the thickness of the depleted layer remains fairly constant, and the dissolution becomes congruent. According to

Chou and Wollast (1985), the rate of this last step in the case of albite is controlled by the decomposition of activated surface complexes whose configurations depend on pH and the concentration of dissolved Al, and probably also on Na concentration especially in alkaline solutions. On the time scale of geologic processes, this step dominates the dissolution of many silicate minerals (feldspars, chain silicates, phyllosilicates). The rate equation for the steady-state dissolution is linear:

$$\Re\ (\text{mol cm}^{-2}\ \text{s}^{-1}) = \frac{dm}{dt} = k \tag{9.108}$$

where k is the rate constant for the forward, dissolution reaction. The zeroth-order kinetics implies that the rate-determining step is surface-controlled.

As in the case of carbonate minerals, the dissolution rates of silicate minerals far from equilibrium (i.e., in highly unsaturated solutions) are a strong function of pH but the variation is not uniform. For example, in the absence of organic ligands, the experimentally determined dissolution rate of albite at 25°C and 1 bar decreases rapidly with increasing pH of the solution at low pH values, is practically independent of pH in the 5 to 8 range, and then increases rapidly at pH > 8 (Fig. 9.13a). As shown schematically in Fig. 9.13b, many other silicate minerals show a similar pattern of pH dependence far from equilibrium (Brady and Walther, 1989. 1990; Carroll and Walther, 1990; Lerman, 1990; Drever, 1994), although at any particular pH the observed values of dissolution rate for silicate minerals at 25°C vary by about five orders of magnitude, as can be seen from the calculation results presented in Table 9.4.

The rate laws consistent with the experimentally determined silicate dissolution rates in the acidic (pH < ~4.5), neutral (pH between ~4.5 and ~8), and alkaline (pH > ~8) regions are (Drever, 1994; Blumm and Stillings, 1995):

$$\Re_{\text{acidic}} = k_{\text{acidic}}\ (a_{H^+})^{m_{\text{acidic}}} \tag{9.109}$$

$$\Re_{\text{neutral}} = k_{\text{neutral}} \tag{9.110}$$

$$\Re_{\text{alkaline}} = k_{\text{alkaline}}\ (a_{H^+})^{m_{\text{alkaline}}} \tag{9.111}$$

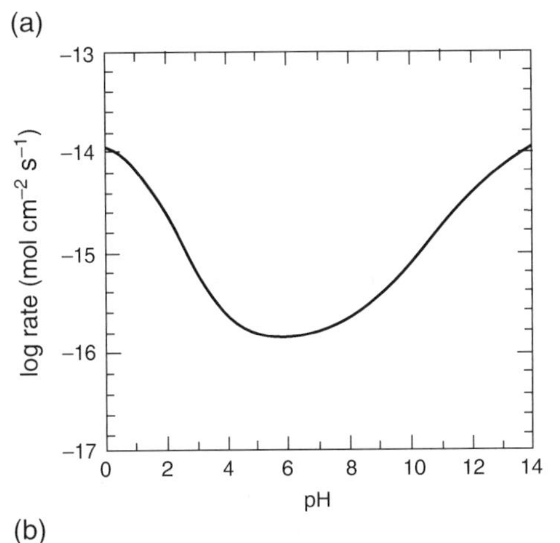

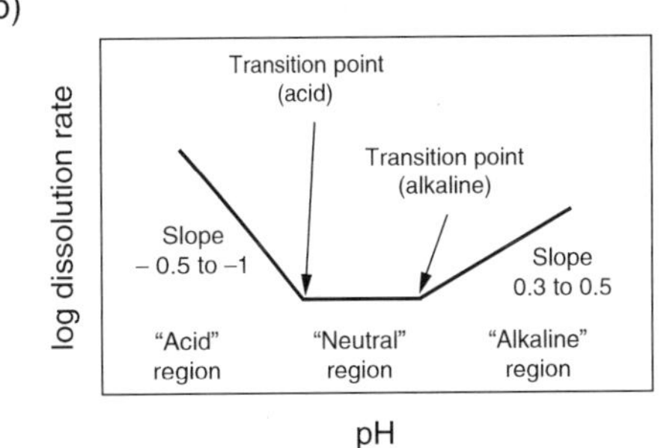

Fig. 9.13 (a) Experimentally determined variation of the rate of steady-state dissolution of albite as a function of solution pH at 25°C far from equilibrium (after Chou and Wollast, 1985, Figure 5, p.977). (b) Schematic relationship between dissolution of silicate minerals and solution pH at 25°C far from equilibrium (after Drever, 1994, Figure 1,p. 2326).

where the exponents m_{acidic} and m_{alkaline} are positive numbers specific to the mineral under consideration, and represent the slopes in a log $\Re$ versus pH plot (see Fig. 9.13b). For most silicate minerals, the value of m_{acidic} lies between 0.5 and –1 and that of m_{alkaline} between +0.3 and +0.5 (Brady and Walther, 1989; Drever, 1994). The fractional exponents suggest that the dissolution rate is controlled by the adsorbed H^+ and OH^- on the silicate mineral surface rather than their concentrations in the solution (Brady and Walther, 1989).

Table 9.4 Mean lifetime (approximate time required for complete dissolution) of a hypothetical 1 mm sphere of selected minerals in dilute aqueous solution at 25°C, pH = 5, and far fro equilibrium.

Mineral	Log rate (mol m^{-2} s^{-1})	Lifetime (yr)	Mineral	Log rate (mol m^{-2} s^{-1})	Lifetime (yr)
Quartz	–13.39	34,000,000	Gibbsite	–11.45	276,000
Kaolinite	–13.28	6,000,000	Enstatite	–10.00	10,100
Muscovite	–13.07	2,600,000	Diopside	–10.15	6,800
Microcline	–12.50	921,000	Forsterite	–9.5	2,300
Albite	–12.26	575,000	Anorthite	–8.55	112
Sanidine	–12.00	291,000	Wollastonite	–8.00	79

Source: compilation by Lasaga et al. (1994).

Comparison with field estimates of silicate dissolution rates

A comparison of weathering rates determined in the laboratory on pure minerals in presumably highly undersaturated solutions and those estimated from mass balance calculations in catchments have shown that the laboratory rates are greater by one to four orders of magnitude (Velbel, 1985; Schnoor, 1990; Burch *et al.*, 1993; Drever, 1994; Drever and Clow, 1995). For example, based on dissolved silica as a conservative tracer of weathering, Schnoor (1990) found laboratory rates on the order of 10^{-12} to 10^{-11} mol m^{-2} s^{-1} compared with field rates on the order of 10^{-14} to 10^{-12} mol m^{-2} s^{-1}.

It is possible that, unlike experimental systems studied at far from equilibrium conditions, natural systems are close to equilibrium and, therefore, involve rate-reducing, backward reactions. However, citing two well-constrained natural systems as examples, Velbel (1985) argued that the lower weathering rates of alkali feldspar in natural systems could not be accounted for by lower thermodynamic affinities of the reactions, because the calculated affinities (see section 9.3) turned out to be several times larger than the critical value below which the affinity term would have a discernible rate-reducing effect. The most likely causes of the relatively higher laboratory rates, according to Velbel (1985), are: artifacts created during sample preparation (such as very fine particles and crystal defects) that accelerate reaction rate; loss of reactive locations, such as crystal defects, in natural systems (as evidenced by etch pits on mineral surfaces); and access to mineral surfaces in natural systems limited by fluid migration through pore and fracture networks. Possible contributing factors for higher rates in natural environments include the orders-of-magnitude longer duration of weathering, complexities of solution chemistry (especially dissolved aluminum concentrations, which inhibit aluminosilicate dissolution) and pH, and organic ligands (Schnoor, 1990; Blum and Stillings 1995; White and Brantley, 1995). An additional complication is that the chemistry of waters involved in the weathering of silicate minerals in the field may be significantly different from the chemistry of stream and other surficial waters on which mass-balance estimates of weathering rates are commonly based.

9.7 Summary

1. For a reaction, a A + b B $\Rightarrow$ c C + d D, the rate-law is of the form

$$\Re = k\ [A]^{n_A}[B]^{n_B}[C]^{n_C}[D]^{n_D}$$

   where $\Re$ is the rate of the reaction, k is the experimentally determined rate constant of the particular reaction (at a specified temperature and pressure), [A], [B], etc., are concentrations, the exponents n_A, n_B, etc., denote the order of the reaction with respect to the species A, B, etc., and $n_{overall} = n_A + n_B + n_C + n_D$. Values of the exponents bear no relationship to the stoichiometric coefficients in the balanced overall reaction and must be determined experimentally.
2. Many overall reactions, whether homogeneous or heterogeneous, consist of several elementary reactions (or steps) that may occur sequentially or concurrently. The rate of an elementary reaction is independent of the concentrations of the products if the reaction only in the forward direction is considered, and the value of the exponent for a reactant is the same as the stoichiometric coefficient of the species in the reaction. In the case of sequential elementary reactions, the slowest of the elementary reactions is the rate-determining step; in the case of concurrent elementary reactions, the fastest branch determines the reaction mechanism.
3. In most geochemical systems of interest the reactions are second-order or lower. The relevant expressions for simple zeroth-order, first-order, and second-order elementary reactions are summarized in Table 9.5.

Table 9.5 Rate laws, their integrated forms, and expressions for half-lives for simple zero-order, first-order, and second-order elementary reactions.

	Zeroth-order	First-order	Second-order
Reaction	$A \xrightarrow{k} \text{Products}$	$A \xrightarrow{k} \text{Products}$	$2A \xrightarrow{k} \text{Products}$
Rate – law expression	$-\frac{d[A]}{dt} = -k$	$-\frac{d[A]}{dt} = -k\ [A]$	$-\frac{d[A]}{dt} = -k\ [A]^2$
Integrated rate equation	$[A] = [A]_0 - kt$	$\ln [A] = \ln [A]_0 - kt$	$\frac{1}{[A]} = \frac{1}{[A]_0} + kt$
Linear plot to determine k	[A] vs. t	ln [A] = ln [A] vs. t	$\frac{1}{[A]}$ vs. t
Half - life ($t_{1/2}$)	$\frac{[A]_0}{2k}$	$\frac{\ln 2}{k}$	$-\frac{1}{[A]_0 k}$

4. For an elementary reaction at equilibrium, the rate in the forward direction is equal to the rate in the reverse direction (the principle of detailed balancing).
5. Factors that affect rates of reactions are: nature of reactants (the more bonds that need to be broken, the slower the reaction rate) concentration of reactants, reaction mechanism, temperature, and the presence of inorganic or organic catalysts.
6. The temperature dependence of the rate constant of a reaction is described by the Arrhenius equation:

$$k = A\, e^{-E_a/RT}$$

where A is a "frequency factor" and E_a the *activation energy* of the reaction.
7. The relationship between activation energy and standard enthalpy change of an elementary reaction is

$$E_{a\,(\mathrm{for})} - E_{a\,(\mathrm{rev})} = \Delta H_r^0$$

8. The relationship between the net rate of a reaction ($\Re_{net}$) and its free energy change (ΔG_r) at the temperature (T) and pressure (P) of interest is

$$\Re_{net} = \Re_+ - \Re_- = \Re_+\,(1 - e^{\Delta G_r/RT})$$

where $\Re_+$ and $\Re_-$ are the forward and reverse rates, and $-\Delta G_r$ the affinity of the reaction.
9. A catalyst is a substance that accelerates the rate of a thermodynamically favorable chemical reaction, without being transformed or consumed by the reaction, by providing an alternative pathway that requires lower activation energy.
10. The advection–diffusion equation (in one dimension) for the transport of solutes in aqueous solutions is:

$$\left(\frac{\partial C_i}{\partial t}\right)_{\mathrm{fluid}} = \underbrace{-v_x\left(\frac{\partial C_i}{\partial x}\right)}_{\mathrm{advection}} \underbrace{- D_i\left(\frac{\partial^2 C_i}{\partial x^2}\right)}_{\mathrm{diffusion}} \pm \underbrace{\left(\frac{\partial C_i}{\partial t}\right)}_{\mathrm{reaction}}$$

where C_i is the concentration, D_i the diffusion coefficient of the *i*th species, and v_x denotes the constant average linear velocity of fluid flow (in the x direction).
11. The temperature dependence of diffusion coefficient can be described by an equation of the Arrhenius form:

$$D_i = A_D\, e^{-E_D/RT}$$

where A_D is the frequency factor and E_D is the activation energy of diffusion for the species i.
12. At fixed temperature and P_{CO_2} the equation for net rate of calcite dissolution/precipitation that satisfies the results of kinetic experimental studies by Plummer *et al.* (1978) is

$$\Re = k_1\, a_{H^+} + k_2\, a_{H_2CO_3^*} + k_3\, a_{H_2O} - k_4\, a_{Ca^{2+}}\, a_{HCO_3^-}$$

where $H_2CO_3^* = H_2CO_{3\,(aq)} + CO_{2\,(aq)}$, a_i represents activity in the bulk fluid, and k_1, k_2, k_3, and k_4 are rate constants. Experimental studies also show that the rate of calcite and aragonite precipitation increases exponentially with increase in Ω_{cal} and Ω_{arg}, respectively, and dissolved Mg retards the precipitation of calcite.
13. The dissolution rates of silicate minerals far from equilibrium are a function of the solution pH but the variation is not uniform. Rates determined in the laboratory are greater by one to four orders of magnitude compared to those estimated from field measurements. The discrepancy may be related to variability in factors such as duration of reaction, surface area of minerals available for reaction, access of solutions to reactive surfaces, and solution composition.

9.8 Recapitulation

Terms and concepts

Activation energy
Advection–diffusion equation
Affinity of a reaction
Arrhenius equation
Activated complex
Diffusion coefficient
Elementary reaction
Eyring equation
Fick's laws of diffusion
Heterogeneous catalysis
Heterogeneous reaction
Homogeneous catalysis
Homogeneous reaction
Half-life of decay
Integrated rate equations
Order of reaction
Parallel reactions
Principle of detailed balancing
Rate laws
Rate constant
Rate-determining step
Reaction intermediate
Reaction mechanism
Reaction quotient
Sequential reactions
Steady state
Transition state theory
Transport- and surface-controlled reactions

Computation techniques

- Graphical technique of determining the order of a reaction.
- Graphical technique for determination of activation energy.
- Variation of rate constant as a function of temperature.
- Variation of activation energy as a function of temperature.

9.9 Questions

1. Consider the degradation of a pollutant A into products P_1 and P_2 by the following two parallel reactions, with rate constants k_1 and k_2,

$$A + B \xrightarrow{k_1} P_1$$
$$A + C \xrightarrow{k_2} P_2$$

Write an expression for $\frac{d[A]}{dt}$, assuming that the reverse reactions are too slow to be of consequence.

2. A mechanism proposed for the reaction between H_2 gas and CO gas to form formaldehyde, H_2CO, is as follows:

$H_2 \Leftrightarrow 2H$ (fast, equilibrium)

$H + CO \Rightarrow HCO$ (slow)

$H + HCO \Rightarrow H_2CO$ (fast)

(a) Write the balanced equation for the overall reaction.
(b) Is this reaction mechanism consistent with the observed rate dependence, which is one-half order in H_2 and first order in CO (i.e., rate law = $k\,[H_2]^{1/2}[CO]$)?

3. The rate constant for a first-order decay reaction is $3.5 \times 10^{-4}\,s^{-1}$. What is its half-life? What percentage of the reactants would be left after 10 min?

4. Species A is involved in a first-order reaction, and [A] is measured as a function of time, as given below:

Hours	0	24	48	96
Molality	1.0	0.905	0.819	0.670.

(a) Determine the value of the rate constant.
(b) Determine the half-life of A.
(c) Calculate [A] when t = 120 h.
(d) Calculate t when [A] is reduced to 10%.

5. The overall reactions for oxidation of Fe^{2+} and the corresponding rate laws are as follows (Langmuir, 1997):

At pH above 2.2 but below 3.5, $Fe^{2+} + 0.25O_{2\,(g)} + 0.5H_2O_{(l)} \xrightarrow{k_1} FeOH^{2+}$

$$\frac{d[Fe^{2+}]}{dt} = -\,k_1(Fe^{2+})\,P_{O_2};\quad k_1 = 10^{-3.2}\ \text{bar}^{-1}\ \text{day}^{-1}\ \text{at } 20^\circ C$$

At pH above 4, $Fe^{2+} + 0.25O_{2\,(g)} + 2.5H_2O_{(l)} \xrightarrow{k_2} Fe(OH)_{3(s)} + 2H^+$

$$\frac{d[Fe^{2+}]}{dt} = -\,k_2\frac{(Fe^{2+})}{(H^+)^2}P_{O_2};$$
$$k_2 = 1.2\times10^{11}\text{mol}^2\text{bar}^{-1}\text{day}^{-1}\ \text{at}\ 20^\circ C$$

(a) Assuming $P_{O_2} = 0.2$ bar, plot a graph of log k_+ (mol day^{-1}) versus pH in the range 2.2 to 7. What can you conclude from this graph?
(b) If [Fe] = 1×10^{-2} mol l^{-1}, what is the rate of oxidation at pH = 5?
(c) How does the half-life of Fe^{2+} change when the pH is increased from 4 to 7? (For a constant value of $P_{O_2} = 0.2$ bar, the rate law becomes pseudo first-order.)

6. For a reaction $A \Rightarrow B + C$ carried out at a particular temperature, the measured molality of the reactant A, m_A, at various intervals during the progress of the reaction were as follows:

Time (h)	m_A(mol kg^{-1})	Time (h)	m_A(mol kg^{-1})
0	10.000	6	1.653
2	5.488	8	0.907
4	3.012	10	0.498

Construct an appropriate graph to show that the reaction was first order. Determine from the graph the rate constant and the half-life of the reaction.

7. Using the reaction rate data given below (selected from Rimstidt and Barnes, 1980, p. 1689), determine the activation energy and frequency factor for the reaction representing the dissolution of quartz:

$$SiO_2\ (\alpha - \text{quartz}) + 2H_2O \Rightarrow H_4SiO_{4\,(aq)}$$

Assume that the activation energy and the frequency factor are independent of temperature.

Temperature (°C)	Rate constant (s^{-1})	Temperature (°C)	Rate constant (s^{-1})
65	3.81×10^{-9}	202	3.21×10^{-6}
105	3.62×10^{-9}	223	2.24×10^{-6}
145	3.03×10^{-8}	265	1.85×10^{-6}
173	9.10×10^{-7}		

8. For a reaction with E_a = 10 kcal mol^{-1} at 25°C, what is the fraction of molecules with kinetic energy greater than E_a? How would the fraction change for a reaction with E_a = 20 kcal mol^{-1}?

9. For the reaction

$$CaCO_3\ (\text{calcite}) + H_2CO_{3\,(aq)} \Rightarrow Ca^{2+} + 2HCO_3^-$$

the rate constant at 25°C is $3.47 \times 10^{-5}\,s^{-1}$ and the activation energy is 41.85 kJ mol^{-1}. Determine the value of the frequency factor for this reaction. What will be your prediction regarding the rate constant of the reaction at 10°C and 50°C?

10. For the first-order gas phase reaction

$$N_2O_5 \Rightarrow NO_2 + NO_3$$

the activation energy is 88 kJ mol^{-1} and the rate constant at 0°C is 9.16×10^{-3} s^{-1}.
    (a) What would be the value of the rate constant at 25°C?
    (b) At what temperature would this reaction have a rate constant = 3.00×10^{-2} s^{-1}?
11. Geochemists often use a rule of thumb that the rate of a reaction approximately doubles with every 10° C rise in temperature. Is this statement correct? Discuss your answer by calculating the increase in the rate constant of a reaction for increases in temperature from (i) 300 to 310 K, (ii) 500 to 510 K, (iii) 700 to 710 K, and (iv) 900 to 910 K, assuming a constant activation energy of 50 kJ mol^{-1} for the reaction.
12. From the dissolution rate data given in Table 9.4, calculate how long it will take to dissolve a 1-mm-thick layer of albite (density 2.62 g cm^{-3}). Assume that the calcite has a planar surface.
13. The reaction $SO_2Cl_2 \Rightarrow SO_{2\,(g)} + Cl_{2\,(g)}$ is first order and has a rate constant of 2.24×10^{-5} s^{-1} at 320°C.
    (a) Calculate the half-life of the reaction.
    (b) What fraction of a sample of the reactant would remain after being heated for 5.00 h?
    (c) Calculate how long a sample of the reactant needs to be maintained at 320°C to decompose 92% of the initial amount.
14. The rate of a particular reaction at 373 K is four times faster than its rate at 323 K. What is the activation energy of the reaction?
15. For a reaction A ⇒ B + C carried out at a particular temperature, the measured molarity of the reactant A, M_A, at various intervals during the progress of the reaction were as follows:

Time (min)	M_A(mol L^{-1})	Time (min)	M_A(mol L^{-1})
0.0	2.000	6.00	0.338
2.00	1.107	8.00	0.187
4.00	0.612	10.00	0.103

    (a) What is the order of this reaction?
    (b) Write the rate-law expression for the reaction.
    (c) What is the value of the rate constant at this temperature?

Part III
Isotope Geochemistry

Although the measurements on which isotope geology is based will probably continue to be made by a small but expanding number of experts, the interpretation of the data should be shared increasingly with geologists who are familiar with the complexities of geological problems.

Faure (1986)

10 Radiogenic Isotopes

The geologic time scale gave a focus to efforts to determine the duration of geologic time. There were a variety of qualitative estimates based on physical and geological processes, as understood in the 1800's. But a truly quantitative approach to measurement of geologic time only became possible with the discovery of radioactivity in 1896, and the eventual understanding of radioactive decay processes in the early 1900's.

Armstrong (1991)

From the perspective of applications to geologic systems, we are interested in only those isotopes that show measurable variations relatable to natural process. Such isotopes may be divided into two distinct groups: *radiogenic isotopes* that are produced by radioactive decay of unstable parent nuclides; and *stable isotopes* that do not undergo radioactive decay. Thus, in a closed system, the abundance of a radiogenic isotope, unlike that of a stable isotope, is a function of the time interval over which the decay process has occurred and is the basis of radiometric dating of geologic samples. A radiogenic isotope itself may be radioactive (e.g., ^{226}Ra, an isotope of radium) or stable (e.g., ^{206}Pb, an isotope of lead).

Both radiogenic and stable isotopes are used extensively for gaining insights into geochemical processes. Reliance on isotope ratios for the interpretation of geochemical processes and products is a relatively recent phenomenon, but helped by the development of instruments and techniques for measurement of ultrasmall amounts of isotopes in earth materials, the study of isotopes has turned out to be a rapidly expanding field of research. The main applications of radiogenic isotopes in geochemistry include:

(1) dating of rocks, minerals, and archeological artifacts (radiometric geochronology);
(2) tracing geochemical processes such as the evolution of the Earth's oceans, crust, mantle, and atmosphere through geologic time, and magmatic differentiation and assimilation (DePaolo and Wasserburg, 1979; DePaolo, 1981);
(3) tracing the sources and transport paths of dissolved and detrital constituents involved in the formation of rocks and mineral deposits (e.g., Hemming *et al.*, 1995, 1996; Capo *et al.*, 1998; Kamenov *et al.*, 2002; Panneerselvam *et al.*, 2006).

In this chapter we will discuss some of these applications of radiogenic isotopes, with emphasis on techniques of radiometric dating. More elaborate treatments of the subject can be found, for example, in Faure (1986, 2001), Dickin (1995), and Banner (2004).

10.1 Radioactive decay

10.1.1 Abundance and stability of nuclides

Only about 430 of approximately 1700 known nuclides occur naturally and of these only about 200 are stable. The stability of a nuclide is a function of the configuration of its nucleus, and the binding energy (per atomic particle), which generally decreases with increasing atomic weight above mass 56. An examination of the atomic structure of the nuclides reveals some interesting general patterns about their stability.

(1) With the exception of only two nuclides, $^1_1$H and $^3_2$He, which have fewer neutrons than protons, stable nuclides are characterized by nearly equal numbers of protons (Z)

Introduction to Geochemistry: Principles and Applications, First Edition. Kula C. Misra.

and neutrons (N) (the so-called *symmetry rule*). Actually, the $N : Z$ ratio increases from 1 to about 3 with increasing values of the mass number (A).

(2) Most of the stable nuclides have even numbers of protons and neutrons; stable nuclides with either odd Z or odd N are much less common, and nuclides with odd Z and odd N are rare (the so-called *Oddo–Harkins rule*).

(3) Nuclides that have one of the so-called "magic numbers" (2, 8, 10, 20, 28, 50, 82, and 126) as their Z or N are relatively more stable.

An unstable nuclide decays spontaneously at a constant rate until it is transformed into an isotope with a stable nuclear configuration. *Radioactivity* is this phenomenon of spontaneous decay of unstable nuclides, which are referred to as radioactive nuclides or *radionuclides*, to daughter isotopes. The phenomenon was discovered by the French physicist Henri Becquerel in 1896 but the term "radioactivity" was actually coined by the Polish scientist and Nobel laureate Marie Curie a couple of years later on the basis of ionizing radiation emitted by radium.

Most of the radionuclides that were produced during stellar evolution (see Chapter 12) have become extinct because of their fast decay rates, although they can be produced artificially through nuclear reactions. For geologic applications we are interested in a small group of radionuclides that continue to exist in geologic materials in measurable quantities for one of the following reasons (Banner, 2004):

(1) their decay rates are too slow relative to the age of the Earth (e.g., $^{87}_{37}$Rb, $^{235}_{92}$U);

(2) they are decay products of long-lived, naturally occurring radioactive parents (e.g., $^{222}_{86}$Rn produced from $^{235}_{92}$U);

(3) they are produced naturally as a result of the bombardment of stable isotopes by cosmic rays (e.g., $^{14}_{6}$C from $^{14}_{6}$N);

(4) they are produced by human activities, such as operation of nuclear reactors and testing of nuclear explosions, and released to the environment (e.g., $^{90}_{38}$Sr, $^{239}_{94}$Pu).

10.1.2 *Mechanisms of radioactive decay*

Radioactive decay occurs by three mechanisms: *beta decay*, *alpha decay*, and *spontaneous nuclear fission*, usually accompanied by emission of energy in the form of gamma rays. The alpha/beta/gamma terminology was adopted in the late 1800s, before physicists understood what each of these emissions was made of.

Beta decay

The process of beta decay depends on the ratio of proton (or atomic) number (Z) to neutron number (N) in the radionuclide; in all cases beta decay is accompanied by radiant energy in the form of gamma rays. Radionuclides with an excess of neutrons (i.e., with $N : Z > 1$) decay spontaneously by emitting from the nucleus a negatively charged beta particle or *negatron* (β^-) that is identical in charge and mass to an extranuclear electron. The *negatron decay* may be regarded as the transformation of a neutron into a proton, an electron, which is expelled from the nucleus as a β^- particle, and an *antineutrino* by the reaction

$$\text{neutron} \Rightarrow \text{proton} + \beta^- + \bar{\nu} + \gamma \quad (10.1)$$

in which energy and mass are conserved as in any chemical reaction. An *antineutrino* ($\bar{\nu}$), an antiparticle of neutrino, is a stable elementary particle that has near zero rest mass and zero electric charge, but its kinetic energy conserves the energy and momentum of the β^- decay nuclear transition. (An *antiparticle* has the same mass as a particle but opposite charge.) The decay produces a new isotope, a *daughter*, in which N is decreased by 1 whereas Z is increased by 1 (Table 10.1). An example is the β^- decay of $^{14}_{6}$C: $^{14}_{6}\text{C} \Rightarrow {}_{7}^{14}\text{N} + \beta^- + \bar{\nu} + \gamma$.

For some radionuclides that are deficient in neutrons (i.e., with $N : Z < 1$), beta decay occurs by emission from the nucleus of a positively charged beta particle or *positron* (β^+), which has the same mass as an electron but carries a positive charge. The *positron decay* may be regarded as the transformation of a proton into a neutron that is accompanied by the emission of a positron and a neutrino according to the reaction:

$$\text{proton} \Rightarrow \text{neutron} + \beta^+ + \nu + \gamma \quad (10.2)$$

A *neutrino* (ν) is a stable elementary particle with variable kinetic energy but with very small (but nonzero) rest mass and zero electric charge; neutrinos conserve the energy and momentum of the positron decay nuclear transition. Neutrinos travel at or close to the speed of light, and are able to pass through ordinary matter almost undisturbed. They are extremely difficult

Table 10.1 Daughter nuclide produced by alpha and beta decay of parent radionuclide.

	Decay mechanism	Atomic number	Neutron number	Mass number
Parent nuclide		Z	N	$A = Z + N$
Daughter nuclide	α^{2+} (alpha) decay	$Z - 2$	$N - 2$	$A - 4$
	β^- (negatron) decay	$Z + 1$	$N - 1$	A
	β^+ (positron) decay	$Z - 1$	$N + 1$	A
	Electron capture decay	$Z - 1$	$N + 1$	A

to detect. The positron decay produces a new isotope in which N increases by 1 whereas Z decreases by 1 (Table 10.1). An example is the β^+ decay of $^{10}_{6}C$: $^{10}_{6}C \Rightarrow {}^{10}_{5}B + \beta^+ + \nu + \gamma$.

A third type of beta decay is called *electron capture*. This process involves the spontaneous capture of an electron orbiting close to the nucleus, and transformation of a proton to a neutron and a neutrino (which is ejected from the atom's nucleus) by the reaction:

$$\text{proton} + e^- \Rightarrow \text{neutron} + \nu + \gamma \quad (10.3)$$

The decay produces a new isotope in which, as in the case of positron decay, N increases by 1 whereas Z decreases by 1 (Table 10.1). An example is the electron capture by $^{11}_{6}C$: $^{11}_{6}C + e^- \Rightarrow {}^{11}_{5}B + \nu$. Electron capture is also called K-capture because the captured electron usually comes from the atom's K-shell.

Alpha decay

A large number of radionuclides undergo decay by emission of *alpha particles* from the nucleus. An alpha particle (α^{2+}) is composed of two neutrons and two protons, identical to the nucleus of helium ($^{4}_{2}He$); so both N and Z for the new nuclide formed by *alpha decay* decrease by 2, resulting in a decrease of the atomic mass (A) by 4 (Table 10.1). An example is the alpha decay of $^{238}_{92}U$ to $^{234}_{90}Th$: $^{238}U \Rightarrow {}^{234}Th + \alpha^{2+}$ + energy.

Nuclear fission

A few of the heaviest isotopes, such as those of uranium and transuranic elements (elements with atomic mass greater than that of uranium), decay by spontaneous *fission*. In this process the nucleus breaks into two or more unequal fragments, accompanied by the release of a large amount of energy. Generally, the fission products have excess neutrons and therefore undergo further decay by β^- and gamma-ray emission until a stable nuclide is formed. The operation of fission nuclear reactors around the world is based on this principle.

Most radionuclides decay by one of the mechanisms described above, the most common being β^- decay. Some isotopes, however, undergo *branched decay* – decay by more than one mechanism simultaneously – each branch producing a different daughter. For example, in a given population of $^{40}_{19}K$ atoms, 88.8% decay to stable $^{40}_{20}Ca$ atoms by negatron emission and 11.2% decay to stable $^{40}_{18}Ar$ atoms by positron emission and electron capture (see section 10.3.5).

10.2 Principles of radiometric geochronology

10.2.1 *Decay of a parent radionuclide to a stable daughter*

As proposed first by Rutherford and Soddy (1902), the rate of decay of a radioactive nuclide at any instant is proportional to the number of atoms of the nuclide remaining at that instant. Denoting the number of radioactive parent atoms remaining at any time t by N (not to be confused with the neutron number mentioned earlier), the mathematical expression for the rate of radioactive decay (dN/dt), a first-order reaction (see section 9.1.3), can be written as

$$-\frac{dN}{dt} \ \alpha \ N \quad \text{or} \quad -\frac{dN}{dt} = \lambda N \quad (10.4)$$

where the negative sign reminds us that N decreases with time, and λ is the proportionality constant, usually called the *decay constant*, which is expressed in units of reciprocal time. The decay constant states the probability that a given atom of the radionuclide will decay within a stated time. Every radioactive decay scheme has its own characteristic numerical value of λ, which is not affected by temperature, pressure, and any chemical changes involving the isotope. The constant rate of radioactive decay over geologic time is the fundamental premise of age dating by radiogenic isotopes.

The integrated form of the rate equation (10.4), as mentioned in section 9.1.3 and elaborated in Box 10.1, is:

Box 10.1 Integrated form of the radioactive decay equation

Integrating equation (10.4), we have

$$\int -\frac{dN}{N} = \int \lambda \, dt = \lambda \int dt$$

$$-\ln N = \lambda t + \text{Integration constant (IC)}$$

Denoting the value of N as N_0 at $t = 0$, $-\ln N_0 = IC$. Substituting for IC,

$$-\ln N = \lambda t - \ln N_0$$

Rearranging, we get, $-\ln (N/N_0) = \lambda t$, which is commonly expressed as

$$N = N_0 \, e^{-\lambda t} \quad (10.5)$$

$$N = N_0 \, e^{-\lambda t} \quad (10.5)$$

where N_0 represents the number of radioactive parent atoms present when the rock or mineral was formed (i.e., at time $t = 0$), t is the time interval (in years) over which radioactive decay has occurred, and N represents the number of radioactive parent atoms remaining after time t. Equation (10.5) describes the basic quantitative relationship among N, N_0, and t for any radioactive decay process of known λ. The decay constant is related to the half-life ($t_{1/2}$) of the decay process – the value of t when $N = 1/2\ N_0$ – by the equation (see section 9.1.3):

$$t_{1/2}(\text{yr}) = \frac{0.693}{\lambda\ (\text{y}^{-1})} \quad (10.6)$$

We can express equation (10.5) in terms of daughter atoms. Let us assume that the radioactive parent produces D^* atoms of a stable daughter during time t. Since the decay of one parent atom produces one daughter atom,

$$D^* = N_0 - N \tag{10.7}$$

Substituting for N (equation 10.5),

$$\boxed{D^* = N_0 - N = N_0 - N_0 e^{-\lambda t} = N_0(1 - e^{-\lambda t})} \tag{10.8}$$

Equation (10.5) describes the time-dependent exponential decay of a known population of parent atoms and equation (10.8) the complementary growth of the stable daughter atoms (Fig. 10.1).

10.2.2 *Basic equation for radiometric age determination*

The direct application of equations (10.5) and (10.8) to natural geochemical systems, such as samples of rocks and minerals, is limited by our lack of knowledge of N_0. We can, however, circumvent this limitation by expressing N_0 in terms of N and D^*, quantities we can actually measure in the laboratory. Substituting for N_0 in equation (10.7), we get the equation for the growth of the daughter,

$$\boxed{D^* = N_0 - N = Ne^{\lambda t} - N = N(e^{\lambda t} - 1)} \tag{10.9}$$

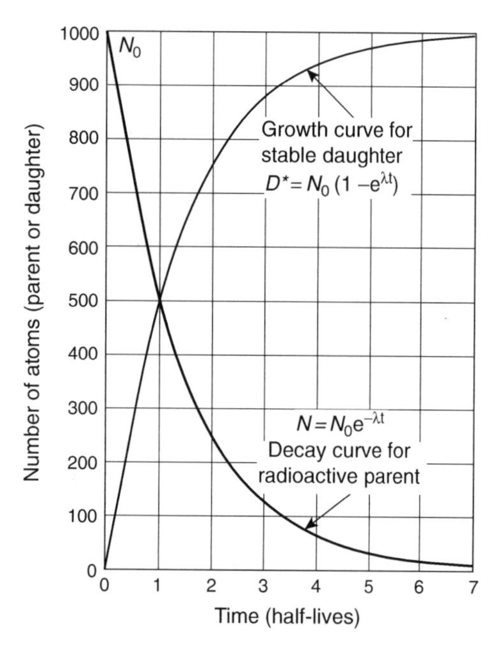

Fig. 10.1 The decay curve for a hypothetical population of 1000 radioactive atoms (N_0) as a function of time expressed in units of half-lives and the complementary growth curve representing the generation of atoms of a stable daughter by radioactive decay (D^*). Both the curves are asymptotic to the time axis.

where t is the geologic age of the sample, and N and D^*, respectively, are the number of parent atoms remaining and the number of the stable daughter atoms generated in time t per unit weight of the sample.

To derive a general equation for isotope-based geochronology, we should also consider the possibility that the system under consideration may have contained some daughter atoms prior to the incorporation of the radioactive parent into the sample. In that case,

$$D = D_0 + D^* \tag{10.10}$$

where D and D_0 represent, respectively, the total number (after time t) and the initial number (at time t_0) of the stable daughter atoms per unit weight of the sample. Substituting for D^* in equation (10.10) we get the basic general equation for age determination of rocks and minerals (Fig. 10.2):

$$\boxed{D = D_0 + N(e^{\lambda t} - 1)} \tag{10.11}$$

which may be rearranged to yield the following expression for t:

$$\boxed{t = \frac{1}{\lambda} \ln \left[\frac{D - D_0}{N} + 1 \right]} \tag{10.12}$$

The quantity D_0 is of interest also because of the information it contains about the history of the daughter before its incorporation into the sample being analyzed.

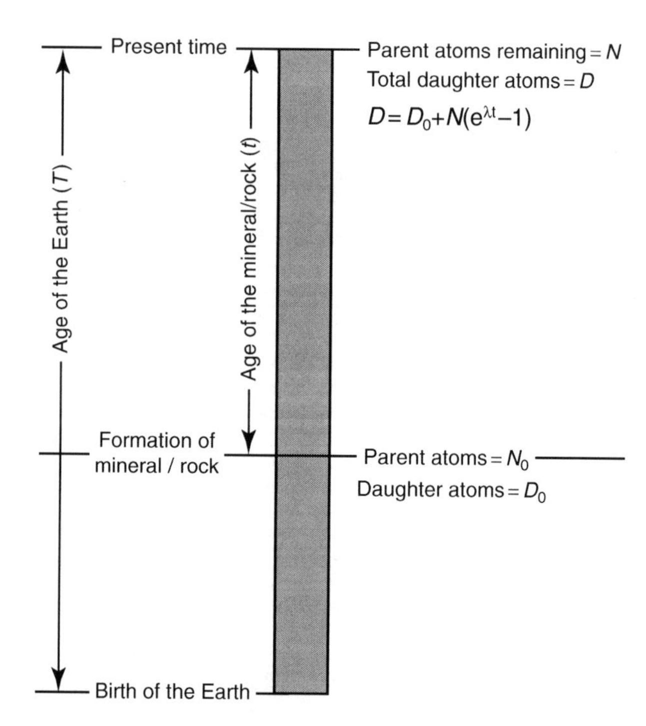

Fig. 10.2 Schematic representation of the geochronologic equation (10.11).

We can use equation (10.12) to calculate t for a mineral or rock sample, provided the following conditions are satisfied (Wetherill, 1956b; Faure, 1986).

(1) The value of λ is known accurately and has not changed over geologic time (for parent nuclides with long half-lives, a small error in λ translates into a large uncertainty in the calculated age).
(2) The measurements of D and N are accurate and representative of the rock or mineral to be dated.
(3) The system has remained closed to the parent, the unstable intermediate daughters, and the stable daughter since the formation of the mineral or rock t years ago; i.e., the D^*/N ratio in the sample has changed since that time only as a result of radioactive decay, not by any gain or loss of either the parent or the daughter. Commonly, this is the most difficult condition to satisfy.
(4) It must be possible to assign a realistic value to D_0. This is accomplished either by analyzing several cogenetic (i.e., with same t and D_0) samples (see Fig. 10.4) or by assuming a value based on experience with the particular decay scheme and samples. The latter is not a particularly desirable approach, but the error introduced due to an incorrect value of D_0 is relatively small if the samples chosen for geochronology are highly radiogenic (i.e., $D^* >> D_0$).

Conditions (3) and (4) define a model for the geologic history of the sample. The date calculated from equation (10.12) is commonly referred to as the *model age* because the determined age is valid only to the extent that the actual geologic history of the sample used conforms to the assumed model defined by conditions (3) and (4).

As ratios of isotope abundances can be measured more accurately in the laboratory than absolute abundances, the basic practical equation for geochronology is obtained by dividing both sides of equation (10.11) by the abundance of a nonradiogenic isotope that has remained constant in the samples of interest through geologic time. For example, for the decay of ${}^{176}_{71}\text{Lu}$ to ${}^{176}_{72}\text{Hf}$ (${}^{176}_{71}\text{Lu} \Rightarrow {}^{176}_{72}\text{Hf} + \beta^- + \bar{\nu}$ + energy), where Q represents the output energy), the equation incorporating the nonradiogenic isotope ${}^{177}_{72}\text{Hf}$ takes the form

$$\frac{{}^{176}\text{Hf}}{{}^{177}\text{Hf}} = \left(\frac{{}^{176}\text{Hf}}{{}^{177}\text{Hf}}\right)_0 + \frac{{}^{176}\text{Lu}}{{}^{177}\text{Hf}}(e^{\lambda t} - 1) \quad (10.13)$$

where $({}^{176}\text{Hf}/{}^{177}\text{Hf})_0$ is the initial ratio, and ${}^{176}\text{Hf}/{}^{177}\text{Hf}$ and ${}^{176}\text{Lu}/{}^{177}\text{Hf}$ are the present-day ratios in the sample of interest. Note that the ratios in this equation (and other geochronologic equations) are atomic ratios.

Radioactive decay schemes commonly used for age dating of earth materials are listed in Table 10.2, and a few of the systems are discussed in section 10.3. In all cases, isotope ratios are measured with a mass spectrometer, an instrument designed to separate charged atoms and molecules of different masses based on their motions in electrical and/or magnetic

Table 10.2 Radioactive decay schemes commonly used for geochronology.

Parent / daughter pair	Decay mechanism	λ (yr^{-1})*	(a) Half-life (yr) (b) Effective range for geochronology (yr)	Typical samples
^{238}U / ^{206}Pb	Decay chain $8\alpha^{2+} + 6\beta^-$	1.55×10^{-10}	(a) 4.47×10^9 (b) $T - 10^7$	Zircon, badellyite, uraninite, monazite, Pb-bearing minerals
^{235}U / ^{207}Pb	Decay chain $7\alpha^{2+} + 4\beta^-$	9.85×10^{-10}	(a) 7.04×10^8 (b) $T - 10^7$	Zircon, badellyite, uraninite, monazite, Pb-bearing minerals
^{232}Th / ^{208}Pb	Decay chain $6\alpha^{2+} + 4\beta^-$	4.95×10^{-11}	(a) 1.40×10^{10} (b) $T - 10^7$	Zircon, badellyite, uraninite, monazite, Pb-bearing minerals
^{87}Rb / ^{87}Sr	β^-	1.42×10^{-11}	(a) 4.88×10^{10} (b) $T - 10^7$	K-feldspar, mica, whole rock
^{40}K / ^{40}Ar	Electron capture	5.81×10^{-11}	(a) 1.19×10^{10} (b) $T - 10^3$	Sanidine, hornblende, plagioclase, mica, whole rock
^{147}Sm / ^{143}Nd	α^{2+}	6.54×10^{-12}	(a) 1.06×10^{11} (b) $T - 0$	Pyroxene, amphibole, feldspars, whole rock
^{14}C / ^{14}N	β^-	1.21×10^{-4}	(a) 5.73×10^3 (b) 70, 000 – 0	Charcoal, wood, peat
^{176}Lu / ^{176}Hf	β^-	1.94×10^{-11}	(a) 3.57×10^{10}	Apatite, zircon, garnet, monazite, whole rock
^{187}Re / ^{187}Os	β^-	1.64×10^{-11}	(a) 4.23×10^{10}	Molybdenite, osmiridium, laurite

*Ages reported in the older literature may have been based on slightly different values of decay constants, and should be recalculated using the decay constants given above. T = age of the Earth.
Sources of data: compilations by Faure (1986), Dickin (1995), Brownlow (1996), and Banner (2004).

fields. The reader may refer to Faure (1986, pp. 56–63) for a succinct discussion of mass spectrometry used in isotope geochemistry.

10.2.3 *Decay series*

The formulations presented in the previous section presume a one-step, direct decay of a parent radionuclide into a stable daughter. This, however, is not always the case. Some radionuclides, such as ^{238}U, ^{235}U, and ^{232}Th, decay to a stable daughter through a series of short-lived, radioactive daughter isotopes. For example, the decay of ^{238}U into the stable daughter ^{206}Pb occurs through 14 intermediate isotopes of 8 different elements (Fig. 10.3). The chain involves both α and β^- decay and branches repeatedly, but all possible paths lead to ^{206}Pb, the stable isotope of lead. One of the decay products in the ^{238}U decay series is radon gas (^{222}Rn), which is colorless and odorless, but radioactive with a half-life of 3.83 days. Radon gas is chemically inert, but it is a potential health hazard because both radon and its intermediate decay products, isotopes of polonium (Po), are radioactive. When humans inhale radon gas or its radioactive decay products, polonium particles stick to the lungs' bronchi, where they decay by emitting α particles in all directions. Some of the α particles hit the cells lining the bronchi, and may cause cell mutation and ultimately lung cancer in some cases.

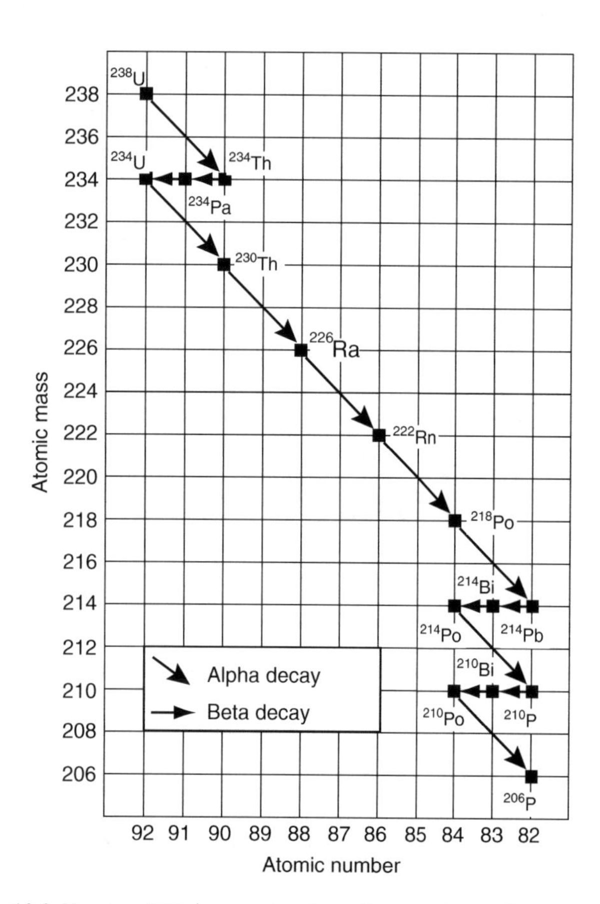

Fig. 10.3 Uranium-238 decay series: the radioactive decay of parent $^{238}_{92}U$ nuclide (half-life = 4.47×10^9 yr) to its stable daughter nuclide $^{206}_{82}Pb$ via a series of intermediate radiogenic daughters, including radium ($^{226}_{88}Ra$; half-life = 1622 yr) and radon gas ($^{222}_{86}Rn$; half-life = 3.83 days).

Consider a decay series comprised of N_A atoms of a parent A decaying to N_B atoms of a radioactive daughter B, which then decays to N_C atoms of a second radioactive daughter C, and so on to N_S atoms of a stable daughter S. Let λ_A, λ_B, λ_C, ... represent the corresponding decay constants. If λ_A is much smaller than the decay constants of the intermediate daughters (i.e., the parent has a much longer half-life compared to the daughters), then it can be shown that (see, e.g., Faure, 1986)

$$\lambda_A N_A = \lambda_B N_B = \lambda_C N_C = \ldots \qquad (10.14)$$

and

$$N_S = N_A(e^{\lambda_A t} - 1) \qquad (10.15)$$

This condition, in which the rate of decay of the daughter nuclide is equal to that of its parent nuclide, is known as *secular equilibrium*. When a decay series has achieved secular equilibrium, the system can be treated as though the initial parent decayed directly to the stable daughter without intermediate daughters. Thus, equation (10.11) (which is identical to equation 10.15) is valid, for example, for a ^{238}U (parent) $\Rightarrow$ ^{206}Pb (stable daughter) system that has achieved secular equilibrium.

10.3 Selected methods of geochronology

10.3.1 *Rubidium–strontium system*

Rubidium has two naturally occurring isotopes – radioactive $^{87}_{37}Rb$ (27.835 atom %) and stable $^{85}_{37}Rb$ (72.165 atom %) – whereas strontium has four – $^{88}_{38}Sr$ (82.58 atom %), $^{87}_{38}Sr$ (7.00 atom %), $^{86}_{38}Sr$ (9.86 atom %), and $^{84}_{38}Sr$ (0.56 atom %), all of which are stable. The Rb–Sr dating method is based on the β^- decay of ^{87}Rb to ^{87}Sr in Rb-bearing minerals and rocks ($\lambda_{Rb} = 1.42 \times 10^{-11}\,yr^{-1}$; $T_{1/2\,(Rb)} = 48.8 \times 10^9$ yr). The relevant equation for Rb–Sr geochronology (see equation 10.11), incorporating ^{86}Sr as the nonradiogenic isotope for normalization, is:

$$\boxed{\frac{^{87}Sr}{^{86}Sr} = \left(\frac{^{87}Sr}{^{86}Sr}\right)_0 + \frac{^{87}Rb}{^{86}Sr}\,(e^{\lambda_{Rb} t} - 1)} \qquad (10.16)$$

where $(^{87}Sr/^{86}Sr)_0$, $^{87}Sr/^{86}Sr$, and $^{87}Rb/^{86}Sr$ denote, respectively, the initial mole ratio, the measured mole ratio, and the present-day mole ratio in the sample, and t is its geologic age.

In principle, equation (10.16) can be solved for t if this is the only unknown variable (see equation 10.12). The $^{87}Sr/^{86}Sr$ ratio in a sample is measured with a suitable mass spectrometer, and the concentrations of Rb and Sr are determined either by X-ray fluorescence or by isotope dilution technique. The ratio of Rb/Sr concentrations is converted to $^{87}Rb/^{86}Sr$ by the following equation (Faure, 1986):

$$\frac{^{87}\mathrm{Rb}}{^{86}\mathrm{Sr}} = \left(\frac{\mathrm{Rb}}{\mathrm{Sr}}\right)\left(\frac{\text{abundance of }^{87}\mathrm{Rb}}{\text{abundance of }^{86}\mathrm{Sr}}\right)\left(\frac{\text{atomic wt of Sr}}{\text{atomic wt of Rb}}\right) \quad (10.17)$$

where $^{87}Rb/^{86}Sr$ is the present atomic ratio of the isotopes in a unit weight of the sample, and Rb/Sr is the ratio of concentrations of these elements in unit weight of the sample. The atomic weight and isotopic composition of rubidium is known to be the same at the present time ($^{87}Rb = 27.8346$ atom %) in all samples of terrestrial, meteoritic, and lunar rubidium, but the abundance of ^{86}Sr and the atomic weight of strontium depend on the abundance of ^{87}Sr and therefore appropriate values must be calculated for each sample from measured strontium isotope ratios.

Example 10–1. Calculation of the atomic weight of Sr in a sample from measured Sr isotope ratios (Faure, 1986)

Consider a rock sample in which the Sr isotope ratios have been measured as:

$^{87}Sr/^{86}Sr=2.5000$ $^{86}Sr/^{88}Sr=0.11940$ $^{84}Sr/^{88}Sr=0.006756$

The masses of Sr isotopes are (in amu):

$^{84}Sr=83.9134$ $^{86}Sr=85.9092$ $^{87}Sr=86.9088$
$^{88}Sr=87.9056$

Let us calculate the atomic weight of Sr in this sample.

$$\frac{^{87}\mathrm{Sr}}{^{88}\mathrm{Sr}} = \frac{^{87}\mathrm{Sr}}{^{86}\mathrm{Sr}} \times \frac{^{86}\mathrm{Sr}}{^{88}\mathrm{Sr}} = 2.500 \times 0.1194 = 0.2985$$

Sum of iotope ratios,

$$\sum \text{ratios} = \frac{^{84}\mathrm{Sr}}{^{88}\mathrm{Sr}} + \frac{^{86}\mathrm{Sr}}{^{88}\mathrm{Sr}} + \frac{^{87}\mathrm{Sr}}{^{88}\mathrm{Sr}} + \frac{^{88}\mathrm{Sr}}{^{88}\mathrm{Sr}}$$
$$= 0.006756 + 0.11940 + 0.2985 + 1.0000 = 1.4246$$

Fractional abundances of the Sr isotopes in the sample:

$$^{84}\mathrm{Sr} = \frac{\frac{^{84}\mathrm{Sr}}{^{88}\mathrm{Sr}}}{\Sigma\text{ratios}} = \frac{0.00675}{1.4246} = 0.004738; \quad ^{86}\mathrm{Sr} = \frac{\frac{^{86}\mathrm{Sr}}{^{88}\mathrm{Sr}}}{\Sigma\text{ratios}} = \frac{0.1194}{1.4246} = 0.08381$$

$$^{87}\mathrm{Sr} = \frac{\frac{^{87}\mathrm{Sr}}{^{88}\mathrm{Sr}}}{\Sigma\text{ratios}} = \frac{0.2985}{1.4246} = 0.20953; \quad ^{88}\mathrm{Sr} = \frac{\frac{^{88}\mathrm{Sr}}{^{88}\mathrm{Sr}}}{\Sigma\text{ratios}} = \frac{1.0000}{1.4246} = 0.70195$$

Abundances of the Sr isotopes expressed as atom %:

$^{84}Sr=0.474$ $^{86}Sr=8.381$ $^{87}Sr=20.953$ $^{88}Sr=70.195$

Atomic weight of Sr in the sample

$$= \frac{(0.474 \times 83.9134) + (8.381 \times 85.9092) + (20.953 \times 86.9088) + (70.195 \times 87.9056)}{100}$$
$$= \frac{39.775 + 720.005 + 1821.000 + 6170.533}{100} = \frac{8751.313}{100} = 87.513$$

The initial ratio, $(^{87}Sr/^{86}Sr)_0$, can be estimated in various ways. If the mineral to be dated is strongly enriched in radiogenic ^{87}Sr, one can assume a value of $(^{87}Sr/^{86}Sr)_0$ based on experience – for example, an initial ratio of 0.704 for uncontaminated mafic volcanic rocks of recent age derived from the upper mantle – because in such cases the age calculated using equation (10.16) is relatively insensitive to the value of $(^{87}Sr/^{86}Sr)_0$. A more reliable approach is the one proposed by Nicolaysen (1962), which takes advantage of the fact that the form of equation (10.16) is that of a straight line. So, a plot $^{87}Sr/^{86}Sr$(y axis) against $^{87}Rb/^{86}Sr$ (x axis) for a suite of comagmatic igneous mineral separates and whole-rocks (i.e., all of them are of the same age and can be assumed to have the same value of initial strontium ratio) that span a wide range of Rb:Sr ratios, will ideally define a straight line with y-axis intercept = $(^{87}Sr/^{86}Sr)_0$, and slope $(m) = e^{\lambda_{Rb} t} - 1$ (Fig. 10.4). We can calculate t, which in this case represents the age of magmatic crystallization, from the slope of the line:

$$\boxed{t = \frac{\ln (m+1)}{\lambda_{\mathrm{Rb}}}} \quad (10.18)$$

The straight line generated in such a plot is called an *isochron* (iso = equal, chron = time), because all samples having the same age (and the same initial ratio) should plot on this line. In practice, a straight line is fitted to the array of points by a statistical method called linear regression, and a correlation

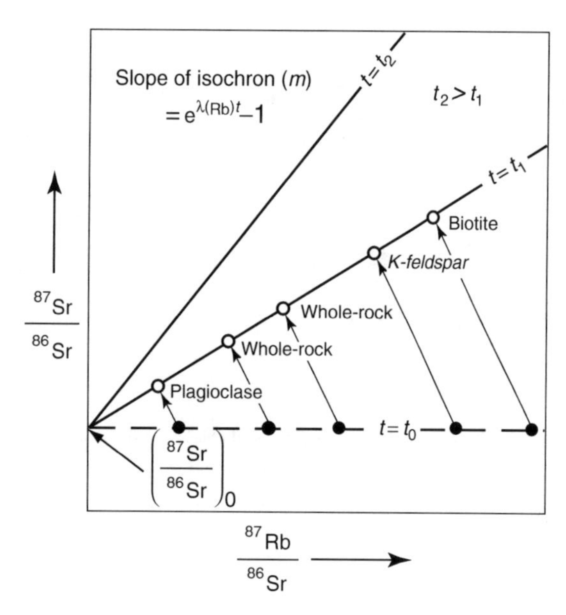

Fig. 10.4 Schematic Rb–Sr isochron diagram for two different suites of comagmatic igneous minerals and whole-rocks. The two suites are of different age (t_1 and t_2) but formed from homogeneous magmas having the same initial strontium ratio. The horizontal line is also an isochron (t_0), representing the time when the minerals and rocks had the same initial ratio and no radiogenic ^{87}Sr. The arrows indicate the generation of radiogenic ^{87}Sr by the decay of ^{87}Rb over the time period t_1. The radiogenic ^{87}Sr is proportional to the ^{87}Rb originally present in the sample. Note that the $(^{87}Sr/^{86}Sr)_0$ of a comagmatic suite can be approximated by the $^{87}Sr/^{86}Sr$ ratio of a sample with a very low concentration of ^{87}Rb.

error is calculated for the isochron (York, 1969; Dickin, 1995). A computer program titled *Isoplot 3.00* (Ludwig, 2003) that runs under Microsoft Excel is also available for such calculations, and can be obtained without charge from Ken Ludwig (kludwig@bgc.org). The opportunity for a statistical treatment of the data to estimate the uncertainty in the determined age is a big advantage of the Rb–Sr and other isochron methods. For example, the age and initial ratio of the Sudbury Nickel Irruptive, the host for the world-famous nickel sulfide deposits, and its initial ratio were determined by Rb–Sr as 1843 ± 133 Ma and 0.7071 ± 0.0005 (Fig. 10.5a).

10.3.2 *Samarium–neodymium system*

Samarium and neodymium each have seven naturally occurring isotopes. The isotopes, along with their abundances in atom % are: $^{144}_{62}Sm$ (3.1%), $^{147}_{62}Sm$ (15.0%), $^{148}_{62}Sm$ (11.2%), $^{149}_{62}Sm$ (13.8%), $^{150}_{62}Sm$ (7.4%), $^{152}_{62}Sm$ (26.7%), and $^{154}_{62}Sm$ (22.8%); and $^{142}_{60}Nd$ (27.1%), $^{143}_{60}Nd$ (12.2%), $^{144}_{60}Nd$ (23.9%), $^{145}_{60}Nd$ (8.3%), $^{146}_{60}Nd$ (17.2%), $^{148}_{60}Nd$ (5.7%), and $^{150}_{60}Nd$ (5.6%). The Sm–Nd method of age dating is based on the alpha decay of $^{147}_{62}Sm$ to $^{143}_{60}Nd$ ($\lambda_{Sm} = 6.54 \times 10^{-12}\,yr^{-1}$; $t_{1/2}$ (Sm) $= 1.06 \times 10^{11}$ yr). The equation for Sm–Nd geochronology is analogous to that for the Rb–Sr system:

$$\frac{^{143}Nd}{^{144}Nd} = \left(\frac{^{143}Nd}{^{144}Nd}\right)_0 + \frac{^{147}Sm}{^{144}Nd}\,(e^{\lambda_{Sm}t} - 1) \tag{10.19}$$

where ^{144}Nd used for normalization is the nonradiogenic isotope of constant abundance through geologic time, $^{143}Nd/^{144}Nd$ is the measured mole ratio in a sample, $(^{143}Nd/^{144}Nd)_0$ is the initial neodymium ratio, $^{147}Sm/^{144}Nd$ is the present-day mole ratio in unit weight of the sample, and t is the geologic age of the sample. In essence, age determination by the Sm–Nd method is similar to that by the Rb–Sr method, and consists of establishing an isochron in coordinates of $^{143}Nd/^{144}Nd$ (y axis) and $^{147}Sm/^{144}Nd$ (x axis) by analyzing a suite of cogenetic mineral separates or whole rocks (or a combination of the two), and calculating the age from the slope of the isochron (m),

$$m = (e^{\lambda_{Nd}t} - 1) \tag{10.20}$$

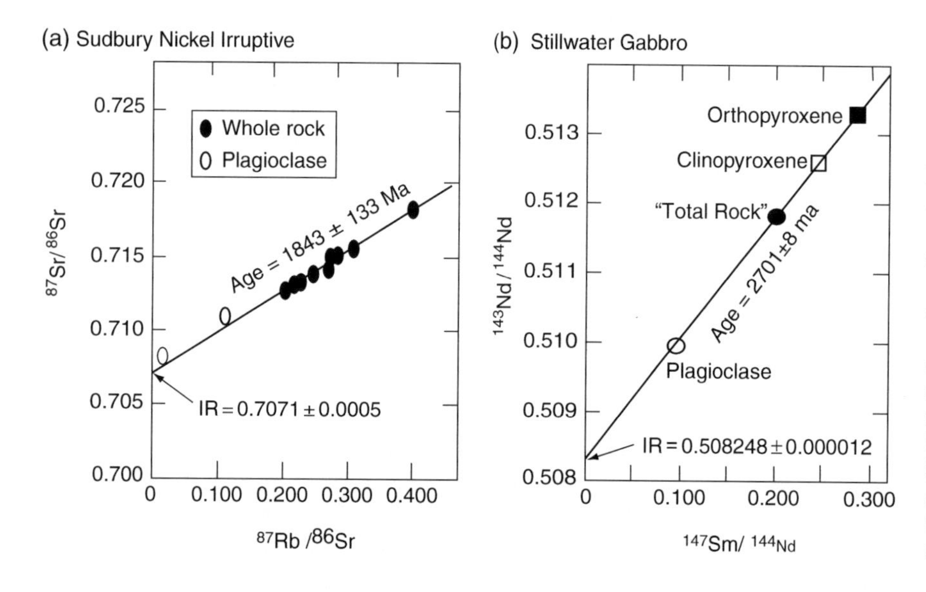

Fig. 10.5 Isochron diagrams. (a) Rb–Sr isochron for the norite unit of the Sudbury Nickel Irruptive (Ontario, Canada) defined by a suite of whole-rock samples of norite and two plagioclase concentrates (Hurst and Farhat, 1977). The age from the slope of the isochron, recalculated to $\lambda(^{87}Rb \Rightarrow {}^{87}Sr) = 1.42 \times 10^{-11}$ by Faure (1986) is 1843 ± 133 Ma, and its y-axis intercept gives an initial $(^{87}Sr/^{86}Sr)$ ratio (IR) of 0.7071 ± 0.0005. (b) Sm–Nd isochron for the Stillwater Layered Complex (Montana, USA). The calculated age from the slope of the isochron is 2701 ± 8 Ma ($\lambda_{Sm} = 6.54 \times 10^{-12}\,yr^{-1}$) and its y-axis intercept gives an initial ratio of 0.508248 ± 0.000012. (Source of data: DePaolo and Wasserburg, 1979.)

or from equation (10.16) after obtaining an estimate of the initial neodymium ratio. For example, the age of the Stillwater Layered Complex was determined as 2701 ± 8 Ma by the Sm–Nd isochron method (Fig. 10.5b; DePaolo and Wasserburg, 1979).

The Sm–Nd method is particularly suitable for dating mafic and ultramafic igneous rocks, whereas the Rb–Sr method is more applicable to dating intermediate and felsic igneous rocks. This is because the Sm : Nd ratios of igneous rocks decrease with increasing degree of differentiation in spite of an increase in the concentrations of both elements. On the other hand, because of the geochemical coherence between Rb and K and between Sr and Ca, the Rb : Sr ratio increases with increasing degree of differentiation. The usefulness of the Sm–Nd system for dating ancient rocks is enhanced by the ability of Sm and Nd to reside in some common rock-forming minerals (e.g., clinopyroxene in komatiites and basalts) and common accessory minerals (e.g., monazite, allanite, titanite, and zircon in granitoids) that are resistant to postcrystallization modification. Thus, the Sm and Nd isotopes remain relatively undisturbed on the whole-rock scale during metamorphism, hydrothermal alteration, or weathering. A limitation of the Sm–Nd method is the lack of a large range of Sm : Nd ratios in natural rocks and minerals owing to the strong geochemical coherence between the two rare earth elements; this limits the precision of absolute age determined from a Sm–Nd isochron.

10.3.3 *Uranium–thorium–lead system*

Naturally occurring isotopes of the U–Th–Pb system include three radioactive isotopes of uranium [$^{238}_{92}U$ (99.2743 atom %), $^{235}_{92}U$ (0.7200 atom %), and $^{234}_{92}U$ (0.0057 atom %)], only one isotope of thorium ($_{90}{}^{232}Th$) that is radioactive, and four stable isotopes of lead [$^{204}_{82}Pb$ (1.4 atom %), $^{206}_{82}Pb$ (24.1 atom %), $^{207}_{82}Pb$ (22.1 atom %), and $^{208}_{82}Pb$ (52.4 atom %)]. The isotope ^{204}Pb has no long-lived radioactive parent, so that its terrestrial abundance may be presumed to have been constant through geologic time. The other three isotopes are radiogenic and their terrestrial abundances have steadily increased throughout geologic time by radioactive decay of parent uranium and thorium nuclides into ^{206}Pb, ^{207}Pb, and ^{208}Pb (Table 10.2). Lead isotopes are commonly measured as ratios relative to nonradiogenic ^{204}Pb, and the corresponding equations for geochronology are:

$$\frac{^{206}Pb}{^{204}Pb} = \left(\frac{^{206}Pb}{^{204}Pb}\right)_0 + \frac{^{238}U}{^{204}Pb}(e^{\lambda_{238}t} - 1) \quad (10.21)$$

$$\frac{^{207}Pb}{^{204}Pb} = \left(\frac{^{207}Pb}{^{204}Pb}\right)_0 + \frac{^{235}U}{^{204}Pb}(e^{\lambda_{235}t} - 1) \quad (10.22)$$

$$\frac{^{208}Pb}{^{204}Pb} = \left(\frac{^{208}Pb}{^{204}Pb}\right)_0 + \frac{^{232}Th}{^{204}Pb}(e^{\lambda_{232}t} - 1) \quad (10.23)$$

where λ_{238}, λ_{235}, and λ_{232} are the decay constants for ^{238}U, ^{235}U, and ^{232}Th, respectively, and the subscript "0" refers to initial ratio. U–Pb age determinations are more accurate when the lead isotopic composition is highly radiogenic because the age calculation then becomes almost independent of the initial lead isotope ratio.

In principle, each of the equations (10.21)–(10.23) can be used to construct an isochron diagram and determine the age of a cogenetic suite of rocks, as we have discussed for the $^{87}Rb - {}^{87}Sr$ method, and the ages so determined should ideally be concordant (that is, $t_{206} = t_{207} = t_{208}$). Smith and Farquhar (1989) applied such an approach (a plot of $^{206}Pb/^{204}Pb$ versus $^{238}U/^{204}Pb$) for direct dating of certain Phanerozoic corals. In practice, however, such an isochron approach does not work for U–Th–Pb systems as the samples analyzed seldom satisfy the critical requirement of having remained closed to U, Th, and Pb (as well as the intermediate daughters) during the lifetime of the system being dated. A way to minimize this problem is to reframe the geochronologic equations in terms of lead isotope ratios (e.g., $^{207}Pb/^{206}Pb$). Ratios are likely to be relatively insensitive to lead loss because of geochemical coherence among the lead isotopes. Three widely used dating methods involving lead isotope ratios are briefly discussed below.

U–Pb Concordia diagram

Rearranging equation (10.21), we get

$$\boxed{\frac{^{206}Pb^*}{^{238}U} = (e^{\lambda_{238}t} - 1)} \quad (10.24)$$

where $^{206}Pb^*$ represents radiogenic ^{206}Pb produced during the time interval t; that is,

$$\frac{^{206}Pb*}{^{238}U} = \frac{\frac{^{206}Pb}{^{204}Pb} - \left(\frac{^{206}Pb}{^{204}Pb}\right)_0}{\frac{^{238}U}{^{204}Pb}} \quad (10.25)$$

Similarly, equation (10.20) can be rearranged to give

$$\boxed{\frac{^{207}Pb*}{^{235}U} = (e^{\lambda_{235}t} - 1)} \quad (10.26)$$

where $^{207}Pb^*$ represents radiogenic ^{207}Pb produced during the time interval t, and can be expressed by an equation similar to equation (10.25). We can calculate t from equations (10.24) and (10.26) by inserting the measured isotopic composition of a U-bearing mineral into the left-hand side of these equations, provided we can estimate $(^{206}Pb/^{204}Pb)_0$ and $(^{207}Pb/^{204}Pb)_0$ or ignore these two terms with justification.

Accessory minerals that have been used for U–Pb dating include zircon ($ZrSiO_4$), badellyite (ZrO_2), monazite ($CePO_4$), apatite [$Ca_5(PO_4)_3(OH, F, Cl)$], titanite ($CaTiSiO_5$), perovskite ($CaTiO_3$), rutile (TiO_2), allanite (a complex Ca–Mn–Fe–Al silicate of the epidote group), and xenotime (YPO_4). Zircon is by far the most preferred mineral as it has a relatively high concentration of U (because U^{4+} substitutes quite readily for Zr^{4+} due to identical charge and similar ionic radius), is widely distributed (present in most felsic to intermediate igneous rocks), and contains very little initial Pb (because Pb^{2+} does not substitute easily for U^{4+} in the zircon crystal structure). Zircon is extremely resistant to hydrothermal alteration, and can survive diagenesis, metamorphism, or weathering that may modify or destroy its host rock (Amelin *et al.*, 1999; Valley *et al.*, 2005). Zircon crystals with heavy radiation damage or post-magmatic alteration can be identified easily and excluded from analysis. Moreover, the newly developed chemical abrasion ("CA–TIMS") technique involving high-temperature annealing, followed by partial HF digestion at high temperature and pressure, can be effective at removing portions of zircon crystals that have lost Pb, without affecting the isotopic systematics of the remaining material (Mattinson, 2005). The small amounts of initial lead isotopes, $(^{206}Pb)_0$ and $(^{207}Pb)_0$, incorporated in a zircon crystal can be estimated from measurements of the amount of ^{204}Pb in the mineral and $^{206}Pb/^{204}Pb$ and $^{207}Pb/^{204}Pb$ ratios of the whole rock (Dickin, 1995). The estimated amounts of $(^{206}Pb)_0$ and $(^{207}Pb)_0$ are subtracted from the present-day ^{206}Pb and ^{207}Pb values to yield the amounts of radiogenic ^{206}Pb and ^{207}Pb in the zircon. Another advantage of using zircon crystals is that they may be zoned (Fig. 10.6), the different zones representing separate events such as magmatic crystallization or metamorphism. In such cases, the zones, which must be treated as different populations, may provide a chronology of successive events experienced by the host rock.

If a uranium-bearing mineral behaves as a closed system, equations (10.24) and (10.26) should, in principle, yield concordant ages. In reality, however, even zircon crystals often yield discordant ages because of some Pb loss from the system of interest. In some such cases it may still be possible to interpret the isotopic data by means of the *concordia* diagram devised by Wetherill (1956a). The concordia curve is constructed by plotting compatible values of $^{206}Pb^*/^{238}U$ (y axis) against $^{207}Pb^*/^{235}U$ (x axis) calculated for chosen values of t (Fig. 10.7), so that every point on the concordia corresponds to a particular value of t, if the system can be presumed to have been closed to U, Pb, and the intermediate daughters. Any U–Pb system that plots on the concordia yields concordant dates, which can be calculated from the coordinates of the point on the concordia; any U–Pb system that does not, yields discordant dates – i.e., the age calculated using equation (10.24) is not the same as that calculated using equation (10.26).

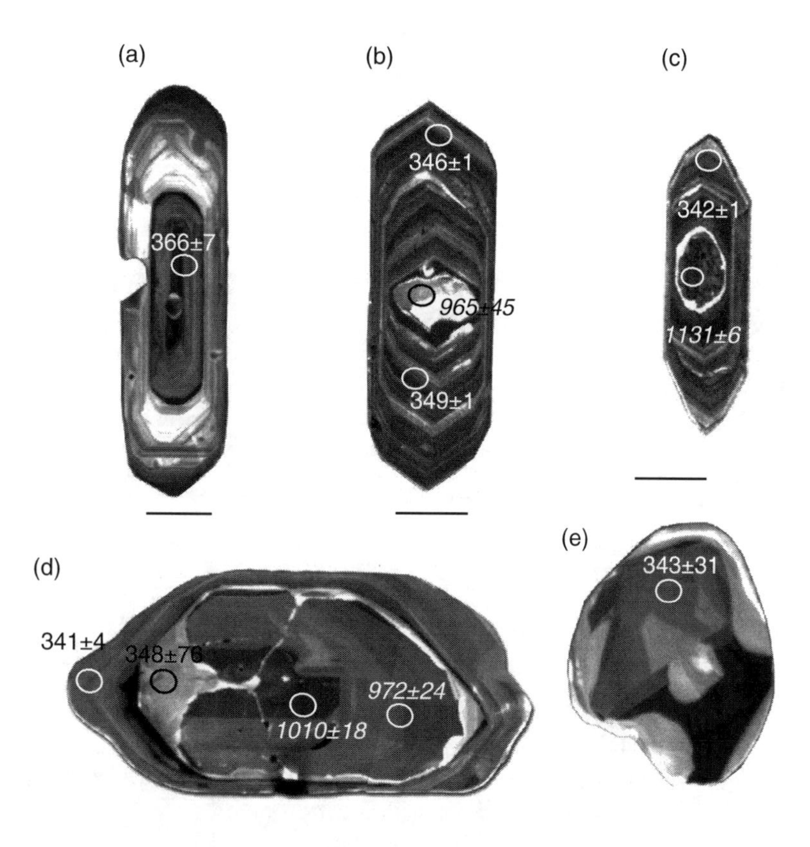

Fig. 10.6 Cathodoluminescence images of zoned zircon crystals showing radiometric ages (SHRIMP-RG analyses) at different spots. Italicized numbers are Pb/Pb ages, all others are $^{206}Pb/^{238}U$ ages. Circular black spots are mineral inclusions. Each scale bar is 100 μm. (a) Magmatic zircon with oscillatory growth zoning (Byars, 2009). (b and c) Magmatic zircons with oscillatory growth zoning and inherited cores (Stahr, 2007). (d) Detrital zircon with faintly zoned, roughly concentric metamorphic rim overgrowths (Merschat, 2009). (e) Soccerball metamorphic zircon with sector zoning (Merschat, 2009). (Compiled by Arthur Merschat, Department of Earth and Planetary Sciences, The University of Tennessee, Knoxville.)

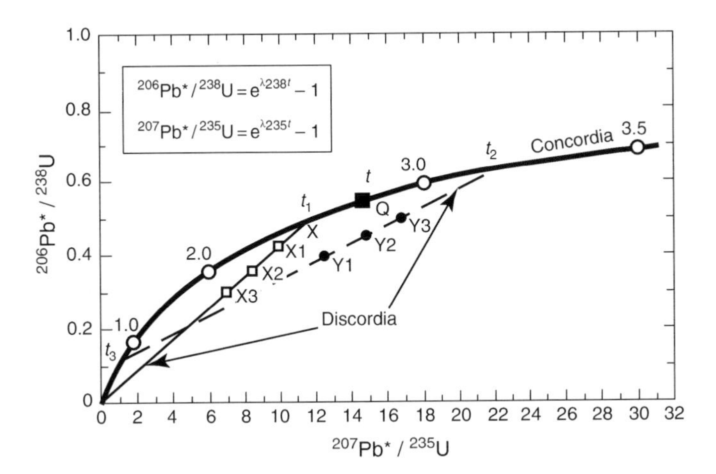

Fig. 10.7 U–Pb concordia diagram illustrating interpretations of lead isotopic data for uranium-bearing minerals such as zircon. The concordia is generated by plotting $^{206}Pb^*/^{238}U$ against $^{207}Pb^*/^{235}U$, which are calculated for selected values of t (a few of which are marked on the concordia), using equations (10.22) and (10.24). For a zircon plotting on the concordia (e.g., point Q), an identical age (t) can be calculated from either of the equations. For discordant zircon crystals that experienced an episode of Pb loss (or U gain) to varying degrees, represented by the points X1, X2, and X3 on the discordia passing through the origin, the crystallization age (t_1) is given by the point X on the Concordia. For the discordia defined by zircons of the same age but variable degrees of Pb loss (points Y1, Y2, and Y3), the upper intersection of the discordia with the concordia gives the age of zircon crystallization (t_2). The lower intersection (t_3) is the age of the lead-loss event if the lead loss was episodic; it has no age significance if the Pb loss was by continuous diffusion over a long period of time.

Let us consider a population of cogenetic zircon crystals of age t_1 whose isotopic composition is represented by point X on the concordia (Fig. 10.7). Suppose that the zircon population experienced an episode of Pb loss resulting from metamorphism or hydrothermal alteration, and that the lead lost had the same isotopic composition as the total lead that was in the mineral (a condition not likely to be satisfied if lead was added to the mineral). It seems that Pb is more easily lost from crystals of small size, crystals containing high concentration of U, or crystals with a high degree of radiation damage. With variable loss of Pb, the changed isotopic compositions of the zircon crystals would lie on a straight line joining point X and the origin of the concordia diagram (the coordinates of any zircon crystal from which all the radiogenic Pb has been lost). Such a chord is called a *discordia* because U–Pb systems lying on it yield discordant dates. Discordant zircons that have lost varying degrees of U as a result of chemical weathering recently may also lie on a discordia that passes through the origin but they should plot above the concordia. In general, this graphical procedure will not be applicable where lead has been added to a U–Pb system unless the isotopic composition of the added lead is known.

In some cases, zircon crystals separated from a single rock (e.g., a granite) plot on a discordia that intersects the concordia at two points (Fig. 10.7). The upper intercept represents the crystallization age of zircon (t_2). Assuming that the Pb loss was instantaneous, the lower intercept is interpreted as the age (t_3) of the event that disturbed the system (Wetherill, 1956b); the intercept has no age significance if the Pb was lost by continuous diffusion over a long period of time. Additional geological information would be required to distinguish between episodic and continuous Pb loss models for a given set of samples. The computer code *Isoplot* (version 3.00) by Ludwig (2003) can be used for constructing the concordia, statistical fitting of discordias to analytical data, and calculating the coordinates of the points of intersection with the concordia.

Isotopic composition of lead in U- and Th-bearing minerals and whole rocks

An equation relating the isotopic composition of lead in U- and Th-bearing minerals to model age can be obtained by combining equations (10.21) and (10.22) and substituting for the present-day $^{238}U/^{235}U$ ratio, which is equal to 137.88:

$$\frac{\dfrac{^{207}Pb}{^{204}Pb} - \left(\dfrac{^{207}Pb}{^{204}Pb}\right)_0}{\dfrac{^{206}Pb}{^{204}Pb} - \left(\dfrac{^{206}Pb}{^{204}Pb}\right)_0} = \frac{^{235}U}{^{238}U}\left(\frac{e^{\lambda_{235}t} - 1}{e^{\lambda_{238}t} - 1}\right) = \frac{1}{137.88}\left(\frac{e^{\lambda_{235}t} - 1}{e^{\lambda_{238}t} - 1}\right) \quad (10.27)$$

Usually, we write this equation in a more compact form as

$$\frac{^{207}Pb^*}{^{206}Pb^*} = \frac{^{235}U}{^{238}U}\left(\frac{e^{\lambda_{235}t} - 1}{e^{\lambda_{238}t} - 1}\right) = \frac{1}{137.88}\left(\frac{e^{\lambda_{235}t} - 1}{e^{\lambda_{238}t} - 1}\right) \quad (10.28)$$

where $^{207}Pb^*/^{206}Pb^*$ refers to the ratio of the two lead isotopes produced by radioactive decay during time t.

Equation (10.27) indicates that a suite of closed U–Pb systems having the same age and the same initial lead isotopic composition should form a straight line array in a plot of $^{207}Pb/^{204}Pb$ (y axis) versus $^{206}Pb/^{204}Pb$ (x axis), the measured present-day ratios in the samples. This straight line, an isochron, will pass through a point whose coordinates are $(^{207}Pb/^{204}Pb)_0$ and $(^{206}Pb/^{204}Pb)_0$, and its slope (m) will be given by:

$$\text{slope } (m) = \frac{1}{137.88}\left(\frac{e^{\lambda_{235}t} - 1}{e^{\lambda_{238}t} - 1}\right) \quad (10.29)$$

Evidently, the slope of the isochron depends only on the age t, which in turn is determined by the isotopic composition of lead in the samples (and does not require any knowledge of the U and Pb concentrations in the samples), and can be determined from the equation of the regression line fitted to the data points. Equation (10.29) is transcendental (i.e. it cannot be solved for t by algebraic methods) and therefore has to be solved by an

Table 10.3 Calculated values of slopes (m) of isochrons for selected values of model age (t), using equation (10.29)

t (Ga)	Slope (m)	t (Ga)	Slope (m)	t (Ga)	Slope (m)
0.1	0.04801	1.6	0.09872	3.1	0.23701
0.2	0.05011	1.7	0.10419	3.2	0.25242
0.3	0.05233	1.8	0.11004	3.3	0.26897
0.4	0.05470	1.9	0.11629	3.4	0.28674
0.5	0.05722	2.0	0.12299	3.5	0.30583
0.6	0.05990	2.1	0.13015	3.6	0.32635
0.7	0.06276	2.2	0.13782	3.7	0.34840
0.8	0.06580	2.3	0.14604	3.8	0.37211
0.9	0.06904	2.4	0.15483	3.9	0.39760
1.0	0.07250	2.5	0.16426	4.0	0.42502
1.1	0.07619	2.6	0.17437	4.1	0.45451
1.2	0.08012	2.7	0.18520	4.2	0.48624
1.3	0.08432	2.8	0.19682	4.3	0.52039
1.4	0.08881	2.9	0.20929	4.4	0.55715
1.5	0.09360	3.0	0.22266	4.5	0.59672

$$\text{slope } (m) = \frac{1}{137.88}\left(\frac{e^{\lambda_{235}t}-1}{e^{\lambda_{238}t}-1}\right); \lambda\,(^{238}\text{U}) = 1.55125 \times 10^{-10}\text{yr}^{-1};$$
$\lambda\,(^{235}\text{U}) = 9.8485 \times 10^{-10}\text{yr}^{-1}$

iterative procedure, preferably with the help of a computer, to obtain a value of t that satisfies the equation with an acceptable level of residual. Approximate solutions, however, can be obtained by preparing a table of calculated slopes (m) corresponding to selected values of t (Table 10.3) and interpolating between appropriate values of t for a particular value of the slope (see Example 10–2). A computer code called *CLEO* (Common Lead Evaluation using Octave), is also available on the web (http://www.iamg.org/CGEditor/index.htm) for calculation of regressions, using the algorithm after York (1969), and the corresponding ^{207}Pb–^{206}Pb age (Gaab *et al.*, 2006).

Example 10–2: Calculation of "207–206" model age of the black shale sequence of the Niutitang Formation, China, from measured lead isotope ratios listed in Jiang et al. (2006)

A plot of ^{207}Pb/^{204}Pb versus ^{206}Pb/^{204}Pb ratios for the black shale whole-rock samples given in Jiang *et al.* (2006) defines an isochron with a slope of 0.0580 (Fig. 10.8), which corresponds to an age (t) between 0.5 Ga (^{207}Pb*/^{206}Pb* = 0.0572) and 0.6 Ga (^{207}Pb*/^{206}Pb* = 0.0599) (Table 10.3). We can calculate an approximate value of t by interpolation between these two limits, using the lever rule:

$$\frac{0.6-0.5}{0.0599-0.0572} = \frac{0.6-t}{0.0599-0.0580}$$

$t = 530$ Ma

An iterative procedure in which the interpolation is made between successively narrower limits would yield an age of

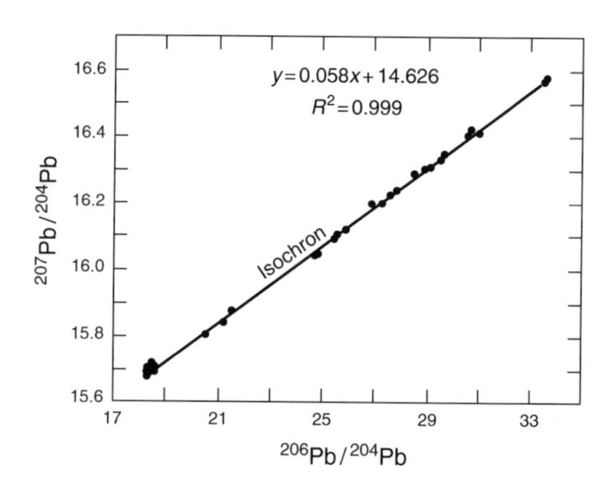

Fig. 10.8 ^{207}Pb–^{206}Pb isochron for the black shale sequence of the Niutitang Formation, China. The isotope ratios are from Jiang *et al.* (2006). The slope of the isochron is 0.0580 and the age calculated by interpolation (Table 10.3) is 530 Ma.

531 Ma. Thus our time-saving approximation in this case results in an error of less than 0.2%.

This dating method works better than the U–Pb and Th–Pb methods (equations 10.19, 10.20, and 10.21) because the "207–206 date" does not depend on the present-day U content of the sample and, therefore, is not affected by recent loss of U, for example, by chemical weathering.

The whole-rock lead method of dating was first applied to meteorite samples. From analysis of lead isotopes in three stony meteorites and two iron meteorites (one of which was the Canyon Diablo meteorite), Patterson (1956) was the first to calculate an age of 4.55 ± 0.05 Ga from a Pb–Pb isochron diagram ($\lambda_{238} = 1.537 \times 10^{-10}\text{yr}^{-1}$; $\lambda_{235} = 9.72 \times 10^{-10}\text{yr}^{-1}$). He also showed that the isotopic composition of terrestrial lead in recent oceanic sediment plots on the isochron and concluded that the age of the Earth is essentially the same as that of meteorites. [This is also corroborated by the fact the isotopic composition of Os of the Earth's mantle fits the Re–Os isochron for iron meteorites and the metallic phase of chondrites (see section 10.4.4). Subsequent studies have shown that the Pb–Pb ages of meteorites lie between 4.55 Ga and 4.57 Ga (Allègre *et al.*, 1995).

Common lead method

This method is particularly useful for dating sulfide mineral deposits, which commonly contain Pb-bearing minerals. *Common lead* refers to the Pb that occurs in minerals whose U : Pb and Th : Pb ratios are so low that its isotopic composition does not change appreciably with time. The preferred common-lead mineral for lead-isotope analysis is galena (PbS), which is widely distributed in mineral deposits. When galena is not available, other base metal sulfide minerals containing trace amounts of Pb and even K-feldspar, in which Pb^{2+} replaces K^+ to a small extent, may be used.

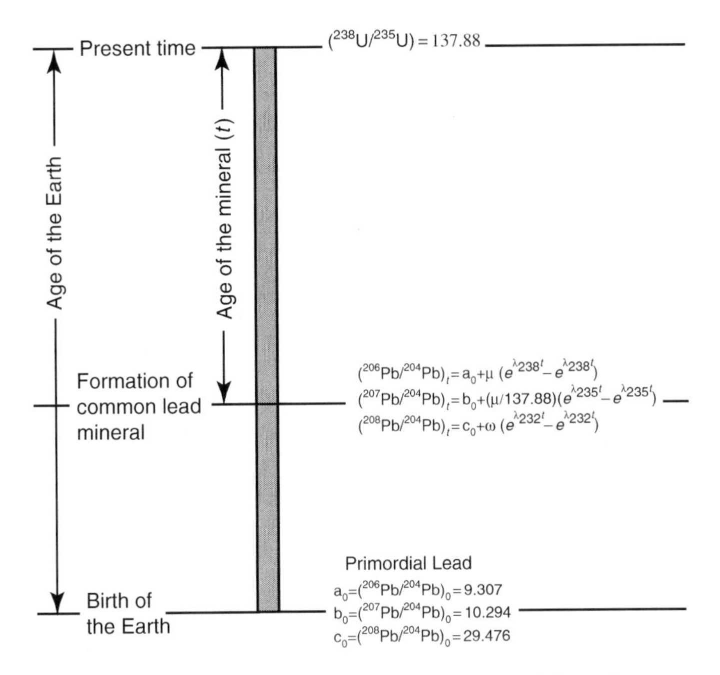

Fig. 10.9 Schematic representation of the formulation of the Holmes–Houtermans single-stage lead evolution model. The lead isotope ratios at time t are assumed to be the same as at the present time.

Single-stage leads. The simplest formulation of the lead isotope evolution is the Holmes–Houtermans single-stage model (Holmes, 1946; Houtermans, 1946), which is based on the premise that present-day isotopic ratios of common lead are the same as those at the time of its formation (Fig. 10.9). The model involves several assumptions, which are as follows (Cannon et al., 1961).

(1) The *primordial lead*, the lead at the birth of the Earth (time T = 4.55 Ga), had a unique and uniform isotopic composition, which is assumed to be the values in the troilite phase (FeS) of the Canyon Diablo meteorite because the Earth and meteorites are believed to have formed at the same time from an isotopically homogeneous solar nebula. The lead in this troilite is the least radiogenic lead ever found. It is virtually free of U and Th, and the isotopic composition of lead in this phase, therefore, can reasonably be assumed to have remained the same since its crystallization. The currently accepted values of the primordial ratios, commonly denoted as a_0, b_0, and c_0, are (Tatsumoto *et al.*, 1973):

$$a_0 = (^{206}Pb/^{204}Pb) = 9.307$$

$$b_0 = (^{207}Pb/^{204}Pb) = 10.294$$

$$c_0 = (^{208}Pb/^{204}Pb) = 29.476$$

(2) The changes in the $^{238}U/^{204}Pb$, $^{235}U/^{204}Pb$, and $^{232}Th/^{204}Pb$ ratios in any given region of the Earth were entirely due to radioactive decay within one or more closed systems, with U/Pb and Th/Pb ratios maintained at constant values in each system.

(3) The formation of galena (or other Pb-bearing minerals) during a mineralization event represents the separation of Pb from a mantle or crustal source; the lead-isotope ratios of the source are incorporated in the mineral without fractionation and are frozen in the mineral. In other words, the present-day lead isotopic ratios ($^{206}Pb/^{204}Pb$, $^{207}Pb/^{204}Pb$, and $^{208}Pb/^{204}Pb$) in a sample of galena are the same as they were at the time of its separation from the source.

(4) The lead source is an infinite reservoir, so that withdrawal of some Pb, U, and Th from the source by magmas, ore-forming fluids, or other mechanisms has not significantly disturbed the ratios involving the radioactive parent isotopes ^{238}U, ^{235}U, or ^{232}Th.

Commonly, the *single-stage lead evolution curve* for a system of given μ (= $^{238}U/^{204}Pb$) is constructed by plotting calculated values of $^{206}Pb/^{204}Pb$ (x axis) and $^{207}Pb/^{204}Pb$ (y axis) corresponding to a series of chosen values of t (geologic age) lying between $t = T$ (the age of the Earth) and $t = 0$ (the present time), using equations (10.30) and (10.31) (see Box 10.2)

$$\left(\frac{^{206}Pb}{^{204}Pb}\right)_t = a_0 + \mu\,(e^{\lambda_{238}T} - e^{\lambda_{238}t}) \quad (10.30)$$

Box 10.2 Equations for the common lead method of geochronology

Notations

$$\frac{^{206}Pb}{^{204}Pb} = a_0 = 9.307 \qquad \frac{^{207}Pb}{^{204}Pb} = b_0 = 10.294 \qquad \frac{^{208}Pb}{^{204}Pb} = c_0 = 29.476$$

$$\frac{^{238}U}{^{204}Pb} = \mu \qquad \frac{^{232}Th}{^{204}Pb} = \omega \qquad \frac{^{232}Th}{^{238}U} = \kappa \qquad \left(\frac{^{238}U}{^{235}U}\right)_{present} = 137.88$$

Equations for single-stage growth curves and primary isochrons
According to the single-stage model, the $^{206}Pb/^{204}Pb$, $^{207}Pb/^{204}Pb$, and $^{208}Pb/^{204}Pb$ ratios of common lead that got separated, without isotope fractionation, from a U–Th–Pb reservoir t years ago are given by:

$$\left(\frac{^{206}Pb}{^{204}Pb}\right)_t = \left(\frac{^{206}Pb}{^{204}Pb}\right)_0 + \frac{^{238}U}{^{204}Pb}(e^{\lambda_{238}T} - 1) - \frac{^{238}U}{^{204}Pb}(e^{\lambda_{238}t} - 1) = a_0 + \mu\,(e^{\lambda_{238}T} - e^{\lambda_{238}t}) \quad (10.30)$$

$$\left(\frac{^{207}Pb}{^{204}Pb}\right)_t = \left(\frac{^{207}Pb}{^{204}Pb}\right)_0 + \frac{^{235}U}{^{204}Pb}(e^{\lambda_{235}T} - 1) - \frac{^{235}U}{^{204}Pb}(e^{\lambda_{235}t} - 1) = b_0 + \frac{\mu}{137.88}(e^{\lambda_{235}T} - e^{\lambda_{235}t}) \quad (10.31)$$

Box 10.2 (Cont'd)

$$\left(\frac{^{208}\text{Pb}}{^{204}\text{Pb}}\right)_t = \left(\frac{^{208}\text{Pb}}{^{204}\text{Pb}}\right)_0 + \frac{^{232}\text{Th}}{^{204}\text{Pb}}(e^{\lambda_{232}T} - 1) - \frac{^{232}\text{Th}}{^{204}\text{Pb}}(e^{\lambda_{232}t} - 1)$$
$$= c_0 + \mu\kappa\,(e^{\lambda_{232}T} - e^{\lambda_{232}t}) \qquad (10.32)$$

Combining equations (10.30) and (10.31), we get

$$\frac{\left(\frac{^{207}\text{Pb}}{^{204}\text{Pb}}\right)_t - b_0}{\left(\frac{^{206}\text{Pb}}{^{204}\text{Pb}}\right)_t - a_0} = \frac{1}{137.88}\left(\frac{e^{\lambda_{235}T} - e^{\lambda_{235}t}}{e^{\lambda_{238}T} - e^{\lambda_{238}t}}\right) \qquad (10.33)$$

which is the equation of a straight line in coordinates of ^{206}Pb/^{204}Pb (x axis) and ^{207}Pb/^{204}Pb (y axis), with origin at (a_0, b_0) and slope (m) given by

$$m = \frac{\left(\frac{^{207}\text{Pb}}{^{204}\text{Pb}}\right)_t - b_0}{\left(\frac{^{206}\text{Pb}}{^{204}\text{Pb}}\right)_t - a_0} = \frac{1}{137.88}\left(\frac{e^{\lambda_{235}T} - e^{\lambda_{235}t}}{e^{\lambda_{238}T} - e^{\lambda_{238}t}}\right) \qquad (10.34)$$

This straight line represents the isochron corresponding to age t. Note that the slope of the isochron is a function of t but is independent of the value of μ, whereas the isotope ratios are functions of both t and μ (equations 10.30–10.32).

Similar equations can be written relating ^{206}Pb and ^{208}Pb, and ^{207}Pb and ^{208}Pb.

$$\left(\frac{^{207}\text{Pb}}{^{204}\text{Pb}}\right)_t = b_0 + \frac{\mu}{137.88}\,(e^{\lambda_{235}T} - e^{\lambda_{235}t}) \qquad (10.31)$$

(Single-stage evolution curves can also be represented in coordinates of ^{206}Pb/^{204}Pb and ^{208}Pb/^{204}Pb, or ^{207}Pb/^{204}Pb and ^{208}Pb/^{204}Pb.) Each point on the single-stage curve marks the theoretical lead isotope ratios corresponding to a particular value of t when that Pb was separated from its source region and incorporated into a common lead mineral such as galena. The straight line joining such a point to the origin is the corresponding *isochron* and its slope (m) is given by

$$m = \frac{1}{137.88}\left(\frac{e^{\lambda_{235}T} - e^{\lambda_{235}t}}{e^{\lambda_{238}T} - e^{\lambda_{238}t}}\right) \qquad (10.34)$$

Note that the slope of the isochron is independent of the value of μ so that all single-stage leads that were removed from sources with different μ but at the same time t must lie on the isochron corresponding to time t. For a source region of known μ, the value of t is defined by the intersection of the corresponding growth curve with the isochron. The isochron representing leads corresponding to $t = 0$ is called the *geochron* because all modern single-stage leads in the Earth and in meteorites must lie on it. For the purpose of illustration, a set of single-stage evolution curves for reservoirs with different assumed values of μ and corresponding isochrons for a few selected values of t are shown in Fig. 10.10.

It may appear that once we have measured the lead isotopic ratios, (^{206}Pb/^{204}Pb)$_t$ and (^{206}Pb/^{204}Pb)$_t$, in a sample,we should be able to calculate t from the slope of the corresponding isochron (equation 10.34). This equation, however, is transcendental and has to be solved the same way as equation (10.29) (discussed earlier) – either by an iterative procedure or by interpolation between appropriate values of t for a particular value of the slope (Table 10.4). After the model age has been calculated, we can use equation (10.30) or (10.31) to solve for μ (see Example 10–2). The computer code *Isoplot* (version 3.00) by Ludwig (2003) can be used for constructing single-stage growth curves and isochrons, and for calculating the age corresponding to any point on the growth curve.

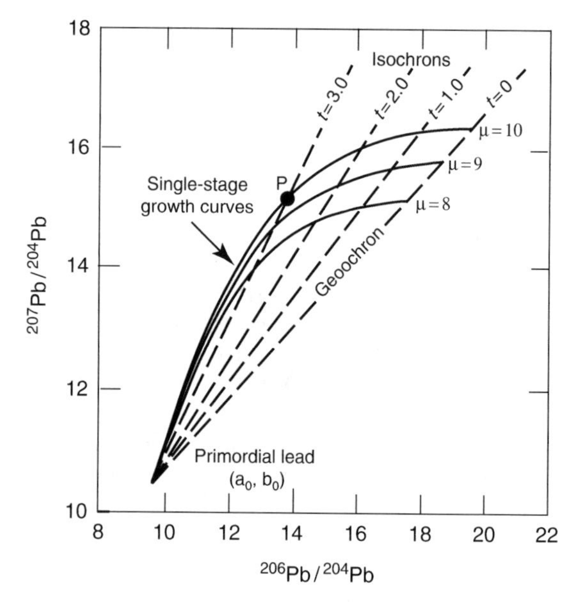

Fig. 10.10 Single-stage lead evolution curves for U–Pb systems with present-day μ (= ^{238}U/^{204}Pb) values of 8, 9, and 10. The curves were constructed by solving equations (10.28) and (10.29) for different chosen values of t, assuming that the age of the Earth (T) = 4.55 Ga. The straight lines are isochrons for $T = 4.55$ Ga, and selected values of t (3.0, 2.0, 1.0, and 0 billion years); the isochron for $t = 0$ is called the geochron. The coordinates of point P represent ^{206}Pb/^{204}Pb and ^{207}Pb/^{204}Pb ratios of a galena lead that was withdrawn 3.0 Ga froma source region of $\mu = 10.0$; the model age of this galena is 3.0 Ga. $a_0 = 9.307$, $b_0 = 10.294$, $\lambda_{238} = 1.55 \times 10^{-10}\,\text{yr}^{-1}$, and $\lambda_{235} = 9.85 \times 10^{-10}\,\text{yr}^{-1}$.

Table 10.4 Calculated values of slopes (m) of isochrons for selected model ages (t), using equation (10.34)

t (Ga)	Slope (m)	t (Ga)	Slope (m)	t (Ga)	Slope (m)	t (Ga)	Slope (m)
0.2	0.6356	1.4	0.7815	2.6	1.0341	3.8	1.5039
0.4	0.6551	1.6	0.8142	2.8	1.0930	4.0	1.6175
0.6	0.6763	1.8	0.8501	3.0	1.1583	4.2	1.7450
0.8	0.6992	2.0	0.8895	3.2	1.2310	4.4	1.8885
1.0	0.7243	2.2	0.9330	3.4	1.3121	4.5	1.9670
1.2	0.7516	2.4	0.9810	3.6	1.0341		

$m = (1/137.88)\,(e^{\lambda_{235}T} - e^{\lambda_{235}t}) / (e^{\lambda_{238}T} - e^{\lambda_{238}t})$

$\lambda(^{238}U) = 1.55125 \times 10^{-10}\,yr^{-1}$; $\lambda(^{235}U) = 9.8485 \times 10^{-10}\,yr^{-1}$; $T = 4.55 \times 10^9\,yr$

Example 10–2: Calculation of the age (t) of a galena sample from its lead isotope composition, assuming that the lead evolution conformed to the Holmes–Houtermans model. Use values of a_0, b_0, λ_{238}, λ_{235}, and T as given in the text

Suppose, the lead isotope ratios of a single-stage galena from a zinc–lead sulfide ore deposit are: $^{206}Pb/^{204}Pb = 14.1855$, and $^{207}Pb/^{204}Pb = 15.3899$. What is the age ($t$) of this galena?

To determine the Holmes–Houtermans model age of this galena sample, we first calculate the slope of the isochron corresponding to age t from the given values of the isotope ratios (equation 10.34):

$$m = \frac{\left(\frac{^{207}Pb}{^{204}Pb}\right)_t - b_0}{\left(\frac{^{206}Pb}{^{204}Pb}\right)_t - a_0} = \frac{15.3899 - 10.294}{14.1855 - 9.307} = \frac{5.0959}{4.8785} = 1.0446$$

This value of slope corresponds to an age (t) lying between 2.6 Ga (slope = 1.0341) and 2.8 Ga (slope = 1.0930) (Table 10.4). We now calculate the value of t by applying the lever rule:

$$t = 2.6 + \frac{(2.8 - 2.6)\,(1.0446 - 1.0341)}{1.0930 - 1.0341} = 2.6 + 0.0356$$
$$= 2.636\ \text{Ga}$$

The error resulting from this approximation is very small – theoretical calculation, as for Table 10.4, would yield a slope of 1.0446 for $t = 2.637$ Ga.

We can also calculate the $^{238}U/^{204}Pb$ ratio (or μ) of the source region of lead for this galena by using equation (10.30):

$$\mu = \frac{\frac{^{206}Pb}{^{204}Pb} - a_0}{e^{\lambda_{238}T} - e^{\lambda_{238}t}} = \frac{14.1855 - 9.307}{e^{(1.5513\times10^{-10})(4.55\times10^9)} - e^{(1.5513\times10^{-10})(2.636\times10^9)}}$$
$$= \frac{4.8785}{e^{0.7058} - e^{0.4089}} = \frac{4.8785}{2.0255 - 0.5203} = \frac{4.8785}{0.5203} = 9.38$$

Anomalous leads. If the isotopic composition of the lead contained in a rock or a mineral deposit can be accounted for by the single-stage evolution model (i.e., the composition plots on the single-stage evolution curve), it is called *single-stage lead* or *ordinary lead*, and its model age gives the time of its formation within experimental error; if not, it is called *anomalous lead*, and its model age may be appreciably different from its actual time of formation. An unmistakable example of anomalous lead is the J-type lead (Russell and Farquhar, 1960), so named after its type occurrence in the galena–sphalerite ores of Joplin (Missouri, USA), which contains excess radiogenic lead and yields model ages that are younger than the age of the ore deposit inferred from other lines of evidence.

Many leads are anomalous in the context of the Holmes–Houtermans model, because their lead isotope ratios have been evaluated with reference to an evolution model that assumes that the μ and ω values of the lead reservoir have changed with time only by radioactive decay and not by other processes such as differentiation and homogenization. Such an assumption is unrealistic in view of the dynamic evolution of the Earth involving repeated interaction between the mantle and the crust. This realization has led to several recently proposed lead evolution models that allow for variation in μ and ω values of the lead reservoir by mixing of leads of different isotopic compositions. For example, Stacey and Kramers (1975) constructed an "average" lead growth curve incorporating a two-stage process that involved a change in the μ and ω values of the reservoir (by geochemical differentiation) at 3.7 Ga and found this to yield good model ages for lead samples of different ages (Fig. 10.11). Their model postulates a first stage of lead evolution from a primordial composition assumed to be that of meteoritic troilite lead ($\mu = 7.19$, $\omega = 32.19$) beginning at 4.57 Ga, and a second stage evolution beginning at 3.7 Ga with ($\mu \approx 9.74$, $\omega \approx 37.19$) in those portions of the Earth that took part in the mixing events, giving rise to "average" lead. Doe and Zartman (1979) discussed the lead isotope data from several geologic environments in relation to four growth curves generated by a dynamic lead evolution model:

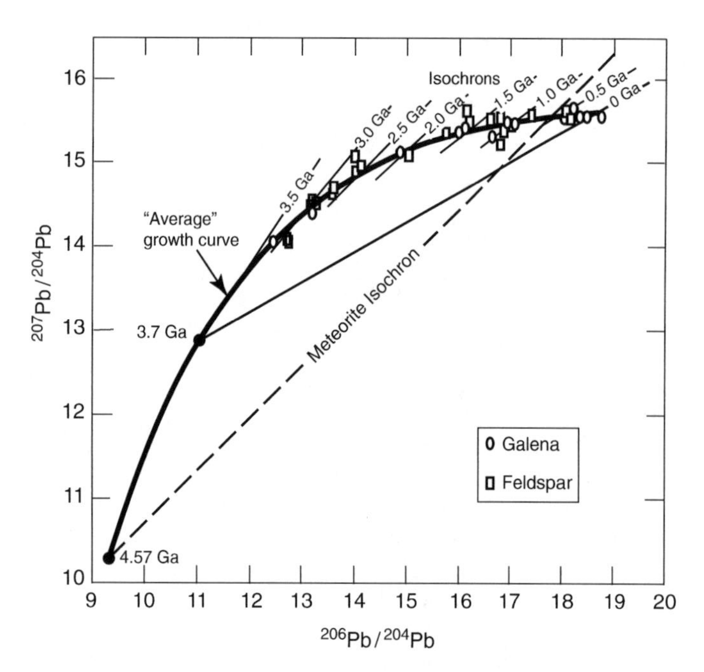

Fig. 10.11 Lead isotopic ratios of 13 galena samples from conformable deposits and 23 least radiogenic feldspar leads in relation to the two-stage lead evolution curve of Stacey and Krammers (1975). For the first-stage evolution $\mu = 7.19$ and for the second stage $\mu = 9.74$; the first stage began at $t = T$ (4.5 Ga), and the second stage at $t = 3.7$ Ga. The difference between the accepted age and the model age obtained from the two-stage lead evolution curve lie within ± 100 Ma for all galena samples (actually within ±50 Ma for all except the two youngest samples) and for 17 out of 23 feldspar samples. (After Stacey and Kramers 1975.)

(i) a curve for the mantle lead ($\mu = 8.92$), (ii) a curve for the "orogene," representing a balance of lead input from the mantle, upper crust, and lower crust sources ($\mu = 10.87$), corresponding to the "average" curve of Stacey and Kramers (1975); (iii) a curve reflecting lead contribution from only the upper crust to the orogene ($\mu = 12.24$); and (iv) a curve reflecting lead contribution from the lower crust only to the orogene ($\mu = 5.89$). When combined with regional and local geologic data, such dynamic models offer significant improvements in the interpretation of anomalous leads, although they often do not lead to unequivocal identification of the lead source.

10.3.4 *Rhenium–osmium system*

Osmium, a platinum-group element (PGE), has seven naturally occurring isotopes – $^{184}_{76}$Os (0.023%), $^{186}_{76}$Os (1.600%), $^{187}_{76}$Os (1.510%), $^{188}_{76}$Os (13.286%), $^{189}_{76}$Os (16.251%), $^{190}_{76}$Os (26.369%), and $^{192}_{76}$Os (40.957%) – all of which are stable. Rhenium has two naturally occurring isotopes – stable $^{185}_{75}$Re (37.398%) and radioactive $^{187}_{75}$Re (62.602%). The basis of age dating by Re–Os isotopes is the *b*-decay of $^{187}_{75}$Re to $^{187}_{76}$Os ($^{187}_{75}\text{Re} \Rightarrow {}^{187}_{76}\text{Os} + \beta^- + \bar{\nu}$ + energy).

For the decay of $^{187}_{75}$Re to $^{187}_{76}$Os, the geochronologic equation incorporating the nonradiogenic isotope $^{186}_{76}$Os for normalization takes the form

$$\frac{^{187}\text{Os}}{^{186}\text{Os}} = \left(\frac{^{187}\text{Os}}{^{186}\text{Os}}\right)_0 + \frac{^{187}\text{Re}}{^{186}\text{Os}}\left(e^{\lambda_{\text{Re}}t} - 1\right) \qquad (10.35)$$

where $(^{187}\text{Os}/^{186}\text{Os})_0$ is the initial ratio, ^{187}Os/^{186}Os and ^{187}Re/^{186}Os are the present-day ratios, λ_{Re} is the decay constant of ^{187}Re, and t is the geologic age of the sample. There is some uncertainty about the value of λ_{Re}. As discussed by Dickin (1995), the most appropriate value for use in geologic age studies is 1.64×10^{-11} yr^{-1} (half-life = 42.3×10^9 yr) determined by Lindner *et al.* (1989). Equation (10.35) is analogous to equation (10.15) and the procedure for age determination is the same as described for the Rb–Sr method.

The Re–Os method is particularly attractive for dating mineral deposits containing phases, such as molybdenite (MoS_2) or copper sulfides, that generally carry high Re : Os ratios, although the possible mobility of Re in hypogene and near-surface environments (Walker *et al.*, 1989; McCandless *et al.*, 1993) is a potential problem. A problem in using Os-rich minerals of the platinum-group elements, such as osmiridium (OsIr) and laurite [Ru(Os,Ir)S_2], for geochronology is that a significant fraction of ^{186}Os in the sample could have been generated by the rare long-lived unstable isotope ^{190}Pt, as in copper ores of the Sudbury igneous complex (Walker *et al.*, 1991).

The Os-rich minerals, however, are very useful for tracing how the mantle osmium isotopic composition has evolved as a function of time from its primordial value of 0.807 ± 0.006 (Fig. 10.12a). The consistent association of these minerals with ultramafic rocks derived from mantle sources suggests that the minerals inherited the mantle Os isotopic composition at the time of their formation, and this composition has not changed with time because these minerals incorporated virtually no Re. As shown in Fig. 10.12b, a plot of ^{187}Os/^{186}Os ratios of a suite of these minerals (excluding the laurite sample from the Bushveld Complex) against time define a highly correlated straight line, suggesting that the mantle evolved with a homogeneous Re : Os ratio (despite the formation of the crust by magmatic activity within it) that changed only as a result of ^{187}Re decay. The regression equation for the Os-evolution line is:

$$^{187}\text{Os}/^{186}\text{Os} = 1.040 - 0.050768\ t \qquad (10.36)$$

where t is the geological age of the sample (i.e., the time that has elapsed since the removal of the Os-mineral from its source in the mantle) in units of Ga. This equation can be used as a geochronometer to date common Os minerals (i.e., Os minerals that formed practically without incorporating any Re) if they had formed from a mantle source whose Os isotopic composition was the same as predicted by the Os-evolution line in Fig 10.12b, and which evolved in a system closed to Os and Re. Application of this simple technique, which equires the measurement only of

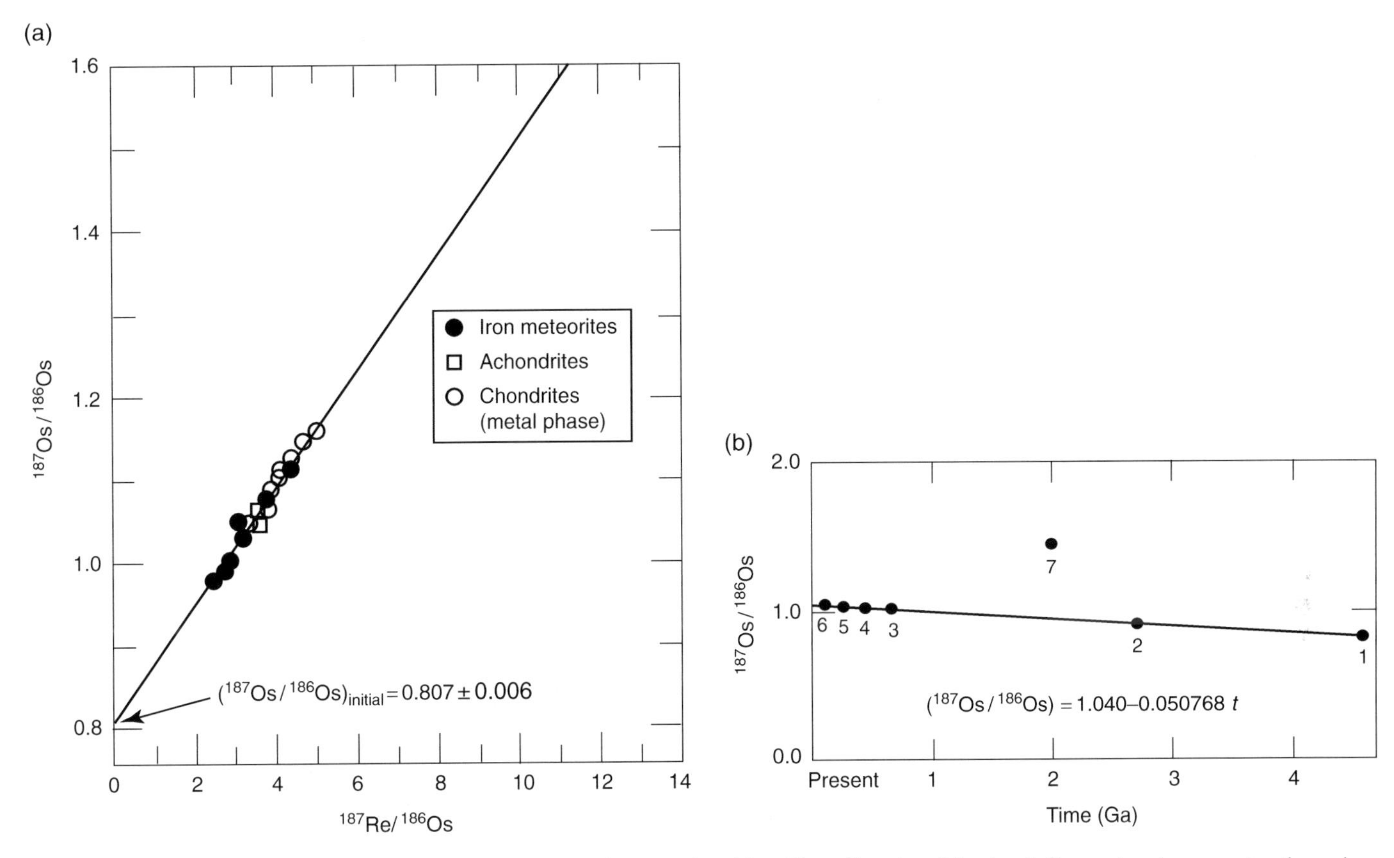

Fig. 10.12 (a) Re–Os isochron for iron meteorites and metallic phases in chondrites. The colinearity of the data indicates that the meteorites formed within a narrow time range from a primordial source that was isotopically homogeneous with respect to Os. The primitive mantle composition included in this figure, $^{187}Re/^{186}Os = 3.34$, and the present $^{187}Os/^{186}Os = 1.040$ are taken from the mantle evolution curve plotted in Fig. 10.12b. The fit of the estimated mantle composition (Luck and Allègre, 1983) to the Re–Os isochron is evidence that the parent bodies of meteorites and the Earth formed at about the same time from the same primordial source. (Sources of data: Allègre and Luck, 1980; Luck and Allègre, 1983.) (b) Evolution of the isotopic composition of osmium in the Earth's mantle based on samples of iron meteorites, osmiridium, and laurit. Data points for the Bushveld Complex (McCandless and Ruiz, 1991) plot above the mantle evolution curve, and probably reflect crustal contamination.

$^{187}Os/^{186}Os$ ratios in samples, is limited by the rarity of Os-rich minerals. Attempts to apply the technique to date Ni–Cu sulfides (e.g., Luck and Allègre, 1984) were not successful, probably because of open-system behavior of the sulfides.

10.3.5 Potassium (^{40}K)–argon (^{40}Ar) method

The radioactive isotope $^{40}_{19}K$, which constitutes only a minute fraction of naturally occurring potassium (effectively a constant value of 0.01167 atom % in all rocks and minerals), undergoes branched decay to $^{40}_{20}Ca$ by β^- emission (88.8% of the ^{40}K nuclides) and to $^{40}_{18}Ar$ by electron capture and β^+ decay (11.2% of the ^{40}K nuclides). The radiogenic ^{40}Ca and ^{40}Ar generated over time t in a K-bearing system closed to K, Ca, and Ar can be expressed as

$$^{40}Ca^* + {}^{40}Ar^* = {}^{40}K(e^{\lambda_K t} - 1) \qquad (10.37)$$

where ^{40}Ca and ^{40}Ar represent the daughters generated over time t, and λ_K is the total decay constant of ^{40}K. Denoting the decay constants of the branched decay of ^{40}K to ^{40}Ca and to ^{40}Ar as λ_β and λ_{ec}, respectively, and using the recommended values of the decay constants given in Steiger and Jäger (1977),

$$\lambda_K = \lambda_\beta + \lambda_{ec} = (4.962 \times 10^{-10}\,yr^{-1}) + (0.581 \times 10^{-10}\,yr^{-1}) = 5.543 \times 10^{-10}\,yr^{-1} \qquad (10.38)$$

The branched decay of ^{40}K results in two potential geochronometers. The decay to ^{40}Ca has been used, for example, by Marshall and DePaolo (1982) to date the Pikes Peak batholith of Colorado, but the method has limited application as a dating tool because in most rock systems the radiogenic ^{40}Ca is overwhelmed by the abundance of nonradiogenic ^{40}Ca, the latter constituting about 97% of the total Ca in terrestrial samples. The decay to ^{40}Ar, on the other hand, has been widely used to date K-bearing minerals, volcanic glass, and whole rocks. Minerals suitable for K–Ar dating include feldspars, feldspathoids, micas, amphiboles, and pyroxenes.

In general, the total number of ^{40}Ar atoms in a rock or mineral sample is the sum of three components:

$$^{40}\text{Ar} = (^{40}\text{Ar})_0 + {}^{40}\text{Ar}^* = (^{40}\text{Ar})_{\text{atm}} + (^{40}\text{Ar})_{\text{excess}} + {}^{40}\text{Ar}^* \quad (10.39)$$

where ^{40}Ar* is the radiogenic Ar produced by decay of ^{40}K in the sample subsequent to its crystallization or since it became a closed system, and $(^{40}\text{Ar})_0$ is the total amount of initial Ar atoms in the sample. The latter is composed of $(^{40}\text{Ar})_{\text{atm}}$, the atmospheric Ar adsorbed onto or contained within the sample, and $(^{40}\text{Ar})_{\text{excess}}$, the *excess argon* that was derived by outgassing of old rocks in the crust and mantle and retained by the sample at the time of its formation.

The fraction of ^{40}K atoms that decay into ^{40}Ar* is $(\lambda_{ec}/\lambda_K)^{40}$K, so that the geochronologic equation for the total amount of ^{40}Ar atoms in a closed K-bearing system can be written as

$$^{40}\text{Ar} = (^{40}\text{Ar})_0 + {}^{40}\text{Ar}^* = (^{40}\text{Ar})_0 + \frac{\lambda_{ec}}{\lambda_K}\,{}^{40}\text{K}\,(e^{\lambda_K t} - 1) \quad (10.40)$$

where t is the geologic age of the sample. As it is very difficult to determine $(^{40}\text{Ar})_0$ in a sample and a mineral is unlikely to trap appreciable amounts of the inert argon gas in the crystal structure at the time of its formation, in the conventional ^{40}K–^{40}Ar method it is assumed that the sample contained no initial argon, that is, $(^{40}\text{Ar})_0 = 0$. The assumption does not introduce significant error in the case of samples highly enriched in ^{40}Ar* relative to $(^{40}\text{Ar})_0$. Equation (10.40), then, simplifies to

$$^{40}\text{Ar} = {}^{40}\text{Ar}^* = \frac{\lambda_{ec}}{\lambda_K}\,{}^{40}\text{K}(e^{\lambda_K t} - 1) \quad (10.41)$$

which can be solved for t if the concentrations of ^{40}K and ^{40}Ar* in the sample are known from measurements:

$$t = \frac{1}{\lambda_K}\ln\left[\frac{^{40}\text{Ar}^*}{^{40}\text{K}}\left(\frac{\lambda_K}{\lambda_{ec}} + 1\right)\right] \quad (10.42)$$

In the conventional ^{40}K–^{40}Ar method, ^{40}K and ^{40}Ar contents are measured on separate aliquots of the sample to be dated. In practice, the total K content is measured on one aliquot by some standard analytical technique (such as X-ray fluorescence, atomic absorption, or flame photometry); the ^{40}K content is obtained by multiplying the total concentration of K in the sample by a factor of 0.0001167, the fractional abundance of ^{40}K. The other aliquot is heated to fusion within an ultra-high vacuum system (to minimize argon contamination from the atmosphere), and the released gas is analyzed in a mass spectrometer, using known quantities of ^{38}Ar as a tracer. The amount of ^{40}Ar* in the sample is calculated by subtracting the nonradiogenic ^{40}Ar liberated during the experiment from the measured value of ^{40}Ar. The nonradiogenic ^{40}Ar is assumed to be atmospheric contamination and to have the same ^{40}Ar/^{36}Ar ratio (= 295.5) as the present atmosphere (Dalrymple and Lanphere, 1969). Thus, the correction is made by measuring the ^{36}Ar content of the argon gas released during the experiment, and multiplying it by 295.5 to get the amount of contaminating ^{40}Ar:

$$^{40}\text{Ar}^* = {}^{40}\text{Ar}_m - {}^{36}\text{Ar}_m(295.5) \quad (10.43)$$

where the subscript "m" means the value measured by the mass spectrometer.

The assumption of no initial argon in a sample is a limitation of the conventional K–Ar dating method described above; it tends to overestimate the age of a sample because of the presence of a small amount of nonradiogenic Ar component (in the order of 10^{-9} cm^3 g^{-1}; Roddick, 1978) within minerals at the time of their crystallization. However, for a group of K-bearing minerals of the same age and identical $(^{40}\text{Ar})_0$, a condition likely to be satisfied if the analysis is restricted to several K-bearing minerals coexisting in the same rock specimen, the age and initial ratio can be determined by the isochron method similar to that discussed earlier for the Rb–Sr and Sm–Nd systems. The equation for the ^{40}K–^{40}Ar isochron method, normalized to the abundance of the nonradiogenic isotope ^{36}Ar, is

$$\frac{^{40}\text{Ar}}{^{36}\text{Ar}} = \left(\frac{^{40}\text{Ar}}{^{36}\text{Ar}}\right)_0 + \frac{\lambda_{ec}}{\lambda_K}\,\frac{^{40}\text{K}}{^{36}\text{Ar}}(e^{\lambda_K t} - 1) \quad (10.44)$$

The age t can be calculated from the slope (m) of the isochron, a straight line, defined in ^{40}Ar/^{36}Ar (y axis) – ^{40}K/^{36}Ar (x axis) space by the relation

$$m = \left(\frac{\lambda_{ec}}{\lambda_K}\right)(e^{\lambda_K t} - 1) \quad (10.45)$$

Equation (10.45) can be rearranged to obtain the following expression for t:

$$t = [\ln\,(m\,\frac{\lambda_K}{\lambda_{ec}} + 1)]/\lambda_K \quad (10.46)$$

The y-axis intercept of the isochron gives the value of $(^{40}\text{Ar}/^{36}\text{Ar})_0$, which represents the sum of the atmospheric component and the excess argon retained by the samples at the time of their formation:

$$\left(\frac{^{40}\text{Ar}}{^{36}\text{Ar}}\right)_0 = \left(\frac{^{40}\text{Ar}}{^{36}\text{Ar}}\right)_{\text{atm}} + \left(\frac{^{40}\text{Ar}}{^{36}\text{Ar}}\right)_{\text{excess}} = 295.5 + \left(\frac{^{40}\text{Ar}}{^{36}\text{Ar}}\right)_{\text{excess}} \quad (10.47)$$

This formulation assumes that all the samples have the same nonradiogenic argon isotope composition, i.e., the same value of $(^{40}Ar/^{36}Ar)_0$.

The ^{40}K–^{40}Ar method is the only major dating method that involves a gaseous daughter product. Argon tends to diffuse out of many minerals to some extent, even at modest temperatures of a few hundred degrees; in such situations, the determined ^{40}K–^{40}Ar age would be erroneously younger.

In reality, a ^{40}K–^{40}Ar date calculated using equation (10.41) represents not the age but rather the time elapsed since the mineral being dated cooled through its *blocking* (or *closure*) *temperature*, the temperature below which Ar loss from a particular mineral by diffusion becomes negligible compared with its rate of accumulation. The blocking temperature of a mineral depends on the cooling rate. For example, sanidine from volcanic rocks is highly retentive because of rapid cooling, but K-feldspar from plutonic rocks is not because of slow cooling. An additional issue for the isochron method (equation 10.44) is that the blocking temperature also depends on the crystal structure and, therefore, varies from mineral to mineral. Among the common minerals in igneous and metamorphic rocks, hornblende, muscovite, and biotite retain Ar fairly well. However, if the temperature during an episode of metamorphism is maintained at a sufficiently high temperature for all the Ar to escape from a rock, the K–Ar system will be reset and the ^{40}K–^{40}Ar date will record the age of metamorphism rather than the age of mineral formation.

10.3.6 *Argon (^{40}Ar)–argon (^{39}Ar) method*

A useful variant of the ^{40}K–^{40}Ar method is the ^{40}Ar–^{39}Ar technique that is based on the conversion of a known fraction of the $^{39}_{19}K$ in a K-bearing sample to $^{39}_{18}Ar$ by irradiation with fast neutrons in a nuclear reactor. The efficiency of irradiation to produce ^{39}Ar is determined by irradiating a standard of known ^{39}K concentration and age (commonly referred to as the flux monitor) simultaneously with the unknown sample. The sample is heated under ultrahigh vacuum (as in the case of the conventional K–Ar approach) after irradiation, and the released gas is analyzed by mass spectrometry to obtain the $^{40}Ar^*/^{39}Ar$ ratio and to calculate the age of the sample. The calculation includes appropriate corrections for argon isotopes introduced from the atmosphere and produced within the sample by undesirable neutron reactions with calcium and potassium (Dalrymple and Lanphere, 1971, 1974; Hanes, 1991).

Since the ^{39}Ar produced is proportional to the ^{39}K content of the sample, and the ratio of ^{39}K to ^{40}K is a constant of known value, a measure of ^{39}Ar in the sample yields a measure of the ^{40}K in the same sample. Thus, the $^{40}Ar^*/^{39}$ Arratio is proportional to the $^{40}Ar^*/^{40}K$ ratio, and hence the age of the sample. The relevant age equation in this case turns out to be

$$t = \frac{1}{\lambda_K} \ln\left(\frac{^{40}Ar^*}{^{39}Ar} J + 1\right) \tag{10.48}$$

where J, a measure of the efficiency of conversion of ^{39}K to ^{39}Ar, is given by

$$J = \frac{e^{\lambda_K t_s} - 1}{(^{40}Ar^*/^{39}Ar)_s} \tag{10.49}$$

In equation (10.49), t_s is the known age of the standard and $(^{40}Ar^*/^{39}Ar)_s$ the measured value of this ratio in the standard. An advantage of the $^{40}Ar^*/^{39}Ar$ method is that we measure a ratio of argon isotopes, which is inherently more precise than the measurement of absolute abundances of K and Ar.

The $^{40}K-^{39}Ar$ method has two variations: the *total fusion technique*, in which the sample is heated to complete fusion after irradiation, and the age is calculated from the $^{40}Ar^*/^{39}Ar$ ratio of all of the Ar released in a single experiment; and the *incremental heating* or *age spectrum* technique in which the sample is heated in incremental steps (starting at a few hundred degrees centigrade and finishing with fusion of the sample) and an apparent age is calculated from the $^{40}Ar^*/^{39}Ar$ ratio of the released argon gas corresponding to each step. This latter variation is useful in special cases, for example, where some Ar has been lost in a post-crystallization thermal event (Dalrymple and Lanphere, 1971). Because the ^{39}Ar in the sample is due to atom-per-atom conversion of ^{40}K, it is possible to liberate argon in stages from different domains of the sample and still recover full age information from each step (Dickin, 1995). If the sample has been closed to K and Ar* since the time of initial cooling, the ratio of $^{40}Ar^*$ to reactor-produced ^{39}Ar will be the same for each increment of gas released and each heating step will give the same apparent ^{40}Ar–^{39}Ar age. The result will be an age spectrum that will plot essentially as a horizontal line (or a "plateau") in a plot of apparent ^{40}Ar–^{39}Ar ages against the fraction of ^{39}Ar released, from which a meaningful age can be extracted (Fig. 10.13a). On the other hand, if Ar* was lost from only some crystallographic sites but not others, then the apparent age calculated for each heating step will be different and the age spectrum will take a more complicated form. The age of such a sample and perhaps some of its geologic history might still be inferred from the release pattern, even though its conventional $^{40}K-^{40}Ar$ age would be erroneous.

The incremental heating data can also be used to construct a $^{40}Ar/^{36}Ar$ versus $^{39}Ar/^{36}Ar$ isochron diagram (Fig. 10.13b), similar to those discussed earlier for other isotope systems. The relevant equation is

$$\left(\frac{^{40}Ar}{^{36}Ar}\right)_m = \left(\frac{^{40}Ar}{^{36}Ar}\right)_0 + \left(\frac{^{40}Ar^*}{^{39}Ar}\right)_k \left(\frac{^{39}Ar}{^{36}Ar}\right)_m \tag{10.50}$$

where the subscripts "m" and "k" refer, respectively, to measured ratios and to argon produced by potassium in the sample. Equation (10.50) defines a straight line (an isochron) in coordinates of $(^{40}Ar/^{36}Ar)_m$ and $(^{39}Ar/^{36}Ar)_m$ whose slope is the

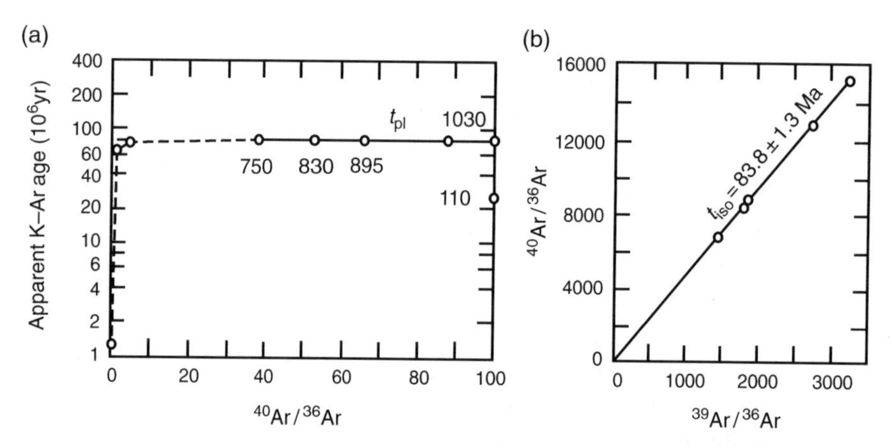

Fig. 10.13 Age of a relatively "undisturbed" muscovite sample by the ^{40}Ar–^{39}Ar method: (a) $^{40}Ar/^{39}Ar$ age spectrum diagram and (b) isochron diagram. The weighted average plateau age (t_{pl}) and the isochron age (t_{iso}) are not significantly different from each other or from the age determined by the conventional ^{40}K–^{40}Ar method (80.6 ± 1.0 Ma) at the 95% confidence level). The number by the side of each data point marks the temperature (in degrees Celsius) of the heating step, and the steps included in the calculation of the plateau age are connected by solid lines. The first three heating steps yield low apparent ages, which may indicate a small (1% or less) loss of radiogenic argon, but the error in age is indistinguishable from statistical uncertainty. On the isochron diagram, the intercept representing $(^{40}Ar/^{39}Ar)_0$ is 50 ± 128 Ma for the muscovite sample. (After Dalrymple and Lanphere, 1974.)

$^{40}Ar^*/^{39}Ar$ ratio that is related to the age of the sample by equation (10.48) and whose intercept is the initial $^{40}Ar/^{36}Ar$ ratio.

10.3.7 *Carbon-14 method*

Cosmogenic radionuclides are produced by the collision of cosmic rays, which contain high-energy protons and other particles, with nuclei of atoms in the atmosphere and on the Earth's surface. Table 10.5 presents a list of cosmogenic radionuclides that are important for geologic applications. In this chapter we will briefly discuss only the use of carbon-14 isotope for dating samples containing biogenic carbon; for a more detailed discussion of radiocarbon geochronology, the reader is referred to Trumbore (2000). The radiocarbon dating technique was introduced by Willard F. Libby (1908–1980) and his colleagues at the University of Chicago in 1949, an achievent that brought Libby the 1960 Nobel Prize for chemistry.

Carbon has two stable isotopes, $^{12}_{6}C$ (98.89 atom %) and $^{13}_{6}C$ (1.11 atom %), and a radioactive isotope, $^{14}_{6}C$, produced in the upper atmosphere by the interaction between cosmic-ray neutrons and the nucleus of stable $^{14}_{7}N$ atoms:

$$\text{neutron} + ^{14}_{7}N \Rightarrow ^{14}_{6}C + \text{proton} \qquad (10.51)$$

Most of ^{14}C is quickly oxidized to CO_2 or incorporated into CO and CO_2 molecules through exchange reactions with ^{12}C

Table 10.5 Some cosmogenic radionuclides and their applications

Parent/stable daughter	Half – life (yr)	λ (yr⁻¹)	Principal applications
$^{10}_{4}Be$ / $^{10}_{5}Be$	1.5×10^6	0.462×10^{-6}	Dating marine sediment, Mn-nodules, glacial ice, quartz in rock exposures, terrestrial age of meteorites, and petrogenesis of island-arc volcanics
$^{14}_{6}C$ / $^{14}_{7}N$	5730 ± 40	0.1209×10^{-3}	Dating of biogenic carbon, calcium carbonate, terrestrial age of meteorites; carbon cycling among the Earth's carbon reservoirs
$^{26}_{13}Al$ / $^{26}_{12}Mg$	0.716×10^6	0.968×10^{-6}	Dating marine sediment, Mn-nodules, glacial ice, quartz in rock exposures, terrestrial age of meteorites
$^{32}_{14}Si$ / $^{32}_{16}Si$	276 ± 32	0.251×10^{-2}	Dating biogenic silica, glacial ice
$^{36}_{17}Cl$/ $^{36}_{16}S$, $^{36}_{18}Ar$	0.308×10^6	2.25×10^{-6}	Dating glacial ice, exposures of volcanic rocks, groundwater, terrestrial age of meteorites
$^{39}_{18}Ar$ / $^{39}_{19}K$	269	0.257×10^{-2}	Dating glacial ice, groundwater
$^{53}_{25}Mn$ / $^{53}_{24}Cr$	3.7×10^6	0.187×10^{-6}	Terrestrial age of meteorites, abundance of extraterrestrial dust in ice and sediment
$^{81}_{36}Kr$ / $^{81}_{35}Br$	0.213×10^6	3.25×10^{-6}	Dating glacial ice, cosmic-ray exposure age of meteorites
$^{182}_{72}Hf$ / $^{182}_{74}W$	9.0×10^6	7.7×10^{-8}	Dating of the formation of planetesimals and planets, and the initiation and duration of core formation in terrestrial planets

Sources of data: Faure (1986), Halliday and Lee (1999)

and ^{13}C, and dispersed rapidly through the atmosphere and the hydrosphere. The radioactive $^{14}_{6}C$ isotope decays to stable $^{14}_{7}N$ by emission of β^- particles (decay constant $\lambda_{C-14} = 1.209 \times 10^{-4}\,yr^{-1}$; half-life = 5730 ± 40 yr). A balance between the rates of production and decay of ^{14}C is responsible for maintaining a steady equilibrium concentration of ^{14}C in the atmosphere.

Plants acquire ^{14}C through consumption of $^{14}CO_2$ molecules during photosynthesis, and by absorption through the roots; animals acquire ^{14}C by eating plant material and by absorbing CO_2 from the atmosphere and the hydrosphere. As a result of rapid cycling of carbon between the atmosphere and living biosphere, the concentration (more correctly, the activity) of ^{14}C in living organisms is maintained at a constant level, which is approximately equal to that of the atmosphere (once the $^{14}C/^{12}C$ ratio has been corrected for mass-dependent isotope fractionation effects). When an organism dies it no longer absorbs $^{14}CO_2$ from the atmosphere, and the concentration of ^{14}C in the organism decreases as a function of time because of radioactive decay. Thus, the present concentration of ^{14}C in dead plant or animal material is a measure of the time elapsed since its death and is the principle underlying the carbon-14 method of dating. For the radiocarbon age of an organic matter sample to correspond with is actual age, the ratio of carbon isotopes in the measured material must not have changed – except by the radioactive decay of ^{14}C – since the death of the organism.

Radioactive decay of ^{14}C is a first-order reaction and its integrated rate equation for a sample of carbon extracted from plant or animal tissue that died t years ago is given by (see section 9.1.3; Fig. 9.2):

$$[A]_t = [A]_0\, e^{-\lambda_{C-14} t} \tag{10.52}$$

where $[A]_t$ is the concentration of ^{14}C in a sample in equilibrium with the atmosphere, and $[A]_0$ is the concentration of ^{14}C in the same sample at the time of death of the animal or plant (commonly referred to as *specific activity* of ^{14}C), both measured in units of disintegrations per minute per gram of carbon (dpm $g^{-1}$of C), and λ_{C-14} is the dissociation constant for ^{14}C decay. Note that, unlike other radiometric dating methods, the ^{14}C method does not require the ratio of parent to daughter, or even the concentration of ^{14}N, in the sample. Using $\lambda_{C-14} = 1.209 \times 10^{-4}\,yr^{-1}$ and converting natural logarithm to the base 10, we obtain the following expression for t, the ^{14}C age of the sample:

$$t\ (\text{years}) = \frac{1}{\lambda_{C-14}} \ln \frac{[A]_0}{[A]_t} = 19.035 \times 10^3 \log \frac{[A]_0}{[A]_t} \tag{10.53}$$

To determine $[A]_t$ in a sample, it is first converted into CO_2 or another carbon compound, and then the concentration of ^{14}C is measured by counting the disintegrations with an ionization chamber or a scintillation counter. The value of $[A]_0$ is assumed to be known and to have been constant during the past 70,000 years, regardless of the species of plant or animal whose dead tissues are being dated and their geographic location. The best estimate of $[A]_0$ is 13.56 ± 0.07 dpm g^{-1} of C (Libby, 1955), but the value should be corrected for systematic variation of the ^{14}C content of the atmosphere in the past (Taylor, 1987; Faure, 1998).

Materials commonly used for ^{14}C dating include wood, charcoal, peat, nuts, seeds, grass, and paper. The rapid decay of ^{14}C limits the useful range of the method to about 12 half-lives or about 70,000 years (more realistically, 50,000 to 60,000 years) and, thus, its application in geology, such as dating volcanic ash and glacial events, but it is very useful for dating archeological samples, groundwater, and organic samples relevant to environmental geochemistry. A problem with dating groundwater is that it may contain carbon from dissolution of carbonates; this carbon does not contain any ^{14}C and, therefore, the groundwater may appear much older than it really is. Samples younger than about 1860 cannot be dated by the ^{14}C method owing to profound changes in the ^{14}C content of the atmosphere since the industrial revolution. Burning of fossil fuels has released large amounts of ^{14}C-absent CO_2 that has diluted the ^{14}C concentration in the atmosphere, thereby decreasing the amount of ^{14}C incorporated by organisms. Testing of nuclear devices, on the other hand, has added significant amounts of ^{14}C to the atmosphere since about 1945.

Example 10–3: Calculation of the maximum ^{14}C age of a wooden carving, given that measured ^{14}C activity of the wood is 12.37 dpm g^{-1} of C

As mentioned above, $[A_0]$ = 13.56 dpm g^{-1} of C; given, $[A_t]$ = 12.37 dpm g^{-1} of C. Substituting for $[A_0]$ and $[A_t]$ in equation (10.53),

$$t\ (\text{yr}) = 19.035 \times 10^3 \log \frac{[A]_0}{[A]_t} = 19.035 \times 10^3 \log \frac{13.56}{12.37}$$
$$= 760\ [= 19.035 \times 10^3 \log(1.096)$$
$$= 19.035 \times 10^3 \times 0.0399 = 0.76 \times 10^3 = 760]$$

Since wood for the carving could not have been obtained before the death of the source tree, the maximum age of the carving is 760 years.

10.4 Isotope ratios as petrogenetic indicators

The usefulness of radiogenic isotopes in the petrogenetic interpretation of igneous rocks lies in the fact that the isotopes of an element, because of their geochemical coherence, are not fractionated during crystal–liquid equilibria such as partial melting of source rocks or crystallization of magma. Thus, magma generated by partial melting inherits the isotopic composition of its source as its own initial composition and, assuming no contamination due to interaction with isotopically distinct wall rocks or other batches of magma, passes on this initial isotopic composition to the products as the magma crystallizes. Such products would have the same initial ratio as the source but different present-day ratios depending on their age. It follows that the isotopic characteristics of mantle source regions can be inferred from studies of uncontaminated oceanic volcanic rocks. Conversely, it should be possible to estimate the degree of contamination experienced by a suite of igneous rocks if we know the isotopic composition(s) of the source as well as that of the contaminant(s) (DePaolo, 1981; Powell, 1984).

10.4.1 *Strontium isotope ratios*

Papanastassiou and Wasserburg (1969) analyzed seven basaltic achondrites (stony meteorites) and established an isochron that gave a model age of 4.39 ± 0.26 Ga (using the old value of $\lambda_{Rb} = 1.39 \times 10^{-11} yr^{-1}$) and an initial $^{87}Sr/^{86}Sr$ ratio of 0.69899 ± 0.00005, which they referred to as *Basaltic Achondrite Best Initial* (BABI). The basaltic achondrites are appropriate samples for this purpose for two reasons: they seem to have crystallized from silicate melts like terrestrial igneous rocks; and their extremely low Rb/Sr ratios suggest that their $^{87}Sr/^{86}Sr$ ratios have remained practically unchanged since crystallization. Subsequent studies have shown that most stony meteorites formed within the relatively narrow time interval of 4.5 ± 0.1 Ga, and the BABI value should be adjusted to 0.69897 ± 0.00003. Thus, we can conclude that the Earth was born 4.5 ± 0.1 Ga ago with a primordial $^{87}Sr/^{86}Sr$ ratio of approximately 0.699; this is now accepted as the $^{87}Sr/^{86}Sr$ ratio for a hypothetical "uniform reservoir" (UR) of planetary material from which the Earth's mantle and all its derivative products were formed.

The BABI value of 0.699 provides a reference point for tracing the evolution of terrestrial Rb–Sr systems. Sometime after its birth, the Earth developed an inhomogeneous continental crust (see section 12.2.3) that was dominated by granitic rocks enriched in Rb (which substitutes easily for K in silicate minerals), whereas Sr (which substitutes easily for Ca in silicate and carbonate minerals) was preferentially retained in the residual, depleted mantle. Consequently, crustal rocks have evolved with elevated Rb/Sr ratios and have developed much higher present-day $^{87}Sr/^{86}Sr$ ratios, whereas the depleted mantle (i.e., depleted in Rb/Sr ratio) has evolved with much lower Rb/Sr ratios and is characterized by much lower $^{87}Sr/^{86}Sr$ ratios (Fig. 10.14a). This difference allows us to use initial $^{87}Sr/^{86}Sr$ ratios to discriminate between mantle and crustal components in igneous rocks.

The present-day $^{87}Sr/^{86}Sr$ ratios of the mantle, as inferred from analyses of young oceanic basalts and large gabbroic intrusives, which are believed to have originated in the mantle and not experienced significant crustal contamination, lie in the range of 0.702 to 0.706 (Fig. 10.15), with an average of about 0.704 ± 0.002, indicating that the mantle is not entirely homogeneous. It can be calculated that to produce a ratio of 0.704 from the BABI value of 0.699 over 4.55 billion years (the Earth's age) would require an average Rb/Sr ratio of 0.027. The Rb/Sr ratio of the mantle must have decreased as a function of time because of preferential enrichment of Rb in the partial melts extracted from the mantle, and a reasonable estimate of the $^{87}Sr/^{86}Sr$ ratio of the mantle at any time in the past can be obtained by linear interpolation between 0.699 and 0.704. The evolution of strontium isotopes in continental crust is more complicated because of its heterogeneity in terms of the ages and Rb/Sr ratios of crustal rocks, but rocks having initial $^{87}Sr/^{86}Sr$ ratios greater than 0.706 indicate a crustal component.

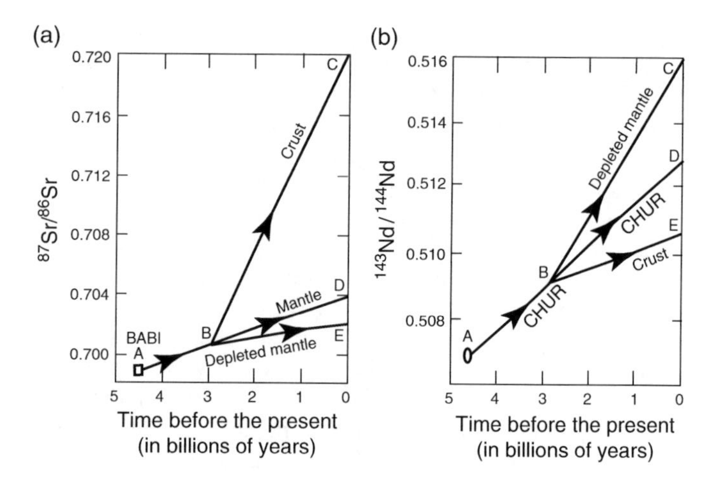

Fig. 10.14 (a) Evolution of terrestrial $^{87}Sr/^{86}Sr$ ratio over geologic time. Line ABD represents the hypothetical, closed-system evolution of mantle material, with Rb/Sr ratio of 0.027, to a present $^{87}Sr/^{86}Sr$ ratio of 0.702; in reality, this should be a slightly convex-upward curve because of the time-dependent decrease in the Rb/Sr of the upper mantle due to extraction of crustal material by partial melting. Line BC represents the closed-system evolution of a batch of crustal material with a Rb/Sr ratio of 0.15, extracted from the mantle 2.9 Ga. Line BE represents the evolution of the depleted mantle, depleted in Rb/Sr ratio resulting from extraction of the crustal material. (b) Evolution of terrestrial $^{143}Nd/^{144}Nd$ ratio over geologic time. Point A represents a $^{143}Nd/^{144}Nd$ ratio of 0.5066 for a hypothetical chondritic uniform reservoir (CHUR) 4.6 Ga. . Line ABD represents the hypothetical, closed-system evolution of the mantle material to a present $^{143}Nd/^{144}Nd$ ratio of 0.5126. Line BC represents closed-system evolution of a batch of crustal material extracted from the mantle by partial melting 2.9 Ga. Line BE represents the evolution of the depleted mantle, depleted in Sm/Nd ratio resulting from extraction of the crustal material.

10.4.2 *Neodymium isotope ratios*

Compared with Sr, the rare-earth elements, including Sm and Nd, are relatively immobile during postdepositional alteration (Banner, 2004). As a result, Nd isotope ratios in rocks are likely to be modified less than Sr isotope ratios in the same rocks.

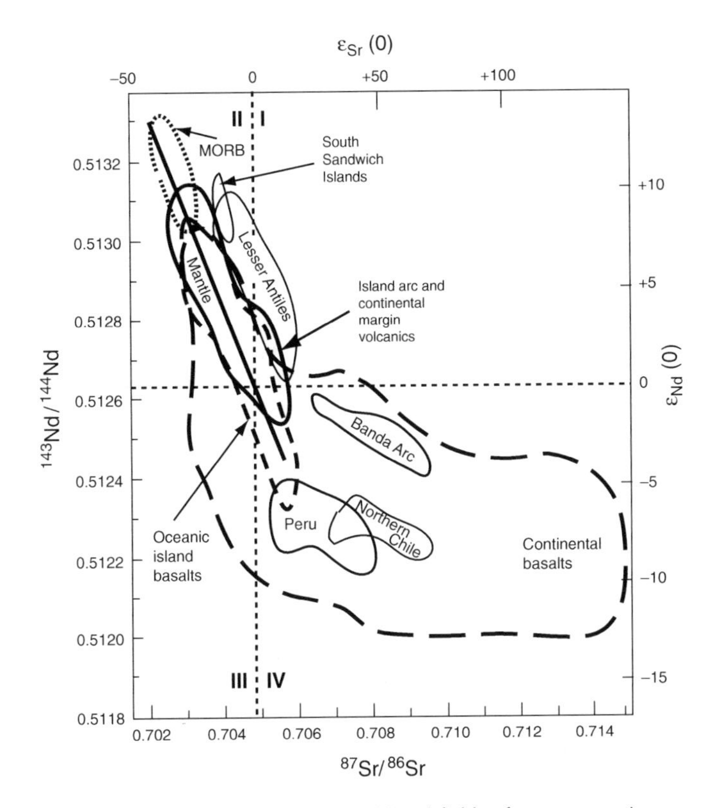

Fig. 10.15 Nd and Sr isotope compositional fields of young oceanic basalts (MORB and OIB), subduction-related basalts (island arcs formed by subduction of oceanic crust under oceanic lithosphere and continental margins adjacent to subduction zones), and continents (intraplate continental flood basalts and continental rift zone basalts). The elevated $^{87}Sr/^{86}Sr$ ratios of island arc basalts from the Lesser Antilles and the South Sandwich Islands are due to Sr contamination arising from interaction with seawater. The relatively higher $^{87}Sr/^{86}Sr$ ratios and lower $^{143}Nd/^{144}Nd$ ratios of basalts and andesites from Peru, North Chile, and the Banda Arc (Indonesia) are due to Sr and Nd contamination from crustal rocks. (Source of data: compilation by Faure (1986) from published literature.)

Chondritic meteorites, primitive planetary objects composed of condensates from the solar nebula, are considered to be the building blocks from which terrestrial planets were constructed. The terrestrial evolution of Nd isotopes is described in terms of an initial "chondritic uniform reservoir" (CHUR; DePaolo and Wasserburg, 1976) whose Sm/Nd ratio = 0.31 and whose Nd isotope ratios were uniform and the same as those of chondrites. Because the chondrites have not been affected by magmatic processes, they provide an estimate of the bulk-earth values of Sm/Nd and $^{143}Nd/^{144}Nd$. From analysis of several chondrites and an achondrite, Jacobsen and Wasserburg (1980) determined an isochron age of 4.6 Ga for the meteorites and an initial $^{143}Nd/^{144}Nd$ ratio of 0.505828 ± 0.000009 for CHUR (or 0.506609 ± 0.000008 when normalized to a $^{146}Nd/^{144}Nd$ ratio of 0.7219). The present-day values of the $^{143}Nd/^{144}Nd$ ratio (relative to a $^{146}Nd/^{144}Nd$ ratio of 0.7219) and the $^{147}Sm/^{144}Nd$ ratio of CHUR are 0.512638 and 0.1967, respectively, based on analyses of stony meteorites. The $^{143}Nd/^{144}Nd$ ratio of CHUR at any other time t (years before the present) can be calculated using the equation

$$\left(\frac{^{143}\text{Nd}}{^{144}\text{Nd}}\right)_{t\text{ (CHUR)}} = \left(\frac{^{143}\text{Nd}}{^{144}\text{Nd}}\right)_{0\text{ (CHUR)}} - \left(\frac{^{147}\text{Sm}}{^{144}\text{Nd}}\right)_{0\text{ (CHUR)}} \left(e^{\lambda_{\text{Sm}} t} - 1\right) \qquad (10.54)$$

where $(^{143}Nd/^{144}Nd)_{t\text{ (CHUR)}}$ is the value of this ratio in CHUR at time t in the geologic past (and would be the initial ratio of a rock derived from CHUR at time t), $(^{143}Nd/^{144}Nd)_{0\text{ (CHUR)}}$ is the present-day value of this ratio in CHUR (= 0.512638), and $(^{147}Sm/^{144}Nd)_{0\text{ (CHUR)}}$ is the value of this ratio in CHUR at the present time (= 0.1967).

The terrestrial evolution of the $^{143}Nd/^{144}Nd$ ratio from its CHUR value of 0.505828 at 4.6 Ga (Fig. 10.14b) is analogous to that of the $^{87}Sr/^{86}Sr$ ratio from its BABI value of 0.699 in UR. The $^{143}Nd/^{144}Nd$ ratio has increased as a function of the duration of ^{147}Sm decay, both in the bulk earth and in its differentiated products, but at a faster rate in the depleted mantle compared with the continental crust. The higher enrichment of ^{143}Nd in the depleted mantle is because of preferential partitioning of Nd, relative to Sm, into felsic differentiates. As a result, compared with the depleted mantle (with an elevated Sm/Nd ratio), the crust has evolved with lower Sm/Nd ratios and developed significantly lower present-day $^{143}Nd/^{144}Nd$ ratios. This difference allows us to use initial $^{143}Nd/^{144}Nd$ ratios to discriminate between mantle and crustal components in igneous rocks. If the initial $^{143}Nd/^{144}Nd$ ratio for a cogenetic igneous suite is significantly higher than the corresponding $^{143}Nd/^{144}Nd$ ratio of CHUR at the time of crystallization of the rocks, the magma most probably was derived by partial melting of the mantle; if the initial ratio is significantly lower, the magma most probably had a crustal source or at least a crustal component. Note that interpretations based on high and low initial $^{87}Sr/^{86}Sr$ ratios are just the opposite.

10.4.3 *Combination of strontium and neodymium isotope ratios*

A combination of Nd and Sr isotope ratios provide a better appreciation of the constraints on petrogenesis of igneous rocks and the heterogeneity of the mantle as reflected by the isotopic composition of magmas generated in the mantle. The variation in measured $^{207}Pb/^{204}Pb$ and $^{206}Pb/^{204}Pb$ ratios of young oceanic basalts also attest to the heterogeneity of the mantle (Wilson, 1989).

In general, differences in the Nd isotope ratios between the rocks of interest and CHUR are quite small, and it is convenient to represent the comparison by the "epsilon parameter" introduced by DePaolo and Wasserburg (1976):

$$\varepsilon_{\text{Nd}} = \left[\frac{(^{143}\text{Nd}/^{144}\text{Nd})_{\text{meas}}}{(^{143}\text{Nd}/^{144}\text{Nd})_{\text{p (CHUR)}}} - 1\right] \times 10^4 \qquad (10.55)$$

where $(^{143}Nd/^{144}Nd)_{meas}$ is the measured value of this ratio in a sample and $(^{143}Nd/^{144}Nd)_{p\,(CHUR)}$ the present value of this ratio in CHUR (= 0.512638, normalized to $^{146}Nd/^{144}Nd$ = 0.7219). Calculation of this parameter by normalization of all data to CHUR also circumvents the problem of different normalization procedures followed by different laboratories. We often consider ε_{Nd} values together with ε_{Sr} values, the latter defined in an analogous way as

$$\varepsilon_{Sr} = \left[\frac{(^{87}Sr/^{86}Sr)_{meas}}{(^{87}Sr/^{86}Sr)_{p\,(UR)}} - 1 \right] \times 10^4 \qquad (10.56)$$

where UR represents "uniform reservoir", equivalent to bulk earth, and $(^{87}Sr/^{86}Sr)_{p\,(UR)}$ is the present value of this ratio in the uniform reservoir (= 0.7045). The bulk earth reference reservoir for Sr, however, is not well defined (White and Hofmann, 1982), and many authors prefer to use measured $^{87}Sr/^{86}Sr$ ratios instead of calculated ε_{Sr} values.

A value of zero for either ε_{Nd} or ε_{Sr} would indicate that the corresponding measured isotopic ratio of the igneous rock is same as that of the CHUR or UR at present. A positive ε_{Nd} value implies that the magma was derived from a depleted mantle source; a negative ε_{Nd} value implies that the magma was derived from, or contaminated by, a source with a lower Sm/Nd ratio, such as old crustal material that had separated from CHUR (see Fig. 10.14a). The implications of positive and negative ε_{Sr} values are just the opposite of each other; for example, a crustal source of magma or its contamination by a crustal source is indicated by positive ε_{Sr} values (see Fig. 10.14b).

Figure 10.15 is a plot of measured $^{87}Sr/^{86}Sr$ versus $^{143}Nd/^{144}Nd$ ratios of selected young oceanic, subduction-related, and continental basalts. The most conspicuous feature of the plot is the negative correlation between ε_{Nd} and ε_{Sr} for mid-oceanic ridge basalts (MORBs) and oceanic island basalts (OIBs), often referred to as the *Mantle Array*. The MORBs and most of the OIBs fall in quadrant II of this plot ($\varepsilon_{Nd} > 0$, $\varepsilon_{Sr} < 0$) and are interpreted to have originated from "depleted" mantle sources having higher Sm/Nd ratios than CHUR and lower Rb/Sr ratios than UR (see Fig. 10.14). Some OIBs, however, plot in quadrant IV ($\varepsilon_{Nd} < 0$, $\varepsilon_{Sr} > 0$), indicating the presence of "enriched" magma sources (i.e., with lower Sm/Nd ratios than CHUR and higher Rb/Sr ratios than UR) in the mantle under the ocean basins. The wide range of Nd and Sr isotope ratios marking the Mantle Array has been interpreted variously as the result of mixing of magmas from "depleted" and "enriched" sources (DePaolo and Wasserburg, 1979), partial melting of a heterogeneous mantle source (Zindler *et al.*, 1979), incorporation of subducted oceanic or continental crust into the magma (Hawkesworth *et al.*, 1979a), and magma generation by disequilibrium melting of a mantle with variable phlogopite contents (Flower *et al.*, 1975). A detailed discussion of the isotopic characteristics of oceanic basalts is given by Cousens and Ludden (1991).

Volcanic rocks from island arcs (resulting from subduction of oceanic crust under oceanic lithosphere) plot largely in quadrant II along the Mantle Array, but their lower Nd and Sr ratios are not consistent with magmas formed by remelting of an oceanic crust composed only of MORBs. Volcanic rocks from continental margins, such as those from South America, plot in quadrant IV; these rocks represent magmatic activity resulting from subduction of oceanic crust under continental crust and contamination with Nd and Sr derived from felsic crustal rocks. As expected, crustal contamination is more pronounced in the case of continental basalts. Some continental basalts, however, plot in quadrant II, and may represent sources under the continent similar to those for OIBs.

The paucity of rocks falling in quadrant I ($\varepsilon_{Nd} > 0$, $\varepsilon_{Sr} > 0$) and in quadrant III ($\varepsilon_{Nd} < 0$, $\varepsilon_{Sr} < 0$) is to be expected because simultaneous enrichment or depletion of both Nd and Sr is incompatible with their geochemical behavior. The Sr enrichment in some of the Lesser Antilles rocks is most likely due to seawater contamination (Hawkesworth *et al.*, 1979b) and the unusual Nd and Sr of some continental basalts (such as the Tertiary volcanic province of northwest Scotland) plotting in quadrant III is believed to be due to contamination from basement granulites (Carter *et al.*, 1978).

10.4.4 *Osmium isotope ratios*

The Re–Os system is unique compared to Rb–Sr, U–Pb, and Sm–Nd systems for which all parent–daughter pairs are incompatible in mantle phases, whereas Re and Os display marked fractionation between mantle and crustal systems. Osmium is a highly compatible element in mantle phases, and is strongly retained in the mantle during most magmatic processes that produce crust from mantle. Rhenium, on the other hand, is an incompatible element in mantle phases ($K_{D(Re)} < 0.01$; Walker *et al.*, 1988) and readily partitioned into crustal material relative to Os. The recycling of crustal material into the mantle may complicate the interpretation of intra-mantle processes when using other isotopic systems. The Re–Os sysem, when used together with other isotopic systems may provide a unique insight into the evolution of the mantle and the crust over geologic time. The Re–Os sysem in the mantle is likely to have been insensitive to the incorporation of crustal material because of the similar Re abundances and much lower Os abundances in the crust relative to the mantle (Walker *et al.*, 1988).

Because of the decay of ^{187}Re, which is preferentially partitioned into the crust, the crustal osmium is more radiogenic compared to that in the mantle, the $^{187}Os/^{186}Os$ ratio increasing with decreasing geologic age. Thus, osmium isotopic composition is a powerful tracer of crustal contaminaton of mantle-derived magmas. For example, the inhomogeneity of the initial $^{187}Os/^{186}Os$ ratio (ranging from the mantle value of 0.9 corresponding to an age of 2.66 Ga to as high as 1.15) for chromite bands in the

Stillwater Layered Complex in Montana has been interpreted to reflect contamination of mantle-derived mafic magmas by an enriched crustal compoent (Lambert *et al.*, 1989). The high $^{187}Os/^{186}Os$ ratios of laurite from chromitite units (UG1 and UG2) and the Merensky Reef of the Rustenburg Layered Suite (2.05 Ga), Bushveld Complex (Fig. 10.12b), may also be due to a large crustal component in the related magma (McCandless and Ruiz, 1991). Alternatively, the high radiogenic osmium may be indicative of a hydrothermal origin of the Mrensky PGE ore, as proposed by many authors (see, e.g., Boudreau and McCallum, 1986). Crustal contamination increases the activity of SiO_2 in a mafic magma, and thereby decreases the solubility of sulfur in the magma (Haughton *et al.*, 1974). This is an important consideration for the formation of Ni–Cu sulfide deposits hosted in mafic–ultramafic igneous bodies.

10.5 Summary

1. Radioactivity is the phenomenon of spontaneous decay of unstable nuclides, which are referred to as radionuclides. Radioactive decay occurs by three mechanisms: beta decay – emission of a negatively charged beta particle (β^-), or emission of a positively charged beta particle (β^+), or electron capture; alpha decay, emission of an alpha particle (α^{2+}) composed of two neutrons and two protons; and spontaneous nuclear fission, involving the break up of the nucleus and release of a large amount of energy.
2. The rate of radioactive decay is written as:

$$-\frac{dN}{dt} = \lambda N$$

where N denotes the number of parent atoms remaining at any time t, and λ is the *decay constant* (expressed in units of reciprocal time), which is characteristic for a particular radionuclide and is assumed to have remained constant during Earth's history.
3. The decay constant is related to the half-life ($t_{1/2}$) of the decay process by the equation:

$$t_{1/2} = \frac{0.693}{\lambda}$$

4. Radiogenic isotopes, isotopes produced by radioactive decay of radionuclides, have wide application to geochronology, mantle evolution, igneous petrogenesis, chemical stratigraphy, provenance studies, and the study of temporal changes in Earth surface processes.
5. The basic general equation used for age determination of rocks and minerals in closed systems is:

$$D = D_0 + N\,(e^{\lambda t} - 1)$$

where D and D_0, respectively, are the total number of daughter atoms after time t and the number of daughter atoms present initially per unit weight of the sample. The equation for Rb–Sr dating is:

$$\frac{^{87}Sr}{^{86}Sr} = \left(\frac{^{87}Sr}{^{86}Sr}\right)_0 + \frac{^{87}Rb}{^{86}Sr}\,(e^{\lambda_{Rb} t} - 1)$$

A plot of $^{87}Sr/^{86}Sr$ (y axis) against $^{87}Rb/^{86}Sr$ (x axis) for a suite of comagmatic samples will ideally define a straight line, an isochron, with the y-axis intercept = $(^{87}Sr/^{86}Sr)_0$, the initial ratio, and slope $(m) = e^{\lambda_{Rb} t} - 1$ from which we can calculate t. The isochron method of age determination for other systems, such as ^{147}Sm / ^{143}Nd, ^{176}Lu / ^{176}Hf ^{187}Re / ^{187}Os, and ^{40}K / ^{40}Ar, is based on the same principles and equations of the same form.
6. Three widely used approaches for geochronology based on U–Th–Pb systems are: (i) the U–Pb concordia diagram, which is useful for providing age information in spite of lead loss from the system; (ii) the "207–206 date," which does not depend on the present-day U content of the sample and, therefore, is not affected by recent loss of U; and (iii) the common lead method, which is particularly suitable for dating sulfide deposits.
7. The ^{40}Ar–^{39}Ar method of geochronology involves the conversion of a known fraction of the $^{39}_{19}K$ in a K-bearing sample to $^{39}_{18}Ar$ by irradiation with fast neutrons in a nuclear reactor, and then measuring the $^{40}Ar^*/^{39}Ar$ ratio in the gas released by heating the sample to fusion either in one step or in several incremental steps.
8. A method used for dating relatively young (<70,000 years) samples containing biogenic carbon is based on the decay of ^{14}C (half-life = 5730 ± 40 yr):

$$t = \frac{1}{\lambda_{C-14}} \ln \frac{[A]_0}{[A]_t} = 19.035 \times 10^3 \log \frac{[A]_0}{[A]_t}$$

where $[A]_t$ is the concentration of ^{14}C in a sample in equilibrium with the atmosphere and $[A]_0$ the concentration of ^{14}C in the same sample at the time of death of the animal or plant (commonly referred to as specific activities of ^{14}C).
9. Sr and Nd isotopes provide insight into the evolution of the mantle and genesis of igneous rocks because the isotopes, owing to their geochemical coherence, are not fractionated during crystal–liquid equilibria such as partial melting or magma crystallization processes.
10. The mantle evolution curve for osmium isotopes is very useful for assessing crustal contamination of mantle-derived magmas of known age. The mantle evolution curve can also be used for age determination

just from measured $^{187}Os/^{186}Os$ ratios of Os-rich samples such as osmiridium and laurite ($^{187}Os/^{186}Os$ = 1.040 − 0.050768 t).

10.6 Recapitulation

Terms and concepts

Alpha decay
Anomalous lead
Basaltic achondrite best initial (BABI)
Beta decay
Blocking temperature
Branched decay
Chondritic uniform reservoir (CHUR)
Common lead
Concordant dates
Concordia
Cosmogenic nuclides
Daughter (product)
Decay constant
Decay series
Depleted mantle
Discordia
Electron capture
Epsilon parameters (ε_{Nd}, ε_{Sr})
Excess argon
Geochron
Geochronology
Half-life
Initial ratio
Isochron
Mantle Array
Model age
Nuclear fission
Ordinary lead
Primordial lead
Radioactivity
Radioisotope
Radionuclide
Single-stage lead
Uniform reservoir (UR)

Computation techniques

- Calculation of age using geochronologic equations for various systems.
- Graphical techniques for determination of geologic age and intial ratios.
- Construction of concordia diagrams.
- Construction of single-stage growth curves and isochrons for lead isotopes.
- Construction of $^{40}Ar/^{39}Ar$ age spectrum diagram.

10.7 Questions

The relevant decay constants are as listed in Table 10.2. The use of spreadsheets, as in the computer program Excel, is strongly recommended.

1. Show that the decay constant (λ) for a radionuclide is related to its half-life ($t_{1/2}$) by the equation:

$$t_{1/2}(\text{years}) = \frac{0.693}{\lambda\ (y^{-1})}$$

2. The isotope ^{176}Lu undergoes β^- decay to ^{176}Hf with a long half-life. From a plot of $^{176}Hf/^{177}Hf$ (y axis) versus $^{176}Lu/^{177}Hf$ (x axis) for a suite of meteorites (eucrites), Patchett and Tatsumoto (1980) obtained a 10-point isochron with a slope of 0.0934 and an initial $^{176}Hf/^{177}Hf$ ratio of 0.27973. The age of the meteorites is known to be 4550 Ma. Calculate the half-life of ^{176}Lu decay.
3. How many half-lives would it take for a given population of plutonium atoms ($^{239}_{94}Pu$) to be reduced to less than 0.1% of the original? Half-life of $^{239}_{94}Pu = 2.44 \times 10^4$ yr.
4. Calculate the $^{147}Sm/^{144}Nd$ ratio of a rock containing 1.83 ppm Sm and 5.51 ppm Nd. The atomic weight of Sm is 150.36 and that of Nd is 144.24. The isotopic percentage abundance of ^{147}Sm is 15.0 and that of ^{144}Nd is 23.954.
5. For a sample of volcanic rock, the measured lead isotope ratios are as follows:

   $^{204}Pb/^{206}Pb = 0.143$; $^{206}Pb/^{206}Pb = 1.000$; $^{207}Pb/^{206}Pb = 12.95$; and $^{208}Pb/^{206}Pb = 21.96$.

   Recalculate the abundance of the ratios in terms of atom percent and calculate the atomic weight of lead in this sample (Faure, 1986, p. 304). The masses of the isotopes (in amu) are:

   $^{204}Pb = 203.970$; $^{206}Pb = 205.9744$; $^{207}Pb = 206.9759$; and $^{208}Pb = 207.9766$.

6. Determine the model age and the initial ratio of the Sudbury Nickel Irruptive norite by constructing an isochron diagram using the whole-rock Rb–Sr isotope ratios given below (Gibbins and McNutt, 1975). What is the initial ratio and what is its significance?

Sample	$^{87}Rb/^{86}Sr$	$^{87}Sr/^{86}Sr$	Sample	$^{87}Rb/^{86}Sr$	$^{87}Sr/^{86}Sr$
W101	0.3243	0.7155	W118	0.1129	0.7097
W102	0.3852	0.7172	W120	0.1213	0.7098
W103	0.3736	0.7173	W124	0.2259	0.7129
W104	0.4228	0.7181	W129	0.3212	0.7153
W110	0.2953	0.7144	W134	0.2866	0.7143
W113	0.2374	0.7129	W136	0.3011	0.7148
W116	0.1158	0.7100			

7. The Sm–Nd isotopic data given below are for samples of mafic–ultramafic lavas from the Norseman–Wiluna greenstone belt in the Kambalda area, Western Australia (Claoué-Long *et al.*, 1984).

Sample	$^{147}Sm/^{144}Nd$	$^{143}Nd/^{144}Nd$	Sample	$^{147}Sm/^{144}Nd$	$^{143}Nd/^{144}Nd$
1	0.1436	0.511610	7	0.2592	0.514110
2	0.1804	0.512413	8	0.2034	0.512997
3	0.1520	0.511832	9	0.2053	0.512942
4	0.1923	0.512634	10	0.1976	0.512762
5	0.2254	0.513375	11	0.2002	0.512843
6	0.2352	0.513601			

Fit the data to an isochron by least-squares regression, determine the initial $^{143}Nd/^{144}Nd$ ratio from the graph, and calculate the age based on the slope of the regression line. What is the significance of the initial ratio for this suite of lavas?

8. The Rb–Sr data given below are for whole-rock samples from amphibolite metamorphic facies of the Lewisian gneiss complex in Scotland (Moorbath *et al.*, 1975), and the Sm-Nd data are for whole-rock samples from granulite metamorphic facies of the same complex (Hamilton *et al.*, 1979). Do the data yield concordant Rb–Sr and Sm–Nd dates for the complex? If not, suggest a reasonable explanation for the discordance. Assuming that the complex is of igneous origin, what can you say about the source region for the magma from the initial ratios?

Sample	$^{87}Rb/^{86}Sr$	$^{87}Sr/^{86}Sr$	Sample	$^{147}Sm/^{144}Nd$	$^{143}Nd/^{144}Nd$
1	0.133	0.7085	1	0.1960	0.512756
2	0.177	0.7090	2	0.2346	0.513518
3	0.086	0.7040	3	0.1707	0.512195
4	0.240	0.7107	4	0.1082	0.511006
5	0.501	0.7211	5	0.0741	0.510376
6	0.594	0.7224	6	0.0875	0.510634
7	0.191	0.7096	7	0.1030	0.510962
8	0.104	0.7049	8	0.0775	0.510501
9	0.400	0.7153	9	0.1127	0.511143
10	1.710	0.7668	10	0.0871	0.510638
11	0.368	0.7152	11	0.0999	0.510902
12	0.049	0.7032			
13	0.298	0.7132			

9. Listed below is a set of measured lead–uranium isotope ratios for a suite of monazite and uraninite samples from Zimbabwe (Rhodesia) as reported by Ahrens (1955b):

Sample	$^{206}Pb^{*}/^{238}U$	$^{207}Pb^{*}/^{235}U$
(1) Monazite (Manitoba)	0.634	14.75
(2) Monazite (Ebonite)	0.507	12.45
(3) Monazite (Jack Tin)	0.420	10.10
(4) Monazite (Irumi)	0.383	9.02
(5) Uraninite (Manitoba)	0.270	5.85
(6) Monazite (Antsirabe)	0.241	5.16

Plot a concordia diagram for this suite of samples. What is a reasonable interpretation of the concordia diagram? What are the assumptions for your interpretation?

10. Calculate the "^{207}Pb–^{206}Pb age" of the Lake Johnston batholith in the Kalgoorlie–Norseman area, Western Australia from the measured $^{206}Pb/^{204}Pb$ and $^{206}Pb/^{204}Pb$ ratios of K-feldspar (K-fe), plagioclase (Plag) and whole-rock (WR) samples listed below (Oversby, 1975):

Sample	$^{206}Pb/^{204}Pb$	$^{207}Pb/^{204}Pb$	Sample	$^{206}Pb/^{204}Pb$	$^{207}Pb/^{204}Pb$
1 (K-fe)	14.437	15.121	7 (Plag)	23.202	16.674
2 (K-fe)	15.321	15.238	8 (Plag)	20.958	16.233
3 (K-fe)	14.319	15.079	9 (WR)	29.003	17.696
4 (K-fe)	15.196	15.275	10 (WR)	28.106	17.497
5 (K-fe)	16.519	15.445	11 (WR)	29.230	17.772
6 (K-fe)	15.330	15.256	12 (WR)	24.172	16.829

11. A sample of galena from a sulfide deposit has the following isotopic composition: $^{206}Pb/^{204}Pb = 14.97$; $^{207}Pb/^{204}Pb = 15.30$; and $^{208}Pb/^{204}Pb = 34.82$. Calculate the model age and the $^{238}U/^{204}Pb$ ratio (μ) of the source region of the lead, using the single-stage Holmes–Houtermans model of lead evolution and the values of the parameters as given in the text. How would you test if the Pb actually had a single-stage history?

12. Construct a single-stage growth curve for common lead in $^{206}Pb/^{204}Pb$ (x axis) – $^{207}Pb/^{204}Pb$ (y axis) space, using the values of a_0, b_0, λ_{238}, and λ_{238} as given in the text. Assume $\mu = 10$. Interpret the lead isotopic data (Brown, 1969) given below for samples of galena from an ore deposit in southeast Missouri.

Sample	$^{206}Pb/^{204}Pb$	$^{207}Pb/^{204}Pb$	Sample	$^{206}Pb/^{204}Pb$	$^{207}Pb/^{204}Pb$
1	20.15	15.79	8	20.72	15.94
2	20.34	15.87	9	20.87	15.96
3	20.41	15.92	10	20.96	15.99
4	20.82	15.93	11	20.80	15.96
5	20.81	15.91	12	20.90	15.96
6	20.76	15.99	13	21.08	16.07
7	20.79	15.97	14	21.33	16.06

13. Analyses of volcanic rock samples from McLennans Hills yielded the following K–Ar isotope ratios (McDougall *et al.*, 1969):
Determine the age and initial ratio of the volcanic suite using an isochron plot. How much excess argon does this suite contain?

Sample	$^{40}Ar/^{36}Ar$	$^{40}K/^{36}Ar$ ($\times 10^6$)	Sample	$^{40}Ar/^{36}Ar$	$^{40}K/^{36}Ar$ ($\times 10^6$)
1	331.83	5.483	6	297.52	0.856
2	335.57	5.533	7	309.96	4.222
3	318.12	4.317	8	319.81	4.112
4	342.38	7.642	9	338.95	6.980
5	347.02	8.454			

14. The data listed below pertains to the $^{40}Ar-^{39}Ar$ dating of Apollo 15 lunar anorthosite rock 15415,9 by the incremental heating method (Husian *et al.*, 1972). Calculate the apparent age for each heating step and determine the "plateau age" of the sample from the appropriate graph. For this problem, $J = 0.0983$.

Temperature (°C)	Cumulative fraction ^{39}Ar	Radiogenic $^{40}Ar/^{39}Ar$	Temperature (°C)	Cumulative fraction ^{39}Ar	Radiogenic $^{40}Ar/^{39}Ar$
800	0.03	58.14	1500	0.79	83.32
900	0.10	61.34	1650	1.00	79.80
1000	0.27	72.77	800 – 1650	1.00	77.20
1200	0.61	80.15	1000 – 1650	0.90	79.00

15. The specific ^{14}C activity of a sample of charcoal is determined to be 12.47 dpm g^{-1} of carbon. What is the age of the sample if the initial activity of ^{14}C in this sample was 13.56 dpm g^{-1} of carbon? What would be the error if the initial activity were 15.30 dpm g^{-1} of carbon?
16. The Sr and Nd isotopic data for the young (< 1 Ma) basaltic rocks of the Azores Islands (Atlantic Ocean) presented below are from the compilation by White and Hofmann (1982). SM = Sao Miguel Island; F = Faial Island.

Sample	$^{87}Sr/^{86}Sr$	$^{143}Nd/^{144}Nd$	Sample	$^{87}Sr/^{86}Sr$	$^{143}Nd/^{144}Nd$
SM-2a	0.70337	0.512926	F- 2	0.70386	0.512871
SM-12	0.70514	0.512707	F-14	0.70394	0.512870
SM- 6	0.70430	0.512812	F-21	0.70384	0.512943
SM- 6	0.70429	0.512810	F-29	0.70392	0.512877
SM-28	0.70525	0.512732	F-33	0.70393	0.512843

Construct a ε_{Sr} (x axis) versus ε_{Nd} (y axis) plot and discuss the possible reasons for the spread of the isotopic data.

11 Stable Isotopes

Stable isotope geochemistry has become an important discipline within the earth sciences. It is a discipline that is applicable to almost all studies, from those concerning the paleoecology of extinct life forms to those that seek answers about the origin of the earth and solar system.

Kyser (1987c)

In this chapter we will discuss stable isotopes whose abundances in geologic materials vary not as a function of time (because they are not daughter atoms of radionuclides) but because of separation of light from heavy isotopes as a result of various equilibrium and kinetic processes. A large number of stable isotopes occur naturally, but modern stable isotope geochemistry has traditionally been involved with isotopes of a few light elements such as H, C, N, O, and S. Each of these elements has a lighter, more abundant isotope and one or more heavier, less abundant isotopes, the ratios of which vary differentially in natural substances (Table 11.1). The characteristics that render stable isotopes of these elements useful for interpretation of some geochemical and biological processes are as follows (O'Neil, 1986a).

(1) These elements are the main components of most minerals, rocks, and fluids; they also are the basic constituents of most forms of life.
(2) The relative difference in mass between heavy and light isotopes of each of these elements is fairly large (Table 11.1), much more pronounced than in the case of isotopes of heavier elements (e.g., ^{238}U and ^{235}U). The difference ranges from a high of 100% for hydrogen (^{1}H and ^{2}H) to a low of 6.25% for sulfur (^{32}S and ^{34}S). The relative mass difference is an important consideration for *mass-dependent fractionation* of stable isotopes. Hydrogen isotope fractionations, for example, are about ten times larger than those of the other elements of interest.
(3) The occurrence of more than one oxidation state, as with C, N, and S, or of very different types of bonds as in H–O, C–O or Si–O (ranging from ionic to highly covalent) enhances the mass-dependent isotopic fractionation of these elements.
(4) For each of the five elements of interest, the abundance of the least common isotope is sufficiently high (a few tenths of a percent to a few percent) to allow for high precision measurements. Depending on the instrument used, the analytical error of deuterium analysis is up to ten times larger than those of heavier elements because of the low abundance of deuterium (about 160 ppm) in nature.

In contrast, fractionation of isotopes of the elements useful in radiogenic isotope geochemistry (e.g., Sr, Nd, Hf, Os, Pb, etc.) are quite small and can generally be ignored. This is because these elements form dominantly ionic bonds, generally exist in one oxidation state, and are characterized by only small differences in mass between heavy and light isotopes of interest.

Variations of stable isotope ratios in minerals, rocks, and fluids have provided valuable insights into a wide spectrum of geologic issues, ranging in scale from the formation of individual

Introduction to Geochemistry: Principles and Applications, First Edition. Kula C. Misra.

Table 11.1 Abundances of selected stable isotopes and reference standards used for measurement of isotope ratios.

Element	Isotope	Relative abundance (%)	Primary reference standard	Comments
Hydrogen	^1H (protium)	99.9844	V–SMOW* (Vienna Standard Mean Ocean Water)	$(D/H)_{SMOW}=155.76\times10^{-6}$
	^2H (D)	0.0156		$(D/H)_{SMOW}=1.050\ (D/H)_{NBS-1}$
Carbon	^{12}C	98.89	PDB (Pee Dee Belemnite)	$(^{13}C/^{12}C)_{PDB}=1123.75\times10^{-5}$
	^{13}C	1.11		
Nitrogen	^{14}N	99.63	N_2 in the atmosphere	$(^{15}N/^{14}N)_{atm}=361.3\times10^{-5}$
	^{15}N	0.37		
Oxygen	^{16}O	99.763	V–SMOW* (Vienna Standard Mean Ocean Water)	$(^{18}O/^{16}O)_{SMOW}=2005.2\times10^{-6}$
	^{17}O	0.0375		$(^{18}O/^{16}O)_{SMOW}=1.008\ (^{18}O/^{16}O)_{NBS-1}$
	^{18}O	0.1995	V–PDB** (Pee Dee Belemnite)	$(^{18}O/^{16}O)_{PDB}=2067.2\times10^{-6}$
Sulfur	^{32}S	95.02	CDT (Canyon Diablo Troilite)	$(^{34}S/^{32}S)_{CDT}=449.94\times10^{-4}$
	^{33}S	0.75		
	^{34}S	4.21		
	^{36}S	0.02		

* Original supplies of SMOW and PDB have been exhausted; V-SMOW and V-PDB are identical to SMOW and PDB, respectively, within limits of analytical errors.
** $\delta^{18}O_{V\text{-}SMOW}=1.03091\ \delta^{18}O_{PDB}+30.91$; $\delta^{18}O_{PDB}=0.97002\ \delta^{18}O_{SMOW}-29.98$.
Source of data: compilations by Kyser (1987a, p. 2) and O'Neil (1986b).

minerals to the evolution of the Earth and its biosphere through time. Some notable applications of these isotopes include: geothermometry; recognition and quantification of crustal assimilation by magmas and mixing of hydrothermal fluids; inferences about the sources of sulfide, sulfate, carbonate, reduced carbon, and metals in rocks and ore deposits; estimation of water : rock ratios in water–mineral reactions; and tracing the evolution of the atmosphere–biosphere system. In this chapter we will focus on a few such applications of oxygen–hydrogen and sulfur isotopes. More detailed treatments of stable isotope geochemistry can be found in Faure (1986, 2001), Valley *et al.* (1986), Kyser (1987c), Criss (1999), Hoefs (2004), and Anbar and Rouxel (2007).

Studies of stable isotope geochemistry, however, have not been limited to the few traditional elements mentioned above (O, H, S, C, and N). With the advent of advanced analytical instruments such as multicollector, inductively coupled plasma mass spectrometers (MC–ICP–MS), capable of attaining precisions of 0.05 to 0.2 per mil (‰) for many isotope systems, large portions of the Periodic Table are now accessible to stable isotope studies. The isotopic geochemistry of a number of nontraditional stable isotopes of elements, such as Li, Mg, Cl, Ca, Cr, Fe, Cu, Zn, Se, and Mo, is being seriously investigated, and has been reviewed recently in Johnson *et al.* (2004a). Some innovative applications of nontraditional isotopes include reconstruction of the paleochemistry of the oceans (Ca, Fe, Mo), fingerprinting the source of contaminants in groundwater (Cr, Cu, Zn, Cd), and the monitoring of metal transfer processes in the human body (Fe, Ca). As the brief discussion of Fe isotopes included in this chapter would illustrate, nontraditional stable isotopes are governed by the same general principles as the traditional stable isotopes.

11.1 Isotopic fractionation

The partitioning of isotopes of an element between two coexisting substances resulting in different isotopic ratios in the two substances is called *isotopic fractionation*. The isotopes of an element have nearly identical electronic structure and, therefore, very similar chemical properties, but the mass difference causes the stable isotopes to fractionate during certain reactions. Biological and most geological processes produce such mass-dependent fractionation.

11.1.1 Causes of isotopic fractionation

The energy of a molecule can be described in terms of the interactions among the electrons as well as different kinds of motions of atoms in a molecule: vibrational (the stretching and compression of the chemical bonds between the atoms), rotational, and translational (linear motion). Isotopic fractionation arises largely because such motions are associated with energies that are mass dependent. By far the largest contribution comes from vibrational motion, the only mode of motion available to atoms in a solid substance. For a molecule, $E_{vibrational} = 1/2\ h\upsilon$, where E=virational energy, ν=vibrational frequency, which is inversely proportional to the square root of its mass, and h=Planck's constant. In the case of hydrogen and deuterium, differences in the rotational component are also important. Isotopic fractionation between two chemical species results in redistribution of the isotopes of interest to minimize the energy of the system. So, when a heavier isotope of an element replaces a lighter isotope, the vibrational energy of the molecule is decreased and the molecule tends to be more stable compared with a molecule containing the lighter isotope. As a general rule, between coexisting molecules the heavier isotope is preferentially partitioned into the one in which it can form the stronger bond, that is, the bond associated with lower vibrational frequency. For example, in a liquid–vapor system, such as H_2O (liquid)–H_2O (vapor), the liquid phase with stronger bonds is enriched in the heavier isotopes (^{18}O and ^2H), whereas the vapor phase with weaker bonds preferentially concentrates the lighter isotopes (^{16}O and ^1H).

Another reason for mass-dependent fractionation of isotopes is the kinetic energies of molecules. The kinetic energy of a molecule is a function of its mass ($E_{kinetic} = 1/2\,mv^2$, where m = mass and v=velocity), and is the same for all ideal gases at a given temperature. Differences in the velocities of isotope molecules lead to isotopic fractionations in a variety of ways. Isotopically light molecules, such as $^{12}C^{16}O$, can preferentially diffuse out of a system and leave the residual reservoir enriched in the heavy isotope $^{12}C^{18}O$. In the case of evaporation of water, the greater average translational velocity of the lighter $^1H_2{}^{16}O$ molecules allows them to break through the liquid surface preferentially, resulting in an isotopic fractionation between vapor and liquid (O'Neil, 1986a).

11.1.2 *Mechanisms of isotopic fractionation*

The main mechanisms for fractionation of stable isotopes are (Faure, 1986; Hoefs, 2004):

(1) isotopic exchange reactions, which involve the redistribution of isotopes of an element among different molecules containing that element, without any change in the chemical make-up of the reactants or products;
(2) unidirectional reactions, such as bacterially mediated reduction of sulfate species to sulfide species, in which reaction rates depend on isotopic compositions of the reactants and products; and
(3) physical processes such as evaporation and condensation, melting and crystallization, adsorption and desorption, and diffusion of ions or molecules due to concentration or temperature gradients, in which mass differences come into play.

11.1.3 *Fractionation factor*

Irrespective of mechanism, the degree of isotopic fractionation between two coexisting phases A and B in isotopic equilibrium is expressed by a ratio called the *fractionation factor*, α_{A-B}, which is defined as:

$$\boxed{\alpha_{A-B} = \frac{R_A}{R_B}} \tag{11.1}$$

where R is an atomic ratio and by convention is always written as the ratio of heavy (less abundant) isotope to the light (more abundant) isotope (such as D/H, $^{13}C/^{12}C$, $^{15}N/^{14}N$, $^{18}O/^{16}O$, or $^{34}S/^{32}S$) in the phase indicated by the subscript. For example, the fractionation factor describing the fractionation of the carbon isotopes ^{13}C and ^{12}C between calcite and CO_2 is defined as:

$$\alpha_{calcite-CO_2} = \frac{\left(^{13}C/^{12}C\right)_{calcite}}{\left(^{13}C/^{12}C\right)_{CO_2}} \tag{11.2}$$

which has an empirically determined value of 1.0098.

Isotopic fractionation factor is analogous to the distribution coefficient and is the most important quantity used in evaluating stable isotope variations observed in nature. Factors influencing the sign and magnitude of fractionation factors are (O'Neil, 1986a): (i) temperature; (ii) chemical composition; (iii) crystal structure and chemical bonds; and (iv) pressure. The variation of fractionation factors as a function of temperature can be quite pronounced in some systems and is the basis of stable isotope geothermometry (discussed later). In general, bonds to ions with low atomic mass and high ionic potential are associated with high vibrational frequencies, and such ions tend to incorporate the heavy isotope preferentially in order to lower the free energy of the system. Thus, molecules containing the heavier isotope (e.g., ^{13}C–H, ^{34}S–O) have higher dissociation energies and are more stable than those containing the lighter isotope (e.g., ^{12}C–H, ^{32}S–O). In natural equilibrium assemblages, quartz in which the oxygen is bonded to the small, highly charged Si^{4+} ion is considerably enriched in ^{18}O, whereas magnetite in which the oxygen is bonded to the relatively large and divalent Fe^{2+} ion is ^{18}O-deficient. Similarly, carbonate minerals are ^{18}O-rich because the oxygen in these minerals is bonded to the small, highly charged C^{4+} ion. Effects of crystal structure on the fractionation factor are commonly minor in importance compared to those arising from chemical bonding, but can be large in some cases. For example, the ^{18}O fractionation between aragonite and calcite is only about 1.0‰ at 25°C, but that between H_2O and D_2O, the latter with a more ordered structure, is more than 15‰ at 25°C. The effect of pressure on fractionation in solid phases is negligible, no more than 0.1‰ over

20 kbar, because the change in molar volumes of solids on isotopic substitution is very small, typically hundreds to tenths of a percent; the effect may, however, be significant in mineral–fluid systems.

Fractionation factors are obtained in three ways (O'Neil, 1986a): (i) semi-empirical calculations using spectroscopic data and the methods of statistical mechanics; (ii) laboratory calibration studies; and (c) measurements of natural samples whose formation conditions are well known or highly constrained. Experimentally or empirically determined fractionation factors for a large number of systems are listed in O'Neil (1986a), Kyser (1987a), and Campbell and Larsen (1998).

11.1.4 *The delta (δ) notation*

It is easier and more precise to measure the difference in absolute isotope ratios between two substances than the value of R in each substance. As such stable isotope ratios are normally reported as delta (δ) values in units of parts per thousand or per mil (‰) relative to appropriate reference standards. For a sample, δX_{sample} is defined as:

$$\boxed{\delta X_{\text{sample}}(\text{per mil}) = \frac{\left(R_{\text{sample}} - R_{\text{std}}\right)}{R_{\text{std}}} \times 10^3} \tag{11.3}$$

where R_{sample} is the atomic ratio of the heavy to the light isotope in the sample, R_{std} is the corresponding ratio in the standard, and δX stands for δD, $\delta^{18}O$, $\delta^{34}S$, $\delta^{13}C$, or $\delta^{15}N$. The oxygen isotope ratio of a carbonate sample, for example, is reported as

$$\delta^{18}O_{\text{sample}}(\text{per mil}) = \frac{(^{18}O/^{16}O)_{\text{carbonate}} - (^{18}O/^{16}O)_{\text{std}}}{(^{18}O/^{16}O)_{\text{std}}} \times 10^3 \tag{11.4}$$

Carbon dioxide is the most suitable gas for analysis of oxygen isotope ratios, which are measured directly with an isotope ratio mass spectrometer. The isotope composition of oxygen in water is determined by analysis of CO_2 that was equilibrated isotopically with the water sample at a known temperature. The isotopic composition of hydrogen in water is measured by analysis of H_2 gas that was prepared by reacting the water sample with metallic uranium at about 750°C. The isotopic composition of oxygen in a carbonate sample is determined from CO_2 gas obtained by reacting the sample with 100% phosphoric acid, commonly at 25°C. In an actual mass spectrometer run, the isotope ratio of the unknown sample gas is compared repeatedly with the isotope ratio of a standard gas derived from the appropriate reference standard by rapidly switching back and forth between sample and standard by means of a magnetically operated valve system. This ensures that the two gases are measured under identical spectrometer conditions. The precision of $\delta^{18}O$ values is of the order of ±0.2% or better, including instrumental, analytical, and sampling errors; the corresponding precision of δD values is ±1%.

The convention of expressing δX in per mil (instead of percent) is to avoid dealing with very small numbers. It follows from the definition of δX that for a standard $\delta X = 0$. A positive value of δ means enrichment of the heavier isotope in the sample relative to the standard, a negative value indicates relative depletion of the heavier isotope in the sample. For example, a sample with $\delta^{18}O = -10‰$ is depleted in $^{18}O/^{16}O$ ratio by 10.0‰ (or 1%) relative to the standard. Typical ranges of oxygen, sulfur, and carbon isotope ratios in natural geologic systems are summarized in Fig. 11.1.

Isotope laboratories use various working standards for the measurement of δ values, but to facilitate interlaboratory comparisons, the δ values are scaled to internationally accepted standards (Table 11.1). Troilite (FeS) from the Canyon Diablo iron meteorite is the standard used for reporting $\delta^{34}S$ values. The standard commonly used for reporting $\delta^{18}O$ and δD values is SMOW (Standard Mean Ocean Water), which was defined by Craig (1961) in terms of a National Bureau of Standards (USA) reference water, NBS-1, as follows:

$$(^{18}O/^{16}O)_{\text{SMOW}} = 1.008\ (^{18}O/^{16}O)_{\text{NBS-1}}$$
$$(D/H)_{\text{SMOW}} = 1.050\ (D/H)_{\text{NBS-1}} \tag{11.5}$$

The universally employed reference standard for reporting $\delta^{13}C$ values is the Chicago PDB standard (Belemnite from the Cretaceous Peedee Formation, South Carolina), which was the laboratory working standard used at the University of Chicago during the time that the oxygen isotope paleotemperature scale was being developed. The original SMOW and PDB standards have been exhausted. V-SMOW and V-PDB, whose scales are identical to those of SMOW and PDB within the limits of analytical uncertainty, refer to standards available from the International Atomic Energy Commission (IAEA) in Vienna. A commonly employed silicate standard at present is NBS-28 quartz, which has $\delta^{18}O = +9.60‰$ on the SMOW scale. A relatively new working standard for carbon isotopes is the carbonate reference standard NBS-19, which is scaled as:

$$\delta^{13}C_{\text{NBS-19/PDB}} = 1.95‰ \tag{11.6}$$

11.1.5 *Calculation of the fractionation factor from δ values*

Isotopic fractionation factors can be approximated in terms of δ values. For two species, A (reactant) and B (product), coexisting in isotopic equilibrium, the relationship is (see Box 11.1 for derivation):

$$\boxed{1000 \ln \alpha_{A-B} \approx \delta_A - \delta_B = \Delta_{A-B}} \tag{11.7}$$

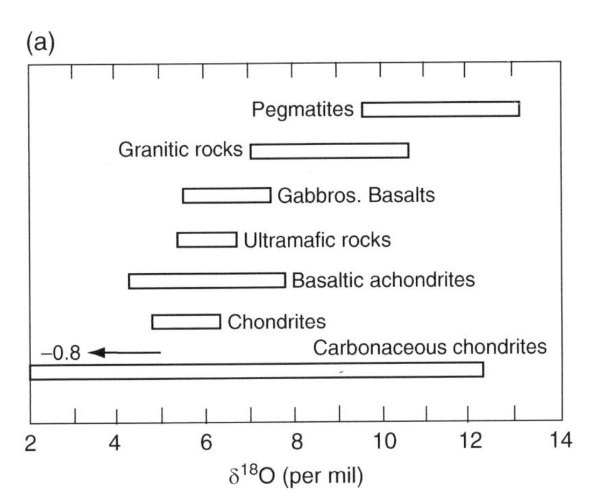

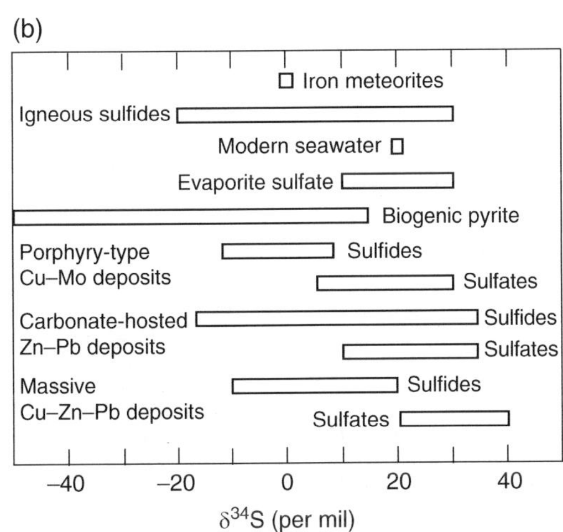

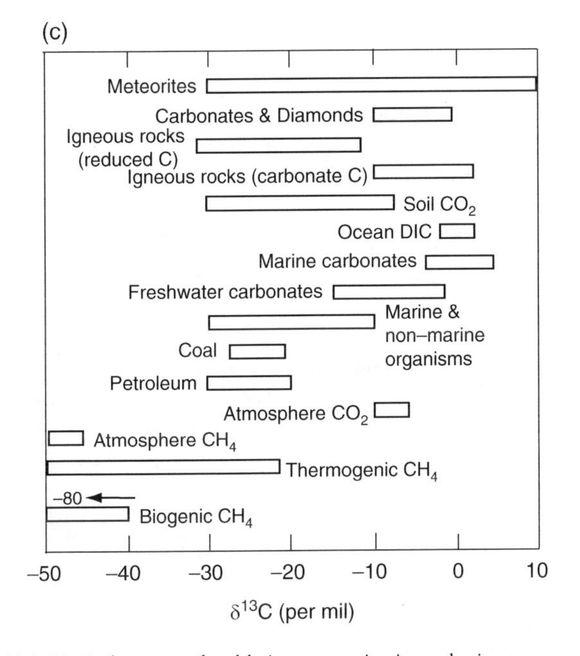

Fig. 11.1 Typical ranges of stable isotope ratios in geologic systems: (a) Oxygen; (b) Sulfur; (c) Carbon. (Sources of data: compilations by Faure (1986), Hoefs (1997), and Misra (2000).)

Box 11.1 Relationship between isotopic fraction factor and δ values for two species A and B coexisting in equilibrium ($A \Leftrightarrow B$)

From equation (11.3),

$$\delta_A = \left(\frac{R_A}{R_{std}} - 1\right) \times 10^3; \quad \delta_B = \left(\frac{R_B}{R_{std}} - 1\right) \times 10^3 \tag{11.8}$$

which can be rearranged to obtain expressions for R_A, R_B, and R_A/R_B in terms of δ_A and δ_B:

$$R_A = \left(\frac{\delta_A + 1000}{1000}\right) R_{std}; \quad R_B = \left(\frac{\delta_B + 1000}{1000}\right) R_{std} \tag{11.9}$$

$$\boxed{\alpha_{A-B} = \frac{R_A}{R_B} = \frac{\delta_A + 1000}{\delta_B + 1000}} \tag{11.10}$$

From equation (11.1),

$$\alpha_{A-B} - 1 = \frac{R_A}{R_B} - 1 = \frac{R_A - R_B}{R_B} \tag{11.11}$$

Substituting for R_A and R_B in equation (11.11),

$$\alpha_{A-B} - 1 = \frac{\delta_A - \delta_B}{\delta_B + 1000} \tag{11.12}$$

Because δ_B is very small compared to 1000, we make the approximation that $\delta_B + 1000 \approx 1000$, and simplify equation (11.12) to:

$$1000\,(\alpha_{A-B} - 1) \approx \delta_A - \delta_B \tag{11.13}$$

Values of α for the light stable isotope pairs (except for the D–H pair) are very close to unity and typically vary in the third decimal place. Most values, therefore, are of the form 1.00X, where X is rarely greater than 4, so that $1000\ln(1.00X) \approx X$. For example, for X=4 (i.e., α=1.004), $1000\ln(1.004) = 1000\,(0.00399) \approx 4$. This is a useful relationship because experimental studies have shown that $1000\ln\alpha$ is a smooth, and in some cases a linear, function of temperature for many minral–mineral and mineral–fluid pairs (see section 11.3).

Also, for small values of X, $1000\,(1.00X - 1) \approx 1000\ln(1.00X)$. Using these approximations,

$$1000\ln\alpha_{A-B} = 1000\,(\alpha_{A-B} - 1) \tag{11.14}$$

Substituting in equation (11.13),

$$\boxed{1000\ln\alpha_{A-B} \approx \delta_A - \delta_B = \Delta_{A-B}} \tag{11.7}$$

The isotopic fractionation factor between any two species A and B can also be calculated from either of the following relationships (which can easily be deduced from first principles):

$$\boxed{\alpha_{A-B} = \frac{\alpha_{A-C}}{\alpha_{B-C}}} \tag{11.15}$$

$$\boxed{\Delta_{A-B} = \Delta_{A-C} - \Delta_{B-C}} \tag{11.16}$$

where C denotes a third coexisting species.

where Δ_{A-B} represents the isotopic fractionation between A and B. Fractionation factors obtained by merely subtracting δ values will be in excellent agreement with 1000 ln α, within limits of analytical error, for values of both Δ and δ_i up to about ±10‰ (O'Neil, 1986b, p. 563). For per mil fractionations or δ_i values significantly greater than |10|, as often is the case, for example, in the hydrogen system, calculations should use equations (11.10) or (11.12). Experimentally determined fractionation factors between mineral pairs or mineral–fluid pairs yield smooth curves in 1000 ln α versus $1/T^2$ plots (T in Kelvin), generally above 150°C.

Example 11–1: Calculation of isotopic fractionation factor from δ values

Given that δ values of liquid water (lw) and water vapor (wv) in equilibrium at 10°C are: $\delta^{18}O_{lw}$=−0.80‰, and $\delta^{18}O_{wv}$=−10.79‰, what are the values of Δ_{lw-wv} and the fractionation factor α_{lw-wv} at 10°C?

Using equation (11.7),

$$\Delta_{lw-wv} = \delta_{lw} - \delta_{wv} = -0.80-(-10.79) = +9.99‰$$

$$\ln \alpha_{lw-wv} = (\delta_{lw} - \delta_{wv})/1000 = 9.99/1000 = 0.00999$$

$$\alpha_{lw-wv} = e^{0.00999} = 1.01004$$

Note that using equation (11.12), we get

$$\alpha_{lw-wv} = \frac{\delta_{lw} - \delta_{wv}}{\delta_{wv} + 1000} + 1 = \frac{9.99}{-10.79 + 1000} + 1 = 0.0101 + 1 = 1.0101$$

which is almost identical to the value of α_{lw-wv} calculated above.

11.2 Types of isotopic fractionation

Effects of isotopic fractionation are generally evaluated as: (i) equilibrium isotopic effects, produced by equilibrium isotope exchange reactions, which are independent of the pathways or mechanisms involved in the achievement of equilibrium; and (ii) kinetic isotopic effects, produced by unidirectional processes or unequilibrated chemical reactions (e.g., evaporation, diffusion, bacteria-mediated generation of methane from organic matter), which depend on reaction mechanisms and possible intermediate products.

11.2.1 Equilibrium isotope effects

If CO_2 containing only ^{16}O is mixed with water containing only ^{18}O (although such molecules with only one isotope of oxygen actually do not exist in nature), the two substances will exchange ^{16}O and ^{18}O until the system achieves equilibrium. For ease of mathematical manipulations, isotope exchange reactions are usually written in such a way that only one atom of the isotope is exchanged. For example, the oxygen isotope exchange between $CO_{2(g)}$ and $H_2O_{(l)}$ may be represented as

$$\frac{1}{2}C^{16}O_{2(g)} + H_2{}^{18}O_{(l)} = \frac{1}{2}C^{18}O_{2(g)} + H_2{}^{16}O_{(l)} \qquad (11.17)$$

Isotope exchange equilibrium may be viewed as a special case of general chemical equilibrium in which the isotopes of a single element are exchanged between two different substances, but without any net reaction. The equilibrium constant for this exchange reaction, $K_{CO_2-H_2O}$, is written in terms of concentrations as

$$K_{CO_2-H_2O} = \frac{(C^{18}O_2)^{\frac{1}{2}}(H_2{}^{16}O)}{(C^{16}O_2)^{\frac{1}{2}}(H_2{}^{18}O)} = \frac{(^{18}O/^{16}O)_{CO_2}}{(^{18}O/^{16}O)_{H_2O}} \qquad (11.18)$$

and it can be calculated using methods of statistical mechanics. Concentrations are used rather than activities or fugacities because ratios of activity or fugacity coefficients for isotopically substituted molecules are equal to unity (O'Neil, 1986a).

If the isotopes are randomly distributed over all possible crystallographic sites in the species involved in the exchange reaction, the fractionation factor (α) is related to the equilibrium constant (K) of the reaction by the relation

$$\alpha_{CO_2-H_2O} = K^{1/n}_{CO_2-H_2O} \qquad (11.19)$$

where n is the number of atoms exchanged. For the reaction under consideration, n=1, and the equilibrium constant is identical to the fractionation factor:

$$K_{CO_2-H_2O} = \frac{(^{18}O/^{16}O)_{CO_2}}{(^{18}O/^{16}O)_{H_2O}} = \frac{R_{CO_2}}{R_{H_2O}} = \alpha_{CO_2-H_2O} \qquad (11.20)$$

The experimentally determined value of $\alpha_{CO_2-H_2O}$ at 25°C (298.15 K) and 1 bar is 1.0412 (Kyser, 1987a), which means that this exchange reaction at equilibrium should result in CO_2 being enriched in ^{18}O by 41‰ relative to H_2O (and H_2O being depleted in ^{18}O by 41‰ relative to CO_2).

The standard-state (298.15 K, 1 bar) Gibbs free energy change for the reaction is a small quantity:

$$\Delta G^1_{r,\,298} = -RT \ \ln K_{CO_2-H_2O}$$
$$= -(8.314\text{J mol}^{-1}\text{K}^{-1})\ (298\ \text{K})\ (\ln 1.0412)$$
$$= -100.03\ \text{J mol}^{-1}$$

In fact, the Gibbs free energy changes for typical isotope exchange reactions in nature are in the order of $-100\,J\,mol^{-1}$ or less, not large enough to initiate chemical reactions. That is why it is not necessary to consider isotopic compositions of the species involved in chemical equilibria. On the other hand, conditions that can break and reform Al–O and S–O bonds to allow equilibrium distribution of ^{18}O among the minerals should be sufficient to establish chemical equilibrium in the system. Thus, oxygen isotope equilibrium among minerals in a rock is strong evidence that the minerals are also in chemical (mineralogical) equilibrium (O'Neil, 1986a).

11.2.2 *Kinetic isotope effects*

Biochemical reactions such as bacterially mediated reduction of sulfate (SO_4^{2-}) to H_2S or generation of methane (CH_4) from organic matter are unidirectional because the rate of backward reaction is much slower than that of the forward reaction, and the reaction does not attain equilibrium. Some chemical reactions such as evaporation, diffusion, and dissociation are also unidirectional. The instantaneous isotopic fractionation accompanying such a unidirectional reaction can be considered in terms of rate constants for the component reactions.

In principle, the exchange of sulfur isotopes ^{32}S and ^{34}S between sulfate (SO_4^{2-}) and its reduced product H_2S can be represented by the exchange reaction

$$^{32}SO_4^{2-} + H_2{}^{34}S = {}^{34}SO_4^{2-} + H_2{}^{32}S \qquad (11.21)$$

The calculated theoretical value of the equilibrium constant for this reaction, $K_{SO_4^{2-}-H_2S}$, defined as

$$K_{SO_4^{2-}-H_2S} = \frac{(^{34}SO_4^{2-})\,(H_2\,^{32}S)}{(^{32}SO_4^{2-})\,(H_2\,^{34}S)} = \frac{(^{34}S/^{32}S)_{SO_4^{2-}}}{(^{34}S/^{32}S)_{H_2S}} = \alpha_{SO_4^{2-}-H_2S} \qquad (11.22)$$

is 1.075 at 25°C (Tudge and Thode, 1950). Thus, if this exchange takes place, although no mechanism is yet known, it should lead to the enrichment of sulfate in ^{34}S by amounts up to 75‰ relative to H_2S (or the enrichment of H_2S in ^{32}S by amounts up to 75‰ relative to sulfate).

In reality, bacterially mediated reduction of sulfate to H_2S involves two competing isotopic reactions with different mass-dependent rate constants:

$$^{32}SO_4^{2-} \xrightarrow{k_1} H_2\,^{32}S \quad \text{and} \quad ^{34}SO_4^{2-} \xrightarrow{k_2} H_2\,^{34}S \qquad (11.23)$$

For this process, the ratio of rate constants for the two forward reactions, k_1/k_2, is the *kinetic isotopic effect* that measures the extent of isotopic fractionation; i.e., $\alpha_{SO_4^{2-}-H_2S} = k_1/k_2$. The isotopic fractionation in this case occurs because one of the isotopes reacts more rapidly than the other. In general, molecules containing the lighter isotope (e.g., $^{32}SO_4^{2-}$) have the faster reaction rate compared to molecules containing the heavier isotope (e.g., $^{34}SO_4^{2-}$) because of smaller bonding energy, so that the product (H_2S) tends to be enriched in the lighter isotope (^{32}S) and depleted in the heavier isotope (^{34}S) relative to the parent SO_4^{2-} (see section 11.6).

The magnitude of equilibrium fractionation factor between two species is independent of pathway; it depends only on temperature (the effect of pressure is negligible at pressures less than about 10 kbar). The kinetic isotope effect, on the other hand, depends on the reaction pathway. Other factors that influence the magnitude of kinetic isotope effect are those that affect the rates of chemical reactions and the metabolic activity of sulfate-reducing bacteria (such as *Desulfovibrio desulfuricans*). These factors include concentration of the reactants, temperature, pressure, pH, catalysis, the organisms involved (Detmers *et al.*, 2001), and availability of nutrients such as phosphate and nitrate. A detailed discussion of the rate-controlling mechanisms of bacterial sulfate reduction is provided by Rees (1973).

The magnitude of kinetic fractionation, which can be calculated using statistical mechanics or determined from experiments, is generally smaller than that of equilibrium fractionation, and kinetic fractionation effects become smaller with increasing rates of overall reaction. Harrison and Thode (1957) determined a value of 1.024 for the k_1/k_2 ratio, which means that the H_2S produced at any instant by the reduction of SO_4^{2-} should be enriched in ^{32}S, or depleted in ^{34}S, by 24‰ relative to the remaining SO_4^{2-}. Subsequent studies have shown that the kinetic $\Delta_{SO_4^{2-}-H_2S}$ value accompanying bacterial sulfate reduction varies from 15 to 25‰ at relatively high rates of sulfate reduction to more than 60‰ at low rates of sulfate reduction (Goldhaber and Kaplan, 1975).

11.3 Stable isotope geothermometry

The isotopic fractionation factor, like the equilibrium constant, is a function of temperature and generally approaches unity at very high temperatures. Experimentally determined temperature dependence of α_{A-B} can usually be described by the following expressions:

$$\text{High } T \text{ (above about 200°C)} \qquad 1000 \ln \alpha_{A-B} \approx C_1 + \frac{C_2}{T^2}$$

$$\text{Low } T \text{ (below about 200°C)} \qquad 1000 \ln \alpha_{A-B} \approx C_3 + \frac{C_4}{T} \qquad (11.24)$$

where C_1, C_2, C_3, and C_4 are empirical constants and T is in Kelvin. The temperature dependence of isotopic fractionation factors, determined experimentally, empirically, or theoretically, and expressed in the form of equations or calibration graphs, is the basis of stable isotope geothermometry. Usually, experimentally determined calibrations are more reliable, but the attainment of equilibrium in experiments at low temperatures, the most useful range for stable isotope geothermometry, is a potential limitation. On the other hand, the magnitude of isotopic fractionations decrease rapidly with increasing temperature because of the dependence of the fractionation factor on $1/T^2$ and are generally too small at temperatures in excess of about 800°C for reliable geothermometry. A unique advantage of isotope geothermometry is that the effect of pressure on isotopic fractionation is negligible in most systems, especially at pressures less than 10 kb. This is because the volumes of isotopically substituted molecules and crystals are not sensitive to pressure changes.

The best isotope geothermometers are those for which (i) the temperature dependence of the fractionation factor is large, (ii) the fractionations between the substances under consideration are large relative to the experimental errors of isotopic analysis, and (iii) it is unlikely that the isotopic compositions of the substances have changed after their formation. Hydrogen isotope fractionations are usually large, but their relatively small temperature dependence in mineral–H_2O systems, especially below 400°C, and the ease of isotope exchange between fluids and minerals even at low temperatures preclude their use in geothermometry. Similarly, carbon isotopic fractionation among carbonate minerals is too insensitive to temperature variations to be useful in geothermometry.

Commonly used isotope geothermometry involves oxygen and sulfur isotopes. An assemblage of *N* oxygen-bearing or sulfur-bearing minerals in equilibrium for which calibration curves or thermometric equations are available should give, in theory, *N*-1 oxygen or sulfur isotopic temperatures, although all of the possible pairs will not be suitable for geothermometry. For example, the quartz–feldspar pair, a very common assemblage in felsic igneous and metamorphic rocks, is not a useful geothermometer because $\Delta^{18}O_{qz\text{–}feldspar}$ values are small and rather insensitive to temperature variation, and feldspar is particularly susceptible to post-formational isotope exchange with hydrothermal fluids.

An essential condition for reliable isotopic geothermometry is that isotopic equilibrium was achieved between phases of interest at the time of their formation and has been maintained since then. A close agreement of temperatures determined from different pairs of minerals increases the reliability of the estimated temperature and provides evidence for isotopic equilibrium among the minerals in the assemblage, indicating their formation from the same fluid at fairly uniform temperature. A lack of concordance of the temperatures usually implies a lack of isotopic equilibrium among the minerals or noncontemporaneity of one or more minerals in the assemblage, assuming that the calibrations are accurate. Another indication of a lack of isotopic equilibrium is a reversal of the commonly observed order of isotopic enrichment among coexisting phases of an assemblage. For example, the order of decreasing ^{18}O-enrichment in a granite should be quartz > feldspar > biotite > magnetite; thus, a higher $\delta^{18}O$ value for feldspar compared to quartz would indicate a disequilibrium assemblage. Generally, oxygen isotopic equilibrium among minerals in a rock is strong evidence for chemical equilibrium in the rock. The energy required to break Al–O and Si–O bonds in silicate minerals and allow rearrangement towards isotopic equilibrium is sufficient to effect chemical equilibrium (O'Neil, 1986a).

11.3.1 *Oxygen isotope geothermometry*

One of the earliest applications of oxygen isotope thermometry was the estimation of temperatures of ancient oceans from measurements of oxygen isotope ratios in carbonate minerals (calcite or aragonite) that constitute remains of marine organisms or limestones of known geologic age (Urey, 1947; Craig, 1965). The basis of this thermometry is that the fractionation of oxygen isotopes between calcium carbonate minerals and ocean water at any given time is a function of the water temperature. Although sound in principle, this technique has not worked well except for post-Miocene $CaCO_3$ sediments. The main limitation is that, although the $\delta^{18}O$ value of ocean water has fluctuated significantly over geologic time, the variation is not a function of temperature only. The carbonate paleothermometry is further complicated by variations in the pH of seawater (Zeebe, 2001; Royer *et al.*, 2004), post-deposition changes in the isotopic composition of the carbonate minerals, and the biochemical fractionation of oxygen by shell-forming organisms.

Silicate and oxide minerals in igneous and metamorphic rocks equilibrate at relatively higher temperatures and therefore are not saddled with complications arising from kinetic isotopic effects that occur at lower temperature equilibration typical of sedimentary rocks. As reviewed by Taylor (1997), equations describing oxygen isotopic fractionation between many mineral–H_2O pairs and mineral–mineral pairs as a function of temperature can be found in a number of publications (e.g., Clayton and Kieffer, 1991; Mathews, 1994; Zheng, 1993a,b). A set of equations (called *reduced partition functions*) of the form

$$\begin{aligned} f_{\text{mineral–H}_2\text{O}} &= 1000 \ln \alpha_{\text{mineral–H}_2\text{O}} \\ &= \text{A}_{\text{mineral}}x - \text{B}_{\text{mineral}}x^2 + \text{C}_{\text{mineral}}x^3 \end{aligned} \quad (11.25)$$

for several major rock-forming minerals relative to H_2O is listed in Table 11.2. In these equations, A, B, and C are mineral-specific constants, and $x = 10^6/T^2$ (*T* in Kelvin). The equations are based on a combination of laboratory experiments and statistical thermodynamic calculations, and they yield good approximations for temperatures greater than 400 K.

At a given temperature, the equilibrium oxygen isotope fractionation between any two minerals included in Table 11.2 can be computed as the algebraic difference between the two corresponding $f_{\text{mineral–H}_2\text{O}}$ functions. For example, for the mineral pair quartz (qz)–magnetite (mt), which is widely

Table 11.2 Equations ($f_{mineral-H_2O}$) for selected individual minerals for oxygen isotope geothermometry.

$f_{mineral-H_2O} = 1000 \ln \alpha_{mineral-H_2O}$
$= A_{mineral} X - B_{mineral} X^2 + C_{mineral} X^3$
$x = 10^6/T^2$ (T in Kelvin)

Mineral	A	B	C
Quartz	12.116	0.370	0.0123
Calcite	11.781	0.420	0.0158
Albite	11.134	0.326	0.0104
Jadeite	10.319	0.287	0.0088
Muscovite	10.221	0.282	0.0086
Anorthite	9.993	0.271	0.0082
Zoisite	9.983	0.269	0.0081
Diopside	9.237	0.199	0.0053
Garnet	8.960	0.182	0.0047
Fosterite	8.326	0.142	0.0032
Rutile	6.960	0.088	0.0017
Magnetite	5.674	0.038	0.0003

Source of data: Taylor (1997, a compilation of data from Clayton and Kieffer, 1991; Mathews, 1994; O'Neil and Taylor, 1967, 1969).
The value of 1000 ln α for any mineral–mineral pair at a given temperature is the difference between the $f_{mineral-H_2O}$ values of the two minerals. For example, for the pair quartz (qz)–magnetite (mt), $D_{qz-mt} = 1000 \ln \alpha_{qz-mt} = f_{qz-H_2O} - f_{mt-H_2O}$. At $T = 1000$ K, using given values of the mineral-specific constants, $1000 \ln \alpha_{qz-mt} = 11.7583 - 5.6363 = 6.122$.

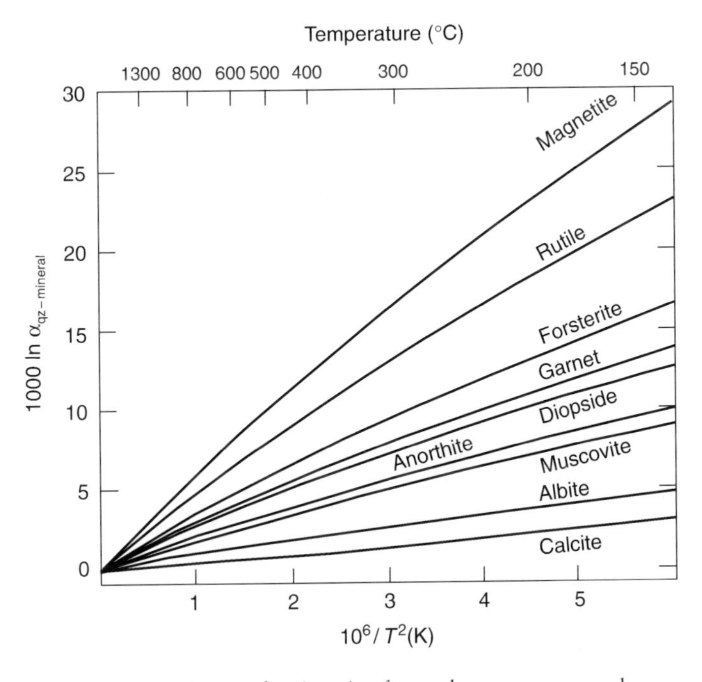

Fig. 11.2 Oxygen isotope fractionation factors between quartz and various silicate and oxide minerals as a function of temperature. The fractionation factors are all positive because at equilibrium ^{18}O is preferentially partitioned into quartz relative to all the other minerals. Calibration curves for mineral pairs were calculated directly from the respective $f_{mineral}$ functions given in Table 11.2, as illustrated for the rutile–magnetite pair in Example 11–2.

distributed in rocks and frequently used for geothermometry (see equation 11.16),

$$\Delta_{qz-mt} = 1000 \ln \alpha_{qz-mt} = 1000 \ln \alpha_{qz-H_2O} - 1000 \ln \alpha_{mt-H_2O} \quad (11.26)$$

which can be framed in terms of $f_{mineral-H_2O}$ functions as:

$$\begin{aligned} 1000 \ln \alpha_{qz-mt} &= 1000 \ln f_{qz-H_2O} - 1000 \ln f_{mt-H_2O} \\ &= (A_{qz}x - B_{qz}x^2 + C_{qz}x^3) - (A_{mt}x - B_{mt}x^2 + C_{mt}x^3) \end{aligned} \quad (11.27)$$

where A, B, and C are mineral-specific constants, and $x = 10^6/T^2$ (T in Kelvin). For the purpose of illustration, a set of such calculated calibration curves for several minerals relative to quartz is presented in Fig. 11.2. In principle, each such calibration curve can be used to determine graphically the isotopic equilibration temperature of the mineral pair from their measured oxygen isotope ratios.

Example 11–2: Calculation of oxygen isotope thermometric equation and calibration curve for the rutile (rut)–magnetite (mt) pair based on data in Table 11.2

For the rutile (rut)–magnetite (mt) pair,

$$1000 \ln \alpha_{rut-mt} = 1000 \ln \alpha_{rut-H_2O} - 1000 \ln \alpha_{mt-H_2O} \quad (11.28)$$

Expressing in terms of $f_{mineral-H_2O}$, and incorporating the values of the coefficients as listed in Table 11.2, we obtain the thermometric equation for the rutile–magnetite pair as

$$\begin{aligned} 1000 \ln \alpha_{rut-mt} &= (A_{rut}x - B_{rut}x^2 + C_{rut}x^3) - (A_{mt}x - B_{mt}x^2 + C_{mt}x^3) \\ &= (6.960x - 0.088x^2 + 0.0017x^3) \\ &\quad - (5.674x - 0.038x^2 + 0.0003x^3) \\ &= 1.286x - 0.05x^2 + 0.0014x^3 \end{aligned} \quad (11.29)$$

where $x = 10^6/T^2$ (T in Kelvin). To construct a $1000 \ln \alpha_{rut-mt}$ versus temperature calibration curve (Fig. 11.3) for the rutile–magnetite geothermometer, we calculate $1000 \ln \alpha_{rut-mt}$ for assumed values of temperature as listed below:

$x=10^6/T^2$	T (K)	$1000 \ln \alpha_{rut-mt}$	$x=10^6/T^2$	T (K)	$1000 \ln \alpha_{rut-mt}$
1	1000	5.63	4	500	22.11
2	707.11	11.20	5	447.21	27.46
3	577.35	16.69	6	408.25	32.74

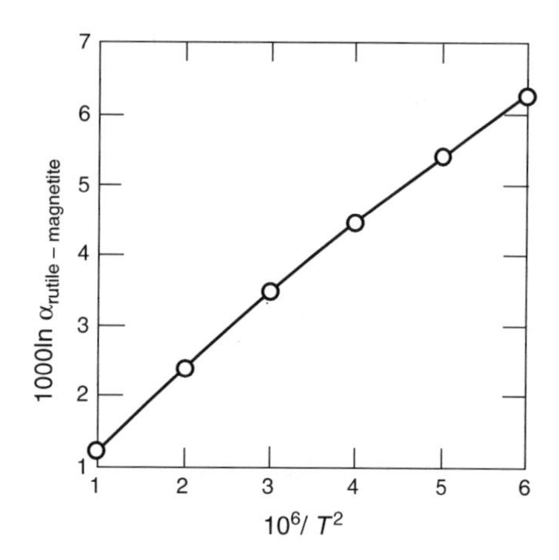

Fig. 11.3 Rutile–magnetite oxygen isotope geothermometer calculated using equations for $f_{\text{rutile-H}_2\text{O}}$ and $f_{\text{rutile-H}_2\text{O}}$ given in Table 11.2.

11.3.2 Sulfur isotope geothermometry

Equations for the commonly used sulfur isotope geothermometers, derived on the basis of a critical evaluation of the available theoretical and experimental data, are given in Table 11.3. The pyrite–galena pair is the most sensitive sulfide geothermometer, but temperatures obtained involving pyrite may show an appreciable spread because of the common occurrence of several generations of pyrite in a mineral assemblage. Sphalerite–galena is the most commonly used pair and in many cases it yields temperatures consistent with those obtained from fluid inclusion microthermometry. Inconsistent temperatures obtained from this geothermometer in some studies have been attributed to isotopic disequilibrium (Botinikov *et al.*, 1995). Even in an equilibrated assemblage, the temperature estimated may not represent the temperature of sulfide formation because sulfide minerals are prone to relatively rapid reequilibration.

11.4 Evaporation and condensation processes

Consider the fractionation of isotopes X1 and X2 in a system consisting of two substances A and B. The change in the isotopic composition of A and B with increasing fractionation will depend on the magnitude of the fractionation factor and whether the system is closed or open to A and B. To illustrate this, let us consider the fractionation of oxygen and hydrogen isotopes during evaporation of ocean water and condensation of the water vapor so formed.

Hydrogen has two stable isotopes (^{1}H and ^{2}H or D) and oxygen has three (^{16}O, ^{17}O, and ^{18}O). Thus, there are nine possible combinations for isotopically distinct water molecules, with masses ranging from 18 for $^1H_2{}^{16}O$ to 22 for $D_2{}^{18}O$. A water molecule containing the lighter isotope (^{1}H or ^{16}O) has a higher vibrational frequency and, therefore, weaker bonds compared to one containing the heavier isotope (D or ^{18}O) (see section 11.1.1). This results in a greater tendency of the light water molecules to escape into the vapor phase above a body of liquid water (i.e., $^1H_2{}^{16}O$ has a significantly higher vapor pressure than $D_2{}^{18}O$). Consequently, water vapor formed by evaporation of ocean water is enriched in ^{16}O and ^{1}H relative to the ocean water, whereas raindrops formed by condensation of water vapor are enriched in ^{18}O and D relative to the water vapor.

11.4.1 Evaporation of ocean water

Because ocean water (ow) is well mixed, it is considered to have a relatively uniform isotopic composition ($\delta D = +5$ to $-7‰$, $\delta^{18}O = +0.5$ to $-1.0‰$), with mean values very close to the defined values of the SMOW standard ($\delta D = 0‰$,

Table 11.3 Sulfur isotope geothermometers.

	Equation (T in Kelvin)	Uncertainties in the calculated temperature*	
Mineral pair in equilibrium (M1–M2)	$\Delta^{34}S = \delta^{34}S_{M1} - \delta^{34}S_{M2}$	**1**	**2**
Pyrite – galena	$T = \frac{(1.01 \pm 0.04) \times 10^3}{\sqrt{\Delta^{34}S}}$	± 25	± 20
Sphalerite (or pyrrhotite)– galena	$T = \frac{(0.85 \pm 0.03) \times 10^3}{\sqrt{\Delta^{34}S}}$	± 20	± 25
Pyrite – chalcopyrite	$T = \frac{(0.67 \pm 0.04) \times 10^3}{\sqrt{\Delta^{34}S}}$	± 35	± 40
Pyrite – sphalerite (or pyrrhotite)	$T = \frac{(0.55 \pm 0.04) \times 10^3}{\sqrt{\Delta^{34}S}}$	± 40	± 55

*1 = Uncertainty in estimated temperature from uncertainty in the equation (at 300°C).
2 = Uncertainty from the analytical uncertainty of ± 0.2‰ for Δ values (at 300°C).
Source of data: Ohmoto and Rye (1979).

$\delta^{18}O=0‰$). Larger variations exist, but only in special situations, such as seas in arid regions with restricted access to the open ocean.

For all practical purposes, the ocean is an infinite reservoir of water with constant $\delta^{18}O$ and δD values, which do not change because of evaporation of a very small fraction of the reservoir to water vapor (wv) or addition of a relatively small volume of meteoric water by rivers and precipitation. Thus, the ocean water is considered to be an open system with respect to liquid water. The system is also open with respect to water vapor because it escapes to the atmosphere as soon as it is formed instead of staying in equilibrium with water.

The values of isotopic fractionation factors for evaporation of water under equilibrium conditions at 25°C (298.15 K) are (Dansgaard, 1964):

$$\alpha^{ox}_{ow-wv} = \frac{(^{18}O/^{16}O)_{ow}}{(^{18}O/^{16}O)_{wv}} = 1.0092 \quad (11.30)$$

$$\alpha^{hy}_{ow-wv} = \frac{(D/H)_{ow}}{(D/H)_{wv}} = 1.074 \quad (11.31)$$

That the fractionation factor for both oxygen and hydrogen isotopes is > 1 is consistent with our expectation that the ocean water would be enriched in the heavier isotopes relative to water vapor. The values of $\delta^{18}O_{wv}$ and δD_{wv} at 25°C (298.15 K can be calculated using equation (11.10):

$$\alpha^{ox}_{ow-wv} = \frac{\delta^{18}O_{ow} + 1000}{\delta^{18}O_{wv} + 1000} = \frac{0 + 1000}{\delta^{18}O_{wv} + 1000}$$

$$\delta^{18}O_{wv} = \frac{1000}{\alpha^{ox}_{ow-wv}} - 1000 = \frac{1000}{1.0092} - 1000 = -9.12‰ \quad (11.32)$$

Similarly,

$$\delta D_{wv} = \frac{1000}{\alpha^{hy}_{ow-wv}} - 1000 = \frac{1000}{1.074} - 1000 = -68.90‰ \quad (11.33)$$

Deviations from these values in the open oceans are usually attributed to kinetic isotopic effects (Craig and Gordon, 1965).

11.4.2 *Condensation of water vapor*

The process of condensation of water vapor is the reverse of evaporation of ocean water. In contrast to ocean water, however, a mass of water vapor in the atmosphere is not a reservoir of constant isotopic composition. As condensation is dominantly an equilibrium process, the isotopic composition of the first raindrops falling out of a new mass of water vapor ($\delta^{18}O_{wv}=-9.12$; $\delta D_{wv}=-68.90$) over the ocean is very close to that of ocean water ($\delta^{18}O_{ow} \approx 0$; $\delta D_{ow} \approx 0$). However, the continuing preferential removal of ^{18}O and D from a mass of water vapor into liquid water causes the remaining water vapor to be progressively enriched in ^{16}O and ^{1}H (and depleted in ^{18}O and D). This, in turn, causes the rainwater in equilibrium with the water vapor to become progressively depleted in ^{18}O and D as well.

Thus, the water vapor (wv)–rainwater (rw) system is closed to the parent (wv), but open to the product species (rw) in the sense that it is continuously removed from the system. Isotopic fractionation in such a system can be modeled by the *Rayleigh distillation equation* (see also section 12.3.3):

$$R = R_0\, f^{(\alpha_{rw-wv} - 1)} \quad (11.34)$$

where R_0 is the initial $^{18}O/^{16}O$ or D/H ratio of the water vapor, R is the $^{18}O/^{16}O$ or D/H ratio of the remaining water vapor after precipitation, f is the atomic fraction of the water vapor remaining in the system with respect to the original amount at any particular extent of evolution of the system (i.e., $f_{wv}=1$ at the beginning condensation and $f_{wv}=0$ when all the water vapor has rained out), and α_{rw-wv} is the rainwater (product)–water vapor (parent) fractionation factor (i.e., $\alpha_{rw-wv}=R_{rw}/R_{wv}$).

Equations for calculating the oxygen and hydrogen isotopic compositions of the remaining water vapor and the rainwater in equilibrium with that vapor as a function of f are (see Box 11.2):

For oxygen isotopes,

$$\delta^{18}O^{remaining}_{wv} = (\delta^{18}O^{initial}_{wv} + 1000) f_{wv}^{(\alpha_{rw-wv} - 1)} - 1000 \quad (11.35)$$

$$\delta^{18}O_{rw} = (\delta^{18}O^{remaining}_{wv} + 1000)\ \alpha_{rw-wv} - 1000 \quad (11.36)$$

For hydrogen isotopes,

$$\delta D^{remaining}_{wv} = (\delta D^{initial}_{wv} + 1000) f_{wv}^{(\alpha_{rw-wv} - 1)} - 1000 \quad (11.37)$$

$$\delta D_{rw} = (\delta D^{remaining}_{wv} + 1000)\ \alpha_{rw-wv} - 1000 \quad (11.38)$$

Calculations of $\delta^{18}O^{remaining}_{wv}$ and $\delta^{18}O_{rw}$ as a function of f_{wv} are presented in Example 11-3 and the results are shown in Fig. 11.4.

Box 11.2 Equations for the oxygen and hydrogen isotopic compositions of rainwater and remaining water vapor as a function of progressive condensation at constant temperature

In essence, the exercise involves expressing R and R_0 in equation (11.34) in terms of δ values. For oxygen isotopes, expressions for R and R_0 from equation (11.9) are:

$$R_{0\,(wv)} = \left(\frac{\delta^{18}O_{wv}^{initial} + 1000}{1000}\right) R_{std};\ R_{wv} = \left(\frac{\delta^{18}O_{wv}^{remaining} + 1000}{1000}\right) R_{std} \quad (11.39)$$

Dividing,

$$\frac{R_{(wv)}}{R_{0\,(wv)}} = \frac{\delta^{18}O_{wv}^{remaining} + 1000}{\delta^{18}O_{wv}^{initial} + 1000} = f_{wv}^{(\alpha_{rw-wv} - 1)}$$

where f_{wv} is the fraction of the water vapor remaining, and α_{rw-wv} is the rainwater (rw)–water vapor (wv) fractionation factor (i.e., $\alpha_{rw-wv} = R_{rw}/R_{wv}$). Rearranging, we get the equation to calculate the $\delta^{18}O$ value of the remaining water vapor for a given value of f_{wv}:

$$\delta^{18}O_{wv}^{remaining} = (\delta^{18}O_{wv}^{initial} + 1000) f_{wv}^{(\alpha_{rw-wv} - 1)} - 1000 \quad (11.35)$$

The corresponding $\delta^{18}O$ value of rainwater can be obtained from the definition of α_{rw-wv}:

$$\alpha_{rw-wv} = \frac{R_{rw}}{R_{wv}^{remaining}} = \frac{\delta^{18}O_{rw} + 1000}{\delta^{18}O_{wv}^{remaining} + 1000}$$

Rearranging the terms, we get

$$\delta^{18}O_{rw} = (\delta^{18}O_{wv}^{remaining} + 1000)\ \alpha_{rw-wv} - 1000 \quad (11.36)$$

Equations for $\delta D_{wv}^{remaining}$ (equation 11.37) and δD_{rw} (equation 11.38) can be derived in a similar manner.

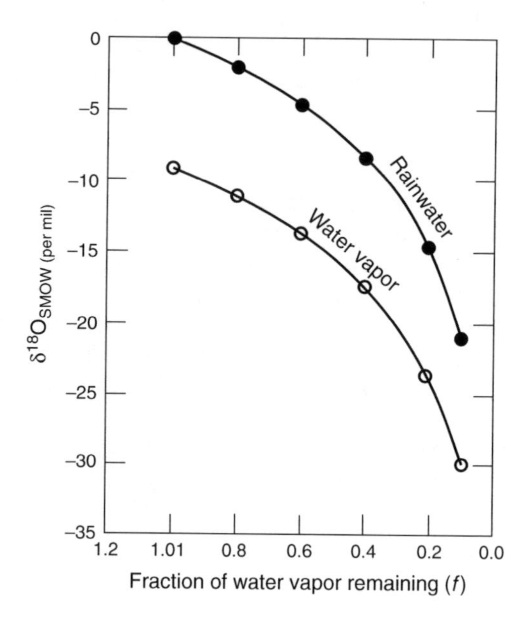

Fig. 11.4 Variation of $\delta^{18}O$ in course of condensation of water vapor ($\delta^{18}O_{wv}^{initial} = -9.1$) at 25°C, according to Rayleigh distillation equation. $\alpha_{rw-wv} = 1.0092$. The $\delta^{18}O$ value of the first condensate is taken as 0.0. The $\delta^{18}O$ value of rainwater is negative irrespective of what fraction of the water vapor condenses, and it becomes increasingly more negative as the condensation progresses.

Example 11–3: Calculation of the progressive change in d18O (water vapor) and in the corresponding d18O (rainwater) at 25°C (298.15 K)

Let us accept that the initial $\delta^{18}O$ value of the water vapor is 9.1‰ and $\alpha_{rw-wv} = 1.0092$ at 25°C. For any chosen value of f_{wv} ($f_{wv} \leq 1$), we can calculate $\delta^{18}O_{wv}^{remaining}$ using equation (11.35) and $\delta^{18}O_{rw}$ using equation (11.36). The results are tabulated below (Table 11.4) and presented in the form of graphs in Fig. 11.4.

Note that the $\delta^{18}O$ value of rainwater is negative irrespective of what fraction of the water vapor condenses, and it becomes increasingly more negative with increased condensation, which accounts for the fact that $\delta^{18}O$ values of meteoric waters (water recently involved in atmospheric circulation)

Table 11.4 Progressive change in $\delta^{18}O$ (water vapor) and in the corresponding $\delta^{18}O$ (rainwater) at 25°C (298.15 K) with increasing precipitation.

Fraction of water vapor remaining (f_{wv})	$\delta^{18}O_{wv}^{remaining}$ (per mil)	$\delta^{18}O_{rw}$ (per mil)	Fraction of water vapor remaining (f_{wv})	$\delta^{18}O_{wv}^{remaining}$ (per mil)	$\delta^{18}O_{rw}$ (per mil)
1.0	−9.1	0.0	0.5	−15.4	−6.4
0.9	−10.1	−1.0	0.4	−17.4	−8.4
0.8	−11.2	−2.1	0.3	−20.0	−11.0
0.7	−12.4	−3.3	0.2	−23.7	−14.7
0.6	−13.8	−4.7	0.1	−29.9	−21.0

are negative compared with ocean water. Similar calculations will show that δD values of meteoric waters are also negative relative to ocean water.

11.4.3 *Meteoric water line*

Because isotopic fractionation increases with decreasing temperature, δD and $\delta^{18}O$ values of meteoric water vary with latitude and elevation: the higher the latitude or the elevation, the more depleted is the meteoric water in ^{18}O and D. $\delta^{18}O$ values of less than −50‰ and δD values of less than −450‰ have been measured at the South Pole. The D/H fractionation, however, is proportional to the $^{18}O/^{16}O$ fractionation because condensation of H_2O in the Earth's atmosphere is essentially an equilibrium process. Thus, in a δD versus $\delta^{18}O$ plot, all meteoric waters around the globe fall close to a straight line (Fig. 11.5), called the *global meteoritic water line* (GMWL), which is defined by the equation (Craig, 1961):

$$\delta D = 8\,\delta^{18}O + 10 \tag{11.40}$$

The intercept of +10‰ for the GMWL is a consequence of the nonequilibrium evaporation process and is the reason why ocean water does not fall on the GMWL line. If evaporation were an equilibrium process, the intercept would be zero. Surface waters that have been affected by evaporation, hydration, or interaction with silicate and carbonate minerals may deviate from the global meteoric water line. For any location on the globe, we can establish a *local meteoric water line* (LMWL) by analyzing the isotope ratios of local precipitation (rain and snow) over a period of time. Equation (11.40), however, satisfactorily describes the slope and intercept of the meteoric water line on a global scale. The value of the intercept can vary substantially from one locality to another, mainly because of variability in evaporation; the slope of the meteoric water line is very close to 8.0 for all precipitation that has not experienced isotopic fractionation subsequent to condensation.

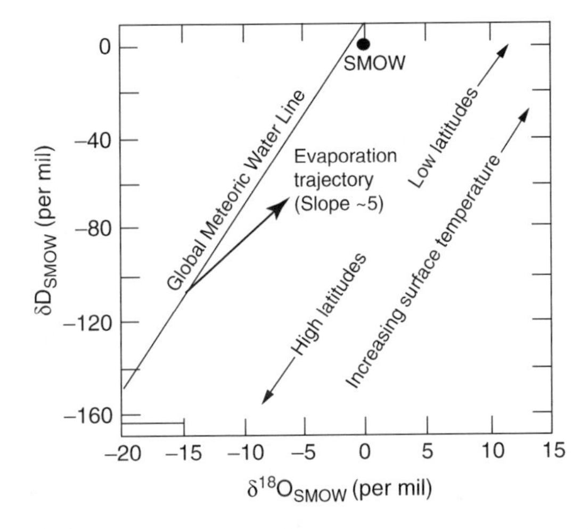

Fig. 11.5 A δD versus $\delta^{18}O$ plot showing the global meteoric water line (GMWL) as defined by equation (11.40) and a typical trajectory for deviations from the GMWL.

11.5 Source(s) of water in hydrothermal fluids

The term "hydrothermal fluid" refers to naturally occurring aqueous fluids involved in the dissolution, precipitation, and alteration of minerals. The waters of hydrothermal fluids may be classified into four end-member categories: (i) *meteoric* – water recently involved in atmospheric circulation (e.g., rainwater, water in rivers and lakes, ocean water); (ii) *connate* – formation water trapped in pores of sediments deposited in an aqueous environment; (iii) *metamorphic* – water produced by metamorphic dehydration reactions or water that equilibrated with metamorphic rocks at temperatures above about 300°C; and (iv) *magmatic* – water derived from magma, regardless of the ultimate source of the water.

A combination of $^{18}O/^{16}O$ and D/H ratios has proved very useful in tracing the origin and evolution of hydrothermal fluids, including fluids responsible for the formation of ore deposits (Taylor, 1974, 1997; Sheppard, 1986). $^{18}O/^{16}O$ and D/H ratios are affected somewhat differently in water–rock systems. Because of the extremely small amount of hydrogen in rocks (usually < 2000 ppm), the D/H ratio of a fluid undergoes negligible change by mineral–water exchange reactions. In contrast, the $^{18}O/^{16}O$ ratio of the fluid may be modified significantly unless water : rock ratios are very high or the isotopic exchange reactions are ineffective for kinetic or other reasons. The magnitude of the shift, which is generally to higher $\delta^{18}O$ values because of the water evolving towards oxygen isotopic equilibrium with ^{18}O–rich silicate and carbonate minerals, is determined by: (i) the ratio of the amount of oxygen in the exchangeable minerals to that in the fluid; (ii) the magnitude of mineral–H_2O fractionation factors (which takes into account the temperature of isotopic exchange); and (iii) the initial isotopic composition of the reacting minerals. Thus, the D/H ratio of a hydrothermal fluid is a better indicator of the source of the fluid, whereas its $^{18}O/^{16}O$ ratio may provide information about its subsequent evolution.

Two methods are used to determine D/H and $^{18}O/^{16}O$ ratios of hydrothermal fluids. Direct measurements are possible only in cases where the fluids can be sampled directly, such as connate brines, geothermal waters, and fluid inclusions. Isotopic compositions of fluids that are not amenable to direct sampling, as in the case of igneous and metamorphic rocks, are determined indirectly from mineral assemblages presumed to have been in equilibrium with the fluid. In essence, the indirect method involves the following steps: isotopic analysis of mineral assemblages, estimation of temperature of formation utilizing appropriate geothermometers, and calculation of D/H and $^{18}O/^{16}O$ ratios of fluids in equilibrium with assemblages at temperatures of their formation using appropriate mineral–H_2O fractionation factors. The uncertainty in δD and $\delta^{18}O$ values of fluids estimated by the indirect method arises largely from uncertainties in the temperatures of mineral deposition and in the values of isotopic fractionation factors. The indirect method is more commonly used, especially for $^{18}O/^{16}O$ ratios, because it is technically much easier. Because of the paucity of hydrogen-bearing minerals in many ore deposits, fluid inclusion analysis usually is the only direct method for obtaining information on the δD values of ore-forming fluids. The accuracy of determination of δD is typically an order of magnitude worse than for $\delta^{18}O$ (typically ±1.0‰ versus ±0.1‰). However, the natural variations in D/H are also much larger than for $\delta^{18}O$; so a 10‰ variation generally represents a very large $\delta^{18}O$ change but only a small δD change.

If the oxygen and hydrogen isotopic compositions and temperature of ancient ocean waters were comparable to present-day values, then ancient meteoric waters would be expected to follow a relation similar to GMWL. The variation of ocean water isotopic composition through geologic time is not known with any certainty, but indications are that their composition has been essentially the same at least during the Cenozoic and the Mesozoic (±1‰ for $^{18}O/^{16}O$ and ±10‰ for D/H) and possibly as far back as the Lower Paleozoic (Knauth and Roberts, 1991). $^{18}O/^{16}O$ ratios of ancient cherts suggest that Precambrian ocean water was depleted in ^{18}O by up to 20‰ (Perry 1967), but other studies have concluded that the $\delta^{18}O$ values of the ocean waters have remained relatively constant throughout much of the Proterozoic (Holmden and Muehlenbachs 1993), and even during the Archean (Gregory and Taylor, 1981; Beaty and Taylor, 1982; Gregory, 1991). Discussing the various approaches that have been used to resolve this question, Sheppard (1986) proposed that the isotopic compositions of ancient ocean waters, at least since about 2500 Ma, have stayed within narrow limits of about 0 to −3‰ for $\delta^{18}O$ and about 0 to −25‰ for δD, not radically different from their present-day values. Also, modeling of the interaction between oceanic crust and ocean water, the dominant process for regulating the oxygen isotope composition of the oceans (Muehlenbachs and Clayton, 1976), indicates that the $\delta^{18}O$ value of ocean water may vary by ±2‰, with a time response ranging from 5 to 50 Myr for mid-oceanic ridge expansion rates of 110 cm per year (Lécuyer and Allemand, 1999). Thus, it appears reasonable to use the present-day GMWL as a reference for evaluating fossil hydrothermal systems.

The isotopic compositions of different kinds of natural waters are summarized in Fig. 11.6. Most volcanic and plutonic igneous rocks have very uniform $\delta^{18}O$ values (+ 5.5 to +10.0‰) and δD values (−50 to −85‰); the field shown as "primary magmatic water" corresponds to the calculated isotopic composition of water in equilibrium with such normal igneous rocks at $t \geq 700°C$. The field of metamorphic water also represents calculated values for waters in equilibrium with metamorphic minerals at $t=300°$ to 600°C. Evidently, the fields for magmatic and metamorphic waters shown in Fig. 11.6 are approximate. The *kaolinite line* in Fig 11.6 represents the locus of isotopic data points for pure kaolinites from weathering zones formed in approximate equilibrium with meteoric water at surface temperatures (Savin and Epstein, 1970). It is parallel to the meteoric

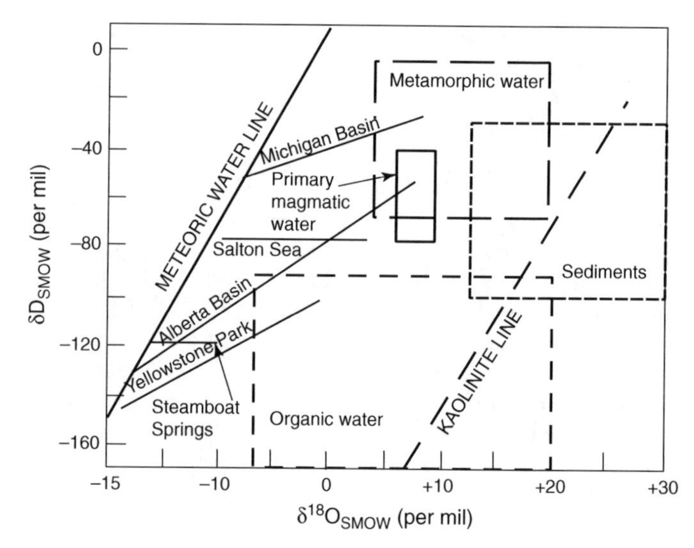

Fig. 11.6 Oxygen and hydrogen isotopic compositions of various hydrothermal waters relative to the SMOW standard. The global meteoric water line (GMWL) defines the compositions of meteoric water; the kaolinite line represents the isotopic compositions of kaolinites formed by chemical weathering. Also shown are trends for the Michigan Basin, the Alberta Basin, and the Yellowstone Park (Wyoming) geothermal brines, and the $\delta^{18}O$ shifts of the geothermal waters of Steamboat Springs (Nevada) and Salton Sea (California) relative to the isotopic composition of present-day meteoric water in these areas. (Sources of data: Taylor, 1974; White, 1974; Sheppard, 1986.)

water line, shifted in $\delta^{18}O$ and δD value by about 27‰ and 30‰, respectively, from the local meteoric water. The kaolinite line is potentially useful for discriminating clay minerals formed by chemical weathering from those formed by hydrothermal alteration. However, the interpretation of variations in clay mineral isotopic data is rather complicated because clay minerals may be composed of a mixture of detrital and authigenic components, and because particles of different ages may have exchanged isotopes to varying degrees (Hoefs, 1997).

Analyses of waters from active geothermal systems around the world have shown that these waters essentially are of local meteoric derivation. Some examples are shown in Fig. 11.6. In almost all cases, the hot water or steam shows a characteristic shift to higher $\delta^{18}O$ values as a result of isotopic exchange with silicate and carbonate minerals of the country rocks. In contrast, because of the extremely small initial hydrogen content of the rocks relative to the amounts of water involved, the δD values either remain almost identical to local meteoric waters (e.g., Steamboat Springs, Nevada, USA) or show a linear increase compared with local meteoric waters because of evaporation (e.g., Yellowstone Park, Wyoming, USA). Meteoric water is considered to be the main, if not the only, source of recharge for the Salton Sea geothermal brines, California (Craig, 1966; Clayton *et al.*, 1968). The 1.5 to 2‰ higher $\delta^{18}O$ in these waters relative to the meteoric water line is probably because of evaporation of meteoric waters in the desert environment prior to infiltration (White, 1974). On the basis of δD versus $\delta^{18}O$ trends, oilfield brines (formation waters), once thought to represent connate waters (seawater trapped in sediments), are now believed to contain a major component of circulating meteoric groundwater. Similar to surface meteoric waters, oilfield brines show a general increase in δD and, to a lesser extent, in $\delta^{18}O$ toward lower latitudes (e.g., the Michigan Basin brines versus the Alberta Basin brines, Fig. 11.6). Within a given sedimentary basin, however, the δD values generally increase with increasing $\delta^{18}O$ (and salinity). According to Taylor (1974) this trend is a result of one or more of the following processes: mixing of meteoric waters with waters of connate or other origin; isotopic exchange with clay minerals in the rocks; fractionation effects by processes such as membrane filtration; and reactions involving petroleum hydrocarbons.

11.6 Estimation of water: rock ratios from oxygen isotope ratios

Interaction of rocks with hydrothermal fluids results in a shift in $\delta^{18}O$ values, which can be used to estimate the water: rock ratio in the system during the alteration. The basis of this exercise is the mass balance equation pertaining to oxygen isotopes (Taylor, 1977, 1979):

$$w\delta^{i}_{\text{water}} + r\delta^{i}_{\text{rock}} = w\delta^{f}_{\text{water}} + r\delta^{f}_{\text{rock}} \tag{11.41}$$

The superscripts i and f refer, respectively, to the initial state (i.e., before water–rock interaction) and the final state (i.e., after interaction), w is the atom percent of meteoric water oxygen in the rock + water system, and r is the atom percent of exchangeable rock oxygen in the rock + water system.

Let us first consider the simplest model, a "closed" system in which the water is recyled over and over again through the rock until the rock attains the equilibrium value δ^{f}_{rock} at a certain temperature and then expelled from the system, and $\delta^{f}_{\text{water}}$ is determined by isotopic equilibration with the rock. The material-balance water: rock ratio (w/r) for this ideal "closed" system, integrated over its lifetime, can be calculated from the following relationship derived by rearranging equation (11.41):

$$\left(\frac{w}{r}\right)_{\text{closed}} = \frac{\delta^{f}_{\text{rock}} - \delta^{i}_{\text{rock}}}{\delta^{i}_{\text{water}} - (\delta^{f}_{\text{rock}} - \Delta_{r-w})} \tag{11.42}$$

where $\Delta_{r-w} = \delta^{f}_{\text{rock}} - \delta^{f}_{\text{water}}$ at the temperature of equilibration. This is the theoretical w/r ratio; the actual w/r ratio may be different depending on the efficiency of the exchange reaction.

At the other extreme is the "open" system model in which each infinitesimal packet of water makes only a single pass through the system and does not recycle. The equation for the integrated w/r ratio in this case is (Taylor, 1977):

$$\left(\frac{w}{r}\right)_{\text{open}} = \ln\left[\frac{\delta^{i}_{\text{water}} + \Delta_{r-w} - \delta^{i}_{\text{rock}}}{\delta^{i}_{\text{water}} - (\delta^{f}_{\text{rock}} - \Delta_{r-w})}\right] = \ln\left[1 + \left(\frac{w}{r}\right)_{\text{closed}}\right] \tag{11.43}$$

Note that for a given set of initial conditions and constant w/r ratio, δ^{f}_{rock} is determined solely by Δ_{r-w}, which is a function of temperature only. Conversely, for a given set of initial values and constant temperature, δ^{f}_{rock} is determined only by the w/r ratio. For both closed and open systems, we can calculate the w/r ratios as a function of δ^{f}_{rock} and temperature (for more details, see Taylor, 1977; Gregory *et al.*, 1989; Larson and Zimmerman, 1991; Nabelek, 1991). Both the models give minimum values of w/r ratio, because an appreciable amount of water may have moved through fractures in the rock without exchanging oxygen isotopes.

Uncertainties in the results obtained using equations (11.42) and (11.43) arise from the fact that several of the variables in them cannot be measured and, therefore, have to be estimated. δ^{i}_{rock} is estimated either on the basis of experience with similar rock types or from measured values of a sample far removed from the effects of hydrothermal alteration; $\delta^{i}_{\text{water}}$ is estimated either from the paleogeographic reconstruction of meteoritic water values or from a range of assumed values; and Δ_{r-w} from temperature of equilibration determined by some method of geothermometry. Moreover, because of the common lack of isotopic equilibrium among the various minerals of a rock, especially at low temperatures, we should calculate a separate w/r ratio for each mineral, and then combine these with modal abundances to calculate a bulk w/r ratio for the rock. A common practice for the sake of convenience is to assume that $\delta^{18}O^{f}_{\text{rock}}$ at equilibrium is equal to the $\delta^{18}O^{f}$ value of plagioclase feldspar, and calculate $\Delta^{18}O_{r-w}$ utilizing the plagioclase feldspar–H_2O geothermometer (O'Neil and Taylor, 1967):

$$1000 \ln \alpha_{\text{plag-water}} = \Delta^{18}O_{\text{plag-water}}$$
$$= -3.41 - 0.41\ \text{An} + (2.91 - 0.76\ \text{An})\left(\frac{10^6}{T^2}\right) \tag{11.44}$$

where An denotes the anorthite content of the plagioclase expressed as mole fraction, and T is the temperature of equilibration in Kelvin. This is a reasonable simplification because plagioclase is an abundant mineral in most rocks, and it exhibits the greatest rate of ^{18}O exchange with an external fluid phase.

Because the hydrogen concentration of a rock is only a fraction of the oxygen concentration, interaction with hydrothermal solution affects the isotopic composition of hydrogen much more strongly than that of oxygen. Even for very low w/r ratios, the δD values of the hydrous minerals change rapidly, while $\delta^{18}O$ values remain essentially unchanged. Appreciable change in $\delta^{18}O$ values occurs only when w/r ratios become quite high, by which time hydrogen in the hydrous phases would have completely equilibrated with the infiltrating fluid.

Example 11–4: Calculation of the water : rock ratio for a hydrothermally altered igneous body from the given oxygen isotope data, assuming that $\delta^{18}O^f_{rock}$ at equilibrium is equal to the $\delta^{18}O$ value of plagioclase (An_{30}) and 500°C is the temperature of isotopic equilibration.
$\delta^{18}O^i_{rock}$=+ 6.5‰; $\delta^{18}O^f_{rock}$=−4‰; and $\delta^{18}O^i_{water}$=−14 ‰

Substituting values in the plagioclase (An_{30})–H_2O geothermometer of O'Neil and Taylor (1967), equation (11.44) yields:

$$\Delta^{18}O_{plag-water}(\text{at } 500°C) = -3.41 - (0.41 \times 0.30) + [2.91 - (0.76 \times 0.30)]\left(\frac{10^6}{(500+273)^2}\right) = -3.533 + (2.682 \times 1.674) = 0.956$$

Assuming $\delta^{18}O_{plag} \approx \delta^{18}O^f_{rock}$, $\Delta^{18}O_{rock-water} = \Delta^{18}O_{plag-water} = 0.956$

For closed system:

$$\left(\frac{w}{r}\right)_{closed} = \frac{\delta^{18}O^f_{rock} - \delta^{18}O^i_{rock}}{\delta^{18}O^i_{water} - (\delta^{18}O^f_{rock} - \Delta^{18}O_{r-w})} = \frac{-4.0 - 6.5}{-14 - (-4 - 0.956)} = 1.16$$

For open system: $$\left(\frac{w}{r}\right)_{open} = \ln\left[1 + \left(\frac{w}{r}\right)_{closed}\right] = \ln[1 + 1.16] = 0.77$$

$$\delta^{18}O^f_{water} = \delta^{18}O^f_{rock} - \Delta^{18}O_{r-w} = -4 - 0.956 = -4.956 ‰$$

Thus, a decrease in $\delta^{18}O_{rock}$ from +6.5‰ to −4‰ is compensated by an increase in $\delta^{18}O_{water}$ from −14‰ to −4.96‰ for maintaing mass balance in the rock–water system.

11.7 Sulfur isotopes in sedimentary systems

There are two major reservoirs of sulfur on the Earth that have uniform sulfur isotopic compositions: the mantle, which has $\delta^{34}S \approx 0$‰ and in which sulfur is present primarily in the reduced form as sulfide (H_2S); and seawater, which has $\delta^{34}S$=+ 20‰ (at the present time) and in which sulfur is present in the oxidized form as dissolved sulfate (SO_4^{2-}). The main source of H_2S for the precipitation of sulfides in sediments is the reduction of dissolved seawater SO_4^{2-} by organic matter. Dissolved SO_4^{2-} may also be derived from connate water, and/or dissolution of sulfate minerals (mainly gypsum and anhydrite). The major sources of organic matter in sediments are crude oil, thermogenic natural gas and/or gas condensates, and microbial methane. A generalized, overall reaction for the reduction of sulfate by organic matter, which is accompanied by the oxidation of organic carbon, may be written as:

$$SO_4^{2-} + 2CH_2O \Rightarrow H_2S + 2HCO_3^- \quad (11.45)$$

where the organic matter is approximated as CH_2O. The reducing agent could also be H_2 or CH_4 generated by the breakdown of organic matter:

$$SO_4^{2-} + 2H^+ + 4H_2 \Rightarrow H_2S + 4H_2O \quad (11.46)$$

$$SO_4^{2-} + CH_4 \Rightarrow H_2S + CO_3^{2-} + H_2O \quad (11.47)$$

The sulfate reduction is either catalyzed by anaerobic, sulfate-reducing bacteria at relatively low temperatures and is referred to as *bacterial sulfate reduction* (BSR), or it occurs at relatively elevated temperatures without the involvement of bacteria and is referred to as *thermochemical sulfate reduction* (TSR). In either case, the product H_2S becomes enriched in ^{32}S relative to the parent SO_4^{2-} and the parent in ^{34}S relative to H_2S. The major reaction products and byproducts by BSR and TSR are identical. However, as discussed by Machel *et al.* (1995), petrographic, isotopic, and compositional data of these products and byproducts may permit identification of a BSR versus a TSR origin of the H_2S.

In most marine sediments, 75% to 90% of the H_2S (or HS^-, depending on the pH) produced by SO_4^{2-} reduction is reoxidized by reaction with dissolved oxygen in overlying porewater, or with Fe(III) or Mn(IV) within the sediment. The rest is incorporated in the sediment as sulfide minerals, predominantly as pyrite (FeS_2) by way of iron "monosulfide" precursors (with stoichiometry close to FeS), during early diagenesis when the sediments are still within about 1 m of the seawater–sediment interface, as long as the sediments contain enough reactive iron. When the supply of reactive iron is exhausted, H_2S (or HS^-) combines with organic matter to form organosulfur compounds. In typical shales, 10 to 30% of the reduced sulfur in sediments is fixed as organic sulfur and the rest as pyrite; in Fe-poor hosts, such as carbonate sediments, organically bound sulfur usually exceeds pyrite sulfur by a factor of more than two (Dinur *et al.*, 1980). The $\delta^{34}S$ value of organic sulfur tends to be higher than the coexisting pyrite by 1 to 10‰, and organic matter within a given stratigraphic unit may have a distinct $\delta^{34}S$ signature that would permit the use of sulfur species in identifying the source rocks and the migration paths of petroleum (Dinur *et al.*, 1980; Thode, 1981). In general, bacterial sulfate reduction is limited by the availability of metabolizable organic carbon, which typically amounts to about 15–45% of the total organic carbon in the sediment. Commonly, this relationship is reflected by a positive correlation between organic carbon and pyrite-bound sulfur in normal marine sediments (Goldhaber and Kaplan, 1975; Berner, 1985):

$$\text{Pyrite S (wt\%)} = 0.37 \times \text{organic carbon (wt\%)} \quad (11.48)$$

If the observed S : C mass ratio in the sediment is much greater than 1, diagenetic sulfide precipitation may not be the only or even the dominant source of sulfur in the sediment.

The $\delta^{34}S$ value of H_2S generated by sulfate reduction is determined by several factors: (i) the initial isotopic composition of the seawater sulfate; (ii) the mechanism of reduction (BSR or TSR) and factors affecting such reduction; (iii) the kind of sedimentary system (open or closed to SO_4^{2-} and/or H_2S); and (iv) oxidation of sulfides. Fractionation of sulfur isotopes associated with pyrite formation from dissolved sulfide in the laboratory has been reported to be less than 1‰ (Price and Shieh, 1979). Thus, for all practical purposes, sulfide minerals precipitated by reaction of metal complexes with H_2S, an important process in the formation of sulfide deposits in sedimentary environments, inherit the sulfur isotopic composition of H_2S.

11.7.1 *Bacterial sulfate reduction (BSR)*

Sulfur is an essential, although minor, constituent of living cells. A large and diverse group of microorganisms assimilate sulfur from the environment, most commonly as sulfate (but sometimes as sulfide or even as thiosulfate, polythionates and elemental sulfur), which is then reduced to sulfide for incorporation into the principal organosulfur compounds within cells. In this process, usually H_2S is not produced from sulfate in detectable amounts, except as a transient intermediate. This metabolic pathway, which requires input of energy, is known as *assimilatory sulfate reduction*, a complex biochemical process that generally is associated with small sulfur isotope fractionations, $\Delta^{34}S_{sulfate—organic\ sulfur}$ (= 1000 ln $\alpha_{sulfate—organic\ sulfur} \approx \delta^{34}S_{sulfate} - \delta^{34}S_{organic—S}$) being in the order of 3‰ (Canfield, 2001b). A much smaller, specialized group of microorganisms (prokaryotes), who gain energy for their growth by catalyzing thermodynamically favorable reactions such as (11.45) and (11.46), reduces sulfate in great excess of nutritional requirements and produces H_2S depending on the amount of organic material or H_2 dissimilated. This process of reduction of sulfate to sulfide is termed *dissimilatory sulfate reduction*, which is associated with large kinetic fractionations, the $\Delta^{34}S_{sulfate—sulfide}$ ($\approx \delta^{34}S_{sulfate} - \delta^{34}S_{sulfide}$) values ranging up to 46‰ (see section 11.2.2).

Sulfate-reducing bacteria, which utilize sulfate as a terminal electron acceptor to oxidize organic matter and produce H_2S, appear to have been active as early as ~3.47 Ga (Shen *et al.*, 2001). The bacteria are an ecologically diverse group and tend to develop wherever sulfate is present and the supply and decomposition of organic matter are sufficient to create anaerobic conditions (Trudinger *et al.*, 1985). The organisms, or evidence of their activities, have been detected in anoxic environments ranging in temperatures from −1.5°C to just over 100°C and ranging in salinities from freshwater to near halite saturation. Most of the described sulfate-reducers are mesophilic bacteria, which can tolerate temperatures from near 0°C to a maximum of about 40° to 60°C. At low temperatures (< 60°C), BSR is perhaps the only effective mechanism of sulfate reduction. The upper temperature limit of BSR has not been established experimentally, but empirical evidence indicates an upper limit of about 60° to 80°C, above which almost all sulfate-reducing bacteria cease to metabolize (Machel *et al.*, 1995). In general, individual sulfate reducers can metabolize only within a narrow temperature range of 20° to 40°C, and organisms with different temperature adaptation fractionate similarly. The wide range of temperatures over which BSR occurs in nature results from the involvement of a variety of organisms with overlapping temperature adaptations (Canfield, 2001a).

The major organisms known to be dissimilatory sulfate reducers are species of *Desulfovibrio*, especially *Desulfovibrio desulfuricans*, and certain Clostridia. Studies on sulfate reducers, largely based on *Desulfovibrio desulfuricans*, have led to the following general observations pertaining to sulfur isotope fractionations by BSR (Canfield, 2001b; Canfield *et al.*, 2006).

(1) When organisms utilize organic electron donors, the extent of fractionation tends to vary inversely as the specific rate of sulfate reduction. According to Goldhaber and Kaplan (1975), the ratio k_1/k_2 (see reaction 11.23) increases from 1.015–1.025 at relatively high rates of sulfate reduction to more than 1.065 at low rates of sulfate reduction. The value of 1.065 is close to the equilibrium value of 1.075 at 25°C, suggesting that the kinetic isotope effects are small in sedimentary environments with a slow rate of sulfate reduction. Recently, Canfield *et al.* (2006) also reported a positive correlation between isotope fractionation and sulfate reduction rate. Detmers *et al.* (2001) and Shen and Buick (2004), however, found no evidence for such correlation in pure cultures or in natural populations.
(2) Within an organism's growth range, specific rates of bacterial sulfate reduction increase with increasing temperature – by two orders of magnitude from about 0° to 40°C – and by the availability of oxidizable organic substrates (i.e., digestible foods) (Ohmoto and Goldhaber, 1997). Experimental study by Canfield *et al.* (2006), however, showed that the relationship is nonlinear: the largest fractionation occurred at low and high temperatures, and the lowest fractionation in the intermediate temperature range.
(3) Fractionation tends to be lower when H_2 is the electron donor, particularly at low specific rates of sulfate reduction.
(4) Fractionation is influenced by the concentration of sulfate. In studies of pure cultures and natural populations, and with abundant sulfate (> 1 millimolar), fractionation has been measured in the range of 5–46‰ (Shen and Buick, 2004), largely independent of absolute sulfate abundance. At low sulfate concentrations, the rate at which sulfate enters the cell is reduced, thereby reducing also the rate of sulfate reduction, and fractionations become very small (≤ 4‰) when sulfate concentration drops below about 1 mM. According to Habicht *et al.* (2002), the transition between high and low fractionation is sharply defined at around 200 μM sulfate concentration.

The generally accepted isotopic signatures of BSR are large spreads in $\delta^{34}S$ values of sulfides (dominantly pyrite) with abundant negative values, and large isotopic offsets between coeval sulfate and sulfide. Fractionation resulting from BSR can be up to −65‰ (i.e., the H_2S or the metal sulfides that originate from BSR could be 65‰ lighter than their parent sulfate) because of the formation of intermediate compounds, such as thiosulfate or polysulfides, as well as disproportionation of these compounds (Jorgensen *et al.*, 1990).

11.7.2 *Thermochemical sulfate reduction (TSR)*

There exists reliable thermometric data indicating that many sedimentary sulfide deposits formed at temperatures that might have been above the tolerance limits of sulfate-reducing bacteria. In such cases, nonbacterial, chemical reduction of sulfate by methane or associated organic matter might have provided H_2S required for the precipitation of sulfides (Barton, 1967).

The kinetic sulfur isotope fractionation by TSR, which is much less compared to that by BSR, decreases with increasing temperature (e.g., ~20‰ at 100°C, ~15‰ at 150°C, ~10‰ at 200°C). On the basis of available experimental data and theoretical considerations, Kiyosu and Krouse (1990) proposed the following equation to describe the temperature dependence of TSR:

$$1000\,(\alpha_{\text{sulfate-sulfide}} - 1) = 3.32\,(10^6/T^2) - 4.91 \qquad (11.49)$$

where $\alpha_{\text{sulfate-sulfide}}$ is the fractionation factor in per mil and T is the temperature in Kelvin. In some natural systems, $\delta^{34}S$ values of sulfides and source sulfates are consistent with those obtained by using equation (11.49) (see, for instance, Machel *et al.*, 1995), but in some deep sour gas reservoirs the parent sulfate and the associated sulfides have nearly the same $\delta^{34}S$ values. For example, the very positive $\delta^{34}S$ values of H_2S in many natural gas fields (about +20‰ or more; Thode and Monster, 1965), approaching that of the source evaporite sulfate, probably reflect the almost complete reduction of sulfate by organic matter at less than 200°C in a system closed to sulfate but open to H_2S (Ohmoto 1986; Krouse *et al.*, 1988).

The lower temperature limit of TSR is not well defined. Available experimental data suggest that nonbacterial reduction of sulfates takes place at sufficiently high rates in the laboratory at temperatures above about 175°C, although it may occur at temperatures as low as about 100°C; as has been interpreted, for example, for the Pine Point (Canada) lead–zinc deposit (Powell and Macqueen 1984). According to Machel *et al.* (1995), TSR is a late diagenetic process, which generally requires temperatures greater than 100°C, anoxic conditions, and the presence of hydrocarbons.

11.7.3 *Sulfur isotopic composition of seawater sulfate through geologic time*

Evaluation of the sulfur isotope data for marine sedimentary sulfides requires knowledge of the isotopic composition of the source sulfate, the dissolved sulfate in coeval seawater. The isotopic fractionation during inorganic precipitation of minerals from seawater is negligible, in the order of 0 to +2.4‰ only. This is the reason why the isotopic compositions of ancient evaporite (gypsum and anhydrite) deposits have been used to estimate the $\delta^{34}S$ values of coeval seawater sulfate through geologic time. Claypool *et al.* (1980) were the first to present a compilation of marine evaporite data spanning the Phanerozoic and Late Precambrian (Fig. 11.7). According to their estimate, the $\delta^{34}S$ value of ancient seawater sulfate fluctuated between approximately +10 and +35‰ and averaged around +17‰ over the past 1000 Myr. The main features of the sulfur isotope age curve are: a pronounced maximum ($\delta^{34}S \approx +30‰$) at the end of the Proterozoic; gradual decrease to a Permian minimum ($\delta^{34}S \approx +10‰$), with a minor excursion during the Devonian ($\delta^{34}S \approx +25‰$); and then an increase towards the value for modern oceanic sulfate ($\delta^{34}S \approx +21‰$). This compilation has been refined by addition of new evaporite

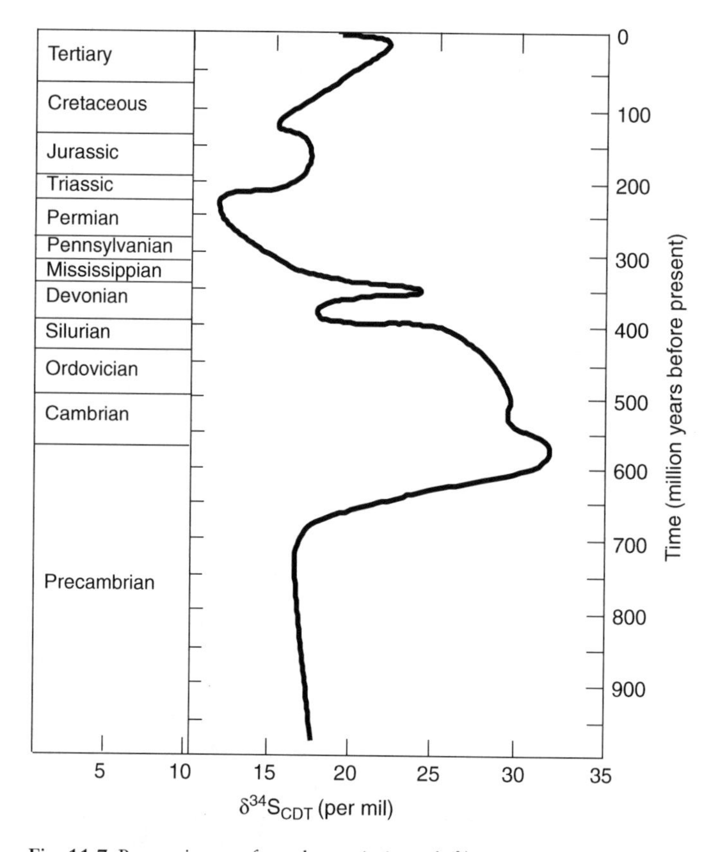

Fig. 11.7 Best estimate of secular variation of $\delta^{34}S_{CDT}$ of seawater based on evaporite derived sulfate–sulfur. The Proterozoic portion of the curve is based on very few data points, and there is a lack of data for the Archean because of the paucity of evaporite deposits. (After Bottrell and Newton, 2006, modified from Claypool *et al.*, 1980.)

data (e.g., Strauss, 1997, 1999), but the overall picture remains essentially the same. The secular variation in $\delta^{34}S$ has been attributed to fluctuations in the rate of production and burial of ^{34}S-depleted sedimentary sulfides, and the consequent enrichment of seawater sulfate in ^{34}S to maintain the overall mass balance of sulfur isotopes.

The evaporite record, however, suffers from some fundamental limitations. The rock successions do not contain a continuous record of evaporite deposits, either due to non-deposition or lack of preservation, and without fossils for biostratigraphic correlation the known evaporite deposits are not always well constrained in geologic age. An alternative approach followed recently by many workers (e.g., Kampschulte *et al.*, 2001; Lyons *et al.*, 2004; Kah *et al.*, 2004) uses trace sulfates (ranging in concentration from several hundred to a few thousand ppm) in biogenic limestones as a proxy for the seawater sulfate. Precipitation of calcium sulfate during precipitation of evaporites is associated with a very small isotopic fractionation ($\delta^{34}S = 0$ to +2‰; Holser and Kaplan, 1966), and the sulfate incorporation into the carbonate mineral lattice is not associated with any isotopic fractionation (Kampschulte *et al.*, 2001). Moreover, the carbonate-associated sulfate (CAS) gets excluded from the lattice of carbonate minerals during recrystallization without any significant sulfur isotope effect. Thus, CAS preserves the sulfur isotopic composition of contemporaneous seawater sulfate. CAS offers a better time resolution of the isotopic data because carbonate sequences provide a more continuous record than marine evaporite sulfate deposits and they also generally contain fossils that allow chronostratigraphic correlation. The CAS approach, for example, has resulted in a more reliable reconstruction of the isotopic composition of seawater sulfate during the Proterozoic for which evaporite sulfate data points are very limited (Bottrell and Newton, 2006).

11.7.4 *Open versus closed sedimentary systems with respect to sulfate and sulfide*

We adopt here the definitions of *open* and *closed* systems with respect to sulfate (SO_4^{2-}) and sulfide (H_2S) as discussed by Ohmoto and Goldhaber (1997). A sedimentary system can be considered open with respect to H_2S produced by reduction of SO_4^{2-} if the H_2S escapes to the overlying oxygenated sediments and water column or is fixed as iron sulfides and organic sulfur that, as a first order approximation, may be assumed to be nonreactive to SO_4^{2-}. In other words, in a system open to H_2S, the H_2S is removed from the system as soon as it is formed; if not, the system is closed with respect to H_2S. A system is considered open with respect to SO_4^{2-} when the supply rates of SO_4^{2-} are much faster than the rates of its reduction, irrespective of the absolute rate of sulfate reduction, so that the isotopic composition of SO_4^{2-} is maintained at a constant level for all practical purposes. Examples of open sulfate systems are a bioturbation zone (typically 0 to ~20 cm depth zone) in normal marine sediments, which is stirred by feeding activities of benthonic organisms, and the anoxic water body in euxinic basins. A system is considered closed with respect to SO_4^{2-} when the rate of sulfate reduction exceeds that of sulfate supply, so that the isotopic composition of the remaining sulfate changes as the reduction progresses and all sulfate may eventually be consumed within the system. Examples of closed sulfate systems are sedimentary sections beneath the bioturbation zone under oxic seawater and the sediments underlying the anoxic water column in euxinic basins.

In a system that is open to both SO_4^{2-} and H_2S, $\delta^{34}S_{SO_4}$ and $\delta^{34}S_{H_2S}$ values remain constant, despite variations in the sulfide content of the system, and so does the value of $\Delta_{SO_4-H_2S}$ (see equation 11.7). For example, assuming an average $\alpha_{SO_4-H_2S} = 1.025$ and $\delta^{34}S_{SO_4} = 20‰$ for seawater, bacterial isotopic fractionation would result in H_2S and precipitated pyrite having $\delta^{34}S \approx 5‰$. The relationship between $\delta^{34}S$ values of SO_4^{2-} and H_2S in a system that is closed to both and in which the two sulfur species are in equilibrium is shown in Fig. 11.8a. For a system closed to SO_4^{2-}, but open to H_2S, the fractionation can be described by the Rayleigh distillation model (equation 11.34) and is shown in Fig. 11.8b. Sulfides precipitated as a result of Rayleigh fractionation tend to be characterized by a large spread in $\delta^{34}S$ values, abundant negative values, and a progressive increase in $\delta^{34}S$ values toward the stratigraphic top in a sedimentary sequence (although the

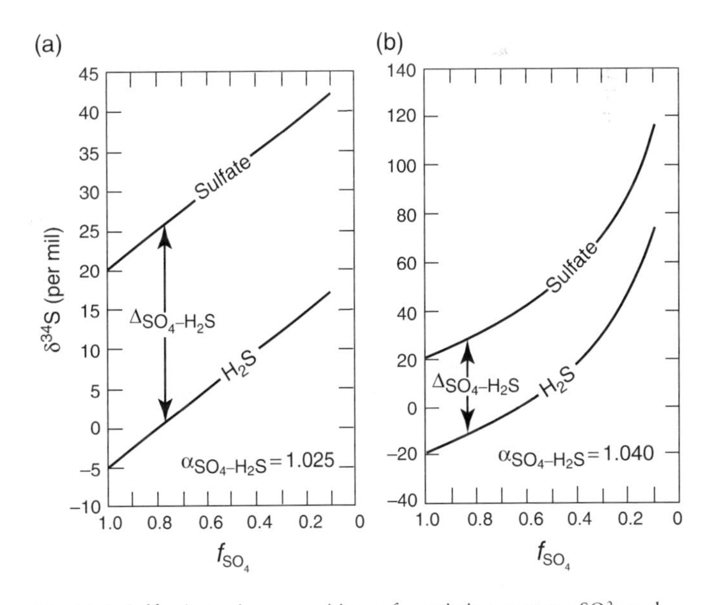

Fig. 11.8 Sulfur isotopic compositions of coexisting aqueous SO_4^{2-} and H_2S: (a) system closed to both SO_4^{2-} and H_2S, and in which the two sulfur species are in equilibrium; and (b) system closed to SO_4^{2-}, but open to H_2S (Rayleigh fractionation). f_{SO_4} denotes the atomic fraction of the initial SO_4^{2-} source remaining at a given instant of time. Note that in both cases, the SO_4^{2-} and H_2S are progressively enriched in ^{34}S relative to the starting sulfate composition as the reduction progresses toward completion, with a constant $\Delta_{SO_4-H_2S}$ in the case of (a) and decreasing $\Delta_{SO_4-H_2S}$ in the case of (b).

Box 11.3 Equations for sulfur isotope fractionation in selected sedimentary systems at constant temperature

(a) System open with respect to both SO_4^{2-} and H_2S

$\delta^{34}S_{SO4}$ value remains the same as the initial value and $\delta^{34}S_{H2S}$ attains a constant value at equilibrium given by (see equation. 11.10):

$$\delta^{34}S_{H2S} = \frac{\delta^{34}S_{SO4}^{initial} + 1000}{\alpha_{SO4-H2S}} - 1000 \quad (11.50)$$

(b) System closed with respect to both SO_4^{2-} and H_2S

As the fractionation progresses, values of both $\delta^{34}S_{SO4}$ and $\delta^{34}S_{H2S}$ change, which can be calculated using the following equations (Ohmoto and Goldhaber, 1997), but $\Delta_{SO4-H2S}$ remains constant at equilibrium:

$$\delta^{34}S_{H2S} = \delta^{34}S_{SO4}^{initial} - f_{SO4}(1000 \ln \alpha_{SO4-H2S}) \quad (11.51)$$

$$\delta^{34}S_{SO4} = \frac{\delta^{34}S_{SO4}^{initial} - \delta^{34}S_{H2S}(1 - f_{SO4})}{f_{SO4}} \quad (11.52)$$

(c) System closed with respect to SO_4^{2-} but open with respect to H_2S (Rayleigh fractionation; see Box 11.2)

As the fractionation progresses, values of both $\delta^{34}S_{SO4}$ and $\delta^{34}S_{H2S}$ change, which can be calculated using the following equations:

$$\delta^{34}S_{SO4}^{remaining} = (\delta^{34}S_{SO4}^{initial} + 1000) f_{SO4}^{(1/\alpha_{SO4-H2S}-1)} - 1000 \quad (11.53)$$

$$\delta^{34}S_{H2S} = (\delta^{34}S_{SO4}^{remaining} + 1000)(1/\alpha_{SO4-H2S}) - 1000 \quad (11.54)$$

In the above equations, $\delta^{34}S_{SO4}^{initial}$ represents the isotopic composition of the initial source sulfate, f_{SO4} the atomic fraction of SO_4^{2-} remaining in the system at any particular extent of the system's evolution relative to the original amount, $\delta^{34}S_{SO4}^{remaining}$ and $\delta^{34}S_{H2S}$ the corresponding isotopic compositions for any particular value of f_{SO4}, and $\alpha_{SO4-H2S}$ the fractionation factor between SO_4^{2-} and H_2S.

$\delta^{34}S$ value of the total sulfide mass would tend to be identical to the $\delta^{34}S$ value of the initial SO_4^{2-}). Equations to calculate $\delta^{34}S_{SO4}$ and $\delta^{34}S_{H2S}$ describing the relationships in the idealized systems mentioned above are given in Box 11.3. In natural sediments, however, there are no simple "open" or "closed" systems, and the interpretation of sulfur isotopic ratios is more complicated. Variable but typically negative $\delta^{34}S_{sulfide}$ values of sedimentary pyrites have generally been regarded as evidence of BSR (and Rayleigh fractionation), but this qualitative conclusion is insufficient to explain the observed variability in $\delta^{34}S_{sulfide}$ distribution in some individual stratigraphic units.

11.7.5 *Sulfur isotope ratios of sulfides in marine sediments*

A compilation of $\delta^{34}S$ values of sulfides in modern and ancient marine sediments over geologic time (Fig. 11.9) reveals several interesting features (Canfield and Teske, 1996; Habicht and Canfield, 2001a; Lyons *et al.*, 2004; Shen and Buick, 2004; Zahnle *et al.*, 2006):

(1) The $\delta^{34}S$ values of sulfides in modern and ancient marine sediments cover a wide range, from about −50‰ to about +40‰ (mostly to about +30‰).
(2) The sulfides of Archean age (> 2.5 Ga) are characterized by very low $\delta^{34}S$ values, within ±5‰ of the contemporaneous seawater sulfate (Habicht and Canfield, 1996, p. 342; Canfield *et al.*, 2000), but there is a sudden, marked increase at ~2.3−2.4 Ga, close to the beginning of the Paleoproterozoic.
(3) The next major increase in sulfur isotope fractionation occurs in the Neoproterozoic, at 0.7–0.8 Ga (Canfield, 2005). Phanerozoic sulfides are highly fractionated, the observed $\delta^{34}S$ values for the entire Phanerozoic ranging from about +40‰ to about −50‰. Modern sediments contain the most ^{34}S-depleted sulfides, with $\delta^{34}S$ values as low as −50‰ corresponding to a fractionation of about 70‰ relative to the present-day seawater sulfate ($\delta^{34}S$=+ 21‰). The 2.3–2.4 Ga and 0.7–0.8 Ga transitions in the sulfur

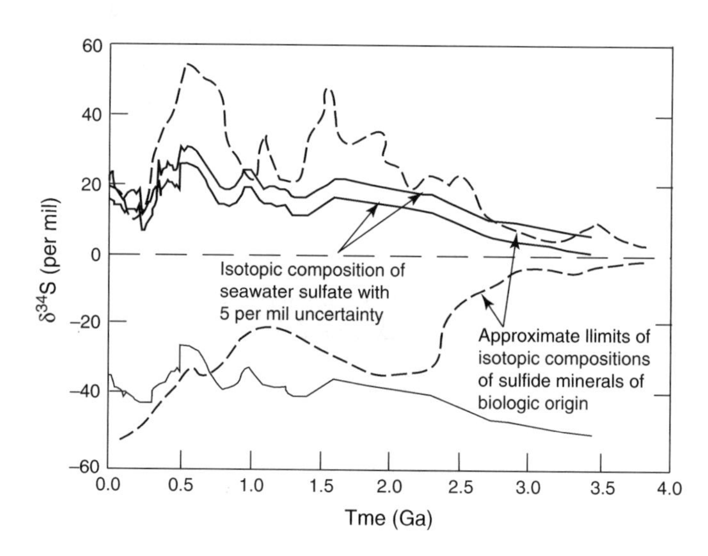

Fig. 11.9 The secular trends in the isotopic composition of seawater sulfate and sulfide minerals of likely biological origin over geological time. The band (double line) in the upper part of the figure represents the inferred isotopic composition of seawater sulfate (mostly anhydrite and gypsum), with 5‰ uncertainty, through geologic time. The single line in the lower part of the figure represents the isotopic composition of sulfate displaced by 55‰ to more ^{34}S-depleted values. Before 2.2 Ga the isotope composition of sulfate is determined mostly from barites, and constraints on the isotopic composition of seawater sulfate before 1.7 Ga are sparse. (Sources of data: Canfield, 1998; Canfield and Raiswell, 1999; Shen *et al.*, 2001.)

isotope record of sedimentary sulfides (Fig. 11.9) are broadly consistent with a two-step oxygenation of the Proterozoic atmosphere (see Chapter 13).

A combination of the secular variation in $\delta^{34}S$ of seawater sulfate Fig. 11.9 and the likely range of kinetic isotope effect (1.015–1.065) should result in a large range of fractionation (with mean $\delta^{34}S$ values in the range of approximately −50 to +15‰) in sedimentary sulfides produced by BSR. However, it appears that BSR cannot account for ^{34}S depletion (the difference between the isotopic composition of sulfate and sulfide) greater than about 46‰, the maximum ^{34}S depletion (the difference between the isotopic composition of the sulfate and sulfide) that has been observed in pure bacterial cultures and in natural populations from highly active microbial mats (Canfield and Teske, 1996; Habicht and Canfield, 1997). The large ^{34}S depletion found in some Phanerozoic sedimentary sulfides (up to 71‰; Canfield and Teske, 1996) is thought to have resulted from an initial fractionation by sulfate–reducing bacteria followed by one or more cycles of bacterially mediated oxidation of H_2S to intermediate sulfur species such as elemental sulfur (S^0), sulfite (SO_3^{2-}), and thiosulfate ($S_2O_3^{2-}$), and their disproportionation to ^{34}S-depleted SO_4^{2-} and H_2S (Jorgensen, 1990; Canfield and Teske, 1996; Habicht *et al.*, 1998). The *disproportionation reactions* (chemical reactions in which an element in one oxidation state in the reactants is converted into two or more oxidation states in the products) can be represented as follows:

$$S_2O_3^{2-} + H_2O \Rightarrow H_2S + SO_4^{2-} \quad (11.55)$$

$$4SO_3^{2-} + 2H^+ \Rightarrow H_2S + 3SO_4^{2-} \quad (11.56)$$

$$4S^o + 4H_2O \Rightarrow 3H_2S + SO_4^{2-} + 2H^+ \quad (11.57)$$

In bacterial cultures, the largest observed isotope fractionation has been found to be associated with the disproportionation of $S_2O_3^{2-}$–^{34}S depletion of 20–37‰ in H_2S and ^{34}S enrichment of 7–12‰ in sulfate (Habicht *et al*, 1998).

Sulfur isotope fractionation by BSR is reduced at low sulfate concentrations and high temperatures. Ohmoto and coworkers (e.g., Ohmoto and Felder, 1987; Kakegawa and Ohmoto, 1999) have argued that the very low $\delta^{34}S$ values of the Archean marine sulfides can be explained by high rates of bacterial sulfate reduction in a warmer ocean (temperature ~30° to ~50°C) that contained appreciable amounts of sulfate (> 1 mM kg^{-1} H_2O), with $\delta^{34}S_{SO_4}$ values of ~ +3‰. However, ^{34}S depletion observed in sulfides in modern natural microbial mats and in experimental runs – under conditions of abundant sulfate (up to 65 mM kg^{-1} H_2O), high rates of sulfate reduction (≥ 10 μmol cm^{-3} d^{-1}), and elevated temperatures (up to 85°C) – range from 13 to 28‰ (Habicht and Canfield, 1996; Canfield *et al.*, 2000), much higher than the ^{34}S depletion observed in Archean sedimentary sulfides. The very small sulfur isotope fractionation in Archean sedimentary sulfides could be the consequence of a nonbiogenic source of sulfur, if sulfur-reducing bacteria had not yet evolved, but sulfate-reducing bacteria were active in the ocean by at least about 3.47 Ga (Shen *et al.*, 2001). The favored interpretation is that BSR in Archean seawater occurred in an environment of low concentration of dissolved sulfate (< 1 mM compared to the present value of 28 mM) that mirrored a lack of oxidative chemical weathering of sulfides (mainly pyrite), the main source of dissolved SO_4^{2-} in seawater, because of an oxygen-deficient atmosphere. Canfield *et al.* (2000) calculated that the complete oxidation of 100 μm pyrite grains in soils (sedimentary pyrites are typically of this size or smaller) should occur with water containing > 1 μM O_2, which is the amount of dissolved O_2 that would be in equilibrium with 0.4% PAL (the *present atmospheric level*) at 25°C. Thus, the sulfur isotope record is consistent with very low atmospheric O_2 concentrations (< 0.4% PAL, as a rough estimate) before ~2.4 Ga. It is also consistent with a lack of sulfate–evaporite deposits in the Archean.

The jump in sulfur isotope fractionation at 2.4–2.3 Ga marked the rise in seawater sulfate concentration to > 1 mM, approximately coincident with the timing of the first stage of a rise in Earth's atmospheric oxygen, often referred to as the *Great Oxidation Event* (Holland, 2002). The fractionations during the interval between 2.4–2.3 Ga and 0.8–0.7 Ga are consistent with 40–45‰ as the upper limit of fractionation via BSR in an ocean containing > 1 mM dissolved sulfate. The dramatic increase in fractionation after ~0.8–0.7 Ga was probably caused by additional fractionations produced by disproportionation of sulfur-containing species (such as S^0, SO_3^{2-}, $S_2O_3^{2-}$) due to greater oxygen availability in the ocean–atmosphere system, leading to evolutionary development of nonphotosynthetic sulfide-oxidizing bacteria and enhanced biotic oxidation of sulfides to intermediate sulfur species (Canfield and Teske, 1996; Lyons *et al.*, 2004). According to Canfield and Teske (1996), atmospheric O_2 in excess of 5–18% PAL would have been high enough to substantially oxidize the surface seafloor in nearshore settings, thus accelerating the formation of intermediate sulfur compounds and the disproportionation process (Canfield, 2005). The two-step oxygenation of the Proterozoic atmosphere, corresponding to the two major transitions in the sulfur isotope record, is compatible with the $\delta^{13}C$ chemistratigraphy of sedimentary carbonates and organic matter (Des Marais *et al.*, 1992; Karhu and Holland, 1996; Kump and Arthur, 1999; Kah *et al.*, 2004). The evolution of the Earth's atmosphere and oceans will be discussed in more detail in Chapter 13.

11.8 Mass-independent fractionation (MIF) of sulfur isotopes

So far our discussion has been restricted to mass-dependent equilibrium and kinetic isotope fractionations of sulfur isotopes, fractionations that are governed primarily by the relative mass differences between isotope species. For sulfur isotopes, this dependence is indicated by highly correlated linear arrays on plots of $\delta^{33}S$ versus $\delta^{36}S$ and $\delta^{36}S$ versus $\delta^{34}S$ for terrestrial sulfide samples younger than 2000 Ma ($\delta^{33}S \approx 0.515\ \delta^{34}S$; $\delta^{36}S \approx 1.91\ \delta^{34}S$; Hulston and Thode, 1965) (Fig. 11.10a). *Mass-independent fractionation* (MIF), on the other hand, is characterized by isotopic compositions that do not plot on the mass-dependent fractionation line (Fig. 11.10b). The strategy for distinguishing between mass-dependent and mass-independent fractionation is similar for other stable isotope systems, such as iron isotopes (Beard and Johnson, 2004).

Strong evidence for an anoxic atmosphere during the Archean is provided by the fact that MIF of sulfur isotopes has been reported in sediments of Archean and Early Proterozoic age (> 2.45 Ga) but not in younger sediments (Farquhar *et al.*, 2000; Farquhar and Wing, 2003; Mojzsis *et al.*, 2003; Papineau and Mojzsis, 2006). Farquhar *et al.* (2001) showed experimentally that MIF-S occurs via gas-phase photochemical reactions such as *photolysis* (see section 13.1) of SO_2 and SO in the stratosphere (see Fig. 13.2) by UV radiation in the 190–220 nm spectral region, the region of wavelengths at which UV radiation is strongly absorbed by ozone and oxygen. Thus, the penetration of such shortwave radiation deep into the atmosphere to cause photolysis of SO_2 could have happened only when the column abundances of ozone and oxygen in the Earth's atmosphere were much lower than at present. Using a one-dimensional photochemical model, Pavlov and Kasting (2002) showed that in an essentially anoxic atmosphere with O_2 concentrations $<10^{-5}$ PAL (i.e., < 2 ppmv), sulfur would be removed from the atmosphere in a variety of oxidation states (from −2 as in H_2S to +6 as in H_2SO_4), each with its own distinct isotopic signature, and incorporated into sediments. On the other hand, in an atmosphere with O_2 concentrations $\geq 10^{-5}$ PAL, any signature of atmospheric MIF-S would be lost because of the oxidation of sulfur-bearing species to H_2SO_4 and subsequent rehomogenization of the sulfur isotopes in the oceanic reservoir (e.g., by mixing with magmatic sulfur with $\Delta^{33}S = 0$ exhaled through submarine volcanism) before being transferred into sediments.

Figure 11.11 is a plot of $\Delta^{33}S$ as a function of age of sedimentary sulfide and sulfate minerals, where the quantity $\Delta^{33}S$ reflects the deviation of measured $\delta^{33}S$ and $\delta^{34}S$ from the $\delta^{34}S - \delta^{33}S$ mass-dependent fractionation array (see Fig. 11.10a), the mass-dependent reference value, and is calculated as (Farquhar and Wing, 2003):

$$\Delta^{33}S = \delta^{33}S - 1000\left(\left(1 + \frac{\delta^{34}S}{1000}\right)^{0.515} - 1\right) \quad (11.58)$$

The $\Delta^{33}S$ record divides the Earth's history into three stages that reflect fundamental changes in the Earth's atmosphere as well as in its sulfur cycle through time (Farquhar and Wing, 2003). Stage I, characterized by a large range of $\Delta^{33}S$ values, represents an anoxic atmosphere. A low-O_2 atmosphere, $<10^{-5}$ PAL according to modeling by Pavlov and Kasting (2002), is necessary for the production of nonzero $\Delta^{33}S$ values, because ozone shields UV radiation at high O_2 levels, inhibiting the photochemical reactions that produce mass-independent fractionation. Low O_2 levels are also necessary for the preservation of nonzero $\Delta^{33}S$ values during the transfer of both oxidized and reduced sulfur species to the surface (Pavlov and Kasting, 2002) and for the retention of the $\Delta^{33}S$ differences between them in surface environments (Farquhar *et al.*, 2000). Stage II is marked

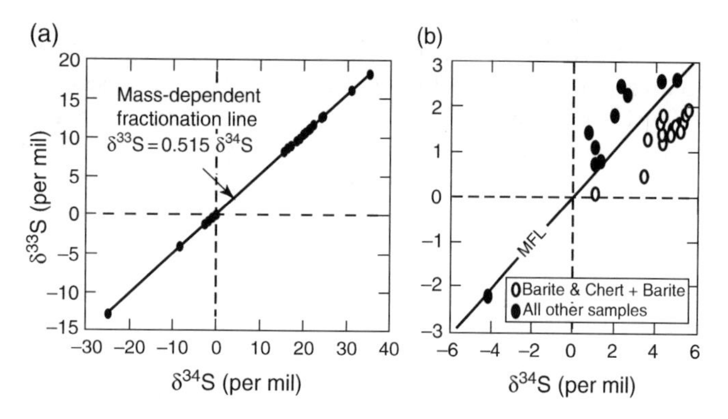

Fig. 11.10 Plot of $\delta^{33}S$ versus $\delta^{34}S$ for terrestrial sulfide (pyrite) and sulfate (barite) samples: (a) samples younger than 2000 Ma, which form a tightly constrained linear array, the mass dependent fraction line (MFL), defined by the relation $\delta^{33}S \approx 0.515\ \delta^{34}S$; and (b) samples older than 3000 Ma, which do not plot on the MFL and represent mass independent fractionation (MIF). (Sources of data: Farquhar *et al.*, 2000; Mojzsis *et al.*, 2003; Hu *et al.*, 2003. After Farquhar *et al.*, 2000, Fig. 3(B), p. 758; Farquhar *et al.*, 2001, Figure 1, p. 32830.)

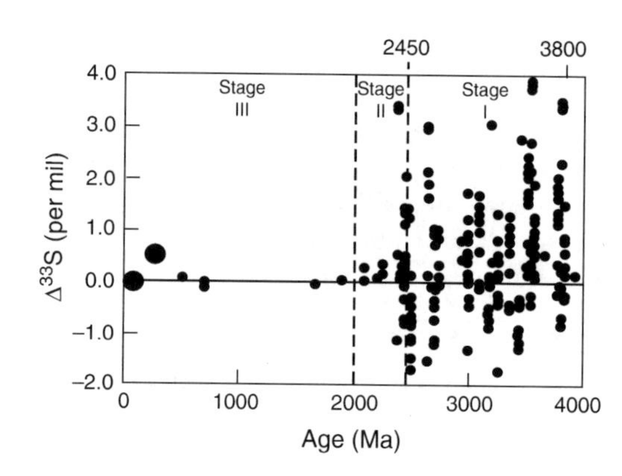

Fig. 11.11 Plot of $\Delta^{33}S$ versus age for terrestrial sulfide (pyrite) and sulfate (barite) samples. The large filled circle represents hundreds of analyses of samples younger than 2.0 Ga. The change from Stage I to Stage II is attributed to a change in the Earth's atmospheric chemistry. Photolysis reactions involving SO_2 and SO in the Earth's early atmosphere, coupled with an efficient transfer of the signature to the Earth's surface, produced this record. The smaller Stage II record may reflect the onset of oxidative weathering or it may reflect stabilization of atmospheric oxygen to intermediate levels (10^{-5}–10^{-2} PAL). (Sources of data: Farquhar *et al.*, 2000, 2001; Heymann *et al.*, 1998; Mojzsis *et al.*, 2003. After Farquhar and Wing, 2003 Figure 1, p. 3.)

by a smaller range of $\Delta^{33}S$ variability (−0.2 to +0.5‰) and is interpreted to indicate the beginning of a dramatic increase in the concentration of free O_2 in the atmosphere around 2.45 Ga. Stage III is characterized by near-zero $\Delta^{33}S$ values (generally much smaller than ±0.2‰) and represents a record of dominantly mass-dependent fractionation over the past 2 Gyr under an oxic atmosphere with increased concentration of atmospheric O_2, probably to > 1% PAL. The oxidation of Fe^{2+} in paleosols and the lack of detrital uraninite in sediments are consistent with atmospheric O_2 concentration of $\geq 10^{-2}$ PAL sometime after 2.3 Ga (Holland, 1984,1994) (see Chapter 13). Syngenetic pyrite from organic-rich shales of the 2.32 Ga Rooihoogte and Timeball Hill formations, South Africa, show no evidence of MIF-S, and large MIF-S signals do not reappear thereafter, indicating that the rise of atmospheric oxygen had occurred by 2.32 Ga (Bekker *et al.*, 2004). Thus the MIF-S record points to a significant rise in atmospheric O_2 between 2.45 and 2.32 Ga, and the transition to an oxygen-rich terrestrial atmosphere according to Farquhar and Wing (2003) was complete by 2 Ga.

A different interpretation was offered by Zahnle *et al.* (2006), who attributed the loss of MIF-S signature in sediments to a collapse of the atmospheric methane in the late Archean, not to a rise in the atmospheric oxygen. A key player in the MIF-S story was elemental sulfur (S_8), a major product of dissociation of SO_2 gas by solar UV radiation in an anoxic atmosphere. The insoluble S_8 particles rained out of the atmosphere, were incorporated in sediments, and preserved the MIF-S signal. Conditions conducive to the formation of S_8 are a source of sulfur (such as volcanic eruptions, at least as large as the volcanic SO_2 source today), an anoxic troposphere, and sufficient amount of a reduced gas (such as CH_4). According to Zahnle *et al.* (2006), a decrease in the atmospheric CH_4 from hundreds of ppmv to below 10 ppmv would eliminate the MIF-S signal. The methane collapse was driven by the increasing importance of dissolved sulfate in the oceans and the increasing competitive advantage of sulfate reducers over methanogens. This scenario would be consistent with the onset of low-latitude glaciation ("Snowball Earth") around ~2.4 Ga. The rise of oxygen to geologically detectable levels occurred after the loss of MIF-S, and was facilitated by generally low levels of atmospheric CH_4 characteristic of the ice ages.

11.9 Iron isotopes: geochemical applications

Transition metals are potentially well suited to be used as proxies for assessing the redox state of ancient oceans because their chemical transformations and speciation are closely coupled to the availability of oxygen in the environment. A promising element in this connection is iron because the oxidation of Fe(II) to Fe(III) in aqueous solutions is typically accompanied by significant fractionation of iron isotopes. A number of laboratories around the world are now investigating the application of iron isotopes to a variety of topics: the study of human blood, pathways of Fe cycling in modern oceans, redox cycling in ancient Earth, and differentiation of the Solar System. Our current understanding of the iron isotope geochemistry has been summarized in a number of recent review papers such as Anbar (2004), Beard and Johnson (2004), Anbar and Rouxel (2007), and Johnson *et al.* (2008a,b).

11.9.1 *Fractionation of iron isotopes*

Iron, the fourth most abundant element in the Earth's crust, has four naturally occurring stable isotopes: ^{54}Fe (5.84%), ^{56}Fe (91.76%), ^{57}Fe (2.12%), and ^{58}Fe (0.28%). The different behavior of Fe(II) compared to Fe(III), and the significant isotopic fractionations (1‰ or more in $^{56}Fe/^{54}Fe$ ratio) that are associated with redox conditions, suggest that Fe isotopes should be useful for tracing the geochemical cycling of Fe.

Iron isotope data are typically reported as $\delta^{56}Fe$:

$$\delta^{56}\text{Fe (per mil)} = \left(\frac{(^{56}\text{Fe}/^{54}\text{Fe})_{\text{sample}}}{(^{56}\text{Fe}/^{54}\text{Fe})_{\text{Ign Rocks}}} - 1 \right) 10^3 \quad (11.59)$$

which is of the same form as equation (11.3). $(^{56}Fe/^{54}Fe)_{Ign\ Rocks}$ is defined by the average of a wide spectrum of terrestrial igneous rocks that have a near-constant Fe-isotopic composition of $\delta^{56}Fe = 0.00 \pm 0.05$ (1σ)‰, irrespective of age, location, composition, and tectonic setting (Beard *et al.*, 2003). A standard commonly used for facilitating interlaboratory comparison of data is the Fe-metal isotope standard IRMM-014; the two reference frames for $\delta^{56}Fe$ are related as:

$$\delta^{56}\text{Fe}_{\text{Ign Rocks}} = \delta^{56}\text{Fe}_{\text{IRMM-014}} - 0.09‰ \quad (11.60)$$

Between two Fe-bearing species, A and B, the mass-dependent isotopic fractionation factor (α_{A-B}) and the isotopic fractionation ($\Delta^{56}Fe_{A-B}$), which may reflect equilibrium or kinetic isotope effects or a combination of the two, are related the same way as other stable isotopes:

$$\Delta^{56}\text{Fe}_{\text{A-B}} = 1000 \ln \alpha_{\text{A-B}} \approx \delta^{56}\text{Fe}_{\text{A}} - \delta^{56}\text{Fe}_{\text{B}} \quad (11.61)$$

Natural mass-dependent variations of $\delta^{56}Fe$ values, whether generated by abiotic or biotic processes, are small and cover a range of only ~4‰ (from ~ +1.0 to −3.0‰; Fig. 11.12), an order of magnitude smaller than the variations observed for $\delta^{34}S$. Fortunately, modern multiple collector inductively coupled plasma mass spectrometers (MC–ICP–MS) are capable of routinely measuring $\delta^{56}Fe$ values to a reported external precision of ±0.1‰ (2σ) in samples weighing < 1 μg, which is adequate to examine iron isotope fractionation in nature (Anbar, 2004, p. 225).

The most important controls on iron isotope fractionations in natural, low-temperature systems are the oxidation state of the species and the strength of their bonds. Molecular vibration theory predicts that iron isotopes may be fractionated during equilibrium (reversible) and disequilibrium (irreversible) abiotic

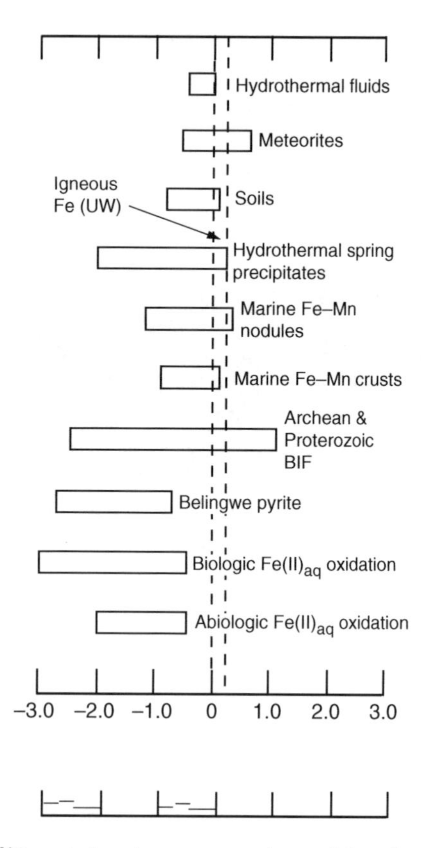

Fig. 11.12 $\delta^{56}Fe$ variations in some natural materials and experimental studies of Fe^{2+}_{aq} oxidation. The box for each category represents the maximum observed range. For California coastal samples, the Fe^{3+}/Fe_{total} ratio exceeds 0.3, reflecting an important influence by Fe^{3+}-reducing bacteria. (Sources of data: Archer and Vance (2006); compilations by Johnson *et al.* (2003), Anbar (2004), and Johnson and Beard (2006).)

reactions between minerals and fluids due to the relative stability of the heavier isotopes in more strongly bonded minerals and aqueous species. Under equilibrium conditions, aqueous Fe^{3+} or compounds that are composed entirely of Fe(III) (e.g., ferrihydride, represented in this discussion as $Fe(OH)_{3(s)}$, which eventually crystallizes to hematite, Fe_2O_3, without any further Fe isotope fractionation) have higher $\delta^{56}Fe$ values than those of mixed Fe^{3+}–Fe^{2+} oxidation state (e.g., magnetite, $Fe^{2+}O.Fe_2^{3+}O_3$), and aqueous species or minerals in which the iron is entirely Fe^{2+} (e.g., siderite, $FeCO_3$) tend to have the lowest $\delta^{56}Fe$ values. Exceptions to these general trends in equilibrium fractionation include aqueous species or minerals with covalently bonded Fe, such as pyrite, which has relatively high $\delta^{56}Fe$ values. As is the case with isotopes of other bioreactive elements (such as carbon, nitrogen, sulfur, and oxygen), Fe isotopes can be fractionated by microbes during both dissimilatory Fe(III) reduction (DIR), in which Fe(III) acts as an electron acceptor for respiration, and assimilatory iron metabolism, which involves uptake and incorporation of iron into biomolecules. Both processes result in depletion of the product in ^{56}Fe relative to ^{54}Fe.

Iron isotope variations observed in some natural materials and determined experimentally are shown in Fig. 11.12. Most of the Fe in the crust, including most igneous rocks as well as their alteration and weathering products, have a $\delta^{56}Fe$ value near zero (Johnson *et al.*, 2004b). So, measured nonzero $\delta^{56}Fe$ values for minerals and rocks have been produced by interaction with Fe-bearing fluids such as the hydrothermal fluids vented at mid-oceanic ridges. Modern MOR fluids have $\delta^{56}Fe$ values between −0.8‰ and −0.1‰, but the values were probably close to zero during the Archean because of higher heat flow (Johnson *et al.*, 2008a).

As summarized by Beard and Johnson (2004) and Johnson and Beard, (2006), experimental investigations have shown that the maximum fractionation between $Fe(OH)_{3(s)}$, the initial Fe^{3+}-precipitate formed by the two-step abiotic reaction, $Fe^{2+}_{aq} \Rightarrow Fe^{3+}_{aq} \Rightarrow Fe(OH)_{3(s)}$, and Fe^{2+}_{aq} (i.e., $\Delta_{Fe(OH)3(s)-Fe^{2+}_{aq}}$) is +3.0‰ at room temperature (i.e., $\alpha_{Fe(III)-Fe(II)}$ ~1.0030), reflecting a Fe^{3+}_{aq}–Fe^{2+}_{aq} fractionation of +2.9‰ and a very small $Fe(OH)_{3(s)}$–Fe^{3+}_{aq} fractionation of +0.1‰. The equilibrium $Fe(OH)_{3(s)}$–Fe^{3+}_{aq} fractionation is very small because the reaction does not involve a change in the oxidation state or coordination of Fe. The $Fe(OH)_{3(s)}$–Fe^{2+}_{aq} fractionation observed in nature is much less, generally between ~ +0.9 to +1.6‰, representing the sum of an equilibrium fractionation of +2.9‰ and a kinetic fractionation of −1.3‰ to −2.0‰ upon precipitation depending on precipitation kinetics. Thus, partial oxidation would produce ferric oxide/hydroxide precipitates that have positive $\delta^{56}Fe$ values, controlled by the net $Fe(OH)_{3(s)}$–Fe^{2+}_{aq} fractionation factor and the extent of reaction, whereas complete oxidation would produce no net change in the $\delta^{56}Fe$ value relative to that of Fe^{2+}_{aq}.

11.9.2 *Abiotic versus biotic precipitation of Fe minerals in banded iron formations*

Significant variations in Fe-isotope compositions of rocks and minerals appear to be restricted to chemically precipitated sediments, such as the Precambrian banded iron formations (BIFs; see Box 13.3). The Fe-isotope ratios in BIFs span almost the entire range yet observed on Earth because some of the largest Fe-isotope fractionations occur between Fe^{2+} and Fe^{3+} species involved in the precipitation of iron oxide and carbonate minerals in BIFs. The precipitation of hematite (Fe_2O_3) involves hydrothermal transport of dissolved Fe^{2+}, oxidation of Fe^{2+}_{aq} to Fe^{3+}_{aq}, and subsequent precipitation of Fe^{3+}_{aq} as $Fe(OH)_{3(s)}/Fe_2O_3$. The formation of magnetite (Fe_3O_4) and siderite ($FeCO_3$) requires reduction of Fe^{3+}_{aq} to Fe^{2+}_{aq}. Both the oxidation of Fe^{2+}_{aq} and the reduction of Fe^{3+}_{aq} can occur by abiotic or biotic processes (see Box 13.3).

Based on the hypothesis that Fe isotope fractionation by bacterially catalyzed kinetic processes would produce much larger effects compared to inorganic fractionation, particularly at equilibrium, much of the initial research on Fe isotopes was prompted by the expectation that Fe-isotope compositions of Fe-bearing minerals could be used to differentiate between biotic and abiotic precipitation of $Fe(OH)_{3(s)}/Fe_2O_3$. So far, this approach has not met with conclusive results because the overall $Fe(OH)_{3(s)}/Fe_2O_3$–Fe^{2+}_{aq} fractionation, as determined by experiments (Bullen *et al.*, 2001; Croal *et al.*, 2004; Balci *et al.*, 2006; Staton *et al.*, 2006), appears to be similar regardless of the oxidative pathway involved. Thus, Fe isotopes, by themselves, do not appear to provide a clear discrimination between biotic and abiotic pathways that involve oxidation.

Detailed experimental studies indicate that bacterial iron reduction (BIR) produce low $\delta^{56}Fe$ values, $\Delta^{56}Fe_{Fe^{2+}_{aq}-Fe(OH)3(s)} = -3.0‰$, whereas $\Delta^{56}Fe_{Fe3O4(s)-Fe^{2+}_{aq}} = +1.3‰$ and $\Delta^{56}Fe_{FeCO3(s)-Fe^{2+}_{aq}} = -0.5‰$ (Johnson *et al.*, 2005, 2008b). Thus, negative $\delta^{56}Fe$ values of magnetite or siderite probably are a signature of biotic reduction of $Fe(OH)_{3(s)}/Fe_2O_3$ to Fe^{2+}_{aq}. A combination of low $\delta^{56}Fe$ values and covariation of $\delta^{56}Fe$ and $\delta^{34}S$, as has been documented for the sedimentary pyrite from the Belingwe greenstone belt, Zimbabwe (2.7 Ga), would provide strong evidence in favor of microbial Fe(III) reduction. This is because such covariation is most readily explained in terms of coupled Fe(III) and SO_4^{2-} reduction during microbial organic matter degradation, as occurs in modern anoxic sediments (Archer and Vance, 2006).

11.10 Summary

1. Some of the important applications of stable isotope geochemistry to natural geochemical systems include: geothermometry; evaluation of the sources of the stable isotopes in minerals, rocks, and fluids; recognition of bacterially mediated reactions in oxidation–reduction reactions, and evolution of the crust–ocean–atmosphere system through geologic time.
2. Stable isotope ratios of samples are reported as δ values, in per mil (‰), relative to appropriate reference standards: For example, $\delta^{18}O_{sample}$ is defined as:

$$\delta^{18}O_{sample}(\text{per mil}) = \frac{(R_{sample} - R_{std})}{R_{std}} \times 10^3$$

where R_{sample} is the atomic ratio of the heavy to the light isotope in the sample (i.e., $^{18}O/^{16}O$), and R_{std} is the corresponding ratio in the reference standard.
3. The fractionation factor (α_{A-B}), the degree of isotopic fractionation between two coexisting phases A and B (Δ_{A-B}) in isotopic equilibrium, is defined as $\alpha_{A-B}=R_A/R_B$, and can be calculated from the equation

$$\alpha_{A-B} = \frac{R_A}{R_B} = \frac{\delta_A + 1000}{\delta_B + 1000}$$

or from the approximate relationship

$$1000 \ln \alpha_{A-B} \approx \delta_A - \delta_B = \Delta_{A-B}$$

if values of both Δ and δ_i are about ±10‰ or less. Between coexisting molecules the heavier isotope is preferentially partitioned into the one in which it can form stronger bonds.
4. Effects of isotopic fractionation are generally evaluated as: (i) equilibrium isotopic effects, produced by equilibrium isotope exchange reactions, which are independent of the pathways or mechanisms involved in the achievement of equilibrium; and (ii) kinetic isotopic effects, produced by unidirectional processes or unequilibrated chemical reactions (e.g., bacterially mediated reduction of SO_4^{2-} to H_2S), which depend on reaction mechanisms and possible intermediate products.
5. Stable isotope geothermometry is based on the variation of isotopic fractionation as a function of temperature; the effect of pressure is negligible in most systems, especially at < 10 kbar.
6. The progress of isotopic fractionation with time in a system depends on whether the system is closed or open to the parent and product species under consideration. For example, isotopic fractionation in a system such as water vapor (wv)–rainwater (rw), which is closed to the parent (wv) but open to the product species (rw), can be modeled by the Rayleigh distillation equation:

$$R = R_0 f^{(\alpha-1)}$$

7. The sulfate reduction of SO_4^{2-} to H_2S is either catalyzed by anaerobic, sulfate-reducing bacteria at relatively low temperatures (bacterial sulfate reduction, BSR), or it occurs at relatively elevated temperatures without the involvement of bacteria (thermochemical sulfate reduction, TSR). In either case, the product H_2S becomes enriched in ^{32}S relative to SO_4^{2-}.
8. The available record of $\delta^{34}S$ values of sulfides over geologic time appears to broadly correlate with a two-step oxygenation of the Proterozoic atmosphere at ~2.2–2.3 Ga and ~0.7–0.8 Ga.
9. Strong evidence for an anoxic atmosphere during the Archean is provided by the fact that mass-independent fractionation (MIF) of sulfur isotopes has been reported in sediments of Archean and Early Proterozoic age (> 2.3 Ga) but not in younger sediments.
10. Fe isotopes do not seem to provide a clear discrimination between biotic and abiotic pathways involving oxidation for the precipitation ferric hydroxides and oxides in banded iron formations (BIFs). Negative values of $\delta^{56}Fe$ for magnetite and siderite, on the other hand, appear to indicate biologic iron reduction (BIR).

11.11 Recapitulation

Terms and concepts

Assimilatory sulfate reduction
Anoxic atmosphere
Bacterial sulfate reduction (BSR)
Delta notation
Disproportionation reaction
Dissimilatory sulfate reduction
Great oxidation event
Isotope geothermometry
Isotopic effects (equilibrium and kinetic)
Isotopic fractionation (mass-dependent and mass-independent)
Isotopic fractionation factor
Kaolinite line
Meteoric water line (global and local)
Open and closed systems
Oxic atmosphere
Rayleigh distillation equation
SMOW
Thermochemical sulfate reduction (TSR)

Computation techniques

- Calculation of fractionation factor from δ values, and of $\delta_{product}$ from δ_{parent}, using fractionation factor.
- Calculations pertaining to isotopic stable isotope geothermometry (oxygen and sulfur).
- Calculations for fractionations in open and closed systems, including calculations using the Rayleigh distillation equation.

11.12 Questions

1. Hydrogen has two stable isotopes and oxygen has three. List the combinations for isotopically distinct water molecules, along with their masses.
2. The relative atomic abundance of oxygen and carbon isotopes are:

   ^{16}O: ^{17}O: ^{18}O = 99.759: 0.0374: 0.2039; and ^{12}C: ^{13}C = 98.9:1.1

   Calculate the relative abundances of (i) masses 44.0, 45.0 and 46.0 in CO_2, and (ii) masses 28.0, 29.0, and 30.0 in CO.
3. Calculate the $\delta^{13}C$ value of $CaCO_3$ precipitated in equilibrium with atmospheric CO_2 gas at 20°C, given that $\alpha_{CaCO_3-CO_2} = 1.01017$ and $\delta^{13}C_{CO_2} = -7.0‰$. What would be the $\delta^{13}C$ value of calcite precipitated at 20°C in isotopic equilibrium with atmospheric CO_2 if the $\delta^{13}C$ value of atmospheric CO_2 decreases to −12‰ due to increased burning of fossil fuels?
4. Consider a hypothetical sediment composed of 15 mol% of detrital quartz ($\delta^{18}O$=+12‰) and 85 mol% marine calcite ($\delta^{18}O$=+ 28‰). Determine the oxygen isotopic composition of each mineral after metamorphic equilibration in a closed system at 500°C. $\alpha_{SiO_2-CaCO_3} = 1.01017$ at 500°C.
5. An air mass whose initial δD value is −71.4‰ at 25°C starts to condense into rain. Plot graphs showing the progressive change in the δD value of water vapor and the δD value of rainwater in equilibrium with that water vapor at 25°C. Assume $\alpha_{rainwater-water\ vapor}$=1.074.
6. The fractionation of oxygen isotopes for the muscovite (mus)–water system as a function of temperature in the 500–800°C range can be described as (Bottinga and Javoy, 1973, p. 257):

   $$1000 \ln \alpha_{musc-water} = 1.90\ (10^6/T^2) - 3.10$$

   What is the $\delta^{18}O$ value of muscovite in equilibrium with water having $\delta^{18}O$=−9‰ at 500°C?
7. A mineralized quartz vein contains rutile (rut) in equilibrium with quartz. Derive a quartz–rutile geothermometry equation and calculate the quartz–rutile equilibration temperature if $\delta^{18}O_{qz} - \delta^{18}O_{rut}$=−4.60 per mil. Equations describing the variation of mineral–water fractionation factors with temperature are (Friedman and O'Neil, 1977):

   $$1000 \ln \alpha_{qz\text{-}water} = 2.51\ (10^6/T^2) - 1.46\ (500\text{–}750°C)$$

   $$1000 \ln \alpha_{rut\text{-}water} = 4.10\ (10^6/T^2) + 1.46\ (575\text{–}775°C)$$
8. Plot graphs showing the variaton of $\delta_{mineral-water}$ as a function of temperature (t=0 to 225°C) for calcite, quartz, and phosphate (calcite–water data relative to PDB standard; quartz–water and phosphate–water data relative to SMOW standard):

   (a) calcite–water: $t\ (°C) = 16.9 - 4.2\ (\delta_{calcite} - \delta_{water}) + 0.13\ (\delta_{calcite} - \delta_{water})^2$

   (b) quartz–water: $t\ (°C) = 169 - 4.1\ \{(\delta_{quartz} - \delta_{water}) + 0.5\}$

   (c) phosphate–water: $t\ (°C) = 111.4 - 4.3\ \{(\delta_{phosphate} - \delta_{water}) + 0.5\}$

   Comment on (i) the temperature dependence of the fractionation for all the three minerals, and (ii) the suitability of these mineral pairs as paleothermometers.
9. Given below are $\delta^{18}O$ and δD values of rainwater and snowmelt samples collected over a period of time at a certain locality in western USA. Plot the data and determine the linear regression equation for the local meteoric water line. How does it compare with the global meteoric water line of Craig (1961)?

δ18O ‰	δD ‰	δ18O ‰	δD ‰	δ18O ‰	δD ‰	δ18O ‰	δD ‰
−8.4	−55	−5.57	−23	−5.7	−21	−18.1	−120
−6.8	−44	−5.7	−21.5	−7.7	−43	−21.75	−163
−8.3	−55	−7.75	−42.5	−5.5	−25	−16.05	−116
−7.9	−52	−5.5	−25	−15.8	−111	−19.7	−128
−7.7	−51	−11.7	−78	−23.2	−151	−18.8	−137
−8.2	−56	−5.6	−23	−22.95	−182.3	−17.6	−129

10. Using the equations for $1000 \ln \alpha_{quartz-H_2O}$ and $1000 \ln \alpha_{magnetite-H_2O}$ given in Table 11.2, construct a calibration curve for $1000 \ln \alpha_{quartz-magnetite}$ as a function of temperature (in Kelvin) in the range $10^6/T^2=0$ to $10^6/T^2=6$ (x coordinate). What is the isotopic equilibrium temperature corresponding to $1000 \ln \alpha_{quartz-magnetite}=18.213$?
11. Calculate w/r ratios as a function of assumed $\delta^{18}O$ values for an altered igneous body, $\delta^{18}O^f_{rock}$, at different equilibraton temperatures (400°C, 500°C, and 600°C), and for "closed" and "open" system models of Taylor (1977). Assume that, at equilibrium, $\delta^{18}O^f_{rock}$ is equal to the $\delta^{18}O$ value of plagioclase (An_{30}) in the rock, and use the plagioclase (An_{30})–water geothermometer of O'Neil and Taylor (1967) to calculate $\Delta^{18}O_{plag-water}$: $\delta^{18}O^i_{rock}=+6.5$‰ and $\delta^{18}O^i_{water}=-14$‰. Show your results on a binary diagram, with x axis$=w/r$ ratios (atom% oxygen) and y axis$=\delta^{18}O^f_{rock}$. What generalized conclusions can you infer from your results?
12. The $\delta^{34}S$ value of the present-day seawater sulfate is 21‰. Calculate the value of $\alpha_{SO_4-H_2S}$ that would produce pyrite with a $\delta^{34}S$ value of −50‰ by sulfate reduction, assuming the system to be open with respect to both SO_4^{2-} and H_2S? What are the factors that determine the $\delta^{34}S$ values of marine sedimentary pyrites? What are the possible causes for the observation that the $\delta^{34}S$ value of pyrites in Archean marine sediments is only about −5‰?
13. Coexisting sphalerite (ZnS)–galena (PbS) pairs in a lead–zinc sulfide deposit in Providencia, Mexico, have $\delta^{34}S$ values (per mil) as listed below (Rye, 1974):

Sample number	$\delta^{34}S$ (galena)	$\delta^{34}S$ (sphalerite)
60-H-67	−1.15	+0.95
60-H-36-57	−1.11	+0.87
62-S-250	−1.42	+1.03
63-R-22	−1.71	+0.01

Using the appropriate equation from Table 11.3, estimate the temperature of formation of the sulfide ore. What are the assumptions involved in your estimate?

14. Consider the bacterial reduction of dissolved SO_4^{2-} to H_2S in a Permian marine basin in which the dissolved SO_4^{2-} had an initial $\delta^{34}S$ value of +15‰. The basin, after its formation, had no connection to the ocean but was "open" to the H_2S produced. Calculate and plot:
(a) variations in $\delta^{34}S_{SO_4}$ and the corresponding $\delta^{34}S_{H_2S}$ as the reduction progresses to completion, using the Rayleigh distillation model;
(b) corresponding variation in $\delta^{34}S$ value of the total H_2S molecules removed from the system since the beginning of fractionation, using the equation (Ohmoto and Goldhaber, 1997)

$$\delta^{34}S_{\Sigma H_2S} = (\delta^{34}S^{initial}_{SO_4} - f_{SO_4}\ \delta^{34}S^{remaining}_{SO_4})/(1-f_{SO_4})$$

where f_{SO_4} is the atomic fraction of the SO_4^{2-} remaining in the system with respect to the original amount at any particular extent of evolution of the system. Assume constant temperature and $\alpha_{SO_4-H_2S}=1.025$.

Part IV
The Earth Supersystem

In order to gain a good working knowledge of the Earth and its processes, we need to understand the interaction not only between systems but also between and among the various scales of activity of the many systems.

Schneider (2000)

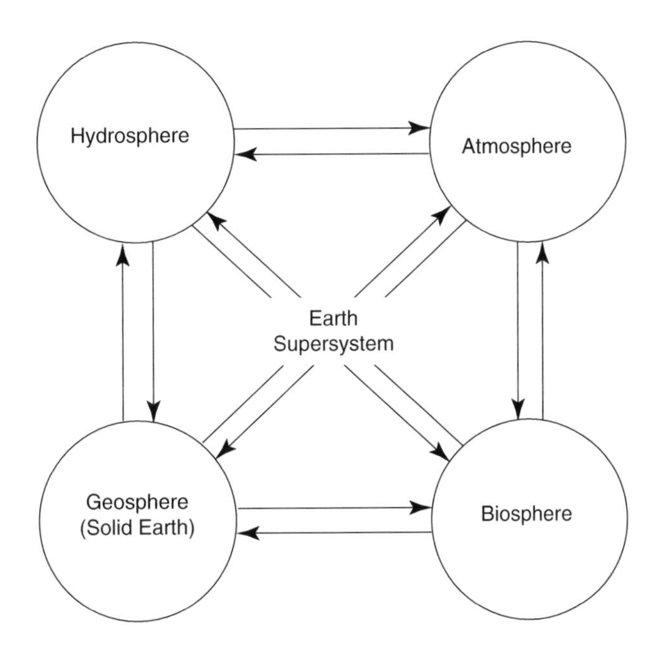

12 The Core–Mantle–Crust System

The questions of origin, composition and evolution of the Earth require input from astronomy, cosmochemistry, meteoritics, planetology, geology, petrology, mineralogy, crystallography, materials science and seismology, at a minimum. To a student of the Earth, these are artificial divisions, however necessary they are to make progress on a given front.

Anderson (1989)

A system is defined as a group of components (parts) that interact and function as a whole. A component can be a reservoir of matter (described by its mass and volume), a reservoir of energy (described by temperature, for example), an attribute of the system (such as body temperature or pressure), or a subsystem (such as the cardiovascular system, one of the interlinked subsystems of the human body) (Kump *et al.*, 2004). The Earth may be viewed as a supersystem composed of four interacting systems: (i) the *geosphere* (or solid Earth), which is divisible into three major units – the core, mantle, and crust (see section 12.2.1); (ii) the *hydrosphere*, which comprises the various reservoirs of water, including ice; (iii) the *atmosphere*, the thin envelope of gases that surround the Earth; and (iv) the *biosphere*, which consists of all living organisms. In this chapter and the next we apply the principles we have learned in the previous chapters to understand some of the geochemical processes that have shaped the evolution of the various Earth systems to their current states.

We will start at the very beginning when the universe suddenly came into existence, and then discuss the origin of elements, the birth of the Solar System (including planet Earth), the differentiation of the Earth into mantle and core, the formation of the atmosphere and the oceans, and later addition of the crust. As geochemical interactions between mantle and crust occur largely through magmatism, we will briefly dwell on the processes of partial melting that generate magmas and crystallization of magmas that produce igneous rocks, and some techniques that can be used to infer the tectonic settings of ancient basaltic volcanic suites.

12.1 Cosmic perspective

The Earth is a very small part of the Solar System and only a tiny fraction of the universe, but it is a very special celestial body. In addition to being the planet we live on and the only entity in the universe with known life, the Earth has a unique composition. One of the fundamental factors that govern the Earth's geochemical processes and products is its composition on different scales. The unique composition of the Earth is a product of three sets of processes: the synthesis of elements (*nucleosynthesis*); the creation of the Solar System; and the formation and evolution of the Earth itself.

12.1.1 *The Big Bang: the beginning of the universe*

How and when did the universe come into existence? The most popular model at present for the origin of the universe is the *Big Bang theory*, which attributes the creation of the universe to very rapid (almost instantaneous), exponential expansion of an infinitely small, very hot, and infinitely dense phase. An analogy may the instantaneous expansion of an infinitely small balloon to the size of the universe, not an explosion like the bursting of a balloon, a common misconception. This event, in

Introduction to Geochemistry: Principles and Applications, First Edition. Kula C. Misra.

cosmological parlance, is known as a "singularity," a term used to describe the mind-boggling concept that an infinitely large amount of matter was gathered at a single point in space–time. Theoretical calculations by different authors indicate that this event occurred about 13 to 15 billion years ago. Prior to this event, nothing existed – not space, or time, or matter, or energy – and suddenly there was our universe, which had all of the mass and energy it contains today! We do not know where this dense starting material came from or why this "big bang" occurred. All that we can argue is that the model is compatible with three key astronomical observations as described below.

(1) Galaxies (massive, gravitationally bound systems consisting of stars, an interstellar medium of gas and dust, and dark matter) appear to be receding from the Earth. (The familiar Milky Way galaxy, to which our Solar System belongs, is only one of some 10^{11} galaxies, each ranging in size from 80,000 to 150,000 light years in diameter and containing in the order of 10^{11} stars. 1 light year is the distance light travels in a year $\approx 10^{13}$ km.) This expansion of the universe is documented by red-shifting of the light waves emitted from the galaxies, a consequence of the *Doppler effect*, the change in frequency and wavelength of a wave as perceived by an observer moving relative to the source of the wave (see Box 12.1).

Box 12.1 The Doppler effect

It is a familiar experience that the pitch of an ambulance's siren becomes higher as the vehicle moves towards you and then becomes lower as it moves past you. This change in pitch results from a shift in the frequency of sound waves as perceived by an observer when the source of the waves is moving relative to the observer. The phenomenon is known as the *Doppler effect* in honor of the Austrian mathematician–physicist Christian Doppler (1803–1853) who discovered it. As the ambulance approaches, the sound waves from its siren are compressed towards the observer and the intervals between waves diminish, which translates into an increase in frequency or pitch. As the ambulance moves away, the sound waves are stretched relative to the observer, causing the siren's frequency to decrease. This Doppler shift can also be described in terms of wavelength because of the inverse relationship between frequency and wavelength. It is possible to determine if the ambulance is approaching you or speeding away from you by the change in the siren's frequency (or wavelength), and to estimate the ambulance's speed from the rate of change of the frequency (or wavelength).

The Doppler effect is also recorded in the spectra of moving astronomical bodies that emit electromagnetic radiation. Light coming from stars and galaxies is *red-shifted* (i.e., it appears more red because the wavelengths are shifted towards the red end of the visible spectrum), indicating that they are receding from us. If the galaxies were moving towards us, the light emitted would be *blue-shifted* (i.e., it would appear more blue because of the shifting of the wavelengths towards the blue end of the visible spectrum). If they were stationery, there would be no shifting of wavelengths.

The recessional velocity of a galaxy, v (in km s^{-1}), can be calculated using *Hubble's Law*, so named after the American astronomer Edwin Hubble (1889–1953), who established the following linear relationship:

$$v = H_0 d \tag{12.1}$$

where d is the distance of the galaxy from the Earth in 10^6 light years, and H_0 is the *Hubble constant* (expressed in units of km s^{-1} per megaparsec; 1 megaparsec (Mpc) = 3.26 million light years). The direct proportionality between v and d – i.e., the greater the distance to a galaxy, the greater the red-shift in its spectral lines – indicates that farther galaxies are receding with higher velocities. The importance of the Hubble constant is that it expresses the rate at which the universe is expanding, from which it is possible to calculate the time when all galaxy trajectories would converge to the initial condition – the age of the universe. The value of H_0 is still being debated, but the accepted range lies between 64 to 72 (km/s)/Mpc. For H_0 = 68 (km/s)/Mpc, the age of the universe works out to be 13.4×10^9 years (Lineweaver, 1999), and for H_0 = 71 (km/s)/Mpc, the age is 13.7×10^9. How long the universe will continue to expand cannot be predicted without a more detailed knowledge about the objects in the universe.

(2) There is a pervasive presence of *cosmic background radiation* in the universe. This radiation, with a maximum wavelength of about 2 mm, which is in the microwave range of the electromagnetic spectrum, has all the characteristics of radiant heat, the same kind of radiant heat as emitted by a blackbody with a temperature of about 3 K. Some theoretical physicists had predicted the pervasive existence of such radiation if the universe began with a hot Big Bang. In 1965, Arno Penzias and Robert Wilson of the Bell Telephone Laboratories in New Jersey confirmed the prediction by documenting the existence of background cosmic radiation, which led to their sharing the Nobel prize in Physics in 1978.

By about 300,000 years after the Big Bang, the temperature of the universe had dropped sufficiently for protons and electrons to combine into hydrogen atoms ($p^+ + e^- \Rightarrow {}^1_1H$), and the interaction between radiation and the background gas became ineffective; this background radiation has propagated freely over the entire universe ever since. The background radiation has been losing energy because its wavelength is stretched by the expansion of the universe, and its temperature has dropped to about 3 K, and will eventually cool to 0 K.

(3) The abundance of the light elements such as hydrogen and helium in the observable universe is consistent with the Big Bang model (see section 12.1.3).

12.1.2 Nucleosynthesis: creation of the elements

Nucleosynthesis is the process of creating new atomic nuclei from preexisting nucleons (protons and neutrons). Our understanding of nucleosynthesis is based on three sets of observations: (i) the abundance of elements (and their isotopes) in the cosmos; (ii) experimental data on the probability of specific nuclear reactions occurring under specified conditions; and (iii) inferences about the existence of appropriate sites for nucleosynthesis. The nucleosynthesis is believed to have occurred in four phases: (i) *Big Bang* (or *cosmogenic*) *nucleosynthesis* (BBN), which occurred immediately after the Big Bang; (ii) *stellar nucleosynthesis* in the interior of larger stars during the process of stellar evolution; (iii) *explosive nucleosynthesis* in supernovae; and (iv) *galactic nucleosynthesis* (*cosmic ray spallation*) in interstellar space. These processes are described below briefly; a detailed treatment of the topic can be found, for example, in Meyer and Zinner (2006).

Big Bang nucleosynthesis

Initially, the universe was very, very hot ($>10^{27}$ K). As the universe expanded, it cooled, and eventually matter formed and coalesced together by gravity to form various celestial bodies. After about 10^{-32} s of the Big Bang, the universe had expanded to 30 cm in diameter, but the pressure and temperature (10^{27} K) were still so high that matter existed only as a quark–gluon plasma, consisting of (almost) free quarks and gluons, which are the basic building blocks of matter. In 10^{-26} s, the universe had attained a size comparable to that of the Solar System, temperature had cooled to about 10^{13} K, and a radical transition led to the formation of protons and neutrons (within the first second after the Big Bang), which became the common state of matter. The temperature at this stage was too high for the formation of more complex matter.

Free neutrons and protons are less stable than helium nuclei. So they have a strong tendency to form $^{4}_{2}$He, but the formation of $^{4}_{2}$He requires the formation of deuterium ($^{2}_{1}$H or D) as an intermediate step. At very high temperatures, the mean energy per particle is greater than the binding energy of $^{2}_{1}$H, so that any $^{2}_{1}$H nuclei formed would be immediately destroyed. However, the temperature of the universe dropped to about 10^{9} K within 3 min of its beginning, and protons (p) and neutrons (n) began to react with each other to form deuterium nuclei(^{2}H or D):

$$p + n \Rightarrow {}^{2}_{1}\text{H} + \gamma \tag{12.2}$$

It took about 300,000 years after the Big Bang for the universe to cool to about 3000 K, cool enough for electrons to be bound to nuclei, forming whole H and He atoms, and for photons to separate from matter causing light to burst forth for the first time. Further reactions such as those listed below produced tritium ($^{3}_{1}$H), the radioactive isotope of hydrogen, along with the isotopes of helium, $^{3}_{2}$He and $^{4}_{2}$He:

$${}^{2}_{1}\text{H} + n \Rightarrow {}^{3}_{1}\text{H} + \gamma \tag{12.3}$$

$${}^{3}_{1}\text{H} + p \Rightarrow {}^{4}_{2}\text{He} + \gamma \tag{12.4}$$

$${}^{2}_{1}\text{H} + p \Rightarrow {}^{3}_{2}\text{He} + \gamma \tag{12.5}$$

$${}^{3}_{2}\text{He} + n \Rightarrow {}^{4}_{2}\text{He} + \gamma \tag{12.6}$$

Although $^{4}_{2}$He continues to be produced by other mechanisms (such as stellar fusion and α decay), and trace amounts of $^{1}_{1}$H by *spallation* (an atomic nucleus splitting into three or more fragments due to interaction with energetic particle) and certain types of radioactive decay (proton emission and neutron emission), the abundances of these light isotopes ($^{1}_{1}$H, $^{2}_{1}$H, $^{3}_{2}$He, and $^{4}_{2}$He) in the universe are believed to have been determined primarily by the Big Bang nucleosynthesis. A few helium nuclei also combined into heavier nuclei, producing trace amounts (on the order of 10^{-10}%) of Li, Be, and B (Meyer and Zinner, 2006). No heavier elements could be formed because of the very short duration of the Big Bang nucleosynthesis, which came to an end when, due to continued expansion and cooling of the universe, its temperature and density fell below what was required for nuclear fusion. Any free neutrons decayed away because neutrons outside the nucleus are unstable (having a half-life of only 10 min). Thus, the Big Bang created only two elements in any abundance, hydrogen and helium, and trace amounts of three very light elements, lithium, beryllium, and boron. The agreement between the abundance observed and that predicted by theoretical calculations is good for $^{4}_{2}$He, and even better for $^{3}_{2}$He and $^{2}_{1}$H (whose abundances can be measured with greater accuracy). The predicted and observed values, however, differ by a factor of two for $^{7}_{3}$Li, most likely because of the uncertainties in the reconstruction of the primordial $^{7}_{3}$Li abundance. Elements formed during this time were in the plasma state and did not cool to the state of neutral atoms until much later. Carbon ($^{12}_{6}$C) and elements with atomic numbers greater than that of carbon came into existence only after the formation of stars by various nuclear processes in their hot interiors.

Stellar nucleosynthesis

Stellar nucleosynthesis occurs in stars during the process of stellar evolution. Inhomogeneities in the hot gas, which was the state of the universe for quite some time after the Big Bang, led to the formation of proto-galaxies (perhaps about 0.5 Gyr after the Big Bang) through a process of gravitational attraction and collapse, and collapse due to instabilities within the proto-galaxies led to the formation of stars.

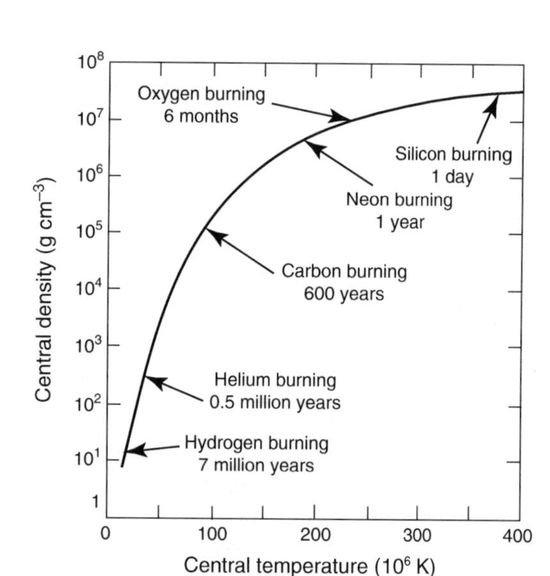

Fig. 12.1 Evolutionary path of the core of a massive star of 25 solar masses (after Bethe and Brown, 1985). The period spent in each step of the nuclear fusion process depends on the mass of the star; more massive stars would evolve more rapidly. For most of the star's lifetime the primary energy source is the fusion of hydrogen nuclei to form helium. The final stage of silicon fusion yields a core of iron, from which no further energy can be extracted by nuclear reactions. The iron core cannot resist gravitational collapse, leading to a supernova explosion.

Subsequent nucleosynthesis of elements (from carbon to nickel) occurred primarily within the core regions of stars that pre-dated the Solar System, either by nuclear fusion or by nuclear fission operating in different environments and at different times, with hydrogen as the sole starting material. This view of nucleosynthesis, called the *polygenetic hypothesis*, which can account for the observed abundances of the elements in the Solar System and in nearby stars, was first proposed by E.M. Burbridge, G.R. Burbridge, W.A. Fowler, and F. Hoyle (1957) in a classic paper commonly referred to as B^2FH. A strong argument in favor of this hypothesis is the fact that such nuclear transformations are currently taking place inside stars in the course of their evolution. The successive nuclear fusion processes, each of which occurs inside stars at specific temperatures, are referred to as *hydrogen burning*, *helium burning*, *carbon burning*, *neon burning*, *oxygen burning*, and *silicon burning* (Fig. 12.1; Table 12.1), the ashes from one burning stage providing the fuel for the next and the conversion of gravitational energy into thermal energy providing the automatic temperature rise. Several of these reactions may occur simultaneously in the core and outer shell of a massive star, resulting in a core of different composition compared to its outer shell. Moreover, every star does not experience all of these nuclear reactions; that is why other stars in the Milky Way galaxy do not have the same composition as the Sun. The above mentioned fusion processes are able to create

Table 12.1 Successive stages of stellar nucleosynthesis.

Process	Approximate duration	Fuel for the process	Important products
Hydrogen burning	7×10^6 yr	H	^4_2He, $^{14}_7N$, $^{17}_8O$, $^{26}_{13}Al$
Helium burning	0.5×10^6 yr	He	$^{12}_6C$, $^{16}_8O$, $^{18}_8O$, $^{22}_{10}Ne$, $^{25}_{12}Mg$, *s-process* isotopes*
Carbon burning	600 yr	C	$^{20}_{10}Ne$, $^{23}_{11}Na$, $^{24}_{12}Mg$
Neon burning	1 yr	Ne	$^{16}_8O$, $^{24}_{12}Mg$
Oxygen burning	6 months	O	$^{28}_{14}Si$, $^{31}_{15}P$, $^{32}_{16}S$
Silicon burning	1 day	Mg to Si	Fe–peak isotopes; $^{56}_{26}Fe$, $^{54}_{26}Fe$

*Isotopes formed by nucleosynthesis involving slow capture of neutrons.
Source of data: Bethe and Brown (1985) and Meyer and Zinner (2006).

elements up to iron ($Z = 26$) and nickel ($Z = 28$), the region of the isotopes having the highest binding energy per nucleon. Heavier elements can be synthesized within stars by processes involving neutron capture. The products of stellar nucleosynthesis are generally distributed into the universe as planetary nebulae or through the *solar wind* (streams of high-energy atomic particles).

Hydrogen burning. In hydrogen burning, which starts when the gravitational collapse of a star has proceeded to the point where density reaches 6 g cm^{-3} and temperature reaches 10×10^6 to 20×10^6 K, fusion of protons forms ^4_2He (see Box 12.2). Much of the helium produced is consumed in later stages of stellar evolution or is locked up in stars of lower masses that never reach the temperature required for helium burning. In stars of 1.2 or less solar masses (M_0), this process occurs via the proton–proton chain, the only source of nuclear energy for the first-generation stars that formed from the primordial mixture of hydrogen and helium after the Big Bang. Some stars run through their evolution relatively rapidly (in millions of years) and produce nearly all the heavier elements by various nuclear processes in their hot interiors. The life of such stars ends in a gigantic explosion as a supernova when it runs out of nuclear fuel. The core collapses in milliseconds, and the subsequent "bounce"of the core generates a shock wave so intense that it blows off most of the star's mass (Bethe and Brown, 1985). The newly formed elements within the star are expelled into the interstellar medium, where they are mixed with interstellar gas and ultimately incorporated into next-generation stars. The material in the Milky Way galaxy underwent several such cycles before our Solar System was born.

In more massive, second and higher generation stars, in which the central temperatures exceed 2×10^7 K and which contain carbon ($^{12}_6C$) synthesized by ancestral stars, hydrogen burning is also accomplished via another process, the

Box 12.2 Nuclear reactions associated with hydrogen burning

The net result of hydrogen burning (or fusion of protons) is the production of helium, which can be accomplished by the following two mechanisms:

(1) Proton–proton chain (Burbridge *et al.*, 1957)

$${}^{1}_{1}H + {}^{1}_{1}H \Rightarrow {}^{2}_{1}H + \beta^{+} + \nu \text{ (neutrino)} + 0.422\,\text{MeV} \quad (12.7)$$

$$\beta^{+} + \beta^{-} \Rightarrow 1.02\,\text{MeV (annihilation)} \quad (12.8)$$

$${}^{2}_{1}H + {}^{1}_{1}H \Rightarrow {}^{3}_{2}He + \gamma + 5.493\,\text{MeV} \quad (12.9)$$

$${}^{3}_{2}He + {}^{3}_{2}He \Rightarrow {}^{4}_{2}He + {}^{1}_{1}H + {}^{1}_{1}H + 12.859\,\text{MeV} \quad (12.10)$$

(2) CNO cycle (Bethe, 1968)

$${}^{12}_{6}C + {}^{1}_{1}H \Rightarrow {}^{13}_{7}N + \gamma \quad (12.11)$$

$${}^{13}_{7}N \Rightarrow {}^{13}_{6}C + \beta^{+} + \nu \quad (12.12)$$

$${}^{13}_{6}C + {}^{1}_{1}H \Rightarrow {}^{14}_{7}N + \gamma \quad (12.13)$$

$${}^{14}_{7}N + {}^{1}_{1}H \Rightarrow {}^{15}_{8}O + \gamma \quad (12.14)$$

$${}^{15}_{8}O \Rightarrow {}^{15}_{7}N + \beta^{+} + \nu \quad (12.15)$$

$${}^{15}_{7}N + {}^{1}_{1}H \Rightarrow {}^{12}_{6}C + {}^{4}_{2}He \quad (12.16)$$

The end result is that four ${}^{1}_{1}H$ nuclei fuse to form one nucleus of ${}^{4}_{2}He$ and release a large amount of energy. Note that the nucleus of ${}^{12}_{6}C$ works as a sort of catalyst, so it can be used again and again for starting fresh CNO cycles. The C-cycle reaction also provides a mechanism for the production of ${}^{14}_{7}N$ in the star.

carbon–nitrogen–oxygen (CNO) cycle (Bethe, 1968; see Box 12.2). Most stars in the Milky Way galaxy, including the Sun, are at least second-generation stars because only the very small first-generation stars could have survived in this very old galaxy. At present, the Sun is burning only hydrogen, but it contains heavier elements inherited from earlier-generation stars. The energy released as a result of the conversion of hydrogen to helium is the source of the Sun's temperature and luminosity.

When the hydrogen fuel is exhausted in the core region of the star, hydrogen fusion is switched to a shell outside the core. This is the *red giant* phase of a star's evolution. Helium produced by hydrogen fusion cannot be burned immediately due to the larger nuclear charge of helium, which produces a much higher Coulomb barrier against fusion. So, the helium accumulates in the core, increasing its density and causing it to contract and get hotter. The increase in the core temperature results in an expansion of the envelope and a lowering of the surface temperature, which causes the color to turn red. How soon this would happen depends on the mass of the star.

Helium burning. Helium burning begins in a star if it is massive enough for temperatures to reach 10^{8}K and the core density to reach 10^{4}g cm^{-3}. The critical reaction in this stage is the triple-alpha process in which three ${}^{4}_{2}He$ nuclei (i.e., three α particles) fuse to form the nucleus of ${}^{12}_{6}C$ through ${}^{8}_{4}Be$ as a very unstable intermediate product (half-life of ${}^{8}_{4}Be = 10^{-16}$ s):

$${}^{4}_{2}He + {}^{4}_{2}He \Rightarrow {}^{8}_{4}Be \quad (12.17)$$

$${}^{8}_{4}Be + {}^{4}_{2}He \Rightarrow {}^{12}_{6}C + \gamma \quad (12.18)$$

Another important product of helium burning is ${}^{16}_{8}O$ by the reaction:

$${}^{12}_{6}C + {}^{4}_{2}He \Rightarrow {}^{16}_{8}O \quad (12.19)$$

At this point in a star's evolution, the ${}^{1}_{1}H$ has been converted to ${}^{4}_{2}He$, and the ${}^{4}_{2}He$ into ${}^{12}_{6}C$ and ${}^{16}_{8}O$ with the accumulation of some ${}^{14}_{7}N$. Other isotopes that can be produced by helium burning include ${}^{18}_{8}O$, ${}^{22}_{10}Ne$, and ${}^{25}_{12}Mg$. Helium burning is the key to the synthesis of all elements beyond helium, and probably is responsible for much of the cosmic abundance of ${}^{12}_{6}C$, although in more massive stars the later burning stages consume some of this ${}^{12}_{6}C$.

In contrast to massive stars that became supernovae in a few millions of years, stars with mass less than ~8 M_0 take hundreds of millions to billions of years to burn the H and He in their interior to C and O. When available helium is exhausted, the evolution of low-mass stars, such as the Sun, ends after the red giant phase because the gravitational force is not sufficient to overcome Coulomb repulsion of electrons. Nuclear reactions cease, the star simply contracts and its exterior heats up. The star becomes extremely hot, a *white dwarf* (commonly composed of C and O). In the absence of further fusion reactions, a white dwarf eventually cools off to become a *black dwarf* (a burned out chunk of carbon!). White dwarfs comprise approximately 6% of all known stars in the solar neighborhood. It is believed that over 97% of all stars in our galaxy would eventually turn into white dwarfs. This is the likely fate of our Sun in about 5 Gyr.

Carbon burning. In stars with an initial mass of at least 8 solar masses, gravitational contraction of the star's core can generate temperatures in excess of 5×10^{8} K. At such high temperatures it is possible to overcome the Coulomb repulsion barrier between two ${}^{12}_{6}C$ nuclei and fuse them to form ${}^{24}_{12}Mg^{*}$ (an excited nucleus of ${}^{24}_{12}Mg$, which has higher energy compared to its energy in the ground state, the state of minimum energy):

$${}^{12}_{6}C + {}^{12}_{6}C \Rightarrow {}^{24}_{12}Mg^{*} \quad (12.20)$$

This represents the carbon burning stage in the evolution of a star. $^{24}_{12}Mg^*$ decays mostly to $^{20}_{10}Ne$ (Meyer and Zinner, 2006, p. 73):

$$^{24}_{12}Mg^* \Rightarrow \, ^{20}_{10}Ne + \, ^{4}_{2}He \qquad (12.21)$$

Thus, carbon burning is the major producer of $^{20}_{10}Ne$ in the universe. A number of other less abundant nuclei (including those of Na, Al, P, S, and K) are also synthesized at this stage, and in the subsequent process.

Neon burning. Neon burning, which follows carbon burning and typically occurs at temperatures near 1.5×10^9 K, lasts for a very short duration. It involves disintegration of $^{20}_{10}Ne$ ($^{20}_{10}Ne + \gamma \Rightarrow \, ^{16}_{8}O + \, ^{4}_{2}He$) and recapture of the a particle so liberated to produce $^{24}_{12}Mg$ ($^{20}_{10}Ne$ + 24He $\Rightarrow \, ^{24}_{12}Mg + \gamma$). Thus, the net reaction in neon burning is

$$^{20}_{10}Ne + \, ^{20}_{10}Ne \Rightarrow \, ^{16}_{8}O + \, ^{24}_{12}Mg \qquad (12.22)$$

Oxygen burning. The neon burning stage is followed by oxygen burning at temperatures in excess of about 10^8 K. The main products of oxygen burning are $^{28}_{14}Si$, $^{31}_{15}P$, and $^{32}_{16}S$ via $^{32}_{16}S^*$ (an excited nucleus of $^{32}_{16}S$) as an intermediate product:

$$^{16}_{8}O + \, ^{16}_{8}O \Rightarrow \, ^{32}_{16}S^* \qquad (12.23)$$

$$^{32}_{16}S^* \Rightarrow \, ^{28}_{14}Si + \, ^{4}_{2}He \Rightarrow \, ^{31}_{15}Si + \, ^{1}_{1}H \Rightarrow \, ^{32}_{16}S + \gamma \qquad (12.24)$$

Nuclei of masses up to $A = 40$ may be produced by oxygen fusion through the capture of proton, neutron, and α particles, but at the end of the oxygen burning, the matter in the stellar core is dominated by $^{28}_{14}Si$ and $^{32}_{16}S$.

Silicon burning. Silicon burning commences when the temperatures and densities inside the star exceed about 3×10^9 K and 10^7 g cm^{-3}, respectively. It lasts for a week or less, depending on the mass of the star. The mechanism for silicon burning is more complicated than that for the previous stages because the high charges of the Si and S isotopes prevent them from interacting directly. Photodisintegration of $^{28}_{14}Si$ nuclei and other intermediate-mass nuclei around $A = 28$ produces an abundant supply of protons, neutrons, and α particles, which are captured by the nuclei left from previous burning stages to produce new nuclei up to $A = 56$ (i.e., $^{56}_{28}Ni$). However, $^{56}_{28}Ni$ is unstable and it decays quickly to $^{56}_{27}Co$ (half-life of $^{56}_{28}Ni$ = 6.075 days), and then to stable $^{56}_{26}Fe$ (half-life of $^{56}_{27}Co$ = 77.27 days), which has the highest binding energy per nucleon. This is the reason for the relatively high solar abundances of the iron group elements (see Fig. 12.4). The relevant reactions are (Meyer and Zinner, 2006):

$$^{28}_{14}Si + \, ^{28}_{14}Si \Rightarrow \, ^{54}_{26}Fe + 2\,^{1}_{1}H \qquad (12.25)$$

$$^{28}_{14}Si + \, ^{28}_{14}Si \Rightarrow \, ^{56}_{28}Ni \qquad (12.26)$$

$$^{56}_{28}Ni \Rightarrow \, ^{56}_{27}Co + e^+ + \nu \qquad (12.27)$$

$$^{56}_{27}Co \Rightarrow \, ^{56}_{26}Fe + e^+ + \nu \qquad (12.28)$$

Nuclear statistical equilibrium. Iron ($^{56}_{26}Fe$) is something of a dead end of stellar fusion in the sense that fusion reactions that build elements up to iron release energy, whereas fusion reactions for the formation of elements with atomic numbers higher than that of iron actually consume energy. Thus, whereas small amounts of elements heavier than iron (e.g., gold, silver, uranium, etc.) are formed in stars, their abundances in the Solar System are very low. When increasing temperature at the core of a star brings the chain of nuclear fusion reactions to this dead end, disintegrative as well as constructive nuclear reactions begin to occur among the iron group nuclei ($^{51}_{23}V$, $^{52}_{24}Cr$, $^{55}_{25}Mn$, $^{56}_{26}Fe$, $^{59}_{27}Co$, $^{59}_{28}Ni$), leading to a *nuclear statistical equilibrium;* the relative abundances of iron group elements reflect this steady state.

Heavier elements can be synthesized within stars by a neutron capture process known as the *s process* or in the explosive environments, such as supernovae, by a number of processes. Some of the more important of these include: the *r process*, which involves rapid captures of neutrons; the *rp process*, which involves rapid captures of protons; and the *p process*, which involves photodisintegration of existing nuclei.

s process. Nuclides heavier than the iron group form by the capture of neutrons, the product of earlier helium burning reactions in stellar interiors (the main neutron source reactions being $^{13}_{6}C + \, ^{4}_{2}He \Rightarrow \, ^{16}_{8}O + n$ and $^{22}_{10}Ne + \, ^{4}_{2}He \Rightarrow \, ^{25}_{12}Mg + n$) by the iron group seed nuclei ($^{51}_{23}V$, $^{52}_{24}Cr$, $^{55}_{25}Mn$, $^{56}_{26}Fe$, $^{59}_{27}Co$, $^{59}_{28}Ni$). Neutrons being electrically neutral particles are not affected by the Coulomb repulsion that inhibits reactions involving charged particles. The capture of one neutron by a nucleus increases its neutron number by one unit, producing the next heavier isotope of the original element. If the resulting nucleus is unstable, it tends to undergo β^- decay (i.e., it emits an electron, and one neutron is effectively transformed into a proton) to form a more stable nucleus (an isobar). The neutron number stays the same as that of the original element, but the number of protons (i.e., the atomic number, Z), increases by one because of the transformation of the captured neutron into a proton (see Table 10.1). If the resulting nucleus is stable, there will be no immediate β^- decay. However, when successive neutron captures producing successively heavier isotopes of the original element result in too high a neutron: proton ratio in the nucleus, it would undergo β^- decay to form a more stable nucleus (an isobar) as in the previous case. The two dominant tracks of neutron addition are the slow *s* process and the rapid *r* process (Fig. 12.2), which occur during the final stages of the evolution of red giants.

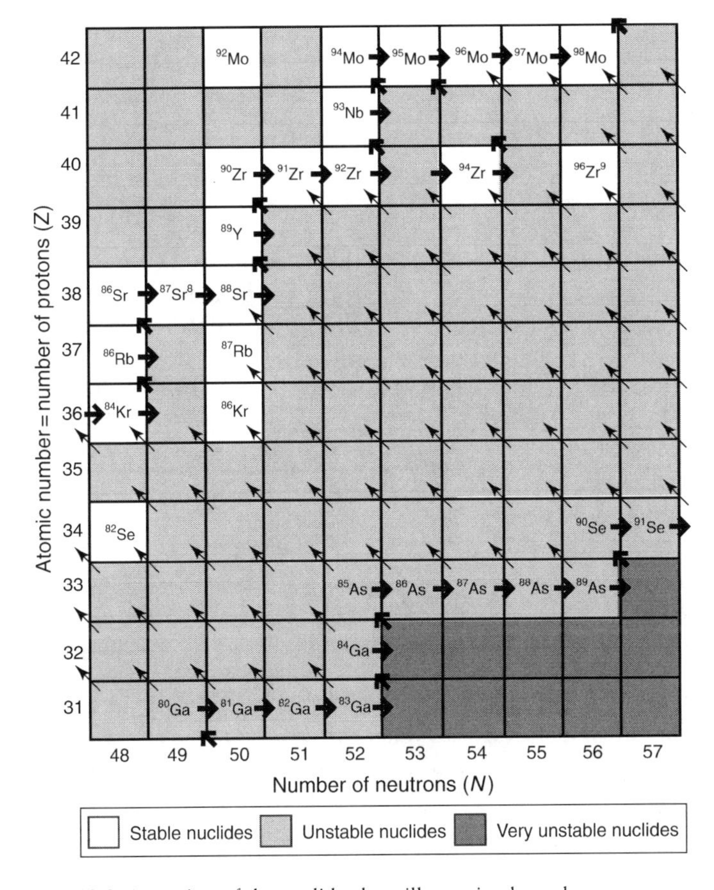

Fig. 12.2 A portion of the nuclide chart illustrating how the *s* process (upper set of arrows) and *r* process (lower set of arrows) form new heavy nuclides by neutron capture and subsequent β decay. In the *s* process, neutron additions (indicated by thick horizontal arrows) move the nucleus rightward on the chart until it is transformed into a neutron-rich unstable nuclide. The unstable nuclide undergoes β decay (indicated by thick diagonal arrows), which shifts it up and to the left, producing an isotope of a different chemical element. The nucleus can then capture additional neutrons until it β decays again, and so forth. In the *r* process, rapid addition of neutrons (rapid compared to the half-lives of nuclides represented by the light gray boxes) creates a chain of neutron-rich unstable nuclides. When neutron addition moves a nucleus into one of the dark gray boxes (very unstable nuclides), it β decays on a time scale even shorter than the duration of a supernova event. When the supernova neutron addition ends, the unstable nuclides left along the *r* process track quickly β decay (thin diagonal arrows) until they become stable nuclides (represented by white boxes). (After J.A. Wood, *The Solar System*, 1st edition, Copyright © 1979, Figure 6-20, p. 152. Used with permission of Pearson Education, Inc., Upper Saddle River, NJ.)

The *s* process occurs at relatively low neutron density and intermediate temperature conditions in stars. It involves slow (relative to the rate of radioactive β^- decay) addition of neutrons until an excess number of neutrons produces an unstable nuclide, which then undergoes β^- decay to form a stable, heavier nuclide. For example, the nucleus of $^{62}_{28}$Ni absorbs a neutron changing it to an excited nucleus of $^{63}_{28}$Ni, which then gets deexcited by emitting a γ-ray, and the radioactive $^{63}_{28}$Ni decays to stable $^{63}_{29}$Ni by emitting a β^- particle (Faure, 1998):

Neutron capture by $^{62}_{28}$Ni: $\quad ^{62}_{28}\text{Ni} + n \Rightarrow\, ^{63}_{28}\text{Ni} + \gamma$

Beta decay of radioactive $^{63}_{28}$Ni:

$$^{63}_{28}\text{Ni} \Rightarrow\, ^{63}_{29}\text{Cu} + \beta^- + \bar{\nu} + 0.0659\,\text{MeV} \tag{12.29}$$

The net result is the production of $^{63}_{29}$Cu from $^{62}_{28}$Ni by neutron capture. $^{63}_{29}$Cu, a stable isotope of copper, can then capture another neutron to form $^{64}_{29}$Cu, and repetition of the process described above would produce other heavy elements. Given enough neutrons, the *s* process can synthesize the majority of the isotopes in the range $23 \leq A \leq 46$ and a considerable proportion of the isotopes in the range $63 \leq A \leq 209$ (α decay prevents further nucleosynthesis by neutron capture), and account for abundance peaks at $A = 90$, 138, and 208. Note that a helium burning star producing heavy elements by the *s* process would have to be at least a second generation star with the iron group nuclides inherited from an earlier-generation star.

Explosive nucleosynthesis

A *supernova* (an exploding massive star) begins with the collapse of a stellar core, from a radius of several thousand kolometers to a radius of 100 km or so in a few tenths of a second. When matter in the central region of the core is compressed beyond the density of nuclear matter (3×10^{14}g cm^{-3}), it rebounds, sending a massive shock wave outward. The temperature increase that results from the compression produces a breakdown of existing nuclei by photodisintegration, for example,

$$^{56}_{26}\text{Fe} + \gamma \Rightarrow 13\, ^{4}_{2}\text{He} + 4n \tag{12.30}$$

$$^{4}_{2}\text{He} + \gamma \Rightarrow 2\, ^{1}_{1}\text{H} + 2n \tag{12.31}$$

The destruction of nuclei that had taken millions of years to build may be viewed as a backward step, but the large number of neutrons and protons produced by photodisintegration leads to three additional processes of nucleosynthesis: the *r* process (rapid process of neutron capture); the *rp* process (rapid capture of free protons); and the so-called *p* process. As the shock wave traverses the outer layers of the star, it compresses and heats them. This initiates explosive helium burning, oxygen burning, and silicon burning in these stellar shells, thereby modifying the composition established in the presupernova burning phase. These processes produce isotopes of S, Cl, Ar, Ca, Ti, and Cr, and some Fe.

The *r* process occurs when a large burst of neutrons, such as that released during the 1 to 100 s of peak temperature during a supernova explosion, allows rapid neutron capture, and immediate transformation of nuclides by β decay. At the end of the supernova event, nuclides formed by the *r* process transform more slowly by successive β decays to more stable nuclides. An example is the rapid capture of 5 neutrons in succession by ${}^{65}_{29}Cu$ to form ${}^{70}_{29}Cu$, which then decays by β^- emission to stable ${}^{70}_{29}Cu$:

Rapid capture of 5 neutrons by ${}^{65}_{29}Cu$: ${}^{65}_{29}Cu + 5n \Rightarrow {}^{70}_{29}Cu + 5\gamma$

Beta decay of radioactive ${}^{70}_{29}Cu$: ${}^{70}_{29}Cu \Rightarrow {}^{70}_{30}Zn + \beta^- + \bar{\nu} + \sim 7.2\,MeV$ (12.32)

The *r* process is responsible for the production of a large number of isotopes in the range $70 \le A \le 209$, and also for the synthesis of uranium and thorium, as well as of the most neutron-rich isotopes of each heavy element. Many of these isotopes can also be synthesized by the *s* process, but some isotopes, such as ${}^{86}Kr$, ${}^{87}Rb$, and ${}^{96}Zr$, can form only by the *r* process.

A number of proton-rich heavy isotopes (e.g., ${}^{74}_{32}Se$, ${}^{195}_{78}Pt$, ${}^{173}_{70}Yb$) cannot be created by neutron capture on any scale. There are two ways to increase the ratio of protons to neutrons in an atomic nucleus: either add protons or subtract neutrons. The *rp* process consists of consecutive proton addition to seed nuclei, and requires a H-rich environment (i.e., a large proton flux) as well as a very high temperature ($> 10^9$ K) to overcome the large Coulomb barrier. The process is inhibited by α decay, which puts an upper limit of ${}^{105}Te$ as the end point.

The *p* process, as it is now understood, is a misnomer because it has nothing to do with proton capture, as the term seems to imply (Meyer and Zinner, 2006, p. 78). Actually, the so-called *p* process operates through the mechanism of photo-disintegration, which occurs when γ-rays strip off first neutrons and then protons and α particles from preexisting *r*- and *s*-process isotopes. This is why the *p* process is sometimes referred to as the γ process. Proton capture, which must overcome Coulomb repulsion, is less probable than neutron capture and thus less effective. The *p* process is significant only for those nuclei that cannot be produced by other mechanisms, and the abundances of elements created by the *p* process only (e.g., ${}^{168}_{70}Yb$, ${}^{174}_{72}Hf$, ${}^{180}_{73}Ta$) tend to be low.

Cosmic ray spallation

Stellar and explosive nucleosynthesis do not produce the lighter elements such as Li, Be, and B; these elements are produced continually by cosmic ray spallation. Cosmic ray spallation (CRS) refers to the fragmentation of carbon, oxygen, and nitrogen nuclei in the interstellar medium by *cosmic rays* (fast-moving high energy particles, such as α particles (${}^{4}_{2}He$) and protons (${}^{1}_{1}H$), which are abundant in the interstellar gas). These reactions occur at high energies (higher than the Big Bang and stellar interiors), but at low temperatures, and are responsible for all of some of the lightest isotopes (such as ${}^{3}_{2}He$, ${}^{6}_{3}Li$, ${}^{9}_{4}Be$, ${}^{10}_{5}B$, ${}^{11}_{5}B$, and some of ${}^{7}_{3}Li$, but not ${}^{2}_{1}H$).

12.1.3 The Solar System

The Solar System is part of the Milky Way galaxy, which contains probably more than 10^{11} stars and has a lens-like form with a diameter of about 100,000 light years and a thickness of 1000 light years. The Solar System is comprised of the Sun, which contains > 99.86% of the mass of the system and those celestial objects bound to it by gravity, all of which were formed from the collapse of a much larger molecular cloud approximately 4.6 Ga. The temperature in the Sun ranges from 15.6×10^6 K at the core to 5800 K at the surface, and the pressure at the core is 250 billion atm. It is composed almost entirely of hydrogen (70%) and helium (28%).

The celestial bodies bound to the Sun by gravity include: 8 planets – which, in order of increasing distance from the Sun, are Mercury, Venus, Earth, Mars, Jupiter, Saturn, Uranus, and Neptune (Table 12.2) – and their satellites; at least five dwarf planets; an asteroid belt between Mars and Jupiter; the Kuiper belt containing a large number of small celestial bodies; meteoroids; comets; and interplanetary dust. A fundamental

Table 12.2 Some data about the Sun and the planets in the Solar System.

Planet/ Sun	Planet type	Equatorial diameter (km)	Distance from the Sun (10^6 km)	Relative mass* (kg)	Average density (g cm^{-1})	Composition	Principal atmospheric constituents
Mercury	Terrestrial	4,878	57.9	0.06	5.4	Iron/rocky	No atmosphere
Venus		12,104	108.2	0.82	5.2	Rocky	CO_2, N_2, Ar
Earth		12,756	149.6	1.00	5.5	Rocky	N_2, O_2, Ar
Mars		6,794	227.9	0.11	3.9	Rocky	CO_2, N_2, Ar
Jupiter	Jovian	142,984	778.3	317.87	1.3	Gassy	H, He
Saturn		120,536	1,426.9	95.14	0.7	Gassy	H, He
Uranus		51,118	2,871.0	14.56	1.2	Gassy/icy	H, He, CH_4
Neptune		49,528	4,497.1	17.21	1.7	Gassy/icy	H, He, CH_4
Sun		1.39×10^6		1.99×10^{30}	1.4	H 70% He 28%	No atmosphere

*Normalized to the mass of the Earth = 6.0×10^{24} kg.

difference between the stars, such as the Sun, and planets, such as the Earth, is that stars have a high-temperature core capable of generating their own energy by nuclear reactions, whereas planets have no such self-sustaining energy source. As a result, a star is seen because it generates its own light, but a planet is visible by virtue of the light from a nearby bright star reflected off its surface.

The planets may be divided into two major groups in terms of composition: *terrestrial planets* (Mercury, Venus, Earth, and Mars); and *Jovian planets* (Jupiter, Saturn, Uranus, and Neptune). The terrestrial planets (also classified as inner planets) are characterized by relatively small size (< 13,000 km in equatorial diameter), proximity to the Sun, high-density (3.9–5.5 g cm^{-3}), and earth-like rocky composition (about 90% composed of Fe, Si, Mg, and O). The Jovian planets (also classified as *outer planets* or *gas giants*), in contrast, are large bodies (> 49,000 km in equatorial diameter) of much lower density (0.7–1.7 g cm^{-3}), composed of gases (mostly H and He, like the Sun) and positioned at much greater distances from the Sun. The Earth is unique among the planets in having an oxygen-rich atmosphere, plenty of liquid water, and life. Pluto, previously considered a planet, is now classified as a *dwarf planet* belonging to the Kuiper belt, which is a region of the Solar System beyond the planets, extending from the orbit of Neptune (at 30 AU) to approximately 55 AU from the Sun, and appears to be composed of left-over material from planetary formation. (1 AU (*astronomical unit*) is a unit of length based on the mean distance between the Earth and the Sun = 149.6×10^6 km). A dwarf planet is a celestial body orbiting the Sun and massive enough to acquire a spherical shape by its own gravity, but which has not cleared its neighboring region of planetesimals and is not a satellite. The International Astronomical Union (IAU) currently recognizes only five dwarf planets – Pluto, Eris, Ceres, Haumea, and Makemake – but many more are suspected to exist in the Kuiper belt.

The asteroids comprise a group of relatively small celestial bodies that orbit around the Sun. They range up to about 1000 km in diameter, but most of the 50,000 or so asteroids that have been observed are only about 1 km across, and are fragments of once larger bodies. Asteroids are believed to have accreted shortly after the beginning of the Solar System's history and represent remnants of the proto-planetary disc in the region where perturbations by Jupiter prevented the accretion of planetary materials into planets or a large planet. The Kuiper belt is similar to the asteroid belt, although it is 20 times as wide and 20–200 times as massive and located outboard of the planets. Also, whereas the asteroid belt is composed primarily of rock and metal, the Kuiper belt objects are composed largely of frozen volatiles (usually referred to as "ices") such as methane, ammonia, and water.

Meteoroids are sand- to boulder-sized particles of debris, smaller than asteroids but larger than interplanetary dust. Comets, the most spectacular and unpredictable bodies in the solar system, measuring a few kilometers to tens of kilometers across, are composed of frozen gases (ammonia, methane, carbon dioxide, and carbon monoxide) that hold together pieces of rocky and metallic materials. Comets orbit the Sun and, when close enough to the Sun, exhibit a visible "coma" (or atmosphere) and/or a tail – both primarily from the effects of solar radiation upon the comet's nucleus.

A viable model for the origin of the Solar System must account for its characteristics listed below.

(1) All the planets revolve in the same direction (counterclockwise) around the Sun in elliptical (but near-circular) orbits, and these orbits all lie within a nearly flat disc called the *ecliptic plane*, which is the apparent path that the Sun traces out in the sky. The exception is Mercury, whose orbit is inclined at 7° to the ecliptic plane. The common plane of rotation and the near-circular orbits suggest *in situ* formation of the planets, simultaneously with formation of the Sun, from a disk of material.
(2) All the planets, except Venus and Uranus, rotate on their axis in a counterclockwise direction. Venus rotates slowly to the right and Uranus is tilted on its side with its north pole pointing towards the Sun, possibly due to a catastrophic collision early in the evolution of the Solar System.
(3) More than 99.86% of the mass of the Solar System is concentrated in the Sun at the center of the system, but 98% of its angular momentum resides in the planets at the edges. (Angular momentum is the measure of motion of objects in curved paths, including both rotation and orbital motion.)
(4) There is an orderly change in the composition of the planets with increasing distance from the Sun. The terrestrial planets have high densities (3.9 to 5.5 g cm^{-3}) and no or only a few moons, and are primarily composed of metals and silicate rocks. In contrast, the Jovian planets have lower densities (0.7 to 1.76 g cm^{-3}) and many moons, and are composed of gases – the two largest planets, Jupiter and Saturn, of hydrogen and helium, and the two outermost planets, Uranus and Neptune, largely of ices (such as methane, ammonia, and water vapor). The anomalously high density of Mercury is best explained by the stripping of much of its silicate mantle by collision with a massive planetesimal.
(5) All the components of the Solar System have the same age – about 4.56 Ga based on radiometric measurements on meteorites (Patterson *et al.*, 1956; Allégre *et al.*, 1995).
(6) The planets have a regular spacing – the mean distances of the planets from the Sun are related to a simple mathematical progression of numbers (*Bode's Law*). (This formulation led Bode to predict the existence of another planet between Mars and Jupiter, what we now recognize as the asteroid belt.)

The model that best accounts for these characteristics envisages the creation of the Solar System from a nebula, the Solar Nebula, about 4.56 billion years ago (see section 12.1.6).

12.1.4 Meteorites

A *meteorite* is a rock that was formed elsewhere in the Solar System, was orbiting the Sun or a planet for a long time (a *meteoroid*), was eventually captured by the Earth's gravitational field, and fell to the Earth as a solid object. When a meteoroid passes through the atmosphere, its exterior is heated to incandescence, producing a visible streak of light called a *meteor* or, more commonly a "falling star" or a "shooting star." Most of these objects are very small and burn up in the atmosphere; parts of a few larger ones survive the journey and land on the Earth as meteorites. It is estimated that about 10^6 to 10^7 kg of meteorites, predominantly smaller than 1 mm in size, rain down on the Earth each year; occasionally they have been large enough to produce impact craters such as the Meteor Crater in Arizona. Our database on meteorites is built on the recovered ones, which represent only a small fraction of the population actually received by the Earth.

Most meteorites (~99.8%) are from the asteroids. The catalogued meteorites also include a few martian meteorites (the SNC group, so named for its principal meteorites Shergotty, Nakhla, and Chassigny) and a few lunar meteorites ("lunaites"), which are interpreted to have been ejected from the surface of Mars and the Moon, respectively. Martian meteorites represent the only known samples of another planet available for research. Lunar meteorites are of major scientific importance because most of them probably originated from areas on the far side of the Moon that were not sampled by the Apollo or Luna missions.

The most primitive meteorite specimens are believed to be samples of the most primitive (pre-planetary) material in the Solar Nebula. Some meteorites are so primitive that they contain traces of interstellar dust, which survived thermal processing in the Solar Nebula. Thus, meteorites provide information about the chemical and physical properties at different locations within the Solar Nebular disk, as well as constraints about time scales and physical processes involved in the formation and evolution of objects in the Solar System.

Meteorites show a range of compositions and textures because their parent bodies came into existence in different regions of the Solar Nebula. The most commonly used parameters for classification of meteorites are petrology (texture, mineralogy, and mineral compositions), whole-rock chemical composition, and O-isotopic composition. The abridged classification presented in Table 12.3 is based on petrology. More detailed treatments can be found, for example, in Wasson (1974) and Weisberg *et al.* (2006), and on the Worldwide Web (http://www.meteorite.fr/en/classification).

Of particular interest in the present context are chondrites, which are considered to be samples of pre-planetary material from the Solar Nebula, the volatile-rich carbonaceous chondrites being the most primitive of all meteorites. The parent bodies of chondrites are small to medium sized asteroids that were never part of any celestial body large enough to undergo melting and planetary differentiation. Many of them, however, reached high enough temperatures for thermal metamorphism in their interiors. These metamorphosed chondrites are represented by the so-called *equilibrated chondrites*. The source of the heat most likely was the decay of short-lived radioisotopes, especially ^{26}Al and ^{60}Fe, although some contribution may have come from the heat released during impacts onto the asteroids. The iron meteorites, in contrast, represent the cores of differentiated asteroids, the melting and metal–sulfide–silicate fractionation occurring effectively at the beginning of geologic time (T_0) (Taylor, 1992).

Conspicuous among the constituents present in almost all chondrites are *chondrules*. Carbonaceous chondrites commonly contain ~50% chondrules by volume, whereas ordinary and enstatite chondrites contain up to 80%. Chondrules are spherical bodies ranging in size from ~0.01 to 10 mm across and composed of quenched crystals of ferromagnesian olivine and pyroxene, Fe–Ni metal, troilite (FeS), and glassy or microcrystalline mesostasis (the last-formed interstitial material between the larger mineral grains). The vast majority of chondrules represent molten droplets formed by flash heating of "cool" (< 650 K) nebular dust at peak temperatures of 1770–2120 K (McSween *et al.*, 2006) followed by very rapid recondensation, possibly over the span of a few minutes (Hewins, 1997), in the Solar Nebula. Evidence for this conclusion includes relict grains that apparently survived the melting, and the porphyritic texture of most chondrules, which require the presence of residual solid nuclei to facilitate the formation of large crystals. Many chondrules appear to have experienced multiple heating episodes, but most were not completely melted. The coexistence of metallic Fe–Ni with generally Mg-rich olivine and pyroxene in many chondrules reflects a reducing environment during chondrule formation. If the chondrules had been melted in an oxidizing environment, their metallic iron would have been oxidized to FeO, which would have been incorporated by the olivine and pyroxene that crystallized upon cooling. The transient heating episodes were localized because there is no evidence for silicate melting on a nebular scale. However, none of the mechanisms proposed to account for the localized heating – collisional impact, lightning, heating by the Sun, and passage of shock waves – is entirely satisfactory, but thermal processing of particles in nebular shock waves appears to meet most constraints imposed by chondrule properties (Ciesla and Hood, 2002).

Calcium–aluminum-rich inclusions (CAIs) constitute another important, but minor, component of carbonaceous chondrites. These irregularly shaped or round inclusions range in size from microscopic to 5–10 cm. They are enriched, commonly by a factor of 20 or so, in refractory elements such as Ca, Al, Ti, Re, and the noble metals (Ru, Rh, Pd, Ag, Os, Ir, Pt, and Au), and depleted in Fe, Mg, and Si relative to bulk chondrites. The main minerals of CAIs are corundum (Al_2O_3), hibonite ($CaAl_{12}O_{19}$), grossite ($CaAl_4O_7$), perovskite ($CaTiO_3$), spinel ($MgAl_2O_4$), anorthite ($Ca_2Al_2Si_2O_8$), melilite (solid solution of $Ca_2MgSi_2O_7$ and $Ca_2Al_2SiO_7$), and Al–Ti-pyroxene. The origin of CAIs is controversial. Earlier they were thought to be high-temperature equilibrium condensates of the Solar Nebula gas, commonly

Table 12.3 Simplified classification of meteorites (with percentage that falls to the Earth).

Classification	Characteristics
Iron meteorites (5.7%) (Fig. 12.3a, b)	Consist essentially (average 98%) of a Ni–Fe alloy (Ni usually between 4 and 20%, rarely greater), most with a characteristic structure known as Widmanstätten pattern, which consists of lamellae of kamacite (a Ni–Fe alloy containing about 6% Ni) bordered by taenite (another Ni–Fe alloy containing about 30% Ni); this pattern is typical of exsolution from a high-temperature alloy by very slow cooling. Accessory minerals include troilite (FeS), schreibersite $(Fe,Ni,Co)_3P$, graphite (C), daubreelite $(FeCr_2S_4)$, cohenite (Fe_3C), and lawrencite $(FeCl_2)$.
Stony–iron meteorites (1.5%)	Composed of Ni-Fe alloys and silicate minerals in approximately equal amounts.
Pallasites (Fig. 12.3c)	Consists of abundant olivine crystals (often of gem quality) in a continuous base of Ni–Fe alloy, rendering the meteorite a distinctive texture. Regarded as samples of core–mantle boundary material from differentiated asteroids.
Mesodiorites	The base of Ni–Fe alloy is discontinuous and olivine, if present, is an accessory phase. The silicate portion is heavily brecciated and consists mostly of pyroxene and plagioclase. Have a complex formation history.
Stony meteorites	Composed predominantly of silicate minerals (olivine and pyroxene). Other constituents include variable amounts of refractory inclusions, particles rich in Fe–Ni and sulfides, and interstellar dust.
Chondrites (85.7%)	Stony meteorites that have not been modified due to melting or differentiation of the parent body. Some were thermally metamorphosed (recrystalized without melting) in the Solar Nebula; these are referred to as equilibrated chondrites. Most chondrites contain chondrules. Other constituents of chondrites are Fe–Ni metal and sulfide grains, ameboid olivine aggregates (AOAs), isolated grains of silicate minerals, and very fine-grained dust with embedded presolar grains that originated elsewhere in the galaxy. The various classes of chondrites differ in their proportions of chondrules, calcium–aluminum inclusions (CAIs), and metal.
Carbonaceous chondrite (Fig. 12.3d)	Consists largely of hydrated ferromagnesian silicates (serpentine or chlorite); up to 10% of complex organic compounds, which are not of biologic origin but may represent precursors of biologic organisms; and CAIs, which are among he first solids condensed from the cooling protoplanetary disk. Some contain veins of carbonates and sulfates. The most volatile rich among meteorites. Chemistry matches that of the Sun more closely than any other class of chondrites. Parent bodies represent some of the most primitive matter that formed in oxygen-rich regions of early Solar System.
Ordinary chondrites (Fig. 12.3e)	Most common class of stony meteorites. Composed mostly of olivine and pyroxene. Contain both volatile and oxidized elements, and are believed to have formed in the inner asteroid belt.
Other chondrites	Less common chondrites of variable petrology and chemical composition. At one end of the spectrum are the highly reduced enstatite or E chondrites, containing enstatite $(MgSiO_3)$ and almost all of their iron in the metallic form; these must have formed in the inner solar system. At the other end are the highly oxidized rumurutiite or R chondrites, containing large amounts of Fe-rich olivine and oxidized iron, suggesting formation farther from the Sun
Achondrites (7.1%) (Fig. 12.3f)	Do not contain chondrules, have lower abundances of volatile and moderately volatile (e.g., the alkalis) elements, and are usually much more coarsely crystalline than chondrites. Many resemble terrestrial basaltic rocks in composition and texture. They crystallized from silicate melts or are residues of partial melting, and represent pieces of differentiated asteroids that have fragmented due to collision. Subclasses include: primitive achondrites (residues from partial melting that took place on small parent bodies having chondritic bulk compositions); meteorites from the asteroid Vesta (HED Group); other evolved achondrites; lunar meteorites (LUN Group); and martian meteorites (SNC Group).

modified by melting, recrystallization, and alteration; they are now generally interpreted to be mixtures of condensate materials and refractory residues of materials vaporized by transitory and local heating events, although some CAIs are probably condensates.

The U–Pb absolute age of CAIs in the Allende chondrite has been determined to be 4.566 ± 0.002 Ga (Allègre *et al.*, 1995), almost identical to the age, 4.5672 ± 0.0006 Ga, obtained for CAIs in the Efremovka chondrite (Amelin *et al.*, 2002), which make them the oldest material in the Solar System sampled and dated so far. Using ^{26}Al–^{26}Mg isotope systematics, it has been shown that many chondrules formed within 1 to 2 Myr of CAIs, and that high-temperature nebular processes, such as CAI and chondrule formation, lasted for about 3–5 Myr (Russell *et al.*, 2006). The Earth and the Moon formed about 50 to 100 Myr later. Carbonaceous chondrites are also notable for their "primitive" chemical composition, "primitive" in the sense that abundances of most chemical elements do not differ by more than a factor of two from those in the Sun's *photosphere* (the region of a star's surface, composed largely of neutral gas

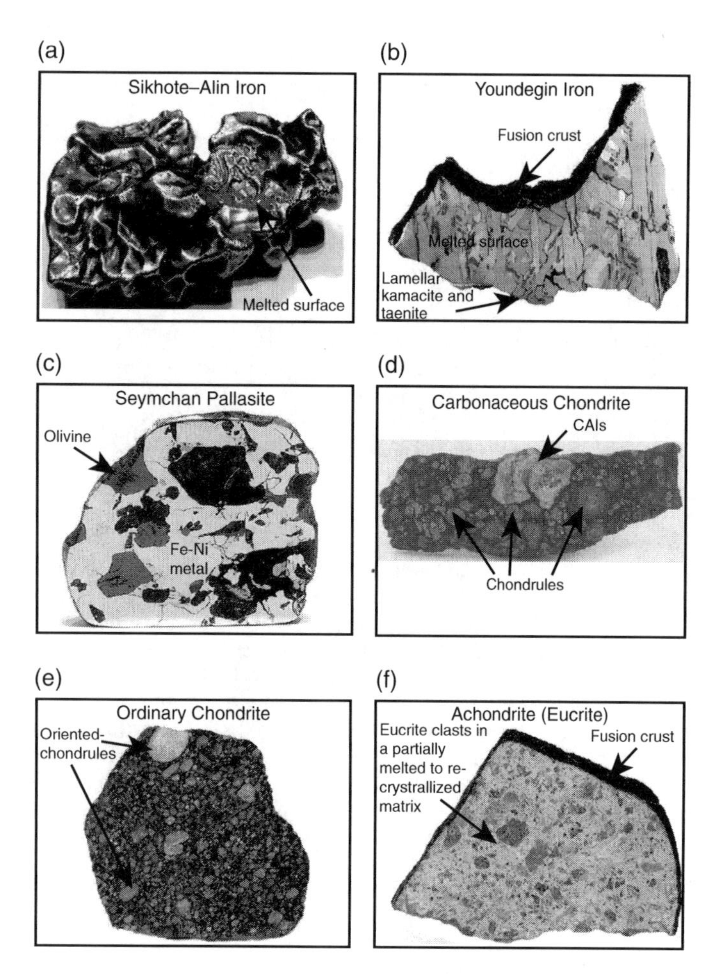

Fig. 12.3 Fragments of selected meteorites (Meteorites Australia Collection). (a) Iron meteorite (Sikhote-Alin; 83 g) with a melted surface. (b) Iron meteorite (Youndegin; 76.1 g complete slice) composed almost entirely of lamellar Fe–Ni alloys kamacite and taenite. (c) Stony iron meteorite (Seymchan pallasite; 15.5 g complete slice) showing olivine crystals in a Fe–Ni metal matrix. (d) Carbonaceous chondrite (NWA 2140; 1.31 g slice) with irregular shaped, light colored Ca–Al-rich inclusions (CAIs) and well preserved chondrules. (e) Ordinary chondrite (NWA 5507; 5.99 g complete slice) containing an abundance of chondrules, some of which show a preferred orientation. (f) Achondrite (NWA 1836 /Eucrite; 7.1 g crusted slice), a melted cumulate rock. (Source: http://www.meteorites.com.au/collection) (Copyright © 2002–2011, Meteorites Australia. Reproduced with permission of Jeff Kuyken, Meteorites Australia.)

atoms at a temperature of ~5700 K, from which radiation escapes), which in turn is believed to define the composition of our Solar System. This is the reason why chondrite compositions are often used as a standard for assessing the degree of chemical fractionation experienced by materials formed throughout the Solar System, including terrestrial rocks (see section 12.2.5). A comparison of the element abundances (normalized to 10^6 Si atoms) in the extensively studied Allende meteorite, a relatively primitive carbonaceous chondrite, with those in the Sun's photosphere (derived from spectral lines of

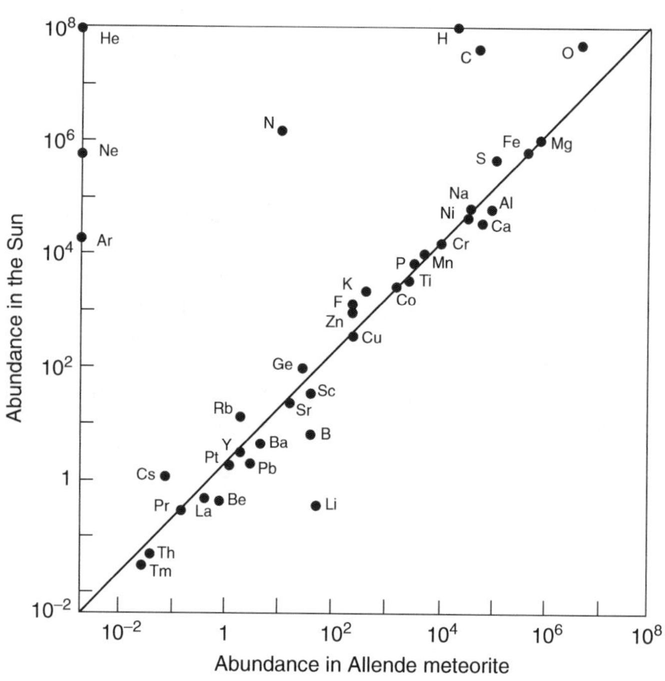

Fig. 12.4 Comparison of element abundances in the Allende meteorite, a carbonaceous chondrite, with those in the solar photosphere. Both axes are normalized to 10^6 silicon atoms. The diagonal line represents perfect correspondence. The scale is logarithmic, but the abundances are within a factor of two of each other, except for the highly volatile elements (H, C, N, O, and inert gases). (From *Geochemistry: Pathways and Processes* by H.Y. McSween, S.M. Richardson, and M.E. Uhle, Figure 15.5, p. 320; Copyright © 2003; Columbia University Press, New York. Used with permission of the publisher.)

the light emitted from the photosphere) illustrates the close match between them (Fig. 12.4), except for depletion of extremely volatile elements (H, He, C, N, O, and noble gases) and enrichment of a few very light elements (Li, Be, and B). It is assumed that the element abundances in the Sun's photosphere are representative of the Solar Nebula. This is a reasonable assumption because nuclear synthesis during the Sun's evolution should have affected only the composition of the deep interior (with the exception of Li, Be, and B, which are destroyed during hydrogen burning and so have been depleted near the Sun's surface).

12.1.5 *Solar System abundances of the elements*

The Solar System abundances of the elements, as compiled by Anders and Grevesse (1989), are listed in Appendix 9 and graphically represented in Fig. 12.5. The abundances are based mainly on the analyses of CI chondrites (a group of carbonaceous chondrites that lack chondrules and CAIs, but represent samples of the most undifferentiated Solar System matter available to us), except for the incompletely condensed elements (H, C, N, and O) and noble gases, for which solar photosphere and other astronomical data were used. The values are in terms of

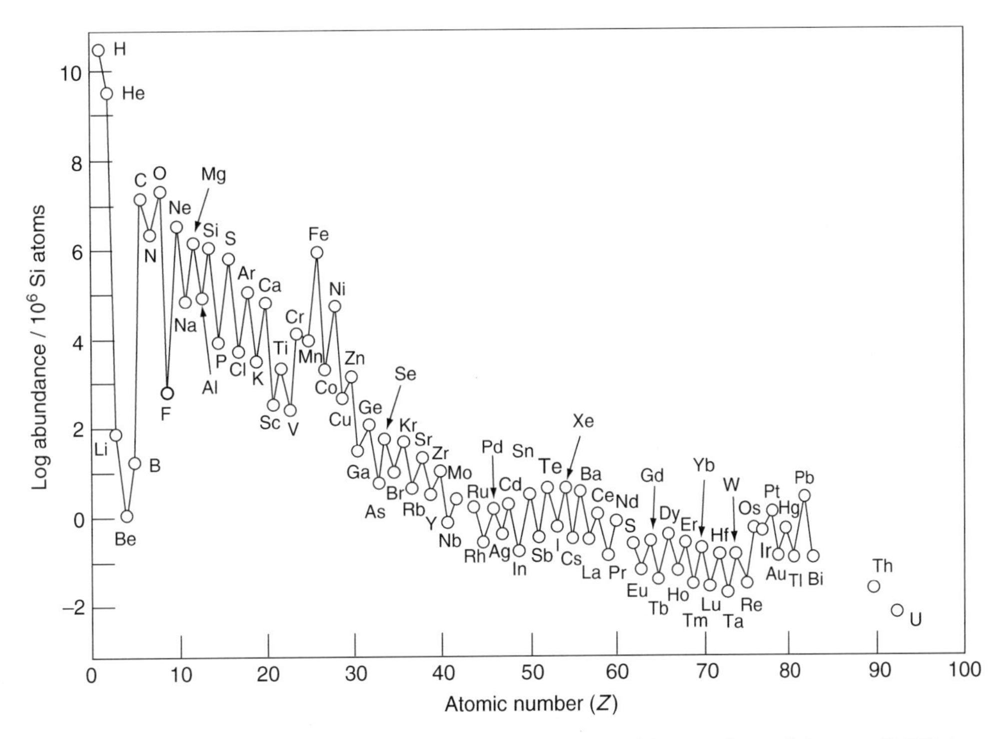

Fig. 12.5 Solar System abundances of the elements normalized to 10^6 silicon atoms. (Source of data: Anders and Grevesse (1989).)

number of atoms, normalized to 10^6 silicon atoms to keep these numbers from being very large. This abundance pattern is often referred to as the *cosmic abundance* of the elements, although it does not necessarily represent the composition of the universe because of variations in the compositions of the stars.

Noteworthy features of the abundance pattern revealed in Fig. 12.5 are as follows.

(1) Hydrogen and helium are, by far, the most abundant elements in the Solar System because these were the primary constituents of the primitive universe; the atomic H : He ratio is about 10.25: 1.

(2) The overall pattern is one of rapidly decreasing abundance with increasing atomic number (although the trend flattens for elements with $Z > 50$), reflecting decreasing nucleosynthesis in the more advanced burning cycles.

(3) Exceptions to the above trend are the abnormally low abundances of Li, Be, and B, which are depleted due to subsequent nuclear burning, and the higher abundance of the iron group relative to its neighbors (for reasons mentioned earlier). The high abundance of Fe makes it an inescapable candidate for being a major constituent of planetary cores.

(4) Elements having even atomic numbers (such as $^{12}_{6}C$) carbon are more abundant than their immediate neighbors having odd atomic numbers (such as $^{14}_{7}C$), which is known as the *Oddo–Harkins* rule. This is because elements with odd atomic numbers have one unpaired proton and are more likely to capture another, thus increasing their atomic number. In elements with even atomic numbers, protons are paired, with each member of the pair offsetting the spin of the other, thus enhancing the stability of the atom. A notable exception to this postulate is $^{1}_{1}H$, the most abundant element in the universe.

12.1.6 *Origin of the Solar System: the planetesimal model*

The most widely accepted model for the origin of the Solar System is that it started about 5 Ga with the formation of a nebula (an interstellar, rotating molecular cloud) from a larger molecular gas cloud some 30 to 40 light years across and 2 to 10 times the mass of the present Solar System. About 10,000 years after the Big Bang, the temperature had fallen to such an extent that the energy density of the universe began to be dominated by massive particles, rather than the light and other radiation that had predominated earlier. Thus, besides the hydrogen and helium gas and plasma generated during the Big Bang, the Solar Nebula consisted of microscopic dust (solid particles) and the ejected matter of long-dead stars. The occurrence of excess amounts of the isotope $^{24}_{12}Mg$ in CAIs of chondrites, attributed to the radioactive decay of the short-lived, now-extinct radionuclide $^{26}_{13}Al$ (half-life = 0.73×10^6 yr), suggests that the Solar Nebula formed very soon after nucleosynthesis. The neutron-rich, short-lived isotope $^{60}_{26}Fe$ (which decays to $^{60}_{28}Ni$ with a half-life of 1.5×10^6 yr only) is not formed in abundance by spallation reactions, but is formed in core collapse supernovae. The confirmation of live $^{60}_{26}Fe$ isotopes in the early Solar System is compelling evidence that the Solar System formed near a massive star that eventually went supernova, injecting elements already synthesized within the massive star into the Solar Nebula (Hester and Desch, 2005).

The Solar Nebula, estimated to have been composed essentially of H_2 and He (99.98%) with a $H_2O:H_2$ ratio of 5.4×10^{-4} (Lodders, 2003), collapsed under gravitational attraction, probably triggered by shock waves from the nearby supernova explosion. The collapsed material began a counterclockwise rotation, much of it gravitating to the center of rotation, and conservation of angular momentum caused the nebula to spin faster and faster. The balance between gravity and rotation led to a flattened disk (often referred to as the *proto-planetary disk*) with about 90% of the material concentrated in the central region, forming the proto-Sun that eventually developed into the present Sun. The planets formed as a byproduct of the evolution of the Solar Nebula, constrained by conservation of angular momentum contained in the system. That is why the planets revolve and rotate in counterclockwise direction, and their solar orbits lie nearly on the ecliptic plane. According to the law of conservation of angular momentum, the Sun should be rotating faster than it actually does (25 days per a 360° rotation); the relatively slow rotation is attributed to magnetic flow lines of the Sun interacting with ionized gases in the Solar Nebula.

The formation of the planets involved two important processes: *condensation* of the hot nebular gases into solid grains (dust); and *accretion* of the dust grains into progressively larger solid bodies, which acquired most of their materials locally. The temperature of the proto-planetary disk varied from ~2000K at the center to ~40K at approximately 50 AU (=7.5×10^9 km) from the proto-Sun. The pressure ranged from less than 0.1 atm at the center to about 10^{-7} atm near the edge of the disk (Cameron and Pine, 1973). According to most astrophysical models, the bulk of the pre-solar solid matter in the interior of the disk vaporized and then condensed as the nebula cooled after the formation of the proto-Sun. The condensation sequence was controlled by *P–T* conditions in various parts of the disk (see, e.g., Grossman and Larimer, 1974). No liquids were produced during condensation because the pressure in the Solar Nebula was too low. Refractory constituents of low volatility such as compounds of magnesium, calcium, iron, and silicon, and Fe–Ni metal, were largely left in planetary masses close to the Sun, where the terrestrial planets formed subsequently by accretion. Not much gas could accumulate in the inner Solar System because of the high temperature. Violent solar flares and solar wind, known as T-Tauri wind, carried hydrogen, helium, the noble gases, and many of the volatile elements outward, beyond 4 to 5 AU, where they accreted to the giant gassy planets, but with rocky, metallic cores. The predominantly gaseous composition of the outer planets is because of the much greater abundance of gases (mainly hydrogen and helium) in the original interstellar material. Water was able to condense as ice in the nebula at a temperature of ~160K at 4–5 AU and be retained in the distant planets and the satellites of the giant planets. This scenario explains the depletion of the terrestrial planets and meteorites in inert gases and volatile elements (such as K, Pb, and Rb). Cameron (1995) provides a detailed discussion of the nebula stage of the Solar System, which lasted about 1 Myr or less.

In the accretional stage of the planets, solid particles in the rotating proto-planetary disk began to coalesce together, possibly in regions of localized higher gravity such as the mid-plane of the accretion disk, to form a hierarchy of solid bodies of progressively larger sizes, starting with small grains and reaching dimensions of the order of 1 km. The kilometer-sized solid bodies are called *planetesimals*. Experiments have shown that irregular dust grains can stick together if they collide at speeds up to a few tens of meters per second. It is estimated by computer simulation that planetesimals with diameters of 5 km are achievable after a few thousand years. Once accretion produced kilometer-sized bodies, they were no longer controlled by nebular gas drag, and the planetesimals grew rapidly into *planetary embryos* of the order of 1000 km in diameter in roughly 10^5–10^6 yr by swallowing smaller objects available in their respective "feeding zone" (a roughly annular region of the order of 0.01 AU in width), a process referred to as "runaway growth" (Chambers, 2004). The ultimate size of a planet depended upon the amount of material available in its feeding zone.

The final stage of planetary accretion occurred over a timeframe of 10^7 years, and involved the collision of a few dozen planetary embryos comparable in size to the Moon or Mars (0.01–0.1 Earth masses). Numerical simulations suggest that the accretion of the Earth may have occurred in as little as ~5 Myr in the presence of a significant amount of nebular gas or in as long as ~100 Myr without the gas, although most of the planetary mass would probably be accreted within the first 10 Myr (Wetherill, 1986). It is not clear how much nebular gas was present during these early growth stages, but the near-circular orbits of the planets can be explained by the dampening of eccentricities due to the presence of even a small amount of nebular gas (Halliday, 2006). A small amount of dust and gases that never accreted into larger bodies still remain in interplanetary space. Smaller bodies not accreted into planets, except those retained in the asteroid belt, were lost from the Solar System.

The final planets were a mixture of material from a broad region of the inner Solar System, but most of it for any planet came from its own locale, the variations in the composition of the planets arising from the difference in the suites of planetesimals they acquired. The formation of the planets, including the Earth, was complete by 4.56 Ga, which is considered the beginning of geologic time (T_0).

12.2 Evolution of the Earth

12.2.1 *The internal structure of the Earth*

The present-day internal structure of the Earth is a composite of its original accretionary growth and subsequent modifications by physical and chemical processes such as heating, convective circulation, geochemical differentiation, etc. In essence, the Earth's interior consists of three concentric shells of markedly different composition and density (Fig. 12.6a): the *core*, the *mantle*, and the *crust*. This interpretation is based on multiple lines of evidence (Fig. 12.6b): (i) sharp breaks in density, which generally increases with depth; (ii) sharp breaks in the velocities of P and S seismic waves, which depend on the density and elastic constants of the medium; (iii) differentiation of the parent bodies of meteorites, as reflected in the spectrum of meteorite compositions; and (iv) mineralogy and estimated *P–T* conditions of equilibration of rock samples from the Earth's interior brought up through magmatism.

The crust

The volumetrically insignificant crust (only 0.7% of the Earth's mass) is composed almost entirely of silicate minerals, only eight elements – O, Si, Al, Fe, Ca, Mg, Na, K – making up about 99% of the total. The crust is of two distinct types: *oceanic crust* and *continental crust*. The low-lying oceanic crust, which covers approximately 70% of the Earth's surface area, is relatively thin (from ~4 km at some mid-ocean ridges to more than 10 km in some oceanic volcanic plateaus produced from profuse outpourings of basaltic lava; average thickness ~7 km), young (≤ 200 Ma), fairly homogeneous on a large scale, and composed of relatively dense rock types such as basalt (enriched in Ca, Mg, and Fe; Table 12.4). The continental crust, on the other hand, is topographically higher, thicker (from 20 to 80 km, average thickness ~40 km), much older (>3.8 Ga at places), and composed of diverse lithologies that yield an average intermediate igneous rock or "andesitic" bulk composition (enriched in Si, Na, K, volatiles, and radioactive elements). The continental crust is significantly enriched incompatible elements (with about 1% of the mass of the primitive mantle, it contains up to 50% of the primitive mantle's budget for these elements); it has a high La/Nb ratio, a low Ce/Pb ratio, and a subchondritic Nb/Ta ratio (Rudnick and Gao, 2004). The presence of a chemically evolved continental crust is one of the Earth's unique features compared to other rocky planets in our Solar System.

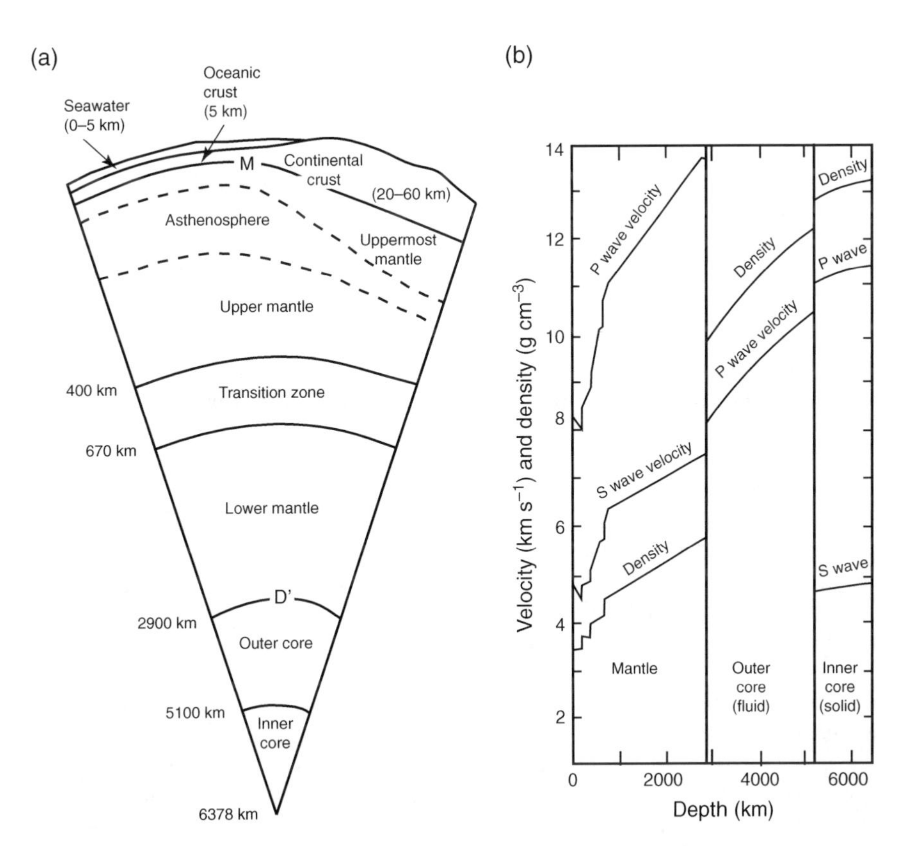

Fig. 12.6 (a) Schematic section of the Earth. Details of the crust and mantle are not to scale. M = Mohorovicic Discontinuity (commonly referred to as "Moho"); D" = the core–mantle boundary (CMB) (Wiechert–Gutenberg discontinuity?). The average densities of the various units are as follows: continental crust – 2.7 to 3.0; oceanic crust – 3.0 to 3.3; mantle – 3.3 to 5.7; outer core (liquid) – 9.9 to 12.2; inner core (solid) – 12.6 to 13.0. (b) Variation of density and velocities of seismic body waves (P- and S-waves) as a function of depth. Note that S-waves do not pass through the liquid outer core. (Sources of data: Wilson (1989); Ernst (2000).)

The mantle

Compared to the crust, the mantle, which represents about 83% of the volume and 68.3% of the mass of the Earth, is enriched in Mg, contains less Fe and much less Si, and is depleted in volatile and radioactive elements. The mantle is divided into three regions on the basis of major discontinuities in seismic velocities (Fig. 12.6a). The *upper mantle* extends from the *Mohorovicic Discontinuity* (or *Moho*), which marks a jump in compressional seismic wave (P-wave) velocity from ~7 km s^{-1} to ~8 km s^{-1} and is interpreted to mark the crust–mantle boundary, to a major seismic discontinuity occurring near a depth of 400 km. In some regions the Moho is transitional rather than a sharp discontinuity, and the exact location of the crust–mantle boundary is debatable. The upper mantle includes the lower part of the *lithosphere*, the strong outer layer of the Earth, including the crust, which reacts to stresses as a brittle solid, and the underlying weaker *asthenosphere* that readily deforms by creep. The asthenosphere is a low-velocity zone in which seismic waves are attenuated strongly, indicating the possible presence of a partial melt phase. This is also the zone in the upper mantle in which convective motions are most likely to occur.

A second major seismic discontinuity occurs near a depth of 650–670 km, and the region between the two seismic discontinuities is referred to as the *transition zone*. The *lower mantle* comprises the large region between the "650 km discontinuity" and the core, the outer limit of which is encountered at a depth of 2900 km. Constrained by the observed P-wave velocities, the compositions of ophiolite complexes and xenoliths recovered from alkali basalts and kimberlites, and the requirement that the mantle material must be capable of producing the large volumes of terrestrial basaltic magmas by partial melting, the upper mantle is inferred to be composed of ultramafic rocks such as peridotite (olivine + pyroxene) and eclogite (garnet + pyroxene). The composition of the upper mantle under oceanic crust is thought to be somewhat different from that under the continental crust. In both cases, however, the peridotite immediately below the Moho has been depleted in incompatible elements because of the extraction of copious volumes of basaltic magmas throughout geologic time, and the depleted peridotite is underlain by primitive, undepleted peridotite (Fig. 12.7). Clark and Ringwood (1964) proposed that the overall composition of this primitive source rock, which is capable of yielding basaltic magma on partial melting, corresponds to a hypothetical mixture of one part basalt to three parts dunite, and called it "pyrolite" (pyroxene–olivine rock) (Table 12.4). The depleted peridotite under the continental crust is thought to contain pockets of eclogite segregations; that beneath the oceanic crust is believed to be composed of an upper layer of strongly depleted harzburgite (olivine + orthopyroxene) due to extraction of mid-oceanic ridge basalts, and a lower layer of less depleted lherzolite (olivine + orthopyroxene + clinopyroxene) (Ringwood, 1991).

Most geophysical measurements are at least consistent with the upper and lower mantle, as well as the transition zone, having broadly the same bulk chemical composition. The mineralogy of the mantle, however, varies as a function of depth (Fig. 12.8) because of progressive increase in load pressure. High-pressure experimental data indicate that the transition zone is characterized by denser polymorphs of olivine: β-olivine, which has a distorted spinel structure (compared to orthorhombic α-olivine, whose structure consists of independent SiO_4 tetrahedra linked by divalent aoms in sixfold coordination); and, at still higher pressures, γ-olivine, which has a spinel structure. At a depth of about 300 km, the pyroxene components begin to form a solid solution with the garnet, and at about 450 km depth, the Si-rich garnet solid solution [termed *majorite*, empirical formula $Mg_3Fe_{1.2}^{3+}Al_{0.6}Si_{0.2}(SiO_4)_3$] becomes the only stable aluminous phase. The mineral assemblage adopted by pyrolite in the lower mantle is believed to comprise three dense phases – Mg-perovskite ($MgSiO_3$ with perovskite structure), Ca-perovskite ($CaSiO_3$ with perovskite structure), and magnesiowüstite ((Mg, Fe)O).

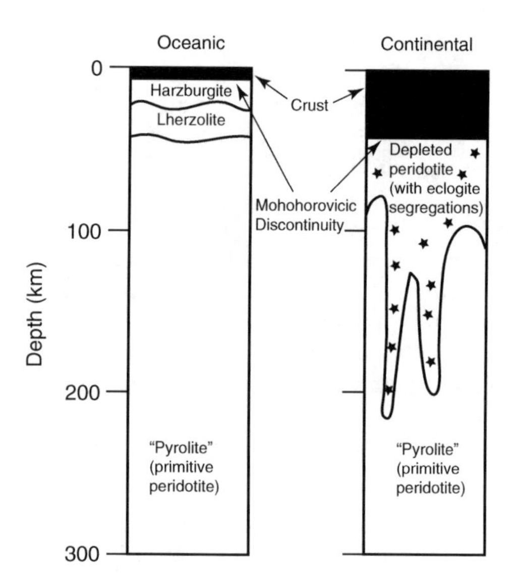

Fig. 12.7 Composition of the upper mantle according to the pyrolite model. (After Ringwood, 1991, Figure 1, p. 2326.)

The core

The Earth's core comprises about 16% of the Earth's volume and 31% of the Earth's mass. The average density of the Earth is 5.515 g cm^{-3}, making it the most dense planet in the Solar System. Since the average density of the crust is only 2.7 to 3.3 g cm^{-3} and that of the mantle is 3.3 to 5.7 g cm^{-3}, the Earth's core must be made of high-density elements, an inference supported by seismic data. If we assume chondritic meteorites to be the building blocks of terrestrial planets, the much less concentrations of siderophile elements in the Earth's upper mantle implies that these elements reside mainly in the core. The Earth's core is believed to be composed of Fe–Ni alloy (primarily Fe, with about 4–5% Ni and 0.2% Co), probably with small amounts of lighter elements such as S, C, O, and Si (Stevenson, 1981; Poirier, 1994; Kargel and Lewis, 1993; Allègre *et al.*, 1995). Model calculations by Kargel and Lewis (1993) yielded

Table 12.4 Estimated model compositions (major elements).

Oxide (wt%)	Primitive mantle† (1)	Primitive mantle (2)	Pyrolite (3)	Continental crust (4)	Oceanic crust (5)	"Normal" MORB (6)
SiO_2	45.0	49.9	45.16	59.7	49.4	50.45
TiO_2	0.20	0.16	0.71	0.68	1.4	1.61
Al_2O_3	4.45	3.65	3.54	15.7	15.4	15.25
Cr_2O_3	0.384	0.44	0.43			
MgO	37.8	35.15	37.47	4.3	7.6	7.58
FeO	8.05*	8.0*	8.04	6.5*	7.6	10.43*
Fe_2O_3			0.46		2.7	
CaO	3.55	2.90	3.08	6.0	12.5	11.30
MnO	0.135	0.13	0.14	0.09	0.3	
NiO	0.25	0.25	0.20			
CoO	0.013		0.01			
Na_2O	0.36	0.34	0.57	3.1	2.6	2.68
K_2O	0.029	0.022	0.13	1.8	0.3	0.09
P_2O_5	0.021	0.06	0.06	0.11	0.2	

(1) McDonough and Sun (1995), based on compositions of peridotite, komatiites, and basalts; (2) Taylor and McLennan (1985), based on CI carbonaceous chondrites; (3) Ringwood (1966); (4) Condie (1997); (5) Ronov and Yaroshevsky (1976); (6) Hofmann (1988).
*Total Fe as FeO.
†Primitive mantle (or bulk silicate Earth) = Earth's mantle immediately after the core formation (i.e., the present mantle plus crust).

the following composition for the core (in wt%): Fe – 85.55, O – 5.18, Ni – 4.88, S – 2.69, Cr – 0.45, Mn – 0.41, P – 0.35, Co – 0.22, Cl – 0.07, and possibly K – 0.02; modeling by Allègre *et al.* (1995) produced similar numbers for O (4 wt%) and S (2.3 wt%), but also 7.3 wt% Si. The outer core is interpreted to be molten (at a temperature of ~5000°C) because of its low viscosity and inability to transmit S-waves, whereas the inner core is solid, with the crystals probably exhibiting a large degree of common orientation. The liquid state of the outer core can be rationalized by the disposition of the melting curve for metallic iron (and minor nickel) relative to the Earth's present geothermal gradient (Fig. 12.9). The Earth's magnetic field owes its origin to fluid flow within the outer core, and the differential motion between the outer liquid, metallic core and the base of the solid, silicate mantle. One effect of this magnetic field is that it deflects the solar wind from the Earth, and therefore prevents certain molecules (e.g., water vapor) in the atmosphere from being swept away into space.

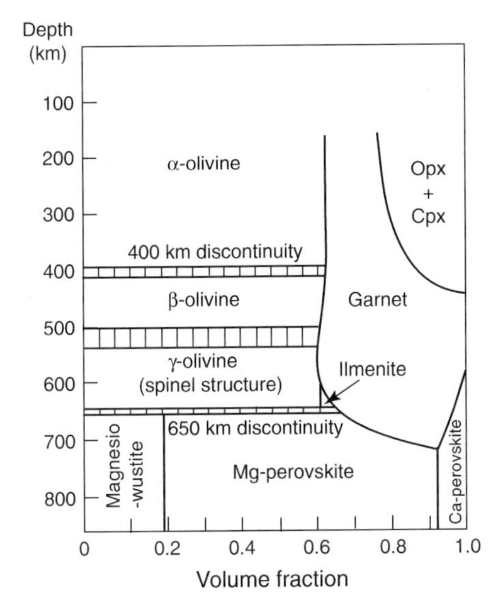

Fig. 12.8 Distribution of mineral assemblages in the upper mantle, assuming a pyrolite compositional model (after Ringwood, 1991). It is assumed that the temperature at 400 km depth is near 1400°C and at 650 km depth is near 1600°C. Opx = orthopyroxene; Cpx = clinopyroxene. (After Ringwood, 1991, Figure 4, p. 2090.)

12.2.2 *Bulk Earth composition*

The ratios of various elements and isotopes in the Earth roughly match those of undifferentiated chondrites. It appears reasonable, therefore, to infer that the bulk composition of the Earth itself must be close to the average composition of chondritic meteorites, the presumed building blocks of the Earth. The Earth's bulk composition can then be computed by distributing the metal and silicates of meteorites so as to satisfy the requirements of density and elastic properties deduced from the behavior of seismic waves. Mason (1966), for example, calculated the bulk composition of the Earth assuming that (i) the core has the composition of the average for iron–nickel in chondrites, and includes the average amount (5.3%) FeS in these meteorites; and (ii) the composition of the *bulk silicate Earth* (BSE), the residual silicate portion of the Earth after core formation, now represented by modern mantle plus crust, is the same as the silicates (plus small amounts of phosphates and oxides) of the "average" chondrite. The problem is that there is no consensus as to what kind of meteorite should be chosen as

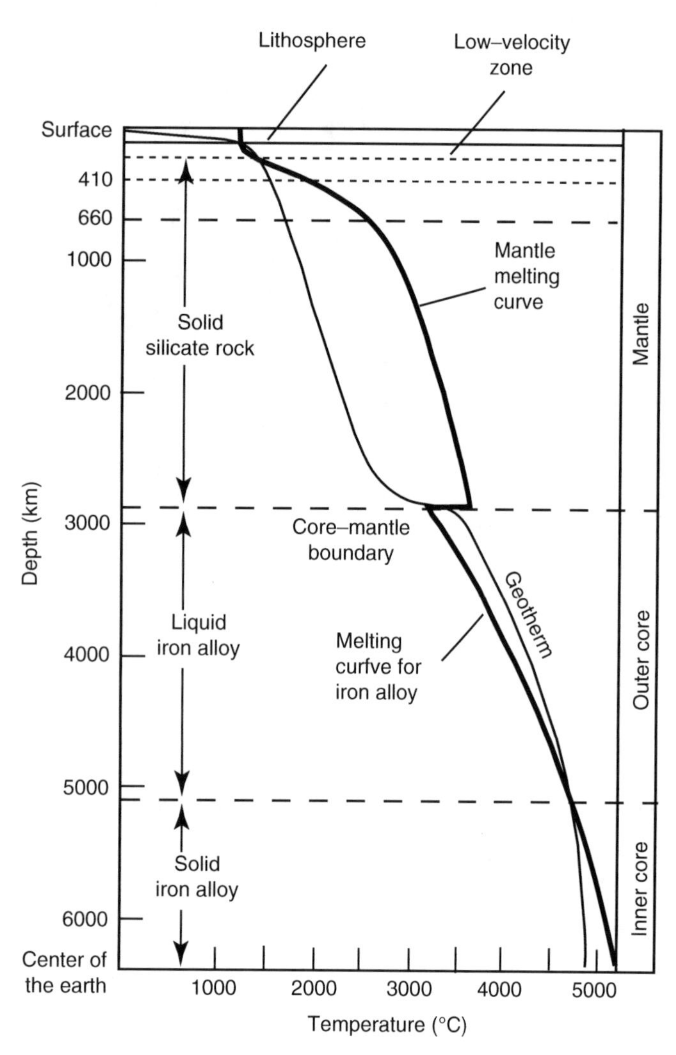

Fig. 12.9 Explanation for the liquid state of the outer core based on the estimated relative positions of the geotherm (which describes the increase in temperature with depth) and the melting curve for iron alloy. The geotherm lies below the mantle melting curve throughout most of the mantle (solid) and below the melting curve for iron alloy in the inner core (solid), but it lies above the iron melting curve in the outer core (liquid). The intersection of the geotherm with the mantle melting curve at the base of the lithosphere is consistent with the partially molten low-velocity zone. (After Grotzinger and Jordan, 2010 , Figure 14.10,p. 379.)

the best representative of an average composition or how meteorite compositions should be weighted to get an average composition.

There is even a more fundamental argument against the above approach. Drake and Righter (2002) compared Mg/Si and Al/Si ratios (Fig. 12.10), oxygen-isotope ratios, osmium-isotope ratios, and D/H, Ar/H_2O and Kr/Xe ratios of material from the Earth, Mars, comets, and various meteorites, and concluded that no primitive material similar to the Earth's mantle is currently represented in our meteorite collections. It appears that the "building blocks" of the Earth, at least in part, were composed of chondrites or achondrites

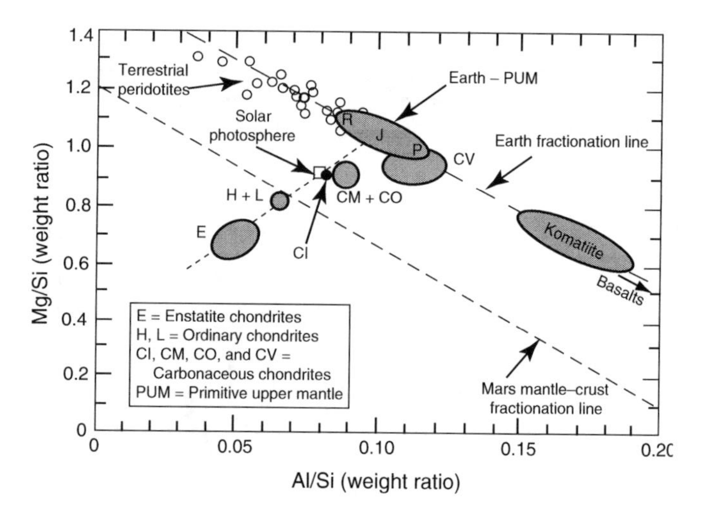

Fig. 12.10 Al/Si versus Mg/Si ratios of primitive material in the inner Solar System define an unexplained trend that is very different from the Earth's fractionation line defined by peridotites, komatiites, and basalts. The intersection of the two trends gives Si/Al ~0.11 and Mg/Si ~1, which are considered as the most plausible values for the bulk silicate Earth (BSE; Drake and Righter, 2002). The letters R, J, and P refer to estimates of the BSE composition by Ringwood (1979), Jagoutz et al.(1979), and Palme and Nickel (1986), respectively. Martian fractionation line is defined by Chassigny, shergottites, and martian soils and rocks from the Viking and Pathfinder mission (Dreibus et al.,1998). (Modified from *Meteorites and the Early Solar System, II*, edited by D. Lauretta, and H.Y. McSween, Jr. © 2006 The Arizona Board of Regents. Reprinted by permission of the University of Arizona Press.)

yet to be sampled. Zolensky *et al.* (2006) have shown that the flux of material to the Earth's surface changes over time such that material accreting to the Earth today may not be entirely representative of that available in the early Solar System. The estimation of the Earth's bulk composition is a difficult task in the absence of known chondrites that could be considered representative of the undifferentiated primitive Earth.

An alternative approach would be to combine the average compositions of the Earth's various units formed by differentiation according to their relative proportions. The problem is that whereas the compositions of the atmosphere, hydrosphere, and crust can be determined by direct measurements, the compositions of the mantle and core are model dependent. There is no way to calculate the core composition directly from seismic data, which merely constrain the density and physical state, and the composition of the mantle is dependent on the model adopted for the composition of the "primitive mantle" (the *bulk silicate Earth*, BSE).

Element abundances for the whole Earth presented in Table 12.5 were estimated by Kargel and Lewis (1993) following a different approach. First, they established a "best" estimate of the BSE. This estimate was based on analyses of oceanic basalts, continental flood basalts, oceanic and continental alkali basalts, ultramafic components of ophiolites, Archean komatiites, and spinel- and

Table 12.5 Estimated bulk Earth element abundances (in ppm).

Z	Element	Abundance	Z	Element	Abundance	Z	Element	Abundance
1	H	36.9?	32	Ge	10.2	63	Eu	0.1035
2	He		33	As	1.73	64	Gd	0.363
3	Li	1.69	34	Se	3.16	65	Tb	0.0671
4	Be	0.052	35	Br	0.13	66	Dy	0.448
5	B	0.292	36	Kr	?	67	Ho	0.1027
6	C	44?	37	Rb	0.76	68	Er	0.294
7	N	0.59	38	Sr	14.4	69	Tm	0.0447
8	O	316700	39	Y	2.88	70	Yb	0.300
9	F	15.8	40	Zr	7.74	71	Lu	0.0449
10	Ne	?	41	Nb	0.517	72	Hf	0.203
11	Na	2450	42	Mo	1.71	73	Ta	0.0281
12	Mg	148600	43	Tc		74	W	0.171
13	Al	14330	44	Ru	1.71	75	Re	0.0674
14	Si	145900	45	Rh	0.227	76	Os	0.898
15	P	1180	46	Pd	0.831	77	Ir	0.889
16	S	8930	47	Ag	0.099	78	Pt	1.77
17	Cl	264	48	Cd	0.068	79	Au	0.157
18	Ar	?	49	In	0.0049	80	Hg	0.0065
19	K	225	50	Sn	0.34	81	Tl	0.0073
20	Ca	16570	51	Sb	0.061	82	Pb#	0.172
21	Sc	11.1	52	Te	0.39	83	Bi	0.0057
22	Ti	797	53	I	0.036	84	Po	
23	V	104	54	Xe	?	85	At	
24	Cr	3423	55	Cs	0.055	86	Rn	
25	Mn	2046	56	Ba	4.33	87	Fr	
26	Fe	320400	57	La	0.434	88	Ra	
27	Co	779	58	Ce	1.114	89	Ac	
28	Ni	17200	59	Pr	0.165	90	Th	0.0543
29	Cu	82.7	60	Nd	0.836	91	Pa	
30	Zn	47.3	61	Pm		92	U	0.0152
31	Ga	4.42	62	Sm	0.272		Sum	100.02

Nonradiogenic Pb only.
Source of data: Kargel and Lewis (1993).

garnet-bearing xenoliths recovered from basalts and kimberlites (see section 12. 3.1). The BSE incorporates the combined effects of core formation and volatility, presumably including condensation and sublimation in the solar nebula and volatile loss during the accretion and early evolution of the Earth. So, next, they established the Earth's *volatility trend*, which was based on a plot of chondrite-normalized abundances of lithophile elements versus calculated nebular condensation temperatures. Element abundances for the core and the bulk Earth were inferred from BSE abundances and the volatility trend.

Examination of Table 12.5 shows that (i) about 90% of the Earth is made up of only four elements: Fe, O, Si, and Mg; (ii) the abundance of only three other elements – Ni, Ca, and Al – exceed 1%; and (iii) ten elements – Na, P, S, Cl, K, Ti, V, Cr, Mn, and Co – occur in amounts ranging from 0.01 to 1%. Thus 17 elements account for about 99.9% of the bulk Earth chemical composition, the rest of the elements comprising 0.1% or less of the bulk.

12.2.3 *The primary geochemical differentiation of the proto-Earth: formation of the Earth's core and mantle*

Timing of core formation

Short-lived radioactive isotopes, such as ^{26}Al (^{26}Al $\Rightarrow$ ^{26}Mg ; $t_{1/2}$ = 0.73 Myr), ^{60}Fe (^{60}Fe $\Rightarrow$ ^{60}Ni; $t_{1/2}$ = 1.5 Myr), and ^{182}Hf (^{182}Hf $\Rightarrow$ ^{182}W; $t_{1/2}$ = 8.9 Myr), can be used to constrain the timimg of differentiation events that occurred within their lifetimes. The decay of the extinct radionuclide ^{182}Hf to ^{182}W is an ideal isotopic system for tracing the rates of accretion and core formation of the Earth and other inner Solar System objects for the following reasons (Halliday and Lee, 1999).

(1) The half-life of ^{182}Hf (8.9×10^6yr) is ideal for evaluating events that occurred up to ~50–60 Myr after the start of the Solar System.

(2) The initial abundance of ^{182}Hf was relatively high ($^{182}Hf/^{180}Hf = 10^{-4}$), so that the isotopic effects produced are easily resolved.
(3) Both Hf and W are refractory elements, so parent : daughter ratios and W isotope compositions of bulk planets are usually chondritic, and hence well defined.
(4) Both Hf and W are strongly fractionated during core segregation, which is considered to be an early process in planetary evolution. Hf, a strongly lithophile element, is retained in the silicate portion (mantle) of a planet or planetesimal, whereas W, a moderately siderophile element, is largely partitioned into a coexisting metallic phase (core). Consequently, if core formation took place when the Earth still had some ^{182}Hf, the W remaining in the silicate mantle would eventually develop an excess abundance of ^{182}W relative to that of chondrites. Conversely, if core formation was late, after (practically) all ^{182}Hf had decayed, the W isotope composition of the silicate Earth would be identical to that found in chondrites.

The slight excess $^{182}W/^{183}W$ ratio observed in the bulk silicate Earth compared to chondrite samples (Yin *et al.*, 2002; Kleine *et al.*, 2004) implies that segregation of the Earth's core occurred within a relatively short interval after the formation of the Solar System, during the life-time of ^{182}Hf. Modeling of the W-isotope composition of BSE as a function of time indicates that the bulk of metal–silicate separation (i.e., core–mantle differentiation) in the proto-Earth was essentially completed within < 30 Myr of the Solar System formation (Yin *et al.*, 2002; Kleine *et al.*, 2002, 2004). These results are consistent with other lines of evidence for rapid planetary formation (Lugmair and Shukolyukov, 1998), and are also in agreement with dynamic accretion models that predict a relatively short time (~10 Myr) for the main growth stage of terrestrial planet formation (Wetherill, 1986).

Formation of the Moon: the giant impact hypothesis

The last major growth stage of the Earth's core formation is defined by the age of the Moon. The most widely accepted model for the origin of the Moon at present is based on the so-called *giant impact hypothesis* (Hartman and Davis, 1975) – the off-center impact delivered by a Mars-sized planetary body (sometimes referred to as Theia) of approximately chondritic composition with the Earth, when the Earth had acquired about 90% of its final mass.

^{182}Hf—^{182}W isotope data indicate that the Moon formed 30–50 Myr after the start of the Solar System, the exact age being dependent on the model deployed (Halliday and Kleine, 2006). The most commonly quoted date is 45 ± 5 Ga (Halliday, 2003; Kleine *et al.*, 2004), which is almost the same as the Moon's Rb–Sr age, 4.48 ± 0.02 Ga (Halliday, 2008). The impactor and the Earth are assumed to have already differentiated by that time into a metallic core and a metal-poor silicate mantle. Hydrodynamic simulations of the giant impact require that as much as 70–90% of the material forming the Moon came from the debris disk of vaporized silicate mantle of the impactor (not from the terrestrial mantle), and the metallic core of the impactor accreted to the Earth contributing the final ~10% of the Earth's mass (Canup and Asphaug, 2001; Canup, 2004; Pahlevan and Stevenson, 2007). Calculations indicate that the Moon could have formed from the impact debris in 10 yr or less, certainly in no more than 100 yr. The impact model explains the angular momentum of the Earth–Moon system (the spin of each, plus the orbital motion of the Moon around the Earth), the volatile-poor and refractory-rich composition of the Moon, and its iron-depleted bulk composition. The Moon has only a tiny metallic core, and estimates of its iron abundance vary between 8 and 12% compared to about 31% for the Earth (McSween and Huss, 2010).

Wiechert *et al.* (2001) showed that oxygen isotope compositions of lunar samples from Apollo missions define, within analytical uncertainties ($2\sigma = \pm 0.016$), a single mass-dependent fractionation line in a plot of $\delta^{18}O$ versus $\delta^{17}O$, and it is identical to the terrestrial fractionation line within uncertainties. They attributed this congruence to isotopic similarity between the Moon-forming impactor and the proto-Earth. The problem is that most of the Moon material came from the impactor. For the impactor to have the same mix of primordial material as the Earth, we have to invoke the unlikely scenario of the two bodies growing at exactly the same distance from the Sun, being fed from the same part of the Solar Nebula, and having similar histories of core–mantle separation. Perhaps, as argued by Pahlevan and Stevenson (2007), the oxygen isotope composition was homogenized as a result of vigorous mixing between the Earth and the proto-lunar disk in the aftermath of the giant impact.

Calculations indicate that the energy released in such an impact would be enough to melt much of the impact debris, implying that the Moon would have formed very hot, possibly entirely molten (Canup, 2004; Pahlevan and Stevenson, 2007). This hot initial state is consistent with the decades-old idea that the Moon was surrounded by a "magma ocean" when it formed. The highly anorthositic composition of the lunar crust, as well as the existence of KREEP-rich lunar rocks (i.e., rich in K, REE, and P) are consistent with the idea that a large portion of the Moon was once molten, and a giant impact scenario could easily have supplied the energy needed to form such a "magma ocean." It is also likely that the Moon-forming impact, the last time that the Earth was hit by another planet-size body, led to the loss of the Earth's existing atmosphere, large-scale melting and vaporization of some of the Earth's mantle, and the formation of a terrestrial "magma ocean" engulfed in a rock-vapor atmosphere (Koeberl, 2006; Zahnle, 2006; Pahlevan and Stevenson, 2007; Wilson, 2008).

Homogeneous versus heterogeneous accretion models

Over the years, much of the debate about the mechanism of the Earth's accretion and its differentiation into a primitive mantle and core has focused on two end-member models (Fig. 12.11): (i) *homogeneous accretion* and (ii) *heterogeneous accretion.*

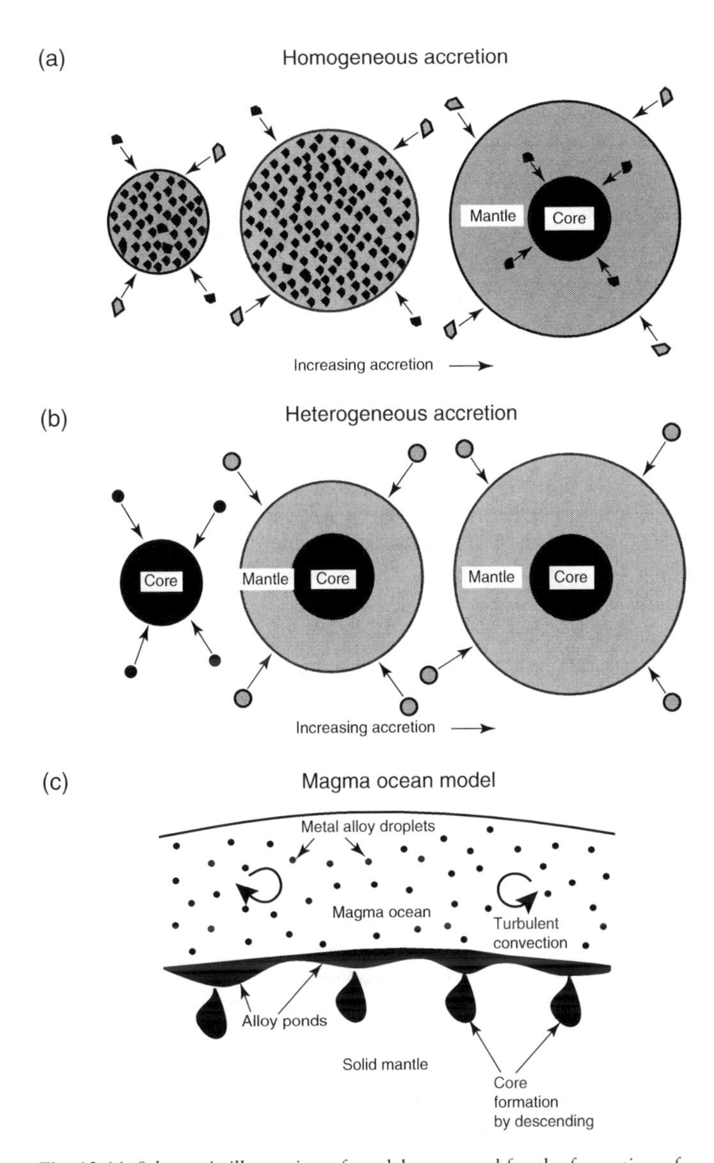

Fig. 12.11 Schematic illustration of models proposed for the formation of the Earth's metallic core and silicate mantle (BSE). (a) Homogeneous accretion, a two-stage process consisting of the accretion of a homogeneous (undifferentiated) proto-Earth and its subsequent differentiation into metallic core and silicate mantle. (b) Heterogeneous accretion, which envisions contemporaneous accretion and differentiation of material that changed from reducing to oxidized with time (followed by the addition of a "late veneer" after the core formation). (c) "Deep magma ocean" model of core formation. Small metal alloy droplets settle rapidly through the strongly convecting silicate liquid, accumulate as ponds of molten alloy at the bottom of the magma ocean, and equilibrate with the surrounding silicate material before descending as diapirs into the growing core.

Both models, neither of them totally satisfactory, assume that the Earth is made of extant meteoritic material that originated in the asteroid belt; the main difference between them lies in the timing and mechanism of metal–silicate separation.

According to the homogeneous accretion model (Ringwood, 1979), condensation of the Earth material from the solar nebula gas was essentially complete before accretion began. The temperature of the nebula is assumed to decrease outward from the Sun, and the composition of the condensate was a function of the distance from the Sun. The solids so formed accreted to form various planets without getting involved further in reactions with the gas; any uncondensed materials were somehow flushed from the system. The Earth accreted as a cool, homogeneous and undifferentiated mixture of silicates and metals, but became molten soon after accretion due to meteorite bombardment, gravitational compression, and radioactive decay of mainly short-lived radionuclides such as ^{26}Al, ^{182}W and ^{142}Nd (especially ^{26}Al), and differentiated internally into the core–mantle structure. The heavier metals (predominantly Fe and Ni) percolated downwards, eventually forming large dense accumulations that sank rapidly towards the center to form the core (Elsasser, 1963; Walter and Tronnes, 2004). Problems with this model, which satisfactorily accounts for planetary mean densities, are as follows (McSween *et al.*, 2003): (i) it is unrealistic to assume that a large planet would accrete only condensates formed at single temperature; and (ii) the planets Venus, Earth, and Mars should have no atmosphere because no condensed volatiles are possible at the condensation temperatures for these planets.

The heterogeneous accretion model, originally proposed by Turekian and Clark (1969), envisages planetary growth by simultaneous condensation and accretion of various compounds as the temperature fell inside an originally hot Solar Nebula. The model is based on the premise that all solids were vaporized in a hot Solar Nebula prior to condensation, and each new condensate was accreted as soon as it formed, preventing its subsequent reaction with the vapor. The planet grew by accretion of chondritic material that became more oxidizing with time. The result was a layered planet – the most refractory material (e.g., Fe + Ni), which would condense first out of a hot nebular gas, accreted forming a single metallic core. This was followed by condensation of less refractory silicates at lower temperatures and addition of this material, in layers, to generate an iron-rich core overlain by the primitive mantle composed of silicates, and finally an atmosphere composed of volatile phases. The heterogeneous accretion hypothesis has been widely accepted, but advances in our understanding of the accretion process have made this hypothesis less viable (Righter, 2003).

Let us examine how the heterogeneous accretion model fits the available geochemical data. It is reasonable to expect that refractory elements were accreted to all planets in cosmic or average solar system abundances (corrected for volatility as described by Newsom and Sims, 1991). The differentiation of a planet into mantle and core would partition trace elements according to the their distribution coefficients under planetary conditions of temperature, pressure, oxidation state, and compositions of the core and mantle, the distribution coefficient of an element i between molten alloy (core) and silicate (mantle) being defined as (see Box 5.6)

$$D_i^{\text{alloy-silicate}} = C_i^{\text{alloy}}/C_i^{\text{silicate}} \qquad (12.33)$$

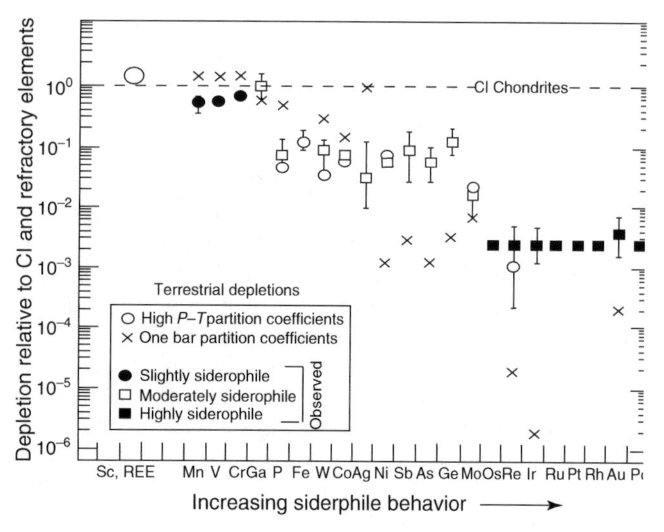

Fig. 12.12 "Stair-step" pattern observed for the depletion of siderophile elements (normalized to CI carbonaceous chondrites and refractory elements) in the Earth's primitive upper mantle. For highly and moderately siderophile elements, the abundances are greater than those calculated using partition coefficients between a peridotite magma and metallic liquid measured at 1 bar pressure and 1200°C–1600°C. This discrepancy is commonly referred to as the "excess siderophile element problem." Calculations using high *P–T* (270 ± 60 kbar; 2250 ± 300 K) partition coefficients overlap the observed depletions for moderately siderophile elements and Re, consistent with metal–silicate equilibrium and homogeneous accretion. The depletions of Ga, Sn, and Cu are due entirely to volatility rather than core formation (After Drake and Righter, 1997, Figure 1, p. 542.)

where C_i^{alloy} and $C_i^{silicate}$ are the concentrations of *i* in weight per cent (or ppm) in the metal alloy liquid and silicate liquid, respectively. However, as shown in Fig. 12.12, the abundances of moderately siderophile (P through Ge) and highly siderophile (Os through Pd) elements in the BSE are too high to be consistent with metal (core)–silicate (BSE) equilibrium based on distribution coefficients measured at low pressure (1 bar) and moderate temperature (1200°C–1600°C). This observation has come to be known as the "excess siderophile element problem" (Ringwood, 1966). Actually, the term is somewhat misleading because some of the moderately and slightly siderophile elements are actually less abundant than predicted by their partition coefficients (Fig. 12.12).

Wänke (1981), a proponent of the heterogeneous accretion model, addressed the stair-step pattern of siderophile elements by suggesting that the material accreting to the Earth changed in composition and oxidation state in time (Fig. 12.11b). The first 80% to 90% of the material was very reducing, so that all elements included in Fig.12.12, except the refractory lithophile elements such as Sc and REE, were quantitatively extracted into the core, and the mantle became almost devoid of Fe^{2+}. The next 20% to 10% or so of the accreting material was more oxidized (such as Fe, Co, Ni, Ga, W, Zn, Ge as oxides) and contained elements of moderately volatile character (such as Na, K, Rb, F) in chondritic abundances. The admixture of a small amount (< 1%) of metallic iron and segregation of this metal was responsible for the extraction of highly siderophile elements (Ir, Os, Au, etc.), and possibly also S and chalcophile elements, present in this material to the core, leaving other elements stranded in the mantle. Icy planetesimals (richer in volatiles such as H_2O, halogens, and CO_2), added the last 1% or so of the accreting material, the so-called "late veneer," after core formation had ceased. This material was too oxidizing to coexist in equilibrium with metallic iron, and all the siderophile elements delivered by the "late veneer" were forced to stay in the mantle, where they were thoroughly homogenized. The term "late veneer," although entrenched in the literature, is misleading, as the last dregs of material accreted to the Earth are well mixed into at least the upper mantle and perhaps the entire mantle, rather than veneering the surface (Drake and Righter, 2002).

The heterogeneous accretion hypothesis makes dynamical sense in that the "feeding zone" of the Earth's accretion must have extended further out from the Sun with the growth of the planet, but the zonal structure of the Earth could not have been predicted from condensation temperatures of earth materials because of the large degree of overlap. Also, according to this model, there should be much more sulfur in the Earth's mantle and Ga (which should have behaved like Ni) in the core than estimated (Drake, 2000).

Magma ocean hypothesis

There is general agreement that the Earth experienced large-scale melting and developed, probably multiple times, one or more magma oceans late in its accretion. The melting was an inevitable consequence of the heating due to collisions among the accreting planetesimals, radioactive decay of short-lived nuclides within the first few million years of Solar System history, and the greenhouse effect of the evolving atmosphere (see section 13.3.3). The last global-scale terrestrial magma ocean probably resulted from the Moon-forming giant impact. The oxygen isotope homogeneity observed in meteorites (angrites, eucrites, and lunar and Martian meteorites) and terrestrial samples is consistent with the involvement of magma oceans in the Earth's evolution (Greenwood *et al.*, 2005). The Earth most likely accreted heterogeneously, with material composition changing with time, but we have not found any evidence for it. It appears that the magma ocean(s) homogenized the pre-existing material, thus obliterating any record of heterogeneous accretion, with the possible exception of the "late veneer" of highly siderophile elements that was added later than core formation (Drake and Righter, 2002).

Accretion followed by magmatic processes in a deep, high *P–T* magma ocean environment offers a reasonable explanation for the Earth's bulk geochemical properties. The essence of the "deep magma ocean" hypothesis (Murthy, 1991; Li and Agee, 1996; Righter and Drake, 1997, 1999; Drake, 2000; Rubie *et al.*, 2003; Wade and Wood, 2005; Righter, 2007) is that droplets (of diameter of about 1 cm and a settling velocity of about 0.5 m s^{-1}, according to Rubie *et al.*, 2003) of metallic liquid descended through a deep magma ocean, equilibrating with the silicate at high temperatures and pressures as they fell. Experimental studies have shown that metal–silicate partition coefficients of siderophile elements, such as Ni and Co, are significantly reduced at elevated temperatures and pressures (Li and Agee, 2003). The liquid metal ponded in pockets at the base of the magma ocean, which extended approximately to the base of the current upper mantle, and subsequently descended rapidly as large diapers to the growing core without further equilibration with the surrounding silicate material (Fig. 12.11c). Modeling suggests a short duration, in the order of 1000 yr, for metal–silicate equilibration and magma ocean crystallization (Rubie *et al.*, 2003), and the short timeframe is consistent with W–Hf dates for Moon formation around 4540 Ma. The principal objection to this model is the improbability that one set of P–T conditions can satisfy the mantle abundances of all elements.

As summarized in Righter (2003), estimates of the temperature and pressure at the base of the magma ocean, based on experimental determination of partition coefficients of many of the excess siderophile elements (Mn, V, Cr, P, Fe, W, Co, Ni, Mo, Re) between metal alloy liquid and a hydrous peridotite magma, range from 2000 K to 4000 K and 25 GPa to 60 GPa, at an oxygen fugacity about 2 log units below that of the iron–wüstite buffer (Li and Agee, 1996; Chabot and Agee, 2003; Righter, 2003; Chabot *et al.*, 2005). Assuming that the core formed by a single-stage process, Wade and Wood (2005) determined that the mantle concentrations of the refractory elements could be matched at a temperature of 3750 K and a pressure of 40 GPa. At 40 GPa, however, the calculated equilibrium temperature turned out to be about 1000 K above the peridotite *liquidus* (the boundary curve on a phase diagram, representing the onset of crystallization of a liquid with lowering temperature). This is an implausible scenario because the base of the magma ocean, where the metal would pond during accretion, must be saturated in crystals and should, therefore, lie at or below the peridotite liquidus. Forcing the temperature to lie on the peridotite liquidus as the Earth grew, Wade and Wood (2005) could match the mantle concentrations of refractory elements provided oxygen fugacity increased during accretion. They proposed that as the Earth continued to grow by accretion, Mg-perovskite (which is stable below 660 km in the present-day Earth and is the principal phase in the lower mantle) started to crystallize at the base of the magma ocean, and the mantle became increasingly more self-oxidizing through cycles of dissolution and reprecipitation of Mg-perovskite as described below.

In peridotite compositions, Mg-perovskite accommodates about 5% Al_2O_3, the charge balance being maintained by the coupled substitution

$$Mg^{2+}\ Si^{4} \Leftrightarrow Fe^{3+}\ Al^{3+} \quad (12.34)$$

This substitution mechanism is so stable that it forces ferrous iron (Fe^{2+}) to disproportionate to ferric iron (Fe^{3+}) and iron metal (Fe^0):

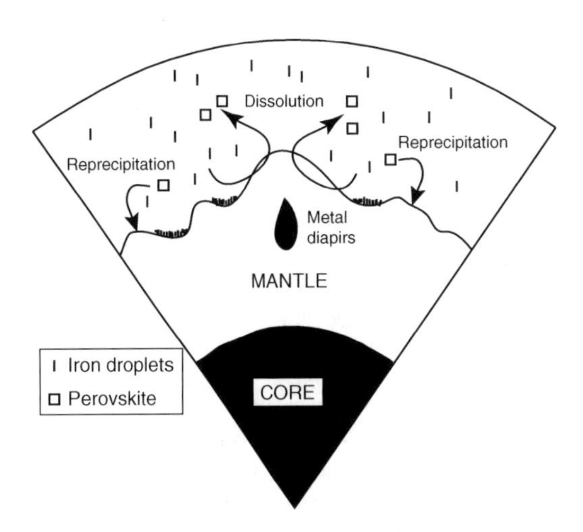

Fig. 12.13 Formation of the Earth's core accompanied by mantle "self-oxidation" as a result of perovskite precipitation at the base of the terrestrial magma ocean. The repeated crystallization and dissolution of Fe^{3+}-containing perovskite (due to fluctuating depth of the magma ocean) acted as a Fe^{3+} "pump," raising progressively the oxygen fugacity of the magma ocean from which the mantle grew by crystallization. (After Wade and Wood, 2005, Figure 10, p. 92.)

$$3Fe^{2+}\text{ (in silicate)} \Leftrightarrow 2Fe^{3+}\text{ (in perovskite)} + Fe^0 \quad (12.35)$$

Stated in terms of oxide components of the lower mantle, perovskite that crystallized from the magma ocean (silicate melt) dissolved Fe^{3+} as the $Fe^{3+}AlO_3$ component and released Fe^0 in the process:

$$3Fe^{2+}O + Al_2O_3 \Leftrightarrow \underset{\text{perovskite}}{2Fe^{3+}AlO_3} + \underset{\text{metal}}{Fe^0} \quad (12.36)$$

The depth of the magma ocean fluctuated continuously, generating fronts of dissolution and precipitation of perovskite (which is stable above 23 GPa) at its lower boundary (Fig. 12.13). The repeated dissolution, which released Fe^{3+} to the magma ocean, and precipitation, which produced more Fe^{3+} (reaction 12.36), resulted in a progressive oxidation of the magma ocean. Later droplets of metal falling through the magma ocean reacted with Fe^{3+}, driving reaction 12.36 to the left and producing more Fe^{2+} that dissolved in the magma ocean and was available for disproportionation to Fe^{3+} and Fe^0. Thus, oxidized iron (Fe^{3+}) content in the magma ocean (mantle) increased as a consequence of perovskite crystallization, raising the oxygen fugacity of core metal–silicate equilibration. In the very final stages of the Earth's accretion, this mechanism may have caused sufficient oxidation to halt metal segregation to the core, thus setting the stage for the "late veneer" to establish the mantle contents of highly siderophile elements.

The deep magma ocean hypothesis is also compatible with partitioning of light elements such as silicon into the metallic core. Gessman *et al.* (2001) and Malavergne *et al.* (2004) have shown that silicon solubility in iron–alloy melt is enhanced at high pressures.

12.2.4 Formation and growth of the Earth's crust

Many questions regarding the formation and evolution of the Earth's crust have not been resolved. When did the first crust form? Was the earliest crust ultramafic, mafic, or felsic? How were continental and oceanic crusts extracted from the mantle? Did the crusts grow continuously or episodically? Here is a brief summary of what we know about answers to these questions.

The earliest crust

We really do not know the nature of the earliest crust, but comparison with other planets suggests that the Earth's first or primary crust was basaltic in composition like modern oceanic crust. At present, new oceanic crust is created predominantly at spreading ridge axes, from basaltic melts generated directly from the upper mantle by partial melting, and recycled into the mantle by subduction at continental margins. The earliest oceanic crust probably formed and was recycled the same way. However, as the Earth's oldest oceanic crust is only about 200 Ma old, and no remnants of Hadean or Archean oceanic crust (e.g., in the form of ophiolites, which are interpreted as obducted oceanic crust) have yet been identified, we have no direct evidence regarding the nature of oceanic crust in the early Earth. It is reasonable to speculate that early oceanic crust probably crystallized from a magma ocean soon after planetary accretion and, like modern oceanic crust, it probably was widely distributed on the Earth's surface. Because of the greater amount of heat in the Archean upper mantle, oceanic crust may have been produced at a considerably faster rate than at present, and thus probably was much thicker than the modern oceanic crust (Condie, 1997). Russell and Arndt (2005) suggested that a high degree of partial melting of hot, dry Hadean (>3.8 Ga) mantle at ocean ridges and in plumes resulted in an oceanic-type crust about 30 km thick, overlain at places by extensive and thick mafic volcanic plateaus.

The continental crust, because of its diversity, is a much more complex system compared to the oceanic crust, but preserved remnants of continental crust (granitic) are much older. The earliest continental crust must have formed by partial melting of the primitive mantle, with incompatible elements (such as Cs, Rb, K, U, Th, and La) mostly segregated into the partial melts. Much of the key evidence regarding the earliest continental crust has been lost owing to repeated erosion, metamorphism, tectonism, and remelting. It has been possible, however, to reconstruct the age of the earliest mantle–crust differentiation by isotopic studies, all of which point to a very early formation of the first crustal rock. The Earth's oldest known *in situ* crustal units, which comprise less than 10% of preserved Archean crust, are mainly of tonalitic gneisses containing fragments of komatiite (high-Mg basalt) and amphibolite (metamorphosed basalt), some of which may be remnants of early oceanic crust. These units have been dated to 3.8–4.0 Ga (e.g., Baadsgaard *et al.*, 1984; Bowring and Williams, 1999), leaving a gap of more than 500 Myr in the crustal record, considering that the surfaces of the Moon, Mars, and Mercury reveal crater-saturated regions of ~4.5 Ga "primordial" crust.

Evidence of crust older than 4.0 Ga – actually as old as ~4.4 Ga – is provided by U–Pb ages of the cores of some isolated, detrital zircon crystals from ~3.0-Ga-old metasedimentary rocks of the Yilgarn Craton, Western Australia (Froude *et al.*, 1983; Compston and Pidgeon, 1986; Wilde *et al.*, 2001; Mojzsis *et al.*, 2001; Valley *et al.*, 2002; Cavosie *et al.*, 2005; Nemchin *et al.*, 2006). The zircon crystals are interpreted to have been derived from granitic terranes, suggesting that at least small amounts of proto-continental crust existed by ~4.4 Ga (only about 200 Myr later than the end of Earth's accretion). Without such buoyant crust, the zircon-bearing sediments would have sunk into the mantle and assimilated. Grieve *et al.* (2006) calculated that pockets of felsic crust could have been produced by differentiation of melts generated by early large-scale impacting of the basaltic crust. Analysis of Lu–Hf isotopes support the view that continental crust had formed by 4.4 to 4.5 Ga and was rapidly recycled into the mantle (Harrison *et al.*, 2005). Oxygen isotope ratios of >4.0-Ga zircon crystals from Jack Hills in the Yilgarn Craton have been interpreted to represent the earliest evidence for oceans on the Earth (Peck *et al.*, 2001) (see section 13.5.1). According to this interpretation, the Earth during the Hadean eon (4.6–3.8 Ga) was most probably characterized by a thick basaltic crust covered by an ocean, with very little dry land (composed mostly of granitic rocks) under an atmosphere probably dominated by CO_2 and/or CH_4.

How did the earliest crust form? Again, we do not know with any certainty, partly because we do not know the nature of the primitive crust. Perhaps, the Earth's earliest crust formed the same way as that of the Moon, by crystallization differentiation and solidification of the deep magma ocean that once covered the entire surface. In lunar magma ocean, the crystallizing plagioclase separated by floating to the top, and is now represented by anorthositic lunar highlands; later, remelting of the olivine- and pyroxene-rich cumulates under reducing conditions produced the mare basalts. The most compelling evidence for this interpretation is provided by the strong europium depletion in mare basalts because of removal of Eu^{2+} (octahedral coordination ionic radius = 1.25 Å) from the magma ocean by easy substitution for Ca^{2+} (octahedral coordination ionic radius = 1.08 Å) in plagioclase. Thus, the negative Eu anomaly indicates that the source of mare lavas was melted under reducing conditions and had experienced plagioclase crystallization (Grove and Krawczynski, 2009).

A problem with invoking an analogous mechanism for the formation of the Earth's cust from a terrestrial magma ocean is the lack of evidence of a terrestrial anorthositic primitive crust. Even without a magma ocean, extensive melting of the upper mantle should have produced large pockets of peridotitic magma and patches of anorthositic crust by crystallization of such magma. Perhaps no anorthositic crust formed on the Earth because the Ca and Al contents of the terrestrial magma(s) were not high enough for plagioclase to be an early crystallization phase, or plagioclase that formed could not float to the top of the hydrous terrestrial magma. If an anorthositic crust did form, then it has been completely destroyed by chemical weathering, leaving no remnants (Warren, 1989). Also, the terrestrial basaltic rocks do not show a Eu anomaly, perhaps because the terrestrial magma ocean was not reducing.

Preserved stratigraphic records in continental terranes as old as 900 Ma provide evidence that the currently operative plate tectonic paradigm has governed the evolution of the lithosphere at least since the late Precambrian (Kröner, 1977). Extending plate tectonics (see section 12.3) to the interpretation of older crustal history is highly controversial because many of the hallmarks of plate tectonics – obducted oceanic crust in the form of ophiolites, tectonic mélanges, high-level thrust belts, accretionary wedges, foreland basins, and low-temperature/high-pressure paired metamorphic belts – are absent in the Archean rock record. Hamilton (2003), for example, has argued that plate tectonics began only at about 2.0 Ga "when continents could stand above oceans and oceanic lithosphere could cool to subduction-enabling density and thickness." On the other hand, many authors (e.g. Moorbath, 1978; Sleep and Windley, 1982; Condie, 1997) contend that some mechanism of plate creation and recycling must have been operative to accommodate the large amounts of heat loss and vigorous convection in the early mantle. Thus, accretion of felsic crust on a limited scale could have occurred shortly after the solidification of the magma ocean. Many kimberlites contain eclogite xenoliths, which are interpreted by some authors to represent metamorphic products of subducted oceanic crust on the basis of their elemental and isotopic compositions (e.g., Helmstaedt and Doig, 1975; Jacob, 2004; Riches *et al.*, 2010). If this interpretation is correct, then the Archean age of some of these xenoliths makes a very powerful case for some form of plate tectonics operating in the late Archean Earth. As Cawood *et al.* (2006) have pointed out, there is mounting evidence in favor of the operation of modern-style subduction processes possibly as far back as the Hadean eon. The evidence includes: the discovery of Archean eclogites in the eastern Baltic Shield; the presence of late Archean subduction-related ("Kuroko-type") volcanogenic massive Cu–Zn–Pb sulfide deposits; the discovery of mid-Archean island arc volcanics; and the discovery of detrital zircon crystals attesting to the existence of at least some amount of proto-continental crust by 4.4 Ga. Some authors (e.g., Watson and Harrison, 2005; Cawood *et al.*, 2006) have suggested that the pattern of modern-style crust production, crust destruction, and sedimentary recycling, similar to the essential processes of plate tectonics, was probably operative as far back as the Hadean eon.

Composition of the continental crust

Matching of seismic wave velocities to lithology supports a threefold subdivision of the continental crust that gets more mafic with depth (Christensen and Mooney, 1995; Rudnick and Gao, 2004): an upper crust (~10 km thick) of granitic composition; a middle crust (~20 km thick) composed of a basalt (amphibolite)–granite mixture; and a lower crust (~10 km thick) of basaltic composition (mafic granulite). All segments of the continental crust are enriched in incompatible elements relative to the primitive mantle, and the upper crust is particularly enriched in LREE (Fig. 12.14). In addition, the upper crust has large negative europium and strontium anomalies that are largely complemented by the positive europium and strontium anomalies of the lower crust; the middle crust has essentially no europium anomaly. These features suggest that the upper granitic crust differentiated from the lower crust through partial melting, crystal fractionation (involving plagioclase in which Eu^{2+} and Sr^{2+} substitute easily for Ca^{2+}), and mixing processes (see section 12. 3). The middle crust has an overall trace-element pattern that is very similar to that of the upper crust, indicating that it too is dominated by the products of intracrustal differentiation (Rudnick and Gao, 2004). Based on observations of the rock types present in the upper continental crust and models of lower-crust composition, the average bulk composition of the continental crust is estimated to be andesitic, with SiO_2 = 57–64% and Mg# (defined as 100 × molar Mg/(Mg + ΣFe)) between 50 and 56 (Rudnick, 1995).

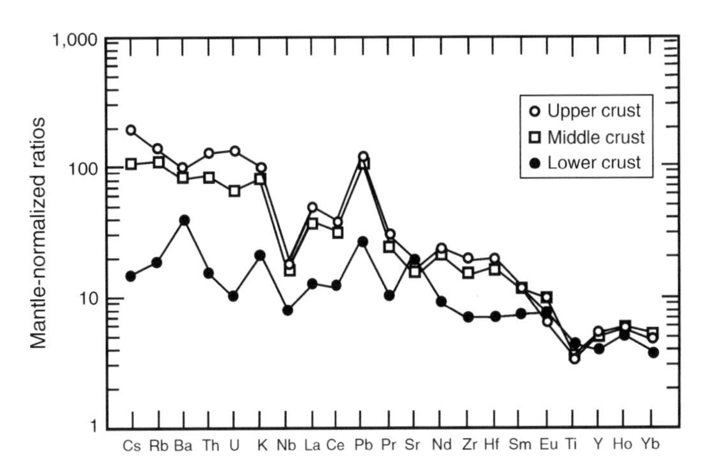

Fig. 12.14 Trace element (including REE) composition of the upper, middle, and lower continental crusts from Rudnick and Gao (2004), normalized to primitive mantle values from McDonough and Sun (1995). (After Rudnick and Gao, 2004, Figure 17b, p. 51.)

Average compositions of the continental and oceanic crusts relative to the primitive mantle composition form surprisingly complementary patterns (Fig. 12.15). In continental crust, concentrations of the most incompatible elements Rb, Ba, Th, and K reach 50 to 100 times the primitive mantle values but < 3 times primitive mantle values in the oceanic crust. In average oceanic crust (represented by "normal" MORB), the maximum concentrations are only about 10 times the primitive mantle values, and they are attained by the moderately incompatible elements Hf, Zr, Ti, Yb, Y, and the intermediate to heavy REE. The patterns cross at element P, and the least incompatible elements Ti, Yb, and Y are more enriched in oceanic crust than in continental crust. Relative depletions in Nb (resulting in characteristic low Nb/La and Nb/Ta ratios),

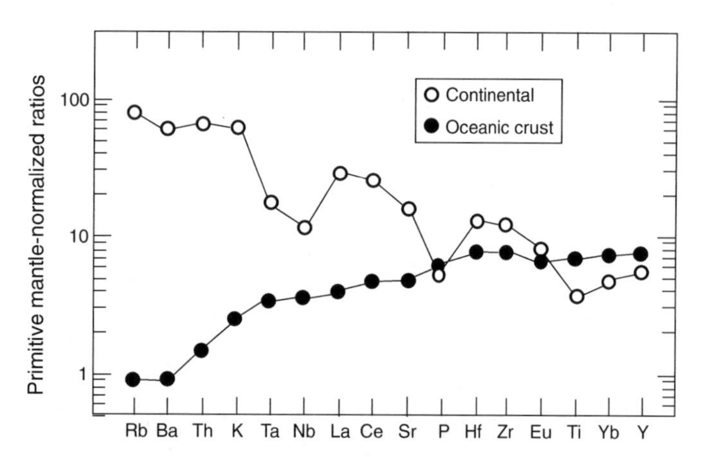

Fig. 12.15 Complementary patterns exhibited by concentrations of elements in the bulk continental crust (Condie, 1997) and "normal-type" mid-ocean ridge basalts (N-MORB) (Hofmann, 1988), both normalized to the estimated values for the primitive mantle (Sun and McDonough, 1989. The continental rust is enriched in incompatible elements and has a high La/Nb ratio, which are characteristic features of convergent margin magmas and rarely observed in intraplate magmas.

P, and Ti, and a strong relative enrichment in Pb are important features of continental crust. This lack of complementary relation between the continental crust and depleted mantle can be balanced by a refractory eclogite reservoir deep in the mantle, which formed by metamorphism of sinking slabs of oceanic crust in subduction zones and is enriched in Nb, Ta, and Ti (Rudnick *et al.*, 2000).

Growth of the continental crust

Hoffman (1988) argued that the compositional relationship between continental crust and oceanic crust could be explained by a simple (employing equilibrium partial melting), two-stage model of first extracting continental crust from the primitive mantle, and then extracting oceanic crust by remelting of the residual (depleted) mantle. Yet most high-pressure single-stage melting experiments conducted on mantle peridotite compositions have yielded magmas that are considerably less evolved (e.g., basalts and picrites) than the bulk continental crust (andesitic). Several models of crustal growth have been proposed to explain this paradox (see reviews in Rudnick, 1995; Kelemen, 1995; Rudnick and Gao, 2004), all of which require the return of mafic-to-ultramafic lithologies to the convecting mantle, and thus assume crustal recycling to have been operative throughout the Earth's history.

The similarity between the composition of high-Mg# andesite (HMA) lavas (e.g., negative Nb–Ta anomalies) found in many contemporary magmatic arcs and the continental crust average, coupled with the occurrence of prolific andesite magmatism at continental convergent margin settings (such as the Andes and the Cascades), prompted Taylor (1967) to propose the "andesite model" of continental growth – that continental crust formed as a result of andesitic arc magmatism and was accreted to a preexisting crustal mass. The andesite model, however, cannot explain the higher Cr and Ni contents or Th/U ratio of the continental crust compared to HMA. Moreover, andesite magmatism is uncommon in volcanic sequences emplaced during the Archean when a substantial part of the present crustal mass is inferred to have formed. This led Taylor and McLennan (1985) to suggest that only post-Archean crustal growth was accomplished by island arc accretion, and growth of continental crust during the Archean probably occurred in intraplate settings in response to plume-related and continental rift magmatism. The recent discovery of andesites with negative Nb and Ta anomalies on the Pacific–Antarctic Rise implies that such magmas are not restricted to subduction zones but can form also at plume-influenced mid-oceanic ridge (Haase *et al.*, 2005). However, post-Archean crust of andesitic composition could not have been formed by accretion of intra-oceanic island arcs because they are estimated to have basaltic rather than andesitic bulk compositions. Magmas in continental arc settings, such as the Andes and the Cascades, are andesitic, presumably due to contamination from preexisting continental crust, but the primary magmas are estimated to have been basaltic, similar to those in intraoceanic settings.

A variation of the andesite model envisages the formation of HMA in the mantle beneath arcs through interaction between rising basaltic melts from mantle peridotite and silicic melts derived from subducted oceanic crust, and the direct addition of the HMA to continental crust. This process is likely to have been more prevalent in a hotter, Archean Earth and involve extensive silicic-melt–peridotite reaction as the slab melts traversed the mantle wedge (Kelemen, 1995). The model, however, suffers from the same shortcomings as the original andesite model – that intra-oceanic island arcs appear to have basaltic bulk compositions, and andesites are uncommon constituents of Archean-aged crust.

An alternative model (Arndt and Goldstein, 1989) proposes "delamination" of the lower continental crust to account for the andesitic composition of the continental crust. High-Mg# andesite may have formed by crustal differentiation of mantle-derived basaltic magmas, and the mafic/ultramafic residue of differentiation forming the lower crust transformed into eclogite during subduction of continental crust as a result of arc–continent collision. Being denser than upper mantle peridotite, the eclogite might have then sunk into the mantle ("delaminated"), leaving only andesitic differentiates in the crust (Kelemen, 1995).

Crustal growth rate

There are two diametrically opposing hypotheses to account for the growth of continental crust to its present dimensions. The first may be called a steady-state hypothesis, which proposes that the present mass of the crust formed almost immediately after the Earth's formation, after which the system reached a steady state of negligible crustal growth through a

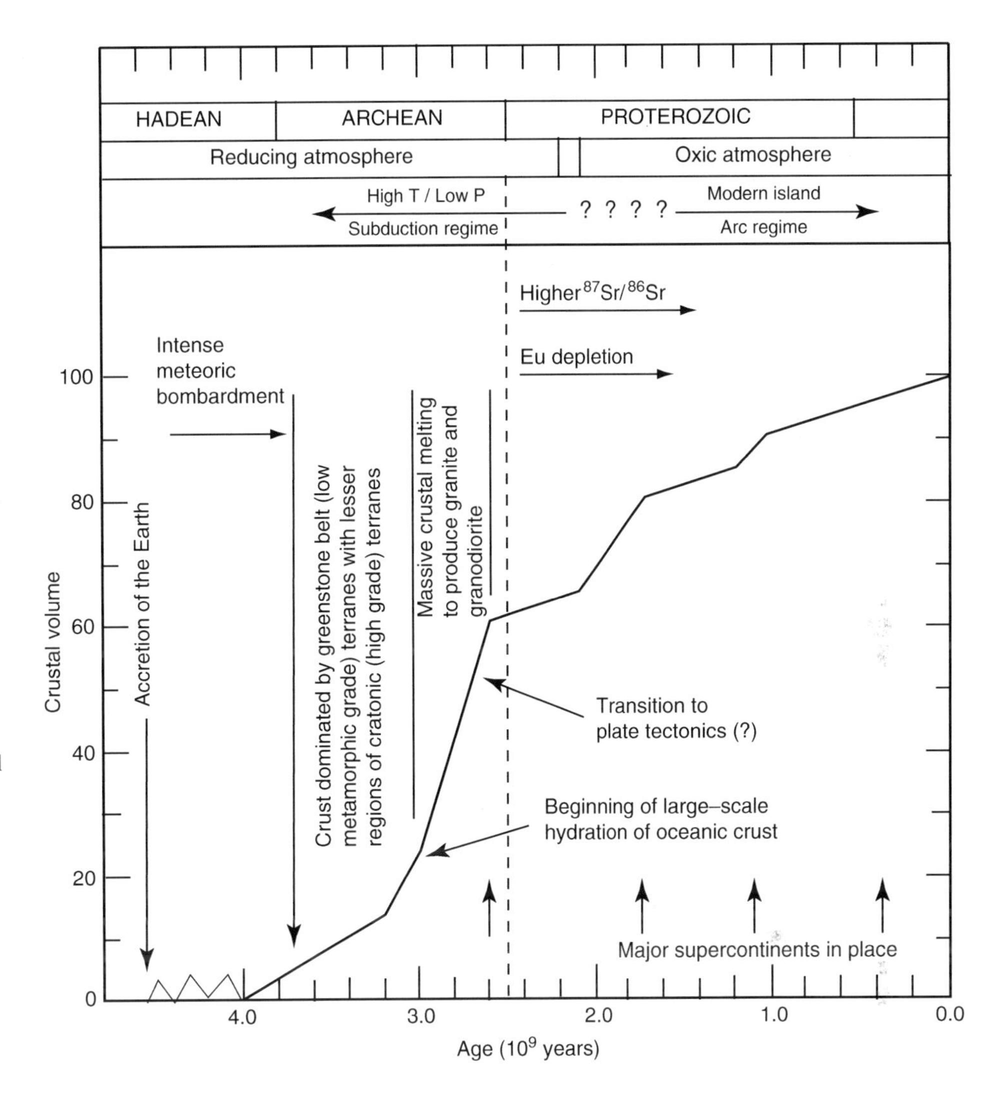

Fig. 12.16 Schematic model for the growth and evolution of the continental crust on a global scale as proposed by Taylor and McLennan (1995). The actual value of crust present at any given time is not well constrained, but a value of 50% crust by about 2.5 Ga is a likely minimum value. Although major global episodes of crustal growth and differentiation are well documented during the late Archean and at about 2.1–1.7 Ga, it is less clear if crustal growth is episodic on a global scale during younger times. In general, there appears to be a relationship between crustal growth episodes and assembly phases of supercontinents. (After Taylor and McLennan, The geochemical evolution of the continental crust, Rev. Geophysics and Space Physics, v. 33, p. 241-265, 1995. Copyright 1995 American Geophysical Union. Used by permission of American Geophysical Union.)

balance between new additions from the mantle and losses by erosion and by recycling via subduction (Armstrong, 1991; Bowring and Housh, 1995). The opposing hypothesis, which is more popular and more consistent with geochemical evidence, argues for episodic growth of continental crust over geologic time (Taylor and McLennan, 1981, 1995; Deming, 2002).

The episodic growth model of Taylor and McLennan (1995) presented in Fig. 12.16 shows that major pulses of continental growth occurred at about 3.8–3.5, 2.9–2.7, 2.0–1.7, 1.3–1.1, and 0.5–0.3 Ga, which appear to correlate reasonably well with phases of supercontinental assembly. During supercontinental assembly, subduction of old oceanic crust would be at a maximum, leading to an increase in arc volcanism. In addition, the continental collisions that terminated the assembly phases would promote crustal thickening, intracrustal melting and differentiation, granulite facies metamorphism, continental underplating, and, accordingly, preservation of such unsubductible crust in the geologic record.

Two striking features of the episodic growth model are particularly well constrained: the absence of significant amounts of continental crust prior to 3.0 Ga; and the most rapid growth of continental crust in the late Archean (2.9–2.7 Ga). At least 60% of the crust was emplaced by 2.7 Ga. There are at least three independent lines of geochemical evidence for this conclusion (Deming, 2002): (i) an abrupt increase in the $^{87}Sr/^{86}Sr$ ratio of ocean water at ~2.5 Ga, which is interpreted to represent the emergence of large continental land masses; (ii) the scarcity of zircons older than 3.0 Ga, and the abundance of 2.5 to 3.0 Ga zircons; and (iii) the absence of europium depletion in rocks older than 2.5 Ga.

The late Archean pulse corresponds to the Archean–Proterozoic transition. This transition was marked by widespread "cratonization" of the crust that involved massive intracrustal melting, resulting in the emplacement of large amounts of K-rich granitic rocks in the upper crust, transfer of heat-producing radioactive elements (K, U, and Th) to the upper crust, and stabilization of the crust. The Archean–Proterozoic boundary can also be correlated with many other geologic events. As summarized by Taylor and McLennan (1995), these include: the widespread occurrence of uranium deposits in basal Proterozoic sediments because of enrichment of the upper crust in incompatible elements due to intracrustal melting; the dramatic increase in ^{87}Sr in the K-rich granites emplaced in the upper crust; and the proliferation of stromatolites and banded iron formations in the Proterozoic, attesting to the development of stable continental shelves.

12.3 Generation and crystallization of magmas

On the basis of tectonic setting, we can define four distinct plate-tectonic environments in which magmas may be generated (Fig. 12.17; Table 12.5): (i) constructive plate margins (spreading centers); (ii) destructive plate margins (subduction zones); (iii) oceanic intraplate settings; and (iv) continental intraplate settings.

The Earth's mantle is composed essentially of solid rocks, except perhaps a very small amount (< 1 wt%) of a melt phase in the asthenosphere. Partial melting resulting from heating of surrounding cold rocks by basaltic underplating may generate magmas locally in the continental crust and lithosphere, but *decompression melting* (pressure-release melting) and lowering of the *solidus* temperature (i. e., the temperature at which it will start to melt) by volatiles appear to be the major mechanisms of partial melting in the mantle. Partial melting beneath mid-oceanic ridges occurs in response to adiabatic decompression of ascending mantle peridotite in the zone of upwelling. The decompression-melting model is also applicable to mantle plumes that are believed to be responsible for intraplate magmatism. In subduction zones (located at convergent plate margins) magmas are generated by partial melting of the upper portion of the subducting oceanic crust and the overlying mantle wedge. As the slab of cold oceanic crust descends into the mantle by subduction, it is progressively heated by conduction of heat from the surrounding mantle and also possibly by frictional heating at the surface of the slab. The increasing temperature and pressure results in dehydrating an originally hydrous mineral assemblage in the slab and releasing aqueous fluids, which depress the melting points of silicate minerals and induce partial melting of the overlying mantle wedge (Fig. 12.18).

12.3.1 *Geochemical characteristics of primary magmas*

It is generally accepted that partial melting of upper mantle material produces magmas dominantly of mafic (basaltic) and ultramafic (picritic and komatiitic) compositions in most tectonic settings. Such magmas are called *primary magmas*, in contrast to *evolved* (or *derivative*) *magmas* generated through differentiation or mixing of primary magmas. Primary magmas appear to be generated within a very restricted depth range, within the upper 100–200 km of the mantle. Diamond-bearing kimberlites probably represent the deepest terrestrial magmas, originating from depths greater than 200–250 km. The geochemical characteristics of primary magmas depend upon parameters such as the source composition and mineralogy, and the depth, degree, and mechanism of partial melting – factors that vary from one tectonic setting to another. The wide spectrum of terrestrial igneous rocks is due partly to the varying composition of the primary magmas and partly to subsequent differentiation and assimilation processes. These

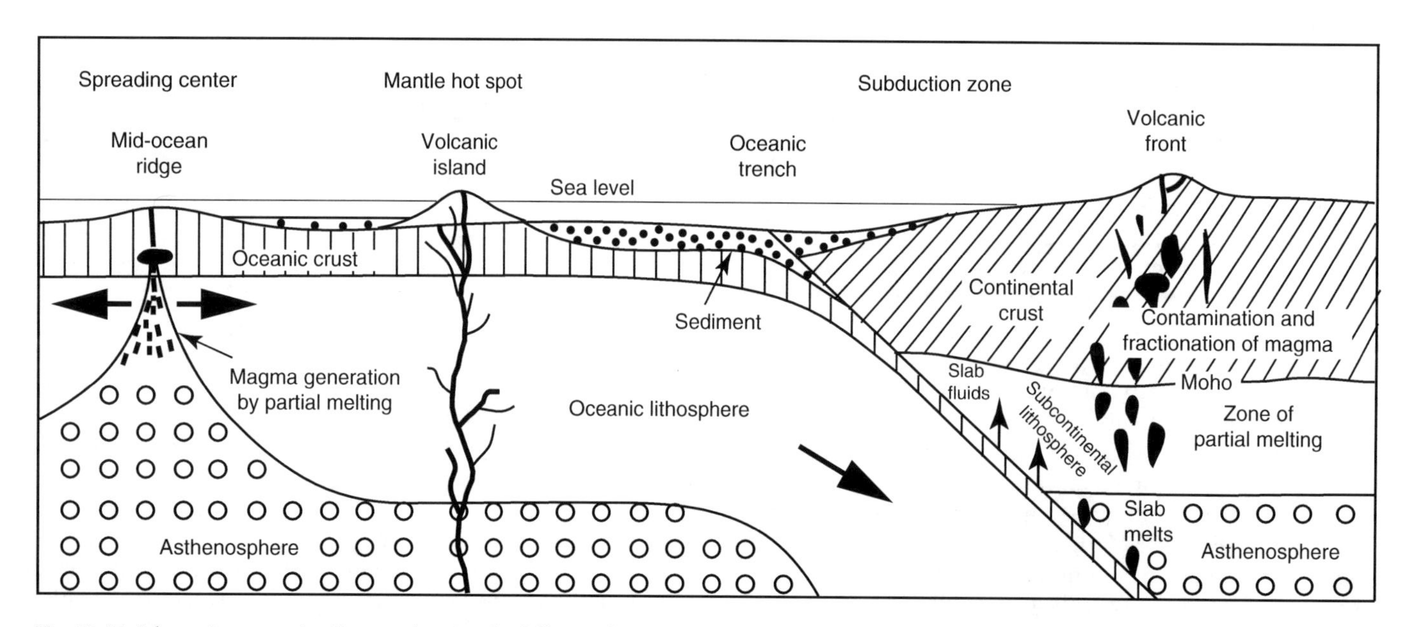

Fig. 12.17 Schematic composite diagram showing the different plate-tectonic environments of magma generation: (a) divergent plate margin (spreading center); (b) oceanic-plate–continental-plate convergent margin (subduction zone); and (c) oceanic intraplate setting (oceanic islands).

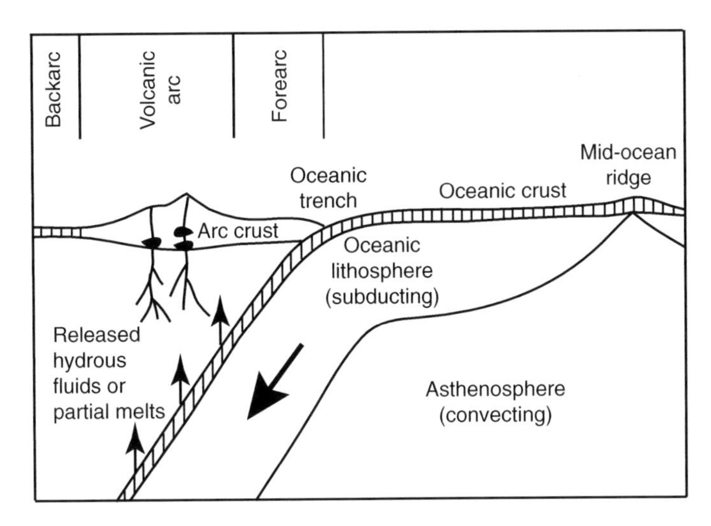

Fig. 12.18 Model of magma generation in a subduction zone marked by the convergence of two oceanic plates. The aqueous fluids released from the subducting oceanic crust due to metamorphic dehydration reactions enable partial melting of the overlying mantle wedge.

include fractional crystallization, magma mixing, and crustal contamination, either during the ascent of the magmas to higher levels or concurrently with crystallization (the so-called assimilation and fractional crystallization (AFC) process; DePaolo, 1981).

Basaltic rocks (SiO_2 = 44 – 53.5 wt%) are the most abundant type of volcanic rocks found within the accessible continental crust and are the overwhelmingly dominant volcanic rocks in the oceanic domain. On the basis of major element chemistry, primary basaltic magmas (and the resulting rocks) are divisible into three major *magma series* (e.g., Middlemost, 1975) – *tholeiitic*, *calc-alkaline*, and *alkalic* – each magma series comprising a genetically related spectrum of mafic to felsic rocks, as evidenced by the trends in their chemistry and mineralogy. Volcanic rocks of the alkalic group are silica-undersaturated, and can be distinguished from subalkalic rocks (which include tholeiitic and calc-alkaline magma series) by their higher alkali (Na_2O and K_2O) contents (Fig. 12.19a and b). The very low-K_2O, subalkalic basalts (Fig. 12.19b) are also depleted in other *large ion lithophile* (LIL) elements (such as Rb, Ba, U, Th, Pb, and LREEs). Some basalts plot within the alkalic field on the SiO_2–Na_2O diagram but within the subalkalic field on the SiO_2–K_2O diagram; such basalts are termed *transitional*. Komatiitic basalts (SiO_2 < 44 wt %) are low-K subalkalic basalts characterized by low Al_2O_3 and high MgO contents. Komatiitic basalts, however, can be readily separated from the low-K basalts of the present ocean floor, as komatiitic basalts normally contain more MgO than Al_2O_3, whereas the "ocean-floor basalts" normally contain more Al_2O_3 than MgO.

The subalkalic basaltic rocks, the most dominant type of volcanic rocks found in both the continental and oceanic domains, can be subdivided into a *tholeiitic series* and a *calc-alkaline* or high-alumina series (Fig. 12.19c). These two series can also be distinguished in terms of their differentiation trends on an AFM diagram (Fig. 12.20) – tholeiitic suites show a strong trend of iron enrichment because of crystallization of Fe–Ti oxides in the early stages of differentiation, which is followed by alkali enrichment. Calc-alkaline suites also show an alkali enrichment trend but no iron enrichment trend. Calc-alkaline basalts may be further subdivided into high-K, medium-K, and low-K types on a plot of SiO_2 versus K_2O (Fig. 12.19d).

At the present time, the calc-alkaline series is restricted to subduction-related tectonic settings, whereas alkalic basalts and their differentiates are commonly found in intraplate tectonic settings such as oceanic islands and continental rifts. The characteristic magma series associated with each of the tectonic settings in which primary magmas are generated are summarized in Table 12.6. Note that the distinction between tholeiitic and alkalic basalts, although important petrologically, does not appear to be correlated with tectonic setting.

12.3.2 *Behavior of trace elements during partial melting of source rocks*

Since the same kind of volcanic rocks occur in more than one tectonic setting (Table 12.6), the tectonic settings of primary basaltic magmas cannot be discriminated on the basis of major elements. We have to, therefore, depend on distributions of trace elements (including REEs) and Sr–Nd–Pb isotopic signatures for such discrimination.

Compatible and incompatible elements

The partitioning of a trace element *i* between a mineral and a melt at equilibrium is described by the *Nernst distribution coefficient* (see Box 5.6):

$$K_{D(i)}^{\text{mineral-melt}} = \frac{C_i^{\text{mineral}}}{C_i^{\text{melt}}} \qquad (12.37)$$

where C_i^{mineral} and C_i^{melt} represent concentrations (e.g., in wt% or ppm) of *i* in the two phases.

Mineral–melt distribution coefficients are determined commonly from the analysis of minerals and their glassy matrix in rapidly cooled volcanic rocks, or from experiments at specified conditions in which synthetic or natural starting materials are doped with the trace element of interest. Values of distribution coefficients depend on a number of variables: composition of the melt, temperature, pressure, oxygen fugacity (especially for Eu), and crystal chemistry of the minerals. Published data on mineral–melt distribution coefficients are rather limited, and often we have to use approximate values for modeling calculations. Useful compilations of mineral–melt distribution coefficients can be found, for example, in Rollinson (1993), Lodders and Fegley (1998), and Salters and Stracke (2004). A compilation of

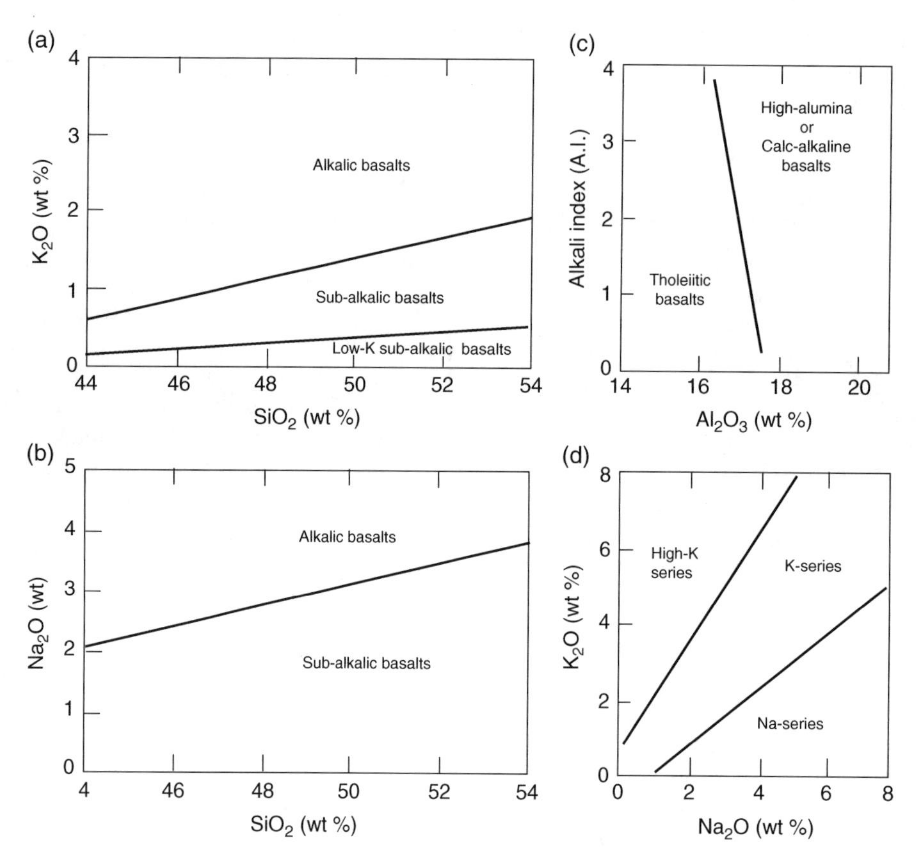

Fig. 12.19 Classification of primary basaltic magmas (after Middlemost, 1975; Wilson, 1989): (a) SiO_2 (wt%) versus K_2O (wt%); (b) SiO_2 (wt%) versus Na_2O (wt%); (c) alkali index (AI) versus Al_2O_3 (wt%); (d) SiO_2(wt%) versus K_2O (wt%). Alkali Index = $(Na_2O + K_2O) / [(SiO_2 - 43) \times 0.17]$.

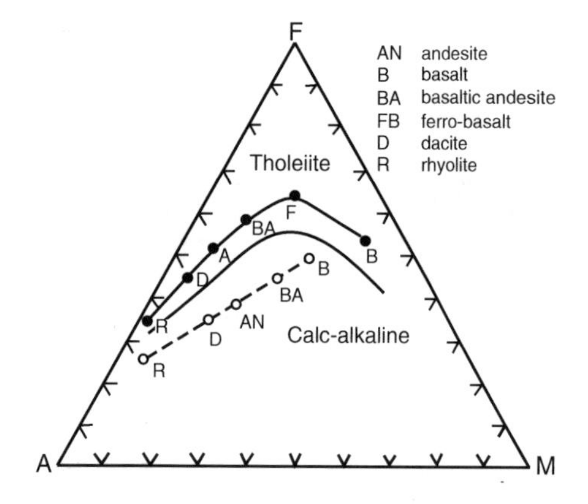

Fig. 12.20 AFM diagram showing typical tholeiitic and calc-alkaline trends of differentiation. A = $Na_2O + K_2O$; F = $FeO + Fe_2O_3$; M = MgO.

mineral–melt distribution coefficients appropriate for very Mg-rich (komatiitic) melts and useful for modeling the evolution of the lunar magma ocean and crust is given in Snyder *et al.* (1992). For the purpose of illustration, a few mineral–mafic melt distribution coefficients for REEs at specific temperature, pressure and oxygen fugacity conditions are listed in Table 12.7.

During partial melting and crystallization, trace elements with $K_{D(i)}^{mineral-melt} > 1$ are preferentially concentrated in the solid phases (minerals) and are called *compatible elements*, whereas trace elements with $K_{D(i)}^{mineral-melt} < 1$ are preferentially concentrated in the melt phase and are called *incompatible elements*. Some trace elements, however, exhibit variable behavior. For example, phosphorus behaves as an incompatible element during partial melting of mantle rocks and prefers the melt phase, but behaves as a compatible element during crystallization of granitic melts and is accommodated in the structure of the minor phase apatite.

Depending on the *field strength* (or *ionic potential*), which is defined as the charge : ionic radius ratio of an ion, incompatible elements may be divided further into two groups: *high field strength* (HFS) elements (such as lanthanides, Sc, Y, Th, U, Pb, Zr, Hf, Ti, P, Nb, and Ta) with ionic potential >2; and *low field strength* (LFS) elements (such as Cs, Rb, K, Ba, Sr, Pb, Eu^{2+}) with ionic potential <2. The LFS elements, along with the light rare earth elements La and Ce, are also referred to as LIL elements. The LIL elements are particularly concentrated in late-stage products of magmatic crystallization. Elements having small ionic radius and a relatively low charge (such as Ni, Cr, Cu, W, Ru, Rh, Pd, Os, Ir, Pt, and Au) tend to be

Table 12.6 Characteristic magma series associated with specific tectonic settings.

	Plate margin		Within-plate	
Tectonic setting	Divergent (constructive)	Convergent (destructive)	Intra-oceanic	Intracontinental
Volcanic feature	Mid-oceanic ridges; backarc spreading centers	Island arcs; active continental margins	Oceanic islands	Continental rift zones; continental flood-basalt provinces
	(Ocean-floor basalts)	(Volcanic-arc basalts)	(Ocean-island basalts)	(Continental basalts)
Characteristic magma series	Tholeiitic	Tholeiitic (low K_2O)	Tholeiitic	Tholeiitic
	–	Calc-alkaline (high Al_2O_3)	–	–
	–	Alkalic	Alkalic	Alkalic
Rock type	Basalts	Basalts and differentiates	Basalts and differentiates	Basalts and differentiates

Sources of data: Pearce and Cann (1973) and Wilson (1989).
Definitions of the three magma series included in this table are illustrated in Fig. 12.17.

Table 12.7 Nernst distribution coefficients, $K_{D(i)}$, for rare earth elements between some minerals and mafic silicate melt.

	Mineral									
Melt	La	Ce	Nd	Sm	Eu	Gd	Dy	Er	Yb	Lu
Diopside [T = 1265°C, P = 1 bar; $-\log f_{O_2}$ = 0.679 (for Eu, $-\log f_{O_2}$ = 10–16) (Grutzeck *et al.*, 1974)]										
	0.69	0.098	0.21	0.26	0.31	0.30	0.33	0.30		0.28
Olivine [T = 1200°C, P = 1 bar; $-\log f_{O_2}$ = 12.4 (Kennedy *et al.*, 1993)]										
	0.000028	0.000038	0.00020	0.00062	0.00015	0.00099	0.0039	0.0087	0.017	0.020
Plagioclase [T = 1150–1400°C, P = 1 bar; $-\log f_{O_2}$ = 0.679 (for Eu, $-\log f_{O_2}$ = 12.5 (Drake and Weil, 1975)]										
	(1)	(2)	(3)	(4)	(5)	(6)	(7)	(8)	0.035	(9)
Perovskite [T = 1150–1400°C, P = 1 bar; $-\log f_{O_2}$ = 0.679 (for Eu, $-\log f_{O_2}$ = 12.5 (Nagasawa *et al.*, 1980)]										
	2.62			2.70	2.34	2.56			0.488	0.411

(1) $\ln K_{D\,(La)} = -6.40 + 7000/T$; (2) $\ln K_{D\,(Ce)} = -5.21 + 4600/T$; (3) $\ln K_{D\,(Nd)} = -4.22 + 2920/T$; (4) $\ln K_{D\,(Sm)} = -4.13 + 2340/T$; (5) $\ln K_{D\,(Eu)} = -3.65 + 1560/T$; (6) $\ln K_{D\,(Gd)} = -3.09 + 240/T$; (7) $\ln K_{D\,(Dy)} = -1.54 - 2360/T$; (8) $\ln K_{D\,(Er)} = -0.12 - 4620/T$; (9) $\ln K_{D\,(Lu)} = -5.40 - 3200/T$.
T in Kelvin

compatible with minerals formed early in the crystallization process. Fig. 12.21 illustrates the distribution trends of a few trace elements in an island-arc tholeiite suite of volcanic igneous rocks formed by magmatic differentiation. The concentrations of compatible elements (e.g., Sr, Cr, Ni) generally decrease, whereas the concentrations of incompatible elements (e.g., Rb, Ba, Zr, Y) generally increase, with increased differentiation as indexed by increasing whole-rock SiO_2 content. Cr^{3+} (ionic radius 0.615 Å) and Ni^{2+} (0.69 Å) substitute easily for Mg^{2+} (0.72 Å) in early formed Mg-minerals (olivines and pyroxenes), and Sr^{2+} (1.18 Å) substitutes for Ca^{2+} (1.0 Å) in plagioclase. Rb^{+} (1.52 Å) and Ba^{2+} (1.35 Å) substitute for K^{+} (1.38 Å) in K-feldspars and, therefore, tend to be concentrated, along with K, in late differentiates. Zr^{4+} (0.72 Å) and Y^{3+} (0.70 Å) are concentrated at the felsic end of a differentiated series because their ionic size and charge make their substitutution for any major cations in common silicate minerals rather difficult. These elements are segregated into the late residual solutions, and if present in appreciable amounts they may form minerals of their own (e.g., zircon, $ZrSiO_4$).

Partial melting models

Depending on mineral–melt distribution coefficients, concentrations of trace elements in a mantle-derived partial melt, unlike those of the major elements, may be quite sensitive to the type of melting process and the degree of partial melting of the mantle source. There are two ideal or end-member models for the partial melting process.

(1) *Equilibrium partial melting* (or *batch melting*), in which the partial melt forms continually and equilibrates thermodynamically with the solid residue at the site of melting until mechanical conditions allow the melt to be removed as a single "batch." Up to this time the bulk composition of the source-rock–melt system remains the

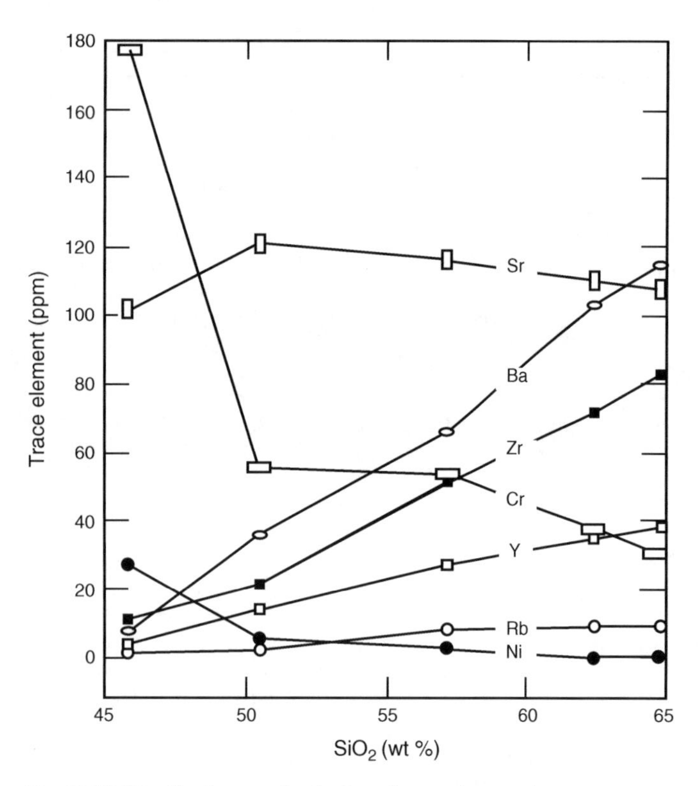

Fig. 12.21 Distribution trends of selected trace elements in a mafic-to-felsic differentiated suite of volcanic igneous rocks from South Sandwich island arc. The rock types of the suite are: basalt 1 (SiO_2 = 45.8%); basalt 2 (SiO_2 = 50.49%); basaltic andesite (SiO_2 = 57.1%); andesite (SiO_2 = 62.39%); and dacite (SiO_2 = 64.81%). In this system, Cr, Ni, and Sr are compatible elements, whereas Ba, Rb, Zr, and Y are incompatible elements. (Source of data: Luff, 1982.)

same as at the start, although the compositions of the partial melt and the solid residue change progressively with partial melting.

(2) *Fractional partial melting* (or *Rayleigh melting*), in which the small amount of partial melt formed at any instant may be in equilibrium with the solid residue, but is immediately and continuously removed from the source-rock–melt system, thereby preventing further interaction with the solid residue. For this process the bulk composition of the system changes continuously along with the compositions of the partial melt and the solid residue.

Which partial melting process is appropriate in a particular situation depends on the permeability threshold of the source? Which determines the ability of a partial melt to segregate from its source region? Because of the requirement of immediate melt removal, fractional melting can occur only if the mantle becomes permeable at very low degrees of partial melting (generally in the order of 1% or less, but may be up to about 2–3%). Thus, fractional melting may be an appropriate model for generation of some mafic melts, whereas batch melting may be the appropriate process for generating felsic melts, which are more viscous and, therefore, have a higher permeability threshold. Also, considering that the removal of infinitesimal amounts of melt is physically unlikely, batch melting may be the more geologically realistic model for the formation of large bodies of magma of uniform composition. An alternative model would be to allow thorough mixing of melt increments collected in a common reservoir, in which case the results for fractional melting would be indistinguishable from batch melting for highly incompatible trace elements.

For both melting processes, the concentration of a trace element i in the partial melt depends on whether the melting is *modal* or *nonmodal*, the degree of partial melting, and the appropriate *bulk distribution coefficient* of the residual solid at the moment when melt is removed from the system. The underlying assumption in all geologically relevant melting models is that the distribution coefficients remain constant throughout each melting event irrespective of changing melt composition and temperature.

Modal and nonmodal melting

In modal melting, the minerals melt according to their modal proportions in the source assemblage so that their weight fractions, and therefore the bulk distribution coefficient, ($D^{\text{rock}}_{0(i)}$), remain the same as for the source rock at the onset of melting, until one of the minerals is completely consumed as melting progresses. For example, in the partial melting of mantle rocks, minor phases such as garnet, amphibole, biotite, and clinopyroxene would, in general, be consumed much earlier than olivine and orthopyroxene. $D^{\text{rock}}_{0(i)}$ is the weighted mean of the individual mineral distribution coefficients for the element i, and is calculated as

$$D^{\text{rock}}_{0\,(i)} = W_0^{\alpha}\; K_{\text{D}\,(i)}^{\alpha\text{–melt}} + W_0^{\beta}\; K_{\text{D}\,(i)}^{\beta\text{–melt}} + W_0^{\gamma}\; K_{\text{D}\,(i)}^{\gamma\text{–melt}} + \dots = \sum_{\alpha} W_0^{\alpha}\; K_{\text{D}\,(i)}^{\alpha s\text{–melt}} \tag{12.38}$$

where W_0^{α} is the weight fraction of the mineral α in the source rock, and $K_{D(i)}^{\alpha\text{-melt}}$ is the α-melt distribution coefficient for the element i, and so on for the other phases in the source rock undergoing partial melting. For example, the bulk distribution coefficient of Ni in a hypothetical source rock containing 30 wt% olivine, 20 wt% orthopyroxene, and 50 wt% clinopyroxene, would be

$$D^{\text{rock}}_{0(\text{Ni})} = 0.3K_{D(i)}^{\text{ol-melt}} + 0.2K_{D(\text{Ni})}^{\text{opx-melt}} + 0.5K_{D(\text{Ni})}^{\text{epx-melt}}$$

Evidently, $D^{\text{rock}}_{0(i)}$ must be recalculated every time one or more of the minerals in the initial assemblage gets consumed completely by melting.

In the more realistic case of nonmodal melting (which is also more difficult to quantify), the proportions of the minerals entering the melt change continuously in the course of partial melting. For a particular value of F, the weight fraction of melt formed (equivalent to the degree of partial melting), the bulk distribution coefficient, D_i^{bulk}, is (Wilson, 1989)

$$D_i^{\text{bulk}} = \frac{D_{0\,(i)}^{\text{rock}} - FP}{1 - F} \tag{12.39}$$

If $p^\alpha, p^\beta, p^\gamma, \ldots$ represent the weight fractions of phases entering the melt, then

$$\begin{aligned} P &= p^\alpha K_{\text{D}\,(i)}^{\alpha\text{-melt}} + p^\beta K_{\text{D}\,(i)}^{\beta\text{-melt}} + p^\gamma K_{\text{D}\,(i)}^{\gamma\text{-melt}} + \ldots \\ &= \sum_\alpha p^\alpha K_{\text{D}\,(i)}^{\alpha\text{-melt}} \end{aligned} \tag{12.40}$$

Note that the value of P, and therefore the concentration of i in the melt, is unaffected by any change in the mineralogy of the source rock that may have occurred prior to the instant under consideration.

Equations for batch and fractional melting

Equations for calculating the concentration of a trace element i in the partial melt, C_i^{melt}, relative to that in the original source rock, $C_{0(i)}^{\text{rock}}$, are as follows (Wood and Fraser, 1976; Allègre and Minster, 1978):

Batch melting (Fig. 12.22)

Modal
$$\frac{C_i^{\text{melt}}}{C_{0\,(i)}^{\text{rock}}} = \frac{1}{D_{0\,(i)}^{\text{rock}} + F(1 - D_{0\,(i)}^{\text{rock}})} \tag{12.41}$$

Nomodal
$$\frac{C_i^{\text{melt}}}{C_{0\,(i)}^{\text{rock}}} = \frac{1}{D_{0\,(i)}^{\text{rock}} + F(1 - P)} \tag{12.42}$$

Fractional melting

Modal
$$\frac{C_i^{\text{melt}}}{C_{0\,(i)}^{\text{rock}}} = \frac{1}{D_{0\,(i)}^{\text{rock}}} (1 - F)^{\left(\frac{1}{D_{0\,(i)}^{\text{rock}}} - 1\right)} \tag{12.43}$$

Nomodal
$$\frac{C_i^{\text{melt}}}{C_{0\,(i)}^{\text{rock}}} = \frac{1}{D_{0\,(i)}^{\text{rock}}} \left(1 - \frac{PF}{D_{0\,(i)}^{\text{rock}}}\right)^{\left(\frac{1}{P} - 1\right)} \tag{12.44}$$

F is the weight fraction of melt produced in the case of batch melting, and the weight fraction of the melt produced and already removed from the source in the case of fractional melting.

In each of the above cases, the concentration of i in the residual solid can be obtained by solving an appropriate mass balance equation (Allègre and Minister, 1978). Equations for more complex scenarios of partial melting have been discussed, for example, by Hertogen and Gijbels (1976) and Allègre and Minster (1978).

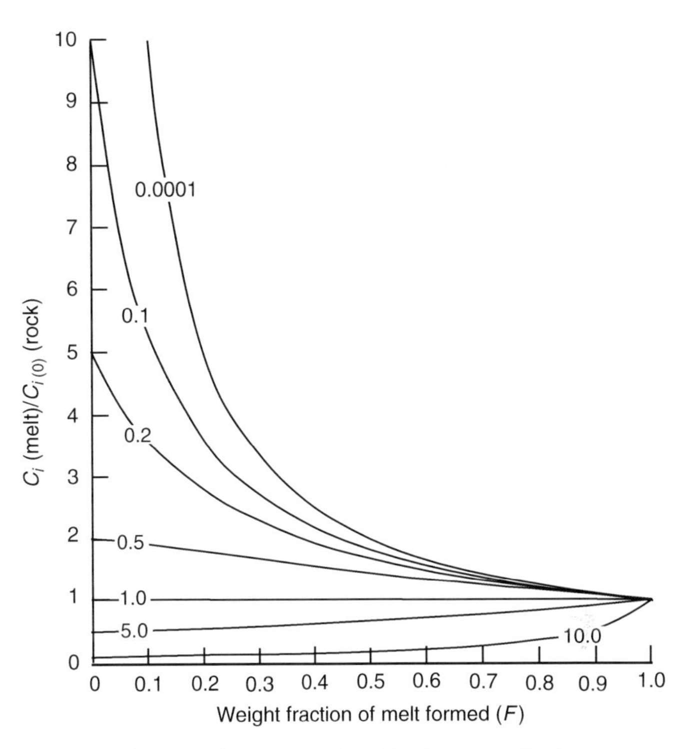

Fig. 12.22 Change in the concentration of an incompatible element, relative to its initial concentration in the source rock, as a function of the weight fraction of melt formed (F) during modal equilibrium (batch) partial melting. The curves are for arbitrarily assumed values (0.0001, 0.2, 0.5, 1.0, 5.0, 10.0) of the bulk distribution coefficient.

In all the four cases stated above, the concentration of a given trace element in the melt relative to the source increases with decreasing value of bulk distribution coefficient for a given degree of partial melting and decreases with increasing degree of partial melting for a given value of the bulk distribution coefficient. In the case of modal batch melting, the simplest case, for very small values of $D_{0(i)}^{\text{rock}}$, equation (12.41) reduces to $C_i^{\text{melt}}/C_{0(i)}^{\text{rock}} = 1/F$, which marks the limit of enrichment of i in the melt for any given degree of batch melting. For very small values of F, equation (12.41) reduces to $C_i^{\text{melt}}/C_{0(i)}^{\text{rock}} = 1/D_{0(i)}^{\text{rock}}$, which represents the maximum possible enrichment of an incompatible element in the melt (or maximum possible depletion of a compatible element in the melt). The concentrations of incompatible elements in the partial melt approach those of the source rock with increasing degree of partial melting and should be the same as in the source rock for 100% melting. The results for modal fractional melting are very similar to that of modal batch melting. A noticeable difference is that, in the range 0–10% partial melting, the changes in element concentrations relative to the original source are more extreme in fractional melting compared to batch melting, although the limiting value of $1/D_{0(i)}^{\text{rock}}$ is the same.

Example 12–1: Calculation of trace element concentrations in a partial melt generated by modal and nonmodal batch and fractional melting

Suppose we are interested in the concentration of the rare earth element Ce, a highly incompatible trace element, in the partial melt formed from an ultramafic mantle source rock composed of three minerals in the following proportion (by weight) – diopside (di) : enstatite (en) : forsterite (fo) = 0.4:0.2:0.4. Phase equilibria in the system di–en–fo indicate that the phases would melt in the proportion 0.7 (di) : 0.2 (en): 0.1 (fo) in the case of nonmodal melting. Our task is to calculate the value of $C_{\text{Ce}}^{\text{melt}}/C_{0(\text{Ce})}^{\text{rock}}$ for modal and nonmodal equilibrium and fractional partial melting for $F = 0.3$, given that the distribution coefficients for Ce for the minerals are: di – 0.10; en – 0.003; and fo – 0.001 (Wilson, 1989, p. 65).

$$D_{0(i)}^{\text{rock}} = W_0^{\text{di}} K_{\text{D(Ce)}}^{\text{di-melt}} + W_0^{\text{en}} K_{\text{D(Ce)}}^{\text{en-melt}} + W_0^{\text{fo}} K_{\text{D(Ce)}}^{\text{fo-melt}}$$

$$= (0.4)\ (0.10) + (0.2)\ (0.003) + (0.4)\ (0.001)$$

$$= 0.04 + 0.0006 + 0.0004 = 0.041$$

$$P = p^{\text{di}} K_{\text{D(Ce)}}^{\text{di-melt}} + p^{\text{en}} K_{\text{D(Ce)}}^{\text{en-melt}} + p^{\text{fo}} K_{\text{D(Ce)}}^{\text{fo-melt}}$$

$$= (0.7)\ (0.10) + (0.2)\ (0.003) + (0.1)\ (0.001)$$

$$= 0.07 + 0.0006 + 0.0001 = 0.0707$$

Now let us calculate the $C_{\text{Ce}}^{\text{melt}}/C_{0(\text{Ce})}^{\text{rock}}$ values for $F = 0.3$.
Batch melting:

Modal
$$\frac{C_{\text{Ce}}^{\text{melt}}}{C_{0\ (\text{Ce})}^{\text{rock}}} = \frac{1}{D_{0\ (\text{Ce})}^{\text{rock}} + F(1 - D_{0\ (\text{Ce})}^{\text{rock}})}$$
$$= \frac{1}{0.041 + (0.3)\ (1 - 0.041)} = 3.04$$

Nomodal
$$\frac{C_{\text{Ce}}^{\text{melt}}}{C_{0\ (\text{Ce})}^{\text{rock}}} = \frac{1}{D_{0\ (\text{Ce})}^{\text{rock}} + F(1 - P)}$$
$$= \frac{1}{0.041 + (0.3)\ (1 - 0.0707)} = 3.13$$

Fractional melting:

Modal
$$\frac{C_{\text{Ce}}^{\text{melt}}}{C_{0\ (\text{Ce})}^{\text{rock}}} = \frac{1}{D_{0\ (\text{Ce})}^{\text{rock}}}\ (1 - F)^{(\frac{1}{D_{0\ (\text{Ce})}^{\text{rock}}} - 1)} = \frac{1}{0.041}(1 - 0.3)^{(\frac{1}{0.041} - 1)}$$
$$= 0.0058$$

Nomodal
$$\frac{C_{\text{Ce}}^{\text{melt}}}{C_{0\ (\text{Ce})}^{\text{rock}}} = \frac{1}{D_{0\ (\text{Ce})}^{\text{rock}}}\ (1 - \frac{PF}{D_{0\ (\text{Ce})}^{\text{rock}}})^{(\frac{1}{P} - 1)}$$
$$= \frac{1}{0.041}\ (1 - \frac{(0.0707)\ (0.3)}{0.041})^{(\frac{1}{0.0707} - 1)}$$
$$= 0.0017$$

To predict the change in the melt composition with continued partial melting, we would have to consider the stabiliy relations among the three phases as depicted in an appropriate phase diagram for the di–en–fo system.

12.3.3 *Behavior of trace elements during magmatic crystallization*

Differentiation of primary magmas on cooling occurs mainly by two processes: (i) liquid immiscibility and (ii) crystallization. *Liquid immiscibility* is the phenomenon of separation of a cooling magma into two or more liquid phases of different composition in equilibrium with each other. Separation of a sulfide liquid in a mafic or ultramafic magma, for example, is believed to be the dominant mechanism in the formation of magmatic nickel–copper sulfide deposits hosted by ultramafic–mafic complexes. Model equations for calculating the concentration of a trace element in the immiscible sulfide liquid have been discussed by Campbell and Naldrett (1979) and Naldrett *et al.* (1984), and summarized in Misra (2000).

Analogous to partial melting, crystallization of magma may be viewed in terms of two end-member models.

(1) *Equilibrium crystallization*, which assumes continuous thermodynamic reequilibration of the crystallized product with the residual (depleted) magma producing crystals without zoning. It is assumed that the cooling is sufficiently slow and the diffusion sufficiently rapid so that the interior of crystals maintain equilibrium with the melt during crystallization.
(2) *Fractional crystallization* (or *Rayleigh fractionation*), which precludes equilibration between the products of crystallization and magma, either because of continuous removal of the products from magma as soon as they are formed or because diffusion in crystals is much slower than in magma. This process is best described by the Rayleigh distillation equation.

Consider the crystallization of a mineral assemblage containing a trace element i from a finite magma reservoir. Assuming that the mineral–magma distribution coefficients remain constant during the crystallization process, equations that may be used to calculate the changing concentration of i in the residual magma relative to the original magma are as follows (Wood and Fraser, 1976; Allègre and Minster, 1978):

Equilibrium crystallization
$$\boxed{\frac{C_i^{\text{res magma}}}{C_{0\ (i)}^{\text{magma}}} = \frac{1}{D_i^{\text{cryst}} + F * (1 - D_i^{\text{cryst}})}} \qquad (12.45)$$

Fractional crystallization
$$\boxed{\frac{C_i^{\text{res magma}}}{C_{0\ (i)}^{\text{magma}}} = F^{*(D_i^{\text{cryst}} - 1)}} \qquad (12.46)$$

where $C_{0(i)}^{\text{magma}}$ is the initial concentration of i in the magma, $C_i^{\text{res magma}}$ is its concentration in the residual magma, F^* is the weight fraction of the original magma remaining (a measure of the degree of crystallization), and D_i^{cryst} is the bulk distribution coefficient of i for the crystallizing mineral assemblage. The calculation of the bulk distribution coefiicient, D_i^{cryst}, is analogous to that of $D_{0(i)}^{\text{rock}}$ (equation 12.30):

$$D_i^{\text{cryst}} = \sum_{\alpha} W_0^{\alpha}\, K_{\text{D}\,(i)}^{\alpha\text{—magma}} \qquad (12.47)$$

where W_0^{α} is the weight fraction of crystallizing mineral α and $K_{\text{D}(i)}^{\alpha\text{-magma}}$ its crystal–magma distribution coefficient for the element i. For either crystallization model, the smaller the value of the bulk distribution coefficient of a trace element, the greater will be its enrichment in the melt for a given degree of crystallization (Fig. 12.23). A comparison of equilibrium and fractional crystallization for the same element would show that the enrichment of the residual magma in incompatible trace elements is very similar in the two cases up to about $F = 0.25$, but at $F > 0.25$ for incompatible elements and for highly compatible elements the two models differ significantly. In a differentiated suite of igneous rocks, the late differentiates are commonly considerably enriched in highly incompatible elements (such as Li, Ba, Rb, and Pb) and depleted in markedly compatible elements (such as Cr and Ni).

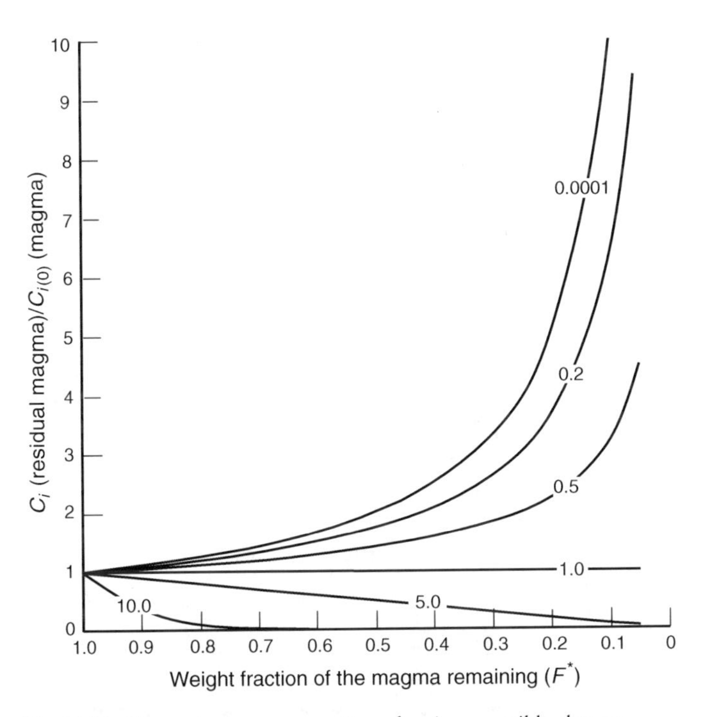

Fig. 12.23 Change in the concentration of an incompatible element, relative to its initial concentration in the source rock, as a function of the weight fraction of melt remaining (F^*) during fractional crystallization. The curves are for arbitrarily assumed values (0.0001, 0.2, 0.5, 1.0, 5.0, 10.0) of the bulk distribution coefficient.

Gravitative settling of crystals to the base of the magma chamber (or flotation to the top, depending on the density of crystals relative to the magma) is no longer considered the dominant mechanism of crystal separation, except perhaps in the case of ultramafic magmas. The evidence now appears to favor *in situ* growth of crystals on the floor and walls of the magma chamber and, to facilitate further crystallization, removal of the depleted magma from contact with the growing crystals by a combination of diffusive and convective processes (Sparks *et al.*, 1984). Equations that describe trace element distributions during *in situ* crystallization have been discussed by Langmuir (1989).

In principle, it should be possible to invert the above equations to calculate the weight proportion of phases crystallizing from a melt for a given degree of fractional crystallization (Allègre and Minister, 1978). At present such calculations are limited because of the lack of high-quality mineral-melt partition coefficient data for many of the elements of interest. In addition, trace element modeling of natural systems is constrained by the problems inherent in the analysis of very small concentrations and the dependence of K_{D} values on the compositions of the phases involved in the crystallization process.

Example 12–2. Crystallization differentiation of oceanic tholeiitic basalts

The average Rb contents of oceanic tholeiitic basalts and oceanic alkalic basalts are 1 and 18 ppm, respectively. The bulk distribution coefficient for Rb in tholeiitic basalt is 0.02, and it is assumed to be constant throughout the crystallization process. Based on the distribution of Rb, could alkalic basalts be a fractional crystallization product of tholeiitic magmas (after Wood and Fraser, 1976)?

The enrichment factor for Rb in alkali basalts (residual magma) relative to tholeiites (original magma) = 18.0 ppm/1 ppm = 18. Let us calculate the degree of crystallization necessary to produce such a residual magma.

Applying the Rayleigh crystallization model, the most favorable condition for enrichment of incompatible elements in the residual magma, the extent of fractional crystallization necessary in this case can be calculated as follows:

$$\frac{C_{\text{Rb}}^{\text{alk basalt magma}}}{C_{\text{Rb}}^{\text{tholeiite magma}}} = F^{*(D_i^{\text{cryst}}-1)} = F^{*(0.02-1)} = F^{*(-0.98)} = 18$$

$$F^* = 0.0524$$

Equivalent percent crystallization = (1 – 0.0524) × 100 = 94.76

Applicaton of the equilibrium crystallization yields a similar result:

$$\frac{C_{\text{Rb}}^{\text{alk basalt magma}}}{C_{\text{Rb}}^{\text{tholeiite magma}}} = F^{*(D_i^{\text{cryst}}-1)} = F^{*(0.02-1)} = F^{*(-0.98)} = 18$$

$$\frac{C_i^{\text{alk basalt magma}}}{C_{0\,(i)}^{\text{tholeiite magma}}} = \frac{1}{0.02 + F^* (1 - 0.02)} = 18$$

$F^* = 0.037$

Equivalent percent crystallization = (1 − 0.037) × 100 = 96.3

It is difficult to imagine any combination of phases that could be removed from a tholeiitic melt so as to achieve 95–96% crystallization without substantially altering the major element chemistry of the residual liquid. It is, therefore, unlikely that alkali basalts are equilibrium or fractional crystallization products of tholeiitic magmas.

12.3.4 *Chemical variation diagrams*

A *chemical variation diagram* is a graphical display of compositional variation in a presumably genetically related suite of rocks. Variation diagrams are plotted in terms of abundances or ratios of major or trace elements (or a combination of both) that are relevant to the rocks being investigated. Trends in such diagrams are particularly useful for the interpretation of crystal–liquid fractionation processes, by either partial melting or fractional crystallization, in a cogenetic suite of volcanic igneous rocks. One of the commonly used types in igneous petrology is the *Harker diagram*, in which constituent oxides (in wt%) are plotted against SiO_2 (in wt%), the latter chosen as a measure of the degree of magmatic differentiation. Because SiO_2 tends to be mobile even in weakly altered rocks, other less mobile constituents, such as MgO or Zr, are often preferred as the differentiation index. For a cogenetic differentiated series of rocks, Harker-type variation diagrams may display continuous linear trends or coherent segmented trends marking the so-called *liquid line of descent*, which represents the changing magma composition with fractional crystallization. In general, the inflection point between two adjacent segmented trends marks the onset of crystallization of a new mineral or a group of new minerals. This principle is illustrated in Fig. 12.24 with a hypothetical example. Basaltic magmas, in which the number of crystallizing phases is normally small, frequently display strong inflected trends; intermediate volcanic suites, on the other hand, typically display inflection-free trends in major element variation diagrams, despite wide variation in the crystallizing assemblage (Wilson, 1989).

Variation diagrams with trace elements as variables may also provide useful information about fractional crystallization. For example, trends of decreasing Ni, Cr, and Ti with differentiation through a rock series indicate fractionation of olivine, clinopyroxene, and Fe–Ti oxides, respectively. In such diagrams Zr is often used as a differentiation index.

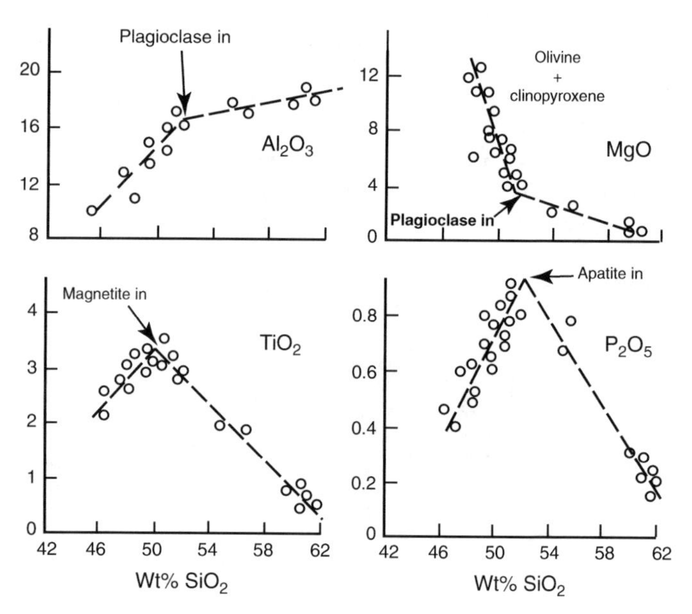

Fig. 12.24 Harker variation diagrams for a suite of cogenetic volcanic rocks related by fractional crystallization of olivine, clinopyroxene, plagioclase, magnetite, and apatite. (From *Igneous Petrogenesis* by Marjorie Wilson, 1989, Fig. 4.7, p. 85, Springer, The Netherlands. Originally published by Chapman and Hall. Reproduced with kind permission from Science+Business Media B.V.)

12.3.5 *Rare earth elements*

The REEs comprise a coherent group of 15 elements – La (Z = 57), Ce, Pr, Nd, Pm, Sm, Eu, Gd, Tb, Dy, Ho, Er, Tm, Yb, and Lu (Z = 71) – characterized by very similar physical and chemical properties. They all form 3+ cations. Eu also exists as Eu^{2+} in igneous systems, substituting mainly for Ca^{2+} in plagioclase, the Eu^{3+} :Eu^{2+} ratio becoming higher with increasing oxidizing condition, and Ce may be tetravalent under oxidizing conditions.

The REEs in both chondrites and basalts exhibit a saw-tooth pattern because of the *Oddo–Harkins effect* (elements with even–odd atomic numbers having higher concentrations than their immediate neighbors with odd atomic number). Normalization of the REE abundances for basalts relative to the abundances in CI chondrite meteorites, believed to be the Earth's parental material, gives smoothed REE patterns that can be compared directly to primordial Earth. The normalization is done by dividing the concentration of each REE by the concentration of the same REE in the chondrite standard. A few examples of chondrite-normalized REE patterns are shown in Fig. 12.25.

Magmas formed by partial melting of upper mantle rocks incorporate REEs depending on their abundances in the source rock and the $D^{\text{mineral–magma}}$ values for the various minerals in the source rock (see section 12.3.2). Therefore, the overall shape of the REE patterns and anomalies associated with individual elements are useful for constraining the source of magma or the participation of specific minerals in

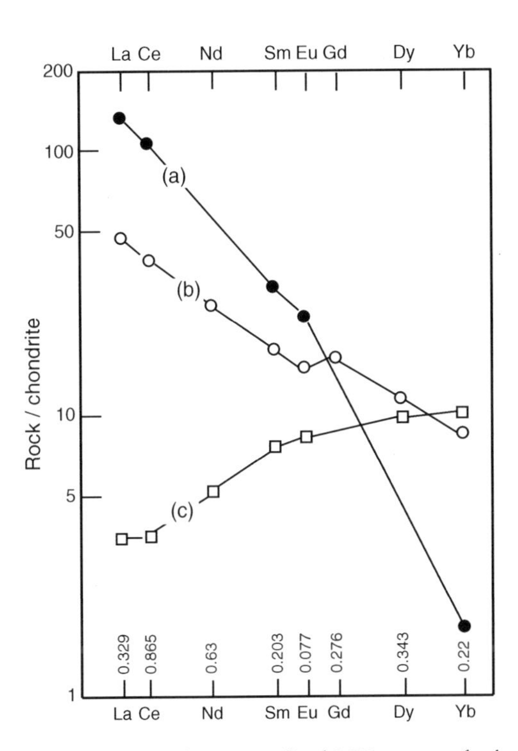

Fig. 12.25 Schematic chondrite-normalized REE patterns for basaltic igneous rocks formed by partial melting of upper mantle rocks. (a) Basalt strongly enriched in light REE, suggesting very small degrees of partial melting or a light-REE enriched source, and very low concentrations of heavy REE, suggesting the presence of residual garnet in the source. (b) Basalt with high concentrations of heavy-REE, suggesting the absence of garnet in the source rock, and a slight negative Eu anomaly indicative of plagioclase crystallization or equilibrium of the partial melt with a plagioclase-bearing mantle source. (c) Basalt showing strong light-REE depletion, suggesting derivation from a light-REE depleted garnet-free source. Chondrite values for normalization, as indicated on the diagram, are from Nakamura (1974). (From *Igneous Petrogenesis* by Marjorie Wilson, 1989, Fig. 2.3, p. 18, Springer, The Netherlands. Originally published by Chapman and Hall. Reproduced with kind permission from Science+Business Media B.V.)

its evolution. For example, garnet has very low $D^{\text{mineral-magma}}$ values for the light REEs and increasingly larger values for heavy REEs (Table 12.7), so that magma generated by partial melting of a garnet-bearing source rock would be depleted in the heavy REEs. On the other hand, $D^{\text{mineral-magma}}$ values for all REEs are less than 0.1 in the case of olivine; thus its presence in the source rock leads to essentially equivalent enrichment of all the REEs. The compatibility of Eu^{2+} with plagioclase, in contrast to low $D^{\text{mineral-magma}}$ values for all other REEs, gives rise to the negative europium anomaly either because Eu^{2+} is retained by plagioclase in the source rock or because of crystallization and removal of plagioclase from the partial melt (Fig. 12.25).

Rare earth elements typically have low solubility in water. So clastic sediments commonly inherit the REE composition of their eroded parents. Consequently, the REE distribution pattern of clastic sediments is a good indication of the composition of the crystalline rock from which the sediments were derived. The interpretation gets complicated if sediments were contributed by more than one parent.

12.4 Geochemical discrimination of paleotectonic settings of mafic volcanic suites

12.4.1 *Tectonomagmatic discrimination diagrams*

Several studies have shown that modern basaltic rocks erupted in different tectonic settings carry different trace element signatures. For example, island arc basalts are characterized by a selective enrichment in incompatible elements, whereas continental tholeiites contain higher concentrations of K, Rb, Ba, and Th compared to MORB. It is, therefore, reasonable to expect that the tectonic settings of ancient basaltic rocks may be deciphered from their trace element geochemistry, a particularly useful exercise in those cases where the paleotectonic setting is not clear from field relations and petrology. The principle behind all the proposed "*tectonomagmatic discrimination diagrams*" is the comparison of selected trace element concentrations in an ancient basaltic suite with their concentrations in present day analogs from known plate-tectonic settings. Some discrimination diagrams use ratios of incompatible elements for such comparisons in order to overcome the effects of fractionation that occur during partial melting or fractional crystallization. The effectiveness of discrimination diagrams utilizing major elements is limited by the extensive overlap in major element chemistry among MORB, backarc basin tholeiites, and volcanic-arc basalts, and the relative mobility of the major elements during alteration and metamorphism.

Trace elements that can be used for these empirical discrimination diagrams must be relatively immobile during post-crystallization alteration and metamorphism. Elements that commonly best satisfy this requirement are: the REEs (with the possible exception of Ce); HFS elements such as Sc, Y, Zr, Hf, Th, Ti, P, Nb, and Ta; and some other elements such as Cr, Ni, Co, and V. Low field strength elements (Cs, Sr, K, Rb, Ba) are relatively mobile, even during low to moderate temperature alteration by seawater; even HFS elements may be somewhat mobile in fluids at lolw pH and with high F contents (Jiang *et al.*, 2005). In addition, the transition metals Mn, Zn, and Cu tend to be mobile, particularly at high temperatures. Such generalizations are normally valid, but many exceptions have been documented. The samples analyzed must not be much altered, as evidenced by other tests, for example, a high degree of correlation between MgO and Ni as expected in unaltered basaltic rocks. The samples should also be free of phenocrysts and be large enough to overcome the effects of small-scale inhomogeneities.

Since the pioneering work of Pearce and Cann (1973), who proposed a Ti–Zr–Y ternary diagram with distinct

Table 12.8 Tectonomagmatic discrimination diagrams for determining paleotectonic settings of ancient basaltic suites.

Tectonomagmatic discrimination diagram	Tectonic settings to be discriminated	Reference	Comments	
Bivariate	Ti–Zr	OFB–LKT–CAB	Pearce and Cann (1973)	Continental (within-plate) tholeiitic basalts must be screened out before using this diagram for altered volcanics (Floyd and Winchester, 1975)
	Cr–Ti	OFB–LKT	Pearce (1975)	
	(Y/Nb)–TiO_2 (Fig. 12.26d)	Oceanic tholeiites–oceanic alkali basalts–continental tholeiites–continental alkali basalts	Floyd and Winchester (1975)	Useful for discriminating between tholeiitic basalts (Y/Nb > 1; horizontal trend) and alkali basalts (Y/Nb > 1; vertical trend)
	Ni–(Ti/Cr)	OFT–IAT (low-K)	Beccaluva *et al.* (1979)	Only tholeiitic rocks with SiO2 between 40 and 56% were used for this diagram
	Y–Cr (Fig. 12.26b)	MORB–WPB–VAB	Pearce *et al.* (1981)	Effectively discriminates between MORB and VAB
	Zr–Ti (Fig. 12.26a)	MORB–WPB–IAB	Pearce *et al.* (1981)	Considerable overlap among the three fields
	(Nb/Y)–(Ti/Y)	MORB–IAB–WPB	Pearce (1982)	Successfully separates MORB from WPB
	Zr–(Zr/Y) (Fig. 12.26c)	MORB–IAB–WPB	Pearce and Norry (1979)	Effectively discriminates between MORB, IAB, and WPB
Ternary	(Ti/100)–Zr–(Yx3) (Fig. 12.26e)	OFB–LKT–CAB–WPB	Pearce and Cann (1973)	Supposed to be very effective for discriminating WPB, but not according to Holm (1982) The OFB field should also represent all tholeiitic basalts formed in a rifting environment (Floyd and Winchester, 1975)
	(Ti/100)–Zr–(Sr/2)	OFB–LKT–CAB	Pearce and Cann (1973)	Should not be used unless Sr can be shown to be immobile
	(Hf/3)–Th–Ta (Fig. 12.26h)	N-MORB–P-MORB–WPB–DPMB	Wood (1980)	Recognizes different types of MORB, and is supposed to be effective at identifying volcanic-arc basalts. Does not work well according to Thompson *et al.* (1980)
	TiO_2–(MnOx10)–(P_2O_5x10) (Fig. 12.26f)	MORB–IAT–OIT–CAB–OIA	Mullen (1983)	Applicable to basaltic andesites and basalts (45–54% SiO2), and their zeolite to green schist facies metamorphosed products, of oceanic regions. Continental tholeiites overlap portions of all five oceanic fields
	(Nbx2)–(Zr/4)–Y (Fig. 12.26g)	N-MORB–P-MORB WPBA–WPBT–VAB	Meschede (1986)	VAB plot within both WPBT and N-type MORB fields and thus cannot be distinguished from these types. Ancient continental tholeiites which plot in the T–Zr–Y diagram within the MORB/VAB field are clearly distinguished from N-type MORB
Major elements	Statistical analysis of major elements	OFB–LKT–CAB–SHO–WPB	Pearce (1976)	Does not successfully discriminate between ocean-island and continental basalts
	FeO*–MgO–Al_2O_3	Ocean floor and ridge–continental tholeiites–ocean island basalts–orogenic–spreading center islands	Pearce *et al.* (1977)	Based on analysis of subalkaline basaltic–andesitic rocks only (51–56% SiO2), mainly Cenozoic in age. Claims to differentiate between oceanic and continental basalts

CAB = calc-alkaline basalts; CON = continental basalts; DMPB = destructive plate margin basalts; IAB = island-arc basalts; IAT/LKT = island-arc (low-K) tholeiites; MORB = mid-oceanic ridge basalts; N-MORB = "normal" mid-oceanic ridge basalts; P-MORB = plume-type enriched MORB; OFB = ocean-floor basalts; OFT = ocean-floor tholeiites; SHO = shoshonites; OIA = oceanic-island alkalic basalts; OIB = ocean-island basalts; OIT = ocean-island tholeiites; VAB = volcanic-arc basalts; WPB = within-plate basalts (oceanic and continental); WPBA = within-plate basalts alkalic; WPBT = within-plate basalts tholeiitic.
Shoshonite is a volcanic-arc basaltic (trachy-andesitic, alkaline) rock composed of olivine and augite phenocrysts in a groundmass of labradorite with orthoclase rims, olivine, augite, small amount of leucite, and some dark colored glass.

fields for WPB, CAB, IAT, and IAT+MORB, a number of tectonomagmatic discrimination diagrams have been proposed by various authors. A selection of useful discrimination diagrams is summarized in Table 12.8, and some of the more commonly used diagrams are presented in Fig. 12.26. Rollinson (1993) provides a comprehensive discussion of tectonomagmatic discrimination diagrams for basaltic and granitic rocks.

It must, however, be mentioned that this approach by itself is unlikely to provide unequivocal interpretation of the paleotectonic settings, especially for highly altered rocks. The tectonomagmatic discrimination diagrams should be used as a possible clue but not as the conclusive evidence.

12.4.2 *Spider diagrams*

Another method of inferring the tectonic setting of an ancient basaltic suite is to compare its pattern of selected trace element abundances with those of their present-day analogs, both normalized to a common reference material, in known tectonic settings. The elements chosen for generating such patterns, referred to as "*spider diagrams*" (Thompson, 1982), behave incompatibly (i.e., $D_i^{\text{mineral—melt}} < 1$) with respect to a small percentage of melt generated by partial melting. Exceptions to this are Sr, which may be compatible with plagioclase, Y and Yb with garnet, and Ti with magnetite. The order in which the elements are arranged in a spider diagram is somewhat arbitrary, commonly designed to give a smooth pattern for average MORB. Generally, the order is one of decreasing incompatibility from left to right in a four-phase lherzolite (clinopyroxene–orthopyroxene–olivine–garnet) undergoing partial melting.

Normalization concentrations adopted for spider diagrams range from chondritic abundances (e.g., Thompson, 1982; Sun and McDonough, 1989) because of their nearly cosmic composition, to a hypothetical primitive mantle (BSE) composition (e.g., Taylor and McLennan, 1985; Kargel and Lewis, 1993; McDonough and Sun, 1995), to an average MORB composition (Pearce, 1983; Bevins *et al*., 1984). Some of the commonly used normalization values are listed in Table 12.9. Normalization to chondritic values, which have been measured directly, may be preferable to the primordial mantle composition, which is estimated. The normalization values (in ppm) used for Fig. 12.26 are chondritic values from Thompson (1982).

MORB-normalized patterns are particularly useful for comparing the trace element characteristics of different types of basalts. The elements used for MORB-normalized spider diagrams include two groups based on their relative mobility in aqueous fluids: Sr, K, Rb, and Ba are mobile and plot at the left side of the pattern; the remaining elements are immobile. The elements are arranged such that the incompatibility of the mobile and immobile elements increases from the outside to the center of the pattern (Wilson, 1989, p. 21). The shape of these patterns, according

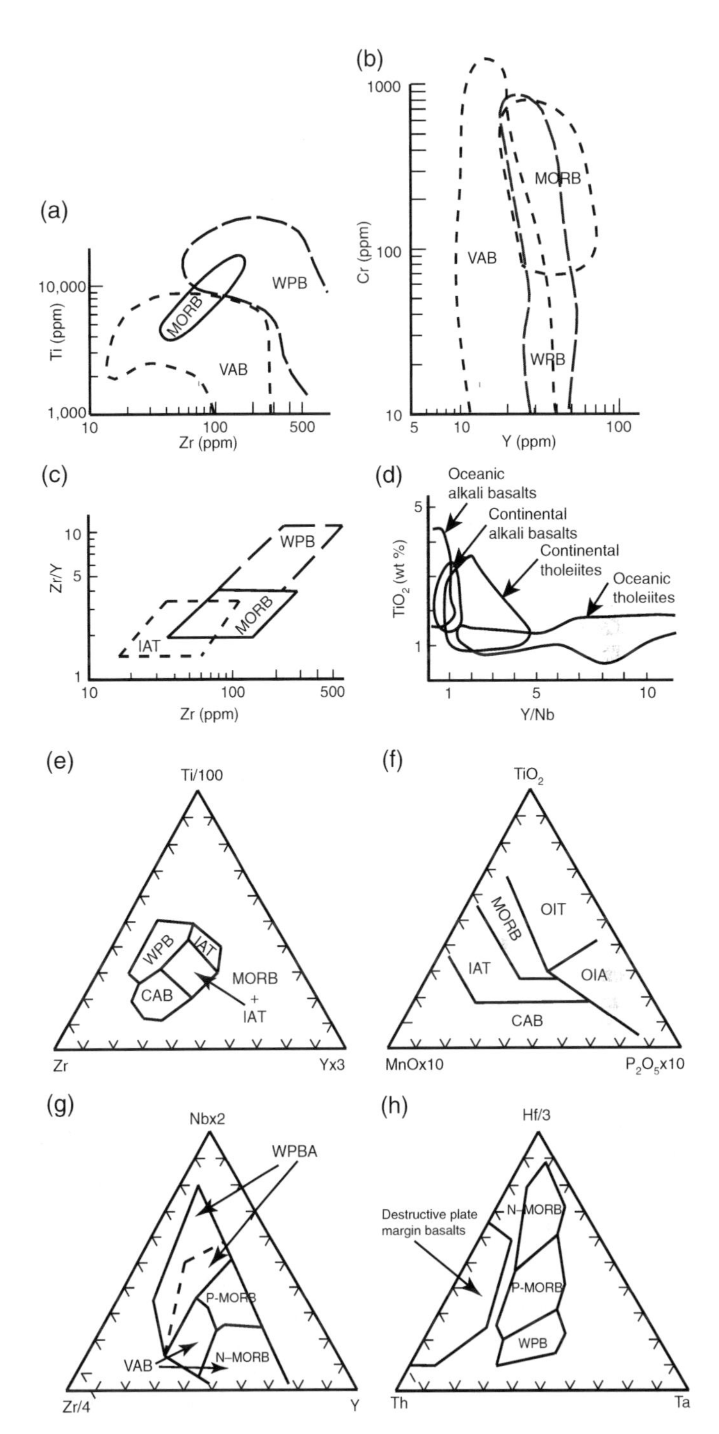

Fig. 12.26 Some commonly used tectonic discrimination diagrams for mafic volcanic rocks: (a) Ti/100–Zr–Yx3 (Pearce and Cann, 1973); (b) TiO_2–MnOx10–P_2O_5x10 (Mullen, 1983); (c) Nbx2–Zr/4–Y (Meschede, 1986); (d) Hf/3–Th–Ta (Wood *et al*., 1979); (e) Zr–Ti (Pearce *et al*., 1981); (f) Y–Cr (Pearce *et al*., 1981); (g) Zr–Zr/Y (Pearce and Norry, 1979); (h) Y/Nb–TiO_2 (Floyd and Winchester, 1975). OFB = ocean-floor basalts; OFT = ocean-floor tholeiites; N-MORB = "normal" mid-oceanic ridge basalts; P-MORB = plume-type enriched MORB; VAB = volcanic arc basalts; IAT/LKT = island arc (low-K) tholeiites; CAB = calc-alkaline basalts; IAB = island arc basalts; SHO = shoshonites. WPB = within-plate basalts (oceanic and continental); CON = continental basalts; WPBA = within-plate basalts alkalic; WPBT = within-plate basalts tholeiitic; OIB = ocean-island basalts; OIT = ocean island tholeiites; OIA = oceanic island alkalic basalts.

Table 12.9 Normalizing values (in ppm) commonly used in the calculation of "spider diagrams."

	Primitive mantle						MORB	
Trace element	**(1)**	**(2)**	**(3)**	**Trace element**	**Chondrite (4)**	**Trace element**	**(5)**	**(6)**
Cs	0.018	0.0131	0.021	Ba	6.90	Sr	120	136
Rb	0.550	0.598	0.600	Rb	0.350	K_2O(%)	0.15	0.15.
Ba	5.1	6.41	6.6	Th	0.042	Rb	2	1
Th	0.064	0.0782	0.0795	K	120	Ba	20	12
U	0.018	0.022	0.023	Nb	0.350	Th	0.20	0.20
K	180	232.4	240	Ta	0.020	Ta	0.18	0.17
Ta	0.040	0.0412	0.037	La	0.329	Nb	3.50	2.50
Nb	0.560	0.765	0.658	Ce	0.865	Ce	10	10
La	0.551	0.622	0.648	Sr	11.800	P_2O_5(%)	0.12	0.12
Ce	1.436	1.592	1.675	Nd	0.630	Zr	90	88
Sr	17.8	20.7	19.9	P	46	Hf	2.40	2.50
Nd	1.067	1.175	1.250	Sm	0.203	Sm	3.50	
P	n.d.	79.4	90	Zr	6.840	TiO_2(%)	1.50	1.50
Hf	0.270	0.300	0.283	Hf	0.200	Y	30	35
Zr	8.3	11.47	10.5	Ti	620	Yb	3.40	3.50
Sm	0.347	0.360	0.406	Tb	0.052	Sc	40	
Ti	960	1180	1205	Y	2	Cr	250	290
Tb	0.087	0.0955	0.099	Tm	0.034			
Y	3.4	3.91	4.30	Yb	0.220			
Pb	0.120	0.149	0.150					

Sources of data: (1) Taylor and McLennan (1985); (2) Kargel and Lewis (1993); (3) McDonough and Sun (1995); (4) Thompson (1982); (5) Pearce (1983); (6) Bevins *et al.* (1984).

to Pearce (1983), is not likely to be modified much by variable degrees of partial melting or fractional crystallization, and the patterns may provide constraints on source characteristics.

As an example, spider diagrams for some typical present-day basaltic rocks from known tectonic settings are presented in Fig. 12.27. Note that the patterns are quite distinct. The oceanic basalts (both tholeiitic and alkalic) are characterized by the relatively smooth, convex upward patterns and high degrees of incompatible element enrichment, especially in Nb–Ta, suggesting that the source may be enriched in incompatible elements. The island arc calc-alkaline basalt, on the other hand, is strongly spiked, with a pronounced Nb–Ta trough and a lesser dip for the relatively less incompatible elements. The positive spikes are mostly a consequence of materials added to the mantle source of the basalts by subduction-zone fluids. In contrast, the MORB has a relatively smooth pattern, with the less incompatible elements (such as Zr, Hf, Ti, and Y) more enriched than the more incompatible elements (such as Ba, Rb, Th, and K). There is no known mechanism for producing such a pattern in the large volumes of partial melt generated at mid-oceanic ridges (Nory and Fitton, 1983). The most logical explanation for the MORB pattern is a large degree of partial melting of a source depleted in the more incompatible elements such as K and Rb and, to a lesser extent, Nb and La. The depletion is probably complementary to the formation of continental crust throughout geologic time.

The major limitation of the reliance on spider diagrams to interpret paleotectonic settings is the sensitivity of the patterns to compositional variability within any magma series due to variations in the partial melting process as well as due to external factors such as magma mixing, crustal contamination, and alteration. For instance, the distinctive Nb–Ta trough for calc-alkaline basalt (Fig. 12.27) is also a typical feature of other types of basalts that have experienced crustal contamination.

Spider diagrams, which may be viewed as an extension of REE diagrams (Fig. 12.25), also provide valuable petrogenetic information concerning crystal–magma equilibria. Consider, for example, the pattern for MORB in Fig. 12.27. The enrichment in less incompatible elements such as Zr, Ti, and Y in the magma indicates high degrees of partial melting of the upper mantle source, which can be reconciled with the much lesser enrichment in the highly incompatible elements such as Ba, Rb, and K only if the source of MORB is depleted in these elements. The complementary trace element patterns of the oceanic crust and continental crust (Fig. 12.15) are consistent with the argument that this depletion of the source has resulted from the formation of the continental crust throughout geological time.

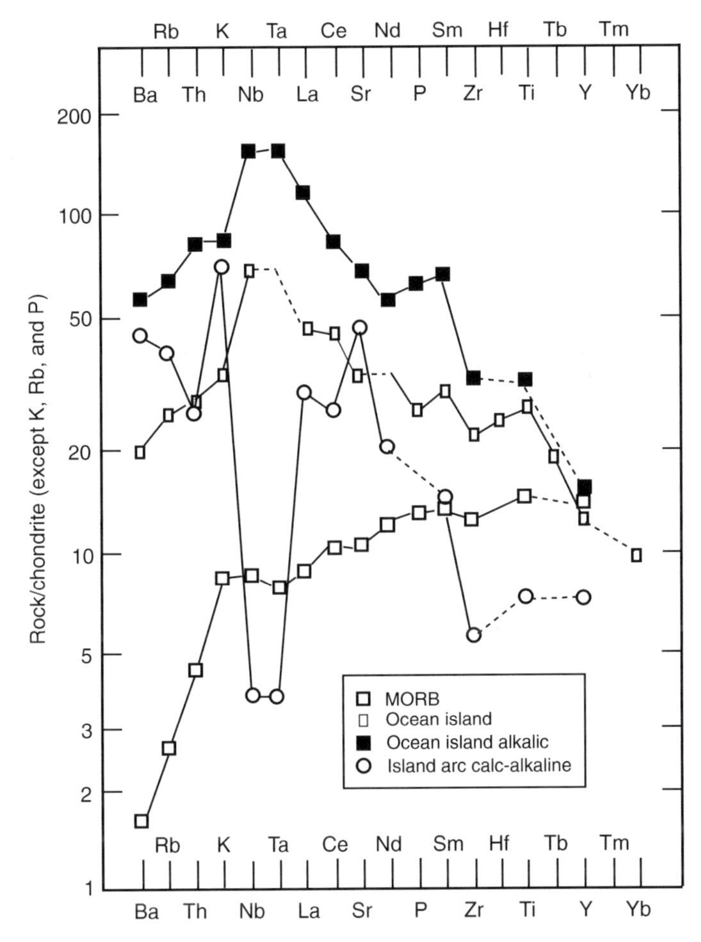

Fig. 12.27 Spider diagrams for typical mid-ocean ridge basalt (MORB), ocean island alkalic basalt, ocean island tholeiite, and island arc calc-alkaline basalt (Basaltic Volcanism Study Project, 1981). The trace elements are arranged in order of decreasing incompatibility. The chondrite normalization factors are from Thompson *et al.* (1984).

12.5 Summary

1. The most popular model for the origin of the universe is the Big Bang theory, which attributes the creation of the universe about 15 Ga to very rapid, exponential expansion of infinitely small, very hot, and infinitely dense matter.
2. The Big Bang created only two elements, hydrogen and helium, in any abundance, and trace amounts of three very light elements – lithium, beryllium, and boron.
3. Elements up to $A = 56$ were formed mostly inside stars by a succession of nuclear fusion processes referred to as hydrogen burning, helium burning, carbon burning, neon burning, oxygen burning, and silicon burning. Heavier elements could be synthesized within stars by a slow neutron capture process known as the *s* process or in the explosive environments, such as supernovae, by a number of processes. Some of the more important of these include: the *r* process, which involves rapid neutron captures; the *rp* process, which involves rapid proton captures; and the *p* process, which involves photodisintegration of existing nuclei.
4. The Solar System formed from a nebula. The formation of the planets within he Solar Nebula involved two important processes: condensation of the nebular gases into solid grains (dust); and accretion of the dust grains into progressively larger solid bodies, reaching dimensions of the order of 1 km (planetesimals). The largest of the planetesimals grew rapidly into planetary embryos of the order of 1000 km in diameter in 0.1–1 million years, and collision of a few dozen of the embryos formed the planets.
5. Age-dating of meteorites indicates that the formation of the planets, including the Earth, occurred at about 4.56 Ga, which is considered the beginning of geologic time.
6. The Earth's interior consists of a series of concentric shells of differing composition and density: the core (outer liquid core and solid inner core), the mantle, and the crust (continental and oceanic).
7. The Earth's differentiation into primitive mantle (bulk silicate Earth, BSE) and metallic core is best explained by magmatic processes operating in a deep high *P–T* magma ocean environment.
8. The oldest oceanic crust is only about 200 Ma old. The earliest crust probably was basaltic in composition (with significant amounts of komatiite?) formed by magmatic processes and was recycled the same way as the modern oceanic crust, but there are no remnants of the earliest oceanic crust. The oldest continental crust material identified so far is in the form of ~4.4-Ga-old detrital zircon crystals in metamorphosed sediments. The earliest continental crust must have formed by partial melting of the primitive mantle, with incompatible elements (such as Cs, Rb, K, U, Th, and La) segregated into the crust-forming partial melts. The continental crust appears to have grown episodically, with at least 60% of the crust emplaced by 2.7 Ga.
9. Primary magmas generated by partial melting of upper mantle rocks, dominantly resulting from decompression, comprise three major magma series – tholeiitic, calc-alkaline, and alkalic – each of which spans a genetically related spectrum of mafic to felsic rocks.
10. Trace element characteristics of primary magmas are governed by the composition of the parent material, degree of partial melting, mineral–melt bulk distribution coefficients of the elements, and whether the melting was an equilibrium or fractional (Rayleigh) process. The concentration of a given trace element in the melt relative to the source increases with decreasing value of bulk distribution coefficient for a given degree of partial melting and decreases with increasing degree of partial melting for a given value of the bulk distribution

coefficient. Behavior of trace elements during crystallization can also be modeled as an equilibrium or fractional process.

11. Crystal–liquid fractionation proceses, by either partial melting or crystallization, in a cogenetic volcanic suite can be interpreted from trends in chemical variation diagrams.
12. Incompatible and immobile trace elements, including REEs, in unaltered (or only slightly altered) and genetically related basaltic suites are useful for recognition of their paleotectonic settings. Tools commonly used for this purpose include empirical tectonomagmatic discrimination diagrams and "spider diagrams."

12.6 Recapitulation

Terms and concepts

Asthenosphere
Big Bang theory
Bulk distribution coefficient
Bulk silicate Earth (BSE)
Chemical variation diagram
Compatible elements
Continental crust
Core
Doppler effect
Enrichment factor
Equilibrium crystallization
Equilibrium (batch) partial melting
Fractional crystallization
Fractional partial melting
Giant impact hypothesis
Harker diagram
Heterogeneous accretion
Ionic potential
Liquidus
Incompatible elements
In situ crystallization
Homogeneous accretion
LIL elements
Liquid immiscibility
Liquid line of descent
Lithosphere
Magma ocean
Magma series
Mantle
Meteorites
Modal partial melting
Moho
Neutron capture
Nonmodal partial melting
Planetesimals
Plate tectonics
p process
Primitive mantle
Nucleosynthesis
Oceanic crust
Pyrolite
Rayleigh equation
r process
rp process
s process
Solar Nebula
Solar System
Solidus
Spider diagram
Tectonomagmatic discrimination diagram

Computation techniques

- Concentration of incompatible trace elements in melts formed by partial melting or in melts remaining during crystallization of a melt (magma).
- Harker-type variation diagrams.
- REE patterns.
- Tectonomagmatic discrimination diagrams.
- Chondrite-normalized spider diagrams.

12.7 Questions

1. Construct a chart showing the importants events, in chronological order, related to the evolution of the Earth during the first billion years or so after its birth.
2. In the Sun's core, each second about 700,000,000 tons of hydrogen are converted to about 695,000,000 tons of helium, and the excess mass is converted into energy. Calculate the energy output per second by this process of thermonuclear fusion, using the famous Einstein equation $E = mc^2$.
3. Show that the predominant form of hydrogen gas at the Sun's surface ($T = 5800\,K$, $P = 1$ bar) is atomic hydrogen (H), not molecular hydrogen (H_2). The value of the equilibrium constant for the reaction $H_2 \Leftrightarrow H + H$ is given by $\log K_T = 6.16 - 23{,}500/T$ (T in Kelvin) over the temperature range 1000–4000 K. Assume the gases to be ideal, and $P_H = P_{H_2} = 1$ bar.
4. Extract from Appendix 5 the standard free energy data for the oxides and sulfides of the following elements:

   Ca, Mg, Al, Ti, Mn, Si, Mn, K, Na, Sn, Zn, U, W, Mo, Co, Cu, Ni, Pb

   Can you separate the lithophile elements from the chalcophile elements on the list based on free energy of formation? Use an appropriate plot (or plots) to illustrate your conclusion.

5. Consider equilibrium partial melting of a gabbroic source rock composed of 50% plagioclase, 35% clinopyroxene, and 15% olivine by volume. Assume the melting to be modal and the densities of the minerals (in g cm^{-3}) as follows: olivine – 3.6; clinopyroxene – 3.4; and plagioclase – 2.7. Using the distribution coefficient values given in Table 12.9. Construct a graph showing the variation of $C_i^{melt}/C_{0\,(i)}^{rock}$ in the partial melt for the incompatible trace element Rb as a function of F. What is the enrichment factor for Rb in the melt if the degree of partial melting is 1%? [Hint: Remember to convert volume proportion of the minerals to weight fractions].
6. Construct graphs for bulk distribution coefficient values of 0, 0.1, 0.5, 1.0, and 2.0 to show the change in the concentration of a species *i* in the partial melt relative to the source rock ($C_i^{melt}/C_{0\,(i)}^{rock}$) as a function of the fraction of melt remaining (*F*) for (i) equilibrium crystallization and (ii) fractional crystallization. Comment on the differences, if any, between the two cases.
7. If weighted mean bulk crystal–melt distribution coefficients (D_0) for eclogite (garnet–clinopyroxene) assemblages are taken as 2.0 for Yb and 0.1 for K^+, then:
   (a) How much crystallization of eclogite would be necessary to increase the K^+ content of residual liquids by a factor of two, assuming Rayleigh fractionation?
   (b) How much would the Yb content of the melts have decreased during this process?
   (c) Repeat the calculations assuming equilibrium crystallization.
   (After Wood and Fraser, 1976.)
8. Calculate the REE patterns, normalized to chondrite, for 10% modal batch melting of a garnet peridotite (typical mantle rock) with the following mineralogic composition (in wt%): olivine – 55; orthopyroxene – 20; clinopyroxene – 20; and garnet – 5. Assume that the REE abundances in the garnet peridotite are twice the abundances in chondrites. The chondrite abundances (in ppm) are: La – 0.315; Ce – 0.813; Nd – 0.597; Sm – 0.192; Eu – 0.772; Dy – 0.325; Yb – 0.208; Lu – 0.0323. The mineral–melt distribution coefficients of the REE are listed below:

Element	Orthopyroxene	Clinopyroxene	Plagioclase	Olivine	Garnet
La	0.002	0.066	–	0.002	0.20
Ce	0.003	0.100	0.023	0.003	0.28
Nd	0.006	0.500	0.023	0.003	0.068
Sm	0.014	1.700	0.024	0.003	0.29
Eu^{2+}	0.023	1.600	0.393	0.005	0.49
Yb	0.110	0.715	0.006	0.009	11.5
Lu	0.110	0.610	0.005	0.009	11.9

13 The Crust–Hydrosphere–Atmosphere System

Atmosphere, ocean, and biosphere write their history in the rocks and move on, renewing themselves in a few years or millennia. Their record is preserved in those parts of the lithosphere, growing scarcer as time goes on, that escape the ravages of Zeus and Pluto. Much of this record is chemical; reading it requires the development of chemical hypotheses and models and the formulation of chemical questions to test them.

Gregor, Garrels, Mackenzie, and Maynard (1988)

The Earth has many unique features compared to other planets in the Solar System: abundant (liquid) water on the surface (the oceans cover ~71% of the Earth's surface); an atmosphere with abundant free oxygen and very little CO_2 or CH_4; a chemically evolved crust; and the presence of life (including what we consider to be advanced forms of life). In this chapter we will review the composition of the present atmosphere and oceans, and how they have evolved over the geologic time. Our current knowledge of the evolution of the atmosphere and hydrosphere is quite fragmentary. The conclusions summarized in this chapter are based on models constructed by different authors, each model limited by its own assumptions.

13.1 The present atmosphere

The atmosphere (or air) refers to the envelope of gases that surrounds the Earth and is retained by the Earth's gravity. The Earth's atmosphere has no outer boundary; it gradually becomes thinner with increasing altitude and just fades into space. An altitude of 120 km (75 miles) is where atmospheric effects become noticeable during reentry of spacecrafts. The commonly accepted boundary between the Earth's atmosphere and outer space lies at an altitude of 100 km (62 miles) and is called the *Kármán line*, so named after the Hungarian–American engineer and physicist (1881–1963) who was the first to calculate that around this altitude the Earth's atmosphere becomes too thin for aeronautical purposes (that is, any vehicle above this altitude would have to travel faster than orbital velocity in order to derive sufficient aerodynamic lift from the atmosphere to support itself). The average mass of the total atmosphere is about 5.2×10^{18} kg (Andrews *et al.*, 2004). About 50% of the atmosphere's mass is contained within 5.6 km of the Earth's surface, about 75% within 11 km of the Earth's surface, and about 99.99997% within 100 km of the Earth's surface. The dense part of the atmosphere (pressure > 0.01 bar), accounting for ~97% of mass of the atmosphere, extends to about 30 km from the Earth's surface.

13.1.1 *Temperature and pressure distribution in the atmosphere*

The temperature of the atmosphere is a direct or indirect consequence of the absorption of solar radiation by resident gas molecules and atoms in the atmosphere. Solar radiation consists predominantly of visible light and the more energetic (shorter wavelengths) ultraviolet (UV) radiation of the electromagnetic spectrum (Fig. 13.1). The absorption by different constituents of the atmosphere occurs at specific wavelengths. The major absorbers are N_2(g), O_2(g), O_3(g), NO_2(g), and particulates (Table 13.1).

On the basis of temperature distribution, the atmosphere can be divided into several layers (Fig. 13.2): *troposphere*, *stratosphere*, *mesosphere*, and *thermosphere*. The troposphere

Introduction to Geochemistry: Principles and Applications, First Edition. Kula C. Misra.

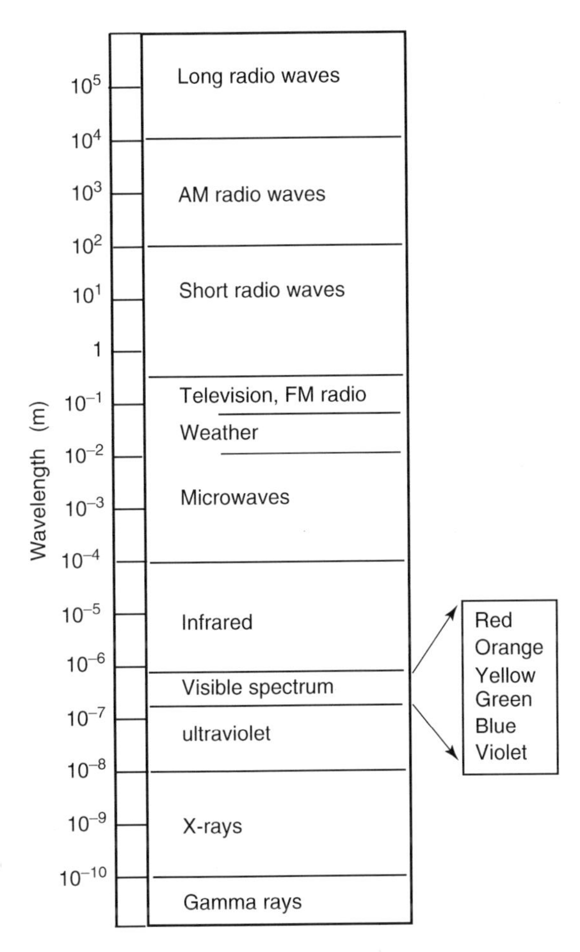

Fig. 13.1 The electromagnetic spectrum. 1 m = 10^6 μm.

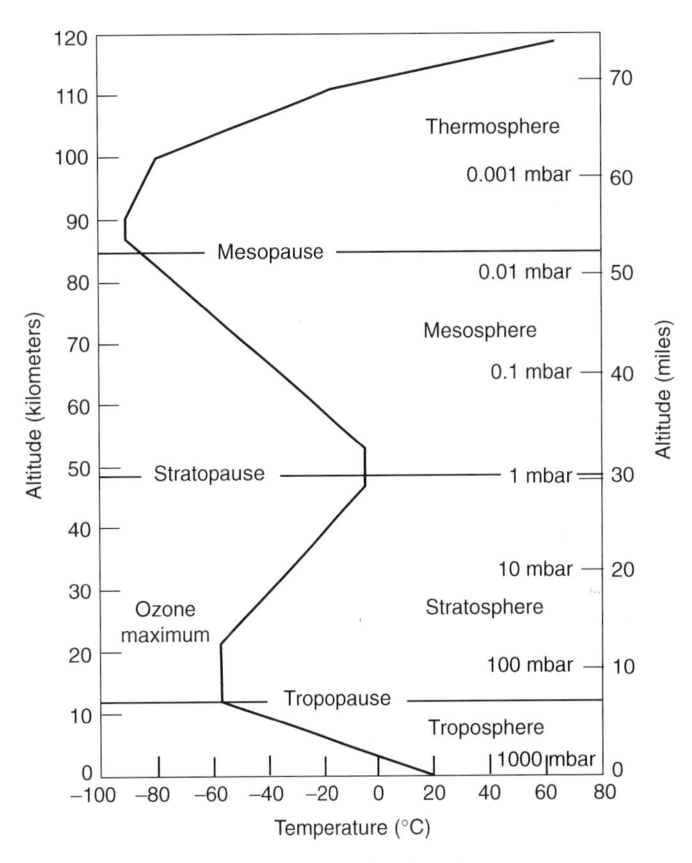

Fig. 13.2 Layers of the Earth's atmosphere based on temperature distribution. Also shown is the exponential decrease in the atmospheric pressure in millibars (mbar) with increasing altitude. Source: http://www.srh.noaa.gov/srh/jetstream/atmos/atmprofile.htm.

Table 13.1 Major absorbers of ultraviolet (UV) radiation in the Earth's atmosphere.

Radiation	Wavelength (μm)	Dominant absorbers	Location of absorption
Far-UV	0.01–0.25	N_2 (g)	Thermosphere, mesosphere
		O_2 (g)	Thermosphere, mesosphere, stratosphere
Near-UV			
UV-C	0.25–0.29	O_3 (g)	Stratosphere
UV-B	0.29–0.32	O_3 (g)	Stratosphere, troposphere
		Particulates	Polluted troposphere
UV-A	0.32–0.38	NO_2 (g)	Polluted atmosphere
		Particulates	Polluted troposphere

1 μm (micrometer) = 10^{-6}m (meter). Source of data: Jacobson (2002)

("tropos"=turning), the lowest layer of the atmosphere, extends to between 7 km at the poles and 17 km at the equator, with an average thickness of 11 km. It has a great deal of vertical mixing by convection due to solar heating, and its temperature decreases with increasing elevation because it is heated by radiation reflected back from the Earth's surface. The troposphere contains about 80–90% of the mass of the atmosphere and almost all of the water vapor in the atmosphere ($\leq$4% H_2O); this is the layer where most of the phenomena that we call weather (clouds, rain, snow, storms) occur. The boundary between the troposphere and the stratosphere, the layer above, is called the *tropopause*.

The stratosphere ("stratos" = layered) extends to about 50 km altitude. It contains the "*ozone layer*," which provides a protective shield from the Sun's ultraviolet rays. At its peak concentration in the stratosphere, around an altitude of 20–25 km, the ozone (O_3) density is typically about 5×10^{12} molecules cm^{-3}, which translates to a concentration of only a few parts per million (see Box 13.1). Temperature increases with elevation within the stratosphere because the ozone absorbs solar ultraviolet radiation and then emits it as infrared (heat) radiation. The pressure at the upper boundary of the staratosphere, the *stratopause*, is 1/1000th of that at sea level.

The mesosphere ("meso" = middle) extends to an altitude of about 80–85 km. This is the zone where most meteors burn up

Box 13.1 Mixing ratio

The abundance of a gas in air is commonly expressed as the *mixing ratio by volume*, the fraction of the volume of air occupied by the gas. For example, a mixing ratio of 10 parts per million by volume (ppmv) in air means 10 units of volume of the gas in 106 units of volume of air. This is a convenient way of expressing the abundance of a gas (e.g., a gaseous pollutant in the atmosphere) because we know from the ideal gas equation that volumes occupied by different ideal gases at the same temperature and pressure are proportional to the numbers of molecules of the gases.

(1) *Conversion of parts per million by volume (ppmv) to number of molecules.*
Assuming ideal behavior, the number of molecules per unit volume of any gas, the so-called *Loschmidt's number* (n_0), can be obtained from the ideal gas equation written as (Hobbs, 2000):

$$P = n_0\, k\, T \qquad (13.1)$$

where P is the pressure of the gas in pascals, n_0 the number of molecules per air, k the Boltzman's constant = 1.381×10^{-23} J deg^{-1} molecule^{-1}, and T the temperature in Kelvin. At P = 1 atm = 1013 × 102 Pa, and T = 273.15 K,

$$n_0 = \frac{P}{kT} = \frac{1013\times 10^2}{(1.381\times 10^{-23})\,(273.15)} = 2.69\times 10^{25} \text{ molecules m}^{-3}$$

Suppose that air contains a gaseous species i with a mixing ratio of C_i (ppmv). Since the volumes occupied by gases at the same temperature and pressure are proportional to the number of molecules in the gases,

$$\frac{\text{no. of molecules of } i \text{ in 1 m}^3 \text{ air}}{\text{no. of molecules in 1 m}^3 \text{ air } (= n_0)} = \frac{\text{volume occupied by molecules of } i \text{ in air}}{\text{volume occupied by air}} = C_i$$

So, the number of molecules of i in 1 m^3 air at 273.15 K and 1 atm = C_i (ppmv) n_0

For example, the mixing ratio of methane in dry air is 1.79 ppmv.
Applying the above equation, the number of molecules of methane in 1 m^3 air at 1 atm and 273 K

$= 1.79$ (ppmv) $\times n_0 = 1.79 \times 2.69 \times 10^{25} \approx 5 \times 10^{25}$ molecules

Let us also calculate the partial pressure exerted by atmospheric methane, $P_{methane}$.

C_i (ppmv) = mole fraction of methane in air = 1.79
So, $P_{methane} = P_{total} \times C_i = 1 \times 1.79 \times 10^{-6} = 1.79 \times 10^{-6}$ atm.

(2) *Conversion of parts per million by volume (ppmv) to μg m^{-3}.*
At 1 atm pressure and T (K) temperature, the relationship between ppmv and the concentration expressed as μg m^{-3} (micrograms of a gas species per cubic meter of air) is:

$$\mu\text{g m}^{-3} = \frac{(\text{ppmv})\,(12{,}187)\,(\text{molecular mass of the gas species})}{T\ (\text{K})} \qquad (13.2)$$

For example, a mixing ratio of 10×10^{-3} ppmv of the pollutant SO_2 in air at 1 atm and 25°C

$$= \frac{(\text{ppmv})\,(12{,}187)\,(\text{molecular mass of SO}_2)}{(273.15+25)}$$

$$= \frac{(10\times 10^{-3})\,(12{,}187)\,(64.06)}{298.15} = 26.18\ \mu\text{g m}^{-3}$$

upon entering the atmosphere. In the absence of ozone, there is no source of internal heating, and the temperature decreases with elevation in this layer, and reaches a minimum of around −100°C at the *mesopause*, the upper boundary of mesosphere and the coldest place on the Earth.

Temperature increases with height in the thermosphere from the mesopause to the *thermopause*, the upper boundary of the thermosphere, and may rise to as much as 1500°C. The height of the thermosphere varies with solar activity and may extend up to about 800 km. The International Space Station orbits in the thermosphere, between 320 and 380 km. The heat generated by the absorption of very energetic UV radiation (λ < 0.12 μm) above approximately 100 km altitude by atoms and molecules of oxygen and nitrogen is the cause of high temperatures in the thermosphere. For example, the UV radiation ionizes oxygen molecules (O_2 + UV photon $\Rightarrow O_2^+ + e^-$), but many of the released electrons can recombine with the oxygen ions producing a large amount of heat energy ($O_2^+ + e^- \Rightarrow O$ + O + heat energy). The UV radiation with wavelengths in the 0.12–0.30 μm range is absorbed mainly in the mesosphere and stratosphere by molecular oxygen (O_2) and ozone (O_3); nitrogen gas (N_2) is not an effective absorber of these wavelengths. The UV radiation with larger wavelengths (λ > 0.30 μm) penetrates into the troposphere. The very outermost fringes of the atmosphere (from 500–1000 km up to 10,000 km), where the gas is so tenuous that collisions between molecules become infrequent, is called the *exosphere*.

The temperature-based atmospheric layers overlap with those based on other characteristics. For example, the *ionosphere*, the part of the atmosphere that is ionized by solar radiation and influences propagation of radio waves to distant places, includes parts of both the mesosphere and the thermosphere.

The atmospheric pressure at any location is a direct consequence of the weight of the overlying air column at that location. The pressure varies with location and time. The average pressure on the Earth's surface is 14.356 lbs per square inch (psi), which is about 2.5% lower than the standard definition of one atmosphere (atm) pressure (14.696 psi or 101.325 kPa) and corresponds to the mean pressure not at sea level but at the mean base of the atmosphere as contoured by the Earth's terrain. The pressure decreases exponentially with increasing altitude (Fig. 13.2), by about a factor of 10 for every 16 km increase in altitude (the *barometric law*). The rapid drop in atmospheric pressure with increasing altitude is the reason why a supplemental supply of oxygen is necessary for climbing very high mountains such as Mount Everest (8.8 km elevation) and why passenger jets (which commonly fly at an altitude of about 10 km) need to have pressurized cabins.

13.1.2 *Photochemical reactions in the atmosphere*

A *photochemical reaction* is a reaction driven by the interaction of a *photon* of electromagnetic radiation of appropriate wavelength and a molecule, and *photolysis* (or *photodissociation*) refers to the dissociation of a molecule by a photochemical reaction. A photon is the smallest discrete amount of energy that can be transported by an electromagnetic wave of a given frequency. Whether a molecule can be involved in a photochemical reaction depends on the probability of the molecule absorbing a photon of sufficient energy to cause its dissociation.

Photochemical reactions play a key role in many aspects of atmospheric chemistry. An important aspect of such reactions is the production of *free radicals*. A free radical (also referred to simply as *radical*) is an atom or a group of atoms containing an unpaired electron in an outer shell. Free radicals are usually very reactive because of the unpaired electron and therefore short-lived. One of the most important reactive species in the atmosphere is the neutral hydroxyl free radical, $OH^\bullet$ (not to be confused with the negatively charged hydroxide ion, OH^-), which is formed by the following photochemically induced reaction sequence, starting with the photodissociation of ozone:

$$O_{3(g)} + \text{UV photon} \Rightarrow O_{2(g)} + O_{(g)} \tag{13.3}$$

$$O_{(g)} + H_2O_{(g)} \Rightarrow 2OH^\bullet_{(g)} \tag{13.4}$$

Many other photochemical reactions (e.g., the breakdown of $NO_{2(g)}$ and $O_{2(g)}$, and photosynthesis) will be discussed in the following sections.

Example 13–1: Calculation of the minimum wavelength of radiation capable of splitting an oxygen molecule into oxygen atoms

The photodissociation reaction for splitting an oxygen molecule is:

$O_{2(g)}$ + UV photon $\Rightarrow O_{(g)} + O_{(g)}$ (endothermic; $\Delta H^1_{r,\,298.15} = +\,498.4\,\text{kJ mol}^{-1}$)

So, the minimum energy required to split 1 mole of oxygen molecules = 498.4 kJ mol^{-1}

The minimum energy required per molecule of oxygen (E)

$$= \frac{498.4\ \text{kJ mol}}{\text{Avogadro's number}} = \frac{498.4\ \text{kJ mol}^{-1}}{6.022\times10^{23}\ \text{molecules mol}^{-1}}$$
$$= 8.276\times10^{-19}\ \text{J molecule}^{-1}$$

The relationship between E and the wavelength of the emitted radiation (λ) is given by the famous Einstein equation (see equation 2.7),

$$\lambda = \frac{hc}{E} \tag{13.5}$$

where c = speed of electromagnetic waves, and h = Planck's constant = 6.626×10^{-34} J s. Substituting the values of E, c, and h,

$$\lambda = \frac{hc}{E} = \frac{(6.626\times10^{-34}\ \text{J s})\,(2.998\times10^{8}\ \text{m s}^{-1})}{8.276\times10^{-19}\ \text{J molecule}^{-1}}$$
$$= 2.4002\times10^{-7}\text{m} = 0.240\ \mu\text{m}$$

Thus, the minimum radiation wavelength for splitting an oxygen molecule is 0.240 μm.

13.1.3 *The Ozone layer in the stratosphere*

Chapman mechanism

The bulk of the atmospheric ozone occurs in the stratosphere, where it is produced mostly by interaction of oxygen with UV radiation. In 1930, British geophysicist Sydney Chapman (1888–1970) proposed a series of photochemical reactions, now called the *Chapman mechanism* (or *Chapman reactions*), for production and destruction of ozone, thus for maintaining a steady-state concentration of ozone gas in the stratosphere. At an altitude of approximately 50 km, solar UV radiation of $\lambda < 0.24\,\mu$m has sufficient energy to dissociate O_2 into two atoms of oxygen ($O_{(g)}$), without ionizing it. Each of the two $O_{(g)}$ atoms then combines with $O_{2(g)}$ to form $O_{3(g)}$. The $O_{3(g)}$ molecule cannot accommodate the energy released by the latter reaction without breaking a bond; so for the $O_{3(g)}$ molecule to be a stable product, the reaction needs a catalyst molecule in the neighborhood to absorb the energy. The catalyst molecule commonly is $O_{2(g)}$ or $N_{2(g)}$ because these two molecules comprise 99% of the atmosphere. The reactions involved in the production of $O_{3(g)}$ may be represented as follows (Hobbs, 2000; Eby, 2004):

$$O_{2(g)} + \text{UV photon} \Rightarrow O_{(g)} + O_{(g)} \quad (\text{endothermic}; \; \Delta H^1_{r,\,298.15} = +498.4\,\text{kJ mol}^{-1}) \quad (13.6)$$

$$O_{(g)} + O_{2(g)} + M \Rightarrow O_{3(g)} + M \quad (\text{exothermic}; \; \Delta H^1_{r,\,298.15} = -106.5\,\text{kJ mol}^{-1}) \quad (13.7)$$

$$O_{(g)} + O_{2(g)} + M \Rightarrow O_{3(g)} + M \quad (\text{exothermic}; \; \Delta H^1_{r,\,298.15} = -106.5\,\text{kJ mol}^{-1}) \quad (13.8)$$

$$3O_{2(g)} + \text{UV photon} \Rightarrow 2O_{3(g)} \quad (\text{net reaction}) \quad (13.9)$$

where UV photon refers to solar radiation in the 0.2 to 0.22 μm and 0.185 to 0.2 μm wavelength regions, and M represents a catalytic molecule, usually N_2 or O_2, the predominant species in the stratosphere. Reaction (13.7) also occurs in the troposphere, but the oxygen in this case originates from NO_2 photolysis ($NO_2 + h\nu \Rightarrow NO + O$), not from O_2 photolysis.

Ozone is destroyed naturally in the stratosphere (and also in the troposphere) by *photolysis*, which occurs by absorption of solar radiation in the ~0.230 to 0.320 μm wavelength region, leading to regeneration of O_2:

$$O_{3(g)} + \text{UV photon} \Rightarrow O_{2(g)} + O_{(g)} \quad (\text{endothermic}; \; \Delta H^1_{r,\,298.15} = +386.5\,\text{kJ mol}^{-1}) \quad (13.10)$$

$$O_{(g)} + O_{3(g)} \Rightarrow 2O_{2(g)} \quad (\text{exothermic}; \; \Delta H^1_{r,\,298.15} = -391.1\,\text{kJ mol}^{-1}) \quad (13.11)$$

$$2O_{3(g)} + \text{UV photon} \Rightarrow 3O_{2(g)} \quad (\text{net reaction}) \quad (13.12)$$

Note that when ozone reacts with atomic oxygen, the ozone is destroyed permanently. On the other hand, every molecule of ozone consumed by UV photon produces an oxygen atom, which is free to combine with another $O_{2(g)}$ molecule and form $O_{3(g)}$. Also, $O_{3(g)}$ can be photolyzed all the way down to the Earth's surface, whereas $O_{2(g)}$ can be photolyzed only above about 20 km altitude. This is because all short-wavelength UV radiation required to split $O_{2(g)}$ is absorbed above this altitude.

Two factors that control the production and destruction of ozone are the availability of high-energy (UV) photons, which decreases towards the Earth's surface, and the density of the atmosphere (i.e., the number of gas molecules per unit volume of the atmosphere), which increases towards the Earth's surface. The optimum combination of these two variables acting in opposite directions occurs at an altitude of approximately 15–35 km (at the equator), the region of the so-called "ozone layer," which contains about 90% of our planet's ozone with peak values of only ~10 ppmv, compared to 0–0.07 ppmv in the unpolluted air near the Earth's surface. The maximum ozone concentration is predicted and found at an altitude of approximately 23 km; the maximum altitude varies with time of year and latitude on the Earth (van Loon and Duffy, 2000). The ozone in a given air parcel in the ozone layer is being continuously destroyed by photolysis (equation 13.12) when the parcel is in sunlight, but from the perspective of Chapman reactions the ozone concentration should stay fairly constant over time scales of weeks to months because of regeneration of ozone (equation 13.9) at almost the same rate.

Other reactions for Ozone destruction

Calculations based only on the Chapman oxygen–ozone cycle ($3O_{2(g)} \Rightarrow 2O_{3(g)} \Rightarrow 3O_{2(g)}$) generate the observed vertical profile for $O_{3(g)}$, with the peak $O_{3(g)}$ concentration at the expected altitude, but the predicted concentrations are too high compared with the measured values. This is because there are other pathways that destroy ozone. Most of the catalytic reactions that have been proposed for the removal of stratospheric ozone are of the form

$$X_{(g)} + O_{3(g)} \Rightarrow XO_{(g)} + O_{2(g)} \quad (13.13)$$

$$XO_{(g)} + O_{(g)} \Rightarrow X + O_{2(g)} \quad (13.14)$$

$$O_{3(g)} + O_{(g)} \Rightarrow 2O_{2(g)} \; (\text{net reaction}) \quad (13.15)$$

where X represents the catalyst and XO the intermediate product. The important point about such a reaction chain involving a catalyst is that, unless there is an appreciable sink for X, a few molecules of X have the potential to eliminate very large numbers of ozone molecules and atomic oxygen. In anthropologically undisturbed stratosphere, the important catalysts are free radicals of three categories (van Loon and Duffy, 2000): HO_x($H^\bullet$, $OH^\bullet$, $HO_2^\bullet$); NO_x($NO^\bullet$, $NO_2^\bullet$); and ClO_x($Cl^\bullet$, $ClO^\bullet$). Modeling of the ozone chemistry shows that the Chapman mechanism, $NO_{x(g)}$, $HO_{x(g)}$, and $ClO_{x(g)}$ account, respectively, for 20–25%, 31–34%, 16–29%, and 19–20% of O_3 destruction in the stratosphere (Lodders and Fegley, 1998). Let us consider the reactions involving $NO_{(g)}$ (nitric oxide) that dominates in the lower stratosphere and those involving OH that dominates in the upper atmosphere. The destruction of ozone by $ClO_{x(g)}$ is discussed in section 13.3.1.

The major sources of $NO_{(g)}$ in the troposphere are emissions from combustion of fossil fuels and synthesis from atmospheric $O_{2(g)}$ and $N_{2(g)}$ by lightning, and the major sources of $NO_{(g)}$ in the stratosphere are transport from the troposphere and the breakdown of $N_2O_{(g)}$ (nitrus oxide; laughing gas), which is emitted during denitrification by anaerobic bacteria in soils, by bacteria in fertilizers and sewage, and during biomass burning, automobile combustion, and combustion of aircraft fuel. The $N_2O_{(g)}$ molecule is not a free radical; it is a stable species with a tropospheric residence time estimated to be 160 yr. The principal mechanism for its removal from the troposphere is by migration into the stratosphere where it is converted into $NO^{\bullet}_{(g)}$ by reaction with excited-state oxygen:

$$N_2O_{(g)} + O^{*}_{(g)} \Rightarrow NO^{\bullet}_{(g)} + NO^{\bullet}_{(g)} \quad (13.16)$$

Then, $NO^{\bullet}_{(g)}$ catalyzes the destruction of $O_{3(g)}$:

$$NO^{\bullet}_{(g)} + O_{3(g)} \Rightarrow NO^{\bullet}_{2(g)} + O_{2(g)} \quad (13.17)$$

$$NO^{\bullet}_{2(g)} + O_{(g)} \Rightarrow NO^{\bullet}_{(g)} + O_{2(g)} \quad (13.18)$$

$$O_{3(g)} + O_{(g)} \Rightarrow 2O_{2(g)} \text{ (net reaction)} \quad (13.19)$$

It has been estimated that in the upper atmosphere ~10^5 molecules of O_3 are destroyed before one molecule of $NO_{x(g)}$ is removed from this catalytic ozone destruction cycle; the chain length decreases to ~10 in the lower stratosphere (Jacobson, 2002). Tropospheric $NO^{\bullet}_{2(g)}$ is unlikely to be transferred into the stratosphere because it is readily removed as HNO_3 by $OH^{\bullet}$ ($NO^{\bullet}_{2(g)} + OH^{\bullet}_{(g)} \Rightarrow HNO_{3(g)}$).

Ozone destruction by the highly reactive $OH^{\bullet}$ (which is produced in the stratosphere by the reaction of $O^{*}_{(g)}$ with $H_2O_{(g)}$, $CH_{4(g)}$, and $H_{2(g)}$) involves the free radical $HO^{\bullet}_{2(g)}$ (hydroperoxy radical) as an intermediate product, and can be represented by reactions such as:

$$OH^{\bullet}_{(g)} + O_{3(g)} \Rightarrow O_{2(g)} + HO^{\bullet}_{2(g)} \quad (13.20)$$

$$HO^{\bullet}_{2(g)} + O_{3(g)} \Rightarrow OH^{\bullet}_{(g)} + 2O_{2(g)} \quad (13.21)$$

$$O_{3\,(g)} + O \Rightarrow 2O_{2\,(g)} \text{ (net reaction)} \quad (13.22)$$

As will be discussed later (see section 13.3.1), stratospheric ozone loss has been enhanced in recent years by the anthropogenic infusion of chlorofluorocarbon (CFC) compounds.

13.1.4 Composition of the atmosphere

Approximately 99.9% (by volume) of the Earth's atmosphere is accounted for by nitrogen (78.1%), oxygen (20.9%), and argon (0.9%). There is a nearly uniform mixture of these three gases from the Earth's surface up to ~80 km altitude (including the troposphere, stratosphere, and mesosphere) because turbulent mixing dominates over molecular diffusion at these altitudes. Abundant atmospheric O_2 distinguishes the Earth from all other planets in the solar system. A number of other gaseous species, with concentration levels in parts per million (ppm) to parts per trillion (ppt), make up the rest 0.1% of the atmosphere (Table 13.2).

The composition of the atmosphere reflects a balance among volcanic activity, sedimentation, and life processes, but are the atmospheric gases in thermodynamic equilibrium? We can answer this question by checking if the partial pressures of gases in the atmosphere are consistent with the partial pressures they should have in a system in equilibrium. Let us consider the reaction that links CO_2 and CH_4 in the atmosphere and assume that the gases behave ideally:

$$CH_{4(g)} + 2O_{2(g)} \Leftrightarrow CO_{2(g)} + 2H_2O_{(g)}\ ; \quad K_{eq} = \frac{P_{CO_2}\,(P_{H_2O})^2}{P_{CH_4}\,(P_{O_2})^2} \quad (13.23)$$

Table 13.2 Average composition of the Earth's unpolluted lower atmosphere (up to an altitude of 25 km).

Gas	Volume%	Source	Estimated residence time
N_2	78.084	Biologic	10^6–10^7 yr
O_2	20.946	Biologic	3000–10,000 yr
Ar	0.934	Radiogenic	Forever
CO_2	0.0383	Biologic, geologic, anthropogenic	2–10 yr
Ne	0.00182	Earth's interior	Forever
He	0.000524	Radiogenic	~10^6 yr
CH_4	0.00017	Biologic, geologic, anthropogenic	2–10 yr
Kr	0.000114	Radiogenic	Forever
H_2	0.000055	Biologic, chemical	4–8 yr
N_2O	0.00003	Biologic, anthropogenic	5–200 yr
Xe	0.000009	Radiogenic	Forever
NO_2	0.0000002	Biologic, anthropogenic	0.5–2 days
O_3	0 to 0.000007	Chemical	100 days

Water vapor (H_2O) is ~0.40% by volume for the whole atmosphere, typically 1 to 4% near the surface. Other species present in the atmosphere at parts per trillion level concentrations include: NO, SO_2, CO, NH_3, H_2S, CS_2, (carbonyl sulfide), CH_3SCH_3 (dimethyl sulfide), methyl chloride (CH_3Cl), methyl bromide (CH_3Br), methyl iodide (CH_3I), hydrogen chloride, CCl_3F (CFC-11 Freon), CCl_2F_2 (CFC-12 Freon), and carbon tetrachloride (CCl_4).

Sources of data: Hobbs (2000); Railsback (2006).

Box 13.2 Residence time of a substance in a reservoir

The *(mean) residence time* (or *mean lifetime*) of a substance (i) in a specified reservoir is defined as the ratio of the mass of the substance in the reservoir (M_i) to either input flux (F_{in}) or the output flux (F_{out}), expressed in terms of mass per unit time, at steady state:

$$\text{Residence time } (\tau) = \frac{M_i}{F_{in} \text{ (or } F_{out})} \tag{13.24}$$

In a steady-state system, the reservoir concentration or mass of a substance does not change with time, so that $F_{in} = F_{out}$, and we can use either of the fluxes to calculate the residence time.

For purpose of illustration, let us calculate the residence time of H_2O in the atmosphere from the following data: amount of H_2O in the atmosphere = 0.13×10^7 kg; amount of H_2O added by evaporation to the atmosphere per year = 4.49 kg (0.63×10^{17} kg from lakes and rivers + 3.86×10^{17} kg from the oceans).

The average residence time of H_2O in the atmosphere

$$= \frac{M_{H_2O}}{F_{in}} = \frac{0.13 \times 10^{17} \text{kg}}{4.49 \times 10^{17} \text{kg yr}^{-1}} = 0.029 \text{ yr} = 10.6 \text{ days}$$

Thus, the average lifetime of a H_2O molecule in the atmosphere, assuming that it is well mixed and in a steady state with respect to H_2O, is about 11 days only!

What is the residence time of H_2O in the oceans? The rivers supply approximately 36×10^3 km^3 of H_2O per year and the volume of the ocean basins is approximately 1350×10^6 km^3. Thus, the average residence time of H_2O in the oceans (or the time that would be taken to "fill" the ocean basins)

$$= \frac{1350 \times 10^6 \text{ km}^3}{36 \times 10^3 \text{ km}^3} = 37{,}500 \text{ yr}$$

The average residence times of dissolved constituents in the oceans vary from millions of years for the highly soluble species (e.g., Na) to approximately 100,000 yr for a soluble but highly bioactive element such as P that is intensely cycled by marine organisms, to a few hundred years or less for relatively insoluble elements such as the REEs. Substances with long residence times can accumulate to relatively high concentrations in a reservoir compared with those with shorter residence times. However, highly reactive substances, even with short residence times, can yield products that cause environmental problems. Residence time is a useful concept because it enables us to evaluate how long a reservoir would take to recover if its chemistry was perturbed, for example, by the addition of a pollutant.

What happens when we perturb a steady-state system, for example, by changing the output flux? Assuming a first-order kinetic model (i.e., the change in the amount of i with time, dM_i/dt, is proportional to its original mass in the reservoir), the integrated rate equation can be written as

$$[M_i]_t = [M_i]_0 \, e^{-kt} \tag{13.25}$$

where k is the rate constant, and $(M_i)_0$ and $(M_i)_t$ denote the amounts of i in the reservoir at time $t = 0$ and some specified time t, respectively. Note that in this model the new steady state for a particular reservoir is approached exponentially.

From the free energy of formation data for the various species (Appendix 5), the equilibrium constant (K_{eq}) for this reaction can be calculated as ~10^{140}, indicating that CH_4 should oxidize readily to CO_2. We can calculate the equilibrium concentration of CH_4 from the concentrations (in vol%) of the other gases in the atmosphere (Table 13.2): O_2 = 20.95%; CO_2 = 0.0383%; and H_2O = 0.4%. Assuming that the volume fractions of these gases are approximately equal to their partial pressures,

$$P_{CH_4} = \frac{P_{CO_2}}{K_{eq}} \frac{(P_{H_2O})^2}{(P_{O_2})^2} = \frac{(3.83 \times 10^{-4})\,(4 \times 10^{-3})^2}{10^{140}\,(0.2095)^2}$$
$$= \frac{(15.32)\,(10^{-4})\,(10^{-6})}{(10^{140})\,(0.0439)}$$
$$= 349 \times 10^{-10} \times 10^{-140} = 3.5 \times 10^{-148} \text{ atm}$$

Thus, the calculated equilibrium concentration of CH_4 is orders of magnitude less than its actual concentration in the atmosphere (1.7×10^{-6} atm), indicating that the atmospheric gas mixture is not in thermodynamic equilibrium. Most of the gases in the atmosphere are in a *steady state* (i.e., the input of each such gas into the atmosphere is balanced by its output into other reservoirs) with *residence times* (see Box 13.2) ranging from 1–10 Myr for oxygen to a few days or less for some trace constituents (Table 13.2).

Chemical reactions among gaseous species in the atmosphere are dominated by intermolecular collisions, and many are catalyzed by aerosol particles. An *aerosol* is a suspension of fine solid particles and/or liquid droplets in a gas (the term "sol" evolved to differentiate suspensions from solutes in a liquid solution). Examples are smoke, haze over the oceans, volcanic emissions, and smog. Natural aerosol particles include mineral dust released from the surface, sea salt, biogenic emissions from the land and oceans, and dust particles in volcanic eruptions, but dust in the atmosphere has increased significantly due to human activities such as surface mining and industrial processes. Atmospheric aerosol particles that stay aloft for a while (by Brownian motion) are mostly in the 0.2–2 μm size range, and they serve as nuclei for condensation of water vapor to droplets of water (cloud particles). High concentrations of aerosol particles, most of which are large enough to cause incoherent scattering of visible light, produce a haze in the atmosphere, limiting visibility. Aerosol particles also play a critical role in the formation of smog (see section 13.3.2).

13.2 Evolution of the Earth's atmosphere over geologic time

13.2.1 *Origin of the atmosphere*

The origin of the terrestrial atmosphere is one of the most puzzling enigmas in the planetary sciences. The Earth probably had a *primary (*or *primordial) atmosphere*, comprised of gases captured gravitationally from the solar nebula. This *solar-type atmosphere*, if it existed, must have been composed, like the Sun itself, of hydrogen and helium. However, marked depletion of the present terrestrial atmosphere in noble gases, by factors of ~10^{-4} to 10^{-11} compared with their abundances in the solar nebula suggests that the solar-type atmosphere dissipated, probably almost completely, when the Solar System was ≤ 1 Myr old (Walker, 1977; Zahnle, 2006).

The Earth's primordial atmosphere was replaced within ~100 Myr by a *secondary atmosphere* (Walker, 1977). Earlier, it was believed that most of the Earth's volatiles were trapped originally in the cold interior of the planet during accretion, and the atmosphere accumulated subsequently when heating of the Earth's interior by radioctive decay gradually released the volatiles through volcanic degassing. More recent models of planetary accretion advocate that the secondary atmosphere formed by shock release and vaporization of chemically bound volatiles contained in the accreting planetesimals. Both qualitatively and quantitatively, the most important of the impact-induced degassed volatiles probably was water vapor. Much of the water accreted by the Earth probably existed in the form of hydrous silicates or was adsorbed onto the dust grains in planetesimals, and entered directly into the atmosphere due to degassing of planetesimals by impact erosion, mainly by dehydration of hydrous silicate minerals (see section 13. 5.1).

Transient steam atmosphere

According to Abe *et al.* (2000), the formation of a transient steam atmosphere started when the mass of the proto-Earth had reached about 0.01 $M_\oplus$ ($M_\oplus$ = Earth mass = 5.97 × 10^{24} kg). Hydrogen likely was a major constituent of such an atmosphere as a consequence of reduction of H_2O by infalling, metallic iron-rich planetesimals (Kasting, 1993). Indirect evidence for the presence of a H_2-rich steam atmosphere is provided by the isotopic composition of noble gases in the atmosphere, some of which appear to have been mass fractionated by the drag created by rapid, hydrodynamic escape of H_2, which carried off the lighter isotopes more easily than the heavy ones. This is believed to be the best explanation for the higher ^{22}Ne:^{20} Ne ratio in the atmosphere than both in the solar wind and the mantle (Zahnle *et al.*, 1988). Xenon ($^{131}_{54}$Xe) is the exception to this trend. Xenon ($^{131}_{54}$ Xe) is depleted in the terrestrial atmosphere by a factor of 4.8 × 10^4 relative to solar composition, whereas krypton ($^{84}_{36}$Kr) is depleted by a factor of only 3.3 × 10^4. Also, Xe isotopes are fractionated by 38% amu relative to solar, whereas Kr isotopes are fractionated by only 7.6% amu. This so-called "missing xenon paradox" – the greater depletion and fractionation of the heavier xenon compared to the lighter krypton in the atmosphere, although the abundances of Kr and Xe in the meteorites are about the same – has been attributed by Dauphas (2003) to the terrerstrial atmosphere being formed by contribution from two sources: fractionated nebular gases, and accreted cometary volatiles having low Xe : Kr ratio and unfractionated isotope ratios.

The evolution of an impact-generated steam atmosphere surrounding the accreting Earth, assuming no remnants of the primordial atmosphere, no explosive removal of a portion of the proto-atmosphere by a large impact, and an average of 0.1 wt% H_2O in the accreting planetesimals has been modeled by Matsui and Abe (1986a,b). The surface temperature of an accreting planet depends on the mass and composition of its atmosphere, and the average heat flux from the planet. The latter corresponds to the sum of the net solar flux and, more importantly, gravitational energy flux released from the surface due to accretion. Water vapor plays a key role in controlling the Earth's surface temperature because its strong absorption bands for infrared radiation prevent the impact energy (heat) from escaping into interplanetary space. When the radius of the planet exceeded 40% of its final radius, the blanketing by water vapor (roughly 100 bar), perhaps with some contribution from radioactive decay of short-lived nuclides, would foster surface temperatures high enough to melt the outer portion of the planet and form a magma ocean containing a predictable amount of dissolved H_2O.

Once a surface magma ocean has formed, the mass of the steam atmosphere and the surface temperature and pressure (estimated to have been about 1500 K and 100 bar) would not vary much during further accretion (Matsui and Abe, 1986a). This is because an increase in the surface temperature would increase the degree of melting, thus increasing the size of the magma ocean and causing more H_2O from the atmosphere to be dissolved in it. A decrease in the H_2O content of the atmosphere would decrease the efficiency of the blanketing effect, thereby decreasing the surface temperature, which would then allow the surface to re-solidify and increase the H_2O in the atmosphere because of outgassing. The exchange of H_2O between the atmosphere and the magma ocean would maintain the abundance of H_2O in the atmosphere just above the critical level required for sustaining a surface magma ocean (Abe *et al.*, 2000).

The mass of H_2O left in the proto-atmosphere would be limited by the solubility of H_2O in the silicate melt (which predominantly is a function of temperature), and be practically independent of the H_2O content of the source planetesimals. Considering that the solubility of H_2O in silicate melt is ~1 wt% (under a 100 bar steam atmosphere), comparable to or greater than the concentration of H_2O in the source materials (estimated to be ~0.1 to 1 wt%), almost all of the H_2O would dissolve in the magma ocean. On the basis of thermal evolution models of the growing Earth, Matsui and Abe (1986a,b) predicted that the maximum final mass of H_2O in an impact-generated proto-atmosphere at the end of accretion would be ~10^{21} kg, corresponding to a pressure of ~3×10^7 Pa (or ~3×10^2 bar), which is rather insensitive to variations in the input parameters. They considered the apparent coincidence of the H_2O abundance in the proto-atmosphere with the present mass of the Earth's oceans (~1.4×10^{21} kg) as evidence for an impact origin of the atmosphere and hydrosphere.

Eventually, the sum of accretional and solar heating would drop below the runaway greenhouse threshold, and the steam in the atmosphere would condense and rain out to form a hot proto-ocean of liquid water on the Earth's surface. The average temperature of this ocean, as estimated from oxygen isotope ratios of Archean metasedimentary rocks, would be ~600 K (Matsui and Abe, 1986b). The remaining atmosphere would be dominated by carbon and nitrogen compounds, mainly CO_2, CO, and N_2. The partial pressure of CO_2 in this atmosphere would be 60 to 80 bar if all of the crustal carbon (estimated at ~10^{23} g) were present in the atmosphere as CO_2, or ~10 bar if only about 15% of this carbon resided in the atmosphere before continents began to grow and carbonate rocks began to accumulate (Kasting, 1993). Modeling by Liu (2004) concluded that after accretion and solidification of a "magma ocean," the proto-atmosphere of the Earth would be composed of 560 bar of H_2O and 100 bar of CO_2, and oceans would start to grow when the temperature of the Earth cooled to below approximately 300–450°C.

Venus, which is comparable in size and mass to the Earth, should also have developed an impact-induced proto-atmosphere with approximately the same mass of H_2O. However, whereas H_2O in the Earth's proto-atmosphere could condense to form a hot ocean, it remained in gaseous state in the proto-Venusian atmosphere. The blanketing by atmospheric CO_2 probably resulted in a runaway greenhouse effect on Venus. Oceans did not form or could not stay around because water was converted to water vapor, the increased water vapor in the atmosphere enhanced the greenhouse effect, which in turn caused a further rise of temperature and more evaporation of water, and so on. A possible explanation for the strongly fractionated D : H ratio of the Venusian atmosphere is the photodissociation of H_2O into hydrogen and oxygen by solar UV radiation, and subsequent preferential escape of H_2 to space through a hydrodynamic process (Matsui and Abe, 1986b; Drake, 2005).

In the later stage of accretion, evolution of the atmosphere would depend on the model of accretion envisaged. If the accreting planetesimals remained small, a hot atmosphere and magma ocean would be sustained. On the other hand, any intermittent impact by a large planetesimal (or proto-planet) would likely repeat the steamy atmosphere–magma ocean–water ocean sequence. It is quite conceivable that the ocean was successfully vaporized and rained out with each major impact during the final stages of accretion.

Effect of the Moon-forming giant impact

The effect of the Moon-forming giant impact on the Earth's pre-existing volatiles inventory is not fully resolved. Ahrens (1993) suggested that the preexisting atmosphere was blown off completely by the impact, whereas Peppin and Porcelli (2002) argue that retention of a fraction of the primary, pre-impact atmosphere is an important reqirement for explaining the present abundance and isotopic composition of Xe in the Earth's atmosphere. It is generally agreed that the volatiles were lost from the side of the Earth that was impacted. The loss of volatiles from the other side would depend on whether there was a deep liquid-water ocean on the surface. Model calculations by Genda and Abe (2005) suggest that for a given impact, the fraction of atmosphere lost from a protoplanet covered with a liquid-water ocean would always be larger than that from one without such an ocean, and the loss would increase with decreasing value of R_{mass} (R_{mass} = atmospheric mass/ocean mass). A thin atmosphere above a thick water ocean would be largely expelled, otherwise the atmosphere would be retained. In either case, the entire liquid-water ocean would survive the impact and the water would be retained (Zahnle, 2006, p. 220).

The Moon-forming giant impact melted most of the mantle, forming a deep magma ocean, and vaporized some of the mantle (Canup, 2004), forming an atmosphere of mostly rock vapor topped by ~2500 K silicate clouds (Ahrens, 1993; Peppin and Porcelli, 2002; Zahnle, 2006). The rock vapor condensed into silicates and rained out at a rate of about 1 m per day, and the volatiles became increasingly important constituents of the atmosphere. Much of the Earth's water remained at first in the molten mantle, which later degassed as the mantle froze. By the end of the main accretionary and core-forming events, the residual atmosphere was dominated by a steamy mixture of N_2, CO_2, and H_2O. It also contained minor amounts of reducing gases such as CO and H_2, and trace amounts of noble gases (Des Marais, 1994; Zahnle, 2006). The Earth's surface temperature would depend on how much CO_2 was present and for how long. Model calculations suggest that, as long as most of the Earth's CO_2 remained in the atmosphere, the surface temperature would have been

~500 K. However if carbonates, the main sink of CO_2 in the pre-biotic Hadean eon, could subduct into the mantle, it could have taken as little as 10 Myr to remove 100 bar of CO_2 from the proto-atmosphere (Zahnle, 2006). The surface heat flux dwindled at the end of the Earth's main accretionary phase, and the steam atmosphere rained out to form an ocean. By this account, water has been on the Earth's surface almost from the beginning. Mantle temperatures in the Hadean eon were much higher than today due to greater concentration of radioactive elements and due to thermal energy released during the impact of accretionary bodies. As a result, the Hadean mantle was drier than its modern counterpart, and most of the Earth's water eventually became trapped in the oceans. As we will elaborate later (section 13.5.1), it is unlikely that any known source could have delivered an ocean of water to the Earth after the Moon-forming impact.

The lunar impact-cratering record indicates that the Earth–Moon system continued to be bombarded by large impactors (> 100-km in diameter) until at least 3.8 Ga, and it is likely that the atmosphere gained some H_2O, CO, and NO during this "heavy bombardment period" (4.5–3.8 Ga) if some of the impactors were of cometary or carbonaceous chondritic composition (Chyba, 1987; Kasting, 1993). CO would have been produced by reduction of ambient, atmospheric CO_2 by iron-rich impactors or by oxidation of organic carbon in carbonaceous impactors. NO gas would have been generated by shock heating of atmospheric CO_2 and N_2. The atmosphere during the Hadean eon (4.6–3.8 Ga), however, was essentially devoid of free (molecular) oxygen.

13.2.2 *A warm Archean Earth: the roles of carbon dioxide and methane*

Except for the absence of O_2 and the presence of elevated levels of greenhouse gases such as CO_2 or CH_4, the atmosphere at the beginning of the Archean probably was not very different from what it is now. CH_4 would be a better choice if the Archean was biologically productive, whereas CO_2 would be the better choice if the Earth was lifeless or nearly so (Zahnle, 2006). The Sun's luminosity has been increasing at the rate of ~5% per billion years as a consequence of the gradual conversion of hydrogen into helium inside the Sun's core. The much lower solar luminosity at 4.6 Ga, estimated to be only ~30% of that at present should have caused the Earth's mean surface temperatures to dip below the freezing point of water, resulting in widespread glaciation. The earliest firm evidence for continental glaciation in North America (the *Huronian glaciation* event in eastern Canada) is found in rocks that are 2.2–2.5 Gyr old; the age of the oldest glaciation event, however, may be ~2.9 Ga as recorded in the Archean Mozaan Group of South Africa (Young *et al.*, 1998). The other well-documented Proterozoic glacial period, which actually consists of a series of glacial episodes of global extent and is collectively termed the *Late Precambrian glaciations*, lasted from about 0.8 to 0.6 Ga. The lack of glaciation on the Earth for more than 1.5 Gyr after its birth is credited by most modelers to a much higher partial pressure of the greenhouse gas CO_2, estimated to have been 1–10 bar at 4.5 Ga and decreasing to 0.03–0.3 bar around 2.5 Ga. Mean surface temperature of an atmosphere so rich in CO_2 would be ~85°C, high enough to prevent a very cold Earth in spite of relatively low solar luminosity at that time (Kasting, 1987, 1993; Zahnle, 2006).

Catling *et al.* (2001) cited several reasons why high P_{CO_2} during the Archean is not the most likely solution to the "faint young Sun (FYS) problem," although CO_2 probably played a contributory role. Rye *et al.* (1995) reported that iron lost from the tops of >2.5-Gyr-old paleosol profiles was reprecipitated lower down as silicate minerals, rather than as iron carbonate (siderite), indicating that atmospheric P_{CO_2} must have been $<10^{-1.4}$ atm, about 100 times today's level of 360 ppm and at least five times lower than that required to compensate for the lower solar luminosity at that time. The mineralogy of the Archean banded iron formations (see Box 13.3) suggests that P_{CO_2} was less than 0.15 bar at 3.5 Ga.

Marine limestones are abundant in the Archean sedimentary record, whereas the opposite should be the case because calcite solubility increases with increasing P_{CO_2} (see Fig. 7.7). Also, there is no evidence for accelerated levels of acid-induced weathering during the Archean that should have resulted from high P_{CO_2}. Catling *et al.* (2001) argued that a relatively high concentration of atmospheric CH_4, not CO_2, was the most likely cause for Archean greenhouse warming. An Archean methanogenic biosphere is suggested also by extremely negative $\delta^{13}C$ values (between −33‰ and −51‰) in a 2.7-Gyr-old paleosol from Western Australia (Rye and Holland, 2000).

It is likely that the Archean Earth contained at least a small amount of CH_4 (1–10 ppmv) formed abiotically from impacts and from serpentinization of ultramafic rocks on the seafloor. At the present time, CH_4 is produced by methanogenic bacteria, most of which metabolize either CO_2 and H_2 from the atmosphere or acetate (represented here as acetic acid) formed from fermentation of photosynthetically produced organic matter:

$$CO_2 + 4H_2 \Rightarrow CH_4 + 2H_2O \qquad (13.26)$$

$$CH_3COOH \Rightarrow CH_4 + CO_2 \qquad (13.27)$$

Once methanogens appeared, they would have converted most of the available H_2 into CH_4. Photochemical modeling suggests a CH_4 concentration of 1000 ppmv or higher for the Late Archean/Paleoproterozoic atmosphere prior to the drastic rise of O_2 around 2.3 Ga (Kasting, 2005). The surface temperature resulting from such high levels of atmospheric CH_4 would have been enough to prevent a global-scale glaciation. Model-based calculations have shown that an increase in CH_4 from the present-day 1.7 ppmv to 100 ppmv would result in a 12 K increase in surface temperature from 283 to 295 K (Pavlov *et al.*, 2003).

The rise of O_2 around 2.3 Ga (see section 13.2.4) eliminated most of the methane ($CH_4 + O_2 \Rightarrow CO_2 + 2H_2$; $CH_4 + OH\bullet \Rightarrow CH_3 + H_2O$) and probably triggered glaciation during the Paleoproterozoic. Concentrations of CH_4 may have remained much higher (10–100 ppmv, compared to the present-day level of 1.7 ppmv) throughout much of the Proterozoic as a consequence of low concentrations of dissolved O_2 and sulfate in the deep oceans and a corresponding enhancement of organic matter recycling by fermentation and methanogenesis (Kasting, 2005).

13.2.3 *Oxygenation of the atmosphere*

The oxidation state of the atmosphere – oxic or anoxic – is a measure of the balance between the sources and sinks of oxygen (Box 13.3). The earliest mechanism of oxygen production in the atmosphere probably was photolysis of water vapor and CO_2 by solar UV radiation, which may be expressed as the following sequence of essentially irreversible reactions:

$$H_2O_{(g)} + \text{UV photon} \Rightarrow H^\bullet_{(g)} + OH^\bullet_{(g)} \quad (13.28)$$

$$CO_{2(g)} + \text{UV photon} \Rightarrow CO_{(g)} + O_{(g)} \quad (13.29)$$

$$O_{(g)} + OH^\bullet_{(g)} \Rightarrow O_{2(g)} + H^\bullet_{(g)}(\uparrow) \quad (13.30)$$

The escape of the very light H atoms produced in these reactions to space would result in a net addition of O_2 to the atmosphere; without hydrogen escape H_2O would be re-formed. This process, however, did not contribute much to the build-up of O_2 in the atmosphere. Almost all of the O_2 so produced must have been quickly consumed by chemical weathering (after the formation of continental crust), by oxidation of reduced gases (H_2, CO, and H_2S) from volcanoes by reactions such as

$$2H_2 + O_2 \Rightarrow 2H_2O \quad (13.31)$$

$$2CO + O_2 \Rightarrow 2CO_2 \quad (13.32)$$

and by the formation of iron oxides and hydroxides utilizing the Fe^{2+} derived from seawater–oceanic-basalt interaction. Typical atmospheric O_2 levels at this stage were probably on the order of 10^{-14} PAL (*present atmospheric level* = 0.2 atm) or lower (Kasting *et al.*, 1992).

Under the Earth's early, anoxic atmosphere, before the development of an ozone shield, organisms most likely lived underground or in water to avoid exposure to lethal UV radiation, and relied on the conversion of organic or inorganic material to obtain energy. At some point, certain anaerobic bacteria, called *phototrophs*, developed the ability to obtain energy from sunlight by a new process called *anoxygenic photosynthesis*, a process in which reduced chemical species (such as H_2S and H_2) and sunlight provide energy for synthesis of organic substances from CO_2, while more oxidized sulfur compounds or water (but no O_2) are released as byproducts. Examples of such reactions are:

$$CO_{2(g)} + 2H_2S_{(g)} + \text{sunlight} \Rightarrow CH_2O_{(aq)} + H_2O_{(l)} + 2S \quad (13.34)$$

$$CO_{2(g)} + 2H_{2(g)} + \text{sunlight} \Rightarrow CH_2O_{(aq)} + H_2O_{(l)} \quad (13.35)$$

where $CH_2O_{(aq)}$ represents a generic carbohydrate dissolved in water and S represents elemental sulfur. Evidently, anoxygenic photosynthesis did not contribute to the accumultion of free oxygen in the atmosphere.

A marked increase in O_2 production occurred with the advent of *oxygenic photosynthesis*, when cyanobacteria (green algae) appeared on the Earth. Oxygenic photosynthesis is the process in which an organism uses sunlight, water, and CO_2 to synthesize organic matter (e.g., carbohydrates), releasing O_2 as a waste product:

$$CO_{2(g)} + H_2O_{(l)} + \text{sunlight} \Rightarrow CH_2O\ (\text{carbohydrate}) + O_{2(g)} \quad (13.36)$$

The effective addition of molecular O_2 to the atmosphere by photosynthesis depends on how much of the organic carbon is protected from reoxidation by burial in sediments, and how much of the O_2 is consumed by other oxidative reactions such as chemical weathering and oxidation of reduced volcanic gases. At present, respiration and decay reverse this reaction on a time scale of ~10^2 yr, consuming > 99% of the O_2 produced by photosynthesis. A small fraction (0.1–0.2%) of organic carbon does escape oxidation through burial in sediments, resulting in an estimated contribution of $10.0 \pm 3.3 \times 10^{12}$ moles of O_2 per year (Holland, 2002; Catling and Claire, 2005).

Box 13.3 What defines an oxic versus an anoxic atmosphere?

Whether an atmosphere is oxic or anoxic can be described by an oxygenation parameter, K_{oxy}, which is defined as (after Catling and Claire, 2005):

$$K_{oxy} = \frac{F_{source}}{F_{sink}} = \frac{F_{source}}{F_{metamorphic} + F_{volcanic} + F_{weathering}} \quad (13.33)$$

where F_{source} = net flux of O_2 generated (in 10^{12} moles per year), and F_{sink} = O_2 consumed (in 10^{12} moles per year). Net sources of oxygen are: (i) photolysis coupled with loss of H to space (reactions 13.28–13.30); and (ii) oxygenic photosynthesis (reactions 13.35 and 13.38), provided the rate of oxygen production > the rate of organic carbon burial (at present the two rates are equal). O_2 is consumed by the reductants derived from volcanic outgassing and metamorphic reactions (mainly hydrogen-bearing gases such as H_2, H_2S, and CH_4), as well as by oxidative chemical weathering and animal respiration (and, in modern times, by the burning of fossil fuels). The atmosphere is oxic when $K_{oxy} > 1$, and anoxic when $K_{oxy} < 1$. For the present atmosphere, $K_{oxy} \approx 6$, which means that the atmosphere is quite oxic.

The atmospheric O_2 budget is further complicated by the response of other redox-sensitive elements, particularly iron and sulfur. During chemical weathering, sulfur in sulfide minerals, such as pyrite, is oxidized to soluble sulfate (SO_4^{2-}), which is carried by rivers to the ocean, where bacteria reduce sulfate and Fe^{3+} to pyrite (FeS_2), and pyrite burial in sediments is balanced by O_2 production. The reactions involved may be represented as (Canfield, 2005):

$$16H^+ + 16HCO_3^- \Rightarrow 16CH_2O + 16O_2 \quad (13.37)$$

$$8SO_4^{2-} + 16CH_2O \Rightarrow 16HCO_3^- + 8H_2S \quad (13.38)$$

$$2Fe_2O_3 + 8H_2S + O_2 \Rightarrow 8H_2O + 4FeS_2 + 15O_2 \quad (13.39)$$

$$\text{(net reaction) } 2Fe_2O_3 + 8SO_4^{2-} + 16H^+ \Rightarrow 8H_2O + 4FeS_2 + 15O_2 \quad (13.40)$$

At present, pyrite burial contributes $7.8 \pm 4.0 \times 10^{12}$ moles of O_2 per year (Holland, 2002). The burial of organic matter and, to a lesser degree, of pyrite accounts for essentially all the oxygen released to the atmosphere.

13.2.4 *The Great Oxidation Event (GOE)*

When exactly molecular oxygen first started to accumulate in the atmosphere is a controversial subject. In one scenario, originally proposed by Dimroth and Kimberley (1976) and later promoted by Ohmoto and his collaborators (Towe, 1990; Ohmoto, 1996, 1997; Watanabbe *et al.*, 1997; Beukes *et al.*, 2002), the atmospheric P_{O_2} level has essentially been constant, probably within ±50% of PAL (0.2 atm), since the advent of oxygenic photosynthesis in the Archean. According to the much more popular scenario, first proposed by Cloud (1972) and subsequently championed by a large number of authors (e.g., Walker *et al.*, 1983; Holland, 1984, 1994, 2002, 2006a,b; Kasting, 1993, 2001, 2006; Des Marais, 1994; Karhu and Holland, 1996; Canfield, 1998; Canfield *et al.*, 2000; Kump *et al.*, 2001; Pavlov and Kasting, 2002; Anbar and Knoll, 2002; Shen and Buick, 2004; Catling and Claire, 2005), the atmosphere was essentially devoid of molecular oxygen during the Hadean and Archean eons, but experienced a dramatic rise in the oxygen level during the Paleoproterozoic, around 2.3 Ga (2.4–2.2 Ga). This irreversible transition from fundamentally reducing to oxidizing conditions on the Earth's surface is often referred to as the *Great Oxidation Event* (GOE) (Holland, 1994). It is estimated that the GOE marked a dramatic increase in P_{O_2} from $<10^{-5}$ PAL (Pavlov and Kasting, 2002) to $>10^{-2}$ PAL and possibly to $>10^{-1}$ PAL (Holland, 1984, 2006a). A chemical equilibrium model by Krupp *et al.* (1994), involving minerals, seawater, and atmospheric gases, yielded a similar scenario: an oxygen-free atmosphere ($P_{O_2} < 10^{-10}$ bar) before ~2.35 Ga, comprised mainly of CO_2($P_{CO_2} \leq 1$ bar) and N_2, with minor amounts of H_2 and H_2S. Sverjensky and Lee (2010) have suggested that the rise in atmospheric oxygen resulted in an explosive growth in the diversification of minerals after the GOE because many elements could then occur in more than one oxidation state in minerals formed in near-surface environments. Another major oxidation event occurred at ~0.8–0.7 Ga, more than 1 Gyr after the GOE (see section 11.7.5), just before the appearance of large animals in the fossil record.

Evidence for the Great Oxidation Event

The hypothesis of a marked increase in the atmospheric P_{O_2} at 2.4–2.2 Ga is supported by the following lines of evidence preserved in the rock record.

(1) A large number of paleosols older than about 2.2–2.3 Ga have lost significant amounts of iron, whereas paleosols younger than 1.9 Ga have retained it (Holland, 1984; Rye and Holland, 1998; Yang *et al.*, 2002). The Fe-poor paleosols can be explained by weathering under low oxygen conditions prior to 2.2 Ga (< 0.001 PAL to 0.01 PAL; Rye and Holland, 1998). Iron contained in igneous and metamorphic rocks is dominantly Fe^{2+}. Under an anoxic atmosphere, the iron released during weathering remained as soluble Fe^{2+}, and was carried away by surface water and groundwater. The level of atmospheric O_2 was high enough after 1.9 Ga to retain the iron in the soil by oxidizing Fe^{2+} to insoluble Fe^{3+}.

(2) Continental red beds (sediments with a red coloration due to ferric oxide and hydroxide coatings on quartz grains), which attest to the oxidation of iron, appear for the first time in the rock record only around 2.2 Ga and become abundant after 2.0 Ga. The oldest, reasonably well-dated red beds, closely preceeding the major interval of red bed deposition after 2.0 Ga, occur in the 2224–2090-Myr-old Rooibergg Group in South Africa.

(3) Detrital uraninite ($U^{4+}O_2$) occurs only in rocks older than ~2.3 Ga, as in the uranium deposits hosted by conglomorates in South Africa (Witwatersrand district, ~2.7 Ga) and Canada (Elliot Lake district, ~2.3 Ga), indicating that the atmospheric O_2 was too low during the weathering of the host rock to oxidize uraninite into soluble U^{6+} -bearing species (Smith and Minter, 1980; Schidlowski, 1981; Smits, 1984). Very little or no O_2 in the Archean atmosphere is also indicated by the presence of detrital uraninite, pyrite, and, locally, siderite in 3250–2750 Ma fluvial sandstones and conglomorates in the Pilbara Craton, Australia (Rasmussen and Buick, 1999). Most estimates of O_2 concentration for the survival of detrital uraninite lie between ≤ 0.01 PAL (Holland, 1984) and 0.005 PAL (Kump *et al.*, 2004).

(4) Banded iron formations (BIFs) (Fig. 13.3), are abundant in the sedimentary record prior to 2.4 Ga, but are largely absent between 2.4 and 2.0 Ga (Fig. 13.4). Since Fe(II) species (such as Fe^{2+} and $FeOH^+$) are relatively soluble in

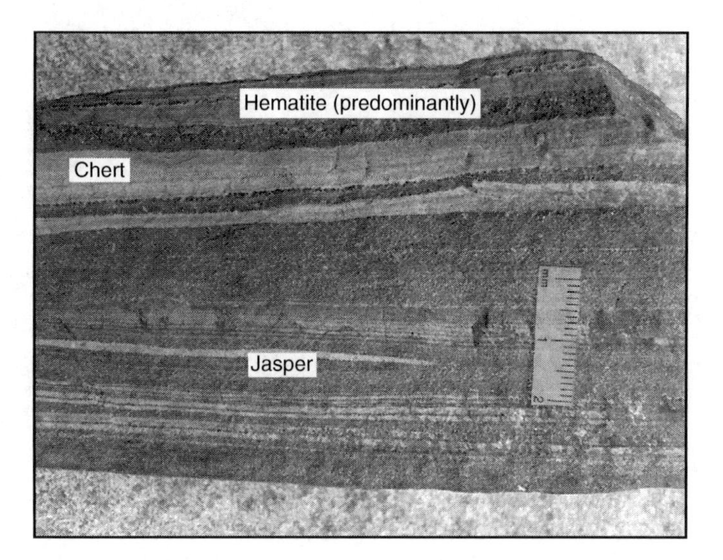

Fig. 13.3 A hand specimen of Archean (~3.0 Ga) banded iron formation from the Buhwa Greenstone Belt, Zimbabwe Craton. (Courtesy Dr Chris Fedo, Department of Earth and Planetary Sciences, The University of Tennessee, Knoxville, USA.)

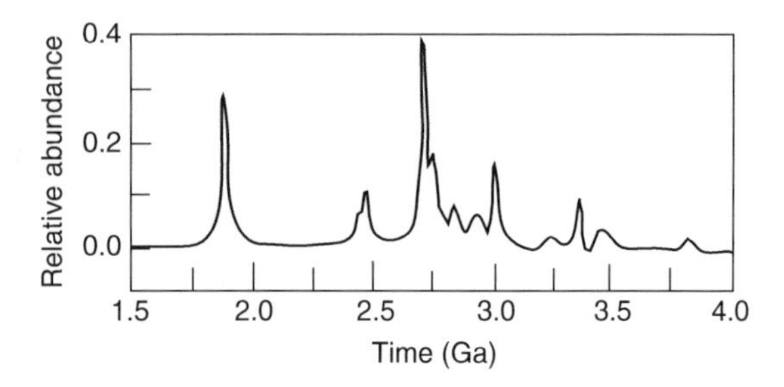

Fig. 13.4 Best estimate of relative abundance of banded iron formations (BIFs) as a function of time. (After Isley and Abbott, Plume-related mafic volcanism and the deposition of banded iron formation, Journal of Geophysical Research, v. 104, p. 15461–15467, 1999. Copyright 1999 American Geophysical Union. Used by permission of American Geophysical Union.)

ocean water, whereas Fe(III) species are not, most models for the formation of BIFs require an anoxic ocean to sustain large quantities of dissolved Fe(II) and some mechanism to precipitate the iron as Fe(III) compounds (see Box 13.4). Because the oxygen in BIFs could have come from several sources, we cannot use BIFs directly to make a quantitative estimate of the O_2 content of the atmosphere, but the time-restricted distribution of BIFs does point to a sudden rise in atmospheric O_2 at ~2.4 Ga.

(5) $\delta^{34}S$ values of sedimentary sulfides (pyrite in fine-grained siliciclastic sediments) show a marked jump to higher values at ~2.4–2.3 Ga. The favored interpretation is that the jump reflects a transition from low concentrations of dissolved SO_4^{2-} (< 1 mM kg^{-1} H_2O) in the oceans under an anoxic atmosphere (O_2 < 0.004 PAL) to higher concentrations of dissolved SO_4^{2-} (>1 mM kg^{-1} H_2O) when the atmosphere turned oxic (see section 11.7.5). This upper limit of O_2 concentration before ~2.4 Ga is in broad agreement with that indicated by the paleosol data.

(6) Mass-independent fractionation of sulfur isotopes (MIF-S) occurs in sediments of Archean and early Proterozoic age (> 2.45 Ga), but not in younger sediments. The loss of the MIF signal is attributed to an increase in the atmospheric oxygen to >10^{-5} PAL between 2.45 and 2.32 Ga (see section 11.7).

(7) Organic carbon is fixed from dissolved inorganic carbon (DIC) by autotrophic processes. Since organic carbon is depleted in ^{13}C, the remaining DIC is enriched in ^{13}C. Carbonate minerals precipitating from ocean water incorporate this DIC, providing a record of $\delta^{13}C$ of marine DIC through time and the intensity of organic carbon burial. Modeling of the intensity of organic carbon burial, the major source of atmospheric O_2, as a function of geologic time by Bjerrum and Canfield (2004) shows a marked transition to high values between 2.4 and 2.2 Ga (Fig. 13.5).

The oxygen jump

What triggered the GOE remains a hotly debated issue. If life existed before 3.8 Ga, it was subjected to intense meteorite bombardment and possibly annihilation. Cyanobacteria, the first oxygenic photosynthesizers, have been identified from organic biomarkers in sediments dated at 2.7 Ga (Brocks *et al.*, 1999), and may have existed since as early as 3.8–3.5 Ga (Buick, 1992; Schopf, 1993; Mojzsis *et al.*, 1996; Shen and Buick, 2004; Tice and Lowe, 2004). Why, then, was there a gap of at least 400 Myr between the emergence of cyanobacteria and the jump in atmospheric oxygen? Some suggested explanations are discussed below.

A rise in the atmospheric O_2 would require an increase in the O_2 source or a decrease in the O_2 sink. An obvious mechanism for an increasing O_2 source would be a long-term increase in organic carbon burial rates, perhaps because of the growth of continental shelves (Godderis and Veizer, 2000), but the carbon isotope data (Fig. 13.5) do not show a systematic increase in the burial rate of organic carbon over geologic time. The atmospheric evolution model of Goldblatt *et al.* (2006) suggests that a mere 3% increase in organic carbon burial would have been enough to trigger the GOE, but such a change is far too small to be detected in the carbon-isotope record.

Another possible cause of the delay in oxygen build-up could be the consumption of oxygen by reduced volcanic gases. At the present time, the O_2 sink provided by the oxidation of volcanic gases accounts for ~25% of the O_2 generated by organic carbon burial; the other ~75% is consumed by oxidative weathering of reduced species in rocks, predominantly of organic carbon, pyritic sulfur, and Fe(II) in shales (Kump *et al.*, 2001). The volcanic sink was certainly much larger during the Archean because of much greater volcanic activity, but the larger O_2 sink would have been offset by a larger release of volcanic CO_2. This, together with the approximately constant $\delta^{13}C$ composition of carbonates (Fig. 13.5), reflecting a

Box 13.4 Banded iron formations (BIFs)

Banded iron formation (BIF), typically, is an iron-rich (~20–40 wt% Fe), siliceous (~40–60 wt% SiO_2), laminated or banded (on scales ranging from < 1 mm to tens of centimeters) sedimentary rock that consists of alternating, millimeter-to-decimeter thick layers of iron-rich minerals (such as magnetite, Fe_3O_4; hematite, Fe_2O_3; and siderite, $FeCO_3$) and silica (SiO_2 as chert or jasper) (Fig. 13.3). Banded iron formations are practically devoid of detrital input, as reflected in their very low Al_2O_3 contents (0–1.8 wt%). They are widely distributed around the world, often as huge deposits, and account for about 90% of the global supply of iron ores. Banded iron formations are important in the context of the evolution of the atmosphere–ocean system because of their time-restricted distribution, and because deposition of BIFs served as a large sink for the atmospheric O_2.

The oldest known BIFs are those of the Isua supracrustal belt (3760 ± 70 Ma; Moorbath *et al.*, 1973). Banded iron formations are quite common in the rock record between ~3.5 and 2.4 Ga, but largely absent from the geologic record between 2.4 and 2.0 Ga. They reappear for a brief interval (2.0–1.8 Ga) in the Paleoproterozoic and then disappear from the rock record quite abruptly after 1.8 Ga (Fig. 13.4). The late Proterozoic (0.85–0.54 Ga) iron formations (Rapitan-type BIFs, which occur in Yukon and NWT of Canada and in the Urucum area of Brazil) are very different from typical BIF in terms of their hematite-dominated mineralogy and stratigraphic settings, especially association with glaciomarine deposits. The Rapitan-type BIFs are thought to be the result of anoxic conditions that arose from stagnation of the oceans beneath a near-global ice cover (referred to as "Snowball Earth").

Most aspects of the genesis of Archean–early Proterozoic BIFs – sources of iron and silica, rhythmic banding of iron-rich and silica-rich layers, precipitation mechanisms for the various facies of BIFs (oxide, silicate, carbonate, and sulfide), spatial and temporal distribution – are controversial and not satisfactorily resolved (James, 1992; Beukes and Klein, 1992; Klein and Beukes, 1992; Morris, 1993; Misra, 2000; Klein, 2005). A comprehensive discussion of the genesis of BIFs is beyond the scope of this book; we will touch only on a few aspects of the story germane to atmospheric oxygen evolution.

The most abundant type of BIF is the Superior-type iron formations, so named because of their abundance in the Lake Superior region of USA and Canada. These predominantly early Proterozoic deposits with well-developed and laterally extensive banding are interpreted to have been deposited on stable continental shelves at depths below the storm wave base (which is at least 200 m in the modern oceans). Prompted by the lack of volcanic association and paucity of detrital components in the Superior-type BIFs, earlier genetic models (e.g., Drever, 1974; Holland, 1984) had invoked a continental source of iron and silica. But the overall REE patterns, positive Eu and negative Ce anomalies, and Nd isotopic data for BIFs ranging in age from 3.8 Ga to 1.9 Ga show that much of the iron and silica were brought from the mantle by mid-oceanic ridge (MOR)-type hydrothermal fluids (Derry and Jacobsen, 1990). As Fe(II) is much more soluble in seawater than Fe(III), the formation of voluminous quantities of BIF requires that large portions of the deep ocean were anoxic to enable the transport of dissolved Fe(II). The dissolved iron was brought to the near-surface regions, probably by wind-induced upwelling of the type that occurs in some modern coastal settings (Klein and Beukes, 1989) or by hydrothermal plumes (Isley, 1995; Isley and Abbott, 1999), and precipitated (at pH > 4) as Fe(III) compounds (here represented by ferric hydroxide, $Fe(OH)_{3(s)}$), believed to be precursors of hematite and magnetite in oxide-facies BIFs. The precipitation occurred by one of the following types of oxidation reactions in the stratified Precambrian ocean.

(1) Indirect biological oxidation, utilizing the dissolved O_2 produced by oxygenic photosynthesis (reaction 13.36) in the *photic zone* of the ocean (the zone, about 100 m for calm and clear water, that is penetrated by sufficient sunlight for photosynthesis to occur), probably in localized "oxygen oases" associated with cyanobacteria blooms,

$$3Fe^{2+}_{(aq)} + O_2 + 7H_2O \Rightarrow 3Fe(OH)_{3(s)} + 5H^+ \quad (13.41)$$

(2) Photochemical oxidation of dissolved Fe(II) in the photic zone of the ocean by solar UV radiation (Cairns-Smith, 1978; François, 1986),

$$Fe^{2+}_{(aq)} + 3H_2O + \text{UV photon} \Rightarrow Fe(OH)_{3(s)} + 2H^+ + 0.5H_2 \quad (13.42)$$

(3) Direct biological oxidation of Fe(II), the electron donor, catalyzed by anoxygenic phototrophic Fe(II) oxidizing bacteria, which convert CO_2 into biomass using light energy (Widdel *et al.*, 1993; Kappler *et al.*, 2005),

$$4Fe^{2+} + CO_2 + 11H_2O + \text{sunlight} \Rightarrow CH_2O + 4Fe(OH)_{3(s)} + 8H^+ \quad (13.43)$$

Although Fe(II) can be oxidized photochemically in simple aqueous systems, such oxidation has not been reported in more complex environments such as seawater at approximately neutral pHs. Also, in the absence of biological silica secretion, the contemporaneous oceans were likely saturated with H_4SiO_4, causing the formation of amorphous Fe-silicate gels and thus limiting the effects of photochemical oxidation. In contrast, both freshwater and marine anoxygenic phototrophs readily oxidize Fe(II) (Widdel *et al.*, 1993; Ehrenreich and Widdel, 1994; Kappler and Newman, 2004) and can account for most, if not all, of the iron in BIFs even if the cell densities of the iron-oxidizing bacteria were much less than found in modern Fe-rich aqueous environments (Konhauser *et al.*, 2002).

An important point in favor of indirect biological oxidation is that primitive O_2-producing photosynthetic bacteria lacked suitably advanced oxygen-mediating enzymes (Cloud, 1973) and, in order to survive, required Fe(II) as an oxygen acceptor. Consequently, these microorganisms would have flourished during periods of abundant supply of Fe(II), leading to $Fe(OH)_3$ precipitation (Crowe *et al.*, 2009). The intervals of limited $Fe(OH)_3$ precipitation, then, may be ascribed to a decrease in Fe(II) supply, causing dwindled population of photosynthetic bacteria.

Whether the oxidation was biotic or abiotic is still unresolved (see section 11.8.1), but it is reasonable to conclude that the intervals of abundant BIFs coincided with an anoxic atmosphere that sustained anoxic conditions over large portions of the deep oceans, thereby permitting the transport of large quantities of iron as dissolved Fe(II) species. Thus, the BIF record provides indirect evidence that the atmosphere turned oxic after ~2.4 Ga. This interpretation implies that a lower level of atmospheric O_2, possibly as low as that before 2.4 Ga, ushered the return of BIF deposition during 2.0–1.8 Ga, which is consistent with a decrease in the calculated fraction of total carbon buried as organic carbon (*f* ratio) as

Box 13.4 (cont'd)

shown in Fig. 13.5. The disappearance of BIFs from the geologic record after 1.8 Ga was perhaps the consequence of very limited supply of dissolved Fe(II) when the deep ocean became oxygenated, a scenario consistent with an oxic atmosphere. Canfield (1998), however, suggested that the deep ocean remained anoxic until the Neoproterozoic, with the Fe(II) being removed by the precipitation of pyrite. In any case, the cessation of BIF deposition eliminated an effective O_2 sink and, thus, contributed to the transition of the atmosphere to an oxygenated state.

A further complication in the BIF story arises from the fact that the major Fe-bearing phases in oxide-facies BIFs, which have not experienced supergene alteration or metamorphism, is magnetite, not hematite or goethite. Conversion of precursor $Fe(OH)_{3(s)}$ to magnetite and siderite at or below the sediment–water interface occurred via several pathways (Johnson *et al.*, 2008a): reactions with seawater Fe^{2+}_{aq} ($2Fe(OH)_{3(s)} + Fe^{2+}_{aq} \Rightarrow Fe_3O_{4(s)} + 2H_2O + 2H^+$; $Fe^{2+} + CO_2 + H_2O \Rightarrow FeCO_3 + 2H^+$); bacteria-catalyzed dissimilatory Fe^{3+}_{aq} reduction (DIR) under Fe^{2+}_{aq}-limited conditions; and DIR in the presence of excess Fe^{2+}_{aq}.

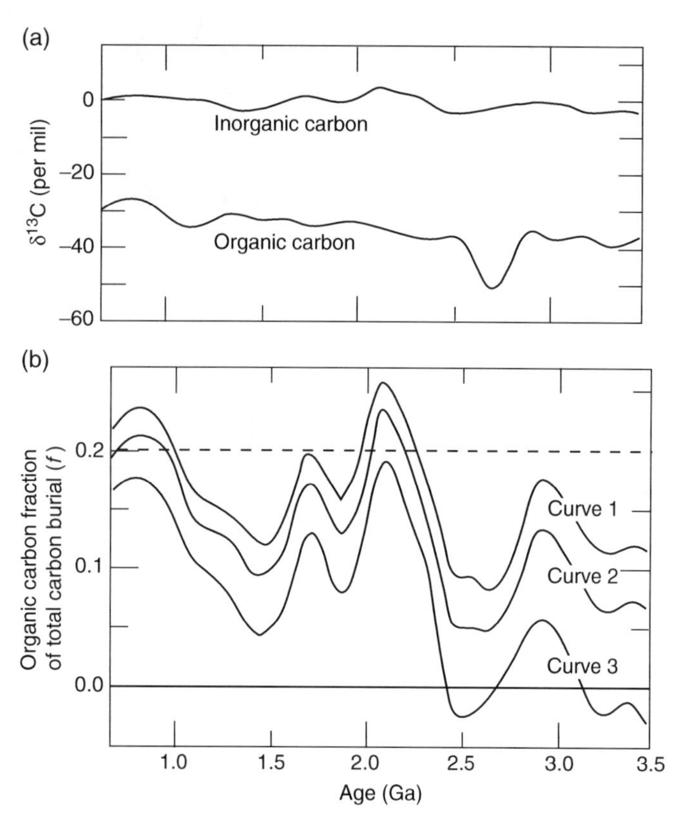

Fig. 13.5 (a) A summary of the isotopic composition of inorganic carbon and organic carbon through geologic time (3.5–0.7 Ga). (b) Reconstructions of the significance of *f* ratio (organic carbon fraction of total carbon burial) through geologic time. Curve 1 does not consider ocean crust carbonatization (OCC). Curves 2 and 3 are based on an isotope gradient of 2% and 5%, respectively, between the surface and deep oceans. (After Bjerrum and Canfield, New insights into the burial history of organic carbon on the early Earth, Geochem. Geophys. Geosyst., v. 5, Q08001, doi:10.1029/2004GC000713 (online), 2004. Copyright 2004 American Geophysical Union. Used by permission of American Geophysical Union.)

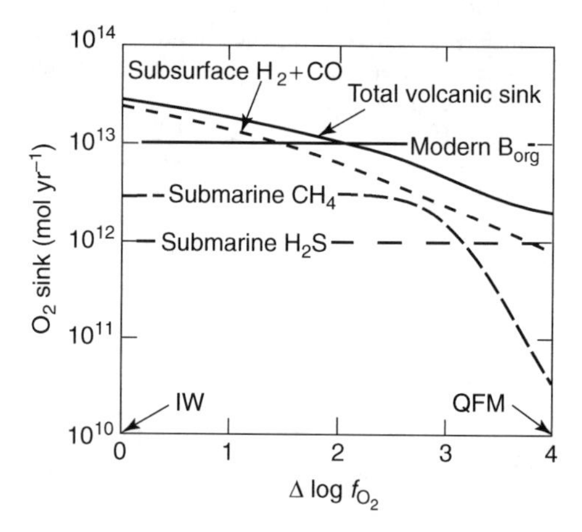

Fig. 13.6 Magnitude of the volcanic O_2 sink as a function of the oxygen fugacity of mantle source regions. B_{org} is the organic carbon burial rate and the presumed rate of O_2 production if most of the organic matter came from oxygenic photoautotrophs. $\Delta\log f_O$ is relative to the iron–wustite (IW) oxygen buffer. Calculations were performed at $T = 1500$ K and $P = 1$ bar for subaerial volcanic gases, and $T = 900$ K and $P = 400$ bar for subaqueous volcanoes. QFM = quartz–fayalite–magnetite oxygen buffer. (After Kump et al., Rise of atmospheric oxygen and the "upside-down" Archean mantle. Geochem. Geophys. Geosyst., v. 2, 1025, doi:1029/2000GC000114 (online), 2001. Copyright 2001 American Geophysical Union. Used by permission of Amerivcan Geophysical Union.)

roughly constant proportion of organic carbon burial, requires that the burial rate of organic carbon was proportionately larger. If oxygenic photosynthesis dominated the production of O_2 during the Archean, then we had an O_2 source that was enhanced in proportion to the volcanic sink. Thus, a higher rate of volcanic outgassing cannot by itself account for the delay in rise of atmospheric O_2 (Kump *et al.*, 2001).

A more plausible explanation for the change in the oxidation state of the atmosphere involves a gradual decrease in the scavenging of atmospheric O_2 due to a corresponding change in the oxidation state of volcanic gases, dominated by H_2 and CO during the early Archean, and escape of H_2 into space (Kasting *et al.*, 1993; Kump *et al.*, 2001; Catling *et al.*, 2001; Holland, 2002; Canfield, 2005).

What caused the change in the oxidation state of volcanic gases? The oxidation state of volcanic gases depends on the oxidation state of upper mantle rocks, where the gases originate: the more oxidized the upper mantle, the more oxidized the volcanic gases. So, for this hypothesis to be viable, the upper mantle must have become progressively more oxidized because of volcanic outgassing of H_2 (and other H-bearing gases, such as CH_4, H_2S, H_2O) by reactions such as (Catling and Claire, 2005)

$$3FeO + H_2O \Rightarrow Fe_3O_4 + H_2 \Rightarrow Fe_3O_4 + 2H\ (\uparrow \text{space}) \quad (13.44)$$

According to Kump *et al.* (2001) the photosynthetic O_2 source exceeded the total volcanic sink when the average f_{O_2} of volcanic gases, no longer dominated by H_2 and CO, rose to a value of 2 log units below the QFM buffer (Fig. 13.6). This marked the transition from an anoxic atmosphere, in which the O_2 sink exceeded the O_2 source, to an oxic atmosphere, in which the O_2 source exceeded the O_2 sink. In a recent article, Kump and Barley (2007) proposed that the anoxic–oxic transition was a consequence of terrestrial volcanism changing from being predominantly submarine during the Archean to being predominantly subaerial at the beginning of the Proterozoic. This is because subaerially erupted volcanic gases have equilibrated with magmas at high temperatures and low pressures, and are dominated by oxidized gases (H_2O, CO_2, and SO_2). Volcanic gases from submarine eruptions, on the other hand, erupt at lower temperatures, and contain higher concentrations of reduced species (H_2, CO, CH_4, and H_2S).

A problem with the above scenario is that the concentrations of redox-sensitive elements, such as Cr and V, in several Archean high-Mg lava flows (komatiites) indicate an oxidation state comparable to, or more oxidizing than, that of the present-day oceanic basalts (Canil, 1997). This apparent discrepancy may be rationalized in several ways: (i) the source regions of komatiites represented pockets of reducing environment in a compositionally inhomogeneous mantle; (ii) the required increase in the f_{O_2} of volcanic gases was actually < 0.5 log units, which is in keeping with the limits set by the Cr and V contents of Archean komatiites (Holland, 2002); and (iii) the oxygen released by oxygenic photosynthesis (reaction 13.36) was largely consumed by the oxidation of continental crust, which, in turn, caused a change in the oxidation state of gases released by metamorphic reactions, independent of the mantle oxidation state (Catling *et al.*, 2001; Catling and Claire, 2005).

Since even a small excess of hydrogen over a steady-state abundance of ~0.1% H_2 (or its equivalent, 0.05% CH_4) could keep the atmosphere in an anoxic state, the anoxic to oxic transformation of the atmosphere must have involved getting rid of most of its H_2 (and CH_4). As suggested by Catling *et al.* (2001), the hydrogen escape rate was probably greatly enhanced by the build up of biogenic methane in the Archean atmosphere and the subsequent photolysis of CH_4. The pertinent methanogenesis reactions are reduction of CO_2 to CH_4 by mantle-derived volcanic H_2,

$$CO_{2(g)} + 4H_{2(g)} \Rightarrow CH_{4(g)} + 2H_2O_{(l)} \quad (13.45)$$

and decomposition of photosynthetically (anoxygenic or oxygenic) produced organic matter to CH_4,

$$2CH_2O_{(aq)} \Rightarrow CH_{4(g)} + CO_{2(g)} \quad (13.46)$$

The photochemistry of CH_4 is complicated, but may be represented by the reactions

$$CH_{4(g)} + \text{UV photon} \Rightarrow C_{(s)} + 4H_{(g)}(\uparrow \text{space}) \quad (13.47)$$

$$C_{(s)} + O_{2(g)} \Rightarrow CO_{2(g)} \quad (13.48)$$

which promoted rapid escape of hydrogen into space. Thus, the combined effect of oxygenic photosynthesis (equation 13.36), methanogenesis (equations 13.45 and 13.46), and hydrogen escape (equations 13.47 and 13.48) resulted in a net gain of O_2.

The availability of phosphorous, a nutrient for cyanobacteria, in the oceans may also have played a role in maintaining an anoxic atmosphere until the disappearance of BIFs. As Fe(III) precipitates strongly adsorb PO_4^{3-} at pH <9, the very low P content of BIFs translates into very low concentrations of P in BIF-depositing waters of Archean and early Proterozoic oceans, only ~10–25% of present-day concentrations. It is likely that low P availability significantly reduced rates of photosynthesis and organic carbon burial, thus contributing to the prolonged anoxic condition of the atmosphere (Bjerrum and Canfield, 2002, p.159). Marine phosphorite deposits made their first appearance during the GOE, after the cessation of BIF deposition (Holland, 2006a).

According to the conceptual model of Goldblatt *et al.* (2006), the advent of oxygenic photosynthesis gave rise to two simultaneously stable steady states for atmospheric oxygen: a low-oxygen state (< 10^{-5} PAL) and a high-oxygen state (> 5×10^{-3} PAL). In a low-oxygen atmosphere, the lack of an effective ozone shield allowed oxygen to be rapidly consumed by a UV-catalyzed reaction with biogenic methane ($CH_{4(g)} + 2O_{2(g)} + \text{UV photon} \Leftrightarrow CO_{2(g)} + 2H_2O_{(g)}$). As the oxygen level increased, the shielding of the atmosphere from UV radiation due to the increased level of ozone impeded oxygen consumption, and the atmosphere attained a higher level of oxygen. Thus, the atmospheric budget could change even if other sources and sinks of oxygen remained constant. The model predicts the persistence of a reducing atmosphere for at least 300 Myr after the onset of oxygenic photosynthesis. Goldblatt *et al.* (2006) also predicted that a much larger perturbation would be required to cause a high-oxygen atmosphere to revert back to a low-oxygen state. This might explain why a relatively high level of oxygen persisted following the GOE.

A yo-yo Archean atmosphere?

As discussed earlier, the presence of MIF-S in many sedimentary rocks older than ~2.4 Ga, and its absence in younger rocks, is considered the strongest evidence for a dramatic change from an anoxic atmosphere to an oxic atmosphere around 2.4 Ga. This is because photolysis of volcanic SO_2 gas by ultraviolet radiation in an oxygen-poor atmosphere is the only mechanism known to produce MIF-S. This interpretation, however, is not consistent with the observation recently reported by Ohmoto *et al.* (2006) that the MIF-S signal is absent throughout ~100 m sections of 2.76-Gyr-old lake sediments and 2.92-Gyr-old marine shales in the Pilbara craton, Western Australia. This suggests that these late Archean rocks preserve a hint of oxic conditions, at least locally, in which rehomogenization of MIF took place. Alternatively, elevated concentrations of other gases in the atmosphere may also have depressed the MIF-S signal during the 2.8–3.0 Ga period (Holland, 2006a).

Examining the MIF-S record for pre-1.6 Ga sedimentary rocks in Australia, Canada, South Africa, and the USA, Ohmoto *et al.* (2006) proposed an Archean "yo-yo atmosphere," which fluctuated from anoxic (< 10^{-5} PAL) before ~3.0 Ga, to oxic (> 10^{-5} PAL or even > 10^{-2} PAL) between ~3.0 and ~2.75 Ga, to anoxic between ~2.75 and 2.4 Ga, and back to oxic after ~2.4 Ga. Anbar *et al.* (2007) reported that the late Archean (2501 ± 8 Ma) Mount McRae Shale in Western Australia is enriched in redox-sensitive transition metals such as Mo and Re. Correlation with organic carbon indicates that these metals were derived from contemporaneous seawater. Mo and Re were probably supplied to Archean oceans by oxidative weathering of crustal sulfide minerals, pointing to the presence of a "whiff of oxygen" in the environment more than 50 Myr before the start of the GOE. The model of Goldblatt *et al.* (2006) mentioned earlier explicitly predicts that the atmosphere was bistable for some time before the GOE. The validity of the "yo-yo atmosphere" hypothesis is still being debated in the scientific community (see discussions by Kasting, 2006; Knauth, 2006; Ohmoto *et al.*, 2006).

13.2.5 A model for the evolution of the atmosphere

Many models have been proposed for the evolution of the atmosphere, but all of them are controversial. The simple conceptual model discussed below is a three-box framework of the atmosphere–ocean system originally proposed by Walker *et al.* (1983) and subsequently refined by Kasting *et al.* (1992) and Kasting (1993). The model does not incorporate the concept of a yo-yo Archean atmosphere. The three "boxes" in this model represent three distinct reservoirs of oxygen: the atmosphere, the surface ocean (the uppermost 75 m or so of water that is stirred rapidly by the action of the wind and, thus, is well mixed), and the deep ocean. The evolution of the atmosphere and the ocean is conceptualized to have progressed in three stages (Fig. 13.7).

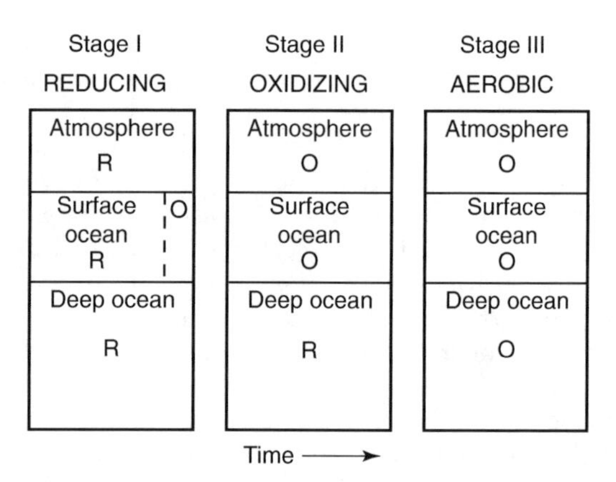

Fig. 13.7 The three–box model of the atmosphere–ocean system pertaining to atmospheric evolution over geologic time proposed by Kasting *et al.* (1992). O = oxidizing, R = reducing. Note that during Stage I, the surface ocean was oxidizing locally.

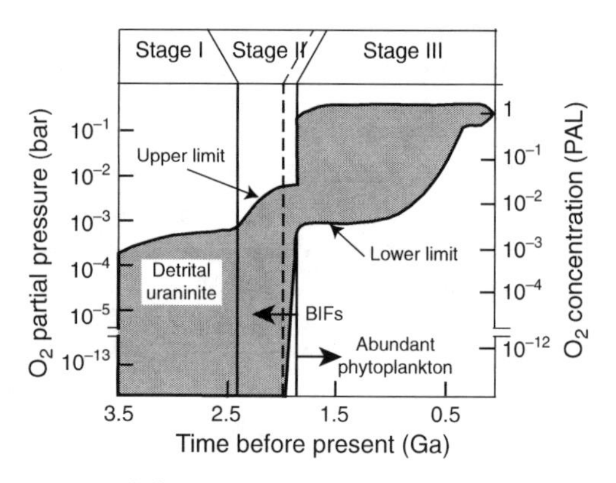

Fig. 13.8 Estimated change in atmospheric O_2 levels through geologic time. The upper and lower limits are constrained by detrital uraninite, paleosols, and the fossil record. (Simplified from J.F. Kasting, H.D. Holland, and L.E. Kump, 1992, *Atmospheric Evolution: the Rise of Oxygen*, Fig. 4.6.1, Cambridge University Press. Used with permission of the publisher.)

Estimated change in the atmospheric O_2 level over geologic time, based on the three-stage model of Kasting *et al.* (1992) and some geologic and biologic data that constrain the model, is presented in Fig. 13.8. During Stage I (prior to the GOE), termed "reducing," both the atmosphere and the ocean were essentially devoid of O_2, although after the arrival of cyanobacteria localized patches of the surface ocean ("oxygen oases") within the photic zone, characterized by high microbial productivity, may have developed substantial concentrations of dissolved O_2. Such "oxygen oases" could have provided an environment conducive to BIF deposition. The deeper oceans were certainly anoxic as evidenced by the occurrence of an abundance of Archean-age BIFs. The occurrence of large marine manganese deposits of Archean age (e.g., the ~3.0-Gyr-old Iron Ore Group, Orissa, India) also points to reducing conditions in the deeper oceans, where Mn^{2+} could accumulate, then upwell into the shallow oceans, and be precipitated there as a constituent of Mn-carbonates or Mn-oxides (Roy, 1997; Holland, 2006a). As mentioned earlier, the atmospheric O_2 level at this stage was generally $<10^{-5}$ PAL (Pavlov and Kasting, 2002). Independent calculations based on the survival of detrital uraninite imply an atmospheric O_2 level between ≤ 0.01 PAL (Holland, 1984) and 0.005 PAL (Kump *et al.*, 2004).

Stage II (from ~2.4 to ~1.85 Ga), termed "oxidizing," represents a period in which small amounts of O_2 were present in the atmosphere and surface ocean, but the deep ocean remained anoxic, as evidenced by the occurence of large marine manganese deposits of Paleoproterozoic age. Models for BIF deposition suggest an upper limit of ~0.03 PAL for atmospheric O_2 during Stage II (Kasting, 1993), whereas paleosol data yield O_2 values of less than 0.15 PAL (Klein and Beukes, 1992). Holland (2006a) estimated that the oxygen level during this stage reached about 0.02–0.04 atm. The reason for the lack of BIFs in the rock record between 2.4 and 2.0 Ga is not known, but the reappearance of BIFs between 2.0 and 1.8 Ga implies a return to lower oxygen levels during the later part of the Paleoproterozoic.

The transition from Stage II to Stage III occurred when the deep ocean became oxic, and it is marked by the disappearance of the Superior-type BIFs around 1.85 Ga (Kasting *et al.*, 1993). Possible causes of this transition include a decrease in the supply of Fe^{2+} to the deep sea (probably due to reduced oceanic ridge hydrothermal activity or a transition to an oxygenated ocean), removal of Fe^{2+} as Fe-sulfide precipitates in a sulfidic ocean (Canfield, 1998; Anbar and Knoll, 2002), an increase in the rate of oxygenic photosynthesis, and an increase in the burial rate of organic carbon. A growing body of evidence suggests that precipitation of Fe-sulfides was probably responsible for removing Fe from deep ocean waters (Canfield, 1998; Anbar and Knoll, 2002; Shen *et al.*, 2003; Arnold *et al.*, 2004).

Stage III (1.85–0.85 Ga), termed "aerobic," represents a situation, somewhat similar to the present atmosphere, in which oxygen pervaded the entire system and P_{O_2} was high enough to support aerobic respiration. Calculations suggest that the atmospheric O_2 must have risen to at least 0.02 PAL for the transition to Stage III; otherwise the deep ocean would have remained anoxic as a consequence of the influx of reductants from hydrothermal vents (Kasting, 1993).

The three-box model does not account for the late Proterozoic (0.85–0.54 Ga) Rapitan-type BIFs, which constitute a very different type of BIF and should not be interpreted as necessarily signaling a return to Stage II conditions. The O_2 concentration in the atmosphere and the sulfate concentration in seawater during 0.85–0.54 Ga rose to levels that were probably not much lower than those of the present day (Holland, 2006a).

Calcium sulfate minerals (gypsum or anhydrite), or their pseudomorphs, are scarce before ~1.7–1.6 Ga (Grotzinger and Kasting, 1993), indicating a low concentration of SO_4^{2-} in the pre-1.7 Ga oceans, and therefore a low concentration of oxygen in the contemporaneous atmosphere. On the basis of $\delta^{34}S$ values in marine sedimentary sulfides, Canfield and Teske (1996) argued that atmospheric P_{O_2} values remained at <5–18% PAL until the late Proterozoic transition. The second line of evidence comes from $\delta^{34}S$ values of carbonate-associated sulfate (CAS), trace amounts of sulfate that substitute for CO_3^{2-} ion in marine carbonate minerals. The amount of CAS increases with increasing sulfate concentration in the solution and, thus, can be used as a proxy for the dissolved sulfate concentration of the contemporaneous ocean. Moreover, CAS trapped in marine carbonates is isotopically buffered against appreciable diagenetic overprint and, therefore, can record the $\delta^{34}S$ of the contemporaneous ocean water (Lyons *et al.*, 2004). Calculations by Kah *et al.* (2004) suggest that marine sulfate concentrations remained low, < 35% of modern values, for nearly the entire Proterozoic. Since lower oceanic sulfate concentrations imply lower rates of pyrite oxidative weathering and, therefore, lower levels of atmospheric P_{O_2}, a significant rise in O_2 may not have occurred until the latest Neoproterozoic (0.54 Ga), just before the Cambrian biological explosion, when sulfate levels may have reached 20.5 mM kg^{-1} H_2O or 75% of present-day levels. The Neoproterozoic jump in atmospheric O_2, probably to not much less than PAL (Holland, 2006a), broadly correlates with a marked increase in organic carbon burial (Fig. 13. 5b).

13.2.6 *The Phanerozoic atmosphere*

The level of atmospheric O_2 probably reached close to the modern value by the Cambrian biologic explosion. A number of studies (e.g., Berner and Canfield, 1989; Berner, 2001, 2003) have concluded that atmospheric O_2 must have fluctuated significantly during the Phanerozoic, but quantitative modeling has been unsatisfactory in the absence of simple indicators of the O_2 level. Berner and Canfield (1989) have constructed a mathematical model that enables calculation of the level of atmospheric O_2 across the Phanerozoic. In this model, the burial rates of organic carbon and pyrite sulfur, which control the accumulation of atmospheric O_2, are not calculated from $\delta^{13}C$ and $\delta^{34}S$ values of ancient oceans as recorded by sedimentary $CaCO_3$ and $CaSO_4$, respectively. Instead, the burial rates are calculated from an assumed constant worldwide clastic sedimentation rate and the relative abundance and C and S contents of three rock types: marine sandstones and shales, coal basin sediments, and nonmarine clastics (red beds, arkoses).

The "best estimate" of the variation of atmospheric O_2 during the Phanerozoic obtained by Berner and Canfield (1989) is presented in Fig. 13.9. The most striking feature of this plot is

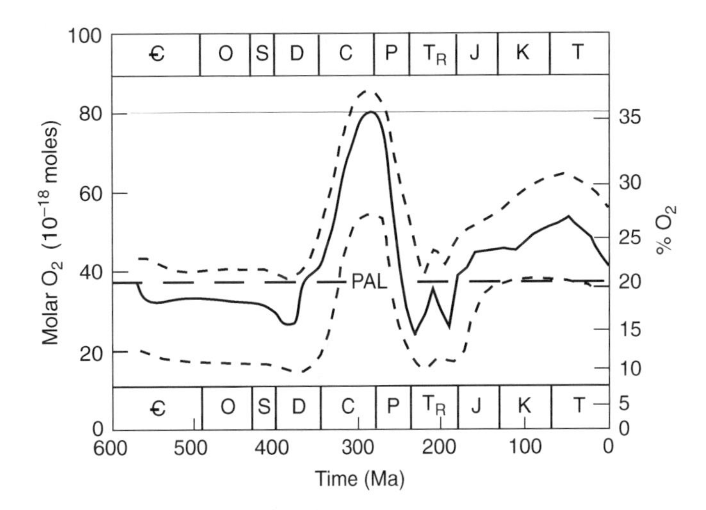

Fig. 13.9 "Best estimate" of atmospheric O_2 during the Phanerozoic as modeled by Berner and Canfield (1989). The values of O_2 shown here should be used in only a qualitative to semiquantitative sense. The dashed lines represent crude error limts, based on their sensitivity analysis. PAL = present atmospheric level of oxygen. (After Berner and Canfield, 1989, Fig. 13, p. 357.)

the pronounced excursion, reaching concentration levels as high as 35% O_2, during the Permo-Carboniferous and its rapid plunge at the end of Permian. The excursion may be ascribed to the evolution of vesicular land plants, which was a new source of organic carbon, and the development of vast lowland swamps where vesicular plants could flourish and then be buried to form coals (Berner, 2004).

Most studies incorporating all known climate forcings implicate CO_2 as the primary driver for the most recent rise in global temperatures (Mann *et al.*, 1998; Crowley, 2000; Mitchell *et al.*, 2001). The revised GEOCARB model for CO_2 levels during the Phanerozoic (GEOCARB III; Berner and Kothavala, 2001) suggests a similar story – a broad correspondence between the atmospheric CO_2 concentration and glaciation. As shown on Fig. 13.10, CO_2 was low (< 500 ppm) during periods of long-lived and widespread continental glaciations (Permo-Carboniferous, 330–260 Ma, and late Cenozoic, past 30 Myr) and high (>1000 ppm) during intervening warmer periods (Royer *et al.*, 2004). The late Ordovician (~440 Ma) represents the only interval during which glacial conditions apparently coexisted with a CO_2-rich atmosphere.

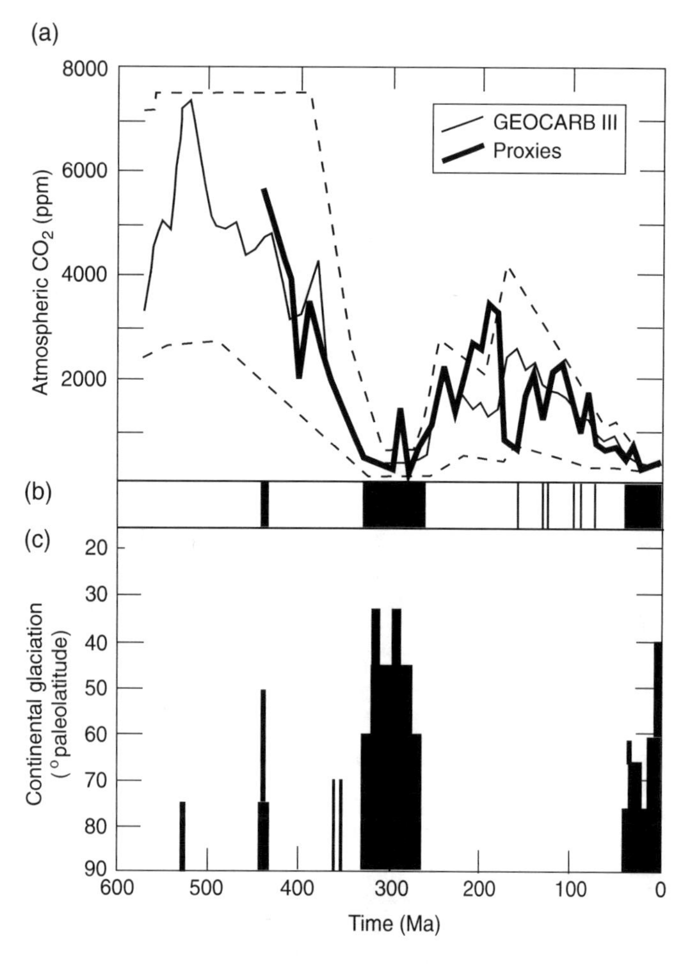

Fig. 13.10 CO_2 and the Phanerozoic climate. (a) Evolution of atmospheric CO_2 over Phanerozoic time according to the GEOCARB III model of Berner and Kothavala (2001). The outer dashed lines represent an estimate of errors in the GEOCARB III model. Also shown for comparison is the graph reconstructed from multiple proxies. (b) Intervals of glacial (solid black) and cool (thin lines) climates. (c) Latitudinal distribution of direct evidence of glaciation, such as tillites and striated bedrock, etc. (Crowley, 1998). (After Royer *et al.*, 2004.)

Pavlov *et al.* (2003) have proposed that the long period of warm climate between the Paleoproterozoic (2.5–1.6 Ga) and Neoproterozoic (0.9–0.542 Ga) glaciations was a consequence of the greenhouse effect of the Proterozoic atmosphere bolstered by 100–300 ppm of CH_4. Their calculations for the estimated CH_4 flux in the Proterozoic assume that (i) most of the organic carbon (CH_2O) was initially produced by photosynthesis, and (ii) consumption of CH_4 by methanotrophic bacteria was ineffective in the Proterozoic ocean, which was anoxic at depth and impoverished in sulfate. The Neoproterozoic glaciation could have been initiated by a decrease in CH_4 flux resulting from the second oxygenation event toward the end of the Proterozoic.

Calculations by Berner (2006), using average nitrogen : carbon ratio (N/C) of sedimentary organic matter, coal, and volcanic/metamorphic gases, and existing models of the carbon cycle indicate that the concentration of atmospheric N_2 changed very little (by <1% of the present value) over Phanerozoic time. This means that significant variations of the mass of O_2 during the Phanerozoic resulted in corresponding variations in the total atmospheric pressure.

13.3 Air pollution: processes and consequences

Air pollution refers to the presence in the atmosphere of substances that are potentially harmful to human beings, animals, vegetation, property, or the environment. Air pollutants can be grouped into three broad classes: (i) inorganic gases such as oxides of nitrogen (N_2O, NO, NO_2), sulfur (SO_2, SO_3), and carbon (CO, CO_2), and other inorganic gases (O_3, NH_3, H_2S, HF, HCl, Cl_2, Rn); (ii) volatile organic compounds (e.g., CH_4, higher hydrocarbons, and chlorofluorocarbon compounds such as CFC-11 and CFC-12); and (iii) suspended particulate matter (particularly small particles of solid or liquid substances <10 μm in diameter, commonly denoted as PM-10, which are not very effectively removed by rain droplets). Air pollutants may be primary or secondary. *Primary pollutants* are the direct products of combustion and evaporation (e.g., CO, CO_2, NO_x, SO_x, and volatile organic compounds (VOC)). *Secondary pollutants* are generated in the atmosphere by reaction with primary pollutants (e.g., O_3 formed by photochemical oxidation of O_2, as discussed earlier). The main components of air pollution in the USA are listed in Table 13.3.

13.3.1 *Depletion of stratospheric ozone – the "ozone hole"*

The ozone in the stratosphere absorbs about 99% of the UV radiation in the 0.23–0.32 μm wavelength region (sometimes referred to as UV-B radiation), thus providing an effective

Table 13.3 The main components of air pollution in the USA, 2008 (fires and dust excluded).

Primary air pollutant	Annual emissions (million tons)	Major anthropogenic sources	Adverse health and environmental effects
Particulates		Fuel combustion, agriculture, unpaved roads, forest fires, oceans (sea salts)	Aggravates respiratory and cardiovascular problems; impairs visibility
PM 10	2		
PM 2.5	1		
Sulfur dioxide (SO_2)	11	Fuel combustion (especially high-sulfur coal; utilities and industrial plants); cement production; volcanoes	Causes sulfurous smog that can aggravate respiratory disease; contributes to acid rain and dry acid deposition; damages materials and vegetation
Carbon dioxide (CO_2)	1592 (carbon)	Fuel combustion (vehicles, power stations), forest fires	Possible global warming
Carbon monoxide (CO)	78	Fuel combustion (especially vehicles)	Reduces the ability of blood to carry oxygen to body tissues; aggravates cardiovascular disease
Nitrogen oxides (NO_x)	16	Fuel combustion (especially vehicles); fertilizers; biomass burning	Causes photochemical smog that can aggravate respiratory problems; contributes to acid rain, dry acid deposition, and ozone and particle formation
Volatile organic compounds (VOC)	17	Fuel combustion, solvents, paints	Increases risk of cancer; contributes to ozone formation

Sources of data: CO_2 data from Carbon Dioxide Information Analysis Center, Oak Ridge National Laboratory website **http://cdiac.ornl.gov**; the rest from US Environmental Protection Agency, Air Quality Trends website http://www.epa.gov/airtrends/aqtrends.html.

shield against the penetration of UV radiation that is harmful to most organisms. Calculations show that an effective ozone shield started to form at an O_2 level of 10^{-3} PAL and became firmly established at an O_2 level of ~10^{-2} PAL (Kasting *et al.*, 1992). In fact, life could not have been sustained on the Earth until there was enough oxygen in the atmosphere to develop an ozone shield for protection from the solar UV radiation. It has been estimated that the incidence of skin cancer in human beings would increase by 2% for every 1% decrease in the concentration of ozone in the stratosphere. Kasting *et al.* (1989) suggested that elemental sulfur vapor (S_8) produced photochemically from volcanogenic SO_2 and H_2S could have played a similar protective role in an anoxic, ozone-free, primitive atmosphere. Some of the earliest organisms were probably thermophillic bacteria that metabolized elemental sulfur.

The stratospheric ozone concentration near the equator, the region of maximum UV photon flux density on the Earth, is fairly constant around 250–300 Dobson units (DU), and it decreases toward the poles but with large seasonal fluctuations at higher latitudes. (The Dobson unit is a unit of measurement of atmospheric ozone columnar density, specifically ozone in the stratospheric ozone layer. It is named after the English physicist and meteorologist G.M.B. Dobson (1899–1976) who in 1928 invented the photoelectric spectrophotometer used in measuring the total ozone over an area from the ground. A Dobson unit for a gas is the amount of gas contained in a layer of the pure gas that is 10^{-5} m in thickness at 1 atm and 0°C; 1 DU = 4.4615×10^{-5} mole m^{-2}). As discussed earlier (Box 13.1), the concentration of O_3 in the stratosphere should be in a steady state because of the balance between production and destruction of O_3, but it is not so because the balance has been destroyed by anthropogenic input of ozone depleting chemicals (ODCs) into the atmosphere.

Significant loss of ozone in the lower atmosphere was first noticed over Antarctica in the 1970s by a research group from the British Antarctic Survey (BAS). In the fall of 1985, BAS scientists reported a much more dramatic depletion in total O_3 column abundance above Antarctica in the austral springs since 1977 – about a 30% decrease relative to pre-1980 measurements – a phenomenon commonly referred to as the "Antarctica ozone hole" (ozone concentration less than 220 DU, which is chosen as the baseline value for an "ozone hole" because total ozone values < 220 DU were not found over Antarctica prior to 1979). The story goes that the measured values were so low and unanticipated that the scientists suspected at first that their instruments were faulty, but measurements with replaced instruments confirmed the earlier results. Many subsequent studies have revealed significant stratospheric ozone depletion in many parts of the world. The most severe depletion has been observed over Antarctica (up to about 70% during September–November, the austral spring), which has now grown to a size greater than that of North America, and to a lesser extent over the Arctic. The "Antarctica ozone hole" is a seasonal event; it typically decreases by 20–25% during January–March.

Box 13.5 Chlorofluorocarbon compounds (CFCs)

Chlorofluorocarbons (CFCs) comprise a group of compounds containing fluorine, chlorine, and carbon atoms. The five main CFCs are CFC-11 (trichlorofluoromethane, $CFCl_3$), CFC-12 (dichlorodifluoromethane, CF_2Cl_2), CFC-113 (trichlorotrifluoroethane, $C_2F_3Cl_3$), CFC-114 (dichlorotetrafluoroethane, $C_2F_4Cl_2$), and CFC-115 (chloropentafluoroethane, C_2F_5Cl). "Freon" is a DuPont trade name for a group of CFCs (including CFC-11 and CFC-12), which are used primarily as refrigerants, but also in fire extinguishers and as propellants in aerosol cans.

Chloroflurocarbon compounds were first synthesized in 1928 by DuPont scientists during a search for a new, nontoxic substance that could serve as a safe refrigerant, replacing ammonia as the standard cooling fluid at that time. Chlorofluorocarbons filled this requirement very well because they are chemically inert (and, therefore, highly stable), nonflammable, nontoxic compounds that are superb solvents. They were considered miracle compounds and soon were in demand for a variety of applications: as coolants in refrigeration and air conditioners; as solvents in cleaners, particularly for electronic circuit boards; and as propellants in aerosols.

Scientists remained unaware of the possible impact of CFCs on the atmosphere until 1970 when the British scientist James Lovelock detected CFC-11 (at concentration levels in parts per trillion) not only in every air sample that passed over Ireland from the direction of London, but also in air samples directly off the North Atlantic uncontaminated by recent urban population. A few years later, in 1974, Mario Molina and Sherwood Rowland demonstrated in their laboratory that CFCs breakdown to form free chlorine radicals ($Cl^{\bullet}_{(g)}$), and predicted that these very reactive radicals could cause environmental problems by catalyzing the destruction of ozone in the stratosphere. Now we know how true that prediction was!

In the 1950s, David Bates and Marcel Nicolet had presented evidence that $OH^{\bullet}$ and $NO^{\bullet}$ could catalyze the recombination reaction $O_{3(g)} + O_{(g)} \Rightarrow 2O_{2(g)}$ (equation 13.19), thereby reducing the amount of stratospheric ozone. Support for this mechanism of ozone depletion came from the suggestion of Prof. Paul Crutzen in 1970 that N_2O gas produced by soil bacteria, fertilizers, and automobile engines with catalytic converters is stable enough to reach the stratosphere, get converted to NO by reaction with atomic oxygen, and catalyze the destruction of ozone (equations 13.10, 13.11, and 13.12).

Ozone can also be removed by reaction with natural chlorine, which is transferred from marine and terrestrial biological sources to the stratosphere mainly as methyl chloride (CH_3Cl), but the natural source accounts for only 25% of the chlorine that is transported across the tropopause. In 1974, Sherwood Rowland, Professor of Chemistry, University of California at Irvine, and his postdoctoral associate Mario J. Molina suggested that long-lived anthropogenic organic halogen compounds, such as chloroflurocarbons (CFCs) and bromofluorocarbons (see Box 13.5), might also catalyze the breakdown of ozone. The prediction turned out to be correct, and Crutzen, Molina, and Rowland were awarded the 1995 Nobel Prize in Chemistry in recognition of their work on stratospheric ozone.

The breakdown of ozone by CFC compounds in the stratosphere (at an altitude of 12–20 km and higher) is a complicated process. The CFC compounds absorb UV radiation in the 190–220 μm range, which results in photodissociation reactions. For example:

$$CF_2Cl_{2(g)}\ (CFC-12) + UV\ photon \Rightarrow Cl^{\bullet}_{(g)} + CClF^{\bullet}_{2(g)} \quad (13.49)$$

$$CFCl_{3(g)}\ (CFC-11) + UV\ photon \Rightarrow Cl^{\bullet}_{(g)} + CFCl^{\bullet}_{2(g)} \quad (13.50)$$

A small amount of chlorine is also generated by the breakdown of methyl chloride (CH_3Cl), which is produced biogenically from the oceans, from the burning of vegetation, and from volcanic emissions ($CH_3Cl + h\nu \Rightarrow CH_3 + Cl^{\bullet}$).

Once released from their parent compounds, the highly reactive $Cl^{\bullet}_{(g)}$ can destroy $O_{3(g)}$ molecules through a variety of catalytic cycles, the simplest being a reaction with $O_{3(g)}$ to form an intermediate compound, chlorine monoxide ($ClO^{\bullet}_{(g)}$), and finally $O_{2(g)}$:

$$Cl^{\bullet}_{(g)} + O_{3(g)} \Rightarrow ClO^{\bullet}_{(g)} + O_{2(g)} \quad (13.51)$$

$$ClO^{\bullet}_{(g)} + O_{(g)} \Rightarrow Cl^{\bullet}_{(g)} + O_{2(g)} \quad (13.52)$$

$$O_{3(g)} + O_{(g)} \Rightarrow 2O_{2(g)}\ (net\ reaction) \quad (13.53)$$

It is estimated that a single $Cl^{\bullet}_{(g)}$ atom participating in the above catalytic cycle is potentially capable of reacting with 100,000 $O_{3(g)}$ molecules, attesting to the large possible impact of even a small amount of CFC introduced into the stratosphere. A similar catalytic chain reaction would apply to bromine released from synthetic bromofluorocarbon compounds (e.g., CF_2ClBr), and from methyl bromide (CH_3Br) produced by marine plankton, agricultural pesticide (fumigant), and biomass burning. Actually, on an atom-per-atom basis, bromine is even more efficient than chlorine in destroying ozone, but its contribution to ozone destruction is minimal because there is much less bromine in the atmosphere at present. Laboratory studies have shown that F and I atoms participate in analogous catalytic cycles. However, in the Earth's atmosphere, F atoms react rapidly with water and methane to form strongly bound HF, whereas organic molecules that contain I react so rapidly in the lower atmosphere that they do not reach the stratosphere in significant quantities.

Most of the $Cl^{\bullet}_{(g)}$ and $ClO^{\bullet}_{(g)}$ released into the atmosphere (at all latitudes) are quickly tied up in chlorine reservoirs such as chlorine nitrate ($ClONO_{2\,(g)}$) and hydrochloric acid ($HCl_{(g)}$) by reactions such as:

$$ClO^{\bullet}_{(g)} + NO_{2(g)} + M \Rightarrow ClONO_{2(s)} + M \quad (13.54)$$

$$Cl^{\bullet}_{(g)} + CH_{4(g)} \Rightarrow HCl_{(g)} + CH_{3(g)} \quad (13.55)$$

Removal of the chlorine and nitrogen species arrests the destruction of ozone, but not for long. During the austral winter, when the temperature of the Antarctica stratosphere drops below −77°C (−196 K), nitric acid trihydrate (NAT) crystals ($HNO_3.3H_2O$) start to crystallize on small sulfuric acid–water aerosol particles, and $ClONO_{2(g)}$ and HCl are consumed by the following reactions that occur on the surface of the NAT solid particles:

$$ClONO_{2(g)} + H_2O_{(water-ice)} \Rightarrow HOCl_{(g)} + HNO_{3(adsorbed)} \quad (13.56)$$

$$ClONO_{2(g)} + HCl_{(adsorbed)} \Rightarrow Cl_{2(g)} + HNO_{3(adsorbed)} \quad (13.57)$$

However, nitric acid trihydrate cannot survive the increased solar UV radiation during the austral spring, and the catalytic destruction of ozone continues through regeneration of $Cl^{\bullet}_{(g)}$ with the arrival of spring season:

$$HOCl_{(g)} + \text{UV photon} \Rightarrow OH^{\bullet}_{(g)} + Cl^{\bullet}_{(g)} \quad (13.58)$$

$$Cl_{2(g)} + \text{UV photon} \Rightarrow Cl^{\bullet}_{(g)} + Cl^{\bullet}_{(g)} \quad (13.59)$$

It turns out that reactions (13.51) and (13.52) alone cannot account for the very large depletion of O_3 that characterizes the Antarctica ozone hole. A cycle catalyzed by $ClO^{\bullet}_{(g)}$ that appears capable of explaining about 75% of the observed O_3 depletion associated with the ozone hole, without involving atomic oxygen (which is in short supply in the stratosphere), is as follows (Hobbs, 2000; van Loon and Duffy, 2000):

$$ClO^{\bullet}_{(g)} + ClO^{\bullet}_{(g)} + M \Rightarrow ClOOCl_{(g)} + M \quad (13.60)$$

$$ClOOCl_{(g)} + \text{UV photon} \Rightarrow Cl^{\bullet}_{(g)} + ClOO^{\bullet}_{(g)} \quad (13.61)$$

$$ClOO^{\bullet}_{(g)} + M \Rightarrow Cl^{\bullet}_{(g)} + O_{2(g)} + M \quad (13.62)$$

$$2Cl^{*}_{(g)} + 2O_{3(g)} \Rightarrow 2ClO^{*}_{(g)} + 2O_{2(g)} \quad (13.63)$$

$$2O_{3(g)} + \text{UV photon} \Rightarrow 3O_{2(g)} \quad \text{(net reaction)} \quad (13.64)$$

It is now well established that CFC compounds are the chief culprits of stratospheric ozone depletion. The best evidence for this conclusion is the strong inverse correlation between ClO and ozone concentrations in the atmosphere (Fig. 13.11). CFC molecules are much heavier than nitrogen or oxygen but they can reach the stratosphere like other heavy gases, such as argon and krypton, because of wind turbulence that keeps the atmospheric gases well mixed, and because they are insoluble and highly stable. Natural sources of chlorine in the troposphere (mainly HCl from volcanic eruptions, and NaCl sprayed from the oceans) are four to five orders of magnitude larger than anthropogenic sources, but NaCl is highly soluble and quickly removed from the atmosphere by rain water, and measurements have shown that volcanic HCl amounts to only ~3% of the stratospheric chlorine. It is estimated that anthropogenic sources account for about 80% of the stratospheric chlorine.

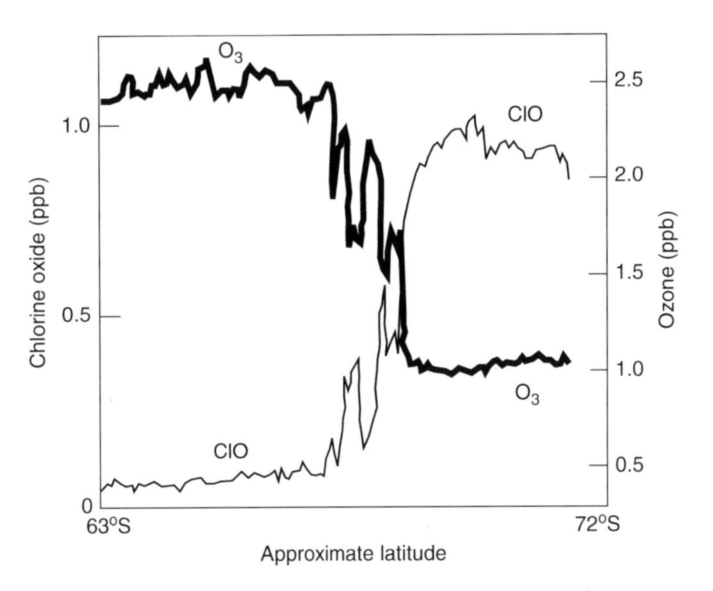

Fig. 13.11 Inverse correlation between chlorine oxide (ClO) and ozone concentrations in the atmosphere over the Antarctic polar region on September 16, 1987. Note that in the high latitudes of Antarctica, the very low ozone concentrations are matched by the relatively high concentrations of ClO produced by the reaction of ozone with Cl. The curves have been smoothed for simplicity. (After Anderson *et al.*, 1991.)

In a report released in 1976, the US National Academy of Sciences concluded that the ozone depletion hypothesis was strongly supported by scientific evidence. Scientists had calculated that if CFC production continued to increase at the going rate of 10% per year until 1990 and then remain steady, the global ozone loss would amount to 5–7% by 1995 and 30–50% by 2050. Responding to this prediction, the USA, Canada, Norway, and Sweden banned the use of CFCs in aerosol spray cans in 1978. Subsequently, as a result of international agreements (Montreal Protocol, 1987; London meeting, 1990; Copenhagen meeting, 1992), various countries have enacted laws for phased reduction and eventual ban on the production and use of CFCs. Promising substitutes, which have practically no ozone destroying potential, include hydrochlorofluorocarbons (HCFCs) and hydrofluorocarbons (HFCs) such as HFC-134a (CH_2FCF_3), HCFC-123 ($CHCl_2CF_3$), HFC-125 (CHF_2CF_3), and HFC-22 (CH_2F_2). The Antarctic "ozone hole," however, is expected to continue to exist for decades because of the longevity of CFCs (between 40 and 150 years) already present in the stratosphere. It is estimated that ozone concentrations in the lower stratosphere over Antarctica will actually increase by 5–10% by 2020 and then return to pre-1980 levels by about 2060–2075.

13.3.2 *Smogs*

The term "*smog*" was introduced in 1905 by Harold Antoine Des Voeux, a member of the Coal Smoke Abatement Society in London, to describe the combination of smoke and fog

Table 13.4 Comparison of general characteristics of sulfurous smog and photochemical smog.

Characteristics	Sulfurous smog	Photochemical smog
Primary pollutants	$SO_{2(g)}$, CO, carbon particulates	$NO_{(g)}$, $NO_{2(g)}$, CO, organics
Secondary pollutants	H_2SO_4, SO_4^{2-} aerosol	O_3, PAN and other organics, acids, aerosols
Principal sources	Industrial plants, households	Motor vehicles
Temperature	Cool (< 35°F)	Hot (> 75°F)
Relative humidity	High, usually foggy (> 85%)	Low, usually hot and dry (< 70%)
Type of temperature inversion	Radiation inversion	Subsidence inversion
Peak occurrence	Winter months early morning	Summer months around mid-day
Effect on human health	Lung and throat irritation	Eye and respiratory irritation

Sources of data: Raiswell *et al.*, 1980; Andrews *et al.*, 2004; Eby, 2004.

[sm(oke and f)og], which was then a common feature in several cities throughout Great Britain. *Smoke* is an aerosol composed of solid and liquid particulates and gases, which are emitted when a material undergoes combustion, together with the air that is entrained or otherwise mixed into the mass. The composition of smoke depends on the nature of the combusted material and the condition of combustion, mainly temperature and oxygen supply. Smoke inhalation, the primary cause of death in victims of indoor fires, kills by a combination of thermal damage, poisoning, and pulmonary irritation caused by CO, hydrogen cyanide and other combustion products.

Fog begins to form when water vapor (a colorless gas) condenses into tiny water droplets in the air. Aerosol particles in the air provide the nuclei required for such condensation. Fog normally occurs at a relative humidity near 100% (that is, when the air is saturated with moisture), which is achieved either by adding moisture to the air or by lowering the ambient air temperature. When the air is saturated with moisture, additional moisture tends to condense rather than stay in the air as water vapor.

The floating solid particles and liquid droplets in the smog, dominantly in the 0.2–2 μm size range, cause incoherent scattering of visible light. This results in reduced visibility, which is commonly referred to as *haze*. In addition, smogs are potentially harmful to human health, vegetation, and the environment if they contain pollutants (such as CO, NO_x, O_3, H_2SO_4, VOC, etc.).

There are two main types of smog: (i) *sulfurous smog* (also referred to as London-type or classical-type smog) and (ii) *photochemical smog* (also referred to as Los Angeles-type smog). A comparison of the general characteristics of the two types of smog is presented in Table 13.4.

Sulfurous smog

Most of the world's energy comes from the combustion of hydrocarbons, an oxidation reaction. With plenty of oxygen around, the hydrocarbon (represented here as "CH") is oxidized to CO_2 gas:

$$4CH + 5O_{2(g)} \Rightarrow 4CO_{2(g)} + 2H_2O_{(g)} \tag{13.65}$$

but this poses no direct health threat because CO_2 gas is not toxic. If enough oxygen is not available for combustion, the main product is CO gas:

$$4CH + 3O_{2(g)} \Rightarrow 4CO_{(g)} + 2H_2O_{(g)} \tag{13.66}$$

CO gas is toxic because hemoglobin in our blood takes up CO much faster than it takes up oxygen. With even less oxygen, the product is "smoke" containing particulates of solid carbon that can damage our lungs:

$$4CH + O_{2(g)} \Rightarrow 4C_{(s)} + 2H_2O_{(g)} \tag{13.67}$$

Additional problems arise because of impurities in the hydrocarbon fuels, the most important of which in the context of air pollution is sulfur, which on combustion is converted into SO_2 gas. Sulfurous smog results from the accumulation of smoke from fuels with high sulfur content, such as coal (~0.2 to 7.0 wt% S) and fuel oils (~0.5 to 4.0 wt% S) used in boilers, furnaces, domestic fireplaces, steam turbines, and thermal power stations. Where the atmosphere is humid, the unburnt carbon particles may serve as nuclei for condensation of water droplets. The water droplets become acidic with H_2SO_4 by dissolving atmospheric SO_2 gas, and form an irritating fog. Thus, sulfurous smog is associated with a high concentration of unburned carbon soot as well as elevated levels of atmospheric sulfur dioxide.

Sulfurous smog was a common phenomenon in the mid-20th century in cities such as London (UK.), but it reached a crisis level in the first week of December, 1952, due to the convergence of a series of conditions conducive to the formation of heavy sulfurous smog. On December 4, there was a cloud cover over the city that did not allow much of the incoming solar radiation to penetrate. Consequently, there was no appreciable warming of the lower layers of air, and the temperature at ground level fell precipitously. The air was stagnant, and the relative humidity climbed to 80%, which was sufficiently high to form a dense fog within the polluted city air during the night. Surface temperatures were close to freezing as the next day commenced.

The fog did not dissipate as the Sun rose the next morning. The high albedo of the fog layer limited the penetration of

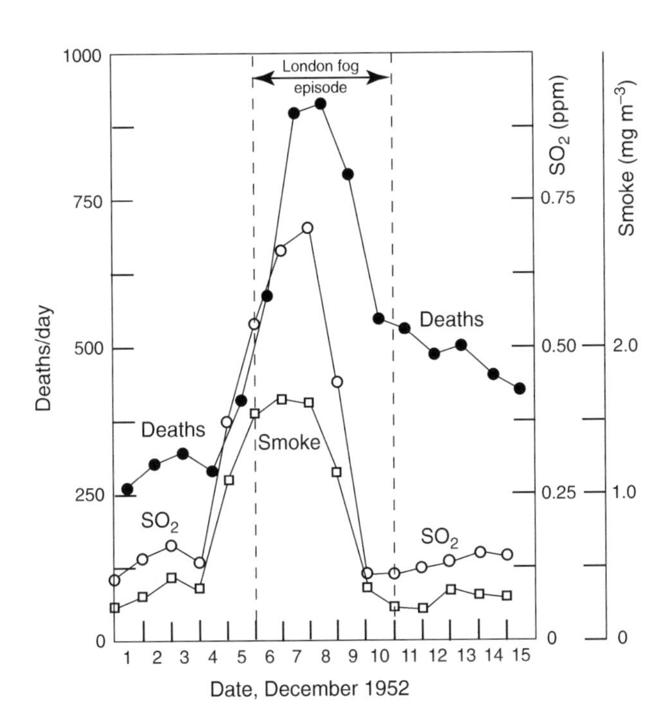

Fig. 13.12 Death toll resulting from the catastrophic London fog episode of December, 1952. Note the high degree of correlation among SO_2, smoke, and number of deaths. (Modified from Wilkins, 1954, Figure 4, p. 10.)

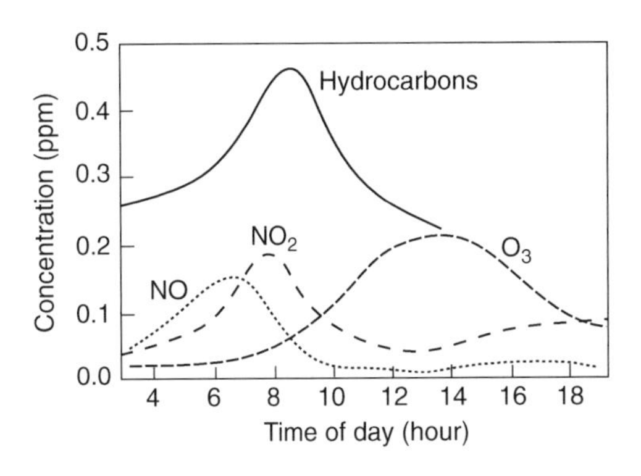

Fig. 13.13 Typical variations in the abundances of some important pollutants, on a 24-h cycle, produced during a photochemical smog event. (After van Loon and Duffy, 2000.)

sunlight to the lower layers of air, and the fog lingered within the Greater London Basin because of exceptionally light winds. By December 6, the noontime temperature had dropped to −2°C (28°F) and the relative humidity had increased to 100%. The cold temperatures and dampness increased the demand for indoor heating in homes and factories; ash, SO_2 gas, and soot from the burning of coal, the primary fuel used in heating, as well as pollutants from automobile exhausts, filled the stagnant air. By December 8, SO_2 reached peak concentrations of ~0.7 ppm (compared to typical annual mean concentration of ~0.1 ppm in polluted cities with large coal usage), and the peak smoke concentrations rose to ~1.7 mg m^{-3}. The resulting smog made breathing very difficult for the Londoners, and in many cases resulted in death. Meteorological conditions improved on December 9 when a breeze developed to remove the stagnant air conditions. The following day a cold front passed through, bringing with it fresh air from the North Atlantic and an end to the smog episode. It is estimated that the so-called "Killer Fog" killed ~4000 people between December 4 and 10 (Fig. 13.12), although it is not certain that the high concentrations of SO_2 and smoke were the sole reasons for the deaths. The low temperatures were probably a significant contributing factor. Sulfurous smog of this magnitude is unlikely to happen again because of much reduced usage of coal as a heating fuel and regulations regarding emissions of SO_2 gas from industrial plants (e.g., UK Clean Air Act of 1956; US Clean Air Act of 1970), but sulfurous smog is still fairly common in early winter mornings in cities with a very large population, a high concentration of coal-burning industrial plants, and humid stagnant air.

Photochemical smog

Photochemical smog was first noticed in Los Angeles during the Second World War; hence, the name Los Angeles-type of smog. It is characterized by high concentrations of a large variety of pollutants such as particulates, nitrogen oxides, ozone, carbon monoxide, hydrocarbons, aldehydes (molecules with a CHO functional group), and sometimes sulfuric acid. (Actually, the term photochemical smog is a misnomer because it involves neither smoke nor fog.) The reactions producing the pollutants are ultimately driven by photochemistry and hence have a diurnal cycle. The effects of the smog are most prominent at mid-day, when solar intensity is highest, and the smog tends to recede at night because of the short half-life of many of the compounds (Fig. 13.13).

Photochemical smog has its origin in the nitrogen oxides and hydrocarbon vapors (e.g., ethylene and butane) emitted by automobiles due to evaporation from fuel tanks and as unburned species in the exhaust, and by industrial plants. These emissions undergo photochemical reactions initiated by sunlight, generating a variety of secondary pollutants, including ozone. Together these products result in smog that pollutes the atmosphere, reduces visibility, and even causes breathing problems. Whereas ozone in the stratosphere plays a very beneficial role, ozone at the ground level is harmful to vegetation and human health, and is the pollutant of primary concern in photochemical smog. The chemical reactions that lead to photochemical smog are extremely complex, and still not completely understood. Outlined below are some of the major chemical reactions considered to be important in the present context.

Although N_2 gas is generally considered as inert, at high temperatures (e.g., in an internal combustion engine) molecular oxygen and nitrogen in the air dissociate into their atomic states and participate in a series of reactions. The three principal reactions (all reversible) for forming NO are:

$$O_{(g)} + N_{2(g)} \Rightarrow NO_{(g)} + N_{(g)} \quad (13.68)$$

$$N_{(g)} + O_{2(g)} + \Rightarrow NO_{(g)} + O_{(g)} \quad (13.69)$$

$$N_{(g)} + OH_{(g)} + \Rightarrow NO_{(g)} + H_{(g)} \quad (13.70)$$

Note that the oxygen atom produced by reaction (13.69) can enter reaction (13.68) and promote a chain of reactions that produce NO. The net reaction obtained by adding 13.68 and 13.69 is:

$$N_{2(g)} + O_{2(g)} \Rightarrow 2NO_{(g)} \quad (13.71)$$

When the NO gas is exhausted to the open atmosphere, it is oxidized to $NO_{2(g)}$ by the following reactions:

$$2NO_{(g)} + O_{2(g)} \Rightarrow 2NO_{2(g)} \quad (13.72)$$

$$NO_{(g)} + O_{3(g)} \Rightarrow NO_{2(g)} + O_{2(g)} \quad (13.73)$$

$$RO_2^{\bullet} + NO_{(g)} \Rightarrow RO^{\bullet} + NO_{2(g)} \quad (13.74)$$

where $RO^{\bullet}$ and $RO_2^{\bullet}$ are referred to as peroxyl radicals in which R represents the root organic molecule.

The $NO_{2(g)}$ produced by the above reactions undergoes photodissociation at wavelengths below 0.38 μm into $NO_{(g)}$ and $O_{(g)}$, leading to rapid formation of $O_{3(g)}$:

$$NO_{2(g)} + \text{UV photon} \Rightarrow NO_{(g)} + O_{(g)} \quad (13.75)$$

$$O_{2(g)} + O_{(g)} + M \Rightarrow O_{3(g)} + M \quad (13.76)$$

$$\text{(net reaction) } NO_{2(g)} + O_{2(g)} + \text{UV photon} \Rightarrow NO_{(g)} + O_{3(g)} \quad (13.77)$$

where M is a catalytic molecule, most likely N_2 or O_2, that absorbs the excess energy released dring the formation of the ozone molecle. This is the main mechanism by which O_3 is formed *in situ* in the troposphere. However, not much ozone is accumulated directly by reaction (13.77) because it is depleted by the rapid reaction that regenerates NO_2 at the expense of ozone:

$$NO_{(g)} + O_{3(g)} \Rightarrow NO_{2(g)} + O_{2(g)} \quad (13.78)$$

Assuming that the ozone budget is determined by reactions (13.75), (13.76), and (13.78), the steady state concentration of O_3 in urban polluted air should be only ~0.03 ppmv, but typical values are well above this concentration and can reach 0.5 ppmv (Hobbs, 2000). Therefore, there must be other chemical reactions, such as the oxidation of hydrocarbon vapors released through the combustion of petroleum fuels, which also convert NO to NO_2 without consuming O_3. The reactions involving hydrocarbons are quite complex and the ubiquitous $OH^{\bullet}$ radical plays a prominent role in promoting these reactions (e.g., $O^* + H_2O \Rightarrow 2OH^{\bullet}$; $NO_2 + H_2O \Rightarrow NO + 2OH^{\bullet}$). As an example, if the petroleum vapor from vehicles is represented by CH_4, then the the net reaction may be written as

$$CH_{4(g)} + 2O_{2(g)} + 2NO_{(g)} + \text{sunlight} \Rightarrow H_2O_{(g)} + HCHO_{(g)} + 2NO_{2(g)} \quad (13.79)$$

The $OH^{\bullet}$ radical does not appear in the net reaction because it acts as a kind of catalyst. Aldehydes, such as formaldehyde (HCHO) formed by the above reaction, are eye irritants and, at high concentrations, tend to be carcinogens.

Conditions necessary for the development of photochemical smog are a high density of automobiles emitting a large amount of N_2O gas, large concentrations of reactive hydrocarbons (RH) from automobile exhausts and industrial plants, plenty of sunlight (high level of UV radiation), and stagnant air due to thermal inversion (i.e., temperature increasing, instead of decreasing, with altitude). Thermal inversion occurs when a layer of warm air above traps the ground-level cooler air, and prevents the cooler air from rising and dispensing the pollutants. A city which meets these conditions is, for example, Los Angeles, and it is no wonder that photochemical smog is a common phenomenon in this city. Photochemical smog, characterized by the reddish-brown hue of NO_2, is now a common feature of many major cities around the world (e.g., Athens, Bangkok, Mexico City, Beijing, and Tokyo).

13.3.3 Acid deposition

Acid deposition, popularly referred to as "acid rain," is the deposition of acidic gases, aerosol particles, or raindrops on the Earth's surface. The acidity in these deposits result from the formation of sulfuric acid (H_2SO_4), nitric acid (HNO_3), or hydrochloric acid (HCl) in the atmosphere. The main sources of these gases are volcanoes, biological processes on the Earth's surface, and human activities. Summarized below are the reactions pertaining to H_2SO_4, the most important contributor to acid deposition; simlar reactions can be written for HNO_3 and HCl.

Gas-phase transformation of sulfur dioxide gas

The gas-phase transformation mechanism involves the following steps (Jacobsen, 2002): (i) gas-phase oxidation of $SO_{2(g)}$ to $H_2SO_{4(g)}$; (ii) condensation of $H_2SO_{4(g)}$ and water vapor onto aerosol particles or cloud drops to produce a $H_2SO_{4(aq)}$–$H_2O_{(aq)}$ solution; and (iii) dissociation of $H_2SO_{4(aq)}$ to SO_4^{2-} and H^+ in the solution, thereby increasing its acidity. The relevant reactions are presented below:

$$SO_{2(g)} + OH^{\bullet}_{(g)} + M \Rightarrow HSO^{\bullet}_{3(g)} + M \text{ (catalyst)}$$
$$HSO^{\bullet}_{3(g)} + O_{2(g)} \Rightarrow HO^{\bullet}_{2(g)} + SO_{3(g)}$$
$$SO_{3(g)} + H_2O_{(l)} \Rightarrow H_2SO_{4(aq)}$$
$$H_2SO_{4(aq)} \Rightarrow 2H^+ + SO_4^{2-} \quad (13.80)$$

Aqueous-phase transformation of sulfur dioxide

When clouds are present, $SO_{2(g)}$ is consumed at a faster rate than can be explained by gas-phase chemistry alone. This is because of reactions in the liquid water droplets. The aqueous-phase transformation of $SO_{2(g)}$ involves the following steps (Jacobsen, 2002): (i) dissolution of $SO_{2(g)}$ into liquid water drops to produce $SO_{2(aq)}$; (ii) conversion (within the water drops) of $SO_{2(aq)}$ to $H_2SO_{3(aq)}$ and dissociation of $H_2SO_{3(aq)}$ to HSO_3^- and SO_3^{2-}; and (iii) oxidation of HSO_3^- and SO_3^{2-} to SO_4^{2-} (within the water drops). The relevant reactions are listed below:

$$
\begin{aligned}
&SO_{2(g)} \Leftrightarrow SO_{2(aq)} \\
&SO_{2(aq)} + H_2O_{(l)} \Leftrightarrow H_2SO_{3(aq)} + H^+ + HSO_3^- \\
&HSO_3^- \Rightarrow H^+ + SO_3^{2-} \\
&HSO_3^- + H_2O_{2(l)} + H^+ \Rightarrow SO_4^{2-} + H_2O_{(aq)} \\
&\qquad\qquad + 2H^+ \text{ (at pH} \le 6) \\
&SO_3^{2-} + O_{3(aq)} \Rightarrow SO_4^{2-} + O_{2(aq)} \text{ (at pH} > 6)
\end{aligned} \tag{13.81}
$$

The last reaction is written in terms of SO_3^{2-} and SO_4^{2-} because the HSO_3^-–O_3 reaction is relatively slow and, at pH > 6, most S(VI) exists as SO_4^{2-}.

Gas-phase transformation, which involves condensation onto raindrops, is the dominant process, whereas aqueous phase transformation, which involves dissolution in cloud drops and rain drops, is the more rapid process. The aqueous H^+ produced by these processes is the reason for the acidity of "wet" deposition. The acidity may be partly neutralized by the formation of sulfates, for example, $(NH_4)_2SO_4$ and $(NH_4)HSO_4$ if ammonia is present in the polluted air:

$$2NH_3 + 2H^+ + SO_4^{2-} \Rightarrow (NH_4)_2SO_4 \tag{13.82}$$

$$NH_3 + 2H^+ + SO_4^{2-} \Rightarrow (NH_4)HSO_4 \tag{13.83}$$

Besides deposition in precipitation ("wet" deposition, predominantly rainfall), sulfur is also transferred from the atmosphere to the Earth's surface by "dry" deposition. Dry deposition occurs when SO_2 gas is absorbed or dissolved directly on land by standing water, vegetation, soil, and other surfaces. Also, sulfate particles may settle out of the air or be trapped by vegetation (dry fallout). Dry deposition of SO_2 gas over land depends on the characteristics of the surface, small-scale meteorological effects, and the atmospheric concentration of SO_2. Dry deposition is greater near the emission sources of the gas and decreases rapidly with distance from the source. In some areas, dry acid deposition may be responsible for as much as 25% of total acid deposition (Berner and Berner, 1996).

Whether wet or dry, acid deposition is harmful to aquatic life, vegetation, structures such as buildings and monuments, and human health.

13.3.4 Greenhouse gases and global warming

The energy received by the Earth as a consequence of solar radiation is the source of the Earth's surface temperature, and this is the energy that sustains life on the Earth, and drives many of its processes such as the circulation of air and the hydrologic cycle. The annual mean global energy balance of the Earth–atmosphere system, expressed in terms of 100 arbitrary units of energy associated with the incoming solar radiation at the top of the atmosphere (TOA), is shown in Fig. 13.14. This radiation has wavelengths in the visible light and ultraviolet region of the electromagnetic spectrum. Out of the 100 units, 49 units are absorbed by the Earth, 20 units are absorbed by the atmosphere, and the rest (31 units) are reflected back into space by the Earth's surface (9 units) and atmosphere (22 units). The Earth receives an additional 95 units of energy resulting from absorption by atmosphere and reradiation to the Earth (greenhouse effect), raising the total to 144 units of energy. Thermal equilibrium at the surface of the Earth is achieved by the transfer of 144 units back to the atmosphere: 114 units as longwave (infrared) radiation, 23 units as evapotranspiration, and 7 units by heat fluxes associated with convection, turbulence, etc. Note that at TOA, the incoming 100 units of solar energy is balanced by 31 units of reflected and backscattered radiation and 69 units of outgoing longwave radiation, which ensures thermal equilibrium for the Earth–atmosphere system as a whole.

Greenhouse gases

The Earth's average annual surface temperature is about 15°C, whereas calculations show that, without an atmosphere, the Earth's average surface temperature at present would be a relatively chilly −18°C (0°F) (Kump *et al.*, 2004). This difference of 33°C is attributed to the warming of the atmosphere by absorption of the IR radiation (95 units, Fig. 13.14) returned from the Earth. The warming of the Earth's lower atmosphere due to natural gases that are transparent to most incoming solar radiation, but selectively absorb and reemit a portion of the outgoing thermal-IR radiation, is the natural *greenhouse effect* of the Earth's atmosphere, and the gases responsible for this effect are referred to as *greenhouse gases*. The phenomenon is analogous to the warmth retained inside a greenhouse in our gardens. The glass exterior of a greenhouse, like the greenhouse gases, allows the penetration of incoming solar radiation, but prevents the escape of a portion of the outgoing thermal-IR radiation, resulting in a net increase in the temperature of the interior (much like what happens when a car with closed windows is left in the sun for a while).

The major natural greenhouse gases are CO_2, CH_4, and N_2O. H_2O (water vapor) is also a strong absorber of infrared radiation, but is usually not considered to be a greenhouse gas because human activities have only a small direct influence on the amount of atmospheric water vapor. As mentioned earlier (section 13.2.2), the most plausible explanation for the

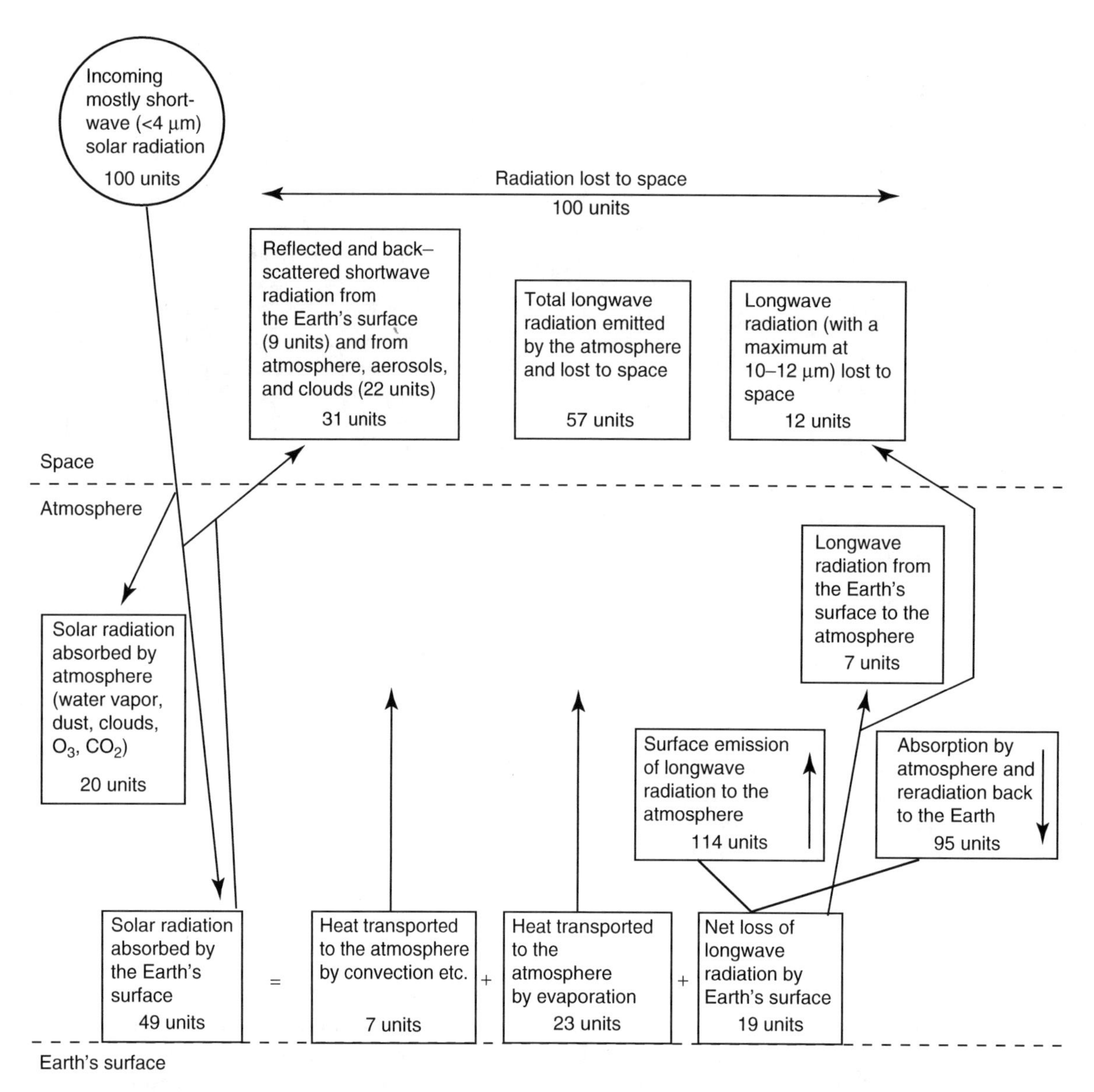

Fig. 13.14 Current estimate of the annual mean global energy balance of the Earth–atmosphere (unpolluted) system expressed in terms of 100 units of incoming solar energy (which represents 342 W m^{-2}) received at the top of the Earth's atmosphere (TOA). (Source of data: Houghton *et al.* (1996).)

absence of glaciation during the Archean, when solar luminosity was 20 to 30% lower than today, is the greenhouse warming due to CH_4 and CO_2 in the atmosphere (Pavlov *et al.*, 2000). Now we are threatened with the likelihood of too high a surface temperature because of sharp increases in the atmospheric concentrations of these gases since the Industrial Revolution, especially since the 1950s (Fig. 13.15). In addition, the atmosphere now contains trace amounts of a number of gases of anthropogenic origin, such as chlorofluorocarbons (CFCs) and hydrofluorocarbons (HFCs), which are exceptionally good absorbers of infrared radiation (in addition to being instrumental in ozone depletion in the stratosphere). The potential increase in the Earth's surface temperature due to increased concentrations of greenhouse gases (and aerosols) in the atmosphere is referred to as *global warming*, which can force global-scale climate changes. Currently recognized warning signs of global warming include an approximately 0.7°C (or 1.3°F) increase in the surface temperature during the 20th century, retreating mountain glaciers, significant melting of the Antarctica ice cap at the edges, and a rise in the sea level by at least 10 cm over the past 100 yr.

The *global warming potential* (GWP) of a greenhouse gas is calculated as the ratio of global warming from one unit mass of the greenhouse gas to one unit mass of carbon dioxide over

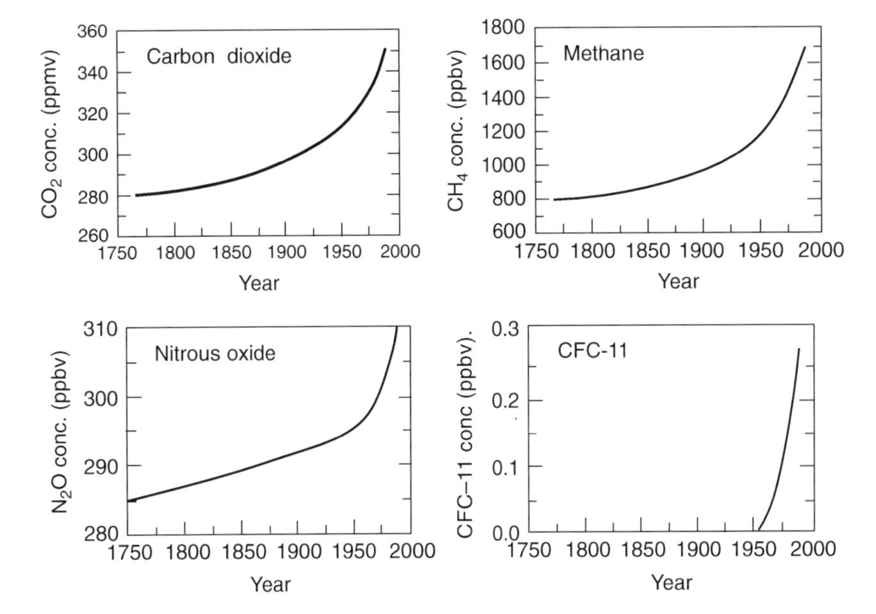

Fig. 13.15 Increases in the concentration of greenhouse gases in the atmosphere since circa 1750. Chlorofluorocarbons (represented by CFC–11), which are entirely anthropogenic in origin, became significant only after 1950. The concentrations of the other gases, which were relatively constant until the 1700s, have increased sharply since then due to human activities, especially combustion of fossil fuels. (After Intergovernmental Panel on Climate Change, 1990.)

Table 13.5 Radiative forcing and global warming potential of greenhouse gases.

Greenhouse gas	Concentration 1990 (ppmv)	Radiative forcing	Total radiative forcing (%)	Lifetime (yr)	Global warming potential (100 yr)
CO_2	354	1.5	61	50–200	1
CH_4	1.72	0.42	17	12	21
H_2O		0.14	6		
N_2O	0.310	0.1	4	120	310
CFC ± 11	0.00028	0.062	2.5	65	3400
CFC ± 12	0.000484	0.14	6	130	7100
Other CFCs		0.085	3.5		

Source of data: Berner and Berner (1966), Intergovernmental Panel on Climate Change (1966), Rodhe (1990), van Loon and Duffy (2000).

a particular time period. The Intergovernmental Panel on Climate Change (IPCC) recommends that GWP be calculated on a 100-year basis. Although CO_2 is considered to be the Earth's main greenhouse gas because of its much higher abundance relative to other greenhouse gases, complex molecules such as CH_4, N_2O, O_3, and the fluorocarbons have much higher global warming potential (Table 13.5).

Radiative forcing

The impact of a factor, such as a greenhouse gas, that can cause climate change is often evaluated in terms of its *radiative forcing*, which is a measure of how the energy balance of the Earth–atmosphere system is influenced when the factor under consideration is altered. Radiative forcing is usually quantified as the rate of energy change per unit area of the globe as measured at the top of the atmosphere, and is expressed in units of Watts per square meter (W m^{-2}). Positive forcings lead to warming of climate, and negative forcings to cooling. The word 'radiative' reminds us that we are looking at the balance between incoming solar radiation and outgoing infrared radiation within the atmosphere, and the word 'forcing' emphasizes that the Earth's radiative balance is being forced out from its normal state.

The principal components of the radiative forcing of climate change over the period from circa 1750 (the beginning of the industrial era) to 2005 are presented in Fig. 13.16. About 60% of the positive radiative forcing is caused by CO_2 (Table 13.4); this is the reason for global concern about the increasing emission of CO_2 to the atmosphere from anthropogenic sources. The greenhouse gases CH_4, N_2O, and halocarbons, despite their combined concentration of ~2 ppmv, have been significant contributors to radiative forcing because of their high global warming potential. Most of the negative forcing is caused by the aerosol particles in the atmosphere, both directly through reflection and absorption of solar and infrared radiation and indirectly through the changes they cause in cloud properties.

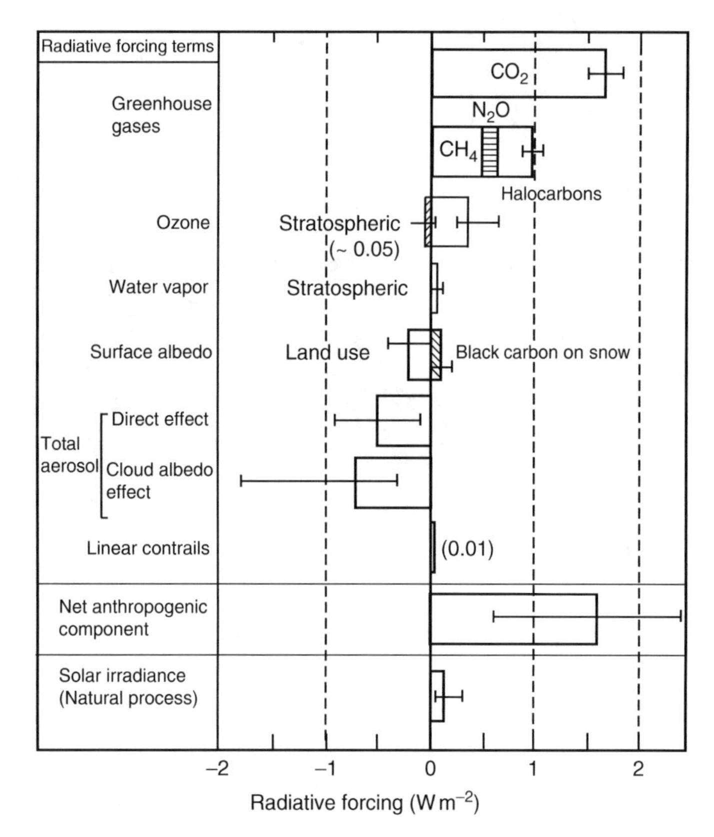

Fig. 13.16 Global average radiative forcing estimates and ranges in 2005, relative to 1750 (the start of the industrial era), for anthropogenic greenhouse gases and other important agents and mechanisms. The thin black line attached to each bar represents the range of uncertainty for the respective value. Positive forcings lead to warming of climate, and negative forcings lead to cooling. Albedo is a measure of the reflectivity of a surface, expressed as the percentage of incoming light that is reflected. Linear contrails refer to linear trails of condensation produced by aircrafts. The only increase in natural forcing of any significance between 1750 and 2005 occurred in solar irradiance. (Adopted from Intergovernmental Panel on Climate Change, 2007.)

The average global temperature has risen by about 0.7°C over the past 100 yr, and the warming trend is likely to continue without substantive changes in global patterns of fossil fuel consumption. It has been estimated that the global average temperature would increase between 1.4 and 5.8°C over the next 100 yr as a result of increases in concentrations of atmospheric CO_2 and other greenhouse gases. Melting of polar ice caps in response to this magnitude of global temperature increase could cause significant rise in average sea level (0.09 to 0.88 m), exposing low-level coastal cities or cities located by tidal rivers such as New Orleans to increasingly frequent and severe floods. In addition, increased concentrations of CO_2 could have a significant impact on the global distribution of vegetation because some plant species respond more favorably to increases in CO_2 than others.

13.4 The hydrosphere

Hydrosphere is the component of the Earth supersystem that includes the various reservoirs of water and ice on or close to the Earth's surface. The global ocean, covering about 71% of Earth's surface to an average depth of 3800 m (12,5000 ft), is the predominant feature of the hydrosphere. The oceans contain ~1370 × 10⁶ km³ of water and account for 97% of Earth's water; groundwater, rivers, lakes, glaciers and ice caps account for the rest of the hydrosphere.

13.4.1 Composition of modern seawater

The oceans constitute a complex geochemical system of amazingly constant bulk composition. The major dissolved constituents, commonly assumed to be present as ionic species (free ions plus ion pairs formed with ions of opposite charge), are almost the same ones encountered in continental waters: Na^+, Ca^{2+}, Mg^{2+}, K^+, Cl^-, SO_4^{2-}, and HCO_3^- (Table 7.1). These seven constituents account for more than 99% of the dissolved

Table 13.6 Selected minor and trace dissolved constituents of seawater (excluding constituents < 1 mm).

Constituent	Concentration		Constituent	Concentration	
	μg kg^{-1} (ppb)	μM (micromoles L^{-1})		μg kg^{-1} (ppb)	μM (micromoles L^{-1})
Br^-	66,000–68,000[a]	840–880	H_4SiO_4	<30–5000	<0.5–180
H_3BO_3	24,000–27,000[a]	400–440	NO_3^-	<60–2400	1–40
Sr^{2+}	7700–8100[a]	88–92	NO_2^-	<4–170	<0.1–4
F^-	1000–1600[a]	50–85	NH_4^+	<2–40	<0.1–2
CO_3^{2-}	3,000–18,000	50–300	Orthophosphate	<10–280	<0.1–3
O_2	320–9600	10–300	Organic carbon	300–2000	–
N_2	9500–19,000	300–600	Organic nitrogen	15–200	–
CO_2	440–3520	10–80	Li^+	180–200	26–27
Ar	360–680	9–17	Rb^+	115–123	1.3–1.4

[a]For a salinity of 35%.
Source of data: compilation by Berner and Berner (1996).

constituents in both seawater and river water. The concentrations of ions as shown on Table 7.1 may vary by about ±10% corresponding to changes in salinity (the total dissolved salt content, which varies from 33‰ to 38‰ in the open ocean), but their ratios vary by less than 1% (Wilson, 1975). On the other hand, the concentration ratios of dissolved minor and trace elements relative to chloride do show significant variations from place to place (Table 13.6), especially the constituents such as phosphates and nitrates, which are nutrients for plankton and reflect the metabolic activities of the organisms that are abundant only in the well-lit, upper few tens or hundreds of meters.

Dissolved constituents in seawater are of three kinds: *conservative*, *recycled*, and *scavenged* (Walther, 2009). The concentrations of conservative constituents (such as Br^-, Cl^-, Mg^{2+}, Na^+) relative to other conservative constituents remain constant. This results in a constant concentration versus depth profile for each conservative element, unless its absolute abundance is modified by a change in the volume of seawater. Such constituents usually have long residence times (> 10^6 years) in the ocean. In contrast, recycled and scavenged constituents are nonconservative and their concentrations vary significantly from place to place. Recycled constituents refer to those (such as organic C, H_4SiO_4, Ca^{2+}, NO_3^-, and PO_4^{3-}) that are recycled through the involvement of organisms. Organisms use these constituents as nutrients and, when they die and start sinking toward the ocean floor, their decomposition and dissolution return the constituents to the seawater. Concentrations of such constituents tend to increase with increasing depth. Scavenged constituents (such as Pb, Mn, Sn) are those that are adsorbed onto the surfaces of solid articles; the concentration of a scavenged constituent is strongly influenced by equilibrium and it typically decreases with increasing depth in the ocean. Broecker and Peng (1982) classified the dissolved chemical constituents in seawater as: *biolimiting constituents*, which are almost totally depleted in surface water due to interaction with organisms (examples: NO_3^-, Zn^{2+}, Cd^{2+}, HPO_4^{-2}, H_4SiO_4, and H_4GeO_4); *biointermediate constituents*, which are partially depleted in surface water (examples: Ca^{2+}, Mg^{2+}, C, etc.); and *biounlimited constituents*, which show no measurable depletion in surface water (examples:Na^+, K^+, etc.).

The composition of modern seawater can be accounted for by steady-state mixing of the two major contributors to ocean chemistry, river water (RW) and mid-ocean ridge (MOR) hydrothermal brines, coupled with precipitation of solid $CaCO_3$ and SiO_2 phases (Spencer and Hardie, 1990). The major compositional gradients in the oceans are vertical, driven by biological cycling processes. The main process is removal of constituents from surface seawater by organisms, and destruction of the organism-produced particles after downward movement. Thus, the most dramatic change occurs in the surface water, and deep water is enriched in most constituents relative to suface water. The only major exception is dissolved oxygen. An example from the northern Pacific Ocean

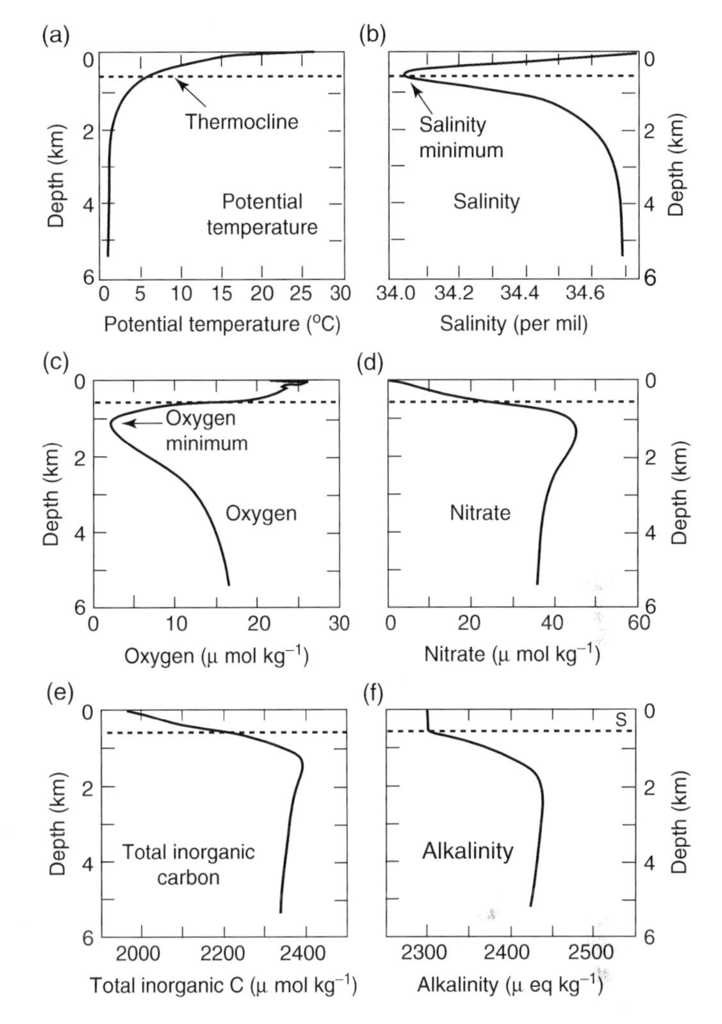

Fig. 13.17 Depth profiles at GEOSECS station 214 in the North Pacific Ocean (32°N, 176°W): (a) potential temperature, (b) salinity, (c) oxygen, (d) nitrate, (e) total dissolved inorganic carbon ($\Sigma CO_2 = CO_2 + HCO_3^- + CO_3^{2-}$), and (f) alkalinity. The dashed line in each figure represents the depth of the salinity minimum. The thermocline is a region (about 1 km thick) of steeply decreasing temperature gradient that separates the surface water (the top 50–300 m that is stirred by the wind and is well mixed) from the deep water. (Modified from Broecker and Peng, 1982, Figure 1-1, p. 4 and Figure 1-2, p. 5.)

(32°N, 176°W) based on data gathered by the Geochemical Ocean Sections program (GEOSECS) during the 1970s is presented in Fig. 13.17. Solar energy is absorbed almost completely within only a few tens of meters of seawater, giving rise to a warm surface zone below which lies a cold ocean, with a nearly constant *potential temperature* of 2°C for most of the deep ocean (Fig. 13.17a). (Water temperatures at depth are generally reported as "potential temperatures," which refer to measured *in situ* temperatures from which the small temperature increase due to compression exerted by the water column has been deducted.) The transitional region between these two zones is called the *thermocline*. As expected, the position of the thermocline varies with distance from the equator and the time

Table 13.7 Influx of dissolved constituents by rivers into the oceans and the time reqauired for these constituents to reach oceanic accounts.

	Concentration				
Dissolved constituent	Average river water[a] (ppm)	Average seawater[b] (ppm)	Mass in the oceans[c] ($\times 10^{20}$ g)	Annual flux from rivers[d] ($\times 10^{14}$ g)	Time for river fluxes to attain oceanic amounts ($\times 10^{6}$ yr)
Cl^-	7.8	19,000	260.3	2.53	102.9
Na^+	6.3	10,500	143.8	2.05	70.2
Mg^{2+}	4.1	1300	17.8	1.33	13.4
SO_4^{2-}	11.2	2650	36.3	3.64	10.0
Ca^{2+}	15.0	400	5.5	4.87	1.1
K^+	2.3	380	5.2	0.75	6.9
HCO_3^-	58.4	140	1.9	18.98	0.1
Br^-	–	65	0.9	–	–
CO_3^{2-}	–	18	0.2	–	–
Sr^{2+}	–	8	0.1	–	–
SiO_2	13.1	6	0.1	4.26	0.02
Organic C	9.6	0.5	0.01	3.12	0.002

[a]Livingstone (1963); [b]Goldberg (1957).
[c]Total mass of ocean water used in the calculation = 1.37×10^{21} kg.
[d]Total river discharge into the oceans per year used in the calculation = 374×10^{14} kg.

of year, and it exercises a major control on the vertical concentration profile of nonconservative dissolved constituents.

The *salinity* of seawater (the total dissolved salts, usually expressed in per mil, ‰) varies from ~30‰, where major rivers discharge into the ocean, to ~40‰ in shallow waters of arid areas, where evaporation is faster than mixing, the average being about 35‰. The vertical salinity profile shows a salinity minimum at a depth of about 600 m, above which the salinity increases toward the surface because of increased evaporation, and a fairly constant salinity in the cold deep ocean (Fig. 13.17b).

Concentrations of dissolved gases in seawater (except for gases such as O_2, CO_2, H_2S, and N_2, which are involved in metabolic processes of organisms) are largely determined by their solubility, which is controlled by the atmospheric partial pressure of the gas. As shown in Fig. 13.17c, the dissolved O_2 content of the surface water in contact with the atmosphere is slightly in excess of its equilibrium value of 0.25 mmol kg^{-1} at 25°C and 1 bar. This is due to the contribution of local photosynthesis to the near-surface O_2 budget. Generally, an oxygen minimum zone develops below the minimum salinity depth because O_2 is consumed during oxidation of organic matter raining down from above. The O_2 level increases in the deep ocean because of the influx of cold oxygenated water that sinks to the bottom of the oceans at the Earth's north and south poles. The present oceans are aerobic because the amount of organic matter is not enough to consume all of the available dissolved O_2. Nitrate is produced as O_2 is consumed, so that the nitrate profile (Fig. 13.17d) has an inverse relationship with that for O_2. The alkalinity of the seawater changes (Fig. 13.17f) mainly because of extraction of Ca by organisms to form $CaCO_3$ skeletal material. Much the $CaCO_3$ skeletal material dissolves after falling to the deep ocean; this accounts for the higher concentration of inorganic carbon ($\Sigma CO_2 = CO_2 + HCO_3^- + CO_3^{2-}$) in the deep ocean (Fig. 13.17e).

The pH of seawater at a given P_{CO_2} is buffered by the complex equilibrium,

$$H_2O + CO_{2(aq)} \Leftrightarrow H_2CO_{3(aq)} \Leftrightarrow H^+ + HCO_3^- \Leftrightarrow 2H^+ + CO_3^{2-} \quad (13.84)$$

The pH varies within the narrow range of ~7.8 to 8.4 because of the relatively small variation in the $HCO_3^-:CO_3^{2-}$ ratio (Skirrow, 1975).

13.4.2 *Mass balance of dissolved constituents in seawater*

Rivers are the major sources of dissolved and particulate materials in the oceans, and it was once thought that the much higher salinity of the oceans (~35‰ compared to ~< 0.2‰ for average river water) is the result of incremental accumulation of the river load over geologic time. However, the compositions of sedimentary rocks and the nature of the fossil record do not indicate any radical change in seawater composition, at least since the Cambrian Period, and possibly over the past 2 Gyr (Holland, 1972, 1984). The only changes for which there is good evidence are a decrease in the concentration of dissolved silica as siliceous organisms developed in the Precambrian, and fluctuations in the oxidation state of the oceans (see section 13.5.2). At the present rate of salt addition by rivers, it would require only a few tens of million years at best to account for the mass of ocean salt. Moreover, when the times required to accumulate the individual dissolved constituents in seawater are calculated, the storage times are found to be only a few million years (Table 13.7). It appears that the seawater composition has maintained some sort of steady-state balance over a long time. Then, the questions we are faced with are: how was river water composition modified to seawater composition that is so different (Table 13.7); and how has the steady-state been maintained over geologic time?

Processes capable of modifying the chemistry of seawater, besides addition of constituents by rivers, fall into six categories (Drever *et al.*, 1988; Berner and Berner, 1996).

(1) Biological processes such as the synthesis of soft tissues or organic matter, secretion of skeletal hard parts, and the bacterial decomposition of organic matter upon death (see Berner and Berner, 1996, for a list of major processes of organic matter decomposition in marine sediments). Such processes exert major controls on the concentration of recycled constituents (Ca^{2+}, HCO_3^-, SO_4^{2-}, H_4SiO_4, CO_2, O_2, NO_3^-, and PO_4^{3-}).
(2) Seafloor volcanics–seawater interaction, both high temperature (200–400°C) reactions at mid-oceanic ridge axes and off-axis low temperature reactions.
(3) Reaction of seawater with detrital solids (silicate minerals and clay minerals) transported from the continents. Such reactions may involve an entire detrital mineral, forming generally a more cation-rich mineral (referred to as "reverse weathering," Mackenzie and Garrels, 1966), or affect only the mineral surface by ion exchange, especially of Ca^{2+} for Na^+. Ion exchange is particularly important for trace elements.
(4) Ocean-to-air transfer of cyclic salts (sea-salt particles blown from the ocean surface onto the land).
(5) Precipitation of evaporite and sulfide minerals, an important mechanism for the removal of Na^+, Ca^{2+}, Cl^-, and SO_4^{2-} from seawater.
(6) Special processes that generally affect only one or two constituents. These include porewater burial of Cl^- and Na^+, and a number of processes unique to the nitrogen and phosphorous cycles (nitrification, denitrification, N_2 fixation, adsorption of phosphate on ferric oxides, and authigenic apatite formation).

Input and output processes for several important dissolved constituents have been discussed by Berner and Berner (1996) and are summarized in Table 13.8. It is likely that these processes have been affected by changes in global parameters such as the rate of tectonic uplift, the rate of seafloor spreading, oxic and anoxic events, and biologic evolution. Perhaps, the apparent constancy of the seawater composition through the geologic record was a consequence of the coupled effects of such changes.

Estimates of inputs and outputs reveal that many of the dissolved constituents in seawater are markedly out of balance (Berner and Berner, 1996). For example, the concentrations of Na^+, Cl^-, and SO_4^{2-} have been increasing in the modern era, although they appear to be in balance on a time scale of millions of years. The present imbalance is due to a lack of evaporite basins for the precipitation of halite (NaCl) and gypsum ($CaSO_4.2H_2O$), and because Cl^- and SO_4^{2-} are being added to seawater through anthropogenic pollution. Over the long term (the past 25 Myr), the input and output of Ca^{2+} have been in balance (within the errors of estimation), but the removal of Ca^{2+} from the present oceans (almost entirely by biogenic $CaCO_3$ deposition) considerably exceeds its input, and this is caused by excessively rapid deposition of $CaCO_3$ on the continental shelves as a result of the rapid post-Ice Age rise in sea level.

13.5 Evolution of the oceans over geologic time

13.5.1 *Origin of the oceans*

Age of the oceans

The occurrence of chemical sediments (such as BIFs) in the 3.8-Gyr-old Isua sequence in west Greenland (Nutman *et al.*, 1997; Whitehouse *et al.*, 1999) is indisputable evidence that liquid water was present on the Earth by that time. A much earlier presence of liquid water on the Earth has been inferred from the ^{18}O-enriched oxygen isotope composition of >3900-Myr-old magmatic zircon crystals recovered from the Jack Hills metasedimentary belt of Western Australia (Mojzsis *et al.*, 2001; Peck *et al.*, 2001; Cavosie *et al.*, 2005). Cavosie *et al.* (2005) reported that preserved magmatic $\delta^{18}O_{V\text{-}SMOW}$ values for individual Jack Hills zircon crystals (4400–3900 Ma) range from 5.3 to 7.3%, increasingly deviating from the mantle range of 5.3 ± 0.3% as the crystals decrease in age from 4400 to 4200 Ma; elevated values up to 6.5% occur as early as 4325 Ma and up to 7.3% by 4200 Ma. A compilation of $\delta^{18}O_{V\text{—}SMOW}$ values for zircons ranging in age from 4.4 Ga to 0.2 Ma by Valley *et al.* (2005) shows that $\delta^{18}O_{V\text{—}SMOW}$ values >7.5% occur only in zircons younger than 2500 Ma.

Oxygen isotope ratios of magmatic zircon crystals reflect the $\delta^{18}O$ of parental magma and, because of the slow rate of oxygen diffusion in zircon in most crustal environments, unaltered zircon crystals preserve the oxygen isotope ratio attained at the time of crystallization. With rare exceptions, the mantle is a remarkable homogeneous oxygen isotope reservoir, and magmatic zircons in high-temperature equilibrium with mantle magmas have average $\delta^{18}O_{V\text{-}SMOW}$ = 5.3 ± 0.6‰ (Valley, 2003), and this mantle value has remained nearly constant (±0.2‰) over the past 4.4 Gyr (Valley *et al.*, 1998; Mojzis *et al.*, 2001). Closed-system fractional crystallization can result in higher $\delta^{18}O_{V\text{-}SMOW}$(whole rock) values, up to ~6.3‰,

Table 13.8 Major processes affecting the concentration of specific dissolved constituents in seawater numbered in the order of approximate decreasing importance, and the status of their mass balance. (Modified from Berner and Berner, 1996.)

			Mass balance	
Constituent	**Input processes**	**Output processes**	**Long-term budget**	**Present-day budget**
Chloride (Cl^-)	1. Rivers (including pollution)	1. Evaporative NaCl deposition 2. Net ocean–air transfer 3. Porewater burial	B	NB
Sodium (Na^+)	1. Rivers (including pollution)	1. Evaporative NaCl deposition 2. Net ocean–air transfer 3. Cation exchange 4. Volcanics–seawater reaction 5. Porewater burial	B	NB
Sulfate (SO_4^{2-})	1. Rivers (including pollution) 2. Polluted rain and dry deposition	1. Evaporative $CaSO_4$ deposition 2. Biogenic pyrite formation 3. Net ocean–air transfer	B	NB
Magnesium (Mg^{2+})	1. Rivers	1. Volcanics–seawater reaction 2. Biogenic Mg-calcite deposition 3. Net ocean–air transfer	B	?
Potassium (K^+)	1. Rivers 2. High-temperature volcanics–seawater reaction	1. Low-temperature volcanics–seawater reaction 2. Fixation on clay minerals near river mouths 3. Net ocean–air transfer	B	?
Calcium (Ca^{2+})	1. Rivers 2. Volcanics–seawater reaction 3. Cation exchange	1. Biogenic $CaCO_3$ deposition 2. Evaporitic $CaSO_4$ deposition	NB	NB
Bicarbonate (HCO_3^-)	1. Rivers (including pollution) 2. Biogenic pyrite formation	1. $CaCO_3$ deposition	NB	NB
Silica (H_4SiO_4)	1. Rivers 2. Volcanics–seawater reaction	1. Biogenic silica precipitation	?	NB
Phosphorus (PO_4^{3-}, HPO_4^{2-}, $H_2PO_4^-$, organic P)	1. Rivers (including pollution) 2. Rain and dry fallout	1. Burial of organic P 2. $CaCO_3$ deposition 3. Adsorption on volcanogenic ferric oxides 4. Phosphorite formation	B	NB
Nitrogen (NO_3^-, NO_2, organic N)	1. N_2 fixation 2. Rivers 3. Rain and dry deposition	1. Denitrification 2. Burial of organic N	NB	?

B = balanced; NB = not balanced.

in more silicic magmas. $\delta^{18}O_{V-SMOW}$ values of magmatic zircons significantly higher than ~6.3‰ have been interpreted to require that the host magma incorporate a significant component of recycled continental crust that had been enriched in ^{18}O by interaction with liquid water under surface or near-surface low-temperature conditions during weathering or diagenesis (Mojzsis *et al.*, 2001; Peck *et al.*, 2001).

The above interpretation is consistent with the presence of a hydrosphere of some extent interacting with the crust by ~4.3 Ga, possibly even by ~4.4 Ga, very soon after the formation of the Earth, but it has been questioned by some authors. Hoskin (2005) proposed that the high $\delta^{18}O$ signature could be the result of localized exchange with a high $\delta^{18}O$ (6–10‰ or higher) fluid. Nemchin *et al.* (2006) determined that younger oscillatory zoned Jack Hill zircon crystals (uniquely interpreted as representing crystallization from a melt), including oscillatory zoned cores in complex grains, have $\delta^{18}O$ values lower than 6.5‰, which are within the range for zircons in high-temperature equilibrium with the normal mantle rocks of $\delta^{18}O = 5.3 \pm 0.6‰$. Nevertheless, as discussed earlier (section 13.2.1), theoretical models indicate that the Earth's atmosphere and oceans had formed within a few tens of millions of years after the end of the main accretion period. The early appearance of oceans, a prerequsite for the origin and evolution of life, has been instrumental in the extensive sequestration of atmospheric CO_2 in the form of sedimentary calcium carbonate minerals (such as aragonite and calcite), thus preventing the development of a runaway greenhouse atmosphere, the likely cause of the lack of a hydrosphere on planet Venus.

Source(s) of water

Where did the Earth's water come from and how did it accumulate to form the oceans? We do not know for sure, but let us discuss some possibilities. Water can be made *in situ* by oxidizing H_2 ($2H_2 + O_2 \Rightarrow 2H_2O$) or organic molecules ($CH_2O + O_2 \Rightarrow CO_2 + H_2O$), but neither works well for the Earth. The former gives a D/H isotopic ratio that is too low, and the latter gives a C/H ratio that is too high (Zahnle, 2006). Besides, there was little free oxygen in the proto-atmosphere for such oxidation reactions on a large scale. Other suggested sources of water in the terrestrial planets include: (i) comets; (ii) hydrous asteroids; (iii) phyllosilicates migrating from the asteroid belt; and (iv) accreting "wet" planetesimals. As discussed below, none of these sources offers a completely satisfactory option.

Comets. For a while, the idea of post-accretion addition of water to an anhydrous proto-Earth from exogenous sources, such as comets and hydrous asteroids, had gained ground because of two reasons. First, it was widely believed that the inner Solar System was too hot for hydrous phases to be thermodynamically stable and be available for accretion. Second, the Earth (and other terrestrial planets) appear to have experienced one or more magma ocean events that would have led to degassing of any existing water acquired in the course of accretion. However, a cometary source of the water (e.g., Deming, 2002; Lunine, 2006) is untenable because of a large discrepancy in hydrogen isotopic ratios. The D : H ratios in the three comets that have been studied so far – Halley, Hyakutake, and Hale-Bopp, all from the Oort Cloud (a spherical shell of cometary nuclei whose average radius is 50,000 AU and which probably has 10^{11} comets) – fall within the range $3.16 \pm 0.34 \times 10^{-4}$, essentially twice the value of 1.56×10^{-4} in Standard Mean Ocean Water (SMOW) (Drake, 2005; Owen and Bar-Nun, 2000). Assuming that these three comets are representative of all comets (a shaky claim as we have not yet sampled any comets from the Kuiper Belt), a late-accreting veneer of volatile-rich material delivered by comets would be limited, variously estimated as < 20% (Abe *et al.*, 2000) to ~35% (Owen and Bar-Nun, 2000) to a maximum of 50% (Drake, 2005) of the water in the Earth's oceans; the rest had to be picked up from some other source, such as the solar nebula, with D : H ratio less than that in SMOW. Moreover, Dauphas and Marty (2002) calculated that comets contributed <0.7–2.7×10^{22} g of material to the accreting Earth, which is less than 1% of the mass of water in one Earth ocean (a minimum mass of 1.4×10^{24} g; Matsui and Abe, 1986a). A limitation of the argument against a cometary source of water is the possibility that we are yet to sample representative comets.

Hydrous asteroids. The mass of the Earth continues to grow at present with the addition of 10^6 to 10^7 kg of meteorites each year, and such additions have certainly occurred throughout geologic time. During accretion, the mass flux of material added to the Earth through meteorites is estimated to have been about 2×10^9 times greater than at present. This contribution dropped five to ten orders of magnitude during the first 100 Myr and has continued to decline ever since, except during a resurgence called the *Late Heavy Bombardment* (LHB), an event of intense bombardment by planetary bodies from a dominantly asteroid reservoir between 3.85 and 3.9 Ga (Koeberl, 2006).

The occurrence of up to ~9 wt% water in carbonaceous chondrites raises the possibility that hydrated asteroids could have been the source of Earth's water. From model computations, Morbidelli *et al.* (2000) proposed that the bulk of the water presently on the Earth was carried by a few planetary embryos, originally formed in the outer asteroid belt and accreted by the Earth at the final stage of its formation. Finally, bombardment of comets from the Uranus–Neptune region and from the Kuiper belt delivered no more than 10% of the present water mass. This is the so-called "late veneer," as postulated for the heterogeneous accretion model (see section 12.2.2). However, the $^{187}Os : ^{186}Os$ ratio of the Earth's primitive upper mantle (~0.1285–0.1305) is significantly higher than that of carbonaceous chondrites (~0.125–0.127), and overlaps that of anhydrous ordinary chondrites (~0.127–0.131), effectively ruling out asteroidal bodies as the source of the "late veneer" (Drake, 2005). It is, however, possible that ordinary chondrites are derived from the metamorphosed outer parts of hydrous asteroids in which case impact of a bulk asteroid could deliver large amounts of water to the accreting Earth.

Phyllosilicates. Ciesla *et al.* (2004) proposed that the source of the Earth's water was phyllosilicate minerals, which formed in the asteroid belt where they were thermodynamically stable, and then migrated into the inner solar system where they were incorporated into the accreting Earth. Since it seems unlikely that phyllosilicates could be decoupled from other minerals and transported into the inner solar system, this hypothesis faces the same objection in terms of Os isotopic composition as stated above for an asteroidal source of water (Drake, 2005).

"Wet" planetesimals. If the accretion disk was too hot for hydrous minerals to form or for phyllosilicate grains migrating from the asteroid belt to survive, the H_2O could have been adsorbed directly onto dust particles prior to their accretion into planetesimals (Drake, 2005). Using a Monte Carlo simulation, Stimpfl *et al.* (2004) have argued that all of the Solar Nebula gases (H_2, He, and H_2O) would interact with the surface of the dust grains, but only H_2O could be adsorbed. Their calculations show that one to three Earth oceans of water could have been adsorbed at temperatures of 500 K to 700 K. Thus, it appears that much or all of the water present on the Earth (as well as on Mars) was indigenous, extracted from the accreting planetesimals. A problem with the "wet" planetisimal hypothesis is that there should be no anhydrous asteroidal bodies in the Solar System, in which case the source of anhydrous chondrites has to be ascribed to metamorphosed, dehydrated parts of hydrous asteroids.

13.5.2 *Oxidation state of the oceans*

As mentioned earlier (section 13.2.5), except for some "oxygen oases" associated with prolific communities of cyanobacteria, the surface oceans were anoxic under an anoxic atmosphere before the Great Oxidation Event (Stage I, Fig. 13.7). The deeper oceans were almost certainly anoxic as evidenced by the occurrence of large Archean-age marine manganese deposits and BIFs. During Stage II (2.45–1.85 Ga), following the GOE, the atmospheric O_2 level had risen to more than 10^{-1} PAL (i.e., > 0.02 atm), perhaps as high as 0.2–0.4 atm (Holland, 2006a). The concentration of O_2 in much of the shallow oceans was probably close to equilibrium with atmospheric O_2, but the deep oceans continued to remain anoxic during much of the interval between 2.45 and 1.8 Ga (Holland, 2006a), as evidenced by the occurrence of large marine manganese deposits of Paleoproterozoic age.

The interval between 1.8 Ga, the end of prolific BIF deposition, and 0.8 Ga, when the BIF deposition resumed, was a period of environmental stability that has been called the "boring billion" (Holland, 2004). Most of the surface oceans were mildly oxygenated, but the oxidation state of the deep oceans during this time interval is controversial. In the model of Kasting *et al.* (1992), the end of BIF deposition is ascribed to the deep oceans becoming oxic. Also, according to Holland (2006a), the absence of major marine manganese deposits in the geologic record of this interval, whereas they are present both before 1.8 Ga (e.g., the Sausar Group, India, ~2.0 Ga) and after 0.8 Ga (e.g., Penganga Group, India, ~800 Ma), suggests that the deep ocean was mildly oxygenated, perhaps because of a very small delivery rate of organic matter to the deep oceans. The limited availability of organic matter is consistent with the absence of marine phosphorite deposits during this entire interval. An alternative interpretation proposed by Canfield (1998) and supported later by other workers (e.g., Anbar and Knoll, 2002; Shen *et al.*, 2003; Poulton *et al.*, 2004) attributes the end of BIF deposition to the development of a euxinic (anoxic and sulfidic) deep ocean. Increased atmospheric O_2 levels enhanced sulfide weathering on land and, thus, the flux of sulfate to the oceans. The increased availability of H_2S through sulfate reduction by organic matter (represented here by CH_2O),

$$2CH_2O + SO_4^{2-} \Rightarrow H_2S + 2HCO_3^- \qquad (13.85)$$

removed the Fe(II) dissolved in ocean water as Fe-sulfide precipitates. The abundance of sediment-hosted lead–zinc sulfide deposits in the Mesoproterozoic may be related to an adequate supply of reduced sulfur in anoxic deep oceans (Goodfellow, 1987). Canfield (1998) argued that atmospheric oxygen did not reach the high levels (within a factor of two or three of PAL) required for the deep oceans to be aerobic until the second oxic event in the Neoproterozoic era, as evidenced by the large increase in $\delta^{34}S$ at 0.6–1.0 Ga (Fig. 11.9), which was also contemporaneous with a significant evolutionary radiation of nonphotosynthetic marine sulfide-oxidizing bacteria (Canfield and Teske, 1996). The model of a globally extensive seafloor anoxia persisting for about 1 Gyr after GOE (often referred to as the "Canfield ocean") is claimed to be consistent with significantly lower $\delta^{97/95}Mo$ values for mid-Proterozoic black shales (the Velkerri Formation, McArthur Basin, Australia; the Wollogorang Formation, northern Australia), compared to values of recent euxinic sediments of the Black Sea (Arnold *et al.*, 2004). Their data may, however, reflect localized euxinic basins rather than a global-scale euxinic ocean.

Compared to the Mesoproterozoic, the Neoproterozoic era (0.8–0.54 Ga) appears to have been a time of pronounced environmental and biological change on a global scale. Major fluctuations in the $\delta^{13}C$ value of marine carbonates were accompanied by several very large glaciation events, the SO_4^{2-} content of seawater rose to values comparable to that of the modern oceans (Horita *et al.*, 2002; Brennan *et al.*, 2004), and the level of atmospheric O_2 probably approached modern values by the time of the biological explosion at the Archean to Proterozoic transition (Holland, 2004).

The Neoproterozoic was a time of abundant marine phosphorite deposits in contrast to their absence from the geologic record over the preceding billion years (1.8–0.8 Ga). The increased downward transport of organic matter required for the formation of phosphorite deposits was apparently sufficient to return anoxic conditions in the deep oceans for at least parts of the Neoproterozoic. The striking association of Rapitan-type BIFs with manganese deposits and glacial deposits suggests that anoxia may have been particularly pronounced during the Neoproterozoic glacial periods (Holland, 2006a).

The surface oceans must have been oxygenated throughout the Phanerozoic for the marine life to fluorish, but the oxidation state of the deeper oceans appears to have fluctuated widely. Oceanic anoxia in shelf and abyssal environments, as evidenced by the extensive deposition of black shales, has been

Box 13.6 Evaporites

Salt deposits believed to have formed by evaporation of surficial brines are called *evaporites*. In a barred basin, deprived of replenishment because of isolation from the ocean, evaporation may increase its salinity to cause precipitation of salts (Hsu, 1972). The sequence of precipited salts, as determined by Usiglio in 1849, would be gypsum (when seawater has evaporated to about 19% of its original volume), followed by halite (when the volume has been reduced to about 10% of the original). Further evaporation would produce a whole series of magnesium and potassium salts known as bitterns (Berner and Berner, 1996).

Marine evaporites preserved in the geologic record appear to be of two different compositional types (Kovalevich *et al.*, 1998): (i) $MgSO_4$-type, characterized by an assemblage of potassium and magnesium sulfates (kieserite, kainite, langbeinite, polyhalite), the type of assemblage that would precipitate out of modern evaporated seawater, although locally chloride minerals (such as sylvite, carnallite, and bischofite) may prevail; and (ii) KCl-type (or $MgSO_4$-poor-type), dominated by chloride minerals (sylvite and carnallite, sometimes bischofite) and free of $MgSO_4$ salts. Gypsum, anhydrite, and halite are common to both types. $MgSO_4$-type evaporites are found in the latest Precambrian (Vendian), Mississippian, Pennsylvanian, Permian, Miocene and Quaternary deposits. KCl-type evaporites are found in Cambrian through Mississippian, and Jurassic through Paleogene deposits (Hardie, 1996, p. 279). Intervals of "calcite seas" are approximately coincident with the deposition of KCl-type evaporites and that of "aragonite seas" with the deposition of $MgSO_4$-type evaporites (Fig. 7.8; Hardie, 1996).

The origin of the KCl-type evaporites is controversial because solubility calculations fail to account for the precipitation of sylvite by evaporation of modern seawater and there is no present-day example of a marine evaporite basin precipitating potassium salts. As the K^+ concentration of the oceans is believed to have been fairly constant over the Phanerozoic (Holland, 2004; Demicco *et al.*, 2005), the evaporation of seawater must be combined with a SO_4^{2-}-depleting process in order to explain the syndepositional precipitation of sylvite instead of Mg-sulfates. In addition to post-depositional changes, proposed mechanisms responsible for sulfate depletion are (Holland *et al.*, 1996): (i) secular changes in the composition of seawater resulting from secular variations in seawater chemistry controlled primarily by fluctuations in the mid-ocean ridge hydrothermal brine flux, which, in turn, have been driven by fluctuations in the production of oceanic crust at mid-ocean ridges (Hardie, 1996; Stanley and Hardie, 1998) because seawater recycling through mid-ocean ridges depletes the water in Mg^{2+}, SO_4^{2-}, and Na^+; and (ii) differences in the extent of dolomitization, followed by anhydrite ($CaSO_4$) or gypsum ($CaSO_4.2H_2O$) precipitation, and in the extent of the albitization of clays during the course of seawater evaporation (Horita *et al.*, 1991, 1996).

a recurring feature of the Earth's history (Holland, 1984; Grotzinger and Knoll, 1995). The result was episodic massive release of H_2S (reaction 13.85), which may have been a contributing factor the in the Late Devonian, Late Permian, and Middle Cretaceous mass extinction events (Kump *et al.*, 2005).

13.5.3 *Composition of the oceans*

Marine evaporite sequences in the rock record and primary fluid (brine) inclusions trapped in primary halite crystals of evaporite beds provide the most important constraints on the composition of ancient seawater (see Box 13.6). Applying our knowledge of the reaction paths for the interpretation of natural evaporite sequences (Harvie *et al.*, 1980), information obtained from brine inclusions in marine evaporite crystals of known age can be used for reconstruction of the contemporary ocean water composition. Evaporation paths for different scenarios can be simulated using, for example, the thermodynamic modeling of the solubility of evaporite minerals by Harvie *et al.* (1984) and the computer program of Sanford and Wood (1991). This approach, however, has some potential limitations: (i) there are no unambiguous criteria for determining marine versus nonmarine origin of evaporites; (ii) the variability among naturally occurring evaporite sequences indicates multiple possible reaction pathways; and (iii) solute concentrations in fluid inclusions may have been significantly modified by diagenesis or brine–rock interactions. Note that there are only a few well-preserved evaporite deposits older than 800 Ma, and so the seawater composition prior to that time is poorly constrained.

In the absence of preserved rocks or fossils, not much can be said about the composition of oceans during the Hadean eon (4.5–3.8 Ga). It is a reasonable guess that Archean oceans contained significant concentrations of Na^+, K^+, Ca^{2+}, Mg^{2+}; Cl^- was certainly the dominant anion in seawater, and the composition and salinity of Archean oceans were not very different from that of present-day seawater (Holland and Kasting, 1992; Holland, 2004). Calcite, aragonite, and dolomite were the dominant carbonate minerals precipitated during the Archean (siderite was a common constituent of BIFs only), implying that the Archean seawater was saturated with respect to $CaCO_3$ and $CaMg(CO_3)_2$. Bedded or massive gypsum/anhydrite formed in evaporitic environments is absent in the Archean (3.8–2.5 Ga) and Paleoproterozoic (2.5–1.8 Ga) record, suggesting a low concentration of SO_4^{2-} in the pre-1.8 Ga oceans. As discussed earlier (section 11.7.5), $\delta^{34}S$ values of marine sedimentary sulfides also indicate low concentration of SO_4^{2-} (<1 mmol kg^{-1}) in the Archean seawater.

During the Proterozoic, seawater continued to be supersaturated with $CaCO_3$, $CaMg(CO_3)_2$, and SiO_2. Its pH certainly rose in response to decreasing atmospheric P_{CO_2}, and it is likely that its salinity did not differ greatly from that of modern seawater (Holland, 1992). The SO_4^{2-} concentration rose to > 1 mmol kg^{-1} H_2O, but stayed well below the present value of 28 mmol kg^{-1} H_2O for most of the Proterozoic. For example, Shen *et al.* (2002) have proposed that the $\delta^{34}S$ values of pyrite in the Paleoproterozoic (1.72–1.73 Ga) sediments of the McArthur Basin (northern Australia) are best explained if SO_4^{2-} concentration in the contemporary seawater was between 0.5 and 2.4 mmol kg^{-1} H_2O.

Bedded marine evaporite deposits of Neoproterozoic age (900–542 Ma) offer the earliest opportunity to reconstruct the composition of contemporary seawater from brine inclusions in marine halite. For example, fluid inclusions in halite from the 545-Ma Ara Group (southern Oman) are Ca-depleted, Mg-rich, Na–K–Mg–Cl–SO_4 brines (saturated with respect to potash minerals and $MgSO_4$), giving the following reconstructed composition (in mmol kg^{-1} H_2O) for the contemporaneous seawater (Brennan *et al.*, 2004): Na^+ = 479; K^+ = 11, Ca^{2+} = 14 (9.5 to 18.5); Mg^{2+} = 52; SO_4^{2-} = 20.5 (16 to 25); Cl^- = 581. These ionic concentrations are similar to those in modern evaporated seawater and fluid-inclusion brines trapped in modern marine halites (Brennan *et al.*, 2004):Na^+ = 485; K^+ = 11, Ca^{2+} = 11; Mg^{2+} = 55; SO_4^{2-} = 29; Cl^- = 565.

A major shift in the ocean chemistry occurred in the Early Cambrian, broadly coincident with the "biologic explosion." Fluid inclusions in marine halite crystals from the Early Cambrian (515 Ma) Angarskaya Formation (eastern Siberia) indicate that, compared to Late Proterozoic, the contemporary seawater had a Ca-depleted, Na–K–Ca–Mg–Cl composition, depleted in SO_4^{2-} and Mg^{2+}, but enriched in Ca^{2+}. The reconstructed Angarskaya concentrations (in mmol kg^{-1} H_2O) – Na^+ = 450; K^+ = 9, Ca^{2+} = 37 (33.5 to 40); Mg^{2+} = 44; SO_4^{2-} = 8 (4.5 to 11); Cl^- = 605 (Brennan *et al.*, 2004) – are similar to those in other early Paleozoic (Cambrian, Silurian, Devonian) fluid inclusions in marine halites (Lowenstein *et al.*, 2001; Horita *et al.*, 2002). The coincidence between the large increase in oceanic Ca^{2+} concentration (an approximately threefold increase compared to the Neoproterozoic) and the Cambrian "biologic explosion" probably reflects the effect of the first major biocalcification event.

Numerous fluid inclusion studies (e.g., Hardie, 1996; Kovalevich *et al.*, 1998; Horita *et al.*, 2002; Lowenstein *et al.*, 2001, 2003; Demicco *et al.*, 2005) have led to the conclusion that the Phanerozoic seawater composition has alternated between two types of brines related to different evaporite associations: a $MgSO_4$-type, rich in SO_4^{2-} but low in Ca^{2+}; and a KCl-type, rich in Ca^{2+} but low in SO_4^{2-}. In marine evaporites, the secular variation has been recorded in the form of time intervals when potash deposits are characterized by $MgSO_4$ salts, such as polyhalite [$K_2MgCa_2(SO_4)_4.2H_2O$], kieserite ($MgSO_4.H_2O$), and langbeinite [$K_2Mg_2(SO_4)_3$], and time intervals when potash deposits are characterized by KCl salts, such as sylvite (KCl), and the absence or paucity of $MgSO_4$ salts (see Box 13.6). It is significant that periods of $MgSO_4$-type evaporites are synchronous with "aragonite seas" and periods of KCl-type evaporites with "calcite seas" (Fig. 7.8). As discussed earlier (section 7.8.4), secular variations in the major ion chemistry of seawater have occurred probably in response to fluctuations in the ratio of MOR brine flux to river water flux as a consequence of changes in the production of oceanic crust at mid-ocean ridges.

13.6 Geosphere–hydrosphere–atmosphere–biosphere interaction: global biogeochemical cycles

The concept of *biogeochemical cycles* (also referred to as *geochemical cycles*) is a comprehensive expression of interactions within the four components of the Earth supersystem – geosphere (solid Earth), hydrosphere, biosphere, and atmosphere – in terms of global reservoirs of materials and transfer of materials from one reservoir to another. Every element has its own unique biogeochemical cycle, and all of these cycles have ben operating simultaneously. Since the composite is too complex to comprehend, it is convenient to consider the cycles for individual elements (or chemical species) separately. Each biogeochemical cycle is generally portrayed in the form of a *box model* in which the reservoirs are represented by boxes (geometric shapes such as reactangles and circles), whereas the processes linking the boxes and the fluxes between them are denoted by arrows. Each cycle comprises two interconnected components: an *exogenic* component, which operates on the surface of the Earth; and an *endogenic* component, which operates in the interior of the Earth. The two components are linked through tectonics.

It is assumed that, in the absence of anthropogenic perturbations, chemical mass transfer among global reservoirs is *cyclic*, which means that the intake of geologically permanent reservoirs is balanced in the long run by their output so that their size and composition remain, within rough limits, constant over long periods of time. Because biogeochemical processes appear to be cyclic, superimposed on the slow secular evolution of the Earth, the modeling of environmental systems boils down to the modeling of cyclic processes or parts of them. The factors involved in the construction of geochemical cycling models include: (i) identification of sources and sinks of the elements; (ii) definition of the boundaries of the reservoirs; (iii) prediction and evaluation of transport paths; (iv) quantitative knowledge of the masses of the substances in reservoirs, and fluxes into and out of reservoirs (often the most difficult parameters to evaluate); and (v) appropriate mathematical models relating the various variables. For the sake of simplicity, however, we will restrict our attention to qualitative aspects of the biogeochemical cycles of carbon (Fig. 13.18), oxygen (Fig. 13.19), nitrogen (Fig. 13.20), sulfur (Fig. 13.21) and phosphorus (Fig. 13.22), which are of prime importance for the biosphere, the most active geochemical realm at present, and the Earth's surface. Quantitative evaluations of these cycles have been discussed, for example, in Gregor *et al.* (1988), and Berner and Berner (1996).

Comprehensive treatments of these biogeochemical cycles can be found in articles by Houghton (2005), Petsch (2005), Galloway (2005), Brimblecome (2005), and Ruttenberg (2005), all included in the *Treatise on Geochemistry*, vol. 8, 2005, edited by W.H. Schlesinger.

13.6.1 The carbon cycle

The carbon cycle (Fig. 13.18) is important for three reasons (Houghton, 2005): (i) carbon is one of the basic elements of the structure of all life on the Earth, making up ~50% of the dry weight of all living things; (ii) the cycling of carbon approximates

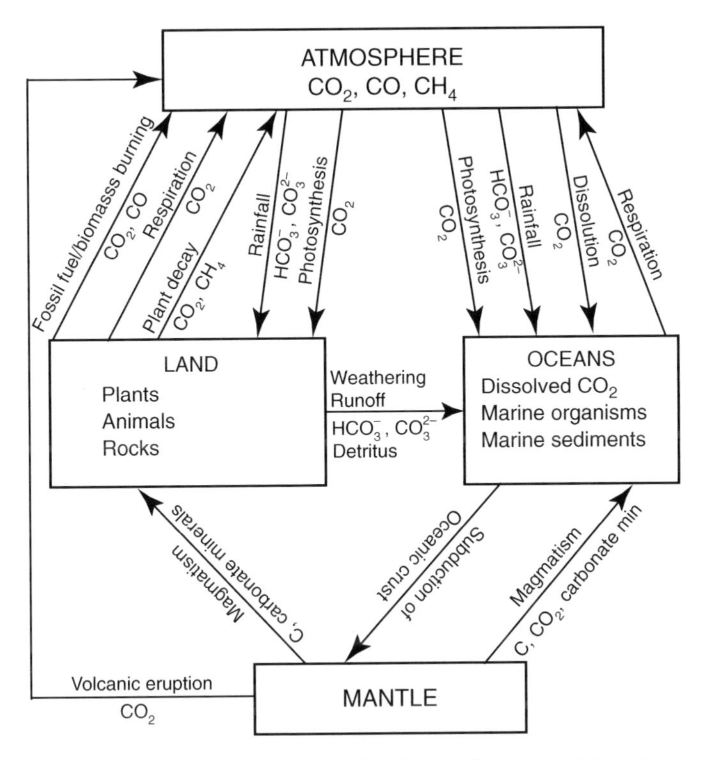

Fig. 13.18 Simplified biogeochemical cycle of carbon. Some CO_2 and CH_4 are returned to the atmosphere via plant decay. The cycle would be in balance if it were not for human interference such as burning of fossil fuels, cement manufacturing, and deforestation.

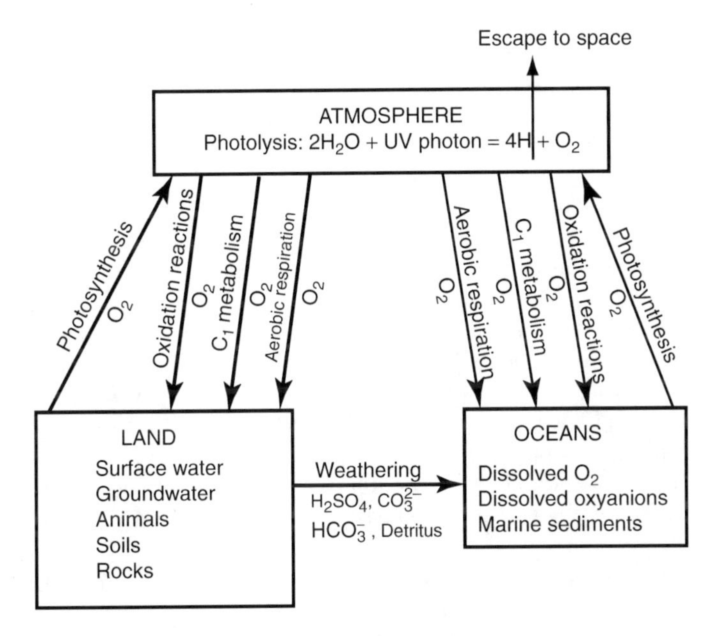

Fig. 13.19 Simplified exogenic biogeochemical cycle of oxygen.

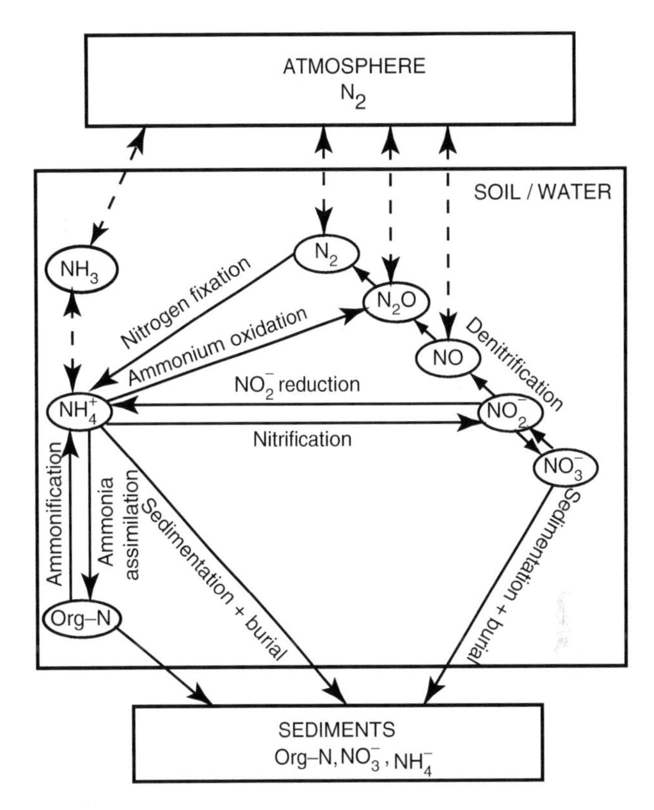

Fig. 13.20 Simplified exogenic biogeochemical cycle of nitrogen. The natural nitrogen cycle has been affected considerably by anthropogenic input of nitrogen gases from various sources.

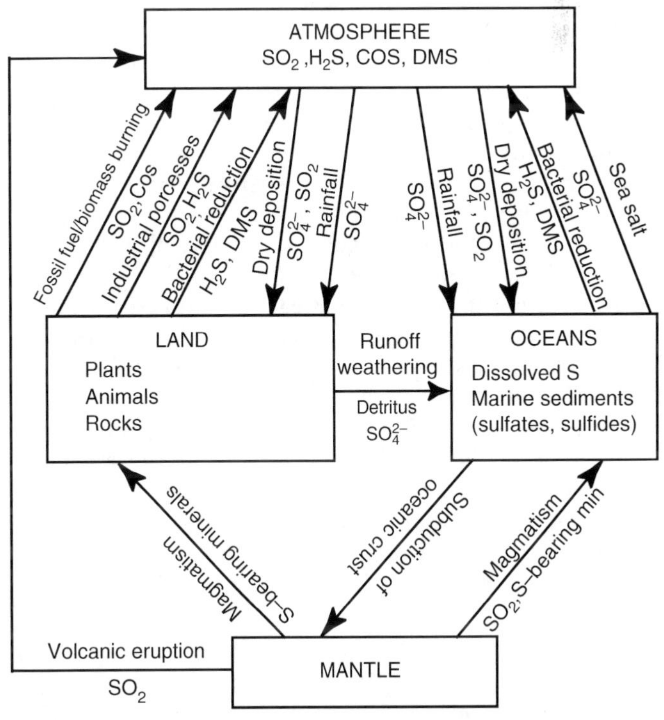

Fig. 13.21 Simplified biogeochemical cycle of sulfur. DMS = Dimethyl sulfide, $(CH_3)_2S$.

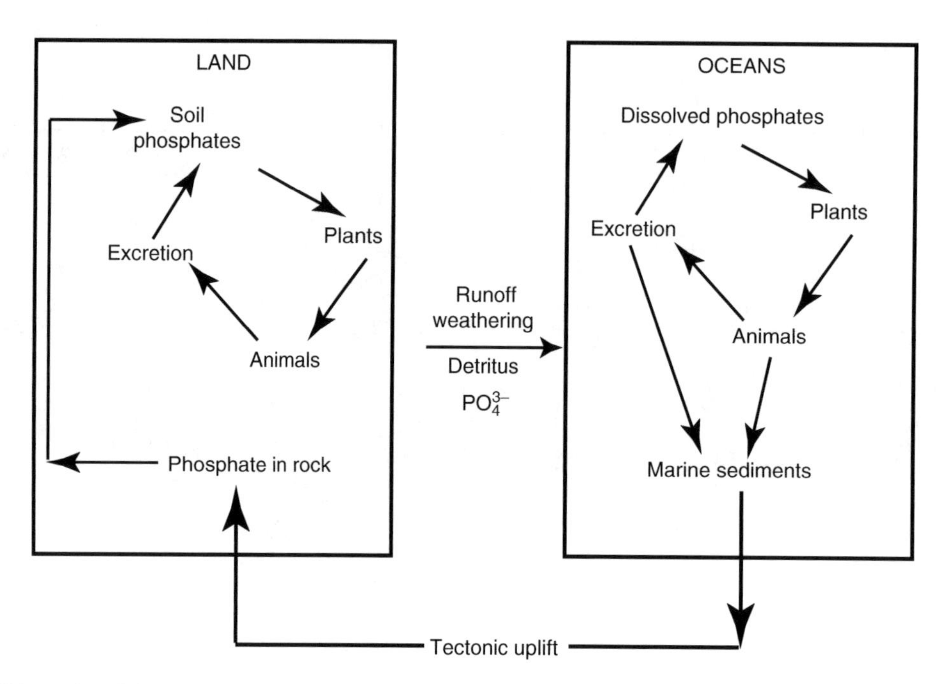

Fig. 13.22 Simplified biogeochemical cycle of phosphorus.

the transfers of energy around the Earth, the metabolism of natural, human, and industrial systems; and (iii) CO_2 and CH_4 are two of the most important greenhouse gases.

The Earth contains ~10^{23} g of carbon, most of it sequestered in carbonate rocks (0.65×10^{23} g) and buried organic matter (0.156×10^{23} g) such as kerogen, oil and natural gas, and coal (Schlesinger, 1997). Only ~0.1% of the carbon (40×10^{18} g) in the Earth's upper crust is cycled throughout active surface reservoirs. The main exogenic reservoirs are the oceans (dissolved organic and inorganic carbon; and marine sediments, mostly in the form of $CaCO_3$), the biosphere (organic matter in plants and animals, with carbon comprising 50% of all living tisues), the land (carbonate rocks and soil carbonate), and the atmosphere (mainly as CO_2 and CH_4, which have played a crucial role in the development of lifeforms and the alteration of Earth's surface environment throughout Earth's history).

Carbon fluxes between the atmosphere and the oceans and between the atmosphere and the continental biosphere are very large. The main processes that control the transfer of carbon among these reservoirs are biological: oxygenic photosynthesis and aerobic respiration by terrestrial plants and phytoplankton (microscopic marine plants that form the base of the marine foodchain) in the oceans (see Chapter 8):

$$\textit{Photosynthesis}: CO_{2\,(g)} + H_2O_{(l)} + \text{sunlight} \Rightarrow CH_2O + O_{2\,(g)} \quad (8.63)$$

$$\textit{Respiration}: CH_2O + O_{2(g)} \Rightarrow CO_{2(g)} + H_2O_{(l)} + \text{energy} \quad (8.66)$$

where CH_2O stands for organic compounds such as carbohydrates. Almost all multicellular life on Earth depends on the production of carbohydrates (the most abundant group of organic compounds that includes sugars, starches, and cellulose) by photosynthesis and breakdown of those carbohydrates by respiration to generate the energy needed for movement, growth, and reproduction of organisms. Most, but not all, of the CO_2 extracted from the atmosphere through photosynthesis is returned to the atmosphere through cell respiration and decay of plants. On an annual basis, the amount of carbon involved in photosynthesis and respiration is about 1000 times greater than the amount that moves through the geological component. Nonbiological processes that release CO_2 to the atmosphere include burning of fossil fuels and biomass, cement manufacturing, decay of organic matter, and breakdown of $CaCO_3$. During the past 150 years (~1850–2000) the atmosphere has registered a 30% increase in the amount of carbon, mostly from the combustion of fossil fuels.

The main geochemical processes involved in the cycling of carbon include weathering of rocks exposed by erosion and tectonic uplift, precipitation of minerals, burial and subduction of rocks and sediments at convergent plate boundaries, and volcanism. As discussed earlier, weathering of rocks containing carbonate and silicate minerals is facilitated by carbonic acid (H_2CO_3) formed by reaction of water with atmospheric CO_2 ($CO_{2\,(g)} + H_2O \Leftrightarrow H_2CO_{3\,(aq)}$; reaction 7.36) and by oxidation of organic matter ($CH_2O + O_2 \Rightarrow CO_2 + H_2O \Rightarrow H_2CO_{3\,(aq)}$; reaction 7.85). Dissolved substances produced by weathering reactions such as

$$\underset{\text{calcite}}{CaCO_3} + H_2O + CO_2 \Rightarrow Ca^{2+} + 2HCO_3^- \quad (13.86)$$

$$CaMg(CO_3)_2 + 2H_2O + 2CO_2 \Rightarrow Ca^{2+} + Mg^{2+} + 4HCO_3^- \quad (13.87)$$
dolomite

$$CaSiO_3 + 3H_2O + 2CO_2 \Rightarrow Ca^{2+} + 2HCO_3^- + H_4SiO_4 \quad (13.88)$$
wollastonite

$$MgSiO_3 + 3H_2O + 2CO_2 \Rightarrow Mg^{2+} + 2HCO_3^- + H_4SiO_4 \quad (13.89)$$
enstatite

are transported to the oceans by rivers, and precipitated as carbonates and silica by reactions such as

$$Ca^{2+} + 2HCO_3^- + Si(OH)_4 \Rightarrow CaCO_3 \text{ (calcite)} + SiO_2 \text{ (silica)} + 3H_2O + CO_2 \quad (13.90)$$

$$Mg^{2+} + 2HCO_3^- + Si(OH)_4 \Rightarrow MgCO_3 \text{ (magnesite)} + SiO_2 \text{ (silica)} + 3H_2O + CO_2 \quad (13.91)$$

The precipitation of calcite (and aragonite) and silica is predominantly biological in the sense that they are precipitated by reef and planktonic organisms to build their skeletons (see sections 7.8.4 and 7.8.5 for abiological and biological precipitation of calcium carbonate). When marine animals and plants die, their remains settle toward the seafloor. Much of this organic debris undergoes decomposition by bacteria and dissolution during their downward journey, replenishing the oceanwater in dissolved CO_2, calcium, silica, and nutrients. The CO_2 is stored in the deeper waters of the oceans for hundreds to a thousand or so years before being returned to the atmosphere by the upwelling of deep ocean waters.

The organic debris that escapes degradation during downward transit accumulates as part of the seafloor sediments. Burial eventually transforms calcite-rich sediments into limestone and the highly altered, finely disseminated organic matter (principally preserved remains of microscopic plant material), called *kerogen*, into oil and gas under appropriate temperature–pressure conditions. Burial of terrestrial plant material, commonly in swampy environments, leads to the formation of different ranks of coal (peat, lignite, bituminous coal, and anthracite) depending on the temperature–pressure conditions experienced by the buried material. The carbon incorporated in sedimentary rocks may be transformed into anthracite or graphite by thermal metamorphism, whereas limestones may recrystalize to form marble.

The carbon cycle continues as seafloor spreading leads to subduction of the seafloor under continental margins. The carbon-bearing material in the subducting slab eventually melts and the carbon is incorporated into magma. Part of the carbon enters crustal igneous rocks formed from this magma and some carbon is released through volcanic eruptions to the atmosphere mostly as CO_2, where the carbon combines wih water and returns to the Earth's surface as H_2CO_3 dissolved in rainwater ($H_2O + CO_2 \Rightarrow H_2CO_3 \Rightarrow H^+ + HCO_3^-$; $HCO_3^- \Rightarrow H^+ + CO_3^{2-}$).

The widespread occurrence of marine carbonate deposits suggests that dissolution of CO_2 in seawater has played an important role in the removal of CO_2 from the Earth's atmosphere. By analogy, the CO_2-rich atmosphere of Mars may be due, at least in part, to the absence of liquid water on the surface. In contrast, the CO_2 atmosphere of Venus may reflect an approximate equilibrium between atmospheric CO_2 and limestone deposits on the surface:

$$CaCO_3 \Leftrightarrow CaO + CO_2 \quad (13.92)$$

13.6.2 *The oxygen cycle*

A unique feature of the Earth among celestial bodies is an abundance of free molecular oxygen (O_2) in its atmosphere. Living organisms depend on oxygen for breathing and for producing energy. Oxygen is also essential for combustion (an oxidation reaction) and it is the ingredient for maintaining the stratospheric ozone shield that provides protection from UV rays. Oxygen is a very reactive element and it readily reacts with most elements of the Periodic Table. Yet, the atmospheric O_2 concenration has been maintained at a fairly constant value (~21% or a partial pressure of oxygen = 0.21 bar) for a long time because of a balance between its production and consumption.

As shown in Fig. 13.19, the main source of atmospheric oxygen is oxygenic photosynthesis by land plants and phytoplankton of the oceans, which produces sugars and oxygen from CO_2 and water in the presence of sunlight:

$$CO_{2(g)} + H_2O_{(l)} + \text{sunlight} \Rightarrow CH_2O \text{ (carbohydrate)} + O_{2(g)} \quad (8.63)$$

A small amount of O_2 is also produced by photolysis of water vapor (reactions 13.28 and 13.30), the net reaction for which can be represented as

$$2H_2O_{(g)} + \text{UV photon} \Rightarrow 4H_{(g)}\uparrow + O_{2\,(g)} \quad (13.93)$$

Atomic oxygen formed by photodissociation of molecular oxygen by UV radiation is the major form of oxygen above ~120 km altitude.

The main processes that result in relatively large fluxes of O_2 from the atmosphere to the Earth's surface include: (i) aerobic respiration, the oxidation of organic substrates with oxygen to yield chemical energy ($CH_2O_{(aq)} + O_{2(g)} \Rightarrow CO_{2(g)} + H_2O_{(l)}$; reaction 8.66); (ii) biologically mediated oxidative metabolism of C_1 compounds such as methane, methanol, and formaldehyde, which are common in soils and sediments as the products of anaerobic fermenation reactions (Petsch, 2005); and (iii) various oxidation reactions. Examples of oxidation reactions that consume O_2 include: combustion of fossil fuels and biomass;

oxidation of sulfide to sulfate (e.g., $HS^- + 2O_{2(g)} \Rightarrow SO_4^{2-} + H^+$; reaction 8.68), oxidation of Fe(II) minerals to Fe(III) minerals (e.g., $2Fe^{2+} CO_{3(s)} + O_{2(g)} + 2H_2O \Rightarrow 2Fe^{3+} OOH + 2HCO_3^-$; reaction 8.67); and oxidation of reduced volcanic gases such as H_2S and CH_4 ($H_2S_{(g)} + O_{2\,(g)} \Rightarrow SO_{2\,(g)} + H_{2\,(g)}$; $CH_{4\,(g)} + O_{2\,(g)} \Rightarrow CO_{2\,(g)} + 2H_{2\,(g)}$). Aerobic respiration is the most important process of oxygen consumption on the Earth.

A small amount of the Earth's oxygen occurs dissolved in the oceans. Oxygen solubility depends on the temperature, pressure, and salinity. At 1 bar pressure and ~35‰ salinity, the solubility decreases from 11.2 mg L^{-1} (= 349 μmol L^{-1}) at 0°C to 4.6 mg L^{-1} (= 146 μmol L^{-1}) at 50°C. By far the largest reservoir of oxygen is the crust–mantle system, which holds ~99.5% of the Earth's oxygen (compared to only 0.49% in the atmosphere and 0.01% in the biosphere), and is recycled through plate tectonics.

13.6.3 *The nitrogen cycle*

Nitrogen is an essential constituent of DNA, RNA, and proteins, the building blocks of life. All organisms, therefore, require nitrogen to live and grow. The availability of nitrogen to sustain life should not be a problem, considering that it constitiutes ~79% (by volume) of the Earth's atmosphere, 2.82×10^{20} moles of N_2 gas and much smaller amounts ($\approx 10^{12}$ moles) of six other molecular species (NH_3, NH, N_2O, NO, NO_2, and HNO_3). The N_2 gas is inert and therefore cannot be utilized directly by plants or animals, except by a few primitive bacteria that are capable of converting N_2 gas to NH_3 or NH_4^+, which is the only way organisms can utilize nitrogen directly from the atmosphere. The other significant reservoirs of nitrogen include sediments and sedimentary rocks and, to a lesser extent, the biosphere and dissolved organic compounds in the oceans. The oceans contain a relatively small amount of nitrogen as dissolved N_2 because of its low solubility in water (~0.03 g kg^{-1} at 0°C to ~0.01 g kg^{-1} at 60°C).

Table 13.9 Oxidation states of nitrogen.

Oxidaton state	Examples
−3	NH_3 (ammonia), NH_4^+ (ammonium ion), NH_2^- (amide ion)
−2	N_2H_4 (hydrazine)
−1-	NH_2OH (hydroxylamine), NH_2Cl (chloramine)
0	N_2 (nitrogen gas)
+1	N_2O (dinitrogen oxide or nitrus oxide), $H_2N_2O_2$ (hyponitrus acid)
+2	NO (nitrogen monoxide or nitric oxide)
+3	N_2O_3 (dinitrogen trioxide), HNO_2 (nitrus acid), NO_2^- (nitrite ion)
+4	NO_2 (nitrogen dioxide), N_2O_4 (dinitrogen tetroxide)
+5	N_2O_5 (dinitrogen pentoxie), HNO_3 (nitric acid), NO_3^- (nitrate ion)

The nitrogen cycle (Fig. 13.20) is a little more complicated than the oxygen cycle because nitrogen exists in both organic and inorganic forms as well as in many oxidation states (Table 13.9). Moreover, nitrogen is not involved to any appreciable extent in mineral dissolution and precipitation; it is strictly a biogenic element that is cycled through the action of microorganisms. Thus, microbial processes have a strong, in many cases controlling, influence on the biogeochemistry of nitrogen. Three classes of microorganisms are important in this connection: those that convert N_2 to NH_4^+; those that convert NH_4^+ to NO_3^-; and those that convet NO_3^- back to N_2, thus completing the cycle.

Nitrogen is cycled through the biosphere, atmosphere, and geosphere by six main natural processes, which transform nitrogen from one chemical form to another (Galloway, 2005): (i) nitrogen fixation ($N_2 \Rightarrow NH_3$); (ii) ammonia assimilation ($NH_3 \Rightarrow$ organic N); (iii) nitrification ($NH_4^+ \Rightarrow NO_3^-$); (iv) assimilatory nitrate reduction ($NO_3^- \Rightarrow NO_2^- \Rightarrow NH_4^+$), (v) ammonification (organic N $\Rightarrow NH_4^+$); and (vi) denitrification ($NO_3^- \Rightarrow N_2$).

Nitrogen fixation is the process by which N_2 is converted to any nitrogen compound in which nitrogen has a nonzero oxidation state. The most common is the reduction of N_2 to NH_3 mediated by certain bacteria (e.g., *Rhizobium*, which live in the root nodules of legumes such as peas and beans, and cyanobacteria):

$$2N_{2(g)} + 6H_2O \Rightarrow 4NH_3 \text{ (wihin the organism)} + 3O_{2(g)} \quad (13.94)$$

The NH_3 is released following the death of the organism, and subsequent hydrolysis of NH_3 produces NH_4^+, which can be assimilated by plants:

$$4NH_3 + 4H_2O \Rightarrow 4NH_4^+ + 4OH^- \quad (13.95)$$

Over geologic history, most reactive nitrogen has been formed by biological mediation. In the latter half of the 20th century, however, the Haber–Bosch process for manufacturing NH_3, which is used to make fertilizers and explosives, has been the dominant process on continents for creating reactive nitrogen. This process uses N_2 from the atmosphere and H_2 from fossil fuels (usually natural gas), and operates at high temperature and pressure with a metallic catalyst (see section 9.4.2):

$$N_{2(g)} + 3H_{2(g)} \Rightarrow 2NH_{3(g)} \quad (13.96)$$

Another reactive form of nitrogen, NO gas, is produced by high-energy natural events such as lightning or high-temperature combustion of fossil fuels (as in automobile engines and thermal power plants):

$$N_{2(g)} + O_{2(g)} + \text{electrical /fossil fuel energy} \Rightarrow 2NO_{(g)} \quad (13.97)$$

Ammonia assimilation is the uptake of NH_3 or NH_4^+ by an organism into its biomass in the form of an organic nitrogen compound. All nitrogen obtained by animals can be traced back to the eating of plants at some stage of the foodchain. We may not realize it, but the nitrogen in our food was fixed initially by nitrogen-fixing bacteria.

Nitrification is the aerobic process by which microorganisms oxidize NH_3 to NO_3^- (nitrate ion) and derive energy from the reaction. Actually, nitrification is a combination of two bacterial processes: oxidation of NH_4^+ to NO_2^- (nitrite ion) by one group of bacteria (e.g., *Nitrosomonas*); and oxidation of NO_2^- to NO_3^- by another group of bacteria (e.g., *Nitrobacter*):

$$2NH_4^+ + 3O_2 \Rightarrow 2NO_2^- + 2H_2O + 4H^+ \quad (13.98)$$

$$2NO_2^- + O_2 \Rightarrow 2NO_3^- \quad (13.99)$$

The conversion of nitrites to nitrates is beneficial for plants because accumulated nitrites are toxic to plant life.

Assimilatory nitrate reduction is the uptake of NO_3^- by an organism and incorporation as biomass through nitrate reduction. It is an important process for input of nitrogen for many plants and organisms.

Ammonification is part of the general process of decomposition that converts reduced organic nitrogen ($R\text{–}NH_2$) to reduced inorganic nitrogen (NH_4^+) by the action of microorganisms.

Denitrification is the reduction of nitrates back to any gaseous nitrogen species, largely N_2:

$$5CH_2O + 4H^+ + 4NO_3^- \Rightarrow 2N_{2(g)} + 5CO_{2(g)} + 7H_2O \quad (13.100)$$

It is an anaerobic process that is carried out mainly in sediments (and occasionally in the water column) by bacterial species such as *Pseudomonas* and *Clostridium*, and the conversion takes place in the following sequence: $NO_3^- \Rightarrow NO_2^- \Rightarrow NO \Rightarrow N_2O \Rightarrow N_2$. Being a gas, N_2 is likely to be lost readily to the atmosphere. Denitrification is the only nitrogen transformation process that removes nitrogen from ecosystems (essentially irreversibly), thereby approximately balancing the nitrogen fixed by the nitrogen fixers described above.

The natural exogenic nitrogen cycle has been affected considerably by anthropogenic input of nitrogen gases from various sources: ammonium-based chemical fertilizers, automobile exhausts, industrial plants, and sewage facilities such as septic tanks and holding tanks. The various forms of nitrogen in our ecosystems contribute to a number of environmental problems such as photochemical smog and acid rain (as discussed earlier), and creation and growth of eutrophic lakes and oceanic dead zones through algal bloom-induced hypoxia (reduced dissolved oxygen content in a body of water).

13.6.4 The sulfur cycle

Sulfur is a very reactive element and combines directly with many elements, especially with hydrogen and oxygen forming H_2S and many oxides of sulfur, the most important of which are SO_2 and SO_3. Sulfur is also a biologically active element and is cycled readily through the foodchain. Some of the earliest organisms on Earth utilized sulfur compounds, particularly through anoxygenic photosynthesis.

A simplified version of the sulfur biogeochemical cycle is presented in Fig. 13.21. Sulfur from the mantle enters the hydrosphere as a result of alteration/weathering of mafic and ultramafic rocks in the oceans and on land, and by emission of SO_2 during volcanic eruptions. In the present oxic state of the atmosphere, any H_2S in the volcanic emissions is rapidly oxidized to SO_2. Volcanic emissions also release SO_2 directly to the atmosphere. Other sources of sulfur in the atmosphere are: biologically produced H_2S, DMS [dimethyl sulfide, $(CH_3)_2S$] and other organic sulfur compounds (e.g., carbonyl sulfide, COS; carbon disulfide, CS_2), which are readily oxidized to SO_2; SO_2 from burning of fossil fuels and biomass, and industrial processes (e.g., roasting of sulfide ores); and SO_4^{2-} from sea-salt aerosol. Most of the SO_4^{2-} flux is redeposited in the ocean as precipitation and dry-fall. The SO_2 combines with H_2O to form H_2SO_4, which leads to acid deposition (see section 13.3.3) on land and in oceans.

Weathering of sulfur-bearing sediments releases the stored sulfur, which is then oxidized into SO_4^{2-}. The SO_4^{2-} is taken up by plants and microorganisms, and converted into organic forms. Animals consume these organic forms through their food, thereby cycling the sulfur through the foodchain. When organisms die and decay, some of the sulfur is again released as SO_4^{2-} and mostly transported as runoff, although some enters the tissues of microorganisms.

The fate of SO_4^{2-} that enters the oceans and lakes from the atmosphere and as runoff from the land is dominated by anaerobic bacteria (such as *Desulfovibrio desulfuricans*) that reduce SO_4^{2-} to H_2S (reaction 8.71). These bacteria can tolerate a large range of pH and salinity conditions, and they occur widely in marine and lacustrine sediments as well as in the overlying water column if it is sufficiently anoxic (Eh < +0.100 V). The H_2S reacts with metals forming sulfides, predominantly iron sulfide (FeS) that accumulates on the seafloor with marine sediments and eventually recrystallizes to pyrite (FeS_2). In addition, sulfate salts (evaporites), such as anhydrite ($CaSO_4$) and gypsum ($CaSO_4.2H_2O$), are precipated episodically in marine and nonmarine evaporite basins. At the present time, sulfate input into the oceans exceeds the output by a large amount, the imbalance being due to excessive inputs of anthropogenic sulfur and the absence of large modern-day evaporite basins. The sulfur incorporated in sedimentary rocks is recycled into the hydrosphere in the form of dissolved SO_4^{2-} when the sedimentary rocks are exposed to chemical weathering.

13.6.5 *The phosphorus cycle*

Phosphorus is an essential element for all organisms and is a key player in fundamental biochemical reacions. It is a critical component of ATP (adenosine triphosphate, $C_{10}H_6N_5O_{13}P_3$), the cellular energy carrier. Adenosine triphosphate contains a large amount of energy stored in its high-energy phosphate bonds; the energy is released when ATP breaks down into ADP (adenosine diphosphate) and utilized for many metabolic processes. Phosphorus, like calcium, is an essential ingradient for vertebrates; in the human body 80% of the phosphorus is found in teeth and bones as organophosphates and in calcium phosphates such as hydrooxyapatite, $Ca_5(PO_4)_3(OH)$, and fluorapatite, $Ca_5(PO_4)_3F$. Organisms cannot directly assimilate phosphorus stored in rocks and soils. Conversion to orthophosphate (PO_4^{3-}), which can be assimilated directly, occurs through geochemical and biochemical reactions at various stages in the global phosphrous cycle (Fig. 13.22).

Very little phosphorus circulates in the atmosphere because at the Earth's normal temperatures and pressures, phosphorus and its compounds are not gases. Thus, the atmosphere is not a reservoir of phosphorus). Sedimentary rocks constitute the largest reservoir of phosphorus. Weathering of continental bedrock results in the dissolution of phosphorus mineals such as apatite ($Ca_{10}(PO_4)_6(OH, F, Cl)_2$) and release of phosphorus as PO_4^{3-} that can be assimilated by organisms. Plants absorb phosphates from soil or water, then bind the phosphate into organic compounds. The phosphorus is passed into animals through successive steps in the foodchain, and eventually returned to the soil or water through the excretion of urine and feces, as well as from the decomposition of plants and animals after death. Much of the phosphates delivered to the oceans (or inland lakes) through runoff, however, are pollutants derived from anthropogenic sources such as mining, leakage of sewage, and use of phosphates for fertilizers, detergents, soft drinks, etc. The phosphorus is transported as soluble PO_4^{3-}, as an adsorbed phase on suspended particles of soil clays and ferric oxides, and as detrital primary apatite and particulate organic compounds. The solubilization/desorption of some of this phosphorus adds to the phosphorus budget of the oceans.

The sole means of phosphorus removal from the ocean or lake water is burial with marine sediments that accumulate on the ocean floor. The phosphatic materials in the sediments include (Berner and Berner, 1996): organic phosphorus compounds (carried to the oceans by rivers, as well as formed in the oceans by photosynthesis) that survive bacterial decay; finely dispersed authigenic calcium phosphate in marine muds of continental margins; phosphate adsorbed on hydrous ferric oxides and incorporated in $CaCO_3$ during growth of calcareous shells. When sediments are stirred up by upwelling, a lot of the phosphorus returns to the surface water and reenters the phosphorus cycle. As is the case with nitrogen, phosphorus undergoes many transfers between deep and surface waters before becoming buried permanently in sediments and locked in the resulting sedimentary rocks. This phosphorus would be released only when the rocks are brought to the surface by tectonic uplift and subjected to weathering.

13.7 Summary

1. On the basis of temperature distribution, the atmosphere can be divided into several layers: troposphere, stratosphere, mesosphere, and thermosphere. The pressure decreases exponentially with altitude, by about a factor of 10 for every 16 km. The stratosphere contains the "ozone layer," which provides a protective shield from the Sun's ultraviolet radiation. The recently discovered depletion of the ozone layer has been caused by the introduction of CFC compounds into the atmosphere from anthropogenic sources.
2. Photochemical reactions play a key role in many aspects of atmospheric chemistry, including the production and destruction of ozone.
3. Three gases – N_2, O_2, and Ar – comprise about 99.9% (by volume) of the atmosphere. Earth is the only planet with abundant atmospheric O_2.
4. The Earth probably had a primary atmosphere composed of H_2 and He captured gravitationally from the solar nebula, but this atmosphere was replaced within ~100 Myr by a secondary atmosphere formed by shock release and vaporization of chemically bound volatiles contained in the accreting planetesimals. Water vapor contained in this steamy atmosphere was the ultimate source of the water in the oceans. A liquid-water ocean similar to modern oceans came into existence at ~4.5 Ga, after the Moon-forming giant impact, by outgassing of a magma ocean when it solidified.
5. The atmosphere during the Hadean (4.5–3.8 Ga) and Archean (3.8–2.5 Ga) was essentially devoid of free (molecular) O_2. Accumulation of free O_2 in the atmosphere had to await the advent of oxygenic photosynthesis, when cyanobacteria appeared on the Earth.
6. Many lines of evidence (BIFs, detrital uraninite, MIF-S, paleosols, etc.) indicate that a dramatic rise in atmospheric P_{O_2}, from $< 10^{-5}$ PAL to $> 10^{-2}$ PAL, occurred during the early Proterozoic, around 2.3 Ga (2.4–2.2 Ga). This is commonly referred to as the Great Oxidation Event (GOE). Another oxidation event occurred at ~0.8–0.7 Ga. Atmospheric O_2 probably reached close to modern values by the Cambrian biologic explosion. The most conspicuous feature of the Phanerozoic scenario was the pronounced excursion, reaching levels as high as 35% O_2, during the Carboniferous–Permian and the rapid decline at the end of Permian.
7. From an environmental perspective, the principal agents of air pollution are: anthropogenic $SO_{2(g)}$ (sulfurous smog, acid deposition); $NO_{x(g)}$ and ground-level ozone

(photochemical smog); $CO_{2(g)}$ and $CH_{4(g)}$ (global warming); and CFC compounds (stratospheric ozone depletion).

8. Theoretical models indicate that the Earth's atmosphere and oceans had formed within a few tens of millions of years after the end of the main accretion period. Much or all of the water present on Earth (as well as on Mars) was indigenous, extracted from the accreting planetesimals.
9. Seven aqueous species – Na^+, Ca^{2+}, Mg^{2+}, K^+, Cl^-, SO_4^{2-}, and HCO_3^- – account for more than 99% of the dissolved constituents in both seawater and river water, and their concentrations are fairly constant (except perhaps that of Ca^{2+}) on a global scale. On the other hand, the concentrations of minor and trace constituents vary significantly. The major compositional gradients in the oceans are vertical, driven by removal of constituents from surface seawater by organisms, and destruction of the organism-produced particles after downward movement.
10. Besides addition of constituents by rivers, several processes are potentially capable of modifying the chemistry of seawater: (i) biological processes; (ii) seafloor volcanics–seawater interaction; (iii) reaction of seawater with detrital silicate and clay minerals transported from the continents; (iv) ocean-to-air transfer of cyclic salts; (v) precipitation of evaporite, sulfide, and phosphate minerals; and (vi) special processes such as porewater burial of Cl^-, nitrification and denitrification, and adsorption of phosphate on ferric oxides.
11. Except for local "oxygen oases," the surface oceans were anoxic under an anoxic atmosphere during the Archean, and the deep oceans continued to remain anoxic until about 1.8 Ga, the end of prolific BIF deposition. The deep ocean was probably sulfidic until 0.8 Ga, when the BIF deposition resumed, and anoxic for at least parts of the Neoproterozoic. The surface oceans must have been oxygenated throughout the Phanerozoic, but the oxidation state of the deeper oceans has fluctuated widely.
12. The seawater composition during the Hadean and Archean was probably not much different from present-day seawater. Seawater continued to be supersaturated with $CaCO_3$, $CaMg(CO_3)_2$, and SiO_2 during the Proterozoic. Its pH certainly rose in response to decreasing atmospheric P_{CO_2}, and it is likely that its salinity did not differ greatly from that of modern seawater. The Phanerozoic seawater composition has alternated between two types of brines related to different evaporite associations: a $MgSO_4$-type, rich in SO_4 but low in Ca; and a KCl-type, rich in Ca but low in SO_4.
13. The biogeochemical cycle of a chemical element (or a chemical species) is a conceptualized representation of its global reservoirs and its transfer from one reservoir to another through chemical and/or biochemical proceses. The biosphere is the most important geochemical realm at present, and the most important biogeochemical cycles are those of carbon, oxygen, nitrogen, sulfur, and phosphorus. The natural cycles of these elements have been significantly affected by input from anthropogenic sources over the past 200 yr.

13.8 Recapitulation

Terms and concepts

Acid deposition
Anoxic
Anoxygenic photosynthesis
Banded iron formation (BIF)
Biogeochemical cycles
Biolimited constituents
Carbon burial
Catalytic reaction chain
Chapman mechanism
Chloroflurocarbon compounds
Euxinic
Evaporites
Exosphere
Free radical
Global warming potential (GWP)
Great Oxidation Event (GOE)
Greenhouse gases
Huronian glaciation
Kármán line
Late Heavy Bombardment
Late Precambrian glaciations
Magma ocean
Mixing ratio
Oxic
Oxygenic photosynthesis
Ozone hole
Ozone layer
Photochemical reaction
Photochemical smog
Photolysis
Photon
Present atmospheric level (PAL)
Primary atmosphere
Primary pollutants
Radiative forcing
Residence time
Salinity
Scavenged constituents
Secondary atmosphere
Secondary pollutants
Snowball Earth stratosphere
Sulfidic ocean
Sulfurous smog
Thermocline
Thermosphere
Troposphere
Ultraviolet radiation

Computation techniques

- Energy required for splitting an atom
- Residence times of constituents in a reservoir
- Calculations pertaining to rates of reactions

13.9 Questions

1. Describe the important events, in chronological order, related to the evolution of the Earth during the first billion years or so after its birth.
2. The concentration of Na^+ in seawater is 460×10^{-3} g kg^{-1} water (Table 7.1) and the mass of seawater = 1.4×10^{24} g
   (a) Calculate the total amount of Na^+ in the oceans.
   (b) Given that the present flux of anthropogenic Na^+ into the oceans is 3.39×10^{12} mol yr^{-1}, how long would it take to increase the concentration of Na^+ in seawater by 10%? Assume no increase in the output flux to counterbalance the increased input.
3. The concentration of ammonia (NH_3) in the atmosphere is 0.456 μg m^{-3} at 0°C and 1 atm. Express the concentration in ppmv and ppbv. The atomic weights of H and N are 1.01 and 14.01, respectively, and the total number of molecules in 1 m^3 of air at 1 atm and 0°C (the *Loschmidt's number*) is 2.69×10^{25}.
4. Using the data given below, calculate the residence times of the solutes in seawater. The amount of water flowing from rivers to oceans = 3.74×10^{19} g yr^{-1}, and the amount of water in oceans = 1.35×10^{24} g.

Solute	Conentration in rivers (ppm)	Concentration in seawater (ppm)	Solute	Conentration in rivers (ppm)	Concentration in seawater (ppm)
Cl	6	19,350	Ca	13	412
Na	5	10,7860	K	1	399
SO_4	8	2712	HCO_3	52	145
Mg	3	1294	Si	10	10

5. What would be the equilibrium pH value of rainwater if atmospheric SO_2 gas were its only dissolved constituent?

$$SO_{2(g)} + H_2O_{(l)} \Leftrightarrow H_2SO_{3(aq)} \quad K_1 = 2.0$$
$$H_2SO_{3(aq)} \Leftrightarrow H_2SO^-_{3(aq)} + H^+ \quad K_2 = 2.0 \times 10^{-2}$$

   Assume that the activity of SO_2 gas in the atmosphere is 5×10^{-9} bar.
6. Assuming that P_{CO_2} in the Earth's early Archean atmosphere was 10^{10} bar, what would be the likely pH of rainwater in equilibrium with the atmospheric P_{CO_2}? Calculate the solubility of calcite in equilibrium with this water. Assume $a_i = m_i$. [Hint: see section 7.5.1.]
7. Climate modeling suggests that the Earth's atmospheric CO_2 concentration by the end of the 21st century will be close to 700 ppmv and global mean surface temperature could be 1.4–5.8°C warmer than in AD 2000. Explain qualitatively the likely impact of this scenario on (i) the pH of the rainwater, (ii) the pH of surface water in equilibrium with calcite, and (iii) the saturation state of surface seawater with respect to calcite.
8. The photolysis of water may be represented by the following reaction series:
   H_2O + UV photon $\Rightarrow$ H + OH
   OH + OH $\Rightarrow$ O + H_2O
   O + OH $\Rightarrow$ O_2 + H
   O + O + M $\Rightarrow$ O_2 + M
   (a) How many water molecules are needed to produce one molecule of O_2 by this reaction series?
   (b) What is a necessary condition for net addition of O_2 by this reaction series?
9. The combination of CO_2 with photolysis reaction series for H_2O (Q. 13.8) yields the following reaction series:
   H_2O + UV photon $\Rightarrow$ H + OH
   CO_2 + UV photon $\Rightarrow$ CO + O
   CO + OH $\Rightarrow$ CO_2 + H
   OH + OH $\Rightarrow$ O + H_2O
   O + O + M $\Rightarrow$ O_2 + M
   Write down the net balanced reaction.
10. Methane gas (CH_4) comprises 7×10^{-5} % by mass of the Earth's atmosphere. What is the residence time of the gas in the atmosphere if it is being lost to the methane sinks at the rate of 4×10^{11} kg yr^{-1}. Estimated mass of the Earth's atmosphere = 5×10^{18} kg.
11. Calculate the residence time of Na^+ in the oceans using the following data: total riverine discharge into the oceans = 37.4×10^{15} L yr^{-1}; average concentration of Na^+ in riverine water = 0.313×10^{-3} mol L^{-1} (Gregor *et al.*, 1988, p. 26); total amount of water in the oceans = 10^{21} kg (Andrews *et al.*, 2004, p. 192); and average concentration of Na^+ in the oceans = 460×10^{-3} mol L^{-1}.
12. For the reaction

$$CO_{2(g)} + 4H_{2(g)} \Leftrightarrow CH_{4(g)} + 2H_2O_{(g)}$$

the equilibrium constants and partial pressures in the primitive atmosphere are as follows:

$K_{eq}(300\ K) = 5.2 \times 10^{19}$; $K_{eq}(500\ K) = 2.7 \times 10^{12}$
$P_{H_2O} = 3.0 \times 10^{-2}$ bar ; $P_{CO_2} = 3.0 \times 10^{-4}$ bar;
$P_{H_2} = 5.0 \times 10^{-5}$ bar

Calculate P_{CH_4} at 300 K and 500 K.

13. A catalytic cycle of reactions that might have contributed to the formation of H_2 from H in the early atmosphere of the Earth is

$$H + CO + M \xrightarrow{k_1} HCO + M$$

$$H + HCO \xrightarrow{k_2} H_2 + CO$$

$2H \Rightarrow H_2$ (net reaction)

Assuming steady state, calculate the concentration of the radical HCO, given the following: concentration of CO = 1.0×10^{12} molecules cm^{-3}; concentration of M = 2.5×10^{19} molecules cm^{-3}; $k_1 = 1.0 \times 10^{-34}\ cm^6\ s^{-1}$ molecule^{-2}; and $k_2 = 3.0 \times 10^{-10}\ cm^3\ s^{-1}$ molecule^{-1}

14. Suppose, the conversion of ozone into molecular oxygen can be achieved by the following elementary reactions:

$O_3 \Leftrightarrow O_2 + O$
$O_3 + O \Leftrightarrow 2O_2$

What is (i) the overall chemical reaction, (ii) the intermediate, (iii) the rate law for each elementary reaction, and (iv) the rate-controlling elementary reaction, if the rate law for the overall reaction is
rate = $k\,[O_3]^2\,[O_2]^{-1}$
where k is the rate constant.

15. Calculate the rate for the reaction $Cl^\bullet + O_3 \Rightarrow ClO^\bullet + O_2$ at 30 km and 235 K. Assume that $[Cl^\bullet] = 5.0 \times 10^{11}$ molecules cm^{-3}, and $[O_3] = 2.0 \times 10^{12}$ molecules cm^{-3}. The rate equation is rate = $k\,[Cl]\,[O_3]$. For the Arrhenius equation, $A = 2.8 \times 10^{12}$, and $E_a = 21\ kJ\ mol^{-1}$.

16. The amount of carbon in the crust is $\sim 10^{20}$ kg. If all this carbon were present in the atmosphere as CO_2, what would be the atmospheric P_{CO_2}? Amount of carbon (inorganic) in the atmosphere = 58×10^{15} moles (Faure, 1991, p. 437).

17. In the upper atmosphere, O_3 is dissociated by ultraviolet radiation by the reaction $O_{3(g)} + UV \Rightarrow O_{2(g)} + O$. What is the energy of a 3400×10^{-8} cm photon that is absorbed in the process? What is the energy absorbed for a mole of these photons?

18. Photosynthesis involves the absorption of light of wavelength 440 nm and emission of radiation of wavelength 670 nm. What is the energy available for photosynthesis by this process?

Appendix 1

Units of measurement and physical constants

Units of measurement

Length	1 centimeter (cm) = 0.01 meter (m) = 0.3937 inch (in) angström (Å) = 10^{-8} cm
Mass	1 gram (g) = 0.0022 pound (lb) = 0.001 kilogram (kg)
Time	second (s)
Temperature	Kelvin (K) = degree Celsius (°C) + 273.15 Degree Fahrenheit (°F) = (°C × $\frac{9}{5}$) + 32 Degree Celsius (°C) = (°F −32) × $\frac{5}{9}$
Pressure	1 bar = 0.987 atmosphere (atm) = 10^6 dynes cm^{-2} = 10^5 pascals (Pa) = 14.504 pounds per square inch (psi) = 1.0197 kg cm^{-2} = 750.06 mm 1 atm = 1.013250 bar = 1.013250 × 10^5 pascal (Pa) = 14.696 lb in^{-2} (psi) = 1.033 kg cm^{-2} = 760 torr = 1.013250 × 106 dyne cm^{-2} 1 Pa = N m^{-2} = J m^{-3} = 10^5 bar
Volume	kJ $kbar^{-1}$
Force	1 newton (N) = 1 kg.m.s^{-2} = 10^5 dynes
Energy/work	1 joule (J) = 1 N.m = 10^7 ergs (= dyne.cm) = 0.2390 calories (cal) 1 electron-volt (ev) = 1.6 × 10^{19} J = 1.6 × 10^{12} ergs
Power	1 watt (W) = 1 J s^{-1}
Energy	1 calorie (cal) = 0.001 kilocalorie (kcal) = 4.184 J = 41.84 cm^3.bar = 41.29 cm^3.atm
Enthalpy (*H*):	kJ
Entropy (*S*):	kJ K^{-1} entropy unit (e.u.) = cal.deg^{-1}.$mole^{-}$
Free Energy (*G*)	kJ
Heat Capacity (*C*)	cal.deg^{-1}

Introduction to Geochemistry: Principles and Applications, First Edition. Kula C. Misra.

Numerical prefixes		Numerical prefixes	
pico (p)	10^{-12}	Kilo (k)	10^{3}
nano (n)	10^{-9}	Mega (M)	10^{6}
micro (μ)	10^{-6}	Giga (G)	10^{9}
milli (m)	10^{-3}	Terra (T)	10^{12}

Physical constants

Absolute zero (of temperature)	–273.15°C
Avogadro's constant (N_A)	6.022094×10^{23} molecules mol^{-1}
Base of natural logarithm (e)	2.71828
Boltzmann constant (k)	1.38063×10^{-23} J K^{-1}
Electron volt (eV)	1.60207×10^{-12} erg $= 1.60207 \times 10^{-19}$ J
Elementary charge (e)	$1.6021892 \times 10^{-19}$ coulomb
Faraday constant (F)	96,487 coulombs $equivalent^{-1}$ = 96.487 kJ $volt^{-1}$ $equivalent^{-1}$ = 23,061 cal $volt^{-1}$ $equivalent^{-1}$
Gas constant (R)	8.314 J mol^{-1} K^{-1} = 1.987 cal mol^{-1} K^{-1} = 82.06 cm^3 atm mol^{-1} K^{-1}
Gravitational constant (G)	6.6732×10^{-11} m^3 kg^{-1} s^{-2} $= 6.6732 \times 10^{-11}$ newton m^2 kg^{-2}
Mass of the electron	9.109×10^{-28} g
Mass of the neutron	1.675×10^{-24}
Mass of the proton	1.673×10^{-24} g
Planck's constant (h)	6.626176×10^{-34} J·s $= 6.62517 \times 10^{-27}$ erg s
Volume of 1 mole of ideal gas:	
At 0°C and 1 bar	22.123 liters (at 1 atm = 22.414 liters)
At 25°C and 1 bar	24.15 liters (at 1 atm = 24.47 liters)
Speed of light (c)	2.998×10^{10} cm s^{-1} $= 2.998 \times 10^{8}$ m s^{-1}
ln x	2.3026 $\log_{10} x$

Dimensions of the Earth

Equatorial radius of the Earth	6.378139×10^{6} m
Polar radius	6.35675×10^{6} m
Mass of the Earth	$5,973 \times 10^{24}$ kg
Mass of the crust	0.024×10^{24} kg
Mass of the mantle	4.02×10^{24} kg
Mass of the outer core	1.85×10^{24} kg
Mass of the inner core	9.7×10^{22} kg
Mass of the atmosphere	5.1×10^{18} kg
Mass of the oceans	1.4×10^{21} kg
Volume of the oceans	1.37×10^{21} L
Mean depth of the oceans	3.8×10^{3} m
Surface area of the continents	1.48×10^{14} m^2
Surface area of the oceans	3.62×10^{14} m^2
Mean density of the whole Earth	5.52 g cm^{-3}
Mean density of the crust	Continental 2.7 g cm^{-3}; Oceanic 3.0 g cm^{-3}
Mean density of the mantle	4.5 g cm^{-3}
Mean density of the core	10.7 g cm^{-3}

Appendix 2

Electronic configurations of elements in ground state

		Electronic configuration														
		1*s*	2*s*	2*p*	3*s*	3*p*	3*d*	4*s*	4*p*	4*d*	4*f*	5*s*	5*p*	5*d*	5*f*	5*g*
Z	Element	*K*	*L*		*M*			*N*				*O*				
1	H	1														
2	He	2														
3	Li	2	1													
4	Be	2	2													
5	B	2	2	1												
6	C	2	2	2												
7	N	2	2	3												
8	O	2	2	4												
9	F	2	2	5												
10	Ne	2	2	6												
11	Na	2	2	6	1											
12	Mg	2	2	6	2											
13	Al	2	2	6	2	1										
14	Si	2	2	6	2	2										
15	P	2	2	6	2	3										
16	S	2	2	6	2	4										
17	Cl	2	2	6	2	5										
18	Ar	2	2	6	2	6										
19	K	2	2	6	2	6		1								
20	Ca	2	2	6	2	6		2								
21	Sc	2	2	6	2	6	1	2								
22	Ti	2	2	6	2	6	2	2								
23	V	2	2	6	2	6	3	2								
24	Cr	2	2	6	2	6	5	1								
25	Mn	2	2	6	2	6	5	2								
26	Fe	2	2	6	2	6	6	2								
27	Co	2	2	6	2	6	7	2								
28	Ni	2	2	6	2	6	8	2								
29	Cu	2	2	6	2	6	10	1								
30	Zn	2	2	6	2	6	10	2								
31	Ga	2	2	6	2	6	10	2	1							
32	Ge	2	2	6	2	6	10	2	2							
33	As	2	2	6	2	6	10	2	3							

Introduction to Geochemistry: Principles and Applications, First Edition. Kula C. Misra.

(*Cont'd*)

Z	Element	Electronic configuration														
		1s	*2s*	*2p*	*3s*	*3p*	*3d*	*4s*	*4p*	*4d*	*4f*	*5s*	*5p*	*5d*	*5f*	*5g*
		K	*L*		*M*			*N*				*O*				
34	Se	2	2	6	2	6	10	2	4							
35	Br	2	2	6	2	6	10	2	5							
36	Kr	2	2		2	6	10	2	6							
37	Rb	2	2	6	2	6	10	2	6			1				
38	Sr	2	2	6	2	6	10	2	6			2				
39	Y	2	2	6	2	6	10	2	6	1		2				
40	Zr	2	2	6	2	6	10	2	6	2		2				
41	Nb	2	2	6	2	6	10	2	6	3		2				
42	Mo	2	2	6	2	6	10	2	6	5		1				
43	Tc	2	2	6	2	6	10	2	6	5		2				
44	Ru	2	2	6	2	6	10	2	6	6		2				
45	Rh	2	2	6	2	6	10	2	6	7		2				
46	Pd	2	2	6	2	6	10	2	6	8		2				
47	Ag	2	2	6	2	6	10	2	6	10		1				
48	Cd	2	2	6	2	6	10	2	6	10		2				
49	In	2	2	6	2	6	10	2	6	10		2	1			
50	Sn	2	2	6	2	6	10	2	6	10		2	2			
51	Sb	2	2	6	2	6	10	2	6	10		2	3			
52	Te	2	2	6	2	6	10	2	6	10		2	4			

Z	Element	Electronic configuration																	
		1s	*2s*	*2p*	*3s*	*3p*	*3d*	*4s*	*4p*	*4d*	*4f*	*5s*	*5p*	*5d*	*5f*	*6s*	*6p*	*6d*	*7s*
		K	*L*		*M*			*N*				*O*				*P*			*Q*
53	I	2	2	6	2	6	10	2	6	10		2	5						
54	Xe	2	2	6	2	6	10	2	6	10		2	6						
55	Cs	2	2	6	2	6	10	2	6	10		2	6			1			
56	Ba	2	2	6	2	6	10	2	6	10		2	6			2			
57	La	2	2	6	2	6	10	2	6	10		2	6	1		2			
58	Ce	2	2	6	2	6	10	2	6	10	2	2	6			2			
59	Pr	2	2	6	2	6	10	2	6	10	3	2	6			2			
60	Nd	2	2	6	2	6	10	2	6	10	4	2	6			2			
61	Pm	2	2	6	2	6	10	2	6	10	5	2	6			2			
62	Sm	2	2	6	2	6	10	2	6	10	6	2	6			2			
63	Eu	2	2	6	2	6	10	2	6	10	7	2	6			2			
64	Gd	2	2	6	2	6	10	2	6	10	7	2	6	1		2			
65	Tb	2	2	6	2	6	10	2	6	10	9	2	6			2			
66	Dy	2	2	6	2	6	10	2	6	10	10	2	6			2			
67	Ho	2	2	6	2	6	10	2	6	10	11	2	6			2			
68	Er	2	2	6	2	6	10	2	6	10	12	2	6			2			
69	Tm	2	2	6	2	6	10	2	6	10	13	2	6			2			
70	Yb	2	2	6	2	6	10	2	6	10	14	2	6			2			
71	Lu	2	2	6	2	6	10	2	6	10	14	2	6	1		2			
72	Hf	2	2	6	2	6	10	2	6	10	14	2	6	2		2			
73	Ta	2	2	6	2	6	10	2	6	10	14	2	6	3		2			
74	W	2	2	6	2	6	10	2	6	10	14	2	6	4		2			
75	Rh	2	2	6	2	6	10	2	6	10	14	2	6	5		2			
76	Os	2	2	6	2	6	10	2	6	10	14	2	6	6		2			
77	Ir	2	2	6	2	6	10	2	6	10	14	2	6	7		2			
78	Pt	2	2	6	2	6	10	2	6	10	14	2	6	9		1			
79	Au	2	2	6	2	6	10	2	6	10	14	2	6	10		1			

(*Continued*)

(*Cont'd*)

		Electronic configuration																	
		1*s*	**2*s***	**2*p***	**3*s***	**3*p***	**3*d***	**4*s***	**4*p***	**4*d***	**4*f***	**5*s***	**5*p***	**5*d***	**5*f***	**6*s***	**6*p***	**6*d***	**7*s***
Z	**Element**	***K***	***L***		***M***			***N***				***O***				***P***			***Q***
80	Hg	2	2	6	2	6	10	2	6	10	14	2	6	10		2			
81	Tl	2	2	6	2	6	10	2	6	10	14	2	6	10		2	1		
82	Pb	2	2	6	2	6	10	2	6	10	14	2	6	10		2	2		
83	Bi	2	2	6	2	6	10	2	6	10	14	2	6	10		2	3		
84	Po	2	2	6	2	6	10	2	6	10	14	2	6	10		2	4		
85	At	2	2	6	2	6	10	2	6	10	14	2	6	10		2	5		
86	Rn	2	2	6	2	6	10	2	6	10	14	2	6	10		2	6		
87	Fr	2	2	6	2	6	10	2	6	10	14	2	6	10		2	6		1
88	Ra	2	2	6	2	6	10	2	6	10	14	2	6	10		2	6		2
89	Ac	2	2	6	2	6	10	2	6	10	14	2	6	10		2	6	1	2
90	Th	2	2	6	2	6	10	2	6	10	14	2	6	10		2	6	2	2
91	Pa	2	2	6	2	6	10	2	6	10	14	2	6	10	2	2	6	1	2
92	U	2	2	6	2	6	10	2	6	10	14	2	6	10	3	2	6	1	2
93	Np	2	2	6	2	6	10	2	6	10	14	2	6	10	4	2	6	1	2
94	Pu	2	2	6	2	6	10	2	6	10	14	2	6	10	6	2	6		2
95	Am	2	2	6	2	6	10	2	6	10	14	2	6	10	7	2	6		2
96	Cm	2	2	6	2	6	10	2	6	10	14	2	6	10	7	2	6	1	2
97	Bk	2	2	6	2	6	10	2	6	10	14	2	6	10	9	2	6		2
98	Cf	2	2	6	2	6	10	2	6	10	14	2	6	10	10	2	6		2
99	Es	2	2	6	2	6	10	2	6	10	14	2	6	10	10	2	6		2
100	Fm	2	2	6	2	6	10	2	6	10	14	2	6	10	10	2	6		2
101	Md	2	2	6	2	6	10	2	6	10	14	2	6	10	10	2	6		2
102	No	2	2	6	2	6	10	2	6	10	14	2	6	10	10	2	6		2
103	Lr	2	2	6	2	6	10	2	6	10	14	2	6	10	10	2	6	1	2

Z = Atomic Number.
Source of data: Cotton and Wilkinson (1980). Also see Eby (2004, pp. 472–473).

Appendix 3

First ionization potential, electron affinity, electronegativity (Pauling scale), and coordination numbers of selected elements

Element (Z)	First ionization potential (eV)[a]	Electron affinity (eV)[a]	Electro-negativity[a]	Approximate ionic character of bond with oxygen[d](%)	Ion	Common coordination numbers[b]	Ionic radius for octahedral coordination[c] (Å)
^{1}H	13.60	0.75	2.20	30			
^{2}He	24.59	*	—	—			
^{3}Li	5.39	0.62	0.98	76	Li^{+}	6	0.76
^{4}Be	9.32	*	1.57	59	Be^{2+}	4	0.27(4)
^{5}B	8.30	0.28	2.04	39	B^{3+}	3,4	0.11(4)
^{6}C	11.26	1.26	2.55	19	C^{4+}	3,4,6	0.15(4)
^{7}N	14.53	*	3.04	4	N^{5+}	3	0.13
^{8}O	13.62	1.46	3.44		O^{2-}		1.40
^{9}F	17.42	3.40	3.98		F^{-}		1.33
^{10}Ne	21.56	*	—	—	—	—	—
^{11}Na	5.14	0.55	0.93	79	Na^{+}	6,8	1.02
^{12}Mg	7.65	*	1.31	67	Mg^{2+}	6	0.72
^{13}Al	5.98	0.44	1.61	55	Al^{3+}	4,6	0.535
^{14}Si	8.15	1.38	1.90	43	Si^{4+}	4	0.26(4)
^{15}P	10.49	0.75	2.19	34	P^{5+}	4	0.17(4)
^{16}S	10.36	2.08	2.58	19	S^{2-}		1.84
					S^{6+}	4	0.12(4)
^{17}Cl	12.97	3.61	3.16	2	Cl^{-}		1.81
^{18}Ar	15.80	*	—	—			
^{19}K	4.34	0.50	0.82	82	K^{+}	8–12	1.38
^{20}Ca	6.11	0.02	1.00	76	Ca^{2+}	6,8	1.00
^{21}Sc	6.56	0.19	1.36	67	Sc^{3+}	6	0.745
^{22}Ti	6.83	0.08	1.54	59	Ti^{3+}	6	0.67
					Ti^{4+}	6	0.605
^{23}V	6.75	0.53	1.63	55	V^{3+}	6	0.64
					V^{4+}	6	0.58
					V^{5+}	4,6	0.54
^{24}Cr	6.77	0.67	1.66	55	Cr^{3+}	6	0.615
^{25}Mn	7.43	*	1.55	59	Mn^{2+}	6	0.83
					Mn^{3+}	6	0.645
					Mn^{4+}	4,6	0.53
^{26}Fe	7.90	0.15	1.83	47	Fe^{2+}	6	0.78
					Fe^{3+}	6	0.643

(Continued)

Introduction to Geochemistry: Principles and Applications, First Edition. Kula C. Misra.

(*Cont'd*)

Element (*Z*)	First ionization potential (eV)[a]	Electron affinity (eV)[a]	Electro-negativity[a]	Approximate ionic character of bond with oxygen[d](%)	Ion	Common coordination numbers[b]	Ionic radius for octahedral coordination[c] (Å)
^{27}Co	7.88	0.66	1.88	47	Co^{2+}	6	0.745
^{28}Ni	7.64	1.16	1.91	43	Ni^{2+}	6	0.69
^{29}Cu	7.73	1.24	1.90	43	Cu^{+}	6,8	0.77
					Cu^{2+}	6	0.73
^{30}Zn	9.39	*	1.65	55	Zn^{2+}	4,6	0.74
^{31}Ga	6.00	0.3	1.81	47	Ga^{3+}	4,6	0.62
^{32}Ge	7.90	1.23	2.01	39	Ge^{4+}	4	0.73
^{33}As	9.79	0.81	2.18	34	As^{3+}	4,6	0.58
					As^{5+}	4,6	0.46
^{34}Se	9.75	2.02	2.55	19	Se^{2-}		1.98
					Se^{6+}	4	0.28(4)
^{35}Br	11.81	3.36	2.96	6	Br^{-}		1.96
^{36}Kr	14.00	*	—	—			
^{37}Rb	4.18	0.49	0.82	82	Rb^{+}	8–12	1.52
^{38}Sr	5.69	0.05	0.95	79	Sr^{2+}	8	1.18
^{39}Y	6.22	0.31	1.22	70	Y^{3+}	6	0.90
^{40}Zr	6.63	0.43	1.33	67	Zr^{4+}	6	0.72
^{42}Mo	7.09	0.75	2.16	34	Mo^{4+}	6	0.65
					Mo^{6+}	4,6	0.59
^{46}Pd	8.34	0.56	2.20	30	Pd^{2+}	6	0.86
^{47}Ag	7.58	1.30	1.93	43	Ag^{+}	8,10	0.94
^{48}Cd	8.99	*	1.69	55	Cd^{2+}	6,8	0.95
^{50}Sn	7.34	1.11	1.96	43	Sn^{2+}	6,8	1.27(8)
					Sn^{4+}	6	0.69
^{51}Sb	8.61	1.05	2.05	39	Sb^{3+}	6	0.80(5)
					Sb^{5+}	4,6	0.60
^{52}Te	9.01	1.97	2.10	34	Te^{2-}		2.21
					Te^{6+}	4,6	0.56
^{53}I	10.45	3.06	2.66	15	I^{-}		2.20
^{54}Xe	12.13	*	2.60	15			
^{55}Cs	3.89	0.47	0.79	84	Cs^{+}	12	1.67
56Ba	5.21	0.15	0.89	82	Ba^{2+}	8–12	1.35
57La	5.58	0.5	1.10	74	La^{3+}	8	1.032
^{58}Ce	5.54		1.12	74	Ce^{3+}	6,8	1.01
^{62}Sm	5.64		1.17	74	Sm	6,8	0.958
^{72}Hf	6.83	~0	1.3	67	Hf^{4+}	6	0.71
^{73}Ta	7.55	0.32	1.5	59	Ta^{5+}	6	0.64
^{74}W	7.86	0.82	1.7	51	W^{6+}	4,6	0.60
^{79}Au	9.22	2.31	2.4	22	Au	8–12	1.37
^{80}Hg	10.44	*	1.9	43	Hg^{2+}	6,8	1.02
^{81}Tl	6.11	0.2	1.8	47	Tl^{+}	8–12	1.50
					Tl^{3+}	6,8	0.67
^{82}Pb	7.42	0.36	1.8	47	Pb^{2+}	6–10	1.19
^{83}Bi	7.28	0.95	1.9	43	Bi^{3+}	6,8	1.03
^{88}Ra	5.28		0.9	79	Ra^{2+}	8–12	1.48
^{90}Th	6.31		1.3	67	Th^{4+}	6,8	0.94
^{92}U	6.19		1.7	51	U^{4+}	6,8	1.00(8)
					U^{6+}	6	0.73

[a]Lide (1998).
[b]Smith (1963).
[c]Krauskopf and Bird (1995); the number in parenthesis indicates a coordination other than tetrahedral.
[d]Calculated based on the ionic character as a function of difference in electronegativity relative to oxygen (Sargent-Welch, 1980; cited in Faure, 1998).
*Elements expected to have electron affinity close to zero on quantum mechanics grounds.
1 ev = 96.485 kJ mol^{-1}

Appendix 4

Thermodynamic symbols

Symbol	Description
A	Hemholtz free energy of a system
A_r	Affinity of a chemical reaction
a_i^{α}	Activity of the ith constituent in the phase α
$a_{\pm}$	Mean ionic activity
°C	Degrees Celsius (temperature unit on Celsius scale)
C	Heat capacity of a substance
$\bar{C}_P$	Molar heat capacity of a substance at constant pressure
$\bar{C}_V$	Molar heat capacity of a substance at constant volume
c	Number of components in a system (the phase rule)
D_i	Diffusion coefficient of constituent i
E	Electromotive force
E_a	Activation energy
Eh	Electromotive force measured relative to standard hydrogen electrode (SHE)
e^-	Electron
$\Im$	Faraday constant
f	Number of degrees of freedom in a system (the phase rule)
f	Fugacity of a nonideal gas
f_i	Fugacity of the ith species in a gas mixture
f_i°	Fugacity of the ith species in a gas mixture at standard state pressure P°
G	Gibbs free energy
$\bar{G}_i$	Gibbs free energy per mole of constituent i in a phase
$\bar{G}_T^{\circ}$	Gibbs free energy per mole of a substance at standard pressure and temperature T (K)
$\bar{G}_T^{P}$	Gibbs free energy per mole of a substance at pressure P and temperature T (K)
$\Delta G_{f,T}^{\circ}$	Standard Gibbs free energy of formation at temperature T
$\Delta G_{r,298.15}^{\circ}$	Gibbs free energy change of a reaction at standard pressure and temperature 298.15 K
$\Delta G_{r,T}^{\circ}$	Gibbs free energy change of a reaction at standard pressure and temperature T (K)
$\Delta G_{r,T}^{P}$	Gibbs free energy change of a reaction at pressure P (bars) and temperature T (K)
G^{EX}	Excess Gibbs free energy (nonideal solution)
H	Enthalpy
$\bar{H}_i$	Enthalpy per mole of constituent i in a phase
$\bar{H}_{298.15}^{\circ}$	Enthalpy per mole of a substance at standard pressure and temperature 298.15 K
$\bar{H}_T^{\circ}$	Enthalpy per mole of a substance at standard pressure and temperature T (K)
$\Delta H_{f,T}^{\circ}$	Standard enthalpy of formation at temperature T (K)
$\Delta H_{r,298.15}^{\circ}$	Enthalpy change of a reaction at standard pressure and temperature 298.15 K
$\Delta H_{r,T}^{\circ}$	Enthalpy change of a reaction at standard pressure and temperature T (K)
H^{EX}	Excess enthalpy (nonideal solution)
$\bar{h}_i^{\alpha}$	Partial molar enthalpy of the ith constituent of phase α
h	Henry's Law constant

(*Continued*)

Introduction to Geochemistry: Principles and Applications, First Edition. Kula C. Misra.

(*Cont'd*)

Symbol	Description
h	Planck's constant
I	Ionic strength of an aqueous solution
IAP	Ion activity product
J_i	Mass flux of the constituent i
K	Degrees Kelvin (temperature unit on Kelvin scale)
K_{eq}	Equilibrium constant of a chemical reaction
K_D	Distribution (or partition) coefficient
k	Boltzmann constant
k_B	Rate constant of a chemical reaction
M_i	Molarity of the ith species in a solution
L	10^{-3} m^3
m_i	Molality of the ith species in a solution
$m_\pm$	Mean ionic molality
N_A	Avogadro's constant
n	Number of moles
n_i	Number of moles of the ith constituent
P	Pressure (total) on a system
P_E	Equilibrium pressure
P_i	Partial pressure of the ith species in a gas mixture
P_i^*	Vapor pressure of the pure ith species
pa_i	$-\log a_i$
pH	A measure of the acidity or basicity of an aqueous solution
p^ε	$-\log a_{e^-}$
Q	Heat exchanged between a system and its surroundings
R	Gas constant
r	Number of restrictions on a system (phase rule)
S	Entropy
$\bar{S}_i$	Entropy per mole of constituent i in a phase
$\bar{S}^o_{298.15}$	Entropy per mole of a substance at standard pressure and temperature 298.15 K
$\bar{S}^o_T$	Entropy per mole of a substance at standard pressure and temperature T (K)
S_o	Residual entropy
$\Delta S^o_{f,T}$	Standard entropy of formation at temperature T
$\Delta S^o_{r,298.15}$	Entropy change of a reaction at standard pressure and temperature 298.15 K
$\Delta S^o_{r,T}$	Entropy change of a reaction at standard pressure and temperature T (K)
$\bar{S}_i^\alpha$	Partial molar entropy of the ith constituent of phase α
s	Solubility
T	Temperature of a system (Kelvin scale)
T_c	Critical temperature
T_E	Equilibrium temperature
U	Internal energy
V	Volume
$\bar{V}_i$	Volume per mole of constituent i in a phase
$\bar{V}^o_{298.15}$	Volume per mole of a substance at standard pressure and temperature 298.15 K
$\bar{V}_T^P$	Volume per mole of a substance at pressure P and temperature T (K)
$\bar{V}_C$	Critical volume per mole
ΔV_r	Volume change for a chemical reaction
V^{EX}	Excess volume (nonideal solution)
$\bar{V}_i^\alpha$	Partial molar volume of the ith species in phase α
W	Work done
$W_{i,j}$	Interaction parameter for components i and j in Margules equations for regular solutions
X_i^α	Mole fraction of the ith constituent in phase α
Y	Any extensive thermodynamic property
$\bar{Y}$	Any extensive thermodynamic property per mole
$\bar{y}_i^\alpha$	Partial molar property of the ith constituent in phase α
α_P	Coefficient of thermal expansion at constant pressure P
β_T	Coefficient of compression at constant temperature T
χ	Fugacity coefficient of a nonideal gas
χ_i	Fugacity coefficient of the ith species in a gas mixture
γ_i	Activity coefficient of the ith constituent (molality basis)
$\gamma_\pm$	Mean activity coefficient (molality basis)

(*Cont'd*)

Symbol	Description
Δ	Difference between the initial and subsequent value of any function
δ_{sample}	Isotopic fractionation of a sample relative to a standard
λ_i	Rational activity coefficient of the *i*th constituent (mole-fraction basis)
ξ	Progress of a chemical reaction
μ_i^α	Chemical potential of the *i*th constituent in phase α
μ_i°	Chemical potential of pure *i* at unit pressure and the temperature of interest
$\mu_\pm$	Mean chemical potential of an aqueous species
v_i	Stoichiometric coefficient of the *i*th constituent
ρ	Density of a substance
ϕ	Number of phases in a system (the phase rule)

Appendix 5

Standard state (298.15 K, 10^5 Pa) thermodynamic data for selected elements, ionic species, and compounds

The most consistent results are obtained if all the thermodynamic data used in a calculation are taken from the same compilation.

	Species or compound	Mineral	$\Delta H^{o}_{f,\,298.15}$ (kJ mol^{-1})	$S^{o}_{298.15}$ (J mol^{-1}K^{-1})	$\Delta G^{o}_{f,\,298.15}$ (kJ mol^{-1})	$V^{o}_{298.15}$ (J bar^{-1})	Source
Elements	Ag metal (c)	Native silver	0	42.55	0	1.0272	1, 17
	Al metal (c)		0	28.35	0	0.999	1
	Al metal (c)		0	28.30	0		4, 10
	Al metal (c)		0	28.30	0	0.999	17
	As metal (c)		0	35.69	0	1.2963	1, 17
	Au metal (c)		0	47.49	0	1.0215	17
	Ba metal (c)		0	62.42	0	3.821	1, 2, 10, 11
	Ba metal (c)		0	62.42	0	3.821	17
	C (c)	Graphite	0	5.74	0	0.5298	1, 4, 10, 17
	C (c)	Graphite	0	5.85	0	0.530	18
	C (c)	Diamond	1.895	2.38	2.900	0.3417	1, 2, 10
	C (c)	Diamond	1.9	2.38	2.9	0.342	17
	C (c)	Diamond	2.07	2.30	3.13	0.342	18
	Ca metal (c)		0	41.63	0	2.6190	1
	Ca metal (c)		0	41.59	0		4, 10
	Ca metal (c)		0	42.90	0	2.619	17
	Cd metal (c)		0	51.80	0	1.3005	1, 17
	Co metal (c)		0	30.04	0	0.6670	1, 17
	Cr metal (c)		0	23.64	0	0.7231	1
	Cr metal (c)		0	23.62	0	0.7231	17
	Cu metal (c)	Native copper	0	33.15	0	0.7113	1
	Cu metal (c)	Native copper	0	33.14	0	0.7113	17
	Fe metal (c)	Native iron	0	27.28	0	0.7092	1
	Fe metal (c)	Native iron	0	27.32	0		10
	Fe metal (c)	Native iron	0	27.09	0	0.7092	17
	Fe metal (c)	Native iron	0	27.32	0	0.709	18
	Hg metal (l)		0	75.90	0	1.4822	1, 17
	K metal (c)		0	64.68	0	4.5360	1, 4, 10
	K metal (c)		0	64.67	0	4.5360	17
	Mg metal (c)		0	32.67	0	1.3996	17
	Mn metal (c)		0	32.01	0	0.7354	1, 17
	Mo metal (c)		0	28.66	0	0.9387	1, 17
	Na metal (c)		0	51.30	0	2.3812	1, 4, 10
	Na metal (c)		0	51.46	0	2.381	17

Introduction to Geochemistry: Principles and Applications, First Edition. Kula C. Misra.

(*Cont'd*)

	Species or compound	Mineral	$\Delta H^{o}_{f,\,298.15}$ (kJ mol^{-1})	$S^{o}_{298.15}$ (J mol^{-1}K^{-1})	$\Delta G^{o}_{f,\,298.15}$ (kJ mol^{-1})	$V^{o}_{298.15}$ (J bar^{-1})	Source
	Ni metal (c)		0	29.87	0	0.6588	1, 17
	P metal (c)		0	22.85	0	1.7200	1
	P metal (c)		0	41.09	0	1.73	17
	Pb metal (c)		0	65.06	0	1.8267	1
	Pb metal (c)		0	64.8	0	1.8267	17
	Pd metal (c)		0	37.82	0	0.8862	1
	Pt metal (c)		0	41.63	0	0.9091	1, 17
	S (c, orthorhombic)		0	31.80	0	1.5511	1
	S (c, orthorhombic)		0	32.05	0		4, 10
	S (c, orthorhombic)		0	32.05	0	1.5511	17
	S (c, monoclinic)		0	37.85	0	?	1
	S (c, monoclinic)		0	33.03	0	?	17
	Sb metal		0	45.52	0	1.8178	1, 17
	Si metal (c)		0	18.81	0	1.2056	1, 4, 17
	Sn metal (c)		0	51.20	0	1.6289	1
	Sn metal (c)		0	51.18	0	1.629	17
	Sr metal (c)		0	55.40	0	3.3921	1
	Sr metal (c)		0	55.70	0		10, 11, 12
	Sr metal (c)		0	55.69	0	3.392	17
	Th metal (c)		0	53.39	0	1.9788	1
	Th metal (c)		0	51.83	0	1.979	17
	Ti metal (c)		0	30.63	0	1.0631	1
	Ti metal (c)		0	30.76	0	1.063	17
	U metal (c)		0	50.29	0	1.2497	1
	U metal (c)		0	50.2	0	1.250	17
	V metal (c)		0	28.91	0	0.8350	1
	V metal (c)		0	28.94	0	0.8350	17
	W metal (c)		0	32.64	0	0.9545	1
	W metal (c)		0	32.65	0	0.9545	17
	Zn metal (c)		0	41.63	0	0.9162	1, 17
	Zr metal (c)		0	38.99	0	1.4016	1
	Zr metal (c)		0	38.87	0	1.4016	17
Gas species	CH_4 (g)		–74.810	186.26	–50.708	*	1, 10
	CH_4 (g)		–74.8	186.26	–50.7	*	17
	CH_4 (g)		–74.81	186.26	–50.66		18
	Cl_2 (g)		0	223.08	0	*	1, 17
	Cl_2 (g)		0	222.972	0		4, 10
	CO (g)		–110.530	197.67	–137.171	*	1
	CO (g)		–110.53	197.551	–137.17		4, 10
	CO (g)		–110.5	197.3	–137.1	*	17
	CO (g)		–110.53	197.67	–137.13		18
	CO_2 (g)		–393.510	213.79	–394.375	*	1
	CO_2 (g)		–393.51	213.676	–394.37		4, 10
	CO_2 (g)		–393.5	213.8	–394.4	*	17
	CO_2 (g)		–393.51	213.70	–394.30		18
	F_2 (g)		0	202.70	0	*	1
	F_2 (g)		0	202.682	0		4, 10
	F_2 (g)		0	202.79	0		17
	H_2 (g)		0	130.68	0	*	1, 17
	H_2 (g)		0	130.571	0		4, 10
	H_2 (g)		0	130.70	0		18
	HCl (g)		–92.312	186.90	–95.229	*	1
	HCl (g)		–92.3	186.9	–95.3	*	17
	HF (g)		–271.1	173.8	–273.2		2
	HF (g)		–273.3	173.670	–275.4		4, 10
	HF (g)		–273.3	173.8	–275.4		17
	H_2O (g)		–241.814	188.83	–228.569	*	1
	H_2O (g)		–241.83	188.73	–228.58		4

(*Continued*)

(*Cont'd*)

	Species or compound	Mineral	$\Delta H^{o}_{f,\,298.15}$ (kJ mol^{-1})	$S^{o}_{298.15}$ (J mol^{-1}K^{-1})	$\Delta G^{o}_{f,\,298.15}$ (kJ mol^{-1})	$V^{o}_{298.15}$ (J bar^{-1})	Source
	H_2O (g)		−241.83	188.726	−228.58		4, 10
	H_2O (g)		−241.8	188.8	−228.6	*	17
	H_2O (g)		−241.81	188.80	−228.54		18
	H_2S (g)		−20.627	205.80	−33.543	*	1
	H_2S (g)		−20.6	205.7	−33.4		2
	H_2S (g)		−20.6	205.696	−33.4		4, 10
	H_2S (g)		−20.6	205.8	−33.4	*	17
	N_2 (g)		0	191.61	0	*	1, 17
	NH_3 (g)		−45.940	192.78	−16.410	*	1
	NH_3 (g)		−46.11	192.5	−16.45		2
	NH_3 (g)		−45.9	192.77	−16.4	*	17
	NO (g)		90.2	210.8	86.6		2
	NO_2 (g)		33.095	240.06	51.25	*	1
	NO_2 (g)		33.1	240.1	51.2	*	17
	N_2O (g)		82.0	219.8	104.2		2
	O_2 (g)		0	205.15	0	*	1, 17
	O_2 (g)		0	205.043	0		4, 10
	O_2 (g)		0	205.20			18
	P_2 (g)		144.3	218.1	103.7	*	2
	S_2(g)		128.49	228.17	79.453	*	1
	S_2(g)		128.6	228.06	79.7		4, 10
	S_2(g)		128.6	228.17	79.7		17
	SO_2 (g)		−296.810	248.22	−300.170	*	1
	SO_2 (g)		−296.8	248.114	−300.1		4, 10
	SO_2 (g)		−296.8	248.2	−300.1	*	17
	SO_3 (g)		−395.722	256.76	−371.046	*	1
	SO_3 (g)		−395.7	256.8	−371.0	*	17
Liquid species	H_2O (l)		−285.830	69.95	−237.141	1.8069	1
	H_2O (l)		−285.83	69.95	−237.14		4, 10
	H_2O (l)		−285.8	70.0	−237.1	1.807	17
	H_2O (l)		−267.43	75.00	----	1.164	18
	H_2SO_4 (l)		−814.0	156.9	−690.0	5.357	17
	Silica (l)		−923.09	11.20	----	2.640	18
	Silica glass		−901.6	48.5	−849.3	2.727	17
Ionic Species	Ag^{+}		105.750	73.38	77.077		1
	Ag^{+}		105.8	73.45	77.1		17
	Ag^{2+}		268.6	88.0	269.0		2, 17
	$AgCl_2^{-}$		−245.2	231.4	−215.4		2
	Al^{3+}		−531.000	−308.00	−489.000		1
	Al^{3+}		−531	−321.7	−485		2
	Al^{3+}		−540	−340	−487.65		3
	Al^{3+}		−540.9	−342	−488.8		10, 14
	Al^{3+}		−538.4	−332	−489.4		17
	$Al(OH)^{2+}$		−778	−204	−696.54		3
	$Al(OH)_2^{+}$		−1000	−16	−901.7		3
	$Al(OH)_3$ (aq)		−1230	108	−1100.6		3
	$Al(OH)_4^{-}$		−1502.5	102.9	−1305.3		2
	$Al(OH)_4^{-}$		−1487	160	−1305.8		3
	$Al(OH)_4^{-}$		−1478.0	127.0	−1307		10, 14
	Ba^{2+}		−537.640	9.60	−560.740		1
	Ba^{2+}		−537.6	9.6	−560.8		2
	Ba^{2+}		−532.5	8.4	−555.36		3
	Ba^{2+}		−532.5	8.4	−555.4		10, 11, 13
	Ba^{2+}		−532.5	8.40	−555.4		17
	CO_2 (aq)		−413.8	117.6	−386.0		2
	CO_2 (aq)		−413.26	119.36	−386.0		4, 10
	CH_4 (aq)		−89.0	83.7	−34.3		2

(*Cont'd*)

Species or compound	Mineral	$\Delta H^{o}_{f,\,298.15}$ (kJ mol^{-1})	$S^{o}_{298.15}$ (J mol^{-1}K^{-1})	$\Delta G^{o}_{f,\,298.15}$ (kJ mol^{-1})	$V^{o}_{298.15}$ (J bar^{-1})	Source
CH_2O (aq)				−129.7		2
H_2CO_3 (aq)		−699.650	187.00	−623.170		1
H_2CO_3 (aq)		−699.09	189.31	−623.14		3
H_2CO_3 (aq)		−699.7	184.7	−623.2		17
HCO_3^- (aq)		−691.900	91.20	−586.850		1
HCO_3^- (aq)		−689.9	98.4	−586.8		4
HCO_3^- (aq)		−689.93	98.4	−586.8		4, 10
HCO_3^- (aq)		−689.9	98.4	−586.8		17
CO_3^{2-} (aq)		−677.140	−56.90	−527.900		1
CO_3^{2-} (aq)		−675.23	−50.0	−527.9		4, 10
CO_3^{2-} (aq)		−675.2	−50.0	−527.0		17
Ca^{2+}		−542.830	−53.1	−553.540		1
Ca^{2+}		−542.8	−53.1	−553.6		2
Ca^{2+}		−543.0	−56.2	−553.6		17
Ca^{2+}		−543.0	−56.2	−552.8		4, 10
$Ca(OH)^+$		−764.4		−718.4		20, 20
$CaCO_3$ (aq)		−1220.0	−110.0	−1081.4		2
$CaSO_4$ (aq)		−1452.1	−33.1	−1298.1		2
Cl^-		−167.000	56.73	−131.270		1
Cl^-		−167.1	56.6	−131.2		4
Cl^-		−167.1	56.6	−131.2		4, 10
Cl^-		−167.1	56.60	−131.2		17
Cd^{2+}		−75.900	−71.20	−77.580		1
Cd^{2+}		−75.900	−72.8	−77.6		17
Co^{2+}		−54.400	−113.00	−58.200		1
Cr^{2+}		−143.51		−176.15		20, 21
CrO_4^{2-}						
Cu^+		71.670	41.00	50.000		1
Cu^+		71.7	40.60	50.0		17
Cu^{2+}		64.770	−99.60	65.520		1
Cu^{2+}		64.9	−98.0	65.1		17
$CuSO_4$ (aq)				−692.2		2
F^-		−335.350	−13.18	−281.705		1
F^-		−332.6	−13.8	−278.8		2
F^-		−335.35	−13.8	−281.5		4
F^-		−335.35	−13.8	−281.5		4, 10
F^-		−335.4	−13.8	−281.5		17
HF (aq)		−320.1	88.7	−296.8		2
HF (aq)		−323.15	88.00	−299.7		10, 11
Fe^{2+}		−89.100	−138.00	−78.870		1
Fe^{2+}		−89.0		−82.88		3
Fe^{2+}		−91.1	−107.1	−90.0		17
Fe^{3+}		−48.500	−316.00	−4.600		1
Fe^{3+}		−48.5		−8.56		3
Fe^{3+}		−49.9	−280.0	−16.7		17
$Fe(OH)^{2+}$		−290.8	−142	−229.4		2
$Fe(OH)^{2+}$		−291.2		−233.20		3
$Fe(OH)_2^+$		−548.9		−450.5		3
$Fe(OH)_3$		−802.5		−648.3		3
$FeSO_4$		−998.3	−117.6	−823.4		2
H^+		0	0	0		1
Hg^{2+}		−171.000	−32.00	164.40		1
Hg^{2+}		−170.2	−36.2	163.5		17
Hg_2^{2+}		172.000	84.50	164.400		1
Hg_2^{2+}		166.9	65.74	153.6		17
$HgCl^+$		−18.8	75	−5.4		2
$HgCl_4^{2-}$		−554.0	293	−446.8		2

(Continued)

(*Cont'd*)

Species or compound	Mineral	$\Delta H^o_{f,\ 298.15}$ (kJ mol^{-1})	$S^o_{298.15}$ (J mol^{-1}K^{-1})	$\Delta G^o_{f,\ 298.15}$ (kJ mol^{-1})	$V^o_{298.15}$ (J bar^{-1})	Source
HgS_2^{2-}				–41.9		2
K^+		–252.170	101.04	–282.490		1
K^+		–252.14	101.2	–282.5		4
K^+		–252.14	101.2	–282.5		4, 10
K^+		–252.1	101.20	–282.5		17
Mg^{2+}		–466.850	–138.00	–454.800		1
Mg^{2+}		–467.0	–237	–455.4		4
Mg^{2+}		–466.8	–138.1	–454.8		2
Mg^{2+}		–467.0	–137	–455.4		4, 10
Mg^{2+}		–467.0	–137	–455.4		17
Mn^{2+}		–220.700	–73.60	–228.00		1
Mn^{2+}		–220.8	–73.60	–228.1		17
NH_3 (aq)		–80.29	111.3	–26.5		2
NH_4^+		–133.260	111.17	–79.457		1
NH_4^+		–132.51	113.4	–79.31		2
NH_4^+		–133.3	111.17	–79.4		17
Na^+		–240.300	58.41	–261.900		1
Na^+		–240.1	59.0	–261.9		2
Na^+		–240.34	58.45	–262.0		4, 10
Na^+		–240.3	58.45	–261.5		17
$NaCO_3^-$		–935.9	–49.8	–792.8		2
$NaHCO_3$ (aq)		–943.9	113.8	–849.7		2
Ni^{2+}		–54.000	–129.00	–45.6		1, 17
Ni^{2+}		–54.0	–128.9	–45.6		2
NO_3^-		–206.9	146.7	–110.8		17
O_2 (aq)		–11.7	110.9	16.4		2
OH^-		–230.025	–10.71	–157.328		1
OH^-		–230.0	–10.9	–157.2		4
OH^-		–230.01	–10.9	–157.2		4, 10
OH^-		–230.0	–10.7	–157.3		17
PO_4^{3-}		–1277.000	–222.00	–1019.000		1
PO_4^{3-}		–1259.6	–222.00	–1001.6		17
HPO_4^{2-}		–1292.1	–33.5	–1089.1		2
$H_2PO_4^-$		–1296.3	90.4	–1130.3		2
H_3PO_4 (aq)		–1288.3	158.2	–1142.5		2
Pb^{2+}		–1.700	10.00	–24.400		1
Pb^{2+}		0.9	18.5	–24.2		17
$PbCl^+$				–164.8		2
$PbCl_2$ (aq)				–297.2		2
$PbCl_3^-$				–426.3		2
Pd^{2+}		149.0		176.5		20
H_2S (aq)		–38.6	126.0			4, 10
HS^-		–16.999	62.80	12.100		1
HS^-		–16.3	67	12.2		2
HS^-		–16.3	67.0			4, 10
HS^-		–16.3	67.0	44.8		17
HSO_4^-		–886.9	131.7	–755.3		2, 4, 10
S^{2-} (aq)		33.000	–15.00	85.800		1
S^{2-} (aq)		33.1	–14.6	85.8		17
SO_3^{2-}		–635.600	–29.00	–486.600		1
SO_3^{2-}		–635.5	–29.0	–486.5		17
SO_4^{2-}		–909.270	20.00	–744.630		1
SO_4^{2-}		–909.34	18.5	–744.0		2
SO_4^{2-}		–909.3	18.5	–744.0		4, 10
SO_4^{2-}		–909.3	18.5	–744.0		17
H_4SiO_4 (aq)		–1460.000	180.00	–1308.000		1
H_4SiO_4 (aq)		–1460.0	180.0	–1307.8		17
H_4SiO_4 (aq)		–1468.6	180	–1316.6		2
H_4SiO_4 (aq)		–1457.3		–1307.9		3

(Cont'd)

	Species or compound	Mineral	$\Delta H^o_{f,\,298.15}$ (kJ mol^{-1})	$S^o_{298.15}$ (J mol^{-1}K^{-1})	$\Delta G^o_{f,\,298.15}$ (kJ mol^{-1})	$V^o_{298.15}$ (J bar^{-1})	Source
	$H_3SiO_4^-$		−1431.7		−1251.8		3
	$H_2SiO_4^{2-}$		−1383.7		−1176.6		3
	Pt^{2+}						
	Sn^{2+}		−8.8	−17	−27.2		2
	Sn^{4+}		−30.5	−117	2.5		2
	Sr^{2+}		−545.800	−33.00	−559.440		1
	Sr^{2+}		−545.8	−32.6	−559.5		2
	Sr^{2+}		−550.9	−31.5	−563.8		17
	Sr^{2+}		−550.90	35.1	−563.83		8
	Sr^{2+}		−550.90	−31.5	−563.86		10, 11, 13
	Th^{4+}		−769.0	−422.6	−705.1		2
	U^{4+}		−613.800	−326.00	−579.100		1
	U^{4+}		−591.2	−410	−531.0		2
					−962.7		2
	UO_2^{2+}		−1019.0	−98.2			4
	UO_2^{2+}		−1019.6	−97.5	−953.5		2
	UO_2CO_3 (aq)				−1537.9		9
	$UO_2(CO_3)_2^{2-}$				−2104.4		9
	$UO_2(CO_3)_3^{4-}$				−2659.2		9
	VO^{2+}		−486.6	−133.9	−446.4		2
	Zn^{2+}		−153.390	−109.60	−147.260		1
	Zn^{2+}		−153.9	−112.1	−147.1		2
	Zn^{2+}		−153.4	−109.8	−147.3		17
	$ZnCl^+$				−275.3		2
Sulfides, sulfosalts	Ag_2S (c)	Acanthite	−32.346	142.84	−40.080	3.419	1
	Ag_2S (c)	Acanthite	−32.6	144.0	−40.7		2
	Ag_2S (c)	Acanthite	−32.0	142.9	−39.7	3.419	17
	AsS (c)	Realgar	−71.340	63.51	−70.230	2.98	1
	AsS (c)	Realgar	−30.9	63.5	−29.6	2.980	17
	As_2S_3 (c)	Orpiment	−169.030	163.60	−168.410	7.051	1
	As_2S_3 (c)	Orpiment	−91.6	163.6	−90.4	7.051	17
	CdS (c)	Greenockite	−149.600	70.29	−145.630	2.9934	1
	CdS (c)	Greenockite	−161.9	64.9	−156.5		2
	CdS (c)	Greenockite	−149.6	72.2	−146.1	2.993	17
	CuS (c)	Covellite	−48.575	66.65	−49.080	2.042	1
	CuS (c)	Covellite	−53.1	66.5	−53.6		2
	CuS (c)	Covellite	−54.6	67.4	−55.3	2.042	17
	Cu_2S (c)	Chalcocite	−80.115	120.75	−86.868	2.7475	1
	Cu_2S (c)	Chalcocite	−79.5	120.9	−86.2		2
	Cu_2S (c)	Chalcocite	−83.9	116.2	−89.2	2.748	17
	$CuFeS_2$ (c)	Chalcopyrite	−194.9	124.9	−195.1	4.392	17
	Cu_5FeS_4(c)	Bornite	−371.6	398.5	−394.7	9.873	17
	FeS_2 (c)	Pyrite	−171.544	52.93	−160.229	2.394	1
	FeS_2 (c)	Pyrite	−178.2	52.9	−166.9		2
	FeS_2 (c)	Pyrite	−171.5	52.9	−160.1	2.394	17
	FeS_2 (c)	Marcasite	−169.450	53.89	−158.421	2.458	1
	FeS_2 (c)	Marcasite	−169.5	53.9	−158.4	2.458	17
	FeS (c)	Troilite	−100.960	60.33	−101.333	1.820	1
	FeS (c)	Troilite	−100.0	60.3	−100.4		3
	FeS (c)	Troilite	−101.0	60.3	−101.3	1.820	173
	$Fe_{0.875}S$ (c)	Pyrrhotite	−97.5	60.7	−98.9	1.749	17
	$Fe_{0.90}S$ (c)	Pyrrhotite	−97.6	63.2	−99.6	1.688	17
	HgS (c)	Cinnabar	−58.155	82.51	−50.645	2.8416	1
	HgS (c)	Cinnabar	−54.3	82.5	−40.7	2.842	17
	HgS (c)	Metacinnabar	−46.735	96.23	−43.315	1.0169	1
	HgS (c)	Metacinnabar	−46.7	96.2	−43.3	3.017	17
	MnS (c)	Alabandite	−213.865	78.20	−218.155	2.146	1
	MnS (c)	Alabandite	−213.9	80.3	−218.7	2.146	17

(Continued)

(*Cont'd*)

	Species or compound	Mineral	$\Delta H^o_{f,\,298.15}$ (kJ mol^{-1})	$S^o_{298.15}$ (J mol^{-1}K^{-1})	$\Delta G^o_{f,\,298.15}$ (kJ mol^{-1})	$V^o_{298.15}$ (J bar^{-1})	Source
	MoS_2 (c)	Molybdenite	−306.269	62.57	−297.421	3.202	1
	MoS_2 (c)	Molybdenite	−271.8	62.6	−262.8	3.202	17
	NiS (c)	Millerite	−84.868	66.11	−86.192	1.689	1
	NiS (c)	Millerite	−82.0	53.0	−79.5		2
	NiS (c)	Millerite	−91.0	53.0	−63.9	1.689	17
	Ni_3S_2 (c)	Hazelwoodite	−202.920	133.90	−197.070	4.095	1
	Ni_3S_2 (c)	Hazelwoodite	−216.3	133.2	−210.2	4.095	17
	PbS (c)	Galena	−97.709	91.38	−96.075	3.149	1
	PbS (c)	Galena	−98.3	91.7	−96.8	3.149	17
	PdS (c)		−75	46	67		2
	PtS (c)	Cooperite	−82.4	55.1	−76.9	2.215	17
	Sb_2S_3 (c)	Stibnite	−174.890	182.00	−173.470	7.341	1
	Sb_2S_3 (c)	Stibnite	−151.4	182.0	−149.9	7.341	17
	WS_2	Tungstenite	−298.320	94.98	−297.945	3.207	1
	WS_2	Tungstenite	−241.6	67.8	−233.0	3.207	17
	ZnS (c)	Sphalerite	−206.900	58.66	−202.496	2.383	1
	ZnS (c)	Sphalerite	−206.0	57.7	−201.3		2
	ZnS (c)	Sphalerite	−204.1	58.7	−199.6	2.383	17
	ZnS (c)	Wurtzite	−203.8	58.8	−199.3	2.385	17
	Al_2O_3 (c)	Corundum	−1675.700	50.92	−1582.228	2.5575	1
Oxides, hydroxides	Al_2O_3 (c)	Corundum	−1675.700	50.92	−1582.3		4, 10
	Al_2O_3 (c)	Corundum	−1675.700	50.820	−1582.199	2.558	16
	Al_2O_3 (c)	Corundum	−1675.7	50.9	−1582.3	2.558	17
	Al_2O_3 (c)	Corundum	−1675.19	50.90	−1581.72	2.558	18
	$Al(OH)_3$ (c)	Gibbsite	−1293.128	68.44	−1154.889	3.1956	1
	$Al(OH)_3$ (c)	Gibbsite	−1293.3	68.4	−1155.1		2
	$Al(OH)_3$ (c)	Gibbsite	−1293.13	68.44	−1154.89		6, 10
	$Al(OH)_3$ (c)	Gibbsite	−1293.1	68.4	−1154.9	3.196	17
	α-AlOOH (c)	Diaspore	−1000.585	35.27	−922.000	1.776	1
	α-AlOOH (c)	Diaspore	−999.4	35.3	−920.9		2
	α-AlOOH (c)	Diaspore	−999.378	35.308	−920.806	1.776	16
	α-AlOOH (c)	Diaspore	−1001.3	35.3	−922.7	1.776	17
	α-AlOOH (c)	Diaspore	−999.40	35.00	−920.73	1.776	18
	γ-AlOOH (c)	Bohemite	−993.054	48.45	−918.400	1.9535	1
	γ-AlOOH (c)	Bohemite	−990.4	48.4	−915.8		2
	γ-AlOOH (c)	Bohemite	−996.39	37.19	−918.4		10, 15
	γ-AlOOH (c)	Bohemite	−996.4	37.2	−918.4	1.954	17
	CaO (c)	Lime	−635.089	38.21	−603.487	1.6764	1
	CaO (c)	Lime	−634.92	38.1	−603.30		10, 16
	CaO (c)	Lime	−635.090	37.750	−603.350	1.676	16
	CaO (c)	Lime	−635.090	37.750	−603.350	1.676	16
	CaO (c)	Lime	−635.1	38.1	−603.1	1.676	17
	$Ca(OH)_2$ (c)	Portlandite	−986.085	83.39	−898.408	3.3056	1
	$Ca(OH)_2$ (c)	Portlandite	−985.2	83.4	−897.5		7
	$Ca(OH)_2$ (c)	Portlandite	−985.16	83.40	−897.5		10, 16
	$Ca(OH)_2$ (c)	Portlandite	−986.1	83.4	−898.0	3.306	17
	Co_3O_4 (c)	Co- spinel	−891.190	102.51	−772.553	3.977	1
	Co_3O_4 (c)	Co- spinel	−891	102.5	−774		2
	$FeCr_2O_4$ (c)	Chromite		146.02		4.401	1
	$FeCr_2O_4$ (c)	Chromite	−1445.5	146.0	−1344.5	4.401	17
	CuO (c)	Tenorite	−157.320	42.63	−129.564	1.222	1
	CuO (c)	Tenorite	−156.1	42.6	−128.3	1.222	17
	Cu_2O (c)	Cuprite	−168.610	93.14	−146.030	2.3437	1
	Cu_2O (c)	Cuprite	−170.06	92.4	−147.8	2.344	17
	FeO (c)		−272.043	59.80	−251.156	1.200	1
	FeO (c)		−272.0	60.6	−251.4	1.200	17
	Fe_2O_3 (c)	Hematite	−824.640	87.40	−742.683	3.0274	1
	Fe_2O_3 (c)	Hematite	−824.7	87.7	−742.8		6

(*Cont'd*)

Species or compound	Mineral	$\Delta H^o_{f,\ 298.15}$ (kJ mol^{-1})	$S^o_{298.15}$ (J mol^{-1}K^{-1})	$\Delta G^o_{f,\ 298.15}$ (kJ mol^{-1})	$V^o_{298.15}$ (J bar^{-1})	Source
Fe_2O_3 (c)	Hematite	–824.718	87.74	–84		10, 2
Fe_2O_3 (c)	Hematite	–825.627	87.437	–743.681	3.027	16
Fe_2O_3 (c)	Hematite	–826.2	87.4	–744.4	3.027	17
Fe_2O_3 (c)	Hematite	–825.73	87.40	–743.73	3.027	18
Fe_3O_4 (c)	Magnetite	–1115.726	146.14	–1012.566	4.4524	1
Fe_3O_4 (c)	Magnetite	–1116.1	146.1	–1012.9		6
Fe_3O_4 (c)	Magnetite	–1116.125	146.138	–1012.931		2, 10
Fe_3O_4 (c)	Magnetite	–1117.403	146.114	–1014.235	4.452	16
Fe_3O_4 (c)	Magnetite	–1115.7	146.1	–1012.7	4.452	17
Fe_3O_4 (c)	Magnetite	–1115.55	146.10	–1012.31	4.452	18
$FeTiO_3$ (c)	Ilmenite	–1236.622	105.86	–1159.170	3.169	1
$FeTiO_3$ (c)	Ilmenite	–1231.947	108.628	–1155.320	3.170	16
$FeTiO_3$ (c)	Ilmenite	–1231.25	108.90	–1154.63	3.169	18
Fe_2TiO_4 (c)	Ulvospinel	–1493.8	180.40	–1399.9	4.682	17
Fe_2TiO_4 (c)	Ulvospinel	–1497.44	175.00	–1401.79	4.682	18
FeOOH (c)	Goethite	–559.330	60.38	–488.550	2.082	1
FeOOH (c)	Goethite	–559.33	60.38	–488.55		1, 10
FeOOH (c)	Goethite	–562.6	60.4	–491.8	2.082	17
FeOOH (c)	Goethite	–561.66	60.40	–490.86	2.082	18
KOH (c)		–424.676	78.91	–378.932	2.745	1
KOH (c)		–424.7	78.9	–378.9	2.745	17
MgO (c)	Periclase	–601.490	26.94	–569.196	1.1248	1
MgO (c)	Periclase	–601.60	26.95	–569.31		7, 10
MgO (c)	Periclase	–601.500	26.951	–569.209	1.125	16
MgO (c)	Periclase	–601.6	26.9	–569.3	1.125	17
MgO (c)	Periclase	–601.65	26.90	–569.34	1.125	18
$Mg(OH)_2$ (c)	Brucite	–924.540	63.18	–833.506	2.463	1
$Mg(OH)_2$ (c)	Brucite	–924.54	63.18	–833.51		2, 10, 15
$Mg(OH)_2$ (c)	Brucite	–925.937	63.064	–834.868	2.468	16
$Mg(OH)_2$ (c)	Brucite	–924.5	63.2	–833.5	2.463	17
$Mg(OH)_2$ (c	Brucite	–924.97	64.50	–834.31	2.463	18
$MgAl_2O_4$ (c)	Spinel	–2299.320	80.63	–2174.860	3.971	1
$MgAl_2O_4$ (c)	Spinel	–2300.313	84.535	–2176.537	3.997	16
$MgAl_2O_4$ (c)	Spinel	–2299.1	88.7	–2176.6	3.971	17
$MgAl_2O_4$ (c)	Spinel	–2300.31	81.50	–2175.64	3.978	18
MnO (c)	Manganosite	–385.220	59.71	–362.896	1.3221	1
Mn_3O_4 (c)	Hausmannite	–1387.830	153.97	–1282.774	4.6950	1
MnO_2 (c)	Pyrolusite	–520.030	53.05	–465.138	1.661	1
MnO_2 (c)	Pyrolusite	–520.3	53.1	–465.14		2
MnO_2 (c)	Pyrolusite	–520.0	52.8	–465.0	1.661	17
NaOH (c)		–425.800	64.43	–379.651	1.878	1
NaOH (c)		–425.8	64.4	–379.6	1.878	17
NiO (c)	Bunsenite	–239.3	38.0	–211.1	1.097	17
SnO_2 (c)	Cassiterite	–580.740	52.30	–519.902	2.155	1
SnO_2 (c)	Cassiterite	–519.6	52.3	–580.7		2
SnO_2 (c)	Cassiterite	–577.6	49.0	–515.8	2.155	17
ThO_2 (c)	Thorianite	–1226.410	65.23	–1168.775	2.6373	1
ThO_2 (c)	Thorianite	–1226.4	65.2	–1169.2	2.637	17
TiO_2 (c)	Rutile	–944.750	50.29	–889.446	1.882	1
TiO_2 (c)	Rutile	–944.750	50.460	–889.497	1.882	16
TiO_2 (c)	Rutile	–944.0	50.6	–888.8	1.882	17
TiO_2 (c)	Rutile	–94419	50.60	–888.92	1.882	18
$CaTiO_3$ (c, ortho)	Perovskite	–1660.630	93.64	–1575.256	3.3626	1
$CaTiO_3$ (c, ortho)	Perovskite	–1660.6	93.6	–1574.8	3.363	17
$FeTiO_3$ (c)	Ilmenite	–1236.622	105.86	–1159.170	3.169	1
$FeTiO_3$ (c)	Ilmenite	–1232.0	108.9	–1155.5	3.169	17
UO_3 (c)		–1223.800	98.62	–1146.461	3.556	1
U_3O_8 (c)		–3574.8	282.55			4
U_3O_8 (c)				–3369.6		9

(*Continued*)

(Cont'd)

	Species or compound	Mineral	$\Delta H^{\circ}_{f,\,298.15}$ (kJ mol^{-1})	$S^{\circ}_{298.15}$ (J mol^{-1}K^{-1})	$\Delta G^{\circ}_{f,\,298.15}$ (kJ mol^{-1})	$V^{\circ}_{298.15}$ (J bar^{-1})	Source
	UO_2 (c)	Uraninite	−1084.910	77.03	−1031.770	2.4618	1
	UO_2 (c)	Uraninite	−1084.9	77.0	−1031.7	2.4622	17
	V_2O_3 (c)	Karelianite	−1218.800	98.07	−1139.052	2.9850	1
	V_2O_3 (c)	Karelianite	−1218.8	98.1	−1139.0	2.985	17
	ZnO (c)	Zincite	−350.460	43.64	−320.477	1.4338	1
	ZnO (c)	Zincite	−348.3	43.6	−318.3		2
	ZnO (c)	Zincite	−350.5	43.2	−320.4	1.434	17
	ZrO_2 (c)	Baddeleyite	−1100.560	50.38	−1042.790	2.115	1
	ZrO_2 (c)	Baddeleyite	−1100.6	50.4	−1042.9	2.115	17
Sulfates, nitrates, carbonates, phosphates, tungstates	$AgNO_3$ (c)		−124.4	140.9	−33.4		2
	$BaCO_3$ (c)	Witherite	−1210.850	112.13	−1132.210	4.581	1
	$BaCO_3$ (c)	Witherite	−1216.3	112.1	−1137.6		2
	$BaCO_3$ (c)	Witherite	−1210.85	112.13	−1132.21		10, 13
	$BaCO_3$ (c)	Witherite	−1210.9	112.1	−1132.2		17
	$BaSO_4$ (c)	Barite	−1473.190	132.21	−1362.186	5.210	1
	$BaSO_4$ (c)	Barite	−1468.3	128.6	−1356.3		10
	$BaSO_4$ (c)	Barite	−1473.6	132.2	−1362.5	5.210	17
	$CaCO_3$ (c)	Calcite	−1207.370	91.71	−1128.842	3.6934	1
	$CaCO_3$ (c)	Calcite	−1207.6	91.7	−1129.07		7
	$CaCO_3$ (c)	Calcite	−1207.6	91.71	−1129.07		7, 10
	$CaCO_3$ (c)	Calcite	−1206.819	91.725	−1128.295	3.690	16
	$CaCO_3$ (c)	Calcite	−1207.4	91.7	−1128.5	3.693	17
	$CaCO_3$ (c)	Calcite	−1207.54	92.50	−1128.81	3.689	18
	$CaCO_3$ (c)	Aragonite	−1207.430	87.99	−1127.793	3.415	1
	$CaCO_3$ (c)	Aragonite	−1206.4	93.9	−1128.3		7
	$CaCO_3$ (c)	Aragonite	−1206.4	93.91	−1128.25		7, 8, 10
	$CaCO_3$ (c)	Aragonite	−1207.4	88.0	−1127.4	3.415	17
	$CaCO_3$ (c)	Aragonite	−1207.65	89.50	−1128.03	3.415	18
	$CaMg(CO_3)_2$ (c)	Dolomite	−2324.480	155.18	−2161.672	6.434	1
	$CaMg(CO_3)_2$ (c)	Dolomite	−2326.3	155.2	−2163.4		2
	$CaMg(CO_3)_2$ (c)	Dolomite	−2325.248	154.890	−2162.354	6.432	16
	$CaMg(CO_3)_2$ (c)	Dolomite	−2324.5	155.2	−2161.3	6.434	17
	$CaMg(CO_3)_2$ (c)	Dolomite	−2324.56	156.00	−2161.51	6.434	18
	$CaSO_4$ (c)	Anhydrite	−1434.110	106.69	−1321.696	4.594	1
	$CaSO_4$ (c)	Anhydrite	−1435.5	106.5	−1321.98		7
	$CaSO_4$ (c)	Anhydrite	−1434.5	106.5	−1321.98		7, 10
	$CaSO_4$ (c)	Anhydrite	−1434.4	107.4	−1321.8	4.601	17
	$CaSO_4.2H_2O$ (c)	Gypsum	−2022.628	194.14	−1797.197	7.469	1
	$CaSO_4.2H_2O$ (c)	Gypsum	−2022.92	193.9	−1797.36		7
	$CaSO_4.2H_2O$ (c)	Gypsum	−2022.92	193.9	−1797.36		7, 10
	$CaSO_4.2H_2O$ (c)	Gypsum	−2023.0	193.8	−1797.0	7.469	17
	$Ca_5(PO_4)_2$ (c)	Whitlockite	−4120.8	236.0	−3883.6	9.762	17
	$Ca_5(PO_4)_3OH$ (c)	Hydroxyapatite	−6721.600	390.37	−6338.434	15.960	1
	$Ca_5(PO_4)_3OH$ (c)	Hydroxyapatite	−6738.5	390.4	−6338.5		2
	$Ca_5(PO_4)_3F$ (c)	Fluroapatite	−6872.200	387.86	−6508.119	15.756	1
	$Ca_5(PO_4)_3F$ (c)	Fluroapatite	−6872	387.9	−6491.5		2
	$Ca_5(PO_4)_3F$ (c)	Fluroapatite	−6872.0	387.9	−6489.7	15.756	17
	$CaWO_4$ (c)	Scheelite	−1645.150	126.40	−1538.361	4.705	1
	$CaWO_4$ (c)	Scheelite	−1645.2	126.4	−1538.0	4.705	17
	$Cu_2(CO_3)(OH)_2$ (c)	Malachite	−1054.0	166.3	−890.2	5.486	17
	$Cu_3(OH_2)(CO_3)_2$ (c)	Azurite	−1632.2	254.4	−1391.4	9.101	17
	$FeCO_3$ (c)	Siderite	−736.985	105.0	−666.698	2.9378	1
	$FeCO_3$ (c)	Siderite	−753.8		−673.05		3
	$FeCO_3$ (c)	Siderite	−755.9	95.5	−682.8	2.938	17
	$FeCO_3$ (c)	Siderite	−761.50	95.00	−688.16	2.938	18
	H_3PO_4 (c)		−1266.920	110.54	−1112.290	4.8520	1
	KNO_3 (c)	Niter	−494.5	133.1	−394.6	4.804	17
	$MgCO_3$ (c)	Magnesite	−1113.280	65.09	−1029.480	2.8018	1

(Cont'd)

	Species or compound	Mineral	$\Delta H^o_{f, 298.15}$ (kJ mol^{-1})	$S^o_{298.15}$ (J mol^{-1}K^{-1})	$\Delta G^o_{f, 298.15}$ (kJ mol^{-1})	$V^o_{298.15}$ (J bar^{-1})	Source
	$MgCO_3$ (c)	Magnesite	–1095.8	65.7	–1012.1		2
	$MgCO_3$	Magnesite	–1113.636	65.210	–1029.875	2.803	16
	$MgCO_3$	Magnesite	–1113.3	65.1	–1029.5	2.802	17
	$MgCO_3$	Magnesite	–1111.59	65.10	–1027.74	2.803	18
	$MnCO_3$ (c)	Rhodochrosite	–889.270	100.00	–816.047	3.1073	1
	$MnCO_3$ (c)	Rhodochrosite	–894.1	85.8	–816.7		2
	$MnCO_3$ (c)	Rhodochrosite	–892.9	98.0	–819.1	3.107	17
	$MnCO_3$ (c)	Rhodochrosite	–891.06	98.00	–817.22	3.107	18
	$NaCO_3$ (c)		–1129.2	135.0	–1045.3	4.160	17
	NH_4NO_3 (c)		–365.560	151.08	–183.803	4.649	1
	NH_4NO_3 (c)		–365.6	151.1	–183.8	4.649	17
	$NiSO_4$ (c)		–872.9	92	–759.9		2
	$NiSO_4$ (c)		–873.2	101.3	–762.7	3.857	17
	$PbCO_3$ (c)	Cerussite	–699.150	130.96	–625.337	4.059	1
	$PbCO_3$ (c)	Cerussite	–699.2	131.0	–625.5	4.059	17
	$PbSO_4$ (c)	Anglesite	–919.940	148.57	–813.026	4.795	1
	$PbSO_4$ (c)	Anglesite	–920.0	148.5	–813.1	4.795	17
	$SrCO_3$ (c)	Strontianite	–1218.680	97.07	–1137.640	3.901	1
	$SrCO_3$ (c)	Strontianite	–1220.1	97.1	–1140.1		2
	$SrCO_3$ (c)	Strontianite	–1225.8	97.2	–1144.73		8
	$SrCO_3$ (c)	Strontianite	–1225.8	97.2	–1144.73		8, 10
	$SrCO_3$ (c)	Strontianite	–1218.7	97.1	–1137.6	3.901	17
	$SrSO_4$ (c)	Celestite	–1453.170	118.00	–1340.970	4.625	1
	$SrSO_4$ (c)	Celestite	–1453.1	117	–1340.9		2
	$SrSO_4$ (c)	Celestite	–1456.9		–1345.7		3
	$SrSO_4$ (c)	Celestite	–1455.9	128.3	–1345.7		10
	$SrSO_4$ (c)	Celestite	–1453.2	117.0	–1339.6	4.625	17
	$ZnCO_3$ (c)	Smithsonite	–812.780	82.42	–731.480	2.8275	1
	$ZnCO_3$ (c)	Smithsonite	–817.0	81.2	–735.3	2.828	17
	$ZnSO_4$ (c)	Zinkosite	–982.820	110.46	–871.530	4.157	1
	$ZnSO_4$ (c)	Zinkosite	–980.1	110.5	–868.7	4.157	17
Halides	AgCl	Chlorargyrite	–127.070	96.23	–109.819	2.5727	1
	AgCl	Chlorargyrite	–127.1	96.2	–109.8	2.573	17
	$BaCl_2$ (c)		–858.6	123.7	–810.4		2
	CaF_2 (c)	Fluorite	–1229.260	68.87	–1176.920	2.4542	1
	CaF_2 (c)	Fluorite	–1219.6	68.9	–1167.3		2
	CaF_2 (c)	Fluorite	–1228.0	68.51	–1175.57		7, 10
	CaF_2 (c)	Fluorite	–1228.0	68.9	–1175.3	2.454	17
	Hg_2Cl_2 (c)	Calomel	–132.610	96.23	–105.415	3.2939	1
	Hg_2Cl_2 (c)	Calomel	–132.7	95.8	–105.2	3.294	17
	KCl (c)	Sylvite	–436.470	82.59	–408.554	3.7524	1
	KCl (c)	Sylvite	–436.5	82.6	–408.6	3.752	17
	NaCl (c)	Halite	–411.260	72.12	–384.212	2.7015	1
	NaCl (c)	Halite	–411.15	72.1	–384.14		2
	NaCl (c)	Halite	–411.3	72.02	–384.2	2.702	17
Silicates	Al_2SiO_5 (c)	Kyanite	–2591.730	83.76	–2441.276	4.409	1
	Al_2SiO_5 (c)	Kyanite	–2594.220	82.430	–2443.370	4.412	16
	Al_2SiO_5 (c)	Kyanite	–2593.8	82.8	–2443.1	4.415	17
	Al_2SiO_5 (c)	Kyanite	–2593.13	83.50	–2442.59	4.414	18
	Al_2SiO_5 (c)	Andalusite	–2587.525	93.22	–2439.892	5.153	1
	Al_2SiO_5 (c)	Andalusite	–2589.972	91.434	–2441.806	5.147	16
	Al_2SiO_5 (c)	Andalusite	–2589.9	91.4	–2441.8	5.152	17
	Al_2SiO_5 (c)	Andalusite	–2588.77	92.70	–2440.97	5.153	18
	Al_2SiO_5 (c)	Sillimanite	–2585.760	96.11	–2438.988	4.990	1
	Al_2SiO_5 (c)	Sillimanite	–2586.091	95.930	–2439.265	4.983	16
	Al_2SiO_5 (c)	Sillimanite	–2586.1	95.4	–2439.1	4.986	17
	Al_2SiO_5 (c)	Sillimanite	–2585.89	95.50	–2438.93	4.986	18

(Continued)

(*Cont'd*)

Species or compound	Mineral	$\Delta H^o_{f,\ 298.15}$ (kJ mol^{-1})	$S^o_{298.15}$ (J mol^{-1}K^{-1})	$\Delta G^o_{f,\ 298.15}$ (kJ mol^{-1})	$V^o_{298.15}$ (J bar^{-1})	Source
$Al_2SiO_4F_2$ (c)	Topaz	–3084.5	105.4	–2910.6	5.153	17
$Al_2Si_2O_5(OH)_4$ (c)	Kaolinite	–4120.114	203.05	–3799.364	9.952	1
$Al_2Si_2O_5(OH)_4$ (c)	Kaolinite	–4119.6	205.0	–3799.7		2
$Al_2Si_2O_5(OH)_4$ (c)	Kaolinite	–4133		–3785.8		3
$Al_2Si_2O_5(OH)_4$ (c)	Kaolinite	–4120.327	203.700	–3799.770	9.952	16
$Al_2Si_2O_5(OH)_4$ (c)	Kaolinite	–4119.0	200.4	–3797.5	9.934	17
$Al_2Si_2O_5(OH)_4$ (c)	Kaolinite	–4122.33	203.70	–3801.72	9.934	18
$Al_2Si_2O_5(OH)_4$ (c)	Dickite	–4118.840	197.07	–3796.305	9.93	1
$Al_2Si_2O_5(OH)_4$ (c)	Dickite	–4118.3	197.1	–3795.9		2
$Al_2Si_2O_5(OH)_4$ (c)	Dickite	–4118.5	197.1	–3796.0	9.856	17
$Al_2Si_2O_5(OH)_4$ (c)	Halloysite	–4101.480	203.00	–3780.713		1
$Al_2Si_2O_5(OH)_4$ (c)	Halloysite	–4101.2	203.3	–3780.5		2
$Al_2Si_2O_5(OH)_4$ (c)	Halloysite	–4101.5	203.0	–3780.7		17
$Al_2Si_4O_{10}(OH)_2$ (c)	Pyrophyllite	–5643.300	239.40	–5269.384	12.782	1
$Al_2Si_4O_{10}(OH)_2$ (c)	Pyrophyllite	–5642.0	239.4	–5268.1		2
$Al_2Si_4O_{10}(OH)_2$ (c)	Pyrophyllite			–5273.3		3
$Al_2Si_4O_{10}(OH)_2$ (c)	Pyrophyllite	–5640.781	239.400	–5266.865	12.76	16
$Al_2Si_4O_{10}(OH)_2$ (c)	Pyrophyllite	–5640.0	239.4	–5266.2	12.81	17
$Al_2Si_4O_{10}(OH)_2$ (c)	Pyrophyllite	–5640.85	239.40	–5266.87	12.810	18
$CaAl_2Si_2O_8$ (c)	Anorthite	–4243.040	199.30	–4017.266	10.079	1
$CaAl_2Si_2O_8$ (c)	Anorthite	–4227.8	199.3	–4002.2		2
$CaAl_2Si_2O_8$ (c)	Anorthite	–4228.730	200.186	–4003.221	10.075	16
$CaAl_2Si_2O_8$ (c)	Anorthite	–4234.0	199.3	–4007.9	10.079	17
$CaAl_2Si_2O_8$ (c)	Anorthite	–4233.48	200.00	–4007.51	10.079	18
$Ca_2Al_2Si_3O_{12}$ (c)	Grossular	–6656.700	255.50	–6294.919	12.530	1
$Ca_2Al_2Si_3O_{12}$ (c)	Grossular	–6632.859	255.150	–6270.974	12.538	16
$Ca_2Al_2Si_3O_{12}$ (c)	Grossular	–6640.0	260.1	–6278.5	12.528	17
$Ca_2Al_2Si_3O_{12}$ (c)	Grossular	–6644.07	255.00	–6280.94	12.535	18
$Ca_2Al_3Si_3O_{12}(OH)$ (c)	Zoisite	–6889.488	297.576	–6494.148	13.588	16
$Ca_2Al_3Si_3O_{12}(OH)$ (c)	Zoisite	–6901.1	295.9	–6504.5	13.65	17
$Ca_2Al_3Si_3O_{12}(OH)$ (c)	Zoisite	–6898.57	297.00	–6502.25	13.575	18
$Ca_2Al_3Si_3O_{12}(OH)$ (c)	Clinozoisite	–6894.968	287.076	–6496.497	13.673	16
$Ca_2Al_3Si_3O_{12}(OH)$ (c)	Clinozoisite	–6898.11	301.00	–6502.98	13.360	18
$CaFeSi_2O_6$ (c)	Hedenbergite	–2839.9	174.2	–2676.3	6.795	17
$CaFeSi_2O_6$ (c)	Hedenbergite	–2843.45	175.00		6.795	18
$Ca_2Fe_2Si_3O_{12}$ (c)	Andradite	–5771.0	316.4	–5427.0	13.204	17
$Ca_2Fe_2Si_3O_{12}$ (c)	Andradite	–5768.09	318.00	–5424.33	13.204	18
$CaMgSi_2O_6$ (c)	Diopside	–3210.760	143.09	–3036.554	6.609	1
$CaMgSi_2O_6$ (c)	Diopside	–3206.2	142.9	–3032.0		2
$CaMgSi_2O_6$ (c)	Diopside	–3200.583	142.500	–3026.202	6.620	16
$CaMgSi_2O_6$ (c)	Diopside	–3201.5	142.7	–3026.8	6.609	17
$CaMgSi_2O_6$ (c)	Diopside	–3202.54	142.70	–3027.80	6.619	18
$Ca_2Mg_5Si_8O_{22}(OH)_2$ (c)	Tremolite	–12355.08	548.90	–11627.91	27.292	1
$Ca_2Mg_5Si_8O_{22}(OH)_2$ (c)	Tremolite	–12360	548.9	–11631		2
$Ca_2Mg_5Si_8O_{22}(OH)_2$ (c)	Tremolite	–12319.7	548.9	–11592.6		5
$Ca_2Mg_5Si_8O_{22}(OH)_2$ (c)	Tremolite	–12305.578	551.150	–11578.55	27.268	16
$Ca_2Mg_5Si_8O_{22}(OH)_2$ (c)	Tremolite	–12303.0	548.9	–11574.6	27.290	17
$Ca_2Mg_5Si_8O_{22}(OH)_2$ (c)	Tremolite	–12309.72	550.00	–11581.42	27.270	18
$CaSiO_3$ (c)	Wollastonite	–1635.220	82.01	–1549.903	3.993	1
$CaSiO_3$ (c)	Wollastonite	–1631.500	81.810	–1546.123	3.983	16
$CaSiO_3$ (c)	Wollastonite	–1634.8	81.7	–1549.0	3.990	17
$CaSiO_3$ (c)	Wollastonite	–1634.04	82.50	–1548.47	3.993	18
$CaTiSiO_5$ (c)	Sphene (titanite)	–2601.400	129.20	–2459.855	5.565	1
$CaTiSiO_5$ (c)	Sphene (titanite)	–2596.652	129.290	–2455.134	5.565	16
$CaTiSiO_5$ (c)	Sphene (titanite)	–2596.6	129.2	–2454.6	5.574	17
$CaTiSiO_5$ (c)	Sphene (titanite)	–2595.55	131.20	–2454.15	5.565	18
Fe_2SiO_4	Fayalite	–1479.360	148.32	–1379.375	4.639	1
Fe_2SiO_4	Fayalite	–1479.360	150.930	–1380.154	4.630	16

(*Cont'd*)

Species or compound	Mineral	$\Delta H^o_{f,\ 298.15}$ (kJ mol^{-1})	$S^o_{298.15}$ (J mol^{-1}K^{-1})	$\Delta G^o_{f,\ 298.15}$ (kJ mol^{-1})	$V^o_{298.15}$ (J bar^{-1})	Source
Fe_2SiO_4	Fayalite	−1478.2	151.0	−1379.1	4.631	17
Fe_2SiO_4	Fayalite	−1478.22	151.00	−1378.98	4.631	18
$FeSiO_3$	Ferrosilite	−1194.375	95.882	−1117.472	3.296	16
$FeSiO_3$	Ferrosilite	−1195.2	94.6	−1118.0	3.300	17
$FeSiO_3$	Ferrosilite	−2388.75	190.60	−2234.53	6.592	18
$Fe_3Al_2Si_3O_{12}$ (c)	Almandine	−5265.502	339.927	−4941.728	11.511	16
$Fe_3Al_2Si_3O_{12}$ (c)	Almandine	−3864.8	342.6	−4942.0	11 532	17
$Fe_3Al_2Si_3O_{12}$ (c)	Almandine	−5263.65	340.00	−4939.80	11.511	18
$KAlSi_3O_8$ (c)	Microcline	−3967.690	214.20	−3742.30	10.872	1
$KAlSi_3O_8$ (c)	Microcline	−3681.1	214.2	−3742.9		2
$KAlSi_3O_8$ (c)	Microcline	−3970.791	214.145	−3745.415	10.869	16
$KAlSi_3O_8$ (c)	Microcline	−3974.6	214.2	−3749.3	10.872	17
$KAlSi_3O_8$ (c)	Microcline	−3975.05	216.00	−3750.19	10.892	18
$KAlSi_3O_8$ (c)	High sanidine	−3959.560	232.90	−3739.776	10.905	1
$KAlSi_3O_8$ (c)	Sanidine	−3959.704	229.157	−3738.804	10.896	16
$KAlSi_3O_8$ (c)	Sanidine	−3965.6	232.8	−3745.8	10.905	17
$KAlSi_3O_8$ (c)	Sanidine	−3964.90	230.00	−3744.21	10.900	18
$KAl_3Si_3O_{10}(OH)_2$ (c)	Muscovite	−5976.740	306.40	−5600.671	14.071	1
$KAl_3Si_3O_{10}(OH)_2$ (c)	Muscovite	−5984.4	305.3	−5608.4		2
$KAl_3Si_3O_{10}(OH)_2$ (c)	Muscovite	−5976.740	293.157	−5596.723	14.087	16
$KAl_3Si_3O_{10}(OH)_2$ (c) (Al/Si disordered)	Muscovite	−5974.8	306.4	−5598.8	14.081	17
$KAl_3Si_3O_{10}(OH)_2$ (c) (Al/Si ordered)	Muscovite	−59990.0	287.7	−5608.4	14.081	17
$KAl_3Si_3O_{10}(OH)_2$ (c)	Muscovite	−5984.12	292.00	−5603.71	14.083	18
$Kfe_3AlSi_3O_{10}(OH)_2$	Annite	−5149.3	415.00	−4798.3	15.43	17
$KMg_3AlSi_3O_{10}(OH)_2$ (c)	Phlogopite		319.66		14.991	1
$KMg_3AlSi_3O_{10}(OH)_2$ (c)	Phlogopite			−5831.8		2
$KMg_3AlSi_3O_{10}(OH)_2$ (c)	Phlogopite	−6207.342	334.158	−5827.224	14.977	16
$KMg_3AlSi_3O_{10}(OH)_2$ (c) (Al/Si disordered)	Phlogopite	−6226.0	334.6	−5846.0	14.965	17
$KMg_3AlSi_3O_{10}(OH)_2$ (c) (Al/Si ordered)	Phlogopite	−6246.0	315.9	−5860.5	14.965	17
$KMg_3AlSi_3O_{10}(OH)_2$ (c)	Phlogopite	−6219.44	3288.00	−5837.42	14.964	18
$Mg_2Al_4Si_5O_{18}$	Cordierite	−9158.727	417.970	−8651.517	23.311	16
$Mg_2Al_4Si_5O_{18}$	Cordierite	−9161.5	407.2	−8651.1	23.322	17
Mg_2SiO_4	Forsterite	−2170.370	95.19	−2051.325	4.379	1
Mg_2SiO_4	Forsterite	−2174.0	95.1	−2055.1		2
Mg_2SiO_4	Forsterite	−2175.7	95.2	−2056.7		5
Mg_2SiO_4	Forsterite	−2174.420	94.010	−2055.023	4.366	16
Mg_2SiO_4	Forsterite	−2173.0	94.1	−2053.6	4.365	17
Mg_2SiO_4	Forsterite	−2171.85	95.10	−2052.75	4.366	18
$MgSiO_3$ (c)	Enstatite (orthorhombic)	−1546.8	67.8	−1459.9		5
$MgSiO_3$ (c)	Enstatite (orthorhombic)	−1545.552	66.170	−1458.181	3.133	16
$MgSiO_3$ (c)	Enstatite	−1545.6	66.3	−1458.3	3.131	17
$MgSiO_3$ (c)	Clinoenstatite	−1547.750	67.86	−1460.883	3.147	1
$MgSiO_3$ (c)	Clinoenstatite	−1549.0	67.7	−1462.1		2
$MgSiO_3$ (c)	Clinoenstatite	−1545.926	66.325	−1458.601	3.131	16
$MgSiO_3$ (c)	Clinoenstatite	−1545.0	67.9	−1458.1	3.128	17
$Mg_7Si_8O_{22}(OH)_2$(c)	Anthophyllite	−12069.032	535.195	−11342.582	26.560	16
$Mg_7Si_8O_{22}(OH)_2$(c)	Anthophyllite	−12070.0	534.5	−11343.4	26.54	17
$Mg_7Si_8O_{22}(OH)_2$(c)	Anthophyllite	−12068.59	536.00	−11342.22	26.540	18
$Mg_3Si_2O_5(OH)_4$ (c)	Chrysotile	−4361.660	221.30	−4034.024	10.850	1
$Mg_3Si_2O_5(OH)_4$ (c)	Chrysotile	−4365.6	221.3	−4037.8		2
$Mg_3Si_2O_5(OH)_4$ (c)	Chrysotile	−4363.356	220.134	−4035.373	10.720	16

(*Continued*)

(*Cont'd*)

Species or compound	Mineral	$\Delta H^o_{f,\ 298.15}$ (kJ mol^{-1})	$S^o_{298.15}$ (J mol^{-1}K^{-1})	$\Delta G^o_{f,\ 298.15}$ (kJ mol^{-1})	$V^o_{298.15}$ (J bar^{-1})	Source
$Mg_3Si_2O_5(OH)_4$ (c)	Chrysotile	–4360.0	221.3	–4032.4	10.750	17
$Mg_3Si_2O_5(OH)_4$ (c)	Chrysotile	–4358.46	221.30	–4030.75	10.746	18
$Mg_{48}Si_{34}O_{85}(OH)_{62}$	Antigorite	–71364.156	3602.996	–66076.529	174.246	16
$Mg_3Si_4O_{10}(OH)_2$ (c)	Talc	–5915.900	260.83	–5536.048	13.625	1
$Mg_3Si_4O_{10}(OH)_2$ (c)	Talc	–5903.3	260.8	–5523.7		2
$Mg_3Si_4O_{10}(OH)_2$ (c)	Talc	–5893	260.7	–5527.1		3
$Mg_3Si_4O_{10}(OH)_2$ (c)	Talc	–5897.387	261.240	–5517.657	13.610	16
$Mg_3Si_4O_{10}(OH)_2$ (c)	Talc	–5900.0	20.8	–5520.2	13.620	17
$Mg_3Si_4O_{10}(OH)_2$ (c)	Talc	–5896.92	260.80	–5516.73	13.625	18
$Mg_3Al_2Si_3O_{12}$ (c)	Pyrope	–6284.620	222.00	–5920.856	11.327	1
$Mg_3Al_2Si_3O_{12}$ (c)	Pyrope	–6286.548	266.359	–5936.009	11.316	16
$Mg_3Al_2Si_3O_{12}$ (c)	Pyrope	–6285.0	266.3	–5934.5	11.312	17
$Mg_3Al_2Si_3O_{12}$ (c)	Pyrope	–6284.23	266.30	–5933.62	11.318	18
$Mg_5Al_2Si_3O_{10}(OH)_8$ (c)	Chlorite (clinochlore)	–8857.4	465.3	–8207.8		5
$Mg_5Al_2Si_3O_{10}(OH)_8$ (c)	Chlorite (clinochlore)	–8909.590	435.154	–8250.546	21.147	16
$Mg_5Al_2Si_3O_{10}(OH)_8$ (c)	Chlorite (clinochlore)	–8919.0	421.0	–8255.8	21.100	17
$Mg_5Al_2Si_3O_{10}(OH)_8$ (c)	Chlorite (clinochlore)	–8929.86	410.50	–8263.35	21.090	18
$Mg_4Si_6O_{15}(OH)_2.6H_2O$(c)	Sepiolite	–10116.9	613.4	–9251.6		5
$MnSiO_3$ (c)	Rhodonite	–1319.350	102.50	–1243.081	3.516	1
$MnSiO_3$ (c)	Rhodonite	–1320.9	100.5	–1240.5		2
$MnSiO_3$ (c)	Rhodonite	–1321.6	100.5	–1244.7	3.494	17
$MnSiO_3$ (c)	Rhodonite	–1321.72	100.50	–1244.76	3.494	18
$NaAlSi_3O_8$ (c)	Albite	–3921.618	224.412	–3703.293	10.083	16
$NaAlSi_3O_8$ (c)	Albite	–3921.618	224.412	–3703.293	10.083	16
$NaAlSi_3O_8$ (c)	Albite	–3935.0	207.4	–3711.6	10.007	17
$NaAlSi_3O_8$ (c)	Low albite	–3935.120	207.40	–3711.722	10.007	1
$NaAlSi_3O_8$ (c)	Low albite	–3935.100	207.443	–3711.715	10.043	16
$NaAlSi_3O_8$ (c)	High albite	–3921.618	224.412	–3703.293	10.083	16
$NaAlSi_3O_8$ (c)	High albite	–3924.84	223.40	–3706.12	10.109	18
$NaAlSi_2O_6.H_2O$ (c)	Analcime	–3309.839	234.43	–3091.730	9.749	1
$NaAlSi_2O_6.H_2O$ (c)	Analcime	–3300.8	234.3	–3082.6		2
$NaAlSi_2O_6.H_2O$ (c)	Analcime	–3310.1	227.7	–3090.0	9.74	17
$NaAlSi_2O_6.H_2O$ (c)	Analcite	–3309.90	232.00	–3090.97	9.740	18
$NaAlSi_2O_6$ (c)	Jadeite	–3029.400	133.47	–2850.834	6.04	1
$NaAlSi_2O_6$ (c)	Jadeite	–3025.118	133.574	–2846.482	6.034	16
$NaAlSi_2O_6$ (c)	Jadeite	–3029.3	133.5	–2850.6	6.040	17
$NaAlSi_2O_6$ (c)	Jadeite	–3027.83	133.50	–2849.10	6.040	18
$NaAl_3Si_3O_{10}(OH)_2$ (c)	Paragonite	–5563.572	277.699	–5563.572	13.216	16
$NaAl_3Si_3O_{10}(OH)_2$ (c) (Al/Si disordered)	Paragonite	–5933.0	295.8	–5555.7	13.21	17
$NaAl_3Si_3O_{10}(OH)_2$ (c) (Al/Si ordered)	Paragonite	–5949.3	277.1	–5568.5	13.21	17
$NaAl_3Si_3O_{10}(OH)_2$ (c)	Paragonite	–5946.33	276.00	–5565.09	13.211	18
$NaSi_7O_{13}(OH)_3$ (c)	Magadiite	–241.83	188.73	–6651.9		3
SiO_2 (c)	α-Quartz	–910.700	41.46	–856.288	2.2688	1
SiO_2 (c)	α-Quartz	–910.7	41.5	–856.3		4, 10
SiO_2 (c)	α-Quartz	–910.700	41.460	–856.288	2.269	16
SiO_2 (c)	α-Quartz	–910.7	41.5	–856.3	2.269	17
SiO_2 (c)	α-Quartz	–910.88	41.50	–856.46	2.269	18
SiO_2 (c)	β-Quartz	–909.07	43.54		2.367	18
SiO_2	Chalcedony	–905.4	–169.0	–855.0		10
SiO_2 (amorphous)		–903.200	47.40	–850.559	2.727	1
SiO_2 (amorphous)		–899.7		–849.1		3
SiO_2 (amorphous)		–899.6	–166.0	–850.2		10
SiO_2 (c)	α–Crystobalite	–908.346	43.40	–854.512	2.5739	1

(Cont'd)

Species or compound	Mineral	$\Delta H^o_{f,\,298.15}$ (kJ mol^{-1})	$S^o_{298.15}$ (J mol^{-1}K^{-1})	$\Delta G^o_{f,\,298.15}$ (kJ mol^{-1})	$V^o_{298.15}$ (J bar^{-1})	Source
SiO_2 (c)	α–Crystobalite	−909.5	42.7	−855.4		2
SiO_2 (c)	α-Crystobalite	−907.753	43.394	−853.918	2.587	16
SiO_2 (c)	Crystobalite	−908.4	43.4	−854.6	2.574	17
SiO_2 (c)	Crystobalite	−906.04	46.50	−853.12	2.610	18
SiO_2 (c)	Tridymite	−907.488	43.93	−851.812	2.653	1
SiO_2 (c)	Tridymite	−909.1	43.5	−855.3		2
SiO_2 (c)	Tridymite high	−907.045	45.524	−853.844	2.737	16
SiO_2 (c)	Tridymite low	−907.750	43.770	−854.026	2.675	16
SiO_2 (c)	Tridymite low	−907.5	43.9	−853.6	2.653	17
SiO_2 (c)	Tridymite	−906.73	46.10	−853.69	2.700	18
$USiO_4$ (c)				−1861.9		9
Zn_2SiO_4 (c)	Willemite	−1636.530	131.38	−1522.936	5.242	1
Zn_2SiO_4 (c)	Willemite	−1636.7	131.4	−1523.2		2
Zn_2SiO_4 (c)	Willemite	−1636.7	131.4	−1523.1	5.242	17
$ZrSiO_4$ (c)	Zircon	−2033.400	84.03	−1918.890	3.926	1
$ZrSiO_4$ (c)	Zircon	−2034.2	84.0	−1919.7	3.926	17
$ZrSiO_4$ (c)	Zircon	−2031.85	84.03	−1917.35	3.926	18

c = crystalline; l = liquid; g = gas; aq = aqueous species; all charged ionic species are considered to be in aqueous solution (standard state, m = 1).
* volume at 298.15 K and 1 bar = 2478.97 J/bar
Sources: (1) Robie *et al.* (1978); (2) Wagman *et al.* (1982); (3) Drever (1997); (4) Cox *et al.* (1989); (5) Helgeson *et al.* (1978); (6) Hemingway and Sposito (1990); (7) Garvin *et al.* (1987); (8) Busenberg *et al.* (1984); (9) Brookins (1988); (10) Nordstrom and Munoz (1994); (11) Grenthe *et al.* (1992); (12) Hemingway (1990); (13) Busenberg and Plummer (1986); (14) Palmer and Wesolowski (1992); (15) Hemmingway *et al.* (1991); (16) Berman (1988); (17) Robie and Hemmingway (1995); (18) Holland and Powell (1998), (19) Krauskopf and Bird (1995); (20) Weast et al. (1986); (21) Garrels and Christ (1965).

Appendix 6

Fugacities of H_2O and CO_2 in the range 0.5–10.0 kbar and 200–1000°C

	Pressure (kbar)	Temperature (°C)								
		200	300	400	500	600	700	800	900	1000
$RT \ln f(H_2O)$	0.5	11.48	20.93	30.10	37.36	43.35	49.04	54.57	60.02	65.40
	1.0	12.48	22.06	31.44	39.52	46.83	53.60	60.06	66.34	72.50
	1.5	13.45	23.14	32.66	41.03	48.84	56.20	63.23	70.04	76.69
	2.0	14.40	24.18	33.81	42.38	50.46	58.17	65.57	72.75	79.75
	2.5	15.34	25.20	34.92	43.64	51.92	59.86	67.53	74.98	82.25
	3.0	16.26	26.19	35.99	44.83	53.26	61.39	69.26	76.91	84.40
	3.5	17.17	27.16	37.03	45.97	54.53	62.80	70.83	78.65	86.31
	4.0	18.06	28.11	38.04	47.07	55.73	64.12	72.29	80.26	88.06
	4.5	18.93	29.03	39.02	48.13	56.89	65.39	73.66	81.76	89.69
	5.0	19.80	29.95	39.99	49.17	58.01	66.60	74.98	83.17	91.22
	5.5	20.66	30.85	40.93	50.18	59.09	67.77	76.23	84.53	92.67
	6.0	21.50	31.74	41.87	51.17	60.15	68.90	77.45	85.82	94.05
	6.5	22.34	32.61	42.78	52.15	61.19	70.00	78.62	87.08	95.38
	7.0	23.17	33.48	43.69	53.10	62.20	71.08	79.77	88.29	96.67
	7.5	23.99	34.34	44.58	54.05	63.20	72.13	80.88	89.47	97.91
	8.0	24.80	35.19	45.57	54.97	64.18	73.16	81.97	90.62	99.12
	8.5	25.61	36.03	46.34	55.89	65.14	74.17	83.03	91.74	100.30
	9.0	26.41	36.86	47.20	56.79	66.09	75.17	84.08	92.83	101.45
	9.5	27.20	37.68	48.06	57.69	67.02	76.15	85.10	93.91	102.58
	10.0	27.99	38.50	48.90	58.57	67.94	77.11	86.11	94.97	103.68
$RT \ln f(CO_2)$	0.5	23.33	29.27	34.97	40.52	45.99	51.40	56.77	62.11	67.42
	1.0	26.51	33.19	39.64	45.94	52.14	58.27	64.34	70.37	76.37
	1.5	29.07	36.13	42.99	49.72	56.35	62.90	69.39	75.84	82.25
	2.0	31.37	38.69	45.83	52.85	59.77	66.62	73.41	80.16	86.87
	2.5	33.52	41.04	48.39	55.63	62.77	69.85	76.87	83.86	90.80
	3.0	35.58	43.26	50.78	58.18	65.51	72.77	79.98	87.15	94.29
	3.5	37.57	45.38	53.04	60.59	68.06	75.48	82.85	90.18	97.48
	4.0	39.51	47.43	55.21	62.88	70.49	78.04	85.54	93.00	100.44
	4.5	41.41	49.43	57.31	65.09	72.81	80.47	88.09	95.68	103.23
	5.0	43.27	51.38	59.35	67.23	75.05	82.81	90.54	98.23	105.89
	5.5	45.07	53.26	61.31	69.28	77.19	85.04	92.86	100.65	108.41
	6.0	46.81	55.07	63.21	71.25	79.24	87.18	95.08	102.96	110.81
	6.5	48.52	56.85	65.05	73.17	81.23	89.25	97.23	105.19	113.12
	7.0	50.18	58.58	66.85	75.03	83.16	91.25	99.31	107.34	115.35
	7.5	51.83	60.28	68.61	76.86	85.05	93.21	101.33	109.43	117.51
	8.0	53.44	61.95	70.34	78.64	86.90	95.12	103.30	111.47	119.61

Introduction to Geochemistry: Principles and Applications, First Edition. Kula C. Misra.

(Cont'd)

Pressure (kbar)	Temperature (°C)								
	200	300	400	500	600	700	800	900	1000
8.5	55.03	63.59	72.03	80.40	88.71	96.99	105.23	113.46	121.66
9.0	56.61	65.22	73.71	82.12	90.49	98.82	107.12	115.40	123.67
9.5	58.16	66.82	75.36	83.82	92.24	100.62	108.97	117.31	125.63
10.0	59.70	68.40	76.98	85.50	93.96	102.39	110.80	119.18	127.55

Source: Holland and Powell (1998). Additional fugacity data in the range 10–120 kbar and 400–1400°C can be found on the Worldwide Web (http://www.esc.cam.ac.uk/astaff/holland/ds5/gases/H2O.html).

Analytical expressions used for calculations of $RT \ln f_{H_2O}$ or $RT \ln f_{CO_2}$: $a + b\,T + c\,T^2$, where T is the temperature in Kelvin, and a, b, and c are polynomials in P (pressure in kbars):

$$a = a_1 + a_2P + a_3P$$

$$b = b_1 + \frac{b_2}{P} + \frac{b_3}{P^2} + \frac{b_4}{\sqrt{P}}$$

$$c = c_1 + c_2P + \frac{c_3}{P} + \frac{c_4}{\sqrt{P}}$$

The values of the constants are as follows:

	H_2O	CO_2
a_1	−3.8610E+1	−9.4290E+0
a_2	1.3854E+0	2.6209E+0
a_3	−1.2280E−3	−1.1704E−2
b_1	1.2716E−1	1.2772E−1
b_2	3.9770E−2	1.1587E+4
b_3	−7.8802E−3	−2.7250E−2
b_4	−4.1436E−2	1.4323E−1
c_1	−9.8194E−6	−6.8060E−7
c_2	6.5419E−8	−2.3744E−7
c_3	1.3849E−6	0.0000E+0
c_4	−1.3900E−5	5.6770E−6

Appendix 7

Equations for activity coefficients in multicomponent regular solid solutions (after Mukhopadhyay *et al.*, 1993)

G^{EX} = molar excess free energy
X_i = mole fraction of i
λ_i = activity coefficient of i
W_{ij} = binary interaction free energy parameter
W_{ijk} = ternary interaction free energy parameter
T = temperature (K)
R = gas constant

Symmetric regular solutions

Taylor series expansion truncated after the second power. Solution with 1-site mixing.

$W_{ij} = W_{ji}$

No ternary interaction free energy parameter.

Expressions for $RT \ln \lambda_2$, $RT \ln \lambda_3$, and $RT \ln \lambda_4$ can be obtained by cyclically rotating the subscripts in the expression for $RT \ln \lambda_i$. For example, in the quaternary solution ($c = 4$), $RT \ln \lambda_2$ can be obtained by substituting 1 by 2, 2 by 3, 3 by 4, and 4 by 1:

$$RT \ln \lambda_2 = X_3W_{23} + X_4W_{24} + X_1W_{21} - G^{EX} = X_1W_{12} + X_3W_{23} + X_4W_{24} - G^{EX}$$

Number of components (c)	$G^{EX} = \sum_{i<j}^{c} X_iX_jW_{ij}$	$RT \ln \lambda_i = \sum_{j=1, j\neq i}^{c} X_jW_{ij} - G^{EX}$
2 (1, 2)	$X_1X_2W_{12}$	$RT \ln \lambda_1 = X_2W_{12} - G^{EX}$
3 (1, 2, 3)	$X_1X_2W_{12} + X_1X_3W_{13} + X_2X_3W_{23}$	$RT \ln \lambda_1 = X_2W_{12} + X_3W_{13} - G^{EX}$
4 (1, 2, 3, 4)	$X_1X_2W_{12} + X_1X_3W_{13} + X_1X_4W_{14} + X_2X_3W_{23} + X_2X_4W_{24} + X_3X_4W_{34}$	$RT \ln \lambda_1 = X_2W_{12} + X_3W_{13} + X_4W_{14} - G^{EX}$

Introduction to Geochemistry: Principles and Applications, First Edition. Kula C. Misra.

Asymmetric regular solutions

All constituent binaries are characterized by symmetric Margule-type parameters
Taylor series expansion truncated after the third power.
Solution with 1-site mixing.

$$W_{ij} \neq W_{ji}$$

$$C_{ijk} = \tfrac{1}{2}\left[W_{ij} + W_{ji} + W_{ik} + W_{ki} + W_{jk} + W_{kj}\right] - W_{ijk} \text{ for all } i, j$$

Number of components (c)	$G^{EX} = \sum_{i<j}^{c} \sum_{<j}^{c} X_i X_j \left[X_j W_{ij} + X_i W_{ji}\right] + \sum_{i}^{c} \sum_{<j}^{c} \sum_{<k}^{c} X_i X_j X_k C_{ijk}$	$RT \ln \lambda_i = 2 \sum_{j=1, j\neq i}^{c} X_i X_j W_{ji} + \sum_{j=1, j\neq i}^{c} X_j^2 W_{ij} + \sum_{j, j\neq i}^{c} \sum_{<k, k\neq i}^{c} X_j X_k C_{ijk} - 2G^{EX}$
2 (1, 2)	$X_1X_2[X_2W_{12} + X_1W_{21}]$	$RT \ln \lambda_1 = 2X_1X_2W_{21} + X_2^2W_{12} - 2G^{EX}$
3 (1, 2, 3)	$X_1X_2[X_2W_{12} + X_1W_{21}] + X_1X_3[X_3W_{13} + X_1W_{31}] + X_2X_3[X_3W_{23} + X_2W_{32}] + X_1X_2X_3C_{123}$	$RT \ln \lambda_1 = 2[X_1X_2W_{21} + X_1X_3W_{31}] + X_2^2W_{12} + X_3^2W_{13} + X_2X_3C_{123} - 2G^{EX}$
4 (1, 2, 3, 4)	$X_1X_2[X_2W_{12} + X_1W_{21}] + X_1X_3[X_3W_{13} + X_1W_{31}] + X_1X_4[X_4W_{14} + X_1W_{41}] + X_2X_3[X_3W_{23} + X_2W_{32}] + X_2X_4[X_4W_{24} + X_2W_{42}] + X_3X_4[X_4W_{34} + X_3W_{43}] + X_1X_2X_3C_{123} + X_1X_2X_4C_{124} + X_1X_3X_4C_{134} + X_2X_3X_4C_{234}$	$RT \ln \lambda_1 = 2[X_1X_2W_{21} + X_1X_3W_{31} + X_1X_4W_{41}] + X_2^2W_{12} + X_3^2W_{13} + X_4^2W_{14} + X_2X_3C_{123} + X_2X_4C_{124} + X_3X_4C_{134} - 2G^{EX}$

Expressions for $RT \ln \lambda_2$, $RT \ln \lambda_3$, and $RT \ln \lambda_4$ can be obtained by cyclically rotating the subscripts in t the expression for $RT \ln \lambda_i$. For example, in the quaternary solution ($c = 4$), $RT \ln \lambda_2$ can be obtained by substituting 1 by 2, 2 by 3, 3 by 4, and 4 by 1:

$$\begin{aligned} RT \ln \lambda_2 &= 2[X_2X_3W_{32} + X_2X_4W_{42} + X_2X_1W_{12}] + X_3^2W_{23} + X_4^2W_{24} + X_1^2W_{21} + X_3X_4C_{234} + X_3X_1C_{231} + X_4X_1C_{241} - 2G^{EX} \\ &= 2[X_1X_2W_{12} + X_2X_3W_{32} + X_2X_4W_{42}] + X_1^2W_{21} + X_3^2W_{23} + X_4^2W_{24} + X_1X_3C_{123} + X_1X_4C_{124} + X_3X_4C_{234} - 2G^{EX} \end{aligned}$$

Appendix 8

Some commonly used computer codes for modeling of geochemical processes in aqueous solutions

Nordstrom (2004) provides an informative review of computer codes available for modeling various low-temperature geochemical processes: mineral dissolution and precipitation, aqueous inorganic speciation and complexation, solute adsorption and desorption, ion exchange; oxidation–reduction, phase transformations, gas uptake or production, organic matter speciation and complexation; evaporation, dilution, water mixing, reactions involving biotic interactions, and photoreactions (Nordstrom, 2004, p. 38).

Listed below are some of the codes in current and common use.

Code	Details
WATEQ4F	This code, originally developed by Truesdell and Jones (1974) of US Geological Survey (USGS) to aid in the interpretation of water quality data, has been updated (Ball and Nordstrom, 1991). The most recent version of the code is available in the Survey's Open-file Report 91-183 (189 pp), and at <http:water.usgs.gov/software/wateq4f.html>.
SOLMINEQ.88	A code called SOLMNEQ was developed by Kharaka and Barnes (1973) of USGS, in parallel with WATEQ, to calculate reaction equilibria above 100°C. The revised version of the code, SOLMINEQ.88 (Kharaka *et al*., 1988; Perkins *et al*., 1990), which covers the temperature range of 0–350°C and the pressure range of 1–1000 bar, and includes mass-transfer options, is particularly applicable to deep sedimentary basins. The code, with complete program documentation, and sample input and output examples, is available from the authors.
PHREEQC	Parkhurst *et al*. (1980) of USGS developed the code PHREEQE to compute, in addition to aqueous speciation, mass transfer and reaction paths. The latest revised version of the code, PHREEQC (version 2) includes ion exchange, evaporation, fluid mixing, sorption, solid–solution equilibria, kinetics, one-dimensional transport, and inverse modeling (Parkhurst and Appelo, 1999). It is available at <http://www.brr.cr.usgs.gov/projects/GWC_coupled/phreeqc/index.html>. Also, a web-implementation version, WEB–PHREEQ, by Bernhardt Saini-Eidukat is available at <http://www.ndsu.nodak.edu/webphreeq/>. In this version the user has the option of choosing the PHREEQC, MINTEQ, or WATEQ4F data base for calculations.
PHRQPITZ	This code, similar to PHREEQE, was developed to simulate geochemical reactions in brines and other electrolyte solutions to high concentrations, using the Pitzer methods (Plummer *et al*., 1988; Plummer and Parkhurst, 1990). It is available at <http://www.brr.cr.usgs.gov/projects/GWC_coupled/phrqpitz>. However, the code is now obsolete, and the current version of PHREEQC contains a complete implementation of the Pitzer specific ion interaction approach.
MINTEQA2/PRODEFA2	Felmy *et al*. (1984) produced the code MINTEQ to calculate aqueous geochemical equilibria by combining the WATEQ3 database (Ball *et al*., 1981) with the computer program MINEEQL (Westall *et al*., 1976). MINTEQ evolved into the USEPA-supported code MINTEQA2 (Allison *et al*., 1991). The more recent upgrade is MINTEQA2/PRODEFA2 (version 4), which is available from Scientific Software Group, P.O. Box 708188, Sandy, Utah 84070. It contains code revisions, updates in thermodynamic data, and modifications to minimize nonconvergence problems, to improve titration modeling, to minimize phase rule violations, to enhance execution speed, and to allow selected results to to be transferred to a spreadsheet. PRODEFA2 is an ancillary program that produces MINTEQA2 input files using an interactive preprocessor (Nordstrom, 2004, p. 49). Visual MINTEQ (version 3.0), built on USEPA's MINTEQA 2 software and maintained by John Peter Gustafsson at KTH, Sweden since 2000, is available at <http://www.lwr.kth.se/english/OurSoftware/Vminteq>. This code can simulate the chemical composition in solutions in contact with gases, solid compounds, and particle surfaces, but it does not contain any transport model, and it cannot handle partitioning of nonpolar organic compounds, nonequilibrium processes, and problems involving high temperature/pressure.

Introduction to Geochemistry: Principles and Applications, First Edition. Kula C. Misra.

(*Cont'd*)

Code	Details
EQ3/6 (Wolery)	This is a software package, developed by Wolrey (Wolrey, 1992a, b; Wolrey and Daveler,1992), for geochemical modeling of aqueous systems. The package consists of four principal components: (i) the EQ3NR code, which computes a static thermodynamic model of an aqueous solution, given inputs such as pH and analytical concentrations; (ii) the EQ6 code, a reaction-path program, which typically is used to compute the consequences of reacting an aqueous solution, with a set of specified "reactants;" (iii) the EQLIB library; and (iv) an appropriate thermodynamic database. The latest version of the software package (EQ3/6 V7.2B) can be ordered from: Energy Science and Technology Software Center (ESTSC – <http://www.osti.gov/estc/orders.jsp>).
CHILLER/SOLVEQ	The CHILLER/SOLVEQ program, developed and maintained by Mark Reed of the University of Oregon (Reed, 1982, 1998), focuses on complex problems in hydrothermal geochemistry, including combinations of boiling, condensation, fluid–fluid mixing, oxidation–reduction, and variations in water–rock reactions. The program can be obtained from Mark Reed.
THE GEOCHEMIST'S WORKBENCH	The GWB software package is a commercial product offered by C. Bethke (1996). The package comes in three tiers. *GWB Essentials* is a set of software tools for solving water quality problems, including those encountered in environmental protection and remediation, the petroleum industry, and mine wastewater planning. *GWB Standard* contains all the components of GWB Essentials plus a full suite of reaction path modeling tools. *GWB Professional* contains, in addition, a full suite of one-dimensional and two-dimensional reactive transport tools. Details for ordering the package are available at <http://www.rockware.com/product/overview.php?id=187.

Appendix 9

Solar system abundances of the elements in units of number of atoms per 10^6 silicon atoms

Z	Element	Abundance	Z	Element	Abundance	Z	Element	Abundance
1	H	2.79×10^{10}	32	Ge	1.19×10^{2}	63	Eu	9.73×10^{-2}
2	He	2.72×10^{9}	33	As	6.56×10^{0}	64	Gd	3.30×10^{-1}
3	Li	5.71×10^{1}	34	Se	6.21×10^{1}	65	Tb	6.03×10^{-2}
4	Be	7.3×10^{-1}	35	Br	1.18×10^{1}	66	Dy	3.942×10^{-1}
5	B	2.12×10^{1}	36	Kr	4.5×10^{1}	67	Ho	8.89×10^{-2}
6	C	1.01×10^{7}	37	Rb	7.09×10^{0}	68	Er	2.508×10^{-1}
7	N	3.13×10^{6}	38	Sr	2.35×10^{1}	69	Tm	3.78×10^{-2}
8	O	2.38×10^{7}	39	Y	4.64×10^{0}	70	Yb	2.479×10^{-1}
9	F	8.43×10^{2}	40	Zr	1.14×10^{1}	71	Lu	3.67×10^{-2}
10	Ne	3.44×10^{6}	41	Nb	6.98×10^{-1}	72	Hf	1.54×10^{-1}
11	Na	5.74×10^{4}	42	Mo	2.55×10^{0}	73	Ta	2.07×10^{-2}
12	Mg	5.74×10^{4}	43	Tc	0	74	W	1.33×10^{-1}
13	Al	8.49×10^{4}	44	Ru	1.86×10^{0}	75	Re	5.17×10^{-2}
14	Si	1.00×10^{6}	45	Rh	3.44×10^{-1}	76	Os	6.75×10^{-1}
15	P	1.04×10^{4}	46	Pd	1.39×10^{0}	77	Ir	6.61×10^{-1}
16	S	5.15×10^{5}	47	Ag	4.86×10^{-1}	78	Pt	1.34×10^{0}
17	Cl	5.240×10^{3}	48	Cd	1.61×10^{0}	79	Au	1.87×10^{-1}
18	Ar	1.01×10^{5}	49	In	1.84×10^{-1}	80	Hg	3.4×10^{-1}
19	K	3.770×10^{3}	50	Sn	3.82×10^{0}	81	Tl	1.84×10^{-1}
20	Ca	6.11×10^{4}	51	Sb	3.09×10^{-1}	82	Pb	3.15×10^{0}
21	Sc	3.42×10^{1}	52	Te	4.81×10^{0}	83	Bi	1.44×10^{-1}
22	Ti	2.400×10^{3}	53	I	9.0×10^{-1}	84	Po*	≈ 0
23	V	2.93×10^{2}	54	Xe	4.7×10^{0}	85	At*	≈ 0
24	Cr	1.35×10^{4}	55	Cs	3.72×10^{-1}	86	Rn*	≈ 0
25	Mn	9.550×10^{3}	56	Ba	4.49×10^{0}	87	Fr*	≈ 0
26	Fe	9.00×10^{5}	57	La	4.460×10^{-1}	88	Ra*	≈ 0
27	Co	2.250×10^{3}	58	Ce	1.136×10^{0}	89	Ac*	≈ 0
28	Ni	4.93×10^{4}	59	Pr	1.669×10^{-1}	90	Th	3.35×10^{-2}
29	Cu	5.22×10^{2}	60	Nd	8.279×10^{-1}	91	Pa	≈ 0
30	Zn	1.260×10^{3}	61	Pm	0	92	U	9.00×10^{-3}
31	Ga	3.78×10^{2}	62	Sm	2.582×10^{-1}	93	Np*	≈ 0
						94	Pu*	≈ 0

*The abundances of the radioactive daughters of uranium and thorium are very low and aretherefore indicated as ≈ 0. The transuranium elements [$Z = 95$ (Am) to $Z = 103$ (Lr)] do not occur in the Solar System and, therefore, are not included in this table. The elements Tc and Pm do not occur in the Solar System because all of their isotopes are unstable and decay rapidly.
Source of data: Anders and Grevesse (1989).

Introduction to Geochemistry: Principles and Applications, First Edition. Kula C. Misra.

Appendix 10

Answers to selected chapter–end questions

Chapter 1

3. Molarity of NaCl = 0.98
4. Mole fraction of ethanol = 0.477
   Mole fraction of water = 0.523
5. Mole fraction of ethanol = 0.419
   Mole fraction of water = 0.581
6. Molality of $Na^{=}$ = 0.0096
   Concentration of $Na^{=}$ = 220.7 ppm
7. Molality of K_2SO_4 = 0.8197
   Molarity of K_2SO_4 = 0.7767
   Mole fraction of solvent = 0.9855
8. Moles of $NaAlSi_3O_8$ = 0.021
   Moles of Si = 0.063
9. Molality of Ca = 2.25
   Molarity of Ca = 2.13

Chapter 2

2. U = 238.0289 amu; Pb = 207.2377 amu
3. Dissolved Ag Cl = 0.0019 g L^{-1}
4. 5×10^{17} H_2O molecules per second
5. Mine the second deposit
6. (a) 18; (b) 2; (c) 14; and (d) 50
7. Fe: $1s^2 2s^2 2p^6 3s^2 3p^6 3d^6 4s^2$
9. $E_{UV}/E_{yellow} = 51.86$

Chapter 3

1. Series (a) and (c)
3. $U_L(KCl) \approx 675$ kJ mol^{-1}
   $E_{cryst}(Cl) \approx -675$ kJ mol^{-1}
4. $U_L(LiCl) \approx 858$ kJ mol^{-1}
5. Paramagnetic
6. Na–Cl: 53.1%; Cl–O: 4.8%; Zn–O: 39.8%

Chapter 4

1. $K_{eq} = 1.1 \times 10^{-5}$
2. $K_{eq} = 10^{-3.7}$ $K_{eq} = 10^{-3.7}$
3. $P_{O_2} = 10^{-87.4}$ atm
   9.8×10^{-66} oxygen molecules per liter
4. Heat required = 313.5 kJ
5. Specific heat capacity of the metal = 0.385 J g^{-1} °C^{-1}
6. $\Delta H_f^1(CH_4) = -74.8$ kJ mol^{-1}
7. Keq = 2.3
8. Difference in work = $R(T_2 - T_1) \ln (P_2 - P_1)$
10. Volume change = 3.266 J bar^{-1} mol^{-1}
11. (a) $\bar{H}_{800} = -3801.9$ kJ; $\bar{S}_{800} = 0.462$ kJ
    $\bar{G}_{800} = -4171.5$ kJ
    (b) the same as (a)
12. $\Delta \bar{V}_{r,500}^{1\ kbar} = -0.349$ J bar^{-1};
    $\Delta \bar{G}_{r,500}^{1\ kbar} = -6.55$ kJ
13. (a) Exothermic; (b) products more stable than reactants; (c) reactants will be favored by increase in temperature, products by increase in pressure

Chapter 5

2. $K_{eq} = 3.47 \times 10^{41}$
4. (a) 856 K; (b) ~50 K; (c) decrease
5. (a) $a_{fo}^{ol} = 0.644$; $a_{fa}^{ol} = 0.297$
   (b) $a_{alm}^{gt} = 0.2780$; $a_{gr}^{gtl} = 0.0161$
   (c) $a_{diop}^{cpx} = 0.190$; $a_{jd}^{cpxl} = 0.272$
6. (a) $a_{Ca_3Al_2Si_3O_{12}}^{grossular} = 3.5 \times 10^6$
   (b) $a_{Ca_3Al_2Si_3O_{12}}^{grossular} = 1$
12. $K_{eq} = 10^{-1.46}$
13. 8.32 mg L^{-1}
14. (a) 19.3 bar; (b) 11.8 bar
15. (a) 4.99×10^{-4} mol L^{-1} at 25°C
    (b) 3.78×10^{-4} mol L^{-1} at 50°C

Introduction to Geochemistry: Principles and Applications, First Edition. Kula C. Misra.

Chapter 6

2. $\ln K_{300} = -42.04$; $\ln K_{700} = -2.38$
3. GAPES: 7.9 kbar; GADS: 7.7 kbar
4. $T = 992°C$; $\log f_{O_2} = -12.1$
5. 1228°C (Green); 1235°C (Krogh)
6. $T = 521°C$
7. $P = 5150$ bars; $T = 800$ K
8. (a) $\log K_{eq} = 2.717$; (b) $P = 2.5$ kbar

Chapter 7

2. $m_{HCO_3^-} = 10^{-4.35}$
3. $K_{sp}(Ag_2CrO_4) = 8.99 \times 10^{-12}$
   Solubility of $Ag_2CrO_4 = 1.31 \times 10^{-4}\,m$
4. Solubility of $Ag_2CrO_4 = 4.99\,g\,kg^{-1}$
5. (a) Seawater is undersaturated in NaCl
   (b) Evaporation by a factor > ~3.5
6. (a) Alkalinity = 0.0024
7. The less soluble barite replaced gypsum
8. Solution with added HCl is a good buffer
11. $CaCo_3 + H_2O \Leftrightarrow Ca^{2+} + HCO_3^- + OH^-$
12. Solubility = $10^{-2.917}$ mol kg^{-1}; pH = 7.6
16. (a) pH= 5.48; (b) pH = 8.1
17. $\Delta H_r^0 = 23.2$ kJ

Chapter 8

2. Oxidation-reduction reactions: (b) and (d)
3. $K_{eq} = 10^{-10.14}$
4. $Eh = +0.76$ volts; $pe = 12.8$
5. $a_{Mn^{2+}} = 10^{3.4}$
6. Dominant sulfur species: H_2S
7. $E^0 = -1.56$ volts
8. $Eh_{H_2O}^{\text{Acid mine drainage}} = 0.98$ volts;
   $Eh_{H_2O}^{\text{Ocean water}} = 0.73$ volts
9. Siderite is not stable in the presence of water
11. Dissolved $N_2 = 15.97$ ppm
12. $\Delta G_{f,\,1000K}^{1\text{ bar}} = -149.22$ kJ mol^{-1}

Chapter 9

1. $d[A]/dt = -[A])k_1[B] + k_2[c])$
2. Reaction mechanism is consistent with the observed rate law
6. Rate constant = -0.3 h^{-1}
   Half-life = 2.31 h
9. Rate constant at 10°C = $10^{-4.85}$ s^{-1}
   Rate constant at 50°C = $10^{-3.89}$ s^{-1}
10. (a) Rate constant = 0.23 s^{-1}; (b) 9°C
12. 933,000 yr
13. (a) Half-life = 3.09×10^4 s
    (b) Fraction remaining = 0.668
    (c) 31.3 h
14. Activation energy = 27.8 kJ mol^{-1}
15. (a) The reaction is first order
    (b) Rate law: Rate constant × [A]
    (c) Rate constant = 0.3 min

Chapter 10

2. Half-life = 3.53×10^{10} yr
3. 9.977 half-lives
4. $^{147}Sm/^{144}Nd = 0.199$
5. Atomic weight of lead = 207.55
6. Age = 1910 Ma; initial ratio = 0.7066
7. Age = 3252 Ma; initial ratio = 0.5085
8. Rb–Sr age: 2613 Ma; initial ratio = 0.7015
   Sm–Nd age: 2922 Ma; initial ratio = 0.5089
9. Age: ~2700 Ma; lead loss: ~500 Ma
10. Age = 2634 Ma
11. Age = 1.97 Ga;$\mu = 8.48$
12. J-type anomalous leads
13. Age = 119,376 yr; initial ratio = 290.1
14. Age: ~4.09 Ga
15. Age (sample with 13.56 dpm g^{-1}) = 693 yr
    Age (sample with 15.30 dpm g^{-1}) = 1690 yr

Chapter 11

2. CO_2: mass 40.0, 98.424%; mass 45.0 1.169%; mass 46.0, 0.403%
3. 3.10 per mil; –1.95 per mil
4. qz: 26.46 per mil; cal: 25.45 per mil
6. $\delta^{18}O_{mus} = -8.9$ per mil
7. $1000 \ln a$ (qz–rut) $= -1.59\,(10^6/T^2) - 2.92$
   T = 1028 K
9. LMWL: $\delta D = 8.0188\,\delta^{18}O + 16.987$
10. $T = 548$ K
12. $\alpha_{SO_4-H_2S} = 1.457$
13. 270 to 375°C

Chapter 12

3. H is the dominant species
7. (a) 54% crystallization would be required
   (b) X_{Yb} in remaining melt / X_{Yb} in original melt = 3.9×10^{-7}
   (c) 56% equilibrium crystallization; Yb depletion by a factor of 0.64

Chapter 13

2. (a) 6.44×10^{20} g Na^+
   (b) 0.83×10^6 yr
4. Residence times (million years): Cl, 116.4; Na, 7.7; SO_4, 12.2; Mg, 15.6; Ca, 1.1; K, 14.4; HCO_3, 0.1; Si, 0.4
5. pH = 4.85
6. pH = 3.4
8. (a) 2 molecules of water
   (b) escape of the hydrogen
9. $2H_2O$ + UV photon = $4H + O_2$

10. 9 yr
11. Residence time = 39×10^6 yr
12. P (CH_4) at 300 K = ~108 bar
13. 8.3×10^6 molecules per cm^3
14. Rate laws: step (i), $k_1[O_3]$; step (ii), $k_2[O_3][O]$
15. Rate = 6.0×10^{-7} molecules $cm^{-3}\ s^{-1}$
17. 352.3 kJ per mole of photon
18. 1.55×10^{-19} J per photon

References

Aaronson, T. (1971) The black box: fuel cells. *Environment*, 13, 10–18.

Abe, Y., Ohtani, E., Okuchi, T., Righter, K., & Drake, M. (2000) Water in the early Earth. In *Origin of the Earth and Moon* (eds R.M. Canup & K. Righter), pp. 413–443. The University of Arizona Press, Tucson, AZ.

Adler, S.F. (2001) Biofiltration – a primer. *Chemical Engineering Progress*, 97, 33–41. (April 2001).

Ahrens, L.H. (1955), Implications of the Rhodesia age pattern. *Geochimica Cosmochimica Geochim. Cosmochim Acta*, 8, 1-1–15.

Ahrens, T.J. (1993) Impact erosion of terrestrial planetary atmospheres. *Annual Review of Earth and Planetary Sciences*, 21, 525–555.

Ai, Y. (1994) A revision of the garnet–clinopyroxene Fe^{2+}–Mg exchange geothermometer. *Contributions to Mineralogy and Petrology*, 115, 467–473.

Allègre, C.J, and Luck, J.-M. (1980) Osmium isotopes as petrogenetic and geological tracers. *Earth and Planetary Science Letters*, 48, 148–154.

Allègre, C.J, and Minster, J.F. (1978) Quantitative models of trace element behavior in magmatic processes. *Earth and Planetary Science Letters*, 38, 1–25.

Allègre, C.J., Manhes, G., & Gopel, C. (1995) The age of the Earth. *Geochimica Cosmochimica Acta*, 59, 1445–1456.

Allègre, C.J., Poirier, J.P., Humler, E., & Hofmann, A.W. (1995) The chemical composition of the Earth. *Earth and Planetary Science Letters*, 134, 515–526.

Allison, J.D., Brown, D.S., & Novo-Gradac, K.J. (1991) MINTEQA2/PRODEFA2. *A Geochemical Assessment Model for Environmental Systems, Version 3.0 User's Manual*. U.S. Environmental Agency (EPA/600/3-91/021).

Altermann, W., Kazmierczak, J., Oren, A., & Wright, T. (2006) Cyanobacterial calcification and its rock-building potential during 3.5 billion years of Earth history. *Geobiology*, 4, 147–166.

Amelin, Y., Lee, D.C., Halliday, A.N., & Pidgeon, R.T. (1999) Nature of the Earth's earliest crust from hafnium isotopes in single detrital zircons. *Nature*, 399, 252–255.

Amelin, Y., Krot, A.N., Hutcheon, I.D., & Ulyanov, A.A. (2002) Lead isotope ages of chondrules and calcium–aluminum-rich inclusions. *Science*, 297, 1678–1683.

Anbar, A.D. (2004) Iron stable isotopes: beyond biosignatures. *Earth and Planetary Science Letters*, 217, 223–236.

Anbar, H.D., & Knoll, A.H. (2002) Proterozoic ocean chemistry and evolution: a bioinorganic bridge? *Science*, 297, 1137–1142.

Anbar, H.D., & Rouxel, O. (2007) Metal stable isotopes in paleooceanography. *Annual Review of Earth and Planetary Sciences*, 35, 717–746.

Anbar, A.D., Duan, Y., Lyons, T.W., *et al.* (2007) A whiff of oxygen before the Great Oxidation Event. *Science*, 317, 1903–1906.

Anders, E., & Grevesse, N. (1989) Abundances of the elements: meteoritic and solar. *Geochimica Cosmochimica Acta*, 53, 197–214.

Anderson, A.J., Mackenzie, F.T., and Ver, L.M. (2003) Solution of shallow-water carbonates: an insignificant buffer against rising atmospheric CO_2. *Geology*, 31, 513–516.

Anderson, D.L. (1989) *Theory of the Earth*. Blackwell Scientific Publications, Boston, 366 pp.

Anderson, D.J., & Lindsley, D.H. (1981) The correct Margules formulation for an asymmetric ternary solution: revision of the olivine-ilmenite thermometer, with applications. *Geochimica Cosmochimica Acta*, 45, 847–853.

Anderson, D.J., & Lindsley, D.H. (1988) Internally consistent solution models for Fe–Mg-Mn-Ti oxides: Fe-Ti oxides. *American Mineralogist*, 73, 714–726.

Anderson, G.M., & Crerar, D.A. (1993) *Thermodynamics in Geochemistry – the Equilibrium Model*. Oxford University Press, New York, 608 pp.

Anderson, J.G., Toohey, D.W., & Brune, W.H. (1991) Free radicals within the Antarctic vortex: the role of CFCs in the Antarcic ozone loss. *Science*, 251, 39–46.

Andrews, J.E., Brimblecombe, P., Jickells, T.D., Liss, P.S., & Reid, B.J. (2004) *An Introduction to Environmental Chemistry*, 2nd edn. Blackwell Science, Oxford, 296 pp.

Anovitz, L.M., & Essene, E.J. (1987) Phase equilibria in the system $CaCO_3$–$MgCO_3$–$FeCO_3$. *Journal of Petrology*, 28, 389–414.

Appelo, C.A.J., & Postma, D. (1995) *Geochemistry, Groundwater and Pollution*, 2nd edn. Rotterdam: Balkema, 649 pp.

Aranovich, L.Y., & Berman, R.G. (1997) A new garnet–orthopyroxene thermometer based on reversed Al_2O3 solubility in FeO–Al_2O_3–SiO_2 orthopyroxene. *American Mineralogist*, 82, 345–353.

Aranovich, L.Y., & Newton, R.C. (1999) Experimental determination of CO_2–H_2O activity–composition relations at 600–1000°C and 6–14 kbar by reversed decarbonation and dehydration reactions. *American Mineralogist*, 84, 1319–1332.

Introduction to Geochemistry: Principles and Applications, First Edition. Kula C. Misra.

Archer, C., & Vance, D. (2006) Coupled Fe and S isotope evidence for Archean microbial Fe(III) and sulfate reduction. *Geology*, 34, 153–156.

Armstrong, R.L. (1991) The persistent myth of crustal growth. *Australian Journal of Earth Sciences*, 38, 613–630.

Arndt, N.T., & Goldstein, S.L. (1989) An open boundary between lower continental crust and mantle: its role in crust formation and custal recycling. *Tectonophysics*, 161, 201–212.

Arnold, G..L., Anbar, A.D., Barling, J., & Lyons, T.W. (2004) Molybdenum isotope evidence for widespread anoxia in mid-Proterozoic oceans. *Science*, 304, 87–90.

Atkinson, A.B. (2002) *A model for the PTX properties of H_2O–NaCl.* Unpublished MS thesis, Virginia Technical, Blacksburg, VA, 133 pp.

Baadsgaard, H., Nutman, A.P., Bridgwater, D., Rosing, M., McGregor, V.R., & Allaart, J.H. (1984) The zircon chronology of the Akilia association and Isua supercrustal belt, West Greenland. *Earth and Planetary Science Letters*, 68, 221–228.

Baas Becking, L.G.M., Kaplan, I.R., & Moore, D. (1960) Limits of the natural environment in terms of pH and oxidation–reduction potentials. *Journal of Geology*, 68, 243–284.

Balci, N., Bullen, T.D., Witte-Lien, K., Shanks, W.C., Motelica, W.C., & Mandernack, K.W. (2006) Iron isotope farctionation during microbially stimulated Fe(II) oxidation and Fe(III) precipitation. *Geochimica Cosmochimica Acta*, 70, 622–639.

Ball, J.W., & D.K. Nordstrom (1991) User's manual for WATEQ4F, with revised thermodynamic data base and test cases for calculating speciation of major, trace, and redox elements in natural waters. *U.S. Geological Survey Open-File Report*, 91-183, 189 pp.

Ball, J.W., Jenne, E.A., & Cantrell, M.W. (1981) WATEQ3 – A geochemical model with uranium added. *U.S. Geological Survey Open–File Report,* 811183, 81 pp.

Banner, J.L. (2004. Radiogenic isotopes: systematics and applications to earth surface processes and chemical stratigraphy. *Earth-Science Reviews*, 65, 141–194.

Barnes, H.L. 1979. Solubilities of ore minerals. In *Geochemistry of Hydrothermal Ore Deposits* (ed. H.L. Barnes), 2nd edn, pp. 404–460. J. Wiley & Sons, New York.

Barrett, T.J., & Anderson, G.M. (1988) The solubility of sphalerite and galena in 1–5 *m* NaCl solutions to 300°C. *Geochimica Cosmochimica Acta*, 52, 813–820.

Barton, P.B. (1967) Possible role of organic matter in the precipitation of the Mississippi Valley ores. In *Genesis of Stratiform Lead–Zinc–Barite–Fluorite Deposits* (ed. J.S. Brown), pp. 371–378. Economic Geology Publishing Company, Lancaster.

Basaltic Volcanism Study Project (1981) *Basaltic Volcanism on the Terrestrial Planets.* Pergamon Press, New York, 1286 pp.

Bazylinski, D.A., & Frankel, R.B. (2000) Biologically controlled mineralization of magnetic iron minerals by magnetotactic bacteria. In *Environmental Microbe–Mineral Reactions* (ed. D.R. Lovley), pp. 109–144. ASM Press, Washington, DC.

Beard, B.L., & Johnson, C.M. (2004) Fe isotope variation in the modern and ancient Earth and other planetary bodies. *Reviews in Mineralogy and Geochemistry*, 55, 319–357.

Beard, B.L., Johnson, C.M., Skulan, J.L., Nealson, K.H., Cox, L., & Sun, H. (2003) Application of Fe isotopes to tracing the geochemical and biological cycling of Fe. *Chemical Geology*, 195, 87–117.

Beaty, D.W., & Taylor, H.P., Jr. (1982) Some petrologic and oxygen isotope relationships in the Amulet mine, Noranda, Quebec, and their bearing on the origin of Archean massive sulfide deposits. *Economic Geology*, 77, 95–108.

Becccaluva, L., Ohnenstetter, D., & Ohnenstetter, M. (1979) Geochemical discrimination between ocean-floor and island-arc tholeiites – application to some ophiolites. *Canadian Journal of Earth Sciences*, 16, 1874–1882.

Bekker, A., Holland, H.D., Wang, P.L., Rumble, D., Stein, H., Hannah, J.L., Coetzee, L.I., & Beukes, N.J. (2004) Dating the rise of atmospheric oxygen. *Nature*, 427, 117–120.

Berman, R.G (1988) Internally consistent thermodynamic data for stoichiometric minerals in the system K_2O–Na_2O–CaO–MgO–FeO–Fe_2O_3–Al_2O_3–SiO_2–TiO_2–H_2O–CO_2. *Journal of Petrology*, 29, 445–522.

Berman, R.G. (1990) Mixing properties of Ca–Mg–Fe–Mn garnets. *American Mineralogist*, 75, 328–344.

Berman, R.G., & Aranovich, L.Y. (1996) Optimized standard state and solution properties of minerals. I. Model calibration for olivine, orthopyroxene, cordierite, garnet, and ilmenite in the system FeO–MgO–CaO–Al_2O_3–TiO_2–SiO_2. *Contributions to Mineralogy and Petrology*, 126, 1–24.

Berman, R.G., Aranovich, L.Y., & Pattison, D.R.M. (1995) Reassessment of the garnet–clinopyroxene exchange thermometer: II. Thermodynamic analysis. *Contributions to Mineralogy and Petrology*, 119, 30–42.

Berner, E.K., & Berner, R.A. (1987) *The Global Water Cycle: Geochemistry and Environment.* Prentice Hall, Englewood Cliffs, NJ, 376 pp.

Berner, R.A. (1971) *Principles of Chemical Sedimentology.* McGraw Hill, New York.

Berner, R.A. (1975) The role of magnesium in the crystal growth of calcite and aragonite from seawater. *Geochimica Cosmochimica Acta*, 39, 489–504.

Berner, R.A. (1976) The solubility of calcite and aragonite in seawater at one atmosphere and 34.5 parts per thousand. *American Journal of Science*, 276, 713–730.

Berner, R.A. (1981) A new geochemical classification of sedimentary environments. *Journal of Sedimentary Petrology*, 51, 359–365.

Berner, R.A. (1985) Sulfate reduction, organic matter decomposition and pyrite formation. *Philosophical Transactions of the Royal Society*, A315, 25–38.

Berner, R.A. (1995) Chemical weathering and its effect on atmospheric CO_2 and climate. In *Chemical Weathering Rates of Silicate Minerals* (eds A.F. White & S.L. Brantley). *Reviews in Mineralogy*, 31, 565–583.

Berner, R.A. (2001) Modeling atmospheric over Phanerozoic time. *Geochimica Cosmochimica Acta*, 65, 685–694.

Berner, R.A. (2003) Phanerozoic atmospheric oxygen. *Annual Reviews of Earth and Planetary Sciences*, 31, 105–134.

Berner, R.A. (2004) *The Phanerozoic Carbon Cycle: CO_2 and O_2.* Oxford University Press, Oxford, 150 pp.

Berner, R.A. (2006) Geological nitrogen cycle and atmospheric N_2 over Phanerozoic time. *Geology*, 34, 413–415.

Berner, E.K., & Berner, R.A. (1996) *Global Environment; Water, Air, and Geochemical Cycles.* Prentice Hall, Upper Saddle River, NJ, 376 pp.

Berner, R.A., & Canfield, D.E. (1989) A new model for atmospheric oxygen over Phanerozoic time. *American Journal of Science*, 289, 333–361.

Berner, R.A., & Kothavala, Z. (2001) GEOPCARB III: a revised model of atmospheric CO_2 over Phanerozoic time. *American Journal of Science*, 301, 182–204.

Berner, R.A., & Morse, J.W. (1974) Dissolution kinetics of calcium carbonate in sea water. IV. Theory of calcite dissolution. *American Journal of Science*, 274, 108–134.

Berner, R.A., & Westrich, J.T. (1985) Bioturbation and the early diagenesis of carbon and sulfur. *American Journal of Science*, 285, 193–206.

Berner, R.A., Lasaga, A.C., & Garrels, R.M. (1983) The carbonate–silicate geochemical cycle and its effect on atmospheric carbon dioxide over the past 100 million years. *American Journal of Science*, 283, 641–683.

Bertrand, P., & Mercier, J.-C.C. (1985) The mutual solubility of coexisting ortho- and clinopyroxene: toward an absolute geothermometer for the natural system. *Earth and Planetary Science Letters*, 76, 109–122.

Bethe, H.A. (1968) Energy production in stars. *Science*, 161, 541–547.

Bethe, H.A., & Brown, G.E. (1985) How a supernova explodes. *Scientific American*, 252 (5), 60–68.

Bethke, C.M. (1996) *Geochemical Reaction Modelling: Concepts and Applications*. Oxford University Press, New York, 397 pp.

Beukes, N.J., & Klein, C. (1992) Models for iron-formation deposition. In *The Proterozoic Biosphere: A Multidisciplinary Study* (eds J.W. Schopf & C. Klein), pp. 147–151. Cambridge University Press, New York.

Beukes, N.J., Dorland, H., Gutzmer, J., Nedachi, M., & Ohmoto, H. (2002) Tropical laterites, life on land, and the history of atmospheric oxygen in the Paleoproterozoic. *Geology*, 30, 491–494.

Bevins, R.E., Kokelaar, B.P., & Dunkley, P.N. (1984) Petrology and geochemistry of lower to middle Ordovician rocks in Wales: a volcanic arc to marginal basin transition. *Proceedings of the Geologists' Association*, 95, 337–347.

Bhattacharya, A., Mohanty, L., Maji, A., Sen, S.K., & Raith, M. (1992) Non-ideal mixing in the phlogopite-annite binary: constraints from experimental data on Mg–Fe partitioning and a reformulation of the biotite–garnet geothermometer. *Contributions to Mineralogy and Petrology*, 111, 87–93.

Birch, F. (1966) Compressibility; elastic constants. In *Handbook of Physical Constants* (ed. S.P. Clark). *Geological Society of America Memoir*, 97, 97–173.

Bischoff, J.L., & Dickson, F.W. (1975) Seawater–basalt interaction at 200°C and 500 bars: implications for origin of sea-floor heavy metal deposits and regulation of seawater chemistry. *Earth and Planetary Science Letters*, 25, 385–397.

Bjerrum, C.J., & Canfield, D.E. (2002) Ocean productivity before about 1.9 Gyr ago limited by phosphorus adsorption onto iron oxides. *Nature*, 417, 159–162.

Bjerrum, C.J., & Canfield, D.E. (2004) New insights into the burial history of organic carbon on the early Earth. *Geochemistry Geophysics Geosystems*, 5, Q08001, doi:10.1029/2004GC000713 (online).

Blowes, D.W., Ptacek, C.J, Jambor, J.L., & Weisener, C.G. (2003) The geochemistry of acid mine drainage. In *Treatise in geochemistry*, eds, H.D. Holland and K.K. Turekian, 9, ed., 149–204. Oxford: Elsevier-Pergamon.

Blum, A.E., & Stillings, L.L. (1995) Feldspar dissolution kinetics. In *Chemical Weathering Rates of Silicate Minerals* (eds A.F. White & S.L. Brantley). *Reviews in Mineralogy*, 31, 291–351.

Bodnar, R.J. (2003) Introduction to aqueous-electrolyte fluid inclusions. In *Fluid Inclusions: Analysis and Interpretation* (eds I. Samson, A. Anderson, & D. Marshall). *Mineralogical Association of Canada Short Course Series*, 32, 81–100.

Bodnar, R.J., & Vityk, M.O. (1994) Interpretation of microthermometric data for H_2O–NaCl fluid inclusions. In *Fluid Inclusions in Minerals: Methods and Applications* (eds B. De Vivo & M.L. Frezzotti), pp. 117–130. Virginia Technical, Blacksburg, VA.

Bohlen, S.R., & Lindsley, D.H. (1987) Thermometry and barometry of igneous and metamorphic rocks. *Annual Review of Earth and Planetary Sciences*, 15, 397–420.

Bohlen, S.R., & Liotta, J.J. (1986) A barometer for garnet amphibolites and garnet granulites. *Journal of Petrology*, 27. 1025–1034.

Bohlen, S.R., Montana, A., & Kerrick, D.M. (1991) Precise determinations of the equilibria kyanite = sillimanite and kyanite = andalusite and a revised triple point for Al_2SiO_5 polymorphs. *American Mineralogist*, 76, 677–680.

Bohlen, S.R., Wall, V.J., & Boettcher, A.L. (1983a) Experimental investigation and application of garnet granulite equilibria. *Contributions to Mineralogy and Petrology*, 83, 52–61.

Bohlen, S.R., Wall, V.J., & Boettcher, A.L. (1983b) Experimental investigations and geological applications of equilibria in the system $FeO–TiO_2–Al_2O_3–SiO_2–H_2O$. *American Mineralogist*, 68, 1049–1058.

Bohr, N. (1913a) Binding of electrons by positive nuclei. *Philosophical Magazine*, 6, 26 (151), 1–25.

Bohr, N. (1913b) Systems containing only a single nucleus. *Philosophical Magazine*, 6, 26 (153), 476–507.

Bosak, T., & Newman, D.K. (2003) Microbial nucleation of calcium carbonate in the Precambrian. *Geology*, 31, 577–580.

Boswell, J. (2002) Understand the capabilities of bio-oxidation. *Chemical Engineering Progress*, December 2002.

Botinikov, N.S., Dobrovol'skaya, M.G., Genkin, A.D., Naumov, V.B., & Shapenko, V. (1995) Sphalerite–galena geothermometers: distribution of cadmium, manganese, and the fractionation of sulfur isotopes. *Economic Geology*, 90, 155–180.

Bottinga, Y., & Javoy, M. (1975) Oxygen isotope partitioning among the minerals in igneous and metamorphic rocks. *Reviews in Geophysics and Space Physics*, 13, 401–418.

Bottrell, S.H., & Newton, R.J. (2006) Reconstruction of changes in global sulfur cycling from marine sulfate isotopes. *Earth-Science Reviews*, 75, 59–83.

Boudreau, A.E., & McCallum, (1992) Concentration of platinum-group elements by magmatic fluids in layered intrusions. *Economic Geology*, 87, 1830–1848.

Bowen, N.L. (1940) Progressive metamorphism of siliceous limestone and dolomite. *Journal of Geology*, 48, 225–274.

Bowring, S,A., & Housh, T. (1995) The Earth's early evolution. *Science*, 269, 1535–1540.

Bowring, S.A., & Williams, I.S. (1999) Priscoan (4.00–4.03 Ga) orthogneisses from northwestern Canada. *Contributions to Mineralogy and Petrology*, 134, 3–16.

Brady, P.V., & Walther, J.V. (1989) Controls on silicate dissolution rates in neutral and basic pH solutions at 25°C. *Geochimica Cosmochimica Acta*, 53, 2823–2830.

Brady, P.V., & Walther, J.V. (1990) Kinetics of quartz dissolution at low temperatures. *Chemical Geology*, 82, 253–264.

Brennan, S.T., Lowenstein, T.K., & Horita, J. (2004) Seawater chemistry and the advent of biocalcification. *Geology*, 32, 473–476.

Brey, G.P., & Kohler, T. (1990) Geothermometry in four-phase lherzolites II. New thermobarometers, and practical assessment of existing thermobarometers. *Journal of Petrology*, 31, 1353–1378.

Brierly, C.L. (1997) Mining biotechnology: research to commercial development and beyond. In *Biomining: Theory, Microbes and*

Industrial Processes (ed. D.E. Rawlings), pp. 3–17. Springer-Verlag, Berlin.

Brimblecombe, P. (2005) The global sulfur cycle. In *Biogeochemistry* (ed. W.H. Schlesinger), Vol. 8, *Treatise on Geochemistry* (eds H.D. Holland & K.K.Turekian), pp. 425–472. Elsevier–Pergamon, Oxford.

Brocks, J.L., Logan, G.A., Buick, R., & Summons, R.E. (1999) Archean molecular fossils and the early rise of eukaryotes. *Science*, 285, 1033–1036.

Broecker, W.S., & Oversby, V.M. (1971) *Chemical Equilibria in the Earth*. McGraw- Hill, New York, 318 pp.

Broecker, W.S., & Peng, T-H. (1982) *Tracers in the Sea*. Lamont–Doherty Geological Observatory, Columbia University, Palisades, NY.

Brookins, D.G. (1978) Eh–pH diagrams for elements from $Z = 40$ to $Z = 52$. Applications to the Oklo natural reactor. *Chemical Geology*, 23, 324–342.

Brookins, D.G. (1988) *Eh–pH Diagrams for Geochemistry*. Springer-Verlag, Berlin, 176 pp.

Brown, J.S. (1969) Isotopic zoning of lead and sulfur in southeast Missouri. In *Genesis of Stratiform Lead–Zinc–Barite–Fluorite Deposits (Mississippi Valley Type Deposits)* (ed. J.S. Brown) *Society of Economic Geologists' Monograph*, 3, 410–426.

Brown, N.E., & Navrotsky, A. (1994) Hematite–ilmenite (Fe_2O_3–$FeTiO_3$) solid solutions: the effects of cation ordering on the thermodynamic mixing. *American Mineralogist*, 79, 485–496.

Brownlow, A.H. (1996) *Geochemistry*, 2nd edn. Prentice Hall, Upper Saddle River, NJ, 580 pp.

Buddington, A.F., & Lindsley, D.H. (1964) Iron–titanium oxide minerals and synthetic equivalents. *Journal of Petrology*, 5, 310–357.

Buening, D.K., & Buseck, P.R. (1973) Fe–Mg lattice diffusion in olivine. *Journal of Geophysical Research*, 78, 6852–6862.

Buick, R. (1992) The antiquity of oxygenic photosynthesis: evidence from stromatolites in sulfate-deficient Archaean lakes. *Science*, 255, 74–77.

Bullen, T.D., White, A.F., Childs, C.W., Davidson, V.V., & Schulz, M.S. (2001) Demonstration of significant abiotic iron isotope fractionation in nature. *Geology*, 29, 699–702.

Burbridge, E.M., Burbridge, G.R., Fowler, W.A., & Hoyle, F. (1957) Synthesis of the elements in stars. *Reviews of Modern Physics*, 29, 547–560.

Burch, T.E., Nagy, K.L., & Lasaga, A.C. (1993) Free energy dependence of albite dissolution kinetics at 80°C and pH 8.8. *Chemical Geology*, 105, 137–162.

Burnham, C.W., Holloway, J.R., & Davis, N.F. (1969) Thermodynamic properties of water to 1000°C and 10 000 bars. *Geological Society of America Special Paper*, 132, 96 pp.

Burns, R.G. (1973) The partitioning of trace transition elements in crystal structures: a provocative review with appliacations to mantle geochemistry. *Geochimica Cosmochimica Acta*, 37, 2395–2403.

Burns, R.G. (1993) *Mineralogical Applications of Crystal Field Theory*, 2nd edn. Cambridge University Press, New York, 551 pp.

Burns, R.G., & Fyfe, W.S. (1964) Site preference energy and selective uptake of transtion–metal ions from a magma. *Science*, 144, 1001–1003.

Burns, R.G., & Fyfe, W.S. (1966) The behavious of nickel during magmatic crystallization. *Nature*, 210, 1147–1148.

Burns, R.G., & Fyfe, W.S. (1967) Crystal-field theory and the geochemistry of transition elements. In *Researches in Geochemistry* (ed. P.H. Abelson), Vol. 2, pp. 259–285. J. Wiley & Sons, New York.

Burton, E.A., & Walter, L.M. (1987) Relative precipitation rates of aragonite and Mg calcite from seawater: temperature or carbonate ion control? *Geology*, 15, 111–114.

Busenberg, E., & Plummer, L.N. (1986) A comparative study of the dissolution and crystal growth kinetics of calcite and aragonite. In Studies in Diagenesis, Mumpton, F.A., ed., *U.S. Geological Survey Bulletin* 1578, 139–2168.

Busenberg,E., Plummer, L.N., & Parker, V.B. (1984) The solubility of $BaCO_3$ (cr) (witherite) in CO_2–H_2O solutions between 0 and 90°C, evaluation of the associated constants of $BaHCO_3^+$(aq) and $BaCO_3^o$ (aq) between 5 and 80°C, and a preliminary evaluation of the thermodynamic properties of Ba^{2+}(aq). *Geochimica Cosmochimica Acta*, 50, 2225–2234.

Byars, H.E. (2009) *Tectonic evolution of the west–central portion of the Newton window, North Carolina Inner Piredmont: Timing and implications for the emplacement of the Paledozoic Vale charnockites, Walker Top granite,and mafic complexes*. Unpublished MS thesis, The University of Tennessee, Knoxville, 248 pp.

Cairns-Smith, A.G. (1978) Precambrian solution photochemistry, inverse segregation, and banded iron formation. *Nature*, 276, 807–808.

Cameron, A.G.W. (1995) The first ten million years in the solar nebula. *Meteoritics*, 30, 133–161.

Cameron, A.G.W., & Pine, M.R. (1973) Numerical models of the primitive solar nebula. *Icarus*, 18, 377–406.

Campbell, A.R., & Larsen, P. (1998) Introduction to stable isotope applications in hydrothermal systems. *Reviews in Economic Geology*, 10, 173–193.

Campbell, I.H., & Naldrett, A.J. (1979) The influence of silicate:sulfide ratios on the geochemistry of magmatic sulfide deposits. *Economic Geology*, 74, 1503–1506.

Canfield, D.E. (1998) A new model for Proterozoic ocean chemistry. *Nature*, 396, 450–453.

Canfield, D.E. (2001a) Biogeochemistry of sulfur isotopes. *Reviews in Mineralogy and Geochemistry*, 43, 607–636.

Canfield, D.E. (2001b) Isotope fractionation by natural populations of sulfate-reducing bacteria. *Geochimica Cosmochimica Acta*, 65, 1117–1124.

Canfield, D.E. (2005) The early history of atmospheric oxygen: homage to Robert M. Garrels. *Annual Review of Earth and Planetary Sciences*, 33, 1–36.

Canfield, D.E., & Raiswell, R. (1999) The evolution of the sulfur cycle. *American Journal of Science*, 299, 697–723.

Canfield, D.E., & Teske, A. (1996) Late Proterozoic rise in atmospheric oxygen concentration inferred from phylogenetic and sulfur-isotope studies. *Nature*, 382, 127–132.

Canfield, D.E., Habicht, K.S., & Thamdrup, B. (2000) The Archean sulfur cycle and the early history of atmospheric oxygen. *Science*, 288, 658–661.

Canfield, D.E., Olesen, C.A., & Cox, R.P. (2006) Temperature and its control of isotope fractionation by a sulfate-reducing bacterium. *Geochimica Cosmochimica Acta*, 70, 548–561.

Canil, D. (1997) Vanadium partitioning and the oxidation state of Archean komatiites magmas. *Nature*, 389, 842–845.

Cannon, R.S., Pierce, A.P., Antweiler, T.C., & Buck, K.L. (1961) The data of lead isotope geology related to problems of ore genesis. *Economic Geology*, 56, 1–38.

Canup, R.M. (2004) Origin of terrestrial planets and the Earth–Moon system. *Physics Today*, April 2004, 56–62.

Canup, R.M., & Asphaug, E. (2001) Origin of the Moon in a giant impact near the end of the Earth's formation. *Nature*, 412, 708–712.

Capo, R.C., Stewart, B.W., & Chadwick, O.A. (1998) Strontium isotopes as tracers of ecosystem processes: theory and methods. *Geoderma*, 82 (197–225.

Capellos, C., & Bielski, B.H.J. (1972) *Kinetic Systems: Mathematical Description of Chemical Kinetics in Solution*. Wiley-Interscience, New York, 138 pp.

Capobianco, C.J., Jones, J.H., & Drake, M.J. (1993) Metal–silicate thermochemistry at high temperature: magma oceans and the "excess siderophile element" problem of the Earth's upper mantle. *Journal of Geophysical Research*, 98, 5433–5443.

Carlson, W.D., & Lindsley, D.H. (1988) Thermochemistry of pyroxenes on the join $Mg_2Si_2O_6$–$CaMgSi_2O_6$. *American Mineralogist*, 73, 242–252.

Carmichael, I.S.E. (1967a) The mineralogy of Thingmuli, a tertiary volcano in eastern Iceland. *American Mineralogist*, 52, 1815–1841.

Carmichael, I.S.E. (1967b) The iron–titanium oxides of salic volcanic rocks and their associated ferromagnesian silicates. *Contributions to Mineralogy and Petrology*, 14, 36–64.

Carmichael, I.S.E. (1991) The redox state of basaltic and silicate magmas: a reflection of their source regions? *Contributions to Mineralogy and Petrology*, 106, 129–141.

Carroll, S.A., & Walther, J.V. (1990) Kaolinite dissolution at 25°, 60°, and 80°C. *American Journal of Science*, 290, 797–810.

Carter, S.R., Evensen, N.M., Hamilton, P.J., & O'Nions, R.K. (1978) Neodymium and strontium isotope evidence for crustal contamination of continental volcanmics. *Science*, 202, 743–747.

Catling, D.C., & Claire, M.W. (2005) How Earth's atmosphere evolved to an oxic state: a status report. *Earth and Planetary Science Letters*, 237, 1–20.

Catling, D.C., Zahnle, K.J., & McKay, C.P. (2001) Biogenic methane, hydrogen escape, and the irreversible oxidation of early life. *Science*, 293, 839–843.

Cavosie, A.J., Valley, J.W., Wilde, S.A., & E.I.M.F. (2005) Magmatic $\delta^{18}O$ in 4400–3900 Ma detrital zircons: a record of the alteration and recycling of crust in the Early Archean. *Earth and Planetary Science Letters*, 235, 663–681.

Cawood, P.A., Kröner, A., & Pisarevsky, S. (2006) Precambrian plate tectonics: criteria and evidence. *GSA Today*, 16, 4–11.

Chabot, N.L., & Agee, C.B. (2003) Core formation in the Earth and Moon: new experimental constraints from V, Cr, and Mn. *Geochimica Cosmochimica Acta*, 67, 2077–2091.

Chabot, N.L., Draper, D.S., & Agee, C.B. (2005) Conditions of core formation in the Earth: constraints from nickel and cobalt partitioning. *Geochimica Cosmochimica Acta*, 69, 2141–2151.

Chatterjee, N.D., & Johannes, W. (1974) Thermal stability and standard thermodynamic properties of synthetic $2M_1$-muscovite, $KAl_2[AlSi_3O_{10}(OH)_2]$. *Contributions to Mineralogy and Petrology*, 48, 89–114.

Cheng, W., & Ganguly, J. (1994) Some aspects of multicomponent excess free energy models with Chermak, J.A., & Rimstidt, J.D. (1990) The hydrothermal transformation of kaolinite to muscovite/illite. *Geochimica Cosmochimica Acta*, 54, 2979–2990.

Chermak, J.A., & Rimstidt, J.D. (1990) The hydrothermal transformation of kaolinite to muscovite/illite. *Geochimica Cosmochimica Acta*, 54, 2979–2990.

Chou, L., & Wollast, R. (1985) Steady-state inetics and dissolution mechanism of albite. *American Journal of Science*, 285, 963–993.

Chou, L., Garrels, R.M., & Wollast, R. (1989) Comparative study of the kinetics and mechanisms of dissolution of carbonate minerals. *Chemical Geology*, 78, 269–282.

Christensen, N.I., & Mooney, W.D. (1995) Seismic velocity structure and composition of the continental crust: a global view. *Journal of Geophysical Research*, 100, 9761–9788.

Chyba, C.F. (1987) The cometary contribution to the oceans of primitive Earth. *Nature*, 330, 632–635.

Ciesla, F.J., Hood, L.L. (2002) The nebular shock wave model for chondrule formation: shock processing in a particle–gas suspension. *Icarus*, 158, 281–293.

Ciesla, F.J., Lauretta, D.S., & Hood, L.L. (2004) Radial migration of phyllosilicates in the solar nebula (abstract # 1219). *35th Lunar and Planetary Science Conference*. CD-ROM.

Clark, S.P., & Ringwood, A.E. (1964) Density distribution and constitution of the mantle. *Rev. Geophys.*, 2, 35–88.

Claoué-Long, J.C., Thirlwall, M.F., & Nesbitt, R.W. (1984) Revised Sm–Nd systematics of Kambalda greenstones, Western Australia. *Nature*, 307, 697–701.

Claypool, G.E., Holser, W.T., Kaplan, I.R., Sakai, H., & Zak, I. (1980) The age curves for sulfur and oxygen isotopes in marine sulfate and their mutual interpretation. *Chemical Geology*, 28, 199–260.

Clayton, R.N., & Kieffer, S.W. (1991) Oxygen isotopic thermometer calibrations. In *Stable Isotope Geochemistry: A Tribute to Samuel Epstein* (eds H.P. Taylor Jr., J.R. O'Neil, & I.R. Kaplan). *Geochemistry Society Special Publication*, 3, 3–10.

Clayton, R.N., Muffler, L.J.P., & White, D.E. (1968) Oxygen isotopic study of calcites and silicates of the River Ranch No. 1 well, Salton Sea geothermal field, California. *American Journal of Science*, 266, 968–979.

Cloud, P.E. (1972) A working model of the primitive Earth. *American Journal of Science*, 272, 537–548.

Cloud, P.E. (1973) Paleoecological significance of banded iron formation. *Economic Geology*, 68, 1135–1143.

Cohen, L.H., & Klement, W., Jr. (1967) High-low quartz inversion: determination to 35 kilobars. *Journal of Geophysical Research*, 72, 4245–4251.

Companion, A.L. (1964) *Chemical bonding*. New York: McGraw Hill, 155 pp.

Compston, W., & Pidgeon, R.T. (1986) Jack Hills, evidence of more very old detrital zircons in Western Australia. *Nature*, 321, 766–769.

Condie, K.C. (1997) *Plate tectonics and crustal evolution*. Oxford, MA: Butterworth and Heinemann, 282 pp.

Cotton and Wilkinson (1980)

Cox, J.D., D.D. Wagman, and V.A. Medvedev (1989) *CODATA Key Values for Thermodynamics*. Washington, D.C.: Hemisphere, 271 pp.

Cousens, B.L., & Ludden, J.N. (1991) Radiogenic isotope studies on oceanic basalts: a window into the mantle. In *Applications of Radiogenic Isotope Systems to Problems in Geology* (eds L. Heaman & J.N. Ludden). *Mineralogical Association of Canada Short Course Handbook*, 19, 225–255.

Craig, H. (1961) Standard for reporting concentrations of deuterium and oxygen-18 in natural waters. *Science*, 133, 1833–1834.

Craig, H. (1965) The measurement of oxygen isotope paleotemperatures. In *Stable Isotopes in Oceanographic Studies and Paleotemperatures*, Spoleto, July 26–27, 1965. Consiglio Nazionale delle Ricerche, Laboratorio di Geologia Nucleare, Pisa, pp. 1–22.

Craig, H. (1966) Isotopic composition and origin of the Red Sea and Salton Sea geothermal brines. *Science*, 154, 1544–1548.

Craig, H., & Gordon, L.I. (1965) Deuterium and oxygen-18 variations in the ocean and in the marine atmosphere. In *Stable Isotopes in Oceanographic Studies and Paleotemperatures*, Spoleto, July 26–27, 1965. Consiglio Nazionale delled Ricerche, Laboratorio di Geologia Nucleare, Pisa, pp. 1–22.

Criss, R.E. (1999) *Principles of Stable Isotope Distribution*. Oxford University Press, New York.

Croal, L.R., Johnson, C.M., Beard, B.L., & Newman, D.K. (2004) Iron isotope fractionation by Fe(II)-oxidizing photoautotrophic bacteria. *Geochimica Cosmochimica Acta*, 68, 1227–1242.

Crowe, S.A., Joners, C.A., Katsev, S., *et al.* (2009) Photoferrotrophs thrive in an Archean ocean. http://www.pnas.org/content/105/41/15938.full (on line).

Crowley, T.J. (2000) Causes of climate change over the past 1000 years. *Science*, 289, 870–872.

Currie, K.L. (1971) The reaction 3cordierite = 2garnet + 4sillimanite + 5quartz as a geological thermometer in the Poinicon Lake region, Ontario. *Contributions to Mineralogy and Petrology*, 33, 215–226.

Dahl, P.S. (1980) The thermal-compositional dependence of Fe^{2+}–Mg distribution between coexisting garnet and pyroxene: applications to geothermometry. *American Mineralogist*, 65, 852–866.

Dalrymple, G.B., & Lanphere, M.A. (1969) *Potassium Argon Dating*. W.H. Freeman, San Francisco: 258 pp.

Dalrymple, G.B., & Lanphere, M.A. (1971) $^{40}Ar/^{39}Ar$ technique of K/Ar dating: a comparison with conventional technique. *Earth and Planetary Science Letters*, 12, 300–308.

Dalrymple, G.B., & Lanphere, M.A. (1974) $^{40}Ar/^{39}Ar$ age spectra of some undisturbed terrestrial samples. *Geochimica Cosmochimica Acta*, 38, 715–738.

Dansgaard, W (1964) Stable isotopes in precipitation. *Tellus*, 16(4), 436–468.

Dasgupta, S., Sengupta, P., Guha, D., & Fukuoka, M. (1991) A refined garnet–biotite Fe–Mg exchange geothermometer and its application in amphibolites and granulites. *Contributions to Mineralogy and Petrology*, 109, 130–137.

Davidson, P.M., & Lindsley, D.H. (1985) Thermodynamic analysis of quadrilateral pyroxenes. *Contributions to Mineralogy and Petrology*, 91, 390–404.

Davidson, P.M., & Lindsley, D.H. (1989) Thermodynamic analysis of pyroxene-olivine-quartz equilibria in the system CaO–MgO–FeO–SiO_2. *American Mineralogist*, 74, 18–30.

Dauphas, N. (2003) The dual origin of the terrestrial atmosphere. *Icarus*, 165, 326–339.

Dauphas, N., & Marty, B. (2002) Inference on the nature and msass of Earth's late veneer from noble metals and gases. *Journal of Geophysical Research*, 107, E12-11–E12-7.

Davies, C.W. (1962) *Ion Association*. Butterworths, Washington, DC, 190 pp.

Deer, W.A., Howie, R.A., & Zussman, J. (1966) *An Introduction to the Rock-Forming Minerals*. J. Wiley & Sons, 696 pp.

Deer, W.A., Howie, R.A., & Zussman, J. (1996) *An Introduction to the Rock-Forming Minerals*. J. Wiley & Sons, New York, 696 pp.

Demicco, R.V., Lowenstein, T.K., Hardie, L.A., & Spencer, R.J. (2005) Model of seawater composition for the phanerozoic. *Geology*, 33, 877–880.

Deming, D. (2002) Origin of the ocean and continents: a unified theory of the earth. *International Geology Review*, 44, 137–152.

DePaolo, D.J. (1981) Trace element and isotopic effects of combined wallrock assimilation and fractional crystallization. *Earth and Planetary Science Letters*, 53, 189–202.

DePaolo, D.J., & Wasserburg, G.J. (1976) Nd isotopic variations and petrogenetic models. *Geophysical Research Letters*, 3, 249–252.

DePaolo, D.J., & Wasserburg, G.J. (1979) Sm–Nd age of the Stillwater complex and the mantle evolution curve for neodymium. *Geochimica Cosmochimica Acta*, 43, 999–1008.

Derry, L.A., & Jacobsen, S.B. (1990) The chemical evolution of Precambrian seawater: evidence from REEs in banded iron formations. *Geochimica Cosmochimica Acta*, 54, 2965–2977.

Des Marais, D.J. (1994) The Archean atmosphere: its composition and fate. In *Archean Crustal Evolution* (ed. K.C. Condie), pp. 505–523. Elsevier, Amsterdam.

Des Marais, D.J., Strauss, H., Summons, R.E., & Hayes, J.M. (1992) Carbon isotope evidence for the stepwise oxidation of the Proterozoic environment. *Nature*, 359, 605–609.

Detmers, J., Brüchert, V., Habicht, K.S., & Kuever, J. (2001) Diversity of sulphur isotope fractionations by sulphate-reducing prokaryotes. *Applied Environmental Microbiology*, 67, 888–894.

Dickin, A.P. (1995) *Radiogenic Isotope Geology*. Cambridge University Press, Cambridge, 490 pp.

Dickson, J.A.D. (2002) Fossil echinoderms as monitor of the Mg/Ca ratio of Phanerozoic oceans. *Science*, 298, 1222–1224.

Dimroth, E., & Kimberley, M.M. (1976) Precambrian atmospheric oxygen: Evidence in the sedimentary distribution of carbon, sulfur, uranium and iron. *Canadian Journal of Earth Sciences*, 13, 1161–1185.

Dinur, D., Spiro, B., & Aizenshtat, Z. (1980) The distribution and isotopic composition of sulfur in organic-rich sedimentary rocks. *Chemical Geology*, 31, 37–51.

Doe, B.R., & Zartman, R.E. (1979) Plumbotectonics 1, the Phanerozoic. In *Geochemistry of Hydrothermal Ore Deposits* (ed. H.L. Barnes), 2nd edn, pp. 22–70. J. Wiley & Sons, New York.

Drake, M.J. (2000) Accretion and primary differentiation of the Earth: a personal journey. *Geochimica Cosmochimica Acta*, 64, 2363–2370.

Drake, M.J. (2005) Origin of water in the terrestrial planets. *Meteoritics & Planetary Science*, 40, 1–9.

Drake, M.J., & Righter, K. (2002) Determining the composition of the Earth. *Nature*, 416, 39–44.

Dreibus, G., Ryabchikov, I., Rieder, R., *et al.* (1998) Relationships between rocks and soil at the Pathfinder landing site and the martian meteorites (abstract). In *Lunar and Planetary Science XXIX*, Abstract 1348. Lunar and Planetary Institute Houston (CD-ROM).

Drever, J.J. (1974) Geochemical model for the origin of Precambrian banded iron formations. *Geological Society of America Bulletin*, 85, 1099–1106.

Drever, J.I. (1994) The effect of land plants on weathering rates of silicate minerals. *Geochimica Cosmochimica Acta*, 58, 2325–2332.

Drever, J.I. (1997) *The Geochemistry of Natural Waters: Surface and Groundwater Environments*, 3rd edn. Prentice Hall, Saddle River, New Jersey, 436 pp.

Drever, J., & Clow, D.W. (1995) Weathering rates in catchments. In *Chemical Weathering Rates of Silicate Minerals* (eds A.F. White & S.L. Brantley). *Reviews in Mineralogy*, 31, 463–483.

Drever, J.I., Li, Y.-H, and Maynard, J.B. (1988) Geochemical cycles: the continental crust and the oceans. In *Chemical cycles in the evolution of the Earth* (eds C.B Gregor, R.M. Garrel, F.T. Mackenzie, & J.B. Maynard), pp. 17–53. J. Wiley & Sons, New York.

Droop, G.T.R. (1987) A general equation for estimating Fe^{3+} concentrations in ferromagnesian silicates and oxides from microprobe analyses, using stoichiometric criteria. *Mineralogical Magazine*, 51, 431–435.

Duan, Z., Moeller, N., & Weare, J.H. (1996) A general equation of state for supercritical fluid mixtures and molecular dynamics simulation of mixture *P–V–X–T* properties. *Geochimica Cosmochimica Acta*, 60, 1209–1216.

Duke, J.M. (1976) Distribution of the Period Four transition elements among olivine, calcic pyroxene and mafic silicate liquid: experimental results. *Journal of Petrology*, 17, 499–521.

Dunn, T., & Sen, C. (1994) Mineral/matrix partition coefficients for orthopyroxene, plagioclase, and olivine in basaltic to andesitic systems: a combined analytical and experimental study. *Geochimica Cosmochimica Acta*, 58, 717–733.

Eby, G.N. (2004) *Principles of Environmental Geochemistry*. Thomson Brooks/Cole, California, 514 pp.

Eckert, J.O., Jr., Newton, R.C., & Kleppa, O.J. (1991) The DH of reaction and recalibration of garnet–pyroxene–plagioclase–quartz geobarometers in the CMAS system by solution calorimetry. *American Mineralogist*, 76, 148–160.

Ehrenreich, A., & Widdel, F. (1994) Anaerobic oxidation of ferrous iron by purple bacteria, a new type of phototrophic metabolism. *Applied and Environmental Microbiology*, 60, 4517–4526.

Ehrlich, H.L. (1996) How microbes influence mineral growth and dissolution. *Chemical Geology*, 132, 5–9.

Ehrlich, H.L. (2002) *Geomicrobiology*. Marcel Dekker, Inc., New York, 768 pp.

Ehrlich, H.L., & Brierley, C.L. (eds) (1990) *Microbial Mineral Recovery*. McGraw-Hill, New York, 454 pp.

Elkins, L.T., & Grove, T.L. (1990) Ternary feldspar experiments and thermodynamic models. *American Mineralogist*, 75, 544–559.

Ellis, D.J., & Green, D.H. (1979) An experimental study of the effect of Ca upon gaqrnet–clinopyroxene Fe–Mg exchange equilibria. *Contributions to Mineralogy and Petrology*, 71, 13–22.

Elsasser, W.M. (1963) Early history of the Earth. In *Earth Science and Meteoritics* (eds J. Geiss & E. Goldberg), pp. 1–30. North Holland, Amsterdam.

Enders, M.S., Knickerbocker, C., Titley, S.R., & Southam, G. (2006) The role of bacteria in the supergene environment of the Morenci porphyry copper deposity, Greenlee County, Arizona. *Economic Geology*, 101, 59–70.

Engi, M. (1983) Equilibria involving Al–Cr spinel: Mg–Fe exchange with olivine. Experiments, thermodynamic analysis, and consequences for geothermometry. *American Journal of Science*, 283-A, 29–71.

Ernst, W.G. (ed.). (2000) *Earth Systems: Processes and Issues*. Cambridge University Press, Cambridge, 566 pp.

Essene, E.J. (1982) Geologic thermometry and barometry. In *Reviews of Mineralogy*, Vol. 10, 153–206. Mineralogical Society of America, Washington, DC.

Essene, E.J. (1989) The current status of thermobarometry. In *Evolution of Metamorphic Belts* (eds J.S. Daly, R.A. Cliff, & B.W.D. Yardley), pp. 1–44. The Geological Society, London.

Eugster, H.P., & Wones, D.R. (1962) Stability relations of ferruginous biotite, annite. *Journal of Petrology*, 13, 147–179.

Evans, R.C. (1966) *An Introduction to Crystal Chemistry*. Cambridge University Press, Cambridge, 410 pp.

Farquhar, J., & Wing, B.A. (2003) Multiple sulfur isotopes and the evolution of the atmosphere. *Earth and Planetary Science Letters*, 213, 1–13.

Farquhar, J., Bao, H., & Thiemens, M. (2000) Atmospheric influence of Earth's earliest sulfur cycle. *Science*, 289, 756–758.

Farquhar, J., Savarino, J., Airieau, S., & Thiemens, M.H. (2001) Observation of wavelength-sensitive mass-dependent sulfur isotopes effects during SO_2 photolysis: implication for the early Earth atmosphere. *Journal of Geophysical Research*, 106, 32829–32839.

Faure, G. (1986) *Principles of Isotope Geology*, 2nd edn. J. Wiley & Sons, New York, 589 pp.

Faure, G. (1991) *Principles and Applications of Inorganic Geochemistry: A Comprehensive Textbook for Geology Students*. Macmillan Publishing Co., New York, 626 pp.

Faure, G. (1998) *Principles and Applications of Inorganic Geochemistry: A Comprehensive Textbook for Geology Students*, 2nd edn. Prentice Hall, Upper Saddle River, NJ, 559 pp.

Faure, G. (2001) *Origin Of Igneous Rocks: The Isotopic Evidence*. Springer-Verlag, Berlin, 496 pp.

Fei, Y., Saxena, S.K., & Eriksson, G. (1986) Some binary and ternary silicate solid solution models. *Contributions to Mineralogy and Petrology*, 94, 221–229.

Felmy, A.R., Girvin, D.C., & Jenne, E.A. (1984) *MINTEQ – A Computer Program for Calculating Aqueous Geochemical Equilibria*. US Environmental Protection Agency (EPA–600/3–84–032).

Ferry, J.M. (1980) A comparative study of geothermometers and geobarometers in pelitic schists from south-central Maine. *American Mineralogist*, 65, 720–732.

Ferry, J.M., & Spear, F.S. (1978) Experimental calibration of the partitioning of Fe and Mg between biotite and garnet. *Contributions to Mineralogy and Petrology*, 66, 113–117.

Finnerty, A.A., & Boyd, F.R. (1987) Thermobarometry for garnet peridotites: basis for the determination of thermal and compositional structure of the upper mantle. In *Mantle Xenoliths* (ed. P.H. Nixon), pp. 381–402. J. Wiley & Sons, New York.

Fletcher, P. (1993) *Chemical Thermodynamics for Earth Scientists*. J. Wiley & Sons, New York, 464 pp.

Flower, M.F.J., Schmincke, H.U., & Thompson, R.N. (1975) Phlgopite stability and the $^{87}Sr/^{86}Sr$ step in basalts along the Reykjanes Ridge. *Nature*, 254, 404–406.

Floyd, P.A., & Winchester, J.A. (1975) Magma type and tectonic setting discrimination using immobile elements. *Earth and Planetary Science Letters*, 27, 211–218.

Folk, R.L. (1974) The natural history of crystalline calcium carbonate: effect of magnesium content and salinity. *Journal of Sedimentary Petrology*, 44, 40–53.

François, L.M. (1986) Extensive deposition of banded iron formations was possible without photosynthesis. *Nature*, 320, 352–354.

Frost, B.R. (1991) Introduction to oxygen fugacity and its petrologic importance. *Reviews in Mineralogy*, 25, 1–9.

Froude, D.O., Ireland, T.R., Kinny, P.D., *et al.* (1983) Ion microprobe identification of 4,100–4,200 Myr-old terrestrial zircons. *Nature*, 304, 616–618.

Fuhrman, M.L., & Lindsley, D.H. (1988) Ternary-feldspar modeling and thermometry. *American Mineralogist*, 73, 201–215.

Fyfe, W.S. (1964) *Geochemistry of Solids*. McGraw–Hill Book Co., New York, 199 pp.

Gaab, A.S., Todt, W., & Poller, U. (2006) CLEO: Common lead evaluation using Octave. *Computers & Geosciences*, 32, 993–1003.

Galloway, J.N. (2005) The global nitrogen cycle. In *Biogeochemistry* (ed. W.H Schlesinger), Vol. 8, *Treatise on Geochemistry* (eds H.D. Holland & K.K. Turekian), pp. 557–584. Elsevier–Pergamon, Oxford.

Gamow, G. (1961) *The Atom and its Nucleus*. Prentice Hall, Englewood Cliffs, NJ, 153 pp.

Ganguly, J. (1973) Activity-composition relation of jadeite in omphacite pyroxene: theoretical deductions. *Earth and Planetary Science Letters*, 19, 145–153.

Ganguly, J., & Kennedy, G.C. (1974) The energetics of natural garnet solid solution: 1. Mixing of the aluminosilicate end-members. *Contributions to Mineralogy and Petrology*, 48, 137–148.

Ganguly, J., & Saxena, S.K. (1984) Mixing properties of aluminosilicate garnets: constraints from natural and experimental data, and applications to geothermobarometry. *American Mineralogist*, 69, 88–97.

Ganguly, J., & Tazzoli, V. (1994) Fe^{2+}–Mg interdiffusion in orthopyroxene: retrieval from the data on intercrystalline exchange reaction. *American Mineralogist*, 79, 930–937.

Ganguly, J., Cheng, W., & Tirone, M. (1996) Thermodynamics of aluminosilicate garnet solid solution: new experimental data, an optimized model, and thermometric applications. *Contributions to Mineralogy and Petrology*, 126, 137–151.

Garrels, R.M., & Christ, R.L. (1965) *Solutions, Minerals, and Equilibria*. Harper & Row, New York, 450 pp.

Garvin, D., Parker, V.B., & White, H.J. (1987) *CODATA Thermodynamic Tables – Selections for Some Compounds of Calcium and Related Mixtures: a Prototype Set of Tables*. Hemisphere, Washington, DC.

Gasparik, T. (1984a) Experimental study of subsolidus phase relations and mixing properties of pyroxene in the system $CaO–Al_2O_3–SiO_2$. *Geochimica Cosmochimica Acta*, 48, 2537–2545.

Gasparik, T. (1984b) Two-pyroxene thermometry with new experimental data in the system $CaO–MgO–Al_2O_3–SiO_2$. *Contributions to Mineralogy and Petrology*, 87, 87–97.

Gaylarde, C.C., & Videla, H.A. (eds), *Bioextraction and Biodeterioration of Metals*. Cambridge University Press, Cambridge, 372 pp.

Genda, H., & Abe, Y. (2005) Enhanced atmospheric loss on protoplanets at the giant impact phase in the presence of oceans. *Nature*, 433, 842–844.

Gessmann, C.K., Spiering, B., & Raith, M. (1997) Experimental study of the Fe–Mg exchange between garnet and biotite: constraints on the mixing behavior and analysis of the cation-exchange mechanisms. *American Mineralogist*, 82, 1225–1240.

Gessmann, C.K., Wood, B.J., Rubie, D.C., & Kilburn, M.R. (2001) Solubility of silicon in liquid metal at high pressure: implications for the composition of the Earth's core. *Earth and Planetary Science Letters*, 184, 367–376.

Ghent, E.D., & Stout, M.Z. (1981) Geothermometry and geobarometry and fluid compositions of metamorphosed calc-silicates and pelites, Mica Creek, British Columbia. *Contributions to Mineralogy and Petrology*, 76, 92–97.

Ghent, E.D., Robbins, D.B., & Stout, M.Z. (1979) Geothermometry, geobarometry, and fluid compositions of metamorphosed calc-silicates and pelites, Mica Creek, British Columbia. *American Mineralogist*, 64, 874–885.

Ghiorso, M.S. (1984) Activity/composition relations in the ternary feldspars. *Contributions to Mineralogy and Petrology*, 87, 282–296.

Gibbins, W.A., & McNutt, R.H. (1975) The age of the Sudbury Nickel Irruptive and the Murray Granite. *Canadian Journal of Earth Science*, 12, 1970–1989.

Gibbs, G.V., Spackman, M.A., & Boisen, M.B., Jr. (1992) Bonded and promolecule radii for molecules and crystals. *American Mineralogist*, 77, 741–750.

Gill, R (1996) *Chemical Fundamentals of Geology*, 2nd edn. Chapman & Hall, London, 289 pp.

Given, R.K., & Wilkinson, B.H. (1985) Kinetics control of morphology, composition, and mineralogy of abiotic sedimentary carbonates. *Journal of Sedimentary Petrology*, 55, 109–119.

Gize, A.P., & Barnes, H.L. (1987) The organic geochemistry of two Mississippi Valley-type lead–zinc deposits. *Economic Geology*, 82, 457–470.

Godderis, Y., & Veizer, J. (2000) Tectonic control of chemical and isotopic composition of ancient oceans: the impact of continental growth. *American Journal of Science*, 300, 434–461.

Goldberg, E.D. (1957) Biogeochemistry of trace metals. In *Treatise on Marine Ecology and Paleoecology* (ed. J.W. Hedgepeth), Vol. 1. *Geological Society of America Memoir*, 67, 345–357.

Goldblatt, C., Lenton, T.M., & Watson, A.J. (2006) Bistability of atmospheric oxygen and the Great Oxidation. *Nature*, 443, 683–686.

Goldhaber, M.B., & Kaplan, I.R. (1975) Controls and consequences of sulfate reduction rates in recent marine sediments. *Soil Science*, 119. 42–55.

Goldschmidt, V.M. (1911) *Die Kontaktmetamorphose in Kristianiagebiet*. Kristiania Vidensk. Skr., I, Math-Naturv. Klasse.

Goldschmidt, V.M. (1937) The principles of distribution of chemical elements in minerals and rocks. *Journal of the Chemistry Society of London*, 1937, 665–673.

Goldschmidt, V.M. (1954) *Geochemistry*. Clarendon Press, Oxford, 730 pp.

Goldsmith, J.R., & Newton, R.C. (1969) P–T–X relations in the system $CACO_3–MgCO_3$ at high temperatures and pressures. *American Journal of Science*, 267-A, 160–190.

Goldstein, R.H., & Reynolds, T.J. (1994) Systematics of fluid inclusions in diagenetic minerals. *Society of Sedimentary Geology Short Course*, 31, 199 pp.

Goodfellow, W.D. (1987) Anoxic stratified oceans as a source of sulfur in sedimenthosted stratiform Zn–Pb deposits (Selwyn basin, Yukon, Canada). *Chemical Geology*, 65, 359–382.

Gottschalk, M. (1997) Internally consistent thermodynamic data for rock forming minerals. *European Journal of Mineralogy*, 9, 175–223.

Grambling, J.A. (1981) Kyanite, andalusite, sillimanite, and related mineral assemblages in the Truchas Peaks region, New Mexico. *American Mineralogist*, 66, 702–722.

Green, N.L., & Udansky, S.I. (1986) Ternary-feldspar mixing relations and thermobarometry. *American Mineralogist*, 71, 1100–1108.

Green, T.H., & Adam, J. (1991) Assessment of the garnet–clinopyroxene Fe–Mg exchange thermometer using new experimental data. *Journal of Metamorphic Geology*, 9, 341–347.

Green, T.H., & Pearson, N.J. (1985) Experimental determination of REE partition coefficients between amphibole and basaltic to andesitic liquids at high pressure. *Geochimica Cosmochimica Acta*, 49, 1465–1468.

Greenwood, N.N. (1970) *Ionic Crystals, Lattice Defects, and Nonstoichiometry*. Chemical Publishing Co., 194 pp.

Greenwood, R.C., Franchi, I.A., Jambon, A., Buchanan, P.C. (2005) Widespread magma oceans on asteroidal bodies in the early solar system. *Nature*, 435, 916–918.

Gregor, C.B., Garrels, R.M., Mackenzie, F.T., & Maynard, J.B. (eds) (1988) *Chemical Cycles in the Evolution of the Earth*. J. Wiley & Sons, New York, 276 pp.

Gregory, R.T. (1991) Oxygen isotope history of seawater revisited: timescales for boundary event changes in the oxygen isotope composition of seawater. In *Stable Isotope Geochemistry: A Tribute to Samuel Epstein* (eds H.P. Taylor Jr., J.R. O'Neil, & I.R. Kaplan). *Geochemistry Society Special Publication*, 3, 65–76.

Gregory, R.T. and Taylor, H.P., Jr. (1981) An oxygen isotope profile in a section of Cretaceous oceanic crust, Samail ophiolite, Oman: evidence for $\delta^{18}O$ buffering of the oceans by deep (> 5 km) seawater–hydrothermal circulation at Mid-Ocean Ridges. *Journal of Geophysical Research*, 86, 2737–2755.

Gregory, R.T., Criss, R.E., & Taylor, H.P., Jr. (1989) Oxygen isotope exchange kinetics of mineral pairs in closed and open systems: applications to problems of hydrothermal alteration of igneous rocks and Precambrian iron formations. *Chemical Geology*, 72, 1–42.

Grenthe, I., Fuger, J., Lemire R.J., *et al.* (1992) *Chemical Thermodynamics of Uranium*. Elsevier, 715 pp.

Grieve, R.A.F., Cintala, M.J., & Therriault, A.M. (2006) Large-scale impacts and the evolution of the Earth's crust: the early years. *Geological Society of America Special Paper*, 405, 23–31.

Grossman, L., & Larimer, J.W. (1974) Early chemical history of the solar system. *Reviews of Geophysics and Space Physics*, 12, 71–101.

Grotzinger, J.P., & James, N.P. (2000) Precambrian carbonates: evolution of understanding. In *Carbonate Sedimentation and Diagenesis in the Evolving Precambrian World* (eds J.P. Grotzinger & N.P. James). *Society of Economic Palaeontologists and Minerologists Special Publication*, 67, 3–20.

Grotzinger, J., & Jordan, T. (2010) *Understanding Earth*, 6th edn. W.H. Freeman and Company, New York, 653 pp.

Grotzinger, J.P., & Kasting, J.F. (1993) New constraints on Precambrian ocean composition. *Journal of Geology*, 101, 235–243.

Grotzinger, J.P., & Knoll, A. H. (1995) Anomalous carbonate precipitates: is the Precambrian the key to the Permian?. *Palaios*, 10, 578–596.

Grove, T.L., & Krawczynski, M. (2009) Lunar mare volcanism: where did the magmas come from? *Elements*, 5, 29–34.

Grover, J. (1977) Chemical mixing in multicomponent systems: an introduction to the use of Margules and other thermodynamic excerss functions to represent non-ideal behavior. In *Thermodynamics in Geology* (ed. D.G. Fraser), pp. 67–97. Reidel, Boston, MA.

Grutzeck, M., Kridelbaugh, S., & Weil, D. (1974) The distribution of Sr and REE between diopside and silicate liquid. *Geophysical Research Letters*, 1, 273–275.

Gunnarsson, I., & Arnórsson, S. (1999) New data on the standard Gibbs energy of H_4SiO_4 and its effect on silicate solubility. In *Geochemistry of the Earth's Surface* (ed. H. Àrmannson), pp. 449–452. Balkema, Rotterdam.

Haase, K.M., Stroncik, N.A., Hékinian, R., & Stoffers, P. (2005) Nb-depleted andesites from the Pacific–Antarctic Rise as analogs for early continental crust. *Geology*, 33, 921–924.

Habicht, K.S., & Canfield, D.E. (1996) Sulfur isotope fractionation in modern microbial mats and the evolution of the sulfur cycle. *Nature*, 382, 342–343.

Habicht, K.S., & Canfield, D.E. (1997) Sulfur isotope fractionation during bacterial sulfate reduction in organic-rich sediments. *Geochimica Cosmochimica Acta*, 61, 5351–5361.

Habicht, K.S., & Canfield, D.E. (2001) Isotope fractionation by sulphate-reducing natural populations and the isotopic composition of sulphides in marine sediments. *Geology*, 29, 555–558.

Habicht, K.S., Canfield, D.E., & Rethmeier, J. (1998) Sulphur isotope fractionation during bacterial sulphate reduction and disproportionation of thiosulphate and sulfite. *Geochimica Cosmochimica Acta*, 62, 2585–2595.

Habicht, K.S., Gade, M., Thamdrup, B., Berg, P., & Canfield, D.E. (2002) Calibration of the sulfate levels in the Archean ocean. *Science*, 298, 2372–2375.

Hackler, R.T., & Wood, B.J. (1989) Experimental determination of Fe and Mg exchange between garnet and olivine and estimation of Fe–Mg mixing properties in garnet. *American Mineralogist*, 74, 994–999.

Halliday, A.N. (2003) The origin and earliest history of the Earth. In *Treatise on Geochemistry, 1: Meteorites, Comets and Planets* (ed. A.M. Davies), pp. 509–557. Elsevier, Oxford.

Halliday, A.N. (2006) The origin of the Earth: what is new? *Elements*, 2, 205–210.

Halliday, A.N. (2008) A young Moon-forming impact at 70–110 million years accompanied by late-stage mixing, core formation and degassing of the Earth. *Philosophical Transactions of the Royal Society*, A366, 4163–4181.

Halliday, A.N., & Kleine, T. (2006) Meteorites and the timing, mechanisms, and conditions of terrestrial planet accretion and early differentiation. In *Meteorites and the Early Solar System II* (eds D.S. Lauretta, & H.Y. McSween Jr), pp. 775–801. University of Arizona Press, Tucson, AZ.

Halliday, A.N., & Lee, D-C. (1999) Tungsten isotopes and the early development of the Earth and Moon. *Geochimica Cosmochimica Acta*, 63, 4157–4179.

Hamilton, P.J., Evensen, N.M. O'Nions, R.K., & Tarney, J. (1979) Sm–Nd systematics of Lewisian gneisses: implications for the origin of granulites. *Nature*, 277, 25–28.

Hamilton, W.B. (2003) An alternative Earth. *GSA Today*, November 2003, 4–12.

Hanes, J.A. (1991) K–Ar and $^{40}Ar/^{39}Ar$ geochronology: methods and applications. In *Applications of Radiogenic Isotope Systems to Problems in Geology* (eds L. Heaman & J.N. Ludden). *Mineralogical Association of Canada Short Course Handbook*, 19, 27–57.

Hanor, J. S. (1997). Controls on the solubilization of lead and zinc in basinal brines. *Soc. Economic Geology Spec. Publ.* 4, 483–500.

Hardie, L.A. (1996) Secular variation in seawater chemistry: an explanation for the coupled secular variation in the mineralogies of marine limestones and potash evaporites over the past 600 m.y. *Geology*, 24, 279–283.

Hardie, L.A. (2003) Secular variations in Precambrian seawater chemistry and the timing of Precambrian aragonite seas and calcite seas. *Geology*, 31, 785–788.

Harker, R.I., & Tuttle, O.F. (1955) The thermal dissociation of calcite, dolomite and magnesite. *American Journal of Science*, 253, 209–224.

Harley, S.L. (1984) An experimental study of the partitioning of Fe and Mg between garnet and orthopyroxene. *Contributions to Mineralogy and Petrology*, 86, 359–373.

Harlov, D.E., & Newton, R.C. (1993) Reversal of the metastable kyanite + corundum + quartz and andalusite + corundum + quartz equilibria and the enthalpy of formation of kyanite and andalusite. *American Mineralogist*, 78, 594–600.

Harrison, A.G., & Thode, H.G. (1957) Mechanisms of the bacterial reduction of sulfate from isotope fractionation studies. *Faraday Society Transactions*, 53, 84–92.

Harrison, T.M., Blichert-Toft, J., Müller, W., Albarede, F., Holden, P., & Mojzsis, S.J. (2005) Heterogeneous Hadean hafnium: evidence of continental crust at 4.4 to 4.5 Ga. *Science*, 310, 1947–1950.

Hart, S.R., & Davis, K.E. (1978) Nickel partitioning between olivine and silicate melt. *Earth and Planetary Science Letters*, 40, 203–219.

Hartman, W.K., & Davis, D.R. (1975) Satellite-sized planetesimals and lunar origin. *Icarus*, 24, 504–515.

Harvie, C.E., Weare, J.H., Hardie, L.A., & Eugster, H.P. (1980) Evaporation of seawater: calculated mineral sequences. *Science*, 208, 498–500.

Harvie, C.E., Møller, N., & Weare, J.H. (1984) The prediction of mineral solubilities in natural waters: the Na–K–Mg–Ca–H–Cl–SO_4–OH–HCO_3–CO_3–CO_2–H_2O system to high ionic strengths at 25°C. *Geochimica Cosmochimica Acta*, 48, 723–751.

Haughton, D.R., Roeder, P.L., & Skinner, B.J. (1974). Solubility of sulfur in mafic magmas. *Economic Geology*, 69, 451–467.

Hawkesworth, C.J., Norry, M.J., Roddick, J.C., & Vollmer, R. (1979a) $^{143}Nd/^{144}Nd$ and $^{87}Sr/^{86}Sr$ ratios from the Azores and their significance in LIL-element enriched mantle. *Nature*, 280, 28–31.

Hawkesworth, C.J., O'Nions, R.K., & Arculus, R.J. (1979b) Nd and Sr isotope geochemistry of island arc volcanics, Grenada, Lesser Antilles. *Earth and Planetary Science Letters*, 45, 237–248.

Helgeson, H.C., & Kirkham, D.H. (1974) Theoretical prediction of the thermodynamic behavior of aqueous electrolytes at high pressures and temperatures: II. Debye–Hückel parameters for activity coefficients and relative partial molal properties. *American Journal of Science*, 274, 1099–1198.

Helgeson, H.C., Delaney, J.M., Nesbitt, H.W., & Bird, D.K. (1978) Summary and critique of the thermodynamic properties of rock-forming minerals. *American Journal of Science*, 278-A, 1–229.

Helmstaedt, H., & Doig, R. (1975) Eclogite nodules from kimberlite pipes of the Colorado plateau – samples of subducted Franciscan-type oceanic lithosphere. *Physics and Chemistry of the Earth*, 9, 95–111.

Hemingway, B.S., 1990, Thermodynamic properties for bustenite, NiO, magnetite, Fe_3O_4, and hematite, Fe_2O_3 with comments on selected oxygen buffers. *Am. Mineral.*, 75, 781-790.

Hemingway, B.S., & Sposito, G. (1989) Inorganic aluminum bearing solid phases. In *The Environmental Chemistry of Aluminum*, (Ed. G. Sposito), pp. 55–85. CRC Press, Boca Raton, FL.

Hemingway, B.S., Robie, R.A., Evans, H.T., Jr., & Kerrick, D.M. (1991) Heat capacities and entropies of sillimanite, fibrolite, andalusite, kyanite, and quartz and the Al_2SiO_5 phase diagram. *American Mineralogist*, 76, 1597–1613.

Hemley, J.J., Cygan, G.L., Fein, J.B., Robinson, G.R., & D'Angelo, W.M. (1992) Hydrothermal ore-forming processes in the light of studies in rock-buffered systems: I. Iron–copper–zinc–lead sulfide solubility relations. *Economic Geology*, 87, 1–22.

Hemming, S.R., McLennan, and Hanson, G.N. (1995) Geochemical and Nd/Pb isotopibc evidence for the provenance of the Early Proterozoic Virginia Formation, Minnesota. Implications for the tectonic setting of the Animikie Basin. *Journal of Geology*, 103, 147–168.

Hemming, S.R., McDaniel, D.K., McLennan, S.M., & Hanson, G.N. (1996) Pb isotope constraints on the provenance and diagenesis of detrital feldspars from the Sudbury Basin, Canada. *Earth and Planetary Science Letters*, 142, 501–512.

Henderson, P. (1982) *Inorganic Geochemistry*. Pergamon Press, Oxford, 353 pp.

Hensen, B.J., & Green, D.H. (1971) Experimental study of cordierite and garnet in pelitic compositions at high pressures and temperatures. I. Compositions with excess alumino-silicate. *Contributions to Mineralogy and Petrology*, 33, 309–330.

Hertogen, J., & Gijbels, R. (1976) Calculations of trace element fractionation during partial melting. *Geochimica Cosmochimica Acta*, 40, 313–322.

Hester, J.J., & Desch, S.J. (2005) Understanding our origins: Star formation in HII Region environments. In *Chondrites and the Protoplanetary Disk* (eds A.N. Krot, E.R.D. Scott, & Bo Reipurth). *Astronomical Society of the Pacific Conference Series*, 341, 107–130.

Hewins, R.H. (1997) Chondrules. *Annual Review of Earth and Planetary Sciences*, 25, 61–83.

Heymann, D., Yancey, T.E., Wolbach, W.S., *et al.* (1998) Geochemical markers of the Cretaceous–Tertiary boundary event at Brazos River, Texas, USA. *Geochimica Cosmochimica Acta*, 62, 173–181.

Hiroi, Y., & Kobayashi, E. (1996) Origin of andalusite–kyanite–sillimanite aggregates in the Nishidohira pelitic rocks in the southernmost part of the Abukuma Plateau, Northeast Japan, and the *P–T* path. *J. Mineral. Petrol. and Economic Geology*, 91, 220–234.

Hirschmann, M. (1991) Thermodynamics of multicomponent olivines and the solution properties of $(Ni,Mg,Fe)_2SiO_4$ and $(Ca,Mg,Fe)_2SiO_4$ olivines. *American Mineralogist*, 76, 1232–1248.

Hobbs, P.V. (2000) *Introduction to Atmospheric Chemistry*. Cambridge University Press, Cambridge, 262 pp.

Hodges, K.V., & Crowley, P.D. (1985) Error estimation and empirical geothermobarometry for politic systems. *American Mineralogist*, 70, 702–709.

Hodges, K.V., & McKenna, L.W. (1987) Realistic propagation of uncertainties in geologic thermobarometry. *American Mineralogist*, 72, 671–680.

Hodges, K.V., & Spear, F.S. (1982) Geothermometry, geobarometry and the Al_2SiO_5 triple point at Mt. Moosilauke, New Hampshire. *American Mineralogist*, 67, 1118–1134.

Hoefs, J. (1997) *Stable Isotope Geochemistry, 4th edition*. Springer-Verlag, Berlin, 212 pp.

Hoefs, J. (2004) *Stable Isotope Geochemistry, 5th edition*. Springer-Verlag, Berlin.

Hofmann, A.W. (1988) Chemical differentiation of the Earth: the relationship between mantle, continental crust, and oceanic crust. *Earth and Planetary Science Letters*, 90, 297–314.

Hoisch, T.D. (1990) Empirical calibration of six geobarometers for the mineral assemblage quartz + muscovite + biotite + plagioclase + garnet. *Contributions to Mineralogy and Petrology*, 104, 225–234.

Hoisch, T.D. (1991) Equilibria within the mineral assemblage quartz + muscovite + biotite + garnet + plagioclase, and implications for the mixing properties of octahedrally-coordinated cations in muscovite and biotite. *Contrib., Mineral. Petrol.*, 108, 43–54.

Holdaway, M.J. (1971) Stability of andalusite and the aluminum silicate phase diagram. *American Journal of Science*, 271, 97–131.

Holdaway, M.J. (2000) Application of new experimental and garnet Margules data to the garnet–biotite geothermometer. *American Mineralogist*, 85, 881–892.

Holdaway, M.J. (2001) Recalibration of the GASP geobarometers in light of recent garnet and plagioclase activity models and versions of the garnet–biotite geothermometer. *American Mineralogist*, 86, 1117–1129.

Holdaway, M.J., & Lee, S.M. (1977) Fe–Mg cordierite stability in high-grade pelitic rocks based on experimental, theoretical and natural observations. *Contributions to Mineralogy and Petrology*, 63, 175–198.

Holdaway, M.J., & Mukhopadhyay, B. (1993) A reevaluation of the stability relations of andalusite: thermochemical data and phase diagram for the aluminum silicates. *American Mineralogist*, 78, 298–315.

Holdaway, M.J., Mukhopadhyay, B., Dyar, M.D., Guidotti, C.V., & Dutrow, B.L. (1997) Garnet–biotite geothermometry revised: new Margules parameters and a natural specimen data set from Maine. *American Mineralogist*, 82, 582–595.

Holdren, G.R., Jr., & Berner, R.A. (1979) Mechanism of feldspar weathering – I. Experimental studies. *Geochimica Cosmochimica Acta*, 43, 1161–1171.

Holland, H.D. (1972) Granites, solutions and base metal deposits. *Economic Geology*, 67, 281–301.

Holland, H.D. (1984) *The Chemical Evolution of the Atmosphere and Oceans*. Princeton University Press, Princeton, NJ, 582 pp.

Holland, H.D. (1992) Distribution an paleoenvironmental interpretation of Proterozoic paleosols. In *The Proterozoic Biosphere: A Multidisciplinary Study* (eds J.W. Schopf & C. Klein), pp. 153–155. Cambridge University Press, New York.

Holland, H.D. (1994) Early Proterozoic atmospheric change. In *Early Life on Earth* (ed. S. Bengston), pp. 237–244. Columbia University Press, New York.

Holland, H.D. (2002) Volcanic gases, black smokers, and the Great Oxidation Event. *Geochimica Cosmochimica Acta*, 66, 3811–3826.

Holland, H.D. (2004) Geologic history of seawater. In *Treatise on Geochemistry* (ed. H. Elderfield), Vol. 6, pp. 583–625. Elsevier, Amsterdam.

Holland, H.D. (2006a) The oxygenation of the atmosphere and oceans. *Philosophical Transactions of the Royal Society London, Series B Biological Sciences*, 361, 903–915.

Holland, H.D. (2006b, When did the Earth's atmosphere become oxic? A reply. *Geochemical News*, 108, 20–22.

Holland, H.D., & Kasting, J.F. (1992) The environment of the Archean earth. In *The Proterozoic Biosphere: A Multidisciplinary Study* (eds J.W. Schopf & C. Klein), pp. 21–24. Cambridge University Press, New York.

Holland, T.J.B. (1979) Reversed hydrothermal determination of jadeite–diopside activities. *Eos (Transactions of the American Geophysical Union)*, 60, 405.

Holland, T.J.B. (1980) The reaction albite = jadeite + quartz determined experimentally in the range 600–1200°C. *American Mineralogist*, 65, 129–134.

Holland, T.J.B., & Powell, R. (1990) An enlarged and updated internally consistent thermodynamic dataset with uncertainties and correlations: the system K_2O-Na_2O-CaO–MgO–MnO–FeO–Fe_2O_3–Al_2O_3–TiO_2–SiO_2–C–H_2–O_2. *Journal of Metamorphic Geology*, 8, 89–124.

Holland, T.J.B., & Powell, R. (1991) A Compensated–Redlich–Kwong (CORK) equation for volumes and fugacities of CO_2 and H_2O in the range 1 bar to 50 kbar and 100–1600°C. *Contributions to Mineralogy and Petrology*, 109, 265–273.

Holland, T.J.B., & Powell, R. (1992) Plagioclase feldspars: activity-composition relations based upon Darken's quadratic formalism and Landau theory. *American Mineralogist*, 77, 53–61.

Holland, T.J.B., & Powell, R. (1998) An internally consistent thermodynamic data set for phases of petrological interest. *Journal of Metamorphic Geology*, 16, 309.

Holland, H.D., Horita, J., & Seyfried, W.E. (1996) On the secular variations in the composition of Phanerozoic marine potash evaporites. *Geology*, 24, 993–996.

Hollister, L.S., & Crawford, M.L. (eds) (1981) *Short Course in Fluid Inclusions: Applications to Petrology*. Mineralogical Association of Canada, Calgary.

Holloway, J.R. (1977) Fugacity and activity of molecular species in supercritical fluids. In *Thermodynamics in Geology* (ed. D.G. Fraser), pp. 161–181. Reidel, Boston, MA.

Holloway, J.R. (1981) Compositions and volumes of supercritical fluids in the Earth's crust. In *Fluid Inclusions: Applications To Petrology* (eds L.S. Hollister & M.L. Crawford), pp. 13–38. Mineralogical Association of Canada, Calgary.

Holm, P.E. (1982) Non-recognition of continental tholeiites using the Ti–Y–Zr diagram. *Contributions to Mineralogy and Petrology*, 79, 308–310.

Holmden, C.E., & Muehlenbachs, K. (1993) The $^{18}O/^{16}O$ ratio of 2-billion-year-old seawater inferred from ancient oceanic crust. *Science*, 259, 1733–1736.

Holmes, A. (1946) An estimate of the age of the earth. *Nature*, 157, 680–684.

Holser, W.T, & Kaplan, I.R. (1966) Isotope geochemistry of sedimentary sulphates. *Chemical Geology*, 1, 93–135.

Horita, J., Friedman, T.J., Lazar, B., & Holland, H.D. (1991) The composition of Permian seawater. *Geochimica Cosmochimica Acta*, 55, 417–432.

Horita, J., Weinberg, A., Das, N., & Holland, H.D. (1996) Brine inclusions in halite and the origin of the Middle Devonian Prairie evaporites of western Canada. *Journal of Sedimentary Research*, 66, 956–964.

Horita, J., Zimmermann, H., & Holland, H.D. (2002) Chemical evolution of seawater during the Phanerozoic: implications from the record of marine evaporates. *Geochimica Cosmochimica Acta*, 66, 3733–3756.

Hoskin, P.W.O. (2005) Trace-element composition of hydrothermal zircon and the alteration of Hadean zircon from the Jack Hills, Australia. *Geochimica Cosmochimica Acta*, 69, 637–648.

Houghton, J.T., Filho, L.G.M., Callander, B.A., *et al.* (Eds) (1996) *Climate Change 1995: the Science of Climate Change. Contribution of Working Group I to the Second Assessment Report of the Intergovernmental Panel on Climate Change*. Cambridge University Press, Cambridge, 564 pp.

Houghton, R.A. (2005) The contemporary carbon cycle. In *Biogeochemistry* (ed. W.H Schlesinger), Vol. 8, *Treatise on Geochemistry* (eds H.D. Holland & K.K. Turekian), pp. 425–472. Elsevier–Pergamon, Oxford.

Houtermans, F.G. (1946) Die Isotopenhäufigkeiten im natürlichen Blei und das Alter des Urans. *Naturwissenschaften*, 33, 185–186, 219.

Hsu, K.T. (1972) Origin of saline giants: a critical review after the discovery of the Mediterranean evaporite. *Earth-Science Reviews*, 8, 371–396.

Hu, G., Rumble, D., & Wang, P.L. (2003) An ultraviolet laser microprobe for the *in situ* analysis of multisulfur isotopes and its use in measuring Archean sulfur isotope mass-independent anomalies. *Geochimica Cosmochimica Acta*, 67, 3101–3117.

Huebner, J.S. (1971) Buffering techniques for hydrostatic systems at elevated pressures. In *Research Techniques for High Pressure and High Temperature* (ed. G.C. Ulmer), pp. 123–178. Springer-Verlag, Berlin.

Huizenga, J.-M. (2001) Thermodynamic modeling of C–O–H fluids. *Lithos*, 55, 101–114.

Hulston, J.R., & Thode, H.G. (1965) Variations in the ^{33}S, ^{34}S, and ^{36}S contents of meteorites and their relation to chemical and nucledar effects. *Journal of Geophysical Research*, 70, 3475–3484.

Hurst, R.W., & Farhat, J. (1977) Geochronologic investigations of the Sudbury Nickel Iruptive and the Superior Province granites north of Sudbury. *Geochimica Cosmochimica Acta*, 41, 1803–1815.

Husian, L.O., Schaeffer, O.A., & Sutter, J.F. (1972) Age of lunar anorthosite. *Science*, 175, 428–430.

Hutchins, S.R., Davidson, M.S., Brierley, J.A., & Brierley, C.L. (1986) Microorganisms in reclamation of metals. *Ann. Rev. Microbiol.*, 40, 311–336.

Indares, A., & Martignole, J. (1985) Biotite–garnet geothermometry in the granulite facies: the influence of Ti and Al in biotite. *American Mineralogist*, 70, 272–278.

Intergovernmental Panel on Climate Change (IPCC) (1990) *Climate Change. The IPCC Assesment* (eds J.T. Houghton, G.J. Jenkins, and J.J. Ephraums). Cambridge University Press, Cambridge.

Intergovernmental Panel on Climate Change (IPCC) (1996) *Climate Change 1995. The Science of Climate Change – Contribution of Working Group I to the Second Assessment Report of the IPCC.* Cambridge University Press, Cambridge, 584 pp.

Intergovernmental Panel on Climate Change (IPCC) (2007) Summary for policymakers. In *Climate Change 2007: The Physical Science Basis. Contribution of Working Group I to the Fourth Assessment Report of the Intergovernmental Panel on Climate Change* (eds S.D. Solomon, M., Qin, Z. Manning, *et al.*). Cambridge University Press, Cambridge.

Irvine, T.N. (1975) Crystallization sequences in the Muskox intrusion and other layered intrusions – II. Origin of chromitite layers and similar deposits of other magmatic ores. *Geochimica Cosmochimica Acta*, 39, 991–1020.

Isley, A.H. (1995. Hydrothermal plumes and the delivery of iron to banded iron formation. *Journal of Geology*, 103, 169–185.

Isley, A.H., & Abbott, D.H. (1999) Plume-related mafic volcanism and the deposition of banded iron formation. *Journal of Geophysical Research*, 104, 15461–15467.

Jacob, D.E. (2004) Nature and origin of eclogite xenoliths from kimberlites. *Lithos*, 77, 295–316.

Jacobsen, S.B., & Wasserburg, G.J. (1980) Sm–Nd isotopic evolution of chondrites. *Earth and Planetary Science Letters*, 50, 139–155.

Jacobson, M.Z. (2002) *Atmospheric Pollution: History, Science, and Regulation*. Cambridge University Press, New York.

Jagoutz, E., Palme, H., Baddenhausen, H., *et al.* (1979) The abundances of major, minor and trace elements in the Earth's mantle as derived from primitive ultramafic nodules. *Proceedings of the Lunar and Planetary Science Conference No. 10. Geochimica Cosmochimica Acta*, Supplement 11, 2031–2050.

James, H.L. (1992) Precambrian iron formations: nature, origin, and mineralogic evolution from sedimentation to metamorphism. In *Diagenesis III, Developments in Sedimentology* (eds K.H. Wolf & Chilingarian), pp. 543–589. Elsevier, Amsterdam.

Jamieson, H.E., & Roeder, P.L. (1984) The distribution of Mg and Fe^{2+} between olivine and spinel at 1300°C. *American Mineralogist*, 69, 283–291.

Jiang, S–Y., Wang, R-C., Xu, X-S., & Zhao, K-D. (2005) Mobility of high field styrength elements (HFSE) in magmatic-, metamorohic-, and submarine-hydrothermal systems. *Physics and Chemistry of the Earth*, 30, 1020–1029.

Jiang, S-Y., Chen, Y-Q., Ling, H-F., Yang, J-H., Feng, H-Z., & Ni, P. (2006) Trace- and rare-earth element geochemistry and Pb–Pb dating of black shales and intercalated Ni–Mo–PGE–Au sulfide ores in Lower Cambrian strata, Yangtze Platform, South China. *Mineralium Deposita*, 41, 453–467.

Johnson, C.M., & Beard, B.L. (2006) Fe isotopes: an emerging technique for understanding modern and ancient biogeochemical cycles. *GSA Today*, 16, 4–10.

Johnson, C.M., Beard, B.L., Beukes, N.J., Klein, C., & O'Leary, J.M. (2003) Ancient geochemical cycling in the Earth as inferred from Fe isotope studies of banded iron formations from the Transvaal Craton. *Contributions to Mineralogy and Petrology*, 144, 523–547.

Johnson, C.M., Beard, B.L., & Albaréde, F. (2004a) Overview and general concepts. *Reviews in Mineralogy and Geochemistry*, 55, 1–24.

Johnson, C.M., Beard, B.L., & Albaréde, F. (eds) (2004b) *Geochemistry of Non-Traditional Stable Isotopes. Reviews in Mineralogy and Geochemistry*, 55, 454 pp.

Johnson, C.M., Roden, E.E., Welch, S.A., & Beard, B.L. (2005) Experimental constraints on Fe isotope fractionation during magnetite and Fe carbonate formation coupled to dissimilatory hydrous ferric oxide reduction. *Geochimica Cosmochimica Acta*, 69, 963–993.

Johnson, C.M., Beard, B.L., & Roden, E.E. (2008a) The iron isotope fingerprints of redox and biogeochemical cycling in modern and ancient Earth. *Annual Reviews in Earth and Planetary Sciences*, 36, 457–493.

Johnson, C.M., Beard, B.L., Klein, C., Beukes, N.J., & Roden, E.E. (2008b) Iron isotopes constrain biologic and abiologic processes in banded iron formation genesis. *Geochimica CosmochimicaActa*, 72, 151–169.

Jones, P., Haggett, M.L., & Longridge, J.L. (1964) The hydration of carbon dioxide. *Journal of Chemical Education*, 41, 610–612.

Jorgensen, B.B. (1990) A thiosulfate stunt in the sulfur cycle of marine sediments. *Science*, 249, 643–645.

Jorgensen, B.B., Zawacki, L.X., & Jannasch, H.W. (1990) Thermophilic bacterial sulfate reduction in deep-sea sediments at the Guaymas Basin hydrothermal vent site. *Deep-sea Research*, 37, 695–710.

Kah, L.C. (2000) Depositional $\delta^{18}O$ signatures in Proterozoic dolostones: constraints on seawater chemistry and early diagenesis. In *Carbonate Sedimentation and Diagenesis in the Evolving Precambrian World* (eds J.P. Grotzinger & N.P. James). *Society of Economic Palaeontologists and Minerologists Special Publication*, 67, 345–360.

Kah, L.C., Lyons, T.W., & Frank, T.D. (2004) Low marine sulphate and protracted oxygenation of the Proterozoic biosphere. *Nature*, 431, 834–838.

Kakegawa, T., & Ohmoto, H. (1999) Sulphur isotope evidence for the origin of 3.4–3.1 Ga pyrite at the Princeton gold mine, Barberton Greenstone Belt. *South African Precambrian Research*, 96, 209–224.

Kamenov, G., Macfarlane, A.W., & Riciputi, L. (2002) Sources of lead in the San Cristobal, Pulacayo, and Potsi mining districts, Bolivia, and a reevaluation of regional ore lead isotope provinces. *Economic Geology*, 97, 573–592.

Kampschulte, A., Brukschen, P., & Strauss, H. (2001) The sulfur isotopic composition of trace sulfates in Carboniferous brachiopods: implications for coeval seawater, correlation with other geochemical cycles and isotope stratigraphy. *Chemical Geology*, 175, 149–173.

Kappler, A., & Newman, D.K. (2004) Formation of Fe(III)-minerals by Fe(II)-oxidizing photoautotrophic bacteria. *Geochimica Cosmochimica Acta*, 68, 1217–1226.

Kappler, A., Pasquero, C., Konhauser, K.O., & Newman, D.K. (2005) Deposition of banded iron formations by anoxygenic phototrophic Fe(II)-oxidizing bacteria. *Geology*, 33, 865–868.

Kargel, J.S., & Lewis, J.S. (1993) The composition and early evolution of the Earth. *Icarus*, 105, 1–25.

Karhu, J.A., & Holland, H.D. (1996) Carbon isotopes and the rise of atmospheric oxygen. *Geology*, 24, 867–870.

Kasting, J.F. (1987) Theoretical constraints on oxygen and carbon dioxide concentrations in the Precambrian atmosphere. *Precambrian Research*, 34, 205–229.

Kasting, J.F. (1993) Earth's early atmosphere. *Science*, 259, 920–926.

Kasting, J.F. (2001) The rise of atmospheric oxygen. *Science*, 293, 819–820.

Kasting, J.F. (2005) Methane and climate during the Precambrian era. *Precambrian Research*, 137, 119–129.

Kasting, J.F. (2006) Ups and downs of ancient oxygen. *Nature*, 443, 643–644.

Kasting, J.F., Zahnle, K.J., Pinto, J.P., & Young, A.T. (1989) Sulfur, ultraviolet radiation, and the early evolution of life. *Origins of Life*, 19, 95–108.

Kasting, J.F., Holland, H.D., & Kump, L.E. (1992) Atmospheric evolution: the rise of oxygen. In *The Proterozoic Biosphere: A Multidisciplinary Study* (eds J.W. Schopf & C. Klein), pp. 159–163. Cambridge University Press, New York.

Kasting, J.F., Eggler, D.H., & Raeburn, S.P. (1993) Mantle redox evolution and the oxidation state of the Archean atmosphere. *Journal of Geology*, 101, 245–257.

Kawasaki, T., & Ito, E. (1994) An experimental determination of the exchange reaction of Fe^{2+} and Mg^{2+} between olivine and Ca-rich clinopyroxene. *American Mineralogist*, 79, 461–477.

Kawasaki, T., & Matsui, Y. (1983) Thermodynamic analyses of equilibria involving olivine, orthopyroxene and garnet. *Geochimica Cosmochimica Acta*, 47, 1661–1679.

Keir, R.S. (1980) The dissolution kinetics of biogenic calcium carbonates in seawater. *Geochimica Cosmochimica Acta*, 44, 241–252.

Kelemen, P.B. (1995) Genesis of high Mg# andesites and the continental crust. *Contributions to Mineralogy and Petrology*, 120, 1–9.

Kennedy, A.K., Lofgren, G.E., & Wasserburg, G.J. (1993) An experimental study of trace element partitioning between olivine, orthopyroxene and melt in chondrules: equilibrium values and kinetic effects. *Earth and Planetary Science Letters*, 115, 177–195.

Kern, R., & Weisbrod, A. (1967) *Thermodynamics for geologists*. Freeman, Cooper & Co., San Francisco, 304 pp.

Kerrick, D.M. (ed.) (1990) *The Al_2SiO_5 Polymorphs*. *Reviews in Mineralogy*, 22, 406 pp.

Kerrick, D.M., & Darken, L.S. (1975) Statistical thermodynamic models for ideal oxide and silicate solid solutions, with applications to plagioclase. *Geochimica Cosmochimica Acta*, 39, 1431–1442.

Kerrick, D.M., & Jacobs, G.K. (1981) A modified Redlich–Kwong equation for H_2O, CO_2, and H_2O–CO_2 mixtures at elevated pressures and temperatures. *American Journal of Science*, 281, 735–767.

Kerrick, D.M., Lasaga, A.C., & Raeburn, S.P. (1991) Kinetics of heterogeneous reactions. *Reviews in Mineralogy*, 26, 583–671.

Kharaka, Y.K., & Barnes, I. (1973) SOLMNEQ: Solution–mineral Equilibrium Computations. *National Technology Information Service Report*, PB214–899, 82 pp.

Kharaka, Y.K, Gunter, W.D., Aggarwal, P.K., Perkins, E.H., & DeBraal, J.D. (1988) SOLMINEQ.88: A computer program for geochemical modeling of water-rock interactions. *U.S. Geological Survey Water Resources Inestigation. Report*, 88-4227, 430 pp.

Kitano, Y., Park, K., & Hood, D.W. (1962) Pure aragonite synthesis. *Journal of Geophysical Research*, 67, 4873–4874.

Kiyosu, Y (1980) Chemical reduction and sulfur-isotope effects of sulfate by organic matter under hydrothermal conditions. *Chemical Geology*, 30, 47–56.

Kiyosu, Y., & Krouse, H.R. (1990) The role of organic acid in the abiogenic reduction of sulfate and the sulfur isotope effect. *Geochem. J.*, 24, 21–27.

Kleeman, P.B. (1995) Genesis of high Mg# andesites and the continental crust. *Contributions to Mineralogy and Petrology*, 120, 1–19.

Kleemann, U., & Reinhardt, J. (1994) Garnet–biotite thermometry revisited: the effect of Al^{VI} and Ti in biotite. *European Journal of Mineralogy*, 6, 925–941.

Klein, C. (2005) Some Precambrian banded iron formations from around the world: Theie geological setting, mineralogy, metamorphism, geochemistry, and origin. *American Mineralogist*, 90, 1473–1499.

Klein, C., & Beukes, N.J. (1989) Geochemistry and sedimentology of a facies transition from limestone to iron–formation deposition in the Early Proterozoic, Transvaal Supergroup, South Africa. *Economic Geology*, 84, 1733–1774.

Klein, C., & Beukes, N.J. (1992) Time distribution, stratigraphy, and edimentologic setting,, and geochemistry of Precambian iron–formations. *The Proterozoic Biosphere: A Multidisciplinary Study* (eds J.W. Schopf & C. Klein), pp. 139–146. Cambridge University Press, New York.

Kleine, T., Münker, C., Mezger, K., & Palme, H. (2002) Rapid accretion and early core formation on asteroids and the terrestrial planets from Hf–W chronometry. *Nature*, 418, 952–955.

Kleine, T., Mezger, K., Palme, H., & Munker, C. (2004) The W isotope evolution of the bulk silicate Earth: Constraints on the timing and mechanisms of core formation and accretion. *Earth and Planetary Science Letters*, 228, 109–123.

Kleypas, J.A., Budddemeier, R.W., Archer, D., Gattuso, J.P., Langdon, C., & Opdike, B.N. (1999) Geochemical consequences of increased atmospheric carbon dioxide on coral reefs. *Science*, 284, 118–120.

Klotz, I.M., & Rosenberg, R.M. (1994) *Chemical Thermodynamics, 5th edn*. J. Wiley & Sons, New York, 533 pp.

Knauth, L.P. (2006) Signature required. *Nature*, 442, 873–874.

Knauth, L.P., & Roberts, S.K. (1991) The hydrogen and oxygen isotope history of the Silurian–Permian hydrosphere as determined by direct measurement of fossil water. In *Stable Isotope Geochemistry: A Tribute to Samuel Epstein* (eds H.P. Taylor Jr., J.R. O'Neil, & I.R. Kaplan). *Geochemistry Society Special Publication*, 3, 91–103.

Koeberl, C. (2006) Impact processes on the early Earth. *Elements*, 2, 211–216.

Kohn, M.J., & Spear, F. (1989) Empirical calibration of geobarometers for the assemblage garnet + hornblende + plagioclase + quartz. *American Mineralogist*, 74, 77–84.

Konhauser, K.O., Hamade, T., Raiswell, R., *et al.* (2002) Could bacteria have formed the Precambrian banded iron formations? *Geology*, 30, 1079–1082.

Kovalevich, V.M., Peryt, T.M., & Petrichenko, O.I. (1998) Secular variation in seawater chemistry during the Phanerozoic as indicated by brine inclusions in halite. *Journal of Geology*, 106, 695–712.

Koziol, A.M. (1989) Recalibration of the garnet–plagioclase–Al_2SiO_5–quartz (GASP) geobarometer and application to natural parageneses. *Eos (Transactions of the American Geophysical Union)*, 70, 493.

Koziol, A.M., & Newton, R.C. (1988) Redetermination of the anorthite breakdown reaction and improvement of the plagioclase–garnet–Al_2SiO_5–quartz geobarometer. *American Mineralogist*, 73, 216–223.

Krauskopf, K.B., & Bird, D.K. (1995) *Introduction to Geochemistry*, 3rd edn. McGraw–Hill, New York, 647 pp.

Krogh, E.J. (1988) The garnet–clinopyroxene geothermometer – a reinterpretation of existing experimental data. *Contributions to Mineralogy and Petrology*, 99, 44–48.

Kröner, A. (1977) Precambrian mobile belts of southern and eastern Africa – ancient sutures or sites of ensialic mobility? A case for crustal evolution towards plate tectonics. *Tectonophysics*, 40, 101–135.

Krouse, H.R., Viau, C.A., Eliuk, L.S., Ueda, A., & Halas, S. (1988) Chemical and isotopic evidence of thermochemical sulfate reduction by light hydrocarbon gases in deep carbonate reservoirs. *Nature*, 333, 415–419.

Krupp, R.T., Oberthür, T., and Hirdes, W. (1994) The early Precambrian atmosphere and hydrosphere: thermodynamic constraints from mineral eposits. *Economic Geology*, 89, 1581–1598.

Kump, L.R., & Arthur, M.A. (1999) Interpreting carbon-isotope excursions: carbonates and organic matter. *Chemical Geology*, 161, 181–198.

Kump, L.R., & Barley, M.E. (2007) Increased subaerial volcanism and the rise of oxygen 2.5 billion years ago. *Nature*, 448, 1033–1036.

Kump, L.R., Kasting, J.F., & Barley, M.E. (2001) Rise of atmospheric oxygen and the "upside-down" Archean mantle. *Geochemistry Geophysics Geosystems*, 2, 1025, doi:1029/2000GC000114 (online).

Kump, L.R., Kasting, J.F., & Crane, R.G. (2004) *The Earth System*, 2nd edn. Prentice Hall, Saddle Reef, NJ, 419 pp.

Kump, L.R., Pavlov, A., & Arthur, M.A. (2005) Massive release of hydrogen sulfide to the surface ocean and atmosphere during intervals of oceanic anoxia. *Geology*, 33, 397–400.

Kyser, T.K. (1987a) Equilibrium fractionation factors for stable isotopes. *Mineralogical Association of Canada Short Course Handbook*, 13, 1–84.

Kyser, T.K. (1987b) Standards and techniques. *Mineralogical Association of Canada Short Course Handbook*, 13, 446–452.

Kyser, T.K. (1987c) Stable isotope geochemistry of low temperature processes. *Mineralogical Association of Canada Short Course Handbook*, 13, 452 pp.

Lambert, D.D., Morgan, J.W., Walker, R.J., *et al.* (1989) Rhenium–osmium and samarium–neodymium isotopic systematics of the Stillwater Complex. *Science*, 244, 1169–1174.

Langdon, C., Takahashi, T., Sweeney, C., *et al.* (2000) Effect of calclium carbonate saturation state on the calcification rate of an experimental coral reef. *Global Biogeochemical Cycles*, 14, 639–654.

Langmuir, C.H. (1989) Geochemical consequences of *in situ* crystallization. *Nature*, 340, 199–205.

Langmuir, D. (1997) *Aqueous Environmental Geochemistry*. Prentice Hall, Upper Saddle River, NJ, 600 pp.

Langmuir, D., & Mahoney, J. (1984) Chemical equilibrium and kinetics of geochemical processes. In *Groundwater Studies. First Canadian/American Conference on Hydrogeology*, June 24–26 (eds B. Hitchon & E.I. Wallick), pp. 69–95. National Water Well Association, Worthington, Ohio.

Larson, P.B., & Zimmerma, B.S. (1991) Variations in $\delta^{18}O$ values, water/rock ratios, and water flux in the Rico paleothermal anomaly, Colorado. In *Stable Isotope Geochemistry: A Tribute to Samuel Epstein* (eds H.P. Taylor Jr., J.R. O'Neil, & I.R. Kaplan). *Geochemistry Society Special Publication*, 3, 169–185.

Lasaga, A.C. (1981) *Rate Laws of Chemical Reactions. Reviews in Mineralogy*, 8, 1–68.

Lasaga, A.C. (1984) Chemical kinetics of water–rock interactions. *Journal of Geophysical Research*, 89, 4009–4025.

Lasaga, A.C. (1998) *Kinetic Theory in Earth Sciences*. Princeton University Press, Princeton, NJ, 728 pp.

Lasaga, A.C., Soler, J.M., Ganor, J., Burch, T.E., & Nagy, K.L. (1994) Chemical weathering rate laws and global geochemical cycles. *Geochimica Cosmochimica Acta*, 58, 2361–2386.

Lécuyer, C., & Allemand, P. (1999) Modelling of the oxygen isotope evolution of seawater: implications for the climate interpretation of the $\delta^{18}O$ of marine sediments. *Geochimica Cosmochimica Acta*, 63, 351–361.

Lee, H.Y., & Ganguly, J. (1988) Equilibrium compositions of coexisting garnet and orthopyroxene: experimental determinations in the system $FeO–MgO–Al_2O_3–SiO_2$, and applications. *Journal of Petrology*, 29, 93–113.

Leighton, P.A. (1961) *Photochemistry of Air Pollution*. Academic Press, New York, 300 pp.

Lerman, A. (1990) Transport and kinetics in surfacial processes. In *Aquatic Chemical Kinetics: Reaction Rates of Processes in Natural Waters* (ed. W. Stumm), pp. 505–534. Wiley-Interscience, New York.

Lewis, G.N., & Randall, M. (1925) *Thermodynamics and the Free Energy of Chemical Substances*. McGraw-Hill, New York, 653 pp.

Lewis, G.N., & Randall, M. (1961) *Thermodynamics*, revised by K.S. Pitzer and L. Brewer. McGraw–Hill, New York, 723 pp.

Li, J., & Agee, C.B. (1996) Geochemistry of mantle–core differentiation. *Nature*, 381, 686–689.

Li, J., & Agee, C.B. (2001) The effect of pressure, temperature, oxygen fugacity and composition on partitioning of nickel and cobalt between liquid Fe-Ni–S alloy and liquid silicate: implications for the Earth's core formation. *Geochimica Cosmochimica Acta*, 65, 1821–1832.

Li, Y.–H., & Gregory, S. (1974) Diffusion of ions in sea water and in deep–sea sediments. *Geochimica Cosmochimica Acta*, 38, 703–714.

Libby, W.F. (1955) *Radiocarbon Dating*, 2nd edn. University of Chicago Press, Chicago, 175 pp.

Lide, D.R. (ed.) (1998) *CRC Handbook of Chemistry and Physics*, 79th edn. CRC Press, Boca Raton, FL.

Lide, D.R. (ed.) (2001) *CRC Handbook of Chemistry and Physics*, 81st edn. CRC Press, Boca Raton, FL.

Linder, M., Leich, D.A., Borg, R.J., *et al.* (1989) Direct laboratory determination of the ^{187}Re half-life. *Nature*, 320, 246–248.

Lindsley, D.H. (1983) Pyroxene thermometry. *American Mineralogist*, 68, 477–493.

Lineweaver, C.H. (1999) A younger age for the universe. *Science*, 284(5419), 1503–1507.

Linkson, P.B., Phillips, B.D., & Rowles, C.D. (1979) Computer methods for the generation of Eh–pH diagrams. *Minerals Science and Engineering*, 11, 65–79.

Liu, L.-G. (2004) The inception of the oceans and CO_2-atmosphere in the early history of the Earth. *Earth Planetary Science Letters*, 227, 179–184.

Livingstone, D.A. (1963) Chemical composition of rivers and lakes. *U.S. Geological Survey Professional Paper*, 440-G.

Lodders, K. (2003) Solar system abundances and condensations temperatures of the elements. *Astrophysical Journal*, 591, 1220–1247.

Lodders, K., & Fegley, B., Jr. (1998) *The Planetary Scientist's Companion*. Oxford University Press, New York, 371 pp.

Loucks, R.R. (1996) A precise olivine–augite Mg–Fe-exchange geothermometer. *Contributions to Mineralogy and Petrology*, 125, 140–150.

Lowenstam, H.A. (1981) Minerals formed by organisms. *Science*, 211, 1126–1130.

Lowenstam, H.A., & Weiner, S. (1989) *On Biomineralization*. Oxford University Press, New York.

Lowenstein, T.K., Timofeeff, M.N., Brennan, S.T., Hardie, L.A., & Demicco, R.V. (2001) Oscillations in Phanerozoic seawater chemistry: evidence from fluid inclusions. *Science*, 294, 1086–1088.

Lowenstein, T.K., Hardie, L.A., Timofeeff, M.N., and Demicco, R.V. (2003) Secular variation in seawater chemistry and the origin of calcium chloride basinal brines. *Geology*, 31, 857–860.

Luck, G.M., & Allègre, G.J. (1983) ^{187}Re–^{187}Os systematics in meteoriters and cosmological consequences. *Nature*, 302, 130–132.

Luck, G.M., & Allègre, G.J. (1984) ^{187}Re–^{187}Os investigation in sulphide from Cape Smith komatiites. *Earth and Plânetary Science Letters*, 68, 205–208.

Ludwig, K.R. (2003) *Isoplot/ex, Version 3.00 a Geochronological Toolkit for Microsoft Excel*. Special Publications No. 4, Berkeley Geochronological Center.

Lugmair, G.W., & Shukolyukov, A. (1998) Early solar system time-scales according to ^{53}Mn–^{53}Cr systematics. *Geochimica Cosmochimica Acta*, 62, 2863–2886.

Lunine, J.I. (2006) Origin of water ice in the Solar System. In *Meteorites and the Early Solar System II* (eds D.S. Lauretta & H.Y. McSween, Jr), pp. 309–319. University of Arizona Press, Tucson, AZ.

Lyons, T.W., Kah, L.C., & Gellatly, A.M. (2004) The Precambrian sulphur isotope record of evolving atmospheric oxygen. In *The Precambrian Earth: Tempos and Events* (eds P.G. Eriksson, W. Altermann, D.R. Nelson, W.U. Mueller, & O. Catuneanu). *Developments in Precambrian Geology*, 12, 421–440.

Machel, H.G., Krouse, H.R., & Sassen, R. (1995) Products and distinguishing criteria of bacterial and thermochemical reduction. *Applied Geochemistry*, 10, 373–389.

MacGregor, I.D. (1974) The system $MgOAl_2O_3$–SiO_2: solubility of Al_2O_3 in enstatite for spinel and garnet peridotite compositions. *American Mineralogist*, 59, 110–119.

MacKenzie, F.T., & Garrels, R.M. (1966) Chemical mass balance between rivers and oceans. *American Journal of Science*, 264, 507–525.

Mackenzie, F.T., Ver, L.M., & Lerman, A. (2000) Coastal zone biogeochemical dynamics under global warming. *International Geological Review*, 42, 193–206.

Mäder, U.K., & Berman, R.G. (1991) An equation of state for carbon dioxide to high pressure and temperature. *American Mineralogist*, 76, 1547–1559.

Malavergne, V., Siebert, J., Guyot, F., *et al.* (2004) Si in the core? New high pressure and high temperature experimental data. *Geochimica Cosmochimica Acta*, 68, 4201–4211.

Mann, S. (1983) Mineralization in biological systems. *Structural Bonding*, 54, 125–174.

Mann, M.E., Bradley, R.S., & Hughes, M.K. (1998) Globalscale temperaturepatterns and climate forcing over the past six centuries. *Nature*, 392, 779–787.

Marshall, B.D., & DePaolo, D.J. (1982) Precise age determination and petrogenetic studies using the K–Ca method. *Geochimica Cosmochimica Acta*, 53, 917–922.

Mason, B. (1966) *Principles of Geochemistry*. J. Wiley & Sons, New York, 329 pp.

Mathews, A., (1994) Oxygen isotope geothermometers for metamorphic rocks. *Journal of Metamorphic Petrology*, 12, 211–219.

Matsui, T., & Abe, Y. (1986a) Evolution of an impact-induced atmosphere and magma ocean on the accreting Earth. *Nature*, 319, 303–305.

Matsui, T., & Abe, Y. (1986b) Impact-induced atmospheres and oceans on Earth and Venus. *Nature*, 322, 526–528.

Mattinson, J.M. (2005) Zircon U–Pb chemical abrasion ("CA-TIMS") method: combined annealing and multi-step partial dissolution analysis for improved precision and accuracy of zircon ages. *Chemical Geology*, 220, 47–66.

McCandless, T.E., & Ruiz, J. (1991) Osmium isotopes and crustal sources for platinum–group mineralization in the Bushveld Complex, South frica. *Geology*, 19, 1225–1228.

McCandless, T.E., Ruiz, J., & Campbell, A.R. (1993) Rhenium behaviour in molybdenite in hypogene and near–surface environments: implications for Re–Os geochronometry. *Geochimica Cosmochimica Acta*, 57, 889–905.

McDonough, W.F., & Sun, S.-S. (1995) The composition of the Earth. *Chemical Geology*, 120, 223–253.

McDougall, I., Polach, H.A., & Stipp, J.J. (1969) Excess radiogenic argon in young subaerial basalts from the Auckland volcanic field, New Zealand. *Geochimica Cosmochimica Acta*, 33, 1485–1520.

McSween, H.Y., Jr., & Huss, G.R. (2010) *Cosmochemistry*. Cambridge University Press, Cambridge, NY, 549 pp.

McSween, H.Y., Jr., Richardson, S.M., & Uhle, M.E. (2003) *Geochemistry: Pathways and Processes*, 2nd edn. Columbia University Press, New York, 363 pp.

McSween, H.Y., Jr., Lauretta, D.S., & Leshin, L.A. (2006) Recent advances in meteoritics and cosmochemistry. In *Meteorites and the Early Solar System II* (eds D.S. Lauretta & H.Y. McSween, Jr), pp. 53–66. University of Arizona Press, Tucson, AZ.

Medaris, L.G. (1968) Partitioning of Fe^{++} and Mg^{++} between coexisting synthetic olivines and orthopyroxenes. *American Journal of Science*, 267, 945–968.

Merschat, A.J. (2009) *Assembling the Blue Ridge and Inner Piedmont: Insight into the nature and timing of terrane accretion in the southern Appalachian orogen from geologic mapping,stratigraphy, kinematic analysis,petrology, geochemistry, and modern geocgronology*. Unpublished MS thesis, The University of Tennessee, Knoxville, 455 pp.

Meschede, M. (1986) A method of discriminating between different types of mid-ocean ridge basalts and continental tholeiites with the Nb–Zr–Y diagram. *Chemical Geology*, 56, 207–218.

Meyer, B.S., & Zinner, E. (2006) Nucleosynthesis. In *Meteorites and the Early Solar System II* (eds D.S. Lauretta & H.Y. McSween, Jr), pp. 69–108. University of Arizona Press, Tucson, AZ.

Middlemost, E.A.K. (1975) The basalt clan. *Earth-Science Reviews*, 11, 337–364.

Misra, K.C. (2000) *Understanding mineral deposits*. Kluwer Academic Publishers, Dordrecht, The Netherlands, 845 pp.

Misra, K.C., Anand, M., Taylor, L.A., & Sobolev, N.V. (2004) Multi-stage metamorphism of diamondiferous eclogite xenoliths from the Udachnaya kimberlite pipe, Yakutia, Siberia. *Contributions to Mineralogy and Petrology*, 146, 696–714.

Mitchell, J.F.B., Karloy, D.J., Hegerl, G.C., Zweirs, F.W., Allen, M.R., & Marengo, J. (2001) Detection of climate change and attribution of causes. In *Climate Change 2001: the Scientific Basis. Contribution of*

Working Group I to the Third Assessment Report of the Intergovernmental Panel on Climate Change. Cambridge University Press, Cambridge, 892 pp.

Moecher, D.P., Essene, E.J., & Anovitz, L.M. (1988) Calculation and application of clinopyroxene–garnet–plagioclase–quartz geobarometers. *Contributions to Mineralogy and Petrology*, 100, 92–106.

Mojzsis, S.J., Arrhenius, G., McKeegan, K.D., & Harrison, T.M. (1996) Evidence for life on Earth before 3800 million years ago. *Nature*, 384, 55–69.

Mojzsis, S.J., Harrison, T.M., & Pidgeon, R.T. (2001) Oxygen-isotope evidence from ancient zircons for liquid water at the Earth's surface 4,3000 years ago. *Nature*, 409, 178–181.

Mojzis, S.J., Coath, C.D., Greenwood, J.P., McKeegan, and Harrison, T.M. (2003) Mass-independent isotope effects in Archean (2.5 to 3.8 Ga) sedimentary sulfides determined by ion microprobe analysis. *Geochimica Cosmochimica Acta*, 67, 1635–1658.

Moorbath, S. (1978) Age and isotope evidence for the evolution of the continental crust. *Philosophical Transactions of the Royal Society London, Series A*, 288, 401–413.

Moorbath, S., O'Nions, R.K., & Parkhurst, R.J. (1973) Early Archaean age for the Isua iron formation, West Greenland. *Nature*, 245, 138–139.

Moorbath, S., Powell, J.L., & Taylor, P.N. (1975) Isotopic evidence for the age and origin of the "grey gneiss" complex of the southern Outer Hebrides, Scotland. *Journal of Geology Society of London*, 131, 213–222.

Morbidelli, A., Chambers, J., Lunine, J.I., *et al.* (2000) Source regions and time scales for the delivey of water to the Earth. *Meteoritics and Planetary Science*, 35, 286–306.

Morel, F.M.M., & Hering, J.G (1993) *Principles and Applications of Aquatic Chemistry*. J. Wiley & Sons, New York.

Mori, T, and Green, D.H. (1978) Laboratory duplication of phase equilibia observed in natural garnet lherzolites. *Journal of Geology*, 86, 83–97.

Morris, R.C. (1993) Genetic modeling for banded iron-formationof the Hamersley H Group, Pilbara craton, Western Australia. *Precambrian Research*, 60, 243–286.

Morse, J.W. (1978) Dissolution kinetics of calcium carbonate in sea water: VI. The near-equilibrium dissolution kinetics of calcium carbonate-rich deep sea sediments. *American Journal of Science*, 278, 344–353.

Morse, J.W. (1983) The kinetics of calcium carbonate dissolution and precipitation. *Reviews in Mineralogy*, 11, 227–264.

Morse, J.W., & Berner, R.A. (1979) Chemistry of calcium carbonate in deep oceans. In *Chemical Modeling in Aqueous Systems: Speciation, Sorption, Solubility, and Kinetics* (ed. E.A. Jenne), *American Chemistry Society Symposium Series*, 93, 537–573.

Morse, J.W., Wang, Q., & Tsio, M.Y. (1997) Influences of temperature and Mg:Ca ratio on $CaCO_3$ precipitates from seawater. *Geology*, 25, 85–87.

Morse, J.W., Gledhill, D.K., & Millero, F.J. (2003) $CaCO_3$ precipitation kinetics in waters from the Great Bahama Bank: Implications for the relationship between Bank hydrochemistry and whitings. *Geochimica Cosmochimica Acta*, 67, 2819–2826.

Motoyama, T., & Matsumoto, T. (1989) The crystal structures and the cation distributions of Mg and Fe in natural olivines. *Mineralogical Journal*, 14, 338–350.

Mottl, M.J., Holland, H.D., & Carr, F.R. (1979) Chemical exchange during hydrothermal alteration of basalt by seawater, II. Experimental results for Fe, Mn, and sulfur species. *Geochimica Cosmochimica Acta*, 43, 869–884.

Mucci, A., & Morse, J.W. (1983) The incorporation of Mg^{2+} and Sr^{2+} into calcite overgrowths: influence of growth rate and solution composition. *Geochimica Cosmochimica Acta*, 47, 217–233.

Muehlenbachs, K., and Clayton, R.N. (1976) Oxygen isotope composition of the oceanic crust and its bearing on seawater. *Journal of Geophysical Research*, 81, 4365–4369.

Mukhopadhyay, B., Basu, S., & Holdaway, M.J. (1993) A discussion of Margules-type formulations for multicomponent solutions with a generalized approach. *Geochimica Cosmochimica Acta*, 57, 277–283.

Mukhopadhyay, B., Holdaway, M.J., & Koziol, A.M. (1997) A statistical model of thermodynamic mixing properties of $Ca–Mg–Fe^{2+}$ garnets. *American Mineralogist*, 82, 165–181.

Mullen, E.D. (1983) $MnO/TiO_2/P_2O_5$: a minor element discriminant for basaltic rocks of oceanic environments and its implications for petrogenesis. *Earth and Planetary Science Letters*, 62, 53–62.

Murr, L.E. (1980) Theory and practice of copper sulphides leaching in dumps and *in-situ*. *Minerals Science and Engineering*, 12, 121–189.

Murthy, V.R. (1991) Early differentiation of the Earth and the problem of mantle siderophile elements: a new approach. *Science*, 253, 303–306.

Nabelek, P.I. (1991) Stable isotpe monitors. In *Contact Metamorphism* (ed. D.M. Kerrick), Vol. 6, pp. 395–435. Mineralogical Society of America, Chelsea, MI.

Nagasawa, H., Schreiber, H.D., & Morris, R.V. (1980) Experimental mineral/liquid partition coefficients of the rare earth elements (REE), Sc and Sr for perovskite, spinel and melilite. *Earth and Planetary Science Letters*, 46, 431–437.

Nakamura, N. (1974) Determination of REE, Ba, Fe, Mg, Na, and K in carbonaceous and ordinary chondrites. *Geochimica Cosmochimica Acta*, 38, 757–773.

Naldrett, A.J., Duke, J.M., Lightfoot, P.C., & Thompson, J.F.H. (1984) Quantitative modfelling of the segregation of magmatic sulphides: an exploration guide. *Canadian Institute of Mining and Metallurgy Bulletin*, 77, 46–56.

Nemchin, A.A., Pidgeon, R.T., & Whitehouse, M.J. (2006) Re-evaluation of the origin and evolution of >4.2 Ga zircons from the Jack Hills metasedimentasry rocks. *Earth and Planetary Science Letters*, 244, 218–233.

Newsom, H.E., & Sims, K.W.W. (1991) Core formation during early accretion of the Earth. *Science*, 252, 926–933.

Newton, R.C. (1983) Geobarometry of high-grade metamorphic rocks. *American Journal of Science*, 283-A, 1–28.

Newton, R.C., & Haselton, H.T. (1981) Thermodynamics of the garnet–plagioclase–Al_2SiO_5–quartz geobarometer. In *Thermodynamics of Minerals and Melts* (eds R.C. Newton, A. Navrotsky, & B.J. Wood), Vol. 1, pp. 129–145. Springer-Verlag, New York.

Newton, R.C., & Perkins, D. III (1982) Thermodynamic calibration of geobarometers for charnockites and basic granulites based on the assemblages garnet–plagioclase–orthopyroxene (clinopyroxene)–quartz with applications to high-grade metamorphism. *American Mineralogist*, 67, 203–222.

Newton, R.C., Charlu, T.V., & Kleppa, O.J. (1980) Thermochemistry of the high structural state plagioclases. *Geochimica Cosmochimica Acta*, 44, 933–941.

Nicolaysen, L.O., Burger, A.J., & Liebenberg, W.R. (1962) Evidence for the extreme age of certain minerals from the Dominion Reef conglomerates and the underlying granite in the Western Transvaal. *Geochimica Cosmochimica Acta*, 26, 15–23.

Nordstrom, D.K. (2004) Modeling low-temperature geochemical processes. In *Treatise on Geochemistry* (eds H.D. Holland, K.K. Turikian & J.I. Drever), Vol. 5, pp. 37–72. Elsevier–Pergamon, Oxford.

Nordstrom, D.K., & Alpers, C.N. (1999) Geochemistry of acid mine waters. *Reviews in Economic Geology*, 6A, 133–160.

Nordstrom, D.K., & Munoz, J.L. (1994) *Geochemical Thermodynamics*, 2nd. edn. Blackwell Scientific Publications, Oxford, 493 pp.

Nordstrom, D.K., & Southam, G. (1997) Geomicrobiology of sulfide mineral oxidation. *Reviews in Mineralogy*, 35, 361–390.

Norry, M.J., & Fitton, J.G. (1983) Compositional differences between oceanic and continental basic lavas and their significance. In *Continental Basalts and Mantle Xenoliths* (eds C.J. Hawkesworth, & M.J. Norry), pp. 5–19. Shiva, Nantwich.

Nutman, A.P., Mojzsis, S.J., & Friend, C.R.L. (1997) Recognition of ≥ 3850 Ma water-lain sediments in West Greenland and their significance for the early Archean Earth. *Geochimica Cosmochimica Acta*, 61, 2475–2484.

Ohmoto, H. (1986) Stable isotope geochemistry of ore deposits. *Reviews in Mineralogy*, 16, 491–560.

Ohmoto, H. (1996) Evidence in pre–2.2 Ga paleosols for the early evolution of atmospheric oxygen and terrestrial biota. *Geology*, 24, 1135–1138.

Ohomoto, H. (1997) When did the Earth's atmosphere become oxic? *The Geochemical News*, 93, 12–22.

Ohmoto, H., & Felder, R.P. (1987) Bacterial activity in the warmer, sulphate-bearing, Archaean oceans. *Nature*, 328, 244–246.

Ohmoto, H., & Goldhaber, M.B. (1997) Sulfur and carbon isotopes. In *Geochemistry of Hydrothermal Deposits* (ed. H.L. Barnes), 3rd edn, pp. 517–611. J. Wiley & Sons, New York.

Ohmoto, H. and Rye, R.O (1979) Isotopes of sulfur and carbon. In *Geochemistry of Hydrothermal Ore Deposits* (ed. H.L. Barnes), 2nd edn, pp. 517–612. J. Wiley & Sons, New York.

Ohmoto, H., Watanabe, Y., Ikemi, H., Poulson, S.R., & Taylor, B.E. (2006) Sulfur isotope evidence for an oxic Archean atmosphere. *Nature*, 442, 908–911.

Olson, G.J. (1991) Rate of pyrite bioleaching by *Thiobacillus ferrooxidans* – results of an interlaboratory comparison. *Applied and Environmental Microbiology*, 57, 642–644.

O'Neill, H. St. C., & Wood, B.J. (1979) An experimental study of Fe–Mg partitioning between garnet and olivine, and its calibration as a geothermometer. *Contributions to Mineralogy and Petrology*, 70, 59–70.

O'Neil, J.R. (1986a) Theoretical and experimental aspects of isotopic fractionation. *Reviews in Mineralogy*, 16, 1–40.

O'Neil, J.R. (1986b) Appendix: terminology and standards. *Reviews in Mineralogy*, 16, 561–570.

O'Neil, J.R., & Taylor, H.P., Jr. (1967) The oxygen isotope and cation exchange chemistry of feldspars. *American Mineralogist*, 52, 1414–1437.

O'Neil, J.R., & Taylor, H.P., Jr. (1969) Oxygen isotope equilibrium between muscovite and water. *Journal of Geophysical Research*, 74, 6012–6022.

O'Neill, P. (1985) *Environmental Chemistry*. Allen & Unwin, London, 232 pp.

Oversby, V.M. (1975) Lead isotopic systematics and ages of Archaean acid intrusives in the Kalgoorlie–Norseman area, Western Australia. *Geochimica Cosmochimica Acta*, 39, 1107–1125.

Owen, T.C., & Bar-Nun, A. (2000) Volatile contributions from icy planetesimals. In *Origin of the Earth and Moon* (eds R.M. Canup & K. Righter), pp. 459–471. The University of Arizona Press, Tucson, AZ.

Pagenkopf, G.K. (1978) *Introduction to Natural Water Chemistry*. Marcel Dekker, Inc., New York, 272 pp.

Pahlevan, K., & Stevenson, D.J. (2007) Equilibration in the aftermath of the lunar-forming giant impact. *Earth and Planetary Science Letters*, 262, 438–449.

Palme, H., & Nickel, K.G. (1986) Ca/Al ratio and composition of the Earth's primitive upper mantle. *Geochimica Cosmochimica Acta*, 49, 2123–2132.

Pan, P., & Wood, S.A. (1994) Solubility of Pt and Pd sulfides and Au metal in bisulfide solutions. II. Results at 200°–350°C and at saturated vapor pressure. *Mineralium Deposita*, 29, 373–390.

Panneerselvam, K., Macfarlane, A.W., & Salters, V.J.M. (2006) Provenance of ore metals in base and precious metal deposits of central Idaho as inferred from lead isotopes. *Economic Geology*, 101, 1063–1078.

Papanastassiou, D.A., & Wasserburg, G.J. (1969) Initial strontium isotopic abundances and the resolution of small time differences in the formation of planetary objects. *Earth and Planetary Science Letters*, 5, 361–376.

Papineau, D., & Mojzsis, S.J. (2006) Mass-independent fractionation of sulfur isotopes in sulfides from the pre–3770 Ma Isua Supracrustal Belt, West Greenland. *Geobiology*, 4, 227–238.

Parkhurst, D.L. (1990) Ion-association models and mean activity coefficients of various salts. In *Chemical Modelling of Aqueous Systems II* (eds D.C. Melchior & R.L. Bassett), *American Chemistry Society Symposium Series*, 416, 30–43.

Parkhurst, D.L. (1995) Users guide to PHREEQC – a computer program for speciation, reaction-path, advective transport, and inverse geochemical calculations. *U.S. Geological Survey Water Resources Investigation Report* 95–4227.

Parkhurst, D.L. & Appelo, C.A. (1999) User's guide to PHREEQC (version 2) – a computer program for speciation, batch-reaction, one–dimensional transport, and inverse geochemical calculations. *U.S. Geological Survey Water-Resources Investigation Report*, 80-96, 195 pp.

Parkhurst, D.L., Plummer, L.N., & Thorstenson, D.C. (1980) PHREEQE – a computer program for geochemical calculations. *U.S. Geological Survey Water-Resources Inverstigation Report*, 80-96, 195 pp.

Patchett, P.J., & Tatsumoto, M. (1980) Lu–Hf total-rock isochron for the eucrite meteorites. *Nature*, 288, 571–574.

Patterson, C.C. (1956) Age of meteorites and the earth. *Geochimica Cosmochimica Acta*, 10, 230–237.

Pattison, D.R.M., & Newton, R.C. (1989) Reversed experimental calibration of the garnet–clinopyroxene Fe–Mg exchange thermometer. *Contributions to Mineralogy and Petrology*, 101, 87–103.

Pavlov, A.A., & Kasting J.F. (2002) Mass-independent fractionation of sulfur isotopes in Archean sediments: strong evidence for an anoxic Archean atmosphere. *Astrobiology*, 2, 27–41.

Pavlov, A.A., Kasting, J.F., Brown, L.L., Rages, K.AS., & Freedman, R. (2000) Greenhouse warming by CH_4 in the atmosphere of early Earth. *Journal of Geophysical Research*, 105, 11981–11990.

Pavlov, A.A., Hurtgen, M.T., Kasting, J.F., & Arthur, M.A. (2003) Methane-rich Proterozoic atmosphere? *Geology*, 31, 87–90.

Pearce, J.A. (1975) Basalt geochemistry used to investigate past tectonic environments on Cyprus. *Tectonophysics*, 25, 41–67.

Pearce, J.A. (1976) Statistical analysis of major element patterns in basalts. *Journal of Petrology*, 17, 15–43.

Pearce, J.A. (1982) Trace element characteristics of lavas from destructive plate margins. In *Andesites: Orogenic Andesites and Related Rocks* (ed. R.S. Thorpe), pp. 525–548. J. Wiley & Sons, New York.

Pearce, J.A. (1983) The role of sub-continental lithosphere in magma genesis at destructive plate margins. In *Continental Basalts and Mantle Xenoliths* (eds C.J. Hawkesworth, & M.J. Norry), pp. 230–249. Shiva, Nantwich.

Pearce, J.A., & Cann, J.R. (1973) Tectonic setting of basic volcanic rocks determined using trace element analyses. *Earth and Planetary Science Letters*. 19, 290–300.

Pearce, J.A., & Norry, M.J. (1979) Petrogenetic implications of Ti, Zr, Y and Nb variations in volcanic rocks. *Contributions to Mineralogy and Petrology*, 69, 33–47.

Pearce, T.H., Gorman, B.E., & Birkett, T.C. (1977) The relationship between major element chemistry and tectonic environment of basic and intermediate volcanic rocks. *Earth and Planetary Science Letters*, 36, 121–132.

Pearce, J.A., Alabaster, T., Shelton, A.W., & Searle, M.P. (1981) The Oman ophiolite as a Cretaceous arc-basin complex: evidence and implications. *Philosophical Transactions of the Royal Society, Series A*, 300, 299–317.

Peck, W.H., Valley, J.W., Wilde, S.A., & Graham, C.M. (2001) Oxygen isotope ratios and rare earth elements in 3.3 and 4.4 Ga zircons: ion microprobe evidence for high $\delta^{18}O$ continental crust and oceans in the Early Archean. *Geochimica Cosmochimica Acta*, 65, 4215–4229.

Pepin, R., & Porcelli, D. (2002) Origin of noble gases in the terrestrial planets. *Reviews of Mineralogy and Geochemistry*, 47, 191–246.

Perchuk, L.L., & Lavrent'eva, I.V. (1983) Experimental investigation of exchange equilibria in the system cordierite–garnet–biotite. In *Kinetics and Equilibrium in Mineral Reactions* (ed. S.K. Saxena). *Advances in Physics and Geochemistry*, Vol. 3, pp. 199–239. Springer-Verlag, New York.

Perkins, D. (1979) *Application of new thermodynamic data to mineral equilibria*. PhD thesis, University of Michigan, Ann Arbor, 214 pp.

Perkins, D., & Chipera, S.J. (1985) Garnet–orthopyroxene–plagioclase–quartz barometry: refinement and application to the English River subprovince and the Minnesota River Valley. *Contributions to Mineralogy and Petrology*, 89, 69–80.

Perkins, D., & Vielzeuf, D. (1992) Experimental investigation of Fe–Mg distribution between olivine and clinopyroxene: implications for mixing properties of Fe–Mg in clinopyroxene and garnet–clinopyroxene thermometry. *American Mineralogist*, 77, 774–783.

Perkins, E.H., Kharaka, Y.K., Gunter, W.D., & DeBraal, J.D. (1990) Geochemical modeling of water-rock interactions usin SOLMINEQ.88. In *Chemical Modeling of Aqueous Systems II* (eds C. Melchoir, & R.L.Bassett). *American Chemistry Society Symposium Series*, 416, 117–127.

Perry, E.G. (1967) The oxygen isotope chemistry of ancient cherts. *Earth and Planetary Science Letters*, 3, 62–66.

Petsch, S.T. (2005) The global oxygen cycle. In *Biogeochemistry* (ed. W.H Schlesinger), Vol. 8, *Treatise on Geochemistry* (eds H.D. Holland & K.K. Turekian), pp. 515–556. Elsevier–Pergamon, Oxford.

Pigage, L.C. (1976) Metamorphism of the Settler Schist, southwest of Yale, British Columbia. *Canadian Journal of Earth Sciences*, 13, 405–421.

Pitzer, K.S. (1973) Thermodynamics of electrolytes I: Theoretical basis and general equations. *Journal of Physical Chemistry*, 77, 268–277.

Pitzer, K.S. (1979) Theory: ion interaction approach. In *Activity Coefficients in Electrolyte Solutions* (ed. R.M. Pytkowicz), Vol. 1, pp. 157–208. CRC Press, Boca Raton, FL.

Pitzer, K.S. (1980) Electrolytes. From dilute solutions to fused salts. *Journal of the American Chemistry Society*, 102, 2902–2906.

Plummer, L.N., & Busenberg, E. (1982) The solubilities of calcite, aragonite and vaterite in CO_2-H_2O solutions between 0 and 90°C, and an evaluation of the aqueous model for the system $CaCO_3$–CO_2–H_2O. *Geochimica Cosmochimica Acta*, 46, 1011–1040.

Plummer, L.N., & Parkhurst, D.L. (1990) Application of the Pitzer equations to the PHREEQE geochemical model. In *Chemical Modeling of Aqueous Systems II* (eds C. Melchoir, & R.L.Bassett). *American Chemistry Society Symposium Series*, 416, 128–137.

Plummer, L.N., Wigley, T.M.L., & Parkhurst, D.L. (1978) The kinetics of calcite dissolution in CO_2–water systems at 5° to 60°C and 0.0 to 1.0 atm CO_2. *American Journal of Science*, 278, 179–216.

Plummer, L.N., Parkhurst, D.L., & Wigley, T.M.L. (1979) Critical review of the kinetics of calcite dissolution and precipitation. In *Chemical Modeling in Aqueous Systems: Speciation, Sorption, Solubility, and Kinetics* (ed. E.A. Jenne), *American Chemistry Society Symposium Series*, 93, 537–573.

Plummer, L.N, Parkhurst, D.L., Fleming, G.W., & Dunkle, S.A. (1988) A computer program (PHRQPITZ) incorporating Pitzer's equations for calculation of geochemical reactions in brines. *U.S. Geological Survey Water Resources Investigation Report* 88–4153, 310 p.

Podvin, P. (1988) Ni–Mg partitioning between synthetic olivines and orthopyroxene: application to geothermometry. *American Mineralogist*, 73, 274–280.

Poirier, J-P. (1994) Light elements in the Earth's outer core: a critical review. *Physics of the Earth and Planetary Interiors*, 85, 319–337.

Poulton, S.W., Canfield, D.E., & Fralick, P. (2004) The transition to a sulphidic ocean ~1.845 billion years ago. *Nature*, 431, 173–177.

Powell, M., & Powell, R. (1974) An olivine–clinopyroxene geothermometer. *Contributions to Mineralogy and Petrology*, 48, 249–263.

Powell, R. (1978) *Equilibrium Thermodynamics inPpetrology: An Introduction*. Harper & Row, London, 284 pp.

Powell, R. (1984) Inversion of the assimilation and fractional crystallization (AFC) equations: characterization of contaminants from isotope and trace element relationships in volcanic suites. *Journal of Geology Society of London*, 141, 447–452.

Powell, R., & Holland, T. (1985) An internally consistent thermodynamic dataset with uncertainties and correlations. *Journal of Metamorphic Geology*, 3, 327–342.

Powell, R., & Holland, T. (1994) Optimal geothermometry and geobarometry. *American Mineralogist*, 79, 120–133.

Powell, R., Condliffe, D.M., & Condliffe, E. (1984) Calcite–dolomite geothermometry in the system $CaCO_3$–$MgCO_3$–$FeCO_3$: an experimental calibration. *Journal of Metamorphic Geology*, 2, 33–42.

Powell, T.G., & Macqueen, R.W. (1984) Precipitation of sulfide ores and organic matter–sulfate reactions at Pine Point, Canada. *Science*, 224, 63–66.

Pownceby, M.I., Wall, V.J., & O'Niel, H.S.C. (1987) Fe–Mn partitioning between garnet and ilmenite: experimental calibration and applications. *Contributions to Mineralogy and Petrology*, 97, 116–126.

Price, F.T., & Shieh, Y.N. (1979) Fractionation of sulfur isotopes during laboratory synthesis of pyrite at low temperatures. *Chemical Geology*, 27, 245–253.

Råheim, A., & Green, D.H. (1974) Experimental determination of the temperature and pressure dependence of the Fe–Mg partition coefficient for coexisting garnet and clinopyroxene. *Contributions to Mineralogy and Petrology*, 48, 179–203.

Railsback, L.B. (2006) *Some Fundamentals of Mineralogy and Geochemistry*. http://www.gly.uga.edu/railsback/FundamentalsIndex.html (online).

Rankama, K., & Sahama, T.G. (1950) *Geochemistry*. University of Chicago Press, Chicago, 912 pp.

Rasmussen, B., & Buick, R. (1999) Redox state of the Archean atmosphere: evidence from detrital heavy minerals in ca. 3250–2750 Ma sandstones from the Pilbara Craton, Australia. *Geology*, 27, 115–118.

Rawlings, D.E. (1997) Mesophilic, autotrophic bioleaching bacteria: description, physiology and role. In *Biomining: Theory, Microbes and Industrial Processes* (ed. D.E. Rawlings), pp. 229–245. Springer-Verlag, Berlin.

Rawlings, D.E. (2002) Heavy metal mining using microbes. *Annual Review of Microbiology*, 56, 65–91.

Rawlings, D.E., & Woods, D.R. (1995) Development of improved biomining bacteria. In *Bioextraction and Biodeterioration of Metals* (eds C.C. Gaylarde & H.A. Videla), pp. 632–684. Cambridge University Press, Cambridge.

Reddy, M.M. (1986) Effect of magnesium ions on calcium carbonate nucleation and crystal growth in dilute aqueous solutions at 25°C. In *Studies in Diagenesis* (eds F.A. Mumpton). *U.S. Geological Survey Bulletin*, 1578, 169–182.

Reddy, M.M., Plummer, L.N., & Busenberg, E. (1981) Crystal growth of calcite from calcium bicarbonate solutions at constant P_{CO2} and 25°C: a test of the calcite dissolution model. *Geochimica Cosmochimica Acta*, 45, 1281–1291.

Redlich, O. & Kwong, J.N.S. (1949) The thermodynamics of solutions. V. An equation of state. Fugacities of gaseous solutions. *Chemistry Reviews*, 44, 233–244.

Reed, M. H. (1982) Calculation of multicomponent chemical equilibria and reaction processes in systems involving minerals, gases, and an aqueous phase. *Geochimica Cosmochimica Acta*, 46, 513–528.

Reed, M, H. (1998) Calculation of simultaneous chemical equilibria in aqueous mineral–gas systems and its application to modeling hydrothermal processes. In *Techniques in Hydrothermal Ore Deposits Geology* (eds J. Richards & P. Larson). *Reviews in Economic Geology*, 10, 109–124.

Rees, C.E. (1973) A steady-state model for sulphur isotope fractionation in bacterial reduction processes. *Geochimica Cosmochimica Acta*, 37, 1141–1162.

Reinsch, D. (1977) High-pressure rocks from Val Chiusella (Sesia–Lanzo Zone), Italian Alps. *Neues Jahrbuch für Mineralogie-Abhandlungen*, 130, 89–102.

Richardson, S.M., & McSween, H.Y., Jr. (1989) *Geochemistry: Pathways and Processes*. Prentice Hall, Englewood Cliffs, NJ, 488 pp.

Riches, A.J.V., Liu, Y., Day, M.D., Spetsius, Z.V., & Taylor, L.A. (2010) Subducted oceanic crust as diamond hosts revealed by garnets of mantle xenoliths from Nyurbinskaya, Siberia. *Lithos*, 120, 360–378.

Richet, P., Bottinga, Y., Denielou, L., Petitet, J.P., & Tequi, C. (1982) Thermodynamic properties of quartz, cristobalite and amorphous SiO_2: drop calorimetry measurements between 1000 and 1800 K and a review from 0 to 2000 K. *Geochimica Cosmochimica Acta*, 46, 2639–2658.

Ridgwell, A., & Zeebe, R.E. (2005) The role of global carbonate cycle in the regulation and evolution of the Earth system. *Earth and Planetary Science Letters*, 234, 299–315.

Riding, R., & Liang, L. (2005) Seawater chemistry control of marine limestone accumulation over the past 550 million years. *Revista Española de Micropaleontologia*, 37, 1–11.

Ries, J.B. (2004) Effect of ambient Mg/Ca ratio on Mg fractionation in calcareous marine invertebrates: a record of the oceanic Mg/Ca ratio over the Phanerozoic. *Geology*, 32, 981–984.

Righter, K. (2003) Metal-silicate partitioning of siderophile elements and core formation in the early Earth. *Annual Review of Earth and Planetary Sciences*, 31, 135–174.

Righter, K. (2007) Not so rare Earth? New developments in understanding the origin of the Earth and Moon. *Chemie der Erde*, 67, 179–200.

Righter, K., & Drake, M.J. (1997) Metal/silicate equilibrium in a homogeneously accreting Earth: New results for Re. *Earth and Planetary Science Letters*, 146, 541–553.

Righter, K., & Drake, M.J. (1999) Effect of water on metal–silicate partitioning of siderophile elements: a high pressure and temperature terrestrial magma ocean and core formation. *Earth and Planetary Science Letters*, 171, 383–399.

Righter, K., & Drake, M.J. (2000) Metal–silicate equilibrium in the early Earth: new constraints from volatile moderately siderophile elements Ga, Sn, Cu, and P. *Geochimica Cosmochimica Acta*, 64, 3581–3597.

Righter, K., Drake, M.J., & Scott, E.R.D. (2006) Compositional relationships between meteorites and terrestrial planets. In *Meteorites and the Early Solar System II* (Eds D. Lauretta & H.Y. McSween, Jr), pp. 803–828. The University of Arizona Press, Tucson, AZ.

Rimstidt, J.D., & Barnes, H.L. (1980) The kinetics of silica water reactions. *Geochimica Cosmochimica Acta*, 44, 1683–99.

Ringwood, A.E. (1955) The principles governing trace element distribution during magmatic crystallization, I, The influence of electronegativity. *Geochimica Cosmochimica Acta*, 7, 189–202.

Ringwood, A.E. (1966) Chemical evolution of the terrestrial planets. *Geochimica Cosmochimica Acta*, 30, 41–104.

Ringwood, A.E. (1979) *Origin of the Earth and Moon*. Springer-Verlag, New York.

Ringwood, A.E. (1991) Phase transformations and their bearing on the constitution and dynamics of the mantle. *Geochimica Cosmochimica Acta*, 55, 2083–2100.

Robie, R.A., & Hemingway, B.S. (1995) Thermodynamic properties of minerals at 298.15 K and 1 bar (10^5 Pascals) pressure and at higher temperatures. *U.S. Geological Survey Bulletin*, 2131.

Robie, R.A., Hemingway, B.S., & Fisher, J.R. (1978) Thermodynamic properties of minerals and related substances at 298.15 K and 1 bar (105 pascals) pressure and at higher temperatures. *U.S. Geological Survey Bulletin* 1452, 456 pp. (Reprinted with corrections, 1979.)

Roddick, J.C. (1978) The application of isochron diagrams in $^{40}Ar–^{39}Ar$ dating: a discussion. *Earth and Planetary Science Letters*, 41, 233–244.

Rodhe, H. (1990) A comparison of the contribution of various gases to the greenhouse effect. *Science*, 248, 1217–1219.

Roedder, E. (1984) *Fluid Inclusions*. *Reviews in Mineralogy*, 12, 646 pp.

Roeder, P.L., & Emslie, R.F. (1970) Olivine–liquid equilibrium. *Contributions to Mineralogy and Petrology*, 29, 275–289.

Rollinson, H.R. (1993) *Using geochemical data: evaluation, presentation, interpretation*. Longman/J. Wiley & Sons, New York, 352 pp.

Ronov, A.B., & Yaroshevsky, A.A. (1976) A new model for the chemical structure of the Earth's crust. *Geochemistry International*, 13, 89–121.

Rouxel, O.J., Bekker, A., & Edwards, K.J. (2005) Iron isotope constraints on the Archean and Paleoproterozoic ocean redox state. *Science*, 307, 1088–1091.

Roy, S. (1997) Genetic diversity of manganese deposition in the terrestrial geologic record. In *Manganese Mineralization: Geochemistry and Mineralogy of Terrestrial And Marine Deposits* (eds K. Nicholson, J.R. Hein, B. Buhn, & S. Dasgupta). *Geological Society of London Special Publication*, 119, 5–27.

Royer, D.L., Berner, R.A., Montañe, I.P., Tabor, N.J., & Beerling, D.J. (2004) CO_2 as a primary driver of Phanerozoic climate. *GSA Today*, 14(3), 4–10.

Ruaya, J.R. (1988) Estimation of instability constants of metal chloride complexes in hydrothermal solutions up to 300°C. *Geochimica Cosmochimica Acta*, 52, 1983–1996.

Rubie, D.C., Melosh, H.J., Reid, J.E., Liebske, C., & Righter, K. (2003) Mechanisms of metal–silicate equilibration in the terrestrial magma ocean. *Earth and Planetary Science Letters*, 205, 239–255.

Rudnick, R. L. (1995) Making continental crust. *Nature*, 378, 571–578.

Rudnick, R.L., & Gao, S. (2004) The composition of the continental crust. In *Treatise oneochemistry*, v. 3, *The Crust*, Rudnick, R.L., ed., 1–64. Amsterdam: Elsevier.

Rudnick, R.L., Barth, M., Horn, I., & McDonough, W.F. (2000) Rutile-bearing refractory eclogites: missing link between continents and depleted mantle. *Science*, 287, 278–281.

Rumble, D. (1977) Configurational entropy of magnetite–ulvospinel$_{ss}$ and hematite–ilmenite$_{ss}$. *Carnegie Institute of Washington Year Book*, 76, 581–584.

Russell, M.J., & Arndt, N.T. (2005) Geodynamic and metabolic cycles in the Hadean. *Biogeosciences*, 2, 97–111.

Russell, R.D., & Farquhar, R.Mc. (1960) *Lead Isotopes in Geology*. Wiley Interscience, New York, 243 pp.

Rutherford, E., & Soddy, F. (1902) The cause and nature of radioactivity. Part I. *Philosophical Magazine, Series* 6(4), 370–396.

Ruttenberg, K.C. (2005) The global phosphorus cycle. In *Biogeochemistry* (ed. W.H Schlesinger), Vol. 8, *Treatise on Geochemistry* (eds H.D. Holland & K.K. Turekian), pp. 585–644. Elsevier–Pergamon, Oxford.

Ryan, C.G., Griffin, W.L., & Pearson, N.J. (1996) Garnet geotherms: pressure–temperature data from Cr–pyrope garnet xenocrysts in volcanic rocks. *Journal of Geophysical Research*, 101, 5611–5625.

Rye, R.O. (1974) A comparison of sphalerite–galena sulfur isotope temperatures with filling temperatures of fluid inclusions. *Economic Geology*, **69**, 26–32.

Rye, R., & Holland, H.D. (1998) Paleosols and the evolution of atmospheric oxygen: a critical review. *American Journal of Science*, 298, 621–672.

Rye, R., & Holland, H.D. (2000) Life associated with a 2.76 Ga ephemeral pond? Evidence from Mount Roe #2 paleosol. *Geology*, 28, 483–486.

Rye, R., Kuo, P.H., & Holland, H.D. (1995) Atmospheric carbon dioxide concentrations before 2.2 billion years ago. *Nature*, 378, 603–605.

Sack, R.O., & Ghiroso, M.S. (1991) Chromian spinels as petrogenetic indicators: thermodynamics and petrological applications. *American Mineralogist*, 76, 827–847.

Salters, V.J.M., & Stracke, A. (2004) Composition of depleted mantle. *Geochemistry Geophyics. Geosystems*, 5, 1–27, doi:10.1029/2003GC000597 (on line).

Sandberg, P.A. (1983) An oscillating trend in Phanerozoic nonskeletal carbonate mineralogy. *Nature*, 305, 19–22.

Sanford, W.E., & Wood, W.W. (1991) Brine evolution and mineral deposition in hydrologically pen evaporite basins. *American Journal of Science*, 291, 687–710.

Sarin, M.M., Krishnaswami, S., Dilli, K., Somayajulu, B.L.K., & Moore, W.S. (1989) Major ion chemistry of the Ganga-Bramaputra river systems, India. *Geochimica Cosmochimica Acta*, 53, 997–1009.

Savin, S.M., & Epstein, S. (1970) The oxygen and hydrogen isotope geochemistry of clay minerals. *Geochimica Cosmochimica Acta*, 34, 43–64.

Saxena, S.K., & Fei, Y. (1988) Fluid mixtures in the C–H–O system at high pressure and temperature. *Geochimica Cosmochimica Acta*, 52, 505–512.

Saxena, S.K., & Ribbe, P.H. (1972) Activity–composition relations in feldspars. *Contributions to Mineralogy and Petrology*, 37, 131–138.

Schidlowski, M. (1981) Uraniferous constituents of the Witwatersrand conglomerates: ore microscopic observations for the Witwatersrand metallogeny.. *U.S. Geol. Survey Prof. Ppaper* 1161, N1-N29.

Schlesinger, W.H. (1997) *Biochemistry: An Anaylsis of Global Change, 2nd edn*. Acadmic Press, San Diego, 558 pp.

Schneider, S.H. (2000) Why study Earth systems science? In *Earth Systems: Processes and Issues* (ed. W.G. Ernst), pp. 5–12. Cambridge University Press, Cambridge.

Schnell, H.A. (1997) Bioleaching of copper. In *Biomining: Theory, Microbes and Industrial Processes* (ed. D.E. Rawlings), pp. 21–143. Springer-Verlag, Berlin.

Schnoor, J.L. (1990) Kinetics of chemical weathering: a comparison of laboratory and field weathering rates. In *Aquatic Chemical Kinetics: Reaction Rates of Processes in Natural Waters* (ed. W. Stumm), pp. 475–504. Wiley-Interscience, New York.

Schopf, J.W. (1993) Microfossils of the early Archean Apex Chert; new evience of the antiquity of life. *Science*, 260, 640–646.

Sen, S.K, and Bhattacharya, A. (1984) An orthopyroxene–garnet thermometer and the application to the Madras charnockites. *Contributions to Mineralogy and Petrology*, 88, 64–71.

Sengupta, P., Dasgupta, S., Bhattacharya, P.K., & Mukherjee, M. (1990) An orthopyroxene-biotite geothermometer and its application in crustal granulites and mantle-derived rocks. *Journal of Metamorphic Geology*, 8, 191–197.

Senko, J.M., Campbell, B.S., Henriksen, J.R., Elshahed, M., Dewers, T.A., & Krumholz, L.R. (2004) Barite deposition resulting from phototrophic sulfide-oxidizing bacterial activity. *Geochimica Cosmochimica Acta*, 68, 773–780.

Shaw, D.M. (1953) The camouflage principle and trace element distribution in magmatic minerals. *Journal of Geology*, 61, 142–151.

Shen, Y., & Buick, R. (2004) The antiquity of microbial sulfate reduction. *Earth And Planetary Science Letters*, 64, 243–272.

Shen, Y., Buick, R., & Canfield, D.E. (2001) Isotopic evidence for microbial sulfate reduction in the early Archean era. *Nature*, 410, 77–81.

Shen, Y., Canfield, D.E., & Knoll. A.H. (2002) Middle Proterozoic ocean chemistry: evidence from the McArthur Basin, Northern Australia. *American Journal of Science*, 302, 81–109.

Shen, Y., Knoll, A.N., & Walter, M.R. (2003) Evidence for low sulphate and anoxia in a mid–Proterozoic marine basin. *Nature*, 423, 632–635.

Shepherd, T.J., Rankin, A,H., & Alderton, D.H.M. (1985) *A Practical Guide to Fluid Inclusion Studies*. Blackie, Glasgow, 239 p.

Sheppard, S.M.F. (1986) Characterization and isotopic variations in natural weaters. *Reviews in Mineralogy*, 16, 163–183.

Singer, P.C., & Stumm, W. (1970) Acidic mine drainage: the rate-determining step. *Science*, 167, 1121–1123.

Sjöberg, E.L. (1976) A fundamental equation for calcite dissolution kinetics. *Geochimica Cosmochimica Acta*, 40, 441–447.

Sjöberg, E.L., & Rickard, D.T. (1984) Temperature deperndence of calcite dissolution kinetics between 1 and 62°C at pH 2.7 to 8.4 in aqueous solutions. *Geochimica Cosmochimica Acta*, 48, 485–493.

Skinner, B.J. (1966) Thermal expansion. In *Handbook of Physical Constants* (ed. S.P. Clark). *Geological Society of America Memoir*, 97, 75–96.

Skirrow, G. (1975) The dissolved gases – carbon dioxide. In *Chemical Oceanography*, 2nd edn (Eds J.P. Riley & G. Skirrow), Vol. 2, pp. 245–300. Academic Press, London.

Sleep, N.H., & Windley, B.F. (1982) Archean plate tectonics: constraints and inferences. *Journal of Geology*, 90, 363–380.

Smith, D. (1999) Temperatures and pressures of mineral equilibration in peridotite xenoliths: review, discussion, and implications. In *Mantle Petrology: Field Observations and High Pressure Experimentation: A Tribute to Francis R. (Joe) Boyd* (eds Y. Fei, M. Bertka, & B.O. Mysen). *Geochemistry Society Special Publication*, 6, 171–188.

Smith, F.G. (1963) *Physical Geochemistry*. Addison-Wesley, Reading, MA.

Smith, N.D., & Minter, W. (1980) Sedimentological controls of gold and uranium in Witwatersrand paleoplacers. *Economic Geology*, 75, 1–17.

Smith, P.E., & Farquar, R.M. (1989) Direct dating of Phanerozoic sediments by the ^{238}U–^{206}Pb method. *Nature*, 341, 518–521.

Smits, G. (1984) Some aspects of the uranium mineralsin Witwatersrand sediments of the early Proterozoic. *Precambrian Research*, 25, 37–59.

Snyder, G.A., Taylor, L.A., & Neal, C.R. (1992) A chemical model for generating the sources of mare basalts: Combined equilibrium and fractional crystallization of the lunar magmasphere. *Geochimica Cosmochimica Acta*, 56, 3809–3823.

Songrong, Y., Jiyan, X., Guanzhou, Q., & Yuchua, H. (2002) Research and application of bioleaching and biooxidation technologies in China. *Minerals Engineering*, 15, 361–363.

Southam, G., & Saunders, J.A. (2005) The geomicrobiology of ore deposits. *Economic Geology*, 100, 1067–1084.

Sparks, R.S.J., Hupert, H.E., & Turner, J.S. (1984) The fluid dynamics of evolving magma chambers. *Philosophical Transactions of the Royal Society London*, A310, 511–534.

Spear, F.S. (1989) Petrologic determination of metamorphic pressure–temperature–time paths. In *Metamorphic Pressure–Temperature–Time Paths* (eds F.S. Spear & S.M. Peacock). *Short Course in Geology*, pp. 1–55. American Geophysical Union, Washington, DC.

Spear, F. (1993) *Metamorphic Phase Equilibria and Pressure–Temperature–Time Paths*. Monograph Series, Mineralogical Society of America, Washington, DC, 799 pp.

Spear, F.S., & Kimball, K.L. (1984) RECAMP – a FORTRAN IV program for estimating Fe^{3+} contents in amphiboles. *Computers in Geology*, 10, 317–325.

Spencer, K.J., & Lindsley, D.H. (1981) A solution model for coexisting iron–titanium oxides. *American Mineralogist*, 66, 1189–1201.

Spencer, R.J., & Hardie, L.A. (1990) Control of seawater composition by mixing of river waters and mid–ocean ridge hydrothermal brines. In *Fluid–Mineral Interactions: A Tribute to H.P. Eugster* (eds R.J. Spencer, & I.-M. Chou), Special Publication 2, pp. 409–419. Geochemical Society, San Antonio, Texas.

Stacey, J.S., & Kramers, J.D. (1975) Approximation of terrestrial lead isotope evolution by a two–staged model. *Earth and Planetary Science Letters*, 26, 207–221.

Stahr, D.W., III. (2007) *Tecnomagmatic evaluation of the eastern Blue Ridgre: Differentiating multiple Paleozoic orogenic pulses in the Glenville and Big Ridge quadrangles, southwestern North Carolina*. Unpublished MS thesis, The University of Tennessee, Knoxville, 263 pp.

Stanley, S.M., & Hardie, L.A. (1998) Secular oscillations in the carbonate mineralogy of reef-building and sediment-producing organisms driven by tectonically forced shifts in seawater chemistry. *Paleogeography Paleoclimatology Paleoecology*, 144, 3–19.

Stanley, S.M., Ries, J.B., & Hardie, L.A. (2002) Low-magnesium calcite produced by coralline algae in seawater of Late Cretaceous composition. *Proceedings of the National Academy of Science USA*, 99, 15323–15326.

Steiger, R.H., & Jäger, E. (1977) Subcommission on geochronology: Convention on the use of decay constants in geo– and cosmochronology. *Earth and Planetary Science Letters*, 36, 359–362.

Steuber, T., & Veizer, J. (2002) Phanerozoic record of plate tectonic control of seawater chemistry and carbonate sedimentation. *Geology*, 30, 1123–1126.

Stevenson, D.J. (1981) Models of the Easrth's core. *Science*, 214, 611–619.

Stimpfl, M., Lauretta, D.S., & Drake, M.J. (2004) Adsorption as a mechanism to deliver water to the Earth (abstract). *Meteoritics and Planetary Science*, 39, A99.

Stormer, J.C., Jr. (1975) A practical two-feldspar geothermometer. *American Mineralogist*, 60, 667–674.

Stormer, J.C., Jr. (1983) The effects of recalculation on estimates of temperature and oxygen fugacity from analyses of multicomponent iron–titanium oxides. *American Mineralogist*, 68, 586–594.

Strauss, H. (1997) The isotopic composition of sedimentary sulfur through time. *Paleogeography Paleoclimatology Paleoecology*, 132, 97–118.

Strauss, H. (1999) Geological evolution from isotope proxy signals – sulfur. *Chemical Geology*, 161, 89–101.

Stumm, W., & Morgan, J.J. (1981) *Aquatic Chemistry: An Introduction Emphasizing Chemical Equilibria in Natural Waters*, 2nd edn. J. Wiley & Sons, New York, 780 pp.

Stumm, W., & Wieland, E. (1990) Dissolution of oxide and silicate minerals: rates depend on surface speciation. In *Aquatic Chemical Kinetics: Reaction Rates of Processes in Natural Waters* (ed. W. Stumm), pp. 367–400. Wiley-Interscience, New York.

Summer, D.Y., & Grotzinger, J.P. (2000) Late Archean aragonite precipitation: petrography, facies associations, and environmental significance. In *Carbonate Sedimentation and Diagenesis in the Evolving Precambrian World* (eds J.P. Grotzinger & N.P. James). *Society of Economic Palaeontologists and Minerologists Special Publication*, 67, 123–144.

Sun, S.S., & McDonough, W.F. (1989) Chemical and isotopic systematics of ocean basalts: Implications for mantle composition and processes. In *Magmatism in the Ocean Basins* (eds A.D. Saunders &

M.J. Norry). *Geological Society of London Special Publication*, 42, 313–345.

Sverjensky, D.A., & Lee, N. (2010) The Great Oxidation Event and mineral diversification. *Elements*, 6, 31–36.

Tatsumoto, M., Knight, R.J., & Allègre, C.J. (1973) Time differences in the formation of meteorites as determined from the ratio of lead-207 to lead-206. *Science*, 180, 1279–1283.

Taylor, B.E. (1987) Stable isotope geochemistry of ore-forming fluids. *Mineralogical Association of Canada Short Course Handbook*, 13, 337–445.

Taylor, H.P., Jr. (1974) The application of oxygen and hydrogen isotope studies to problems of hydrothermal alteration and ore deposition. *Economic Geology*, 69, 843–883.

Taylor, H.P., Jr. (1977) Water/rock interactions and the origin of H_2O in granitic batholiths (30th William Smith Lecture). *Journal of the Geology Society of London*, 13, 509–558.

Taylor, H.P., Jr. (1979) Oxygen and hydrogen isotope relationships in hydrothermal mineral deposits. In *Geochemistry of Hydrothermal Ore Deposits*, 2nd edn (ed. H.L. Barnes), pp. 236–277. J. Wiley & Sons, New York.

Taylor, H.P., Jr. (1997) Oxygen and hydrogen isotope relationships in hydrothermal mineral deposits. In *Geochemistry of Hydrothermal Ore Deposits* (ed. H.L. Barnes), 2nd edn, pp. 229–302. J. Wiley & Sons, New York.

Taylor, L.A., Onorato, P.I.K., & Uhlmann, D.R. (1977) Cooling rate estimates based on kinetic modeling of Fe–Mg diffusion in olivine. *Proceedings of the 8th Lunar Science Conference*, Houston, TX, March 4–18, pp. 1581–1592.

Taylor, S.R. (1967) The origin and growth of continents. *Tectonophysics*, 4, 17–34.

Taylor, S.R. (1987) The unique lunar composition and its bearing on the origin of the Moon. *Geochimica Cosmochimica Acta*, 31, 1297–1306.

Taylor, S.R. (1992) The origin of the Earth. In *Understanding the Earth* (eds G. Brown, C. Hawkesworth, & C. Wilson), pp. 25–43. Cambridge University Press, Cambridge.

Taylor, S.R., & McLennan, S.M. (1981) The composition and evolution of the continental crust: Rare earth element evidence from sedimentary rocks. *Philosophical Transactions of the Royal Society of London*, A301, 381–399.

Taylor, S.R., & McLennan, S.M. (1985) *The Continental Crust: Its Composition and Evolution*. Blackwell, Oxford, 312 pp.

Taylor, S.R., & McLennan, S.M. (1995) The geochemical evolution of the continental crust. *Reviews of Geophysics and Space Physics*, 33, 241–265.

Taylor, W.R. (1998) An experimental test of some geothermometer and geobarometer formulations for upper mantle peridotites with application to the thermobarometry of fertile lherzolite and garnet websterite. *Neues Jahrbuch für Mineralogie-Abhandlungen*, 172, 381–408.

Thode, H.G. (1981) Sulfur isotope ratios in petroleum research and exploration: Wilson Basin. *American Association of Petroleum Geologists Bulletin*, 65, 1527–1537.

Thode, H.G., & Monster, J. (1965) Sulfur isotope geochemistry of petroleum, evaporites and ancient seas. *American Association of Petroleum Geologists Memoir*, 4, 367–377.

Thompson, A.B. (1976) Mineral reactions in pelitic rocks. II. Calculation of some P–T–X (Fe–Mg) phase relations. *American Journal of Science*, 276, 425–454.

Thompson, J.B., Jr. (1967) Thermodynamic properties of simple solutions. In *Researches in Geochemistry* (ed. P.H. Abelson), Vol. 2, pp. 340–361. J. Wiley & Sons, New York.

Thompson, R.N. (1982) Magmatism of the British Tertiary volcanic province. *Scottish Journal of Geology*, 18, 49–107.

Thompson, R.N., Morrison, M.A., Hendry, G.I., & Parry, S.J. (1984) An assessment of the relative roles of a crust and mantle in magma genesis: an elemental approach. *Philosophical Transactions of the Royal Society London, Series A*, 310, 549–590.

Tice, M.M., & Lowe, D.R. (2004) Photosynthetic microbial mats in the 3,416-Myr-old ocean. *Nature*, 431, 549–55.

Tolstikhin, I.N., & Kramers, D. (2008) *The Evolution of Matter: from the Big Bang to the Present Day*. Cambridge University Press, New York, 521 pp.

Towe, K.M. (1990) Aerobic respiration in the Archean. *Nature*, 348, 54–56.

Trudinger, P.A., Chambers, L.A., & Smith, J.W. (1985) Low temperature sulfate reduction: biological versus abiological. *Canadian Journal of Earth Science*, 22, 1910–1918.

Truesdell, A.H., & Jones, B.F. (1974) WATEQ, a computer program for calculating chemical equilibria of natural waters. *U.S. Geological Survey Journal of Research*, 2, 233–248.

Trumbore, S.E. (2000) Radiocarbon geochronology. In *Quaternary Geochronology: Methods and Applications* (Eds J.S. Noller, J.M. Sowers, & W.R. Lettis), AGU Reference Shelf 4, pp. 41–60. American Geophysical Union, Washington, DC.

Tudge, A.P., & Thode, H.G. (1950) Thermodynamic properties of isotopic compounds of sulphur. *Canadian Journal of Resources*, 28B, 567–578.

Turekian, K., & Clark, S.P. (1969) Inhomogeneous accretion of the earth from primitive solar nebula. *Earth and Planetary Science Letters*, 6, 346–348.

Urey, H.C. (1947) The thermodynamic properties of isotopic substances. *Journal of the Chemistry Society*, 1947, 562–581.

Valley, J.W. (2003) Oxygen isotopes in zircon. *Reviews in Mineralogy and Geochemistry*, 53, 343–385.

Valley, J.W., Taylor, H.P., Jr., & O'Neil, J.R. (eds) (1986) *Stable Isotopes in High Temperature Geological Processes. Reviews in Mineralogy*, 16, 570 pp.

Valley, J.W., Kinny, P.D., Schulze, D.J., & Spicuzza, M.J. (1998) Zircon megacrysts from kimberlite; oxygen isotope variability among mantle melts. *Contributions to Mineralogy and Petrology*, 133, 1–11.

Valley, J.W., Peck, W.H., King, E.M., & Wilde, S.A. (2002) A cool early Earth. *Geology*, 30, 351–354.

Valley, J.W., Lackey, J.S., Cavosie, A.J., *et al.* (2005) 4.4 billion years of crustal maturation: oxygen isotope ratios of magmatic zircon. *Contributions to Mineralogy and Petrology*, 150, 561–580.

Van Loon, G.W., & Duffy, S.J. (2000) *Environmental Chemistry*. Oxford University, 492 pp.

Velbel, M.A. (1985) Geochemical mass balances and weathering rates in forested watersheds of the southern Blue Ridge. *American Journal of Science*, 285, 904–930.

Von Seckendorff, V., & O'Neill, H. St. C. (1993) An experimental study of Fe–Mg partitioning between olivine and orthopyroxene at 1173, 1273 and 1423 K and 1.6 GPa. *Contributions to Mineralogy and Petrology*, 113, 196–207.

Wade, J., & Wood, B.J. (2005) Core formation and the oxidation state of the Earth. *Earth and Planetary Science Letters*, 236, 78–95.

Wagman, D.D., Evans, W.H., Parker, V.P., *et al.* (1982) The NBS tables of chemical thermodynamic properties: selected values for inorganic and C1 and C2 organic substances in SI units. *Journal of Physics and Chemistry Reference Data*, 11 (Suppl. 2), 1–392.

Walker, J.C.G. (1977) *Evolution of the Atmosphere*. Mcmillan Publishing Co., New York, 318 pp.

Walker, J.C.G., Klein, C., Schidlowski, M.,, J.W., Stevenson, D.J., & Walter, M.R. (1983) Environmental evolution of the Archean–Early Proterozoic Earth. In *Earth's Earliest Biosphere: Its Origin and Evolution* (ed. J.W. Schopf), pp. 260–290. Princeton University Press, Princeton, NJ.

Walker, R.J., Shirey, S.B., & Stecher, O. (1988) Comparative Re–Os, Sm–Nd and Rb–Sr isotope and trace element systematics for Archean komatiites flows from Munro Township, Abitibi Belt, Ontario. *Earth and Planetary Science Letters*, 87, 1–12.

Walker, R.J., Shirey, S.B., Hanson, G.N., Rajamani, V., & Horan, M.F. (1989) Re–Os, Rb–Sr, and O isotopic systematics of the Archean Kolar schist belt, Karnataka, India. *Geochimica Cosmochimica Acta*, 53, 3005–3013.

Walker, R.J., Morgan, J.W., Naldrett, A.J., & Li, C. (1991) Re–Os isotopic systematics of Ni–Cu sulfide ores, Sudbury Igneous Complex, Ontario: evidence for a major crustal component. *Earth and Planetary Science Letters*, 105, 416–429.

Walter, L.M. (1986) Relative efficiency of carbonate dissolution and precipitation during diagenesis: a progress report on the role of solution chemistry. In *Roles of Organic Matter in Sedimentary Diagenesis* (ed. D.I. Gautier). *Society of Economic Palentologists and Mineralogists Special Publication*, 38, 1–11.

Walter, M.J., & Tronnes, R.G. (2004) Early Earth differentiation, *Earth and Planetary Science Letters*, 225, 253-

Walther, J.V. (2009) *Essentials of Geochemistry*, 2nd edn. Jones and Bartlett Publications, Sudbury, MA, 797 pp.

Walther, J.V., & Wood, B.J. (1984) Rate and mechanism in prograde metamorphism. *Contributions to Mineralogy and Petrology*, 88, 246–259.

Walther, J.V., & Wood, B.J. (1986) Mineral-fluid reaction rates. In *Fluid–Rock Interactions During Metamorphism* (eds J.V. Walther & B.J. Wood), *Advances in Physical Geochemistry*, Vol. 5, 194–211. Springer-Verlag, New York.

Wänke, H. (1981) Constitution of terrestrial planets. *Philosophical Transactions of the Royal Society London, Series A*, 303, 287–302.

Warren, P.H. (1989) Growth of the continental crust: a planetary–mantle prospective. *Tectonophysics*, 161, 165–199.

Wasson, J.T. (1974) *Meteorites*. Springer-Verlag, Berlin, 316 pp.

Watanabe, K., Naraoka, H., Wronkiewicz, D.J., Condie, K.C., & Ohmoto, H. (1997) Carbon, nitrogen, and sulfur geochemistry of the Archean and Proterozoic shales from the Kaapvaal Craton, South Africa. *Geochimica Cosmochimica Acta*, 61, 3441–3459.

Watson, E.B., & Harrison, T.M. (2005) Zircon thermometer reveals minimum melting conditions on earliest Earth. *Science*, 308, 841–844.

Weast, R.C., Astle, M.J., & Beyer, W.H. (eds), 1986) CRC *Handbook of Chemistry and Physics*. CRC Press, Boca Raton, FL.

Weisburg, M.K., McCoy, T.J., & Krot, A.N. (2006) Systematics and evaluation of meteorite classificton. In *Meteorites and the Early Solar System II* (eds D.S. Lauretta & H.Y. McSween, Jr), pp. 19–52. University of Arizona Press, Tucson, AZ.

Westall, J., Zachary, J.L., & Morel, F.M.M. (1976) *MINEQL, a Computer Program for the Calculation of Chemical Equilibrium Composition of Aqueous Systems*. Technical Note 18, Department of Civil Engineering, Massachussets Institute of Technology 91 pp.

Wetherill, G.W. (1956a) An interpretation of the Rhodesia and Witwatersrand age patterns. *Geochimica Cosmochimica Acta*, 9, 290–292.

Wetherill, G.W. (1956b) Discordant uranium–lead ages. *Transactions of the American Geophysical Union*, 37, 320–327.

Wetherill, G.W. (1986) Accumulation of terrestrial planets and implications concerning lunar origin. In *Origin of the Moon* (eds W.K. Hartman, R.J. Phillips, & G.J. Taylor), pp. 519–550. Lunar Planetary Institute, Houston.

White, A.F., & Brantley, S.L (eds) (1995) *Chemical Weathering Rates of Silicate Minerals*. Mineralogical Society of America, Washington, DC.

White, D.E. (1974) Diverse origins of hydrothermal ore fluids. *Economic Geology*, 69, 954–973.

White, W.M. (2002) *Geochemistry*. http://www.geo.cornell.edu/geology/classes/geo455/Chapters.HTML (on line).

White, W.M., & Hofmann, A.W. (1982) Sr and Nd isotope geochemistry of mantle evolution. *Nature*, 296, 821–825.

Whitehouse, M.J., Kamber, B.S., & Moorbath, S. (1999) Age significance of U–Th–Pb zircon data from early Archean rocks of west Greenland: a reassessment of continental crust and oceans on the Earth 4.4 years ago. *Nature*, 409, 175–178.

Whitney, D.L. (2002) Coexisting andalusite, kyanite, and sillimanite: sequential formation of three Al_2SiO_5 polymorphs during progressive metamorphism near the triple point, Sivrihisar, Turkey. *American Mineralogist*, 87, 405–416.

Whitney, J.A., & Stormer, J.C., Jr. (1977) The distribution of $NaAlSi_3O_8$ between coexisting microcline and plagioclase and its effect on geothermometric calculations. *American Mineralogist*, 62, 687–691.

Widdel, F., Schnell, S., Heising, S., Ehrenreich, A., Assmus, B., & Schink, B. (1993) Ferrous iron oxidation by anoxygenic phototrophic bacteria. *Nature*, 362, 834–836.

Wiechert, U., Halliday, A.N., Lee, D.-C, Snyder, G.A., Taylor, L.A., & Rumble, D. (2001) Oxygen isotopes and the Moon–forming giant impact. *Science*, 294, 345–348.

Wilde, S.A., Valley, J. W., Peck, W.H., & Graham, C.M. (2001) Evidence from detrital zircons for the existence of continental crust and oceans on the Earth 4.4 Gyr ago. *Nature*, 409, 175–178.

Wilkins, E.T. (1954) Air pollution and the London Fog of December, 1952. *Journal of the Royal Sanitary Institute*, 74, 1–21, DOI: 10.1177/146642405407400101.

Williamson, S.J. (1973) *Fundamentals of Air Pollution*. Addison-Wesley, Reading, MA.

Wilson, M. (1989) *Igneous Petrogenesis*. Unwin Hyman, London, 466 pp.

Wilson, M. (2008) Isotope-ratio measurements reveal a young Moon. *Physics Today*, 61(2,) 16–17.

Wilson, T.R.S. (1975) Salinity and the major elements of seawater. In *Chemical Oceanography* (eds J.P. Riley & G. Skirrow), 2nd edn, Vol. 1, pp. 365–413. Academic Press, London.

Wolery, T.J. (1992a) *EQ3/6, a Software Package for Geochemical Modeling of Aqueous Systems: Package Overview and Installation Guide*, Version 7.0. UCRL–MA–110662 PT I, Lawrence Livermore National Laboratory, Livermore, CA, 66 pp.

Wolery, T.J. (1992b) *EQ3NR, a Computer Program for Geochemical Aqueous Speciation–Solubility Calculations: Theoretical, Manual, User's Manual, User's Guide, and Related Documentation*, Version 7.0. UCRL–MA–11062 PT III, Lawrence Livermore National Laboratory, Livermore, CA, 246 pp.

Wolery, T.J., & Daveler, S.A. (1992) *EQ6, a Computer Program for Reaction Path Modeling of Aqueous Geochemical Systems: Theoretical Manual, User's Guide, and Related Documentation*, Version 7.0. UCRL–MA–110662 PT IV, Lawrence Livermore National Laboratory, Livermore, CA, 338 pp.

Wollast, R. (1990) Rate and mechanism of dissolution of carbonates in the system $CaCO_3$–$MgCO_3$. In *Aquatic Chemical Kinetics: Reaction Rates of Processes in Natural Waters* (ed. W. Stumm), pp. 431–446. Wiley-Interscience, New York.

Wollast, R., & Chou, L. (1984) Kinetic study of the dissolution of albite with a continuous flow-through fluidized bed reactor. In *Chemistry of Weathering.*, (ed. J.I. Drever), pp. 75–96. D. Reidel Publishing Co., Dordrecht.

Wood, B.J. (1987) Thermodynamics of multicomponent systems containing several solid solutions. *Reviews in Mineralogy*, 17, 71–95.

Wood, B.J., & Banno, S. (1973) Garnet–orthopyroxene and orthopyroxene-clinopyroxene relationships in simple and complex systems. *Contributions to Mineralogy and Petrology*, 42, 109–124.

Wood, B.J., & Fraser, D.G. (1976) *Elementary Thermodynamics for Geologists.* Oxford University Press, Oxford, 303 pp.

Wood, D.A. (1980) The application of Th–Hf–Ta diagram to problems of tectonomagmatic classification and to establishing the nature of crustal contamination of basaltic lavas of the British Tertiary volcanic province. *Earth and Planetary Science Letters*, 50, 11–30.

Wood, D.A., Joron, J.L., & Treuil, M. (1979) A re-appraisal of the use of trace elements to classify and discriminate between magma series erupted in different tectonic settings. *Earth Planetary Science Letters*, 45, 326–336.

Wood, J.A. (1979) *The Solar System.* Prentice Hall, Englewood Cliffs, NJ, 196 pp.

Wood, S.A. (1994) Calculation of activity–activity and log fO_2–pH diagrams. *Reviews in Economic Geology*, 10, 81–96.

Wood, S.A., & Samson, I.M. (1998) Solubility of ore minerals and complexation of ore metals in

Wood, S.A., Pan, P., Zhang, Y., & Mucci, A. (1994) Solubility of Pt and Pd sulfides and Au metal in aqueous sulfide solutions. I. Results at 25°–90°C and 1 atm. *Mineralium Deposita*, 29, 373–390.

Woodland, A.B., & Wood, B.J. (1994) Fe_3O_4 activities in Fe–Ti spinel solid solutions. *European Journal of Mineralogy*, 6, 23–37.

Yang, W., Holland, H.D., & Rye, R. (2002) Evidence for low or no oxygen in the late Archean atmosphere from the ~2.76 Ga Mt. Roe #2 paleosol, Western Australia: Part 3. *Geochimica Cosmochimica Acta*, 66, 3707–3718.

Yin, Q., Jacobsen, S.B., Yamashita, K., Blichert-Toft, J., & Telouk, P. (2002) A short time scale for terrestrial planet formation from Hf–W chronometry of meteorites. *Nature*, 418, 949–952.

Young, G.M., Von Brunn, V., Gold, D.J.C., & Minter, W.E.L. (1998) Earth's oldest reported glaciation: pysical and chemical evidence from the Archean Mozaan Group (~2.9 Ga) of South Africa. *Journal of Geology*, 106, 523–538.

York, D. (1969) Least squares fitting of a straight line with correlated errors. *Earth and Planetary Science Letters*, 5, 320–324.

Zahnle, K.J. (2006) Earth's earliest atmosphere. *Elements*, 2, 217–222.

Zahnle, K.J., Kasting, J.F., & Pollack, J.B. (1988) Evolution of a steam atmosphere during Earth's accretion. *Icarus*, 74, 62–97.

Zahnle, K., Claire, M., & Catling, D. (2006) The loss of mass-independent fractionation of sulfur due to a Paleoproterozoic collapse of atmospheric methane. *Geobiology*, 4, 271–284.

Zeebe, R.E. (2001) Seawater pH and isotopic paleotemperatures of Cretaceous oceans. *Paleogeography, Paleoclimatology, Paleoecology*, 170, 49–57.

Zheng, Y.F. (1993a) Calculation of oxygen isotope fractionations in anhydrous silicate minerals. *Geochimica Cosmochimica Acta, 57*, 1079–1091.

Zheng, Y.F. (1993b) Calculation of oxygen isotope fractionations in hydroxyl-bearing silicates. *Earth and Planetary Science Letters*, 120, 247–263.

Zhong, S., & Mucci, A. (1989) Calcite and aragonite precipitation from seawater solutions of various salinities: precipitation rates and overgrowth compositions. *Chemical Geology*, 78, 283–299.

Zindler, A., Hart, S.R., & Frey, F.A. (1979) Nd and Sr isotope ratios and rare earth element abundances in Reykjanes Peninsula basalts: evidence for mantle heterogeneity beneath Iceland. *Earth and Planetary Science Letters*, 45, 249–262.

Zolensky, M., Bland, P., Brown, P., & Halliday, I. (2006) Flux of extraterrestrial materials. In *Meteorites and the Early Solar System II* (eds D.S. Lauretta & H.Y. McSween, Jr), pp. 869–888. University of Arizona Press, Tucson, AZ.

Index

Introduction to Geochemistry: Principles and Applications, First Edition. Kula C. Misra.

Printed in the United States
By Bookmasters